Hydraulicians in the USA 1800–2000

IAHR Monograph

Series editor

Peter A. Davies
Department of Civil Engineering,
The University of Dundee,
Dundee,
United Kingdom

The International Association for Hydro-Environment Engineering and Research (IAHR), founded in 1935, is a worldwide independent organisation of engineers and water specialists working in fields related to hydraulics and its practical application. Activities range from river and maritime hydraulics to water resources development and eco-hydraulics, through to ice engineering, hydroinformatics and continuing education and training. IAHR stimulates and promotes both research and its application, and, by doing so, strives to contribute to sustainable development, the optimisation of world water resources management and industrial flow processes. IAHR accomplishes its goals by a wide variety of member activities including: the establishment of working groups, congresses, specialty conferences, workshops, short courses; the commissioning and publication of journals, monographs and edited conference proceedings; involvement in international programmes such as UNESCO, WMO, IDNDR, GWP, ICSU, The World Water Forum; and by co-operation with other water-related (inter)national organisations. www.iahr.org

Hydraulicians in the USA 1800–2000

A biographical dictionary of leaders in hydraulic engineering and fluid mechanics

Willi H. Hager

VAW, ETH-Hönggerberg, Zürich, Switzerland
hager@vaw.baug.ethz.ch

CRC Press
Taylor & Francis Group
Boca Raton London New York Leiden

CRC Press is an imprint of the
Taylor & Francis Group, an **informa** business

A BALKEMA BOOK

Published by: CRC Press/Balkema
P.O. Box 11320, 2301 EH Leiden, The Netherlands
e-mail: Pub.NL@taylorandfrancis.com
www.crcpress.com – www.taylorandfrancis.com

© 2015 Taylor & Francis Group, London, UK
CRC Press/Balkema is an imprint of the Taylor & Francis Group, an informa business

No claim to original U.S. Government works

ISBN 13: 978-0-367-57560-1 (pbk)
ISBN 13: 978-1-138-02828-9 (hbk)

Visit the Taylor & Francis Web site at
http://www.taylorandfrancis.com

and the CRC Press Web site at
http://www.crcpress.com

Typeset by V Publishing Solutions Pvt Ltd., Chennai, India

Library of Congress Cataloging-in-Publication Data

To our three daughters Olivia, Mirjam and Caren

Preface

When I started with this project in 2009, I thought to include all individuals of relevance outside from Europe. It became soon evident that the United States of America USA had such a large number of persons to be considered that restriction was made to this country. As in the previous two volumes, one single page is attributed to each individual selected. In addition, biographical references attract the reader's attention to specialized sources. To trail the main contributions of a person, bibliographical data are added to each person. Each page includes thus the following information: (1) Both the dates and locations of birth and death, respectively; (2) Professional career, starting with university grades, followed by professional achievements; (3) Main contributions to fluid engineering and awards obtained from academic and professional institutions; (4) Key references including bibliographical information, as also biographical sources, to aid their retrieval.

A biography is incomplete without the portrait denoted in that book with P. This was again the toughest part of the work. The portrait is a very personal item that is often hard to find. I was once more unable to find the portraits of all the persons, either because he or she passed away too long ago, or the family was impossible to contact. The largest help certainly was provided by ETH Library supporting my constant demand in seeing still other journals and books over the years. It is evident that the quality of a portrait cannot be excellent in all cases, either because the person has passed away long time ago so that time has had its effect on the portrait quality, or because the portrait found was not reproduced in the current quality. In cases no portrait was found, a typical photograph relating to the person's project, or to the title page of a book authored by this person, was selected. I would like to draw the reader's attention to the Acknowledgements.

Based mainly on a personal selection, this biographical work intends to include noteworthy individuals having worked in fluid flow. These may either be persons in physics, mathematics or even chemistry, or engineers with a background in water-related sciences, comprising dam, river, and environmental engineers, scientists in aerodynamics and meteorology, or even in hydrology and groundwater flow. All included in this dictionary have passed away, and their origin or main working site is related to the USA, except for Victor L. Streeter, who has passed 100 years of age. With few exceptions these persons have passed away between 1800 and 2000. This period was selected because it corresponds roughly to the time during which modern engineering was developed into the present day's knowledge. I was unable to find more than two women, highlighting the enormous future need for women in this field of knowledge. As observed during scientific congresses, for example, women are by now much more present in our professional world than decades ago, but still many 'disappear' once they have graduated as a PhD. Given that several hydraulicians with a close relation to the USA are included in the former two books, these persons are stated in a short overview Relation to volumes 1&2, so that their retrieval is easily accomplished.

When looking back to the past two decades or so, I note changes in the access to the engineering biographies. Before the Internet started in the 1990s, most information was sought by the telephone and letters addressed to professional societies or individuals. The USA is certainly a country advanced currently in terms of on-line searches, given the vast information available. In contrast, I would state that it was in many cases easier to find a biography of a person a century ago than today: When looking at the two large American Societies, namely ASCE and ASME, or even AGU, it can be stated that writing an obituary for a colleague having passed away appears to be completely outdated. Last year, only one obituary was published in the ASCE's Transactions, whereas there were typically almost 100 annually around 1900. We obviously do no more feel responsible in this task, resulting in a loss of identity and contact in our profession. This is a dark outlook into the future, because what has been attempted in this work will become so difficult that no more memory of our colleagues will be available, because of a lack of basic information. I can only hope that future generations of our profession will come back to where we once were, to highlight not only the engineering art, but also the masters behind it. From this perspective, it should also be reconsidered whether authors of professional papers be invited to submit with a paper also their portrait, as was done by various journals in the past. This source of information is for instance available from the Journal of Hydraulic Research over almost two decades, and proves to be a valuable basis of portrait retrieve. Overall, the USA has a comparatively excellent state as compared with many other countries allowing for the preparation of biographies. Therefore, I can only invite all to add not only to our profession with research, but also with information on these who have successfully worked in a field, so that we are able to document both in words and images who the great developers and designers of engineering work have been as individuals.

The present work would like to introduce outstanding or notable Americans having contributed to our profession. This work should be regarded as an initiative to advance the knowledge of our professional fathers. I invite anybody to contribute in updating the information presently available by adding to the data. May this book be a contribution to the extraordinary work of the US hydraulicians during the past two centuries!

Willi H. Hager
Zurich, February 2015

Acknowledgements

Mr. Walter Thürig, VAW, scanned the portraits included in this book and assisted me in additional work. Dr. David Vetsch, VAW, helped in the final book presentation and the setting up of the book index, for which I thank. Mrs. Carol Reese, formerly at ASCE, Reston VA, was often consulted for details mainly relating to the bio-data of individuals considered in this book. She was always helpful, and thus added to details of this work. I am particularly indebted to Dr. Hank T. Falvey, Colorado, and to Dr. Jack J. Cassidy, California. I further acknowledge the International Association of Hydro-Environment Engineering and Research (IAHR) for the good collaboration in finalizing this third volume on the hydraulicians, particularly its Secretary General, Dr. Christopher George. Finally, I acknowledge VAW for having given me the chance to prepare this book.

WHH

About the editor

Willi H. Hager (*1951) was educated at ETH Zurich, from where he obtained the civil engineering degree in 1976, the PhD degree in 1981, the habilitation degree in 1994, and he is professor of hydraulics since 1998 at the same institute.

He was research associate at ETH Lausanne from 1983 to 1988, returning to Versuchsanstalt für Wasserbau, Hydrologie und Glaziologie (VAW) in 1989 as head of the scientific staff. During the past 35 years of his professional activities, he was interested in hydraulic structures, wastewater hydraulics, high-speed flows, impulse waves, scour and erosion, and in the history of hydraulics, mainly in biographies of notable individuals of water engineering. He has published numerous works on the above mentioned topics both in peer-review journals as well as in national journals and at scientific congresses. He also authored books on these topics, among which are Dam Hydraulics (1998), Constructions hydrauliques (2009), Wastewater Hydraulics (2010) and two volumes on Hydraulicians in Europe. He was awarded the 1997 Ippen Lecture from the Intl. Association for Hydro-Environment Engineering and Research (IAHR) becoming an Honorary Member in 2013. He also was the recipient of the Best Technical Note Award, and the Hydraulic Structures Medal of the American Society of Civil Engineers (ASCE), among other distinctions.

Homepage: http://www.vaw.ethz.ch/people/hy/hagerw

Copyright statements

Anonymous (1905). Henry L. Abbot. *Engineering News* 54(11): 260. *P*

Anonymous (1936). Early presidents of the society, Julies W. Adams. *Civil Engineering* 6(9): 605–606. *P*

Anonymous (1948). Maurice L. Albertson. *Civil Engineering* 18(1): 45. *P*

Anonymous (1950). William Allan, J.C. Stevens Award winner. *Civil Engineering* 20(10): 676. *P*

Anonymous (1950). Charles M. Allen dies: Authority in hydraulics. *Engineering-News Record* 145(Aug. 24): 24. *P*

Anonymous (1930). K. Allen, sanitary engineer, dies. *Engineering News-Record* 105(11): 433. *P*

Anonymous (1946). Harry H. Ambrose. *Engineering News-Record* 137(Aug. 15): 212. *P*

Anonymous (1961). Alvin G. Anderson. *Civil Engineering* 30(10): 81. *P*

Anonymous (1964). Norval E. Anderson. *Civil Engineering* 34(11): 15. *P*

Anonymous (1969). Engineer is BuRec's new boss. *Engineering News-Record* 183(Nov. 13): 32–37. *P* (relating to Ellis L. Armstrong)

Anonymous (1968). BuRec promotes Arthur to associate chief engineer. *Engineering News-Record* 181(Jul. 25): 44. *P* (relating to Harold G. Arthur)

Anonymous (1941). A.H. (Gus) Ayres. *Engineering News-Record* 126(Apr. 24): 591. *P*

Anonymous (1968). James W. Ball. *Civil Engineering* 38(10): 74. *P*

Anonymous (1956). California fills a prize engineering post: Harvey O. Banks. *Engineering News-Record* 156(Jun. 14): 83–86. *P*

Anonymous (1940). Barrows to retire from MIT. *Engineering News-Record* 124(20): 681. *P* (relating to Harold K. Barrows)

Anonymous (1955). William J. Bauer. *Civil Engineering* 25(10): 713. *P*

Anonymous (1981). Leo R. Beard. *Civil Engineering* 51(10): 84. *P*

Anonymous (1976). Paul C. Benedict. *Civil Engineering* 46(10): 103. *P*

Anonymous (1947). A.W.K. Billings new Honorary Member. *Civil Engineering* 17(1): 43. *P*

Anonymous (1951). Chas. A. Bissell. *Engineering News-Record* 147(Oct. 18): 69. *P*

Anonymous (1910). New chief of Engineers, Col. W.H. Bixby. *Engineering News* 63(18): 524. *P*

Anonymous (1916). Col. William M. Black. *Engineering News* 75(10): 483. *P*

Anonymous (1950). F.W. Blaisdell receives J.W. Rickey Medal. *Civil Engineering* 20(1): 45. *P*

Anonymous (1952). Clarence E. Blee. *Engineering News-Record* 149(Jul. 24): Frontispiece. *P*

Anonymous (1950). Walter E. Blomgren. *Engineering News-Record* 144(May 11): 64. *P*

Anonymous (1981). Don E. Bloodgood. *Civil Engineering* 51(10): 72. *P*

Anonymous (1948). Carl A. Bock. *Engineering News-Record* 140(Mar. 25): 466. *P*

Anonymous (1920). John Bogart dead. *Engineering News-Record* 84(18): 887. *P*

Anonymous (1943). James S. Bowman. *Engineering News-Record* 130(Apr. 29): 652. *P*

Anonymous (1976). Donald C. Bondurant, Hilgard Prize. *Civil Engineering* 46(10): 103. *P*

Anonymous (1958). Nicholls W. Bowden. *Civil Engineering* 28(5): 21. *P*

Anonymous (1950). C.E. Bowers, Collingwood Prize winner. *Civil Engineering* 20(10): 675. *P*

Anonymous (1967). M. Clifford Boyer. *Civil Engineering* 37(11): 79. *P*

Anonymous (1955). Joseph N. Bradley. *Civil Engineering* 25(10): 713. *P*

Anonymous (1965). Charles L. Bretschneider. *Civil Engineering* 35(12): 73. *P*

Anonymous (1942). Earl I. Brown. *Civil Engineering* 12(1): 50. *P*

Anonymous (1964). Fred R. Brown. *Civil Engineering* 34(5): 65. *P*

Anonymous (1897). Charles Benjamin Brush. *Engineering News* 37(June 10): 359. *P*

Anonymous (1939). Members of Lower Rio Grande Project. *Civil Engineering* 9(9):576. *P* (relating to Joseph L. Burkholder)

Anonymous (1942). Thomas R. Camp. *Civil Engineering* 12(1): 50. *P*

Anonymous (1964). Frank B. Campbell. *Civil Engineering* 34(5): 65. *P*

Anonymous (1965). Thomas Carmody. *Civil Engineering* 35(6): 83. *P*

Anonymous (1960). Rolland W. Carter. *Civil Engineering* 30(10): 88. *P*

Anonymous (1953). Harry R. Cedergren. *Civil Engineering* 23(12): 856. *P*

Anonymous (1965). Wallace L. Chadwick. *Civil Engineering* 35(11): 104. *P*

Anonymous (1963). Dr. N.A. Christensen. *Civil Engineering* 33(4): 72. *P*

Anonymous (1976). J.E. Christiansen. *Civil Engineering* 46(10): 103. *P*

Anonymous (1940). Charles Clark new head of New York water supply. *Engineering News-Record* 124(Apr. 18): 552. *P*

Anonymous (1958). Cleanup man on Ohio River. *Engineering News-Record* 161(Oct. 2): 54–61. *P* (relating to Edward J. Cleary)

Anonymous (1922). John F. Coleman. *Engineering News-Record* 89(8): 330. *P*

Anonymous (1911). Francis Collingwood. *Engineering News* 66(8): 242; 66(9): 269. *P*

Anonymous (1953). Victor M. Cone retires, Corps of Engineers expert. *Engineering News-Record* 151(Oct. 8): 291–292. *P*

Anonymous (1939). Hilton H. Cooper. *Engineering News-Record* 123(Aug. 10): 186. *P*

Anonymous (1949). Glen N. Cox. *Civil Engineering* 19(9): 806. *P*

Anonymous (1938). William P. Craighill. *Civil Engineering* 8(3): 219–220. *P*

Anonymous (1950). William P. Creager. *Civil Engineering* 20(1): 46–47. *P*

Anonymous (1932). Herbert S. Crocker: The new president. *Civil Engineering* 2(2): 117–118. *P*

Anonymous (1938). John James Robertson Croes. *Civil Engineering* 8(10): 702–703. *P*

Anonymous (1942). Prof. Hardy Cross. *Engineering News-Record* 128(Feb. 5): 217. *P*

Anonymous (1972). Crozet, Claude (Claudius). *A biographical dictionary of American civil engineers*: 30–31. ASCE: New York. *P*

Anonymous (1933). Arthur P. Davis, Boulder Dam consultant, dies. *Engineering News-Record* 111(6): 181. *P*

Anonymous (1958). Davis, Calvin. *Engineering News-Record* 161(Sep. 11): 31. *P*

Anonymous (1963). Robert B. Diemer. *Engineering News-Record* 171(Oct. 17): 20. *P*

Anonymous (1953). Oswald H. Dodkin. *Civil Engineering* 23(11): 794. *P*

Anonymous (1943). Robert E. Doherty. *Engineering News-Record* 131(Jul. 1): 3. *P*

Anonymous (1956). Chief engineer for NY water: Stanley M. Dore. *Engineering News-Record*: 157(Jul. 18): 72. *P*

Anonymous (1954). John Wilson Dougherty. *Civil Engineering* 24(7): 497. *P*

Anonymous (1947). Richard E. Dougherty. *Engineering News-Record* 139(Jul. 24): 97–99. *P*

Anonymous (1961). ASCE past president R.E. Dougherty dies. *Civil Engineering* 31(11): 75. *P*

Anonymous (1950). L.R. Douglass. *Engineering News-Record* 145(Sep. 14): 65. *P*

Anonymous (1922). John H. Dunlap: ASCE Secretary. *Engineering News-Record* 89(1): 34. *P*

Anonymous (1967). Leonard B. Dworsky. *Civil Engineering* 37(4): 76. *P*

Anonymous (1952). 25 years ago, the first Freeman scholars. *Engineering News-Record* 149(Jul. 17): 30. *P* (relating to Herbert N. Eaton)

Anonymous (1953). Inventor Wesley Eckenfelder. *Engineering News-Record* 150(May 14): 25. *P*

Anonymous (1937). Eddy dies in Montreal. *Engineering News-Record* 118(Jun. 17): 895. *P*

Anonymous (1940). Eliassen joins NYU faculty. *Engineering News-Record* 125(Sep. 5): 310. *P*

Anonymous (1950). Joseph W. Ellms dies. *Engineering News-Record* 144(Feb. 16): 24. *P*

Anonymous (1936). B.A. Etcheverry. *Civil Engineering* 6(8): 533. *P*

Anonymous (1954). Fair of Harvard: Master of the House, and much of what he surveys. *Engineering News-Record* 153(Dec. 2): 52–56. *P*

Anonymous (1941). George H. Fenkell. *Civil Engineering* 11(1): 58. *P*

Anonymous (1897). Albert Fink. *Engineering News* 37(April 8): 215. *P*

Anonymous (1974). Hugo B. Fischer. *Civil Engineering* 44(10): 95. *P*

Anonymous (1926). Desmond Fitzgerald dead. *Engineering-News Record* 97(14): 555. *P*

Anonymous (1938). Early presidents of the Society: Desmond Fitzgerald. *Civil Engineering* 8(8): 557–558; 8(11): 759–760. *P*

Anonymous (1945). Don M. Forester. *Engineering News-Record* 134(Feb. 8): 178. *P*

Anonymous (1949). E.P. Fortson. *Engineering-News Record* 142(Jun. 16): 32. *P*

Anonymous (1937). The center-vent Francis wheel. *Civil Engineering* 7(2): 142–143. (*P*)

Anonymous (1976). Danny L. Fread. *Civil Engineering* 46(10): 104. *P*

Anonymous (1958). Francis S. Friel: Large Dam Congress planned. *Engineering News-Record* 160(Mar. 20): 28. *P*

Anonymous (1975). Albert S. Fry, retired TVA official, dies. *Civil Engineering* 45(3): 94–95. *P*

Anonymous (1962). E. Montford Fucik. *Engineering News-Record* 168(May 10): 65. *P*

Anonymous (1934). George W. Fuller dies. *Engineering News-Record* 112(Jun. 21): 821. *P*

Anonymous (1924). Roy W. Gausmann. *Engineering News-Record* 92(3): 130. *P*

Anonymous (1928). Gen. G.W. Goethals dies. *Engineering News-Record* 100(4): 167. *P*

Anonymous (1976). Alfred R. Golzé. *Civil Engineering* 46(10): 95. *P*

Anonymous (1936). Ernest P. Goodrich. *Civil Engineering* 10(6): 699–700. *P*

Anonymous (1958). Harold B. Gotaas. *Civil Engineering* 28(10): 777. *P*

Anonymous (1954). Richard H. Gould. *Civil Engineering* 24(6): 402. *P*

Anonymous (1921). Samuel M. Gray. *Engineering News-Record* 87(21): 872. *P*

Anonymous (1963). Where seniority is a tradition. *Engineering News-Record* 171(Nov. 21): 53–54. *P* (relating to Samuel A. Greeley)

Anonymous (1931). J.H. Gregory: James Laurie Prize. *Engineering News-Record* 106(4): 164. *P*

Anonymous (2010). Claude Irving 'Pete' Grimm. *Civil Engineering* 80(6): 42. *P*

Anonymous (1949). Benjamin F. Groat. *Civil Engineering* 19(8): 580. *P*

Anonymous (1934). Carl Ewald Grunsky. *Civil Engineering* 4(7): 373. *P*

Anonymous (1924). C.E. Grunsky. *Engineering News-Record* 92(3): 128. *P*

Anonymous (2013): William Guyton. *Engineering News-Record* 270(9): 26. *P*

Anonymous (1946). L. Standish Hall. *Civil Engineering* 16(1): 36. *P*

Anonymous (1959). Eugene E. Halmos. *Civil Engineering* 29(3): 220. *P*

Anonymous (1959). E.E. Halmos, Sr., dies: River, harbour designer. *Engineering New-Record* 162(Feb. 19): 194. *P*

Anonymous (1953). Harry Parker Hammond. *Civil Engineering* 23(12): 868. *P*

Anonymous (1955). Designer of dams retires. *Engineering News-Record* 155(Sep. 1): 59–60. *P* (relating to John J. Hammond)

Anonymous (1916). William P. Hardesty. *Engineering News* 76(2): 55. *P*

Anonymous (1966). D.R.F. Harleman. *Civil Engineering* 36(6): 52. *P*

Anonymous (1940). Harper named chief reclamation engineer. *Engineering News-Record* 125(Jul. 18): 76. *P* (relating to Sinclair O. Harper)

Anonymous (1939). Elmo G. Harris. *Civil Engineering* 9(1): 67. *P*

Anonymous (1912). Benjamin Morgan Harrod. *Engineering News* 68(12): 554. *P*

Anonymous (1953). L.F. Harza dies. *Engineering News-Record* 151(Dec. 3): 25. *P*

Anonymous (1924). Dean E.E. Haskell. *Engineering News-Record* 93(24): 847. *P*

Anonymous (1904). Charles Haynes Haswell. *Engineering News* 51(22): 509–511. *P*

Anonymous (1958). Dam experts coming from 40 nations. *Engineering News-Record* 161(Aug. 14): 27. *P* (relating to Gail A. Hathaway)

Anonymous (1930). Allen Hazendies while on vacation. *Engineering News-Record* 105(5): 189. *P*

Anonymous (1935). D.C. Henny dies in Portland OR. *Engineering News-Record* 115(3): 102. *P*

Anonymous (1965). Henry P. Herbich. *Civil Engineering* 35(11): 14. *P*

Anonymous (1966). John B. Herbich. *Civil Engineering* 36(10): 77. *P*

Anonymous (1918). Rudolph Hering. *Engineering News-Record* 81(6): 274–276. *P*

Anonymous (1950). G.H. Hickox. *Civil Engineering* 20(1): 42–43. *P*

Anonymous (1891). Prof. Julius E. Hilgard. *Engineering News* 25(May 16): 462. *P*

Anonymous (1924). John W. Hill. *Engineering News-Record* 92(16): 647. *P*

Anonymous (1937). Louis C. Hill. *Engineering News-Record* 118(Jan. 21): 108. *P*

Anonymous (1964). Conservative design entails innovation. *Engineering News-Record* 172(Apr. 30): 40–43. *P* (relating to Raymond A. Hill)

Anonymous (1941). Julian Hinds. *Engineering News-Record* 146(Apr. 3): 508. *P*

Anonymous (1946). J.C. Hoyt, former ASCE vice-president, is dead. *Civil Engineering* 16: 366. *P*

Anonymous (1966). P.G. Hubbard. *Civil Engineering* 36(2): 10. *P*

Anonymous (1924). Clarence W. Hubbell. *Engineering News-Record* 93(21): 846. P

Anonymous (1952). Huber nominated for highest ASCE post. *Engineering News-Record* 148(May 8): 80. P (relating to Walter L. Huber)

Anonymous (1932). Charles Warren Hunt dies. *Engineering News-Record* 109(4): 118. P

Anonymous (1972). Hutton, William R. *A biographical dictionary of American civil engineers*: 64–65. ASCE: New York. P

Anonymous (1991). Alfred C. Ingersoll. *Civil Engineering* 61(12): 90. P

Anonymous (1971). Chairman William T. Ingram. *Civil Engineering* 41(4): 79. P

Anonymous (1998). Carl Izzard had major impact on hydraulics in highway design. *Civil Engineering* 68(1): 8. P

Anonymous (1972). Robert B. Jansen. *Civil Engineering* 42(6): 84. P

Anonymous (1979). Joe W. Johnson. *Civil Engineering* 49(10): 89–90. P

Anonymous (1961). Wendell E. Johnson. *Engineering News-Record* 166(Mar. 23): 155. P

Anonymous (1982). Wendell E. Johnson, dam expert, dies. *Civil Engineering* 52(6): 88. P

Anonymous (1945). B.E. Jones. *Engineering News-Record* 135(Aug. 30): 267. P

Anonymous (1978). Joel B. Justin. *Civil Engineering* 48(4):100. P

Anonymous (1950). Director J.D. Justin, power expert, is dead. *Civil Engineering* 20(3): 199. P

Anonymous (1948). A.A. Kalinske. *Civil Engineering* 18(1): 45. P

Anonymous (1945). Marion R. Kays. *Engineering News-Record* 134(Feb. 1): 152. P

Anonymous (1952). C.E. Keefer, Thomas Fitch Rowland Prize. *Civil Engineering* 22(8): 587. P

Anonymous (1952). K.B. Keener. *Civil Engineering* 22(7): 22. P

Anonymous (1914). John C. Kelley and the American water meter industry. *Engineering News* 71(13): 674–675. P

Anonymous (1962). John F. Kennedy. *Civil Engineering* 32(10): 71. P

Anonymous (1952). Karl R. Kennison. *Engineering News-Record* 148(May 29): 68. P

Anonymous (1966). Lewis H. Kessler. *Civil Engineering* 36(7): 13. P

Anonymous (1960). Carl E. Kindsvater. *Civil Engineering* 30(10): 88. P

Anonymous (1951). Horace King, Hon. M. ASCE, is dead. *Civil Engineering* 21(5): 293. P

Anonymous (1913). Brigadier-General Dan C. Kingman, Chief of Engineers, USA. *Engineering News* 70(17): 834. P

Anonymous (1965). S.J. Kline. *Civil Engineering* 35(10): 107. P

Anonymous (1949). Theodore T. Knappen. *Engineering News-Record* 142(Apr. 7): Frontispiece. P

Anonymous (1931). Morris Knowles, Director District 6. *Civil Engineering* 1(6): 559. P

Anonymous (1965). Victor A. Koelzer. *Civil Engineering* 35(4): 50. P

Anonymous (2004). In memoriam: Herman J. Koloseus. *Civil Engineering News* 10(2): 5. P

Anonymous (1956). Ford Kurtz. *Civil Engineering* 26(9): 636–637. P

Anonymous (1963). Emory W. Lane. *Civil Engineering* 33(10): 67–68. P

Anonymous (1979). Henry L. Langhaar. *Civil Engineering* 49(10): 95. P

Anonymous (1953). W.M. Lansford. *Civil Engineering* 23(8): 560. P

Anonymous (1940). Lieut. Col. T.B. Larkin. *Engineering News-Record* 125(Jul. 11): 57. P

Anonymous (1962). Emmett M. Laursen. *Civil Engineering* 32(7): 65. P

Anonymous (1956). Finley B. Laverty, Director, District 11. *Civil Engineering* 26(10): 683. P

Anonymous (1960). WPCF moves consultant into its No. 1 job. *Engineering News-Record* 165(Sep. 29): 52–53. P (relating to Ray E. Lawrence)

Anonymous (1909). Col. Smith S. Leach. *Engineering News* 62(18): 466. P

Anonymous (1950). C.A. Lee, Collingwood Prize winner. *Civil Engineering* 20(10): 675. P

Anonymous (1966). Arno T. Lenz. *Civil Engineering* 36(12): 71. P

Anonymous (1940). George K. Leonard. *Engineering News-Record* 125(Oct. 17): 511. P

Anonymous (1977). Ray K. Linsley. *Civil Engineering* 37(7): 120. P

Anonymous (1962). John Lowe III. *Civil Engineering* 32(9): 23. P

Anonymous (1963). Thomas Maddock, Jr. *Civil Engineering* 33(10): 71. P

Anonymous (1970). Charles Mansur. *Civil Engineering* 40(2): 10. P

Anonymous (1967). Harold M. Martin. *Civil Engineering* 37(7): 66. P

Anonymous (1940). Charles D. Marx dies. *Engineering News-Record* 124(Jan. 4): 29. P

Anonymous (1940). Charles David Marx. *Civil Engineering* 10(2): 121. P

Anonymous (1963). Donald H. Mattern. *Civil Engineering* 33(10): 72. P

Anonymous (1938). Arthur E. Matzke. *Civil Engineering* 8(1): 51. P

Anonymous (1944). Frederic T. Mavis. *Engineering News-Record* 133(Sep. 21): 336. P

Anonymous (1957). Frederic T. Mavis. *Civil Engineering* 27(6): 29. P

Anonymous (1948). William H. McAlpine. *Engineering News-Record* 140(Jan. 22): 95. P

Anonymous (1932). William Jarvis McAlpine. *Civil Engineering* 2(8): 522. P

Anonymous (1950). Harry R. McBirney. *Engineering News-Record* 145(Oct. 19): 63. P

Anonymous (1961). Gerald T. McCarthy. *Engineering News-Record* 166(Mar. 2): 45. P

Anonymous (1962). We need researchers. *Engineering News-Record* 169(Oct. 4): 101–102. P (relating to Jack E. McKee)

Anonymous (1901). Robert Emmet McMath. *Engineering News* 45(3): 44. P

Anonymous (1966). M.B. McPherson, Director for ASCE sewer research project. *Civil Engineering* 36(3): 90. P

Anonymous (1914). Daniel W. Mead. *Engineering News* 73(Jun. 4): 1273. P

Anonymous (1936). Elwood Mead dies at 78. *Engineering News-Record* 116(Jan. 30): 173. P

Anonymous (1924). Mansfield Merriman. *Engineering News-Record* 92(16): 649–650. P

Anonymous (1929). Thaddeus Merriman. *Engineering News-Record* 103(25): 984. P

Anonymous (1939). Thaddeus Merriman. *Civil Engineering* 9(11): 699. P

Anonymous (1944). T.A. Middlebrooks, James Laurie Prize. *Civil Engineering* 14(1): 35. P

Anonymous (1964). Carl R. Miller. *ASCE Panel*: Stable channel design. ASCE: New York. P

Anonymous (1921). Hiram F. Mills dead. *Engineering News-Record* 87(15): 627–628. P

Anonymous (1940). Charles A. Mockmore. *Civil Engineering* 10(1): 55. P

Anonymous (1959). Robert A. Monroe. *Civil Engineering* 29(7): 23. P

Anonymous (1956). Walter L. Moore. *Civil Engineering* 26(12): 850. P

Anonymous (1940). Arthur Morgan returns to Dayton-Morgan company. *Engineering News-Record* 124 (Jun. 6): 768. P

Anonymous (1977). The genius of A.E. Morgan. *Civil Engineering* 47(10): 114–117. P

Anonymous (1942). Samuel B. Morris. *Engineering News-Record* 128(Mar. 19): 447. P

Anonymous (1932). Floyd A. Nagler. *Civil Engineering* 2(1): 57. P

Anonymous (1952). Wesley R. Nelson. *Engineering News-Record* 148(Feb. 14): 163. P

Anonymous (1991). Carl F. Nordin. *Civil Engineering* 61(12): 82. P

Anonymous (1911). Edward P. North. *Engineering News* 66(4): 127. P

Anonymous (1912). North, E.P. *Trans. ASCE* 75: 1167–1176. P

Anonymous (1965). Morrough P. O'Brien. *Civil Engineering* 35(12): 73. P

Anonymous (1942). C.W. Okey. *Civil Engineering* 12(6): 340. P

Anonymous (1921). Excavating the foundation for Hetch Hetchy Dam. *Engineering News-Record* 87(6): 222–224. (P) (relating to Albert E. Paddock)

Anonymous (1955). John C. Page. *Civil Engineering* 25(5): 303. P

Anonymous (1939). G.L. Parker named chief hydraulic engineer. *Civil Engineering* 9(10): 633. P

Anonymous (1938). Theodore B. Parker. *Engineering News-Record* 120(May 19): 700. P

Anonymous (1926). H. de B. Parsons. *Engineering News-Record* 96(3): 132–133. P

Anonymous (1924). William B. Parsons. *Engineering News-Record* 92(16): 643–644. P

Anonymous (1961). Former ASCE Director Carl Paulsen dies. *Civil Engineering* 31(3): 88. P

Anonymous (1953). Civil engineer wins Alfred Noble Prize. *Civil Engineering* 23(12): 849. P (relating to Henry M. Paynter)

Anonymous (1975). Peck receives Medal of Science. *Civil Engineering* 45(11): 93. P

Anonymous (1950). A.J. Peterka. *Civil Engineering* 20(1): 42. P

Anonymous (1968). D. Peterson, Jr., honoured for service to ASCE. *Civil Engineering* 38(3): 81. P

Anonymous (1984). Richard D. Pomeroy. *Civil Engineering* 54(10): 80. P

Anonymous (1958). Chesley J. Posey. *Civil Engineering* 28(10): 778. P

Anonymous (1928). General Potter, Head of Mississippi work, dies. *Engineering News-Record* 101(6): 223. P

Anonymous (1952). E.A. Pratt, consulting engineer. *Civil Engineering* 22(7): 502. P

Anonymous (1941). Preston retires from USBR. *Engineering News-Record* 125(Jan. 9): 67. P (relating to Porter J. Preston)

Anonymous (1961). Reginald C. Price. *Civil Engineering* 31(10): 36. P

Anonymous (1958). Puls succeeds Keener as USBR's chief designer. *Engineering News-Record* 160(Jun. 5): 78. *P*

Anonymous (1939). Reclamation pioneer dies at eighty-nine: John H. Quinton. *Engineering News-Record* 122(May 18): 689. *P*

Anonymous (1920). Isham Randolph dead. *Engineering News-Record* 85(6): 287. *P*

Anonymous (1956). Irrigation expert A.B. Reeves to retire from Reclamation. *Engineering News-Record* 157(Dec. 6): 313. *P*

Anonymous (1927). E.W. Rettger. *Engineering News-Record* 99(5): 195. *P*

Anonymous (1955). Leon B. Reynolds. *Engineering News-Record* 155(Oct. 13): 66. *P*

Anonymous (1985). Thomas J. Rhone. *Civil Engineering* 55(10): 96. *P*

Anonymous (1974). George R. Rich. *Civil Engineering* 44(10): 93. *P*

Anonymous (1961). Everett V. Richardson. *Civil Engineering* 31(10): 84. *P*

Anonymous (1949). Rickey Gold Medal awaits its first award. *Civil Engineering* 19(9): 634. *P*

Anonymous (1938). James W. Rickey. *Engineering News-Record* 120(Jan. 13): 87. *P*

Anonymous (1948). Ross M. Riegel. *Civil Engineering* 18(1): 45–47. *P*

Anonymous (1936). Early presidents of the Society: William Milnor Roberts. *Civil Engineering* 6(12): 833–834. *P*

Anonymous (1913). William T. Rossell: New Chief of Engineers. *Engineering News* 70(7): 321. *P*

Anonymous (1939). Hunter Rouse. *Civil Engineering* 9(1): 53. *P*

Anonymous (1973). William M. Sangster. *Civil Engineering* 43(10): 88. *P*

Anonymous (1946). H.N. Savage, Reclamation Conference 1907. *Civil Engineering* 16(4): 188. *P*

Anonymous (1937). John L. Savage. *Civil Engineering* 7(2): 154. *P*

Anonymous (1942). Thorndike Saville. *Engineering News-Record* 128(Mar. 5): 371. *P*

Anonymous (1970). William W. Sayre. *Civil Engineering* 40(10): 80–81. *P*

Anonymous (1937). Schley is appointed Chief of Engineers. *Engineering News-Record* 119(Sep. 23): 409. *P*

Anonymous (1947). L.A. Schmidt, Jr.: James Laurie Prize. *Civil Engineering* 17(1): 41. *P*

Anonymous (1913). James Dix Schuyler. *Engineering News* 69(16): 785–788. *P*

Anonymous (1942). Fred C. Scobey. *Civil Engineering* 12(12): 696. *P*

Anonymous (1961). James L. Sherard. *Civil Engineering* 31(7): 68. *P*

Anonymous (1947). L.K. Sherman. *Engineering News-Record* 138(Jan. 16): 81. *P*

Anonymous (1954). L.K. Sherman, ASCE Honorary Member, is dead. *Civil Engineering* 24(3): 179. *P*

Anonymous (1963). Samuel Shulits. *Civil Engineering* 33(10): 73. *P*

Anonymous (1991). Silberman. *ASCE News* (12): 75–76. *P*

Anonymous (1960). Daryl B. Simons. *Civil Engineering* 30(12): 78. *P*

Anonymous (1951). Raymond A. Skrinde. *Civil Engineering* 21(10): 605–607. *P*

Anonymous (1927). Charles S. Slichter. *Engineering News-Record* 99(5): 194. *P*

Anonymous (1961). Slichter retires from the Corps. *Engineering News-Record* 166(Feb. 9): 51. *P* (relating to Francis B. Slichter)

Anonymous (1952). Harvey Slocum to direct India dam building. *Engineering News-Record* 148(Apr. 17): 45–46. *P*

Anonymous (1900). Hamilton Smith. *Engineering News* 44(18): 300. *P*

Anonymous (1931). J. Waldo Smith. *Engineering News-Record* 106(Mar. 19): 498. *P*

Anonymous (1971). Waldo E. Smith. *Civil Engineering* 41(3): 6. *P*

Anonymous (1962). Kenneth E. Sorensen. *Engineering News-Record* 168(May 10): 67. *P*

Anonymous (1944). Edward Soucek. *Engineering News-Record* 133(Aug. 31): 257. *P*

Anonymous (1973). G.F. Sowers. *Civil Engineering* 43(2): 39. *P*

Anonymous (1912). Henry Wilson Spangler. *Engineering News* 67(13): 616–617. *P*

Anonymous (1936). Herman Stabler, Director District 5. *Civil Engineering* 6(3): 207. *P*

Anonymous (1934). C. Maxwell Stanley. *Civil Engineering* 4(1): 51. *P*

Anonymous (1952). Owen G. Stanley. *Engineering News-Record* 149(Oct. 23): 74. *P*

Anonymous (1905). Frederic P. Stearns. *Engineering News* 54(11): 263–264. *P*

Anonymous (1961). I.C. Steele. *Civil Engineering* 31(10): 83. *P*

Anonymous (1945). John Cyprian Stevens. *Civil Engineering* 15(2): 97–98. *P*

Anonymous (1931). Charles Storer Storrow. *Engineering News-Record* 107(13): 476. *P*

Anonymous (2000). The engineering genius history forgot. *Civil Engineering* 70(3): 15. *P*

Anonymous (1984). Gen. J.H. Stratton dies. *Civil Engineering* 54(5): 68. *P*

Anonymous (1949). Lorenz G. Straub. *Civil Engineering* 19(5): 328–329. *P*

Anonymous (1938). Victor L. Streeter. *Civil Engineering* 8(1): 51. *P*

Anonymous (1972). Strickland, William. *A biographical dictionary of American civil engineers*: 113–114. ASCE: New York. *P*

Anonymous (1954). E.B. Strowger, Niagara Mohawk Power Co. *Civil Engineering* 24(6): 301. *P*

Anonymous (1924). George F. Swain. *Engineering News-Record* 92(16): 645–646. *P*

Anonymous (1903). Elnathan Sweet. *Engineering News* 49(6): 124–125. *P*

Anonymous (1940). Frederick G. Switzer. *Engineering News-Record* 125(Dec. 12): 792. *P*

Anonymous (1942). A.N. Talbot, research leader, dies. *Engineering News-Record* 128(Apr. 9): 528. *P*

Anonymous (1952). Franklin Thomas. *Engineering News-Record* 149(Sep. 11): 54. *P*

Anonymous (1954). Harold A. Thomas, Jr. *Civil Engineering* 24(9): 609–610. *P*

Anonymous (1945). Paul W. Thompson. *Engineering News-Record* 135(Jul. 12): 90. *P*

Anonymous (1948). R.H. Thomson, Honorary Member ASCE. *Civil Engineering* 18(9): 591. *P*

Anonymous (1944). J.B. Tiffany. *Engineering News-Record* 133(Aug. 24): 207. *P*

Anonymous (1938). Ross K. Tiffany. *Engineering News-Record* 120(Jan. 27): 136. *P*

Anonymous (1956). E. Roy Tinney. *Engineering News-Record* 157(Nov. 22): 53–54. *P*

Anonymous (1958). R.J. Tipton: Scholarship fund. *Engineering News-Record* 160(Jun. 19): 177. *P*

Anonymous (1956). B.E. Torpen. *Engineering News-Record* 156(May 3): 79–80. *P*

Anonymous (1883). John C. Trautwine, C.E. *Engineering News* 10(Sep. 22): 450–453. *P*

Anonymous (1924). John C. Trautwine, Jr. *Engineering News-Record* 92(16): 647. *P*

Anonymous (1965). John C. Trautwine. *Civil Engineering* 35(11): 102. *P*

Anonymous (1954). Dario Travaini. *Civil Engineering* 24(4): 257. *P*

Anonymous (1951). James G. Tripp, Construction Engng. Prize. *Civil Engineering* 21(10): 607. *P*

Anonymous (1965). Top Pentagon Award to Turnbull. *Engineering News-Record* 174(Jun. 17): 213. *P*

Anonymous (1933). Turneaure elected to honorary membership. *Civil Engineering* 3(12): 696. *P*

Anonymous (1943). Boston civil engineers elect Turner president. *Engineering News-Record* 130(Apr. 15): 529. *P*

Anonymous (1950). Vito A. Vanoni. *Civil Engineering* 20(1): 43–44. *P*

Anonymous (1947). N.T. Veatch. *Engineering News-Record* 139(Jul. 17): 89. *P*

Anonymous (1955). John K. Vennard. *Civil Engineering* 25(6): 374. *P*

Anonymous (1989). Warren Viessman, Jr. *Civil Engineering* 59(12): 82. *P*

Anonymous (1966). Prof. James R. Villemonte. *Civil Engineering* 36(12): 71. *P*

Anonymous (1968). TVA chairman A.J. Wagner. *Engineering News-Record* 180(May 23): 50. *P*

Anonymous (1974). William E. Wagner. *Civil Engineering* 44(10): 116. *P*

Anonymous (1921). John F. Wallace. *Engineering News-Record* 87(2): 83–84. *P*

Anonymous (1959). Donald S. Walter. *Civil Engineering* 29(11): 22. *P*

Anonymous (1924). Added reclamation responsibility. *Engineering News-Record* 93(25): 1011. *P* (relating to Raymond F. Walter)

Anonymous (1941). R.B. Ward. *Engineering News-Record* 126(Jan. 9): 65. *P*

Anonymous (1960). How to stopwater pollution. *Engineering News-Record* 164(Mar. 3): 55–58. *P* (relating to Frederick H. Waring)

Anonymous (1954). Walter O. Washington. *Civil Engineering* 24(10): 646. *P*

Anonymous (1947). Paul Weir. *Engineering News-Record* 139(Jul. 10): 170. *P*

Anonymous (1949). Andrew Weiss. *Civil Engineering* 19(1): 43–44. *P*

Anonymous (1937). Ashbel Welch. *Civil Engineering* 7(3): 236–237. *P*

Anonymous (1946). C.E. Wells, with Reclamation Group, 1907. *Civil Engineering* 16(4): 188. *P*

Anonymous (1929). F.E. Weymouth. *Engineering News-Record* 103(25): 984. *P*

Anonymous (1929). Andrew J. Wiley. *Engineering News-Record* 103(25): 984. *P*

Anonymous (1954). Prof. R.B. Wiley to leave Purdue Engineering School. *Engineering News-Record* 152(Jun. 10): 69–70. *P*

Anonymous (1962). 'Harza' means river development in 25 countries. *Engineering News-Record* 168(May 10): 64–67. *P* (relating to Charles K. Willey)

Anonymous (1931). Gardner S. Williams. *Engineering News-Record* 107(27): 1053. *P*

Anonymous (1949). Roy B. Williams retires. *Civil Engineering* 19(12): 878. *P*

Anonymous (1961). Abel Wolman Hon. M. ASCE. *Civil Engineering* 31(10): 80. *P*

Anonymous (1944). James G. Woodburn. *Civil Engineering* 14(11): 489. *P*

Anonymous (1953). ASCE Honorary Member Woodward dies. *Civil Engineering* 23(10): 709. *P*

Anonymous (1943). Sherman M. Woodward. *Engineering News-Record* 130(Jan. 28): 108. *P*

Anonymous (1933). David L. Yarnell. *Civil Engineering* 3(1): 47. *P*

Baker, M.N. (1941). The three Jewells: Pioneers in mechanical filtration. *Engineering News-Record* 126(Jan. 30): 179. *P* (relating to Omar H. Jewell)

Brown, J.L. (2012). Braving the storm: Minot's Ledge Lighthouse. *Civil Engineering* 82(11): 46–49. *P* (relating to Joseph G. Totten)

Brown, J. (2013). The Mississippi Basin in miniature. *Civil Engineering* 83(10): 42–45. (*P*) (relating to Charles B. Patterson)

Buehler, B. (1975). Corps guidelines for dam safety inspection need revamping. *Civil Engineering* 45(1): 74–75. *P*

Clapp, W.B., Murphy, E.C., Martin, W.F. (1908). The flood of March 1907 in the Sacramento and San Joaquin River Basins, California. *Trans. ASCE* 61: 281–376. (*P*) (relating to William B. Clapp)

Eaton, E.C. (1919). Removing algae from a California irrigation canal. *Engineering News-Record* 82(8): 382–383. *P*

Eiffert, C.H., Bennett, C.S. (1938). Sixteen years of flood control in the Miami Valley. *Civil Engineering* 8(5): 343–345. (*P*) (relating to Charles S. Bennett)

Fitzsimons, N. (1966). James Renwick. *Civil Engineering* 36(8): 78. *P*

Fitzsimons, N. (1966). Squire Whipple. *Civil Engineering* 36(5): 48. *P*

Fitzsimons, N. (1966). George Washington Whistler. *Civil Engineering* 36(7): 74. *P*

Fitzsimons, N. (1967). Horatio Allen, Hon. M. ASCE. *Civil Engineering* 37(2): 67. *P*

Fitzsimons, N. (1970). Benjamin Wright: The Father of American Civil Engineering. *Civil Engineering* 40(9): 68. *P*

FitzSimons, N, ed. (1972) Noble, Alfred. *Biographical dictionary of American civil engineers*: 94–95. ASCE: New York. *P*

Fitzsimons, N. (1973). Do-Ne-Ho-Geh-Weh: Seneca Sachem and civil engineer. *Civil Engineering* 43(6): 98. *P*

FitzSimons, N. (1991). Bensel, John A. *A biographical dictionary of American civil engineers* 2: 10–11. ASCE: New York. *P*

Floyd, O.N. (1940). Modern construction methods on earth dams. *Civil Engineering* 10(8): 487–490; 10(9): 586–589. *P*

Foster, H.A. (1939). Standing waves in spillway chutes. *Civil Engineering* 9(8): 499. (*P*)

Gorman, A.E. (1933). Chicago's waterworks: A century of accomplishment. *Engineering News-Record* 110(23): 733–737. *P* (relating to DeWitt C. Cregier)

Gowen, C.S. (1900). The foundations of the New Croton Dam. *Trans. ASCE* 43: 469–565. (*P*)

Griggs, F.E., Jr., ed. (1991). Ockerson, John A. *A biographical dictionary of American civil engineers* 2: 83–84. ASCE: New York. *P*

Gumensky, D.B. (1935). Principles of siphon design for Colorado River Aqueduct. *Engineering News-Record* 114(Jun. 27): 899–903. *P*

Hatton, T.C. (1922). Deposition of sludges resulting from sewage disposal plants. *Trans. ASCE* 85: 448–450. *P*

Howson, G.W. (1917). Discussion of Multiple-arch dams on Rush Creek, California. *Trans. ASCE* 81: 900–901. (*P*)

Hoyt, J.C., Anderson, R.H. (1904). Flood on March 1904. *Engineering News* 51(17): 393–394. (*P*)

Hunt, C.W. (1897). Alfred Wingate Craven. *Historical sketch of the American Society of Civil Engineers*: 23. ASCE: New York. *P*

Johnson, C.F. (1943). Gaging and sampling Louisville's sewage. *Civil Engineering* 13(2): 97–100. *P*

Kolupaila, S. (1960). Early history of hydrometry in the United States. *Journal of the Hydraulics Division* ASCE 86(HY1): 1–51. *P* (relating to Zachariah Allen)

Kolupaila, S. (1960). Murray Blanchard: Early history of hydrometry in the United States. *Journal of the Hydraulics Division* ASCE 86(HY1): 32; 86(HY6): 117–119; 86(7): 33. *P*

Hydraulicians
included in volumes 1 and 2 with
a relation to the USA

VOLUME 1

Surname	First name	Country	Years	Page
Bakhmeteff	Boris A.	Russia	1880–1951	492
Bateman	Harry	UK	1882–1946	640
Bogardi	Janos	Hungary	1909–1998	365
Burgers	Johannes Martinus	The Netherlands	1895–1981	440
Burns	Robert V.	UK	1898–1967	650
Busemann	Adolf	Germany	1901–1986	217
Chapman	Sydney	UK	1888–1970	652
Craya	Antoine	France	1911–1996	120
Einstein	Hans Albert	Switzerland	1904–1973	580
Goldstein	Sydney	UK	1903–1989	674
Hilgard	Karl Emil	Switzerland	1858–1938	592
Hovgaard	George William	Denmark	1857–1950	78
Jacob	Max	Germany	1879–1955	258
Juhasz de	Kalman John	Hungary	1893–1972	367
Karman von	Theodor	Hungary	1881–1963	261
Kolupaila	Stepanos	Lithuania	1892–1964	438
Le Méhauté	Bernard	France	1927–1997	159
Lysne	Dagfinn	Norway	1934–2000	459
Milne-Thomson	Louis-Melville	UK	1891–1974	701
Mises von	Richard	Germany	1883–1953	285
Nemenyi	Paul Felix	Hungary	1895–1952	290
Noetzli	Fred	Switzerland	1887–1933	606
Shields	Albert	USA	1908–1974	322
Spannhake	Wilhelm	Germany	1881–1959	325
Tietjens	Oscar	Germany	1893–1971	335
Tollmien	Walter	Germany	1900–1968	338
Wiest de	Roger	Belgium	1925–1998	60
Zowski	Stanislaw Jan	Poland	1880–1940	467

VOLUME 2

Surname	First name	Country	Years	Page
Baumann	Paul	Switzerland	1892–1982	1499
Biot	Maurice	Belgium	1905–1985	839
Bjerknes	Jacob Aall Bonnevie	Norway	1897–1975	1326
Casagrande	Arthur	Austria	1902–1981	776
Chanute	Octave	France	1832–1910	920
Ericsson	Johan (John)	Sweden	1803–1889	1480

Escoffier	Francis	France	1908–1995	943
Eskinazi	Salamon	Turkey	1922–2003	1539
Feodoroff	Nicholas Vasilievich	Russia	1901–2003	1383
Ferri	Antonio	Italy	1912–1975	1267
Flügge-Lotz	Irmgard	Germany	1903–1974	1059
Fokker	Anthony	The Netherlands	1890–1939	1303
Francis	James Bicheno	UK	1815–1892	1586
Frenkiel	Francois Naftali	Poland	1910–1986	1337
Friedrichs	Kurt	Germany	1901–1982	1062
Fteley	Adolphe	France	1837–1903	951
Guderley	Gottfried	Germany	1910–1997	1075
Gumbel	Emil Julius	Germany	1891–1966	1077
Hermann	Rudolph	Germany	1904–1991	1088
Hoerner	Sighard	Germany	1906–1971	1093
Ippen	Arthur Thomas	UK	1907–1974	1596
Keulegan	Garbis Hvannes	Turkey	1890–1989	1397
Kirkwood	James Pugh	UK	1807–1877	1601
Kramer	Hans	Germany	1894–1957	1111
Kuichling	Emil	Germany	1848–1914	1118
Laufer	Jonas (John)	Hungary	1921–1983	1233
Martinelli	Raymond	Italy	1914–1949	1276
Matthes	Gerard Hendrik	The Netherlands	1874–1959	1309
McNown	John Stephenson	USA	1916–1998	1486
Meyer	Adolf	Switzerland	1894–1978	1524
Moskvitinov	Ivan Iosifovich	Russia	1885–1963	1420
Munk	Max	Germany	1890–1986	1136
Muskat	Morris	Latvia	1906–1998	832
Nelidov	Ivan Manuilovich	Russia	1894–1968	1423
O'Shaughnessy	Michael Maurice	Ireland	1864–1934	1251
Pohlhausen	Karl	Germany	1890–1980	1157
Rand	Walter	Estonia	1914–1994	834
Reissner	Hans	Germany	1874–1967	1165
Robinson	Abraham	Poland	1918–1974	1171
Roebling	Johann (John)	Germany	1806–1869	1172
Rott	Miklos (Nicholas)	Hungary	1917–2006	1237
Schweitzer	Pal (Paul)	Hungary	1893–1980	1239
Spitzglass	Jacob M.	Russia	1869–1933	1445
Terzaghi	Karl	Czech Republic	1883–1963	876
Toebes	Gerrit H.	The Netherlands	1927–1981	1319
Vigander	Svein	Norway	1934–1989	1331
Vogel	Herbert	USA	1900–1974	1210
Wagner	Herbert	Austria	1900–1982	826
Wegmann	Edward	Brazil	1850–1935	1534
Weinstein	Alexander	Russia	1897–1979	1215
Westergaard	Harold Malcolm	Denmark	1888–1950	887
Wislicenus	Georg Friedrich	France	1903–1988	1220
Yevjevich	Vujica	Yugoslavia	1913–2006	1661

USA

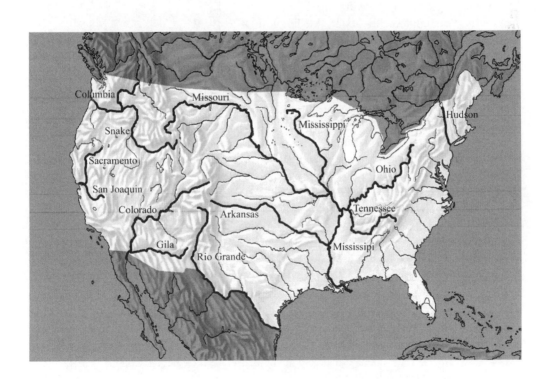

ABBOT

13.08. 1831 Beverly MA/USA
01.10. 1927 Cambridge MA/USA

Henry Larcom Abbot entered military service in 1850, became engineering lieutenant in 1854, joined the North States during the Civil War as Chief Topographical Engineer and finally was a Major General. He was a Member of the Commission for the Mississippi Regulation from 1874, and from 1879 military engineer in the Board of Engineers for Fortification, River and Harbour Improvements. Abbot was a professor of hydraulic engineering at the George Washington University, Washington DC, from 1906.

Abbot's first service from 1857-1861 was assistant to General Humphreys (1810-1883) in an examination of the Mississippi levees. The impressive report on this river included velocity data collected with double-floats, which however were questioned later, as were his equations for river flow. He was later engaged in experimentally establishing a submarine defence for the United States, proceeding to Europe to make contracts for torpedo cables, and observing systems of submarine defence adopted by European States. He was also in charge of devising a permanent plan for reclaiming the alluvial basin of Mississippi River in 1877. From 1885 he reported at what ports fortifications were required. After retirement from active service in 1895, Abbot was a consulting engineer preparing plans for the interior harbour of Manitowoc. From 1897 to 1900 he was a member of the Panama Canal Commission. He was a member of the National Academy of Sciences, or the National Geographic Society. In 1896, he was conferred the degree of honorary doctor by Harvard University. He also published a book on the Panama Canal, where he was greatly involved in the general design features.

Abbot, C.G. (1930). Henry Larcom Abbot. *Biographical Memoirs* 13: 1-101. National Academy of Sciences: Washington DC. *P*

Abbot, H.L. (1902). The Panama Canal, and the regulation of the Chagres River. *Engineering Magazine* 24(12): 329-368.

Anonymous (1905). Henry L. Abbot. *Engineering News* 54(11): 260. *P*

Humphreys, A.A., Abbot, H.L. (1867). Report on the physics and hydraulics of the Mississippi River, upon the protection of the alluvial region against overflow, and upon the deepening of the mouths, based upon surveys and investigations. *Professional Papers* 13. US Army Corps of Engineers: Washington DC.

Poggendorff, J.C. (1898). Abbot, Henry Larcom. *Biographisch-Literarisches Handwörterbuch* 3: 3. Verlag Chemie: Leipzig, with bibliography.

ABERT

22.07. 1828 Philadelphia PA/USA
11.08. 1903 Washington DC/USA

Sylvanus Thayer Abert graduated from Princeton College in 1848, commencing engineering upon the extension of James River and the Kanawha Canal. This canal was designed to connect tidewater at Richmond VA with the Ohio River, but was never completed. Abert was called to Washington in 1849 to make a survey of the Potomac; in 1850 he surveyed the canal entering the East Branch near the Arsenal, recommending a connection with the Ohio and Chesapeake Canal then extending to Washington.

IS A SHIP CANAL PRACTICABLE?

NOTES,

HISTORICAL AND STATISTICAL,

UPON THE PROJECTED ROUTES FOR AN

INTEROCEANIC SHIP CANAL BETWEEN THE ATLANTIC AND PACIFIC OCEANS,

IN WHICH IS INCLUDED

A SHORT ACCOUNT OF THE CHARACTER AND INFLUENCE OF THE CANAL OF SUEZ, AND THE PROBABLE EFFECTS UPON THE COMMERCE OF THE WORLD OF THE TWO CANALS, REGARDED EITHER AS RIVALS, OR AS PARTS OF ONE SYSTEM OF INTEROCEANIC NAVIGATION.

BY

S T. ABERT, C.E.

ILLUSTRATED WITH MAPS.

CINCINNATI:
R. W. CARROLL & CO., PUBLISHERS,
117 West Fourth Street,
1872.

Abert's interesting report on the survey of New River Inlet some 60 km northeast of Wilmington NC in 1856 attracted the US Coast Survey. The mouth formation of New River built with the coast line by storms is reverse of the Mississippi Delta, which extends its channels to the Gulf of Mexico like fingers. Formed by silt of the Mississippi it diverges seaward, whereas the New River Delta formed by sand diverges landward. In the latter river, storms dominate the effects of ebbs. To improve this situation Abert proposed to deepen and widen the channel thereby fixing the channel banks. The report was published in 1856. In 1858 Abert was ordered to the Warrenton Navy Yard FL in charge of works and the ground within the Naval Reservation. He built a foundry, constructed 60 m of sea wall in 10 m deep water by using a diving bell, and prepared plans for a ship basin and graving dock. Its foundation presented special difficulties and it soon became evident that the dock would have to be built in water. After the Civil War, in which he actively participated, he went in 1865 and 1866 to Columbia to improve its infrastructure, namely railways, roads, canals, and water supply. In 1869 he was concerned with the question of the connection of the Great Lakes with the Mississippi, and he conducted a survey of the Arkansas River from Fort Gibson to the mouth of Neosho River to the Mississippi. The distance from Fort Gibson to Little Rock is more than 400 km, with a fall of 70 m. Discharge measurements were taken according to Andrew A. Humphreys (1810-1883) and Henry L. Abbot (1831-1927). Given the high river variability in terms of shape and slope it was recommended to first fix the banks. Abert may thus be considered an early river engineer who added to the practical questions of his country.

Abert, S.T. (1872). *Is a ship canal practicable*? Carroll: Cincinnati. (*P*)
Abert, S.T. (1876). *Report* on a survey of a line to connect the waters of Norfolk Harbor in
        Virginia with the Cape Fear River at or near Wilmington NC. US 44th Congress: 2-38.
Anonymous (1907). Abert, S.T. *Trans. ASCE* 59: 521-530.

ACKERMAN A.J.

21.04. 1901 New Ulm MN/USA
13.03. 1991 Madison WI/USA

Adolph John Ackerman graduated as a civil engineer from the University of Minnesota in 1926. He then joined Fargo Engineering Co. and Commonwealth Power Corp. as assistant engineer on Alcona Dam MI. From 1931 to 1933, he was a chief engineer for Madden Dam, Panama Canal Zone. Upon return to the USA, Ackerman was with the Tennessee Valley Authority TVA in charge of the construction plant facilities and the hydraulic equipment until 1937, when joining *Dravo* Corp. from 1938 to 1945 at Pittsburgh PA, acting as director of engineering for dam construction, tunnels and ship-ways. In parallel he was also a special consultant to the Navy and War Departments from 1943, in charge for harbor design during the Normandy invasion. From 1945 to 1951 Ackerman was in Sao Paolo, Brazil, to investigate inland waterway transportation, a period where he came into contact with Asa White Kenney Billings (1876-1949), the famous hydraulic engineer of Brazil. He there joined also the Brazilian Traction, Light & Power Co. contributing to the hydro-electric power development of the country. From 1952, Ackerman was a private consultant, with projects in India, Peru, and other countries of South America.

Ackerman had chiefly experience on the design and construction of hydro-electric facilities, with aggregate installed capacity of 5.3 Mio HP, or an overall investment of 900 Mio US$. He was the inventor of trust-wheel type cableway towers used for numerous dams, and an automatic spillway gate system used on Calderwood Dam and Bonneville Dam. He was a visiting professor of engineering practice, the University of Illinois, and the author of several books, one of which is a biography on Billings.

Ackerman, A.J. (1931). Gate handling at Calderwood Dam. *Engineering News-Record* 106(19): 754-757.

Ackerman, A.J. (1932). Concreting plant for Madden Dam. *Engineering News-Record* 109(23): 671-675.

Ackerman, A.J., Locher, C. (1940). *Construction, planning and plant*. McGraw-Hill: New York.

Ackerman, A.J. (1953). *Billings and water power in Brazil*. ASCE: New York.

Ackerman, A.J. (1955). Hydro – The answer to Brazil's power needs. *Civil Engineering* 25(5): 272-277.

Anonymous (1964). Ackerman, Adolph J. *Who's who in engineering* 9: 7. Lewis: New York.

http://www.engr.wisc.edu/eday/eday1954.html *P*

ACKERMANN W.C.

07.10. 1913 Sheboygan WI/USA
09.06. 1988 Champaign IL/USA

William (Bill) Carl Ackermann became acquainted with the fundamentals of hydrology at University of Wisconsin, graduating in 1935, and while staying at Tennessee Valley Authority TVA during the next 20 years. He became in 1956 chief of the Illinois State Water Survey, remaining there for another 23 years. His experience as a practicing hydrologist provided him with the sound foundation needed to become a leader in hydrology. He was president of the American Geophysical Union AGU, and of the International Association of Scientific Hydrology IASH. He was also selected by President Truman to the Water Resources Policy Commission and later served as President Kennedy's Water Resources Advisor. He was the recipient of the Robert Horton Medal in 1980.

It was neither research nor scientific papers that brought Ackermann wide recognition. It was rather a combination of three factors, namely his great knowledge, good judgment, and extensive service to technical societies. By 1960, Ackermann had become known as an authority on water issues and a person with common sense and a balanced view. This skill, coupled with service on many committees of AGU and the American Society of Civil Engineers ASCE, brought him great recognition. As head of the Illinois State Water Survey, he saw its role as serving both the state and the nation, with widely diverse research efforts and a state-focused services program. High-quality reports were for him the major product, touching questions on atmospheric chemistry, hydrology and the Midwest weather and climate. He had a strong believe in the research marriage of hydrology and meteorology. The three most challenging achievements related to his interest were the diversion of Lake Michigan waters at Chicago, the development of futuristic water plans for Illinois, and the study of inadvertent weather modification. After retirement in 1989, Ackermann taught hydrology courses at the University of Illinois, Champaign IL, and also served as consultant to state and federal agencies.

Ackermann, W.C. (1973). Sediment trap efficiency of small reservoirs. *Report* TVA/WR/AB-91-14. Tennessee Valley Authority, Knoxville TN.
Ackermann, W.C., White, G.F., Worthington, G.B. (1973). Man-made lakes: Their problems and the environmental effects. *Monographs Series* 17. AGU: Washington DC.
Changnon, S.A. (1989). William C. Ackermann. *Eos* 70(5): 65. *P*
Linsley, R.K., Ackermann, W.C. (1942). Runoff from rainfall. *Trans. ASCE* 107: 825-846.
http://www.agu.org/inside/pastpres_bios_1961-1980.html#ACKERMANN *P*

ADAMS

18.10. 1812 Boston MA/USA
13.12. 1899 Brooklyn NY/USA

Julius Walker Adams was educated in New England academies. At age 18 he was appointed to the US Military Academy, West Point, and there remained 2 years, after which he initiated his career with railways projects, the earliest in the country, then operated with horse power. From 1835, Adams was engaged with engineering works, first as chief engineer for railways companies, then from 1844 as resident engineer for a dock of the US Navy and shortly later for the Boston water works. From 1855 to 1860 he was in charge of the construction of the Brooklyn drainage system, and from 1861 to 1866 chief engineer for the Hudson river bridge at Albany NY. From 1869 to 1878, Adams was chief engineer of the Board of City Works, Brooklyn, remaining in this position until 1889 in the Board of Public Works. He was also involved in Panama Railway Co.

During his long professional career, Adams contributed also to waterworks and sewerage projects. By applying hydraulic principles, including the theory of concentration time that had been proposed by Neville, he reduced the size of sewers and their cost. These installations were so successful that no modifications were needed for decades. He was also actively involved as Colonel in the Civil War. Adams was a founding member of the American Society of Civil Engineers ASCE in 1852; he was elected its president in 1872 and awarded honorary membership in 1888. He owned a large personal library and was a frequent contributor of short articles and editorials to the technical press. In 1852 he was editor of Appleton's Mechanics Magazine and Engineers' Journal, and in the 1880s Editor of Engineering News, then one of the prime journals in the USA. He had a retentive memory, and though he had no diary, he accurately recalled incidents and dates of occurrences, and entered into minute details of engineering works.

Adams, J.W. (1841). *Report of the engineer to the directors of the Lockport and Niagara Falls Railroad Company*, on the proposed extension of their road eastward. Dickson: Boston.
Adams, J.W. (1880). *Sewers and drains for populous districts*, with rules and formulae for the determination of their dimensions under all circumstances. van Nostrand: New York.
Adams, J.W. (1883). Flow of water through pipes. *American Contract Journal* 10(Mar.10): 117.
Adams, J.W. (1899). Stave pipe: Its economic design. *Trans. ASCE* 41: 27-84.
Anonymous (1899). Julius Walker Adams. *Engineering News* 42(Dec.21): 403. *P*
Anonymous (1936). Early presidents of the society. *Civil Engineering* 6(9): 605-606. *P*
http://www.findagrave.com/cgi-bin/fg.cgi?page=gr&GRid=6041406 *P*

AHERN

05.07. 1860 Berkeley CA/USA
19.04. 1944 Dixon CA/USA

Jeremiah Ahern graduated in 1883 from University of California, Berkeley CA. He was engaged by the US Geological Survey as assistant topographer, and from 1889 as topographer, in charge of the Powell Irrigation Survey in Montana. In 1892 he became associated with an engineering company in San Francisco CA, acting as assistant engineer on the South Gila Canal Project in Arizona. He returned in 1895 to the USGS in charge of surveys in Indian Territory. From 1901 he was hydrographer and in charge of hydraulic studies on the Gila River, and Colorado River between Needles CA and Yuma AZ, in collaboration with Joseph B. Lippincott (1864-1942), and William B. Clapp (1861-1911). The story of these studies was published in 1904. One of the six large proposed irrigation projects was Uncompahgre Valley in Colorado, but the water from its river was insufficient to irrigate 700 km^2 of irrigable land, so that it was proposed to supply water by diversion of 35 m^3/s from Gunnison River. The necessary surveys were in 1901 completed by Ahern, including the Grand Canyon of the Gunnison. That portion of the river was so spectacular that it attracted nationwide attention.

The newly established Reclamation Service, under its chief Frederick H. Newell (1862-1932), attracted many engineers from the US Geological Survey, among them also Ahern. He was there first engaged in supervising the DeSmet Project in Wyoming, and then placed in charge of the Shoshone Project, also in Wyoming, including a large storage reservoir for irrigation purposes. The Shoshone Dam was the first high concrete dam of the Reclamation Service, with Hiram N. Savage (1861-1934) responsible for the design. The arch dam is 95 m high, has a 60 m long crest, which has 3 m top width and 33 m base width. The spillway had a discharge capacity of 850 m^3/s. Ahern continued on this project until completion in 1908, resigning then from government service and returning to California, where he had his private practice for the rest of his career. He there was engaged with land surveying and acted as consultant on reclamation and irrigation problems. In 1941 he was seriously injured in a car accident, and passed away three years later due to heart trouble. He was member ASCE since 1904.

Ahern, J. (1904). First Report of the US Reclamation Service. Proc. 1st Conference *Engineers of the Reclamation Service*. Washington DC.
Anonymous (1945). Jeremiah Ahern. *Trans. ASCE* 110: 1638-1641.
http://thoth.library.utah.edu:1701/primo_library/libweb/action/dlDisplay.do *P*

ALBERTSON

30.08. 1918 Hays KS/USA
12.01. 2009 Fort Collins CO/USA

Maurice 'Maurie' Lee Albertson obtained his civil engineering education at State University of Iowa, Iowa City IA, and there submitted his PhD thesis in 1948. He then was an associate professor and later professor of civil engineering, and Head of Fluid Mechanics Research at Colorado State University CSU, Fort Collins CO. From 1857 until retirement, Albertson was in addition Campus Coordinator of the SEATO Graduate School of Engineering. He received the J.C. Stevens Award in 1948, and the Emil Hilgard Award in 1951 from the American Society of Civil Engineers ASCE. He also delivered the 1987 Hunter Rouse Hydraulic Engineering Lecture titled New challenges in water resources engineering. He received honorary membership from ASCE in 2002 'for his global leadership spanning six decades in water resources engineering, international development, and higher education'.

Albertson's professional career included fluid mechanics, water resources systems engineering, hydraulic engineering, water resources research, hydropower, renewable energy resources and international engineering education. He has written numerous papers on alluvial rivers and later on water resources with an outlook into renewable energies. Albertson earned major credit for building CSU's research, graduate and international programs. Of even greater importance were his engineering contributions to international development studies. He played a key role in the establishment of the Peace Corps, and designed and established the SEATO Graduate School of Engineering, now the Asian Institute of Technology, Bangkok Thailand.

Albertson, M.L., Tucker, S., Taylor, D.C. (1970). *Treatise on urban water systems*. Colorado State University: Fort Collins.
Anonymous (1948). Maurice L. Albertson. *Civil Engineering* 18(1): 45. *P*
Anonymous (1958). 1959 Hydraulics Convention. *Civil Engineering* 28(10): 790. *P*
Anonymous (1964). Albertson, Maurice L. *Who's who in engineering* 9: 17. Lewis: New York.
Anonymous (2002). M.L. Albertson. *ASCE News* (12): 14-15. *P*
Rahim Kia, A., Albertson, M.L., eds. (1987). *Design of hydraulic structures*. Colorado State University: Fort Collins CO.
Sayre, W.W., Albertson, M.L. (1959). *The effect of roughness spacing in rigid open channels*. Dept. Civil Engng., Colorado State University: Fort Collins CO.
http://www.today.colostate.edu/story.aspx?id=254 *P*

ALBRIGHT

18.01. 1848 Buchanan VA/USA
20.08. 1931 Buffalo NY/USA

John Joseph Albright graduated from Rensselaer Polytechnic Institute, Troy NY, with the degree in mechanical engineering in 1868. In 1883 he moved to Buffalo after having been active in coal business. In 1896 he became interested in water power on the Madison River, and on the Hudson River in New York State, so that he became associated in 1901 with the Ontario Power Company, on Niagara Falls ON. In 1904 Albright and his associates negotiated on a contract between the Ontario Power Company and the Niagara Power Company for the purchase of Ontario power for distributing electrical power throughout New York State. In 1905 Albright and his associates took a substantial interest in the Niagara, Lockport and Ontario Power Company, to enlarge their scope in hydro-electricity. The charter granted the Company unlimited water supply from Niagara River. A canal would be constructed from the river to a point near Lockport NY to discharge the water in the gorge of Eighteen-Mile Creek, at which point the power plant would be built.

However, after these works were initiated, the Congress passed the Burton Bill to limit the diversion of water to the power plant. In the meantime the power plant on the Canadian side of Niagara Falls was completed. The Burton Act prevented the canal construction, limiting the power production to 60,000 HP instead of the three times larger design value. This legislation struck a staggering blow to this ambitious enterprise. By 1906 the Company's construction had grown to such an extent that it was delivering electric power as far east as Syracuse NY, and in 1913, Buffalo became the controlling factor in the Company when Albright's interests bought out Westinghouse Company's block of stock. The Ontario Company was by then the largest single generating plant. In 1917 the Canadian Government, eager to control its water power resources, negotiated with Albright for the purchase of the Ontario Power Company, but Albright retained his financial interests although he soon retired from active business. His son then became vice-president of the Niagara, Lockport, and Ontario Power Company. One of his father's specialties was the rescue of apparently foundering companies, which he would reorganize until their financial status was beyond reproach.

Anonymous (1931). *John Joseph Albright*. Buffalo Fine Arts Academy: Buffalo NY.
Anonymous (1933). John J. Albright. *Trans. ASCE* 98: 1694-1697.
http://www.albrightknox.org/join-support/planned-giving/the-john-j-albright-society/ *P*

ALDEN

22.04. 1843 Templeton MA/USA
13.09. 1926 Princeton MA/USA

George Ira Alden received his BS diploma from Harvard University in 1868. He was then associated with the Department of Mechanical Engineering of Worcester Polytechnic Institute WPI, Worcester MA until 1896, finally as its head. He further was a consulting engineer for the Norton Wheel Co., and the Plunger Elevator Co. He also chaired the Board of Directors, Weston Co. He was member of both the Worcester School Committee, and the American Society of Mechanical Engineers ASME, serving as vice-president in the term 1891 to 1893.

The Alden Hydraulic Laboratory was founded by Alden in 1894. The original lab, which was originally occupied by a mill, included a 36 in. Venturi meter for discharge measurement. The purpose of this Laboratory was to conduct hydraulic experiments mainly for commercial testing. The laboratory was enlarged in 1910 and in 1926, with a fundamental support from Alden. The main building of this early hydraulic laboratory was 33 m long, 15 m wide and 5 m high. There was also a low-head laboratory. The main projects in 1926 concerned pipe friction laws, measurement of discharge by Pitot tubes and with the salt-velocity method, invented by Charles M. Allen (1871-1950), who was a close associate of Alden and who conducted his first tests in the laboratory. Further work related to differential surge tanks as invented by Raymond D. Johnson (1874-1949), who also was involved in the laboratory. In addition, velocity meters were frequently tested and calibrated in special flumes using the towing method. These meters were then used to determine the discharge at a particular river section. The Alden Hydraulic Laboratory continues its activities as the largest supplier of the National Institute of Standards and Technology for traceable flow meter calibrations.

Alden, G.I. (1877). *Notes on Rankine's Applied mechanics*. CLB: Hartford.

Alden, G.I. (1915). *A comprehensive analogy for the transmission of energy by electricity*. Commonwealth Press: Worcester MA.

Anonymous (1921). Alden, Prof. George Ira. *American men of science* 3: 7. Science Press: New York.

Anonymous (1926). The Alden Hydraulic Laboratory enlarged. *Mechanical Engineering* ASME 48(6): 634-635. *P*

Anonymous (1980). Alden, George I. *Mechanical engineers in America born prior to* 1861: 34. ASME: New York.

Nutt, C. (1919). George I. Alden. *History of Worcester and its people*: 459-460. New York. *P*

ALEXANDER

15.12. 1896 Cuba NY/USA
20.03. 1972 Los Angeles CA/USA

Louis Jessup Alexander graduated in 1922 from the University of Illinois, Urbana-Champaign IL, after having served with the US Navy in World War I. He was then engaged by Joseph B. Lippincott (1864-1942) in California; Alexander began in 1928 his almost forty years-career with Southern California Water Company, where he rose to the position of vice-president and chief engineer. He there paid a prominent part in diversified engineering solutions, which were required for new ways of developing water, designing new facilities, and improving the water treatment and water quality control. He and his group were the first to use a granular-activated carbon filter on the Pacific Coast. Novel design criteria of distribution systems had to be worked out to meet modern requirements. Alexander played a leading role, and in recognition of his efforts received the highest awards of the American Water Works Association AWWA, of which he was member and president in 1955. He was also the recipient of its Fuller Award and a Life Membership. He was in addition member of the American Society of Civil Engineers from 1945, becoming Fellow in 1962. He was president of its Los Angeles Section.

Alexander's greatest contribution to the profession was his work on groundwater control and basin replenishment. He was a leader in the formation of both the West Basin and Central Basin Municipal Water Districts. He was also instrumental in drafting the Replenishment District Act, then creating the Central and West Basin Replenishment District of Los Angeles County. He was from 1963 the representative of the West Basin Municipal Water District on the Metropolitan Water District of the Southern California Board of Directors, serving on three District Board Committees, namely the executive, the engineering and operations, as also the organizational and personnel. His papers are included in the Water resources collection published in 1990. The Louis J. Alexander Award of the Water Replenishment District was installed by Alexander's widow.

Anonymous (1923). Louis J. Alexander. *The 1923 Illio*: 49. University of Illinois: Urbana IL. *P*
Anonymous (1973). Louis J. Alexander. *Trans. ASCE* 138: 615-616.
Bowen, E.R., Alexander, L.J. (1951). *Report* of consulting engineer to the Palm Springs Water
       Company on its water supply and additions to plant. Los Angeles.
Brown, E.G., Banks, H.O. (1959). Investigations of alternative aqueduct systems to serve
       Southern California. *Bulletin* 78. Department of Water Resources: Sacramento.
Kerr, W.H., Alexander, L.J., Baldwin, C.G. (1990). *Water resources collection* 1884-1989.

ALLAN W.

22.05. 1903 Long Island City, Queens NY/USA
28.12. 1989 New York NY/USA

William Allan received education from Polytechnic Institute, Brooklyn, from where he graduated both as BS and MS in civil engineering. He was then at the City College (today's City University CUNY) in New York from 1933, was appointed professor of civil engineering and took over from 1947 to 1970 as Dean of its Technology School. He in addition served as engineering consultant on many hydraulic projects and authored works in fluid mechanics. He was a member of the American Society of Civil Engineers ASCE and winner of its J.C. Stevens Award in 1950 for a discussion on the Panama Canal paper. He was also a co-winner of the 1948 Normal Medal.

The 1946 paper co-authored by Boris A. Bakhmeteff (1880-1951) gives an account of inner processes by which the energy of fluid flow is dissipated by friction. The traditional treatment of this fundamental question was considered by George G. Stokes (1819-1903) and Horace Lamb (1849-1934) yet without elucidating the physical aspects of the problem. It was particularly impossible to reveal the fundamental fact that the loss of energy and its final dissipation into heat do not coincide spatially. The paper attempts to describe the governing processes based on an improved understanding of the mechanisms of turbulent flow. It reveals the consecutive phases of energy dissipation, explaining their significance and the nature of losses involved in the various stages. The processes allow for a better understanding of the fundamental mechanisms of turbulent flow. Once having completed the PhD, Allan moved away from science, serving then in educational and organizational issues of his university. He also made eventually applied research, for example that published in 1949 relating to the Panama Canal. Although it had appeared obvious from the French disaster in the late 19th century that a sea-level project was hardly feasible, this aspect received renewed interest, but so far was never considered in more detail.

Allan, W. (1949). Discussion of Panama Canal: The sea-level project, a symposium. *Trans. ASCE* 114: 841-845.
Allan, W. (1950). Discussion of Aerodynamic theory of bridge oscillations. *Trans. ASCE* 115: 1232
Anonymous (1950). William Allan, J.C. Stevens Award winner. *Civil Engineering* 20(10): 676. *P*
Anonymous (1975). Allan, William. *Who's who in America* 38: 40.
Bakhmeteff, B.A., Allan, W. (1946). The mechanism of energy loss in fluid friction. *Trans. ASCE* 111: 1043-1102.

ALLEN C.M.

12.12. 1871 Walpole MA/USA
15.08. 1950 Holden MA/USA

Charles Metcalf Allen graduated from Worcester Polytechnic Institute in 1894, there received in 1899 the MS degree, and in 1929 the D.Eng. Hon. degree. He was instructor there in mechanical engineering from 1894 to 1902, and from 1906 to 1945 professor of hydraulic engineering. In parallel, he was closely associated with the Alden Hydraulic Laboratory established by George Ira Alden (1843-1926), where he conducted research in power and discharge measurement, and in hydraulic model tests. Being awarded, among other, the 1936 Warner Medal of the American Society of Mechanical Engineers ASME, he received ASME Honorary Membership in 1944 in recognition both of his loyalty to the Society and achievements as an engineer. He was also decorated with the 1949 ASCE John Fritz Medal for 'his exceptional achievement in hydraulic engineering and founder of a notable hydraulic laboratory'.

Allen was an outstanding hydraulic engineer who contributed widely to the profession. He invented the steam engine indicator to rowing, developed current meter rating stations, improved water wheel flow recorders, invented an apparatus for testing the efficiency of gears, and introduced the salt-velocity method for discharge measurement. The latter method was widely used in the first half of the 20th century to determine accurately water discharges where standard instrumentation was not adequate, such as in pipes or in small rivers. Allen published papers notably on water wheel and draft tube testing, on turbine flow recorders, and on the salt-velocity method. After retirement, he continued work at the Alden Hydraulic Laboratory as research director until his death.

Allen, C.M. (1910). The testing of water wheels after installation. *Trans. ASME* 32: 275-309.

Allen, C.M., Winter, I.A. (1923). Comparative tests on experimental draft-tubes. *Proc. ASCE* 49(9): 1813-1845.

Allen, C.M., Taylor, E.A. (1923). The salt velocity method of water measurement. *Trans. ASME* 45: 285-341.

Allen, C.M., Taylor, E.A. (1925). Salt velocity method of water measurement. *Canadian Engineer* 49(5): 195-198.

Allen, C.M. (1930). The salt-velocity method of water measurement. *Mechanical Engineering* 52(4): 375-376. *P*

Anonymous (1950). Charles M. Allen dies: Authority in hydraulics. *Engineering-News Record* 145(Aug.24): 24. *P*

ALLEN H.

10.05. 1802 Schenectady NY/USA
01.01. 1890 Montclair NJ/USA

Horatio Allen entered Columbia College in 1821, graduating in 1823. He first studied law but then decided to enter upon civil engineering and joined the Delaware and Hudson Canal Company. He was appointed in 1824 resident engineer at the canal summit level. Once the first locomotives appeared in the USA, after they were successfully introduced by the British Stephenson in 1825, Allen crossed the Ocean to see the railroad system. In 1829 he was appointed chief engineer of the South Carolina Railroad, and in 1837 chief engineer of the *Croton* Aqueduct. He became in 1844 a member of the firm Stillman, Allen & Co. His last official place was as consulting engineer of the Brooklyn Bridge, New York. He retired from active life in 1870 and was one of the eldest civil engineers when passing away at age 88.

Allen's most noteworthy feature of his life was his participation in the introduction of the locomotive in the USA. He was attracted by Stephenson's experiments and recognized that the steam locomotive had a great future in his country. He bought in Europe two engines and on their arrival at Honesdale PA, Allen made tests himself. In later years, Allen devoted much energies into hydraulic engineering, and he was also associated with the supply of heavy engines to the American steamers before the time of Cunard. In the early 1840s, Allen was instrumental for the Croton Aqueduct supplying drinking water to New York City. Later, he was appointed a Member of the Croton Aqueduct Commission, and he recommended crossing the Harlem River by tunnel rather than by bridge. Some 50 years later, the New Croton Aqueduct was taken into service, including a tunnel across the river. Allen took then interest in naval machinery which was built in his company mainly for the Pacific Mail Steamship Company, and a number of vessels and monitors for the United States during the Civil War. Allen was the most prolific inventor from 1840 to 1880. He served as President ASCE from 1871 to 1873, during which term the ASCE Transactions were issued.

Allen, H. (1832). *Report* to the Board of Directors of the South Carolina Canal and Railroad Co.
Allen, H. (1884). Horatio Allen. *Railroad Gazette* 28(April 4): 253-254. *P*
Anonymous (1890). Horatio Allen. *The Engineering and Building Record* 21(5): 66.
Anonymous (1890). The late Horatio Allen. *Industries* 8(Jan.31): 111. *P*
Anonymous (1936). Early presidents of the Society: Horatio Allen. *Civil Engineering* 6(8): 536. *P*
Fitzsimons, N. (1967). Horatio Allen, Hon. M. ASCE. *Civil Engineering* 37(2): 67. *P*

ALLEN H.C.

10.09. 1864 Newark NJ/USA
05.08. 1932 Syracuse NY/USA

Henry Clayton Allen graduated from Syracuse High
School in 1882. He was then employed there on
surveys, accepting the appointment as assistant city
engineer of Syracuse NY. He was engaged as leveler
for the New Croton Aqueduct Commission, New
York NY, promoted to assistant engineer in 1889.
He returned in 1890 to Syracuse as city engineer,
directing the construction of sewers and municipal
work. He was in 1896 employed by the New York
State in making surveys of Oswego Canal, and then
designed in private practice water supplies for cities
of New York State. In 1897 he further undertook
the reconstruction of a section of the old Erie Canal near Rochester NY. From 1900 he
was resident engineer of the Middle Division, the New York State Canals, including
locks, dams, maintenance and canal repair. He was promoted in 1904 Special Deputy
State Engineer responsible for all new Barge Canals of the State. He resigned from this
position in 1907, to be re-appointed city engineer of Syracuse until 1913, and then once
more from 1916 to 1921, when he opened a private practice as consulting engineer.

From 1907 to 1913 Allen served as member of the Syracuse Intercepting Sewer Board.
In 1918 Syracuse University honored Allen by conferring upon him the honorary MSc
degree in engineering. He was a charter member and past-president of the Syracuse
Technology Club; in 1932 he was elected honorary member in recognition of his
outstanding accomplishments as an engineer, and his long and valued services to the
community. Nearly all his active life was devoted to the interests of the City and the
State. His advice on matters outside his own particular field were continually sought,
and he gave freely of his time and efforts in meeting these demands. The fact that many
knew him during his years of service with the city as 'Straight Line Allen' was a tribute
to his uncompromising honesty of policy and to the directness of his methods as an
engineer. He was a student of the history of the ancient inhabitants of America; he loved
nature, and was particularly interested in his collection of local ferns. He also was a
member of the Archeological Society of Syracuse, and since 1905 a member of the
American Society of Civil Engineers ASCE.

Anonymous (1933). Henry C. Allen. *Trans. ASCE* 98: 1497-1498.
Anonymous (1907). Henry C. Allen. *The Post-Standard* Syracuse NY: 2. *P*
Horton, R.E., Grover, N.C., Hoyt, J.C. (1906). Report of progress of stream measurements.
    Water Supply and Irrigation Paper 166. US Geological Survey: Washington DC.

ALLEN J.R.

23.07. 1869 Milwaukee WI/USA
26.10. 1920 Pittsburgh PA/USA

John Robins Allen graduated in 1892 as mechanical engineer from University of Michigan, Ann Arbor MI, then initiating his professional career with a construction company. After only two years, he started training students in engineering, as instructor and finally as professor of mechanical engineering at University of Michigan. He was engaged as dean in the reorganization of the engineering department, Robert College, Constantinople Turkey, from 1911 to 1913. On his return to the USA, Allen became dean of his *alma mater*. In 1918, he was appointed Director of the Bureau of Research of the American Society of Heating and Ventilating Engineers. He passed away unexpectedly at ago of only 51.

Allen was a prolific author on heating subjects including the books Heat engines and Notes on heating and ventilation. The latter was continued after his death by his co-author Walker, to whom joined also James as the third author. This book was a standard text in heating engineering and in 1946 was in its sixth edition. Allen was president of the American Society of Heating and Ventilating Engineers in 1912, after having been Vice-President of the American Society of Mechanical Engineers ASME. He was also a Member of the British Institution of Heating and Ventilating Engineers and thereby served as consultant for a large building project. Allen possessed a strikingly magnetic personality and favored with the ability to discuss intricate scientific problems in language easy to follow. His death came just as the work at the Research Bureau was beginning to produce important results. It was a satisfaction, however, that Allen expressed shortly before his death that the work was in a shape that it can go on for some time without his directing oversight.

Allen, J.R. (1914). *Heat engines*: Steam, gas, steam turbines and their auxiliaries. McGraw-Hill: New York.

Allen, J.R., Walker, J.H. (1922). *Heating and ventilation*. McGraw-Hill: New York.

Allen, J.R., Walker, J.H., James, J.W. (1946). *Heating and air conditioning*. McGraw-Hill: New York.

Anonymous (1912). Prof. John R. Allen. *The Heating and Ventilating Magazine* 9(2): 35. *P*

Anonymous (1920). Death of John R. Allen. *The Heating and Ventilating Magazine* 17(11): 49. *P*

Cattell, J.M., Brimhall, D.R., eds. (1921). John Robins Allen. *American men of science*: A biographical dictionary 3: 10. The Science Press: New York.

ALLEN K.

06.04. 1857 New Bedford MA/USA
07.09. 1930 White Plains NY/USA

Kenneth Allen graduated with the civil engineering degree in 1879 from Rensselaer Polytechnic Institute RPI, Troy NY. He had been employed from 1875 to 1877 at the Boston Water Works, Sudbury River, Framingham MA. After works with railway firms, he became in 1883 assistant engineer, Philadelphia Water Department, Philadelphia PA, and from 1886 was engaged as superintendent of construction in Kansas City MO. After having served as assistant engineer in charge of public works at Yonkers NY, he was from 1895 to 1900 assistant engineer on the sewerage systems of both Williamsbridge NY, and Baltimore MD. Allen founded then with two colleagues a consulting office there, reporting on water supply and sewerage projects. From 1902 to 1906 he was engineer and superintendent of the Water Department, Atlantic City NJ, and from 1908 to 1914 principal assistant engineer of the Baltimore Sewerage Commission, New York NY.

Allen made in New York studies of NY Harbor in connection with the sewage disposal. He inspected in 1913 similar systems in Great Britain, Holland, France and Germany. He was promoted to engineer of Sewers Planning of NYC in 1915, remaining until his death sanitary engineer of the Board of Estimate and Apportionment, New York City. His chapter in the book Sewage sludge was widely read by the profession; the chapters are I American sewage, II Detritus from grit chambers, III Screenings, IV Sludge from plain sedimentation, V Septic tank sludge, and VI Sludge from Emscher tanks. Most of his papers deal with the clarification of sewage, or the management of biological sludge. Allen was member of the American Society of Civil Engineers ASCE and the American Public Health Association.

Allen, K. (1887). *Water supply*, with special reference to work preliminary to its introduction. Rensselaer Polytechnic Institute: Troy NY.

Allen, K. (1905). *The sanitary protection of water supply*. Franklin Institute: Philadelphia.

Allen, K. (1918). Dissolved oxygen as an index of the pollution of New York Harbor. *American Journal of Public Health* 8(11): 838-842.

Anonymous (1930). K. Allen, sanitary engineer, dies. *Engineering News-Record* 105(11): 433. *P*

Elsner, A., Spillner, F.G., Blunk, H.C., Allen, K., Allen, R.S. (1912). *Sewage sludge*. McGraw-Hill: New York.

FitzSimons, N., ed. (1991). Allen, Kenneth. *A biographical dictionary of American civil engineers* 2: 2. ASCE: New York.

ALLEN Z.

15.09. 1795 Providence RI/USA
17.03. 1882 Providence RI/USA

Zachariah Allen graduated from Brown University, Providence RI, in 1813, where he had acquired a love for knowledge. In 1821 he devised a system to heat several rooms of a house from a single stove with a system of heat-conducting pipes. In 1822 he organized and constructed a woollen mill on the banks of Woonasquatucket River, North Providence RI, constructing a series of dams to provide power to the machinery. The Allendale Mill contained innovative fire-safety features including the first use of heavy fire doors, a sprinkler system, and rotary fire pumps. He invented later the first practical automatic cut-off valve for steam engines, patented in 1833, which was 50 years later proclaimed to be one of the great inventions made in steam engineering. In 1852 Allen purchased a mill near Smithfield RI, rebuilt it and increased its water power by raising the dam height of the millpond. He later added steam power and enlarged the mill further.

Allen deserves credit for the first discharge measurement of a large river in the USA. In the summer of 1841 he visited the Niagara Falls deciding to determine the power of that 'marvel of the world'. First, water discharge was measured. Allen stated: 'Very little attention appears to have been hitherto bestowed on the investigation of the comparative volumes of water discharges by the great rivers of the globe'. A reach was selected below the outlet of Lake Erie, where the flow depth was 9 m. Three cross-sections were considered, 200 m apart, and surface floats were run at 10 locations, resulting in a surface velocity of 3.4 m/s. Using the velocity formula of Johann A. Eytelwein (1764-1848), the discharge was estimated. The power of the Niagara Falls was observed that it was more than 40 times than then used in all Great Britain. Later, it was noted that the true average discharge of Niagara River was about 85% larger than the actual measured discharge, and about 50% larger than the maximum.

Allen, Z., Blackwell, E.R. (1844). On the volume of the Niagara River, as deduced from measurements made in 1841. *American Journal of Science and Arts* 46(1): 67-73.
Kolupaila, S. (1960). Early history of hydrometry in the United States. *Journal of the Hydraulics Division* ASCE 86(HY1): 1-51. *P*
Perry, A. (1883). *Memorial of Zachariah Allen* 1795-1882. Wilson & Son: Cambridge MA.
http://en.wikipedia.org/wiki/Zachariah_Allen *P*
http://library.brown.edu/cds/portraits/display.php?idno=182 *P*

ALLISON J.C.

12.07. 1884 San Diego CA/USA
29.05. 1936 La Jolla CA/USA

Joseph Chester Allison was enrolled for the course in engineering at Stanford University, Stanford CA, but ill health forced him to stop. In 1902 he joined the California Development Company, Calexico CA. It promoted the reclamation of the Imperial Valley Desert by diversion of the water of Colorado River near Yuma AZ. Allison became assistant engineer responsible during the break of the River in 1906, threatening the permanent inundation of Imperial Valley. He then reconstructed the canal structures which were previously installed. He was promoted in 1910 to chief engineer, and advanced to assistant general manager of this development. He first improved Sharp's Heading in Mexico, the main point of diversion for the lateral canal system of Imperial Valley. In 1911 the northern portion of the Imperial Valley irrigation enterprise was taken into service, with Allison as supervisor of the canal system. From 1912 to 1915 he solved the problem of flood water of Colorado River along Volcano Lake, and the water famine due to low-water stage in a scoured river bed below the sill of the Alamo Canal Intake at Hanlon AZ. He succeeded in building a diversion weir across the river by the hydraulic process.

In 1916, when the assets of the California Development Company were acquired by the Imperial Irrigation District, Allison became private consultant at Calexico, acting as the consultant for the District and other companies. He and his associates saw possibilities in land development on the Mexican side of the Colorado River Delta, in Baja California. Irrigation was supplied from the Alamo Canal of the Imperial Irrigation District, entering Mexico near the point of diversion. Allison was the guiding spirit of these engineering features, including the financial and commercial aspects. The project was completed successfully. He was later also involved in the development of the Delta Land Company with its own irrigation system. Both large pumping plants on the Alamo Canal as also a large gravity diversion from the Canal were employed for irrigation, an enterprise which also proved successful. He opened in 1924 an office in Los Angeles CA. A man of exceptional ability, combined with the qualities of an engineer, passed suddenly away at age of only 52. He was member ASCE from 1917.

Allison, J.C. (1916). Control of the Colorado River. *Proc. ASCE* 42(5): 681-709.
Anonymous (1939). Joseph C. Allison. *Trans. ASCE* 104: 1875-1878.
Kerig, D.P. (2001). *El valle de Mexicali y la Colorado River Land Company* 1902-1946: 215.
    Mexicali. *P*

ALLISON W.F.

31.03. 1870 Elgin MN/USA
07.07. 1927 Seattle WA/USA

William Franklin Allison received the BS degree in 1897 from Purdue University, West Lafayette IN. After the Spanish-American War he was sent to the Philippines Islands as engineer officer. Returning to the USA in 1899, Major Allison conducted general engineering in South Dakota, taking then graduate work at South Dakota State College, Brookings SD, receiving in 1904 the civil engineering degree from Cornell University, Ithaca NY. He was appointed instructor of civil engineering then at the Colorado School of Mines, Golden CO, and was promoted to professor of civil engineering in 1905. He acted besides as consultant in projects involving water-works, sewers, and irrigation problems, serving in addition as city engineer of Golden from 1908 to 1911.

Allison entered in 1912 a private practice as civil and sanitary engineer at Portland OR. He was appointed professor of hydraulic and sanitary engineering at the University of Oregon, Corvallis OR, in 1913, but already in 1914 accepted the position of professor of municipal and highway engineering at University of Washington, Seattle WA, which position he held until his death. In parallel he served from 1914 to 1917 the Washington State Board of Health. He went to France with the American Expeditionary Forces from 1917 to 1919, acting among others as engineer within the Peace Commission in Paris. After return to Seattle he served as president of the Western Washington Section of the American Society of Civil Engineers ASCE in 1923, and was member of the special committee on irrigation hydraulics from 1922 to 1927. It was also in this field and in sanitary engineering, in which he published various articles. He further was member of the American Public Health Association APHA, and the Society for the Promotion of Engineering Education. Allison was described as a peace-loving, whole-souled man, loved and respected by all who were privileged to know him.

Allison, W.F. (1922). *Sewage disposal for suburban homes and for isolated institutions*. University of Washington: Seattle WA.
Allison, W.F. (1928). Stream pollution in the Pacific Northwest. *Trans. ASCE* 92: 974-983.
Anonymous (1897). William F. Allison. *Debris*: 29. Purdue University: West Lafayette. *P*
Anonymous (1931). William F. Allison. *Trans. ASCE* 95: 1440-1442.
Anonymous (1988). William F. Allison. *Civil engineering* 1910-1919. University Wash: 29. *P*
Curtiss, R.E., Allison, W.F. (1904). *A determination of the effects of clay in sand used for cement mortars*. Joint Thesis in Civil Engineering. Cornell University: Ithaca NY.

AMBROSE

10.11. 1917 Vicksburg MS/USA
20.05. 1962 Knoxville KY/USA

Harry Harwood Ambrose graduated in 1941 from
Case Institute of Technology, Cleveland OH, with
the BSc degree in civil engineering. He completed
in 1943 a course in naval architecture at University
of Michigan, Ann Arbor MI. After war service, he
returned to his Alma Mater, teaching mechanics and
working for the MSc degree, which he received in
1947. He then joined the faculty of the University of
Tennessee, Knoxville TN, as assistant professor of
civil engineering, securing in 1952 a leave of absence
for advanced study at the University of Iowa, Iowa
City IA, receiving a PhD degree in fluid mechanics.
On returning to University of Tennessee as associate professor, he was appointed there
professor of fluid mechanics in 1958, and head of the Department of Civil Engineering.
He passed away at age of only 45. He was ASCE member from 1959.

Ambrose had research interests in hydraulic engineering and fluid mechanics. His PhD
thesis dealt with the transportation of sand in smooth pipes, a research topic until then
hardly attacked. Whereas the sediment transport in rivers with a sand bed had received
attention from the 1930s, the smooth bed as in a pipe changes the mechanisms of sand
transport. This research was initiated by Hunter Rouse (1906-1996), then director of the
Institute of Hydraulic Research, Iowa State University. Ambrose's interest in these
phenomena arose from his stay at the Waterways Experiment Station, Vicksburg MS, in
1941 and 1942, during which time the research in sediment transport was there initiated.
From the mid-1950s, he turned to questions of fluid resistance and its effect on velocity
distribution. The 1955 discussion describes the differences between wake-interference
and isolated-roughness flows, and its effect on the roughness coefficient.

Ambrose, H.H. (1952). The transportation of uniform sand in a smooth pipe. *PhD Thesis*. Iowa
        Institute of Hydraulic Research. University of Iowa: Iowa City.
Ambrose, H.H. (1955). *The effect of character of surface roughness on velocity distributions
        and boundary resistance*. College of Engineering. University of Tennessee: Knoxville.
Ambrose, H.H. (1955). Discussion of A new concept of flow in rough conduits. *Trans. ASCE*
        120: 402-405.
Ambrose, H.H. (1960). *Turbulent flow in pipes with artificial roughness*. Department of Civil
        Engineering. University of Tennessee: Knoxville.
Anonymous (1946). Harry H. Ambrose. *Engineering News-Record* 137(Aug.15): 212. *P*
Anonymous (1963). Harry H. Ambrose. *Trans. ASCE* 128(5): 113-114.

AMBURSEN

06.02. 1876 Frederickstad/N
17.01. 1953 Pittsburgh PA/USA

Nils Frederick Ambursen was born in Norway. He graduated very young as mechanical engineer from Porsgrund Engineering College, southwest from Oslo and then immediately moved to the United States, where he joined from 1897 to 1899 the International Paper Co. as civil engineer, thereby redesigning the Mill of the Niagara Falls. He founded in 1899 the company Ambursen & Sayles, Watertown NY, and started with the design and construction of paper mills. He designed and built the first reinforced concrete dam in 1903 at Theresa NY. Based on the success of his method, he organized the Ambursen Hydraulic Construction Co., Boston MA.

The first description of the so-called Ambursen Dam was given in 1903. It is built of concrete, reinforced with steel rods and expanded metal. The design consists of a solid concrete toe, a series of solid concrete buttresses upon which rest the inclined upstream sides of the reinforced concrete buttresses anchored to the rock. The weir crest is strengthened with concrete beam. The dam was so constructed that all the pressure is directed to the base, and is therefore entirely a gravity dam. The concrete elements made in sections were prefabricated. The design at Theresa was 3.5 m high and 35 m wide, and built within eighteen working days by ten men, including the removal of the old timber dam and construction of the coffer dams. The maximum overflow depth was 2 m. For overflow dams of considerable height, back faces of the buttresses were added to receive a cover similar to that on the dam front, and thus a properly-shaped spillway. The vacuum formation below the downstream dam side would then also be avoided. Buttress dams in general consist of a wall supported by buttresses on the downstream side. They are also called hollow dams because the buttresses do not form a solid wall stretching across a river. Flat slab buttress dams or Ambursen dams have a flat upstream face, whereas the multiple arch buttress dams feature an upstream face formed by a series of arches, resting on top of the buttresses. These dams require significantly less concrete than gravity dams, but may not be less costly because of the complicated work.

Ambursen & Sayles (1903). A hollow concrete-steel dam at Theresa NY. *Engineering News* 50(Nov.5): 403-404.
Church, W.L., Ambursen, N.F., Sayles, E.W. (1904). *Concrete-steel gravity dams*. Ambursen & Sayles: Boston MA.
Schnitter, N. (2010). Nils Frederick Ambursen. Personal note. *P*

AMOROCHO

03.01. 1920 Bogotá/CO
22.11. 1983 Davis CA/USA

Jaime 'Jim' Amorocho received in 1943 a BS degree from *Universidad Nacional de Colombia*, Bogota Colombia, emigrating later that year to the USA. He received in 1946 his MS in civil engineering from Pennsylvania State University, and his PhD from the University of California, Berkeley CA, in 1961. Soon thereafter, he was appointed assistant professor of hydrology and hydraulics at the University of California, Davis, later being appointed to professor of water science and engineering at this institution.

Amorocho published over 150 papers and reports mainly in the American Geophysical Union AGU. He served as an Associate Editor of Water Resources Research, and as member of committees of the hydrology section. He chaired the Parametric Hydrology Group of the International Association of Scientific Hydrology IASH. Amorocho received the Robert E. Horton Award in 1974, he was a Fellow of the American Society of Civil Engineers ASCE, member of the American Meteorological Society AMS, and of the Academy of Mathematical and Physical Sciences of Colombia. He was a consultant to federal and state agencies, including the Office of Science and Technology. Amorocho is remembered from his laboratory, modifying the famous Cal-State Water Project model, from his consulting office telling a municipal engineer how he ought to design storm drains, from the university conducting his seminars with endless patience for most of his students but short shrift for those judged unworthy. After his death, a Jaime Amorocho Memorial Scholarship was established for deserving civil engineering students in the field of water resources.

Amorocho, J. (1969). *Generalized analysis of small watershed responses*. Water Resources Center, University of California: Davis CA.

Anonymous (1978). Jaime Amarocho. *Eos* 59(6): 527. *P*

Anonymous (1979). Hydrology: president elect Jaime Amarocho. *Eos* 60(35): 640. *P*

Babb, A., Amorocho, J. (1965). *Flow conveyance efficiency of transitions and check structures in a trapezoidal channel*. Dept. Irrigation, University of California: Davis CA.

de Vries, J.J., Amorocho, J., Hartman, W.J. (1980). Sediment modeling for the Sacramento River Diversion to the peripheral canal. *Computer and physical modeling in hydraulic engineering*: 463-474. ASCE: New York.

Lloyd, E.H. (1983). Jaime Amarocho. *Journal Hydrological Sciences* 29(3): 345.

http://content.cdlib.org/xtf/view?docId=hb4d5nb20m&doc.view=frames&chunk.id=div00006&toc.depth=1&toc.id=&brand=calisphere

ANDERSON A.G.

21.04. 1911 Duluth MN/USA
01.07. 1975 New Brunswick NJ/USA

Alvin George Anderson received three university degrees from Minnesota University, BCE in 1933, MS CE in 1935 and the PhD in 1950. He spent his early years with the Soil Conservation Service SCS of the US Department of Agriculture, where he did research on sediment transport in natural streams. He left in 1943 to join the Army, being associated with research on antisubmarine weapons. Returning to University of Minnesota, initially as instructor, he was appointed full professor of civil engineering in 1959, and named director of the St. Anthony Falls Hydraulic Laboratory in 1974. He died only 1 year later following a heart attack suffered while attending a symposium of the American Water Resources Association AWRA at Rutgers University, New Brunswick NJ.

Anderson combined his love for teaching and interest in students with an active research program. In addition to sediment transport, he was also active in research in such areas as flow in bends, culvert hydraulics, air entrainment by high-velocity flow, or hydraulic structures. He was further active as a consultant for both national and international projects. He also took interest in professional associations, including the International Association for Hydraulic Research IAHR, the American Society of Civil Engineers ASCE, or AWRA. He was the recipient of the 1961 ASCE Norman Medal for the benchmark paper Self-aerated flow, co-authored by Lorenz G. Straub (1901-1963), and the 1965 ASCE J.C. Stevens Award for discussing Sediment transportation mechanics.

Anderson, A.G. (1953). The characteristics of sediment waves formed by flow in open channels. Proc. 3rd *Midwestern Conf. Fluid Mechanics* Minnesota: 379-395.
Anderson, A.G. (1961). Sedimentation. *Handbook of fluid dynamics* 18: 1-35, V.L. Streeter, ed. McGraw-Hill: New York.
Anderson, A.G. (1965). Influence of channel roughness on the aeration of high-velocity, open channel flow. 11th *IAHR Congress* Leningrad 1(37): 1-13.
Anonymous (1961). Alvin G. Anderson. *Civil Engineering* 30(10): 81. *P*
Anonymous (2000). CV of Prof. A.G. Anderson. University of Minnesota, Minneapolis.
Killen, J.M., Anderson, A.G. (1969). A study of the air-water interface in air-entrained flow in open channels. 13th *IAHR Congress* Kyoto B(36): 1-9; 5: 241-242.
Marsh, M. (1975). Alvin G. Anderson. *Water Resources Bulletin* 11(5): 1078.
Straub, L.G., Anderson. A.G. (1960). Self-aerated flow in open channels. *Trans. ASCE* 125: 456-481; 125: 485-486.

ANDERSON G.G.

20.04. 1858 Aberdeen/UK
23.12. 1923 Santa Monica CA/USA

George Gray Anderson graduated as civil engineer from the University of Aberdeen, and then served until 1879 at an engineering firm of his native city. He moved in 1880 to the United States, settling in Denver CO, where he became assistant engineer of the Northern Colorado Irrigation Company. He was appointed in 1883 chief engineer of the Platte Land Company, Denver, being involved in the construction of the Platte Valley Canals, including more than 300 km of main canals. Anderson became from 1890 until 1916 consultant of a firm in Denver and then moved to Los Angeles CA continuing as consultant until his death. He was a member of the Institution of Civil Engineers ICE, London UK, the Engineering Institute of Canada, and the American Society of Civil Engineers ASCE.

From 1896 to 1916 Anderson was associated with hydraulic projects, among which were the Irrigation System of the Alberta Railway and Irrigation Company, a study on the Canadian Pacific Irrigation System, Alberta, on the fountain water supply for Pueblo CO, and on the Sacramento Valley Irrigation District. He was involved in the Laramie Valley Irrigation District, Laramie WY, and in the construction of Schaefer Dam and the Sanchez Reservoir in Colorado. He studied the water supply of the Spring Valley Water Company, San Francisco CA, serving as Board member of consulting engineers of the Imperial Irrigation District of California. He contributed papers on these projects in the Transactions ASCE, mainly dealing with irrigation. Anderson was a man of strong individuality, unlimited energy and tireless production. During his long residence in Colorado he earned and enjoyed prestige in his profession. His charming personality gathered to him a host of friends, so that he was remembered with affection and respect by all who knew him. The seriousness with which he accepted his engagements to the profession was illustrated by his insistence two months before his death in rising from the bed in Los Angeles and crossing the continent to attend a meeting in Richmond VA.

Anderson, G.G. (1909). Irrigation in Colorado. *Trans. ASCE* 62: 1-66.
Anderson, G.G. (1910). Some aspects of irrigation development in Colorado. *Proc. Colorado Scientific Society* 9: 273-314.
Anderson, G.G. (1915). The combination of water resources for irrigation and hydro-electric purposes. 2nd *Pan-American Scientific Congress* Washington DC.
Anonymous (1913). G.G. Anderson. *Semi-centennial history of the State of Colorado* 2: 136. *P*
Anonymous (1925). George G. Anderson. *Trans. ASCE* 88: 1338-1340.

ANDERSON N.E.

14.09. 1897 Anna IL/USA
27.11. 1977 La Grange IL/USA

Norval Eugene Anderson was educated at Colorado College, Colorado Springs CO, from 1916 to 1918, and then at the University of Illinois, obtaining in 1920 his BS title in civil engineering. He continued studies at the University of Chicago until 1922 and was from 1934 a professor of civil engineering at the University of Illinois, after having been an assistant engineer for sewage treatment design from 1922 to 1927, and senior engineer until 1931. From 1935 he was also a treatment plant design engineer of the Sanitary District of Chicago, promoted to chief engineer in 1963, shortly before retirement in 1964. Anderson was a member ASCE, as also of today's Central States Water Environment Association CSWEA, from which he was decorated with the 1951 Radebaugh Award.

In his 1950 paper Anderson considers air diffusers for activated sludge treatment on wastewater plants. His paper starts: 'Designers and operators of sewage works learn most from the difficulties encountered', a statement that may still be true today, given the complicated interaction of water flow and solid particles, and air in this particular study, corresponding therefore to a three-phase process. The activated sludge process became popular after World War II with Anderson as one of its promoters in practice. The 1945 paper deals with final settling tanks, in which the density current is identified as the governing process feature. Both local velocities and solids concentrations were measured with a special float developed for these currents, allowing for a discussion of the prominent distributions in a tank, and their improvement for a higher tank efficiency.

Anderson, N.E. (1945). Design of final settling tanks for activated sludge. *Sewage Works Journal* 17(1): 50-65.

Anderson, N.E. (1950). Tests and studies on air diffusers for activated sludge. *Journal Water Pollution Federation* 22(4): 461-476.

Anderson, N.E. (1953). Comprehensive metering of sewage at the world's largest treatment works. *Water and Sewage Works* 100(10): 412-415.

Anderson, N.E., Sosewitz, B. (1971). Chicago industrial waste surcharge ordinance. *Journal Water Pollution Control Federation* 43(8): 1591-1599.

Anonymous (1941). Anderson Norval E. *Who's who in engineering* 5: 36. Lewis: New York.

Anonymous (1964). Norval E. Anderson. *Civil Engineering* 34(11): 15. *P*

Whittemore, L.C., Anderson, N.E. (1937). Design of the sewage treatment works of the Sanitary District of Chicago. *Sewage Works Journal* 9(2): 256-270.

ANDERSON R.H.

02.10. 1877 North Woburn MA/USA
08.08. 1929 Blue Ridge GA/USA

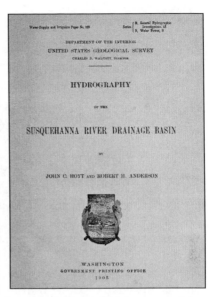

Robert Harlow Anderson was educated at the Robert College, Istanbul, and at Cornell University, Ithaca NY, receiving from the latter in 1898 the degree of civil engineer. He started his career as draftsman at the US Deep Waterway Survey through New York State, serving then as draftsman on a study of the water supply for New York City. In 1900 he was engaged as draftsman on surveys for the Nicaragua Canal, and in 1901 on water power investigations on the Susquehanna River. From 1903 to 1904 he was with the US Geological Survey preparing with John C. Hoyt (1874-1946) a report on its hydrography. He was then appointed assistant engineer with the Isthmian Canal Commission engaged in making studies as one of the first engineers on the hydraulic control of Chagres River.

From 1905 to 1909 Anderson served as engineer for McCall's Ferry Power Company on the Power Dam, Holtwood PA, and he was employed with an irrigation project at Twin Falls ID. His first connection with the Tennessee Electric Power Company was in 1909 when being employed as resident engineer on the construction of two hydro-electric developments for the Power Company at Parksville and Caney Creek, on the Ocoee River in Tennessee State. From 1915 to 1917 he was stationed at Rock Island TN, in charge of the construction of the Great Falls Hydro-Electric Development, and from 1919 to 1921 he was engineer for a company at Rifton NY, during the construction of the dam and the power plant on Wallkill River. Anderson returned to the Tennessee Electric Power Company in 1922 as hydraulic engineer, which position he held until his death. He was first in charge of the construction of additions to the dam at the Great Falls Plant, and subsequently made engineering studies. He was further involved in the construction of an earth storage dam on Toccoa River, a structure 50 m high made up of 43,000 m^3 of earth, with a power installation of 25,000 HP. He left a host of friends both within and without the Tennessee Electric Power Company, who held him in deep admiration for his sterling qualities as a man and a most capable engineer. He was member of the American Society of Civil Engineers ASCE.

Anderson, R.H. (1898). *Ventilation and flow of air currents in sewers*. Cornell University: Ithaca.
Anonymous (1930). Robert H. Anderson. *Trans. ASCE* 94: 1664-1665.
Hoyt, J.C., Anderson, R.H. (1904). Flood on March 1904. *Engineering News* 51(17): 393-394. (*P*)
Hoyt, J.C., Anderson, R.H. (1905). Hydrography of the Susquehanna River Drainage Basin.
　　　General Hydrographic Investigations 13. Government Printing Office: Washington DC.

AREF

28.09. 1950 Alexandria/EG
09.09. 2011 De Land IL/USA

Hassan Aref received his undergraduate degree in 1975 in physics from the University of Copenhagen, Denmark, and earned his doctorate in physics, with a minor in mechanical and aerospace engineering, from Cornell University, Ithaca NY, in 1980. He remained at Cornell for six months as a research associate at its Laboratory of Atomic and Solid State Physics, and spent the summer of 1980 at the Woods Hole Oceanographic Institution. He then joined the engineering faculty at Brown University, Providence RI, as assistant professor, remaining until he joined in 1985 the University of California, San Diego CA, from associate professor to professor of fluid mechanics. In 1992 he was appointed professor of fluid mechanics at the University of Illinois, Urbana-Champaign IL, from where he moved in 2003 as dean to Virginia Tech, Blacksburg VA.

The research interests of Aref included theoretical and computational fluid mechanics, particularly vortex dynamics, the application of chaos to fluid flows, and the mechanics of foams. He was co-editor of two books and author of more than 70 papers in premier journals. He was associate editor of the Journal of Fluid Mechanics from 1984 to 1994, founding editor with David G. Crighton (1942-2000) of the Cambridge Texts in Applied Mathematics, and prior to his death served on the editorial board of Theoretical and Computational Fluid Dynamics and as co-editor of Advances in Applied Mechanics. He was recipient of the G.I. Taylor Award in 2011. He also received the 2000 Otto Laporte Award of the American Physical Society APS, Division of Fluid Dynamics. He was an APS Fellow, Fellow of the American Academy of Mechanics, the Danish Center for Applied Mathematics and Mechanics, and the World Innovation Foundation. Aref was a scholar of high repute, a giant in the field of fluid mechanics, and most of all, a kind and caring mentor.

Anonymous (1994). Aref, Hassan. *American men and women of science* 1: 208. Thomson Gale.
Aref, H. (2007). *Advances in applied mechanics*. Academic Press: London UK.
Aref, H. (2010). Self-similar motion of three point vortices. *Physics of Fluids* 22(5): 057104.
Boyland, P.L., Aref, H., Stremler, M.A. (2000). Topological fluids mechanics of stirring.
    Journal of Fluid Mechanics 403: 277-304.
Nystrom, L.A. (2011). In memoriam: Hassan Aref, Reynolds Metals Professor of Engineering
    Science and Mechanics, and former dean. *Virginia Tech News* (9): 1. P
http://de.wikipedia.org/wiki/Hassan_Aref

ARMSTRONG

30.05. 1914 Cedar City UT/USA
26.01. 2001 Salt Lake City UT/USA

Ellis Leroy Armstrong obtained in 1936 the BS degree in civil engineering from the Utah State University, Logan UT; he did post-graduate work at the Colorado State University, Fort Collins CO. He was from 1936 to 1954 engineer of the US Bureau of Reclamation USBR, engaged on concrete and earth-fill engineering for the Midview and Moon Lake Dams in Utah, from 1938 in charge of the foundation of the Deer Creek Dam UT, and until 1944 with the Anderson Ranch Dam in Southern Idaho, then the highest earth-fill dam of the world. From 1945 to 1948 he was in charge of the earth dam design group, at the chief of engineers office, Denver CO, designing until 1953 26 earth dams, from when he was in charge of Trenton Dam in south-western Nebraska, directing investigations, negotiations, construction, and coordinating designs.

Armstrong was then in 1953 an engineering member of the Egyptian-American Rural Improvement Service, Cairo, formulating long-range plans for the development of the country. He reported to the Government plans for the High Aswan Dam project on the Nile River. From 1954 to 1957 he was project engineer and assistant project manager for an engineering company in New York on the St. Lawrence Power Project, thereby directing earth investigations, from when he was director of the Utah State Highways. He was in 1969 appointed Commissioner of USBR. He regarded his profession as a mission for the good of the mankind; once he decided what needs to be done, he begun working. He succeeded Floyd E. Dominy (1909-2010), who left behind an impressive trail of accomplishments. Armstrong had the technical background that Dominy had at times to replace with a brilliant memory and a sharp tongue. Armstrong reorganized the Bureau during his tenure until 1973. He served also as Lecturer at ColoState on earth dam design, and the development of basic resources. He was a member of the American Society of Civil Engineers ASCE, and the US Committee on Large Dams.

Anonymous (1959). Armstrong, Ellis L. *Who's who in engineering* 8: 66. Lewis: New York.
Anonymous (1969). Engineer is BuRec's new boss. *Engineering News-Record* 183(Nov.13): 32-37. *P*
Armstrong, E.L. (1975). *The St. Lawrence Power and Seaway Project*. University of Florida: Gainesville FL.
Armstrong, E.L., Robinson, M.C., Hoy, S.M. (1978). *History of public works in the United States*, 1776-1976. American Public Works Association: Chicago.

ARNOLDT

18.10. 1820 Heidelberg/D
17.04. 1893 Rochester NY /USA

At his birth the father was officer in the Department of Forestry, District of Heidelberg, so that the son's early associations were with cultured people, and his training imbued his mind with high principles, and strengthened his imperious disposition. Arnoldt became a student at the University of Heidelberg in 1838, passed thence to Karlsruhe, where he entered the Polytechnic School and studied civil engineering. After graduation, he was employed in construction of railroad sections. In the 1848 political eruption agitating Germany he took an active part. With the downfall of the revolution he fled to America.

Landing in New York in 1850, George Arnoldt proceeded to Rochester NY, entering State service in the office of the division engineer of the Erie Canal. During his service he held the positions of second assistant engineer in 1862, second assistant from 1863 to 1866, and assistant from 1867 to 1876. By superior talent he retained his position thus through 27 years, until declining health forced him to retire. The almost 600 km long Erie Canal runs from Albany NY on the Hudson River to Buffalo NY, at Lake Erie, completing a navigable waterway from the Atlantic to the Great Lakes. It contains 36 locks and has a total elevation difference of 170 m. It was constructed from 1817 to 1825. During these times bulk goods were limited to pack animals, given that there was no steamships or railways. The canal fostered a population surge in western New York State, opened regions farther west to settlement, and helped New York City become the chief US port, which was enlarged between 1834 and 1862. In 1918 the western Canal was enlarged to partially become New York State Barge Canal, running parallel to the eastern half and forming its new eastern branch to the Hudson. Erie Canal is currently the east-west route of the New York State Canal System. It was designated in 2000 the Erie Canalway National Heritage Corridor to recognize the national significance of the canal system as the most successful and influential human-built waterway and one of the most important works of civil engineering and construction in North America.

Hill, H.W. (1908). *An historical review of waterways and canal construction in New York State.* Buffalo Historical Society: Buffalo.

Peck, W.F. (1884). *Semi-centennial history of the city of Rochester*, with illustrations and biographical sketches of some of its prominent men and pioneers. Mason: Syracuse NY.

http://mcnygenealogy.com/bios/biographies027.htm P

http://en.wikipedia.org/wiki/Erie_Canal

ARTHUR

23.08. 1914 Lead SD/USA
28.09. 2001 Denver CO/USA

Harold Gilbert Arthur obtained in 1935 the BS degree in civil engineering from the South Dakota School of Mines and Technology, Rapid City SD. He was until 1937 surveyman of the US Bureau of Reclamation USBR at Casper WY, until 1941 junior civil engineer of the US Forest Service, Missoula MT and at Ogden UT, from when he joined USBR as assistant civil engineer in its Earth Dams Section at Denver CO until 1945. After work with an aircraft corporation, he re-joined USBR in the same Section from 1953.

Arthur was promoted in 1968 to associate chief engineer of USBR. He had been chief design engineer since 1965. He previously had served as assistant regional director at Billings MT before going to the USBR headquarters at Denver CO in 1963 as assistant chief design engineer. Arthur was coordinator and editor of USBR's publication Design of small dams, a technical work serving over decades for the design and construction of these hydraulic structures. He also was involved in the description of the Teton Dam Failure, which occurred in 1976 on southeast Idaho, initiated by a large leak on the right dam abutment 40 m below the dam crest. This dam had been designed by USBR and failed as it was completed and filled for the first time. Arthur was awarded by the Secretary of Interior the Distinguished Service Award in 1968, and previously was honored for his liaison work with the International Boundary and Water Commission in the design and construction of Falcon Dam on the Rio Grande. In 1977 he was recipient of the Beavers Engineering Award. He served the Chilean Government in 1960 as adviser on construction of a major earth dam. After his retirement in 1977 he consulted on dams in the Dominican Republic, Ecuador, Greece, the Philippines and many other projects in the USA. He was member of the American Society of Civil Engineers ASCE, and the US Committee on Large Dams.

Anonymous (1959). Arthur, Harold G. *Who's who in engineering* 8: 72. Lewis: New York.
Anonymous (1968). BuRec promotes Arthur to associate chief engineer. *Engineering News-Record* 181(Jul.25): 44. *P*
Arthur, H.G., ed. (1960). *Design of small dams*, 1st ed. Govt. Printing Office: Washington DC.
Arthur, H.G. (1966). *The proposed third powerplant at Grand Coulee Dam*. ASCE: New York.
Arthur, H.G. (1977). *Failure of Teton Dam*. US Dept. of the Interior: Washington DC.
Storey, B.A. (2000). *Oral history interviews*: Harold G. Arthur. USBR: Denver CO.
http://www.geol.ucsb.edu/faculty/sylvester/Teton_Dam/narrative.html

AXTELL

18.08. 1867 Washington IN/USA
14.08. 1929 Port Arthur TX/USA

Frank Foy Axtell received the engineering degree from the Indiana State University, Terre Haute IN in 1891, and was then employed as rodman on a survey of Missouri River. He became recorder in 1892 for a survey of Mississippi River at St. Louis MO, advancing to draftsman in 1894. In 1895 he went to Eagle Pass TX on an irrigation project. After return to the Mississippi River Commission in 1896 he was employed as US surveyor. He was sent by the US Engineer Department to *Calcasieu* Pass LA in 1900 to supervise the extension of its jetties, covering mattress foundation and rip-rap stone work.

Axtell was then transferred back to the US Engineer Office in New Orleans LA, where he advanced to US junior engineer, in charge of the improvement of small streams, including dragging and dredging works. In 1902 he was transferred to Sabine TX, to supervise the improvement of Sabine Pass Harbour and Jetties, and the Sabine and Neches Rivers. In 1906 the US Government acquired title to the Port Arthur Ship Canal, whose supervision was again made by Axtell.

The duties of Axtell from 1902 to 1912 included the (1) topographic and hydrographic surveys of Sabine Pass, and Sabine and Neches Rivers, (2) design of jetty work at Sabine Pass, (3) supervision of canal excavation connecting the Port Arthur Ship Canal with the mouths of Sabine and Neches Rivers, (4) dredging with and repair of the sea-going dredge *Sabine*, and (5) surveys of the Intracoastal Canal from Sabine Pass LA to Galveston Bay. In 1912 Axtell moved to the Texas Company, Port Arthur TX, as construction engineer. In 1913 he was transferred to Tampico MX as chief engineer for hydraulic dredging and the construction of docks. He remained there until 1920, then returning to Port Arthur, working with the Gulf Refining Company, then the largest in the world. Dredging, the filling of a bayou and expansion works were his main tasks. In spite of his failing health, he was actively engaged into these projects until four weeks before his death. He had been a rare personality, loyal to his friends, yet charitable toward all in his judgment. He was unsparing of himself as an engineer. He was member of the American Society of Civil Engineers from 1911.

Anonymous (1895). *Report* of the Secretary of War communicated to the Two Houses of
        Congress: Mississippi River Commission. Govt. Printing Office: Washington DC.
Anonymous (1931). Frank F. Axtell. *Trans. ASCE* 95: 1446-1449.
http://www.google.ch/imgres?q=%22Frank+F.+Axtell *P*

AYRES

01.03. 1883 Concord NH/USA
06.12. 1961 San Francisco CA/USA

Augustine Haines Ayres graduated from Dartmouth College, Hanover NH, with the BSc degree in 1906, and in 1907 with the CE degree from Thayer School of Engineering, Dartmouth NH. He began in 1908 his professional career with design and construction of irrigation projects in Colorado and Alberta. From 1913 to 1920, he was employed by the US Bureau of Reclamation USBR in Montana and Wyoming. He served from 1931 to 1936 as chief engineer of Six Companies Inc., the joint venture of the western contractors that built Hoover Dam. During World War II he was active in naval air base construction of Guam, Ford Island, and Hawaii. He had joined in 1936 the staff of Utah Construction Co., where he remained for seventeen years serving progressively as engineer-estimator, project manager, chief engineer, and vice-president, retiring in 1953. He then served as consultant Bechtel Corporation, Utah Construction Company, Raymond International, and the State of California.

Ayres was a notable hydraulic engineer involved both in the design and the construction of large hydraulic structures. His most important position was certainly as chief engineer of the Hoover Dam, the great pyramid of the American West, whose work was initiated in 1931, and completed in 1936. It is located within a chain of dams along Colorado River, and was a supreme engineering feature of its day, the ultimate expression of machine-age America's ingenuity and technological progress. Currently, Hoover Dam generates enough electric power to serve more than 1 million people in the west. From 1939 to 1949 it was the world's largest hydropower installation, and still now is one of the largest in the USA. The entire power plant base is U-shaped, one wing on Nevada, the other in California. Each power plant wing is 200 m long, rising 100 m above the foundation. There are in total 17 turbines which were installed around 1990 to replace the original installation. The capacity of the power plant is 2,000 MW, based on a top head of 180 m, and a design discharge of 2,000 m^3/s. The plant is maintained by USBR.

Anonymous (1941). A.H. (Gus) Ayres. *Engineering News-Record* 126(Apr.24): 591. *P*
Anonymous (1962). Augustine H. Ayres. *Trans. ASCE* 127(5): 51.
Rogers, J.R. (2010). The new town of Boulder City. *Hoover Dam* 75th Anniversary History Symposium: 40-47.
Stevens, J.E. (1988). *Hoover Dam*: American adventure. University of Oklahoma Press: Norman.
http://www.usbr.gov/lc/hooverdam/faqs/powerfaq.html

BABB

18.06. 1867 Portland ME/USA
02.10. 1937 Granite Falls NC/USA

Cyrus Cates Babb graduated from Massachusetts Institute of Technology MIT in 1890 with a BS. He then joined the US Geological Survey USGS until 1902, staying until 1909 with the US Reclamation Service. From 1910 to 1915 Babb was a member of the Maine State Water Service Commission, and then continued as a private consultant.

Babb's first position within the USGS involved hydrographic reclamation and water power projects. During this period he travelled through the United States and made a trip to Europe. He established a large number of gauging stations, of which one in Montana was named after him, and was still in operation in 1939. He was also instrumental for the preliminary survey for the San Carlos Dam site, on which was built the Coolidge Dam in the 1920s. Following these works, he was in charge of surveys and investigations in Utah and Montana for water supply and irrigation projects, such that he was transferred to the US Reclamation Service once this Act had been accepted. He was concerned with designs of reservoir dams, outlet gates, and canals. During his third career step with the Maine Water Service Commission, Babb became its chief engineer. He investigated the main water resources of the State, examined its water power and storage possibilities, and also prepared the Annual Report of this Commission. He was a member of the Maine Association of Engineers serving as its president in 1911. Finally, as a consultant, his activities included the water supply of towns in Maine, the valuation of water works property, and reports on large hydropower schemes. In 1929, Babb returned to services with the Government in the US Engineer Office, Charleston SC, and later to Norfolk VA as senior hydraulic engineer of the US Corps of Engineers, in charge of water power and flood control investigations. His last position was concerned with a text on Drainage basin problems; he retired in 1936.

Anonymous (1921). Babb, Cyrus Cates. *American men of science*: 24. Science Press: New York.
Anonymous (1939). Cyrus Cates Babb. *Trans. ASCE* 104: 1881-1884.
Babb, C.C. (1892). The hydrography of the Potomac Basin. *Trans. ASCE* 27: 21-33.
Babb, C.C. (1893). Rainfall and flow of streams. *Trans. ASCE* 28: 323-336.
Babb, C.C. (1906). An experiment to determine "n" in Kutter's formula. *Engineering News* 55(Feb.1): 122-123.
Babb, C.C., Hinderlider, M.C., Giles, J.M., Hoyt, J.C. (1907). *Missouri river drainage*. USGS.
Barton, J.D. (1998). *History of Duchesne County*. Utah State Historical Society: Salt Lake City. (*P*)

BABBITT

07.01. 1888 East Orange NJ/USA
10.10. 1970 Seattle WA/USA

Harold Eaton Babbitt was educated at MIT, from where he received the SB degree in 1911, and at the University of Illinois, with an MS degree in 1917. From 1911 to 1913 he was a 'computer' within the Sanitary District of Chicago, then until 1913 an assistant engineer of the Ohio State Board of Health. Successively, Babbitt was an instructor, associate professor and professor at University of Illinois.

Babbitt was the true pacesetter in American sanitary engineering education over four decades. The clarity of his classroom exposition and effectiveness of his investigations in the laboratory identified him as a conspicuously successful teacher and a proficient researcher. Two of his textbooks, Sewerage and sewage treatment, which went through eight editions, and Water supply engineering, dominated this area of engineering textbook literature for years and mirrored his constant search for excellence in education. Upon retiring, Babbitt thrived on the opportunities and challenges awaiting him. He launched into a period of intensive writing that continued for 30 years. Babbitt also prepared the section on Water supply and purification of the widely-used Civil engineering handbook. His research activities continued apace including works of garbage disposal with sewage, the hydraulics of wells and open channel flow of sludge, or removal of radioactive phosphorous from water. Babbitt's professional activities included vigorous participation in numerous technical societies and organizations. He was Honorary Member of the Water Pollution Control Federation WPCF and recipient of the prestigious Fuller Award of the American Water Works Association AWWA.

Anonymous (1937). Babbitt, Harold Eaton. *Who's who in engineering* 4: 46. Lewis: New York.
Anonymous (1971). Harold E. Babbitt. *Journal American Water Works Association* 63(1): 18.
Babbitt, H.E. (1922). Non-uniform flow and significance of drop-down curve in conduits. *Engineering News-Record* 89(25): 1067-1069.
Babbitt, H.E. (1938). Some recent developments in water works practice. *Journal AWWA* 30(3): 453-463.
Babbitt, H.E., Doland, J.J. (1939). *Water supply engineering*. McGraw-Hill: New York.
Babbitt, H.E., Caldwell, D.H. (1940). Turbulent flow of sludges in pipes. *Bulletin* 323. Engineering Experiment Station, University of Illinois: Urbana IL.
Babbitt, H.E. (1952). *Engineering in public health*. McGraw-Hill: New York.
Babbitt, H.E., Baumann, E.R. (1958). *Sewerage and sewage treatment*. Wiley: New York.
http://cee.uiuc.edu/alumni/newsletter/babbitt.aspx *P*

BABCOCK G.H.

17.06. 1832 Unadilla Forks NY/USA
16.12. 1893 Plainfield NJ/USA

George Herman Babcock was born into a family with prolific inventors. He followed the mechanical family tradition, gaining practical experience during the Civil War as draftsman and ship builder. His friendship with the talented mechanical engineer Stephen Wilcox (1830-1893) evolved into a fruitful and long business relationship between the two. In 1857 Wilcox introduced the water-tube boiler, a safe and efficient step in boiler technology. External combustion gases heat thereby tubes through which flowing water is converted into steam and collected in a drum. Because this design permitted higher-pressure operation than earlier versions, the water-tube boiler was readily accepted by the industry.

Boiler explosions were tragically common in the early 19th century. One of the worst occurred in 1850 devastating a manufacturing company in Manhattan. More than 60 workers were killed and 70 injured. Babcock and Wilcox established the firm B&W in 1867 to commercialize boilers based on Wilcox's earlier water-tube principle. The Babcock & Wilcox Non-Explosive Boiler used a sectional tubular design. Whereas fire-tube boilers contain long steel tubes through which the hot gases from the furnace pass and around which the water to be changed to steam circulates, the water-tube boilers use reversed conditions. The company was incorporated in 1881, with Babcock as president and Wilcox as vice-president. B&W boilers were soon powering the nation's first central electrical stations in Philadelphia and New York City. Babcock was active in addition within the American Society of Mechanical Engineers ASME, particularly by issuing the ASME Boiler Testing Code in 1884, providing one of the first standards for the industry. He died only few weeks after his lifelong friend and partner. Today, B&W remains world leader in the power generation industry. In 2012 it was listed the company as one of the nation's 50 most innovative technology companies.

Anonymous (1893). George H. Babcock. *Engineering Record* 29(4): 52.

Anonymous (1894). George H. Babcock. *Trans. ASME* 15: 636-639.

Anonymous (1992). Babcock & Wilcox 1867-1992. *P*

Babcock & Wilcox Company (1922). *Steam*: Its generation and use. B & W: New York.

Babcock, G.H. (1890). Circulation of water in steam boilers. *Scientific American* Supplement 30(745): 11902-11904.

http://www.asme.org/kb/news---articles/articles/boilers/george-herman-babcock

BABCOCK H.A.

23.02. 1917 Rushville IL/USA
23.10. 2003 Monterey VA/USA

Henry Ame Babcock obtained the BS, MS, and PhD degrees from University of Colorado, Boulder CO. He also was a registered engineer in Colorado State. He worked for the US Engineers prior to World War II, and the US Bureau of Reclamation USBR following discharge from the US Navy. He was associated with the Colorado School of Mines CSM, Golden CO, since 1946 for totally 36 years, advancing from instructor to Head of the Basic Engineering Department. He was Honorary Member of CSM from 1982, and enjoyed playing bassoon in an amateur symphony orchestra.

Babcock was interested and taught all through his career engineering subjects. In parallel he was considered a world-authority on the hydraulic transportation of solids both in pipelines and in open channels. His PhD thesis submitted in 1959 dealt also with this problem, whose issues were detected only after World War II, mainly in France and in the USA. The 1956 paper deals, also as an early contribution, with jet flow. The 1967 paper was devoted to the extra head-loss of solid transport in pipes, in addition to the effects of fluid viscosity and boundary roughness. In addition to this research field, Babcock was interested in irrigation engineering, particularly the design of irrigation subdivisions.

Anonymous (1956). H.A. Babcock. Proc. 6th *Hydraulics Conference* Iowa: Frontispiece. *P*

Anonymous (1985). Babcock, Henry A. *Who's who in engineering* 6: 22. AAES: Washington DC.

Arendt, J., Babcock, H.A., Schuster, J.C. (1956). Penetration of a jet into counterflow. *Journal of the Hydraulics Division* ASCE 82(1038): 8-11.

Babcock, H.A. (1959). Tabular solution of open channel flow equations. *Journal of the Hydraulics Division* ASCE 85(HY3): 17-23; 85(HY9): 125-135.

Babcock, H.A. (1967). Head loss in pipeline transportation of solids. Proc. 1st *World Dredging Conf. WODCON* New York 1: 261-289.

Babcock, H.A. (1970). The sliding bed flow regime. Proc. 1st Int. Conf. *Transport of solids in pipes* H1: 1-16. BHRA: Cranfield.

Babcock, H.A. (1971). Heterogeneous flow of heterogeneous solids. *Advances in solid-liquid flow in pipes and its application*: 125-146, I. Zandi, ed. Pergamon Press: Oxford.

Faddick, R.R., Babcock, H.A. (1971). Discussion of Sediment transport mechanics J: Transportation of sediment in pipes. *Journal of the Hydraulics Division* ASCE 97(HY5): 745-748.

BAILEY E.G.

25.12. 1880 Damascus OH/USA
18.12. 1974 Easton PA/USA

Ervin George Bailey obtained his ME degree from Ohio State University in 1903, the Hon. Dr. Eng. degree from Lehigh University in 1937, and in 1943 the Hon. Dr. Sc. degree from Lafayette College. Bailey joined from 1903 to 1907 the staff of the Testing Dept., Consolidation Coal Co. In 1909 he became a partner of the Fuel Testing Co, and was in 1916 the founder and president of the Bailey Meter Co., which he took over in 1944 as chairman. In parallel, Bailey was from 1926 to 1936 president of the Fuller Lehigh Co., and from 1931 to 1951 vice-president of Babcock & Wilcox Co.

Bailey was the 'Dean of combustion engineers' for his revolutionary invention of the Bailey Boiler Meter. This instrument is considered to mark the beginning of the art of automation. Today, both the Bailey Meter and the automation concept it fostered are recognized throughout the world. For almost 50 years Bailey devoted his inventive and executive abilities to improve the science of steam and combustion engineering. Early in his career Bailey saw the need for recording meters to improve combustion efficiency in steam boilers. He developed a meter in which the steam and air flow pens gave an instantaneous indication of steam output and excess combustion air. The meter guided firemen in burning coal, and furnished information to correct faulty operating conditions. It could also be adapted to the firing of oil, gas, and other fuels. The Bailey Boiler Meter is still widely used and has an important factor in improving the fuel burning efficiency of steam boiler plants. His meter proved so successful that Bailey formed his own company to manufacture and sell it. His company began soon to develop other products, including flow meters, combustion control mechanisms, and devices to record pressure, temperature or other parameters of working fluids. Bailey was also a professional leader. He served as president ASME in 1948, and was awarded the ASME Medal in 1942, the society's highest honor. He was an Honorary Member of ASME, ASCE, and the Institution of Mechanical Engineers, London.

Anonymous (1948). E.G. Bailey. *Mechanical Engineering* 70(1): 51. *P*
Anonymous (1954). Bailey, Ervin George. *Who's who in engineering* 7: 89. Lewis: New York.
Anonymous (1975). Ervin George Bailey. *Mechanical Engineering* 97(2): 88-89. *P*
Bailey, E.G. (1916). Steam flow measurement. *Trans. ASME* 38: 775-782.
Bailey, E.G. (1939). Modern boiler furnaces. *Trans. ASME* 61(10): 561-576.
http://www.google.ch/search?q=%22Ervin+george+bailey%22&hl=de&start=0&sa=N

BAILEY G.I.

24.12. 1861 Hempstead TX/USA
28.03. 1908 New York NY/USA

George Irving Bailey entered after graduation from Albany High School in 1880 the Department of the State Engineer. He was a rodman until 1884, then leveller and draftsman until 1892, during which time he was in charge of the lengthening of locks on the Erie Canal. He was then engaged for a storage dam across the Genesee River conducting tests for the tensile and crushing strength of concrete, which counted to the earliest made in dam engineering. He was in 1892 appointed superintendent of the water-works, Albany NY. He in parallel presided the newly formed Board of Water Commissioners.

Bailey was interested in developing the Albany water-works by adding new pumps with a capacity of 700 l/s and 3.3 km of main pipes of 0.80 m diameter. From 1897 filtration of the Hudson River water was considered, which was thought to be the only means to improve the then hopeless situation. Bailey visited various filtration plants, and the matter was finally submitted to Allen Hazen (1869-1930), who recommended slow sand filters to be used. Hazen was then appointed chief engineer under the supervision of Bailey, and the filters were successfully completed. These works were published in the annual reports, with accounts on their operation and the consequent reduction of the typhoid death-rate of the population. The schemes of Bailey and Hazen were considered a definite advance of drinking water treatment, so that visitors from all over the USA were eager to see this engineering feature. In 1902 Bailey was called to New York City to complete defaulted contracts, a work which was satisfactorily completed. He also designed sewers in Boston Road and in Bryant Avenue. Shortly before his death he was engaged with the completion of a filtration plant in Yonkers NY. While apparently in perfect health, he was stricken with heart trouble, passing away the next day. Bailey was member of the American Society of Civil Engineers and of the New England Water Works Association. He published several papers in the Engineering News.

Anonymous (1908). George Irving Bailey. *Trans. ASCE* 61: 556-559.
Bailey, G.I. (1899). The care of fire hydrants in winter. *Journal of the New England Water Works Association* 14(2): 116-123. *P*
Bailey, G.I. (1900). Discussion of The Albany water filtration plant, by A. Hazen. *Trans. ASCE* 43: 296-301.
Bailey, G.I. (1901). The effect of water meters on water consumption in the larger cities of the United States. *Journal of the New England Water Works Association* 15(4): 351-359.

BAKER H.J.M.

14.04. 1878 Lowell MA/USA
02.10. 1943 Seattle WA/USA

Harold James Manning Baker spent his childhood in Hawaii, his boyhood in San Francisco CA, and later in Seattle WA. He graduated from University of Washington, Seattle WA, with the BSc degree in civil engineering, continuing there until 1902 with post-graduate work. In 1902 he became a special student in hydraulic engineering, Cornell University, Ithaca NY, and then was rodman and instrumentman at Port Townsend WA, in charge of topographical surveys at Fort Casey and Fort Worden. In 1903 he began his long engagement with the US Engineer Department, during which Baker spent several years on fortification work. Later he was responsible for all river and harbour work in the Puget Sound, Grays Harbor, Willapa Harbor, Columbia River, and Alaskan areas.

In 1928 Baker was detached from regular river and harbor work, and placed in charge of the preparation of a series of reports covering topics as power, irrigation, flood control, and navigation of the principal rivers in the Seattle Engineer District. The most difficult part of this assignment was the report on the Columbia River, which was widely used by engineers, who produced favorable comments. In referring to the necessity for sound engineering documents prior to government expenditures, Riggs stated in his 1939 paper: 'No better example of such a study can be found than that on the Columbia River made by the Corps of Engineers after four years of study'. Baker was a man of high principles and firm convictions and governed his life according to these standards, quietly and undeviatingly. His relationship with subordinates was friendly, cordial, and courteous. He always found delight in mathematics; a book that he treasured was an autographed copy of The Lowell Hydraulic Experiments, by James B. Francis (1815-1892), which was presented him by Francis' sister when he was student at Cornell University. He was a man of fine spirit and character whose loss was deeply regretted by all his friends. He was member of the American Society of Civil Engineer ASCE from 1908.

Anonymous (1943). Harold J.M. Baker. *Engineering News-Record* 131(Oct.14): 576.
Anonymous (1944). Harold J.M. Baker. *Trans. ASCE* 109: 1561-1562.
Chittenden, H.M., Clapp, J.M., Baker, H.J.M., Clarke, E.L. (1908). *Snohomish River,*
        Washington, from the mouth to Lowell. US Engineer Office: Seattle WA.
Riggs, H.E. (1939). Hazards of uneconomical construction. *Trans. ASCE* 104: 668-689.
https://www.google.ch/search?q=%22Harold+James+Manning+Baker%22& *P*

BALCH

10.03. 1883 Neillsville WI/USA
18.04. 1928 Madison WI/USA

Leland Rella Balch graduated from University of Wisconsin, Madison WI, in 1905 with the BS degree and in 1909 with the civil engineering degree. He started his professional career as engineer assistant with the US Reclamation Service, later as resident engineer in charge of surveys on the Huntley, Sun River and Flathead Irrigation projects in Montana. From 1909 to 1910 he took post-graduate work at his Alma Mater, and was then appointed assistant engineer with the US Reclamation Service on the Shoshone Project in Wyoming. He returned in 1911 to his Alma Mater again as research assistant there conducting hydraulic experiments, resulting in various outstanding technical researches. From 1912 until his death Balch was with the firm of Daniel W. Mead (1862-1948) and Seastone, consulting engineers, Madison WI. There he was responsible for the design and the construction of various works, notably municipal pumping plants, additions to the power plants of the Madison Gas and Electric Power Company, or the Mississippi Valley Public Service Co., Winona MN. He also participated in investigations and the preparation of reports relating to water works, and sanitary works.

During his stay at the University of Wisconsin from 1910 to 1912, Balch studied various hydraulic problems. These included Tests on flash wheels, which were used to lift water from a lower to a higher level. He further made an early study on the jet pump, by which water is again moved by the action of a driver nozzle. Other works include tests of submerged orifices and the flow through submerged tubes. Balch was a person of high ideals, of sterling character and of highest integrity. A man of pleasing personality, he possessed a genial disposition, and a keen sense of humor, endearing himself to all who knew him. He was member of the American Society of Mechanical Engineers ASME and of Civil Engineers ASCE, and the Engineering Society of Wisconsin. He passed away at ago of only 45 for unknown reasons.

Anonymous (1906). Leland R. Balch. *Badger yearbook* 20: 57. University of Wisconsin. *P*
Anonymous (1929). Leland R. Balch. *Trans. ASCE* 93: 1754-1755.
Balch, L.R. (1913). Test of a jet pump. Engineering Series *Bulletin* 7(4): 1-15. Univ. Wisconsin.
Balch, L.R. (1914). Investigation of flow through four-inch submerged orifices and tubes.
    Engineering Series *Bulletin* 8(3): 147-178. University of Wisconsin: Madison WI.
Davis, G.J., Jr., Balch, L.R. (1914). Investigation of flow through large submerged tubes.
    Engineering Series *Bulletin* 7(6): 1-57. University of Wisconsin: Madison WI.

BALDWIN, Jr. L.

16.05. 1780 Woburn MA/USA
30.06. 1838 Boston MA/USA

Loammi Baldwin, Jr., was the son of a prominent civil engineer of the 18th century. He was educated at Westford Academy graduating in 1800 from Harvard College. At age 14 he accompanied his father to the famous canal engineer William Weston (1752-1833) initiating a ten years work of the Baldwin family on Middlesex Canal. From 1807 he devoted his entire energy to hydraulic engineering, traveling to England to inspect public works. Upon return to the USA in 1808 he began engineering practice at Charlestown MA, and was elected in 1810 Fellow, the American Academy of Arts and Sciences.

One of the earliest engineering works of Baldwin was the construction of Strong Fort at Boston Harbor, for defense against the British. In 1819 he completed the construction of Milldam, now that stretch of Beacon Street beyond the Boston Common. From 1817 to 1820 he worked in Virginia, and in 1821 took over as the engineer of the Union Canal in Pennsylvania. Baldwin returned in 1824 to Europe, examining public works mainly in France. In 1827 he was appointed engineer by the United States government which led to the naval dry docks at the Boston Navy Yard in Charlestown and in Norfolk. He was also in charge of the construction of a canal around the Ohio River Falls, or of Harrisburg Canal in Pennsylvania. One year before his death he was stricken by paralysis, a second attack proved fatal. He was then only 58 years of age, but survived under the name 'Father of American civil engineering'. After his death, his brother James Fowle Baldwin, with whom he had often worked, continued his work, especially for the safe water supply of Boston.

Baldwin, L., Jr. (1834). *Report* on the subject of introducing pure water into the city of Boston. Hilliard, Gray Co. Boston.

Baldwin, L., Jr. (1835). *Second Report* made to a committee of the Boston Aqueduct Corporation. Eastburn's Press: Boston.

Schexnadyer, C., Anderson, S. (2011). Construction engineering education: History and challenge. *Journal of Construction Engineering and Management* 137(10): 730-739.

Tower, F.B. (1843). *Illustrations of the Croton Aqueduct*. Wiley & Putnam: New York.

http://en.wikipedia.org/wiki/Loammi_Baldwin,_Jr.

http://www.lib.uchicago.edu/e/scrc/findingaids/view.php?eadid=ICU.SPCL.CRMS203

http://www.google.ch/search?q=%22Loammi+Baldwin%22&hl=de&prmd=imvnso&tbm=isch&tbo =u&source=univ&sa=X&ei=HKTYTrPlJ8jt-ga-g-23Dg&ved=0CG8QsAQ&biw=1044&bih=740 *P*

BALDWIN W.J.

14.06. 1844 Waterford/IE
07.05. 1924 New York NY/USA

William James Baldwin was born on shipboard near Waterford in Ireland. He arrived with his family at Boston MA in 1855, educated in primary schools there and St. Dunstan's High School, Prince Edward Island CA. He became familiar from 1862 in naval architecture, engineering and physics in his father's office, and helped to construct monitors and convert blockade runners in shipyards of East Boston MA. He was from 1866 assistant to Stephen Gates, to design and repair iron ships, and from 1868 worked in general machinery. From 1870 he acted as manager and superintendent at Detroit's Novelty Works. From 1874, he constructed engineering plants for large public buildings. He also was consulting engineer for the Department of Health, New York City, and lectured at the US War College, Washington DC. Baldwin took over in addition from 1880 to 1889 as associate editor the Engineering Record journal, and was Lecturer and professor of thermal engineering at Brooklyn Polytechnic Institute.

The heating of rooms in the 19th century was almost unknown, except by using a fireplace. Baldwin was among those who revolutionized this state by systematically using steam for this purpose. He was a person well set into this position, given his widespread projects achieved until the early 1880s, including works with ships, general machinery, and thermodynamics. Using this then modern equipment improved both the social and hygienic status of those who were in these buildings, mainly in hospitals, schools and large public buildings. Beside a number of books on the topic, he also contributed to the Dictionary of Architecture and Building. Baldwin was a member of the American Society of Mechanical Engineers ASME, and of the American Society of Heating and Ventilating Engineers ASHVE.

Anonymous (1924). William J. Baldwin. *Trans. ASHVE* 30: 404-405.

Baldwin, W.J. (1883). *Steam heating for buildings*. Wiley: New York.

Baldwin, W.J. (1889). *Hot-water heating and fitting or warming buildings by hot-water*: A description of modern hot-water heating apparatus, the methods of construction and the principles. The Engineering & Building Record: New York.

Baldwin, W.J. (1897). *Steam heating data*. New York.

Baldwin, W.J. (1899). *An outline of ventilation and warming*. Baldwin: New York. *P*

Baldwin, W.J. (1908). *Hot water heating and fitting, modern hot water apparatus*: The method of their construction and principles involved. McGraw-Hill: New York.

BALL

20.04. 1905 Larimer County CO/USA
04.07. 2001 Denver CO/USA

James Wesley Ball graduated as civil engineer from Colorado State University. He also held an MS degree in civil engineering from the University of Colorado. Ball joined the US Bureau of Reclamation USBR, Denver CO, for over 30 years, supervising scale-model tests of hydraulic structures. His research was described in a number of technical publications. He was from 1962 a vice-president and managing technical director of the Western Canada Hydraulic Laboratories Inc., directing there hydraulic tests and research related to engineering projects. Later, he taught graduate engineering courses at Colorado State University. In parallel he was a consultant in the Philippines, India, Pakistan and Australia. He was awarded the ASCE Hilgard Hydraulic Prize in 1968.

Ball was an expert in high-speed chute flow, and one of the early hydraulic engineers investigating cavitation and means to reduce damages. His particular concern was that cavitation damages occurred despite the finish quality of the concrete chute surface was excellent, yet extremely small deviations from a perfectly plane surface still persisted, resulting in damages that ultimately led to the loss of the spillway. Ball's classical 1976 paper presents definite numbers under which cavitation damage occurs, thereby guiding hydraulic engineers in terms of quality control. Other research included cavitation observations in a number of hydraulic elements.

Anonymous (1962). James W. Ball. *Civil Engineering* 32(6): 47. *P*
Anonymous (1968). James W. Ball. *Civil Engineering* 38(10): 74. *P*
Ball, J.W. (1963). Construction finishes and high velocity flow. *Journal of the Construction Division* ASCE 92(CO2): 91-110.
Ball, J.W., Tullis, J.P. (1973). Cavitation in butterfly valves. *Journal of the Hydraulics Division* ASCE 99(HY9): 1303-1318; 100(HY4): 620.
Ball, J.W., Sweeney, C.E. (1974). Incipient cavitation damage in sudden enlargements. *Cavitation Symposium* C155: 73-79. Institution of Mechanical Engineers: London.
Ball, J.W. (1976). Cavitation from surface irregularities in high velocity flow. *Journal of the Hydraulics Division* ASCE 102(HY9): 1283-1297; 103(HY4): 469-472; 103(HY8): 945-946; 104(HY8): 1199-1200.
Russell, S.O., Ball, J.W. (1967). Sudden enlargement energy dissipator for Mica Dam outlets. *Journal of the Hydraulics Division* ASCE 93(HY4): 41-56; 94(HY2): 598-600; 94(HY3): 764-766; 94(HY5): 1351-1354; 95(HY2): 705-708.

BALTZER

04.08. 1931 Lansing MI/USA
19.08. 2009 Vienna VA/USA

Robert 'Bob' Arved Baltzer earned his BS and MS degrees from University of Michigan, Ann Arbor MI. After work in practice, he submitted in 1967 his PhD thesis to the Dept. of Civil Engineering, where he had been tutored by Victor L. Streeter (1909-), Ernest F. Brater (1912-2003) and James W. Daily (1913-1991). He then moved to the Surface Water Research Section, Water Resources Division, US Geological Survey USGS, as research engineer remaining there until his retirement.

Baltzer worked all through his career numerically on problems of water resources, open channel hydraulics and environmental hydraulics. The 1961 research report on tidal currents is introduced with a historical review, followed by a discussion on wave motion and the tides. The long-wave equations of continuity and momentum were developed in terms of stage and discharge, thereby neglecting wind and Coriolis forces. Solution methods including Fourier series, power series and the method of characteristics are then presented. A power series technique for discharge is developed using a Taylor series expansion from a location of known boundary and initial conditions to a downstream location of unknown discharge. The difference form of the equations is then used for a computer solution, which is described by flow charts. Four field applications were considered and compared with prototype data, indicating a favorable agreement.

Anonymous (1953). Robert Baltzer. *Michiganensian yearbook*: 167. Ann Arbor MI. *P*

Baltzer, R.A. (1958). *Tidal flow investigations*. US Geological Survey: Washington DC.

Baltzer, R.A., Shen, J. (1961). Flows of homogeneous density in tidal reaches. USGS *Report*.

Baltzer, R.A., Lai, C. (1966). Computer simulation of unsteady flows in rivers and estuaries. ASCE *Hydraulics Division Specialty Conf.* Madison: 1-68. ASCE: New York.

Lai, C., Schaffranek, R.W., Baltzer, R.A. (1992). Frictional resistance treatment in unsteady open-channel flow simulation. *Channel flow resistance*: Centennial of Manning's formula: 409-420, B.C. Yen, ed. Water Resources Publications: Littleton CO.

Regan, R.S., Schaffranek, R.W., Baltzer, R.A. (1996). Time-dependent data system: An interactive program to assemble, manage and appraise input data and numerical output of flow/transport simulation models. *Water Resources Report* 96-4143. USGS.

Rybicki, N.B., Jenter, H.L., Carter, V., Baltzer, R.A., Turtora, M. (1997). Observations of tidal flux between a submerged plant stand and the adjacent channel in the Potomac River near Washington DC. *Limnological Oceanography* 42(2): 307-317.

BANKS F.A.

04.12. 1883 Saco ME/USA
14.12. 1957 Pasadena CA/USA

Frank Arthur Banks graduated as a civil engineer from University of Maine. He went West in 1906 to join the US Reclamation Service, currently the US Bureau of Reclamation USBR. He rose from rodman to construction engineer and district manager of construction and development of the Great Columbia Basin Project. His work in this capacity culminated in the design of Grand Coulee Dam. The Columbia River winds almost 2,000 km from the mountains of British Columbia to the Pacific. As construction and supervising engineer and project manager for USBR from 1933 to 1950, Banks was in charge of the construction of four large earth dams and all irrigation facilities built during the period, in addition to Grand Coulee Dam.

Banks was a consulting engineer after retirement in 1950 to the Province of British Columbia for *Kenny* Dam, and to the Government of India on the construction of *Bhakra* Dam. He received many honors, including the Distinguished Service Award, highest honor of the US Department of the Interior. The citation states that 'his work across the valley of the Columbia will aid the nation as long as water runs downhill'. Banks was a gentle, soft-spoken, white-haired, erudite man who spoke with the trace of a Down East accent which he brought West from Maine. The dams which Banks built in the Great Basin supply water to some 1 million acres. The 30,000 farms for which the dams erected provide water, cover an area of more than 3,000 square miles. In the 1950s, the Grand Coulee Dam was the largest power producing unit of the world. As a scenic attraction its water fall is more than twice the height of Niagara Falls. The Banks Lake is a 40 km long reservoir in central Washington USA, named after Frank Banks.

Anonymous (1939). Frank A. Banks succeeds the late J.D. Ross. *Power Plant Engineering* 43(6): 414. *P*

Anonymous (1939). Banks to be chief at Bonneville. *Engineering News-Record* 122(May 4): 635. *P*

Anonymous (1950). Frank Arthur Banks. *Reclamation Era* 36(7): 130-132. *P*

Anonymous (1958). ASCE Honorary Member F.A. Banks dies. *Civil Engineering* 28(2): 131.

Banks, F.A. (1934). Columbia Basin Project is described by construction engineer. *Southwest Builder and Contractor* 85(Nov.23): 8-9.

McDonald, J.C., Banks, F.A. (1934). Developing the Columbia River Drainage Basin. *Civil Engineering* 4(9): 443-459.

BANKS H.O.

29.03. 1910 Chaumont NY/USA
22.09. 1996 Austin TX/USA

Harvey Oren Banks graduated in 1930 as civil engineer from Syracuse University, Syracuse NY. His moved then as instructor to Stanford University, Stanford CA, where he had also done graduate work in hydraulic and sanitary engineering. Other positions were with the US Soil Conservation Service SCS, the New York Dept. of Public Works, and the city of Palo Alto CA. He joined the California Division of Water Resources in 1938 working on groundwater resources. During World War II he served in the US Army Corps of Engineers, leaving as major. In 1946 Banks left state service to enter private practice as a partner of consulting engineers at Los Angeles CA. One project was a comprehensive plan for the development of the water resources of California's Ventura County.

Banks returned to the Division of Water Resources DWR in 1950, becoming eventually state engineer, its chief in 1955 on the retirement of Arthur D. Edmonston (1886-1957). This job was a civil service position, whereas his new director's position was not. Under his direction, DWR completed its first California Water Plan and initiated the first stage of planning of the California State Water Project SWP. The original SWP design did not call for a creation of a large off-stream storage reservoir along the California Aqueduct, yet Banks and others advocated that the original design be altered. Their plan called for a large multi-use facility near Los Banos CA also tying into the US Federal pre-existing Central Valley Project. This combined facility, later named San Luis Reservoir, offered both projects flexibility. It was under his term as Director that the California Legislature adopted the Burns-Porter Act authorizing construction of SWP in 1959. In 1961 voters approved massive bonds to pay for the construction of the SWP. Banks left the public sector that same year to return to the private sector, where he worked as a consulting engineer until 1996. Banks' papers (1940-1995) are held at the Water Resources Center Archives, University of California, Berkeley CA. The SWP pumping plant at the head of the California Aqueduct near Tracy CA was renamed in 1981 in honor of Banks.

Anonymous (1956). California fills a prize engineering post: Harvey O. Banks. *Engineering News-Record* 156(Jun.14): 83-86. *P*
Banks, H.O. (1952). Utilization of underground storage reservoirs. *Proc. ASCE* 78(114): 1-15.
Banks, H.O. (1955). *Solutions to the water problems.* Committee on Conservation: Sacramento.
Carroll, W.J. (2002). Harvey O. Banks. *Memorial Tributes* 10: 8-13. NAE: Washington DC. *P*
http://en.wikipedia.org/wiki/Harvey_Oren_Banks

BARBAROSSA

07.08. 1915 Boston MA/USA
20.09. 1988 Williamsburg VA/USA

Nicholas Leonard Barbarossa graduated as a civil engineer from Massachusetts Institute of Technology MIT. He joined in 1938 the University of California, Berkeley CA, doing his PhD thesis under Hans A. Einstein (1904-1973) at the hydrodynamics research group. He took over in 1952 as assistant professor of civil engineering at the University of Minnesota, Minneapolis MN. It appears that he preferred a position in engineering practice, so that he joined in the mid-1950s the US Army Engineering Division, responsible for Missouri River, Omaha NE. He was there promoted to chief of its Hydraulics Section in 1959, and later took over as Director of Planning the Missouri River Basin Commission MRBC. He also served from 1962 to 1965 as the first state director of water resources planning for the State of New York, when his responsibilities were enlarged to include regulatory and management tasks as assistant director of the Division of Water Resources. He was a member of the American Society of Civil Engineers.

Barbarossa was involved with Hans Albert Einstein in the re-analysis of river flow resistance. They proposed two divide the total resistance into components made up by roughness drag, and form drag, by which the two were related to the corresponding parameters. The first practical implementation of this concept, the so-called Einstein-Barbarossa method, was used to estimate the relation between flow depth and discharge in alluvial rivers. To validate their procedure, field data from the Missouri River, several of its tributaries and two California rivers were employed. The roughness was estimated using the method of Albert Strickler (1887-1963). This approach was acknowledged milestone in alluvial river mechanics.

Anonymous (1956). N.L. Barbarossa. Proc. 6th *Hydraulics Conference* Iowa: Frontispiece. *P*
Anonymous (1970). Nicholas Leonard Barbarossa. 10th *ICOLD Congress* Montreal 6: 110. *P*
Barbarossa, N.L. (1959). Hydraulic analysis of surge tanks by digital computer. *Journal of the Hydraulics Division* ASCE 85(HY4): 39-78; 86(HY5): 106-112.
Barbarossa, N.L. (1962). *Garrison hydro-power plant transients*. US Army Engineer Division, Corps of Engineers: Omaha NE.
Barbarossa, N.L., Fuhriman, D.K., Maktari, A.M.A. (1977). *Report* on Water resources sector study in the Yemen Arab Republic. USAID: Sanaa.
Einstein, H.A., Barbarossa, N.L. (1952). River channel roughness. *Trans. ASCE* 117: 1121-1146.
Smith, K.V.H. (1970). The Einstein-Barbarossa diagram. *Proc. ICE* 46(2): 169-184.

BARDSLEY

23.04. 1894 St. Louis MO/USA
26.04. 1967 Rockville MD/USA

Clarence Edward Bardsley obtained the BS degree from the Missouri Mining School, St. Louis MO in 1920, the MS degree in civil engineering there in 1924 and the ScD degree in 1926. After a short stay at Northwestern University, he moved from 1928 to 1929 to *Technische Hochschule* Berlin, Germany. Upon return to the USA he was from instructor to professor of civil and hydraulic engineering at his Alma Mater until 1938, from when he became civil engineering professor at Oklahoma Agricultural and Mechanical College, Stillwater OK. From 1942 until retirement, he took over as hydraulic engineer of the Pittsburgh District, US Corps of Engineers, where he again worked on hydraulic issues. He was member of the Missouri and the Oklahoma Academies, and of the International Association of Hydraulic Research IAHR.

Bardsley was engaged during his engineering career with numerous positions, namely as city engineer, county engineer, State representative for the US Coast and Geodesy Survey, railroad engineer and as consultant. He also was engineer for the Skaggs Dam in Texas in the mid-1930s, and only then entered in research when joining Oklahoma College. His reports include a study on the Lower Mississippi River and a typical work on the optimization of hydraulic laboratory studies. It was in these years that hydraulic laboratories became reliable because the basic instrumentation and the fundamental laws of similitude had developed so that model observations could be up-scaled to the prototype. Bardsley in addition wrote two relatively short sketches on the history of hydraulics, one on pipe flow advances, whereas the other deals with the general history of hydraulic sciences.

Anonymous (1943). Bardsley, C.E. *American men of science* 7: 88. Science Press: Lancaster PA.

Bardsley, C.E. (1932). *Investigations of diversion on the Lower Mississippi River*. School of Mines and Metallurgy, University of Missouri: Rolla MO.

Bardsley, C.E. (1939). *Historical resume of the development of the science of hydraulics*. Engineering Expt. Station, Oklahoma Agricultural and Mechanical College, Stillwater.

Bardsley, C.E. (1940). *Historical sketch of flow of fluids through pipes and suggested solutions of pipe flow problems*. Oklahoma Agricultural and Mechanical College. Stillwater OK.

Bardsley, C.E. (1941). *Appurtenances for open channel hydraulics model*. Engineering Experiment Station, Oklahoma Agricultural and Mechanical College. Stillwater OK.

Rouse, H. (1976). Bardsley. *Hydraulics in the USA* 1776-1976: 117. IIHR: Iowa City IA. *P*

BARKER

17.05. 1898 St. Paul MN/USA
08.06. 1959 Knoxville TN/USA

Clifton Thorne Barker was graduated in 1922 from the University of Minnesota, Minneapolis MN, with the degree of engineer of mines. His early experience included appointments with construction companies in the Upper Ohio Valley. The remainder of his life was spent in the service of the federal government. He worked with the US Engineer Department on the Tennessee River Survey at Chattanooga TN from 1924 to 1928, then until 1932 on Missouri River studies for multipurpose projects at Kansas City KS, and on the navigation improvement at Sioux City IA until 1933. He was also transferred to the Wilson Dam AL, to take charge of navigation and power engineering studies.

Barker was with the Tennessee Valley Authority TVA from its establishment in 1933. His knowledge on the design and the economics of inland waterway navigation led to his appointment as chief of the River Transportation Division in 1938, and he was in 1946 made staff specialist in transportation. The US Congress formed TVA in 1933 to rectify poor conditions in Tennessee River Valley, including destructive floods, erosion, deforestation, and periodic river unnavigability. As an independent agency, TVA is governed by a Board of three directors appointed by the US President, who serve for a term of nine years. Among TVA's various projects are dams, of which 26 were built by TVA, electric power, and associated research programs. The headquarters of TVA are at Knoxville KY serving for an area of some 100,000 km^2. Barker was a member of the Society of American Military Engineers SAME, and of the American Society of Civil Engineers ASCE from 1959, whose associate member he was from 1935. The Clifton T. Barker Scholarship at the University of Minnesota is directed to sophomores or juniors.

Anonymous (1922). Clifton T. Barker. *The Gopher* 35: 162. University of Minnesota. *P*
Anonymous (1960). Clifton T. Barker. *Trans. ASCE* 125: 1395-1396.
Barker, C.T. (1938). *TVA Dewey decimal classification system master.* TVA: Knoxville.
Barker, C.T., Spottswood, A.D. (1940). *The initial phase of public-use terminal development* at Chattanooga TN. TVA Knoxville.
Barker, C.T. (1944). *Transportation development and other programs of the Tennessee Valley Authority.* TVA: Knoxville.
Barker, C.T. (1953). *River traffic and industrial growth.* TVA: Knoxville.
Hay, W.W., Barker, C.T. (1942). *Proposed Tennessee River terminals* at Knoxville, Chattanooga, Guntersville, and Decatur. TVA: Knoxville.

BARNARD

19.05. 1815 Sheffield MA/USA
14.05. 1882 Detroit MI/USA

John Gross Barnard graduated in 1833 from the US
Military Academy, West Point NY, and was posted
then as second lieutenant in the US Army Corps of
Engineers. He served on garrisons and fortifications,
participating in the construction of coastal defences
at Fort Columbus/Fort Jay, Fort Hamilton and Fort
Wadsworth in New York City NY, New Orleans LA,
Fort St. Philip LA, and San Francisco CA. During
the American-Mexican War 1846-1848, he headed
construction of American defences at the captured
Mexican port of Tampico. During the Civil War, he
was assigned to the Department of Washington to
defend the capital, becoming chief engineer of the Military District of Washington DC.
President Lincoln appointed Barnard in 1861 brigadier general. He was then engineer
for the Army of the Potomac, participating in the Peninsula Campaign, and directing the
siege works at Yorktown VA. Upon the death of the Chief of the US Army Corps of
Engineers, Brigadier General Joseph Totten (1788-1864), President Lincoln nominated
Barnard as his successor, yet Barnard asked that this nomination be withdrawn. He was
then on the staff of General Grant in the Overland Campaign, and was mustered out in
early 1866.

Barnard was promoted to colonel in the Regular Army and continued his career in the
Army Corps of Engineers until 1881. Soon after the Civil War, he was made president
of the permanent Board of Engineers for Fortifications, River and Harbour Improvements,
a position he held also until 1881. He successfully recast the approach to coastal defense
which was required because of the obsolescence of wooden ships and muzzle loading
guns. He also advocated the successful use of parallel jetties to improve the mouth of
the Mississippi River. He further was a prominent member of the US Lighthouse Board
from 1870 until his retirement in 1881. Barnard was an original member of the Aztec
Club of 1847, as well as the Military Order of the Loyal Legion of the USA.

Abbot, H.L. (1902). Biographical memoir of John Gross Barnard. *National Academy of
        Sciences*: 219-229. *P*
Barnard, J.G. (1859). *Dangers and defences of New York City*. Washington DC.
Barnard, J.G. (1861). *Notes on seacoast defence*. Washington DC.
Barnard, J.G. (1872). *Report on the North Sea Canal of Holland*: The improvement of
        navigation from Rotterdam to the Sea. Government Printing Office: Washington DC.
http://en.wikipedia.org/wiki/John_G._Barnard *P*

BARNES G.E.

17.04. 1898 Washington DC/USA
17.11. 1979 Rancho Palos Verdes CA/USA

George Eric Barnes received education from the Massachusetts Institute of Technology, Cambridge MA, and from the Case School of Applied Science, Cleveland OH, with the CE degree in 1935. He was research engineer at MIT and construction engineer of the Boston Metropolitan District from 1920 to 1923, then until 1929 associate professor of civil engineering at the University of Florida, Gainesville FL. Simultaneously he acted as associate engineer with a company on water supply and sewerage for municipalities, and was design engineer with Metcalf & Eddy, Boston MA, on hydraulic features of dams. He was in addition involved in studies on water supply and irrigation projects in Nicaragua and Honduras. From 1929 to 1932 Barnes joined an engineering company in New York and was engaged with its Department of Sanitation.

Barnes was appointed in 1933 professor of hydraulics and sanitary engineering at Case School, and there also served as head of the Civil Engineering Department. He was simultaneously consulting engineer for the Muskingum Watershed Project, and conducted hydraulic research for 11 dams for flood protection. He was further in charge of the Central Nebraska Irrigation District on Keystone Dam, and associate engineer on waterworks and sewers for various cities in Ohio State. Other projects included the Mahoning Dam for the Pittsburgh Flood Control Project, sewerage works for the cities of Lorain OH, Painesville OH and Cleveland. Barnes was a member of the Board of Inquiry for the Government of Ohio, and was the head technical consultant for the city of Cleveland OH, among many other positions. He was a member of the American Society of Civil Engineers ASCE, presiding its Cleveland Section.

Anonymous (1923). George E. Barnes. *Technique yearbook*: 292. MIT: Cambridge MA. *P*

Barnes, G.E., Kenney, A.A. (1923). *A test of one unit of the new hydro-electric plant of the Amoskeag Manufacturing Company*. Dept. Civil and Sanitary Engineering, MIT: Cambridge MA.

Barnes, G.E. (1935). *Report* on hydraulic model studies for the outlet works and spillway of the Charles Mill Dam near Mansfield, Ohio. US Engineer Office, Zanesville OH.

Barnes, G.E., Jobes, J.G. (1938). Construction and testing of hydraulic models, Muskingum Water-Shed Project. *Trans. ASCE* 103: 227-255.

Barnes, G.E. (1950). Hydraulic models and similitude. *Hydroelectric handbook* ed. 2: 144-147, W.P. Creager, J.D. Justin, eds. Wiley: New York.

BARNES M.G.

17.01. 1867 Reedsburg WI/USA
07.10. 1930 Oak Park IL/USA

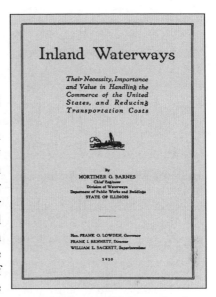

Mortimer Grant Barnes received both the BS and CE degrees from University of Michigan, Ann Arbor MI. He began his civil engineering practice in 1896 in Nebraska. He acted then as assistant to Joseph Riley on the preliminary design of Birmingham Canal from Birmingham AL to the 100 km distant Black Warrior River. As hydraulic engineer he designed a power plant for Sault Ste. Marie MI, and was then engaged as chief engineer with deep waterways surveys in Northern New York State. From 1900 to 1905 he was engaged on the design and the construction of the Illinois and Mississippi Canal, from where he became again assistant of Joseph Riley, now in charge of designs of locks and other hydraulic structures for Panama Canal. Upon completion of the preliminary design and the report, he resigned to accept appointments with the New York Board of Water Supply.

Barnes was a successful canal engineer designing a number of works in the United States. He had a 20-years' experience as civil engineer when adding to the Panama Canal. After resigning these works in 1907 and after having been engaged by the New York Board of Water Supply he formed with a colleague a general hydraulic and construction engineering firm at Albany NY. He was a member of the Advisory Board of Consulting Engineers for Improvement of State Canals from 1907 to 1911, a member of the Board of Consulting Engineers of the New York State Canals from then to 1915, and chief engineer of the Division of Waterways, State of Illinois since 1917. He also was a member of the American Society of Civil Engineers ASCE. He was awarded the degree of master of engineering by the University of Michigan in 1922.

Anonymous (1922). Barnes, Mortimer Grant. *Who's who in engineering* 1: 103. Lewis: New York.

Anonymous (1931). Mortimer G. Barnes. *Trans. ASCE* 95: 1449-1450.

Anonymous (1968). *Starved Rock Locks & Dam*, Illinois Waterway River mile 231, Peru IL. Historic American Engineering Record, Library of Congress, Washington DC.

Barnes, M.G. (1920). *Inland waterways*: Their necessity, importance and value in handling the commerce of the United States, and reducing transportation costs. Barnard & Miller: Chicago. (*P*)

Barnes, M.G. (1922). *The Illinois Deep Waterway*. Collection of Illinois Historical Library: Springfield IL.

Barnes, M.G. (1926). Lockport lock construction plant, Illinois Waterways. *ENR* 96(6): 228-231.

BARROWS

09.11. 1873 Melrose MA/USA
15.03. 1954 Winchester MA/USA

Harold Kilbrith Barrows graduated as civil engineer in 1895 from Massachusetts Institute of Technology MIT. After his work with the Metropolitan Water Board of Boston MA, he was on the Faculty of the University of Vermont for three years. From 1904 to 1908, Barrows was a District Engineer of the US Geological Survey in New England and New York. He then joined the MIT Hydraulic Engineering Department, from where he retired in 1941.

Barrows was both a hydraulics professor for more than 30 years and a consultant, with his office in Boston since 1907. He had an extensive practice in hydraulic engineering, water power, water supply and flood control, and was since 1934 also a regional consultant for the Natural Resources Planning Office. From 1928 to 1930 he acted as a Member of the Advisory Committee of Engineers on flood control for the State of Vermont, and later for the State of New Hampshire. Barrows has written a number of books of which his Water power engineering is certainly most known. This book includes the chapters: (1) Water power development, (2) Hydrology, (3) Stream-flow data, (4) Turbines, (5) Dams, (6) Canals and penstocks, (7-8) Powerhouse, (9) Plant accessories, (10) Speed and pressure regulation, (11) Transmission lines, (12) Cost and value of water power, and (13) Reports and plants descriptions. He also added a chapter to the well-known book of Sherman and Horton on water resources. Additional works include a discussion to one of the first shaft spillways, a hydraulic element that was essentially developed in the United States and has become a standard in hydraulic engineering since the 1960s.

Anonymous (1940). Barrows to retire from MIT. *Engineering News-Record* 124(20): 681. *P*
Anonymous (1954). Harold K. Barrows. *Engineering New-Record* 152(Mar.25): 60.
Barrows, H.K. (1907). *Surface water supply of the United States* 4: St. Lawrence River Basin. Government Printing Office: Washington DC.
Barrows, H.K. (1915). Discussion of The hydro-electric power plant at the Wachusett Dam, Clinton MA. *Journal of the Boston Society of Civil Engineers* 2(3): 111-113.
Barrows, H.K. (1915). Improvement to water supply of the city of Fall River, MA. *Journal New England Water Works Association* 30(9): 46-62.
Barrows, H.K. (1925). Discussion to The hydraulic design of the shaft spillway for the Davis Bridge Dam. *Trans. ASCE* 88: 66-73.
Barrows, H.K. (1927). *Water power engineering*. McGraw-Hill: New York.
Sherman, L.K., Horton, R.E. (1933). Rainfall, runoff and evaporation. *IAHS Bulletin* 20: 22-96.

BATES

10.06. 1777 Morristown NJ/USA
28.11. 1839 Rochester NY/USA

David Stanhope Bates took up the study of survey and mathematics while being a clerk. He was then employed by the owner of large tracts in Oneida County NY to survey his land in 1810. From 1817 he was assistant engineer for the middle division of the Erie Canal, and between 1819 and 1824 there division engineer in charge of works in Irondequoit Valley, thereby designing and constructing the first aqueduct over Genesee River at Rochester NY, which consisted of eleven masonry arches totalling 240 m length. He was also in charge of lock construction at Lockport NY. The almost 600 km long Erie Canal between Lake Erie and Hudson River NY was opened in 1825; it represented a colossal enterprise and the most prestigious project of the young USA, praised then as a civil engineering marvel. This canal allowed for westward migration and within 1840 it turned New York City to the largest sea port of the country. The builders of the Great Western Canal struck out westward from Rome NY in 1817, directed by chief engineer Benjamin Wright (1770-1842). Existing streams and lakes were joined by the canal route. The 130 km middle division was cut in light soil across level terrain, where locks and aqueducts were not required, built by Wright's assistant Bates.

Bates was from 1824 to 1829 the principal engineer of the Ohio Canal system thereby surveying some 1,300 km of canal length and feeders. He acted in parallel as chief engineer of the Louisville and Portland Canal Company, supervising and constructing the canal around the Louisville Falls. From 1828 to 1834 Bates was chief engineer of the Niagara River Hydraulic Company, as also of the Chenango Canal from Utica NY to Binghampton NY. He was commissioned to survey the route of the Genesee Valley Canal and surveyed the line of the Auburn and Rochester Railroad in 1830. He was in 1834 appointed state engineer of Michigan to make surveys for the Erie and Kalamazoo Railroad, yet he resigned in 1835 because of his poor health.

FitzSimons, N., ed. (1972). Bates, David S. *A biographical dictionary of American civil engineers*: 7-8. ASCE: New York.
Herringshaw, T.W. (1858). Bates, D.S. National Library of *American biography*: Washington DC.
Koeppel, G. (2009). *Building the Erie Canal and the American Empire*. Philadelphia. (*P*)
Langmead, D., Garnaut, C. (2001). Erie Canal. *Encyclopaedia of Architectural and engineering feats*: 112-113. ABC-Clio Inc.: Santa Barbara CA.
Vought, J.G., Bates, D.S. (1825). *John G. Vought Letter*. Washington DC.

BAUER

20.09. 1923 Des Moines IA/USA
02.06. 2004 Wheaton IL/USA

William John Bauer obtained his education from the
College of Engineering, University of Iowa, Iowa
City, including the BS in 1947, the MS in 1948, and
the PhD degree in 1951. He then moved as a design
engineer to *Harza* Eng. Co., Chicago IL and in 1955
joined as senior designer the firm John F. Meisner,
Chicago. During this period, he was involved in the
design of improved culverts. In 1959, he founded
the firm Bauer Engineering Inc., Chicago IL, from
where he retired in 1976, and which then changed
the name to Keifer and Associates Inc., presided by
Clint J. Keifer (1920-2007).

Bauer is known in hydraulic engineering for an outstanding paper dealing with the
development of the turbulent boundary layer, which summarizes his PhD findings. This
phenomenon does not only control the head losses along a steep-sloping incline, but
also air entrainment, such that this knowledge is essential for a proper chute design. The
paper describes the variation of the velocity profiles for both smooth and rough chutes
in terms of the chute angle and the unit discharge. The boundary layer thickness, the
coefficient of local resistance, and the parametric shape of the velocity profile were
analyzed as functions of Reynolds number and relative roughness. His research may be
considered an expansion of the boundary layer theory by inclusion of slope and relative
discharge. These results were discussed, among others, by the Frenchman Georges
Halbronn (1920-), who had conducted a similar research at Grenoble University. Few
additions to their work have been advanced until today on this particular aspect of chute
flow. Bauer was also a contributor to the Handbook of applied hydraulics, namely on
Basic Hydraulics, and on Spillway and streambed protection works.

Anonymous (1955). William J. Bauer. *Civil Engineering* 25(10): 713. *P*
Bauer, W.J. (1953). The development of the turbulent boundary layer on steep slopes. *Proc.
    ASCE* 79(281): 1-23.
Bauer, W.J. (1954). Turbulent boundary layer on steep slopes. *Trans. ASCE* 119: 1212-1242.
Bauer, W.J. (1959). Improved culvert performance through design and research studies. *Civil
    Engineering* 29(3): 167-169.
Davis, C.V., Sorensen, K.E., eds. (1969). *Handbook of applied hydraulics*. McGraw-Hill: New
    York.
Koelzer, V.A., Bauer, W.J., Dalton, F.E. (1969). The Chicago area deep tunnel project. *Journal
    of Water Pollution Control Federation* 41(4): 515-534.

BAUMGARTNER

06.04. 1902 Portland OR/USA
22.02. 1959 Tehran/IR

John Albert Baumgartner graduated from Oregon Agricultural College, later the Oregon State College, Corvallis OR, in 1923 with the BSc degree in civil engineering. He began his life-long career with the US Geological Survey USGS in 1925 as engineer in the Water Resources Division, Tucson AZ District. He advanced through the following grades: from 1929 to 1934 he was assistant engineer, from 1935 to 1946 associate engineer, 1946 to 1950 engineer, 1950 to 1956 senior engineer, and finally until his death principal engineer. Baumgartner was from 1930 to 1950 responsible for most of the Arizona flood-peak estimates, giving leadership to the construction of 110 gauging stations in his District. In addition to field work and construction, he was an acknowledged expert in the preparation of stream-flow records. In conjunction with the Point Four-Program, he served as technical adviser to the Iranian government on water resources in addition.

Baumgartner prominently contributed to the regional determination of hydrology and water resources. The 1939 paper involves an improvement of the method of O.P. Stout (1865-1935), drawing a tangent to an existing discharge-stage curve at a point, above which no shift is assumed. The gage correction is expressed in terms of a ratio of the distance between the discharge curve and the tangent, and a graphical interpolation of correction is proposed. Baumgartner's 1943 paper deals with diagrams for the cable length in the air and for a wet line combined on one plate, for discharge measurement in rivers. His 1945 paper is concerned with gage wells driven into Gila River in Arizona using two 20 m railroad rails to 4 m depth to measure discharge. Cable guys were stretched in addition to assure stability of the set-up.

Anonymous (1960). John A. Baumgartner. *Trans. ASCE* 125: 1397.
Baumgartner, J.A. (1939). Discharge determination by the curvature ratio method. *Water Resources Bulletin* (Aug.10): 445-447.
Baumgartner, J.A. (1940). Roughness coefficients for slope-area determinations. *Water Resources Bulletin* (Nov.10): 220-221.
Baumgartner, J.A. (1943). Two-table correction chart. *Water Resources Bulletin* (Nov.10): 158.
Baumgartner, J.A. (1945). Installation of gage wells in silt banks. *Water Resources Bulletin* (May10): 78-79.
Baumgartner, J.A. (1946). Adjustable crane. *Water Resources Bulletin* (Feb.10): 19.
https://www.google.ch/search?q=%22John+A.+Baumgartner%22+Arizona *P*

BAYLEY

30.11. 1821 Saratoga NY/USA
14.12. 1876 New Orleans LA/USA

George Willard Reed Bayley graduated in 1838 from the Rensselaer Polytechnic, Troy NY. He was then employed as railway engineer in his State, but soon after went to sea, spending two years in a tour around the world. After his return, he soon moved to Baton Rouge LA, accepting a position as assistant State engineer until 1852. There he was in charge of hydrographic work of the Mississippi River, which initiated his interest in the physics of large rivers. After work as chief engineer of the New Orleans Rail Road for ten years, he was appointed surveyor of New Orleans LA. Further, he was again in charge of New Orleans Railroad, from where originates his History of railroads in Louisiana, a work of great research, published in issues of the *Picayune* newspaper.

When the contract by Congress was made for the improvement of the South Pass of the Mississippi by jetties, Bayley was selected by James B. Eads (1820-1887) as resident engineer, a position he held until his death. Baylay believed in the ultimate success of this method using jetties of obtaining deep water to the sea. His papers are marked by solidity and research. He wrote an article on the Mississippi. In 1875 he served the Louisiana Levee Company, as engineer and commissioner of levees, also as resident engineer of the jetties. He there prepared an article on levees both in the Johnson's Cyclopedia, and in the ASCE Transactions. He further was for three years member of the Board of Health for New Orleans, coming thereby into contact with epidemics in urban areas; he wrote a paper on the results of the use of disinfectants. He also was from 1874 to 1876 a member of the House of City Representatives, so that by this large accumulation of positions, his health failed, and he passed away at ago of only 55.

Anonymous (1878). G.W.R. Bayley. *Proc. ASCE* 3: 58-60.
Bayley, G.W.R. (1858). *Remarks upon the question of closing the Bayou Plaquemine*. Taylor: Baton Rouge.
Bayley, G.W.R. (1872). *Communication* on construction and management of streets, street gutters, and drainage canals. Board of Health: New Orleans.
Bayley, G.W.R. (1875). Levees as a system for reclaiming lowlands. *Trans. ASCE* 5: 115-146.
Hartley, C. (1983). Sir Charles Hartley and the mouths of the Mississippi. *Louisiana History* 24: 261-287.
Prichard, W., ed. (1947). A forgotten Louisiana Engineer: G.W.R. Bayley. *Louisiana Historical Quarterly* 30(4): 1065-1325. P

BEAN

23.10. 1893 Santa Clara CO/USA
10.06. 1975 Sedona AR/USA

Howard Stewart Bean graduated as a physicist from University of California in 1917. From 1919, Bean was an associate engineer and later chief of the Gage Section, National Bureau of Standards NBS, Washington DC, from where he took over as chief engineer the Capacity, Density and Fluid Meters Section. This Section was formed by consolidation of the Gas Measuring Instrument Section, of which Bean was chief, and the Capacity and Density Section. He was elected Fellow ASME in 1950, received the ASME Worcester Warner Reid Medal in 1955 and retired from the NBS in 1958.

Bean was widely known for his extensive work in the field of fluid measurement and on the application of discharge meters to the fuel gas industry. Both gas and liquid flows were considered. Much of his works included the determination of coefficients upon which the Reynolds number depends. The method and data, for which he was to a great extent responsible, were of basic importance in the steam, oil, and gas industries. As chairman of the Subcommittee on Revision of report, ASME Research Committee on Fluid Meters, he prepared the text and edited the fourth edition of Fluid Meters. From 1936 to 1946, Bean was also chairman of the ASME Committee on flow-nozzle research. A number of committees has benefitted for many years from his work, in which the large number of available data were listed, discussed and analyzed.

Anonymous (1931). Bean, Howard Stewart. *Who's who in engineering* 3: 85. Lewis: New York.
Anonymous (1948) H.S. Bean heads new NBS section. *Mechanical Engineering* 70(11): 933. *P*
Anonymous (1955). Howard Stewart Bean. *Mechanical Engineering* 77(12): 1129. *P*
Anonymous (1975). H.S. Bean. *Mechanical Engineering* 97(10): 97.
Bean, H.S., Buckingham, E., Murphy, P.S. (1929). *Discharge coefficients of square-edged orifices for measuring the flow of air*. US Government Printing Office: Washington DC.
Bean, H.S. (1935). Values of discharge coefficients of square-edged orifices. *American Gas Association Monthly* 17(7): 259-265.
Bean, H.S. (1937). *Flow meters*: Their theory and application. ASME: New York.
Bean, H.S., Beitler, S.R. (1938). Some results from research on flow nozzles. *Trans. ASME* 60(RP-3): 235-244; 61(D-1): 40-42.
Bean, H.S. (1947). Indications of an orifice meter. *American Gas Association Monthly* 29: 337.
Bean, H.S. (1960). *Effect of pipe roughness on orifice meter accuracy*. American Gas Association: New York.

BEARD

06.04. 1917 West Baden IN/USA
21.03. 2009 Austin TX/USA

Leo Roy Beard graduated as a civil engineer in 1939 from Caltech. He worked in water resources development then with the Corps of Engineers until 1972, finally as Director of Hydrologic Engineering Center HEC, Davis CA. Beard was in parallel a professor of civil engineering at the University of Texas, Austin, where he had directed the Center for Research in Water Resources from 1972 to 1980. He was from then associated with *Espey-Huston &* Associates, and the Editor of the journal Water International. Beard was elected to the National Academy of Engineering in 1975. He was made an Honorary Member of the American Water Resources Association in 1980. In 1981 he received the ASCE Julian Hinds Award 'in recognition of his outstanding contributions in the field of water resources planning and management, for leadership and dedication to ASCE and other professional organisations'. In 1993, Beard presented the fourteenth Hunter Rouse Hydraulic Engineering Lecture. He received in 2007 the Ven Te Chow Award from ASCE.

The Ven Te Chow Award sponsored by Chow (1918-1981), and established in 1995, recognizes individuals who in their career in hydrologic engineering made significant contributions through research, education or practice. Beard was decorated for his work in flood hydrograph computations, system techniques for reservoir regulation, statistical methods for streamflow frequency analysis, and the computer-based methods for hydrologic computations. Early in his career, Beard became involved in flood control projects in the Los Angeles and Sacramento areas. The performance of the reservoirs in the Sacramento and San Joaquin River basins was tested during the 1955 floods. He also advocated the use of computer programs, and was credited for the development of the HEC-1 and HEC-3 models. His books have seen widespread use.

Anonymous (1965). Leo R. Beard. *Civil Engineering* 35(4): 50. *P*
Anonymous (1981). Leo R. Beard. *Civil Engineering* 51(10): 84. *P*
Anonymous (1993). Leo R. Beard. *Journal of Hydraulic Engineering* 120(6): v. *P*
Beard, L.R. (1962). *Statistical methods in hydrology*. US Army Engineer: Sacramento CA.
Beard, L.R. (1976). *Flood control by reservoirs*. Hydrologic Engineering Methods: Davis CA.
Beard, L.R., Maxwell, W.H.C., eds. (1982). *Water resources management in industrial areas.*
     Tycooly: Dublin.
http://www.hec.usace.army.mil/misc/LeoRoyBeard/LeoRoyBeard.htm *P*

BEARDSLEY

11.09. 1860 Coventry NY/USA
15.05. 1944 Syracuse NY/USA

James Wallace Beardsley graduated from New York
State Normal School in 1884, obtaining the civil
engineering degree in 1891 from Cornell University,
Ithaca NY. He was from 1892 to 1898 assistant
engineer with the Sanitary District, Chicago IL, in
charge of constructions. The next two years were
spent with the US Board of Engineers on Deep
Waterways, in charge of the St. Lawrence River
Surveys. Similar work then followed until 1902
with the US Army Corps of Engineers for surveys
at the Fox River IL, and the improvements of Sand
Beach Harbor MI. He was from 1902 consulting
engineer to the Philippine Islands Commission after the Spanish-American War, taking
over from 1903 to 1905 as chief of the Philippine Bureau of Engineers, Manila, and
until 1908 as director of Public Works, Philippine Islands.

From 1908 to 1909 Beardsley was consulting engineer for irrigation projects in Java,
then the Dutch Indies, India and Egypt, continuing as chief engineer of the Irrigation
Service of Puerto Rico until 1916. He had from 1916 to 1918 his private consulting
office, and was a consultant in ordinance. Until 1919 he was assistant chief engineer of
the Grand Canal surveys of China, taking over until 1921 as chief engineer and member
of the Junta Central de Caminos, the Republic of Panama. Finally, Beardsley was
engaged by the Public Works Department of the Dominican Republic, Santo Domingo,
from 1926 to 1929, then returning to his consulting office. He was a member of the
American Society of Civil Engineers ASCE, the Western Society of Engineers, and the
Columbos Club of Manila.

Anonymous (1940). Beardsley: Three generations. *Cornell Alumni News* 42(Jan.18): 203. *P*
Anonymous (1945). James W. Beardsley. *Trans. ASCE* 110: 1647-1649.
Beardsley, J.W. (1909). *Preliminary report on irrigation in Java.* Bureau of Public Works:
        Manila.
Beardsley, J.W. (1910). The progress of public works in the Philippine Islands. *Journal of Race
        Development* 1(2): 169-186.
Beardsley, J.W. (1935). Discussion of Water-power development of the St. Lawrence River.
        Trans. ASCE 100: 545-547.
Beardsley, J.W. (1936). Discussion of Flood-stage records of the River Nile. *Trans. ASCE* 101:
        1045-1048.
Beardsley, J.W. (1940). Discussion of Yellow River problem. *Trans. ASCE* 105: 417-418.

BECHTEL

24.09. 1900 Aurora IN/USA
14.03. 1989 Oakland CA/USA

Stephen Davison Bechtel was educated at University of California, Berkeley CA, from where he received in 1954 also an LL.D degree. He joined after studies his father's office, as vice-president from 1925 to 1936, in which time the Hoover Dam was erected, and the more than 3,000 km long Trans-Mountain oil pipeline in Canada was completed. During World War II Bechtel was involved in the California Shipbuilding Corp. and in the 1950s was in charge of various boards dealing with general engineering projects. He was for instance a trustee of the San Francisco Bay Area Council, or director of the Stanford Research Institute. He was an alumnus of the year 1952 of the University of California, a Knight of the Order of St. Sylvester Pope, and among the 50 foremost business leaders in the 1950s. He was a member of the American Society of Civil Engineers ASCE, and recipient of its 1961 John Fritz Medal, and in 1985 elected to its Honorary Member, among many other distinctions.

Bechtel was the creator of the construction empire in the USA. He was president of the Bechtel Corporation from 1936 to 1960, during which time the San Francisco-Oakland Bay Bridge and the first nuclear power plant were completed. Under his aegis, the company became well known as a builder of huge projects all over the world. Works included projects mainly in Saudi Arabia and other parts of the Middle East, in addition to these in the USA. In 1960 his son took over, but the father remained active in the company, holding the title of senior director until his death. At age 84 he helped arrange a joint engineering venture with the government of China. Bechtel's ability to see an entire situation at its reality meant that he could tackle several projects in several fields at once. His analytical ability and business genius won him many awards. By 1980 the Bechtel Corporation was responsible for over half of the power plants in the entire world, and was involved in the construction of large cities and plants in Saudi Arabia. He was not particularly involved in hydraulic engineering or research, but demonstrated outstanding capabilities in engineering for nearly half a century.

Anonymous (1952). Bechtel received Moles construction award. *Civil Engineering* 22(3): 218.
Anonymous (1964). Bechtel, Stephen D. *Who's who in engineering* 9: 114. Lewis: New York.
Anonymous (1989). Stephen D. Bechtel Sr. dies at 88. *Civil Engineering* 59(5): 82.
http://www.bechtel.com/BAC-Stephen-D-Bechtel-Sr.html *P*
http://www.encyclopedia.com/topic/Stephen_Davison_Bechtel.aspx *P*

BEEBE

15.06. 1886 Hampden MA/USA
13.12. 1954 Coeur d'Alene ID/USA

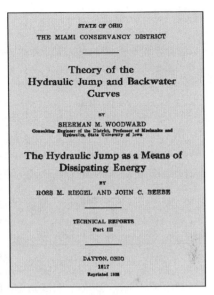

John Cleaveland Beebe graduated as a civil engineer from University of Wisconsin in 1910. He joined first as engineer the Water Resources Branch of the US Geological Survey in Montana, then the US Forest Service in charge of the Prospect Creek Development from 1910 to 1912, and he was from 1913 to 1915 in charge of irrigation projects of the Power and Pump Co., Idaho. Next, Beebe was an assistant engineer of the Miami Conservancy District, Dayton OH. From 1917 to 1921; he was an engineer and manager in Clearmont MT, and then general engineer until 1925 in the states Wyoming and Montana. Further positions through the 1920s and 1930s were with Tibbetts, San Francisco CA, the US War Department, the Pacific Gas & Electric Co., culminating finally as chief division engineer for power-flood control surveys in Washington DC. Beebe retired from this position in 1948.

Beebe is particularly known for hydraulics by his 1917 Report written jointly with Ross Riegel (1881-1966). This work, made for the Miami Conservancy District, deals with hydraulic jumps both in prismatic and in expanding stilling basins. Until then, few experimental works were conducted on this important free surface phenomenon. The main works further were conducted for either small supercritical approach flow Froude numbers, or for extremely small approach flow depths, resulting in scale effects. The results were not readily available but demonstrated that these complex hydraulic processes were amenable by hydraulic modeling. Beebe was also involved in the 1930s in the control of debris flow at Mount Shasta. In 1924, a large mud flow had deposited almost one million m^3 of debris which muddied the Sacramento River. Beebe outlined means to counter future similar scenarios, which had also occurred in the 19th century, and were a constant threat to the Sacramento Valley.

Anonymous (1941). John C. Beebe. *Who's who in engineering* 5: 123. Lewis: New York.

Anonymous (1955). John Cleaveland Beebe. *Civil Engineering* 25(2): 124; 25(2): 128.

Beebe, J.C. (1933). *Report* on investigations for controlling the flow of mud and debris from the Southeast slope of Mount Shasta. US Forest Service: San Francisco CA.

Paul, C.H. (1922). The Flood Control Works of the Miami Conservancy District. *The Military Engineer* 14(2): 141-146.

Riegel, R.M., Beebe, J.C., eds. (1917). The hydraulic jump as a means of dissipating energy. The Miami Conservancy District, *Technical Report* 3. State of Ohio: Dayton OH. (*P*)

BEIJ

10.11. 1893 Hartford CT/USA
26.02. 1986 Franklin NH/USA

Karl Hilding Beij was a junior engineer for the city of Hartford CT from 1914 to 1916, then a surveyor of the Rock Creek & Potomac Parkway Community WA until 1920. After war service, Beij joined as associate engineer the Aero Instrument Section of the National Bureau of Standards NBS, Washington DC, and was there from 1930 a hydraulic engineer in its Hydraulic Laboratory. He was a member of the Philosophical Society of Washington PSW, the American Meteorological Society, the International Association of Hydraulic Research IAHR, and the American Geophysical Union AGU, serving there as Secretary of the Section of Hydrology from 1933 to 1947, and of the General Section from 1947 to 1953.

After Beij had joined the staff of the National Bureau of Standards, he designed original aircraft instruments. In 1942, Beij became assistant of the chief, of NBS Hydraulic Laboratory. His numerous publications include topics as aircraft instruments, copper roofings, and hydraulics. He was also the editor of the 5 volumes of Bibliography in Hydrology for the United States, AGU. In hydraulics, Beij was known for his works relative to flows in pipe bends, spillways, stilling basins, weirs and dry docks. In 1941 he investigated by model tests the thermal effects due to the discharge of large quantities of hot water into a navy yard. He also analyzed the filling systems for dry docks at Pearl Harbor and at Hunter's Point CA. Beij presented one of the first studies on spatially-varied flow which are currently of relevance in side channel spillways. By correctly applying the momentum equation, he was able to validate his experimental observations. He also collaborated in the 1930s with Garbis H. Keulegan (1890-1989).

Anonymous (1948). K. Hilding Beij. *Trans. American Geophysical Union* 29(5): 761. *P*
Anonymous (1954). Beij, Karl Hilding. *Who's who in engineering* 7: 162. Lewis: New York.
Beij-Benoix, B.E. (2004). Karl Hilding Beij. Personal communication. *P*
Beij, K.H. (1934). Flow in roof gutters. *Journal of Research* NBS 12(2): 193-213.
Beij, K.H. (1938). Pressure losses for fluid flow in 90° pipe bends. *Journal of Research* NBS 21(7): 1-18.
Eaton, H.N., Beij, K.H., Brombacher, W.G. (1926). *Aircraft instruments*. Ronald Press: New York.
Keulegan, G.H., Beij, K.H. (1938). Pressure losses for fluid flow in curved pipe bends. *Journal of Research* NBS 18(1): 89-114.

BEITLER

19.03. 1899 Carey OH/USA
18.12. 1994 Cincinnati OH/USA

Samuel Reid Beitler graduated in 1920 as mechanical engineer from Ohio State University, Columbus OH. He then was first an apprentice engineer in various firms and from 1921 an assistant professor of Ohio State University until 1939, when taking over there as associate professor, and from 1944 as professor of hydraulics until retirement. He was from 1929 in parallel involved in the Bailey Meter Co., founded by Ervin George Bailey (1880-1974), and from 1931 member of the ASME Orifice Meter Committee. Beitler was in 1945 nominated for Regional Vice-President ASME, and elected Fellow ASME in 1950.

After practical projects for waterworks, heating and shower-water installations, and canning machinery, Beitler took from 1931 a particular interest into the commercial calibration of large meters. His research added specifically to orifice meters to measure discharge using the pressure difference across a sudden pipeline contraction. This method was quite popular in the first decades of the 20th century, but has been superseded by more accurate and simpler approaches. Beitler was also a consultant on miscellaneous gas-engineering and measurement problems for the Columbia Engineering Corporation and the Ohio Fuel Gas Company. He was further in charge in testing propeller pumps and turbines. He was an expert of fluid flow, particularly in fluid flow measurement. He also published with a colleague a book on Hydraulic machinery. From 1963 to 1974 he was the Executive Director and Secretary of the International Association for Steam Properties.

Anonymous (1941). Beitler, Samuel Reid. *Who's who in engineering* 5: 127. Lewis: New York.

Anonymous (1945). Samuel R. Beitler. *Mechanical Engineering* 67(8): 552-553. *P*

Anonymous (1966). Beitler. *Mechanical Engineering* 88(12): 93. *P*

Bean, H.S., Beitler, S.R., Sprenkle, R.E. (1941). Discharge coefficients of long-range nozzles when used with pipe-wall pressure taps. *Trans. ASME* 63(7): 439-445.

Beitler, S.R., Bucher, P. (1930). The flow of fluids through orifices in six-inch pipes. *Trans. ASME* 52(HYD 7a): 77-87.

Beitler, S.R. (1934). Determination of discharge coefficients of sharp-edge orifices in pipes from one inch to fourteen inches in diameter. *Instruments* 7(1): 3-8.

Beitler, S.R. (1935). The flow of water through orifices. *Bulletin* 89. Engineering Experiment Station. Ohio State University: Columbus OH.

Beitler, S.R., Lindahl, E.J. (1947). *Hydraulic machinery*. Irwin-Farnham: Chicago IL.

BENEDICT

22.09. 1906 Louisburg KS/USA
23.01. 1985 Palo Alto CA/USA

Paul Charles Benedict was educated at Colorado University CU, Boulder CO, where he obtained the BS in civil engineering in 1929. He was then a designer at Boston MA, and in 1930 joined the US Geological Survey USGC Water Branch both in Washington DC and Boise ID until 1941. From then to 1945 Benedict was a district engineer at Iowa City IA, and from 1946 to 1949 was again with the USGS at Lincoln NE as a district engineer, with a promotion to regional engineer from 1950 to 1957, when being appointed chief engineer of the USGS Research Section on Water Quality, Washington DC.

At the time of his retirement in 1967, Benedict had under his immediate supervision 57 USGS research projects. These dealt with a range of subjects as diverse as chemistry and geochemistry of water, aquatic ecology and water quality, estuarine hydrology, sedimentation, glaciology and ice dynamics, geothermal heat transport, aquifer systems, nuclear waste disposal, hydrology, land subsidence, and evapotranspiration. Benedict was nominated for the CU Distinguished Engineering Alumnus Award, for a career that is well summarized by quoting from the citation for Distinguished Service conferred upon him in 1973 "Mr. Benedict has achieved national recognition in the field of fluvial sedimentation and mechanics. Since beginning his career in 1930, he has devoted himself to research of the fundamental laws governing sediment movement in streams'. The award conferred is the highest honor of the Department of the Interior. In 1976 he received ASCE's Hilgard Prize for participation in the preparation of the manual Sedimentation engineering.

Anonymous (1964). Benedict, Paul C. *Who's who in engineering* 9: 127. Lewis: New York.
Anonymous (1976). Paul C. Benedict. *Civil Engineering* 46(10): 103. *P*
Anonymous (1985). Paul C. Benedict. *Eos* 66(14): 153.
Benedict, P.C., Albertson, M.L., Matejka, D.Q. (1955). Total sediment load measured in turbulence flume. *Trans. ASCE* 120: 457-484; 120: 488-489.
Benedict, P.C. (1979). Equipment for investigations of fluvial sediment. *Journal of the Hydraulics Division* ASCE 105(HY3): 163-170.
Vanoni, V.A. (1975). *Sedimentation engineering*. ASCE: New York.
Wolff, R.G., Benedict, P.C. (1964). *Sedimentology*, general introduction and definitions: Fluvial sediment and channel morphology. US Geological Survey: Washington DC.

BENHAM

08.04. 1813 Cheshire CT/USA
01.06. 1884 New York NY/USA

Henry Washington Benham graduated from the US Military Academy in 1837. He was then connected with various government works as member of the US Engineer Corps relating to coast defences and improvements of Savannah River. He served from 1847 to 1848 in the Mexican War, and then was until 1852 superintending engineer of the sea wall for the protection of Great Brewster Island, Boston Harbor, and until 1852 of the Washington DC Navy Yard. He was in charge of the Buffalo Lighthouse and assistant of the US Coast Survey Office until 1856 when sent to Europe in connection with these duties.

After service during the Civil War during which he again was in charge of fortifications at Boston and Portsmouth Harbors, he was mustered out as lieutenant colonel in 1866. President Johnson nominated Benham then for the award of the brevet grade of major general, and to colonel in 1867. In this position he was again responsible for the Boston Harbor sea wall until 1873, and the defenses of New York Harbor from 1877 to 1882. Benham was in addition an expert in the construction of pontoon bridges, thereby inventing the picket shovel and developing rapid construction methods by means of 'simultaneous bays'. Following the Union defeat at Chancellorsville VA, Benham constructed several pontoon bridges across Rappahannock River under fire. This action allowed an Army to retreat from Chancellorsville in 1863. The Henry W. Benham Family Papers encompass his military and engineering career, containing letters and documents from various historical figures. He may be considered both a national military and engineering person whose activities spanned the expansion of the USA in the 19th century. He published the Mexican War memoirs in 1871, and also published papers on his war experiences.

Benham, H.W. (1871). Recollections of Mexico and the Battle of Buena Vista, Mexico, 1847. *Old and New* (6/7): 1-27.
Benham, H.W. (1873). *The West Virginia Campaign* of 1861 by an officer of the US Engineers. Roberts: Boston.
FitzSimons, N. (1991). Benham, Henry W. *A biographical dictionary of American civil engineers* 2: 9-10. ASCE: New York.
http://en.wikipedia.org/wiki/Henry_Washington_Benham *P*
http://www.lib.utexas.edu/taro/utarl/00017/00017-P.html

BENNETT

30.01. 1892 Cincinnati OH/USA
10.03. 1951 Dayton OH/USA

Charles Stuart Bennett obtained the BS degree in civil engineering from Purdue University, Lafayette IN, in 1914. He was then research engineer of the Kentucky Department of Public Roads until 1917, taking over as field engineer until 1922 within the Miami Conservancy District, Dayton OH, on surveys for flood control projects. Until his retirement, he then was in charge of the Office Administration, the Miami Conservancy District, on this large flood control project. He was in 1947 appointed assistant to the chief engineer on the maintenance of this project, and was further in charge of hydraulic data and studies. Bennett was member of the American Society of Civil Engineers ASCE, and of the Dayton Engineering Club.

Bennett had been for 33 years with the Miami Conservancy Project, namely from its initiation in the 1910s to the well-developed status of this notable flood control scheme. He had been responsible for the collection of hydrological and hydraulic data, and the supervision of the flood warning and river stage predictions. The 1938 paper gives an overview on actions completed and planned relating to the Miami Valley. The 1929 paper deals with the 1929 storm, which produced an excessive runoff, attaining more than the rainfall discharge. It was noted that the scheme set up withstood this massive attack, producing only minor damages. Bennett collaborated in this period with Arthur E. Morgan (1878-1975), and Ivan E. Houk (1888-1972).

Anonymous (1948). Bennett, Charles S. *Who's who in engineering* 6: 144. Lewis: New York.
Anonymous (1951). Charles S. Bennett. *Engineering News-Record* 146(Apr.5): 55.
Bennett, C.S. (1929). Miami flood protection works tested by high flood. *Engineering News-Record* 102: 599-601.
Bennett, C.S. (1934). Hydrological data from the Miami Conservancy District. *Engineering News-Record* 113(Nov.1): 556-558.
Bennett, C.S. (1935). Discussion of Determinate stream flow. *Trans. ASCE* 100: 371-373.
Bennett, C.S. (1935). Discussion of The reservoir as a flood control structure. *Trans. ASCE* 100: 918-919.
Bennett, C.S. (1935). Discussion of Practical river laboratory hydraulics. *Trans. ASCE* 100: 168-170.
Eiffert, C.H., Bennett, C.S. (1938). Sixteen years of flood control in the Miami Valley. *Civil Engineering* 8(5): 343-345. (*P*)

BENSEL

16.08. 1863 New York NY/USA
19.06. 1922 Bernardsville NJ/USA

John Anderson Bensel graduated from Stevens Institute of Technology, Hoboken NJ, in 1884 as mechanical engineer, and was the recipient of the Hon. D. Engr. degree in 1921. He was from 1884 a rodman in the Department of Public Works for the New York Aqueduct and the New Croton Dam. He then joined various railways companies as assistant engineer in Pennsylvania; he was from 1889 to 1896 assistant engineer within the Department of Docks NY, in charge of the construction of work on bulkheads and sea-wall docks. From 1896 he took up a private practice for the river wall along a mile of water-front on Delaware River. In parallel he was associated with the Department of Docks and Ferries of New York City, finally as chief engineer until 1906, and in 1907 as commander. From 1908 to 1914 Bensel was president of the Board of Water Supply of NYC, and then continued as New York State Engineer conducting works of public improvement including the Barge Canal.

During World War I Benson commanded an engineering batallion as major general of the US Army Corps of Engineers. After return to his country in 1919, he was again consulting engineer for the New York and New Jersey Bridge and Terminal Commission. He was associated closely all through his engineering career with the American Society of Civil Engineers ASCE, acting as director from 1899 to 1901, as vice-president from 1907 to 1910, taking then over as president. He further was member of the American Society of Mechanical Engineers ASME. He has written papers on his works on the docks, the civil engineering profession, and on floods and their mitigation. Bensel's New Jersey mansion built in 1905 was named Queen Anne Farm. It is currently part of the Morristown National Historic Place. The surrounding grounds contain a five-story stone water tower and a large silver maple tree planted by Bensel in 1906.

Anonymous (1923). John Anderson Bensel. *Trans. ASCE* 86: 1612-1617.
Bensel, J.A. (1905). Wharves and piers: Observations on dock work in New York Harbor. *Trans. ASCE* 54: 1-17.
FitzSimons, N. (1991). Bensel, John A. *A biographical dictionary of American civil engineers* 2: 10-11. ASCE: New York. *P*
Townsend, C.M., Bensel, J.A., Dabney, T.G., Mead, D.W. (1917). Final report of the Special Committee on Floods and Flood Prevention. *Trans. ASCE* 81: 1218-1229; 81: 1303-1310.
h http://en.wikipedia.org/wiki/John_A._Bensel *P*

BIGELOW

13.07. 1814 Watertown MA/USA
15.04. 1862 New Bedford MA/USA

Charles Henry Bigelow graduated in 1835 from the US Military Academy, serving until 1846 in the US Corps of Engineers as an assistant engineer for the construction of Fort Warren and Fort Independence at Boston Harbor. From 1846 to 1857 he was chief engineer of Essex Company in the construction of Lawrence MA. He also supervised the construction of the dam, the North Canal and various mills. He was also associated with the Water Power Company of Lewiston ME, the Niagara Falls Canal, preparing in addition a dam and a canal at Sherbrooke QC in 1856. Later he was engineer of water measurements for the Augusta Manufacturing Co., and the Mills Co. Dam and Canal at St. Anthony Falls, Minneapolis MN. During his last years Bigelow served as engineer of the New Bedford Copper Co.

The Pemberton Mill was a large factory at Lawrence MA, which collapsed without warning in 1860. It was considered the worst industrial accident in Massachusetts history, and one of the worst industrial calamities in American history, resulting in 145 killed and 166 injured workers. The five-storied, 90 m long and 25 m wide Mill was built in 1853 by Bigelow, but it was sold only four years later due to financial panic. The new owners jammed the mill with more machinery, and it ran with great success. The collapse was caused by the high loads, substandard construction, so that this tragedy initiated the improvement of safety standards of industrial buildings. The Niagara Falls Hydraulic Company was to draw water power from Niagara River immediately above the Falls. It included a navigable canal 20 m wide and 3 m deep nearly 1 km upstream from the Falls. Bigelow recommended the use of a water turbine 0.35 m in diameter under a head of 100 m. The Company failed however because the construction cost exceeded the estimates. A new company was formed in 1856 and taken into service in 1862. The canal was completed in 1877, but the entire Company was sold the same year.

Dorgan, M.B. (1913). *Lawrence yesterday and today* 1845-1913. Dick & Trumpold: Lawrence.
FitzSimons, N. (1991). Bigelow, Charles H. *A biographical dictionary of American civil engineers* 2: 12. ASCE: New York.
Washburn, W.D., Bigelow, C.H. (1858). *Water power of the Minneapolis Mill Co*. Minneapolis.
http://www.mikalac.com/tech/pow/niagara2.html
http://lawrence.essexcountyma.net/lawrencedam2.html (*P*)

BILLINGS

08.02. 1876 Omaha NE/USA
03.11. 1949 San Diego CA/USA

Asa White Kenney Billings received his higher education at Harvard University, from where he obtained the degrees AB in 1895, and AM in 1896. His first work was construction of street railways and steam electric power plants at Pittsburgh PA. In 1899 he went to Cuba in charge of similar works. From 1902 to 1906, he was chief engineer for the Havana Central Railroad Co., whereas from 1906 to 1909 he maintained a consulting practice at Havana. The next two years were spent in the USA; from 1912 to 1916 Billings was then in Barcelona, Spain, as manager of construction, during which time he was engaged in building the *Talarn* Dam, then the largest and highest concrete dam in Europe.

After war service and return to Spain, Billings served from 1921 to 1924 as construction manager for the Mexican Light & Power Co., Mexico City, and the Brazilian Traction, Light & Power Co., Rio de Janeiro. For the next 20 years, he was vice-president of the latter company and its several subsidiaries in Brazil, and from 1945 to 1946 there served as president. Notably among Billings achievements in Brazil was the damming up and reversing the flow of streams in the São Paulo area to take advantage of a 600 m drop from the mountain plateau to the coastal plain. In addition to providing cheap power, which has had a significant effect on industrialization, this project has been valuable also for flood control. Billings further was responsible for the design and construction of large hydro-electric installations of the Brazilian Hydroelectric Co., the São Paulo Tramway, Light & Power Co., the São Paulo Electric Co., among other companies. Billings was an associate member ASCE from 1906, and a member from 1908. He was decorated with ASCE Honorary Membership in 1946. He was also a member ASME, and member of the Institution of Civil Engineers, London UK.

Ackerman, A.J. (1953). *Billings and water power in Brazil*. ASCE: New York. *P*
Anonymous (1946). A.W.K. Billings. *Engineering News-Record* 137(Aug.15): 202. *P*
Anonymous (1947). A.W.K. Billings new Honorary Member. *Civil Engineering* 17(1): 43. *P*
Billings, A.W.K., Dodkin, O.H., Knapp, F., Santos, Jr., A. (1933). High-head penstock design. Proc. *ASME Water Hammer* Symposium: 29-61. ASME: New York.
Billings, A.W.K. (1936). Water power in Brazil, with special reference to the São Paulo development. *Journal Institution of Civil Engineers* London 3(8): 677-698.
Billings, A.W.K. (1938). Water power in Brazil. *Civil Engineering* 8(8): 520-523.

BINGER

28.02. 1917 Greenwich NY/USA
21.04. 2008 Chappaqua NY/USA

Wilson Valentine Binger his received education at Harvard University, from where he graduated with an AB degree in 1938, and an MS in civil engineering in 1939. After World War II, he was foundation and soil engineer on dams in Venezuela and Argentina, as also on the Sea Level project of Panama Canal. Later, he was engaged with dams on Missouri River for the US Corps of Engineers. He joined the firm Tippetts-Abbett-McCarthy-Stratton TAMS in 1952, became a partner in 1962, and chairman in 1975. He was in parallel the Secretary of the US Commission of Large Dams USCOLD from 1961 to 1979, and president of the American Institute of Consulting Engineers AICE in 1973. In 1981, Binger was elected president of the International Federation of Consulting Engineers FIDIC. In 1985, he retired from TAMS, where he had spent 33 years of his 46 years' professional career. Binger was elected to the National Academy of Engineering in 1975. His election citation states 'Leadership in the development of large dams, water resources, soil mechanics and foundation engineering'.

TAMS was highly involved in one of the largest building sites of the 1970s and 1980. Tarbela Dam is on the Indus River, Pakistan, about 50 km northwest of Islamabad. The dam is 150 m high with a reservoir of 250 km^3, thus corresponding to the largest earthfill dam worldwide. Reservoir sedimentation there is a great concern, as was also cavitation damage due to high-speed flows. TAMS was acquired by Earth Tech in 2002, and is currently part of AECOM. Binger was from 1978 to 1981 a vice-president of the International Commission of Large Dams ICOLD.

Anonymous (1964). Binger, Wilson V. *Who's who in engineering* 9: 149. Lewis: New York.
Anonymous (1970). Wilson V. Binger. 10th *ICOLD Congress* Montreal: 111. *P*
Binger, W.V. (1948). Analytical studies of Panama Canal slides. Proc. 2nd Intl. Conf. *Soil Mechanics and Foundation Engineering* Rotterdam 2: 54-60.
Binger, W.V. (1978). *Environmental effects of large dams*. ASCE: New York.
Binger, W.V. (1979). United States looks to small hydro. *Water Power & Dam Construction* 31(4): 19. *P*
Binger, W.V. (1989). Jams Hobson Stratton. *Memorial tributes* 3: 326-331. National Academy of Engineering: Washington DC.
Lowe, J. III, Binger, W.V. (1982). *Tarbela dam project*, Pakistan. USCOLD: New York.
http://cache:X7-Sn7zXl14J:enr.construction.com/people/people/archives/080514.asp+%22Binger%22+TAMS *P*

BINGHAM

08.12. 1878 West Cornwall VT/USA
06.11. 1945 Easton PA/USA

Eugene Cook Bingham graduated from Middlebury College in 1899 with an AB, and gained his PhD in 1905 from Johns Hopkins University, to continue then studies at Leipzig, Berlin and Cambridge in 1906. After having been a teacher at Johns Hopkins from 1900 to 1903, he was appointed professor of chemistry at University of Richmond from 1906 to 1916, from where he moved in the same position to Lafayette College for the next 30 years. He received a Certificate of merit from the Franklin Institute. Bingham was a Member of the American Chemical Society, the American Association for Advancement of Science AAAS, and a Honorary Member of the Virginia Chemists' Club.

Bingham is known for the rheological description of fluids, currently referred to as Bingham fluids. His 1916 paper was the start of these researches, though original work was already published in 1906. His main contributions to science were works in viscosity and fluidity, including a definition of fundamental properties, on the precise measurement of viscosity, plasticity and related characteristics, and on the design of instruments for precise measurement of plastic flow. His original type of plastometer was in use for decades. His clear and concise statement of flow phenomena and their inclusion in the science as a generalization of the Newtonian approach, led to what is currently referred to as rheology, a notion proposed by Bingham. It was also him who proposed the term poise as the basic unit of viscosity. More than 100 papers were written on rheology, but additional research was also made in physiology, chemical education, or mensuration. The tributes in a memorial edition from the American scientific societies bear adequate testimony to the high esteem in which Bingham was held as a man of science, to his technical skill, to his pleasant personality and friendly spirit, and to his enthusiasm and work for the Society of Rheology, which he formed.

Anonymous (1946). Prof. E.C. Bingham. *Nature* 158(Sep.14): 370-371.
Anonymous (1964). Bingham, Eugene C. *Who's who in engineering* 9: 111. Lewis: New York.
Anonymous (1979). Bingham. *Applied Mechanics Review* 32(2): 146. *P*
Bingham, E.C. (1922). *Fluidity and plasticity.* McGraw-Hill: New York.
Bingham, E.C. (1924). Plasticity and elasticity. *Journal of Franklin Institute* 197(1): 99-112.
Bingham, E.C. (1932). The (fluidity) mixture law. *Journal of Rheology* 3(1): 95-112.
Poggendorff, J.C. (1925). Bingham, Eugene Cook. *Biographisch-Literarisches Handwörterbuch* 5: 116-117; 6: 226-227; 7b: 396. Verlag Chemie: Leipzig, Berlin, with bibliography.

BIRKHOFF

19.01. 1911 Princeton NJ/USA
22.11. 1996 Water Mill NY/USA

Garrett Birkhoff was the son of George Davis, a noted mathematician of the day. At age 17, Garrett enrolled at Harvard University, graduating from there in 1932. He won a Fellowship to Cambridge University switching his focus from mathematical physics and quantum mechanics to abstract algebra. There, Birkhoff developed his theory of lattices. He published his first book in 1935 on abstract integrals and discovered Dedekind's work on lattices. In 1936 he became an instructor at Harvard University, so that the father's and son's teaching appointments overlapped until 1944.

Throughout his career, Birkhoff aimed to link theory with applications. In the 1950s he served as consultant to Westinghouse Corporation, working with scientific computing and adding to problems of nuclear reactors. He authored numerous papers on applied mathematics, including also the books Hydrodynamics in 1950, and Jets, wakes and cavities in 1957. Birkhoff further considered aspects of homogeneous turbulence and potential flows with free streamlines. He continued as a consultant to military and industrial institutions, including the Los Alamos Science Laboratory, the General Motors Corporation, and the Rand Corporation. In the 1980s, he worked with the Naval Postgraduate School of Monterey CA, conducting research, studying fluid dynamics and teaching. A member of the National Academy of Sciences, the American Academy of Arts and Sciences, and many other academic organizations, he was named George Putnam professor in 1969 at Harvard University until retirement his in 1981. He also delivered the John von Neumann Lecture in 1981.

Anonymous (1980). Birkhoff to deliver von Neumann Lecture. *SIAM News* 13(6): 1. *P*

Birkhoff, G. (1950). *Hydrodynamics*: A study in logic, fact and similitude. University Press: Princeton NJ, 2nd. ed. 1960.

Birkhoff, G., Zarantonello, E.H. (1957). *Jets, wakes and cavities*. Academic Press: New York.

Birkhoff, G., Kampé de Fériet, J. (1958). Kinematics of homogeneous turbulence. *Journal of Mathematics and Mechanics* 7(5): 663-703.

Birkhoff, G. (1961). Calculation of potential flows with free streamlines. *Journal of the Hydraulics Division* ASCE 87(HY6): 17-22; 88(HY2): 187-189; 88(HY3): 223-224; 88(HY4): 293-299; 88(HY4): 315-316; 89(HY2): 147-148.

Davis, P.J. (1997). From where I sat: Garrett Birkhoff. *SIAM News* 30(2): 3. *P*

Halmos, P.R. (1987). G. Birkhoff. *I have a photographic memory*: 518. AMS: Providence. *P*

BIRKINBINE

21.05. 1819 Reading PA/USA
21.04. 1886 Reading PA/USA

Henry Peter Miller Birkinbine operated in 1844 a forge and auger works at Reading PA. He served for 10 years then as chief engineer of the Philadelphia Water Department. In the 1860s he founded a firm, because of his expertise in water supply and sewage treatment, becoming involved in water projects, thereby contributing for improved public health.

As in many cities, the outbreak of typhoid fever and cholera, epidemics caused by contaminated drinking water, led to the erection of water treatment plants. In the mid-1800s, both the Schuylkill and Delaware Rivers were heavily contaminated. The public's demand for improved streams and drinking water initiated environmental protection; the city's ambition to be a prominent economic hub led to the development of the 5,000 km sewage system of Philadelphia today. Birkinbine proposed a water works along the Schuylkill, with a pumping station above the Flat Rock Dam at Shawmont, and reservoirs located higher up the steep river banks, providing water by gravity through distribution mains in the streets. Construction began after the end of the Civil War, with the pumping station at Shawmont completed in 1869. The steam-powered pumps forced water uphill into a reservoir 100 m above city, located at present-day Eva and Dearnley Streets at Roxborough. To increase the capacity of these works and allow for water to flow by gravity, the pumping station on the Schuylkill was expanded in the 1890s, and a larger reservoir was built higher up the ridge along Port Royal Avenue. Besides engineering projects, Birkinbine was interested also in basic hydraulics, including pumps and pumping of water, land drainage, turbines, and fire protection of cities. He thus was an early hydraulic engineer of note.

Birkinbine, H.P.M. (1850). *Montgolfier's hydraulic ram*: A simple and effective machine for forcing a portion of any brook or spring over any required distance and elevation. Philadelphia.
Birkinbine, H.P.M. (1861). *Report* on the experiments with turbine wheels, made 1859-60, at Fairmont Works, Philadelphia. Geddes: Philadelphia PA.
Birkinbine, H.P.M. (1867). Purity of water and hydrographical survey of the Schuylkill River. *Annual Report*. Water Department of the City of Philadelphia: Philadelphia PA.
Birkinbine, H.P.M. (1869). Report to the Water Commissioners upon a *Better supply of water for the City of Harrisburg*. Singerly: Harrisburg. (*P*)
Birkinbine, H.P.M. (1879). *Future water supply of Philadelphia*. Merrihew: Philadelphia.
http://www.phillywatersheds.org/watershed-history-roxborough-water-works

BISHOP

13.01. 1884 Des Moines IA/USA
06.03. 1933 Denver CO/USA

Lyman Edgar Bishop moved in 1887 with his parents to Denver CO. He graduated in 1903 with the BSc degree in civil engineering from Boulder University of Colorado. During the next few years he lived in Denver and there was occupied with engineering work in hydraulics and irrigation. He was involved also in the city water supply. The data acquired on the supply sources were later used by the Water Board, constituting one of the best data record then available. He was also connected as assistant with the Denver Reservoir and Irrigation Company on a large project including four reservoirs. In 1914 he was in private practice locating 30 km of the Grand Valley Canal for the US Reclamation Service. From 1915 to 1917 he was deputy state engineer of Colorado, becoming then construction superintendent on the 36 m high *Costilla* Earth Dam NM. He returned to private practice, designing hydraulic plants, and reporting on the geological conditions for the 30 m high masonry dam on the Bear River, Alexander ID.

Bishop was appointed in 1929 consulting irrigation engineer for the Union of Socialist Soviet Republics, with headquarters at Tashkent, Turkestan. He served in this capacity for two years and was also consulting engineer for the General Cotton Committee. In a letter written home, he described a two weeks' field trip from where he had just returned: 'The Ferghana Valley is egg-shaped, about 100 km wide and 220 km long. There are now some 8,000 km^2 under irrigation and cultivation in this valley. Roughly two-thirds of the area is devoted to cotton. Some of this land has been under irrigation for over 1,200 years'. He returned to the USA in 1931, after his family had made a visit to Tashkent. During his trip, he went by way of India, the Philippines and the Pacific Coast, spending several months to examine irrigation projects. At Denver his former clients had found other consultants, so that his practice had strong financial difficulties: He passed out of life as a result. He had been a man of strong convictions and when he formed an opinion or conclusion, it was difficult for him to compromise. This strong inflexibility of mind was a handicap to him, but he also possessed admirable qualities.

Anonymous (1907). Bishop, Lyman E. *The Coloradoan* 9: 145. *P*
Anonymous (1935). Lyman E. Bishop. *Trans. ASCE* 100: 1603-1605.
Bishop, L.E. (1911). Economic canal location in uniform countries. *Trans. ASCE* 74: 179-204.
Bishop, L.E. (1924). *Report on Laramie Water Company Irrigation System*. Denver CO.
Bishop, L.E. (1926). *South Platte River investigation*. Denver CO.

BISSELL

11.09. 1881 Navasota TX/USA
01.01. 1974 Menlo Park CA/USA

Charles Arthur Bissell obtained in 1902 a BS degree in civil engineering from Austin College, Austin TX, and the civil engineering degree in 1906 from the University of Texas, Austin TX. He was rodman and instrumentman of the Santa Fe Railways until 1908, from when he joined the US Bureau of Reclamation USBR as design engineer of irrigation works for the Sunnyside WA, Elephant Butte NM, and El Paso TX Schemes. From 1914 he was in charge of designs, reports, and technical studies at Washington DC, from when he was there in charge of the Technical Section, the later USBR Chief of Engineering Division. He there was engaged in swamp and overland flows at the Yazoo Delta of Mississippi River, the reclamation and the rural development in the Southern States; in 1929 Bissell was engaged on the Red Bluff Project on Sacramento River CA; and in 1930 on the Central Valley Project in California. Until 1939 he was then office engineer for the Metropolitan Water District of Southern California MWD; he also was in charge of the Colorado River Aqueduct, moving from 1939 to 1944 to the USBR headquarters at Denver CO, becoming regional engineer of Region 3 at Boulder City NV from 1945. He retired in 1952 after 34 years of service for USBR at Boulder City NV.

Bissell had spent almost his entire professional career with USBR, where he had been involved in a number of irrigation, flood control, and dam projects. He had also been chief of the Engineering Division at Washington DC, chief engineer at the USBR Denver office, and Bureau representative, the US Government Interdepartmental Board of Surveys and Maps. The MWD is the largest supplier of treated water in the USA; it is a corporative of 14 cities and 12 municipal water districts providing water to 18 million people in its 13,000 km^2 service area.

Anonymous (1948). Bissell, Charles A. *Who's who in engineering* 6: 172. Lewis: New York.
Anonymous (1951). Chas. A. Bissell. *Engineering News-Record* 147(Oct.18): 69. *P*
Bissell, C.A., ed. (1916). *Irrigation, hydraulics, and water power*. National Research Council: Washington DC.
Bissell, C.A. (1919). *Progress in national land reclamation in the United States*. Smithsonian Institution: Washington DC.
Bissell, C.A., Weymouth, F.E. (1935). Memoir for Arthur P. Davis. *Trans. ASCE* 100: 1582-1591.
Bissell, C.A., ed. (1939). *The Metropolitan Water District of Southern California*: History and First Annual Report. MWD: Los Angeles.

BIXBY F.L.

01.07. 1880 Moorehead MN/USA
13.11. 1955 Reno NV/USA

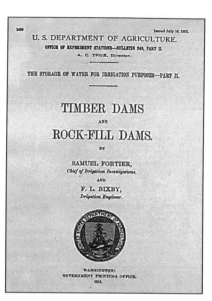

Frederick Louis Bixby received the BSc degree in civil engineering in 1905 from the University of California, Berkeley CA, and the degree of civil engineer from the University of Nevada, Reno NV. He was employed after graduation by the Oregon Short Line Railway in Idaho. He made surveys and investigations for hydro-electric developments in the Salt Lake City UT area in 1907, joining in 1908 the US Department of Agriculture, later serving as office engineer at its headquarters at Washington DC. In 1910 Bixby directed investigations on water resources and water uses in New Mexico, continuing later in this capacity in Nevada. During this period he became a part-time member of the Engineering Faculty, New Mexico College of Agriculture and Mechanic Arts, Las Cruces NM, and also served in a similar capacity in the Civil Engineering Department of the University of Nevada, Reno NV, becoming in 1922 a full-time member of the latter department, and chairman of the Department in 1939 until his retirement in 1948.

Bixby was an expert in irrigation engineering. He has written with Samuel Fortier (1855-1933) a book on irrigation techniques. Part 1 includes the then current knowledge on earth-fill and hydraulic-fill dams, whereas Part 2 deals with timber and rock-fill dams. This work is richly illustrated with photos to show successful designs, and to explain the main aspects of a design. Bixby thereby refers to the two fundamental books in dam design, namely these of Edward Wegmann (1850-1935) and of James D. Schuyler (1848-1912). The 1915 paper deals with the hydraulic aspects of submerged orifices, as often encountered in irrigation problems if the tailwater level is above a certain elevation so that the orifice becomes submerged. Bixby was an ASCE member from 1919.

Anonymous (1933). Bixby, Frederick L. *American men of science* 5: 98. Science Press: NY.
Anonymous (1957). Frederick L. Bixby. *Trans. ASCE* 122: 1234-1235.
Bixby, F.L. (1915). Tests of submerged-orifice head-gates for the measurement of irrigation water. Agricultural Experiment Station *Bulletin* 97. University of Nevada: Reno.
Bixby, F.L., Hardman, G. (1928). The development of water supplies for irrigation in Nevada by pumping from underground sources. Agricultural Experiment Station *Bulletin* 112. University of Nevada: Reno.
Fortier, S., Bixby, F.L. (1912). *The storage of water for irrigation purposes* 2: Timber dams and rock-fill dams. Government Printing Office: Washington DC. (*P*)

BIXBY W.H.

27.12. 1849 Charlestown MA/USA
29.09. 1928 Washington DC/USA

William Herbert Bixby entered the Massachusetts Institute of Technology MIT in 1866, but the sudden death of his father forced him to leave. He took then a position in a mercantile house, and received after 2 years an appointment as cadet at the US Military Academy, West Point NY, from where he graduated in 1873. He was for two years on duty at New York Harbour. From 1875 to 1879, he was teacher at his Alma Mater, then left for France, taking instruction courses at *Ecole des Ponts et Chaussées*, Paris F. After graduation in 1880 he was assigned to special duty studying fortifications in Europe. He returned in 1882 to the USA, on duty at Willets Point NY, preparing a Report on fortifications.

From 1884 to 1891 Captain Bixby was in charge of the US Engineer Office, Wilmington NC, conducting river and harbor improvements in this District. He was then transferred to Newport RI, having charge of the Rhode Island, Massachusetts, and Connecticut Districts. After his stay at the Fourth Lighthouse District, he was transferred in 1897 to the Cincinnati District, having charge of the entire Ohio River improvement work from Cincinnati to Pittsburgh PA. In 1901 he was transferred to Detroit MI, where his work included the Great Lakes from Buffalo NY to Duluth MN, with plans for the third lock at the Sault, including canal improvements. From 1904 to 1907 he was in charge of the Chicago IL District, thereby supervising all work in the Western Great Lakes and Northern Mississippi River Districts, with special works for the deep waterway between Chicago and St. Louis MO. He was transferred in 1907 to St. Louis, having charge as division engineer of the Western Division. He was appointed president of the Mississippi River Commission in 1908, and in 1910 took over the office of Chief of Engineers, so that he acted as head of the Army Corps of Engineers, thereby succeeding William L. Marshall (1846-1920). He retired in 1913. During his long service in the Corps, he was responsible for a large number of tasks requiring research and engineering knowledge.

Anonymous (1910). New chief of Engineers, Col. W.H. Bixby. *Engineering News* 63(18): 524. *P*
Bixby, W.H. (1906). *West fork of south branch of Chicago River*. GPO: Washington DC.
Bixby, W.H. (1910). *Regarding practicability of storage reservoirs* to prevent floods and to benefit navigation on the Ohio and other rivers of the United States. Washington DC.
Bixby, W.H. (1911). *Drainage areas and surface levels of the Great Lakes*. Washington DC.
Bixby, W.H. (1912). *River and harbor improvements*. US Government Printing: Washington DC.
http://en.wikipedia.org/wiki/William_Herbert_Bixby *P*

BLACK

08.12. 1855 Lancaster PA/USA
24.09. 1933 Washington DC/USA

William Murray Black graduated in 1877 from the US Military Academy, and from the Engineering School of Application in 1880. He then was until 1886 instructor at his Alma Mater, and from 1891 to 1895 instructor in civil engineering at Willets Point NY. He was later engaged in practical construction works on the locks and dams on Kanawha and Ohio Rivers, served the Philadelphia Harbor Commission as secretary, and opened the mouth of St. John's River in Florida. He further contributed to the exact knowledge of littoral drift of sand, and the behaviour of bars in tidal estuaries. He then was ordered to the Spanish-American War, serving as chief engineer in Puerto Rico in 1898. From 1899 to 1900 he was chief engineer at Havana, Cuba, in charge of sewer projects and works for the improvement of the ocean front.

From 1903 to 1904 Black observed work of the Isthmian Canal Commission in Panama. He was until 1906 then in charge of rivers, and harbors in Maine, taking over until 1909 the Department of Public Works in Cuba. He was stationed until 1916 in New York NY to improve the East River and Hudson River, receiving in 1912 the ScD degree from the Franklin and Marshall College, Lancaster PA. He also was senior member of the Board of Engineers for Rivers and Harbors, and of the New York and Boston Harbor Line Boulevards. Black was promoted to lieutenant colonel in 1905, to colonel in 1908, and to brigadier general and Chief of Engineers in 1916. He enlarged the US Corps of Engineers more than hundred times during World War I. He served as member of the Committee of Engineers and Education, and chaired the Inland Waterway and Railway Transportation in 1917, retiring as major general in 1919. In 1920 he received the PhD degree in engineering from the Pennsylvania Military College, Chester PA. He finally was a consultant, ASCE member and awarded its Rowland Prize and Wellington Prize.

Anonymous (1916). Col. William M. Black. *Engineering News* 75(10): 483. *P*
Black, W.M. (1893). The improvement of harbours on the South Atlantic Coast of the United States. *Trans. ASCE* 29: 223-276.
Black, W.M. (1895). *The United States public works*. Wiley: New York.
Black, W.M. (1910). *Report* concerning the location of sewer outlets and the discharge of sewage into New York Harbor. Brown: New York.
Black, W.M. (1925). Waterway and railway equivalents. *Trans. ASCE* 88: 538-552.
http://en.wikipedia.org/wiki/William_Murray_Black *P*

BLAISDELL

21.07. 1911 Goffstown NH/USA
19.03. 1998 Minneapolis MN/USA

Fred William Blaisdell received education from the University of New Hampshire, with a BS in 1933, from where he continued at Massachusetts Institute of Technology MIT, with an MS in 1934. He joined the Soil Conservation Service of the Agricultural Research Service ARS at State College PA in 1935, and from 1936 to 1940 the Hydraulic Laboratory of the US Bureau of Standards. He then moved to the Saint Anthony Falls SAF Hydraulic Laboratory, Minneapolis MN, where he was promoted to chief engineer of ARS in 1954, and to the Science and Education Administration, ARS, in 1978. Blaisdell was the recipient of the 1969 J.C. Stevens Award from ASCE, and he was elected ASCE Fellow in the 1980s.

Blaisdell was known as a hard worker and a meticulous experimenter. His 1937 paper on sluice-gate flow compares laboratory and prototype observations, at this time an important issue because so-called scale effects were largely unknown. Researches as these gave confidence to laboratory investigations as an alternative to prototype investigations, which are often extremely limited by practical issues. In 1947, Blaisdell developed the SAF stilling basin, one of the first proposals made, and later followed by similar designs mainly by the US Bureau of Reclamation USBR, Denver CO. Whereas the latter are still in worldwide use, the SAF basin has in the meantime lost interest. Other works of Blaisdell relate to the designs of drop spillways and plunge pools.

Anonymous (1950). F.W. Blaisdell receives J.W. Rickey Medal. *Civil Engineering* 20(1): 45. *P*
Anonymous (1964). Blaisdell, Fred W. *Who's who in engineering* 9: 160. Lewis: New York.
Anonymous (1992). Fred W. Blaisdell and wife Harriet. *ASCE News* (10): 4. *P*
Blaisdell, F.W. (1937). Comparison of sluice-gate discharge in model and prototype. *Trans. ASCE* 102: 544-560.
Blaisdell, F.W. (1948). Development and hydraulic design, Saint Anthony Falls stilling basin. *Trans. ASCE* 113: 483-520; 113: 551-561.
Blaisdell, F.W., Donnelly, C.A. (1956). The Box Inlet Drop spillway and its outlet. *Trans. ASCE* 121: 955-986; 121: 992-994.
Blaisdell, F.W., Manson, P.W. (1963). Loss of energy at sharp-edged pipe junctions in water conveyance systems. *Technical Bulletin* 1283. US Dept. Agriculture: Washington DC.
Blaisdell, F.W., Anderson, C.L. (1991). Pipe plunge pool energy dissipator. *Journal Hydraulic Engineering* 117(3): 303-323; 118(10): 1448-1453.

BLANCHARD

25.07. 1874 Peru IL/USA
03.10. 1961 Rocky River OH/USA

Murray Blanchard received his BS degree from the University of Michigan in 1898, and there his civil engineering degree in 1903. He was from 1898 in parallel a recorder of the US Deep Water Survey, then a junior engineer from 1899 to 1905 with the US Lake Survey, until 1910 then engineer for the hydraulic investigations at J.G. White CO, followed by a position until 1914 with Water Power Ga. Ry. & Pr. CO. He then was with the Sanitary District of Chicago until 1917. After war service, Blanchard was a hydraulic engineer with the State of Illinois until 1930, and finally was both a consultant and with the US Public Works Administration until retirement in 1956. He was a president of the ASCE Illinois Section and published works in ASCE journals.

In his 1902 Report, Blanchard describes discharge measurement of Detroit River at Fort Wayne. Eleven current meters of Haskell type as described by Eugene Edwin Haskell (1855-1933) were used from a catamaran. Single observations lasted for three minutes resulting in vertical velocity plots. Almost two hours were necessary to take eleven profiles with one meter. The numerical data were presented in Tables including average velocity ratios at each 10% flow depth. This work was important in the USA in terms of accurate discharge measurement of large rivers, such that Stepanos Kolupaila (1892-1964) included Blanchard in a list of notable persons in hydrometry with significant observations during the 19th century. The 1920 paper compares records with Haskell and Price meters under various field conditions, stating that the first is more reliable. The 1921 paper proposes the Manning formula for uniform flow predictions. The 1932 paper relates to a set of discharge curves for various falls between two gaging stations.

Anonymous (1925). Murray Blanchard. *Illinois Catholic Historical Review* 7(3): 224. *P*
Anonymous (1959). Blanchard, Murray. *Who's who in engineering* 8: 219-220. Lewis: New York.
Blanchard, M. (1920). Hydraulics of the Chicago Sanitary's District main channel. *Journal of the Western Society of Engineers* 25(13): 471-524; 26(8): 300-303.
Blanchard, M. (1921). Manning's formula better than Kutter's. *Engineering News-Record* 87(3): 96.
Blanchard, M. (1932). A discharge diagram for uniform flow in open channels. *Trans. ASCE* 97: 865-886.
Kolupaila, S. (1960). Murray Blanchard: Early history of hydrometry in the United States. *Journal of the Hydraulics Division* ASCE 86(HY1): 32; 86(HY6): 117-119; 86(7): 33. *P*

BLANEY

19.07. 1892 Los Angeles CA/USA
18.10. 1976 Los Angeles CA/USA

Harry French Blaney, distinguished engineer in water conservation, made key contributions to irrigation research not only in California, but throughout the world. He received in 1915 the BS degree in civil engineering from the University of California, Los Angeles CA, working then for Southern California Gas Company. He began with the US Department of Agriculture in 1917, a career that continued until his retirement in 1962. Until 1973 he was then a research associate at his Alma Mater. Blaney was awarded the 1966 John Deere Medal, the American Society of Agricultural Engineers, the 1966 Royce Tipton Award, the American Society of Civil Engineers ASCE, and he was a Honorary Member of the International Commission of Irrigation and Drainage ICID.

Blaney's career was with the US Department of Agriculture including the following stations: From 1917 to 1918 he was engaged with irrigation and silt studies of Colorado River and at Imperial Valley, from when he was in charge of the Denver Irrigation Field Laboratory, conducting there evaporation and evapotranspiration studies. From 1919 to 1927 he conducted studies on the cost of irrigation water, from when he continued with work on the consumptive use of water by irrigated crops and native vegetation and rainfall disposal until 1934. Until 1941 he then was a water conservation engineer in New Mexico, California and Texas and was in charge of a technical mission to Cuba to investigate the rehabilitation of agriculture, various dam sites, and water supply for rice irrigation. From 1943 to 1951 he was project supervisor in charge of consumptive irrigation studies with the State Engineer of California and the US Soil Conservation Service SCS. Finally from 1954 to 1962, Blaney acted as principal irrigation engineer of the US Agricultural Research Service ARS in Pacific Southwest.

Blaney, H.F., Huberty, M.R. (1930). *Cost of irrigation water in California.* Department of Water Resources: California State Printing Office: Sacramento CA.
Blaney, H.F. (1939). Discussion of A theory of silt transportation. *Trans. ASCE* 104: 1754.
Blaney, H.F. (1951). Consumptive use of water. *Proc. ASCE* 77(Oct., Separate 91): 1-19.
Blaney, H.F. (1958). Evaporation from free water surfaces at high altitudes. *Trans. ASCE* 123: 385-395; 123: 402-404.
Cullen, R.J., Willard, B.R. (1978). *Writings of Harry F. Blaney,* distinguished water conservation engineer. California Water Resources Center, Archives Report No. 24. University of California: Davis CA. *P*

BLEE

12.12. 1886 Santa Ana CA/USA
07.04. 1975 Pasadena CA/USA

Clarence Earl Blee obtained in 1911 the BSc degree in civil engineering from Stanford University, and was then until 1912 assistant engineer in charge of engineering on the construction of large hydraulic fill dams in British Columbia. Until 1914 he was in charge of the engineering and the construction of a concrete dam, the intake and the pipelines of the Sooke Lake Water Supply for Victoria BC, and then was employed as assistant engineer on the water supply of the Pan-Pacific International Exposition. Until 1918 he then was assistant engineer of the California Ore Power Co., again on the construction of a concrete dam and a power plant. After having been involved in the construction of a dry-dock, he was from 1919 to 1920 assistant professor at the Department of Civil Engineering, Stanford University, Stanford CA. Until 1923 Blee was then principal engineer at Vancouver BC on the design, the preparation, and construction supervision of three hydro-electric projects until 1935. He joined then Tennessee Valley Authority TVA as principal engineer until 1936, as project engineer until 1940 in charge of the construction of the Hiwassee Project, until 1943 as project manager on the Fort Loudoun Dam including dam construction, the power plant, and the locks. He was in this period also in charge of TVA's Fontana Dam and Power Plant. From 1943 to 1957 Blee was TVA chief engineer of its planning, design, and construction activities, from when he was a private consultant in southern California.

Blee was in 1952 involved in the largest power expansion of TVA, whose capacity should be doubled within four years. He was described as a keen, wiry man with a ready smile and a habit of choosing his words carefully. He was further characterized as becoming restless at times in the confines of his office at Knoxville KY, stating that 'My chief relaxation is to get out on one of the jobs, and see things actually being done'. He was member of the American Society of Civil Engineers ASCE, and the Society of Professional Engineers of British Columbia.

Anonymous (1952). Clarence E. Blee. *Engineering News-Record* 149(Jul.24): Frontispiece. *P*
Anonymous (1959). Blee, Clarence E. *Who's who in engineering* 8: 221. Lewis: New York.
Blee, C.E. (1929). Tests on model dam determine baffle weir design. *ENR* 103(23): 931-933.
Blee, C.E. (1953). Development of Tennessee River Waterway. *Trans. ASCE* CT: 1132-1146.
Rich, G.R., Blee, C.E., Jessop, G.A. (1955). Discussion of Modernization of the Hales Bar Plant. *Trans. ASCE* 120: 557-561.

BLOMGREN

16.09. 1891 Denver CO/USA
06.09. 1964 Denver CO/USA

Walter Edward Blomgren graduated in 1913 as civil engineer from University of Colorado, Denver CO. He joined the US Reclamation Service as junior and assistant engineer until 1918 for the Boise Project, Idaho, and the *Ochoco* Irrigation District, Oregon. From then to 1920, he was active with the Shoshone Project in Wyoming, continuing with the Fort Hall Irrigation Project in Idaho until 1929. From 1935, Blomgren was chief of the Dam Division, US Bureau of Reclamation USBR, Denver. From 1944 to 1947, he was there chief engineer of the General Design Section. From 1949, he was USBR assistant chief engineer, from where he retired in 1955. He then was a consultant at Denver.

Blomgren was more than 42 years in government service. His career began in 1910 with the USBR's predecessor, the Reclamation Service. After work with various hydropower installations in the West, he contributed to irrigation systems for a consulting engineer, then becoming project engineer in the Office of Indian Affairs' Fort Hall Project. He was involved in a number of large dams, notably Shoshone Dam in the Yellowstone National Park, inaugurated in 1931, then with 100 m the second highest dam in the world, with a top width of 30 m. In the 1930s Blomgren was in charge of design for the tunnels and the outlets of Boulder Dam, today's Hoover Dam, one of the large dams still today. When completed in 1936, it was both the world's largest hydropower generator and the world's largest concrete structure. In the 1940s, Blomgren was then involved in Shasta Dam in California, a curved gravity concrete dam on Sacramento River above Redding, 183 m high and 1 km long with a base thickness of 165 m, based on a design similar to Hoover Dam. Blomgren was awarded the US Interior Department Gold Medal for distinguished services. He was from 1953 to 1957 a chairman of the US National Committee of the International Commission on Irrigation and Drainage ICID.

Anonymous (1950). Walter E. Blomgren. *Engineering News-Record* 144(May11): 64. *P*
Anonymous (1954). Blomgren, Walter E. *Who's who in engineering* 7: 223. Lewis: New York.
Anonymous (1955). Walter E. Blomgren retires. *Engineering News-Record* 154(Feb.17): 182. *P*
Anonymous (1964). Walter E. Blomgren. *Civil Engineering* 34(11): 105.
Blomgren, W.E., Milliken, J.G. (1953). Colorado-Big Thompson Project. *Indian Journal of Power and River Valley Development* 3(2): 89-97. *P*
Blomgren, W.E. (1955). Messstationen für Niederschläge mit Radiomeldung. *Schweizerische Bauzeitung* 73(21): 320.

BLOODGOOD

01.02. 1903 Whitewater WI/USA
19.02. 1985 West Lafayette IN/USA

Don Evans Bloodgood graduated as a civil engineer from University of Wisconsin in 1935. He was then a laboratory assistant of the Milwaukee Sewerage Community and eventually an engineer in charge of activated sludge plants of the Indianapolis Sanitary District. He was from 1943 lecturer and later until retirement in 1971 professor of sanitary engineering at Purdue University. He was the recipient of the 1971 Distinguished Service Award sponsored by the National Clay Pipe Institute, and in 1981 became Honorary Member ASCE for Advancing the concept of reclaiming industrial wastes. Don E. Bloodgood Memorial Award is presented to outstanding students in environmental engineering.

Bloodgood authored a number of highly detailed manuals on the operation of individual wastewater treatment plants. He also initiated as early as in 1944 the Purdue Industrial Waste Conference, an event bringing together experts from the various sectors of the profession, to focus on fields of industrial waste management. These conferences were forerunners of similar events throughout the world. He was also a consultant to many industries and municipalities, and he found time to serve as editor of Industrial Wastes for five years. Once retired, he continued a number of posts such as on advisory committees to the Ohio River Water Sanitation Commission, and the Environmental Protection Agency. Next to ASCE, he was also associated with the major societies devoted to water pollution, including the Water Pollution Control Federation WPCF, the American Public Works Association APWA, the American Water Works Association AWWA, and the Air Pollution Control Association APCA. Among the many honors, he received the Charles Alvin Emerson Medal and the Harrison Prescott Eddy Medal.

Anonymous (1959). Bloodgood, Don E. *Who's who in engineering* 8: 224. Lewis: New York.
Anonymous (1981). Don E. Bloodgood. *Civil Engineering* 51(10): 72. P
Anonymous (1985). Bloodgood, author and educator, dies. *Civil Engineering* 55(4): 72. P
Bloodgood, D.E., Boegly, W.J., Smith, C.E. (1956). Sedimentation studies. *Journal Sanitary Engineering Division* ASCE 82(SA5, 1083): 1-21.
Bloodgood, D.E. (1957). *Sewage treatment practices.* Scranton: Chicago.
Bloodgood, D.E., Teletzke, G.H., Pohland, F.G. (1959). Fundamental hydraulic principles of trickling filters. *Sewage and Industrial Wastes* 31(3): 243-253.
Bloodgood, D.E., Bell, J.M. (1961). Manning's coefficient calculated from test data. *Journal Water Pollution Control Federation* 33(2): 176-183.

BLUNT

01.02. 1823 Portsmouth NH/USA
10.07. 1892 Boston MA/USA

Charles Edward Blunt graduated in 1846 from the US Military Academy, West Point NY, was captain from 1860, lieutenant colonel from 1867, and then colonel from 1882, retiring after more than 40 years in 1887. He served from 1846 to 1854 as assistant engineer in the construction of Fort Winthrop MA, from 1852 to 1853 as superintending engineer of the East Dennis Breakwater and Provincetown Harbour MA, and from 1854 to 1865 in charge of repairs of Fort Montgomery NY. During the Civil War, he was assistant engineer of the defenses of Washington DC, and harbor improvements on Lakes Champlain and Ontario. From 1865 to 1867 he was member of a Board of Engineers on the defenses in the vicinity of Boston MA, from when he was in charge of various river and harbor improvements on Lake Ontario and St. Lawrence River. From 1869 to 1874 he was involved in the construction of Fort Taylor FL, and of Portsmouth Harbor, from when he continued with works on river and harbor improvements on Lake Erie until 1877. He then was ordered for the construction of defenses of the coasts of Maine and New Hampshire until 1883, and there was finally also responsible as supervising engineer for the state river and harbor improvements.

In 1868, Blunt prepared a Letter to the House of Representatives on the surveys for a ship canal to connect Lakes Erie and Ontario. It is stated that 'the importance of an uninterrupted water communication between these two lakes for the largest class of vessels was, as early as 1808, considered worthy of the efficient aid of the general government'. In total five different lines were considered. Preference was given to a route near Niagara River. The Lewiston route would include a system of double locks, which would then have a higher cost as compared with the others. The Eighteen-mile Creek route has the optimum condition of security, and would also be further removed from the frontier, and as a route of commerce, would be shorter. The average cost of the Canal would be about 12 millions of dollars.

Blunt, C.E. (1868). *Report and memoir on proposed lines for a ship canal around Niagara Falls*. Letter from the Secretary of War, 40th Congress, 2nd Session. Washington DC.
Blunt, C.E. (1884). *Results of surveys* of Penobscot River and Bangor Harbor ME, breakwater at mouth of Saco River ME, and near the mouth of Great Bay. GPO: Washington DC.
Powell, W.H. (1890). *Records of living officers of the US Army*: 68-69. Hamersly: Philadelphia.
http://www.arlingtoncemetery.net/ceblunt.htm (*P*)

BOARDMAN

21.01. 1869 Menasha WI/USA
07.01. 1965 Reno NV/USA

Horace Prentiss Boardman received his degree in civil engineering in 1894 from the University of Wisconsin, Madison WI. He was first a topographic surveyor in charge of the Sanitary District, Chicago, was then active for various construction firms until being appointed professor of civil engineering in 1907 at University of Nevada, Reno. From 1921, he was there in addition Director of the Engineering Experiment Station, and served from 1917 to 1921 as acting Dean of the College of Engineering. Since 1927, Boardman chaired the Nevada Snow Surveys, including stream runoff in the Sierra Nevada mountains. One interruption to his consecutive work at Reno came during World War I, when he was on leave of absence in 1918, employed on the engineering force of the US Explosive Plant at Nitro WV. He retired as professor in 1939 continuing as a consultant. He received the Honorary Degree of Doctor of Science from University of Nevada in 1950.

Boardman designed various structures, including the substructure of Glasgow Bridge MO over the Missouri River. Further work also comprised railroad projects, in which hydraulic problems were particularly considered. Another work related to retaining walls and earth pressures, published in 1905. His work in snow surveys might be described as a hobby, though it proved of immense value for estimating and predicting the seasonal water sources for industrial use, especially following disastrous droughts in the 1930s. He was member ASCE, and of the American Geophysical Union AGU in whose journal he published numerous articles on stream flow.

Anonymous (1921). Boardman, Prof. Horace Prentiss. *American men of science* 3: 69. Garrison: New York.
Anonymous (1964). Boardman, Horace P. *Who's who in engineering* 9: 168. Lewis: New York.
Anonymous (1981). Horace P. Boardman. *Trans AGU* 46(3): 599-600.
Boardman, H.P. (1949). Snow surveys for forecasting stream flow in Western Nevada. *Bulletin* 184. Nevada Agricultural Experiment Station: Reno.
Boardman, H.P. (1951). Buttress type dam with curved upstream face. *Civil Engineering* 20(6):346.
Boardman, H.P. (1952). *Truckee River floods and high water years*. Private printing.
Boardman, H.P. (1959). *Some interesting and important facts about Lake Tahoe*. Snow Surveys Committee of State Association of Soil Conservation Districts: Reno.
www.wcc.nrcs.usda.gov *P*

BOCK C.A.

20.04. 1883 West Side IA/USA
01.11. 1966 Pompano Beach FL/USA

Carl August Bock obtained in 1906 the BS degree from Cornell College, Mount Vernon IA, and the CE degree in 1908. He was employed from 1906 to 1912 in the USA, Mexico, Bolivia, and the West Indies with various constructions, then until 1933 he was involved in planning, design and construction of water control projects, as engineer in charge of a consulting engineer. From 1933 to 1940 Bock was with the Tennessee Valley Authority TVA, serving as director of engineering, assistant chief engineer, and chief consulting engineer. He was in parallel consulting engineer of the Puerto Rico Utilization of Water Resources, San Juan PR, since 1935, and consulting engineer of the National Resources Planning Board, and consulting engineer of the Interstate Commission on the Delaware River Basin since 1939. Bock had a general engineering practice from 1940 to 1942, and then was the chief engineer of the Puerto Rico Water Resources. He was member of the American Society of Civil Engineers ASCE.

Bock was the first engineer employed by the TVA when it was organized in 1933. He was soon promoted to assistant chief engineer and was then chief consulting engineer. He had a long connection to the hydro-electricity program in Puerto Rico, mainly after World War II. The *Caonillas* hydro-electric scheme was completed in 1949, constituting then the largest of the power units built under the direction of the Puerto Rico Water Resources Authority, an organization that supplied 15% of the island's power. Caonillas Dam consists of a straight gravity concrete dam 66 m high and 230 m long located in a narrow gorge, creating the largest reservoir of the island, and providing a head of 160 m on the turbines. Bock's contribution to the power supply of Puerto Rico was particularly acknowledged by the consumers.

Anonymous (1939). Carl A. Bock: Resigning TVA engineer. *Engineering News-Record* 123(Nov.30): 721. *P*
Anonymous (1948). Carl A. Bock. *Engineering News-Record* 140(Mar.25): 466. *P*
Anonymous (1948). Bock, Carl A. *Who's who in engineering* 6: 190. Lewis: New York.
Bock, C.A. (1918). *History of the Miami Flood Control Project*. MCD: Dayton OH.
Bock, C.A., Moneymaker, B.C. (1934). *Stratigraphy and structural geology* of the Hiwassee River Basin in vicinity of Coleman dam site. TVA: Knoxville TN.
Pike, A., Morgan, A.E., Bock, C.A. (1934). *Virginia-West Virginia Bandy quadrangle*. US Geological Survey: Washington DC.
http://nationalcalamityeaster1913flood.blogspot.ch/2013/01/morgans-pyramids.html

BODFISH

16.08. 1844 Chicopee MA/USA
17.05. 1894 Washington DC/USA

Sumner Homer Bodfish graduated in 1871 from the US Military Academy, West Point NY, after having served during the Civil War. After graduation, he started his professional career with his father in hydraulic engineering at Langley GA, constructing a dam for the Langley Cotton Mills. In 1872 he went to Washington DC, where he was appointed first assistant engineer of the Board of Public Works. He there had charge of underground works until 1878, when engaged with a large dredging contract on Patapsco River at Baltimore MD. In 1878 he was appointed topographer in the Powell Survey, and from then until 1890 remained in the employ of the US Geological Survey USGS as topographer, and later as irrigation engineer. His first work for the Powell Survey in 1878 was to measure angles in a scheme of triangulation of Grove K. Gilbert (1843-1918). From 1878 to 1880, Bodfish mapped the country around the Grand Canyon.

In 1880, upon the organization of the US Geological Survey, Bodfish was appointed topographer, spending the seasons 1880 to 1881 in topographic work in southern Utah. In 1882 he was engaged on topographic work in the quicksilver mining region of California. Until 1888 he was then on similar work in the neighborhood of Washington DC, and in Massachusetts. Shortly after the creation of the Irrigation Survey, he was appointed irrigation engineer in 1889, and spent the next year in charge of the Colorado Division. He thereby made surveys on optimum reservoir sites, and the modes by which water could be conducted from them to the surrounding arid lands. In 1889 he computed the cost of the dams proposed and the capacities of the reservoirs. In 1890, Bodfish founded a private practice, engaged in numerous works on railways projects. His last project was a survey of the Great Falls of Potomac River for a Boston company, and a determination of the quantity and value of the water power which could be derived from these falls. Ill health forced him in 1893 to leave work, from when he remained an invalid until his death. Bodfish was a careful, painstaking and conscientious worker, an excellent engineer, and a notable surveyor of the 19th century.

Anonymous (1894). Sumner Homer Bodfish. *Proc. ASCE* 20: 96-98.
Evans, R.T., Frye, H.M. (2009). History of the Topographic Branch. *Circular* 1341. US
    Department of the Interior: Washington DC.
Menkes, D. (2007). A man goes west: The 1879 letters of Leonard Herbert Swett. *Utah
    Historical Quarterly* 75(3): 204-219. *P*

BOGART

08.02. 1836 Albany NY/USA
25.04. 1920 New York NY/USA

John Bogart was educated at Rutgers College, New Brunswick NJ, graduating in 1853 with a BA degree; the DSc degree was conferred on him in 1912. After service for the US Corps of Engineers, he became assistant of the Engineering Department, New York State, employed on works of canal reconstructions. He was then engaged until the outbreak of the Civil War with the construction of the Central Park, New York, including water supply and drainage projects. After War, he was appointed Deputy State Engineer, New York NY, staying in office until 1888, thereby being involved in the construction of Washington Bridge over Harlem River, New York, and in tunnels under Hudson River.

In 1890 Bogart became interested in the hydro-electric development. He was appointed consulting engineer of the Cataract Construction Company, of the Niagara Falls Power Company. At this time few knowledge was available on power transmission, so that he visited Europe to study the latest developments. In 1899 he investigated the possibilities of utilizing the power of the St. Lawrence River at Massena NY, designing then the plant of the St. Lawrence Power Company there. His reputation became established as hydraulic engineer, so that he served for instance as engineer of the Cascade Power Company in British Columbia, the Lake Superior Power Company, or the Sault Ste. Marie Power Company. His greatest contribution was the hydro-electric development of the Chattanooga and Tennessee River Power Company, now as chief engineer, a 60,000 HP installation on Tennessee River. This project was particularly difficult on account of the great head variation on the turbines, and in terms of dam foundation. In 1913 Bogart formed a partnership until his death. He was US delegate to the Permanent International Navigation Congresses PIANC held at Düsseldorf in 1902, at Milan in 1905, at St. Petersburg in Russia in 1908, and served as chairman of the Inland Section in the 1912 Philadelphia Congress. He was member of the American Society of Civil Engineers ASCE, serving as director from 1873 to 1875, and as treasurer and secretary.

Anonymous (1920). John Bogart dead. *Engineering News-Record* 84(18): 887. *P*
Anonymous (1925). John Bogart. *Trans ASCE* 88: 1346-1350.
King, M. (1899). John Bogart: Civil engineer, consulting engineer Cataract Construction Co. *Notable New Yorkers* of 1896-1899: A companion volume to King's Handbook of New York City: 412. Orr: New York.
www.wcc.nrcs.usda.gov *P*

BOLTON

05.10. 1856 London/UK
18.02. 1942 New York NY/USA

Reginald Pelham Bolton was educated in England, but moved to the USA to widen his professional experience in 1879. He was assistant to Erasmus D. Leavitt (1836-1916) at Boston MA on the design of mining machinery. Following ill health, Bolton in 1881 returned to England, where he established his own consulting office. During these years he made various inventions, including a hydraulic governor, the Bolton and Hartley air compressor, or an electric hammering apparatus. Bolton returned to the USA in 1894 and first was mainly engaged on the design of heating and ventilating systems of large buildings in New York. Until 1915 he published numerous papers on his varied technical works. He was of course also highly interested in professional activities, and was member of many professional societies, including the American Society of Civil Engineers ASCE, the American Society of Mechanical Engineers ASME, the American Society of Naval Engineers, or the American Society of Heating and Ventilating Engineers ASHVE, of which he was president in 1911, and made honorary member in 1936 'in recognition of the eminent place occupied as a consulting engineer, author, and historian'. He also was member of the Institution of Civil Engineers ICE, London UK, and was awarded its Telford Gold Medal.

Bolton was an active person not only in engineering, but was also interested in the relics of early Indian habitations. His studies of these remains and the research into the life and living of the colonial period was so thorough that he became an authority on the history of New York and Westchester counties. His findings were recorded in pamphlets or publications of the Museum of the American Indian, or the American Scenic and Historic Preservation Society. His devotion to historic matters caused him to be known as the 'Number One Citizen of Washington Heights'. Another part of his developments related to the building and equipment of modern skyscrapers, without which the high office building could never have attained the success it enjoyed in the early 20th century.

Anonymous (1944). Reginald P. Bolton. *Trans. ASCE* 109: 1465-1470.
Bolton, R.P. (1897). Circulation of steam for heating purposes at or below the pressure of the atmosphere. *Trans. ASHVE* 3: 155-175.
Bolton, R.P. (1903). Test of a hydraulic elevator system. *Trans. ASME* 24: 933-944.
Bolton, R.P. (1914). Hydro-electric power compared with steam. *Trans. ASHVE* 20: 374-391.
http://myinwood.net/weekend-archeologists-of-northern-manhattan/ *P*

BONDURANT

11.09. 1908 Charleston MO/USA
14.06. 1985 Heber Springs AR/USA

Donald (Don) Connelly Bondurant graduated with the BSc degree from the University of Missouri, Columbia MO, in 1932. He was in the 1940s Head of the Hydrology and Sediment Section, Albuquerque District, US Army Corps of Engineers, Albuquerque NM, and in the early 1950s then the Head of the Sedimentation Section, Missouri River Division, US Army Corps of Engineers, Omaha NE. He retired in 1972 after 40 years with the Corps, the last 24 of which he spent as head of the sedimentation studies. In addition he was assigned as consultant for other Division offices representing the Chief of Engineers on various interagency committees involved in sedimentation investigations. He was awarded the 1976 Hilgard Prize for his work in the Manual of Sedimentation.

The 1951 paper is concerned with sedimentation problems in reservoirs, involving not only in the loss of storage but also the distribution of storage loss with respect to the many functions of multiple purpose reservoirs, and the possible damage resulting from aggradation above, or degradation below the reservoir. The paper presents data on the form, extent, and type of sediment deposits along with a description of the reservoir, its use, and the contributing drainage. The 1963 Report co-authored by Frederick R. Brown (1912-1999) includes a description of the Nile and its importance for Egypt. Further, the Hydraulic Laboratory, its staff and research plans are reviewed. Another section deals with the degradation of Nile River and possible measures against it.

Anonymous (1946). D.C. Bondurant, New Mexico Section. *Civil Engineering* 16(8): 368. *P*
Anonymous (1976). Donald C. Bondurant, Hilgard Prize. *Civil Engineering* 46(10): 103. *P*
Bondurant, D.C. (1951). Sedimentation studies at Conchas Reservoir in New Mexico. *Trans. ASCE* 116: 1283-1295.
Bondurant, C.D. (1955). *Report* on Reservoir delta reconnaissance. US Army Corps of Engineers: Omaha NE.
Bondurant, D.C., Brown, F.R. (1963). *Review and evaluation* of hydraulic problems at the Hydraulic Research and Experimental Station, Delta Barrage, United Arab Republic. US Army Corps of Engineers: Vicksburg MS.
Bondurant, D.C. (1972). Discussion of River bed degradation after closure of dams. *Journal of the Hydraulics Division* ASCE 98(11): 2051-2052.
Reuss, M. (2004). *Designing the Bayous*: The control of water in the Atchafalaya Basin 1800-1995. A&M University Press: College Station TX.

BORLAND

06.11. 1905 Holyoke CO/USA
02.10. 2001 Peoria AZ/USA

Whitney McNair Borland received a Bachelor's degree from the University of Nebraska and a Master's degree from the University of California, Berkeley CA, both in mechanical engineering. Later he also graduated as a civil engineer.

Borland spent all his professional career with the US Bureau of Reclamation USBR. He joined the Montrose hydraulic laboratory in 1930 testing models for the Imperial, Grand Coulee, and Stewart Mountain Dams. From 1936, he was permanently assigned to the Spillway Design Section in Denver, completing a study on the backwater caused by Grand Coulee Dam on Columbia River. After war service Borland rejoined USBR in 1946 the newly established sedimentation section. In 1950 he took over the responsibility for guiding the sedimentation program until 1970. He thereby supervised the preparation of many reports, forming the basis of technical papers, which were prepared for the journals of ASCE, the American Geophysical Union AGU, and the Federal Interagency Sedimentation Conferences. From 1970 to 1972, Borland prepared a two-volume report on the collection of basic data, for sedimentation and degradation problems associated with the Mekong River in Southeast Asia. After retirement in 1972, he became a consultant on numerous projects. With his expertise in sediment transport and river hydraulics, his work varied from international projects to working directly on river migration and stability studies. He maintained a mutual interest in sedimentation research with Emory W. Lane (1891-1963). Of particular interest to them was the design of stable channels. Lane had developed an approach in the early 1950s. Borland postulated that a practical approach should include in addition to equations also engineering judgment, experience and skill. It resulted in a relation for defining problems with stream morphology.

Borland, W.M., Miller, C.R. (1958). Distribution of sediment in large reservoirs. *Journal of the Hydraulics Division* ASCE 84(HY2): 1-18.
Borland, W.M. (1960). *Stream channel stability*. USBR: Denver.
Borland, W.M. (1963). *Sediment study for Vee Project, Susitna River, Alaska*. USBR: Denver.
Lane, E.W., Borland, W.M. (1951). Estimating bed load. *Trans. AGU* 32(1): 121-123.
Miller, C.R, Borland, W.M. (1963). Stabilization of Fivemile and Muddy Creeks. *Journal of the Hydraulics Division* ASCE 89(HY1): 67-98.
Pemberton, E.L., Strand, R.I. (2005). Whitney M. Borland and the Bureau of Reclamation, 1930-1972. *Journal of Hydraulic Engineering* 131(5): 339-346. *P*

BOUILLON

13.04. 1869 Custinne/B
14.01. 1935 Seattle WA/USA

Alfred Victor Bouillon was born in the Belgian Ardennes. His family moved in 1883 to the USA, eventually settling at Seattle WA. After studies in Canada, he assisted his father, an architect, until 1890. From 1892 he was on the city's employ until 1895, when accepting as assistant chief engineer an offer with the Seattle Power Company. Shortly later he joined the staff of a shipbuilder in Seattle as chief draftsman, where he was involved in design of ships mainly for Alaskan waters where the Klondike gold rush was in full progress. The firm also did ship repair and salvage works, and built a floating dry dock, mainly designed by Bouillon. In 1908, when the firm was closed, he became superintendent of public utilities, Seattle, and chairman of the Board of Public Works.

From 1910 to 1911 Bouillon was engaged as city commissioner of public works and utilities by Edmonton AB, to bring the city-owned utilities out of the chaos that resulted from inexperienced and improper planning. He then returned to Seattle WA, forming a successful partnership with a colleague as consulting civil and mechanical engineers until 1917. He also was during this term expert and assistant to the general manager of the Newport News Shipbuilding and Dry Dock Company on special assignments until 1918. Its head stated: 'I am venturing to call to your attention the very striking ability of Mr. Bouillon; he is industrious and earnest in his work, straightforward and honorable. He is wonderful exact in everything that he does, ingenious in consideration of methods of mathematical demonstration, an absolute master of detail'. In 1918 Bouillon entered government service, becoming the assistant manager of the Contact Division, the US Shipping Board Emergence Fleet Corporation, remaining there until his death. His duties included the study of shipyards, and the development of the Construction Department. He was in 1925 appointed secretary of the American Marine Standards Committee, where he analyzed the design, equipment and operation of vessels, with the objective to simplify and standardize the construction. His combined knowledge and experience relating to the various phases of the marine industry made him exceptionally valuable. He was member of the American Society of Civil Engineers ASCE since 1913, and of the Society of Naval Architects and Marine Engineers SNAME since 1918.

Anonymous (1935). Alfred V. Bouillon. *Trans. ASCE* 100: 1608-1610.
Anonymous (1935). Alfred Victor Bouillon. *Trans. SNAME* 43: 307. (*P*)
http://www.findagrave.com/cgi-bin/fg.cgi?page=gr&GRid=59889456 (*P*)

BOWDEN

23.03. 1888 Tangipahoa LA/USA
07.01. 1972 Knoxville TN/USA

Nicholls White Bowden was educated at Louisiana State University, Baton Rouge LA, from where he graduated with the BS degree in 1908, and the CE degree in 1928. After graduation he was employed by the US Army Corps of Engineers at the Nashville and Chattanooga Districts, where he was engaged in the first survey to locate the site for Wilson Dam at Muscle Shoals AL. He further supervised works for river improvements on Tennessee River, and from 1918 to 1922 was there in charge of dredging works. After other engineering works he completed from 1930 arial photographic mapping survey of rivers, including the route of the proposed Lake Erie - Ohio River Canal along the Beaver and *Mahoning* Rivers in Pennsylvania and Ohio.

In 1936 Bowden accepted the position of principal hydraulic engineer in the General Engineering and Geology Division of the Tennessee Valley Authority TVA. He was first involved in the completion of the report The unified development of the Tennessee River System, forming the basis for the development of the TVA water control system. Later he took over as TVA's first chief river control engineer, responsible to direct the operation of the rapidly growing TVA reservoir system, from which position he retired in 1958. He particularly directed the regulation of the second highest flood on record in 1957 at Chattanooga TN. He was in addition to his TVA assignment also Alternate Member and water consultant for the Water Resources Committee of the National Resources Planning Board. After retirement, Bowden was a consulting engineer. He has written a number of papers on 'open channel' river improvement and on the operation of multi-purpose reservoirs, for which he was awarded by the American Society of Civil Engineers ASCE. He was elected to ASCE Fellowship in 1959.

Anonymous (1958). Nicholls W. Bowden. *Civil Engineering* 28(5): 21. *P*
Anonymous (1964). Bowden, Nicholls W. *Who's who in engineering* 9: 186. Lewis New York.
Anonymous (1972). Bowden, Nicholls White, F. ASCE. *Trans. ASCE* 137: 1053-1055.
Bowden, N.W. (1912). *Regulation of the Hiwassee River near Charleston TE*. Washington DC.
Bowden, N.W. (1942). Operation experiences: Tygart Reservoir. *Trans. ASCE* 107: 1374-1376.
Bowden, N.W. (1950). Multiple-purpose reservoirs: General problems of design and operation. *Trans. ASCE* 115: 803-817.
Christopher, G.S., Bowden, N.W. (1957). Mosquito control in reservoirs by water level management. *Mosquito News* 17(4): 273-277.

BOWERS

03.09. 1919 Hanna WY/USA
22.12. 2008 Goleta CA/USA

Charles Edward Bowers attended the University of Wyoming, Laramie WY, graduating in 1942 as a civil engineer. He then joined the David Taylor Navy Laboratory, Washington DC. After World War II, he moved to Minnesota University to obtain the Master's Degree in 1949. He then joined the Saint Anthony Falls SAF Hydraulic Laboratory and eventually became professor of civil engineering. Bowers was a great teacher and received many awards in the field of education. He moved with his wife in 2000 to California, where he spent the last years. Bowers was the recipient of the 1950 ASCE Collingwood Prize.

While at the Model Basin, Bowers developed facilities and equipment used in model ship testing. It included an investigation of ship performance in restricted channels, the results of which were published in the paper awarded in 1950. From the 1960s, Bowers turned his interest to dam engineering, where he was interested in stilling basins. One of the problems of these hydraulic structures are turbulent pressures by which a structure may get damaged. Bowers investigated in 1969 these for a basic arrangement. Later, he made a notable investigation into the extreme pressure distribution by which stilling basins are subjected, resulting in a maximum pressure head fluctuation of the order of the approach flow kinetic energy head. Bowers conducted studies of dams throughout the world, including an analysis of the southern California *Cachuma* Dam spillway.

Anonymous (1950). C.E. Bowers, Collingwood Prize winner. *Civil Engineering* 20(10): 675. *P*

Bowers, C.E., Tsai, F.Y. (1969). Fluctuating pressures in stilling basins. *Journal of the Hydraulics Division* ASCE 95(HY6): 2071-2079.

Bowers, C.E., Toso, J.W. (1988). Karnafuli project-model studies of spillway damage. *Journal of Hydraulic Engineering* 114(5): 469-483; 116(6): 850-855.

Herbich, J.B., Ziegler, J., Bowers, C.E. (1956). Experimental studies of hydraulic breakwaters. Saint Anthony Hydraulic Laboratory, University of Minnesota: Minneapolis.

Straub, L.G., Bowers, C.E., Tarapore, Z.S. (1959). Experimental studies of pneumatic and hydraulic breakwaters. *Technical Paper* 25, Series B. SAF, University of Minnesota: Minneapolis.

Toso, J.W., Bowers, C.E. (1988). Extreme pressures in hydraulic jump stilling basins. *Journal of Hydraulic Engineering* 114(8): 829-843.

http://www.legacy.com/obituaries/startribune/obituary.aspx?page=lifestory&pid=121907543 *P*

BOWMAN

29.04. 1889 West Liberty IA/USA
28.09. 1962 Knoxville TE/USA

James Schenck Bowman graduated from University of Iowa, Iowa City IA, with the degree of Bachelor of Science in civil engineering in 1913. He continued with graduate studies at University of Wisconsin, Madison WI in 1933. Following graduation, he was first concerned with drainage projects until 1918, becoming then assistant engineer with the Miami Conservancy District, Dayton OH. From 1919, he was irrigation engineer in Wyoming State. From then until 1927, he was a hydraulic engineer with *Fargo* Engineering, Jackson MI, joining until 1932 *Harza* Engineering, Chicago IL. After having been instructor of hydraulic engineering at University of Wisconsin, Madison WI, Bowman began in 1933 service with the Tennessee Valley Authority TVA as hydraulic engineer. By 1941 he was assistant chief water control planning engineer, and in 1943 became chief water control planning engineer, following Sherman M. Woodward (1871-1953). He held this position until retirement from TVA in 1955, and continued to serve TVA as a consultant in matters involving river control.

Prior to his stay with TVA, Bowman was a hydraulic engineer of Norris Dam. Once with the TVA Water Control Planning Department, he took responsibility for preparation of plans for most of TVA's major projects. He contributed the chapter on Spillway crest gates to the 1952 Handbook of applied hydraulics. Bowman was a Member of the American Society of Mechanical Engineers ASME, the Tennessee Historical Society, a Honorary Member of the Knoxville Technical Society, and Past President of the East Tennessee Historical Society. He was elected Associate Member ASCE in 1919, Member ASCE in 1925, and Fellow ASCE in 1959. He was interested also in matters of American history, thereby presiding over the East Tennessee Historical Society.

Anonymous (1943). James S. Bowman. *Engineering News-Record* 130(Apr.29): 652. *P*
Anonymous (1955). Bowman, James S. *Who's who in America* 28: 285. Marquis: Chicago.
Anonymous (1963). James Schenck Bowman. *Trans. ASCE* 128(5): 115-116.
Bowman, J.S., Bowman, J.R. (1952). Spillway crest gates. *Handbook of applied hydraulics* 8: 291-334, C.V. Davis, ed. McGraw-Hill: New York.
Dawson, F.M., Bowman, J.S. (1933). Interior water supply piping for residential buildings. Experiment Station *Report* 77: 1-50. University of Wisconsin: Madison.
Rothrock, M.U., Bowman, J.S. (1946). *The French Broad-Holston Country*: A history of Knox County, Tennessee. East Tennessee Historical Society: Knoxville.

BOYDEN

17.02. 1804 Foxborough MA/USA
17.10. 1879 Boston MA/USA

Uriah Atherton Boyden was educated in country schools and later assisted in blacksmithing and farming. He was then concerned with a first survey for Boston & Providence Railroad, on construction of mills in Lowell MA until 1833, when opening an office of engineering at Boston MA. From 1840 he designed hydraulic works, including an improved Fourneyron-type water turbine of 75 HP at Lowell, operating at 75% efficiency, known as the Boyden turbine in the United States. In 1846 he built three 190 HP turbines. Among improvements were scroll penstocks, suspended top bearing, diffuser showing principles of modern flaring draft tube, and the hook gauge. He proposed the spiral approach flow to a turbine admitting water to turbine at uniform velocity. His works inspired James Francis (1815-1892), who later proposed the Francis tangential inlet turbine. The talents of the two were complimentary. Boyden was abler in mathematical theory, while Francis was the greater experimentalist. Boyden was a shy genius, almost a recluse: He published none of his findings and had no close professional associates apart from Francis. Boyden however had a profound understanding of the defects of the existing hydraulic science and did much to set Francis upon the right course, who had the skill to show how these deficiencies might be overcome. Their insight led each of them to see how a radical improvement in turbine design could be made. Francis was apparently the one who published the results. Thus the Francis turbine rightfully bears his name.

Boyden retired in 1850 to Boston to study pure science, particularly the velocity of sound and light, the compressibility of water, and heat in general. In 1874 he sponsored 1000 US$ to the Franklin Institute to be awarded to a resident of North America who should determine the velocity of light. He also established the Soldiers' Memorial Building, and gave another 1000 US$ for the Boyden Public Library, Foxborough MA. Most of his fortune was left for building observatories on mountain tops.

Anonymous (1887). Boyden, U.A. *Appletons' cyclopaedia of American biography* 1: 341.
Anonymous (1980). Boyden, Uriah Atherton. *Mechanical engineers in America born prior to 1861: A biographical dictionary*: 67-68. ASME: New York. *P*
Pursell, C.W., ed. (1990). *Technology in America*: A history of individuals and ideas. MIT: Cambridge MA.
http://www.rootsweb.ancestry.com/~mafhs/foxuriahboyden.htm *P*

BOYER

02.01. 1908 Boise ID/USA
01.09. 1986 Kennebunk ME/USA

Marion Clifford Boyer graduated in 1928 with the BS degree from the University of Colorado, Denver CO, and as professional engineer in 1935. In 1947 he received the MS degree in hydraulic engineering from the State University of Iowa, Iowa City IA. From 1928 to 1942 he was in the Water Resources Branch of the US Geological Survey at Washington DC and in the Northwest, then in 1943 with the US Bureau of Yards and Docks as an aviation facilities engineer, from 1943 to 1946 on active duty in the US Naval Reserve as a civil Corps officer, from 1947 to 1956 as an associate professor of mechanics and hydraulics at the University of Iowa, and from 1957 to 1962, he was hydraulic engineer with the State of Indiana. Since then he was natural resources coordinator for Fresno County in California.

Boyer was instrumental in the translation of the monthly issued Soviet hydraulics journal *Gidrotekhnicheskoe Stroitelstvo*, namely the Hydrotechnical Construction. This journal is devoted primarily to the design of hydraulic structures, and summarizes the hydraulic, structural, and soil engineering involved in these projects. The first issue was translated in 1967, with a delay between the Russian and English version of less than six months. Boyer served as translation editor, whose skill in Russian was acquired during his stay at Washington DC. He translated scientific Russian articles for the Office of Naval Intelligence since 1958. Boyer had been previously involved in a historical book on the hydraulic developments at the University of Iowa. He also contributed a chapter to the Handbook of applied hydrology in 1964.

Anonymous (1967). M. Clifford Boyer. *Civil Engineering* 37(11): 79. *P*
Boyer, M.C. (1935). *The evaluation of values of C in terms of the mean depth in the computation of flow in open channels*. University of Colorado: Denver.
Boyer, M.C. (1947). *The measurement of velocity of flowing water by electrical methods*. State University of Iowa: Iowa City.
Boyer, M.C., ed. (1949). Third decade of hydraulics. *Bulletin* 33. State University of Iowa: Iowa.
Boyer, M.C. (1961). Irrigation and drainage potential in humid areas. *Trans. ASCE* 126: 184-193.
Boyer, M.C. (1964). Streamflow measurement. *Handbook of applied hydrology* 15: 1-41, V.T. Chow, ed. McGraw-Hill: New York.
McNown, J.S., Boyer, M.C., eds. (1953). Proc. 5th *Hydraulics Conference*. Iowa Institute of Hydraulic Research: Iowa City.

BRACKENRIDGE

15.06. 1858 Richmond Hill NY/USA
29.11. 1929 Santa Barbara CA/USA

THE HARNESSING OF NIAGARA

THE CASSIER MAGAZINE CO.
NEW YORK AND LONDON
1895

William Algernon Brackenridge attended schools at New Brighton, Staten Island NY, and Toms River NJ, commencing practical civil engineering in New York until 1880. He was then transitman in the Survey Corps until 1885. From then to 1889 he was principal assistant engineer of the Long Island Cable Railway. From 1890 he was associated with Clemens Herschel (1842-1930), making plans and reports to the Railroad Commissioners of Connecticut State. He started in 1891 his work with the development of hydro-electricity, accepting an appointment with the Board of Consulting Engineers to develop water power at Niagara Falls, an engagement resulting later in his appointment to engineer in charge of all the construction operations of the Niagara Falls Power Company; this included the main tail-race tunnel and the wheel-pits, the power houses 1 and 2 on the American river side, and the development of the Canadian Niagara Power Company. He superintended these works including the filter plant and the hydraulic pumping plant.

In 1904 Brackenridge was appointed to be one of the five expert civil engineers to act as Advisory Board of consulting engineers on an expenditure of 100,000,000 US$ for the improvement of the New York State Canals, and the construction of the Barge Canal to connect the Hudson River with the Great Lakes. He was engaged in this large project until 1909 when changing his residence from Buffalo NY to Pasadena CA. He had been called by the Southern California Edison Company to advise on the reconstruction of its Kern River No. 1 hydro-electric plant, a work completed in 1909 after a failure of its pressure main and tunnel had suspended its operation. He became vice-president and general manager of the Company, continuing until 1918, when appointed president, a position he held until 1920, from when he was senior vice-president until 1928. Under his direction a large steam installation was made at Long Beach CA, with some 150 km of steel tower transmission line, and also the 32,000 kW hydro-electric Kern River No. 3, including two Francis turbines operating under a head of 240 m. He was identified as important man having developed the American hydro-electric industry.

Anonymous (1895). *The harnessing of Niagara*. Cassier Magazine: New York. (*P*)
Anonymous (1932). William A. Brackenridge. *Trans. ASCE* 96: 1419-1422.
Brackenridge, W.A. (1909). Transmission of electrical energy in Southern California. *Electrical World* 54(18): 1037-1040.
Whitford, N.E. (1921). *History of the Barge Canal of New York State*. Lyon: Albany NY.

BRACKETT

30.11. 1851 Newton MA/USA
26.08. 1915 Brighton MA/USA

Dexter Brackett graduated in 1868 from Brighton High School near Boston MA. Next year he entered the city engineer's office, Boston MA, remaining there for 26 years. He was assigned in 1874 to the supervision of its distribution system of the water works. He was highly instrumental in developing an unusually adequate piping system for the city. He resigned in 1895 in place with the City of Boston to accept the position of engineer of the Distribution Department of the Metropolitan Water Works. He had thereby charge of the construction of the pipe system, pumping stations, and distributing reservoirs within the limits of the District.

From 1907 to the time of his death, Brackett was chief engineer of Boston Metropolitan Water Works. He had always been interested in questions relating to the consumption and waste of water, and had written papers and reports on these subjects. He persistently advocated the use of meters within the Metropolitan District as a means of reducing the waste of water, and had the privilege of seeing his views adopted and the consumption of water thereby reduced. Up to 1907, it had been increasing so rapidly to a point out of necessity for an additional water supply within few years, but the gradual introduction of meters reduced the consumption of water so rapidly that no additional supply was then needed; it reduced to less than two-thirds up to the days of his death. Brackett was member and director of the American Society of Civil Engineers ASCE, he had been president of the Boston Society of Civil Engineers, and the New England Water-Works Association NEWWA. He also was member of the American Water Works Association AWWA. The Dexter Brackett Memorial Medal was instituted by NEWWA in 1916 shortly after Brackett's death. The award is presented annually to a member of the association who authored the most meritorious paper published in its Journal.

Anonymous (1915). Dexter Brackett. *Engineering News* 74(10): 477. *P*
Brackett, D. (1895). Consumption and waste of water. *Trans. ASCE* 34: 185-203.
Brackett, D. (1886). The distribution system of the Boston water works. *Journal of the Association of Engineering Societies* 5(4): 107-115.
Brackett, D. (1886). Rainfall received and collected on the water-sheds of Sudbury River and Mystic Lakes. *Journal of the Association of Engineering Societies* 5(11): 395-401.
Kempe, M. (2006). New England water supplies: A brief history. *Journal of the New England Water Works Association* 120(3): 1-157. *P*

BRADLEY

23.08. 1903 Chicago IL/USA
01.04. 1993 Washington DC/USA

Joseph Newell Bradley graduated as civil engineer from University of Illinois, Urbana IL. He joined the US Bureau of Reclamation USBR, Denver, early in the 1930s staying there until 1955, when joining the Hydraulic Research Section of the Bureau of Public Roads, Washington DC. From the 1980s, Bradley was a private consultant.

Bradley's career fell into the American rise of hydraulics. He states in his 1987 paper: 'The decade beginning in 1930 was the golden age of dam building and hydraulic model testing for USBR. That was during the great depression when the federal government was involved in putting men to work, and wages for construction workers ranged from 50 to 85 cents an hour'. The large dams included Hoover, Grand Coulee, and Shasta. These and many others had a common feature: the storage of water for irrigation in the Western desert lands. Bradley was fully involved into these activities, relating to USBR projects in the west. His first notable research was directed to the control of cavitation damage, as was detected during World War II at Boulder Dam. Then, he proposed a hydraulic design for the morning glory spillway, often used in reservoirs with a side-channel. Further works included the determination of friction factors for tunnels, discharge coefficients for overflow crests, and, most importantly, his works for stilling basins in collaboration with Alvin J. Peterka (1911-1983). Once having left USBR for the Federal Highway Administration, he prepared a notable book on bridge hydraulics, and he was stated to have a major credit for the contribution to the 1975 Circular on energy dissipators.

Anonymous (1955). Joseph N. Bradley. *Civil Engineering* 25(10): 713. *P*

Anonymous (1975). *Hydraulic design of energy dissipators for culverts and channels*. Hydraulic Engineering Circular 14, ed. 1. Federal Highway Administration, Arlington VA.

Bradley, J.N. (1945). *Study of air injection into the flow in the Boulder Dam spillway tunnels*. USBR: Denver.

Bradley, J.N. (1951). *Friction factors for large conduits flowing full*. USBR: Denver.

Bradley, J.N. (1952). *Discharge coefficients for irregular overfall spillways*. USBR: Denver.

Bradley, J.N. (1970). *Hydraulics of bridge waterways*. US Bureau of Public Roads. US. Govt. Printing Office: Washington DC.

Bradley, J.N. (1987). Advances in hydraulic engineering practice: The last four decades and beyond. *Journal of Hydraulic Engineering* 113(7): 955-956.

Falvey, H.T. (2005). Joseph Newell Bradley. Personal communication.

BRATER

07.04. 1912 Saginow MI/USA
29.01. 2003 Ann Arbor MI/USA

Ernest Frederick Brater graduated in 1936 with an MS from the University of Michigan, and there obtained the PhD degree in 1938. He was first engaged with flood prediction within the US Army Corps of Engineers, then was involved in dam design with Ayres Co., Ann Arbor, and after World War II joined the University of Michigan at the Lake Hydraulics Laboratory, where he investigated harbour design and wave forces. He was appointed professor of hydraulic engineering there in 1949, after having served in the staff from 1938.

Brater has authored and co-authored a number of highly popular books in hydraulic education, mainly directed to students and to hydraulic practice. The following chapters are included in the 1996 edition of the Handbook of hydraulics: 1. Fluid properties and hydraulic units, 2. Hydrostatics, 3. Fundamental concepts of flow, 4. Orifices and gates, 5. Weirs, 6. Pipes, 7. Steady uniform flow in open channels, 8. Open channels with non-uniform flow, 9. High-velocity transitions, 10. Wave motion, 11. Spatially variable and unsteady flow, 12. Measurement of flowing water, 13. Advances in hydraulics using computer technology, 14. Applicable computer programs. The second book of note is Hydrology, written in collaboration with C.O. Wisler (1881-1961). Its main chapters are: Hydrograph, Runoff, Precipitation, Infiltration and soil moisture, Groundwater, Water losses, Semiarid regions, Snow, Floods, and Stream-flow records.

Anonymous (1954). Brater, Ernest F. *Who's who in engineering* 7: 274. Lewis: New York.
Brater, E.F. (1950). *Beach erosion in Michigan*: An historical review. Publication 2, Lake Hydraulics Laboratory. University of Michigan: Ann Arbor MI.
Brater, E.F., Sherrill, J.D. (1975). *Rainfall-runoff relations on urban and rural areas*. Environmental Protection Agency: Cincinnati OH.
Brater, E.F. (1976). *Handbook of hydraulics* for the solution of hydraulic engineering problems. McGraw-Hill: New York.
King, H.W., Brater, E.F. (1963). *Handbook of hydraulics* for the solution of hydrostatic and fluid-flow problems. McGraw-Hill: New York.
Wisler, C.O., Brater, E.F. (1959). *Hydrology*. Wiley & Sons: New York.
Woo, D.C., Brater, E.F. (1961). Laminar flow in rough rectangular channels. *Journal of Geophysical Research* 66(12): 4207-4216.
Woo, D.C., Brater, E.F. (1962). Spatially varied flow from controlled rainfall. *Journal of the Hydraulics Division* ASCE 88(HY6): 31-56; 89(HY4): 233-240; 89(HY6): 249-250.

BRETSCHNEIDER

09.11. 1920 Red Owl SD/USA
27.09. 2009 Honolulu HI/USA

Charles Leroy Bretschneider obtained his MS degree from University of California in 1950, and the PhD degree from Texas A&M University in 1959. His study on physical oceanography was awarded with the 1960 ASCE Coastal Engineering Research Prize. Bretschneider had been engaged from 1951 to 1956 with waves and wave forces on pilings, then until 1961 he was concerned as hydraulic engineer with beach erosion as a member of the US Army Corps of Engineers, and directed eastern operations of the National Science Corporation, Washington DC from 1961 to 1964. He was in 1967 appointed professor of physical oceanography and ocean engineering at University of Hawaii, Manoa HI.

Bretschneider during his long career was interested in the large field of oceanographic engineering. He advanced questions relating to the variability, the forecasting, the spectra and the generation and decay of ocean waves. He was further interested in wave forces particularly related to harbour and bridge design, hurricane surges, and the rubble mound breakwater stability. He was known for his excellent 1991 book co-edited by John B. Herbich (1922-2008). This work represents the effort of 14 experts providing valuable assistance in solving practical problems. Chapters include information on ports, offshore structures, marine foundations, sediment processes, wave mechanisms and hydraulic modeling. Bretschneider was member of the American Meteorological Society, the Permanent International Association of Navigation Congresses PIANC, the Society of Naval Architects and Marine Engineers SNAME, ASCE, and AGU.

Anonymous (1965). Charles L. Bretschneider. *Civil Engineering* 35(12): 73. *P*
Anonymous (1975). Bretschneider, Charles Leroy. *Who's who in America* 38: 365.
Anonymous (1994). Bretschneider, Charles L. *American men and women in science* 1: 872.
Bretschneider, C.L. (1952). The generation and decay of wind waves in deep water. *Trans. AGU* 33(3): 381-389.
Bretschneider, C.L. (1958). Hurricane design wave practices. *Journal of the Waterway and Harbour Division* ASCE 83(WW2, Paper 1238): 1-33.
Bretschneider, C.L. (1958). Revisions in wave forecasting: Deep and shallow water. Proc. 6th Conf. *Coastal Engineering*: 30-67.
Bretschneider, C.L., ed. (1969). *Topics in ocean engineering*. Gulf: Houston.
Herbich, J.B., Bretschneider, C.L., eds. (1991). *Handbook of coastal and ocean engineering*. Gulf: Houston.

BRIDGMAN

21.04. 1882 Cambridge MA/USA
20.08. 1961 Randolph MA/USA

Percy Williams Bridgman graduated as a physicist from Harvard College in 1904. He there pursued his career and was initially mainly interested in the properties of fluids at high pressures. One of his first inventions was a self-tightening joint which enabled pressures to be maintained much higher than previously investigated. There, he developed a pressure-measuring device using a primary free-piston gauge and a secondary gauge based upon the variation of electrical resistance of mercury. Here, his skill as a mechanic found full scope, and he was able to measure pressures of up to 13,000 kg/cm^2 with an accuracy of 0.1%. He then measured the thermodynamic properties of liquids between 20 and 80°C, the results being published from 1911 to 1915, providing an immense fund of data relating to compressibility, change of states, and melting curves.

From 1925 to 1927 Bridgman was concerned with research on the changes of viscosity with pressure. He found that liquids, with the exception of water, behave qualitatively alike, their viscosity increasing with pressure rapidly. These results were largely due to Bridgman's personal efforts. He is well known in hydraulics for his contributions to the dimensional analysis, particularly with his 1946 article published in the Encyclopaedia Britannica. Together with Edgar Buckingham (1867-1940), Bridgman must be therefore considered the real promoter of systematic experimentation in physical systems. He was elected assistant professor in 1913, professor in 1919, and finally Hollis professor of Mathematics and Natural Philosophy in 1926. However, he took only little interest in education and administration, but mainly was interested in research. He was appointed Higgins' university professor in 1950. After his retirement in 1954, he continued his experimental work until shortly before his death. He was awarded in 1946 the Noble Prize of physics and elected in 1949 Foreign Member of the Royal Society, London UK, among many other distinctions.

Bridgman, P.W. (1914). High pressures and five kinds of ice. *Franklin Institute* 177: 315-322.
Bridgman, P.W. (1927). *The logic of modern physics*. MacMillan: New York.
Bridgman, P.W. (1946). Dimensional analysis. *Encyclopaedia Britannica* 7: 439-449.
Kemble, E.C., Birch, F. (1970). Percy Williams Bridgman. *Biographical memoirs*, National Academy of Sciences USA: 23-67. Columbia University Press: New York. *P*
Newitt, D.M. (1962). Percy Williams Bridgman. *Biographical memoirs*, Royal Society 8: 26-40. *P*

BRODIE

05.07. 1879 Pittsburgh PA/USA
05.10. 1943 Westerleigh NY/USA

Orrin Lawrence Brodie graduated as civil engineer from Columbia University, New York NY in 1901. He then worked for the city in railroad and street projects. From 1906 he directed studies and designs for the New York City Board of Water Supply, and conducted hydraulic studies on the Catskill Water Supply System. From 1919 to 1927 he was assistant design engineer for the Holland Tunnel organization during which he proposed an important method of tunnel ring stress and silt pressure determination. He then returned to the Board of Water Supply as design engineer, a position he held until his death. He also completed the Grant City model studies concerned with the diversion tunnel of Rondout River at the Merriman Dam. He further assisted Charles M. Allen (1871-1950) at the Worcester Polytechnic Institute, Worcester MA, for the Rondout stilling pool.

Brodie became nationally known for the 1910 book High masonry dam design, an early account on the then proposed design methods. Its second edition was entitled Masonry dam design including high masonry dams, and was published in 1916. At times, Brodie was lecturer and instructor at College of the City of New York on Masonry construction and design, and also served in similar courses at the New York University. At his home he had a complete workshop, where he designed models. He built an exact scale model of the multi-million-dollar diversion tunnel and reservoir spillway, later constructed upstate in connection with the city water supply. It was the prototype of the projected scheme in every detail, and was used in experiments to determine how various features of the actual project would function under all manner of conditions, so that Brodie and his staff were able to decide the best method to be adopted. The taxpayers saved a considerable amount of money through what they learned during these tests. Brodie was described as to live the maxim of never causing pain to any one. He was member of the American Society of Civil Engineers from 1913, of the Association of Engineers of New York, and of the Society of American Military Engineers.

Anonymous (1945). Orrin L. Brodie. *Trans. ASCE* 110: 1665-1667.
Brodie, O.L. (1915). Discussion of The constant-angle arch dam. *Trans. ASCE* 78: 726-727.
Brodie, O.L. (1917). Disc. of Reconstruction of the Stony River Dam. *Trans. ASCE* 81: 1041-1044.
Jackson, R.W. (2011). *Highway under the Hudson*: A history of the Holland Tunnel. New York.
Morrison, C.E., Brodie, O.L. (1910). *High masonry dam design*. Wiley: New York.
http://shootingwithhobie.blogspot.ch/2008_04_01_archive.html *P*

BROWN C.B.

12.11. 1910 Salisbury NC/USA
05.05. 1963 Washington DC/USA

Carl Barrier Brown obtained an undergraduate title from the University of North Carolina, and his MA degree in geology from the University of Cincinnati, Cincinnati OH. He joined in 1934 the Agriculture Department, where he remained until his early death at age 52. His special concerns were sedimentation research and the problems of watershed protection. He had served as Chief of the Water Conservation Division, Assistant Chief of Operations, Director of the Planning Division, and Assistant to the Assistant Administrator for Watersheds. He received in 1959 the Department's Superior Service Award. He was a member of the American Geophysical Union from 1938, and of the American Society of Civil Engineers from 1950, whose Fellow he was from 1959.

Brown became a leading national expert on watershed and flood protection. He helped draft the Watershed Protection and Flood Prevention Act thereby having been one of the main contributors. His legacy in hydraulics was the chapter Sedimentation engineering in the 1950 book edited by Hunter Rouse (1906-1996). The main subchapters are entitled: A Sediment problem, B Sediment properties, C Fundamental principles of movement, D Design of canals and desilting works, E Stabilization and improvement of rivers, F Sedimentation in reservoirs, G Coastal sediment problems, and H Collection and analysis of field data. This chapter clearly indicates the practical background of Brown's work. Shortly later, a more scientific approach to engineering sedimentation problems was provided e.g. by Vito Vanoni (1904-1999). It should be remembered that the interest in these problems initiated in the USA only after World War II, whereas significant European reviews were available from 1900.

Anonymous (1929). Carl B. Brown. *Yackety yearbook* 39: 256. *P*
Anonymous (1963). Carl B. Brown. *Trans. AGU* 44(3): 753.
Brown, C.B. (1941). Factors in control of reservoir silting. *Journal AWWA* 33(6): 1022-1040.
Brown, C.B. (1943). The control of reservoir silting. *Miscellaneous Publication* 521. US Dept. Agriculture: Washington DC.
Brown, C.B. (1945). Sediment complicates flood control. *Civil Engineering* 15(2): 83-86. *P*
Brown, C.B. (1950). Sediment transportation. *Engineering hydraulics* 12: 769-857, H. Rouse, ed. Wiley & Sons: New York.
Eakin, H.M., Brown, C.B. (1939). Silting of reservoirs. *Technical Bulletin* 524. US Department of Agriculture: Washington DC.

BROWN E.I.

13.10. 1874 Carrollton GA/USA
25.01. 1963 Durham NC/USA

Earl Ivan Brown graduated from West Point Military
Academy in 1898. A veteran of World Wars I and II
he served in the US Army Corps of Engineers for
45 years. He first was employed both in Cuba and
the Philippines during and after the 1898 American-
Spanish War, participating in the Moro Expedition
and the Philippine Insurrection before travelling to
the Caribbean for the Cuban Pacification in 1906.
He was member of the American Society of Civil
Engineers ASCE, the Permanent Intl. Association of
Navigation Congress PIANC, and the Wilmington
Engineers Club.

After World War I Brown developed river and harbour improvements in the United
States. He principally worked on the Atlantic and Gulf coasts and on the Ohio River. He
also contributed to the Chesapeake and Delaware Canal, finishing his career in 1938 as
the Division Engineer of the South Atlantic Division. The 1932 paper deals with surge
waves as encountered in canals due to a change of a boundary condition or a rapid
increase or decrease of discharge. Based on a review of then existing theories Brown
stated their limitations. He then developed a new approach by assuming the reflection of
the canal wave from a basin, stating that the degree of reflection is not only governed by
the canal but also by the basin. He also detailed the effect of friction in reducing the
surge height in an infinitely long canal, the time required for the wave crest to reach any
point, the effect of canal width, interference of two opposite waves, and the propagation
of a tide in a canal. The results were applied to the Suez Canal including a successful
comparison with experimental data.

Anonymous (1942). Earl I. Brown. *Civil Engineering* 12(1): 50. *P*
Brown, E.I. (1912). *Guard locks in canals connecting tidal bodies of water*. US Army Corps of
　　　Engineers: Washington DC.
Brown, E.I. (1928). Inlets on sandy coasts. *Proc. ASCE* 54(2): 505-553.
Brown, E.I. (1931). The Chesapeake and Delaware Canal. *Trans. ASCE* 95: 716-765.
Brown, E.I. (1932). The flow of water in tidal canals. *Trans. ASCE* 96: 749-834.
Brown, E.I. (1935). A study of the effect upon navigation and upon the upkeep of the banks and
　　　bed of canals and canalized rivers. 16th *PIANC Congress* Brussels.
Brown, E.I. (1936). *Flow in tidal rivers*. US Engineer School: Fort Belvoir VA.
Brown, E.I. (1939). Beach erosion studies. *Proc. ASCE* 65(1): 29-89; 66(8): 869-919.
http://www.arlingtoncemetery.net/eibrown.htm *P*

BROWN F.R.

15.02. 1912 Peoria IL/USA
30.04. 1999 Vicksburg MS/USA

Frederick Raymond Brown received education from University of Illinois, Urbana IL, from where he obtained the BS degree in civil engineering in 1934. He joined the US Waterways Experiment Station, Vicksburg MS, remaining there all through his career until retirement in 1985. He eventually became chief of the Hydrodynamics Branch, responsible for the conduct of all hydraulic model tests of hydraulic structures, dams, and harbour developments. He has had a large impact on the hydraulic laboratory test procedures in the USA, counting among the experts in this field. Brown was a member of the American Society of Civil Engineers ASCE and elected to its Honorary Member in 1981.

Brown published two notable papers, one on Conformity between model and prototype effects, the other on Models of hydraulic structures. It was known for a long time that each model is unable to reproduce all details of a prototype, because a number of fluid characteristics remain constant in both, resulting in so-called scale effects. These are often negligible but may become excessive if the model scale is incorrectly selected. A typical effect is spray formation, which is hardly noticeable in a scale model, but may become excessive in a prototype. Application limits for scale modelling are thus of concern, and Brown was one who has added significantly to this problem. Other papers and reports deal with laboratory techniques, the optimum instrumentation for a specific problem, and concerns relating to the management and the financing of these works.

Anonymous (1964). Fred R. Brown. *Civil Engineering* 34(5): 65. *P*
Anonymous (1966). Fred R. Brown. *Civil Engineering* 36(12): 71. *P*
Bondurant, D.C., Brown, F.R. (1963). *Review and evaluation of hydraulic problems* at the hydraulic research and experimental station Delta Barrage, UAR. USACE:Vicksburg.
Brown, F.R. (1944). Conformity between model and prototype in hydraulic structures: Grand Tower, Mississippi River. *Trans. ASCE* 109: 100-102; 109: 191-193.
Brown, F.R. (1947). The control of reservoir silting. *Miscellaneous Publication* 521. Washington.
Brown, F.R. (1947). *Models of hydraulic structures*. US Waterways Exp. Station: Vicksburg.
Brown, F.R. (1963). Cavitation in hydraulic structures: Problems created by cavitation phenomena. *Journal Hydraulics Division* ASCE 89(HY1): 99-115; 89(HY5): 141-145.
Tiffany, J.B., Brown, F.R. (1952). Hydraulic experimentation and engineering design. *Hydrodynamics in modern technology*: 64-73. Cambridge University Press: Cambridge.
http://chl.erdc.usace.army.mil/Media/8/5/5/images/Chap6_img_7.jpg *P*

BRUSH C.B.

15.02. 1848 New York NY/USA
03.06. 1897 New York NY/USA

Charles Benjamin Brush graduated in 1867 as a civil engineer from the University of the City of New York. His first professional service was within the Corps of the Croton Aqueduct until 1869, from when he commenced as consulting civil engineer, but in 1874 was appointed adjunct professor of civil engineering at his Alma Mater. In 1888 he became there professor and there was Dean of the School of Engineering. He was elected associate member of the American Society of Civil Engineers in 1871, became member in 1877 and was its director from 1888 to 1891, and vice-president in 1892.

Brush was extensively engaged as a consulting engineer on the design and construction of waterworks all through the USA. Among the more prominent designs were these of Cincinnati, Chicago, Memphis, Jamestown, East New York, Passaic, Easton and Montclair. He also was the chief engineer of the Hoboken Land and Improvement Company, of the North Hudson County Railway Co., and of the Hackensack Water Co. The development of the latter was especially due to his careful study and management. He was for instance called upon for expert examination of the foundation of bridges over Harlem River in New York, and at the Thames River bridge at New London CT. He was engineer for the contractor in the construction of the Washington bridge over Harlem River, or engineer of the Hudson River tunnel. He was further engaged with sewers in North Hudson County, in New Jersey and in Irvington NY. Personally Brush was a man of sterling integrity, painstaking in his professional work, tireless in energy and esteemed by his professional colleagues. He was broadminded and charitable, almost to a fault, though of exceptional business ability.

Anonymous (1897). Charles Benjamin Brush. *Engineering News* 37(June10): 359. *P*
Anonymous (1899). Brush, C.B. *National cyclopaedia of American biography* 9: 33. New York. *P*
Brush, C.B. (1880). The Hudson River Tunnel. *Trans. ASCE* 9: 259-277.
Brush, C.B. (1886). *Insurance maps of Hudson County NJ*. Hoboken NJ.
Brush, C.B. (1888). Some facts in relation to friction, waste and loss of water in mains. *Trans. ASCE* 19: 89-126.
Brush, C.B. (1891). Discussion of The nozzle as an accurate water meter, by J.R. Freeman. *Trans. ASCE* 24: 513-527.
Brush, C.B. (1891). Aeration of a gravity supply. *Proc. American Water Works Association* 10(2): 73-77.

BRUSH W.W.

28.07. 1874 Orange NJ/USA
25.10. 1962 New York NY/USA

William Whitlock Brush graduated in 1893 as civil engineer from New York University, New York NY. He was from 1894 to 1907 engineering assistant and assistant engineer on city water supply projects for Brooklyn NY, until 1910 then engineer with the NY Board of Water Supply on the Catskill System and department engineer of this Board on the design of the City Aqueduct. From 1910 to 1917 he was then deputy chief engineer of the Bureau of Water Supply, Department of Water Supply, Gas & Electricity. He was appointed in 1917 chief engineer, serving from 1919 to 1926 as deputy chief engineer, and taking then again over as chief engineer until his retirement in 1934. Brush was member of the American Society of Civil Engineers ASCE, of the American Water Works Association AWWA, serving as president in 1928, and becoming honorary member in 1937. He also was member and president of the New England Water Works Association, and served as president of the New York University Alumni Association.

Brush started his engineering career as assistant on the Brooklyn water supply system, and subsequently designed considerably of New York's delivery network, including its aqueduct system, from the Catskill Mountains to the city center. After his appointment as chief engineer, he wrote articles on water supply problems and construction projects. After his retirement, he became editor of the Water Works Engineering journal, which was founded in 1847, and so represented one of the first journals in this field. It became in 1964 the Water and Waste Engineering journal.

Anonymous (1947). William W. Brush. *Mueller Record* 34(1): 7. *P*

Anonymous (1959). Brush, William W. *Who's who in engineering* 8: 314. Lewis: New York.

Anonymous (1962). William W. Brush dies: Municipal water expert. *Engineering News-Record* 169(Nov.1): 52.

Brush, W.W. (1914). *Plan for functional organization of the engineering work of the city of New York*. Municipal Engineers: City of New York.

Brush, W.W., French, D.W. (1915). Maintenance of the water supply distribution system of New York City. *Journal of the American Water Works Association* 2(1): 206-237.

Brush, W.W. (1952). *Water works*: Questions and answers; problems in water works operation and maintenance and their solution. Case-Shepperd-Mann: New York.

Hayes, N.J., Dietz, J.J., Smith, M.H., Brush, W.W. (1920). *The municipal water supply system of the City of New York*. New York.

BUCKINGHAM

08.07. 1867 Philadelphia PA/USA
29.04. 1940 Washington DC/USA

Edgar Buckingham obtained a degree as physicist from Harvard University in 1887, and his PhD from University of Leipzig, Germany in 1893. He was an assistant in physics at Harvard University from 1888 to 1892, with a short stay in 1889 at Strasburg University, then Germany, and from 1893 to 1898 he was an assistant professor in physics at the Bryn Mawr College, Bryn Mawr PA. From 1898 to 1901 Buckingham was an instructor at the University of Wisconsin, Madison WI, then a physicist at the US Department of Agriculture from 1902 to 1905, and from then until retirement in 1937 with the staff of the National Bureau of Standards NBS, Washington DC. In parallel, he was Lecturer in thermodynamics at the US Naval Academy, and an associate scientific attaché of the US Embassy in Rome, Italy, in 1918.

Buckingham is best known for his early work on thermodynamics and for his later study of dimensional theory. Attracted to problems that could not be solved by pure calculation but requiring experimentation as well, he demonstrated more clearly than anybody before him how the planning and interpretation of experiments can be facilitated by the method of dimensions, later referred to as dimensional analysis. He pointed out the advantages of dimensionless variables and how to generalize empirical equations. His frequently cited 'π-theorem' serves to reduce the number of independent variables and shows how to experiment on geometrically similar models so as to satisfy the most general requirements of physical as well as dynamic similarity.

Anonymous (1941). Buckingham, Edgar. *Who's who in engineering* 5: 246. Lewis: New York.
Bean, H.S., Buckingham, E., Murphy, P.S. (1929). Discharge coefficients of square-edged orifices for measuring the flow of air. *Journal of Research* NBS 2(RP49): 561-658.
Buckingham, E. (1900). *Theory of thermodynamics*. Macmillan: New York.
Buckingham, E. (1924). Research in heat transmission. *Mechanical Engineering* 46(7): 386-388.
Buckingham, E., Bean, H.S. (1956). Notes on some recently published experiments on orifice meters. *Trans. ASME* 78(2): 379-387.
Hersey, M.D. (1970). Edgar Buckingham. *Dictionary of scientific biography* 2: 565-566. Scribner's: New York.
Poggendorff, J.C. (1925). Buckingham, Edgar. *Biographisch-Literarisches Handwörterbuch* 5: 183; 6: 363; 7: 620. Verlag Chemie: Leipzig, Berlin, with bibliography.
http://soil.scijournals.org/cgi/content/full/69/2/328/FIG1 *P*

BUEHLER

09.05. 1910 Baraboo WI/USA
27.03. 1999 Knoxville TN/USA

Robert 'Bob' Joseph Buehler was a graduate of the University of Wisconsin, Madison WI, with the BS and MS degrees in civil engineering. He then joined the Flood Control Branch of the Tennessee Valley Authority TVA for 41 years until 1975, following a mandatory retirement as branch chief. He from then was consultant and registered engineer in Tennessee. He was elected Fellow ASCE in 1978.

Buehler's 1969 ASCE paper deals with the numerical simulation of transient flows. During the past years major advances were made by TVA using digital computers to solve the basic differential equations for transient flow of water in open channels. The mathematical model employed used an explicit scheme to compute stage, discharge, and velocity transients from known initial and boundary conditions. Finite difference expressions of the governing equations of continuity and momentum were solved. Stage, discharge, local inflows, variable roughness along the channel, and the channel geometry can be used, resulting in discharges and stages along the channel for known initial conditions for both steady and transient flows. The results of this work indicate the diverse range over which the model was applied. There were two reservoirs, a natural river and an uncontrolled channel connection to the two reservoirs. Predictions agreed well with prototype measurements in all four cases considered. Other researches considered similar flow processes in TVA hydraulic structures, and were also involved in the monetary values of life and health.

Buehler, B.J., Price, J.T., Garrison, J.M. (1968). Transient flow investigations for TVA's Browns Ferry generating station. Proc. 7th Conf. *Annual Sanitary and Water Resources Engineering*: 125-138. Technical Report 16. Vanderbilt University: Nashville.

Buehler, B.J., Garrison, J.M., Granju, J.-P., Price, J.T. (1969). Digital computer simulation of transient flows in the TVA system. Proc. 13th *IAHR Congress* Kyoto 1: 345-352; 5: 88.

Buehler, B. (1975). Corps guidelines for dam safety inspection need revamping. *Civil Engineering* 45(1): 74-75. *P*

Buehler, B. (1975). Monetary values of life and health. *Journal of the Hydraulics Division* ASCE 101(HY1): 29-47; 101(HY9): 1297-1300; 101(HY11): 1455-1456; 102(HY3): 418-419.

Buehler, B. (1976). Upstream flood lowering in reservoirs. *Journal of the Hydraulics Division* ASCE 102(HY2): 151-170; 102(HY12): 1784-1785.

Buehler, B. (1977). U.S. floods and their management. *Eos* 58(1): 4-15. *P*

BULL

20.10. 1856 Bergen/N
18.11. 1907 Madison WI/USA

Storm Bull was born in Norway. He graduated as a mechanical engineer in 1877 from the Polytechnic Institute, today's ETH, Zurich. He then travelled extensively in Switzerland, Germany, Belgium, and France, visiting various shops and institutes, returned to Norway and there was for 2 years head draftsman in shipbuilding in the naval yard. In 1879 he moved to the United States, becoming in 1879 instructor in mechanical engineering at University of Wisconsin, Madison WI, and there was promoted in 1884 to assistant professor, then to professor of mechanical engineering in 1886, and in 1891 to professor of steam engineering, a position he held until his premature death due to a cancer of the stomach.

Bull made contributions to the scientific and literary magazines, both in the USA and in Norway; his works showed marked talent and careful study. His main papers were published in the Engineering News Journal, where he was also a frequent contributor to its Engineering Literature Supplement. He authored the book Steam engineering, a widely used text in the USA in this field. During the 1900 Universal Paris Exhibition, he acted as vice-president of the Jury of Awards, whereas during the 1904 Louisiana Purchase Exposition, he was president of the Jury of Awards. He worked in addition with the State on various power and heating plants for institutions around Wisconsin. The Capitol Commission hired him in 1907 to complete plans and drawings for the Capitol heat and power plant equipment. He was member of the American Society of Mechanical Engineers ASME, the Western Society of Engineers, in which he won the Chanute Medal in 1903, and the Society for the Promotion of Engineering Education, acting as vice-president from 1901 to 1902.

Anonymous (1907). Storm Bull, Prof. of steam engineering. *Engineering News* 58(21): 565-566.
Bull, S. (1899). Engineering education as preliminary training for scientific work. *Science* 10: 282.
Bull, S. (1903). Use of superheated steam. *J. of the Western Society of Engineers* 8(6): 691-715.
Bull, S. (1905). *Steam engineering*. University of Wisconsin: Madison.
Hersh, M. (1927). Portraits to be presented at Engineering Society Meeting: Storm Bull. *The Wisconsin Engineer* 31(5): 161. *P*
http://www.rockvillemama.com/dane/bullstorm.txt
http://www.wisconsinhistory.org/whi/fullRecord.asp?id=10681&qstring=http%3A%2F%2Fww
w.wisconsinhistory.org%2Fwhi%2Fresults.asp%3Fsubject_narrow%3DEngineering *P*

BURD

09.10. 1887 Grand Rapids MI/USA
05.11. 1934 Jackson MI/USA

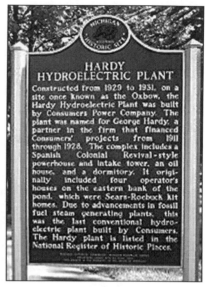

Edward Morris Burd graduated from the University of Michigan, Ann Arbor MI, in 1911 with BS degree in civil engineering. He then became draftsman in the Grand Rapids city office; in 1912 was engaged as inspector for an engineering company at Jackson MI. Until 1915 he was in charge of studies for water power projects. After war service, he formed with a colleague an engineering firm at Grand Rapids doing power plant and municipal engineering work. In 1921 he returned to the Consumers Power Company, Jackson MI, heading its civil-hydraulic engineering department until his death, which occurred tragically while flying his own plane near Jackson crashed to the ground, resulting in his fatal quick death.

During his career, Burd contributed to the engineering community several works and papers, including the Hodenpyl Dam, a 25 m high structure on the Manistee River MI, or the 30 m high Hardy Dam on the Muskegon River MI, including the designs of the spillways, the power hose and the access structures. He further reported on the hydro-electric developments during 1928, or on the Comparative costs of generating electricity by water power and steam. His design experience in hydraulic engineering was extensive, and he was equally competent in structural design. He also had extensive field experience on the construction of large hydro-electric power plants. He was a thinker, accurate in analysis and was acknowledged to have been one of the best students. He possessed a keen sense of humour, was happy when being out of the doors, and loved nature. His outstanding qualities were honesty, courage, with always plenty of ambition and initiative. He was member of the American Society of Civil Engineers ASCE since 1930.

Anonymous (1935). Edward M. Burd. *Trans. ASCE* 100: 1614-1617.
Burd, E.M. (1926). Consumers Power Co. adds Hodenpyl to system. *Power Plant Engineering* 30(Jul.15): 803-806.
Burd, E.M. (1928). Water power in the Middle West. *The Michigan Engineer* 46(9): 12-21.
Burd, E.M. (1932). Hardy Dam provides 40,000 HP for Michigan peak loads. *Power Plant Engineering* 36(Mar.01). 194-198.
Burd, E.M. (1932). Hardy Dam goes on the line. *Electrical World* 99(Feb.27): 412-413.
Burd, E.M. (1934). High dams on pervious glacial drift. *Trans. ASCE* 99: 792-846.
memory.loc.gov/pnp/.../mi/.../mi0439data.pdf (*P*)

BURDEN

22.04. 1791 Dunblane/UK
19.01. 1871 Troy NY/USA

Henry Burden was born in Stirlingshire. He studied mathematics, engineering and drawing at University of Edinburgh UK. He came to America in 1819, and there was first involved in agricultural implements for a firm at Albany NY. He invented a cultivator in 1820, and a hemp and flax machine in 1822. Other inventions include a machine for producing railroad spikes and horseshoes. He in 1833 built a 90 m long iron catamaran. Later he purchased all stock in the Iron and Nail Factory, Troy NY, to become the sole owner in 1848. Burden became a Honorary Member ASCE in 1852.

Burden made significant contributions to the steamboat and water mills. He had a great ambition to build a vessel of minimum water draft. In 1833, he launched the steamboat *Helen*, which rested on two cigar-shaped hulls and had a paddle-wheel amidships of 9 m in diameter. A speed of 29 km/h was made in 1834. Burden was the first advocate of Anglo-Saxon naval architects in the construction of long vessels for ocean navigation. In 1846 he proposed a transatlantic steam-ferry company. Although this company was never organized, its salient features were subsequently imitated by the *Cunard* ocean line. Another accomplishment of Burden was the huge water-wheel *Niagara*. It was built in 1851 as overshot wheel of 1,200 HP, 18 m in diameter and 7 m wide, containing 36 buckets, each 2 m deep. The axis comprised 6 hollow cast-iron tubes, keyed into flanges, from which diverged 264 iron rods of 5 cm thickness, terminating the outer wheel edge. By a lever its revolution and power were controlled.

Anonymous (1972). Burden, Henry. *A biographical dictionary of American civil engineers*: 19-20. ASCE: New York.

Burden, H. (1843). *A brief account on the invention and improvement of the spike machine.* Carroll & Cook: Troy.

Burden Proudfit, M. (1904). *Henry Burden*: His life and a history of his inventions. Troy NY.

Sweeny, F.R.I. (1915). The Burden water-wheel. *Trans. ASCE* 79: 708-726.

Uselding, P.J. (1970). Henry Burden and the question of Anglo-American technological transfer in the nineteenth century. *Journal of Economic History* 30(2): 312-337.

http://209.85.135.132/search?q=cache:Yt5yRKm5zrIJ:en.wikipedia.org/wiki/Henry_Burden+Burden+brief+account+on+the+invention+and+improvement+of+the+spike+machine&cd=1&hl=de&ct=clnk&gl=ch

http://www.archive.org/details/henryburden00prou *P*

BURDICK

06.03. 1874 Chicago IL/USA
17.02. 1955 Chicago IL/USA

Charles Baker Burdick graduated as civil engineer in 1895 from the University of Illinois, Urbana IL. He was first employed by two consultants in Chicago forming in 1902 the partnership Alvord & Burdick until 1922, from when it was Alvord, Burdick & Howson. Burdick contributed significantly to the development of methods for both water and sewage treatment. He also was consulted on the hydraulics of stream flow, floods, and flood relief, authoring with Alvord the 1918 paper on flood prevention and flood protection. He early realized the economic possibilities in the use of super-heated steam in steam turbines and made early installations of that type of equipment in waterworks service. He later converted many steam pumping stations to electric drive.

The advice of Burdick was widely sought in the USA. His outstanding developments included the *Mokelumne* River East Bay City project in San Francisco CA, the Deer Creek Water Development at Salt Lake City UT, the Moffat Filter Plant at Denver CO, and the extensive gallery system at Des Moines IA, which he directed for more than forty years. He also was consultant to the City of Chicago IL on the world's two largest filtration plants, on the Pioneer Mechanical Filtration Plant at Niagara Falls NY with its phenomenal record of typhoid reduction, and to the Long Island Water Company, New York NY. Burdick was also active in affairs relating to the American Society of Civil Engineers ASCE, whose member he was from 1911. He served as Director from 1935 to 1937, and vice-president in 1941 and 1942, and served as chairman of various ASCE Committees. He was elected Honorary Member ASCE in 1946. He was further member of the American Water Works Association AWWA, and the Chicago Engineer's Club, of which he was also honorary members.

Alvord, J.W., Burdick, C.B. (1915). *Report* of the Rivers and Lakes Commission on the Illinois River and its Bottom Lands, with reference to the conservation of agriculture and fisheries and the control of floods. State of Illinois: Chicago.
Alvord, J.W., Burdick, C.B. (1918). *Relief from floods*: The fundamentals of flood prevention, flood protection and the means for determining proper remedies. McGraw-Hill: NY.
Anonymous (1957). Charles B. Burdick. *Trans. ASCE* 122: 1232.
Burdick, C.B. (1913). *Basic principles of ground water collection*. Chicago.
Burdick, C.B. (1923). Water works pumping station design. *Engineering News-Record* 90: 964.
http://www.abhengineers.com/history.html *P*

BURGESS

30.06. 1848 West Sandwich MA/USA
12.07. 1891 Boston MA/USA

Edward Burgess graduated in 1871 from Harvard University, Cambridge MA. He became secretary then of the Boston Society of Natural History. In 1879 he was engaged as instructor in entomology at Harvard until 1883, when travelling to Europe to study naval architecture. At home he designed and built vessels for his own use, but the success was so enormous and a fortune threw him upon his own resources that he turned to the design of sailing yachts. Several of his boats won fame in the eastern USA, and when in 1884 it became necessary to build a large sloop yacht to represent the USA in international races, he was selected by a committee of Bostonians to draw plans for a suitable vessel. His *Puritan* easily defeated the English *Genesta* in the America's Cup of 1885, which was a remarkable triumph because it was the first attempt of an American designer to solve shipbuilding problems to which the English had given their attention for years.

In 1886 the *Mayflower* of Burgess, slightly larger than the *Puritan*, led in the race with the English *Galatea*. In 1888 his fishing schooner *Carrie E. Phillips* distanced four competitors in the Fisherman's Race held at Boston harbour. Burgess' *Volunteer* won the America's Cup against the *Thristle*, the special product of British genius, in the international races of 1887. His other yachts included the *Mariquita* and the *Gossoon*, both remarkably swift sloops designed to counter the success of the Clyde-built cutter *Minerva* by the British in 1888. The 1887 book of the Boston City Council shows the beauty of these yachts and highlights the pride of all involved in Burgess' successes. Burgess was inducted into the America's Cup Hall of Fame in 1994.

Anonymous (1887). *A testimonial to Charles J. Paine and Edward Burgess* from the City of Boston for their successful defence of the America's Cup. City Council: Boston. *P*
Anonymous (1892). Edward Burgess. Proc. *American Academy of Arts and Science* 27(Memoirs): 357-360.
Anonymous (1900). Burgess, Edward. *Appleton's cyclopaedia of American biography* 1: 451.
McVey, A.G. (1892). Edward Burgess and his work. *New England Magazine* 5(1): 49-62. *P*
Rosbe, J.W. (2002). *Maritime Marion Massachusetts*. Arcadia: Charleston SC.
http://en.wikipedia.org/wiki/Edward_Burgess *P*
http://en.wikisource.org/wiki/Collier%27s_New_Encyclopedia_(1921)/Burgess,_Edward
http://www.herreshoff.org/achof/edward_burgess.html

BURKE

16.04. 1938 North Kansas City MO/USA
03.09. 1991 Liberty MO/USA

Thomas Dean Burke graduated in 1960 with the BSc degree in civil engineering from the University of Missouri, School of Mines and Metallurgy, Rolla MO; he did graduate work at University of Missouri, Kansas City KS, until 1969, obtaining the MS degree in 1970 from Colorado State University, Fort Collins CO. He began his professional career with Kansas City District, US Army Corps of Engineers, in 1960 while working on the Ottawa Kansas Flood Control Project. He was then until 1970 a civil engineer in river development, and until 1988 chief of the river development, from when he was until 1991 chief of River and Lake Engineering of the Kansas City District.

Burke was an expert in river engineering, with a particular emphasis on the Missouri River. He received a posthumous citation from the Missouri Department of Conservation for his design of the Missouri River Bank Stabilization and Navigation project, and for the resulting fish and wildlife conservation. Further, the District Engineer from the US Army Corps of Engineers presented the Commanders Award for public service. The Missouri-Arkansas River Basin Association elected Burke Man of the Year 1988. Burke has written several papers relating to Missouri River, including the historical account in 1981, the major river migration characteristics of Kansas River in 1983, and the 1979 paper on the effect of river development on the habitat diversity. It was stated that considerable uncertainty exists on the ecological benefits of river rehabilitation and their long-term performance, mainly for large rivers because of inherent spatial and temporal complexities. He was member of the American Society of Civil Engineers ASCE, and of the Missouri Society of Professional Engineers. He served as an advisor on the Board of Director of Kansas City River Front Inc.

Anonymous (1961). Burke. *Rollamo yearbook*: 217. Missouri University of Science: *P*
Anonymous (1992). Thomas D. Burke. *Trans. ASCE* 157: 501.
Burke, T.D., Robinson, J.W. (1979). River structure modifications to provide habitat diversity. *Mitigation Symposium* National Workshop on mitigation losses of fish and wildlife habitat: 556-561. Fort Collins CO.
Burke, T.D. (1981). Navigation on the Missouri: 70 years in the making. *Water Forum* '81: 456-463. ASCE: New York.
Burke, T.D. (1983). Channel migration of the Kansas River. *River meandering* New Orleans: 250-258. ASCE: New York.

BURKHOLDER

11.10. 1884 Roanoke IL/USA
06.04. 1953 San Diego CA/USA

Joseph Lloyd Burkholder obtained an engineering degree in 1907 from Kansas University, Lawrence KS, which in 1947 bestowed upon him a citation for distinguished service in his profession. After work for a railway company, he became in 1910 engineer of the US Bureau of Reclamation USBR with projects at Boise ID, El Paso TX, and Denver CO. From 1922 to 1924 he was then in charge of the *Barahona* Project, Dominican Republic, from when he was on a project for flood control at Kansas City. From 1926 to 1939 Burkholder was chief engineer of the Middle Rio Grande Conservancy District, Albuquerque NM.

He was then until 1942 assistant general manager of the Mountain Water District of South Carolina on the construction of the Colorado River Aqueduct. Until 1944 he was senior engineer of the American Section, the International Body Commission, of the US & Mexican Operating Manager, on naval fleet and air facilities in the Caribbean Area, stationed at San Juan PR. This service rendered him the Meritorious Civilian Service Award by the Bureau of Yards and Docks. After the War he moved to California as construction engineer at Los Angeles CA, in charge of the Horseshoe Dam on Verde River, 80 km north of Phoenix AZ. From 1945 until his retirement, Burkholder was the general manager and chief engineer of the San Diego County Water Authority.

Burkholder was the leading figure in the building of the San Diego Aqueduct. He led the campaign which resulted in the construction of the second barrel to the Aqueduct, connecting the Los Angeles Metropolitan Water District Aqueduct with Colorado River. The San Diego Aqueduct is a system of four aqueducts supplying about 90% of water to the city. The waters are carried from Colorado River west to the outskirts of the city. The 110 km long First Aqueduct runs from the Colorado River Aqueduct near San Jacinto CA to the San Vicente Reservoir, 25 km north of the city. Burkholder therefore has outstanding merits for having supplied San Diego with water.

Anonymous (1939). Members of Lower Rio Grande Project. *Civil Engineering* 9(9): 576. *P*
Anonymous (1948). Burkholder, Joseph L. *Who's who in engineering* 6: 275. Lewis: New York.
Burkholder, J.L. (1919). Drainage works of Rio Grande irrigation project. *ENR* 83(12): 543-549.
Burkholder, J.L. (1928). *Plan for flood control, drainage and irrigation of the Middle Rio
    Grande Conservancy Project*. The District: Albuquerque NM.
Whitcomb, W.B., Swing, P.D., Burkholder, J.L. (1946). *San Diego County Water Authority*.
    The Authority: San Diego CA.

BUTLER J.S.

25.05. 1872 Rogersville TN/USA
19.10. 1934 Nashville TN/USA

John Soule Butler graduated in 1894 from Vanderbilt University, Nashville TN, with the BSc degree in civil engineering. He was then engaged for the next 23 years in the improvement of rivers and harbours, mostly on Cumberland River, in connection with the Engineer Office, Nashville TN. During this period he laid out work for the construction of temporary buildings at Lock A. He also was responsible for the design of locks, dams, abutments, and retaining walls including steel lock gates. He became in 1912 then assistant engineer visiting the Canal Zone, and was assigned to regular duties at the Nashville Engineer Office, supervising all works on Cumberland River. After war service, Butler was in 1918 commissioned as major, US Army Corps of Engineers, and assigned as chief of Construction Division 2, in charge of constructions at Wilson Dam, Muscle Shoals AL, on Tennessee River. In 1924 Butler was assigned to duty at the Canal Zone, in charge of fortifications of the Panama Canal.

Butler was in 1927 made district engineer at Seattle WA; his selection was a tribute to his ability to plan the development of a great river system. The survey of the Upper Columbia River in Washington from the Snake River to the International Boundary, a distance of some 700 km, was under his jurisdiction. This river is the highest potential power stream in the USA. It was in addition also important in terms of irrigating 5,000 km^2 of potentially rich land. Grand Coulee Dam was later designed for these purposes. In 1931, Butler submitted his famous Report to the Chief of Engineers. To increase low-water flow of Columbia River, he recommended to create the Hungry Horse Dam. It was stated that it required great vision and courage to recommend a dam at Grand Coulee of such a height and an enormous discharge at this time. In 1932 the US Bureau of Reclamation concurred with this vision, undertook modifications, and then started the construction of this dam. Butler was promoted in 1933 to lieutenant-colonel. He was member of the American Society of Civil Engineers ASCE from 1913, and member of the Society of American Military Engineers.

Anonymous (1946). John S. Butler. *Trans. ASCE* 111: 1443-1448.
Butler, J.S. (1931). The Columbia River for irrigation and power: Comprehensive study by Army Engineers. *Civil Engineering* 1(12): 1075-1080.
Butler, J.S. (1931). Columbia River for irrigation and power. *Civil Engineering* 1(9): 1075-1080.
http://www.nps.gov/history/history/online_books/dams/columbia_basin/images/fig45.jpg (*P*)

CAMP

05.11. 1895 San Antonio TX/USA
15.11. 1971 Boston MA/USA

Thomas Ringgold Camp was known affectionately throughout his life as 'Tom Camp'. He graduated in 1916 in architectural engineering from Texas A&M University, College Station TX. After World War I he designed pumping stations, filtration works and sewer systems before enrolling for graduate studies in sanitary engineering at MIT. After receiving an MS degree, he joined in 1925 a firm, specializing in water works and sewerage. In 1929, Camp became instructor at Department of Sanitary Engineering, MIT, where he coupled his education with research for the next 15 years.

Camp was one of the first realizing that sanitary engineering was an interdisciplinary field, which is currently referred to as environmental engineering. He mastered all the associate disciplines. Both theoretical and pragmatic new concepts were generated in water and wastewater treatment, and soon he became a national authority. By 1944, these demands had increased to the point where he decided to enter full-time consulting practice with an associate. In his office, hundreds of collaborators were occupied, dealing with questions in water resources, water and wastewater works, solid waste disposal, water and air pollution abatement, and flood control, both in the United States and abroad. In the private sector, Camp's professional contributions continued. He was president of the New England Water Pollution Control Federation, the New England Water Works Association, and the Boston Society of Civil Engineers. Camp's name is particularly associated with the design of sedimentations basins, for which he developed a design based on the particle fall velocity. He was the recipient of numerous awards, including Honorary Member ASCE in 1965.

Anonymous (1942). Thomas R. Camp. *Civil Engineering* 12(1): 50. *P*
Anonymous (1954). Camp, Thomas R. *Who's who in engineering* 7: 199. Lewis: New York.
Camp, T.R. (1946). Sedimentation and the design of settling tanks. *Trans. ASCE* 111: 895-958.
Camp, T.R. (1952). Water distribution. *Handbook of applied hydraulics* 20: 881-944, C.V.
     Davies, ed. McGraw-Hill: New York.
Camp, T.R. (1963). *Water and its impurities*. Reinhold: New York.
Camp, T.R., Graber, S.D. (1968). Dispersion conduits. *Journal Sanitary Engineering Division*
     ASCE 94(SA1): 31-39; 94(SA4): 762-764; 94(SA6): 1295; 95(SA5): 943-947.
Dresser, H.G. (1971). Dr. Thomas Ringgold Camp. *Journal Boston Society of Civil Engineers*
     58: 298-301. *P*

CAMPBELL F.B.

16.09. 1904 Ironton OH/USA
29.05. 1992 Richardson TX/USA

Frank Bixby Campbell graduated as a civil engineer from Cornell University, Ithaca NY, in 1928. He was then involved in stream gauging with the US Geological Survey. A participant of the 1930 World Power Conference in Berlin, he toured through Europe visiting dams and hydraulic laboratories. From 1930 to 1934 Campbell joined the US Bureau of Reclamation USBR in the design of Boulder Dam. He was on the staff of US Soil Conservation Service from 1935 to 1938. From 1940, Campbell was an associate engineer of the US Engineering Office, Nashville TN, for the Wolf Creek Project on Cumberland River. After war service, he was from 1945 to 1951 a senior engineer with the US Corps of Engineers USACE in the Omaha District, responsible for the Missouri River flood control, navigation and power projects. Finally, Campbell was from 1951 to 1965 a chief engineer within USACE at Waterways Experiment Station, Vicksburg MI.

Campbell authored a number of papers relating to flood defence, hydraulic structures and hydraulic experimentation. A notable work was presented during the 1953 IAHR Congress relating to bottom outlets. It was demonstrated that previous observations were conducted in too small hydraulic models, resulting in scale effects, so that the air entrainment was underestimated. Campbell and Guyton's relation for air entrainment involves as the basic parameter the approach flow Froude number at the contracted section downstream of the gate, in agreement with current knowledge.

Anonymous (1964). Campbell, Frank B. *Who's who in engineering* 9: 263. Lewis: New York.

Anonymous (1964). Frank B. Campbell. *Civil Engineering* 34(5): 65. *P*

Anonymous (1967). Frank B. Campbell. *Civil Engineering* 37(7): 4. *P*

Campbell, F.B., Bauder, H.A. (1940). A rating curve for determining silt-discharge of streams. *Trans. AGU* 21(2): 603-607.

Campbell, F.B., Guyton, B. (1953). Air demand in gated outlet works. 5th *IAHR Congress* Minnesota: 529-533.

Campbell, F.B., Cox, R.G., Boyd, M.B. (1965). Boundary layer development and spillway energy losses. *Journal of the Hydraulics Division* ASCE 91(HY3): 149-163; 91(HY6): 238-245.

Campbell, F.B. (1966). Hydraulic design of rock riprap. *Miscellaneous Paper* 2-777. US Corps of Engineers: Vicksburg MI.

Campbell, F.B., Pickett, E.B. (1966). Prototype performance and model-prototype relationship. *Miscellaneous Paper* 2-857. US Corps of Engineers, Vicksburg MI.

CAMPBELL J.C.

19.08. 1817 Cherry Valley NY/USA
26.03. 1890 New York NY/USA

John Cannon Campbell was at his era one of the most prominent engineers in the United States, who devoted his time primarily to water works. His word was authority on all matters relating to hydraulic engineering. Toward the end of his career, he was interested in important water works in California, but his health was failing so that he returned to New York. He had studied engineering in New York and then joined the Croton Water Works under John B. Jervis (1795-1885). The 1877 report shows a section of Croton Aqueduct including a gauging station at Croton Dam. A hydrographic map shows the dam site including also Campbell.

Campbell acted finally as chief engineer of the Public Works Department of New York City, and thereby was particularly associated with its Croton Water Works. The Old Croton Aqueduct was completed in 1842 and in service until 1965. The water supplied to the city originates from Croton River in its North, which was transported through a 66 km long aqueduct. Its remnants are since 1992 a part of the Historical Landmark. It was a sort of homecoming from the West for Campbell because his main responsibility was the Aqueduct. He was the person who designed and executed the first stage of this outstanding hydraulic project, which was around 1900 taken up in its second stage by Edward Wegmann (1850-1935). Since then, the scheme was steadily improved and currently feeds one of the largest cities with drinking water. Campbell started working under Jervis, showing marked ability for his profession. Later he was involved in a portion of the Hudson River Railroad through the highlands. In 1850 Campbell went to Panama as chief engineer of that railroad, but his failing health obliged him to return home, where he designed soon after a road in Indiana, and subsequently acted there as General Superintendent of a long railroad in Wisconsin. On returning to New York, he was appointed engineer of Croton Reservoir, and became engineer in chief of the Public Works, a position he held for decades then.

Anonymous (1877). John C. Campbell. *Scribner's Magazine* 14(2): 164. *P*
Anonymous (1890). John C. Campbell. *The New York Times* March 27.
Anonymous (1890). John C. Campbell. *Engineering Journal* 64(5): 237-238.
Rideing, W.H. (1877). *Croton water*: New York's water supply. Scribner's: New York.
New York NY Daily Graphic 1875 Aug-Jan 1876 Grayscale – 0180.pdf *P*
search.ancestry.com/.../sse.dll *P*

CAPEN

01.11. 1895 Jersey City NJ/USA
12.03. 1987 Sarasota FL/USA

Charles Herbert Capen graduated as a civil engineer from Cornell University, Ithaca NY, in 1917. He was then an assistant engineer at Pittsburgh PA and New York and in 1919 joined as assistant engineer the New Jersey State Department of Health. From 1925 until retirement in 1955, he was engaged with the North Jersey District Water Supply Commission NJDWSC, finally as chief engineer. In parallel, Capen was from 1927 a consultant to more than 40 municipalities in design, construction and operation of water and sewage facilities, pumping stations, sewer systems, and water works. During his career, Capen was a member, chairman and director of the American Water Works Association AWWA, receiving in 1938 their first G.W. Fuller Award, the 1941 Fuertes Graduate Medal from Cornell University, among others. He was AWWA president in 1952 and Member of the American Society of Civil Engineers.

The NJDWSC is responsible for the operation of the largest water supply system in New Jersey State, including more than 250 km^2 of watershed area, two major reservoirs, two river-diversion pumping stations and a more than 5 m^3/s water filtration plant. At peak, this institution serves the water needs of more than 2 million people in Northern New Jersey. Capen was one of the major collaborators of NJDWSC, thereby supporting its activities in engineering and administration. He guided NJDWSC through the difficult 1930s and finally succeeded in a vital and economic water distributor. He in parallel developed also the design of large diameter pipes in terms of friction loss, thereby accounting for the scientific developments shortly before World War 2. The history of water supply in New Jersey was of further interest to him.

Anonymous (1947). Charles H. Capen. *Journal American Water Works Association* 39(8): 14. *P*
Anonymous (1964). Capen, Charles Herbert. *Who's who in engineering* 9: 370. Lewis: New York.
Anonymous (1967). Charles H. Capen. *Journal American Water Works Association* 79(6): 126. *P*
Capen, C.H. (1936). History of the development of the use of water in northeastern New Jersey. *Journal of the American Water Works Association* 28(8): 973-982.
Capen, C.H. (1937). How much water do we consume? *Journal AWWA* 29(2): 201-212.
Capen, C.H. (1941). Trends in coefficients of large pressure pipes. *Journal of the American Water Works Association* 33(1): 1-83.
Kelly, H.A., Capen, C.H., Buckley, R.C. (1953). *Flood control attainable by land use and treatment*. New Jersey.

CARMODY

06.05. 1928 Bethlehem PA/USA
05.09. 1980 Tucson AZ/USA

Thomas Carmody received education from the State University of Iowa, Iowa City IA, obtaining the PhD degree in 1963 for a work in the field of wake flow behind a disk. He was tutored by Hunter Rouse (1906-1996), who also proposed to Carmody and his colleague Helmut Kobus (1937-) the translation of Daniel Bernoulli's hydraulic text *Hydrodynamica* from Latin to English. Carmody went in the late 1960s as associate professor of civil engineering to the University of Arizona, Tucson AZ, taking over in the 1970s there as professor. He was involved in the 1965 Arizona Hydraulics Division Conference.

The 1967 paper deals with the flow establishment region of an air jet issued with an efflux velocity of some 10 m/s from a 0.30 m diameter nozzle into still air. The data include mean axial and radial velocities, mean static pressures, turbulence intensities, turbulent shear and pressure fluctuations. The fluctuating pressure and the turbulence intensity fields were nearly similar. Using these data, all terms of the integral and differential forms of the momentum and mean energy equations were evaluated, so that results pertain to their variation throughout the flow field observed. The 1980 paper introduces a model for border irrigation predicting the rates of advance and recession, and the duration and depth of infiltration if the slope, length, roughness, inflow and infiltration rates are specified. The model was successfully validated by simulating field runs measured at the University of Arizona.

Anonymous (1965). Thomas Carmody. *Civil Engineering* 35(6): 83. *P*
Carmody, T. (1964). Establishment of the wake behind a disk. *Journal of Basic Engineering* ASME 86(4): 869-882.
Carmody, T., Kobus, H. (1968). *Daniel Bernoulli's Hydrodynamica*. Dover: New York.
Carmody, T. (1980). A critical examination of the 'largest' floods in Arizona: A study to advance the methodology of assessing the vulnerability of bridges to floods, Arizona Department of Transportation. *General Report* 1. University of Arizona: Tucson.
Fonken, D.W., Carmody, T., Laursen, E.M., Fangmeier, D.D. (1980). Mathematical model of border irrigation. *Journal Irrigation and Drainage Division* ASCE 106(IR3): 203-220.
Rouse, H. (1960). Distribution of energy in regions of separation. *La Houille Blanche* 15(3): 221-234; 15(4): 391-403.
Sami, S., Carmody, T., Rouse, H. (1967). Jet diffusion in the region of flow establishment. *Journal of Fluid Mechanics* 27: 231-252.

CARRIER G.F.

04.05. 1918 Millinocket ME/USA
08.03. 2002 Cambridge MA/USA

George Francis Carrier earned in 1939 a mechanical engineer degree, and in 1944 a doctorate in applied mechanics, both from Cornell University, Ithaca NY. From 1946 to 1952 he advanced from assistant professor to professor of engineering at Brown University, Providence RI. He then moved in the same position to Harvard University, Cambridge MA. Carrier's many awards include in 1978 the Timoshenko Medal of the American Society of Mechanical Engineers, the von Karman Medal in 1977 of the American Society of Civil Engineers, the Dryden Medal of the American Institute of Aeronautics and Astronautics, and the National Medal of Science in 1990.

Carrier was interested throughout his long professional career in the solution of key technological problems, including the rolling of plastic material, the technology of viscometry or centrifugal isotope separation. Later, he produced a series of papers in the field of combustion, particularly relevant in the performance of internal combustion engines. Further notable contributions include wind-driven circulation in the ocean, or the mixing of ground and sea water resulting from the diffusion of tides. His oceanographic studies have explained the mechanics of water waves on sloping beaches and in confined harbors. He further analyzed the behaviour of ocean currents in general, and the Gulf Stream in particular. Carrier was an associate editor of the Journal of Fluid Mechanics from 1956 to 1986, and was the editor of Quarterly of Applied Mechanics since 1952.

Abernathy, F.H., Bryson, A.E. (2007). George F. Carrier. *Memorial tributes* 11: 46-51. National Academy of Engineering: Washington DC. *P*

Anonymous (1979). George F. Carrier. *Mechanical Engineering* 101(2): 88. *P*

Anonymous (1991). Carrier wins National Medal of Science. *SIAM News* 24(1): 1. *P*

Anonymous (1991). G.F. Carrier. *Notices American Mathematical Society* 38(4): 277-278. *P*

Carrier, G.F. (1966). Phenomena in rotating fluids. *Applied mechanics*: 69-87. H. Görtler, ed. Springer: Berlin.

Carrier, G.F., Pearson, C.E. (1988). *Partial differential equations*: Theory and techniques, 2nd ed. Academic Press: Boston MA.

Carrier, G.F., Bakshi, P. (1998). *Internal wave generation by a moving object*. Harvard University.

Munk, W.H., Groves, G.W., Carrier, G.F. (1950). Note on the dynamics of the Gulf Stream. *Journal Marine Research* 9(3): 218-238.

CARRIER W.H.

26.11. 1876 Angola NY/USA
10.10. 1950 New York NY/USA

Willis Haviland Carrier graduated in 1901 from Cornell University and accepted a position with the Buffalo Forge Company, New York NY. He was determined to put mechanical engineering on a more rational basis then was the practice at that time. To investigate the principles behind the work of the company's products, Carrier founded the world's first industrial laboratory in his company. With an improved understanding, the Buffalo Forge designers were able to design safer and more efficient products. The laboratory more than paid for itself in its first year of existence.

Carrier's first design job was to develop an air conditioning system for a lithographing company. The problem faced was not heat but humidity, because the paper in the printing plant would shrink or expand depending on the amount of water it absorbed from the air. Carrier met this problem by designing a system that cooled the air in its plant to a constant temperature and thereby reduced the air's humidity. This raised Carrier's interest in the physical properties of air. His first major contribution to the field of air conditioning and refrigeration came in 1904 when discovering that air could be dehumidified by spraying water through it. He knew that the amount of water vapour air can contain depends on its temperature. As the temperature of air drops, the amount of water vapour it can contain also drops. If humid air is then cooled, the amount of water vapour it can contain drops. Despite the extra water being sprayed through it, the amount of water vapour in the air becomes larger than the maximum amount of water vapour the air can contain, resulting in water. The system of air conditioning became of large interest such that the Carrier Air Conditioning Corporation was founded in 1906. It became available for anybody from the 1950s. Carrier was president of the American Society of Refrigeration Engineers in 1927, and awarded Honorary Member of ASME.

Anonymous (1939). W.H. Carrier. *Mechanical Engineering* 61(7): 537. *P*
Anonymous (1943). W.H. Carrier. *Mechanical Engineering* 65(1): 67. *P*
Carrier, W.H. (1911). Rational psychometric formulae. *Trans. ASME* 33: 1005-1050.
Carrier, W.H. (1925). Temperatures of water evaporation into air. *Mechanical Engng.* 47: 327-331.
Koréni, Z., Tolnai, B. (2007). Carrier, Willis Haviland. *Az áramlás-és hőtechnika nagyjai*: 449-450. Műegytemi Kiadó: Budapest (in Hungarian).
Raines, J. (1995). W. Carrier. *Notable 20th century scientists* 1: 318-319. Gale Research: New York.
http://en.wikipedia.org/wiki/Willis_Carrier *P*

CARTER

25.07. 1916 Union County LA/USA
07.11. 2011 Fairfax VA/USA

Rolland William Carter was employed from 1941 by the Surface Water Branch of the US Geological Survey at Alabama GA. In the mid 1950s, he was there the chief of the research station. Later, Carter was a staff member of the US Department of Interior, Washington DC. Around 1970 he was the vice-president of the Commission of Surface Water, of the International Association of Hydrological Sciences IAHS. He was awarded the 1956 and the 1960 Norman Medals from the American Society of Civil Engineers ASCE for papers written with his colleague Carl E. Kindsvater (1913-2002).

Carter authored a number of papers in hydraulics and hydrology. His researches in collaboration with Kindsvater became important. The 1955 paper describes bridge hydraulics for subcritical approach flow. Various parameters were considered, namely the contraction rate, channel shape, pier corner rounding, Froude number, angularity and guide walls, among others. The approach is currently outdated but guided engineers in questions relating to flood defence for decades. Another paper relates to discharge measurement using thin-plate weirs. After detailed experimentation from the 1890's, Kindsvater and Carter's approach accounts for additional parameters including scale effects due to viscosity and surface tension. Following Theodor Rehbock (1864-1950), these were expressed with substitute lengths adjusting both weir width and overflow height. The resulting discharge equation was also used for decades.

Anonymous (1956). Rolland W. Carter. *Civil Engineering* 26(10): 687. *P*
Anonymous (1960). Rolland W. Carter. *Civil Engineering* 30(10): 88. *P*
Anonymous (1972). Rolland Carter. *Eos* 53(12): 1152. *P*
Benson, M.A., Carter, R.W. (1973). *A national study of the streamflow data-collection program.* US Government Printing Office: Washington DC.
Carter, R.W. (1973). Accuracy of current meter measurement. *IAHS Publication* 99. IAHS Press: Wallingford UK.
Kindsvater, C.E., Carter, R.W., Tracy, H.J. (1953). Computation of peak discharge at contractions. *Circular* 284. US Geological Survey: Washington DC.
Kindsvater, C.E., Carter, R.W. (1955). Tranquil flow through open-channel constrictions. *Trans. ASCE* 120: 955-980; 120: 991-992.
Kindsvater, C.E., Carter, R.W. (1959). Discharge characteristics of rectangular thin-plate weirs. *Trans. ASCE* 124: 772-801; 124: 818-822.

CASEY

24.07. 1898 Brooklyn NY/USA
30.08. 1981 White River Junction VT/USA

Hugh John Casey received education from the US Military Academy, West Point NY, until 1918, from the US Engineering School, and from the *Technische Hochschule* Berlin, Germany, from where he obtained the Dr.-Ing. degree in 1935. He also was recipient of the honorary doctoral degree from New York University in 1954. He was professor of military sciences at the University of Kansas, Lawrence KA, from 1922 to 1926, then until 1929 assistant district engineer for river flood control, locks and dams in Pennsylvania, until 1933 executive officer of the Rivers and Harbors Section at the Office of Chief of Engineers, Washington DC, and from 1935 to 1936 chief engineer of Passamaquoddy Tidal Power Project, Eastport ME. Further stations in his professional career included New England flood control surveys, engineering adviser on water power and flood control to the Philippine government, chief engineer to General MacArthur's commands during and after World War II, and the chief engineering position in New York State. He had hoped to become Chief of Engineers, but President Truman passed him over.

Casey is known for his PhD thesis on sediment transport, which he conducted shortly before Albert Shields (1908-1974) presented his fundamental study in 1937 on sediment entrainment at the same institution. Casey also prepared a voluminous report on flood control for the Pittsburgh PA District. He was involved in the design and construction of the Deadman Island Lock and Dam on the Ohio River. In the early part of World War II he was concerned with the enormous wartime construction program. His most notable and lasting achievement was the involvement in the design of The Pentagon, the largest office building in the world.

Anonymous (1964). Casey, Hugh J. *Who's who in engineering* 9: 283-284. Lewis: New York.
Casey, H.J. (1929). Deadman Island Lock and Dam, Ohio River. *The Military Engineer* 21(119): 444-451.
Casey, H.J. (1935). Über Geschiebebewegung. *Mitteilung* 19. Preussische Versuchsanstalt für Wasserbau und Schiffbau: Berlin.
Casey, H.J. (1947). *Engineers of the Southwest Pacific* 1941-45: Reports on operations, US Army Forces in the Far East, Southwest Pacific Area. US Government Printing Office: Washington DC.
Fine, L., Remington, J.A. (1972). *The Corps of Engineers*: Construction in the United States. Washington DC.
http://en.wikipedia.org/wiki/Hugh_John_Casey *P*

CEDERGREN

02.01. 1911 Seattle WA/USA
25.02. 1996 Sacramento CA/USA

Harry Roland Cedergren received his BS degree in civil engineering from Washington University in 1938, and the MS degree in soil mechanics from Harvard University, Cambridge MA in 1939, under Prof. Arthur Casagrande (1902-1981). He was in 1962 senior materials and research engineer, the California Division of Highways, Sacramento CA, and later there was a consultant.

Cedergren was acclaimed for his extensive work in the areas of seepage analysis and drainage design. During a career spanning almost five decades he was active consultant, lecturer and teacher, holding positions with US Corps of Engineers and the California Division of Highways. He was a member of the Transportation Research Board and author of Drainage of highway and airfield pavements. He also noted that engineering works involving water, including dams and levees, can usually be made safe by (1) Keeping the water out of places where it can cause harm by using watertight barriers, and (2) Controlling by drainage methods including filters and drains or relief wells. These basic means have always had their effect both in the past and in the future. The 1962 paper is concerned with filters in hydraulic engineering. The water removing capabilities of two common but distinctly different filter types are analysed by the flow-net technique. Typical solutions and numerical examples are presented to emphasize the importance of boundary conditions and permeability upon the water-removing capacity. The 1985 work is concerned with the design of drainage systems, which can be of great benefit to mankind if properly designed, and often represent the only way to protect these works from the damaging effects of water.

Anonymous (1953). Harry R. Cedergren. *Civil Engineering* 23(12): 856. *P*
Casagrande, A. (1937). Seepage through dams. *Journal of the New England Water Works Association* 51(6): 295-336.
Cedergren, H.R. (1948). Use of flow-net in earth dam and levee design. Proc. 2nd Int. Conf. *Soil Mechanics and Foundation Engineering*: 293-298.
Cedergren, H.R. (1962). Seepage requirements of filters and pervious bases. *Trans. ASCE* 127: 1090-1108; 127: 1111-1113.
Cedergren, H.R. (1975). *Drainage of highway and airfield pavements*. Wiley: New York.
Cedergren, H.R. (1977). *Seepage, drainage and flow nets*. Wiley: New York.
Cedergren, H.R. (1985). *Design of drainage systems for embankments and other civil engineering works*. Metallurgical Society of AIME, Conifer CO.

CERMAK

08.09. 1922 Hastings CO/USA
21.08. 2012 Huerfano CO/USA

Jack Edward Cermak graduated from Colorado State University, Fort Collins CO with the BS degree in 1947, the MS degree in 1948, and from Cornell University, Ithaca NY with the PhD degree in 1959. He then stayed as NATO post-doctoral fellow in 1961 at Cambridge UK. Cermak was a professor of fluid mechanics and wind engineering at his Alma Mater from 1960 to 1985. He also was from 1963 to 1972 chairman of the engineering fluid dynamics and diffusion laboratory, and from 1965 to 1972 president of Colorado State University. From 1985 he was president of Cermak/Peterka and Associates Inc., among many other commitments after retirement from ColoState. He further served on the editorial board of the International Journal of Wind Engineering, and the Archives of Meteorology, Geophysics and Bioclimatology. Cermak was a Fellow of the American Society of Civil Engineers ASCE and the American Academy of Mechanics, and was member of the National Academy of Engineering, among many others. He was awarded the 2010 Flachsbart-Medal from *Windtechnologischer Gesellschaft*, Germany for his outstanding contributions to wind engineering.

Cermak is regarded as one of the fathers of wind engineering. With his laboratory at ColoState he made revolutionary observations relating mainly to the effect of wind on civil engineering structures. He thereby advanced research considerably in this regard but also solved numerous problems of wind engineering dilemmas in practical situations. All through his career he took interest in structural responses to wind, and the atmospheric transport of pollutants, snow, sand and water. He was for instance involved in the wind tunnel testing of the WTC Twin Towers in New York City during their design in the early 1960s. In 2002 ASCE established the Jack E. Cermak Medal to be awarded annually for outstanding contributions to wind engineering.

Anonymous (1970). J.E. Cermak. *Mechanical Engineering* 92(4): 80-81. *P*
Anonymous (1987). Jack E. Cermak. *Who's who in America* 44: 471. Marquis: Chicago.
Cermak, J.E., Sethu Raman, S. (1971). Simulation of density currents. *River mechanics* 31: 1-35, H.W. Shen, ed. Colorado State University: Fort Collins.
McFarland, A.R., Wedding, J.B., Cermak, J.E. (1977). Wind tunnel evaluation of a modified Andersen Impacter and all-weather sampler inlet. *Atmosph. Environment* 11: 535-539.
Poreh, M., Cermak, J.E. (1959). Flow characteristics of circular submerged jet impinging normally on smooth boundary. Proc. 6th Midwestern Conf. *Fluid Mechanics* Austin: 198-212.

CHADWICK

04.12. 1897 Loring KS/USA
05.06. 1996 Claremont CA/USA

Wallace Lacy Chadwick moved as a child to South California. After two years of study at University of Redlands, Redlands CA, his academic career was interrupted by World War I. He later returned to the school, yet to serve for 42 years on its Board of Trustees. In 1965, the University conferred upon him an honorary doctorate in engineering science. After a six-year stint with the Metropolitan Water District, he joined the Southern California Edison Company in 1922, remaining there for almost 40 years. He retired from it in 1962 as vice-president of engineering and construction but went on for another 30 years as consultant to the Edison and Bechtel Corp., San Francisco CA. During his 'second career' Chadwick travelled more than 7 million miles to work on projects such as the San Francisco Bay Area or the Washington DC subway system, hydro-electric power plants in Canada, a power facility in Saudi-Arabia and an official inquiry into the causes of the 1976 failure of Grand Teton Dam.

Chadwick is still known for his 1988 book on the Development of dam engineering in the United States, edited with a colleague. This impressive work evidents the large advances made in dam engineering within 100 years, counting still to the countries with the largest hydropower potential. Chadwick was member of the National Academy of Engineering, and a Fellow of the American Society of Civil Engineers ASCE, of which he also served as president in 1964. During this time he filled in a number of committee assignments including professional conduct, membership qualifications, and water policy and planning. The Society awarded him also its Rickey Medal in 1940. In addition Chadwick belonged to the International Committee on Large Dams ICOLD, and was a Honorary Member ASCE. In 1978 he was named Construction Man of the year by Engineering News-Record for his work on the Grand Teton Dam failure. Although he retired in 1990, he remained professionally active for the rest of his life.

Anonymous (1965). Wallace L. Chadwick. *Civil Engineering* 35(11): 104. *P*
Anonymous (1996). Wallace Chadwick, ASCE's 96th president, dies at age 98. *Civil Engineering* 66(9): 79-80. *P*
Chadwick, W.L. (1974). Chadwick, Wallace L. *Scienziati e tecnologi* 1: 234-235, R. Adams, D. Hodgkin, eds. Mondadori: Milano. *P*
Kollgaard, E.B., Chadwick, W.L. (1988). *Development of dam engineering in the United States.* Pergamon Press: New York.

CHAMBERLIN

25.09. 1843 Mattoon IL/USA
15.11. 1928 Chicago IL/USA

Thomas Chrowder Chamberlin graduated in 1866. He then became a teacher and later principal in a high school. In parallel he spent a year at University of Michigan to strengthen his overall scientific background, for which he had already as a boy been interested. From 1873, he was professor of geology at Beloit Faculty, Beloit WI, and also staff member of geological survey of Wisconsin. In 1876, he was appointed chief geologist, publishing a four volumes book with colleagues on the results of this survey. He also wrote chapters on artesian wells and glacial deposits. The results brought national attention and appointment as head of the glacial division of the National Survey in 1881.

From 1887 to 1892 Chamberlin was president of the University of Wisconsin, Madison. Then he accepted the offer to organize a department of geology at the new University of Chicago, where he remained until retirement in 1918. He thereby created one of the nation's premier departments with a strong research program. He also founded the Journal of Geology and acted as its editor. It was thus at Chicago where Chamberlin fully matured as a leading scholar. From 1904 to 1906 he and a colleague authored the textbook Manual of geology which was of significant influence until after World War II. Besides geology, Chamberlin also significantly contributed to groundwater flow, particularly to artesian wells. As early as in 1885 he published a hydro-geological article urging both geologists and the public to understand and appreciate groundwater. Chamberlin was president of the Geological Society of America in the term 1894, and president of the American Association for the Advancement of Science in 1908. He was a Member of the National Academy of Sciences from 1903, and the first medalist of the Geological Society of America in 1927.

Anonymous (1887). Chamberlin, T.C. *Appletons' Cyclopaedia of American biography* 1: 566.
Back, W. (1996). T.C. Chamberlin, early American hydrogeologist. *Hydrogeology Journal* 4(2): 94-95. *P*
Chamberlin, T.C. (1883). Artesian wells. *Wisconsin Geological Survey* 1: 689-701.
Chamberlin, T.C. (1885). The requisite and qualifying conditions of artesian wells. 5th *Annual Report*: 125-173. US Geological Survey. Washington DC.
Chamberlin, T.C., Salisbury, R.D. (1906). *Manual of geology*. Holt: New York.
http://www.gsajournals.org/perlserv/?request=get-document&doi=10.1130%2F1052-5173(2006)16%5B30%3ARSTCC%5D2.0.CO%3B2&ct=1 *P*

CHASE E.S.

14.07. 1884 Merrimacport MA/USA
08.07. 1969 Boston MA/USA

Edward Sherman Chase graduated in 1907 as a civil engineer from MIT. He was then an assistant at Worcester Polytechnic Institute and from 1909 a research engineer at Reading PA in sewage disposal. Later, Chase became a sanitary engineer of the New York Health Department at Albany NY, and he was from 1920 to 1927 in the staff of Metcalf & Eddy, and partner then until his retirement in 1959, after which he was a consultant to the firm. Chase was a Fellow of the American Society of Civil Engineers ASCE, an Honorary Member of the Institution of Water Engineers UK from 1948, and president of the New England Water Works Association in 1934, among other.

Chase worked from 1920 closely with Harrison P. Eddy (1870-1937), a founder of Metcalf & Eddy. He was best known for the broad field of sanitary engineering, including water supply and the treatment of sewage and industrial waste. His keen perception, sound judgement, quick wit and human understanding led clients and friends to value highly his conclusions and advice. He was also instrumental in the collaboration between British and American developments in wastewater technology, together with Harold Gourley (1886-1956). In 1968 Chase became the first recipient of the Friendship Medal of the Institution for his activities in fostering relations among these engineers. He also served as General Reporter for water supply at the Third International Water Supply Congress in London, 1955. In 1960, the annual E. Sherman Chase Award of the New England Water Pollution Control Association was established.

Anonymous (1931). Chase, E.S. *Who's who in engineering* 3: 226; 7: 409. Lewis: New York.
Anonymous (1934). E. Sherman Chase. *Journal New England Water Works Association* 47(1): Frontispiece. *P*
Anonymous (1952). Chase elected head of sewage group. *ENR* 149(Oct.9): 88. *P*
Anonymous (1955). E. Sherman Chase. *Journal Institution of Water Engineers* 9: Plate 2. *P*
Anonymous (1968). E. Sherman Chase. *Journal Water and Water Engineering* 5(5): 66-67. *P*
Chase, E.S. (1958). Flotation treatment of sewage. *Sewage and Industrial Wastes* 30(6): 783-791.
Chase, E.S. (1962). British and American water works practices, likes, unlikes and dislikes. *Journal British Waterworks Association* 44: 583-591.
Chase, E.S. (1964). Nine decades of sanitary engineering. *Water Work and Waste Engineering* 34(6): 48-49; 34(7): 49-50.
Shaw, A.L. (1969). Edward S. Chase. *Journal Institution Water Engineers* 23(7): 470-472. *P*

CHEPIL

01.01. 1904 Gimli Man MB/CA
06.09. 1963 Manhattan KS/USA

William Steven Chepil was educated at University of Saskatchewan, Saskatoon SK. He joined the US Department of Agriculture within the Wind Erosion Unit in 1947, where he became group leader in 1953. Shortly later, he proposed the wind erosion equation, which in a way paralleled the Universal Soil Loss Equation used for predicting the water erosion. Most of the work on this equation was completed under Chepil, but he died of cancer at age 59 before he could see its first publication.

As soon as the Dust Bowl began in the early 1930s, observers asked why it happened when and why it and what it caused. According to the government, the Dust Bowl was caused by the recent arrival of farmers on the southern plains. Settlers had plowed land unsuitably to crop farming, exposing bare soil to high wind. From then scientists have explored the physics off wind erosion. Chepil's wind erosion equation identifies five factors that contribute to blowing soils, namely climatic forces including precipitation, temperature and wind, soil texture, surface roughness, field length and quantity of vegetation. Knowing these parameters allows for a prediction of wind erodibility. Methods to reduce soil erosion by wind include plowing furrows perpendicular to the prevailing wind direction, corrugating fields by plowing steeper furrows to increase surface roughness, or breaking up long stretches of bare soil with intermittent grass strips.

Chepil, W.S. (1945). Dynamics of wind erosion. *Soil Science* 60(4): 305-320; 60(5): 397-411; 60(6): 475-480; 61(2): 167-177.

Chepil, W.S., Woodruff, N.P. (1954). Estimations of wind erodibility of field surfaces. *Journal of Soil Water Conservation* 9(6): 257-265.

Chepil, W.S. (1957). Sedimentary characteristics of dust storms: Sorting of wind-eroded soil material. *American Journal of Science* 255(1): 12-22.

Chepil, W.S. (1958). *Soil conditions that influence wind erosion*. US Dept. Agriculture: Washington DC.

Chepil, W.S. (1959). Equilibrium of soil grains at the threshold of movement by wind. *Soil Science Society America* 23(6): 422-428.

Chepil, W.S. (1963). Function and significance of wind in sedimentology. *Sedimentation Conference*: 89-94. *Miscellaneous Publication* 970. ARS: Washington DC.

http://www.weru.ksu.edu/new_weru/multimedia/staff/staffimages.htm. *P*
http://www.weru.ksu.edu/new_weru/HistoryWERU.pdf *P*

CHESBROUGH

06.07. 1813 Baltimore MD/USA
19.08. 1886 Chicago IL/USA

Ellis Sylvester Chesbrough was taken from school at age 13 only to become chainman to an engineering party engaged in the preliminary survey of the Baltimore and Ohio Railroad. Later he was engaged on the Alleghany and Portage Railroad to become in 1831 associated with a partner in the construction of the Paterson and Hudson River Railroad. In 1837 he was appointed senior assistant on the building of the Louisville, Cincinnati and Charleston Railroad. He became chief engineer of the Boston waterworks in 1846, planning important structures including the Brookline Reservoir. He was appointed in 1850 sole commissioner in the Boston water department, and during the following year city engineer, having charge of all the waterworks besides being surveyor of the streets and harbour improvements.

Chesbrough became engineer for the Chicago Board of Sewerage commissioners in 1855, and in that capacity planned the sewage system for the city. In 1879 he resigned the office of commissioners of public works. The river tunnels were planned by him and have proven successful despite much criticism. He achieved a high reputation as an authority in the water supply and sewerage of cities, and in this capacity was consulted by the authorities of New York, Boston, Cambridge, Toronto and Detroit. Chesbrough was a Corresponding Member of the American Institute of Architects, and in the term 1877 president of the American Society of Civil Engineers ASCE.

Anonymous (1887). Chesbrough, E.S. *Appletons' cyclopaedia of American biography* 1: 599.
Cain, L.P. (1985). Raising and watering a city: Ellis Sylvester Chesbrough and Chicago's first sanitation system. *Sickness and health in America*: Readings in the History of medicine and public health: 531-541, J.W. Leavitt, R.L. Numbers, eds. University of Wisconsin: Madison.
Chesbrough, E.S., Durant, C.F. (1849). *Letters on hydraulics*: On the physical laws that govern running water. Narine: New York.
Chesbrough, E.S. (1851). *Tabular representation of the present condition of Boston*. Eastburn: Boston.
Chesbrough, E.S. (1858). *Chicago sewerage*. Board of Sewerage Commissioners: Chicago.
Chesbrough, E.S. (1874). Sketch of the plans and progress of the Detroit River tunnel. *Trans. ASCE* 2: 85-91.
http://www.encyclopedia.chicagohistory.org/pages/300017.html *P*

CHICK

26.10. 1896 Limerick ME/USA
16.04. 1973 Warwick RI/USA

Alton Charles Chick received in 1919 his BS degree in mechanical engineering and in 1926 his MS degree in civil engineering from Brown University, Providence RI. In 1922, Chick became principal assistant to John R. Freeman (1855-1932), with whom he remained until 1932. During this ten-year period, he had the opportunity to work on many and varied engineering problems. Chick accepted then the position as an engineer with the Manufacturers Group of six fire insurance companies, who later merged to the Manufacturers Mutual Fire Insurance Company. He was from 1938 until retirement its assistant vice-president and engineer. This work involved application of engineering principles to the prevention of fire and the protection of industrial plants. Chick was also the president of the Providence Engineering Society in 1937.

During his stay with Freeman, Chick compiled the 886 pages book Hydraulic Laboratory Practice and another book on earthquake damage and insurance. The first book gives an overview on hydraulic experimentation at the end of the 1920s, including the main actors. An appendix accounts for the then recent additions in the USA. Both Freeman and Chick finally presented an outstanding volume which has still high value for historical studies. The laboratories, their facilities and instrumentation are highlighted, and recent research is presented. The laboratory directors are also introduced with a biography and their pertinent publications. It appears that such a volume has never been prepared again. Prior to Freeman's death, Chick recomputed a series of experiments on pipe flow and pipe fittings that were made by Freeman in 1892. The results were published in 1941 in a large volume adding to the understanding of these flows.

Anonymous (1943). Alton C. Chick. *Mechanical Engineering* 65(8): 611-612. *P*
Anonymous (1959). Chick, Alton Charles. *Who's who in engineering* 8: 415. Lewis: New York.
Chick, A.C. (1929). Dimensional analysis and the principle of similitude as applied to hydraulic experiments with models. *Hydraulic laboratory practice*: 775-827, J.R. Freeman, ed. ASME: New York.
Chick, A.C. (1929). *Notes on theory of hydraulic experiments with models*. ASME: New York.
Chick, A.C. (1933). *The Long Beach Earthquake* of March 10, 1933, and its effect on industrial structures. Manufacturer's Mutual Fire Insurance Company: Rhode Island.
Freeman, C., Chick, A.C. (1941). *Experiments upon the flow of water in pipes and pipe fittings*. ASME: New York.

CHITTENDEN

25.10. 1858 Yorkshire NY/USA
09.10. 1917 Seattle WA/USA

Hiram Martin Chittenden graduated in 1884 from the US Military Academy, and from the Engineering School of Application in 1887. He was ordered then to Omaha as engineering officer of the Department of Platte, prepared a topographical map of Colorado, Wyoming and Utah two years later, and was from 1889 to 1891 assigned to the improvement of the Missouri River above Sioux City IA. In 1893 he was assigned to duty on the Louisville and Portland Canal bypassing the Falls of the Ohio River near Louisville KY. In 1894 he had charge of the canal survey between Lake Erie and the Ohio River as executive officer, and was promoted to captain in 1895. From 1896 he was secretary of the Missouri River Commission in charge of the improvement of Osage and Gasconade Rivers in Missouri. He served from 1898 as lieutenant-colonel and chief engineer in the Spanish-American War.

Chittenden was from 1899 to 1906 in charge of road construction in Yellowstone, being promoted to major in 1904. Until 1916 he was placed in charge of the Lake Washington Canal to connect Puget Sound WA with lakes in and bordering Seattle WA. During these years, he also was appointed chairman of the Federal Commission on Yosemite Park to consider changes of its boundaries, and was engaged in projects to investigate the Sacramento Flood Control. Chittenden retired in 1910 with the rank of brigadier general, following a stroke causing partial paralysis. He continued in 1912 as consultant for flood control to the Spring Valley Water Company, San Francisco CA, reporting also to the Miami Conservancy District on that problem. Chittenden was a member of the American Society of Civil Engineers ASCE, and the Missouri Historical Society. He published besides papers on technical problems also a notable work on the Yellowstone National Park, and on the American fur trade.

Chittenden, H.M. (1895). *The Yellowstone National Park*. Stewart & Kidd: Cincinnati.
Chittenden, H.M. (1903). *The history of early steamboat navigation on the Missouri River*. Ross & Haines: Minneapolis.
Chittenden, H.M. (1918). Detention reservoirs with spillway outlets as an agency in flood control. *Trans. ASCE* 82: 1473-1540.
FitzSimons, N., ed. (1991). Hiram M. Chittenden. *A biographical dictionary of American Civil Engineers* 2: 21. ASCE: New York.
http://en.wikipedia.org/wiki/Hiram_M._Chittenden *P*

CHRISTENSEN N.A.

19.01. 1903 Provo UT/USA
12.04. 1996 Albuquerque NM/USA

Nephi Albert Christensen graduated in 1928 with a BS as civil engineer from University of Wisconsin, Madison WI, and with the MS in 1934 and the PhD in 1938 from California Institute of Technology, Pasadena CA. He was from 1928 to 1933 professor of exact sciences at Ricks College, Rexburg ID, then until 1938 a hydraulic engineer at Caltech, and then until 1948 Dean of Engineering, Colorado State University CSU, Fort Collins CO. In parallel he was there also Director of its Engineering Division. From 1948 until retirement, Christensen was Director of the School of Civil Engineering, Cornell University, Ithaca NY. From 1954 he was also a member of the New York State Flood Control Commission, and consultant to the Brookhaven National Laboratory of Atomic Energy Commission, and the Ordnance Office of Research, Durham NC.

Christensen was particularly working in engineering education, with a multitude of appointments at American universities. In parallel, he was also interested in hydraulic engineering, as at CSU, where he contributed to the Colorado Agricultural Experiment Station. He submitted a discussion to the noteworthy 1943 ASCE paper on drop structures, in which a first attempt was made to standardize these important elements of hydraulic engineering. The water flow entrains air due to jet formation, thereby complicating the flow analysis. These flows were modeled using the similitude of Froude, provided the model sizes were sufficiently large to avoid scale effects due to surface tension and viscosity. This work also includes two-phase flow of water and sediment at the outlet, which was also investigated to avoid scour. This and other papers had a largely increased value due to discussions, a fact that is currently less observed. On leave of absence from 1942 to 1945, Christensen served successively as chief engineer of the Ballistics Research Laboratory, and chief of the Rocket Research Division at the Ordnance Research and Development Center, Aberdeen MD.

Alger, P.L., Christensen, N.A., Olmsted, S.P. (1965). *Ethical problems in engineering*. Wiley: New York.

Anonymous (1948). Nephi Albert Christensen. *Civil Engineering* 18(9): 612. *P*

Anonymous (1959). Christensen, N.A. *Who's who in engineering* 8: 422. Lewis: New York.

Anonymous (1963). Dr. N.A. Christensen. *Civil Engineering* 33(4): 72. *P*

Christensen, N.A., Gunder, D. (1943). Discussion of Hydraulic design of drop structures for gully control. *Trans. ASCE* 108: 927-930.

CHRISTIANSEN

09.04. 1905 Hyrum UT/USA
28.10. 1989 Logan UT/USA

Jerald Emmett Christiansen was educated at Utah State Agricultural College as agricultural engineer, and at University of California as a civil engineer. He was from 1928 to 1936 a junior agricultural engineer at University of California, Berkeley CA, and promoted then to assistant irrigation engineer. From 1942 to 1946, Christiansen was at the USDA Regional Salinity Laboratory, Riverside CA, when being appointed professor of civil engineering at Utah State Agricultural College, Logan UT, thereby directing also its Engineering Experiment Station. He in parallel served as an irrigation engineer for FAO, UN Montevideo Uruguay, and was a visiting professor to the University of Davis CA. He retired in 1971, but continued as a consultant for the water requirements in Latin America and also counselling graduate students in the Agricultural and Irrigation Engineering Department. Christiansen was the recipient of the 1976 Royce J. Tipton Award for his 'half-century of dedication to worldwide service resulting in significant advancement in irrigation and drainage engineering and salinity control'.

Christiansen mainly worked in agricultural engineering. Whereas his first study relating to the distribution of silt in open channel flow was shortly later generalized by Hunter Rouse (1906-1996), most of his later studies were related to irrigation by sprinklers and sprinkler hydraulics. These included the uniformity of sprinkler irrigation, measuring discharge in the agricultural environment, and the manifold problem of a sprinkler system. Work in the 1960s was concerned with evapo-transpiration and evaporation. Christiansen was one of the frequent participants of the Congresses of the International Commission of Irrigation and Drainage ICID.

Anonymous (1959). Christiansen, J.E. *Who's who in engineering* 8: 423. Lewis: New York.
Anonymous (1971). J.E. Christiansen. *Civil Engineering* 41(1): 37. *P*
Anonymous (1976). J.E. Christiansen. *Civil Engineering* 46(10): 103. *P*
Christiansen, J.E. (1935). Distribution of silt in open channels. *Trans. AGU* 16(2): 478-485.
Christiansen, J.E. (1936). Measuring water for irrigation. *Bulletin* 558. University of California, Berkeley CA.
Christiansen, J.E. (1941). The uniformity of application of water by sprinkler systems. *Agricultural Engineering* 22(3): 89-92.
Christiansen, J.E. (1942). Hydraulics of sprinkling systems for irrigation. *Trans. ASCE* 107: 221-250.

CHURCH A.H.

20.01. 1906 Mauch Chunk PA/USA
10.10. 1978 Hendersonville NC/USA

Austin Harris Church graduated as a mechanical engineer from Cornell University, Ithaca NY, in 1928, and obtained in 1934 the MS degree from New York University. Early in his career he was a junior engineer at Westinghouse Electric Co., where he designed turbines. He was then machine designer at Cooper Union and from 1937 to 1940 engineer with the *De Laval* Steam Turbine Co., Trenton NJ. He was appointed assistant professor of machine design at Cooper Union, New York University, in 1940 because he was competent in the then current turbine design and practice. From 1946 to retirement in 1973, as a professor of mechanical engineering at the same university, he wrote five books and numerous technical publications on machine design, vibration, pumps and compressors. His book Centrifugal pumps and blowers was considered an outstanding text, and was read widely by design engineers and students. After retirement, Church served as a consultant mainly in New York State.

Church was a life Fellow of the American Society of Mechanical Engineers ASME, who was active in society affairs for over 50 years. He further served as chairman of the Machine Design Division and the Pumping Subdivision of the Fluids Engineering Division ASME, thereby greatly advancing the standardization of pumping devices. He was in addition a Member of the Shock and Vibration Committee of the Division of Applied Mechanics, ASME. Church was never a pure scientist, but aimed to combine technological advance with economical and practical issues. He was awarded the title Fellow ASME in 1965. He was also member of the American Society of Engineering Education ASEE, among other societies.

Anonymous (1955). Austin H. Church. *Mechanical Engineering* 77(5): 465. *P*
Anonymous (1975). Church, Austin Harris. *Who's who in America* 38: 561. Marquis: Chicago.
Anonymous (1979). Austin H. Church. *Mechanical Engineering* 101(7): 89.
Church, A.H. (1944). *Centrifugal pumps and blowers*. Wiley: New York.
Church, A.H. (1948). *Elementary mechanical vibrations*. Pitman: London.
Church, A.H., Guillet, G.L. (1950). *Kinematics of machines*. Wiley: New York.
Church, A.H. (1957). *Mechanical vibration*. Wiley: New York.
Church, A.H., Plunkett, R. (1961). Balancing flexible rotors. *Trans. ASME* B 83: 383-389.
Newman, A.B., Church, A.H. (1935). The temperature during the constant-rate heating of a solid circular cylinder. *Trans. ASME* 2E: 96-98.

CHURCH B.S.

17.04. 1836 Belvidere NY/USA
10.12. 1910 New York NY/USA

Benjamin Silliman Church graduated in 1856 from Dartmouth College, Hanover NH, and after a course in civil engineering became a topographer on Croton Aqueduct for the water supply of New York City. This survey was made to locate the optimum sites for storage reservoirs. In 1858 he was transferred to the construction of the new receiving reservoir in Central Park, New York City, having then a capacity of 270 million liters. He was appointed then resident engineer in charge of the Old Croton Aqueduct, the Croton Reservoirs, and the pipe system, continuing in this position until 1880. Later he was Colonel of Engineers in New York State, so that he was known thereafter as Col. Church.

During his long period connected with the Croton Water Works, Church had an excellent opportunity to study the weak points of the scheme, and to investigate plans for the increase of the city's water supply. As early as in 1868, he planned the construction of a large masonry dam across Croton River, 6 km in the tailwater of the Old Croton Dam, whereby a large storage reservoir of 300 km^2 would be formed. In 1880 he was made consulting engineer of the Department of Public Works, assisting Ellis S. Chesbrough (1813-1886) in preparing plans for a large reservoir in Croton watershed, and of a new aqueduct from it to the city. The reservoir was to be formed with a 80 m high masonry dam near the Old Quaker Bridge, so that it was named Quaker Bridge Dam. In 1883 Church was appointed chief engineer of the Aqueduct Commission, holding this position until 1888, when made its consulting engineer. In 1889 he retired from the service of NYC after 33 years. He did then general engineering business as a consultant, his advice being sought in projects for water supply and water power both in the USA and in Canada. In 1897, Church became interested in steam engines, and then continued to work in this field until his death. His endeavor was to produce an engine which, while retaining all characteristics of a water turbine, would provide for the steam flow having a perfect force balance due to the steam pressure and the mechanical stresses. Church also was the inventor of a water meter, and a waste water detector.

Anonymous (1910). Benjamin S. Church. *Engineering News* 64(24): 670; 64(25): 702.
Church, B.S., Fteley, A. (1889). *Design and construction of high masonry dams*. New York.
Church, B.S. (1902). *Concerning the water supply of New York*. Croton Aqueduct Dept.: NYC.
http://digitalgallery.nypl.org/nypldigital/dgkeysearchdetail.cfm?trg=1&strucID=487897&image ID=1213861&parent_id=487785&s *P*

CHURCH I.P.

22.07. 1851 Ansonia CT/USA
08.05. 1931 Ithaca NY/USA

Irving Porter Church graduated in 1873 from Cornell University as a civil engineer and then practiced until 1876 when being appointed assistant professor of civil engineering at his Alma Mater. From 1892 until retirement in 1916 he was there professor of applied mechanics and hydraulics. His long career extended over a period when the study of engineering and mechanics expanded greatly, and he took an important part in shaping the technical courses and in compiling appropriate texts. His students attained eminence in all fields of engineering. Church was an associate member of civil engineering, and a member of the Cornell Association of Civil Engineers. In 1929 the Society for the Promotion of Engineering Education awarded him the Gold Medal for 'accomplishment in technical teaching and actual advancement in the art of technical training'.

Church published three books, which were then compiled into his basic text Mechanics of engineering. This epoch-making work went through a number of editions and became one of the most widely used textbooks of the United States. In addition to his various books, Church also published in technical journals. He was the founder of the technical school as opposed to the trade school. He donated to Cornell University the Fuertes Laboratory, including a 12-inch telescope lens. He further established in 1917 the Irving P. Church Fund for Cornell alumni to purchase books. Painting was one of his favorite diversions in his later years, and his house was filled with his copies of masterpieces.

Anonymous (1905). Church, Irving P. *Who's who in America*: 267. Marquis: New Providence NJ.
Anonymous (1921). Irving P. Church. *American men of science* 3: 126. Science Press: New York.
Anonymous (1980). Church, Irving Porter. *Mechanical engineers in America born prior to* 1860: 89. ASME: New York.
Church, I.P. (1887). *Applied mechanics and hydraulics*. Cornell University: Ithaca NY.
Church, I.P. (1889). *A treatise on hydraulics and pneumatics*. Wiley: New York.
Church, I.P. (1898). *Mechanics of engineering*. Wiley: New York.
Church, I.P. (1902). *Diagrams of mean velocity of water in open channels*. Wiley: New York.
Church, I.P. (1905). *Hydraulic motors* with related subjects, including centrifugal pumps, pipes, and open channels, designed as a text-book for engineering schools. Wiley: New York.
Church, I.P. (1913). Friction head in backwater problems. *Engineering News-Record* 68(6): 168.
Church, I.P. (1915). Discussion of Penstock and surge-tank problems. *Trans. ASCE* 79: 272-277.
http://cdsun.library.cornell.edu/cgi-bin/newscornell?a=d&d=CDS19171109.2.1.15.1&e= *P*

CLAPP

11.04. 1861 Conway MA/USA
27.12. 1911 Pasadena CA/USA

Vol. XXXIV.   FEBRUARY, 1908.   No. 2.

AMERICAN SOCIETY OF CIVIL ENGINEERS.
INSTITUTED 1852.

PAPERS AND DISCUSSIONS.

This Society is not responsible, as a body, for the facts and opinions advanced in any of its publications.

THE FLOOD OF MARCH, 1907,
IN THE SACRAMENTO AND SAN JOAQUIN RIVER
BASINS, CALIFORNIA.*†

BY W. B. CLAPP, M. AM. SOC. C. E., E. C. MURPHY, ASSOC. M. AM.
SOC. C. E., AND W. F. MARTIN, JUN. AM. SOC. C. E.

INTRODUCTION.

The Sacramento and San Joaquin Valleys were visited, in March, 1907, by one of the most destructive floods that have ever occurred in California, the resulting financial loss being unquestionably greater than that from any other flood of which there is record. The greatest damage was done in the valleys of the trunk streams, especially Sacramento Valley. The Lower Sacramento River and its two largest tributaries, Feather and American Rivers, reached the highest stages ever recorded, and record stages were reached by other tributaries of the Sacramento and by the San Joaquin and its tributaries.

The flood was remarkable in many respects. In the first place, it was preceded by a period of heavy precipitation, and consequent flood stages of all the streams, a condition which had prevailed intermittently for many preceding weeks. As a result, the earth was thoroughly

*The data upon which this paper is based were collected by the Water Resources Branch of the United States Geological Survey in co-operation with the State of California, and the paper is published by permission of the Director of the Survey.
Further acknowledgments are due to Mr. J. H. Scarr, the district forecaster of the United States Weather Bureau, and to the engineering department of the Southern Pacific Railway, for data furnished.
†This paper will not be presented at any meeting, but written communications on the subject are invited for publication with it in Transactions.

William Billings Clapp migrated with his parents to Southern California, settling on the Pasadena lands. He acquired his engineering education in the field, becoming in 1882 assistant in public land surveys of Northern California. From 1883 to 1886 he acted as assistant engineer for the construction of the Los Angeles & San Gabriel Valley Railroad, and from then made private work at Seattle WA. He in 1891 returned to Pasadena CA, serving there as its city engineer until 1900, during which period transition from the rural settlement to the modern city was initiated. Clapp was appointed assistant engineer in 1903 of the Geological Survey, from when he was identified until his illness with hydraulic studies of its California Bureau, thereby having advanced to district engineer.

Although Clapp had been in touch with many aspects of engineering, he particularly was inclined to deal with hydraulic problems. He was constantly in touch with water supply problems. Within a short time his reports and opinions were regarded as authoritative and were widely accepted. His principal contribution to the engineering society was the 1908 paper jointly written with Edward C. Murphy (1859-1934) and a colleague on the 1907 flood of the Sacramento River. The analysis of the disastrous flood involved lots of complications. The Sacramento floods have been a perplexing problem throughout California's history. Shortly before the 1907 flood extensive plans were proposed for reclaiming the enormous agricultural values of the Sacramento Basins. Evidence on the accuracy of this work was contained in a report prepared by a Board of Army Engineers. It contains the statement: 'It is thought that the estimates of Messrs. Clapp, Murphy and Martin should be followed closely in determining the necessary channel widths, and their maximum is assumed with certain allowances for flattening of the flood wave in passing down the improved channels'. These works had not yet been begun at Clapp's death, but the future hydraulic developments of that State were guided by his fundamental input. He was described as a person whose best qualities were exposed only to these who achieved his full confidence.

Anonymous (1912). Clapp, W.B. *Trans. ASCE* 75: 1148-1150.

Clapp, W.B., Murphy, E.C., Martin, W.F. (1908). The flood of March 1907 in the Sacramento and San Joaquin River Basins, California. *Trans. ASCE* 61: 281-376. (*P*)

La Rue, E.C., Henshaw, F.F., Clapp, W.B., Martin, W.F., Stevens, J.C. (1910). *Geological survey water-supply papers*. Government Printing Office: Washington DC.

CLARK C.M.

27.02. 1873 Potsdam NY/USA
06.04. 1945 Potsdam NY/USA

Charles Morton Clark was educated at the Potsdam Normal School, and entered the city service in 1893 with the Aqueduct Commission on surveys for the Croton Reservoir. Clark went through all positions from rodman to assistant engineer, in charge of the 33 m high and 360 m long Cross River masonry dam of the Croton Scheme in 1905. In 1908 he was transferred to the Board of Water Supply, and in 1912 made division engineer of the Croton Division, covering the construction of 20 km of the Catskill Aqueduct. 3 years later he was placed in charge of the southern aqueduct department between Peeskill and Hillview Reservoir, except for Kensico Dam. When the city went to the Schoharie watershed for additional supply, Clark served as division engineer from 1918 to 1921. He was then appointed department engineer in charge of the 28 siphons needed to bring the Catskill Aqueduct up to its full capacity. In 1927 he was made department engineer in charge of the 32 km City Tunnel No. 2. When the Delaware-Roundabout system was begun in 1936 he was made deputy chief engineer. He was in 1940 promoted to chief engineer of the New York Water Supply.

The Catskill Aqueduct, part of the New York City water supply system, brings water from the Catskill Mountains to Yonkers, where it connects to other parts of the system. Its construction was commenced in 1907 and completed in 1916. The 262 km aqueduct consists of 89 km of cut and cover aqueduct, 45 km of grade tunnel, 56 km of pressure tunnel, 10 km of steel siphon, and 63 km of conduit. The 67 shafts vary in depth from 50 to 362 m. Water flows by gravity through the aqueduct at 1.2 m/s. Its operational capacity of 2,100,000 m^3/day north of the Kensico Reservoir in Valhalla NY. Clark was shortly taught at school, so that he studied engineering at night over a correspondence course. In 1940, Clarkson College of Engineering, Potsdam NY, conferred on him an honorary degree of Doctor of Engineering.

Anonymous (1940). Charles Clark new head of New York water supply. *Engineering News-Record* 124(Apr.18): 552. *P*

Clark, C.M. (1942). Development of the Catskill supply. *Water Works Engineering* 95(21): 1268-1269; 95(21): 1292.

Clark, C.M. (1942). History of the Delaware supply. *Water Works Engineering* 95: 1276-1277.

http://genealogyandfamilyhistory.yuku.com/topic/470/Obituaries?page=1

http://en.wikipedia.org/wiki/Catskill_Aqueduct

CLARK R.R.

13.12. 1886 Pilot Grove IA/USA
01.06. 1977 Lake Oswego OR/USA

Roy Ross Clark obtained in 1911 the BSc degree in civil engineering from the Oregon State College, Corvallis OR, and in 1912 made in addition special courses at Cornell University, Ithaca NY. He then was from 1913 to 1917 assistant engineer at Portland OR. After war service in France, he was consultant engineer in Portland OR, dealing with over 200 projects in surveys and designs of irrigation systems. Clark was appointed in 1933 design engineer of the Bonneville Project on the Columbia River, in charge of its spillways. In 1939 he became design engineer of the US Engineers on the Willamette Basin Project in Oregon, for which he designed and constructed two flood control dams. During World War II he was head of the Design Department of air bases, and then returned to the design of flood control and multi-purpose dams, in particular contributing to the McNary Dam on the Columbia River, and four Snake River Dams in Idaho and Oregon States. He also served as a vice-president of the Board of Governors, the Oregon State College. He was a member of the American Society of Civil Engineers ASCE, and the Professional Engineers of Oregon.

Clark was mainly a designer and builder of irrigation schemes and dam constructions, so that he has written relatively few papers. Some deal with the optimum constituents of cements for dam design. The 1936 paper is an experimental study on the handling of large lift gates at Bonneville Dam. The Dam was designed for a maximum discharge of 45,000 m^3/s passing through 18 bays, each 16 m wide. The discharge is controlled with roller gates, each consisting of two sections raised or lowered by direct lift from a crane. Each gate section weighs 260 t. A hydraulic model at scale 1:10 was prepared to study the discharge characteristics, the pressure distribution, and the correct operation of the entire set-up.

Anonymous (1948). Clark, Roy R. *Who's who in engineering* 6: 357-358. Lewis: New York.
Clark, R.R. (1936). Spillway gate model supplements calculations. *Engineering News-Record* 116(Mar.05): 343.
Clark, R.R. (1937). *The value of Portland Puzzolan cement in mass concrete*. Oregon State College: Corvallis OR.
Clarke, R.R. (1956). Bonneville Dam stilling basin repaired after 17 years. *Journal of the American Concrete Institute* 27(8): 821-837.
http://records.ancestry.com/Ross_Clark_records.ashx?pid=73480770 *P*

CLAUSER F.H.

25.05. 1913 Kansas MO/USA
03.03. 2013 La Canada CA/USA

Francis Hettinger Clauser obtained his BS degree in 1934, and the PhD degree in aeronautics in 1937 from Caltech. He was for the next ten years engineer of aerodynamic design with Douglas Aircraft Corp., from 1946 to 1960 then professor of aeronautics and department chairman at Johns Hopkins University, Baltimore MD, when joining University of California, Santa Cruz CA, as professor of mechanics until 1964, and as professor of engineering and vice-chancellor until 1969. Then Clauser was until his retirement in 1980 professor of aeronautics and Clark B. Millikan engineering professor at Caltech, Pasadena CA. He was recipient of the Alumni Distinguished Scientific Award from Caltech in 1966, and was a Fellow AIAA, the American Physical Society, the American Academy of Arts and Science, and from 1970 Member of the National Academy of Engineering, with the citation 'For innovations in engineering research and education'.

At Johns Hopkins Clauser established the Department of Aeronautics. He published widely in the fields of aerodynamics, of non-linear mechanics, fluid dynamics and on the reduction of combustion engine emissions. His 1937 PhD thesis deals with the curvature effect on the transition from laminar to turbulent boundary layers, a topic in which his tutor Theodor von Karman (1881-1963) was interested since decades. The result was applied to flows over the upper surface of a wing. The discrepancy between the predicted and the actual transition point was due to the effect of streamline curvature, which may become relatively large due to the relatively small wing curvature radius.

Anonymous (1949). Francis H. Clauser. *Aeronautical Engineering* 8(3): 16. *P*

Anonymous (1954). Clauser, Francis H. *Who's who in engineering* 7: 441. Lewis: New York.

Anonymous (1994). Clauser, Francis H. *American men and women in science* 2: 325.

Clauser, F.H. (1954). Ramjet diffusers at supersonic speeds. *Journal American Rocket Society* 23(2): 79-84.

Clauser, F.H. (1956). The turbulent boundary layer. *Advances in applied mechanics*: 1-51, H.L. Dryden, T. von Karman, eds. Academic Press: New York.

Clauser, F.H. (1957). The structure of turbulent shear flow. *Nature* 179(4550): 60.

Clauser, F.H. (1960). Plasma dynamics. *Aeronautics and Astronautics*: 305-343, N.J. Hoff, W.G. Vincenti, eds. Pergamon Press: Oxford. *P*

Clauser, M., Clauser, F.H. (1937). The effect of curvature on the transition from laminar to turbulent boundary layer. NACA *Technical Note* 613. Washington DC.

CLAUSER M.U.

25.05. 1913 Kansas City MO/USA
26.01. 1980 Monterey CA/USA

Milton Ure Clauser, the twin brother of Francis H. (1913-2013), obtained his BS degree in 1934 and the PhD degree in aeronautics in 1937 from Caltech. He joined until 1938 as aerodynamic design engineer Douglas Aircraft Corp., was then involved in flight tests, power plant designs until joining from 1950 to 1954 Purdue University, West Lafayette IN, as head of its School of Aerodynamics. He also was member of the Propulsion Wind Tunnel Panel ARO Inc., at Tullahoma TN. From 1954 to 1957 he served as Secretary for Research and Development, US Dept. of Defence and in parallel directed the Aeronautical Research Staff of Ramo-Woolridge Corp. From 1956 to 1959 he was a member of the US SAF Scientific Research Board, from when he took over as chairman the Clauser Technical Corp., Torrance CA. He was from 1967 to 1970 Director of the Lincoln Laboratory, Lexington MA, where he was involved in advanced technology to problems of national security. Finally, Clauser accepted the position of Faculty Head at the Naval Postgraduate School, Monterey CA, as academic dean.

Clauser had originally a similar career development as his twin brother, but eventually became more involved in research management, whereas Francis stayed mainly in basic research. Clauser was from the 1950s mainly interested in Magneto-hydrodynamics MHD in which the dynamics of electrically-conducting fluids are studied. Examples include plasmas, liquid metals, and salt or water electrolytes. The word MHD is derived from the magnetic field affecting liquid movement, and was initiated by Hannes Alfvén (1908-1995), a Nobel Prize Awardee in 1970. The fundamental concept of MHD is that magnetic fields induce currents in a moving conductive fluid, which in turn creates forces on the fluid and also charges the magnetic field itself.

Anonymous (1949). Dr. Milton U. Clauser. *Aeronautical Engineering Review* 13(12): 38. *P*
Anonymous (1964). Clauser, Milton U. *Who's who in engineering* 9: 323. Lewis: New York.
Clauser, M., Clauser, F.H. (1937). The effect of curvature on the transition from laminar to turbulent boundary layer. NACA *Technical Note* 613. Washington DC.
Clauser, M.U. (1959). Magneto-hydrodynamics. *Space technology* 18: 1-21, H.S. Seifert, ed. Wiley: New York.
Clauser, M.U. (1960) The magnetic induction plasma engine. *Report* STL/TR-60-0000-00263. Physical Research Lab. Space Technology Laboratories: Los Angeles.
Clauser, M.U., Meyer, R.X. (1964). Magneto-hydrodynamic control system. *US Patent* 3,162,398.

CLEARY

16.06. 1906 Newark NJ/USA
31.03. 1984 Cincinnati OH/USA

Edward John Cleary obtained the BS degree in 1929, the MS degree in 1933, and the CE degree in 1935 from Rutgers University, New Brunswick NJ. He was from 1929 field engineer for an engineering corporation at Chicago IL, moving then to New York NY, taking over in 1932 as research assistant at Rutgers University in its Department of Water and Sewage. He was from 1937 to 1941 special lecturer at the College of Engineering, New York University, and was in parallel until 1949 executive editor of the Engineering News-Record journal. He then took over as executive director and chief engineer the Ohio River Valley Water Sanitation Commission, an interstate pollution control agency at Cincinnati OH. He also was a consultant of the Atomic Energy Commission, and a Lecturer at the College of Medicine, University of Cincinnati. He further was a trustee of the Research Foundation of the American Public Works Association APWA, whose president he was in 1951. Cleary was a member of the Rutgers Engineering Society, the Federation of Sewage and Industrial Wastes Association, the American Water Works Association AWWA, and the American Society of Civil Engineers ASCE.

When the Ohio River Valley Water Sanitation Commission ORSANCO was formed in 1948 to clean up America's dirtiest river valley, many persons were skeptical, because it was improbable that eight states could remain together long enough on a job requiring forthright action. In 1958 these skeptics were confounded as ORSANCO looked toward completion of one of its two big jobs. In 1948, the municipal sewage treatment plants existed only for 1% of the population, whereas in 1958 facilities were in service for 92%. The people who pay for these treatment projects want a river that doesn't smell, flows clearly, has no floating scum or solids, tastes good when treated, and is even suitable for swimming and fishing. It was largely due to the efforts of Cleary that these ambitious goals were attained after 10 years of service. He was a man of boundless enthusiasm and energy, giving new perspectives to his country.

Anonymous (1958). Cleanup man on Ohio River. *Engineering News-Record* 161(Oct.2): 54-61. *P*
Anonymous (1959). Cleary, Edward J. *Who's who in engineering* 8: 443. Lewis: New York.
Cleary, E.J. (1967). *The Orsanco Story*: Water quality management in the Ohio Valley under an Interstate Compact. Johns Hopkins Press: Baltimore MD.
Palange, R.C., Megregian, S., Cleary, E.J., Streicher, L. (1958). Monitoring of stream water quality. *American Water Works Journal* 50(9): 1211-1226.

CLEMENS

29.05. 1900 Marcellus MI/USA
27.01. 1958 Leopoldville/BC

George Reginald Clemens made engineering studies at the University of Michigan, Ann Arbour MI, from where he graduated in 1921. He joined the US Army Corps of Engineers USACE staying there all through his professional career. He was employed in the late 1920s as district engineer, engaged on surveys of Red River and its tributaries to improve the combined purposes of navigation, water power, flood control, and irrigation. He eventually became transportation specialist in river navigation, devoting his knowledge to the Mississippi River. He was in 1952 appointed assistant chief of the USACE Transportation Section. Clemens was a senior engineer of the Mississippi River Commission, killed in Belgian Congo following an airplane accident.

Clemens devoted his professional career almost entirely to the improvement of the Mississippi River. The 1935 paper is concerned with the analysis of flood control by means of reservoirs. The design of a reservoir system is analyzed by discussing the (1) Selection of the method of operation, (2) Test of the method, (3) Transmission of reservoir reduction to area to be protected, and (4) Examples of storage operation. The 1937 paper gives an account of the Memphis Engineer District's flood fight who helped build mud boxes at the Cairo IL flood wall. The 1938 paper deals with the effect of cutoffs on the flooding characteristics along Mississippi River. Measurements made between 1936 and 1938 indicate that the flood crests can be lowered, as was also demonstrated during the Great 1937 Flood, described by Julian L. Schley (1880-1965). During the two years almost 500 km of meandering river reaches were reduced to only 200 km. The paper describes these works and the measures to be taken.

Anonymous (1921). George R. Clemens. *Michiganensian yearbook*: 113. Univ. Michigan. *P*
Anonymous (1929). George R. Clemens. *The Michigan Technic* 42(4): 20.
Anonymous (1958). George R. Clemens. *Civil Engineering* 28(6): 482.
Clemens, G.R. (1935). The reservoir as a flood-control structure. *Trans. ASCE* 100: 879-927.
Clemens, G.R. (1936). Straightening the father of waters. *Engineering News-Record* 116(Feb.20): 269-276.
Clemens, G.R. (1937). The Mississippi meets the 1937 flood. *Civil Engineering* 7(6): 379-383.
Clemens, G.R (1938). Cutoffs lower flood crests. *Engineering News-Record* 121(Nov.17): 608-614.
Clemens, G.R. (1940). Improving Mississippi River navigation. *Civil Engineering* 10(6): 359-361.

CODE

22.11. 1865 Saginaw MI/USA
28.03. 1951 Hollywood CA/USA

William Henry Code received from 1888 to 1890 his engineering education from the University of Michigan, Ann Arbor MI, graduating in 1892 with the BSc degree in civil engineering. He was first assistant state engineer of Wyoming, then went to Arizona as chief engineer of the Consolidated Canal Co., in the Salt River Valley. His experience and knowledge in irrigation projects paired with political connections was particularly valuable. In 1902 the Reclamation Act passed the Congress, providing federal largeness for irrigation projects on federal land throughout the West. From 1902 to 1911 Code was chief engineer of irrigation for the US Indian Service in Arizona, and there also special agent on irrigation investigations for the US Department of Agriculture. In the Gila River Valley, where the Pima Tribe consisting of some 1,000 people had long used the waters of the river, problems with the water arose, because settlers had ignored the water rights. The Pimas were the first recipients of water from a canal, although Code argued for electric well pumps. He believed that the Pima water rights could not be restored. He further did not state that the canal to be built from Roosevelt Dam was charted along a higher elevation than needed for the Pimas.

In 1911 Code was one of the three advisory engineers selected to report on the method of water distribution of the Los Angeles Aqueduct. In 1912 Code formed a partnership of consulting civil engineering at Los Angeles CA. Two years later another partner joined the firm so that its name was Quinton, Code & Hill. Code continued as member of this partnership until 1936 when he retired from active service. During these years the firm handled many projects in Canada and Mexico, as well as in the USA, mainly in the field of hydraulics. Code continued to specialize in irrigation works. Although he resided at Los Angeles, he never lost touch with the irrigation problems of Arizona State. Primarily an engineer, he was also interested in arts and civil affairs. He was socially inclined and his sense of humour was keen. He was member of the American Society of Civil Engineers ASCE from 1907, and became a Life Member in 1936.

Anonymous (1952). William H. Code. *Trans. ASCE* 117: 1296.
web.mst.edu/.../History-Geotech%20Firms%20Los%20Angeles.doc
www.gilariver.com/lessons%5Cwaterlosslesson1...
http://chandlerpedia.com/Exhibits/Chandler_and_the_Greening_of_the_Desert/The_Consolidated_Canal_Company
http://oldhomesoflosangeles.blogspot.ch/2012/10/william-h-code-1729-whitley-ave.html *P*

COLBY B.H.

20.07. 1854 Cherry Valley OH/USA
03.01. 1933 Normal IL/USA

Branch Harris Colby graduated from the University of Michigan, Ann Arbor MI, in 1877 with the BSc degree in civil engineering. He first worked as assistant engineer on the triangulation of the Great Lakes, from when he was for five years US Assistant Engineer with the Mississippi River Commission. From 1884 he engaged in private practice. In 1889 he was appointed by Robert E. McMath (1833-1918), then city engineer of St. Louis MO, for the perfection of its drainage system. Colby thereby designed a slide-rule, which received attention from surveyors in the USA. He also designed and built several large trunk sewers, of which the largest was of horse-shoe shape, 4 m high and 5 m wide. He was further involved in other projects for the improvement of the life conditions at St. Louis, mainly in an improved sewer system.

In 1917 Colby was appointed engineer on steel ship building at Hog Island PA. Here he learned ship construction and then was sent to Chicago IL, to supervise its manufacture. In 1918 he moved to San Francisco CA, to construct the Government Island Shipyard, where it was proposed to build concrete ships. He was then recalled to Philadelphia PA as Board Member to investigate and report on the condition and operation of shipyards along the Atlantic Coast. He was sent as resident engineer to Jacksonville FL in charge of constructing the dry dock, the marine railway, and a repair plant. Finally, Colby was engaged as engineer in 1920 on the sewerage system of Muscle Shoals AL, during which he suffered an almost fatal attack of angina pectoris, after which he was obliged to give up all hope of resuming active practice. During his last 12 years he went several times to New York NY there enjoying a visit to Mackinac Island. His engineering work covered a wide range of experiences, including surveys and triangulation, dredging work, design and construction of sewers, building a shipyard, a dry dock, and not to speak of countless minor activities. He was president of the Engineers' Club of St. Louis, and vice-president of the American Society for Municipal Improvements. He was elected member of the American Society of Civil Engineers ASCE in 1895.

Anonymous (1933). Branch H. Colby. Trans. *ASCE* 98: 1519-1522.
Colby, B.H. (1893). Recent surveys of St. Louis. *J. Association of Engng. Societies* 2(3): 490.
Colby, B.H. (1903). Pollution of streams with special reference to the Chicago Drainage Channel. *Journal of the Association of Engineering Societies* 24(2): 137-145.
Welsch, B. (2013). *Branch H. Colby*. The Engineers' Club: St. Louis MO. P

COLBY B.R.

07.08. 1908 Roseville MN/USA
21.08. 1978 Hubbard County/USA

Bruce Ronald Colby was educated at Minnesota and California Universities as agricultural engineer. He joined in 1934 the US Geological Survey USGS which from its formation rather sought to improve the methods of measurement and analysis on an individual basis than as a concerted agency effort. Stations of his career included positions at St. Paul MN, Boston MA, Iowa City IA, as well as Nebraska State. Three individualists there contributed mainly to the mechanics of sedimentation and sediment transport, namely William W. Rubey (1898-1974), Thomas Maddock, Jr. (1907-1991), and Colby.

When looking at Colby's publications, it becomes obvious that he considered journal papers as unsuitable for his works. Instead, most of his research is published in Reports of the USGS or the US Department of Agriculture. This characteristics was reflected in his personality: A person who did not want to generalize findings but applied these to detailed studies, both in time and in location. Colby also exclusively worked in the narrow field of sediment mechanics, yet his knowledge was so profound that he counted among the then top American experts. He further took interest into the relation of sediments and the surface water chemistry, the effect of sediment on scour, or the effects of sediment transport on the velocity patterns in sand-bed streams.

Anonymous (1930). Bruce R. Colby. *Gopher yearbook*: 84. Minneapolis MN. *P*
Anonymous (1960). Colby, Bruce R. *American men of science* 10: 720. Cattell: Tempe AZ.
Colby, B.R. (1948). Sediment discharge, methods of computation. *Water Resources Bulletin* (Feb.10): 34-39.
Colby, B.R., Hembree, C.H., Rainwater, F.H. (1956). Sedimentation and chemical quality of surface waters in the Wind River Basin, Wyoming. *Water-Supply Paper* 1373. USGS.
Colby, B.R. (1960). Discontinuous rating curves for Pigeon Roost and Cuffawa Creeks in North Mississippi. *Agricultural Research Service* 41-36. US Dept. Agriculture: Washington DC.
Colby, B.R. (1963). *Fluvial sediments*: a summary of source, transportation, deposition, and measurement of sediment discharge. US Government Printing Office: Washington DC.
Colby, B.R. (1964). Scour and fill in sand-bed streams. *Prof. Paper* 462-D. USGS: Washington.
Colby, B.R. (1964). Discharge of sands and mean velocity relationships in sand-bed streams. *Professional Paper* 462-A. USGS: Washington DC.
Colby, B.R., Scott, C.H. (1965). Effects of water temperature on the discharge of bed material. *Professional Paper* 462-G. USGS: Washington DC.

COLE E.S.

29.12. 1871 Washington DC/USA
18.03. 1950 Upper Montclair NJ/USA

Edward Smith Cole studied at University of Illinois from 1890 to 1892, obtaining his MS degree from Cornell University in 1894 and the PhD degree in aeronautics in 1937. He was until 1903 principal assistant in his father's engineering office, Chicago IL. He was then active in charge of studies for the Department of Water-Supply, Gas and Electricity of New York City. In 1905 he was the founder of the *Pitometer* Company, New York, which was in 1920 expanded to the British Pitometer Company Ltd. with Kennedy, London and Kilmarnock UK. Cole was Fellow of the American Society of Mechanical Engineers ASME, received their 1948 Worcester Reid Warner Medal, of the American Society of Civil Engineers, and Honorary AWWA Member.

Cole all through his career developed the Pitot tube to measure air flow velocities. This basic instrument in hydraulics and aerodynamics was developed by Frenchman Henri de Pitot in 1732, bringing him lasting recognition. His device consists of two parallel tubes mounted on a slender frame containing a scale and four petcocks, one of the tubes being straight whereas the other bent by 90° at its lower end. After immersion of the base into flowing water, or into air flow to a certain location, the cocks would be closed, the instrument withdrawn to note the difference in elevation of the two columns of water, to determine the velocity head. Today, the Pitot tube is still widely used in hydraulic laboratories for precise velocity measurement, provided the velocity head is larger than some 5 cm. Many modern airplanes also use Pitot tubes for speed control. Cole was particularly interested in the speed determination of ships.

Anonymous (1941). Edward S. Cole. *Water Works Engineering* 94(Jun.18): 796. *P*
Anonymous (1949). Edward Smith Cole. *Mechanical Engineering* 71(1): 83. *P*
Anonymous (1951). Cole, Edward Smith. *Who's who in America* 26: 525. Marquis: Chicago.
Anonymous (1954). Cole, Edward Smith. *Who's who in engineering* 7: 441. Lewis: New York.
Cole, E.S. (1902). Measurement of flow of water in pipes. *Sibley Journal of Engineering* 16(8): 335-339.
Cole, E.S. (1907). The pitotmeter. *Journal of the Franklin Institute* 164: 425-441.
Cole, E.S. (1935). Pitot tube practice. *Trans. ASME* 57(HYD-8): 281-294.
http://209.85.129.132/search?q=cache:XpNEQREihUoJ:www.findagrave.com/cgi-bin/fg.cgi%3Fpage%3Dgr%26GRid%3D10721+%22Edward+Smith+Cole%22+1871-&cd=1&hl=de&ct=clnk&gl=ch

COLE J.D.

02.04. 1925 Brooklyn NY/USA
17.04. 1999 Albany NY/USA

Julian David Cole graduated as mathematician from Cornell University in 1944, and there took his PhD degree in 1949. He was from 1951 to 1959 assistant professor of aeronautics and applied mechanics at the Caltech, Pasadena CA, then until 1969 professor. From 1969 to 1976 he was professor and chairman of the Department of Applied Mechanics, University of California UCLA, Los Angeles CA, and until 1982 professor of mathematics. From then until his retirement in 1990, he was a professor of applied mathematics, Rensselaer Polytechnic, Troy NY. He was awarded the 1984 von Karman Prize from the Society of Industrial and Applied Mathematics SIAM. He was member of the American Institute of Aeronautics and Astronautics AIAA, receiving its 1992 Fluid and Plasma-dynamics Award.

Cole is remembered professionally for his seminal research and his books Perturbation methods, and Aerodynamics. He bridged mathematics and thereby added significantly to engineering problems. It was a commitment to the scientific formulation and mathematical solution of relevant problems rather than to develop mathematics for its own sake. As his teacher Theodor von Karman (1881-1963), Cole was mostly attracted by nonlinear problems in aerodynamics and transonic flow, having the mathematical tools to devise a solution. Cole found for instance in the 1950ies that weak shocks are described by the current Burgers' equation. He found the exact solution of the Navier-Stokes equation using an ingenious transformation to the linear diffusion equation known as the Cole-Hopf formulation. Students of Cole stated that they were amazed by his ability to understand quickly the essence of a mathematical problem, even on the blackboard, dealing for instance with nonlinear mixed transient flow. All through his career he was an open mind to new approaches, paired with a remarkable intuition.

Anonymous (1995). Cole, Julian D. *Who's who in America* 49: 718. Marquis: Chicago.
Bluman, G., Cook, L.P., Flaherty, J., Kevorkian, J., Malmuth, N., O'Malley, R., Schwendeman, D.W., Tulin, M. (2000). Julian D. Cole. *Notices of the AMS* 47(4): 466-473. *P*
Cole, J.D. (1968). *Perturbation methods in applied mathematics*. Blaisdell: Waltham MA.
Cole, J.D. (1975). Modern developments in transonic flow. *Journal Applied Math.* 29: 763-787.
Cole, J.D., Cook, L.P. (1986). *Transonic aerodynamics*. North-Holland: Amsterdam.
http://209.85.129.132/search?q=cache:XpNEQREihUoJ:www.findagrave.com/cgi-bin/fg.cgi%3Fpage%3
        Dgr%26GRid%3D10721+%22Edward+Smith+Cole%22+1871-&cd=1&hl=de&ct=clnk&gl=ch

COLEMAN J.F.

23.11. 1866 Church Hill MS/USA
03.06. 1944 New Orleans LA/USA

John Francis Coleman started his professional career with works for railroad companies. In 1889 he began an apprenticeship in hydraulic engineering, becoming assistant engineer for the US Engineers in charge of observations on the Mississippi River, an assignment which laid the foundation for his later experience on river control. After various engineering works, he became in 1900 consultant of his firm which was dissolved in 1903, however. At this time there was a change in the transportation facilities on Mississippi River from steamboat to modern means, with New Orleans as the seaport. Coleman was responsible for the design and construction of harbour facilities, which served subsequently as standard for other US ports. He was occupied in the 1910s with varied engineering projects and also became active in the support of the Allied Forces during World War I.

Coleman returned in 1919 to New Orleans LA and there opened with his son the J.F. Coleman Engineering Company. He served there as consultant of various ports on the Atlantic Ocean, and was retained by the Board of Commissioners of the Port of New Orleans as consulting engineer. He was sent to Europe to inspect major ports there to formulate a policy for the administration of the Inner Harbor Navigation Canal at New Orleans, connecting Mississippi River with Lake Pontchartrain. He also was consultant on river hydraulics for the diversion of Brazos River in Texas, the control of Mississippi River at Memphis TN, or the control of Pearl River near Columbia MS. Following the record floods in 1927 on Mississippi River, the US Engineers formulated the Jadwin Plan to mainly protect New Orleans from further hazards. The construction of the Carre Spillway was proposed, with Coleman as the design engineer. Of particular note are also his services toward the American Society of Civil Engineers ASCE. He served as director from 1915 to 1917, as vice-president in 1918, and as president in 1930. He was elected Honorary Member in 1935. Coleman was an engineer who possessed the ability to apply common sense to all problems. He was outspoken in his views and tenacious in his opinions, broad-minded in his judgment, and loyal to his friends.

Anonymous (1922). John F. Coleman. *Engineering News-Record* 89(8): 330. *P*
Anonymous (1945). John F. Coleman. *Trans. ASCE* 110: 1608-1615.
Coleman, J.F. (1923). The port of New Orleans. *Proc. ASCE* 49(6): 1167-1176.
Coleman, J.F. (1924). The river and harbor problems of the Lower Mississippi: The port of New Orleans. *Trans. ASCE* 88: 1015-1024.

COLEMAN N.L.

03.09. 1930 Belvidere IL/USA
09.07. 1994 Oxford MS/USA

Neil Lloyd Coleman received education at Cornell College, Mount Vernon IA, from where he received in 1952 the BS degree, at University of Chicago, Chicago IL, receiving the MS degree in 1957, and the PhD degree in 1960 as a geologist. He was from 1961 associate professor of civil engineering at the University of Mississippi, Oxford MS, directing its Sedimentation Laboratory from 1981 to 1988. From then he was a consulting geologist. Coleman was member of the American Geophysical Union AGU, and of the International Association of Hydraulic Research IAHR. He was awarded the 1983 IAHR Schoemaker Award for the best paper published in its Journal of Hydraulic Research.

Coleman was interested all through his career in soil erosion, and sediment transport and deposition. One of his first important papers written in 1969 with a colleague deals with the data unification of sediment transport using hydraulic similitude. The resulting parameters include the Froude and the Reynolds numbers, and a transport similitude number accounting for the effects of inertia, gravity, and viscous forces. The unified correlation of sediment discharge data originated from numerous laboratory flume studies and a systematic data analysis. In the 1986 paper written with Gary Parker (*1950) a theoretical model of sediment-laden flow is applied to dilute open channel suspensions. The effect of sediment is manifested in terms of a reduced depth and coefficient of resistance, and an increased flow velocity. The theoretical results were found in agreement with observations. Another prediction concerns the turbulence level. The ratio of the power consumed by the flow in holding sediment in suspension to the power supplied to the flow by the work of the weight on the sediment determines a ratio. If this ratio is smaller than unity because of fine sediment, turbulence is intensified.

Coleman, N.L. (1972). The drag coefficient of a stationary sphere on boundary of similar spheres. *La Houille Blanche* 27(1): 17-22.
Coleman, N.L. (1976). *Low sill tested for total sediment-load measurement*. ARS: New Orleans.
Coleman, N.L. (1981). Velocity profiles with suspended sediment. *J. Hydr. Res.* 19(3): 211-229.
Coleman, N.L. (1992). Modified Cebeci-Smith model. *J. Hydr. Res.* 30(4): 555-568. *P*
Parker, G., Coleman, N.L. (1986). Simple model of sediment-laden flows. *Journal of Hydraulic Engineering* 112(5): 356-375.
Willis, J.C., Coleman, N.L. (1969). Unification of data on sediment transport in flumes by similitude principles. *Water Resources Research* 5(6): 1330-1336.

COLES D.E.

08.02. 1924 St. Paul MN/USA
02.05. 2013 Altadena CA/USA

Donald Earl Coles graduated in 1947 as aeronautical engineer from the University of Minnesota, St. Paul MN. In 1948 he went to Caltech, obtaining there the PhD title in 1953. After having worked in its Jet Propulsion Laboratories, he was from 1953 to 1956 a research fellow at Caltech, and there from 1964 professor of aerodynamics until retirement in 1996. In parallel he also was a consultant for the industry from 1954. Coles was a member of the National Commission of Fluid Mechanics Films from 1960, of the American Institute of Aerodynamics and Astronautics AIAA, being the recipient of its 1953 Sperry Award for his PhD thesis, and a member of the American Physical Society APS.

Coles is mainly remembered for his outstanding research in the 1950s relating to the turbulent boundary layer. He was interested in the turbulent skin friction on smooth plates, with a large potential to practical applications in the airplane and naval industries, but also in all other high-speed mobiles from cars and trains to rockets. In the 1960s he became interested in jet noise, a problem that was partially solved only recently, after jet airplanes had produced excessive noise mainly in urban environments. Coles in these years also produced a total of 23 educational films mainly relating to problems in aerodynamics. Later, he was member of the Organizing Committee, Stanford Conf. on Computation of Turbulent Boundary Layers. His major papers then dealt with similarity laws for turbulent flow, shock tube design, transition in circular Couette flow, stalling of airfoils, vortex shedding from cylinders, and coherent structures in turbulence.

Anonymous (1954). Dr. Donald Coles. *Aeronautical Engineering Research* 13(3): 45. *P*
Anonymous (1970). Coles, Donald Earl. *Who's who in America* 36: 442. Marquis: Chicago.
Coles, D. (1954). The problem of the turbulent boundary layer. *Zeitschrift für Angewandte Physik* 5(3): 181-203.
Coles, D. (1954). Measurements of turbulent friction on a smooth plate in supersonic flow. *Journal of the Aeronautical Sciences* 21(7): 433-448.
Coles, D.E. (1956). The law of the wake in the turbulent boundary layer. *Journal of Fluid Mechanics* 1: 191-226.
Coles, D. (1957). Remarks on the equilibrium turbulent boundary layer. *Journal of the Aeronautical Sciences* 24(7): 495-506.
Coles, D. (1963). Estimating jet noise. *The Aeronautical Quarterly* 14(2): 1-16.
http://www.caltech.edu/content/donald-coles-0 *P*

COLES J.S.

03.06. 1913 Mansfield PA/USA
13.06. 1996 Falmouth MA/USA

James Stacy Coles graduated from the Columbia University, New York City NY, with the AM degree in 1936, and the PhD degree in physical chemistry in 1939. He then joined the Middlebury College until 1943 as assistant professor, was research group leader until 1945 of the Underwater Explosives Research Lab, Woods Hole Oceanography Institute of the Massachusetts Institute of Technology MIT, joining then the staff from assistant to associate professor of Brown University, Providence RI until 1952. From then until 1968 he was the president of Bowdoin College, Brunswick ME, and until 1982 the president of the Research Corporation based in New York, a foundation for the advancement of science, from where he retired. During his occupancy, the foundation's assets increased from 11 to 46 million US$. Coles had a large number of honorary degrees including the LLD degrees from Brown University presented in 1955, from Bowdoin College in 1968, and the DSc degree in 1958 from the University of New Brunswick, Fredericton CA.

Under Coles' leadership at Bowdoin College, its science department modernized, programs for gifted students were implemented, and an independent studies program was initiated. Coles' presidency also witnessed a major building program, including the Coles Tower, Wentworth Hall and Chamberlain Hall in 1964. Inaugurated in 1964, Coles Tower was for many years the tallest building in Maine. It is a sixteen-story brick residence housing some 200 students in quad suites. Coles was chairman of the Maine Higher Education Advisory Committee from 1965 to 1967, which helped establish the University of Maine System. He further co-authored the book Physical principles of chemistry. The James Stacey Coles Papers include the personal, business and Brown University and Bowdoin College correspondence. Lecture notes, research material and writings both as manuscripts and published are also included. Coles may therefore be considered an outstanding leader of American education during the 1960s and 1970s.

Anonymous (1994). Coles, James Stacey. *American men and women in science* 2: 397.
Cole, R.H., Coles, J.S. (1965). *Physical principles of chemistry*. Freeman: San Francisco.
Coles, J.S. (1994). *Papers* 1936-1993. Woods Hole Oceanographic Institution: Woods Hole MA.
http://library.bowdoin.edu/arch/archives/jscg.shtml.
http://en.wikipedia.org/wiki/James_S._Coles
http://dla.whoi.edu/manuscripts/node/164641#id2636522 *P*

COLGATE

13.01. 1916 Canyon City CO/USA
11.04. 2011 Longmont CO/USA

Donald (Mike) Ben Colgate obtained the BS degree from Colorado State University, Fort Collins CO in 1942. He joined after war service the US Bureau of Reclamation USBR, Denver CO, staying there all through his professional career, thereby eventually becoming its expert in cavitation.

Cavitation damage is a serious concern in hydraulic engineering because any structure subjected with a high-speed water flow may be damaged and will be ultimately 'lost'. Small streamline curvature along a structural boundary causes a significant reduction of pressure, causing a transition of fluid to vapor flow. Once the boundary returns its curvature, positive pressure establishes, so that the vapor flow abruptly returns to fluid flow, associated with bubble implosion. The resulting pressure peaks may be so large that the structural integrity of the boundary material is attacked, resulting in massive structural damage if the pressure load is sufficiently high or long, as during a flood. Cavitation damage was first observed at Grand Coulee Dam in the late 1940s, where both the discharge and the velocity of the spillway flow are high. Because these loads were of design interest, means had to be developed to counter cavitation damage. After decades of research all over the world, the USBR was among the first to propose flow aeration as a simple but highly-effective approach. Colgate was eminently involved into these developments, although his colleagues Alvin J. Peterka (1911-1983), who indeed made the first aerator tests, and James W. Ball (1905-2001), Colgate's boss, are often credited for these works. Colgate was however the engineer who transferred Peterka's concepts into practice by air supply using specially-designed aerators.

Borden, R.C., Colgate, D., Legas, J., Selander, C.E. (1971). Documentation of operation, damage, repair and testing of Yellowtail Dam Spillway. *Report* ERC-71-23. USBR: Denver.
Colgate, D. (1959). Cavitation damage of roughened concrete surface. *Journal of the Hydraulics Division* ASCE 85(HY11): 1-10.
Colgate, D. (1965). *Resistance of selected protective coatings for concrete to high-velocity water jets*. US Bureau of Reclamation: Denver.
Colgate, D., Legas, J. (1972). Aeration mitigates cavitation in spillway tunnel. Conf. *National Water Resources Engineering* Atlanta: 1571-1605. ASCE: New York.
Falvey, H.T. (2012) Donald Colgate. Personal communication. *P*
Lanning, C.C. (1973). High-head gates and valves in the United States. *Journal of the Hydraulics Division* ASCE 99(HY10): 1727-1775.

COLLES

09.05. 1739 Dublin/IR
05.10. 1816 New York NY/USA

Christopher Colles was orphaned at an early age. His uncle employed him on a canal near Kilkenny in the late 1750s, where he stayed until 1770. Colles left Cork in Ireland with his family, arriving in 1771 at Philadelphia. Despite his bright intellect he had difficulty making a living. In 1773 he was asked to design a steam engine, which he made only with a partial success. Keeping in mind that Watt's patent was only issued in 1769, Colles was credited with the construction of the first American steam engine. Moving to New York just before the Revolution, he designed municipal water supply systems, but was not properly paid due to war actions. After the war he was bursting with ideas for canals and river works. He offered Washington to remove obstructions of River Ohio in 1783. A year later he promoted the development of Mohawk River thereby conducting an engineering survey. Again, he was premature, because works for the Erie Canal were started only 32 years later. During its opening in 1825, Colles' effigy was carried in the parade.

In the early 1790s Colles was engaged to engineer the bypass canals at South Hadley and Turners Falls on Connecticut River. This project survived the 18th century. Colles' most fantastic proposal was a timber flume crossing Northern New Jersey and connect New York Bay with the Delaware River. This work was conceived around 1800 and received publicity, but again little financial support. Before he died, penniless and alone save for few old friends, the brilliant mind of the luckless engineer produced ingenious inventions. He for instance displayed in 1812 the solar microscope. Colles is known also for A survey of the roads of the United States of America, which he published in 1789. The survey consists of 83 sheets covering the main routes from Albany NY to Williamsburgh VA, including sections in Connecticut, Pennsylvania, Maryland and Delaware. Colles lies in an unmarked grave in St. Paul's Episcopal Churchyard in New York City.

Fitzsimons, N. (1967). Christopher Colles, engineer extraordinary. *Civil Engineering* 37(9): 72.
Levitt, A. (2009). Christopher Colles: America's first steam-engine builder. *Intl. Journal for the History of Engineering and Technology* 79(2): 245-279. *P*
Popper, D.E. (2005). Poor Christopher Colles: An innovator's obstacles in early America. *Journal of American Culture* 28(2): 178-190.
http://de.wahooart.com/A55A04/w.nsf/Opra/BRUE-8CEFL7 *P*

COLLINGS

07.07. 1888 Beaver City NE/USA
21.11. 1941 Los Angeles CA/USA

William Tatem Collings Jr. graduated from the high school of his city in 1906. He was then employed by the US Indian Service in Los Angeles CA. From 1904 he worked for the harbour district, and on the Los Angeles Aqueduct. He was city engineer from 1916 at Superior NE, and in 1917 joined the US Reclamation Service on the Rio Grande Project in New Mexico, first on canal location, and then on its construction. He had then for the next two years his private practice at Las Cruces NM, where he acted also as city engineer. He became in 1921 a project engineer for the New Mexico State Highway Dept., but moved in 1923 to Los Angeles as field engineer, engaged in municipal improvement work. In 1927 he accepted the position of engineer with the Imperial Irrigation District, engaged in the installation of a large irrigation system near Niland CA. He later also supervised the construction of the distribution system, and was active in studies of the drainage requirements of the district.

The greatest contribution of Collings in the engineering field was from 1929 to 1938 in connection with the All-American Canal Project. He was employed by the US Bureau of Reclamation USBR in field investigations for the canal. He was then sent to Denver CO as the representative of the Imperial Irrigation District, so that this project became his in the next seven years. A review of his correspondence shows his capacity for work, his understanding of the problems, and his versatility. His reports range from hydrographic studies to financial details for the project. In 1937 he became assistant to the chief engineer at Imperial CA, involved with the construction of the hydro-electric plants of the Canal, but his health began to fail in 1938, so that he retired from this position. In retrospect, it was remembered that throughout the strenuous period devoted to the construction of the Canal, he was suffering from the disease which finally caused his death. He had made a remarkable work both as designer and builder. He was ASCE member from 1931.

Anonymous (1942). William Tatem Collings. *Civil Engineering* 12(1): 70.
Anonymous (1943). William T. Collings, Jr. *Trans. ASCE* 108: 1568-1570.
Collings, Jr., W.T. (1934). Discussion of A problem of soil in transportation in the Colorado River. *Trans. ASCE* 99: 550-555.
http://www.mytrees.com/ancestry/Nebraska/Born-1888/Co/Collings-family/William-Collings-hu000981-40426.html
http://netmole.blogspot.ch/2008/09/scenes-from-bureau-of-reclamation.html (*P*)

COLLINGWOOD

10.06. 1834 Elmira NY/USA
18.08. 1911 Avon-by-the-sea NJ/USA

Francis Collingwood entered at age 17 Rensselaer Polytechnic Institute, Troy NY, graduating with the class of 1855. Among his contemporaries there was Washington A. Roebling, son of John A. Roebling (1806-1869), who was later engineer of Brooklyn Bridge. From 1865 to 1869, Collingwood was city engineer of Elmira NY, and from then until 1883 assistant engineer in the construction of Brooklyn Bridge. During this period he became member of the American Society of Civil Engineers ASCE. He was not only a frequent discusser of papers and often attended ASCE Meetings, but he also wrote various papers in its Transactions. Among these is the 1885 paper entitled Preservation of forests, in which he summarized the forestry operations up to that time and called attention to the necessity of economical methods of lumbering if the forests were to be saved from rapid extinction.

From 1883 Collingwood was engaged to supervise repairs to the Allegheny Suspension Bridge between Pittsburgh PA and Allegheny PA. He was appointed chief engineer in 1887 of the Chesapeake Dry-Dock & Construction Co., Newport News VA, returning in 1888 to New York City. He was there appointed member of the New Croton Aqueduct Commission, which reported on the reservoir and aqueduct work recently completed then. He also served Elmira NY in examining the Chenung River to devise means for protecting the city from floods. Collingwood was in 1891 elected ASCE Secretary and served in this position until 1894. Until his death he was then expert examiner for New York City, Lecturer on foundation techniques at New York University, and consultant. He was awarded the Telford Medal by the Institution of Civil Engineers ICE, London UK, for a paper describing the repairs at Allegheny Bridge, as also the Trevithick Medal. He was director ASCE from 1873 to 1876, and the founder of the Collingwood Prize for Juniors, one of the eldest decorations of ASCE, instituted in 1894.

Anonymous (1911). Francis Collingwood. *Engineering News* 66(8): 242; 66(9): 269. *P*

Collingwood, F. (1872). Experiments on the power of water to transport sand in sluices. *Trans. ASCE* 1: 246-266.

Collingwood, F. (1874). The foundations of the Brooklyn anchorage of the East River Bridge. *Trans. ASCE* 3: 142-146; 4: 205-210.

Collingwood, F. (1876). Notes on the masonry of the East River Bridge. *Trans. ASCE* 6: 7-34.

Collingwood, F. (1890). *The protection of the city of Elmira NY against floods*. Elmira NY.

COLLINS

14.12. 1868 Narragansett RI/USA
19.09. 1937 Clarksburg WV/USA

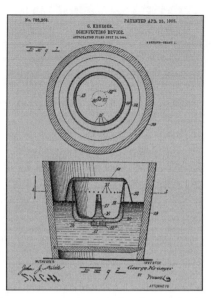

Clarke Peleg Collins completed the civil engineering course at the International Correspondence School, Scranton PA. In 1888 he joined the city engineer of Providence RI, and from 1890 to 1900 was employed as assistant engineer in the Mining Engineering and Survey Department of the Cambria Iron Company, Johnstown PA. He was engaged to study the water supply for its Water Company, and designed the No. 2 Dam on Mill Creek for the Johnstown Water Co., whose capacity was 340,000 m^3. After having been engaged in private practice, railways projects and municipal engineering, he became in 1912 special engineer for a coal mine, designing treatment plants for water supply and the disposal of sewage at Windber PA. From 1914 to 1918 he was employed at Johnstown as sanitary engineer. His plans for the sewage system were approved by Rudolph Hering (1847-1923), serving for a population of 170,000. Until 1920 Collins acted as senior engineer, the US Shipping Board, Washington DC, in charge of sewerage, and sewage disposal. He then moved to Clarksburg WV, engaged in private practice until his death.

In 1920 Collins was engaged by the Mountain Water Supply Company, Philadelphia PA, to study the pollution of its dam on Indian Creek by mine drainage. His report was the first relating to the pollution of streams by mines. Since that time extensive work was done in mine sealing to prevent stream pollution. In 1927 he was retained by Clarksburg to value the plant and holdings of the City Light and Heat Company in presenting its case to protestants before the West Virginia Public Service Commission. In 1933 and 1934, Collins prepared and submitted the data necessary for a large loan from the Public Works Administration for a sewerage system for Clarksburg; this grant was however not accepted. Collins has written several technical papers, namely on The Engineer on Development work, A comparison of the combined and separate systems of sewerage, or on Drainage and sewerage. He was described as accurate, painstaking, possessing a high conception of ethics in his profession. He was ASCE member from 1916.

Anonymous (1938). Clarke P. Collins. *Trans. ASCE* 103: 1766-1770.
Collins, C.P. (1898). A comparison of the combined and separate systems of sewerage. *Municipal Engineering* 15(4): 222-228. (*P*)
Collins, C.P. (1923). Pollution of water supplies by coal mine drainage. *Engineering News-Record* 91(16): 638-641.
Collins, C.P. (1935). *Map of the Metropolitan District* of Clarksburg WV.

CONE

31.08. 1883 Chanute KA/USA
15.07. 1970 Tucson AZ/USA

Victor Mann Cone obtained the civil engineering degree from University of Kansas in 1914. He was engaged from 1906 with flood control, irrigation and drainage works first at the US Department of Agriculture, then with flood control and navigation works on Mississippi River in the Memphis District to eventually become Chief of the Engineering Division, in charge of all investigations, reports, plans and specifications for extensive programs of the US Corps of Engineers until 1937. Until 1940 Cone then joined the Southwestern Division of the Corps at Little Rock AR, and from then until retirement in 1953 was chief of the General Engineering Division and Technical Advice, Nashville TN. He was a member ASCE.

Cone was involved in a number of dam designs for the Corps of Engineers serving for flood control, power development and navigation purposes. From 1929 the Memphis District of the Corps requested his services to carry out the "308" surveys for the St. Francis, White and Arkansas Rivers. Once with the Mississippi River Commission he was involved in a study of a comprehensive reservoir system to control floods. He also invented the Venturi flume which later became known in a modified shape as Parshall Flume, developed by Ralph L. Parshall (1881-1959). Both have a polygonal plan shape, with Cone's flume having a drop at the outlet. He also wrote a number of reports and bulletins dealing with alternative discharge measurement structures as weirs and gates, and presented works on flood control, irrigation and drainage, water power and river navigation.

Anonymous (1953). Victor M. Cone retires, Corps of Engineers expert. *Engineering News-Record* 151(Oct.8): 291-292. *P*
Anonymous (1959). Cone, Victor Mann. *Who's who in engineering* 8: 476. Lewis: New York.
Cone, V.M. (1911). Irrigation in the San Joaquin Valley, California. *Bulletin* 239. US Office of Experiment Stations. Government Printing Office: Washington DC.
Cone, V.M. (1916). New standard irrigation weir. *Engineering News* 76(3): 222-223.
Cone, V.M. (1916). Flow through weir notches with thin edges and full contraction. *Journal of Agricultural Research* 5(23): 1051-1113.
Cone, V.M. (1916). A new irrigation weir. *Journal of Agricultural Research* 5(24): 1127-1143.
Cone, V.M. (1917). Divisors. *Bulletin* 228. Agricultural Experiment Station, Fort Collins CO.
Cone, V.M. (1917). The Venturi flume. *Journal of Agricultural Research* 9(4): 115-129.

CONGER

05.06. 1875 Gouverneur NY/USA
26.05. 1931 Wellesley MA/USA

Alger Adams Conger graduated as a civil engineer from Cornell University, Ithaca NY, in 1897. He was then engaged until 1899 by the US Deep Waterways then directed by Alfred C. Noble (1844-1914). His duties included hydrographic surveys and the design of the proposed ship canal from the Great Lakes to the Atlantic Ocean. He was engaged with a similar work with the Isthmian Canal Commission until 1900, and until 1901 with the New York State Barge Commission at Albany NY. Conger thereby gained a detailed insight into the variety of the engineering problems. From 1901 to 1903 he was in charge of the hydro-electric power development on the canal and power house at Sault St. Marie MI, for the Michigan-Lake Superior Power Company. Until 1906 he was employed again by the New York State Barge Canal in the construction of the canalized river.

Conger joined in 1906 a company in New York as assistant hydraulic engineer, engaged on a plant at West Buxton ME, the *Comerio* Development in Porto Rico, the Ocmulgee Development in Georgia, the La Cross Development in Wisconsin, and the San Joaquin Development in California. He further went to Venezuela preparing plans and estimates for the *Mamo* Hydro-Electric Development. He also was in charge of the design of the hydro-electric power plant at Vernon VT on the Connecticut River, which was the first plant forming the nucleus of the later New England Power Association. When in 1911 this Company was formed, Conger was hydraulic engineer continuing his work until his death. The hydro-electric developments in Massachusetts and Vermont, at Harriman and Sherman on the Deerfield River, and at Bellows Falls on the Connecticut River, which were installed in the 1920s, are examples of the soundness of Conger's designs. He had a mind trained to weigh his problems and make his decisions with common sense, justice and humanity. His associated valued his counsel, and he was a trusted adviser and valued friend to all who knew him. He was remembered by his associated as one who never departed from the highest ideals of his profession. He was one in whom kindness, tact, and sympathy could always be counted on. He did the work that life gave him to do with charity, fidelity, and honor. He was member of the Boston Society of Civil Engineers, and of the American Society of Civil Engineers ASCE.

Anonymous (1931). Alger A. Conger. *Engineering News-Record* 106(25): 1030.
Anonymous (1933). Alger A. Conger. *Trans. ASCE* 98: 1523-1526.
http://radio-timetraveller.blogspot.ch/2011/09/mediumwave-along-erie-canal-part-1.html (*P*)

COOLEY L.E.

05.12. 1850 Canandaigua NY/USA
03.02. 1917 Chicago IL/USA

Lyman Edgar Cooley, brother of Mortimer E. (1855-1944), graduated as a civil engineer from Rensselaer Polytechnic Institute in 1874. From 1879 to 1884 he was assistant engineer on Mississippi and Missouri River improvements. After having been assistant editor of *Engineering News* from 1876 to 1878, he took over as editor the *American Engineer* in 1884. He made a report in favour of the Chicago Sanitary Canal in 1885 and organized the drainage and water supply commission of the Chicago Sanitary District CSD. He was from 1889 its consulting engineer, working on problems of the relation to the lakes and rivers. He was from 1896 member of the Deep Waterways Commission, proposing solutions for the Isthmian Canal. Cooley went in 1897 to Panama and Nicaragua with engineers, becoming advisory engineer for the Erie Canal in 1898. He worked also with water power companies at Keokuk IA, advised on water works for Omaha NE in 1904, was consultant on the optimum location of the Rochester Barge Canal, and contributed to the flood problem in Grand Rapids MI. He was from 1909 consulting engineer for the Lakes-to-the-Gulf Deep Waterway Association, and from 1912 to 1915 chaired the Commission on Sewage Disposal and Water Power Development, CSD.

In the late-19th century, when the Chicago River was a huge sewer, when the sewage of a million population ran straight into the lake, Cooley started to build the Chicago Drainage Canal, then one of the largest engineering works. He set his mind to work as to how Chicago's sewage could be kept out of the water supply. He was the man who figured how and when dirt should fly, and how the earth and rock could be cut to form the canal conveying the sewage of Chicago to the ocean, and on its way be purified by the action of water. This project was one of the achievements of the century.

Cooley, L.E. (1902). Proposed dam and water power on the Mississippi River at Keokuk IA. *Journal of the Western Society of Engineers* 7(1): 10-19.
Cooley, L.E. (1913). *The diversion of the waters of the Great Lakes by way of the Sanitary and ship canal of Chicago*. Chicago Sanitary District: Chicago.
Davenport, F.G. (1973). Lyman Cooley: The sanitation revolution in Illinois. *Journal of the Illinois State Historical Society* 66(3): 306-326. P
Seddon, J.A., Cooley, L.E., Randolph, I. (1900). *A deep waterway from the Great Lakes to the Gulf of Mexico*. Western Society of Engineers: Chicago.
http://genforum.genealogy.com/cooley/messages/2762.html

COOLEY M.E.

28.03. 1855 Canandaigua NY/USA
25.08. 1944 Ann Arbor MI/USA

Mortimer Elwyn Cooley graduated in 1878 from the US Naval Academy, Annapolis MD. He completed in 1879-1880 his sea duty on the North Atlantic with the USS Alliance, and was then assigned to the Bureau of Steam Engineering at Navy Department. He was in 1881 promoted to assistant engineer, assigned to University of Michigan, Ann Arbor MI, teaching steam engineering and iron shipbuilding, and taking there over as professor of mechanical engineering, thereby resigning his commission with the Navy. In addition to university work, he was a consultant in mechanical engineering. He was from 1895 to 1911 chief engineer and officer of Michigan Naval Brigade. He was in 1898 also chief engineer of the US Navy during the Spanish-American War, attached to USS Yosemite. Cooley was a Fellow of the American Association for the Advancement of Science AAAS, a member of the American Society of Mechanical Engineers ASME, of which he was vice-president in 1902, of the US Naval Institute, the US Society of Naval Engineers, and Honorary Member ASCE. The Regents of his university conferred upon him in 1885 the honorary degree of mechanical engineer. He also received the LLD degree from the Michigan Agricultural College in 1907.

Cooley would be currently considered a general engineer, given his widespread interests in engineering. He was for instance involved in the assessment of hydro-electric power plants around 1900, or in railways projects. In parallel his role as the Grand Old Man of the College of Engineering of the University of Michigan was outstanding. He built the College from a rough temporary shop of only some 160 m^2 to three largely equipped buildings of a total of some 50,000 m^2 to hundreds of courses taught by some 160 professors. The original enrollment was below 30, and gradually increased to 1800. He became Dean in 1904, and retired in 1928. Through his long years of service, he was a representative of the whole University.

Cooley, M.E. (1891). *Reports* on the new Gaskill pumping engine, Kalamazoo MI. Ann Arbor.
Cooley, M.E. (1898). *Notes on dynamics of machinery*. Edwards: Ann Arbor.
Cooley, M.E. (1899). *Notes on water wheels*: Water power. Ann Arbor.
Cooley, M.E., Mencken, H.L. (1923). *Dynamics of reciprocating engines*. Wahr: Ann Arbor.
Hinsdale, B.A., Demmon, I.N. (1906). Mortimer E. Cooley. *History of the University of Michigan*: 263-264. University of Michigan Press: Ann Arbor. *P*
Sutin, H.A. (1935). Mortimer E. Cooley. *The Michigan Technic* (3): 103-105.

COOPER A.J.

02.02. 1913 New Orleans LA/USA
12.04. 1989 Knoxville TN/USA

Alfred Joseph Cooper graduated in 1934 with the BSc degree from Tulane University, New Orleans LA, and received in 1951 the MSc degree in civil engineering from University of Tennessee, Knoxville TN. Following graduation, he then first worked as engineer on the construction of the Gretna LA water works, and was in 1934 employed as engineer in the Division of Water Control Planning, the Tennessee Valley Authority TVA, Knoxville TN. He headed three sections within the Division, namely Hydrology and River, Procedures Development, and also River Forecasting, before taking over as assistant chief of the River Control Branch in 1958, and Branch Chief in 1960. Cooper was responsible for directing the impoundment and release of water from the TVA system to achieve maximum benefits in terms of navigation, power production, and flood control in Tennessee River and its tributaries. He also maintained the close corporation between TVA and the US Army Corps of Engineers in regulating the outflows from Tennessee and Cumberland Rivers to reduce flood stages along the Lower Ohio and the Mississippi.

Cooper has written several papers on flood control and reservoir management. His 1976 paper deals with the operation of an integrated river system during a flood. The largest flood experienced on Tennessee River occurred in 1973. TVA, through its organization and equipment, put into effect measures to mitigate flood damages not only on the integrated Tennessee River system but also on the Lower Ohio and Mississippi Rivers. The weather and runoff conditions generating this flood and the available weather services and hydrologic data formed the primary bases for actions taken to control the flood waters. The coordination with public officials, and water and power regulators was essential to successful flood control operations of an integrated system of reservoirs. Despite the unusual areal distribution of the rainfall and resulting runoff, the available control by reservoirs coupled with advance warnings of anticipated rising water levels and crest stages were the key to the success in mitigating flood damages.

Anonymous (1933). Alfred J. Cooper. *The Jambalaya*: 81. Tulane University: New Orleans. *P*
Anonymous (1989). Alfred J. Cooper. *Trans*. ASCE 154: 500-501.
Cooper, A.J., Snyder, W.M. (1956). Evaluating effects of land-use changes on sediment load. *Journal of the Hydraulics Division* ASCE 82(HY1): 1-14.
Cooper, A.J. (1976). Integrated river system operation in a major flood. *Journal of the Power Division* ASCE 102(PO1): 63-80.

COOPER H.H.

18.07. 1913 Quitman GA/USA
13.09. 1990 Leon FL/USA

Hilton Hammond Cooper received education from
the University of Florida with the BS degree in civil
engineering in 1937. He also made post-graduate
studies at Florida State University, Tallahassee FL
in 1952 and at Columbia University, New York NY
in 1960. He was up to 1938 a rodman and draftsman
at the Florida State Road Department, then until
1941 junior engineer at the Water Resources Branch
of the US Geological Survey, from when he became
from assistant to staff engineer of the US Geological
Survey, Florida State University, until retirement.
He was a member of the American Society of Civil
Engineers ASCE, the American Geophysical Union AGU, and the American Water
Works Association AWWA.

Once at the US Geological Survey Cooper began working on fundamental problems of
groundwater flow. After a series of papers on Florida groundwater, he published in
1946 with Charles E. Jacob (1914-1970) an application of the unsteady state theory to
an entire well field. When the water near the Miami well did not follow his theory
Cooper sought the reason by past approaches and later proved the principle of sea water
recirculation as a component of the discharge of fresh groundwater to the coast. He then
considered the mystery of the apparently erratic responses of certain wells to earthquakes
by considering the factors of the well and in the water body determining the magnitude
of the water-level fluctuations, thereby contributing both to groundwater theory and
seismology. Cooper's critical mind perceived the basic errors in the works of respected
groundwater scientists and the resulting confusion. The 1966 paper is the most rigorous
treatment on the basic groundwater flow model, for which he was awarded the 1969
Meinzer Award.

Anonymous (1939). Hilton H. Cooper. *Engineering News-Record* 123(Aug.10): 186. *P*
Anonymous (1964). Cooper, Hilton H. *Who's who in engineering* 9: 361. Lewis: New York.
Cooper, H.H. (1950). *Ground water in Florida*. Geological Survey: Tallahassee.
Cooper, H.H. (1966). The equation of ground-water flow in fixing and deforming coordinates.
    Journal of Geophysical Research 71(20): 4785-4790.
Cooper, H.H., Jr., Bredehoeft, J.D., Papadopulos, I.S. (1967). Response of a finite-diameter well
    to an instantaneous charge of water. *Water Resources Research* 3(1): 263-269.
Theis, C.V. (1972). Presentation of the O.E. Meinzer Award to Hilton H. Cooper, Jr. *Bulletin
    Geological Society of America* 83(4): xxiv-xxv. *P*

COOPER H.L.

28.04. 1865 Sheldon MN/USA
24.06. 1937 Stamford CT/USA

Hugh Lincoln Cooper was 16 years old when he undertook his first engineering project, a 12 m long bridge across a creek that ran through a farm on which he was working. He graduated from Rushford High School and then started work at Chicago and Milwaukee. Although he had no technical training he acquired it through self-study and practical field work. He joined a firm at Dayton OH in 1891, which built water wheels and installed power plants. Eventually he designed power plants in Jamaica, and in New York State, then spending three years with a power-plant company in Brazil.

On returning to the USA, he started his own consulting office. He was able to tackle the seemingly impossible-a plant to harness the raging Horseshoe Rapids above Niagara Falls. Despite skepticism and opposition from the Canadian authorities, Cooper devised a design for Ontario State, which was completed in 1907. He then returned to the USA, designing a hydropower installation across the Mississippi River between Keokuk IA and Hamilton IL. Completed in 1913, the Keokuk Water Power Project marked a significant change in hydropower generation because it employed a wide, slow-moving river to drive its turbines. Cooper's next project was in Egypt to help convert the Aswan Dam into a power source. The work was interrupted by World War I and only resumed in 1920. In the early 1920 Cooper was retained by the Soviet Union to construct the *Dneprostroi* Dam across Dnieper River in the Ukraine, at the time the largest power-plant in Europe. Completed in 1932, its generating capacity was 500 MW, and the dam made the river navigable for the first time by raising its level. Destroyed in World War II the facility was rebuilt in 1947. Cooper's work in Russia was regarded as a model for transfer of industrial skills. It was stated: 'Although Stalin was a powerful change agent, he did not have enough technological knowledge to lead the transfer effort'.

Cooper, H.L. (1912). The water power development of Mississippi River Power Co. at Keokuk, Iowa. *Journal of the Western Society of Engineers* 17(3): 213-226.

Cooper, H.L. (1921). *Comments on the Wooten-Bowden report to the International Joint Commission*. New York.

Dorn, H. (1979). Hugh Lincoln Cooper and the first détente. *Technology and Culture* 20(2): 322-347.

Green, W.P. (1913). Hugh Lincoln Cooper. *Scientific American* 109(Nov.08): 366. *P*

http://www.asce.org/PPLContent.aspx?id=2147487347

COPELAND

11.02. 1815 Coventry CT/USA
05.02. 1895 Brooklyn NY/USA

Charles Wilson Copeland was the son of a builder of steam engines and boilers at Hartford. Under his guidance, the son received lessons in the profession which was to become his life work. These were followed by a course at Columbia College. He was appointed at age 21 superintendant of the West Point Foundry Association, beginning with the design of the machinery of the *Fulton*, the first steam war-vessel to be constructed under the direct supervision of the US Navy Department. In 1839 he received a government appointment under which he signed himself 'Naval Engineer', and was entrusted with the machinery designs for *Mississippi* and *Missouri*, among others. In 1850, this work for the government completed.

Copeland was then appointed superintendent of the *Allaire* Works in New York City, where he designed and supervised a large number of merchant steamers, two of which broke the record of speed in transatlantic travels. During the Civil War, his experience as a marine engineer was used by the government in the adaptation of merchant steamers for service in the Southern blockade. After war, he became the constructing engineer for the US Lighthouse Board, which position he held almost to the time of his death. Copeland was simple and kindly in manner, frugal in personal economy, and untiring in industry. He was a lover and reader of books and a discriminating collector. He was life member of the American Society of Civil Engineers ASCE and maintained an interested and helpful relation to the work of this society. He was also a founding member of the American Society of Mechanical Engineers ASME.

Anonymous (1895). Copeland, Charles W. *Engineering and Mining Journal* 59(Feb.16): 155.
Anonymous (1895). Charles W. Copeland. *Trans. ASME* 16: 1191-1192.
Anonymous (1895). Charles W. Copeland. *J. American Society of Naval Engrs*. 7(1): 201-202.
Anonymous (1930). Copeland, Charles W. *Dictionary of American biography* 4: 423. Scribner: New York.
Bennett, F.M. (1896). Charles W. Copeland. *The steam navy of the United States*: A history of the growth of the steam vessel of war in the US Navy, and of the Naval Engineer Corps: 36-37. Nicholson: Pittsburgh. *P*
Shipman, W.D., Copeland, C.W. (1862). *The Clark patent steam and fire regulator company versus Charles W. Copeland*. Nesbitt: New York.
Thurston, R.H. (1888). *A manual of steam-boilers*. Wiley: New York.

COPPÉE

30.03. 1853 West Point NY/USA
08.05. 1901 Greenville MS/USA

Henry St. Leger Coppée was educated at Lehigh University, Bethlehem PA, where his father was its president. He graduated with the civil engineering degree in 1872, and then was engaged on surveys at San Diego CA. From 1874 to 1875 he acted as an assistant supervisor with the Pennsylvania Railroad Company. In 1878 he became associated with the Mississippi River Commission as an assistant field engineer in charge of gauging operations at Memphis TN, remaining there until 1882, when transferred to Vicksburg MS, where he was placed in charge of improvements at the Office of the Mississippi River Commission. In 1892 he was appointed assistant engineer of the Third Mississippi River District, which assignment he filled until his death. He was thereby actively engaged in the supervision of levee building, bank construction, and river control and improvement along the River and its tributaries in Arkansas, Mississippi and Louisiana States.

Coppée contributed various technical writings based on his experiences and knowledge of river improvements and flood control. The main work was his 1896 paper, for which he was awarded the Thomas Fitch Award of the American Society of Civil Engineers ASCE. The Lower Mississippi spans from Cairo IL to the river mouth as alluvial river, involving a significant sediment transport. The author describes the various kinds of revetments used for bank protection from erosion, including their cost and the results obtained. He received recognition for this paper because its approach to an economical and engineering problem of revetment design was clearly highlighted, as is observed from the various discussions. He also presented his 1898 paper on the Standard levee sections, relating to some 1,600 km of river along which there are almost 2,000 km of levees. Their standard dimensions are 2.5 m crown width, both river and back slopes 3:1; for levees higher than 3.5 m a banquet was added at elevation 2.5 m. Coppée further contributed discussions to papers entitled Dredges and dredging, and the Discharge of the Mississippi River. He had an enviable reputation among colleagues, and was known as a capable engineer of highest integrity. He was elected ASCE Member in 1887.

Anonymous (1932). Henry St. Leger Coppée. *Trans. ASCE* 96: 1444-1445.
Coppée, H.S. (1896). Bank revetment on the Lower Mississippi. *Trans. ASCE* 35: 141-240.
Coppée, H.S. (1898). Standard levee sections. *Trans. ASCE* 39: 191-236.
Coppée, H.S. (1901). Lower Mississippi and its regulation. *Riparian lands of the Mississippi River*: 47-63, F.H. Tompkins, ed. New Orleans. *P*

CORLISS

02.06. 1817 Easton NY/USA
21.02. 1888 Providence RI/USA

George Henry Corliss, a noted mechanical engineer, developed the Corliss steam engine, then a real great improvement over any other stationary steam engine of his time. This engine is considered one of the 19th century engineering achievements, which provided a reliable and efficient source of industrial power. Corliss gained international acclaim for his works during the late 19th century, and is known for the Centennial Engine, the huge centrepiece of the 1876 Centennial Exposition in Philadelphia PA.

Corliss graduated in 1838 from the Academy of Castleton VT. He displayed early signs of his mechanical abilities in 1837, after a flood washed away a bridge over the Batten Kill at Greenwich. He organized local builders in the erection of a replacement structure. In 1838, he established his own general store at Greenwich, where he remained for three years. He moved to Providence RI in 1844 with hopes of finding funding to perfect his sewing machine. However, he abandoned this work to focus on a new endeavor, the improvement of the stationary steam engine, which at the time was regarded as an inefficient alternative to water power. In 1848 he entered a partnership and his company built the first engine utilizing improvements, which was essentially the Corliss steam engine of years later. Corliss and his associates erected a new factory at Providence, where the company expanded greatly in the years to follow. By his death in 1888, the floor space within the plant covered five acres, and the company employed over 1,000 people. In 1849, Corliss was granted an US Patent for his valve gear. In 1856 the Corliss Steam Engine Company CSEC was incorporated with George Corliss as president. The dramatic fuel efficiency of the Corliss engine was a major selling point. CSEC supplied the US Government with machinery during the Civil War. When the USS *Monitor* was constructed in 1861 it was found that a large ring must be made, upon which its turret could revolve, and CSEC was found to be one of the few plants in the country that had the necessary machinery large enough to 'turn' up the large ring. When Corliss found out what the work was for, he worked day and night to get this important ring completed, which was done on time to be placed on the *Monitor*, which later met with the *Merrimac* in the historic Battle of Hampton Roads.

Anonymous (1930). George H. Corliss. *Dictionary of American biography* 4: 441. Scribner: NY.
Wilson, J.G., ed. (1887). George H. Corliss. *Appleton's cyclopedia of American biography* 1:
       740. Appleton: New York.
http://en.wikipedia.org/wiki/George_Henry_Corliss *P*

CORNISH L.D.

20.03. 1877 Lee Center NY/USA
12.05. 1934 Chicago IL/USA

Lorenzo Dana Cornish graduated in 1902 as a civil engineer from Syracuse University, Syracuse NY. He then entered the service of the US Army Corps of Engineers USACE as junior engineer, engaged in the construction of the Locks and Dams No. 2 to 7 on Ohio River. He was ordered to Washington DC in 1905 to study the Panama Canal situation, and later engaged on the design of its locks and dams. In 1907 his office was transferred to Panama, in charge of the execution of these works, remaining there until 1913. In 1914, he went to China as principal assistant engineer to the American National Red Cross Board of Engineers, to report on the Huai River Conservancy Project. In 1915 he entered the US Engineer Office, Cincinnati OH, as principal assistant engineer in charge of the improvements on the Ohio River. On return from war service in France, he was appointed design engineer in the Division of Waterways, State of Illinois. He was engaged on the design of the locks, gates, and dams of the Illinois Waterway. During this period he developed a formula to obtain the spacing of ports from culverts into the lock chamber to prevent water surging during the filling operation. His approach was successful, because there were no damages since completion.

Cornish advanced in 1920 to the assistant chief engineer of the Waterways Division, in recognition of his able work. He was then in charge of flood relief and prevention work along the various rivers in and bounding Illinois State. In 1928, upon the resignation of Mortimer Grant Barnes (1867-1930), Cornish was appointed chief engineer, vigorously carrying on the works on the Illinois Waterway until 1929, when the finances became short, and no additional funds were obtained by the State. The Governor requested the federal government to take over the Waterway, and complete and operate it, who agreed to do so, if Illinois State would rebuild the bridges over the Waterway, which was then finalized by Cornish. Due to ill health, he resigned in 1933 from his position. He had been a member of the American Society of Civil Engineers ASCE since 1910.

Anonymous (1935). Lorenzo D. Cornish. *Trans. ASCE* 100: 1638-1640.
Cornish, L.D. (1916). Design of the lock walls and valves of the Panama Canal. *Trans. of the International Engineering Congress* San Francisco: 65-86.
Jackson, F.E. (1911). Lorenzo D. Cornish. *Makers of the Panama Canal*: 48. P
Mulvihill, W.F., Barnes, M.G., Cornish, L.D. (1928). *Flood control in the Mississippi Valley*. US Army Corps of Engineers: Washington DC.

CORRSIN

03.04. 1920 Philadelphia PA/USA
02.06. 1986 Baltimore MD/USA

Stanley Corrsin obtained in 1940 his BS degree from University of Pennsylvania, Philadelphia PA, his MS degree from Caltech in 1942 and the PhD degree there in 1947. He was from 1940 to 1947 a Caltech research and teaching assistant in aerospace engineering, then until 1951 assistant professor at Johns Hopkins University, Baltimore MD, there continued for the next four years as associate professor of aerodynamics, and from 1955 to 1960 as professor of mechanical engineering. Finally, he was professor of fluid mechanics at Johns Hopkins until his retirement. Corrsin was the recipient of the distinguished alumnus citation from University of Pennsylvania in 1955; he was from 1964 Fellow of the American Physical Society APS, serving as chairman of its Fluid Dynamics Section, and he was awarded its Fluid Dynamics Prize in 1983. Corrsin was further a member of the American Academy of Arts and Sciences, of ASME, and of the National Academy of Engineering. He died of cancer at age 66.

When Corrsin left Caltech in 1947 he was an acknowledged expert in turbulence research. This field with all its manifestations remained his primary interest throughout his career. He thereby contributed successfully to both theoretical and experimental research. He for instance familiarized with diagram techniques to clarify the sequence of nonlinear coupling terms in wave number space. His quest for clarity and precision had just one negative result: He never finished the book he planned on fluid mechanics. His papers deal mainly with dimensional analysis, or the interpretation of the viscous terms in the turbulent energy equation. Corrsin's contribution to the *Handbuch* proof also of his pedagogical interests, which culminated in a total of 25 PhD theses.

Anonymous (1987). Corrsin, Stanley. *Who's who in America* 44: 584. Marquis: Chicago.
Comte-Bellot, G., Corrsin, S. (1971). Simple Eulerian time correlation of full and narrow band velocity signals in grid-generated isotropic turbulence. *J. Fluid Mech.* 48(2): 271-337.
Corrsin, S. (1951). The decay of isotropic temperature fluctuations in an isotropic turbulence. *Journal of the Aeronautical Sciences* 18(6): 417-423.
Corrsin, S. (1963). Turbulence. *Handbuch der Physik* 8(2): 524-590. Springer: Berlin.
Liepmann, H.W. (1994). Stanley Corrsin. *Memorial tributes* 3: 95-98. National Academy of Engineering: Washington DC. P
Lumley, J.L. (1987). Stanley Corrsin. *Physics Today* 40(2): 126.
Lumley, J.L., Davis, S.H. (2003). Stanley Corrsin. *Annual Review of Fluid Mechanics* 35: 1-10.

COWLES

14.09. 1887 Cleveland OH/USA
26.11. 1935 Yountville CA/USA

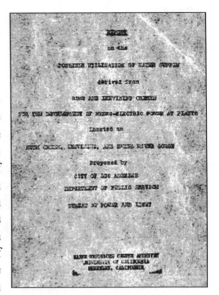

The family of Robert Fulton Cowles moved in 1896 to California. He made an uncompleted study at the Leland Stanford University, San Francisco CA. He had his first engineering assignment in 1908 for Los Angeles Aqueduct. In 1912 he was appointed chief engineer on the construction of the 30 m^3/s main canal for the Casa Grande Valley Water Users in Arizona. In 1913 he was employed at Los Angeles on the design of water and irrigation systems, South Oceanside CA. He was retained by the Bureau of Light and Power, Los Angeles, in 1915, to compile the data of seven potential hydro-electric projects in Owen Valley and the Mono Lake region, which involved stream measurement and run-off studies, as well as the location of the proposed reservoirs, dams, and power plants. In 1916 he was employed by Michael M. O'Shaughnessy (1864-1934) on the Hetch Hetchy project, for the water supply of San Francisco CA.

After service during World War I, from where he returned in 1919, Cowles was engaged in engineering practice at Los Angeles. He was retained for topographical surveys of the Gillespie Project AZ, and hydraulic studies of the Sweetwater Water Company, San Diego CA. In 1921 he had charge of surveys for the Arrowhead Lake Project CA, and was from 1922 chief engineer, the South Coast Land Company, Los Angeles, in charge of the design and the operation of extensive water supply and irrigation systems. He was from 1925 engineer-in-charge of the Bureau of Water Development, San Diego, charged to design and build a 30 km long pipe line 1 m in diameter. He was retained in 1928 by the Campbell Irrigation Company CA for the design of overhead irrigation systems. He then served the Water Department, Los Angeles, as topographer of the Colorado River project. In 1930 he was appointed assistant hydraulic engineer, State Division of Water Resources, California, to investigate the availability and the use of water in the Napa and the Santa Clara Valleys. He finally served as supervising engineer, US Coast and Geodetic Survey, at Kennett Reservoir CA, but his health became poor, so that he passed away at age of only 48. During his career he worked with a high degree of professional ability on several of the then really large hydro-power projects in the West.

Anonymous (1936). Robert F. Cowles. *Trans. ASCE* 101: 1540-1542.
Cowles, R.F. (1914). *Extension of rating curve for Santa Ynez River*. UC Berkeley.
Cowles, R.F. (1916). *Possible utilization of water supply derived from Rush and Leevining Creeks for development of hydroelectric power*. Dept. Public Service: Los Angeles. (P)

COX

21.03. 1903 Keosauqua IA/USA
29.01. 1991 Los Angeles CA/USA

Glen Nelson Cox obtained the BE degree in 1925, the MS degree in 1926 from Iowa State University, Iowa City IA, and the PhD degree from Wisconsin University, Milwaukee WI in 1928. He was then an instructor until 1929 at University of Illinois, Urbana IL, associate professor of hydraulic and mechanical engineering at Louisiana State University, Baton Rouge LA until 1936, from when he took over there as professor of hydraulic engineering until 1948, the last two years serving also as director of the School of Hydraulics there. He was in parallel chief of the hydraulic section, the Department of Public Works in Louisiana from 1944 to 1945. Cox was from 1949 to his retirement professor of hydraulics and mechanical engineering at New York University, New York NY. He was a member of the American Society of Civil Engineers ASCE, the American Society of Engineering Education ASEE, and the American Geophysical Union AGU.

Cox was interested in general hydraulics, among which he was particularly working in weir flow, hydraulics, hydrology, precipitation, runoff and evaporation. His PhD thesis was concerned with submerged weir flow, which is known to be much less stable than free weir flow, but under certain tailwater conditions is the only practicable approach for discharge determination. His MS thesis dealt indeed with the free flow weir arrangement, in which he investigated the effect of roughening the upstream weir face. The 1946 book Fluid mechanics is a common text of the time, in which the Bernoulli theorem, weir and gate flow with the corresponding discharge coefficients, head losses across hydraulic elements or orifice flow are explained. Further the book includes also sections on turbulent pipe flow including the concept of the Reynolds number, and on instruments to measure various parameters of fluid flow.

Anonymous (1925). Glen N. Cox. *Hawkeye yearbook*: 70. University of Iowa: Iowa. *P*
Anonymous (1949). Glen N. Cox. *Civil Engineering* 19(9): 806. *P*
Anonymous (1955). Cox, Glen N. *American men of science* 9: 394. Science Press: Lancaster.
Cox, G.N. (1926). *The flow of water over a rectangular weir as affected by various degrees of roughening of its upstream face*. MS Thesis. State University of Iowa: Iowa City.
Cox, G.N. (1928). *The submerged weir as a measuring device*: A method for making accurate stream flow measurements. University of Wisconsin: Milwaukee.
Cox, G.N., Germano, F.J. (1946). *Fluid mechanics*. Van Nostrand: New York.
Cox, G.N., Plumtree, W.G. (1954). *Engineering mechanics*. Van Nostrand: New York.

CRAIGHILL

01.07. 1833 Charlestown WV/USA
18.01. 1909 Charlestown WV/USA

William Price Craighill graduated in 1853 from the US Military Academy, then joining the US Corps of Engineers, in which he finally rose to brigadier-general and chief of engineers. He was from 1853 to 1856 on duty of the improvement of Savannah River including fortifications and harbour works for Charleston Harbour. He moved then to Washington DC as assistant to the chief of engineers, but was ordered to the US Military Academy as professor of engineering in 1859, where he compiled the Army Officer's pocket companion, a treatise which proved useful for practice.

After service during the Civil War, he was sent in 1865 to the Baltimore District in charge of constructions on the fortifications and of river and harbor works, serving also Andrew A. Humphreys (1810-1883). He remained there until 1895 when promoted to chief engineer. The river and harbor works extended from the Susquehanna to Cape Fear NC, and as far west as the Great Kanawha WV. The improvement of the Cape Fear River presented a notable engineering problem, but the successful closing of the false mouth across the shifting sands along the coast was an important work, in which the log and brush mattress was evolved. Another engineering feature was the canalization of the Great Kanawha River employing movable dams as proposed by the Frenchman Chanoine. These were introduced in the USA after a visit to France in 1878. Craighill was in addition involved in the breakwaters for Delaware Bay, Sandy Bay MA, and San Pedro CA. As chief of engineers he arranged for the prompt construction of new defense works. He retired in 1897; the Washington and Lee Universities conferred on him then the LL.D. degree. He was member of the American Society of Civil Engineers ASCE, was elected to its president in 1894, and Honorary Member in 1896. As president he was exceedingly efficient, combining in a rare degree affability, tolerance, tact, dignity, retentiveness of memory, and swift comprehension of controversies.

Anonymous (1909). William P. Craighill. *Trans.* ASCE 65: 517-524. *P*
Anonymous (1938). William P. Craighill. *Civil Engineering* 8(3): 219-220. *P*
Craighill, W.P., Brunot de Rouvre, P. (1862). *The army officer's pocket companion*: Principally designed for staff officers in the field. Van Nostrand: New York.
http://en.wikipedia.org/wiki/William_Price_Craighill *P*
http://www.lighthousefriends.com/light.asp?ID=416
http://www.mywvhome.com/old/locks.html

CRAMP C.H.

09.05. 1828 Philadelphia PA/USA
06.06. 1913 Philadelphia PA/USA

Charles Henry Cramp, eldest son of William (1807-1879), was famous pioneer shipbuilder of America, who established shipyards on Delaware River near Philadelphia in 1830. The son was educated at the Philadelphia Central High School, after which he was employed in his father's shipyards, making himself master of each detail of ship construction. He showed special aptitude as a naval architect and designer, and after becoming his father's partner in 1849, it was to that working branch that he devoted himself. His inventive capacity and resourcefulness, together with the complete success of his innovations in naval construction soon gave him high rank as an authority in shipbuilding, and made his influence in that industry widely felt.

In the Mexican War he designed surf boats for the landing of troops at Vera Cruz; during the Civil War he designed and built several ironclads for the US Navy, notably the New Ironsides in 1862, and the light-draught monitors used in the Carolina Sounds. After 1887, he entirely or partially designed his own constructions, many of the most powerful ships in the 'new Navy', including the cruisers Columbia, Minneapolis, and Brooklyn, and the battleships Indiana, Iowa, Massachusetts and Brooklyn. In each step of ocean shipbuilding, in the transformation from sail to steam, and from wood to iron and steel, Cramp had a prominent part. His fame as a shipbuilder extended to Europe, so that he built warships for several foreign navies, among others the *Retvizan* and the *Variag* for the Russian government. He also constructed a number of freight and passenger steamers for transatlantic lines. He was considered a captain of industry in all that the term implies, and if the best title of leadership rests upon a knowledge and understanding of industry itself, then he is doubly titled, for he was industry personified. He was a most generous man. He was described as the heart to which even the striker appeals.

Buell, A.C. (1906). *The memoirs of Charles H. Cramp*. Lippincott: Philadelphia. *P*
Cramp, C.H. (1893). *Evolution of the Atlantic Greyhound*. Deutsch: Baltimore MD.
Cramp, C.H. (1894). *American shipbuilding and commercial supremacy*. Moyer & Lesher: Philadelphia.
Cramp, C.H. (1904). *Sixty years of shipbuilding on the Delaware*. Philadelphia.
Cramp, C.H. (1906). *Charles H. Cramp papers*. Archives: Philadelphia.
http://www.thefullwiki.org/Charles_Henry_Cramp

CRAMP W.

22.09. 1807 Kensington PA/USA
06.07. 1879 Atlantic City NJ/USA

William Cramp, father of Charles H. Cramp (1828-1913), studied naval engineering at Philadelphia PA. In 1830 he established his own shipbuilding firm on the Delaware River, first in Kensington, and then in a larger facility at Richmond. Over the next decades, his shipyard grew becoming one of the important in the USA, constructing wood, ironclad, iron, and steel ships. He remained president of the firm for forty-nine years, from its founding until his death. Cramp modernized his equipment to adapt to changes in ship construction. The ability of his firm to adapt to the technology of shipbuilding and ship propulsion through the mid-nineteenth century brought lasting fame. As sail power gave way to steam-powered paddle wheels and then propeller-driven vessels, he developed steam-engine facilities. During the Civil War, he provided the Union Navy steam-propelled ironclads, which played crucial roles in blockade duty and sea engagements. Then iron construction replaced wooden hulls through the 1870s. The firm was ready for the revolution of the 1880s, which brought steel-hulled ships into the naval and merchant fleets. During his lifetime Cramp oversaw the construction of more than 200 ships.

Cramp brought his sons into the business, admitting Charles H. Cramp as a partner. In 1872 he changed the firm name to William Cramp and Sons' Ship and Engine Building Company. The company won contracts to build warships for foreign navies, including those of Russia and Venezuela. The performance of Cramp ships in the Russian fleet in the war between Turkey and Russia (1876-1878) enhanced the reputation of the firm as a constructor of warships. As the US Navy pressed then for expansion of the fleet, the William Cramp and Sons played a role both in advocacy of the new navalism, and in construction of the steel ships of the new navy. It was also a major contributor to the growth of Delaware River shipbuilding, concentrating manpower, skills, technological capability, facilities, and political power. In addition to Cramp's firm and Philadelphia Navy Yard, many shipbuilding companies on both sides of the river converted this 25 km stretch of the Delaware into one of the major shipbuilding centers of the USA.

Taylor, F.H. (1895). *Hand book of the Lower Delaware River*: Ports, tides, quarantine stations, light-house, service life-saving and maritime reporting stations. Harris: Philadelphia.
http://kennethwmilano.com/page/Encyclopaedia/KensingtonPortraitsBiographies/KensingtonBiographies/1stPresbyterianChurchStainglassWindowBiograph/WilliamCramp/tabid/279/Default *P*
http://kennethwmilano.com/page/Encyclopaedia/WilliamCrampShipyard/tabid/188/Default.aspx

CRAVEN A.W.

20.10. 1810 Washington DC/USA
29.03. 1879 Chiswick/UK

Alfred Wingate Craven, son of the naval officer Thomas Tingey, graduated in 1835 from Columbia University as a civil engineer. He was from 1837 associated with large works at Charleston SC, and rapidly rose to the first rank in his profession. At the end of his first years in the profession, he joined a large and dashing corps of young engineers in railroad engineering in South Carolina, where he won many friends by his marked character. After another decade in the railroad business in the Northeast of the United States he found that his family required a change. He became in 1849 engineer commissioner to the Croton Water Board of New York, a position he held until 1868. He moved to England in 1878 and there rapidly passed away due to cerebral softening.

During his stay at Croton water works, Craven designed and carried out the building of the large reservoir in Central Park, New York. He was also engaged in the enlargement of pipes across High Bridge, and the construction of the reservoir in Boyd's Corner, Putnam. He further asked for an accurate survey of Croton Valley, with a view of ascertaining its capacity for furnishing the adequate water supply to the city of New York. He was in addition instrumental in securing the passage of the first law establishing a general sewage system for New York City. Later, he was associated in the design of the underground railway extension along 4th Avenue. He was one of the original Members of the American Society of Civil Engineers ASCE, a director for many years, and its president in the term 1869-1871.

Anonymous (1856). *Report* of the Water Committee of the Common Council of the City of Brooklyn. Spooner: Brooklyn.

Anonymous (1880). Alfred W. Craven. Minutes, *Institution of Civil Engineers* 60: 390-393.

Anonymous (1880). Alfred Wingate Craven, past president ASCE. *Trans. ASCE* 6: 24-26.

Anonymous (1887). Craven, Alfred Wingate. *Appletons' cyclopaedia of American biography* 2: 3. Appleton: New York.

Craven, A.W. (1860). *Answer* of A.W. Craven, chief engineer Croton Aqueduct, to charges made by Fernando Wood, Mayor, New York. Baker & Goodwin: New York.

Finch, J.K. (1929). *Early Columbia engineers*: John Stevens, James Renwick, Horatio Allen, Alfred W. Craven. Columbia University Press: New York.

Hunt, C.W. (1897). Alfred Wingate Craven. *Historical sketch of the American Society of Civil Engineers*: 23. ASCE: New York. *P*

CRAVEN A.

16.09. 1846 Bound Brook NJ/USA
30.09. 1926 Pleasantville NY/USA

Alfred Craven graduated in 1867 from US Military Academy, Annapolis MD, remaining in the service until 1871. He then joined the California Geological Survey, and was of the party who first climbed the Tower Peak. He was engaged on irrigation work in the Sacramento and San Joaquin Valleys, and then went into private practice at Virginia City NV. In this work he was associated with Adolph H. Sutro (1830-1898) in the construction of the Sutro Tunnel. In 1884 he became assistant engineer for the New Croton Aqueduct of the New York water supply. The Aqueduct was almost 50 km long, constructed entirely in rock, and lined with brick, designed for a capacity of 13 m^3/s. From 1885, Craven served there as division engineer. He was in charge of the construction of the Reservoir D, near Carmel NY, and of the completion of Reservoir M on Titicus River, near Purdys NY. These are two of the chain of storage reservoirs built before and since the Croton Basin. Until 1900 he was employed on the building of Jerome Park Reservoir in the Borough of the Bronx.

Under Craven's administration as chief engineer, the great Dual Subway System was designed and built, corresponding to one of the largest municipally constructed projects, comparable in magnitude with Panama Canal. This project may be considered the monument to Craven. He resigned as chief engineer in 1916, and took over then as a consulting engineer of the Commission, until he retired from service in 1920. He had been one of the men, without any apparent effort on his own, who won respect and affection of all who served under him. He never sought popularity: He had but to request that a thing be done and his staff would work cheerfully to accomplish the result. His fairness and straightforward way of thinking appealed to those who had business with him, and his decisions were generally accepted, whether they were for or against the claimants. He was a member of the Society of Municipal Engineers of the City of New York, and of the American Society of Civil Engineers ASCE, serving as director from 1903 to 1905, and as vice-president from 1916 to 1917. In 1908 he was awarded, under an Act of Congress, the Civil War Medal.

Anonymous (1930). Alfred Craven. *Trans. ASCE* 94: 1506-1508.
Craven, A. (1906). Discussion of Changes at the New Croton Dam. *Trans. ASCE* 56: 52-63.
Goldman, J.A. (1997). Building of New York's sewers. Purdue University Press: W. Lafayette *P*
http://whittlesey-whittelseygenealogy.com/showmedia.php?mediaID=137&medialinkID=148 *P*

CREAGER

21.09. 1878 Baltimore MD/USA
04.04. 1953 Buffalo NY/USA

William Pitcher Creager graduated in 1901 as a civil engineer from Rensselaer Polytechnic Institute. He was until 1904 a provincial supervisor for the Philippine Government, then a designer of the New York State Barge Canal until 1906, a draughtsman to the chief hydraulic engineer of the White Engineering Company until 1922, and then chief engineer of the New York Power Corporation. From 1931, Creager was consultant at Baltimore MD. He served as a member of the Advising Engineering Council, Princeton University. He was a member of the American Society of Civil Engineers ASCE, and the American Institute of Consulting Engineers.

Creager is particularly known for his books Engineering for masonry dams published in 1917, and the Hydro-electric handbook published ten years later with Joel Justin (1881-1950). He also published other books and chapters of larger works, besides a number of technical articles in professional journals. Creager's name is known for the spillway surface profile geometry ensuring practically atmospheric pressure conditions under design discharge. Whereas up to 1900, the spillway crest profile was almost abrupt in geometry, Richard Muller (-1944) was the first to propose a crest profile of which the curvature change was continuous, to avoid local sub-pressures resulting in cavitation damage. Currently, Creager's name is retained for this feature because he referred to this effect in his book, thereby overlooking Muller's earlier proposal. Both designs were based on detailed observations of the lower nappe profile of sharp-crested weir flow made by Henry Bazin (1829-1917) in France. Apart from this work, Creager otherwise was a successful general engineer not involved in detailed hydraulic computations. His books may be regarded as excellent examples prior to World War II having had a large influence on the golden years in dam engineering.

Anonymous (1950). William P. Creager. *Civil Engineering* 20(1): 46-47. *P*
Anonymous (1951). Creager, William P. *Who's who in America* 26: 596. Marquis: Chicago.
Anonymous (1953). William P. Creager. *Civil Engineering* 23(5): 329-330. *P*
Anonymous (1953). William P. Creager dies, hydraulics authority. *Engineering News-Record* 150(Apr.16): 106.
Creager, W.P. (1917). *Engineering for masonry dams*. Wiley: New York.
Creager, W.P. (1923). *La construction des grands barrages en Amérique*. Gauthier-Villars: Paris.
Creager, W.P., Justin, J.D. (1927). *Hydro-electric handbook*. Wiley: New York.

CREGIER

01.06. 1829 New York NY/USA
09.11. 1898 Chicago IL/USA

DeWitt Clinton Cregier was employed at age 19 in a machine shop of New York City, later becoming the Morgan Iron Works. In 1853 Cregier moved to Chicago IL to assist in installing the 'Old Sally', the first pumping engine at the new municipally owned waterworks. Serving as chief engineer of the North Side Chicago waterworks for 27 years, his ability as a design engineer was indicated by the many patents he obtained on mechanical devices, pertaining to waterworks use. Many of them were adopted by the City of Chicago without royalty fees. His design of the 400 l/s pumping engine at North Side Pumping Station, installed in 1872, is probably the best known example of his ability. He also was a designer of a fire hydrant used as standard equipment for many years.

It was through Cregier's untiring efforts and ability that, following the Great Fire of 1871 destroying the North Side waterworks, this station was again put back into service after only eight days. When Ellis S. Chesborough (1813-1886) resigned in 1879 as city engineer, Cregier was appointed in his place. Three years later Cregier was appointed commissioner of public works serving in that office until he resigned in 1886, to accept a position as superintendent of a railways company. Chicago was at that time pumping in two average days as much water as all through 1853, the first year of Cregier's service in Chicago. The first pair of Corliss compound-condensing beam engines at the New West Side pumping station were installed in 1876. In the 1930s, the station was entirely electrically operated, with pumping equipment of centrifugal type. In 1889, the year of Cregier's election to city mayor, the city area was increased five times through annexation, with a population over a million. Cregier realized the necessity of a unified water system in the Chicago metropolitan area, and gave as mayor his approval for large extension plans. The progress of the Chicago water supply system within a century was highlighted during the Great International Exposition in 1933. Chesborough and Cregier largely contributed to the water supply of the city, and to that of the USA in general.

Gorman, A.E. (1933). Chicago's waterworks: A century of accomplishment. *Engineering News-Record* 110(23): 733-737. *P*
Pierce, B.L. (1957). *A history of Chicago*: The rise of a modern city 3. University of Chicago Press: Chicago IL.
http://enc.tfode.com/DeWitt_C._Cregier
http://www.firehydrant.org/pictures/i2/DewittCCregier.jpg *P*

CROCKER

20.06. 1867 Haverhill NH/USA
08.03. 1949 Denver CO/USA

Herbert Samuel Crocker obtained the BS degree in civil engineering from University of Michigan, Ann Arbor MI, in 1889, and the ME degree there in 1919. He was awarded the DSc degree in 1934 from the University of Colorado, Denver CO. After works in railways projects and consulting engineering firms, he was bridge engineer until 1896, and until 1901 assistant engineer on the Board of Public Works at Denver CO, joining 1906 then the Western Division of the American Bridge Co., Chicago IL until 1908, when starting with his private practice as consulting engineer at Chicago IL until 1917. After war service he returned to his practice but now to his Denver Office until 1933. Until 1937 he was consulting engineer in charge of construction for collection canals, tunnels, conduits for the transmission diversion of water from the Western Slope of the Rocky Mountains to Denver CO, thereby being a member of the Board of Water Commissioners. Until 1942 Crocker acted as engineering consultant on various projects, including the Twin Lakes transmission diversion, and was further member of the Mississippi Valley Commission. From 1942 he was senior member of the firm Crocker & Ryan, Denver CO, principally engaged as architectural engineer, and consulting engineering to the States of Colorado and New Mexico.

Crocker was a general civil engineer with a specialization mainly in irrigation and water supply projects. He also served on various State Boards to improve river conditions and to control flood flows. He had a particular relation to the American Society of Civil Engineers ASCE, serving as director from 1917 to 1919, vice-president in 1920, acting secretary from 1919 to 1929, and finally ASCE President in 1932 during the difficult early 1930s, becoming honorary member in 1939. He further was member of the ASCE Colorado Section, and the Texas Section, and was member of the Western Society of Civil Engineers, among many others.

Anonymous (1932). Herbert S. Crocker: The new president. *Civil Engineering* 2(2): 117-118. *P*
Anonymous (1948). Crocker, Herbert S. *Who's who in engineering* 6: 432. Lewis: New York.
Anonymous (1949). Herbert S. Crocker dies: Former ASCE President. *Engineering News-Record* 142(Mar.17): 72.
Crocker, H.S. (1935). *Ralston Creek Reservoir Site*: Irrigation Division No. 1, Water Districts No. 6 and 7. Moffat Tunnel Extension Project, Municipal Water Works: Denver CO.
http://www.worldcat.org/title/presentation-at-moffat-filter-plant/oclc/48017835&referer=brief_results *P*

CROES

25.11. 1834 Richmond VA/USA
17.03. 1906 Yonkers NY/USA

John James Robertson Croes graduated as a civil engineer from St. James College, Hagerstown MD in 1853. He started his civil engineering practice in 1856, assisting first James Pugh Kirkwood (1807-1877) in the construction of Ridgewood Reservoir of the Brooklyn Waterworks, then resident engineer of the first high masonry dam in the USA at Boyd's Corner NY from 1865 to 1870, with his 1874 paper being awarded the first Norman Medal from ASCE. He then became topographical engineer for the New York Park Department until 1878. In the 1880s he made expert reports on the Quaker Bridge Dam, a part of New York Aqueduct, and later on the New Croton Dam of the same scheme. From 1903 Croes was a consulting engineer of the New York Health Department. He was a member of the American Society of Civil Engineers ASCE, serving as treasurer from 1877 to 1887, and as ASCE president in 1901. From 1882 he was a member of the Institution of Civil Engineers, London, the American Public Health Association, and the New England Water Works Association.

Croes is known for a large number of reports mainly as an expert of hydraulic schemes. His works related principally to questions on water supply and sewage works. These include the water supply of Newark and Syracuse NY, or the design and construction of waterworks in Indianapolis IN. He also contributed a paper on the History of American water works. He was further associated with the journal Sanitary Engineer, for which he regularly submitted editorials. His 1891 report relates to the dam break at Johnstown PA, resulting in a large loss of lives and goods. ASCE's John James Robertson Croes Medal was established in 1912. It is awarded annually to the Author of a paper published by ASCE journals for its merits as a contribution to engineering science.

Anonymous (1905). Croes, John J.R. *Who's who in America* 4: 338. Marquis: Chicago.
Anonymous (1938). John James Robertson Croes. *Civil Engineering* 8(10): 702-703. *P*
Croes, J.J.R., Craven, A.W. (1871). *Report of the citizen's committee on water*. Halstaed: Syracuse NY.
Croes, J.J.R. (1874). Construction of a masonry dam. *Trans. ASCE* 3: 337-367; 4: 307-309.
Croes, J.J.R. (1889). *Manual of American waterworks*. Engineering News: New York.
Croes, J.J.R. (1891). *The rivers at Johnstown*, Pennsylvania: Report to the Board of Trade of the city of Johnstown. Engineering Press: New York.
Croes, J.J.R. (1901). A century of civil engineering. *Science* 14(342): 83-94.

CROSS

10.02. 1885 Nansemond VA/USA
11.02. 1959 Virginia Beach VA/USA

Hardy Cross graduated in 1908 as a civil engineer from Massachusetts Institute of Technology MIT, obtaining in 1911 his master degree from Harvard University. Cross was from 1908 to 1910 a bridge engineer, then from 1911 to 1918 assistant professor of civil engineering, Brown University, Providence RI, general practice structural engineer for the next three years and from 1921 to 1937 then professor of structural engineering at the University of Illinois, Urbana-Champaign IL, to finally stay as professor of civil engineering at Yale University, New Haven CT. He was awarded Honorary Master of Arts from Yale University in 1937, the 1933 Norman Medal from the American Society of Civil Engineers ASCE, of which he was also a member, and the 1944 *Lamme* Medal from the American Institute of Consulting Engineers AICE.

Cross was not a hydraulic engineer, yet the methods devised for structural engineering were fully adopted in hydraulic engineering. If complicated pipeline systems have to be designed or reassessed, a system of linear equations must be solved. Cross proposed an engineering method to determine the system unknowns, namely discharge and pressure head in branched systems, as are typical in water-supply systems. His original method relates to the moment distribution in structural engineering. It consists of (1) Imagine all joints in the structure held so that they cannot rotate and compute the moments at the member ends for this condition, (2) Distribute at each joint the unbalanced fixed-end moment among the connecting members in production to the stiffness, (3) Multiply the moment distributed to each member at a joint by the carry-over factor, (4) Distribute these moments, (5) Repeat the process until the iteration process is satisfied. Next to a number of bulletins and research papers, Cross has particularly received fame by his 1932 book Continuous frames of reinforced concrete.

Anonymous (1942). Prof. Hardy Cross. *Engineering News-Record* 128(Feb. 5): 217. *P*
Anonymous (1951). Cross, Hardy. *Who's who in America* 26: 605. Marquis: Chicago.
Anonymous (1959). Hardy Cross dies. *Engineering News-Record* 162(Feb.19): 28. *P*
Cross, H. (1926). Temperature deformations in concrete arches. *ENR* 96(5): 190-191.
Cross, H., Morgan, N.D. (1932). *Continuous frames of reinforced concrete*. Wiley: New York.
Cross, H. (1936). Analysis of flow in networks of conduits or conductors. *Bulletin* 286. University of Illinois: Urbana IL.
Cross, H. (1952). *Engineers and ivory towers*. McGraw-Hill: New York.

CROWE

12.10. 1882 Trenholmville Que/CA
26.02. 1946 Redding CA/USA

Francis Trenholm Crowe graduated in 1905 as civil engineer from the University of Maine, Orono ME. He was invited by Frank E. Weymouth (1874-1941) to join him in the West. Crowe's first work with the Reclamation Service was for the Lower Yellowstone irrigation system, and in 1906 he moved to another work involving siphons, sluiceways, and spillways, but returned to the Reclamation Service in 1908 as superintendent of construction of Minidoka Project in Idaho. He was engaged until 1911 on the Snake River storage project, including the construction of the Jackson Lake Dam. Crowe was appointed then assistant superintendent of construction of the Arrowrock Dam in Idaho, with 110 m then the highest concrete dam in the world. It was during this period that Crowe saw the great possibilities for the use of cableways in the construction of concrete dams.

From 1913 to 1915 Crow was in charge of the construction of the Boise Project in Idaho including the building of 80 km of large open drains. In 1916 he was appointed project manager of the Flathead Project in Montana, remaining there until 1920. He left the Reclamation Service to form a partnership at Missoula MT, which was resolved, however, only one year later, so that Crowe returned again to the Reclamation Service as construction engineer of the 70 m high Tieton Dam on the Yakima Project in the State of Washington. On completion of this dam, he was in 1924 general superintendent of construction in the Office of the Chief, US Bureau of Reclamation, Denver CO. He had there general supervision of all construction activities in seventeen western States, so that he became one of the top-ranking men of the Bureau. During these years and his association with Frederick H. Newell (1862-1932), Arthur P. Davis (1861-1933), and David C. Henny (1860-1935), he was determined to build even larger dams. From 1925, when Crowe left the Bureau, he was mainly involved in dam construction, culminating in the construction of Hoover Dam, which was completed two years ahead of schedule. He received as the crowning distinction the honorary ASCE membership in 1944. He had been pre-eminent as a dam builder.

Anonymous (1948). Francis T. Crowe. *Trans. ASCE* 113: 1397-1403.
Dunn, P. (2010). Frank Crowe: General superintendent. Proc. 75th *Hoover Dam Anniversary History Symposium*: 307-317. ASCE: Reston VA.
http://en.wikipedia.org/wiki/Frank_Crowe *P*
http://hooverdamstory.com/crowe.htm *P*

CROZET

01.01. 1790 Villefrance/F
28.01. 1864 Richmond VA/USA

Claude Crozet was born in France, where he was educated at *Ecole Polytechnique*, Paris until 1807, and then received the engineering education at Metz Military Academy until 1809. Crozet served then in Germany and Holland in Napoleon's headquarters until the latter's defeat at Waterloo in 1815. In 1816 he moved to the young United States becoming at Point West assistant professor at the US Military Academy. He was appointed in 1823 state engineer of Virginia State, served there until 1832, and being in 1837 reappointed. In between he was Louisiana State engineer. In Virginia he was in charge of the Board of Public Works and dealt with turnpike projects; later he was involved in railway construction, including the 1.2 km long Blue Ridge Tunnel which at the time was the longest in North America. He also worked on the James River and Kanawha Canal survey in the 1830s.

Crozet was a founding member of the Virginia Military Institute, where he introduced the study of descriptive geometry, a subject greatly developed by his former colleague Jean-Victor Poncelet (1788-1867) at Metz. He left VMI in 1845 serving then for the erection of the Washington Aqueduct from 1857 to 1859, from when he was until his death the principal of the Richmond Academy. Besides having largely contributed to the road network of Virginia State, which was stated to be the best of the country, Crozet was an outstanding civil engineer of America in the first half of the 19th century. The 1989 book on Crozet is described a beautiful piece of scholarship, offering a view of a capable engineer plying his profession in the young USA from the canal to the railroad era. It brings together information from a vast array of sources on an early engineer of the United States.

Anonymous (1972). Crozet, Claude (Claudius). *A biographical dictionary of American civil engineers*: 30-31. ASCE: New York. *P*

Couper, W. (1936). *Claudius Crozet*: Soldier, sailor, educator, engineer. Historical Publication: Charlottesville. *P*

Crozet, C. (1821). *Treatise of descriptive geometry* for the use of the cadets of the US Military Academy. West Point.

Hunter, R.F., Dooley, Jr., E.L. (1989). *Claudius Crozet*: French engineer in America 1790-1864. University of Virginia Press: Charlottesville.

http://www.wvencyclopedia.org/articles/1682 *P*

CURTIS

29.09. 1894 Berlin MD/USA
11.06. 1961 Denver CO/USA

Howard Gray Curtis graduated from the University of Colorado, Boulder CO, with the BSc degree in civil engineering in 1917. From then until 1926 he worked for the water works facility of Nitro VA, a number of irrigation districts, municipal water works, and with consulting engineers, as well as for the St. Paul, Chicago, and Milwaukee Railroad Co. Curtis began his career with the US Bureau of Reclamation USBR in 1926. An authority on pipeline design, he supervised the extensive USBR irrigation systems, including those of the Central Valley Project, and of the All-American Canal System in California. He eventually rose to the Head, the Canals and Pipelines Section, Canals Branch, the Office of the Assistant Commissioner and chief engineer of USBR, Denver CO. Curtis was a member of the American Society for Testing Material, and the American Concrete Agricultural Pipe Association. He was a registered professional engineer of Colorado State. He was further a member of the American Society of Civil Engineers ASCE from 1950, and ASCE Fellow from 1959. He was posthumously awarded in 1962 the highest degree of the Department of the Interior, the Distinguished Service Award.

Curtis was mainly a hydraulic engineer dealing with the design and the construction of large pipeline systems. In addition, he was also involved in the construction of water supply, wastewater, and irrigation systems. Once with the USBR he took charge of canal structures for the Kittitas, Owyhee, and Vale dam schemes, including the design of intake towers, spillways, power plant substructures until 1933. From 1934 he had immediate supervision of the design of canal structures and automatic radial gates. He for instance developed the articulation of large reinforced concrete sections for the All-American, and the Gila Canals.

Anonymous (1917). Howard G. Curtis. *Coloradoan yearbook*: 82. Boulder CO. *P*
Anonymous (1948). Curtis, Howard G. *Who's who in engineering* 6: 445. Lewis: New York.
Anonymous (1963). Howard G. Curtis. *Trans. ASCE* 128(5): 121.
Curtis, H.G. (1937). Progress on All-American Canal. *Engineer's Bulletin* 21(3): 8-10.
Curtis, H.G. (1950). *The Bureau of Reclamation and concrete pipe*. USBR: Denver.
Curtis, H.G. (1952). *The design of concrete pipe irrigation systems*. USBR: Denver.
Slater, W.R., Curtis, H.G., Oliver, P.A. (1950). Flood control works: All-American Canal, Coachella Branch, Boulder Canyon Project. *Joint Summer Convention* of ASCE and EIC Toronto: 1-27.

DAGGETT

06.06. 1867 Foxcraft ME/USA
17.03. 1932 Winthrop MA/USA

Herbert Chapin Daggett graduated in 1891 from the Massachusetts Institute of Technology MIT, with the BSc degree in civil engineering. He then took a position as assistant engineer of the Proprietors of the Locks and Canals, on Merrimack River, Lowell MA, where he tested a variety of water-wheels and their settings, as also transmission methods of the power developed, as at these times there were not yet direct-connected electric generators. He received also training in the discharge measurement through open and closed channels. In 1899 Daggett became chief engineer of the Swain Turbine Manufacturing Company, Lowell MA, where he designed and supervised the manufacture of hydraulic turbines. From 1901 to 1903, he was engineer in the Hydraulic Turbine Division of the Holyoke Machine Company, Worcester MA, where he had similar duties as previously. His work with the two companies served to broaden his experience and his knowledge in the development of hydraulic machinery.

From 1903 to the time of his death, Daggett had charge of the New England Office of the S. Morgan Smith Company, York PA, manufacturers of hydraulic turbines and allied equipment, as engineer and sales manager. He thereby widened his acquaintance with engineers and builders of hydraulic power plants. He also saw the rapid introduction of electrical equipment and the change from the driving of the machinery to the direct connection of generators, and the driving of machinery by motors. He had a thorough knowledge of the various types of turbines, and the particular conditions they were best suited to satisfy the requirements. Few important developments were made in New England without Daggett's advice. Because of his knowledge of the discharge of the various rivers, he was able to determine the possibilities of undeveloped water power. His notable accomplishment was his broad knowledge of the water-power situation in New England, and his acquaintanceship with hydraulic engineers. He was a man of high ideals, having the respect of all with whom he came into contact. He was member of the Boston Society of Civil Engineers, and of the American Society of Civil Engineers.

Anonymous (1895). *Holyoke Machine Co.*, Worcester MA. Blanchard: Holyoke.
Anonymous (1914). *S. Morgan Smith Records* 1890-1914. York PA.
Anonymous (1933). Herbert C. Daggett. *Trans. ASCE* 98: 1526-1527.
Anonymous (1956). *The Hydraulic Turbine Laboratory* of the S. Morgan Smith Co. York PA.
Peckham, B. (1946). *S. Morgan Smith Co*. Story of a dynamic community, York PA: 197. *P*

DAILY

19.03. 1913 Columbia MO/USA
27.12. 1991 Pasadena CA/USA

James Wallace Daily obtained the BA degree from the Stanford University in 1935, and the MS degree from California Institute of Technology, Pasadena, in 1937, from where he also received the PhD title in 1945. From 1936 to 1937 he had been a research assistant in hydraulics at Caltech, then until 1940 a research fellow in hydraulic machinery, and until 1946 an instructor in mechanical engineering. Daily then was appointed assistant professor of hydraulics at the Massachusetts Institute of Technology MIT, was promoted to associate professor in 1949, and was a professor there from 1955 to 1964. From then until 1972, Daily was a professor of engineering mechanics at University of Michigan, Ann Arbor MI, and there professor of fluid mechanics and hydraulic engineering until retirement in 1981. He was a visiting professor to University of Delft, and a visiting scientist to Electricite de France, Paris, in 1971, and a visiting professor to the East China College of Hydraulic Engineering, Nanking, in 1979.

Daily and Donald R.F. Harleman (1922-2005) authored the outstanding textbook Fluid dynamics. This work may be considered an essence of the then available knowledge in basic fluid dynamics, including both potential and viscous flow problems, expanding into turbulence, among other subjects. The second known book of Daily is Cavitation co-authored by Robert T. Knapp (1899-1957) and Frederick G. Hammitt (1923-1989). It may be considered one of the very first books dealing exclusively with this hydraulic phenomenon, which is known for failure of both civil and mechanical engineering elements. Daily was one of the Americans who was working in both branches, and therefore contributed from a generalized point of view to these problems. He was president of the International Association of Hydraulic Research IAHR from 1967 to 1971, and awarded IAHR Honorary Membership, among many other distinctions.

Anonymous (1956). First cavitation seminar. *Mechanical Engineering* 78(1): 116. *P*
Anonymous (1964). Daily, James Wallace. *Who's who in engineering* 9: 407. Lewis: New York.
Anonymous (1987). Daily, James Wallace. *Who's who in America* 44: 636. Marquis: Chicago.
Daily, J.W. (1949). Cavitation characteristics of a hydrofoil section. *Trans. ASME* 71(4): 269-284.
Daily, J.W., Harleman, D.R.F. (1966). *Fluid dynamics*. Addison-Wesley: Reading MA.
Harleman, D.R.F. (1993). James Wallace Daily. *Memorial tributes* NAE 6: 22-24. *P*
Knapp, R.T., Daily, J.W., Hammitt, F.G. (1970). *Cavitation*. McGraw-Hill: New York.
Prins, J.E. (1992). Jim Daily. *IAHR Bulletin* 30(1): 36. *P*

DALRYMPLE

27.06. 1904 Llano TX/USA
02.09. 1996 McAllen TX/USA

Tate Dalrymple received education from University of Texas, Austin TX, obtaining in 1931 the BS degree in civil engineering. He was then until 1935 hydrographer of the Texas Board of Water Engineers from when he joined until retirement the Geological Survey as hydraulic engineer with works in Texas, New Mexico, California and Ohio. From the 1950s he was at the USGS Headquarters at Washington DC, acting from 1951 as chief of the Floods Section, and the Surface Water Branch. He was a member of the American Society of Civil Engineers ASCE, and of the American Geophysical Union AGU.

Dalrymple was an expert in connection with the measurement of floods, preparing various reports on specific floods, as these in Texas in the 1930s. He also developed methods of adapting stream flow data to problems of bridge design, particularly in relation with pier and abutment scour. He authored and co-authored a number of reports on flood frequencies and flood discharges, and the methods of obtaining these data under flooding conditions. He also prepared papers to highway designs. He for instance determined the energy head slope by plotting it against the discharge. Assuming nearly uniform flow, the discharge then was determined from the Gauckler-Manning-Strickler formula. He also discussed various methods for determining flood discharges, among which are current-meter measurements, indirect discharge measurement, the slope-area method, the contraction method, and a method involving culverts.

Anonymous (1929). Dalrymple, Tate. *Cactus yearbook*: 378. University of Texas: Austin. *P*
Anonymous (1964). Dalrymple, Tate. *Who's who in engineering* 9: 409. Lewis: New York.
Benson, M.A., Dalrymple, T. (1967). *General field and office procedures for indirect discharge measurements*. US Geological Survey: Washington DC.
Breeding, S.D., Dalrymple, T. (1944). *Texas floods* of 1938 and 1939. US Government Office: Washington DC.
Dalrymple, T. (1956). Measuring floods. *Symposia Darcy* Dijon 3: 380-404. Association Internationale d'Hydrologie AIH: Louvain.
Dalrymple, T. (1960). *Flood-frequency analyses*. US Geological Survey: Washington DC.
Dalrymple, T. (1963). *Flood-plain mapping activities of the US Geological Survey*. ASCE: New York.
Dalrymple, T., Benson, M.A. (1967). Measurement of peak discharge by the slope-area method. *Techniques of Water-Resources Investigations* 3(A2): 1-12. USGS: Washington DC.

DARRACH

08.09. 1846 Philadelphia PA/USA
01.06. 1927 Philadelphia PA/USA

Charles Gobrecht Darrach graduated in 1866 from the Philadelphia Central High School and began his technical career as assistant engineer in an office of engineering. In 1868 he was assistant engineer with the Reading Railroad Company. He was employed by other American railroad companies from 1870 to 1875. Then he joined as principal assistant engineer the Water Department of Philadelphia until 1885, when becoming a member of an engineering office. From 1888 he designed the waterworks system in the Borough of Ridley Park, Delaware County PA. Darrach was then consultant from 1899 until 1922. He was a member of the Engineers Club of Philadelphia.

Darrach was known as a general engineer who made expert examinations and reports in legal cases, proposed designs for public utilities, water and sewage purification, and heating and ventilation of large buildings. He wrote a paper on the Mechanical installation in the modern office buildings. He was further concerned with designs and improvements of port facilities and the improvement of water supply and sewerage of the city of Philadelphia, and also planned its waterworks, drainage and sewer systems. He was a man of sterling integrity, greatly interested in municipal matters, particularly those involving engineering problems. He was the founder and first president of the Institute of Operating Engineers, and a member of the American Geophysical Society, among others. He was also a member of the American Society of Civil Engineers from 1876. His Collection of Papers archived at the Historical Society of Pennsylvania consists of seven volumes, each pertaining to a different subject. The volume The port of Philadelphia includes a pamphlet on public utility corporations of 1913. The smallest of Darrach's compilations Water supply, Philadelphia, includes notes on water purity, news clippings, and a letter to the Historical Society of Pennsylvania. Also included is an enormous colored map showing a proposed Delaware River Ship Channel.

Anonymous (1928). Charles G. Darrach. *Trans. ASCE* 92: 1681-1682.
Darrach, C.G. (1878). The flow of water in pipes under pressure. *Trans. ASCE* 7: 114-121.
Darrach, C.G. (1879). Die Bewegung des Wassers in Röhren. *Journal für Gasbeleuchtung und Wasserversorgung* 22: 236-241.
Darrach, C.G. (1902). Mechanical installation in modern office building. *Trans. ASCE* 48: 1-16.
Darrach, C.G. (1913). *Valuation of the properties of public utilities corporations*. Sherwood: Philadelphia PA. (*P*)

DAUGHERTY

14.09. 1885 Irvington IN/USA
20.08. 1978 Pasadena CA/USA

Robert Long Daugherty obtained his AB and ME degrees from Leland Stanford University, Stanford CA, in 1909 and 1914, respectively. He was from 1909 to 1910 there an instructor, then until 1916 an assistant hydraulics professor at Cornell University, Ithaca NY, and professor of hydraulic engineering at Rensselaer Polytechnic, Troy NY, from 1916 to 1919. Daugherty was professor of hydraulic and mechanical engineering at Caltech, Pasadena CA, from then until retirement in 1956. In parallel he was consultant of numerous firms and corporations, including the Metropolitan Water District on pumps for the Colorado River Aqueduct, or the Bureau of Reclamation USBR on pumps for the Grand Coulee project on Columbia River from 1938 to 1940. He was a member of the ASME Commission for Revision of Code for Testing of Hydraulic Power Plants.

Daugherty is known for a number of books in mechanical engineering, relating to the design and execution of turbines and pumps. His book Hydraulic turbines was first published in 1913. The third edition includes the following chapters: 1. Introduction, 2. Types of turbines and settings, 3. Water power, 4. Tangential water wheel, 5. Reaction turbine, 6. Turbine governors, 7. General theory, 8. Theory of tangential water wheel, 9. Theory of reaction turbine, 10. Turbine testing, 11. General laws and constants, 12. Turbine characteristics, 13. Selection of type of turbine, 14. Cost of turbines and water power, 15. Design of tangential water wheel, 16. Design of reaction turbine, and 17. Centrifugal pumps. Besides he published books on the various chapters of his previous book. He was closely related with ASME all through his career.

Anonymous (1959). Daugherty, Robert L. *Who's who in engineering* 8: 563. Lewis: New York.
Anonymous (1969). Robert L. Daugherty. *Mechanical Engineering* 91(11): 108. *P*
Anonymous (1978). Robert L. Daugherty. *Mechanical Engineering* 100(11): 109.
Daugherty, R.L. (1915). *Centrifugal pumps*. McGraw-Hill: New York.
Daugherty, R.L. (1920). *Hydraulic turbines*. McGraw-Hill: New York.
Daugherty, R.L. (1930). Pumping machinery. *Mechanical Engineering* 52(4): 381-382. *P*
Daugherty, R.L. (1937). *Hydraulics*: Text on practical fluid mechanics. McGraw-Hill: New York.
Daugherty, R.L. (1938). Centrifugal pumps for the Colorado River Aqueduct. *Mechanical Engineering* 60(4): 295-299.
Daugherty, R.L. (1961). Fluid properties. *Handbook of fluid dynamics* 1: 1-22, V.L. Streeter, ed. McGraw-Hill: New York.

DAVIDIAN

22.05. 1926 New York NY/USA
20.01. 1983 Fresno CA/USA

Jacob Davidian received education from Iowa State University, Iowa City IA. He was then at Georgia Institute of Technology, Atlanta GA, collaborating with Carl R. Kindsvater (1913-2002), receiving the MS degree in 1959. Later he joined as hydraulic engineer the US Geological Survey USGS. Davidian was a member of the American Society of Civil Engineers, and of the American Geophysical Union AGU.

The 1962 Report is a follow-up of previous work of Kindsvater. It describes a laboratory study defining the discharge characteristics, the flow distribution pattern, and the backwater effect of open channel constrictions with a variety of opening geometries. The 1965 paper deals with methods to measure discharge in rivers at gauging stations. The procedure normally accounts for a range of discharges, by which a head-discharge equation may be defined, and expanded for extreme discharges both at the lower and the upper ends. The approach was generalized in the 1968 paper mainly for application by practicing engineers. The 1970 paper details a method to calibrate a number of current meters in hydraulic laboratories using a submerged jet, whose hydraulic characteristics are well defined. The 1961 paper is a classic work in collaboration with Herman J. Koloseus (1919-2004) relating to the definition of open channel flow in the turbulent rough flow regime. Instead of considering natural roughness, a simple and well-defined roughness pattern was selected to obtain fundamental insight in the flow processes.

Anonymous (1956). J. Davidian. Proc. 6th *Hydraulics Conference* Iowa: Frontispiece. *P*

Carter, R.W., Davidian, J. (1965). Discharge ratings at gauging stations. Hydraulic measurements and computation. *Techniques of water resources investigations* 1(12): 1-15. USGS.

Carter, R.W., Davidian, J. (1968). General procedures for gauging streams. *Techniques of Water Resources Investigations* Book 4(A6): 1-13. US Geological Survey: Washington DC.

Davidian, J., Carrigan, Jr., P.H., Shen, J. (1962). Flow through openings in width constrictions. *Water Supply Paper* 1369D. US Geological Survey: Washington DC.

Davidian, J. (1970). Calibration of current meters in a submerged jet. Proc. *Koblenz Symposium*: 99-108.

Davidian, J. (1984). Computation of water-surface profiles in open channels. *Techniques of Water Resources Investigations* Book 3(A15): 1-48. USGS: Washington DC.

Koloseus, H.J., Davidian, J. (1961). Flow in an artificially roughened channel. *Professional Paper* 424-B. US Geological Survey: Washington DC.

DAVIS A.P.

09.02. 1861 Decatur IL/USA
07.08. 1933 Oakland CA/USA

Arthur Powell Davis was a nephew of John Wesley Powell (1834-1902). In 1872 the family moved from Illinois to a farm in Kansas, where Arthur graduated from Kansas Normal School, Emporia. After moving to Washington in 1882, he graduated in 1888 with the BS degree from today's George Washington University. He began his career in 1882 in the US Geological Survey USGS as assistant topographer. In 1884 he was appointed topographer of its Rocky Mountains Division. He was then for five years head of the topographic work in the USGS Southwest Section. Appointed hydrographer in 1896, he had charge of all stream measurements in the United States carried out by the USGS. Further, he was engaged as US hydrographer in charge of the examination of rainfall, stream flow, and flood control under the Isthmian Canal Commission, for both the proposed Nicaragua and Panama Canal routes.

In 1909 Davis was named Member of the Board of Engineers to examine engineering problems pertaining to Panama Canal. In 1915 he was a member of a committee to study the Culebra slides into Panama Canal. In 1911 he was also engaged by the czarist government to investigate the irrigation of the *Kara Kum* Desert in Tajikistan, and in 1914 was sent to China to survey the *Huai* River Conservancy Project. Following the organization of the US Reclamation Service, Davis was appointed in 1903 supervising engineer, becoming in 1908 chief engineer and from 1914 to 1923 its director. During this period, the *Shoshone* and *Arrowrock* dams were built, each then the largest in the world. Davis described these and many other irrigation schemes in his 1917 book. He was also instrumental for the first design of the Boulder Dam. Under his direction, more than one hundred dams were placed. In his 1922 report on problems of the Imperial Valley he summarizes results made during his term.

Anonymous (1933). Arthur P. Davis, Boulder Dam consultant, dies. *ENR* 111(6): 181. *P*
Anonymous (1944). Davis, Arthur Powell. *Dictionary of American biography* 2.1: 224-226. Scribner's: New York.
Davis, A.P. (1899). Nicaragua and Isthmian routes. *National Geographical Magazine*: 247-266.
Davis, A.P. (1899). Rainfall and temperature in Nicaragua. *M. Weather Review* 27(5): 211-212.
Davis, A.P. (1917). *Irrigation works constructed by United States Government.* Wiley: New York.
Davis, A.P., Wilson, H.M. (1919). *Irrigation engineering.* Wiley: New York.
http://www.usbr.gov/history/CommissBios/davis.html *P*

DAVIS C.V.

01.07. 1897 Camden NJ/USA
15.09. 1981 Western Springs IL/USA

Few biographical information is available on Calvin Victor Davis. After graduation to civil engineer, he joined the Ambursen Dam Co. as chief engineer in the late 1920s in New York. In 1947 he then joined Harza Company as vice-president until 1953, then taking over as president of this large engineering firm. From 1963 until retirement in 1968 he was chairman of the Board of Directors.

Davis was on the one hand instrumental for one of the largest engineering companies of the world, and on the other the editor of one of the most significant books in hydropower engineering. His Handbook of applied hydraulics was published first in 1942, with the second edition dating of 1952. It includes 25 chapters on 1250 pages with a large number of then well known experts, including for example James S. Bowman (1889-1961) on Gates, Thomas R. Camp (1895-1971) on Water supplies and Water treatment, George H. Hickox (1903-1986) on Hydraulic models, Julian Hinds (1881-1977) on Canals and pipelines, Ivan E. Houk (1888-1972) on Arch dams and Irrigation, Phillip Z. Kirpich (1914-2008) on Hydrology, Emory W. Lane (1891-1963) on Spillways, John Lowe III (1916-2012) on Earth dams, Lewis F. Moody (1880-1953) on Hydraulic machinery, George R. Rich (1896-1977) on Water hammer and Surge tanks, I. Cleveland Steele (1886-1973) on Rock-fill dams, and John C. Stevens (1876-1970) on Hydroelectric plants. As president of the Harza Company, he widely travelled to assure the contacts to his clients. He for instance was involved in the *Derbendi Khan* Dam in Iran, or Harza was the general consultant of the entire Indus River scheme in Pakistan, where Harza coordinated the activities of the individual consulting firms, preparing a detailed design of the *Mangla* Dam scheme, or the overall project for the settlement of water usage between Pakistan and India.

Anonymous (1958). Davis, Calvin. *Engineering News-Record* 161(Sep.11): 31. *P*
Anonymous (1962). Calvin V. Davis. *Engineering News-Record* 168(May10): 64. *P*
Anonymous (1982). Calvin V. Davis. *Civil Engineering* 52(3): 95. *P*
Davis, C.V. (1932). Stability of dams increased by more economical use of materials. *Engineering News-Record* 107(6): 210-214.
Davis, C.V., ed. (1952). *Handbook of applied hydraulics*, 2nd ed. McGraw-Hill: New York.
Davis, C.V. (1958). Rockfill dams: The Derbendi Khan Dam. *Journal of Power Division* ASCE 84(PO4, 1741): 1-23; 85(PO2): 75-79; 85(PO3): 109.
http://www.usbr.gov/history/CommissBios/davis.html *P*

DAVIS C.H.

16.01. 1807 Boston MA/USA
18.02. 1877 Washington DC/USA

Charles Henry Davis was commissioned as mid-shipman in 1823. He served on a frigate until 1828 in the Pacific, and was promoted to lieutenant in 1834. From 1846 to 1849, he worked for the US Coast Survey, discovering a previously unknown shoal that had caused shipwrecks off the coast from New York. He was also responsible for researching tides and currents and acted as inspector for various shipyards. From 1849 to 1855 he was the first super-intendent of the American Nautical Almanac Office.

Davis was promoted in 1854 to commander and given the command of the *St. Mary's*. In 1859, while commanding it, he was ordered to go to Baker Island in the North Pacific to obtain samples of guano, which was used as fertilizer, becoming possibly the first American to set foot there since it was annexed by the USA in 1857. It was previously thought that this island is inaccessible. During the Civil War, Davis was appointed to the Blockade Strategy Board in 1861, and shortly later promoted to captain. He was made acting Flag Officer in command of the Western Gunboat Flotilla. It fought a short battle with Confederate ships on the Mississippi River at Plum Point Bend TN in 1862. Two of the Union ships were badly damaged and had to run into shoal water to keep from sinking. The Confederate vessels escaped with only minor damage. In summer 1862 with another flotilla in the attack on Vicksburg MS, they were forced to withdraw. Davis then proceeded up the Yazoo River and successfully seized Confederate supplies and munitions. After this excursion he was made chief of the Bureau of Navigation, returning to Washington DC. From 1865 to 1867 he was the superintendent of the US Naval Observatory, from when he was given the command of the South Atlantic Squadron with *Guerriere* as his flagship. In 1869 he returned home and served both on the Lighthouse Board as well as in the Naval Observatory. His son, Commander Charles H. Davis, Jr., served as Chief Intelligence Officer of the Office of Naval Intelligence from 1889 to 1892. Several ships of the US Navy were named in the father's honor, namely the torpedo boat USS *Davis* (TB-12), and the destroyers USS *Davis* (DD-65 and DD-395). Davis, finally a Rear Admiral, advanced therefore the US Coast Survey and was a notable American hydrographer.

Anonymous (1877). C.H. Davis. Proc. *American Academy of Arts and Sciences* 12: 313-320.
Davis, Jr., C.H. (1902). Biographical memoir of Charles Henry Davis. *National Academy of Sciences* 4: 23-55.
http://en.wikipedia.org/wiki/Charles_Henry_Davis *P*

DAWSON

03.09. 1889 Truro NS/CA
23.03. 1963 Iowa City IA/USA

Francis Murray Dawson graduated in 1913 as a civil
engineer from Cornell University. He began as a
construction engineer in 1906, becoming from 1910
to 1912 an instructor in civil engineering and dock
construction at Halifax. Dawson was from 1921 to
1922 assistant professor of hydraulics at Cornell
University, then until 1924 associate hydraulics
professor at University of Kansas, and there full
professor until 1928. Until 1936 he was a professor
of hydraulics and sanitary engineering at University
of Wisconsin, Madison WI, and finally the Dean of
the College of Engineering, University of Iowa IA
until retirement. In parallel, Dawson was also a consultant on engineering projects. He
was a member of the American Society of Civil Engineers ASCE, and of the American
Society for Engineering Education ASEE, which he presided in 1951.

Dawson authored in collaboration with Ernest William Schoder (1879-1968) the basic
text Hydraulics. Its chapters are: 1. Hydrostatics, 2. Total liquid pressure, 3. Stability of
gravity dams, 4. Air and gases, 5. Buoyancy and flotation, 6. Logarithmic plotting, 7.
Flow of liquids through orifices, 8. Converging and diverging flows, Bernoulli's theorem,
9. Flow of water over weirs, 10. Exponential laws of variation, 11. Steady uniform flow
of water in pipes and open channels, 12. Flow of water in pipes, 13. Exact formulas of
flow of water in pipes, 14. Equivalent, compound, looping and branching pipes, 15.
Uniform flow of water in open channels, 16. Viscous flow of oil and water, 17. Nozzle-
type water turbines, 18. Water turbines, and 19. Centrifugal pumps. This book is of 'old
style' both in terms of topics as also in presenting the material. Limited recourse to
laboratory observations is for instance made, despite the authors' activities in this field.

Anonymous (1951). Dawson, Francis Murray. *Who's who in America* 28: 652. Marquis: Chicago.
Anonymous (1964). Francis M. Dawson. *Trans. ASCE* 129: 936-937.
Dawson, F.M., Kalinske, A.A. (1937). *Report* on Hydraulics and pneumatics of plumbing
        drainage systems. University of Iowa: Iowa City.
Dawson, F.M., Kalinske, A.A. (1937). Cross-connections and back-siphonage research. *Tech.
        Bulletin* 1. Natl. Association of Plumbing, Housing, Heating: Washington DC.
Dawson, F.M., Kalinske, A.A. (1939). Methods of calculating water-hammer pressures. *Journal
        of the American Water Works Association* 31(11): 1835-1864.
Schoder, E.W., Dawson, F.M. (1927). *Hydraulics*. McGraw-Hill: New York.
http://www.telusplanet.net/public/jrbaines/v91.html *P*

DEBLER

08.01. 1885 Beatrice NE/USA
02.10. 1976 Denver CO/USA

Erdman Bruno Debler received the BS degree in 1907 from University of Nebraska. After graduation he was employed by railroad companies in Illinois and Nebraska, and by private engineering firms. He joined in 1918 the US Bureau of Reclamation USBR, Denver CO. Starting as an instrument man, he rose to head the Bureau's water resources and project investigation section. He was named in 1943 Director of Project Planning for the Bureau, and in 1944 became Director of the newly established Lower Missouri Region. He there was responsible for reclamation activities in Nebraska, Kansas, East Colorado, south-western Wyoming, and a portion of South Dakota, an area encompassing the southern half of the Missouri River Basin, the upper part of the Arkansas River Basin, and trans-mountain diversions from Colorado River to the Platte and Arkansas Basins.

Debler was a known river engineer, contributing significantly to the development of water resources in the West. His 1949 paper describes the USBR policy relating to multi-purpose reservoirs. Upon passage of the 1902 Reclamation Act, the Reclamation service was organized as a branch of the US Geological Survey and so continued until 1904, when the Reclamation Service was established as an independent branch of the US Department of the Interior. The name of the agency was changed to the current USBR in 1923. The paper highlights the many achievements and future developments to be done, relating to irrigation, power, flood control, navigation and fish and wildlife. Upon retirement from the USBR in 1947, Debler was cited as having had 'an enviable career'. His government had indeed encompassed more than 30 years. His name is engraved in a bronze tablet at Hoover Dam in commemoration of his contribution to its design and construction. He was a member of the American Society of Civil Engineers ASCE from 1925, becoming Fellow ASCE in 1959.

Anonymous (1929). Erdman B. Debler. *New Reclamation Era* 20(2): 31. *P*
Anonymous (1977). Erdman B. Debler. *Trans. ASCE* 142: 562.
Debler, E.B. (1930). *Hydrology of the Boulder Canyon Reservoir*. USBR: Denver.
Debler, E.B. (1932). *Final report on Middle Rio Grande investigations*. USBR: Denver.
Debler, E.B. (1949). Development of policy by the Bureau of Reclamation. Multipurpose reservoirs: A symposium. *Proc. ASCE* 75(3): 295-300.
USBR (2008). *USBR*: History essays from the Centennial Symposium. USBR: Denver.

DELAMATER

30.08. 1821 Rhinebeck NY/USA
07.02. 1889 New York NY/USA

Cornelius Henry Delamater left at age of fourteen school to become an errand boy in a hardware store. At sixteen he entered an office who had established the *Phoenix* Foundry in New York City, and there started as a clerk but he made himself so useful and gained such an insight into the work that he was offered the business. Delamater accepted with a colleague, doing repair work and building boilers and engines for side-wheel steamers. When in 1839 John Ericsson (1803-1889) settled in New York, he was persuaded to give his work to the *Phoenix* Foundry. There he met Delamater and a life-long friendship was formed. Delamater's foundry built, after designs of Ericsson, the first iron boats and the first steam fire-engines in the USA. The 36-inch cast-iron pipe used for the Croton Aqueduct was also made there and before the end of 1840, over fifty propeller steamers were constructed. Delamater then bought property on the North River, New York, and in 1850 there built a large establishment which was later known as the Delamater Iron Works.

Upon the outbreak of the Civil War, the firm entered a new chapter in its history. When the government announced for proposals to build iron-clad vessels, Delamater at once called Ericsson for help. A contract on the basis of Ericsson's designs was received in 1861, the keel of the *Monitor* was laid and the engines were built. *Monitor* left New York in early 1862 and was then engaged in the famous battle with the *Merrimac*. In 1869, the Delamater Works constructed thirty gunboats, each armed with a 100-pound bow-chaser, for the Spanish government. The firm was notable too for its propellers, for air compressors and for the construction of the first successful submarine torpedo boat in 1881. Delamater, the guiding genius, was a man of deep, kind, tolerant sympathies, well known for his disposition, and his warm and lasting friendships. He died suddenly of pneumonia.

Anonymous (1930). Delamater, C.H. *Dictionary of American biography* 5: 211-212. Scribner's: New York.

Delamater, C.H. (1886). *The steam users' manual* for use of owners, engineers and firemen, with rules and directions for the care of boilers. New York.

Hutton, F.R. (1915). *A history of the American Society of Mechanical Engineers* from 1885 to 1915. ASME: New York.

http://en.wikipedia.org/wiki/Cornelius_H._DeLamater *P*

DELAPP

18.07. 1912 Kansas City KS/USA
03.06. 2002 Frazee MN/USA

Warren William DeLapp obtained the BS degree in civil engineering from Kansas State University, Manhattan KS, in 1935. He then joined the United Fruit Co. as a hydraulic engineer and designed and constructed irrigation schemes in Guatemala. In the early 1940s he obtained the MS degree from State University of Iowa, Iowa City IA, and then accepted a teaching position at the University of Minnesota, Minneapolis MN, and there instructed mathematics and mechanics, ultimately receiving the PhD degree. He accepted in 1947 a position at the Department of Civil Engineering of Colorado State University, Fort Collins CO, from where he took a leave of absence from 1959 to 1961 to Peshawar, Pakistan, helping to develop a hydraulic laboratory at the university. He then returned to Colorado, where he worked until his retirement from the University in 1970. He then was engaged by a consulting engineer at Denver CO, and worked on complicated water drainage and water right problems for years. He permanently returned to Minnesota in 1986, where he enjoyed several years of lakeside retirement life.

DeLapp made few but interesting publications during his stay in academia. The 1940 Report is his MS thesis, dealing with sediment transport in upward flow, as is typically observed with wave run-up in the coastal environment. The 1947 PhD thesis deals with a then novel problem of air entrainment in open channel flow. During the following decade, the University of Minnesota in general, and its St. Anthony Falls Hydraulic Laboratory in particular, were heavily involved in this task, culminating in benchmark papers of Lorenz G. Straub (1901-1963) and his colleagues.

Anonymous (1960). De Lapp, Warren W. *American men of science* 10: 908. Cattell: Tempe AZ.
DeLapp, J.R. (1995). *Stories from my early years*: Warren W. DeLapp, with additional
    biographical notes about Warren DeLapp's life. Blurb: San Francisco. *P*
DeLapp, W.W. (1940). *Sediment behaviour in upward flow*. University of Iowa: Iowa City.
DeLapp, W.W. (1943). Discussion of Entrainment of air in water. *Trans. ASCE* 108: 1448-1451.
DeLapp, W.W. (1947). *The high velocity flow of water in a small rectangular channel*.
    University of Minnesota: Minneapolis.
DeLapp, W.W. (1947). *Hydraulic model studies for Chippewa Reservoir Dam*. St. Anthony
    Falls Laboratory. University of Minnesota: Minneapolis.
http://www.123people.com/ext/frm?ti=personensuche%20telefonbuch&search_term=warren%2
0delapp&search_country=

DEMING

21.02. 1817 Berlin CT/USA
10.01. 1894 Salem OH/USA

John Deming established in the 1830s with his elder brother William a general store at New Lyme OH. Before moving to Salem OH, John was involved in wholesale and retail businesses with his father-in-law. Around 1856, he there opened a grocery store, thereby becoming interested in the great success of manufacturers of labor-saving tools for carriage-makers and blacksmiths. He saw how the market for the firm's products continued to grow throughout the country.

Deming Pumps including the old 'Pitcher Spout Pump', on display at the Salem Historical Society Museum, symbolizes the Deming Co. This is the type of pump that could be found in the kitchens of many old farm houses. Also called 'Cistern Pump', it was used most commonly in houses for pumping cistern or well water. Because the cylinder was in the stock of the pump, it was limited in its capacity to withdraw water from a depth of no more than 7 m. When Deming rode into Salem on horseback in 1862, carrying his only possessions in a money belt, it hardly was an auspicious beginning for what today is an internationally-known name in the pump business. He eventually joined the manufacturing business of Dole and Silver, which became the Dole, Silver & Deming Co., and later the Silver & Deming Mfg. Co. It began manufacturing hand and windmill pumps in 1880. In 1890 the firm divided into two separate entities, with Deming specializing in pumps and hydraulic machinery. Up to 1894, the line of pumps manufactured included house force pumps, well pumps, rotary pumps, hydraulic rams, and a small line of spray pumps. The firm grew rapidly, expanding to produce pumps in sizes and capacities from the smallest to the large triplex and deep-well power pumps for use in mines, factories and waterworks. Deming products were sold throughout the world. Its pumps earned the reputation of being the world's best. One of the company's most memorable pumps was built in 1905. It was one of the largest ever manufactured in the USA. The giant pump, shipped to southern California for irrigation purposes, had a capacity of five million gallons a day. In 2006, the Deming Co. closed its doors and left Salem.

Anonymous (1894). John Deming. *Iron Age* 53: 166.
Shaffer, D.E. (2002). *Salem*: A Quaker city history. Arcadia Publishing Co.: Charleston SC.
http://www.salemohiohistory.com/HistoryMakers/John-Deming.aspx *P*
http://vintagemachinery.org/mfgindex/detail.aspx?id=1017&tab=7
http://jimdeming.com/ *P*

DETRA

23.03. 1925 Thompsontown PA/USA
30.06. 1997 Gloucester MA/USA

Ralph William Detra was educated at the University of Pennsylvania, Philadelphia PA until 1944, and at Cornell University, Ithaca NY until 1951, from where he moved to the Swiss Federal Institute of Technology, ETH Zurich, submitting his PhD in 1952. After having been a flight test engineer in the late 1940s, and instructor at Mechanical Engineering Department, Cornell University in 1950, he became after return from Switzerland to the USA supervising aerodynamic researcher, Westinghouse Electric Co., within its Aviation Gas Turbine Division, Kansas City MO until 1955. Until 1959 he was engaged then as principal research scientist at *Avco* Everett Research Laboratory, Everett MA, from where he moved as project manager to Avco Wilmington MA. He was from 1972 to 1981 associated with *energy technol*, acting there as vice-president, and president until his retirement in 1990. He was a member of the American Institute of Aeronautics and Astronautics AIAA.

Detra was from the 1950s interested in hypervelocity flight, and high temperature gas dynamics. His PhD thesis includes the main chapters 1. General considerations, 2. Preliminary theoretical investigations, 3. Experimental investigations, 4. Theory for the secondary flow in curved pipes, 5. Application of the theory, and 6. Losses due to secondary flow. It may be noted that in the early 1950s few works on the topic of secondary flow in bends were available, which is on the one hand astonishing because of the fundamental flow pattern and characteristics, but on the other hand may be explained with the relatively poor instrumentation by then available.

Anonymous (1964). Detra, Ralph W. *Who's who in engineering* 9: 446. Lewis: New York.
Detra, P. (2013). Ralph W. Detra. Personal communication. *P*
Detra, R.W. (1953). *The secondary flow in curved pipes*. ETH Zurich: Zürich.
Detra, R.W., Kemp, N.H., Riddell, F.R. (1957). Heat transfer to satellite vehicles re-entering the atmosphere. *Jet Propulsion* 27(2): 132-137; 27(12): 1256-1257.
Detra, R.W., Hidalgo, H. (1961). Generalized heat transfer formulas and graphs for nose cone re-entry into the atmosphere. *ARS Journal* 31(3): 318-321.
Ehrich, F.F., Detra, R.W. (1954). Transport of the boundary layer in secondary flow. *Journal of Aeronautical Sciences* 21(2): 136-138.
Kemp, N.H., Rose, P.H., Detra, R.W. (1959). Laminar heat transfer around blunt bodies in dissociated air. *Journal of the Aerospace Sciences* 26(7): 421-430.

DEXHEIMER

20.04. 1901 Denver CO/USA
26.11. 1974 Falls Church City VA/USA

Wilbur App Dexheimer was educated at Colorado State College, Fort Collins CO, from where he obtained the BS degree in 1926. He was then assistant city engineer of Fort Collins until 1927, location engineer of the Union Oil Company of California until 1928, from when he joined the US Bureau of Reclamation all through his professional career at Denver CO. He was a field engineer for the Hoover Dam from 1931 to 1936, then for the Bartlett Dam in Arizona until 1938, and for the Shasta Dam in California until 1942. After war service, Dexheimer returned in 1947 to USBR as chief of construction, becoming in 1948 assistant chief construction engineer and in 1953 Commissioner USBR until 1959.

In parallel to his engagements with USBR Dexheimer was also involved in a number of hydropower and irrigation projects as consultant, namely the Taiwan Power Company, the *Yanhee* Electricity Authority in Bangkok, Thailand, the Salto Grande Project of the Argentine-Uruguay Commission, the *Oras* Dam in Brazil, the *Bhakra* Dam in Punjab, India, or the State Electricity Commission of Victoria, Australia. He acted as a member of the Executive Commission of Large Dams USCOLD, and was a member of the Executive Board of the US National Commission of the World Power Conference. He finally was appointed from 1957 to 1960 president of the International Commission on Irrigation and Drainage ICID. He also authored several books on dam engineering, to which he had contributed significantly during his entire career.

Anonymous (1964). Dexheimer, Wilbur A. *Who's who in engineering* 9: 449. Lewis: New York.
Dexheimer, W.A. (1938). *World's highest multiple arch dams*. Bureau of Reclamation: Denver.
Dexheimer, W.A., ed. (1954). *Dams and control works*. Bureau of Reclamation. US
 Government Printing Office: Washington DC.
Dexheimer, W.A., Larson, E.A. (1957). *Vernal Unit, Central Utah Project*. USBR: Denver.
Dexheimer, W.A. (1958). *Friant-Kern Canal:* Technical record of design and construction
 constructed 1945-1951. US Dept. Interior, Bureau of Reclamation: Denver.
Dexheimer, W.A. (1965). Dams. *Encyclopedia Americana* 8: 433-445. Americana Corp.: NY.
Parks, T.L. (2004). *Glen Canyon Dam*. Arcadia: Charleston SC.
Seaton, F.A., Dexheimer, W.A. (1958). *Molokai Project*, Hawaii. USBR: Denver.
http://www.usbr.gov/history/CommissBios/dexheimer.html *P*
http://content.lib.utah.edu/cdm4/item_viewer.php?CISOROOT=/VE_Photos&CISOPTR=327 *P*

DIALOGUE

13.05. 1828 Philadelphia PA/USA
23.10. 1898 Atlantic City NJ/USA

John Henry Dialogue was an industrialist of French-German ancestry. His father, Adam, was also an entrepreneur and inventor, and established himself as manufacturer of riveted fire hose. Dialogue grew up in Philadelphia and was educated at Central High School, graduating in 1846. His uncle taught him machine work and drafting; in 1850, Dialogue moved to Camden NJ.

Dialogue started his business at Camden by repairing locomotives for the Camden & Amboy Railroad Co., as well as working on Camden and Philadelphia and West Jersey Ferry Companies ferryboats, which were then there common at the time, located on the Delaware River. In 1854, he purchased a foundry where his workers performed general machine work, as well as building Corliss stationary engines under a special license for the inventor George H. Corliss (1817-1888). This engine was a new invention, greatly increasing the steam engine's efficiency because of its innovative governor and valve design. In 1862, the Camden National Iron Armor and Shipbuilding Company was founded, which constructed small ships, but the company closed before the Civil War ended. Dialogue then acted as subcontractor for Wilcox and Whiting, which took over the shipyard during the 'weak' economic period from 1865 to 1870. In 1870 Dialogue founded the River Iron Works, Dialogue & Wood, proprietors, building iron ships. Dialogue eventually became partner in the firm with his son, which then became known as John H. Dialogue & Son. This shipyard produced a large number of tugboats, both for civilian use and for the US Navy. The shipyard was innovative, and was one of the first to adopt the compound marine engine and the Scotch boiler. By the late 19th century, the large shipyard had the honor of doing reconstruction work on the famous USS *Constitution*. After trying to restore his ailing heart by resting at his home at Atlantic City, he died, so that his son, John H., Jr., took over the shipyard continuing work before World War I, when the younger Dialogue was forced into bankruptcy. The property was purchased by the Reading Railroad at less than half its appraised value, and the shipyard was demolished and reconstructed into Reading's Camden terminal.

Anonymous (1898). John H. Dialogue. *American Engineer and Railroad Journal* 72: 418.
Anonymous (1898). John H. Dialogue. *Engineering Record* 38: 465.
Anonymous (1898). John H. Dialogue. *Iron Age* 62(Oct.27): 21.
http://en.wikipedia.org/wiki/John_H._Dialogue
http://dvrbs.com/people/CamdenPeople-JohnHDialogue.htm *P*

DIEMER

27.04. 1888 Palmyra MO/USA
27.10. 1966 Los Angeles CA/USA

Robert Bernard Diemer graduated in 1911 from the University of Missouri, Columbia MO, with the BSc degree in civil engineering. Upon his graduation he moved to the North Platte Project in Wyoming and Nebraska States, as principal assistant within the US Bureau of Reclamation USBR. He collaborated later with Frank E. Weymouth (1874-1941) on irrigation projects in Mexico. In 1929 he accepted work with Los Angeles CA on its study of a Colorado River water supply. When the newly formed Metropolitan Water District MWD in 1930 took over the design and construction of this project, Diemer became first assistant and general manager. He successively rose up through all positions becoming in 1950 assistant general manager of MWD, and in 1952 general manager and chief engineer, serving in this position until his retirement in 1961. He was recipient of the 1961 George A. Elliott Award of the American Water Works Association AWWA. He also was honorary member of the American Society of Civil Engineers ASCE.

The purposes of the Colorado River Aqueduct were a continuation of the long trend of population and the industrial growth in Southern California without limitation due to lack of water supply, replenishment of the region's underground water storage basins by spreading or by partial decreases of well pumping, and protection against at least the major effects of droughts. Without the Aqueduct, the unprecedented region's growth would not have occurred. It appeared in the early 1950s that the optimum solution for solving the water supply problem was replenishment. By limiting the pumpage to a safe yield of the basin and supplying all additional requirements from the Colorado River Aqueduct, water levels of wells were effectively restored in spite of years of low rainfall and runoff. The Aqueduct was thus a substantial addition to California, and was at least partially responsible for the great success of the Golden State in the following decades. Diemer and his collaborators have notably contributed to this success. The Robert B. Diemer Treatment Plant at Yorba Linda is one of five in the MWD.

Anonymous (1963). Robert B. Diemer. *Engineering News-Record* 171(Oct.17): 20. *P*
Anonymous (1967). Diemer, Robert B. *Trans. ASCE* 132: 679.
Diemer, R.B. (1953). Expansion of the Colorado River Aqueduct System. *Journal of the American Water Works Association* 45(4): 397-404.
Diemer, R.B. (1959). Colorado River Aqueduct as a source of present and future supplies in Southern California. *Journal of the American Water Works Association* 51(4): 463-470.

DIXON J.W.

01.04. 1905 Burlington IA/USA
04.12. 1978 Arlington VA/USA

John Wesley Dixon was educated at University of Iowa, receiving the BE and ME degrees in 1927 and 1940. He was from 1927 to 1940 a member of the US Corps of Engineers, working at St. Louis MO and Rock Island IL, joining then the Public Works Project Review Section, Washington DC, as chief of the Water Resources Section until 1943. In 1945 he acted as engineering assistant to the Commissioner of the Bureau of Reclamation USBR, Washington DC, and from 1946 to 1954 he was the Director of Project Planning USBR. Dixon was member of the American Society of Civil Engineers, the American Geophysical Union, the Permanent International Association of Navigation Congresses PIANC, and the International Commission of Irrigation and Drainage ICID.

Dixon published during his career numerous reports to the Congress on proposed public works and water resources developments as official federal documents. These include particularly projects sponsored by the World Bank in which he was involved in the design and construction stages. He was also an author of books on environmental engineering related to the economic analysis, and on the relation between dams and the environment, which has been a major topic first in the USA starting in the 1960s, and eventually spreading to the other continents. He acted as chairman of the US Sections of two international engineering boards. He also was a consultant to the Middle East from 1954 to 1956. Further he was assistant to the president of Harza Engineering Company from then, becoming its chief hydraulic engineer and special assistant of this large firm founded by Leroy F. Harza (1882-1953), then taken over by son Richard D. Harza (1923-). Dixon was the recipient of the John Dunlap Memorial Award from the Iowa Engineering Section in 1940.

Anonymous (1927). John W. Dixon. *Hawkeye yearbook*: 71. University of Iowa: Iowa. *P*
Anonymous (1945). Jack W. Dixon. *Engineering News-Record* 134(Mar.8): 154. *P*
Anonymous (1964). Dixon, John A. *Who's who in engineering* 9: 460. Lewis: New York.
Dixon, J.W. (1952). Social and economic implications of irrigation development. *Civil Engineering* 22(8): 689-690; 22(9): 834-835; 22(9): 839.
Dixon, J.W. (1953). Planning an irrigation project today. *Centennial Trans. ASCE*: 357-387.
Hufschmidt, M.M., James, D.B., Meister, A.D., Bower, B.T., Dixon, J.A. (1983). *Environment, natural systems and development*: An economic valuation guide. The Johns Hopkins University Press: Baltimore MD.

DOBBINS

01.03. 1913 Woburn MA/USA
29.12. 1991 Smithtown NY/USA

William Earl Dobbins attended the Massachusetts
Institute of Technology MIT, receiving the degrees
BSc in sanitary engineering in 1934, MS in 1935
and PhD in 1941. The latter, made under Thomas R.
Camp (1895-1971), examined the turbulence effects
on sedimentation in sewage treatment stations. Then
Dobbins held a variety of positions in academia and
consulting. From 1936 to 1939 he taught at Roberts
College in Turkey, and from 1940 to 1950 he held
positions with a number of Boston consulting firms.
From 1943 to 1946 he served with the US Navy
Seabees in Europe, the Middle East, and Northern
Africa. In 1950 he joined the New York University as associate professor of sanitary
engineering, and became professor of civil engineering in 1957, staying there until
1967. It was there where he completed the research for which he is well known.

In 1966 Dobbins co-founded the environmental consulting firm Teetor-Dobbins, Long
Island NY, joining in 1967 the consulting firm as its president. The firm constructed
wastewater treatment plants and other major facilities in New York and Connecticut.
Dobbins continued in this firm until his retirement in 1975. He was a Fellow of the
American Society of Civil Engineers ASCE, a Member of the Water Pollution Control
Federation, and the American Waterworks Association AWWA, and a Diplomate of the
American Academy of Environmental Engineers. He received ASCE's Rudolph Hering
Medal in 1959 and 1965, the 1956 Desmond Fitzgerald Award from the Boston Society
of Civil Engineers, and the Kenneth Allen Memorial Award from the New York Water
Pollution Control Association in 1955. He was known for the pioneering work he led at
New York University in the area of stream dissolved oxygen relations. He published in
1958 with a colleague the widely used formula for stream aeration bearing their names.

Anonymous (1962). Dobbins, William E. *Who's who in American education*: 402. *P*
Anonymous (1992). William E. Dobbins. *Trans. ASCE* 157: 502-503.
Bella, D.A., Dobbins, W.E. (1968). Difference modelling of stream pollution. *Journal of the
Sanitary Engineering Division* ASCE 94(SA5): 995-1016.
Dobbins, W.E. (1944). Effect of turbulence on sedimentation. *Trans. ASCE* 109: 629-678.
Dobbins, W.E. (1964). BOD and oxygen relationships in streams. *Journal of Sanitary
Engineering Division* ASCE 90(SA3): 53-78.
O'Connor, D.J., Dobbins, W.E. (1958). Mechanism of reaeration in natural streams. *Trans.
ASCE* 123: 641-666; 123: 679-684.

DOBSON

27.02. 1881 Philadelphia PA/USA
29.12. 1945 Tulsa OK/USA

Gilbert Colfax Dobson graduated as a civil engineer from the University of Missouri, Kansas MO, in 1909. He was first assistant city engineer at Manila, the Philippines until 1910, then a special agent of the US General Land Office, to continue until 1916 on Panama Canal with field work for the Gatun spillway and locks. The following year was spent with railroad works at Kansas City, and from 1917 to 1918 Dobson was a captain of the US Corps of Engineers in France. Upon return to the USA he was a professor of military sciences and tactics at Colorado School of Mines until 1920, when joining a firm on special hazard insurance reports until 1925. After having been a consultant in Tulsa OK, Dobson joined from 1929 to 1935 the staff of the US Engineering Office, Memphis TN, dealing with flood control studies of Arkansas River and its tributaries. Finally, Dobson was employed with research work in the Soil Conservation Service SCS, Washington DC, thereby advancing to Chief of the Sedimentation Division until retirement in 1943. During his last years he was a private consultant at Tulsa OK.

During his stay with SCS from 1935 to 1943, Dobson served both as field engineer and specialist in mathematical investigations. His most valuable technical contribution was a review of the mathematical processes employed by Paul Du Boys (1847-1924) in his Study of the regime of the Rhone River and the action exercised by the water on an indefinitely shifting bed of gravel, published in 1879. His first position in SCS was as hydraulic engineer, supervising investigations of bed-load movements in natural streams. In 1936 he became Head of the Sedimentation Section, in charge of mainly reservoir sedimentation studies with related laboratory work. He prepared reports and papers with the US Department of Agriculture, most notably the 1940 bulletin.

Anonymous (1939). Dobson, G.C. 1st *Hydraulics Conference* Iowa: Frontispiece. *P*
Anonymous (1941). Dobson, Gilbert C. *Who's who in engineering* 5: 467. Lewis: New York.
Anonymous (1947). Gilbert Colfax Dobson. *Trans. ASCE* 112: 1434-1437.
Dobson, G.C. (1938). Discussion of Turbid water through Lake Mead. *Trans. ASCE* 103: 759-763.
Dobson, G.C., Johnson, J.W. (1940). Studying sediment loads in natural streams. *Civil Engineering* 10(2): 93-96.
Dobson, G.C. (1943). Discussion of Suspended-matter concentration. *Trans. ASCE* 108: 958-959.
Happ, S.C., Rittenhouse, G., Dobson, G.C. (1940). Some principles of accelerated stream and valley sedimentation. *Bulletin* 695. US Department of Agriculture: Washington DC.

DODGE

07.11. 1893 Whitmore Lake MI/USA
08.12. 1971 Ann Arbor MI/USA

Russell Alger Dodge was educated at University of Michigan, Ann Arbor MI, from where he obtained the civil engineering degrees BS and MS in 1916 and 1918, respectively. He was from 1916 to 1917 a teaching assistant there, and an instructor of storm and sanitary sewer constructions. After war service in 1918, Dodge returned to the USA joining the US Reclamation Service for the Rio Grande Project in 1919. In 1920 he was with the Consumers Power Co., Jackson MI, assistant to its hydraulic engineer. From 1921 Dodge was from staff member first to professor of engineering, and from 1953 chairman of the Department of Engineering, University of Michigan. He was a member of the American Society of Civil Engineers ASCE.

Dodge was known for his works in drainage, wastewater treatment and water supply engineering. He has written in 1937 the book Fluid mechanics with a colleague, in which the principles of hydraulics are presented. Its chapters are: 1. Fundamentals, 2. Fluid statics, 3. The flow of an ideal fluid, 4. The flow of a real fluid, 5. Similarity and dimensional analysis, 6. Fluid flow in pipes, 7. Fluid flow in open channels, 8. Fluid measurement, and 9. Flow about immersed objects. Other works deal with hydraulic modeling, particularly relating to sediment hydraulics and the various procedures to be taken into consideration, including for instance scale effects, to guarantee correct up-scaling from the laboratory observations to prototype structures. The 1963 paper deals again with fluvial hydraulics related to jetties.

Anonymous (1964). Dodge, Russell Alger. *Who's who in engineering* 9: 464. Lewis: New York.
Carlson, E.J., Dodge, R.A. (1963). Control of alluvial rivers by steel jetties. *Trans. ASCE* 128(4): 347-375.
Dodge, R.A., Thompson, M.J. (1937). *Fluid mechanics*. McGraw-Hill: New York.
Dodge, R.A., Smith, H.J., Sonnemann, G. (1957). *A study and development of paravanes of the high lift-drag ratio type and the high-lift type*. Engineering Research Institute: University of Michigan: Ann Arbor MI.
Dodge, R.A. (1983). *Model similitude*: Extended for active sediment transport. Water Resources Research Laboratory, USBR: Denver CO.
http://um2017.org/faculty-history/faculty/russell-alger-dodge *P*
deepblue.lib.umich.edu/.../bac9524.0001.001.p... *P*

DODKIN

22.06. 1902 Foxboro MA/USA
07.09. 1953 Sao Paulo/BR

Oswald Hewitt Dodkin received the BSc degree in mechanical engineering in 1923 from Worcester Polytechnical Institute, Worcester MA. He joined in 1926 the Brazilian Traction, Light and Power Co., Sao Paulo BR, as test engineer. After having served in various capacities, he became in 1950 head of that company's Planning Department, which was then organized. Dodkin was also active in civic affairs in the American community of Sao Paulo, and there was member of several technical societies. He further helped to organize the Brazilian Section of the American Society of Civil Engineers ASCE, whose member he was from 1948.

Dodkin was enormously active in the conception and construction of Brazilian hydro-power developments, directing investigations for important projects. For the *Cubatao* Project near San Paulo, he suggested and helped develop the idea of making the pumping plants reversible for peaking purposes, a procedure often applied currently in hydraulic engineering. The scheme was inaugurated in 1927 providing a capacity of 35 MW. It was later expanded before an unstable mountain slide dictated further expansion as an underground scheme. Dodkin also contributed to the knowledge of water hammer effects on unprecedented penstocks for the Cubatao Project. These studies resulted in the 1935 paper authored in collaboration with Asa W.K. Billings (1876-1949) and Fred H. Knapp (1901-1971), which is considered a classic addition to engineering literature. The 1940 paper presents tests on the verification of the reliability of the salt-velocity method for measuring water discharge. The procedure involves cross checks with the method itself, check tests using the pressure-time approach of Norman R. Gibson (1880-1967), and by volumetric tests. The agreement among these methods was found satisfactory.

Anonymous (1953). Oswald H. Dodkin. *Civil Engineering* 23(11): 794. *P*
Anonymous (1953). Oswald Hewitt Dodkin. *Mechanical Engineering* 76(1): 132.
Anonymous (1956). Oswald H. Dodkin. *Trans. ASCE* 121: 1416-1417.
Billings, A.W.K., Dodkin, O.H., Knapp, F., Santos, Jr., A. (1933). High-head penstock design. Proc. *ASME Water Hammer Symposium* Chicago: 29-61.
Dodkin, O.H. (1940). Field checks of the salt-velocity method. *Trans. ASME* 62(11): 663-676.
McDowell, D. (1988). *The Light*: Brazilian Traction, Light, and Power Company Limited 1899-1945. University of Toronto Press: Toronto.

DOHERTY

22.01. 1885 Clay City IL/USA
19.10. 1950 Winter Park FL/USA

Robert Ernest Doherty received his education from University of Illinois, with a BS degree in 1909, and from Union College, Schenectady NY, with the MS degree in 1921. He further received the honorary degrees MA from Yale in 1931, and LL.D from the University of Pittsburgh PA in 1936. He was from 1909 to 1931 employed at the General Electric Co., Schenectady, from when he was professor of electric engineering at Yale University until 1936. Later in his carrier until retirement in 1950, he took over as president the Carnegie Institute of Technology, Pittsburgh PA. He served on the Civilian Advisory Council of the Office of Chief of Ordnance from 1942 to 1945, on the Board of Visitors to the US Naval Academy in 1944, and was in the National Advisory Commission for Aeronautics in 1940.

Doherty was one of the outstanding members of the profession in the USA, combining in excellent proportions the practicing engineer and the teacher. All through his life he was interested in technical education and thereby made a great contribution to its improvement. He supported technical organizations and under his presidency the Engineers' Council for professional development made its greatest advance. Besides he also accepted heavy obligations on behalf of his country. For instance, he was chairman of the Consultative Committee on Engineering, an expert of the Army Specialized Training Division or chaired the OPM Production Planning Board in 1941. President Roosevelt appointed him also to the Board of Naval Academy at Annapolis. He has written the text Mathematics of modern engineering, dealing with the mathematical formulation of relevant engineering problems, basic engineering mathematics, vector analysis, and Heaviside operational calculus. A second volume was published in 1961.

Anonymous (1943). Robert E. Doherty. *Engineering News-Record* 131(Jul.1): 3. *P*
Anonymous (1944). Dr. R.E. Doherty. *The Engineering Journal* 27(3): 166. *P*
Anonymous (1948). Doherty, Robert E. *Who's who in engineering* 6: 519. Lewis: New York.
Doherty, R.E. (1909). *The comparison of gas engines and steam turbines as prime movers*, and the design of a six hundred kW steam turbine plant. University of Illinois: Urbana.
Doherty, R.E., Keller, E.G. (1936). *Mathematics of modern engineering*. Wiley: New York.
Doherty, R.E. (1950). *The development of professional education*: The principles which have guided the reconstruction of education at Carnegie Institute of Technology 1936-1950. University of Illinois: Urbana-Champaign.

DOLAND

01.08. 1890 Denver CO/USA
23.12. 1960 Champaign IL/USA

James Joseph Doland obtained his BS degree in 1914 from University of Colorado, his MS degree from the University of Illinois in 1932, and in 1944 received the Honorary PhD degree from St. John's University, Collegeville MN. He was until 1916 an instructor of engineering mathematics at University of Colorado, until 1923 an engineer at Minneapolis MN, and until 1926 an engineer with the US Bureau of Reclamation USBR, from when he took over as civil engineering professor at University of Illinois, Urbana IL. From 1936 he was in parallel consulting engineer for the National Resources Planning Board. In 1941 Doland was also the principal engineer for USED, and British colonies in the Caribbean for lend-lease airbases. From 1944 he was a consulting hydraulic engineer for the Union Electric Co., St. Louis MO. He was further member of the Joint Council on National Water Policy. Doland was a member ASCE, and of the American Institute of Consulting Engineers AICE.

Doland was at University of Illinois for 32 years. He was recognized as one of the nation's prominent engineers dealing with power development, hydrology and water resources planning. Students were excited of Doland's lectures. One stated: 'He was not only a wonderful teacher, but he was much more than that. He had the unique character of making everybody in his class feel somehow that he was somebody Doland was personally interested in, and he certainly gave that feeling from the start. Hundreds of his former students share similar feeling that I have in that respect'. Doland is known for three books, namely Hydro-power engineering, Water-supply engineering and Low dams. He is also noted for his co-authorship with Hardy Cross (1885-1959) of the Cross-Doland Method on flow analysis in water distribution systems.

Anonymous (1941). Doland, James Joseph. *Who's who in engineering* 5: 472. Lewis: New York.
Anonymous (1959). Doland, James Joseph. *Who's who in America* 28: 708. Marquis: Chicago.
Doland, J.J. (1930). Design of monolithic concrete siphons simplified by use of diagrams. *Engineering News-Record* 104(26): 1047-1052.
Doland, J.J., Chow, V.T. (1952). Discussion of River channel roughness, by H.A. Einstein, N.L. Barbarossa. *Trans. ASCE* 117: 1134-1139.
Doland, J.J. (1954). *Hydro-power engineering*. Ronald Press: New York.
http://209.85.129.132/search?q=cache:a7e7nXpUGSYJ:cee.illinois.edu/about/history/doland+%22James+Joseph+Doland%22&cd=1&hl=de&ct=clnk&gl=ch *P*

DOMINY

04.12. 1909 Hastings NE/USA
20.04. 2010 Boyce VA/USA

Floyd Erin Dominy graduated from the University of Wyoming, Laramie in 1933, with a BA degree in economics and a BS in agricultural engineering. He was then appointed county agricultural agent, and in 1938 became field agent for the Western Division of the Agricultural Administration, Washington DC. In 1944 he was commissioned in the US Naval Reserve serving during the last two war years on Pacific islands. He joined in 1946 the US Bureau of Reclamation USBR as land development specialist responsible for establishing procedures by which the newly irrigated land could be returned to war veterans. He learned the Bureau from the ground, advancing to chief of the Irrigation Division, and to assistant commissioner in 1957, taking over in 1959 as Commissioner the Bureau by appointment of President Eisenhower. He was retired in 1969. He was considered as one of the Nation's giants in water and land development projects and was named in 1966 one of the Public Works Man of the Year.

The 1966 paper deals with intakes in irrigation works and the optimum of desilting the water discharge. It is stated that the economic disposal of sediment requires a thorough knowledge of sediment control devices including cost, construction, operation and maintenance. If the effect of sediment is not considered, then injurious effects on the land and the crop may result. By utilizing hydraulic laboratory techniques, the USBR adopted final prototype designs for the construction of its irrigation system, which are presented in this work. Later, it was stated that Dominy was a man who grew up in the dry lands of Nebraska, and knew first-hand the damage that lack of water can have. It was for this reason he dedicated himself to building dams supplying the society with water. His energy and honesty came through loud and clear.

Dominy, F.E. (1966). Design of desilting works for irrigation systems. *Journal of the Irrigation and Drainage Division* ASCE 92(IR4): 1-26; 93(IR1): 281-287; 94(IR3): 339-341.
Dominy, F.E. (1968). Role of irrigation in the West's expanding economy. *Journal of Irrigation and Drainage Division* ASCE 94(IR4): 401-418; 95(IR4): 607-623; 96(IR2): 223.
Pisani, D.J. (2002). A tale of two commissioners: Frederick H. Newell and Floyd E. Dominy. *History of Bureau of Reclamation*: A symposium: 637-650. Las Vegas NV. *P*
Warne, W.E. (1973). *The Bureau of Reclamation*. Praeger: New York. *P*
http://en.wikipedia.org/wiki/Floyd_Dominy *P*
http://www.legacy.com/obituaries/washingtonpost/obituary.aspx?n=floyd-dominy&pid=142123170#fb

DONNELLY

24.10. 1917 Spartanburg SC/USA
19.02. 1979 Minneapolis MN/USA

Charles 'Chuck' Allen Donnelly joined in 1941 the Soil Conservation Service CSC as engineering aide, and in the early 1950s as hydraulic engineer the St. Anthony Falls Hydraulic Laboratory, University of Minnesota, Minneapolis MN, where he stayed all through his professional career. Few details on his education are currently available. Donnelly has mainly collaborated with Fred W. Blaisdell (1911-1998), with whom he particularly developed various hydraulic structures, notable drop structures and particular intake structures. Donnelly retired at age of only 58 in 1975.

The 1956 paper deals with a particular drop structure. A large number of parameters affecting the flow and scour features was tested. The 1966 report is concerned with the development of a generalized method to determine the free flow capacity of the box-inlet drop spillway, whereas no generalized approach was found for submerged flow conditions. The latter may be estimated by interpolation using submergence curves available for a wide range of flow parameters. The 1975 paper presents experimental results on a rectangular drop inlet of width equal to the barrel diameter, a flat or semi-cylindrical bottom, and a flat, horizontal anti-vortex plate supported above the drop inlet crest, which is referred to as two-way drop inlet because water enters over only the two sides of the rectangular drop inlet. All its dimensions are expressed in terms of the pipe diameter so that the results apply to any structure similar to that tested.

Anonymous (1948). Charles Donnelly. *Gopher yearbook*: 352. University of Minnesota. *P*

Blaisdell, F.W., Donnelly, C.A. (1956). The Box Inlet Drop spillway and its outlet. *Trans. ASCE* 121: 955-986; 121: 992-994.

Blaisdell, F.W., Donnelly, C.A. (1966). Hydraulic design of the box-inlet drop spillway. *Agricultural Handbook* 301. Agricultural Research Service: Washington DC.

Blaisdell, F.W., Donnelly, C.A., Yalamanshili, K., Hebaus, G.G. (1975). The two-way drop inlet self-regulating siphon spillway. Symposium *Design and operation of siphons and siphon spillways* London C(4): 31-53. BHRA: Cranfield UK.

Blaisdell, F.W., Donnelly, C.A. (1975). The hood inlet self-regulating siphon spillway. Symposium *Design and operation of siphons and siphon spillways* London C(111): 137-154. BHRA: Cranfield UK.

Donnelly, C.A., Hebaus, G.G., Blaisdell, F.W. (1974). Hydraulics of closed conduit spillways. Agricultural Research Service *Report* ARS-NC-14. US Dept. Agriculture: Washington.

DORE

19.06. 1898 Somerset MA/USA
22.01. 1987 Clearwater FL/USA

Stanley Milburn Dore obtained in 1920 the BS degree in civil engineering from Brown University, Providence RI. He was from 1922 member of a firm at Boston MA, and engineer of the US Engineers Office at Providence RI, taking there over from 1928 to 1939 as assistant engineer for its Water Supply Board. He was then assistant engineer and assistant to the chief engineer from 1950 for Massachusetts Metropolitan District Water Supply Commission and its successor, on the broad extension of water supply and sewerage systems of Boston, particularly Quabbin Reservoir and the New Pressure Aqueduct.
He was also in charge of the Nut Island sewage treatment plant, the storm overflow conduits, and pumping stations. From 1950 to 1952 Dore was deputy chief engineer of the Allegheny County Sanitary Authority of Pittsburgh PA. He was appointed in 1952 deputy chief engineer in charge of the Design Department of the Board of Water Supply in New York, on additions to the water supply project developed on Delaware River. In 1956 he became its chief engineer, in charge of the design and construction on major water supply projects for the city, thereby succeeding Karl R. Kennison (1866-1977).

The New York City water supply went parallel to the massive increase of population in the city. Around 1900 the city leaders were forced to expand the supply. The watershed land in the Catskill Mountains were purchased and new reservoirs were created by dams so that the Catskill and Delaware Reservoirs System was created, which were then completed in the 1920s, and 1960s, respectively. The current watershed used for water supply covers the size of Delaware State. The water flows to the city through aqueducts. The Croton System has 12 reservoirs and 3 lakes. The largest, the New Croton Reservoir, supplies up to 10% of the city's water supply. Dore's contribution to this scheme was mainly in the technical support. Previously, he had become expert in soil mechanics thereby also having published the 1940 book. He was member of the American Society of Civil Engineers ASCE and in 1939 winner of its Laurie Award. He also was member of the New England Water Works Association, serving as its president in 1952.

Anonymous (1956). Chief engineer for NY water: Stanley M. Dore. *Engineering News-Record*: 157(Jul.18): 72. *P*
Anonymous (1959). Dore, Stanley M. *Who's who in engineering* 8: 644. Lewis: New York.
Dore, S.M. (1937). Permeability determinations, Quabbin Dams. *Trans. ASCE* 102: 682-711.
Plummer, F.L., Dore, S.M. (1940). *Soil mechanics and foundations*. Pitman: New York.

DOUGHERTY J.W.

01.01. 1913 Edmore ND/USA
01.05. 1954 Pasadena CA/USA

John Wilson Dougherty received in 1934 the degree of Bachelor of Science in civil engineering from Oregon State College, Corvallis OR. In 1938 he received the degree of Master of Science from the Carnegie Institute of Technology, Pittsburgh PA. He then taught hydraulics at Carnegie Institute of Technology and at State University of Iowa, Iowa City IA. After war service in Europe, he joined in 1946 the Bechtel Corporation, Los Angeles CA, as chief supervisory engineer. He was associated with the American Society of Civil Engineers ASCE, starting in 1934 as Junior Member, becoming in 1947 Associate Member, and member in 1951. Dougherty passed away due to unknown reasons at age of only 41.

The 1935 paper by Leroy F. Harza (1882-1953) was concerned with the basic principles of flow under dams made of sand. Both analytical methods and the electric analogy were applied, resulting in (1) Hydrostatic pressure along foundation contact, (2) Hydraulic gradient with which the water escapes upward at the dam toe, and (3) Approximate leakage under the dam structure. The Discussion by Charles A. Mockmore (1891-1953) and his student Dougherty presents experiments made at Oregon State University with vertical water flow through beds made up of various granulometry. Generally, the approach of Harza was found valid, but limitations must be stated to correctly apply it by engineers to prototypes. Dougherty's responsibilities with Bechtel Corporation were the design of hydropower installations in the high Sierras. He was also involved in works with flumes, tunnels and inverted siphons. He was the chairman of the Reception Committee of the Los Angeles ASCE Section in 1952, and a member of the Regional Subsection Committee in 1953. He further was a registered professional engineer both in California and Oregon.

Anonymous (1954). John Wilson Dougherty. *Civil Engineering* 24(7): 497. *P*
Anonymous (1954). John Wilson Dougherty. *Engineering News-Record* 152(May20): 124.
Anonymous (1955). John W. Dougherty. *Trans. ASCE* 120: 1562.
Dougherty, J.W., Douglas, J.M. (1940). Mining practice at the Hollinger gold mine. *Mining Technology* 4(2): 20-21.
Harza, L.F. (1935). Uplift and seepage under dams on sand. *Trans. ASCE* 100: 1352-1406.
Mockmore, C.A., Dougherty, J.W. (1935). Discussion of Uplift and seepage under dams on sand. *Trans. ASCE* 100: 1396-1401.

DOUGHERTY R.E.

13.02. 1880 New York NY/USA
29.09. 1961 White Plains NY/USA

Richard Erwin Dougherty graduated in 1911 as a civil engineer from Columbia University, New York NY. For the next year he there was an instructor in engineering, from when he was associated with the New York Central System until 1951. This included until 1918 resident and district engineer, from then to 1924 design engineer, and engineering assistant later to both the vice-president and president. Toward the end of his career, he acted as chairman of the research council. He retired in 1948, yet remained advisory consultant, including the supervision of the Grand Central Terminus real estate until 1951. He was then a private consultant, dealing among others with the Lake Front Dock & Rail Road Terminal Co., Toledo NY. In 1948, Dougherty served as president, the American Society of Civil Engineers ASCE.

Dougherty was mainly a railroad engineer, but also involved in projects of hydraulic engineering. His largest and best known job was the elimination of all of the railroad's grade crossings along New York City's Hudson River waterfront, which permitted the construction of the West Side elevated highway. Both the railroad's pier and wharf construction and its real estate operations were other facets of his responsibilities. Dougherty particularly served the engineering community of the United States, both through commitments toward the city of New York, but particularly through his services in ASCE. These included ASCE directorship from 1928 to 1930, ASCE vice-president from 1944 to 1945 and finally ASCE president in 1948. He held many awards, among them Columbia University's *Egleston* Medal and Citation for distinguished engineering achievements, and the Columbia University Medal for Excellence. He further received the Townsend Harris Medal from the Association of Alumni, City College of New York in 1949.

Anonymous (1947). Richard E. Dougherty. *Engineering News-Record* 139(Jul.24): 97-99. *P*
Anonymous (1948). Inaugurated as 1948 ASCE President. *Civil Engineering* 18(2): 104.
Anonymous (1953). Elected president of the Moles. *Civil Engineering* 23(6): 422.
Anonymous (1959). Dougherty, Richard E. *Who's who in America* 28: 719. Marquis: Chicago.
Anonymous (1961). ASCE past president R.E. Dougherty dies. *Civil Engineering* 31(11): 75. *P*
Anonymous (1961). Noted railroad engineer dies. *Engineering News-Record* 167(Oct.5): 27. *P*
Dougherty, R.E. (1948). The Society's future and its economic limitations: Annual Presidential
    Address. *Trans. ASCE* 113: 1391-1396.

DOUGLASS D.B.

21.03. 1790 Pompton NJ/USA
21.10. 1849 Geneva NY/USA

David Bates Douglass graduated in 1813 from Yale University to join the US Army Corps of Engineers as second lieutenant at West Point. In 1815 he was assigned as assistant professor of natural philosophy to the Military Academy. Due to success he was quickly promoted to the chairs of mathematics and engineering, and in parallel received key outside assignments from the government. He served with surveys of the defenses of Long Island Sound, or the exploration of the Lake Superior region in 1820. Later he acted as consultant of canal and railroad corporations, so that he resigned from the Army to devote his future entirely to engineering.

Douglass took interest in the Morris and Essex Canal of New Jersey, which was under construction in the 1830s. He was particularly interested in the substitution of inclined planes with mechanical lifting power for canal locks. After having directed these works, he was professor of civil engineering at the City University of New York, thereby being engaged in the water supply of New York City. Acting from 1834 to 1836 as engineer for the commissioners, Douglass selected the Croton watershed in preference to two other sources, located the route of the aqueduct, and determined all the essential features of the system, including the crossing of Harlem River on a high bridge. With later enlargements, this system continued to supply New York with water for seventy-five years. Before the actual construction of the Croton Aqueduct had been begun, Douglass was superseded as chief engineer, but his plans were essentially followed. It appeared that incompatibility had developed between him and the chairman of the Board of Commissioners. Douglass then continued in designing cemeteries, and from 1848 was a professor of mathematics at Geneva College, Geneva NY. He died the following year as a result of a paralytic strike.

Anonymous (1887). Douglass, David Bates. *Appletons' cyclopaedia of American biography* 2: 216-217. Appleton: New York.

Anonymous (1930). Douglass, David Bates. *Dictionary of American biography* 5: 405-406. Scribner's: New York.

Jackman, S.W. (1964). David Bates Douglass' journal. *American Neptune* 24(4): 280-293.

Jackman, S.W., Freeman, J.F., eds. (1969). *American voyageur:* The journal of David Bates Douglass. Northern Michigan University Press: Marquette MI.

http://clarke.cmich.edu/detroit/douglass1820.htm *P*

DOUGLASS L.R.

02.03. 1888 Gallup NM/USA
13.01. 1979 Boulder City CO/USA

Louis Rea Douglass obtained the BSc degree in 1928 from the University of Colorado, Boulder CO, the civil engineering degree in 1934, and the MSc degree in 1939. He had previously attended Colorado Agricultural College, Fort Collins CO. He was from 1909 to 1917 rodman and office engineer of a firm at Trinidad CO, and until 1919 in the US Army. From 1919 to 1921 Douglass was engineer in charge of hydroelectric developments. From 1922 to 1925 he was supervising engineer for both the design and the construction of storm sewers at Trinidad CO. After having been engaged in private practice at Denver CO dealing with irrigation projects, Douglass then joined in 1933 the US Bureau of Reclamation USBR, where he was successively employed as civil engineer, chief safety engineer, assistant supervising engineer on the construction of dams, canals and other irrigation structures for water conservation, and utilization projects in the Great Plains area. He was in the mid-1940s promoted to USBR engineer assistant to the commander, and in 1950 to acting director of power for the Boulder Canyon Project.

Douglass became a staff member of the USBR once the large reclamation projects were initiated in 1933. He worked on the design of Hoover Dam, and in 1937 was appointed the first safety engineer of the Bureau. In 1941 he became assistant to the supervising engineer on the construction of water conservation and utilization projects under the Wheeler-Case Program. He was appointed in 1948 consultant to the National Resources Section, Supreme Allied Powers, on land reclamation in Japan. From 1950 to 1954, as director of power, he supervised the installation of three new generating units at the Arizona wing of the Hoover Dam, one of which was the first to have a solid stainless steel turbine runner. At retirement in 1954, Douglass received the US Distinguished Service Citation and Gold Medal 'for outstanding contributions in the field of water conservation and control'. He was member of the American Society of Civil Engineers ASCE, the Colorado Society of Engineers, and the Denver Teknik Club.

Anonymous (1948). Douglass, Louis R. *Who's who in engineering* 6: 529. Lewis: New York.
Anonymous (1950). L.R. Douglass. *Engineering News-Record* 145(Sep.14): 65. *P*
Douglass, L.R. (1916). Irrigation-ditch velocity and discharge. *Engineering News* 76(2): 72-73.
Douglass, L.R. (1947). Davis Dam completes storage regulation of Colorado River below
      Boulder. *Civil Engineering* 17(1): 14-17.
http://en.wikipedia.org/wiki/Louis_R._Douglass *P*

DOUMA

30.05. 1912 Hanford CA/USA
04.10. 2004 Great Falls VA/USA

Jacob Hendrick Douma obtained the BS degree from University of California, Berkeley CA in 1935, and then joined the US Army Corps of Engineer for two years as hydraulic engineer at Vicksburg MS. From 1936 to 1939 he was a staff member of the US Bureau of Reclamation, and then returned to his former position until 1955, finally as chief hydraulic engineer until 1979. Douma then was a consulting hydraulic engineer until 1990 at Washington DC. He was a member of the US Commission of Large Dams, and the Permanent International Association of Navigation Congress PIANC. He was appointed to the National Academy of Engineers, and was a Fellow of the American Society of Civil Engineers ASCE, serving as member in its Executive Committee of the Hydraulics Division ASCE from 1968 to 1974.

Douma was all through his career a hydraulic engineer dealing with water power and the related hydraulic problems. His first works relate to high-speed air-water flows on chutes, the so-called white water flow. After this topic was originally investigated in the 1920s in Austria, the first thorough input was made in the ASCE Symposium on entrainment of air in flowing water, with two main contributions to free surface and pipe flows. Douma participated as a discusser by presenting results relative to the air-water volumetric ratio in terms of a modified Froude number. The larger the latter, the higher the ratio was found. Douma also proposed a limit Froude number for incipient air entrainment of the order of 5. He further analyzed the chute flow features with an adapted backwater relation, and the effects of aerated flow on the energy dissipator.

Anonymous (1973). J.H. Douma. 11th *ICOLD Congress* Madrid 5: 35. *P*

Anonymous (1974). Jacob Douma. *New Civil Engineer* 3(10): 6. *P*

Anonymous (1994). Douma, Jacob H. *American men and women of science* 2: 896.

Douma, J.H. (1943). Discussion of Open channel flow at high velocities, by L. Standish Hall. *Trans. ASCE* 108: 1462-1473.

Douma, J.H. (1955). Engineering problems in US tidal waterways. *Proc. ASCE* 81(789): 1-22.

Douma, J.H. (1983). Hydraulic and hydrologic considerations. *Safety of existing dams*: 71-131, R.B. Jansen, ed. National Academy Press: Washington DC.

Jobes, J.G., Douma, J.H. (1942). Testing theoretical losses in open channel flow. *Civil Engineering* 12(11): 613-615; 12(12): 667-669;

http://140.194.76.129/publications/eng-pamphlets/ep870-1-56/bio.pdf *P*

DRAKE

29.03. 1819 Greenville NY/USA
08.11. 1880 Bethlehem PA/USA

Edwin Laurentine Drake was the driller of the first productive oil well in the USA. Raised on farms in New York and Vermont, he worked as a hotel clerk before becoming agent for the Boston and Albany Railroad. In 1850, he became a conductor on the New York and New Haven Railroad, but few years later had to retire for health reasons. In 1857, while living at New Haven CT, Drake met stockholders of the Pennsylvania Rock Oil Company, claiming a lease on land near Titusville PA, where oil had been gathered from ground-level seepages for medicinal uses. The company hoped to make money selling the oil for lighting, so that the stockholders sent Drake to Titusville to assess the viability of the enterprise. Letters of introduction to businessmen in the area referred to Drake as 'Colonel', so that he was known as Colonel Drake for the rest of his life. After Drake returned to New Haven with a favorable report, the New Haven stockholders formed the Seneca Oil Company, sold some stock to Drake, and sent him back to develop the site.

Drake studied the techniques of drilling salt wells and decided to bore for the oil. He began drilling in 1858 and immediately found it impossible to maintain a borehole in the loose rock and soil just below the surface. He solved the problem by driving pipe sections into the ground until bedrock was struck, and from there the drilling continued until the top of an oil deposit was reached at a depth of 21 m on August 27, 1859. With the spread of this drilling techniques, Titusville and other northwestern Pennsylvania communities became boomtowns. Drake drilled two more wells for the Seneca Co., but failed to patent his drill-pipe methods and never became a success in oil speculation. He worked at various jobs in Titusville, then moved to New York City, Vermont, and New Jersey. In 1870, after years of poverty, he returned to Pennsylvania, where he eventually was awarded a pension by the state legislature. In 1901 an executive of the Standard Oil Company paid to erect a monumental tomb in Drake's honor at Titusville cemetery, to where Drake's body had been moved. In 1946, the Commonwealth of Pennsylvania built a replica of Drake's original oil derrick and engine house at the well site, which subsequently became part of the Drake Well Museum.

Anonymous (1880). Edwin L. Drake. *Scientific American* 43: 344.
http://www.britannica.com/EBchecked/topic/170909/Edwin-Laurentine-Drake
http://cdm16038.contentdm.oclc.org/cdm/ref/collection/p15017coll24/id/472 *P*
http://www.oil150.com/essays/2008/04/_edwin-laurentine-drake-1819-1880_-by-dr-william-r-brice

DRESSLER

04.06. 1920 Philadelphia PA/USA
27.08. 1999 Perkiomenville PA/USA

Robert Franklin Dressler graduated in 1940 from the University of Pennsylvania as a mathematician. From 1954 to 1958, he was chief of mathematics, Physics Division, National Bureau of Standards NBS, Washington DC. Until 1962 he acted as Assistant Director of Research, *Philco* Corp., Philadelphia PA, moving then until 1966 as chief of mathematics analysis to the US Federal Aviation Administration, Washington DC. From then until 1968 Dressler was first the chief scientist of the Swedish Government Aerospace Agency, Stockholm, when joining as professor of mechanical and civil engineering City University, New York NY. From 1976 to 1983 he was manager of the NASA science program, Washington DC, and from then until 1989 Director of Engineering Research, International Water Resources Institute, George Washington University, Washington DC.

Dressler worked in the 1960s on the sonic boom effect in aerodynamics, after having published excellent theoretical and experimental papers on dam break waves. His 1952 paper is a first account on the effect of bottom friction, which is large at the wave front but reduces in the upstream direction. Using a perturbation approach, Dressler was able to obtain the leading wave features, thereby generalizing the classical results of Adhémar Barré de Saint-Venant (1797-1886). The theoretical results were compared in the 1954 paper with laboratory tests, whose size was small, however. The 1959 paper deals with the effect of bottom slope on the dambreak wave, another complication in this hydraulic problem, which was again mathematically solved. Around 1980 Dressler attempted to solve the Boussinesq flow problem relating to streamline curvature effects.

Dressler, I. (2001). Prof. Dr. Robert Franklin Dressler. Personal communication. *P*
Dressler, R.F. (1949). Roll-waves in inclined open channels. *Communications in Pure and Applied Mathematics* 2(2-3): 149-194.
Dressler, R.F. (1952). Hydraulic resistance effect upon the dam-break functions. *Journal of Research* 49(3): 217-225.
Dressler, R.F. (1954). Comparison of theories and experiments for the hydraulic dam-break wave. Proc. Intl. *IUGG Congress* Rome 3: 319-328.
Dressler, R.F. (1958). Unsteady non-linear waves in sloping channels. *Proc. Royal Society* London A 247: 186-198.
Dressler, R.F., Yevjevich, V. (1984). Hydraulic-resistance terms modified for the Dressler curved-flow equations. *Journal of Hydraulic Research* 22(3): 145-156. *P*

DRISKO

11.05. 1906 Winchester MA/USA
03.01. 2004 Nazareth PA/USA

John Bucknam Drisko obtained in 1927 a BS degree from Massachusetts Institute of Technology MIT, Cambridge MA. He was then a Freeman Scholar until 1929, visiting Technische Hochschule Berlin, Germany, becoming assistant and instruction civil engineer at MIT after return until 1933. He was an assistant engineer of the US Bureau of Reclamation USBR, Denver CO until 1935, associate engineer with the Soil Conservation Service SCS, US Dept. of Agriculture in 1936, from when he was engineer with the US Engineering Department until 1942. He was from technical assistant to director, and from 1953 to 1954 assistant director of the Experimental Towing Tank, Stevens Institute of Technology, New York NY. He joined then as principal staff engineer Tippets, Abbett & McCarthy Engineering in New York.

The 1933 Report is concerned with a discussion on the various types of waves, including the solitary, tidal or oscillatory waves, which is followed by a description of experiments made in a small flume. This work covers solitary waves, waves propagated into a channel by tides and surge waves as typically occur in power channels due to a rapid change of discharge. The simultaneous stage-time plots for four points, recorded on a chronograph, accompany this report. The 1933 paper published by the American Geophysical Union discusses surge waves in more detail. The experimental work made by the author supports the historical results of John Scott Russell (1808-1882). Drisko also translated the famous book of Wilhelm Spannhake (1881-1959) in English.

Anonymous (1943). Drisko, John B. *American men of science* 7: 476. Science Press: Lancaster.
Drisko, J.B. (1932). Discussion of Piezometer investigation. *Trans. ASME* 54(1): 11-16.
Drisko, J.B. (1932). Model research in the River Hydraulic Laboratory of the Massachusetts
     Institute of Technology. *Trans. AGU* 13: 384-387.
Drisko, J.B. (1933). *Report on wave studies*. Dept. Civil Engineering MIT: Cambridge MA.
Drisko, J.B. (1933). Wave motion in a channel. *Trans. AGU* 14: 516-518.
Drisko, J.B. (1934). *Centrifugal pumps, turbines, and propellers*: Basic theory and characteristics
     (Translation of German original by W. Spannhake). MIT: Cambridge.
Drisko, J.B. (1942). Discussion of Evaluation of flood losses and benefits. *Trans. ASCE* 107:
     912-914.
Drisko, J.B. (1944). Resistance of V-bottom hulls at speel-length ratios up to 5. *Report* 264.
     Davidson Laboratory: Hoboken NJ.
Rouse, H. (1976). Drisko. *Hydraulics in the USA* 1776-1976: 117. IIHR: Iowa City IA. *P*

DRYDEN

02.07. 1898 Pocomoke City MD/USA
02.12. 1965 Washington DC/USA

Hugh Latimer Dryden received education from Johns
Hopkins University, completing his BA in 1916, his
MA in 1918, and his PhD degree in physics in 1919.
He then joined the staff of the National Bureau of
Standards as chief of its Aerodynamics Section. In
1934 he was named chief of the bureau's Mechanics
and Sound Division, which during World War II
supported the development of guided glide bombs.
In corporation with the US Navy, this section also
developed the BAR radar homing missile. Dryden
became associate NBS director in 1946. In the next
year he joined the National Advisory Committee for
Aeronautics NACA as director of research, to be named NACA director in 1949.
During his tenure NACA became the leading authority on supersonic flight. High-speed
wind tunnel research, flight testing of the X series aircraft, or studies on the critical re-
entry heating problems by missiles were then conducted.

Dryden had become an administrator of scientists and engineers, and devoted his
energies to formulating broad research policy rather than pursuing his personal research
interests. He therefore stated: 'Conventional management procedures are well adapted
to operations in which the product consists of a series of nearly identical items'. But 'A
research laboratory produces ideas and new knowledge verified by experiments. This
cannot be considered as nearly identical in scope, difficulty, or effort required'. Fully
realizing the difficulties, he prepared to play a key role in the US response to the
orbiting of the first artificial satellite, Sputnik I, by the Soviet Union. When President
Eisenhower established the National Aeronautics and Space Administration NASA in
1958, Dryden along with 8,000 employees was transferred to this new civilian agency.
As deputy administrator he participated in the planning of the successful US manned
space program culminating in the landing on the Moon in 1969. He emphasized the
importance of international corporation to space research, and served there from 1962 to
1965, when loosing the battle against cancer.

Anonymous (1956). A salute to Hugh L. Dryden. *Journal of Aeronautical Sciences* 23(1): 1. *P*
Anonymous (1964). Dryden, Hugh Latimer. *Who's who in engineering* 9: 488. Lewis: New York.
Hunsaker, J.C., Seamans, R.C. (1969). Hugh Latimer Dryden. *Biographical memoirs* 40: 35-68.
        Columbia University Press: New York. *P*
Neuman Ezell, L. (1981). Dryden, H.L. *Dictionary of scientific biography* 17(Suppl.2): 243-244.
Smith, R.K., ed. (1974). *The Hugh L. Dryden papers*. Johns Hopkins University: Baltimore MD. *P*

DUDGEON

1819 Tain/UK
09.04. 1895 New York NY/USA

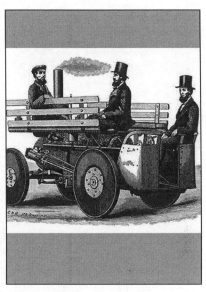

Richard Dudgeon was born in the northern highlands of Scotland, emigrating with his parents when he was a child to Utica NY. He worked as young man in a machinery shop of Manhattan NY, then leading manufacturer of steam engines, providing him an excellent education in mechanical engineering. He opened in 1849 his own machine shop near the East River shipyards, New York NY. Dudgeon invented various devices during his lifetime, including the earliest and most profitable portable hydraulic jack in 1851. When clumsy, inefficient screw jacks were mainly used, the Dudgeon hydraulic jack was small and efficient, making it popular in heavy industry. This mechanical lifting device incorporates an external lever to which force is applied to cause a small internal piston pressurizing usually oil in a chamber. The pressure exerts force on a longer piston, causing it to move vertically upward and raise the bearing plate above it.

Another of Dudgeon's creations was the roller tube expander. The boilers used to power steam engines for steamboats were at his time large water barrels through which snaked iron tubes filled with superheated air, to maximize the surface area of water, that was heated to steam. It was thereby critical that the joints in the tubes be perfectly water-tight, since any leakage caused a steam explosion. His hand-powered roller was much simpler than previously, when a boiler tube was beaten with a hammer, it was also less labor intensive and considerably more precise. He obtained the patent in 1867. Between 1853 and 1867, Dudgeon also designed and manufactured a steam carriage, a self-propelled, steam powered vehicle. The precursor of the automobile was powered by a two-cylinder steam engine connected to a horizontal boiler, a firebox, and a smoke stack. The car body was designed similar to a farm wagon with long bench seats. His steam carriage never interested neither investors nor the public, it was just too far ahead of its time. The 1866 prototype is currently preserved at the Smithsonian Institute. Another innovative design was his steam powered flying machine, which he described in 1877.

Anonymous (1913). The hydraulic jack. *Railway and Locomoitive Engineering* 26(6): 213.
Chanute, O. (1894). *Progress in flying machines*. New York.
Dudgeon, R. (1870). *Maker and patentee* of the hydraulic jack, hydraulic punch, roller tube expander, direct acting steam hammers, rotary steam engines. Phair: New York.
www.glencoveheritage.com/flyingmachine.pdf
http://www.google.ch/imgres?imgurl=http://gorod.tomsk.ru/uploads/34046/1278143545/2.jpg&imgrefurl *P*

DUKLER

05.01. 1925 Newark NJ/USA
12.02. 1994 Houston TX/USA

Abraham Emanuel Dukler obtained his BS from Yale University in 1945, his MS degree in 1950 and PhD degree in 1951 from University of Delaware, Newark DE. From 1945 to 1948 he was employed by an engineering office at Philadelphia PA; he was then until 1952 a research engineer with the Shell Oil Company at Houston TX, and from then joined University of Houston, Houston TX. First he was a faculty member within the Department of Chemical Engineering, from 1963 there professor of chemical engineering, department chairman from 1967 to 1973, and dean of engineering from 1976 to 1983.

In parallel he was also Director, State of Texas Energy Council, from 1973 to 1975, and consultant to a number of firms in the oil industry. He was the recipient of the 1974 Research Award Alpha Chi Sigma, and of the 1989 Kern Research Award. He was from 1978 a Fellow of the American Institute of Chemical Engineers AIChE and was elected to the National Academy of Engineering in 1977.

Dukler was known for his researches in two-phase flows involving gas and liquid, and gas and solid flows. He was also interested in film flow phenomena and in the flows of drops. His 1986 paper for example deals with the shape of the two-dimensional steady Taylor bubble as first described by Geoffrey Ingram Taylor (1886-1975), relating to the bubble rise in vertical channels. The inviscid treatment leads to an infinite set of shapes. Using experimental data, only one particular shape is finally retained. An integral mass and momentum balance analysis between the bubble free surface and the wall takes also viscous effects into account. Matching the two profiles at the nose results in a good description of the phenomenon.

Anonymous (1976). A.E. Dukler. *Mechanical Engineering* 98(11): 86. *P*
Anonymous (1985). Dukler, Abraham. *Who's who in engineering* 6: 171. AAES: Washington DC.
Anonymous (1991). Dukler, Abraham E. *Who's who in America* 46: 886. Marquis: Chicago.
Couët, B., Strumolo, G.S., Dukler, A.E. (1986). Modeling two-dimensional large bubbles in a rectangular channel of finite width. *Physics of Fluids* 29(8): 2367-2372.
Dukler, A.E., Wicks, M. (1963). Gas-liquid flow in conduits. *Modern Chem. Engng.* 1: 349-435.
Dukler, A.E., Hubbard, M.G. (1975). A model for gas-liquid slug flow in horizontal and near horizontal tubes. *Journal of Industrial Engineering and Chemistry* 14(4): 337-347.
Taitel, Y., Lee, N., Dukler, A.E. (1978). Transient gas-liquid flow in horizontal pipes: Modeling the flow pattern transitions. *AIChE Journal* 24(5): 920-934.

DUNLAP F.C.

08.02. 1869 Mullica Hill NJ/USA
25.01. 1928 Philadelphia PA/USA

Frederic Clark Dunlap obtained in 1886 the civil engineering degree from the Polytechnic School, Philadelphia PA. He then entered an engineering office at this city, and was in charge of surveys and the construction of water supply works, tests of pumping machinery, and water power installations. He had his private practice from 1890 to 1893, and then was until 1896 assistant superintendent of the Union Hydraulic Works, Philadelphia PA, engaged on the manufacture of water-work supplies. From 1900 he was an assistant engineer on designs for the Philadelphia Water Supply under John W. Hill (1848-1930). From 1902 to 1905, he was in charge of the construction of the Torresdale Filter Plant, then the largest of the world. He was appointed in 1907 chief of the Bureau of Filtration, remaining in charge until 1912. During this period the Queen Lane Filtration Plant was completed with extensive improvements added in pumping and distribution mains, so that the filtered water was furnished for the first time to every section of the city. He then went West to recover from a chronic rheumatic illness.

Dunlap returned to Philadelphia in 1914 and was employed by the National Dredging Company in the development of its sand and gravel deposits. In 1916 he returned to the city division of sewage disposal, becoming assistant chief engineer of the Department of City Transit. He served from 1917 to 1923 as chief of the Bureau of Highways, when being re-transferred to the former position as chief of the Water Bureau. The situation with regard to the water supply had then reached a point at which it appeared that steps had to be taken to secure an increased supply. Dunlaps's study led him to believe that a more economical plan for an increased supply would be to adopt that proposed many years before to bring water from Delaware River above Trenton NJ to Philadelphia. This plan however aroused opposition from politicians, so that Dunlap resigned from his position in 1924, starting a political campaign by which he convinced the city officials for a division of the water rights along the Upper Delaware. However, he soon passed away, so that he could no more see the fruits of his long struggle. His life demonstrated that of many loyal employees who have given their best to the service of their fellow man frequently receive scant notice and recognition.

Anonymous (1932). Frederic C. Dunlap. *Trans. ASCE* 96: 1461-1463.
Stearns, G.R., Dunlap, F.C. (1909). *Description of the Filtration Works*. Philadelphia PA.
http://www.phillyhistory.org/PhotoArchive/Detail.aspx?assetId=4857 *P*

DUNLAP J.H.

09.09. 1882 Harrisville NH/USA
29.07. 1924 Chicago IL/USA

John Hoffman Dunlap obtained the BA degree in 1905 from Dartmouth College, Hanover NH, and in 1908 the CE degree from Thayer School of Civil Engineering, Hanover. He then was from chainman to levelman at US Reclamation Center, Fallon NE, and field instructor of Thayer School until the mid-1910s. From 1915 he was then from instructor to professor of hydraulics and sanitary engineering at the College of Applied Science, the State University of Iowa, Iowa City IA. He also was employed as sanitary engineer by Iowa State Board of Health. He left Iowa City in 1922 to become Secretary of the American Society of Civil Engineers ASCE, and also opened a private practice as civil and sanitary engineer.

Dunlap was instrumental in erecting the hydraulic laboratory at Iowa State University. Hunter Rouse (1906-1996) states that this was in a way reminiscent of the laboratory at Worcester Polytechnic Institute, Worcester MA, where George I. Alden (1843-1926) had founded one of the first in the 1880s. In 1903 the water rights to a gristmill, located on Iowa River upstream from the campus were deeded to Iowa's School of Applied Science, which later became the College of Engineering. A grant permitted the construction of a new dam just downstream from the campus, and an opening was left for a 3 m gate at the west abutment. Before 1906, a small instructional laboratory had existed in the basement of a building, and in 1914 the newly created Department of Mechanics and Hydraulics undertook the design of a retaining wall next to the dam, including a 3×3×40 m channel, and a 6×6 m laboratory structure. This project was in charge of Dunlap, but taken over after his move to the ASCE by Floyd A. Nagler (1892-1933). Dunlap also invented the diagonal-jet drinking fountain, thereby replacing the less-sanitary vertical-jet fountain before his untimely death following a railroad accident.

Anonymous (1905). John H. Dunlap. *Aegis yearbook*: 70. Dartmouth College: Hanover NH. *P*
Anonymous (1922). John H. Dunlap: ASCE Secretary. *Engineering News-Record* 89(1): 34. *P*
Anonymous (1926). John H. Dunlap. *Trans. ASCE* 89: 1580-1586.
Dunlap, J.H. (1914). *Water works statistics of thirty-eight cities of Iowa*. University of Iowa: Iowa.
Dunlap, J.H. (1918). Small town sewerage works operation: Problems analyzed. *Engineering News-Record* 80(16): 773-775.
Rouse, H. (1976). John H. Dunlap. *Hydraulics in the United States* 1776-1976: 88-89. IIHR: Iowa City. *P*

DURAND W.F.

05.03. 1859 Bethany CT/USA
09.08. 1958 New York NY/USA

William Frederick Durand entered the US Naval Academy, Annapolis MD, in 1876, graduating there in 1880. He joined the Naval Engineering Corps working on problems of marine engineering. He was sent by the Navy to work on a PhD, and in 1888 graduated from Lafayette College, Easton PA. He accepted the post as professor of mechanical engineering at Agricultural and Mechanical College of Michigan, remaining there for three years, when moving to Cornell University, Ithaca NY, to teach marine engineering. In 1904 he moved to Stanford University, Stanford CA, as professor of mechanical engineering. He there became involved in the new technologies of airplanes and began to study the problems of flight. Durand created an aeronautical engineering center at Stanford that became one of the leading of the nation.

The US Government recognized the importance of the aeronautical development by establishing the National Advisory Committee for Aeronautics NACA in 1915. Durand served as Member of the committee until 1933, and from 1941 to 1945. His research team at Stanford led contractors with NACA funded propeller studies. NACA's research on aircraft engines was the first major success and helped develop the Liberty Engine, the major US contribution to aeronautics during World War I. In addition to these works, Durand participated in numerous technical committees and advisory boards. He was for instance also a member of the Boulder Dam project from 1929, given his earlier works in unsteady pipe flows and his interest in high-speed chute flows. Durand received numerous awards, including the Presidential Award in 1946. He died at age 99, just as the space age was dawning.

Anonymous (1925). W.F. Durand, President ASME. *Trans. ASME* 47: Frontispiece. *P*
Anonymous (1958). William Frederick Durand. *Mechanical Engineering* 80(11): 171.
Durand, W.F. (1901). *Practical marine engineering*. Marine Engineering: New York.
Durand, W.F. (1907). *Motor boats*. Marine Engineering: New York.
Durand, W.F. (1912). On the control of surges in water conduits. *Trans. ASME* 34(3): 319-377.
Durand, W.F., ed. (1934). *Aerodynamic theory*. Springer: Berlin.
Durand, W.F. (1940). Flow of water in channels under steep gradients. *Trans. ASME* 62(1): 9-14.
Launius, R.D. (1995). W.F. Durand. *Notable 20th century scientists* 1: 534-535. Gale: New York.
Terman, F.E. (1960). William F. Durand. *Aeronautics and Astronautics*: 4-8. Pergamon Press: Oxford. *P*

DWORSKY

05.01. 1915 Chicago IL/USA
28.03. 2008 Ithaca NY/USA

Leonard Barbara Dworsky obtained the BS degree in civil engineering from the University of Michigan, Ann Arbor MI in 1936, and the MA degree in public administration in 1955 from the American University Washington DC. He worked from 1936 to 1941 as a sanitary engineer with, Illinois Department of Public Health. After war service, he was then commissioned in 1946 officer of the US Public Health Service, retiring after 18 years as naval captain. As a senior administrator, he formulated legislative and policy initiatives which became the basis of the nation's environmental programs for decades.

The 1948 Water Pollution Control Act was notable for recognizing that environmental and political boundaries often differ, and that further formal interstate arrangements are necessary for a successful pollution control. Dworsky supervised the publication of 15 major basin summary reports covering the 226 sub-basins of the USA. He also was a member of the first Federal Interagency Committee's River Basin Committee from 1947; he was the secretary of the Missouri Basin Interagency Committee from 1956, and chairman of the Columbia Basin Interagency Committee from 1959 to 1962. He joined in 1964 the Cornell Faculty as the first director of the Water Resources and Marine Sciences Center, where he was a major player in the development of the field of water resources research. He there studied and taught on river basin management, water quality planning, management of resources in international boundary areas, and conflict resolution, thereby seeking to bridge the gap between social problems and science and technology. After having been an emeritus from 1985, Dworsky presented water policy seminars for another 15 years, and continued to write. He was a recipient of numerous honors, including the 1994 Caulfield Medal for exemplary contributions to the nation.

Anonymous (1967). Leonard B. Dworsky. *Civil Engineering* 37(4): 76. *P*
Dworsky, L.B. (1963). *Problems of water quality management.* US Dept. Health: Washington DC.
Dworsky, L.B. (1965). *Canadian-United States water resources problems and policies.* Water Resources Center, Cornell University: Ithaca NY.
Dworsky, L.B., Allee, D.J., North, R.M. (1991). Water resources planning and management in the US federal system: Long term assessment and intergovernmental issues. *Natural Resources Journal* 31(3): 475-547.
theuniversityfaculty.cornell.edu/.../dworsky.pdf.
http://www.news.cornell.edu/stories/April08/dworsky.obit.aj.html *P*

EADS

23.05. 1820 Lawrenceburg IN/USA
08.03. 1887 Nassau/BS

James Buchanan Eads was from 1838 a purser on a Mississippi River steamboat, where he became fully acquainted with river flow. He had the idea of a diving bell, which he patented in 1842, from when he was entirely engaged in steamboat salvaging. After unsuccessful work, he in 1856 proposed to the Congress to remove all snags and wrecks from the large Midwest rivers and to keep their channels open. In 1861 he was summoned to Washington by President Lincoln to advise him on the best method of utilizing western rivers for attack and defence. Eads proposed a fleet of armour-plated, steam-propelled gunboats and constructed seven of these 600 tons vessels. Within two weeks, over four thousand men were engaged and in forty-five days, *St. Louis*, the first of the seven boats, was launched. In the course of the Civil War, Eads constructed another fourteen gunboats incorporating several new ordnance inventions. Following this heroic undertaking, his health became seriously impaired, yet he was fit again in 1865.

In this year, the Congress asked for the construction of a bridge across the Mississippi at St. Louis MO. The project was thought to be impracticable by the leading engineers, yet Eads proceeded in 1867 with its construction which he successfully completed in 1874. It was of steel and masonry with a center span of 160 m and a pier depth of 50 m below high water resting on bedrock. The conditions encountered during the construction were so extraordinary that only an inventive genius of Eads' caliber could provide the many appliances needed for the subaqueous work and for the superstructure. It is known today as 'Eads Bridge'. Shortly later, Eads proposed to the Congress to open one of the mouths of Mississippi River into the Gulf. This project was attacked by army engineers with works confined to the small South Pass. By a system of jetties so arranged that the river deposited the sediments where he wanted, Eads also successfully accomplished this task in 1879. His name thus was one of the foremost in river engineering.

Anonymous (1887). Eads, James B. *Appletons' cyclopaedia of American biography* 2: 287. *P*
Anonymous (1930). Eads. *Dictionary of American biography* 5: 587-589. Scribner's: New York.
Dorsey, F.L. (1947). *Road to the sea*: The story of James B. Eads and the Mississippi River.
    Rinehart: New York. *P*
How, L. (1970). *James B. Eads*. Libraries Press: Freeport NY.
McHenry, E. (1884). *Addresses and papers of James B. Eads*. Slawson: St. Louis. *P*
http://de.wikipedia.org/wiki/James_Buchanan_Eads *P*

EASTWOOD

27.03. 1857 Minneapolis MS/USA
10.08. 1924 Fresno CA/USA

John Samuel Eastwood was an American engineer who in 1908 built the world's first reinforced concrete multiple-arch dam on bedrock foundation at Hume Lake CA. He attended the University of Minnesota, Minneapolis MS, as civil engineering student; prior to graduation in 1880, he headed west to work on railroad projects in the Pacific Northwest. In 1883, he moved to Fresno CA and established an office as civil engineer, becoming in 1885 Fresno's first city engineer, but apparently was not well suited for the bureaucratic life and soon resigned. He focused for the remainder of his career on work in the private sector or as a consulting engineer.

In 1895, Eastwood became engaged with the Pacific Light and Power Co. as engineer in charge of designing a large hydroelectric project on the South Fork of the San Joaquin River, known as the Big Creek Hydroelectric Project. Eastwood had great hopes for the Project, and he planned storage dams to ensure that a drought could not stop its power production. He therefore had devised an inexpensive type of reinforced concrete dam design minimizing the amount of material required and thus reduced construction cost. In 1908, while waiting for work on Big Creek to begin, he designed and built the Hume Lake Dam. This structure is located in the Sierra Nevada 70 km south of Big Creek. The first of its kind, its completion in 1909 demonstrated the practicality of the multiple-arch design. Shortly later, he received the contract for the design of a multiple-arch dam to supersede the 1884 Big Bear Valley Arch Dam near San Bernardino CA. In the early 20th century, Salt Lake City regularly experienced severe water shortages. To alleviate this deficit, bonds were floated in 1914 to finance the construction of three storage dams, the largest being Mountain Dell Dam in Parley's Canyon, 16 km east of the city, built in two stages from 1914 to 1925. The reputation of Eastwood as dam engineer and author of these projects in the engineering literature grew so that he continued his career in the water power development and dam design. He drowned in 1924 while swimming.

Anonymous (1999). *Inventory of the John S. Eastwood Papers* 1884-1979. University of California: Riverside CA.
Jackson, D.C. (1996). *Building the ultimate dam*: John S. Eastwood and the control of water in the West. University Press of Kansas: Lawrence KS. *P*
http://en.wikipedia.org/wiki/John_S._Eastwood
https://sunsite.berkeley.edu/WRCA/eastwood.html *P*

EATON A.C.

03.08. 1883 Lunenburg MA/USA
06.03. 1934 Worcester MA/USA

Arthur Chester Eaton graduated from the University of Vermont, Burlington VR, in 1907 with the BSc degree in civil engineering. In 1932, his Alma Mater conferred on him in recognition of his professional attainments the honorary degree of civil engineer. His early career was in connection with the Catskill Aqueduct of the New York Board of Water Supply. In 1907 and 1908 he was civil engineering instructor at the University of Vermont, and then engaged on water power projects for the New York State Water Supply Commission. Eaton became associated in 1912 with the Power Construction Company which started water power developments upon the Deerfield River in Southern Vermont and in Western Massachusetts. As assistant engineer of Alger A. Conger (1875-1931), he was connected with this important water power projects of his Company, which expanded into a system of 1,000,000 HP capacity, about one-half of which was water power.

From 1919 Eaton was hydraulic engineer of large water power developments, including the Searsburg, the Harriman, and the Sherman Plants, on Deerfield River, and Bellows Falls, the Fifteen Mile Falls and the McIndoes Plants, on Connecticut River. These were examples of the highest skill and best engineering construction. He was awarded for the 1925 paper the Desmond FitzGerald Medal by the Boston Society of Civil Engineers, of which he was member. His sudden death at age 51 was a great shock to his family and many friends. The great esteem felt by his associates, and the importance of his life work and recognition of his splendid character, were deeply expressed by the Board of Directorate of the New England Power Company. He was acknowledged an authority on hydraulic matters of power developments, and was stated to have been a person of integrity. He was a good friend and loyal worker through all the years during which he was engaged by the Company, providing his service, knowledge and experience. He was member of the American Society of Civil Engineers ASCE since 1920.

Anonymous (1934). Arthur C. Eaton. *Trans. ASCE* 99: 1445-1447.
Anonymous (2013). A.C. Easton. *Vermont Alumni Weekly* (Jul.15): 488. Special Collections. *P*
Eaton, A.C., Collins, E.B. (1923). Davis Bridge Development. *Electrical World* 81(24): 1410-1413.
Eaton, A.C. (1925). The New England Power Co. Davis Bridge Development. *Journal of the Boston Society of Civil Engineers* 12(1): 1-48.
Safford, A.T., ed. (1922). Report of the Committee on Run-off. *Journal of the Boston Society of Civil Engineers* 9(8): 157-212.

EATON E.C.

07.09. 1881 Providence RI/USA
22.02. 1967 Kerrville TX/USA

Eugene Courtlandt Eaton was educated at McGill University, Montreal, Canada, obtaining his BSc degree in 1908. He then was a construction engineer at Calgary AB, later a resident engineer at Boise ID, and at Marblemount WA, until joining the Water Supply Division of the Panama-Pacific International Exposition of San Francisco in 1904. Eaton became in 1905 resident engineer for an irrigation district at Lindsay CA, and Terra Bella CA. He was engineer in charge of dam construction for California State from 1921 to 1927, and later involved in the design of the San Gabriel Dam. He was finally an irrigation engineer of the State of California where he was involved in engineering investigations of water rights for the Central Valley Project. He also was member of the Commission and Supervision of Dams, until being appointed chief engineer for Los Angeles County.

Eaton was a renown irrigation engineer who spent his entire career in the West of the US. Toward the end of his professional career, he served as chief engineer to control floods in Southern California by designing and constructing dams, and by developing water supplies and irrigation systems. A Report deals with the supervision of dams and the safety provisions to be considered for an overall successful engineering design. He has written a number of papers in ASCE journals, the Journal of Western Construction, the Engineering News and the Engineering Record, from 1916 united to the Engineering News-Record. One of his typical papers relating also to flood and erosion control was published in 1936. He was member of the American Society of Civil Engineers ASCE.

Anonymous (1927). Eaton long on hydraulic work. *Engineering News-Record* 98(14): 580.
Anonymous (1954). Eaton, Eugene C. *Who's who in engineering* 7: 691. Lewis: New York.
Eaton, E.C. (1919). Removing algae from a California irrigation canal. *ENR* 82(8): 382-383. *P*
Eaton, E.C., Adams, F. (1927). Irrigation development through irrigation districts. *Trans. ASCE* 90: 773-790.
Eaton, E.C. (1931). *Report on check dams*, Los Angeles County Flood Control District, Forest Service. US Department of Agriculture: Washington DC.
Eaton, E.C. (1931). *Comprehensive plan for flood control and conservation*: Present conditions and immediate needs. Los Angeles County Flood Control District: Los Angeles.
Eaton, E.C. (1932). *San Gabriel Project*: Check dams. County Flood Control District: Los Angeles.
Eaton, E.C. (1936). Flood and erosion control problems and their solution. *Trans. ASCE* 101: 1302-1362.

EATON F.

23.09. 1855 Los Angeles CA/USA
11.03. 1934 Los Angeles CA/USA

Frederick Eaton went to work at age of fifteen for a privately owned Los Angeles waterworks becoming there superintendant but resigned in 1883 to go into private practice, building domestic and irrigation waterworks. In 1886 he was appointed Los Angeles city engineer, initiating its sewerage system. In 1892 he visited Owens Valley to study possibilities of an irrigation project. However, his plan was not accepted, but Eaton realized the immense quantities of water that could be used for water supply of Los Angeles. He was elected mayor of the city in 1899, but resigned two years later. He reconsidered his water supply idea realizing that an aqueduct would be suitable for water supply.

In 1904 the US Reclamation Service began general investigations of all possible major irrigation projects in the West, among which was Owens Valley. Eaton proceeded with his studies, securing options for land ownership. His idea was to organize a water supply company for Los Angeles. In 1904, the city purchased the real estate necessary to this plan. To carry out the municipal plan the Reclamation Service had to abandon its project, and to aid the city. During a conference, it was stated that the project must be municipally owned in whole. Eaton accepted these stipulations, his expenses being refunded, and turned his options over to the city. This public-spirited action was regarded generously. Since 1933 the city owns 270,000 acres in the drainage basin of Owens Valley for water supply purposes. The withdrawal of the US Reclamation Service from Owens Valley was recommended by President Roosevelt stating that this domestication of water was its highest use. The project was executed and is known as the Los Angeles Aqueduct, to which Eaton has largely contributed. He had been always active as a leading citizen in enterprise of importance. He was also active in the scientific organizations of the city.

Anonymous (1913). *Pictorial history of the Aqueduct*. Times Mirror: Los Angeles. *P*
Anonymous (1934). Frederick Eaton. *Engineering News-Record* 112(Mar.22): 399.
Anonymous (1935). Frederick Eaton. *Trans. ASCE* 100: 1645-1647.
Hundley, N. (1992). Fred Eaton. *The great thirst*: Californians and water 1770s – 1990s: 143. University of California Press: Berkeley CA. *P*
Mulholland, W. (1916). *Complete report on construction of the Los Angeles Aqueduct* with introductory historical sketch. Department of Public Service: City of Los Angeles.
http://www.ulwaf.com/LA-1900s/SpecialReports/Mayors/MayorEaton.html *P*

EATON H.N.

04.11. 1892 Auburn MA/USA
18.04. 1970 Hagerstown MD/USA

Herbert Nelson Eaton was educated at the Worcester Polytechnic Institute WPI, from where he received the AB degree in 1916, at Johns Hopkins University with the AM degree in 1923, and at the Technische Hochschule Karlsruhe and Berlin as a Freeman Scholar with degrees in 1928. After having been an instructor at WPI in 1917, he moved from 1919 to 1926 to the National Bureau of Standards at the Aeronautical Instrumentation Section, and then made his trip to Europe. He returned to the Bureau after return to the USA as hydraulic engineer. Eaton was member of the American Society of Mechanical Engineers ASME, the Washington Academy of Sciences, the Washington Academy of Engineers, and the American Geophysical Union AGU.

Eaton was among the few engineering graduates being granted a Freeman Travelling Fellowship for advanced studies in Europe. John Ripley Freeman (1855-1932) initiated this initiative in 1927. A photo taken in the classroom of Richard Winkel (1883-1951) at Hannover Technical University shows Eaton, J.B. Drisco, Frederic T. Mavis (1901-1983), Blake Ragsdale van Leer (1893-1956), K.C. Reynolds, Morrough P. O'Brien (1902-1988), and Lorenz George Straub (1901-1963). These young American engineers have had the opportunity to pass on to other young men the spirit and encouragement they received from Freeman, who supported scientific education all through his career. Eaton first was a scientist working in aeronautical engineering, finally as Chief of its Hydraulics Section, later rather interested in hydraulic problems of large buildings. He for instance studied roof drainage and the implications for air-water flows in vertical shafts.

Anonymous (1937). Eaton, Herbert N. *Who's who in engineering* 4: 387. Lewis: New York.
Anonymous (1952). 25 years ago, the first Freeman scholars. *Engineering News-Record* 149(Jul.17): 30. *P*
Eaton, H.N., Beij, K.H., eds. (1926). *Aircraft instruments*. Ronald Press: New York.
Eaton, H.N. (1932). The National Hydraulic Laboratory. *Mechanical Engineering* 54(4): 263-266; 54(5): 335-340.
Manas, V.T., Eaton, H.N. (1957). *National plumbing code handbook*: Standards and design information. McGraw-Hill: New York.
Wyly, R.S., Eaton, H.N. (1961). *Capacities of stacks in sanitary drainage systems for buildings*. US Department of Commerce, National Bureau of Standards: Washington DC.

ECKENFELDER

15.11. 1926 New York NY/USA
28.03. 2010 Nashville TN/USA

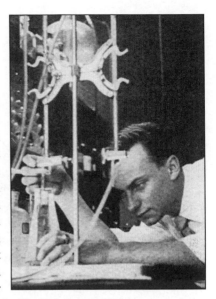

William Wesley Eckenfelder, Jr., was educated at Manhattan College, from where he obtained the BCE degree in 1946, at Penn State University, from where he received the MS degree in 1948, and at New York University obtaining the MCE degree in 1956. He was a sanitary engineer from 1948, then until 1950 a research associate until joining from 1955 to 1965 Manhattan College as assistant and associate professor of civil engineering. From then to 1970 he was a professor of health engineering at the University of Texas, and further until 1989 a distinguished professor of environmental and water resources engineering at Vanderbilt University, Nashville TN. In parallel he was vice-president of Weston, Eckenfelder & Associates from 1952 to 1956, and from then a private consultant. He also was the president of Hydroscience Inc., Nashville TN.

Eckenfelder was a leading scientist in the biological treatment of sewage and industrial wastes, of mass transfer and aeration in waste treatment, in process design of industrial waste treatment plants, and in water quality management. He was deemed the godfather of industrial wastewater treatment by many of his colleagues because he played a key role in the development of wastewater treatment. Through research, publications plus multiple courses, his name became well-known to those who were in his field. He was awarded the honorary degree D.Sc. from Manhattan College in 1990, the Gold Medal of the Synthetic Organic Chemical Manufacturers Association in 1974, the 1981 Camp Medal from the Water Pollution Control Federation, and the 1990 Imhoff-Koch Medal of the International Association of Water Pollution Research and Control.

Anonymous (1953). Inventor Wesley Eckenfelder. *Engineering News-Record* 150(May14): 25. *P*
Anonymous (1974). The new AWWA 6th Federal Convention. *Water* 1(2): 16-17. *P*
Anonymous (1994). Eckenfelder, W. Wesley. *American men and women of science* 2: 1011.
Eckenfelder, Jr., W.W., Melbinger, N. (1957). Settling and compaction characteristics of biological sludges. *Sewage and Industrial Wastes* 29(10): 1115-1122.
Eckenfelder, W.W., Jr., Barnhart, E.L. (1963). Performance of a high rate trickling filter using selected media. *Journal Water Pollution Control Federation* 35(12): 1535-1551.
Eckenfelder, W.W., Jr. (1966). *Industrial water pollution control*. McGraw-Hill: New York.
Eckenfelder, W.W., Jr. (1980). *Principles of water quality management*. CBI: Boston MA.
Eckenfelder, W.W., Jr., Musterman, J.L. (1998). *Activated sludge*: Treatment of industrial wastewater. Taylor & Francis: Oxford UK.

EDDY

29.04. 1870 Millbury MA/USA
15.06. 1937 Montreal/CA

Harrison Prescott Eddy was educated at Worcester Polytechnic Institute, graduating in 1891 with a BS in chemistry. He then was employed in the sewer department of Worcester City, remaining there for sixteen years. During this time he designed over hundred miles of sewers and drains, enlarged the chemical treatment plant, and added intermittent sand filters to the disposal scheme. In parallel he studied sewage treatment and stream pollution. During this period of the pioneering stage of sewage treatment Eddy was appointed in 1906 to the board of engineers preparing the expansion of the sewage system of Louisville KY, marking his first activity as a consultant.

In 1907 Eddy joined the firm of Leonard Metcalf (1870-1926), a civil engineer known in the field of water supply, which proved to be a long and fruitful partnership. With headquarters at Boston MA, they specialized in water supply, sewage treatment and disposal. Metcalf & Eddy began their practice at a time when few communities in the USA provided purified water, and still fewer treated sewage, a time when typhoid fever annually took 35,000 lives. During the next three decades the firm was retained by more than 125 cities, including Fitchburg MA, where Eddy designed one of the first Imhoff trickling filter systems, as proposed by Karl Imhoff (1876-1965) in Germany. Eddy was co-author of the standard three volumes American Sewerage Practice, and of the widely used text Sewerage. He was further an author of technical papers, two of which winning awards, namely the 1913 Desmond Fitzgerald Medal of the Boston Society of Civil Engineers, and the 1925 Normal Medal of ASCE. As member of the ASCE Committee on Public Works, Eddy also played an important role in helping to secure enactment of the Federal Emergency Relief Act. He was further instrumental in setting up ASCE's journal Civil Engineering in 1931. Eddy was fatally stricken during his trip to Montreal, where he should receive Honorary Membership of the Engineering Institute of Canada.

Anonymous (1926). Harrison P. Eddy. *Engineering News-Record* 96(3): 132. *P*
Anonymous (1937). Eddy dies in Montreal. *Engineering News-Record* 118(Jun.17): 895. *P*
Anonymous (1939). Harrison Prescott Eddy. *Trans. ASCE* 104: 1867-1871.
Anonymous (1958). Eddy, Harrison Prescott. *Dictionary of American biography* 22: 169-170.
        Scribner's: New York.
Metcalf, L., Eddy, H.P. (1915). *American sewerage practice*. McGraw-Hill: New York.
Metcalf, L., Eddy, H.P. (1922). *Sewerage and sewage disposal*. McGraw-Hill: New York.

EDMONSTON

12.11. 1886 Ferndale CA/USA
22.02. 1957 San Francisco CA/USA

Arthur Donald Edmonston graduated from Stanford University, Stanford CA, in 1910 with the degree of BA in civil engineering. Until 1924 he was employed on the location, the design, and the construction of irrigation, hydro-electric, and also municipal water projects in California. He entered in 1924 the service of California State, becoming soon after deputy State engineer in charge of studies of its water resources. He was appointed in 1945 assistant state engineer, and in 1950 state engineer, mainly in charge of water resources developments, water rights, the safety of dams, water quality, flood control, and beach erosion. In his position he was member and executive officer of many professional commissions. He was also a member of professional and honorary societies, retiring in 1955.

Edmonston was largely responsible for the formulation of the State Water Plan of 1931, including the Central Valley Project, and originated and directed the studies for the California Water Plan. This Plan included first an inventory of the state water resources. Next, a detailed study of the then employed water utilization, including the irrigable and habitable lands, was made, so that the future water needs could be estimated. The third phase of the study was the formulation of a plan for the full practicable development, control, and utilization of water resources throughout the State. This work included flood control in regions where excess waters were wasted, and the generation of hydro-power to control these waters. Edmonston also conceived and directed the planning of the Feather River Project. The Edmonston Pumping Plant, the biggest, most powerful of its kind at the time, by which water is transported from the north to the south of the State, is named for Edmonston, who directed the early planning of the entire project. He was member of the American Society of Civil Engineers ASCE from 1944.

Anonymous (1958). Arthur D. Edmonston. *Trans. ASCE* 123: 1290.
Edmonston, A.D. (1950). *Kaweah River*: Diversions and service areas. California Division of
     Water Resources: Sacramento CA.
Edmonston, A.D. (1951). *California water resources and the New State Water Plan*. Sacramento.
Edmonston, A.D. (1953). California proposes diversion of Feather River water to Southern
     California. *Civil Engineering* 23(4): 229-231.
Edmonston, A.D. (1955). *The California Water Plan*. Waddell: Sacramento CA.
Tibbetts, F.H., Edmonston, A.D. (1930). *Sacramento Valley*: Duty of water. San Francisco.
http://www.water.ca.gov/swp/history.cfm *P*

EDSON

16.12. 1916 Springfield MA/USA
23.03. 2002 Gainesville FL/USA

Charles Grant Edson was educated at Massachusetts State College, Boston MA, and at the Massachusetts Institute of Technology MIT, graduating in 1943 as civil engineer. He joined the Corps of Engineers in 1938, becoming involved in flood control planning and forecasting of Connecticut River. From 1940 he was engaged in hurricane analysis and levee design on Lake Okeechobee FL, and in meteorology and oceanography at MIT. From 1943 Edson joined the US Army in Europe to forecast there surf, sea and operational weather, completing this task in 1945.

From 1946 Edson was in educational charge in hydraulics and fluid mechanics at the University of Florida, Gainesville FL. He was also a research professional engineer in Florida, and from 1951 consulting engineer on military hydrology within the US Corps of Engineers, Washington DC. He thereby developed codes of communications systems for the flood prediction of Rhine River in World War II, perfecting trigonomic methods for computing great circle distances. He was a member of the American Geophysical Union AGU, American Society of Civil Engineers ASCE, and the Florida Engineering Society. His 1951 paper relates the physical characteristics of a watershed to two distinct effects, namely that runoff is brought to a valley, whereas the other gets a runoff through the mouth. An empirical formula is developed for the peak discharge and the time to peak, based on these two effects. The phenomenon is demonstrated to be so complex that no direct relationship is amenable between basic watershed parameters. This is stated to be the reason that such correlations generally fail. Edson proposed an alternative of which he expected a more successful analysis.

Anonymous (1938). Charles G. Edson. *Olio yearbook*: 74. Amherst MA. *P*

Anonymous (1954). Edson, Charles G. *Who's who in engineering* 7: 700. Lewis: New York.

Edson, C.G. (1951). Parameters for relating unit hydrographs to watershed characteristics. *Trans. AGU* 32(4): 591-596.

Edson, C.G. (1951). Nomograph provides method for comparison of weir formulas. *Civil Engineering* 21(8): 476-477.

Edson, C.G. (1952). Two nomographs developed for trapezoidal channels. *Civil Engineering* 22(2): 148-149.

Edson, C.G. (1954). Hydraulic drop as a function of velocity distribution. *Civil Engineering* 24(12): 814-815; 25(6): 370-371.

Sawyer, W.L., Edson, C.G. (1960). *Engineering mechanics*. University of Florida: Gainesville.

ELDER C.C.

13.06. 1895 Salt Lake City UT/USA
12.11. 1966 Los Angeles CA/USA

Clayburn Combes Elder graduated from University of Utah, Salt Lake City UT, with the BSc degree in civil engineering in 1916. He was then employed as hydrographer for the Twin Falls Land and Water Co., Jerome ID. After service in the US Army, he was from 1919 employed as office engineer for the Idaho State Highway Department, Jerome ID, but in 1920 returned to the Twin Falls Land and Water Co. From 1921 to 1928 he worked for the US Bureau of Reclamation USBR at American Falls ID, and later at Denver CO. He was employed by the National Irrigation Commission of Mexico until 1929, when he left to work for the Los Angeles Department of Water and Power. He was in 1930 transferred to the Metropolitan Water District of Southern California. He was promoted there from senior engineer to hydrographic engineer in 1944, working in this capacity until his retirement in 1965.

Elder served as an expert witness in the District's behalf at innumerable hearings before legislative bodies at Sacramento, California and Washington DC, and on behalf of the State of California and the District in the prolonged litigation Arizona versus California before the US Supreme Court, extending from 1952 to 1964, over the water rights of Colorado River. Because of his extensive knowledge of the Colorado River Basin problems the District's Board of Directors voted upon his retirement to retain him as a consultant. Elder was a member of the American Water Works Association AWWA, the American Meteorological Society AMS, the American Geophysical Union AGU, and the American Mathematical Society AMS. He was from 1940 a member of the American Society of Civil Engineers ASCE, elected ASCE Fellow in 1959, and Life Member in 1961.

Anonymous (1936). Clayburn C. Elder. *The Improvement Era* 39(7): 419. *P*
Anonymous (1967). Clayburn C. Elder. *Civil Engineering* 37(2): 14.
Anonymous (1968). Clayburn C. Elder. *Trans. ASCE* 133: 741-742.
Elder, C.C. (1935). Water supply for construction camps. *Journal of the American Water Works Association* 27(5): 613-626.
Elder, C.C. (1950). Determining the future water requirements. *Journal of the American Water Works Association* 43(2): 124-135.
Elder, C.C., Petersen, E. (1969). *Clayburn C. Elder hydrographic engineer*. University of California: Los Angeles.

ELEVATORSKI

23.03. 1909 New Providence NJ/USA
20.06. 1997 Bonsall CA/USA

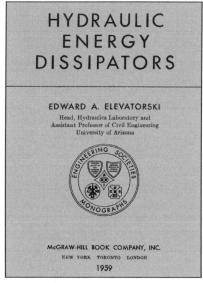

Edward Arthur Elevatorski's professional career is poorly known. He was in the mid-1950s hydraulic engineer at the Bureau of Reclamation USBR, Salt Lake City UT, and in the late 1950s head of the Hydraulic Laboratory and assistant professor of civil engineering, the University of Arizona, Tucson AZ. He was mainly known for his 1959 book on hydraulic energy dissipators, which received mixed comments, however, from the hydraulic community.

The book mentioned includes 20 chapters, namely 1 Functions of energy dissipators, 2 Basic concepts of open channel flow, 3 Elements of hydraulic jump, 4/5 Hydraulic jump in rectangular/ non-rectangular channels, 6 Hydraulic jump in sloping channel, 7 Energy dissipation by the hydraulic jump, 8 Hydraulic-design criteria, 9 Hydraulic-jump-type stilling basins, 10/12 Stilling-basins, 13 Outlet-works control mechanisms, 14 Jet-diffusion and impact stilling basins, 15 Special stilling basins, 16 Ski-jump and arch-dam energy dissipators, 17 Drop-structure stilling basins, 18 Bucket-type energy dissipators, 19 Model tests and hydraulic similitude, 20 Erosion below dams. Apart from the 1935 book of Armin Schoklitsch (1888-1969), Elevatorski's book may be considered the first on the subject topic. It is a basic textbook for hydraulic engineers in which the fundamental concepts of open channel flow are described, including the energy and momentum conservation equations and their applications to practice. The book is excellently illustrated with figures and photographs explaining the concepts described, and it includes a remarkable list of referenced on the topic. The book presents the ten different types of stilling basins as introduced by Alvin J. Peterka (1911-1983) and Joseph N. Bradley (1903-1993), popularizing their work internationally. It also contains a short overview on laboratory modeling of energy dissipators. Later in his career, Elevatorski obviously entered the mining world, dealing mainly with gold and its occurrence in the world.

Anonymous (1958). E.A. Elevatorski. *Desert* College of Engng: 107. Univ. Arizona. Tucson. *P*
Elevatorski, E.A. (1957). Discussion of A study of bucket-type energy dissipator characteristics. *Journal of the Hydraulics Division* ASCE 83(1417): 33-35.
Elevatorski, E.A. (1958). Direct solution for apron elevation. *Civil Engineering* 28(8): 596-597; 28(11): 857.
Elevatorski, E.A. (1959). *Hydraulic energy dissipators*. McGraw-Hill: New York. (*P*)
Elevatorski, E.A. (1982). *Volcanogenic gold deposits*. Minobras: Dana Point CA.
Elevatorski, E.A. (1998). Gold resources of Mexico. Minobras: Dana Point CA.

ELIASSEN

22.02. 1911 Brooklyn NY/USA
14.03. 1997 Palo Alto CA/USA

Rolf Eliassen received his BS degree from MIT in 1932, the MS in 1933, and the PhD title in 1935. He was first for one year design engineer at Pittsburgh PA, then for three years a sanitary engineer at Los Angeles CA, to be appointed in 1940 professor of sanitary engineering at New York City University. He served then as chairman of the Civil Engineering Department at the Biarritz American University, France in 1945. Eliassen was from 1949 to 1960 professor of sanitary engineering and director of the Sedgwick Laboratories in Sanitary Sciences at MIT, Cambridge MA. Until 1973 he joined the staff as professor of civil engineering at Stanford University, Palo Alto CA, from then being Silas H. Palmer emeritus professor. Eliassen was a member of the American Academy of Arts and Sciences, the National Academy of Engineering NAE, the American Water Works Association AWWA, and a Honorary Member ASCE.

Eliassen was a known expert in drinking water and waste disposal engineering. He has written numerous papers in specialized journals, and a notable book as a co-author, dealing with the principles and the management of solid wastes. Besides his academic career, Eliassen was also a consultant. He was a partner of the firm Metcalf & Eddy, Boston MA, from 1961 to 1973, then until 1988 a chairman. He was further a consultant for the International Atomic Energy Agency IAEA, the World Health Organization WHO of United Nations, Executive Office President of the California Water Resources Federal Power Commission, and of the US Senate Commission on Public Works.

Anonymous (1940). Eliassen joins NYU faculty. *Engineering News-Record* 125(Sep.5): 310. *P*
Anonymous (1954). Eliassen, Rolf. *Who's who in engineering* 7: 711. Lewis: New York.
Anonymous (1956). Rolf Eliassen. *Journal American Water Works Association* 48(5): 94. *P*
Anonymous (1995). Eliassen, Rolf. *Who's who in America* 49: 1063. Marquis: Chicago.
Eliassen, R. (1946). Sedimentation and the design of settling tanks. *Trans. ASCE* 111: 947-949.
Eliassen, R., Skrinde, T., William, D. (1958). Experimental performance of ,miracle' water
        conditioners. *Journal American Water Works Association* 50(10): 1371-1385.
McCarty, P.L. (2001). Rolf Eliassen. *Memorial tributes* 9: 64-68. National Academy of
        Engineering: Washington DC. *P*
Tchobanoglous, G., Theisen, H., Eliassen, R. (1977). *Solid waste engineering principles and
        management issues*. McGraw-Hill: New York.
http://news.stanford.edu/pr/97/970319eliassen.html.

ELLET

01.01. 1810 Penn's Manor PA/USA
21.06. 1862 Cairo IL/USA

Charles Ellet, Jr., attended the *Ecole Polytechnique* and the *Ecole des Ponts et Chaussées*, Paris, France, proposing a suspension bridge across Potomac River in 1832. He was then chief engineer of James River and Kanawha Canal. In 1842 he designed and built the first major wire-cable suspension bridge of the USA over the Schuylkill River at Fairmount PA, spanning some 100 m. In 1847 he built a suspension bridge across Niagara River below the falls whose length was more than 200 m.

After having been involved in a number of railroad projects Ellet developed theories for improving flood control and navigation of the mid-western rivers. He had advocated in 1849 the use of reservoirs built in the upper reaches of drainage basins to retain water during the wet season, which was then released during periods of low discharge to improve navigation. This then would also tend to reduce the level of flooding during high flow. The Secretary of War directed Ellet in 1850 to make surveys and reports on the Mississippi and Ohio Rivers to prepare adequate plans for flood prevention and navigation improvement. His report had a considerable influence on later engineering. Ellet was appointed in 1861 colonel of engineers to develop the US Ram Fleet, corresponding to a group of rams on the Mississippi River during the American Civil War. He converted several powerful river towboats, heavily reinforcing their hulls for ramming. These ships had light protection for their boilers, engines and upper works, but later they were fitted with guns. Ellet led his flotilla in action during the Battle of Memphis in 1862, where these rams played a major role in the Union victory against the Confederate River Defense Fleet. However, Ellet died several days later of a wound received at that action. Ellet published the Report of the overflows of the Delta of the Mississippi River, which helped to reshape New Orleans' waterfront. He also noted that an artificial embankment created an overflowing delta. His assertions were taken seriously only decades later and then used in flood control decisions.

Anonymous (1963). Ellet, Charles. *Who was who in America* 1607-1896: 167. Marquis: Chicago.
Ellet, C., Jr. (1845). *The position and prospects of Schuylkill navigation company*. Philadelphia.
Ellet, C., Jr. (1852). *Report on the overflows of the delta of the Mississippi River*. Hamilton.
Ellet, C., Jr. (1855). *Coast and harbour defences*, or the substitution of steam battering rams for ships of war. Clark: Philadelphia.
Lewis, G.D. (1968). *Charles Ellet, Jr.*: The engineer as individualist. Illinois Press: Urbana IL.
http://en.wikipedia.org/wiki/Charles_Ellet,_Jr. P

ELLIOT

06.01. 1910 Somerville MA/USA
09.12. 1983 Knoxville TN/USA

Reed Archer Elliot graduated from Tufts College, Medford MA in 1933, with the BSc degree in civil engineering. He was recruited for employment with the Tennessee Valley Authority TVA in 1935, in the group of Dana M. Wood (1884-1954), head of the Power Studies Branch. This Branch developed plans for the unified development of the TVA water resources, identifying sites which would satisfy the basic tenants of the TVA Act, namely development for flood control, navigation, and power. Initially Elliot was a hydraulic engineering aide. From 1946 to 1954 he efficiently directed the TVA Branch. The proposals for construction of any water control project originated in that group, including the physical features, scope, costs and benefits, and the economic analysis. The results were published in a report, and if approved by the TVA Board, used as a basis for obtaining the first appropriations from the Congress.

In 1955 Elliot was appointed director of the Water Control Planning Division, in which position he directed the work of six Branches, comprising the Division. He supervised these organizations ably throughout the following 18 years, making notable contributions to TVA during the floods of 1957, 1963, and 1973 when reservoir operations have been given credit for reducing enormous flood damages. Elliot retired in 1973, after 38 years of professional career with TVA. He left his mark on the organization, where he was remembered as a highly efficient director. He was Fellow of the American Society of Civil Engineers ASCE, and president of its Knoxville Branch, serving also on various ASCE Committees. He also was member of the Permanent International Association of Navigation Congresses PIANC, and the US Commission on Large Dams, chairing its Study Group in stream Channelization. He further was a registered engineer in the State of Tennessee. He authored a number of papers in the Civil Engineering Journal. He was in 1971 honored for his work, and was named one of the Top Ten Public Works Men of the Year, by the American Public Works Association APWA.

Anonymous (1948). Elliot, Reed A. *Who's who in engineering* 6: 579. Lewis: New York.
Anonymous (1984). Reed A. Elliot. *Trans.* ASCE 149: 361-362.
Elliot, R.A. (1957). TVA develops its rivers. *Civil Engineering* 27(1): 14-17.
Elliot, R.A. (1973). The TVA experience: 1933-1971. *Man-made lakes*: Their problems and environmental effects: 251-258. AGU: Washington DC.
http://toto.lib.unca.edu/findingaids/photo/nfnc/nfnc_set_broadhead/nfnc_broadhead_al.htm *P*

ELLIOTT

08.06. 1850 Lowell IL/USA
14.09. 1926 Washington DC/USA

Charles Gleason Elliott received the BS degree in 1877 from the University of Illinois, Urbana IL, and the civil engineering degree in 1893. He was first engaged in general engineering practice, including drainage engineering in the North Central States, and in geological studies in the West. He was in 1884 appointed sanitary engineer at Indianapolis IN to study its drainage and sewerage schemes. From 1887 to 1889 he was drainage engineer in Iroquois County IL on the drainage of 16,000 acres of land. He was engaged in private practice in a similar position in Illinois State from 1890 to 1901, thereby preparing the first Bulletin on agricultural drainage issues by the US Department of Agriculture USDA. He entered in 1902 government service as drainage expert, later taking over as chief of the Drainage Investigations, in the Office of Experiment Stations USDA, where he continued to work until 1913. He was commissioned by the Secretary of Agriculture in 1908 to visit Europe, to study its land drainage, with a particular reference to the drainage of turf and peat lands in Holland, England and Germany. He also reported on plans for the drainage of Minnesota and prairie lands in Canada, thereby developing the methods then generally used in the arid regions of the West for draining irrigated lands, thus saving many irrigation projects from apparent failure.

After 1913 Elliott was engaged in private practice as a consulting drainage engineer, and thereby was member of the Elliott-Harman Engineering Company, with offices at Peoria IL, Memphis TN, and Washington DC. He authored various books relating to drainage engineering, including Practical farm drainage, Engineering for land drainage, or Drainage of farm lands. He also presented a notable paper at the meeting of the International Engineering Congress at San Francisco CA in 1915 on The drainage as a correlative of irrigation. Elliott therefore was a pioneer in agricultural drainage who was recognized as a leading American authority in that engineering branch. He was member of the American Society of Civil Engineers ASCE, the Illinois Society of Engineers, and of the Washington Society of Engineers.

Anonymous (1926). Charles G. Elliott. *Engineering News-Record* 97(13): 521.
Anonymous (1927). Elliott, Charles G. *Trans.* ASCE 91: 1077-1078.
Elliott, C.G. (1908). *Practical farm drainage*: A manual for farmer and student. Wiley: New York.
Elliott, C.G. (1911). *Engineering for land drainage*. Wiley: New York.
http://www.google.ch/imgres?q=%22Charles+G.+Elliott%22+1850 *P*

ELLIS G.H.

18.07. 1884 Mason MI/USA
24.04. 1975 Butte MT/USA

George Henry Ellis received the degrees BS in 1907 and CE in 1912, respectively, from the University of Michigan, Ann Arbor MI. From 1907 to 1920 he rose from rodman to assistant engineer within the US Reclamation Service. Until 1923 he was then a hydrographer of Montana State, joining then Harza Engineering, Chicago IL, on the construction of Dix Dam and the Lock 7 Power Plant in Kentucky, and Devils River Plants in Texas. He was also engaged with stream measurements on Kentucky River and exploration works of the South Fork damsite. From 1929 Ellis was in charge of the Montana Power Company, Butte MT. He was an associate member of the American Society of Civil Engineers from 1914, and later became Member ASCE.

Ellis was one of the irrigation engineers not only working in the field but also advancing the profession by technical developments. He, as many others, presented tables to facilitate uniform flow computations in circular and horseshoe profiles, he proposed a car to attach to calibrate flow meters in well-designed channels as used by the US Geological Survey from 1911, he described the typical flow features in irrigation channels for uniform flow based on both laboratory and prototype observations, and he investigated the flow features of drops, involving a complex air-water flow. Moreover, he designed and partly constructed dams in the West of the USA, including Shoshone Dam in Wyoming, or Sun River Dam in Montana. The University of Wyoming also holds a scrapbook with poems of Ellis written between 1944 and 1962.

Anonymous (1941). Ellis, George Henry. *Who's who in engineering* 5: 525. Lewis: New York.
Anonymous (2009). G.H. Ellis. American Heritage Center. University of Wyoming: Laramie. *P*
Ellis, G.H. (1915). Tables of circular and horse-shoe conduit sections. *Engineering News* 73(24): 1182-1183.
Ellis, G.H. (1916). The flow of water in irrigation channels. *Trans. ASCE* 80: 1644-1688.
Ellis, G.H. (1917). *Hydraulic and excavation tables*. Department of the Interior, US Reclamation Service. Government Printing Office: Washington DC.
Ellis, G.H. (1920). Car for current-meter gaging stations. *Engineering News-Record* 84(2): 80.
Ellis, G.H. (1920). Hydraulics of the intake to a pipe drop. *Engineering News-Record* 85(12): 565.
Ellis, G.H. (1923). Flood flows or maximum runoffs of Montana streams. *Engineering News-Record* 91(25): 1016-1017.

ELLIS T.G.

25.09. 1829 Boston MA/USA
08.01. 1883 Hartford CT/USA

Theodore Gunville Ellis graduated as civil engineer, was chief engineer of Sackett's Harbour and Saratoga Railroad, and subsequently had charges of silver mines in Mexico to become engineer of Hartford Dyke in 1859. He entered the Federal Army, shortly later being engaged at *Antietam* and *Fredericksburg* during the Civil War, was promoted major in 1863. In 1864 he commanded the camp at Annapolis MD. Ellis was mustered out in 1865, with the brevet rank of brigadier-general. He became surveyor-general for Connecticut in 1867. Then in 1874, he conducted hydraulic experiments at *Holyoke* MA. At the time of his death, he had charge of the government works on the Connecticut River. Ellis was vice-president of the American Society of Civil Engineers ASCE for several years.

The 1876 paper of Ellis relates to orifice flow from large apertures. These were made in Holyoke MA to explore the relation between the discharge coefficient and the approach flow conditions for various square-edged orifice geometries. Ellis added a considerable number of data to this classical topic in hydraulics, yet the effects of fluid characteristics including fluid viscosity and surface tension were not thoroughly accounted for, such that a large number of additional tests were conducted mainly in the 1930s. Noteworthy papers in the same field was previously presented by Jean-Victor Poncelet (1788-1867) and Joseph Lesbros (1790-1860) in 1832 and 1852. One of their and Ellis' main problem was exact discharge measurement by weirs. Their accurate characteristics were only determined in the 1970s, involving the location of upstream flow depth measurement or scale effects. For his 1876 paper, Ellis was awarded the 1877 ASCE Normal Medal.

Anonymous (1887). Ellis, Theodore G. *Appletons' cyclopaedia of American biography* 2: 334.
Ellis, T.G. (1866). *Description of the iron bridge over the Connecticut River*, on the Hartford & New Haven R.R. Brown & Gross: Hartford CT.
Ellis, T.G. (1876). Hydraulic experiments with large apertures at Holyoke, Mass., in 1874. *Trans. ASCE* 5: 19-101; 5: 297-298.
Ellis, T.G. (1878). *Surveys and examinations of Connecticut River*. House of Representatives, Report of the Chief of Engineers, Appendix B14: 248-391. Washington DC.
Ellis, T.G. (1881). Flow of water in open channels. *Engineering News* 8(48): 478-479.
Ellis, T.G. (1892). Fire protection by direct high pressure from pumps in combined pumping and reservoir. *Journal New England Water Works Association* 7(1): 27-34.
http://www.picturehistory.com/product/id/7241 *P*

ELLMS J.W.

04.10. 1867 Ayer MA/USA
07.02. 1950 Cleveland OH/USA

Joseph Wilton Ellms graduated as a civil engineer from Massachusetts Institute of Technology MIT in 1893. He then first joined as assistant chemist the Massachusetts Health Board, and the Louisville Water Company, then both the Brooklyn Board of Health, and the Cincinnati Water Commission as chemist, where he was promoted to test engineer. From 1913 to 1932, Ellms was a private consultant, and in parallel from 1916 also an engineer of the Cleveland Filtration Plants. He also was a member of the committee which developed the Standard methods of water analysis in 1905. He was the recipient of the 1940 ASCE Rudolph Hering Medal for his leadership in sewage disposal, and the Case Institute conferred on him the honorary doctoral degree in 1942.

Ellms worked mainly in the field of water purification by rapid sand filters, but he also was a hydraulic engineer dealing with means to improve water quality. He further was interested in microscopy and the physical characteristics of natural waters, and colloids in relation to water purification. The 1916 paper deals with a high-velocity method to wash sand or mechanical filters, including a description of corrosion effects of a brass wire screen, effects of depth and size of the gravel layers, and the optimum discharge conditions. The 1927 paper deals with the hydraulic jump as a mixing device to improve water quality, one of the few works available in this field. Even currently, only few works are available dealing also with the turbulence generation of hydraulic jumps and the associated mixing capacity. Ellms was among the American pioneers of sewage treatment, with Robert S. Weston (1869-1943) and George W. Fuller (1868-1934).

Anonymous (1936). Joseph W. Ellms. *Water Works Engineering* 89(12): 714. *P*

Anonymous (1937). Ellms, Joseph Wilton. *Who's who in engineering* 4: 403. Lewis: New York.

Anonymous (1950). Joseph W. Ellms dies. *Engineering News-Record* 144(Feb.16): 24. *P*

Anonymous (1953). Joseph Wilton Ellms. *Trans. ASCE* 118: 1244-1245.

Ellms, J.W. (1916). A study on the behavior of rapid sand filters subjected to the high-velocity method of washing. *Trans. ASCE* 80: 1342-1428.

Ellms, J.W. (1918). *Water purification*. McGraw-Hill: New York.

Ellms, J.W. (1937). Advances in water purification during sixty years. *Water Works Engineering* 90(11): 666-669. *P*

Levy, A.G., Ellms, J.W. (1927). The hydraulic jump as a mixing device. *Journal American Water Works Association* 17(1): 1-26.

ELLMS R.W.

24.11. 1903 Cincinnati OH/USA
14.05. 2000 Cleveland OH/USA

Robert Wilton Ellms graduated in 1927 from Case School of Applied Science, Cleveland OH, and he obtained the MS degree in mechanical engineering there in 1930. He started his professional career by designing traveling cranes at Cleveland, becoming subsequently connected with the Pennsylvania Water & Power Company, Baltimore MD, as a hydraulic test engineer. He was in charge of model tests on turbine units installed at Safe Harbor PA. In 1930 he was appointed head of research work on coke crushers in Pittsburgh PA, a work that led to the introduction of new crushing methods. Later in his career he was a staff member of a manufacturing company at Cleveland, with research on surface finishing methods, including brushes and brushing, as well as molding machines and methods for foundries using sand molding. He was a member of the American Society of Mechanical Engineers ASME.

Ellms has published two notable papers. His 1928 work deals with the tailwater depth of a so-called classical jump, and the extension to a hydraulic jump on a sloping channel. From the origins of the interest to this basic hydraulic phenomenon in the early 19th century, few additions were made, particularly as regards the slope effect. Ellms applied the momentum equation to obtain a generalized formula of the well-known Bélanger equation which includes the slope effect. He tested the result with selected laboratory observations and found general agreement. Note that Joseph W. Ellms (1867-1950) was his father, who had previously worked on a similar topic. Following the discussions on this paper, Ellms prepared in 1932 an improved version obtaining again a generalized Bélanger equation. He noted that by increasing the bottom slope the height of the jump decreases, so that in general more energy is dissipated in jumps produced by a sloping channel as compared with the horizontal channel. It was also stated that there exist relatively few hydraulic differences between the two jump types, at least for bottom angles up to 17°. In the discussion of the 1932 paper, additional questions arose so that the topic was reconsidered in the next decades.

Ellms, R.W. (1927). *A study of certain hydraulic jump coefficients*. Case School: Cleveland OH.
Ellms, R.W. (1928). Computation of the tail-water depth of the hydraulic jump in sloping flumes. *Trans. ASME* 50(HYD-5): 1-10.
Ellms, R.W. (1932). Hydraulic jump in sloping and horizontal flumes. *Trans. ASME* 54(HYD-6): 113-121. *P*

EMERSON

26.05. 1882 Saginaw MI/USA
18.02. 1931 Cheyenne WY/USA

Frank Collins Emerson obtained the BSc degree in civil engineering from University of Michigan, Ann Arbor MI, in 1904. He moved to West settling in Wyoming, where he first was engaged in his private practice, including irrigation projects. From 1905 he was employed by the *La Perle* Ditch and Reservoir Company, Douglas WY, and then served a railways company as assistant engineer. In 1906 he worked for the Wyoming State Engineer to locate two large canals on the Shoshone Indian Reservation. Fifteen years later the Wyoming Canal was built along this plan. In 1907 he was appointed chief engineer of the Wyoming Land and Irrigation Company in charge of the location, design, construction and operation of the irrigation projects on Paint Rock in the Big Horn Basin. In 1909 the Shell Canal providing water for 45 km^2 was completed, but water shortage forced to build the Adelaide Lake Reservoir, for both of which Emerson contributed his work.

Emerson was appointed in 1914 superintendent of the Big Horn Canal Association, Worland WY, supervising the reconstruction of the Big Horn Canal. Later he supervised also the Lower Hanover Canal Association. Until 1919 he was in parallel also engaged in general engineering practice, mainly in irrigation and drainage. He was then appointed State Engineer of Wyoming. Until his death he played a leading part in the development of the Wyoming's water resources. He had been appointed Wyoming Commissioner on the Joint Colorado River Commission in 1921, resulting in the signing of the famous Colorado River Compact in 1922. It was mainly through the untiring efforts of Emerson that finally the development of the Boulder Canyon became possible. He was in 1926 elected Governor of Wyoming so that he could also develop the State by political input. However, his health was impaired by hard work during the re-election campaign in 1930. After a strong cold, pneumonia set in resulting finally in his death. He had been a builder, belonging to the long line of pioneers who through unfaltering determination developed the resources of the Nation. He was elected member of the American Society of Civil Engineers ASCE in 1918, and served also in its Irrigation Division.

Anonymous (1932). Frank C. Emerson. *Trans. ASCE* 96: 1470-1474.
Emerson, F.C. (1921). *Irrigation laws* and laws relating thereto of Wyoming State. Cheyenne.
Emerson, F.C. (1925). *Water conservation and control in Wyoming*. Seattle WA.
Tyler, D. (2003). *Western water compacts*. University of Oklahoma Press: Norman OK.
http://www.nndb.com/people/660/000210030/ *P*

ENGER

05.05. 1881 Decorah IA/USA
13.05. 1956 Escondido CA/USA

Melvin Lorenius Enger was educated at University of Minnesota and University of Illinois, where he gained the BS degree in 1906 and the MS degree in civil engineering in 1916. He was from 1908 an instructor at his Alma Mater, from 1911 assistant, from 1917 an associate, and from 1919 professor of theoretical and applied mechanics. Further, from 1934 to 1949 he was the director of its Engineering Experiment Station. Enger was a member of the American Water Works Association AWWA and the recipient of its 1940 Goodell Prize, the Western Society of Engineers WSE, the American Society for Testing Materials ASTM, and the American Society of Engineering Education ASSE, among others.

Enger dealt with a wide range of questions relating to engineering design and research. The 1918 work relates to discharge coefficients of orifices in which a small effect of scale was found provided the head on the orifice or the orifice opening are too small. This finding was in agreement with many others then conducted worldwide. The 1929 paper deals, as one of the first, on manifolds, an element widely used in industries, heating and distribution techniques. The topic has also application in lock design, furnaces or wastewater treatment stations, but the systematic investigation initiated only in the 1950s. Enger and his colleague therefore made an early study on these flows found in collecting or distributing systems. Further research was directed toward flow resistance through locomotive water columns, penstock design, pressure transmission in granular material, or air inlet valves for pipelines.

Anonymous (1940). Melvin L. Enger. *Engineering News-Record* 124(May 2): 614. *P*
Anonymous (1941). Melvin L. Enger. *Water Works Engineering* 94: 566. *P*
Anonymous (1954). Enger, Melvin L. *Who's who in engineering* 7: 725. Lewis: New York.
Anonymous (1956). Melvin L. Enger. *Engineering News-Record* 156(Jun.07): 114.
Enger, M.L. (1908). Relation of waterways to drainage areas. *Engineering Record* 57: 138-139.
Enger, M.L. (1914). Locating leaks by water-hammer diagram. *Engineering News* 71(19): 1040.
Enger, M.L. (1918). The orifice bucket for measuring water. Hydraulic experiments with valves, orifices, hose, nozzles and orifice buckets 4: 27-40. *Bulletin* 105, Engineering Experiment Station. University of Illinois: Urbana IL.
Enger, M.L., Levy, M.I. (1929). Pressures in manifold pipes. *Journal American Water Works Association* 21(5): 659-667.

ERICSON

21.10. 1858 Upland/S
16.04. 1927 Chicago IL/USA

John Ernst Ericson was born in Sweden, graduating in 1880 as civil engineer from the Royal Polytechnic Institute, Stockholm. He came to the USA in 1881, joining as resident engineer a railways company in Illinois, and as bridge engineer at St. Louis MO. In 1883 he was appointed assistant on government surveys for the proposed enlargement of the Illinois and Mississippi River Canal, including Hennepin Canal. From 1884 he was a draftsman in the Water Department in the City of Chicago IL, becoming in 1885 assistant city engineer until 1889, when moving to Seattle WA to develop the new gravity water works system. He returned in 1890 to Chicago IL to develop the Great Drainage Canal System within the Sanitary District, remaining there until 1892, when being appointed assistant engineer in the Bureau of Engineering. In 1893 he was appointed first assistant city engineer of Chicago, taking over as city engineer from 1897 until his death.

Ericson was as first assistant city engineer of Chicago in charge of the design and the construction of additions to the city water supply system, then the second largest city of the USA. As city engineer he was in addition in charge of all bridge designs. He had exceptional opportunities for experiments to determine the elements of water flow in large tunnels, presenting his extensive treatise on this topic in 1911, for which he was awarded to Society's Medal. He published further works and reports on water works, and harbors, always dealing with technical problems related to the city of Chicago. He was president of the Swedish Engineers' Society of Chicago, member of the American Society of Civil, and Mechanical Engineers ASCE and ASME, and of the American Water Works Association AWWA. Ericson was recognized as a leading authority on city improvement in general, and as an engineer who solved problems relating to urban water supply in particular. Various buildings still stand today in Chicago as monuments remembering to his name.

Anonymous (1928). John E. Ericson. *Trans. ASCE* 92: 1686-1687.
Ericson, J.E. (1905). *The water supply system of Chicago*: Its past, present and future. Chicago.
Ericson, J.E. (1911). Investigations of flow in brick lined conduits. *Journal of the Western Society of Engineers* 16(8): 657-705.
Ericson, J.E. (1913). Chicago waterworks. *J. Western Society of Engineers* 18(8): 763-796.
Ericson, J.E. (1925). *The water supply problem in relation to the future Chicago*. Chicago.
http://www.google.ch/imgres?q=%22John+Ernst+Ericson *P*

ERNST

27.06. 1842 Cincinnati OH/USA
21.03. 1926 Washington DC/USA

Oswald Herbert Ernst studied at Harvard University, graduating in 1864 from the US Military Academy. After his war service until 1864 as captain, he was assistant engineer in the construction of fortifications in San Francisco Harbor until 1868, attaining the rank of captain of engineers in 1867. He was until 1871 instructor at Engineering School of Practice, Willit's Point, Long Island NY, and instructor at the Military Academy. From 1878 to 1886 he was in charge of river and harbour improvements of Osage River MO, and of the Mississippi and the Missouri Rivers in Illinois. From 1886 to 1889 he supervised the digging of the deep-sea channel to Galveston Harbor TX. He attained the rank of colonel of engineers in 1893 and was the superintendent of the US Military Academy until 1898, when taking over as brigadier general in the Spanish-American War. He served as Inspector General in Cuba until 1899.

From 1900 Ernst was closely related to the Isthmian Canal as member of the Canal Commission. He visited Europe and Central America in connection with the route proposed. He also was in the commission which determined that Panama Canal should have locks. He was involved in parallel in river and harbor improvements. Ernst was from 1903 to 1906 president of the Mississippi River Commission, chairing the Board of Engineers to survey the route of the 4 m deep waterway between Chicago and St. Louis MO. He was promoted to colonel in 1903 retiring from active service in 1906 as brigadier general. He was appointed to the International Waterways Commission, and chaired the American Section until 1913. From 1907 he was a consulting engineer preparing plans to protect the levees in St. Louis. He presided from 1916 the Board of Consulting Engineers and thereby examined plans to protect the valley of the Miami River in Ohio from floods. He finally was given the rank of major general in 1916.

Ernst, O.H. (1873). *A manual of practical military engineering*, prepared for the use of the cadets of the US Military Academy, and for engineer troops. Van Nostrand: New York.
Ernst, O.H. (1904). Report on Respecting tunnels under the Chicago River. Washington DC.
Ernst, O.H. (1910). *Report* of the Intl. Waterways Commission, on the regulation of Lake Erie, with a discussion of the regulation of the Great Lakes system. Buffalo NY.
FitzSimons, N. (1991). Ernst, Oswald H. *A biographical dictionary of American civil engineers* 2: 33-34. ASCE: New York.
http://www.navsource.org/archives/09/22/22133.htm *P*

ETCHEVERRY

30.06. 1881 San Diego CA/USA
26.10. 1954 New Haven CT/USA

Bernard Alfred Etcheverry obtained his BC degree in civil engineering from University of California, Berkeley CA in 1902. He was appointed associate professor of civil engineering and physics in 1903 at University of Nevada, returned to his Alma Mater as assistant professor of irrigation in 1905, became there associate professor in 1910, taking over as professor of irrigation and drainage engineering in 1917 until retirement in 1951. He was in parallel a consultant for irrigation projects, including the US Department of Agriculture, the Government of the Province of British Columbia, or the Sacramento-San Joaquin Drainage District. He was also a Member of Board of consulting engineers, the State Department of Public Works, to study water resources of California State in the 1920s, or of the Orange County Flood Control District in the 1930s. He was finally a consulting engineer for the Los Angeles County Flood Control District from 1936, and of the Central Valley Project, for the US Bureau of Reclamation USBR.

Etcheverry was an authority on irrigation and known for his role in California's Water Resources Board. Since the latter was set up in 1946, he was a member of committee first and its vice-chairman from 1950. He authored various books in this field, namely Irrigation practice and engineering including the chapters: 1. Soil moisture and plant growth, 2. Disposal of irrigation water applied to the soil, 3. Water requirements of irrigated crops, 4. Results of investigations and irrigation practice regarding proper time to irrigate, 5. Duty of water, 6. Preparation of land for irrigation and method of applying water to the land, 7. Farm ditches and structures for the distribution of irrigation water, and 8. Selection and cost of a small pumping plant. Etcheverry was described as man of extraordinary friendliness. He was highly regarded and deeply esteemed by his students and colleagues. He was loyal to his professional tasks and to his intimate friends.

Anonymous (1936). B.A. Etcheverry. *Civil Engineering* 6(8): 533. *P*
Anonymous (1954). Etcheverry, Bernard A. *Who's who in engineering* 7: 739. Lewis: New York.
Anonymous (1954). B.A. Etcheverry dies. *Engineering News-Record* 153(Nov.4): 26. *P*
Anonymous (1954). B.A. Etcheverry, former director, dies. *Civil Engineering* 24(12): 846. *P*
Etcheverry, B.A. (1912). Units of measurement of irrigation water. *Engineering and Contracting* 38(13): 359-362.
Etcheverry, B.A. (1915). *Irrigation practice and engineering*. McGraw-Hill: New York.
Etcheverry, B.A. (1931). *Land drainage and flood protection*. McGraw-Hill: New York.

EVANS

13.09. 1755 Newport DE/USA
21.04. 1819 New York NY/USA

Oliver Evans was apprenticed as a wheelwright. He invented at age 22 a machine for manufacturing the carding teeth used in textile industry. While working in the flour business he invented the grain elevator, conveyor and the hopper boy, automating the milling process to the point that a mill could be run by only one person. Around 1800 Evans refined the steam engines of the day, developing the first steam engine based on the high-pressure principle. Although he had been working on plans for a steam-powered carriage, he adapted his high-pressure steam engine to further improve the milling process, which had been normally powered by water wheels.

In 1803 the Philadelphia Board of Heat commissioned Evans to build a steam-powered dredger, the first to be used in the USA. It consisted of a small steam engine and the machinery to raise the mud from the Schuylkill River. It was powered to move on land over wheels and in the river by means of a paddle wheel. This amphibious vehicle was the first in the USA in which steam power was used to propel a land carriage. He urged that his idea be adapted to move vehicles on rails of wood or iron and, although he lobbied for a railroad to be built between Philadelphia and New York, the country's first commercial railroad track was not laid until the early 1830s, years after Evans's death. He refined the steam engine throughout his life and initiated innovative manufacturing techniques. While he failed to develop his *Orukter Amphibolos* into a true steam-carriage or a true paddle wheel boat, Evans long maintained that he could have well created these conveyances. He grieved that the credit had gone to other inventors. Evans published in 1797 a book on his early inventions relating to milling techniques, and in 1805 The young engineers guide. Both books were translated into French. He also founded in 1811 the Pittsburgh Steam Engine Company, which in addition to engines made other heavy machinery and castings. The location of the factory in the Mississippi watershed was important in the development of the high-pressure steam engine for the use of river boats. While in New York City in 1819, he was informed that his workshop had burned to the ground in Philadelphia. Evans suffered from a stroke, and died soon after.

Anonymous (1887). Oliver Evans and the steam engine. *Scientific American Suppl.* 24(620): 9896.
Evans, O. (1805). *The young engineer's guide*. Philadelphia.
http://www.madehow.com/inventorbios/26/Oliver-Evans.html
http://en.wikipedia.org/wiki/File:Oliver_Evans_(Engraving_by_W.G.Jackman,_cropped).jpg *P*

EWALD

12.06. 1881 Fairchild WI/USA
17.07. 1953 Mount Lebanon PA/USA

Robert Franklin Ewald was educated at University of Wisconsin, with a BS degree in 1905, and the CE degree in 1915. He accepted in 1906 a position as assistant engineer to Daniel W. Mead (1862-1948), becoming in 1907 engineer within the US Bureau of Reclamation USBR. He joined in 1912 as engineer the Aluminium Company of America, where he then stayed all through his career. Ewald was a member of the American Society of Civil Engineers ASCE, and the Engineers Society of Western Pennsylvania.

Ewald had interest in hydraulic engineering practice all through his career, being involved for instance in the construction of *Cheoah* Dam in Tennessee, and later in the preliminary planning of Fontana Dam, erected later by the Tennessee Valley Authority TVA. The 1931 paper deals with energy dissipation, as from the years after World War I became important due to increased dam heights. These flows contain an enormous hydraulic energy which must be dissipated at the dam toe, to avoid large-scale erosion and scour of the tailwater. Depending on the fall height and the tailwater topography, stilling basins were normally employed involving a hydraulic jump basin in which the hydraulic excess energy is considerably reduced. The paper gives a short overview on these hydraulic structures and presents the results made by a laboratory study. The 1916 discussion adds to the first important study of hydraulic jumps in the USA conducted by Karl R. Kennison (1886-1977). After hydraulic jumps have first technically been described in the early 19th century in Italy, the Frenchman Jean-Baptiste Bélanger (1790-1874) correctly applied the momentum equation in 1838 and determined the so-called sequent depths of the classic hydraulic jump. Kennison was one who checked this result using hydraulic observations, and found agreement also for large so-called approach flow Froude numbers.

Anonymous (1906). Robert F. Ewald. The 1906 *Badger* 20: 70. Univ. Wisconsin: Madison. *P*
Anonymous (1948). Ewald, Robert F. *Who's who in engineering* 6: 608. Lewis: New York.
Anonymous (1953). Robert F. Ewald. *Engineering News-Record* 151(Aug.13): 67.
Dunham, H.F., Ewald, R.F. (1916). Discussion of The hydraulic jump in open channel flow at high velocity. *Trans. ASCE* 80: 390-404.
Ewald, R.F. (1912). *Historical report of office engineer*, 1910 to 1911. USBR: Denver. (*P*)
Ewald, R.F. (1931). Preventing erosion below overfall dams. *Civil Engineering* 1(6): 527-531.
Ewald, R.F. (1948). Discussion of Saint Anthony Falls stilling basin. *Trans. ASCE* 113: 550-551.
http://www.waterhistory.org/histories/strawberry/

EWBANK

11.03. 1792 Durham/UK
16.09. 1870 New York NY/USA

Thomas Ewbank came to the USA in 1819. He there developed methods of tinning lead, patented in 1832, and improved steam safety valves, which he patented in 1831. He was in 1836 able to retire from business and devote himself to studies and writings on mechanics. He travelled in 1845 to Brazil and on his return published the book Life in Brazil. He was a founder and an active member of the American Ethnological Society. He has written a number of books including 'The world a workshop, or the physical relationship of men to the earth' in 1855, or 'Thoughts on matter and force' in 1858.

Ewbank published in 1842 the first of at least four editions of his book on hydraulic and other machines, a five-part tome of more than 550 pages, copiously illustrated with nearly 300 engravings and as garrulous as its title would indicate. Though the Author protested that this would not be the case, it appeared to describe every type of pump that had until then been invented, including a primitive centrifugal unit. He in addition discoursed on innumerable related matters like atmospheric pressure or the ability of flies to walk on ceilings. Ewbank later headed the US Patent Office, and his in 1859 published 'Reminiscences on the patent office' include miscellaneous essays on the philosophy and the history of inventions, giving an account on this activity. He also published various papers in the Transactions of the Franklin Institute. His 'Experiments on marine propulsion, or the virtue of form in propelling blades' was first published in 1860, and then reprinted in Europe. He was a founder of the American Ethnological Society AES in 1842, the oldest professional anthropological association of the USA.

Anonymous (1963). Ewbank, T. *Who was who in America* 1607-1896: 173. Marquis: Chicago.

Ewbank, T. (1842). *A descriptive and historical account of hydraulic and other machines for raising water*, ancient and modern, including the progressive development of the steam engine. Appleton: New York.

Ewbank, T. (1850). On the propulsion of streamers. *Report* of the Commissioner of Patents for the year 1849. Washington DC.

Ewbank, T. (1855). *The world a workshop*: The physical relationship of man to the Earth. Appleton: New York.

Rouse, H. (1976). Thomas Ewbank. *Hydraulics in the United States* 1776-1976: 33-34. Iowa Institute of Hydraulic Research: Iowa City. *P*

http://en.wikipedia.org/wiki/Thomas_Ewbank.

FAIR

27.07. 1894 Burghersdorp/SA
11.02. 1970 Cambridge MA/USA

Gordon Maskew Fair graduated in 1913 from a gymnasium in Berlin, Germany, and then in 1916 with the SB degree from Massachusetts Institute of Technology MIT. Later he studied at Tufts College graduating in 1934 with the degree of honorary MS. He received in 1951 the honorary doctorate from Technische Hochschule Stuttgart, Germany, and honorary fellowship from Imperial College, London. He was a Honorary Member ASCE from 1962.

Fair began his professional career with the Canadian Expeditionary Force CEF as sanitary engineer with research on water disinfection. From 1918 to 1935 he held progressive ranks from instructor to professor at Harvard University. From then to 1965 he was there Gordon McKay professor of sanitary engineering, and from 1946 to 1949 dean of its Faculty of Engineering. He worked as consultant on sanitary engineering for government agencies, industries and foundations, including the National Military Establishment from 1946 to 1953, the International Health Division of the Rockefeller Foundation from 1945 to 1948, or as a commissioner on environmental hygiene from 1949 to 1954. Fair was a chairman of the Environmental Health Study Section from 1952 to 1955, and a panel member on environmental sanitation of the World Health Organization. Volume 1 of his 1966 book includes the chapters: 1. Water in the service of cities, 2. Water system, 3. Wastewater system, 4. Information analysis, 5. Water and wastewater volumes, 6. Elements of hydrology, 7. Rainfall and runoff, 8. Storage and runoff control, 9. Groundwater flow, 10. Groundwater collection, 11. Surface-water collection, 12. Transmission of water, 13. Water distribution, 14. Wastewater flow, 15. Wastewater collection, 16. Machinery and equipment, 17. Optimization techniques, and 18. Engineering projects.

Anonymous (1935). Gordon M. Fair. *Water Works Engineering* 88(May1): 515. *P*
Anonymous (1948). Gordon M. Fair honored by British Water Engineers. *Water Works Engineering* 101(3): 223. *P*
Anonymous (1954). Fair of Harvard: Master of the House, and much of what he surveys. *Engineering News-Record* 153(Dec.2): 52-56. *P*
Anonymous (1970). Gordon Maskew Fair. *Trans. ASCE* 135: 1115-1116.
Fair, G.M., Geyer, J.C. (1954). *Water supply and waste-water disposal*. Wiley: New York.
Fair, G.M., Geyer, J.C., Okun, D.A. (1966). *Water and wastewater engineering*. Wiley: New York.
Imhoff, K., Fair, G.M. (1940). *Sewage treatment*. Wiley: New York.

FANNING

31.12. 1837 Norwich CT/USA
06.02. 1911 Minneapolis MN/USA

John Thomas Fanning studied architecture until 1858 perfecting himself in building construction. He began the general engineering practice in 1862, opening an office at Norwich CT, thereby working for city projects until 1870. Until 1880 Fanning was engaged as chief engineer for water works including Manchester NH, to where he moved his office. In 1880, he was called to New York City to report on the adequate public water supply for all cities in the Hudson Valley. This project included an aqueduct 350 km long with a capacity of one billion gallons of water per day. In 1885 Fanning prepared plans to develop water power of the St. Anthony Falls on the Mississippi River at Minneapolis MN. He was in parallel a consulting engineer of the Upper Red River Valley Drainage Commission, conducting a detailed topographic survey and reporting on the drainage of more than 6,000 km^2 prairie land.

Fanning is still known for a formula describing pipe flow. It is essentially the Darcy-Weisbach formula in which the Fanning resistance coefficient corresponds to one forth of Darcy-Weisbach's. He was awarded in 1883 from the New England Agricultural Society its highest prize for architectural and engineering designs. He also secured patents for a water-wheel, a turbine motor valve, a steam boiler, a steam pumping engine to improve fire-proof building construction, and other designs for hydraulic apparatus. He was a Fellow of the American Association for the Advancement of Science. Besides his still known book, Fanning published reports in technical matters. He died at his home at Minneapolis from an attack of pneumonia after an illness of ten days.

Anonymous (1887). Fanning, J.T. *Appletons' cyclopaedia of American biography* 2: 406-407.
Anonymous (1911). John Thomas Fanning. *Engineering News* 65(7): 214; 65(11): 333.
Anonymous (1911). Col. John T. Fanning. *Engineering Record* 63(7): 74.
Anonymous (1972). Fanning, John T. *A biographical dictionary of American civil engineers*: 41-42. ASCE: New York.
Fanning, J.T. (1877). *A practical treatise on hydraulic and water-supply engineering.* van Nostrand: New York, with re-editions in 1887 and 1902.
Fanning, J.T. (1889). *Report* on the Winnipeg water power of the Assiniboine River in Manitoba. McIntyre: Winnipeg.
Fanning, J.T. (1906). *Water supply engineering.* van Nostrand: Princeton NJ.
Rouse, H. (1976). Fanning. *Hydraulics in the USA* 1776-1976: 57. University of Iowa: Iowa. *P*

FARLEY

30.03. 1863 Fort Plain NY/USA
26.08. 1932 White Plains NY/USA

John Moyer Farley graduated in 1886 from Rutgers University, New Brunswick NJ, with the BS degree, and in 1889 received from there the MS degree in civil engineering. He then designed extensions of a sewerage system at New Brunswick. From 1889 he was engaged as assistant engineer on general canal work for New York State. As principal assistant to John Bogart (1836-1920) he then planned sewerage systems in New York State. In 1890 he made flow measurements for streams feeding into the Genesee Valley, and run-off measurements on the Seneca and Cayuga Lakes watershed. From 1892 to 1900 he had joined a firm for which he superintended the construction of the extension to the sewerage system of White Plains. He thus served as engineer for the Board of Water Commissioners for White Plains from its organization in 1886 until 1911, having charge of the design and construction of all its improvements to the water supply system, including dams, reservoirs and filter plants.

From 1904 to 1906 Farley was in partnership with Edward Wegmann (1850-1935) under the name Wegmann & Farley, engaged in general consulting for reservoirs and dams. Farley had in this period charge of an extension of the sewerage system at Ballston Spa NY. He was employed by the Board of Water Commissioners from 1907 to 1908 to report on a proposed source of water supply for Batavia NY. He also designed the sewer system for Mount Kisco NY, which was completed in 1913, and for Mount Vernon NY, completed in 1910. During this time, as chief engineer, he handled some of the most important water power and water supply cases against New York City in the Ashokan Reservoir condemnation proceedings. He designed the intercepting filter wells and stand-pipe in 1910 for the Board of Water Commissioners at White Plains. From 1913 to 1921 he served as consulting engineer on improvements to the plant and dam of the Water Company, New Castle NY. From 1921 until his death he was engaged in general consulting work on surveys, dams, and reservoir construction. His wide professional knowledge in applied science and engineering was important toward the betterment of life in the many communities in which he worked. He was since 1897 a member of the American Society of Civil Engineers ASCE.

Anonymous (1933). John M. Farley. *Trans. ASCE* 98: 1541-1545.
Farley, J.M. (1930). *City of White Plains and environs*. White Plains NY.
http://www.cccellc.com/History.html *P*

FARMER M.G.

09.02. 1820 Boscawen NH/USA
25.05. 1893 Chicago IL/USA

Moses Gerrish Farmer started studies at Dartmouth College, Hanover NH where he made rapid progress but became seriously ill with typhoid fever. He then spent part of the year 1842 in the office of a civil engineer at Portsmouth NH, and continued for the next years as teacher in private schools. He became also interested in mathematics and engineering, and mainly into experimentation with electricity. He thereby developed an electric fire alarm system, which was included at Boston City. He perfected further a water motor to drive electric dynamos. In 1856, he succeeded in depositing electrolytically aluminium, and founded a prosperous business, but it was wiped out after the panic of 1857, resulting in a large financial loss. Further developments of Farmer included the incandescent electric lamp in 1859, the self-exiting dynamo in 1868 used to light a private residence at Cambridge MA. In the 1870s, Farmer mainly advanced the art of torpedo warfare for the US Navy, but ill health necessitated his resignation in 1881. During his last years, Farmer and his wife had moved to Eliot ME, but he died suddenly at Chicago, where he had gone against the advice of his physician.

Just before 1880 Farmer concentrated his attention to electric power generation and distribution and then acted as consulting electrician for the US Electric Light Company, New York. He has no particular place as a hydraulician, but was greatly responsible for the generation of electricity based on hydropower. Like many other pioneers, Farmer did not profit greatly from his inventive work, but he led the way to many applications of the electric current. Instead of laboring to perfect marketable invention, he would rather lay aside what he has done and proceed in search of the unknown. A friend of his stated: 'He was deserving of more honor than he ever received'. He for instance invented the electric-striking apparatus for fire alarm service in 1848, or in 1858 invented the incandescent electric lamp. He also was from 1872 to 1881 the electrician of the US Torpedo Station, Newport RI.

Anonymous (1931). Farmer, Moses Gerrish. *Dictionary of American biography* 6: 279-281. Scribner's: New York. P
Anonymous (1963). Farmer, Moses G. *Who was who in America* 1807-1896: 176. Marquis: Chicago.
Martin, T.C., Coles, S.L. (1919). *The story of electricity*. Marcy: New York.
Prescott, G.B. (1879). *Farmer on the electric light*. Russel Brothers: New York.

FENKELL

04.02. 1873 Chagrin Falls OH/USA
08.11. 1949 Detroit MI/USA

George Harrison Fenkell obtained in 1895 his civil engineering degree from the University of Michigan, Ann Arbor MI. He joined then the Commissioners of water-works at Erie PA until 1907, had until 1913 a similar position at Detroit MI, and joined its public works department until 1918. From then he was general manager of its Department of Water Supply, and from 1924 chief engineer of the New Water Supply Project, Detroit MI. In the 1920s Fenkell also was a member of the Engineering Board of the Sanitary District of Chicago. He further was a member of the American Society of Civil Engineers ASCE, its director from 1923 to 1926, of the American Water Works Association AWWA and its president in 1930. Fenkell was awarded the 1902 ASCE Norman Medal for his outstanding paper co-authored by Gardner S. Williams (1866-1931).

Fenkell was all through his career attached to the city of Detroit, largely in connection with its water supply. In 1913 he was drafted there as commissioner of public works. He built in this position more than 300 km of sewers and paved nearly 500 km of streets. From 1918 to 1938 Fenkell, then general manager of the Department of Water Supply, undertook a vast program, involving the institution of a meter system, the construction of the then largest rapid sand filtration plant of the world, culminating in the renowned *Springwells* Station in 1933. As ASCE director, he took interest in all his official duties. It was his custom after each Board Meeting to forward his members a circular letter informing them on salient accomplishments. Largely as a result of his attitude, a fine tradition was engendered. He was often heard to say that all relations between superiors and subordinates should be so conducted as to convince those concerned that their associates had been advantageous, both as individuals and as employees.

Anonymous (1924). George H. Fenkell. *Engineering News-Record* 93(21): 846. *P*
Anonymous (1935). George H. Fenkell. *Water Works Engineering* 88(May 1): Frontispiece. *P*
Anonymous (1939). Fenkell, George H. *Who's who in America* 20: 879. Marquis: Chicago.
Anonymous (1941). George H. Fenkell. *Civil Engineering* 11(1): 58. *P*
Anonymous (1949). George H. Fenkell dies. *Water Works Engineering* 102(12): 1104. *P*
Anonymous (1954). George Harrison Fenkell. *Trans. ASCE* 119: 1333-1334.
Fenkell, G.H. (1901). A study in hydraulics. *Association Engineering Societies* 26(3): 155-192.
Williams, G.S., Hubbell, C.W., Fenkell, G.H. (1902). Experiments at Detroit MI, on the effect of curvature upon the flow of water in pipes. *Trans. ASCE* 47: 1-369.

FIELD

16.11. 1873 Montague City MA/USA
27.01. 1941 Montreal/CA

Frederick Elbert Field graduated from Massachusetts Institute of Technology in 1896 with the BSc degree in sanitary engineering. He was appointed assistant engineer in the Department of Water Supply, Boston MA, supervising during five years the installation of water supply pipes under Boston Harbour. He also obtained experience in the design, construction and operation of the city water distribution system. He moved to Philadelphia PA in 1901, serving there as assistant engineer of the Bureau of Filtration. In 1904 he was appointed division engineer of the Bureau of Filtration, Pittsburgh PA, where he was in charge of supervision of the sedimentation basins, and the slow sand filters. He joined in 1910 the Pittsburgh Flood Commission as assistant engineer. He was in addition engaged by Rudolph Hering (1847-1923) and George W. Fuller (1868-1934) on the design of the best means of securing an improved water supply for Montreal CA, then one of the most important shipping centers of North America. Field supervised the construction of the filters; the exceptionally high quality of his work reflected his thorough supervision. In 1915 he became resident engineer on the Filtration and Aqueduct Works, and in 1919, was appointed assistant superintendent of the Water Works Department.

After the completion of the filtration works, Field devoted his attention to the Aqueduct, then under construction from St. Lawrence River to Verdun for water supply and power development. He also served as consulting engineer of St. Hyacinthe QC, in the design and construction of its filter plant, and pumping station to the water works system. In 1920 he was appointed filtration engineer of the Montreal Water Board in charge of the extension of the filtration works. He conducted extended studies on a large reservoir in eastern Montreal. He was member of the American Water Works Association AWWA, and of the American Society of Civil Engineers ASCE since 1910. He published papers on the Montreal Water Works System, the design of the Montreal Filtration Works, and on water purification in general. He was conscientious in his work, and respected by his associates and friends for his ability, integrity and sound principles of conduct.

Anonymous (1941). Frederick E. Field. *Trans. ASCE* 106: 1578-1580.
Field, F.E. (1914). Emergency water supply system for Montreal. *Contract Record* 28: 948-953.
Field, F.E. (1918). Water filtration plant at St. Hyacinthe. *Canadian Engineer* 34(3): 217-220.
Field, F.E. (1927). The water supply system of Montreal. *Trans. ASCE* 91: 538-541.
http://search.ancestry.com/cgi-bin/sse.dll?gl *P*

FINCH

10.10. 1881 Austin TX/USA
09.01. 1972 Austin TX/USA

Stanley Phister Finch graduated from University of Texas, Austin TX, with the BA degree in 1902, and there obtained the civil engineering degree in 1905. He became there instructor in civil engineering and 1906 spent at the University of Wisconsin, Madison WI, doing special work in its hydraulic laboratory. In 1908 he made further studies at Massachusetts Institute of Technology MIT, Cambridge MA, from where he received the MS degree in hydraulics in 1909. Finch then taught continuously at his Alma Mater until his retirement in 1952, becoming adjunct professor in 1913, associate professor in 1919, and professor in 1923. He further served as chairman of the Department of Civil Engineering from 1937 to 1943, and was director of the Bureau of Engineering Research from 1927 to 1938. During his long period on the faculty, Finch served on nearly all College committees and on many of the University committees.

Finch was a conscientious committee member, chairing most controversial committees in a way which satisfied everyone. The College of Engineering particularly recalled his effective administrative work with the Engineering Building Committee. Teaching was the work Finch liked most and in which he really excelled. His early interest was in hydraulics, but in a small faculty he was flexible in the courses he handled. As a specialist in hydraulics became available, Finch willingly shifted more to strength of materials and structures. He had the ability to analyze a problem and resolve it in terms of fundamentals. Memorizing past problem solutions was not what he sought from his students. While everyone regarded him as patient with students who tried, some regarded him as very strict. Many of the latter came back, after few years of engineering, to thank him for insisting on clear thinking. He was member of the American Society of Civil Engineers ASCE from 1930, becoming ASCE Fellow in 1959. The Stanley P. Finch Centennial Professorship in Engineering was established by the University of Texas in 1982, for the benefit of the Cockrell School of Engineering.

Anonymous (1932). Prof. S.P. Finch. *Cactus yearbook*: 43. University of Texas: Austin TX. *P*
Anonymous (1973). Stanley P. Finch. *Trans. ASCE* 138: 624-625.
Finch, S.P. (1909). Test of an Humphrey turbine. *MS Thesis*. Dept. of Civil Engineering. Massachusetts Institute of Technology: Cambridge MA.
Giesecke, F.E., Finch, S.P. (1918). *Physical properties of dense concrete as determined by the relative quantity of cement*. University of Texas: Austin.

FINK

27.10. 1827 Lauterbach/D
03.04. 1897 Sing Sing (Ossining) NY/USA

Albert Fink was born in Eastern Hessen, Germany. He was educated in his native country, moving around 1850 then to the USA, where he entered the service of the Baltimore & Ohio Railroad. After his service as resident engineer at Parkersburg WV, he went to the Louisville & Nashville Railroad as chief engineer and superintendent. He designed and built, among others, the Louisville Bridge across the Ohio River, the Nashville Bridge over the Cumberland, or the Decatur Bridge over Tennessee River. He was elected president of the American Society of Civil Engineers ASCE in 1879, thus leading an important American engineering organisation.

Despite Fink's career was mainly devoted to railroad and bridge engineering, he was also involved in hydraulic problems, particularly as regards the design of piers and abutments of the bridges. The bridges mentioned span partly over wide rivers, so that they had to be resistant particularly during flood conditions. During the Civil War, he was also active in repairing bridges damaged during hostile actions. In 1875, the railroad and steamship association of the South was founded, uniting 25 companies; it corresponded to the first larger association of this kind, and is stated to have been one of the most successful. To a large extent, it was the idea of Fink to combine naval and railroad transportation for the ease of users. He was from 1877 to 1889 general director of a railroad company, which had formed from former forty individual companies. Despite this and other successes, and a high salary of then 25,000 US$, he resigned in favor of a consulting position in New York City

Anonymous (1897). Albert Fink. *Engineering News* 37(April 8): 215. *P*
Anonymous (1899). Albert Fink. *Trans. ASCE* 41: 626-638.
Anonymous (1900). Fink, Albert. *The National cyclopaedia of American biography* 9: 489. White: New York. *P*
Anonymous (1927). *Albert Fink*: A biographical memoir of the father of railway economics and statistics in the United States. Bureau of Railway Economics: Washington DC.
Anonymous (1963). Fink, Albert. *Who was who in America* 1607-1896: 180. Marquis: Chicago.
Faust, A.B. (1909). *The German element in the United States*. Houghton Mifflin: Boston MA.
Lindenthal, D. (1899). Albert Fink. *Mitteilungen des Deutschen Technischen Verbandes* 4(3): 141-157. *P*
http://en.wikipedia.org/wiki/Albert_Fink *P*

FISCHER

16.05. 1937 Lakehurst NJ/USA
22.05. 1983 Bridgeport CA/USA

Hugo Breed Fischer was educated at the California Institute of Technology, Berkeley CA, with the BS degree in 1958, the MS degree in 1963, and the PhD degree in 1966. He was then at his Alma Mater assistant professor of civil engineering until 1970, associate professor until 1974, and professor until his premature death at age 46 following a sail plane accident. Fischer was recipient of the 1966 Straub Award, the ASCE 1969 Croes Medal, the ASCE 1971 Hilgard Prize, and the 1974 ASCE Huber Prize. He was Member ASCE, of the International Association of Hydraulic Research IAHR, and the American Geophysical Union AGU.

Fischer was a notable researcher in environmental engineering. His PhD thesis was supported by the US Geological Survey, enabling him to test his dispersion theories in real rivers. He was later cited 'For his research on fundamentals of dispersion in natural systems and the application of these results to practical problems'. His further research interests included mixing processes, dispersion in estuaries, disposal of sewage effluents and waste heat, and effects of stratification. His 1979 book written jointly with four colleagues was a notable addition to environmental engineering. The Hugo B. Fischer Award, made to honor Fischer's pioneering work on San Francisco Bay-Delta water quality modeling, was endowed in 1995. It is presented annually for (1) development, refinement or innovative application of a computer model, and (2) furtherance of the effective use of models in planning or regulatory functions.

Anonymous (1971). Hugo B. Fischer. *Civil Engineering* 41(10): 55. *P*

Anonymous (1974). Hugo B. Fischer. *Civil Engineering* 44(10): 95. *P*

Anonymous (1981). Hugo B. Fischer. *Who's who in America* 4: 1088. Marquis: Chicago.

Fischer, H.B. (1967). The mechanics of dispersion in natural streams. *Journal of the Hydraulics Division* ASCE 93(HY6): 187-216; 94(HY6): 1548-1559; 95(HY4): 1458-1461.

Fischer, H.B. (1973). Longitudinal dispersion and turbulent mixing in open channel flow. *Annual Review of Fluid Mechanics* 5: 59-78.

Fischer, H.B., List, E.J., Koh, R.C.Y., Imberger, J., Brooks, N.A. (1979). *Mixing in inland and coastal waters*. Academic Press: New York.

Fischer, H.B., ed. (1981). *Transport models for inland and coastal waters*. Academic Press: New York.

http://www.cwemf.org/FischerAwardWinners.htm *P*

FITCH

21.01. 1743 Winsor CT/USA
02.07. 1798 Bardstown KY/USA

John Fitch established a brass shop at East Windsor CT in 1764 and was in charge of the Trenton Gun Factory during the Revolutionary War. He surveyed lands along the Ohio Valley and the North-western Territory from 1780 to 1785. He invented in 1787 a steamboat and in 1788 launched an 18 m long steam paddle-propelled boat used to carry passengers from Philadelphia to Burlington NJ. He received in 1791 French and US patents for a steamboat, but he lost the financial support through inefficient handling of financial affairs, even though he had perfected and constructed four steamboats.

A rival of Fitch was James Rumsey (1743-1792). Fitch thought to discard his original idea of paddle boards on continuous chains for the jet-propulsion idea proclaimed by Benjamin Franklin. But Fitch was persuaded by his mechanic to adhere to the original paddle-board idea, though this was soon changed to a system of crank-mounted paddles driven by a self-designed steam engine. A skiff so propelled made its first short trip to Delaware with two fabricators as passengers in 1786, and an improved craft was publicly demonstrated in the following year. Later, a better-streamlined boat with stern rather than side paddles was built, and by 1790 Fitch operated a passenger and freight service between Philadelphia and Bordentown NJ. Fitch, as Rumsey, sought exclusively patent and operating rights for his steamboat from various states, essential to which was the proof of priority of invention. However, Rumsey had the better claim but a bit of skullduggery seems to have been introduced by his supports if not by Rumsey himself. The model that Washington certified was apparently a mechanical device not utilizing steam at all, as should be apparent from his forthright comments. The other certification evidently describes the 1786 boat, for it agrees in detail with Rumsey's own description of 1788. Though Fitch secured many affidavits correcting such misstatement, Rumsey seems to have won the case. Dissatisfied with the recognition he received in the States, Fitch sought support in France, but with even less success. Plagued by unfortunate personality traits and bad luck, he returned home a bitter man.

Anonymous (1963). Fitch, John. *Who was who in America* 1607-1896: 182. Marquis: Chicago.
Boyd, T. (1972). *Poor John Fitch, inventor of the steamboat*. Ayer: Manchester NH. *P*
Rouse, H. (1976). John Fitch. *Hydraulics in the United States* 1776-1976. Iowa Institute of
        Hydraulic Research: Iowa City. *P*
http://en.wikipedia.org/wiki/John_Fitch_(inventor) *P*

FITZGERALD

20.05. 1846 Nassau/BS
22.09. 1926 Brookline MA/USA

Desmond Fitzgerald was born on the Bahamas Islands, where his father was a captain in the British Army. The family moved to Providence RI when Desmond was a child. Here he became acquainted with engineering in the office of a firm. In the early 1870s Fitzgerald was appointed chief engineer of the Boston & Albany Railroad, but in 1873 began his career as a hydraulic engineer by becoming superintendant of the western division of the Boston water works. Most of his pioneering work was done in connection with the sanitary protection of water supplies, the improvement of reservoirs, and the study of related biology in drinking water. He also built some of the largest and most important storage reservoirs of Boston City during these years.

Fitzgerald was a pioneer in the study of colour in water and of methods of reducing it by swamp drainage, as well as of the sunlight in bleaching stored water. He established a biological laboratory in connection with water supply. Fitzgerald made a long series of tests at Chestnut Hill Reservoir on the evaporation from water area aiding to establish fundamental knowledge on this subject. When the Metropolitan Water Board absorbed the Boston supply works in 1898, he continued in charge of operation until resignation in 1903. As a consultant, he was connected with many projects, including the Chicago Drainage Canal, and the water supplies of Washington DC, San Francisco and Manila. After retirement, Fitzgerald, a distinguished lover of art, erected in Brookline MA his art gallery, which became a center of interest in paintings, and Korean and Chinese pottery. The 1886 and 1892 papers published in the ASCE Transactions were awarded the Norman Medal. He was a Member ASCE and president ASCE in 1899.

Anonymous (1924). Desmond Fitzgerald. *Engineering-News Record* 92(16): 644-645. *P*
Anonymous (1926). Desmond Fitzgerald dead. *Engineering-News Record* 97(14): 555. *P*
Anonymous (1928). Desmond Fitzgerald. *Trans. ASCE* 92: 1656-1661.
Anonymous (1931). Fitzgerald, Desmond. *Dictionary of American biography* 6: 434-435.
    Scribner's: New York.
Anonymous (1938). Early presidents of the Society: Desmond Fitzgerald. *Civil Engineering*
    8(8): 557-558; 8(11): 759-760. *P*
Fitzgerald, D. (1886). Evaporation. *Trans. ASCE* 15: 581-646.
Fitzgerald, D. (1892). Rainfall, flow of streams and storage. *Trans. ASCE* 27: 253-306.
Swain, G.F. (1927). Desmond Fitzgerald. *Proc. American Academy* 62(9): 255-257.

FLAD

30.07. 1824 Rennhof/D
20.06. 1898 Pittsburgh PA/USA

Henry (Heinrich) Flad was born in Bavaria; he graduated in 1846 from the University of Munich and then served in the Bavarian Army during the German revolution of 1848. He fled to the USA in 1849, engaged until 1860 as a railroad engineer. He served from private to colonel in the US Army during the Civil War, becoming then construction assistant to James B. Eads (1820-1887), during which period the Eads Bridge over Mississippi River was completed. Flad was from 1868 to 1876 a member of the Board of the Water Commission of St. Louis MO, during which time the city water-works were completed. He was a member of the American Society of Civil Engineers ASCE, presiding over it in 1866. He also was a founder of the Engineers Club of St. Louis, serving from 1868 to 1880 as its president.

During the Civil War Flad had participated in civil construction. Once serving for Eads he developed measurement methods as a member of the Mississippi River Commission. In connection with James P. Kirkwood (1807-1877), Flad prepared plans for the old waterworks at Bissell's Point, St. Louis, forming a solid basis of the present magnificent system. The construction of Eads' Bridge demonstrated the great ability and fertility of resource to apply scientific principles to the mastery of engineering problems placed him in the estimation of engineers. Flad was elected the first president of the Board of Public Improvements of St. Louis in 1877, and resigned only in 1890 when accepting from the President of the United States the appointment of member of the Mississippi River Commission. Taking into account his outstanding ability, Flad's simplicity of character was grand. He was without guile and deceit, and it was always difficult with him to believe that it existed in others with whom he was brought into contact. Nothing but indisputable evidence of deception brought him to believe that he had been wronged. And then it provoked no animosity, nor would he give vent in strong language of condemnation, but there would come from him expression of sadness, as though something had happened which moved his soul in sorrow.

Anonymous (1963). Flad, Henry. *Who was who in America* 1607-1896: 183. Marquis: Chicago.
FitzSimons, N. (1972). Flad. *Biographical dictionary of American civil engineers*: 45. ASCE: NY.
Rouse, H. (1976). Henry Flad. *Hydraulics in the United States* 1776-1976: 45. Iowa Institute of Hydraulic Research: Iowa City.
http://home.usmo.com/~momollus/CiCmtg/Flad.htm
http://www.google.ch/imgres?q=%22Henry+Flad%22+1824 *P*

FLINN

04.08. 1869 New Berlin PA/USA
14.03. 1937 Scarsdale NY/USA

Alfred Douglas Flinn obtained the BS degree from Worcester Polytechnic Institute, Worcester MA in 1893. He was awarded the Honorary Doctorate of applied science by the University of Louvain, Belgium in 1927. He was from 1895 engineer with the Metropolitan Water Works, Boston MA, from 1902 to 1904 then managing editor of the journal Engineering News, New York, and engineer of the Croton Aqueduct Commission, New York until 1905, from when he took over until 1918 as deputy chief engineer the Board of Water Supply of New York City, namely the Catskill Aqueduct. He was until 1934 secretary of the United Engineering Trustee, Inc. He had previously also served as secretary of the Engineering Council and had been in the early 1920s director of the Engineering Foundation. He was from 1918 to 1923 member of the National Research Council, and chairman of its Engineering Division. Flinn was a Knight of the Order of the White Lion, the Republic of Czechoslovakia. He was a member ASCE.

Flinn devoted his professional career mainly to water supply and water distribution in urban areas. His 1894 paper deals with a trapezoidal weir invented by the Italian Cesare Cipolletti (1843-1908) by which the head-discharge relation becomes particularly simple, so that this hydraulic structure was of relevance in irrigation techniques. The experiments were conducted at the Water Power Company, Holyoke MA, under the direction of Clemens Herschel (1842-1930). It was found that the weir studied has a performance similar to the standard rectangular thin-plate weir. Flinn was an author of the book Waterworks handbook with Robert S. Weston (1869-1943). Its main chapters are 1. Sources of water supply, 2. Collection of water, 3. Transportation and delivery of water, 4. Distribution of water, and 5. Character and treatment of water.

Anonymous (1918). Alfred D. Flinn. *Power* 47(1): 31. *P*
Anonymous (1943). Flinn, Alfred D. *Who was who in America* 1: 407. Marquis: Chicago.
Flinn, A.D., Dyer, C.W.D. (1894). The Cippoletti trapezoidal weir. *Trans. ASCE* 32: 9-33.
Flinn, A.D. (1900). The Wachusett Dam for the Metropolitan Water Supply, Boston MA.
    Engineering News 44(Sep.13): 174.
Flinn, A.D., Weston, R.S., Bogert, C.L. (1916). *Waterworks handbook*. McGraw-Hill: New York.
Flinn, A.D. (1918). New York city's Catskill mountain water supply. *Professional memoirs*. US
    Army Corps of Engineers: Washington DC.
Flinn, A.D. (1909). The world's greatest aqueduct. *Century Magazine* 30(9): 707-721.

FLOYD

12.04. 1878 Hampshire TN/USA
30.11. 1944 Dallas TX/USA

Ozro Nowlin Floyd graduated with the BSc degree in civil engineering in 1905 from the University of Tennessee, Knoxville TN. He was employed by an engineering company at Memphis TN in 1911. The Great Dayton Flood of 1913 gave him opportunity to display his genius in flood control work. He arrived at Dayton OH two years before the Miami Conservancy District was founded, as the principal assistant to Arthur E. Morgan (1878-1975). He was connected with the plan to achieve flood protection using retarding basins and channel enlargements. Floyd remained in the District until its completion in 1922. From 1923 to 1925 he served on the design and construction of two dams for the Wichita Falls TX. He entered then a partnership until 1933 during which a water supply dam for Waco TX, or a storage dam near San Angelo TX were erected.

From 1929, Floyd was retained as consultant to the New Orleans District of the US Army Corps of Engineers. He advised this office on the flowage rights for the Bonnet Carré Spillway. He was later also in charge of the Conowingo power development on Susquehanna River, the design and construction of the comprehensive development of the entire river basin of the Tennessee Valley Authority TVA, the development of an irrigation, flood control, and hydro-electric power project for the Red Bluff Water Power Control District of Pecos TX, or on the design and construction of Denison Dam in Texas, where he was called the father of this scheme. During his career Floyd made numerous projects relating to flood control, including that of Pueblo CO. He also made a study on Madden Dam, the Panama Canal Zone. The complete confidence which he enjoyed from the various agencies was considered a fair measure of his honesty and recognized ability. He was indeed one of the outstanding authorities in the USA in the fields of water conservation, flood control, soil erosion, hydro-power generation, and the construction of large dams. He was member of the American Society of Civil Engineers ASCE, and member of its Executive Committee of the Construction Division.

Anonymous (1946). Ozro N. Floyd. *Trans. ASCE* 111: 1465-1468.

Floyd, O.N. (1935). *Inventory* of the water resources. US National Resources Board: Washington.

Floyd, O.N. (1940). Modern construction methods on earth dams. *Civil Engineering* 10(8): 487-490; 10(9): 586-589. *P*

Fry, A.S., Thompson, R.A., Floyd, O.N. (1923). Fast hydraulic filling of the Wichita Falls Dam. *Engineering News-Record* 91(25): 1004-1008.

FLYNN

12.05. 1838 Tralee/UK
01.06. 1893 Los Angeles CA/USA

Patrick John Flynn was born in Kerry County UK. He received education from Queen's College, Cork, he was sent as a probationer in 1859 to the Civil Engineering College, Roorkee, India. He was there appointed in 1860 assistant engineer in the Public Works Department of the Government of India. In the next ten years of his service he was engaged in the Punjab on the construction of bridges, culverts and embankments on the Grand Trunk Road from Umballa to Rawalpindi.

IRRIGATION CANALS

AND OTHER

Irrigation Works,

INCLUDING

The Flow of Water in Irrigation Canals

AND

OPEN AND CLOSED CHANNELS GENERALLY,

WITH

TABLES

Simplifying and Facilitating the Application of the Formula of
KUTTER, D'ARCY AND BAZIN,

BY

P. J. FLYNN, C. E.

Member of the American Society of Civil Engineers; Member of the Technical Society of the
Pacific Coast; Late Executive Engineer, Public Works Department, Punjab, India.

AUTHOR OF
"Hydraulic Tables based on Kutter's Formula,"
"Flow of Water in Open Channels," etc.

[ALL RIGHTS RESERVED.]

SAN FRANCISCO, CALIFORNIA.
1892.

In 1873 Flynn proceeded to the United States. For the rest of his life he practiced in California as a hydraulic engineer. He carried out works first in connection with the water supply of San Francisco and from 1876 was engaged with the design and the construction of irrigation canals in Central California at Fresno CA. From 1884 he served the municipality of Los Angeles CA designing and partly constructing waterworks at Santa Monica CA. He prepared in 1889 plans for the sewerage scheme of Los Angeles, and also executed important works for the Tulare Irrigation District as engineer. Flynn settled at Los Angeles in 1891 acting there as a consultant. He had achieved a high reputation as a hydraulic engineer and enjoyed an extensive and lucrative practice. Flynn authored a number of technical works, among which are Tables to apply the formula of Wilhelm R. Kutter (1818-1888) for uniform flow, a more general account on this topic in 1886, and an early book on irrigation works published in 1892. Note that toward the end of the 19th century, practicing engineers often applied Tables for their daily problems, and there existed relatively few works in so-called advanced hydraulics directed to engineering application. The irrigation book was certainly a masterly work, given the background of its author.

Anonymous (1894). Patrick J. Flynn. *Minutes* Proc. Institution of Civil Engineers 118: 445-446.
Anonymous (1894). Patrick J. Flynn. *Proc. ASCE* 20: 68-69.
Flynn, P.J. (1883). *Hydraulic tables* for the calculation of the discharge through sewers, pipes and conduits, based on Kutter's formula. Van Nostrand's: New York.
Flynn, P.J. (1885). Hydraulic tables based on the formulae of d'Arcy and Kutter. *Technical Society of the Pacific Coast* 2(1): 37-61.
Flynn, P.J. (1886). *Flow of water in open channels*, pipes, sewers, conduits, with tables based on formulae of Darcy, Kutter, Bazin etc. Van Nostrand: New York.
Flynn, P.J. (1892). *Irrigation canals and other irrigation works*, including the flow of water in irrigation canals, and open and closed channels generally. Spaulding: San Francisco. (*P*)

FOCHT

31.08. 1923 Rockwall TX/USA
22.10. 2010 Houston TX/USA

John Arnold Focht, Jr., received in 1944 the BS degree in engineering from the University of Texas, Houston TX, where his father served on the civil engineering faculty. After army service in Europe, he attended Harvard University, Cambridge MA, studying under both Karl Terzaghi (1883-1963) and Arthur Casagrande (1902-1981), earning the MS degree in 1948. Focht began then work at the US Army Waterways Experiment Station, Vicksburg MS, designing levees, locks and other structures built by the Corps along the Mississippi River, including the *Morganza* Floodway. He moved to an engineering firm at Houston TX then, where he was actively involved in the increase of the office offering a range of geotechnical services to industry and government.

The city of Houston reaped the most benefits from Focht, who designed both the Lake Livingstone and Lake Conroe Dams. He also oversaw much of the original development of NASA's Johnson Space Center, and was instrumental in the design and construction of many southeast Texas and Louisiana refineries, as well as Port of Houston facilities. He also worked in the support of offshore exploration and drilling structures. He was honored for these achievements with the Offshore Energy Center's Pioneer Award. Focht was elected in 1990 ASCE President. The John A. Focht Jr. Citizen Engineer Award, given annually to an engineer for outstanding contributions to the community, was established in 1990. He was elected to the National Academy of Engineering NAE in 1986. The Chi Epsilon civil engineering society made him a National Honor Member in 2000. He died from pancreatic cancer at age 87.

Anonymous (1964). Focht, John A., Jr. *Who's who in engineering* 9: 595. Lewis: New York.

Focht, J.A., Jr. (1978). *An engineer remembers*. Austin TX.

Focht, J.A., Jr. (1986). Investigation of failures. *Terzaghi Lectures* 1974-1982: 223-283. ASCE: New York.

Gemeinhardt, J.P., Focht, J.A., Jr. (1970). Theoretical and observed performance of mobile rig footings on clay. *Second Offshore Technology Conf.* Houston 1: 549-558.

McClelland, B., Focht, J.A., Jr. (1955). *Soil mechanics as applied to mobile drilling structures*. Petroleum Division: Houston TX.

McClelland, B., Focht, J.A., Jr. (1958). Soil modulus for laterally loaded piles. *Trans. ASCE* 123: 1049-1063; 123: 1081-1086.

http://www.asce.org/PPLContent.aspx?id=12884902522 *P*

FOLLANSBEE

07.06. 1879 Minneapolis MN/USA
22.07. 1952 Denver CO/USA

Robert Follansbee obtained in 1902 the degree in civil engineering from Cornell University, Ithaca NY. He was then engineer in a firm at Cleveland OH until 1903, becoming recorder of the US Lake Survey until 1904, from when he was all through his career with the US Geological Survey USGS. In 1906 he studied surface runoff for power, irrigation and flood protection. Later he was in charge of the municipal water supply, and navigation projects in Minnesota, Montana, Colorado, Wyoming, South Dakota and Nebraska. He thereby acted as federal representative of the Idaho-Wyoming Boundary Commission, and was member of the Advisory Commission for State Planning Boards. Follansbee was a member of the American Society of Civil Engineers ASCE, the American Geophysical Union AGU, and the Colorado Society of Engineers.

Follansbee was in the group investigating the Castlewood dam failure in 1933. The old structure of combined rockfill and masonry type was overtopped and half of it was eroded by large floods, releasing almost 5,000 m³/s discharge into a drainage canal of Denver, causing luckily only two fatalities. It was stated that these conditions had to be thoroughly improved to secure flood protection of the city. The 1948 report describes the pioneer efforts of American hydrometric engineers. The Introduction is given by Frederick H. Newell (1862-1932), whereas the evaluation of the years 1913 to 1919 is written by Nathan C. Grover (1868-1957). The part covering the years 1919 to 1947 written by Follansbee consists of three volumes in typewritten form. The 1935 paper deals with stream-flow data of Sacandaga River at Conklingville NY, where gaging stations were compared with power plant records. The plant was rated with the Gibson method; deviations up to nearly 8% resulted during 10 years of comparison.

Anonymous (1933). Castlewood dam failure floods Denver. *Engineering News-Record* 111(6): 174-176.
Anonymous (1942). Robert Follansbee. *Second Hydraulics Conference* Iowa: Frontispiece. *P*
Anonymous (1948). Follansbee, Robert. *Who's who in engineering* 6: 658. Lewis: New York.
Follansbee, R. (1934). Evaporation from reservoir surfaces. *Trans. ASCE* 99: 704-715.
Follansbee, R. (1943). Accuracy of stream-flow records. *Water Resources Bulletin* (Nov.10): 181-183.
Follansbee, R. (1948). *A history of the Water Resources Branch of the US Geological Survey*. Washington DC.

FOLSOM O.H.

10.06. 1909 Los Angeles CA/USA
12.10. 1990 Sun City FL/USA

Oliver Hazard Folsom, brother of Richard Gilman (1907-1996), received education from University of Southern California, Los Angeles CA, obtaining the BS degree in general engineering in 1934. He was then until 1937 chairman of the junior engineering office of the Metropolitan Water District in charge of the Colorado River Project, until 1941 engineer of the US Bureau of Reclamation USBR, in charge of Parker Dam and the Central Valley Project, from where he left to war service. He returned to the USBR in 1945 as engineer until 1952, engaged with Delta Mendota Canal of the Central Valley Project at Sacramento CA, and served at New Delhi, India, from 1955 to 1959 for the River Valley Development, from where he went to Beirut, Lebanon, as chief of the Public Works Division, and there became Director General of the Central Water Authority of the Government of Jordan, on loan from the US AID-State Department. From 1964 to 1966 he was chief mission engineer at Dacca, then East Pakistan and today Bangladesh.

Folsom retired in 1970 becoming a part-time consultant. He for instance moved for the US Army Corps of Engineers from 1971 to 1975 to Vietnam sponsored by the Agency for International Development AID, and later also made business trips to Pakistan, Jordan, Bolivia and Syria. These activities are described for example by Kenneth F. Vernon, originally engineering assistant to the USBR Commissioner and finally director of engineering for the Agency for International Development within the Bureau of Reclamation Oral History Program. Folsom retired with the rank of colonel within the US Army Corps of Engineers. He also was a chartered engineer in India and in the United Kingdom. He was a Fellow of the American Society of Civil Engineers ASCE, the Institution of Engineers, India, and a member of the International Commission of Irrigation and Drainage ICID.

Anonymous (1962). Folsom, Oliver H. *Who's who in America* 32: 1032. Marquis: Chicago.
Anonymous (1985). Folsom, Oliver H. *Who's who in engineering* 6: 205. AAES: New York.
Folsom, O.H. (1968). *Water resources development in Pakistan and India.* Science and the
        human condition in India and Pakistan: 204-208, W. Morehouse, ed. Rockefeller
        University: New York.
Olivier, H. (1975). *Damit.* MacMillan South Africa: Johannesburg.
http://www.asce.org/PPLContent.aspx?id=12884902522 *P*
www.usbr.gov/history/OralHistories/VERNON,KENNETHF.pdf

FOLSOM R.G.

03.02. 1907 Los Angeles CA/USA
11.03. 1996 Napa CA/USA

Richard Gilman Folsom, brother of Oliver Hazard (1909-1990), received education from the California Institute of Technology, Pasadena CA, obtaining the BS degree in 1928, the MS degree in 1929, and the PhD degree in 1932. He was in these years also a teaching assistant at his Alma Mater, then until 1933 engineer of the Water Department, Pasadena CA, moving then to the University of California, Berkeley CA as instructor, assistant and associate professor, and from 1947 to 1953 as professor of mechanical engineering and Department chairman. He accepted in 1953 at the University of Michigan, Ann Arbor MI, the position of Director of Engineering Research, becoming in 1959 president of Rensselaer Polytechnic Institute, Troy NY, from where he retired in 1971. Its new library inaugurated during his presidency was named in his honour.

Folsom lectured and contributed widely to the transactions and the scientific societies on the transportation of sand in pipelines, the design of propeller pumps and fans, the axial adjustment of deep-well turbine pumps, the performance characteristics of deep-well turbine pumps, the code for measurement of water discharges, nozzle coefficients for free and submerged discharge measurement structures, manometer errors, laboratory manuals and jet propulsion engineering. He was a member of the American Rocket Society ARS, the American Society of Engineering Education ASEE, the Aeronautical Society, and a Fellow of the American Society of Mechanical Engineers ASME.

Anonymous (1964). Folsom, Richard G. *Who's who in engineering* 9: 597. Lewis: New York.
Folsom, R.G. (1932). An experimental investigation of the phenomena produced by the highly turbulent flow of water past a series of sharp obstacles. *PhD Thesis*. Caltech: Pasadena.
Folsom, R.G. (1939). Nozzle coefficients for free and submerged discharge. *Trans. ASME* 61(4): 233-238.
Folsom, R.G., Iversen, H.W. (1948). Pipe factors for quantity rate flow measurements with Pitot tubes. *Mechanical Engineering* 70(12): 1019-1020.
Folsom, R.G. (1956). Review of the Pitot tube. *Trans. ASME* 78(7): 1447-1460.
O'Brien, M.P., Folsom, R.G. (1948). The design of current meters. *Trans. AGU* 29(2): 243-250.
Rouse, H. (1976). Richard Folsom. *Hydraulics in the United States* 1776-1976: 131. Iowa Institute of Hydraulic Research: Iowa City. *P*.
http://www.absoluteastronomy.com/topics/Richard_G._Folsom
http://en.wikipedia.org/wiki/Richard_G._Folsom *P*

FOOTE

24.05. 1849 Guilford CT/USA
24.08. 1933 Hingham MA/USA

Arthur DeWint Foote entered Sheffield Scientific School, Yale University, New Haven CT, but left it after one year in 1870. In 1874 he was an assistant engineer on the Sutro Tunnel driven to drain and ventilate the famous Comstock Mines, Virginia City NV. He was then from 1876 to 1877 resident engineer for the New Almaden Mine CA, then the largest quicksilver mine in America. In 1880 and 1881 he made a trip to Mexico to report on mines with the US Geological Survey. He went in 1882 to Idaho as chief engineer of the Idaho Land and Irrigation Company, Boise ID, to build an irrigation system covering 2,500 km². This large engineering project involved storage reservoirs, dams, and two canals. The design discharge of the 48 km long main canal was 110 m³/s, but the Company became bankrupt after one year. The project, which included Arrowrock Dam, was later completed by the US Reclamation Service, as the Boise Project.

In 1893 Foote served as chief engineer on the Snake River Division of the Reclamation Service, locating various canals and reservoirs which then were constructed. He was sent in 1895 to study recent mine-pumping plants driven by electricity along Lake Superior, concluding that compressed air for the transport of power was better in his project. This involved a pipeline with a head of 230 m, a 5 m Pelton turbine and a power house, an innovation in mining engineering. Foote remained there until 1913, serving then as consulting engineer. He was a true pioneer, never afraid of trying something new. His 1909 paper published in the Transactions ASCE deals with his works in the Great Valley of California. Hydraulic mining was stopped in these years because the rivers were filled with so much debris that damages to the valleys were considered too large. He had received the utmost loyalty and esteem from his assistants and employees, who stated of their 'boss': 'All of us who started our mining experience with him hold him in affectionate remembrance. His fine character and high ideals permeated the whole organization'. He was member of the Mining and Metallurgical Society of America MMSA, and the American Society of Civil Engineers ASCE.

Anonymous (1934). Arthur D. Foote. *Trans. ASCE* 99: 1449-1452.
Foote, A.D. (1910). The redemption of the Great Valley of California. *Trans. ASCE* 66: 229-279.
Rickard, T.A. (1922). A.D. Foote. *Interviews with mining engineers*: 171-189. San Francisco. *P*
http://en.wikipedia.org/wiki/Arthur_De_Wint_Foote
http://www.centeredriding.org/newsshow.asp?int_id=26

FORESTER

08.12. 1890 Bell Air IL/USA
25.09. 1969 Monte Vista CO/USA

Don Montell Forester obtained the BSc degree from Georgia School of Technology, todays GeorgiaTech, Atlanta GA, in 1914. He was then until 1916 junior engineer of the Illinois Highway Department, and after war service until 1919 joined the Flood Control Engineering of Mississippi and Alabama States. He was staff member of a construction company from 1926 to 1928 in Mississippi State, and in 1929 an engineer in charge of tests and inspections of floods at Chicago IL. Forester was construction engineer in 1930, and in 1931 chief engineer at Moscow USSR, Berlin D, and Budapest H. He joined in 1932 the US Bureau of Reclamation in charge of water supply for the Boulder Dam, taking over in 1934 as field engineer at Imperial Dam and Desilting Works, the All-American Canal. From 1938 to 1939 he was division engineer of this large Canal, and also engineer in charge of the Rogue River Basin Investigations in Oregon State. From 1941 to 1944 he was project planning engineer of USBR, becoming until 1945 chief of division, USBR Branch of Project Planning, from when he acted as project engineer of the USBR San Luis Valley Project.

The San Luis Valley Project is in the south-central portion of Colorado. The authorized project includes the Conejos Division, which regulates the water supply for 320 km^2 of land in the Conejos Water Conservancy District, and the Closed Basin Division, which will salvage shallow ground water now being lost to evapotranspiration in the Closed Basin of San Luis Valley. The water is delivered to the Rio Grande for beneficial use in accordance with the Rio Grande Compact among the States of Colorado, New Mexico, Texas, and the 1906 Treaty with Mexico. The Conejos Division included construction of Platoro Dam and Reservoir, which was completed in 1951. Forester was member of the American Society of Civil Engineers ASCE.

Anonymous (1945). Don M. Forester. *Engineering News-Record* 134(Feb.8): 178. *P*
Anonymous (1948). Forester, Don M. *Who's who in engineering* 6: 664. Lewis: New York.
Forester, D.M. (1938). Desilting works for the All-American Canal. *Civil Engineering* 8(10): 649-652.
Forester, D.M. (1938). Disc. of Turbid water through Lake Mead. *Trans. ASCE* 103: 755-757.
Forester, D.M. (1957). Discussion of Methods of determining consumptive use of water in irrigation. *Trans. ASCE* 122: 818-819.
http://www.usbr.gov/projects/Project.jsp?proj_Name=San+Luis+Valley+Project

FORRER

06.01. 1793 Harrisburg PA/USA
25.03. 1874 Dayton OH/USA

Samuel Forrer demonstrated a natural aptitude for mechanical pursuits and mill work at young age. In 1814, at age 21, he visited Ohio and returned there three years later, traveling down the river from Pittsburgh on a skiff, settling first at Cincinnati OH, working there as a carpenter. He was engaged soon later as deputy surveyor of Hamilton County. In 1820 he was hired to examine the summit between Scioto and Sandusky Rivers, to determine whether Lake Erie and the Ohio River might be connected with a canal. The result of this survey initiated an examination of the value of such a canal. The canal commissioners appointed James Geddes (1763-1838) as chief engineer, and Forrer advanced soon to assistant engineer in 1822. Two routes were proposed, namely the Erie-Ohio Canal connecting Ohio River at Portsmouth with Lake Erie, and the Miami Canal connecting Cincinnati to Dayton OH. In 1825, both canals were accepted. The Canal Commission appointed Forrer as resident engineer. After marriage in 1826, he lived at Dayton OH, serving until 1831. The Miami Canal was opened in 1829, with 'The Forrer' as the second boat arriving at Dayton, illustrating his importance for the city and the creation of the canal.

In 1832 Forrer was appointed to the Board of Canal Commissioners; he managed during three years the activities of the Miami Extension. In 1836 he was appointed principal engineer of the Miami Canal. In 1839 he agreed to the position of engineer and general superintendent of the turnpikes, including the Dayton and Lebanon Turnpike, The Dayton and Springfield Turnpike, and the Great Miami Turnpike. Due to political changes, Forrer became consultant soon later for public works projects throughout Ohio and the Midwest, including his advice on the proposed Richmond and Brookville Canal in Indiana. In 1846 he travelled east hoping to be hired as contractor of the Chesapeake and Ohio Canal. From then he was mainly engaged in projects relating to canals and to railways in Indiana and Missouri. In the 1860s he was responsible for the entire canal.

Huntington, C.C., McClelland, C.P. (1905). *History of the Ohio Canals*: Cost, use and partial abandonment. Ohio State Archaeological and Historical Society: Columbus OH.
Trevorrow, F.W. (1973). Ohio canal men: Samuel Forrer. *Ohio's canals*: History, description, biography: 16-70. Oberlin OH.
http://lisarickey.wordpress.com/2012/07/19/bio-sketch-samuel-forrer-1793-1874-miami-erie-canal-engineer/ *P*

FORSHEY

18.07. 1812 Somerset County PA/USA
25.07. 1881 Carrollton LA/USA

Caleb Goldsmith Forshey attended from 1833 to 1836 the US Military Academy, West Point NY, but did not graduate. He was professor of mathematics and civil engineering then at Jefferson College, Washington MS, and from 1838 was employed on engineering projects along the Mississippi River. He lived at Vidalia LA until 1848 while serving as city engineer of Natchez MS, across the river from his house. In 1848 he constructed at Carrollton LA a hydrologic station to measure the river discharge until 1855 for the Mississippi Delta Survey.

After work for railroad companies Forshey founded in 1854 the Texas Military Institute, Galveston TX. The school moved two years later to Rutersville TX, where he served until 1861 as superintendent when the school was closed on the onset of the Civil War. He then worked in the Engineering Corps on the defense of the Texas Coast, playing an important role in 1862 in planning the recapture of Galveston. In 1863 he supervised the building of Fort Esperanza on Matagorda Island TX. He from then planned Confederate fortifications near Orange along Sabine River. After the war he was an engineering consultant to the city of Galveston. He published in 1866 a report proposing a system of railroads designed to lead from the port of Galveston into the inferior of Texas. He was in 1870 chairman of a committee that suggested improvements to the channels and harbors of Galveston Bay. He worked along the Red River in the mid-1870s returning eventually to the Mississippi Delta, where he died at age 69.

Anonymous (1881). Caleb Goldsmith Forshey. *Weekly Picayune*. New Orleans LA.

Evans, D.S., Olson, D.W. (1990). Early astronomy in Texas. *Southwestern Historical Quarterly* 93(4): 433-456.

Forshey, C.G. (1850). *Memoir on the physics of the Mississippi River*, and certain internal improvements in the State of Louisiana. Office of the Bee: New Orleans.

Forshey, C.G. (1873) *The delta of the Mississippi*. Wilson & Son: Cambridge.

Forshey, C.G. (1878). *Physics of the Gulf of Mexico and of its chief affluent*. Salem Press: TX.

Geiser, S.W. (1958). *Men of science in Texas* 1820-1880. Southern Methodist University Press.

Kolupaila, S. (1960). Caleb G. Forshey. Early history of hydrometry in the United States. *Journal of the Hydraulics Division* ASCE 86(HY1):12. P

Terreo, J. (1980). *Caleb Goldsmith Forshey diaries*, 1838-1879. State University: Memphis TN.

Waterfield, M. (2003). *Errant rebel*: Caleb Goldsmith Forshey. Waterbend: Maumee OH.

http://www.tshaonline.org/Handbook/online/articles/ffo16

FORTIER

24.04. 1855 Leeds ON/CA
19.08. 1933 Oakland CA/USA

Samuel Fortier obtained from the McGill School, Montreal his BSc, ME and PhD degrees in 1885, 1896 and 1907, respectively. He was first assistant engineer of the Denver Water Company from 1886 to 1890, chief engineer of the Ogden Water Works, and Bear River Canal & Irrigation Company until 1893, when being appointed professor of hydraulic engineering at the Agricultural College of Utah. In parallel he was hydrographer of the US Geological Survey USGS and consulting engineer for irrigation works. From 1899 to 1903 he directed the Montana Experiment Station, Bozeman MT. He then was resident hydrographer of the USGS in Montana, and irrigation engineer of the US Department of Agriculture. From 1903 to 1907 Fortier was in charge of the Pacific Coast District for irrigation investigations of the US Office of Experiment Stations. From then to 1915 he was the chief of irrigation investigations, then taking over as adviser of the Government of British Columbia on irrigation. From 1915 he was chief of the Division of Irrigation, US Office of Public Roads and Rural Engineering, retiring in 1924 as a consultant. He was a member of ASCE, of the American Society of Irrigation Engineers, and of the Canadian Society of Civil Engineers CSCE, which awarded him the *Gowski* Medal in 1896 for a paper on water storage.

Fortier was a well-known expert in irrigation engineering who was greatly in charge of the reclamation projects in the West of the United States. He advanced in collaboration with Fred C. Scobey (1880-1962) a formula to determine the permissible velocities in canals such that erosion was avoided. This topic had been considered already earlier in France, but it was only advanced in the 1930s by Albert Shields (1908-1974), using the relevant dimensionless quantities describing this process for sand beds. Fortier further presented a notable book on the use of water in irrigation.

Anonymous (1917). Fortier, Samuel. *Who's who in America*: 849. Marquis: Chicago.
Anonymous (1933). Samuel Fortier. *Engineering-News Record* 111(Aug.24): 243.
Fortier, S. (1896). Earthen dams. *Bulletin* 46. Utah Agricultural Experiment Station: Logan UT.
Fortier, S. (1915). *Use of water in irrigation*. McGraw-Hill: New York, 3rd ed. in 1926.
Fortier, S., Hoff, E.J. (1920). Defects in current meters and a new design. *Engineering News-Record* 85(20): 923-924.
Fortier, S., Scobey, F.C. (1926). Permissible canal velocities. *Trans. ASCE* 89: 940-956.
Rouse, H. (1976). S. Fortier. *Hydraulics in the USA* 1776-1976: 94. University of Iowa: Iowa. P

FORTSON

16.10. 1906 Washington GA/USA
22.01. 1980 Bay County FL/USA

Eugene Palmer Fortson, Jr., was educated at Texas A&M University, College Station TX, receiving his BS degree in civil engineering in 1932. He then joined the Waterways Experiment Station, Vicksburg MS, as hydraulic model engineer until 1940, from when he served in the US Army under the Persian Gulf Command. Upon return to the USA, Fortson rejoined the Experiment Station, now as chief of its Hydraulics Division. He was from the 1940s member of the American Society of Civil Engineers ASCE.

Fortson was a hydraulic engineer dealing mainly with laboratory studies within the US Army Waterways Experiment Station. His 1963 paper reports on visits to three European hydraulic laboratories, namely the Hydraulic Research Station HRS, the British Dept. of Scientific and Industrial Research at Wallingford UK, then the *Laboratoire National d'Hydraulique* LNH of Electricité de France EdF at Chatou, France, and *Laboratoire Central d'Hydraulique de France* LCH, Paris, France. Despite the then leadership of the USA in many fields of hydraulics and hydraulic engineering, Fortson and others made these visits to obtain information on alternative methods for studying fluid flow. In the early 1960s, the United Kingdom and France counted among the leaders in these fields in Europe. The 1970 Report deals with questions of hydraulic laboratory modeling, mainly with the model similitude and scale effects by which up-scaling from model data to prototype dimensions becomes incorrect because of additional effects of mainly viscosity and surface tension. The photo of the 1949 Report includes notable American individuals of hydraulic engineering, namely Robert T. Knapp (1899-1957), Hunter Rouse (1906-1996), or Arthur T. Ippen (1907-1974), together with Fortson.

Anonymous (1948). Eugene P. Fortson, Jr. *Who's who in engineering* 6: 666. Lewis: New York.
Anonymous (1949). E.P. Fortson. *Engineering-News Record* 142(Jun.16): 32. *P*
Anonymous (1957). Eugene P. Fortson, chief Hydraulics Division. *Civil Engineering* 27(6): 28. *P*
Fortson, Jr., E.P. (1944). Engineering foreign service. *Civil Engineering* 14(3): 127-128.
Fortson, Jr., E.P., Fenwick, G.B. (1961). Navigation model studies of New Ohio River Locks. *Trans. ASCE* 126: 171-183.
Fortson, Jr., E.P. (1963). *Visits to hydraulic laboratories in England and France.* US Army Engineer Waterways Experiment Station: Vicksburg MS.
Fortson, Jr., E.P. (1970). Capabilities of hydraulic models. *Miscellaneous Paper* H-70-5. US Army Waterways Experiment Station: Vicksburg MS.

FOSS J.H.

07.01. 1879 Loleta CA/USA
05.01. 1946 Paia HI/USA

John Harrison Foss graduated from Leland Stanford Junior University, Stanford CA, in 1903. His first work was then on the construction and development of a collection system for mountain water, Haleakala HI, to irrigate the fertile isthmus between east and west Maui. At this time Michael M. O'Shaughnessy (1864-1934) was employed to build Koolau Ditch, an aqueduct extending the collection system to a distance of 50 km from its use. To bring the water to the Maui Agricultural Company's fields, a new Ditch was built simultaneously. Foss rose within a short time there to the engineer in charge, completing 20 km of tunnels, siphons, and open ditch. After completion of this scheme, Foss moved to the plantation as civil engineer, extending the irrigation system, building reservoirs, flumes, weirs, and regulating works. He was in 1907 drawn back to Stanford for 12 years as assistant professor of civil engineering, and as head of its Civil Engineering Department, demonstrating clarity of thought, a human approach to his students, and patience as teacher. He was closely associated with the growth of Maui, returning in 1919 to plan and largely build the 13 km long *Honolula* Ditch, remaining there until his death.

The professional life of Foss was a summary of the progress of his community in Maui. The *Wailoa* Ditch, involving 16 km of concrete-lined tunnels, and a discharge capacity of 8 m^3/s completed in 1923 was his most important work from the technical standpoint, because it is not only the largest in capacity of Hawaiian aqueducts but, in its use of automatic gates, concrete flumes, and self-supporting steel bridge pipes, is a milestone in engineering progress. The development of hydraulic energy as a by-product of the irrigation project was the contribution of Foss to the economy of the plantation. As a consultant, his services in the reconstruction and completion of the Alexander Dam in 1932 contributed largely to the successful methods of internal drainage of a hydraulic-fill construction. His personal qualities were a guide to the entire engineering profession of Hawaii. His integrity, accuracy and dependability, and his comprehensive study of complex problems were an inspiration. His passing left a gap in the civil engineering profession of Hawaii. He was member of the American Society of Civil Engineers ASCE from 1910.

Anonymous (1947). John H. Foss. *Trans. ASCE* 112: 1570-1572.
Siddall, J.W., ed. (1921). John H. Foss. *Men of Hawaii* 2: 153. Star: Bulletin: Honolulu. *P*

FOSS W.E.

04.09. 1868 Sharon, Norfolk County MA/USA
03.03. 1949 Hillside TS/USA

William Everett Foss was most probably educated at the Massachusetts Institute of Technology MIT, Cambridge MA. He had been from assistant to chief engineer of the waterworks of Boston MA. He was thereby responsible for all engineering works. Later he became chief engineer of the Metropolitan Water Works, Boston MA. From 1920, an annual report was published dealing extensively with matters relating to the advance of this important waterworks.

The Water Board of Boston established in 1889 a laboratory to systematically study the biology of the water-supply sources. The Chestnut Hill Laboratory was the country's first, operating under Desmond Fitzgerald (1846-1926). Foss, and George C. Whipple (1866-1924) also worked in this lab. Foss would later also collaborate with Frederick P. Stearns (1851-1919). The 1894 paper deals with a typical problem of the time, the determination of the discharge in a pipe or a channel. The paper first introduces a then recently developed formula by the Frenchman Alfred A. Flamant (1839-1915), who had analyzed data of various experimenters of the 19th century. Foss thus proposed power-law equations for a variety of pipe conditions, including cast-iron pipes and riveted sheet-iron pipes. He then compares his proposal with these of Wilhelm R. Kutter (1818-1888) and Robert Manning (1816-1897), finding that the latter agrees well with his proposal for so-called turbulent rough flow conditions. The 1919 paper describes the Wachusett Power Station located just downstream of Wachusett Dam, Clinton MA. The problems encountered with one of the turbines are particularly highlighted.

Fitzgerald, D., Foss, W.E. (1894). The color of water. *Franklin Institute* 188(11): 400-412.

Fitzgerald, D. (1895). *A short description of the Boston Waterworks*. Rockwell and Churchill: Boston.

Foss, W.E. (1894). New formulas for calculating the flow of water in pipes and channels. *Journal of the Association of Engineering Societies* 13(6): 295-309.

Foss, W.E. (1919). Break in No. 2 hydraulic turbine at Wachusett Power Station, Clinton MA. *Journal of the New England Water Works Association* 33(6): 143-152. (*P*)

Foss, W.E. (1921). Report of the Director and Chief Engineer of Water Division. *First Annual Report*: 87-144. Metropolitan District Commission: Boston.

Foss, W.E. (1932). *Description of the Metropolitan Water Works* 1846-1932. Metropolitan District Commission: Boston MA.

Rogers, J.R. (2012). William E. Foss. Personal communication.

FOSTER

20.09. 1891 Geneva NY/USA
13.12. 1963 New York NY/USA

Henry Alden Foster obtained in 1913 the BS degree in civil engineering from the University of Arizona, Tucson AZ, and the civil engineering degree in 1916 from Cornell University, Ithaca NY. He was then a draftsman for a company at Newark NJ and served from 1917 to 1919 in the US Army. From 1919 to 1925 he was assistant engineer for an engineering company in New York City, joining then until 1933 as associate engineer on the New York Water Power Investigations the firm Parsons, Klapp, Brinkerhoff & Douglas, New York NY. He made in 1934 a study on flood frequencies for the US Geological Survey USGS, and continued as associate engineer with the New York firm. From 1936 to 1938, now as research engineer, he conducted another study on hydro-electricity and water storage projects of the North Platte Basin in Nebraska. He continued work with Parsons, Klapp, Brinkerhoff & MacDonald until 1947 in various positions, and he was appointed in 1947 principal associate. From 1949 until 1957 he was in parallel adjunct professor of hydrology at the Brooklyn Polytechnic Institute. Foster was member of the American Society of Civil Engineers ASCE, and the American Geophysical Union AGU. He was awarded the Fuertes Gold Medal in 1924 from Cornell University.

Foster suggested in 1924 that a previous work on extrapolation of theoretical frequency curves could be improved by using historical hydrological data. He applied the various curves proposed for this analysis, stating that the problem was to determine the form of curves resulting in the most logical representation of the original series. The 1945 paper deals with the use of flow nets for potential flows to construct solutions in engineering hydraulics. The examples discussed include ditches, dikes, dams and other engineering structures. Shortly before his death he conducted studies of improvements for Panama Canal including the possibility of converting it to a sea-level canal. Foster's death was caused by an accident with a car while crossing a street in South Orange NY.

Anonymous (1959). Foster, Henry A. *Who's who in engineering* 8: 819-820. Lewis: New York.
Anonymous (1964). Henry Alden Foster. *Princeton Alumni Weekly* 64(15): 15.
Foster, H.A. (1924). Theoretical frequency curves and their application to engineering problems. *Trans. ASCE* 87: 142-173.
Foster, H.A. (1939). Standing waves in spillway chutes. *Civil Engineering* 9(8): 499. (*P*)
Foster, H.A., Howe, J.W., Jarvis, C.S. (1945). Construction of the flow net for hydraulic design. *Trans. ASCE* 110: 1237-1252. (*P*)

FRANCIS C.

10.08. 1842 Lowell MA/USA
29.04. 1914 Davenport IA/USA

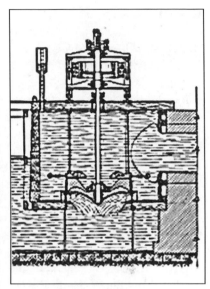

Charles Francis was the third son of James Bicheno Francis (1815-1892), one of the most famous US hydraulic engineers. Charles entered in 1860 Harvard University, graduating after war service in 1864. He remained for some time at Lowell MA, learning the trade of machinist, as a preliminary to following his father's profession as a hydraulic engineer. He was assistant to his father on the Turner Falls Dam across Connecticut River, and the Provincetown MA Dike. In 1871 he went to California where he was engaged in hydraulic mining operations for 9 years. In 1880 he was with the engineering staff of the Mexican Central Railroad in Mexico. In 1883 he returned to Lowell, being associated with his father until 1889. Charles Francis then went to Davenport IA as engineer in charge of the construction of Rock Island Dam for the US Government Arsenal at Rock Island IA. After completion of this work he continued to make Davenport his home, carrying on private practice as an engineer and contractor. He was for 7 years member of the State Board of Health, and he also served for two years as commissioner of Public Works, Davenport IA. Francis was member of the American Society of Civil Engineers ASCE.

The first inward-discharge reaction wheel was patented in 1838. James B. Francis in 1848 made its redesign building the center-vent wheel with an efficiency of almost 80%. This design was the foundation on which the modern wheel was developed. An account on most of Francis' experimental work is contained in the 1871 book, marking the start of a new era in the literature of hydraulic engineering, and was recognized as a standard authority for a long time. Previously, in 1852, Francis refined the use of tube-shaped floats, weighted at one end to ride vertically and thus provide a measure of the average velocity along a vertical element in a stream. This scheme required correction factors, and therefore another set of accurate measurements of discharge, using weirs. Therefore large weirs up to 3 m wide were used, resulting in the Francis weir formula, which was used up to the early 20th century. During his stay at the Lowell Hydraulic Laboratory, Charles Francis was involved in these developments, thereby contributing to his father's work. Once the father retired from the Laboratory, Charles was not enough engaged in the future of hydraulic developments, assuming other responsibilities.

Anonymous (1914). Charles Francis. *Engineering News* 71(19): 1051-1052.
Anonymous (1937). The center-vent Francis wheel. *Civil Engineering* 7(2): 142-143. (*P*)
Francis, J.B. (1871). *Lowell hydraulic experiments*. Van Nostrand: New York.

FRANCIS J.

30.03. 1840 Lowell MA/USA
01.12. 1898 Lowell MA/USA

James Francis was the second son of James Bicheno Francis (1815-1892). His education was mainly in private schools at Lowell MA and Boston MA. At age 20 he entered the Lowell Machine Shop. At the outbreak of the Civil War, he was with the Army of the Potomac as second lieutenant, rising in 1865 to the rank of lieutenant-colonel. He then entered the engineering department of the Hoosac Tunnel being engaged in surveys. Returning to Lowell in 1866 he became assistant engineer of locks and canals on Merrimac River. He thereby studied the hydraulic works of his father. Up to the latter's retirement in 1885, Francis became agent of the company in which position he managed the water power of Merrimac River and all other company's issues.

A notable work was his scheme for the power regulation of Concord River at Whipple's Falls, Lowell. He designed a system to regulate the power, which was in operation since 1894, to the satisfaction of all parties involved. Francis' dealings with corporations or individuals were marked by absolute impartiality, and his honesty of thought and purpose were apparent in all his social relations. He was severely hurt during work on the Pawtucket Canal in 1888, both legs being broken. He recovered from these injuries, but a few months before his death was thrown from a carriage and never regained his former vigor after this accident, losing steadily strength until his death. Francis served as a member of the Lowell Board of Aldermen in the mid-1880s, and was director of the Lowell Gas Light Company. He also was a member of the American Society of Civil Engineers ASCE since 1893, of the American Society of Mechanical Engineers ASME, the Institution of Civil Engineers ICE, London UK, and of the Massachusetts Military Historical Society. The document Finding aid for the James Francis letter books contains his large correspondence both professionally and private all over the world. These are kept at Lowell National Historical Park, Lowell MA.

Anonymous (1899). James Francis. *Journal of the Association of Engineering Societies* 22: 12-14.
Anonymous (1901). Francis, James. *Trans. ASCE* 45: 627-628.
Francis, J.B. (1871). *Lowell hydraulic experiments*. Van Nostrand: New York.
French, J.A. (1986). *Boston's water resource development*. ASCE: New York.
Malone, P.M. (2009). *Waterpower in Lowell*: Engineering and industry in nineteenth-century America. Johns Hopkins University Press: Baltimore MD. *P*
www.nps.gov/lowe/.../J-Francis-Letter-Bks-Finding-Guide-15582.pdf

FREAD

17.07. 1938 Tuscola IL/USA
05.02. 2009 Huntingdon PA/USA

Danny Lee Fread worked a plethora of jobs to pay his way through his Liberal Arts and Engineering degrees from Carthage College, Kenosha WI, and the University of Missouri-Rolla, Rolla MO. After working for Texaco, he returned to his Alma Mater to continue his education, receiving the PhD degree in civil engineering in 1971. He joined the Office of Hydrology at the National Weather Service then in Silver Spring MD as a research hydrologist. Inspired by the tragedy of the failure of Grand Teton Dam in 1976, his research focused on developing computer models to forecast the flow of flooding rivers and dam failures. His computer models were used around the globe. Fread received national awards for his work, including the Department of Commerce Gold Medal, the Huber Research Prize from the American Society of Civil Engineers ASCE, its 1976 J.C. Stevens Award, and the Association of State Dame safety Officials National Award of Merit. He also was a Fellow of the American Meteorological Society. He ended his career as Director of the Office of Hydrology. Following his retirement, he moved with his wife to Pennsylvania to be near their daughter and family.

The 1973 ASCE paper presents a conceptual model to alleviate flood damages due to overtopping failures of future small earthfill dams including the erosion pattern. The potential reduction in the reservoir release due to the proposed erosion retarding layer is also investigated. A method to determine the optimum layer location is provided so as to minimize the maximum possible reservoir release due to a gradually-breached earth dam. The transient reservoir flow is simulated by a numerical model based on the solution of the one-dimensional Saint-Venant equations, which are solved by the method of characteristics subjected to appropriate boundary conditions. The numerical simulation provides the reduction in release discharge in terms of various parameters.

Anonymous (1976). Danny L. Fread. *Civil Engineering* 46(10): 104. *P*
Fread, D.L. (1972). Dynamic flood routing in rivers with major tributaries. *Spring National Meeting* Washington DC: 1-29. American Geophysical Union: Washington DC.
Fread, D.L., Harbaugh, T.E. (1973). Transient hydraulic simulation of breached earth dams. *Journal of the Hydraulics Division* ASCE 99(HY1): 139-154.
Fread, D.L. (1973). Effects of time step size in implicit dynamic routing. *Water Resources Bulletin* 9(2): 338-351.
Holly, F., Jr. (2009). Dr. Danny Lee Fread. *Hydrolink* IAHR (2): 31. *P*

FREEMAN J.R.

27.07. 1855 West Bridgton ME/USA
06.10. 1932 Providence RI/USA

John Ripley Freeman graduated as a civil engineer from MIT in 1876. He then joined a water power company at Lawrence MA, and became assistant to Hiram F. Mills (1836-1921). Hydraulic tests were conducted relating to water power. In 1886 Freeman became engineer for a fire insurance company at Boston MA. He conducted experiments toward the improvement of fire prevention, notably on water jets and the effect of nozzle design. He was awarded the 1890 and 1891 ASCE Norman Medals for the two papers. During the next two decades Freeman was consultant both in fire prevention and improved building standards, as also in hydraulic engineering. Engineering was his recreation, and he practiced it with an enthusiasm and thoroughness that made him early a prominent figure in the engineering world.

Freeman was greatly interested in the application of hydraulic laboratory research to problems of river and harbor flow. He wrote two papers on the history and the need of a National Hydraulic Laboratory, and shortly later presented his Hydraulic Laboratory Practice, summarizing the main European facilities. To further the education of young Americans in hydraulic engineering, he provided grants known as Freeman scholarships starting in 1927. He was an Honorary Member ASCE and ASME of which he was president in 1922 and 1905, respectively, the Boston Society of Civil Engineers, the New England Water Works Association, among many other learned societies. With his death, a notable figure both in hydraulic engineering and in the civil engineering profession had left behind a large vacuum.

Anonymous (1904). John R. Freeman, President ASME. *American Machinist* 27(Dec.24): 1637. *P*
Anonymous (1932). John Ripley Freeman. *Mechanical Engineering* 54(11): 781-783. *P*
Anonymous (1933). John Ripley Freeman. *Trans. ASCE* 98: 1471-1476.
Bush, V. (1935). John Ripley Freeman. *Biographical memoirs* 8: 171-187. National Academy of Sciences: Washington DC, with bibliography. *P*
Freeman, J.R. (1889). Experiments relating to the hydraulics of fire streams. *Trans. ASCE* 21: 303-482.
Freeman, J.R. (1922). Address at the annual convention. *Trans. ASCE* 85: 1601-1630.
Freeman, J.R. (1924). The need of a National Hydraulic Laboratory for the solution of river problems. *Trans. ASCE* 87: 1033-1097.
Freeman, J.R. (1929). *Hydraulic laboratory practice*. ASME: New York.

FREEMAN R.M.

20.07. 1892 Winchester MA/USA
21.01. 1925 New York NY/USA

Roger Morse Freeman, the son of John R. (1855-1932), graduated from the Massachusetts Institute of Technology MIT, Cambridge MA, as electrical engineer, and he had made additional studies at the University of Charlottenburg, Berlin D. He then was during one year at his father's engineering office, working on water power development studies. He further was engaged as construction engineer by the Waterbury Manufacturing Company, Waterbury CT. He was further involved in the construction of large steel works following the entrance of USA in World War I. After the Armistice, the US Navy officials proposed a much larger Navy, so that an armor-plate plant was built at Charleston WV, serving research, and to check the actual cost on bids submitted by steel works.

Following his natural talent, Freeman had hoped to continue in metallurgy, but as no openings were found in the depressed conditions of the industry, he opened an office in New York NY, returning to construction in hydro-electricity. His experience and ability, his friendliness with builders of hydraulic and electrical machinery, and his unbounded patience quickly brought to successful completion a plant of 12,000 HP, on Tippecanoe River near Monticello IN. His second design of a 15,000 HP hydro-electric plant near Oakdale IN, with more than 500 men employed, was a more difficult task in terms of construction, particularly during the winter. The confidence of his employers may be judged from the fact that the formal contract for his services had not been signed at his death, although a draft had been prepared. The reservoir created by these works was subsequently named Freeman Lake in his honor. His death at 33 years of age was caused by heart failure following an operation for appendicitis. He was member of the American Institute of Electrical Engineers AIEE, the American Society of Civil Engineers ASCE, the American Society of Mechanical Engineers ASME, and of the American Association of Engineers AAE. The memoir in the ASCE Transactions was written by his father John Ripley Freeman, the famous hydraulic engineer then in his country.

Anonymous (1925). Death of Roger M. Freeman. *Mechanical Engineering* 47(3): 230-231. *P*
Anonymous (1925). Roger M. Freeman dies. *Engineering News-Record* 94(5): 208.
Anonymous (1931). Roger M. Freeman. *Trans. ASCE* 95: 1485-1488.
Freeman, R.M. (1920). The armor-plate and gun-forging plant of the US Navy Department at
        South Charleston WV. *Trans. ASME* 42: 983-1032.
http://specialcollections.tulane.edu/archon/?p=collections/

FRIEDKIN

18.10. 1909 Brooklyn NY/USA
14.01. 2008 Miranda CA/USA

Joseph Frank Friedkin was educated at the Texas College of Mines, from where he received the BS degree in 1932. He accepted then the position of engineer at the International Boundary and Water Commission at El Paso, where he was engaged with stream gaging, hydrologic and runoff studies for the design of reservoirs for irrigation and flood control. After war service from 1942 to 1945, where he rose to major, he was assigned to the Mississippi River Commission, Vicksburg MS, in charge of channel stabilization. From 1946 to 1950 he was a hydraulic engineer of the San Diego Office again of the Intl. Boundary and Water Commission for flood control, irrigation and drainage works along the Lower Colorado and the Tijuana River. He finally became principle engineer of his organization from 1952 until retirement, during which time he was engaged with river regulations, flood control and drainage aspects of all major rivers along the American-Mexican border. Friedkin was a Fellow of the American Society of Civil Engineers and received in 1990 honorary membership of ASCE.

Friedkin became known for his report on river meandering, a research topic basically attacked only after World War II. It was observed that meandering occurs for rivers at moderate slopes, as compared to straight river flow and river braiding for smaller and larger slopes, respectively. From the 1960s these questions were much more rigorously considered using computational and experimental approaches. From the mid-1940s he was concerned with the large rivers of the southwestern USA, where he had to deal with all aspects of successful river engineering, including hydraulics, groundwater flow, sedimentology, salinity, environmental concerns and economy, so that he was an expert of his field and often asked for professional advice.

Friedkin, J.F. (1945). *A laboratory study of the meandering of alluvial rivers*. US Waterways Experiment Station: Vicksburg MS.
Friedkin, J.F. (1987). International water treaties. *Water resources policy for Asia* 25. Boston.
Friedkin, J.F. (1988). The international problem with Mexico over the salinity of the Lower Colorado River. *Water and the American West*: 31-52, D.H. Getches, ed. University of Colorado: Boulder.
Jordan, D.H., Friedkin, J.F. (1967). The International Boundary and Water Commission: United States and Mexico. 5th Intl. Conf. *Water for Peace* Washington DC: 192-203.
http://www.google.ch/imgres?q=%22Joseph+Friedkin *P*

FRIEL

24.02. 1894 Queenstown MD/USA
11.02. 1964 Bryn Mawr PA/USA

Francis de Sales Friel obtained the civil engineering degree in 1916 from Drexel Institute of Technology, Philadelphia PA, and there was awarded the D.Eng. degree in 1949. He joined in 1917 the later Albright & Friel Inc., Consulting Engineers, Philadelphia PA, which was established in 1890 by John J. Albright (1848-1931), specializing in water supply, sewage and industrial wastes, power plants, flood control, and dams, among other engineering projects. During his long engineering career, Friel was involved in more than 2,000 projects, involving a construction cost of more than 1.3 milliards of US$ in 1960. He was the winner of the 1948 Medal of the American Public Works Association APWA, and designated in 1956 Engineer of the Year by the Engineering Society of Philadelphia. He was member of the American Society of Civil Engineers ASCE, serving as national vice-president from 1956 to 1958. He also was member of the International Committee of Large Dams ICOLD, taking over from 1955 to 1958 as chairman the US Committee. He was further member of the International Association of Hydraulic Research IAHR, and of the International Council of Soil Mechanics and Foundation Engineering.

ICOLD was founded in 1933, and is still the major association of dam engineers, with its headquarters in Paris F. ICOLD Congresses are held all three years. The 1958 Congress was held in New York NY, during which four questions, a standard of ICOLD, were discussed. These were devoted to the heightening of existing dams, deformations and stresses in dam foundations, compaction methods on earth and rock-fill dams, and the use of admixtures and pozzolanic materials in dam concrete. Also as a standard, three study tours were organized following a Congress to visit dams in the Southeast, the Midwest, and the Northwest regions of the USA. Following this successful congress, and his outstanding engineering career, Friel became President ASCE in 1959.

Anonymous (1958). Francis S. Friel: Large Dam Congress planned. *Engineering News-Record* 160(Mar.20): 28. *P*
Anonymous (1959). Friel, Francis de S. *Who's who in engineering* 8: 843. Lewis: New York.
Anonymous (1964). ASCE Past President dies. *Civil Engineering* 34(3): 76.
Friel, F.S. (1927). Sewage from Lower Merion pumped to Philadelphia System. *Engineering News-Record* 98(4): 160-161.
Friel, F.S. (1959). *Addresses of Francis S. Friel while president of ASCE.* ASCE: New York.
Friel, F.S. (1959). Current challenges in civil engineering. *Trans. ASCE* 124: 1038-1048.

FRIZELL

13.03. 1832 Barford QC/CA
04.05. 1910 Dorchester MA/USA

Joseph Palmer Frizell taught himself in mathematics and engineering after basic schooling. He joined in 1850 a cotton mill in Manchester NH and entered there the office of the city engineer in 1854. From 1856 he was engineer assistant of the proprietors of the locks and canals on the Merrimack River under James B. Francis (1815-1892), the foremost hydraulic engineer then of the United States. Francis, having just completed and published The Lowell hydraulic experiments, was carrying out additional work, and under his tutelage, Frizell collaborated from 1857 to 1861, and from 1866 to 1867. During the Civil War, he was an assistant civil engineer of the US Army, engaged with fortifications along the Gulf Coast.

WATER-POWER.

AN OUTLINE OF THE DEVELOPMENT AND APPLICATION OF THE ENERGY OF FLOWING WATER.

BY

JOSEPH P. FRIZELL,
HYDRAULIC ENGINEER,
Member of the American Society of Civil Engineers,
Member of the Boston Society of Civil Engineers.

THIRD EDITION, ENLARGED.
FIRST THOUSAND.

NEW YORK:
JOHN WILEY & SONS.
LONDON: CHAPMAN & HALL, LIMITED.
1903.

Frizell was engaged from 1870 to 1878 with consulting engineering at Boston MA, and in parallel patented an air compressor utilizing the direct action of falling water. He then went West again as an assistant civil engineer to the US Engineers Department and was concerned with hydraulic investigations on the Mississippi headwater. From 1890 to 1892 he was chief engineer of public works at Austin TX, but then returned to Boston. In 1901 he published the results of researches on Water power, which was the first practical book of its kind in the USA. In later life Frizell contributed a number of papers in the similar field, after having retired from his consulting office in 1903. He then lived at Dorchester MA. Frizell was described as a most able member of engineers, largely self-taught, who established the basis on which the modern science of hydraulics rests.

Anonymous (1911). Joseph Palmer Frizell. *Trans. ASCE* 73: 501-503.
Anonymous (1931). Frizell, Joseph Palmer. *Dictionary of American biography* 6.1: 39-40. Scribner's: New York.
Frizell, J.P. (1892). Mr. Fanning's report upon the dam at Austin TX. *Engineering News* 28(Sep.29): 303-304; 28(Nov.10): 440; 28(Dec.22): 592; 29(Jan.12): 88.
Frizell, J.P. (1893). The old-time water-wheels of America. *Trans. ASCE* 28: 237-249.
Frizell, J.P. (1898). Pressure resulting from changes of velocity of water in pipes. *Trans. ASCE* 39: 1-18.
Frizell, J.P. (1901). *Water power*: An outline of the development and application of the energy of flowing water. Wiley: New York. (*P*)
Reynolds, T.S. (1983). *Stronger than a hundred men*: A history of the vertical water wheel. Johns Hopkins University Press: Baltimore MD.

FRY

13.03. 1892 LeClaire IA/USA
23.12. 1974 Knoxville TN/USA

Albert Stevens Fry graduated with the BS in civil engineering from University of Illinois, Urbana IL in 1913, receiving the professional degree in 1918. He was first employed by engineering companies at Memphis TN and Dallas TX from 1913 to 1931 on the planning, design and construction of large-scale flood control, drainage, municipal and irrigation projects in the Midwest of the USA. Fry began as a junior engineer and advanced to head engineer in 1928. He designed and supervised from 1931 to 1933 flood control and navigation works along Mississippi River for the US Corps of Engineers at Memphis TN, and Cairo IL.

Fry joined in 1933 Tennessee Valley Authority TVA to head its General Engineering and Geology Division. His responsibilities not only included hydrology and hydraulics but also geological explorations of dam sites. Under him, TVA set up its engineering laboratory and began to forecast flood levels for Tennessee River and its tributaries. From 1936, following an internal reorganization, Fry headed the TVA Hydraulic Data Division until retirement in 1961. From then, he was a consultant of the US Geological Survey. In his long career Fry became internationally known both in hydraulics and hydrology. He directed pioneering work in watershed hydrology, recorded reservoir sedimentation and developed automatic radio rain gages and stream gages. He was in parallel active in the American Geophysical Union AGU, the Soil Conservation Society SCS, the International Association of Hydraulic Research IAHR, among others. He was also instrumental in the reorganization of ASCE's Hydraulics Division in 1950, chairing it, and conceived the idea of holding Specialty Conferences. He joined ASCE in 1918, became a Member in 1926, a Fellow in 1959 and a Honorary Member in 1962.

Anonymous (1962). Albert S. Fry. *Civil Engineering* 32(10): 67. *P*
Anonymous (1975). Albert S. Fry, retired TVA official, dies. *Civil Engineering* 45(3): 94-95. *P*
Anonymous (1975). Albert Stevens Fry. *Trans. ASCE* 140: 567-569.
Fry, A.S. (1941). Big waters on little streams. *Agricultural Engineering* 22: 424-426.
Fry, A.S. (1942). *Flood control for upper French Board River and tributaries*. TVA: Knoxville.
Fry, A.S. (1945). *Report on initial phases*: Chestuee watershed project. TVA: Knoxville TN.
Fry, A.S. (1945). Hydrometeorology. *Handbook of meteorology*. McGraw-Hill: New York.
Fry, A.S. (1948). Recent developments in hydrology with respect to stream flow forecasting.
    IAHS Congress Oslo: 143-151.

FUCIK

25.01. 1914 Chicago IL/USA
06.04. 2010 Lake Forest IL/USA

Edward Montford Fucik was educated at Princeton University, from where he gained the BS degree in civil engineering in 1935. He was then until 1937 an assistant in civil engineering at Harvard University, Cambridge MA, earning the MS degree, and designer of flood control works for the US Engineering Dept., Ithaca NY. He was engaged from 1938 to 1940 by Harza Company, Chicago IL, on the Santee-Cooper hydro-electric project, Moncks Corner SC, thereby supervising the construction of all earth dams. Next, until 1943 he was engineer and senior engineer at the Panama Canal, in charge of the Soil Mechanics Section on the Third Locks Project, and there also remained for war service until 1945. He then returned to Harza as vice-president until 1953 taking over as executive vice-president until retirement. He was a member of the Princeton Engineering Association, and Fellow and National Director of the American Society of Civil Engineers ASCE. He was winner of the ASCE Thomas Fitch Rowland Prize and the ASCE Rickey Medal. He also was elected member of the National Academy of Engineering NAE.

Harza Engineering Company, then led by Leroy F. Harza (1882-1953), counted to the largest firms world-wide. Fucik's projects included works in hydro-electric power plants, dams and related structures. In addition to these works he was retained by the Panama Canal as special consultant on foundations and related problems, which had caused significant troubles from the original design until today. He also was concerned with studies to the methods of increasing safety and capacity of the Canal. He served as a consultant to the St. Lawrence Seaway Development in Canada. Fucik was elected Chicago Engineer of the Year and actively participated in matters of the International Commission of Large Dams ICOLD.

Anonymous (1962). E. Montford Fucik. *Engineering News-Record* 168(May10): 65. *P*
Anonymous (1964). Fucik, Edward Montford. *Who's who in engineering* 9: 627.
Fucik, E.M. (1941). *Canal Zone*: Third Locks Project. Special Engineering Division: Balboa.
Fucik, E.M. (1962). *Spillways for high dams*. Harza Engineering Company: Chicago.
Fucik, E.M. (1964). Consulting engineering in foreign countries. *Journal of Professional Practice* ASCE 90(1): 37-43.
Weaver, K.D., Bruce, D.A. (2007). *Dam foundation grouting*. ASCE Press: Reston VA.
http://www.legacy.com/obituaries/chicagotribune/obituary.aspx?n=edward-montford-fucik&pid=141974515&fhid=4182. *P*

FUERTES E.A.

10.05. 1838 San Juan PR/USA
16.06. 1903 Ithaca NY/USA

Estevan Antonio Fuertes was born on the former Spanish island Porto Rico, since 1898 an American possession. He received education from Salamanca University, Spain, and at the Rensselaer Polytechnic Institute, Troy NY. From 1861 to 1863 he acted as assistant engineer in the Department of Public Works of Puerto Rico, and subsequently served as director of public works for the western district of the island. From 1864, he was successively assistant engineer and engineer to the Croton Aqueduct Board, thereby preparing a report on the connection of the Croton Water Supply with New York City. From 1870 to 1871 he was chief engineer of the American Isthmian Canal expeditions to Nicaragua to investigate the practicability of a ship canal connecting the Caribbean and the Pacific. He was appointed in 1873 dean of the Civil Engineering Department, Cornell University, directing from 1890 to 1902 its College of Civil Engineering. He became professor of astronomy in 1902 there supervising the construction of the A.C. Barnes Observatory.

Fuertes developed the hydraulic laboratory of Cornell University, where many of the leading hydraulic engineers of the country were then to study. In 1880 the university trustees appropriated £100,000 to equip certain departments of the university, so that Fuertes revisited Europe to study pedagogical laboratory practice, indicating the interest in these activities also in the USA. A more systematic visiting activity from the West to the East initiated in the late 1920s with the Freeman Scholarship, which was inverted then after World War II with the move of many Europeans to the USA. Fuertes was further interested in activities in hydraulic engineering, as is evidenced from several discussions to papers published in the Transactions ASCE. This interest originated from Fuertes' laboratory activities in connection with the Croton Dam Project.

Anonymous (1963). Fuertes, E.A. *Who was who in America* 1607-1896: 430. Marquis: Chicago.
Fuertes, E.A. (1875). Education of civil engineers. *Trans. ASCE* 3: 259-263.
Fuertes, E.A. (1888). Discussion of High masonry dams. *Trans. ASCE* 19: 190-192.
Fuertes, E.A. (1892). Discussion of Notes on the Holland dikes. *Trans. ASCE* 26: 686-689.
Fuertes, E.A. (1900). Discussion of Flow of water over dams. *Trans. ASCE* 44: 339-340.
Poole, M.E. (1916). A story historical of Cornell University with biographies of distinguished Cornellians. Cayuga Press: Ithaca.
Rouse, H. (1976). Estevan A. Fuertes. *Hydraulics in the United States* 1776-1976: 72-73. Iowa Institute of Hydraulic Research: Iowa City. *P*
http://en.wikipedia.org/wiki/Estevan_Antonio_Fuertes *P*

FUERTES J.H.

10.08. 1863 Ponce PR/USA
30.01. 1932 New York NY/USA

James Hillhouse Fuertes, son of Estevan Antonio (1838-1903), graduated with the BS degree in civil engineering from Cornell University, Ithaca NY, in 1883. From then until 1926, when failing health limited strenuous activities, he was mainly associated with sanitary and hydraulic engineering. Until 1891 he served successively as assistant and engineer various cities, and was engaged in private practice at Camden AR, designing sewers and water works. He then moved to New York City continuing in private practice designing drainage, sewerage, and sewage disposal works for more than forty cities in the USA, Canada and Latin America. Since 1907 he was involved in works for flood protection of Harrisburg PA. He was member of the Miami Conservancy Commission for flood regulation works in the Miami River Valley. In 1909 he reported on a dam at Harrisburg PA, on the improvement of Susquehanna River, and on sanitary works in Manhattan. In 1924 he was a member of the Board of Review of the Sanitary District of Chicago, investigating the relation of the Great Lake levels to Chicago problems.

Fuertes was during his career associated with Rudolph Hering (1847-1923) and Allen Hazen (1869-1930) on important sanitary works. His services were so satisfactory that he was often retained by clients, who insisted on receiving his opinion. He authored the books Water and public health, Water filtration works, and the European Sanitary Engineering Series published in 1897 in the Journal Engineering Record. The 1901 book includes the chapters 1. Introductory, 2. Intakes, sedimentation, and settling tanks, 3. Purification of water by slow sand-filters, 4. Design, construction and operation of slow sand-filters, 5. Purification of water by rapid sand-filtration, 6. Construction and operation of rapid sand-filters, 7. Conclusions, and 8. Filtered-water reservoirs. He was of studious disposition, with keen perception, sound judgment, broad vision, and unusual ability to analyze conditions and solve problems. His drawings, often made by himself with great skill, were unusually complete and clearly detailed. He was member of the American Society of Civil Engineers ASCE since 1895.

Anonymous (1932). James H. Fuertes. *Trans. ASCE* 96: 1482-1485.
Fuertes, J.H. (1897). *Water and public health*. Wiley: New York.
Fuertes, J.H. (1901). *Water filtration works*. Wiley: New York.
Fuertes, J.H., Marilley, A.L., Dougherty, J.H. (1906). *Waste of water in New York* and its reduction by meters and inspection. Merchants Association: New York. *P*

FULLER G.W.

21.12. 1868 Franklin MA/USA
15.06. 1934 New York NY/USA

George Warren Fuller graduated in 1890 from the Massachusetts Institute of Technology in chemistry, and continued studies at the University of Berlin, Germany. Upon return to the USA he started a more than forty years lasting career in sanitary engineering. Until 1895 he stayed with the Massachusetts State Board of Health, one of the leading organisations in this field. He was primarily concerned with water and sewage investigations conducted at the Lawrence Experiment Station. He in addition lectured at MIT on biology and bacteriology. In 1895 he conducted in Louisville KY basic studies on water purification, particularly with the so-called rapid sand filtration, which are described in his classic 1898 Report.

Fuller opened in 1899 a private consulting office in New York NY, counting then to the experts in water supply, water purification and sewerage. In 1901 he became a partner with Rudolph Hering (1847-1923), an association which lasted until 1911, when Fuller returned to independent practice. He was awarded the Rowland Prize by ASCE for his paper published in 1903. He was a member of a large number of professional societies, including next to ASCE also the Institution of Civil Engineers, Great Britain, Verein Deutscher Ingenieure VDI, Germany, and Association Générale des Hygienistes et Techniciens Municipaux, France. Fuller was largely responsible for the development and widespread adoption of the standard methods of water and sewage analyses. The 1925 book on water works practice was largely based on his initiative.

Anonymous (1915). George W. Fuller. *Engineering Record* 71(21): Frontispiece. *P*
Anonymous (1924). George W. Fuller. *Engineering News-Record* 93(21): 847. *P*
Anonymous (1931). Fuller, George Warren. *Dictionary of American biography* 21 Suppl. 1: 325-327. Scribner's: New York.
Anonymous (1934). George W. Fuller dies. *Engineering News-Record* 112(Jun.21): 821. *P*
Fuller, G.W. (1898). Report on the investigations into the purification of the Ohio River water at Louisville KY. van Nostrand: New York.
Fuller, G.W. (1903). The filtration works of the East Jersey Water Company at Little Falls NJ. *Trans. ASCE* 50: 394-472.
Fuller, G.W. (1912). *Sewage disposal*. McGraw-Hill: New York.
Wolman, A., ed. (1925). *Manual of American water works practice*, published under the auspices of the American Water Works Association. Williams & Wilkins: Baltimore.

FULLER W.E.

27.07. 1879 Phillips ME/USA
22.06. 1935 New York NY/USA

Weston Earle Fuller obtained the civil engineering degree from Cornell University, Ithaca NY, in 1900. He there served as instructor of civil engineering, and continued in 1902 as assistant engineer of the Ithaca Water Company, during which time he tested a mechanical filter plant, and came into contact with Allen Hazen (1869-1930). Fuller went as resident engineer for Hazen & Whipple to Watertown NY on the construction of its mechanical filtration plant, and was transferred to Poughkeepsie NY then to superintend a slow sand filtration plant. As a result of the excellent work done, he continued at the New York Office as chief draftsman, becoming in 1907 junior partner of Hazen & Whipple, whom included his name in 1915, because of the special value of his services. In 1922 Fuller resigned his position, accepting the civil engineering professorship at Swarthmore College, Swarthmore PA.

Although successful as a professor, the increasing demand of his services as expert led him in 1929 to resign this position, devoting himself from then entirely to private practice at Swarthmore. The largest and most interesting work was the taking over of the properties of the Consolidated Water Company, Passaic NJ. His services resulted in an award from the Condemnation Commission. From 1928 to 1930 Fuller represented the State of Massachusetts in the Massachusetts-Connecticut Case on the Connecticut River. In 1930 he was retained to represent the water power interests for New York State. In 1931 he formed a partnership, which continued after Fuller's death. During this partnership the activities were mainly with the Passaic Valley Water Commission. In his 1914 paper Fuller introduced the term Return period, as the likelihood of an event as a flood, measuring statistically a data set denoting the average recurrence interval over a certain time period. Fuller was a member of the American Water Works Association AWWA, and the American Society of Civil Engineers ASCE since 1912.

Anonymous (1930). Weston E. Fuller. *The Halcyon*: 28. Swarthmore College: Swarthmore PA. *P*
Anonymous (1936). Weston E. Fuller. *Trans. ASCE* 101: 1556-1560.
Fuller, W.E. (1908). *Report* on proposed filtration plant and other improvements. Board of
      Water Commissioners: Ogdensburg.
Fuller, W.E. (1913). Loss of head in bends. *J. New England Water Works Assoc.* 27(4): 509-521.
Fuller, W.E. (1914). Flood flows. *Trans. ASCE* 77: 564-694.
Fuller, W.E. (1914). Disc. of Determination of storm-water run-off. *Trans. ASCE* 78: 1193-1198.

FULTON

14.11. 1765 Fulton PA/USA
24.02. 1815 New York NY/USA

Robert Fulton constructed at age of thirteen paddle-wheels, which he applied with success to a fishing boat. In 1786 Fulton went to London to study there, and actively engaged in a project to improve canal navigation in 1793. In 1796 he made plans for a cast-iron aqueduct which was later built for crossing River *Dee*. Further works related to a passage boat, a despatch boat and a trader to be used on canals. He also published in 1796 his Treatise. During his visit to France in 1797, Fulton experimented on *Seine* River with a boat for submarine navigation, and in 1801 he conducted experiments in Brest with an improved design, yet the boat failed to blow up British ships that sailed along the coast. On return to England, the government thought that Fulton's torpedo was of value, but these bursted harmlessly beside the French ships. The first successful trial occurred in 1805, resulting in the complete destruction of a warship.

In 1806 Fulton returned to the United States, continued experimentation with torpedoes, but never realized these for the Navy. From then he turned his attention to steam navigation. In spring 1807 the first boat with an engine of Watt & Boulton, England, navigated on Hudson River. A voyage from Hudson to Albany lasted then 32 hours. The first attempt to connect a steam-engine with a screw propeller was made by Joseph Bramah (1748-1814) in 1795. Whatever may have been Fulton's honor as to his invention, he undoubtedly deserves the credit of first bringing into practical use the steamboat as a conveyance for passengers and freight. The success of his *Clermont* was followed by a rapid increase of steamboats.

Anonymous (1901). Unveiling Fulton Memorial. *American Machinist* 24(De.28): 1381-1382. *P*

Fulton, R. (1796). *A treatise on the improvement of canal navigation*. Taylor: London.

Fulton, R. (1806). *Letters on submarine navigation*. London.

Fulton, R. (1812). *De la machine infernale maritime, ou de la tactique offensive et defense de la torpille*. Paris.

McNeil, I. (1964). Robert Fulton: Man of vision. *Chartered Mechanical Engineer* 11(1): 16-21. *P*

Philip, C.P. (1985). Robert Fulton: An American Leonardo da Vinci. *Mechanical Engineering* 107(11): 44-54; 108(5): 79-85. *P*

Preble, G.H. (1881). *History of steam navigation*. Hamersly: Philadelphia.

Tyler, D.B. (1946). Fulton's steam frigate. *American Neptune* 6(10): 253-274.

http://de.wikipedia.org/wiki/Robert_Fulton *P*

GAILLARD

04.09. 1859 Fulton SC/USA
05.12. 1913 Baltimore MD/USA

David Du Bose Gaillard graduated in 1884 from the US Military Academy. He was then instructor at the Engineering School of Application, Willet's Point NY until 1887, from when he was engaged in charge of harbour improvements in Florida until 1891. Until 1898 he was member of the International Boundary Commission between Mexico and the USA, and involved in the Washington DC Aqueduct and water supply. During the Spanish-American War he acted as chief engineer in Cuba, and from 1899 he was assistant to the engineer commissioner of District of Columbia. From 1901 to 1903 he was placed in charge of river and harbour improvements on Lake Superior at Duluth MN, promoted to major in 1904. After further studies at the Army War College he was in the general staff in Cuba until 1907.

From 1907 Gaillard was engaged with dredging excavation of Panama Canal, organizing the Chagres Division by excavating almost 40 km between Gamboa and Gatun. Next he was in charge of the Central Division for the 35 km long reach between the Atlantic and Pacific Locks including the excavation through the Continental Divide by the Culebra Cut. He was promoted to lieutenant colonel in 1909, collapsed on the job in 1913, and died five months later. The name of the Culebra Cut changed to Gaillard Cut, and the army post at Culebra was named by President Wilson Camp Gaillard in 1914. The Culebra Cut runs through the backbone of the American Continent. Gaillard dug hill after hill, found its angle of repose, and checked for landslides. He never knew what next morning was to bring because over the night, mountains moved covering their deposits, the tracks and the cars for removal of the earth. His collaborators noted that he worked 12 hours a day on the Cut, then taking his share in the general administration.

FitzSimons, N., ed. (1991) Gaillard, D.D. *A biographical dictionary of American civil engineers* 2: 39-40. ASCE: New York. *P*

Gaillard, D.D. (1896). Gigantic earthwork in New Mexico. *American Anthropologist* A9(9): 311.

Gaillard, D.D. (1904). *Wave action in relation to engineering structures*. GPO: Washington DC.

Gaillard, D.D. (1905). Harbors on Lake Superior, particularly Duluth-Superior Harbor. *Trans. ASCE* 54: 263-296.

Gaillard, D.D. (1912). Culebra Cut and the problems of slides. *Scientific American* 107: 388-390.

http://www.arlingtoncemetery.net/ddgaillard.htm

http://en.wikipedia.org/wiki/David_du_Bose_Gaillard *P*

GALLOWAY

13.10. 1869 San Jose CA/USA
10.03. 1943 Berkeley CA/USA

John Debo Galloway graduated in 1889 from Ross Polytechnic Institute, Terre Haute IN, from where he returned to California, and had his headquarters at San Francisco CA. From 1892 to 1896 he was in charge of harbour works, sewers, and foundation work for a construction company. Until 1899, he was then instructor in drawing and mechanics at the California School of Mechanical Arts, San Francisco, where he met his later business partner. From 1900 to 1906, just prior the San Francisco earthquake, he designed and supervised the hydraulic construction of hydro-electric plants, which were later absorbed by the Pacific Gas and Electric Company. He became one of the first advocates of earthquake-resistant design. In 1907 he served with Charles D. Marx (1857-1939) on the Board of Advisory Engineers on Stanislaus Hydroelectric Power Plant, Strawberry CA. From 1908 he was associated as partner with the above mentioned colleague, and continued himself from 1920 until his death.

The activities of the partnership included a hydro-electric power plant in the Yosemite National Park, the Butte and Tehama Power Company, or the Santa Barbara Gas and Electric Company. Galloway also studied the hydro-electric power plants of Treadwell Mines in Alaska, the proposed power development on Potomac River at Harpers Ferry WV, or the power developments on Vancouver Island BC. He was further involved in irrigation systems, namely the Turloch Irrigation District, the Fresno Canal, the Tulare Lake Basin Water Storage District, or the Sutter Butte Canal. Together with Joseph B. Lippincott (1864-1942), Bernard A. Etcheverry (1881-1954), and Frederick H. Tibbetts (1882-1938), he was member of the Board of Advisory Engineers on the Sacramento Valley District of the Central Valley Project in California. He was thus involved in the development of Shasta Dam, or in hydro-power plants on the Hetch Hetchy Aqueduct. He thereby wrote authoritative texts on these topics. He was Honorary Member of the American Society of Civil Engineers ASCE and awarded its 1941 T.F. Rowland Prize.

Anonymous (1944). John D. Galloway. *Trans. ASCE* 109: 1451-1456. *P*
Galloway, J.D. (1915). The design of hydro-electric power plants. *Trans. ASCE* 79: 1000-1035.
Galloway, J.D. (1922). Hetch Hetchy Water Supply for San Francisco. *Proc. ASCE* 48: 1846-1858.
Galloway, J.D. (1923). Hydro-electric developments on Pacific Coast. *Trans. ASCE* 86: 803-815.
Galloway, J.D. (1939). The design of rock-filled dams. *Trans. ASCE* 104: 1-92.
http://library.ucr.edu/wrca/collections/portraits/galloway.html *P*

GASKILL

19.01. 1845 Royalton NY/USA
01.04. 1889 Lockport NY/USA

Harvey Freeman Gaskill moved in 1861 with his parents to Lockport NY, where he studied for two years at the College and then entered a Commercial College, from where he graduated in 1866. He then joined his uncle's law office to study business law, and was engaged with the manufacture of clocks. Later he was interested in a planning-mill combined with a sash-and-blind factory. In both branches he applied his inventive genius mainly to improve the mechanical equipment. In 1873 he joined a firm at Lockport as a draftsman where pumping machinery for waterworks was manufactured. He immediately turned attention to steam-pumps. At that time, water-works pumping machinery was made in the United States principally at his and another firm. Competition between the two for the supremacy and business was keen.

Gaskill attempted for higher steam economy and larger pumping capacity, launching in 1882 the Gaskill pumping engine. It was the first crank and fly-wheel high duty engine built as a standard for waterwork service. It gave a fairly high steam economy, had a large pumping capacity, was extremely compact and convenient, and was lower in the cost than the preceding types. This engine was quickly accepted nationally and gave his firm advantage, until the Worthington high-duty engine of Henry Rossiter Worthington (1817-1880) appeared. Gaskill was superintendent of his company from 1877, vice-president from 1885, and would eventually have become president but for his untimely death. The automatic cut-off gear of the Holly pumping engine, by which the point of cut-off was varied by the pressure, was also his invention. He was a member of the American Society of Mechanical Engineers ASME.

Anonymous (1889). Harvey F. Gaskill. *Trans. ASME* 10: 833.
Anonymous (1931). Gaskill, Harvey Freeman. *Dictionary of American biography* 6.1: 177-178. Scribner's: New York.
Anonymous (1996). Gaskill, Harvey Freeman. *Biographical dictionary of the history of technology*: 282, L. Day, I. McNeil, eds. Routledge: London.
Gaskill, H.F. (1882). Vertical pumping engine. US Patent Office: Washington DC.
Gaskill, H.F. (1884). *Steam and hydraulic pumping engine.* US Patent Office: Washington DC.
Gaskill, H.F. (1888). *Patent* of Duplex engine. Canada Patent CA 29859.
Hague, C.A. (1907). *Pumping engines for water works.* McGraw-Hill: New York. *P*
http://jerseyman-historynowandthen.blogspot.com/2011_07_01_archive.html

GAUSMANN

01.09. 1882 Brooklyn NY/USA
01.06. 1974 Newtown CT/USA

Roy Warner Gausmann was educated at Columbia University, New York NY, as a civil engineer. He was from 1902 to 1925 in charge of the New York water supply up to division engineer, except for the war period. He also served from assistant to section engineer on the construction of dams, dikes, bridges and other appurtenances of the Ashokan and Gilboa Reservoirs, New York. He was further engaged with the construction of the lower half Shandaken Tunnel, New York State. From 1925 he was general manager of all construction of the new water supply of the Greek capital Athens, and the final completion of reclamation works in Macedonia. From 1941 he was assistant general manager of the US Overseas Division, organizing a repair and maintenance force for the Ordnance Department in the Near East. He was further involved in hydraulic works at Aruba Island, at Lake Ontario, harbour studies in Liberia, and hydro-electric plants in Ecuador, Columbia, and Turkey. He was member of the American Society of Civil Engineers ASCE, the US Military Engineers and the American Water Works Association AWWA. He was awarded the title Commander of the Saviour, Greece.

The 1923 paper is concerned with notable hydraulic model experiments of the Gilboa Dam in Schoharie County NY. Of particular importance was the determination of the spillway capacity using a stepped spillway. To test the hydraulic similitude, models of scales between 1:50 and 1:8 were considered, thereby representing one of the very first scale family. Special gauges were developed to measure both the lower and upper jet surfaces. Another point of relevance was the proper crest section as the transition from the reservoir to the spillway of constant bottom slope. The best results were obtained by smaller steps from the crest to the tangency point, from where the spillway had constant bottom slope and constant step height, as is often adopted presently. A primitive scaling according to the current Froude law served as scale family indicator. This outstanding paper was awarded the 1923 ASCE James Laurie Prize.

Anonymous (1924). Roy W. Gausmann. *Engineering News-Record* 92(3): 130. *P*
Anonymous (1948). Gausmann, Roy W. *Who's who in engineering* 6: 713. Lewis: New York.
Gausmann, R.W., Madden, C.M. (1923). Experiments with models of the Gilboa Dam and spillway. *Trans. ASCE* 86: 280-319.
Gausmann, R.W. (1927). Shandaken Tunnel. *Proc. ASCE* 53(5): 681-706.
Gausmann, R.W. (1933). Athens builds modern water works. *Civil Engineering* 3(1): 1-5.

GEDDES

22.07. 1763 Carlisle PA/USA
17.08. 1838 Geddes NY/USA

James Geddes was the son of a Scottish farmer who worked on his parents farm and taught school before moving to Kentucky. He arrived at Syracuse NY in 1793 to manufacture salt. He was not allowed to buy land next to the salt springs, however, so that he returned to his home in Carlisle to form a business to manufacture salt. Later he studied brine springs, setting up a salt works at Geddesburgh, now Solvay. He was hired in 1797 by the State of New York to survey salt springs.

Geddes became involved in canal building in the first decade of the 1800s. He was a self-trained engineer and surveyor, and an early supporter of a proposed canal to the Great Lakes. He was appointed by the NY State Surveyor General to explore possible routes for a canal. He determined that only two routes had the necessary water sources to support a canal, namely these becoming the Ohio and Erie Canal, and the Miami and Erie Canal. A canal commission was formed in 1810, with Geddes among the five engineers selected in 1816 to supervise construction of the Erie Canal. This canal was only 12 m wide and 1.2 m deep, with a total length of some 600 km and an elevation difference of some 150 m, so that it was not historically record-breaking, but a most noteworthy achievement for the young country. Many constructional innovations, including stump removing or earth-moving equipment, came to be utilized that were unheard-of even in England, where hand shovels, picks and wheelbarrows were still in common use. Above all, the operation is often said to have rivalled West Point in its production of engineers, the Erie School, it is frequently called, who learned on-the-job and soon came to be in demand in other parts of the country. Geddes also was appointed chief engineer of the Ohio and Erie Canal. After completion of these two large projects, Geddes helped survey the Chesapeake and Ohio Canal, working in this capacity until 1828, when accepting a position with Pennsylvania State to design the state's canal system. In the late 1820s he also surveyed a canal in Canada. He was also a member of the New York State Assembly in 1804 and 1822, and was elected as Federalist to the 13th US Congress, holding the office then from 1813 to 1815.

Geddes, J. (1823). *Canal report*. Board of Canal Commissioners: Ohio.
Hawley, M.S. (1872). The Erie Canal: *The question of the origin of the Erie Canal*, considered in reference to Gouverneur Morris, Joshua Forman, James Geddes, and Jesse Hawley.
Rouse, H. (1976). James Geddes. *Hydraulics in the United States*: 24-24. IIHR: Iowa City. P
http://en.wikipedia.org/wiki/James_Geddes_(engineer) P

GELDERN von

06.09. 1852 Berlin/D
17.02. 1932 San Francisco CA/USA

Otto von Geldern was brought as child to the USA, spending his youth at Sonoma CA, where he was also educated. From 1872 he served as an assistant engineer on the construction of the Mare Island Dry Dock, of which he wrote a paper many years later. He was then consulted on the raising of the earth dam of the Chabot Water Works, Vallejo CA, and had jobs as hydrographic assistant in the US Coast and Geodetic Survey. From 1880 he was assistant engineer on the Cascade Locks on Columbia River, remaining then in Oregon in charge of surveys and investigations of proposed river and harbour works.

Because of his experience on the harbours of the Pacific Coast, von Geldern became familiar with ebb and tide flows, so that he was involved as an authority on tide-marsh and ocean-front ownerships on a study relating to the delimitation of shore areas. He was appointed in 1890 secretary of the Technical Society of the Pacific Coast, eventually becoming its soul and heart, continuing in contact with its members long after its further activity had been rendered unnecessary. It was to him that reference was made when specific information about a fellow engineer was desired. As regards his hydro-graphic work, he made surveys of the Columbia River Bar, of the entrance to the San Diego Harbour, or the various rocks in San Francisco Bay. He was an adviser to Sutter County on flood control problems, and consulting engineer to Oakland CA on matters pertaining to the determination of tidal planes as water front boundaries. He planned head-works, and supervised their construction for the Modesto Irrigation District. He also invented a new and practical method of adjusting a modern coast defence gun, an account on which was published in 1905. Von Geldern was remembered as an able engineer who held the ethics of their calling above gain. His life was full, and not restricted to the attainment of his own selfish ambitions. He was an asset to his State. He was a member of the California Academy of Sciences, serving as vice-president and council member from 1915; he was also member of the American Society of Civil Engineers ASCE.

Anonymous (1933). Otto von Geldern. *Trans. ASCE* 98: 1655-1657.
Cumming, J.M. (1907). Otto von Geldern. Theory made practice. *Sunset Magazine* 19(1): 46. *P*
Geldern von, O. (1894). The La Grange Dam. *Engineering Record* 30(23): 372.
Geldern von, O. (1890). Notes on the dry dock and coffer-dam of Navy Yard, Mare Island CA.
    Trans. Technical Society of the Pacific Coast 7(1): 9-28.
Geldern von, O. (1928). *Proposed rehabilitation of hydraulic mining in California*. Yuba CA.

GERIG

25.03. 1866 Ashland MO/USA
03.04. 1944 Arkadelphia AR/USA

William Lee Gerig graduated as a civil engineer in 1885 from the University of Missouri, Columbia MO, which conferred on him in 1937 the honorary degree of Doctor of Laws. He began his professional career as rodman on railroad construction in Arkansas, and then was engaged for 15 years with the Mississippi River Commission, three years as division engineer on the Panama Canal, then many years again with railroad projects, and another 15 years as consulting engineer in the Office of the Chief of Engineers, US Army, retiring with the title of head engineer.

Gerig was employed in 1899 by the Mississippi River Commission. He became noted authority in hydraulics, revetments, and dredging. He designed a number of dams, and was given credit for an improved usage of hydraulics in dredging works. He assisted also in the development of fascine river mats for flood control. He was further involved as assistant engineer in the Chicago Drainage Canal. Given his successful projects, he was retained in 1905 as division engineer on the Panama Canal. Chief Engineer John F. Stevens (1853-1943) asked Gerig for two dams along the Canal instead of three, which was finally adopted. Gerig was then transferred to the Atlantic side of the Canal where he served as division engineer on the Gatun Locks and Dam. He left Panama in 1909 to return to railroad construction as vice-president and general manager of the Pacific and Eastern Railway, Portland OR. In 1917, he began, as assistant to the chief engineer, the construction of the railroad in Alaska for the federal government. In 1923 he was called to Washington DC as head consulting engineer in the Office of the Chief of Engineers, US Army, and remained there until he was retired in 1938. As consultant on dredging, and river and harbour work, he was a member of the board that assisted the preparation of the Jadwin Flood Control Plan, adopted by Congress in 1928, for Mississippi River. As authority on dams and earthworks, he contributed to the design of Fort Peck Dam in Montana, or Conchos Dam in New Mexico. He contributed many papers to the Military Engineer, and to Dredge design and operation. He was considered an eminent American engineer, a master builder, and a man of vision and courage.

Anonymous (1945). William L. Gerig. *Trans. ASCE* 110: 1708-1710.
Billington, D.P., Jackson, D.C. (2006). *Big dams of the New Deal Era*. University of Oklahoma.
Gerig, W., Jadwin, E. (1927). *Rainfall* Mississippi drainage basin. US Army Corps of Engineers.
http://www.loc.gov/search/?q=&fa=Subject%3APanama+Canal+(Panama) *P*
http://www.encyclopediaofarkansas.net/encyclopedia/entry-detail.aspx?entryID=1202

GIBSON J.E.

19.12. 1874 Pine Bluff AR/USA
15.10. 1947 Charleston SC/USA

James Edwin Gibson graduated from the University of Arkansas, Fayetteville AK, in 1894 with the BSc degree in mechanical engineering. He was employed then at Philadelphia PA, assisting numerous boiler trials. He became in 1897 assistant engineer to John W. Ledoux (1860-1932) of the American Pipe and Construction Company, Philadelphia, manufacturing pipes for water systems. During Gibson's stay of 20 years, his activities covered a wide range, including the design and construction of hydro-electric plants, and the design of water works. He further served as manager of Waukesha WI water works, supervising the sinking of two 500 m artesian wells. Another large project was the development of the water supply system for the main lines of the Pennsylvania RR Company. This work embraced every type of water supply, including many impounding dams, steel tanks, standpipes, and more than 800 km of pipes.

In 1903 Gibson made his first acquaintance with Charleston SC, making studies for the proposed Goose Creek supply for the city. The unique plant was built in 1904, with a tidal stream dammed by a low earth dike to impound fresh water upstream, and keep out salt water downstream. Climatic conditions at times caused the evaporation from the reservoir surface to reach about four times the city's requirements. Once the water work system was purchased by Charleston, Gibson was in charge of the plant operation until his death. He was a leader in the use of cement lining for cast-iron pipes, contributing greatly to their development. The pipelines installed in 1922 at Charleston were so successful that no more other lining type was used. Another outstanding achievement was the design of the Goose-Creek aqueduct for water supply of the city. With his death, an outstanding water works engineer had passed away, and a citizen of highest standards. He was a man of greatest integrity, who never failed to exert himself for the best interests of the community he served for 30 years. It was stated of him: 'If you wish to see his monument, look around you'. His Alma Mater conferred on him in 1941 the Honorary Doctor of Science degree, and he was bestowed the George W. Fuller Award by the American Water Works Association. He was member of the American Society of Civil Engineers ASCE from 1910.

Anonymous (1948). James E. Gibson. *Trans. ASCE* 113: 1458-1461.
Gibson, J.E. (1922). Cement-lined cast-iron pipe at Charleston SC. *ENR* 89(10): 387-390.
http://www.charlestonwater.com/water_history_part2.htm *P*

GIESECKE

28.01. 1869 Latium TX/USA
27.06. 1953 New Braunfels TX/USA

Frederick Ernest Giesecke was educated at the Texas A.&M. College, from where he graduated with the ME degree in 1890, at MIT with the BS degree in 1904, and Technische Hochschule Berlin, Germany. He further submitted his PhD thesis in 1924 to the University of Illinois, Urbana IL. He was at his Alma Mater professor of drawing from 1888-1906, professor of architectural engineering from 1906 to 1912, head of the engineering research division until 1927, and finally from then to 1939 director of the Experiment Station. After retirement, Giesecke was a consultant. He was awarded the 1942 F. Paul Anderson Medal of the American Society of Heating and Ventilating Engineers ASHVE, whose president he was in 1940. He also was Member ASCE and ASME.

Giesecke published various papers relating to loss coefficients of bends in pressurized flow systems. Few data were by then available, yet his own data may be subjected by significant scale effects, given the relatively small pipe diameters he used. Even today, bend flow is not fully understood, given the complexities by the highly spatial flow configuration. Giesecke also authored a book on hot-water heating systems in which the previously mentioned works are important. These systems were by then not generally tested in terms of hydraulic optimum, and many improvements were added only after World War II, once the mechanics of manifold flows were investigated. A manifold is composed of a number of main pipes distributing or collecting a fluid from a tank. They are widely used in hydraulic practice, including in heating elements, ventilation systems or furnaces. Giesecke was also interested in general heating methods for water, panel heating and heat transmission.

Anonymous (1940). F.E. Giesecke. *Heating and Ventilating* 37(2): 49. *P*
Anonymous (1941). Giesecke, Frederick E. *Who's who in engineering* 5: 658. Lewis: New York.
Anonymous (1945). Dr. F.E. Giesecke. *Heating and Ventilating* 42(10): 130. *P*
Anonymous (1953). Frederick Ernest Giesecke. *Civil Engineering* 23(9): 644.
Giesecke, F.E. (1917). The friction of water in iron pipes and elbows. *Trans. American Society of Heating and Ventilating Engineers* 23(455): 499-509.
Giesecke, F.E. (1926). Friction of water in elbows. *Trans. ASHVE* 32(754): 303-314.
Giesecke, F.E. (1926). The art of heating buildings by gravity-circulation hot-water heating systems. *PhD Thesis*. University of Illinois: Urbana IL.
Giesecke, F.E. (1947). *Hot-water heating and radiant heating and radiant cooling*. Austin TX.

GILBARG

17.09. 1918 Boston MA/USA
20.04. 2001 Palo Alto CA/USA

David Gilbarg graduated in 1937 from City College, New York NY, completing his PhD thesis in 1941 at the Indiana University, Bloomington IN, tutored by a then eminent algebraic number theorist. During World War II Gilbarg served at the National Bureau of Standards, Washington DC, leading the Fluid Dynamics and Theoretical Mechanics Division at the Naval Ordnance Laboratory. Returning after the war to the University of Indiana as an assistant professor he also was a visiting professor to Stanford University, Stanford CA. He accepted in 1957 the position of mathematics professor there, which he held until retirement in 1989, but continued to be active in his profession until his death.

Gilbarg's work focused on fluid dynamics and non-linear partial differential equations. His 1977 book treats elliptic partial differential equations, which was one of the most read text in this field of mathematics. His work in fluid dynamics, particularly his papers on compressible subsonic flow, established him a leader. He was considerably ahead of his time in seeing trends, and his enthusiasm provided important impetus. Gilbarg's tenure as chairman of the Mathematics Department was stated as 'He was a very good chairman, well organized, with good judgement, always looking out for the members of the Department, with high standards for research and teaching. He was helpful and made people feel at home right away. He had a wonderful memory'.

Finn, R., Gilbarg, D. (1957). Asymptotic behavior and uniqueness of plane subsonic flows. *Communications on Pure and Applied Mathematics* 10(1): 23-63.

Gilbarg, D., Serrin, J. (1950). Free boundaries and jets in the theory of cavitation. *Journal of Mathematics and Physics* 29(1): 1-12.

Gilbarg, D. (1951). The existence and limit behavior of the one-dimensional shock layer. *American Journal of Mathematics* 73(2): 256-274.

Gilbarg, D. (1952). Unsteady flow with free boundaries. *ZAMP* 3(3): 34-43.

Gilbarg, D. (1960). Jets and cavities. *Handbuch der Physik* 9, 311-445, C. Truesdell, ed. Springer: Berlin.

Gilbarg, D., Trudinger, N.S. (1983). *Elliptic partial differential equations of second order*. Springer: Berlin.

histsoc.stanford.edu/pdfmem/GilbargD.pdf.

http://upload.wikimedia.org/wikipedia/commons/thumb/0/08/David_Gilbarg.jpg/220px-David_Gilbarg.jpg. *P*

GILBERT

06.05. 1843 Rochester NY/USA
01.05. 1918 Jackson MI/USA

Grove Karl (Carl) Gilbert graduated in 1862 from Rochester University. He was particularly attracted by his professor in geology, and therefore entered his employ after studies. Gilbert's duties consisted mainly in preparing and arranging collections in natural history for teaching and museum purposes. This was a good training for it gave him a wide range of knowledge of materials and forms in the organic and inorganic world. After five years, Gilbert was appointed volunteer on the Geological Survey of Ohio, marking his entrance in his career as a geologist. He was in parallel also engaged at Columbia College, New York, where he came into contact with the eminent geologists of the USA. Gilbert further was appointed to the newly established survey west of the 100th meridian to train field work.

In 1874, Gilbert entered in contract with John Wesley Powell (1834-1902), Director of Geological Surveys, where he remained until his death. In 1879, the two published a notable report on the extinct Lake Bonneville, in Nevada and Utah. From 1889 to 1892, Gilbert was chief geologist of the US Geological Survey, but this position had little liking. He therefore turned his interests to Niagara River and recent earth movements in the region of the Great Lakes. His last publication related to the transportation of debris by water flow, and had particular reference to the results of hydraulic mining in California. Gilbert disliked controversy and rarely entered upon sensational fields. He was certainly one of the best balanced and most philosophical American geologists. He was a member of the National Academy of Sciences NAS, among many other learned societies.

Anonymous (1931). Gilbert, Grove Karl. *Dictionary of American biography* 6.1: 268-269. Scribner's: New York.

Baulig, H. (1958). La leçon de Grove Karl Gilbert. *Annales de Géographie* 77(7-8): 289-307.

Cole, G.A.J. (1918). Dr. G.K. Gilbert. *Nature* 101(Jul.11): 370-371.

Davis, W.M. (1926). Grove Carl Gilbert. *Memoirs* National Academy of Sciences 21: 1-303. *P*

Gilbert, G.K. (1914). The transportation of debris by running water. *Professional Paper* 86. US Geological Survey: Washington DC.

Poggendorff, J.C. (1898). Gilbert, G.K. *Biographisch-Literarisches Handwörterbuch* 3: 516; 4: 498; 5: 425. Barth: Leipzig, with bibliography.

Rouse, H. (1977). G.K. Gilbert. *Hydraulics in the United States* 1776-1976: 81. IIHR: Iowa IA.

GILCREST

27.10. 1906 Hamilton OH/USA
01.08. 1973 Cincinnati OH/USA

Bruce Robert Gilcrest received engineering education at the Colorado State University, Fort Collins CO. During his entire career he was associated with the US Army Corps of Engineers. After war service, he remained during the late 1940s at the Office of the Division Engineer, Cincinnati OH. Later, he moved to the head-office at Vicksburg MI, where he was involved mainly in flood control projects.

Gilcrest was known for his chapter on flood routing in the book Engineering hydraulics of Hunter Rouse (1906-1996). The chapter is divided into four parts, namely 1. Introduction, in which the general flood routing problems are addressed, 2. Mathematics of flood routing, treating the differential equations of unsteady flow, the travel rate of flood waves, wave attenuation, discharge characteristics of unsteady flow, approximations based on neglecting the momentum equation, and a third approximation based on two differential equations, 3. Routing of floods through reservoirs, including the reservoir-storage characteristics, the use of storage factors, and the various factors affecting storage, and 4. Flood routing in open channels including the stage-discharge-storage relations, the attenuation of flood waves, the description of the Muskingum method, the analysis of wedge storage, and the complete method of solution routing problems either by the method of characteristics, or mechanical and electrical devices for flood routing. This chapter was written before computers were available and may be considered an excellent summary of the methods available in the pre-computer era. The other works of Gilcrest deal equally with the varied aspects of unsteady open-channel flows, and with flooding management for example in the Ohio River Valley, in which the plan developed by the US Army Corps of Engineers is described.

Anonymous (1935). Gilcrest, Bruce. *Silver Spruce Yearbook*: 121. Colostate: Fort Collins. *P*
Gilcrest, B.R., Marsh, L.E. (1941). Channel-storage and discharge-relations in the Lower Ohio River Valley. *Trans. AGU* 22(3): 637-649.
Gilcrest, B.R. (1950). Flood routing. *Engineering hydraulics*: 635-710, H. Rouse, ed. Wiley: New York.
Gilcrest, B.R., Schuleen, E.P., Landenberger, E.W. (1957). Flood control plan for the Ohio River basin. *Proc. ASCE* 83(WW1, Paper 1209): 1-16.
Gilcrest, B.R., Moors, A.J., Bierhorst, J.W.J. (1961). *Spillway for Markland Locks and Dam*, Ohio River, Kentucky and Indiana: Hydraulic model investigation. Defense Technical Information Center: Fort Belvoir.

GILES

29.03. 1902 Somerset/UK
31.01. 1984 Delray Beach FL/USA

Ranald Victor Giles graduated in 1922 as a civil engineer from the Wesleyan University, Middletown CT. He was from 1934 professor of hydraulics at the Drexel Institute of Technology, Philadelphia PA. He was awarded the National Science Foundation Award in 1959 to study hydraulic engineering at the University of Delft, Delft NL.

Giles became known for his 1962 basic textbook on fluid mechanics. It is noted in the Preface that it is primarily designed to supplement standard texts. It is based on the author's conviction that understanding of the basic principles is accomplished best by means of illustrative examples. The book includes 12 chapters and various appendices, namely 1. Properties of fluids, 2. Hydrostatic force on surfaces, 3. Buoyancy and flotation, 4. Translation and rotation of liquid masses, 5. Dimensional analysis and hydraulic similitude, 6. Fundamentals of fluid flow, 7. Fluid flow in pipes, 8. Equivalent, compound, looping and branching pipes, 9. Measurement of flow of fluids, 10. Flow in open channels, 11. Forces developed by moving fluids, and 12. Fluid machinery. Each chapter gives a short account on the fundamental knowledge, and then includes a large section of 'Solved problems', in which the computations are demonstrated in detail. Further, there are in each chapter also 'Supplementary problems' to be solved by the readers. The relative success of Giles' book is certainly based on these problems, because few books are available in which these are solved as in his books. On the other hand, these books have become a kind of outdated, given the basic book character. It is also typical that in Giles' book there are no references at all, so that the reader can rely only on this text. It appears that the entire Schaum's Outline Series is based on this concept, which was successful in the 1950s and 1960s, but hardly would be accepted from present book publishers.

Anonymous (1961). Giles, R.V. *Lexerd yearbook*: 134. Drexel University: Philadelphia PA. *P*
Giles, R.V. (1936). A model of a rapid sand filter plant. *Civil Engineering* 6(11): 779.
Giles, R.V. (1952). *Mechanics of liquids*. Drexel Institute of Technology Press: Philadelphia PA.
Giles, R.V. (1961). *An American's impressions of life and study in the Netherlands*: A report on educational and personal experiences while on a year's leave of absence from Drexel Institute of Technology under the auspices of a National Science Foundation Science Faculty Fellowship. Drexel: Philadelphia PA.
Giles, R.V. (1962). *Theory and problems of fluid mechanics and hydraulics*. Schaum's Outline Series. McGraw-Hill: New York.

GILMAN

31.03. 1921 Portland ME/USA
20.09. 1983 State College PA/USA

Stanley Francis Gilman obtained the BS degree in 1943 from Maine University, Orono ME, the MS degree from the University of Illinois, Urbana IL in 1948, and the PhD degree in mechanical engineering in 1953. He was co-partner of Gilman Furnace Co. until 1947, then research assistant until 1949 in air conditioning research at Urbana, and then assistant professor of mechanical engineering there until 1953 from when he headed the air conditioning section, Carrier Corp., Syracuse NY. He was member of the American Society of Heating and Ventilation Engng. ASHVE, presiding in 1971 over the later American Society of Heating, Refrigerating and Air-conditioning Engineers ASHRAE.

The 1953 Report deals with takeoffs for air duct system. Based mainly on previous results obtained at the Technical University of Munich, Germany, under Dieter Thoma (1881-1942), the head-loss coefficients of more complex dividing flow elements were determined experimentally. The purpose was thereby to define optimum geometrical shapes of these elements, and at the same time would be simple enough in construction. The pressure losses of all elements were found to vary exclusively with the ratio of branch to the approach flow pipes. The 1955 paper deals with a critical review of the nature of pressure losses in divided-flow fittings. The available data for the straight-though branch is found to deviate from new observations. Their behaviour is explained using a simple hydraulic model, resulting in a good agreement with the test data.

Anonymous (1955). Gilman, S.F. *American men of science* 9: 689. Science Press: Lancaster.
Anonymous (1983). Obituary of Dr. Stanley Gilman, ASHRAE President 1971-72. *ASHREA Journal* 25(11): 70. P
Anonymous (1989). Gilman, Stanley F. *Who was who in America* 9: 135. Marquis: Wilmette.
Ashley, C.M., Gilman, S.F., Church, R.A. (1956). Branch fitting performance at high velocity. *Trans. ASHAE* 62(1571): 279-294.
Gilman, S.F. (1955). Pressure losses of divided-flow fittings. *Heating, Piping and Air Conditioning* 61(4): 141-147.
Gilman, S.F., ed. (1977). *Solar energy heat pump systems for heating and cooling buildings*. Pennsylvania State University Press: University Park PA.
Konzo, S., Gilman, S.F., Holl, J.W., Martin, R.J. (1953). Investigation of the pressure losses of takeoffs for extended-plenum type air conditioning duct systems. *Bulletin Series* 415. University of Illinois: Urbana.

GOETHALS

29.06. 1858 Brooklyn NY/USA
21.01. 1928 New York NY/USA

George Washington Goethals graduated from West Point Military Academy in 1880. He then joined the Army Engineer Corps and there rose up in degrees from lieutenant to colonel, for works of Ohio River improvements, assistant of civil engineering at the Academy, improvements on the Cumberland and Tennessee Rivers, completion of the Muscle Shoals Canal, assistant to the chief of engineers, then chief engineer in Porto Rico, river and harbour works from Block Island to Nantucket, and finally from 1907 to 1914 construction of the Panama Canal.

In 1880, Ferdinand de Lesseps (1805-1894) started construction of the sea-level canal across the Isthmus of Panama, basically as the Suez Canal which had been successfully opened in 1869. Until 1898, when the project was finally abandoned, the French had struggled against tropical disease, administrative mismanagement and poor technical organisation. The United States then investigated the feasibility of a ship canal across Central America. In 1904, a treaty with the Republic of Panama was ratified, making the project of Panama Canal possible. In 1907, after the two previous chief engineers had retired, the president of the USA appointed Goethals to the post of chief engineer for the Panama Canal Zone. All responsibility for either success or failure from then rested on him. Many difficulties confronted Goethals and his colleagues. Approximately 30,000 employees of a variety of nationalities had to be kept in health in a climate which had been considered the worst in the world. A special sanitation department was erected to eliminate the sources of yellow fever and malaria. By persistent and intelligent study and labour the gigantic engineering problems mainly relating to the Culebra Cut were resolved. However, Goethals considered problems in relation with the human element as the most difficult. The Panama Canal was practically completed in 1913, and opened to the commerce of the world in August 1914.

Anonymous (1907). Major George W. Goethals. *Engineering News* 57(11): 290. *P*
Anonymous (1914). Colonel George W. Goethals. *Engineering News* 71(11): 584. *P*
Anonymous (1928). Gen. G.W. Goethals dies. *Engineering News-Record* 100(4): 167. *P*
Anonymous (1930). George W. Goethals. *Mechanical Engineering* 52(4): 300. *P*
Anonymous (1931). Goethals, George Washington. *Dictionary of American biography* 6.1: 355-357. Scribner's: New York.
Goethals, G.W. (1915). Panama slides. *Engineering News* 74(22): 1009-1015.
Hennig, R. (1923). George W. Goethals. *Buch berühmter Ingenieure*: 204-227. Neufeld: Berlin. *P*

GOLDMARK

15.06. 1857 New York NY/USA
15.01. 1941 New York NY/USA

Henry Goldmark graduated from Harvard College, Cambridge MA, with the BA degree in 1878, and completed his technical training at *Polytechnische Schule*, Hannover D. In 1880 he moved West, but he returned in 1882. From 1884 he was engaged by a railroad company and was thereby also involved in dam engineering. From 1893 he was in charge of steel dam construction for Ogden Pioneer Electric Company. During these years he published papers on topics including continuous bridge design, cost of sewers, the distortion of riveted pipe by backfilling, or pressures resulting from velocity changes in pipes. He was awarded the 1898 ASCE Thomas Fitch Rowland Prize for the 1897 paper.

Goldmark accepted in 1897 an appointment as design engineer with the US Board of Engineers on Deep Waterways, to study the US Ship Canal from the Great Lakes to salt water. He joined in 1899 the Missouri River Commission working there on the history of locks for ship canals. After work for a railroad company, and his knowledge of locks, he became design engineer for the Isthmian Canal Commission, stationed at Culebra, the Canal Zone. He there designed and superintended the erection and installation of 94 mitering lock gate leaves and various movable caisson emergency dams. This excellent work was incentive for a paper series on the various phases of the work. Upon return from Panama, Goldmark started his consulting practice in New York City, remaining active until 1928. He was associated with George W. Goethals (1858-1928) on the New Orleans Inner Navigation Canal. From 1923 to 1926 Goldmark studied the hydraulics of the proposed Santee Cooper hydro-electric project in South Carolina. He was described as mild in his manner, exact and courtly in speech, and abstemious in habit. He was a member of the American Society of Civil Engineers ASCE since 1888, the Institution of Civil Engineers, London UK, and the Engineering Institute of Canada.

Anonymous (1941). Henry Goldmark. *Trans. ASCE* 106: 1588-1593.
Goldmark, H. (1897). The power plant, pipe line and dam of the Pioneer Electric Company at Ogden UT. *Trans. ASCE* 38: 246-305.
Goldmark, H. (1899). *Locks and lock gates for ship canals*. Cornell University: Ithaca.
Goldmark, H. (1928). Emergency dam on Inner Navigation Canal at New Orleans LA. *Trans. ASCE* 92: 1589-1645.
Goldmark, H., Feld, J. (1930). *Steel dams and gates*. New York.
Jackson, F.E. (1911). Henry Goldmark. *The makers of the Panama Canal*: 34. New York. *P*

GOLZÉ

06.07. 1905 Washington DC/USA
23.02. 1987 Sacramento CA/USA

Alfred Rudolph Golzé obtained the BS degree in 1930 from University of Pennsylvania, and the civil engineering degree in 1940. After having served as a design engineer at Washington DC, he joined in 1933 the US Bureau of Reclamation USBR in charge of the Civilian Conservation Camps until 1943, was then until 1947 assistant director of operations and maintenance, director of programs and finance until 1953, from when he was the chief of program coordination until 1958, then until 1961 assistant commissioner. Finally Golzé was chief engineer of the California Department of Water Resources until 1967, then department director until 1971, and chief of the water resources engineering from 1973. Golzé further involved in dam building activities in Turkey and East Pakistan. He further was a member of the executive committee of the US Commission of Large Dams, and the US Commission of Irrigation and Drainage. Golzé was awarded the 1962 *Toulmin* Medal from the Society of American Military Engineers, and was a Fellow and from 1976 an Honorary Member of the American Society of Civil Engineers.

Golzé designed and constructed the Oroville Dam and the high-lift Tehachapi Crossing during his years with the California Department of Water Resources. As chief engineer of the state he directed the design, construction and initial operation of the massive State Water Project, one of the largest ever undertaken. He was further extremely busy for ASCE: He has served both the National Capital and Sacramento Sections as president, and was member of several task forces and special committees. He was member of various international organisations, including the International Commission of Large Dams ICOLD or the International Commission of Irrigation and Drainage ICID. He was the author of a reference book on reclamation projects in the USA.

Anonymous (1966). Golzé. *Mechanical Engineering* 88(2): 97. *P*
Anonymous (1975). Golzé, Alfred R. *Who's who in America* 38: 1174. Marquis: Chicago.
Anonymous (1976). Alfred R. Golzé. *Civil Engineering* 46(10): 95. *P*
Golzé, A.R. (1959). Multipurpose investigation on the Blue Nile. *Civil Engineering* 29(10): 695-697. *P*
Golzé, A.R. (1961). *Reclamation in the United States.* Caxton: Caldwell ID.
Golzé, A.R. (1967). Power from Oroville. *Water Power* 19(3): 91-100.
Golzé, A.R. (1971). Edward Hyatt (Oroville) underground power plant. *Journal of the Power Division* ASCE 97(PO2): 419-434.

GOODIER

17.10. 1905 Preston/UK
05.11. 1969 Stanford CA/USA

James Norman Goodier was educated at Cambridge University, Cambridge UK, from where he obtained the degrees of MA and PhD. He moved in 1929 to the United States obtaining the ScD degree from the University of Michigan, Ann Arbor MN, in 1931. He had been sponsored by the Commonwealth Fund from 1929 to 1931, was then a research fellow of the Ontario Research Foundation until 1938, when being appointed professor of engineering mechanics at Cornell University, Ithaca NY. In 1947 Goodier moved to Stanford University in the same position until retirement. He was a member of the American Society of Mechanical Engineers ASME, the American Mathematical Society AMS, among others. He was the recipient of the first Westinghouse Award in 1946, from the American Society of Engineering Education. Goodier has written a number of books and numerous papers.

Goodier was an internationally renowned scholar and researcher in elasticity, plasticity, thermal stresses, wave propagation in solids, the mechanics of crack propagation and the dynamics of elastic bodies. He was the first to receive the George Washington Award for excellence in teaching, and in 1961 was recipient of the ASME Timoshenko Medal for his contributions to research and teaching. He was elected to Fellow ASME in 1964. Goodier was married a daughter of Stephen Timoshenko, famous researcher in elasticity. He was a shy person, gentle, non-aggressive, soft-spoken, considerate of others, and above all, a dedicated man in the company of university scholars. The course of his lifelong philosophy is beautifully expressed by George Eliot: 'O may I join the coir invisible, of those immortal dead who live again, in minds made better by their presence, so to live is heaven…'.

Anonymous (1956). J.N. Goodier. *Mechanical Engineering* 78(11): 1075. *P*
Anonymous (1959). Goodier, James N. *Who's who in engineering* 8: 926. Lewis: New York.
Anonymous (1970). James N. Goodier. *Mechanical Engineering* 92(1): 111. *P*
Goodier, J.N., Hodge, P.G., Jr. (1958). *Elasticity and plasticity*. Wiley: New York.
Goodier, J.N., Florence, A.L. (1966). Localized thermal stress at holes, cavities, and inclusions disturbing uniform heat flow: Thermal crack propagation. *Applied mechanics* Munich: 562-568, H. Görtler, ed. Springer: Berlin.
Timoshenko, S., Goodier, J.N. (1951). *Theory of elasticity*. McGraw-Hill: New York.
histsoc.stanford.edu/pdfmem/GoodierJN.pdf

GOODING

01.04. 1803 Bristol NY/USA
04.03. 1878 Lockport IL/USA

William Gooding was an important early settler at Lockport IL. From 1836 he served as engineer of the Canal. He determined that Lockport should be the canal headquarters because of its potential for water power. He felt that this town would be a large future manufacturing center. After canal completion, he became secretary to the Canal trustees, serving until 1871, when the Canal paid its debts and the State regained control of the waterway. Gooding also served as one of the supervisors for the US Army Corps of Engineers from 1869 to 1871 survey for the canal enlargement. He was further appointed surveyor for Oregon State, but he turned it down in favour of staying at Lockport.

Gooding was trained on the job, as most canal engineers of the 19th century. From 1800 to 1860 these men knew each other, designing and constructing a series of canals that connected the Hudson River to the Great Lakes and the Ohio River to the east coast and the Great Lakes. Gooding design his Canal connecting the Great Lakes to Mississippi River via the Illinois River, referred to as the Illinois and Michigan Canal. He was hired by the Canal Commissioners in 1836 because of his wide experience in this field. As the construction of the canal proceeded, towns along it as Lockport, La Salle or Morris were erected. Gooding's principle interest was in water power development, which was not incidental to the construction but a central part of the entire project. This concept led to the establishment of Lockport, because there was the largest fall along the canal. Gooding proposed there a basin fed by water from Lake Michigan, thereby increasing the value of State land and give income to the canal from water rental. Despite works had started well, financial difficulties slowed down the construction in 1839. In 1847 Gooding offered to resign, but the works were then almost completed. The Canal was inaugurated in 1848, enabling navigation across Chicago Portage and helped establish Chicago as the transportation hub of the USA before railroads arrived. Its function was replaced in 1900 by the wider and shorter Chicago Sanitary and Ship Canal until the transportation operations ceased in 1956.

Lamb, J. (1982). *William Gooding*: Chief engineer of the I. and M. Canal. Illinois Canal Society 5: 131-143. *P*

Lamb, J. (1999). *Lockport Illinois*: The old canal town: 12. Arcadia Publishing: Charleston SC. *P*

http://www.lewisu.edu/imcanal/JohnLamb/section_29.pdf
http://en.wikipedia.org/wiki/Illinois_and_Michigan_Canal

GOODRICH

07.05. 1874 Decatur MI/USA
07.10. 1955 Brooklyn NY/USA

Ernest Payson Goodrich obtained the BS degree in 1898 from University of Michigan, Ann Arbor MI. He then joined the US Navy as a civil engineer, was a chief engineer for the Bush Terminal Co., New York, from 1903 to 1907, when starting as a private consultant designing harbours, the regional plan of New York City and public works for other cities. He was port engineer and consultant to the Albany Port Commission from 1924 to 1933, and from 1933 deputy commissioner as well as chief engineer of the Department of Sanitation, New York City. He was from 1934 professor of engineering economics at New York University. In 1943 Goodrich became member of the Prize Adjustment Board for the US Navy. He was a president of the Institute of Traffic Engineers, and a trustee of the Brooklyn Polytechnic Institute. He was awarded the honorary doctorate from his Alma Mater in 1935, and posthumously by the Brooklyn Polytechnic Institute.

Goodrich was a general engineer dealing with the various daily engineering tasks. He was for instance the inventor of the progressive system of electric light signal street traffic control. One of his special interests concerned municipal personnel and the organisation problems on which he has been called to advise more than a hundred cities. He was a Fellow of the American Association for the Advancement of Science AAAS, member and director of the American Society of Civil Engineers ASCE, being awarded the 1905 Collingwood Prize for his 1904 paper, and becoming in 1951 president of the American Institute of Consulting Engineers AICE.

Anonymous (1936). Ernest P. Goodrich. *Civil Engineering* 10(6): 699-700. *P*
Anonymous (1937). E.P. Goodrich. *Engineering News-Record* 118(Apr.29): 614. *P*
Anonymous (1955). Goodrich, Ernest P. *Who's who in America* 28: 1017. Marquis: Chicago.
Anonymous (1955). Ernest P. Goodrich. *Civil Engineering* 25(11): 824. *P*
Goodrich, E.P. (1904). Lateral earth pressure and related phenomena. *Trans. ASCE* 53: 272-321.
Goodrich, E.P., Ford, G.B. (1913). *Report* of suggested plan of procedure for the City Plan Commission, City of New Jersey NJ.
Goodrich, E.P. (1916). Varied uses of valuation: A survey of the bases, upon which physical appraisal may be employed for rate-irking, taxing, purchase, and the control of security issues. *Analyst* 7(Apr.3): 428-429.
Goodrich, E.P. (1946). Traffic engineering reminiscences. Proc. *Annual Meeting*, Institute of Traffic Engineers, New Haven CT.

GORRINGE

11.08. 1841 Bridgetown/BB
07.07. 1885 New York NY/USA

Henry Honeychurch Gorringe was born in the West Indies. He came to USA at an early age and entered the merchant marine service. He served through the Civil War with distinction, rising in rank from a common sailor to lieutenant commander in 1868. He commanded the sloop Portsmouth in the South Atlantic, and from 1876 to 1878 the Gettysburg in the Mediterranean. When the Egyptian government presented the obelisk to the USA, Gorringe was given charge of transporting it to America. William H. Vanderbilt paid for the expense of its removal more than 100,000 US$. Gorringe dug it out of the old location by removing 1,300 m^3 of earth. Then, by an ingenious device of his own invention, lowered it to a horizontal position and cut a hole in the iron steamer Dessoug, purchased from the Egyptian government, through which the obelisk was placed in the hold. The 20 m shaft, which was erected by Thothmes III at Heliopolis about 1600 B.C., was removed to Alexandria in 22 B.C. It has been claimed that the stone has masonic markings on its base. It was finally erected in 1881 at Central Park, New York NY. He wrote Egyptian obelisks dealing with the expedition to retrieve the obelisk, and a study of the other obelisks in Paris and London.

In 1862 Gorringe enlisted in the Union Navy as able-bodied seaman, and was attached to the Mississippi squadron three months later. By 1865 he had risen through successive promotions for gallantry to the rank of acting-volunteer lieutenant. He was promoted to Lieutenant Commander in 1868, and from 1869 until 1871 was the commander of the South Atlantic Squadron's sloop USS Portsmouth. He was engaged in the hydrographic office at Washington DC from 1872 to 1876. Subsequently Gorringe criticized naval matters in public, and when called to account, offered his resignation, which was accepted. He then entered the shipbuilding field, but his venture failed. He was member of Anglo-Saxon Lodge No. 137 of New York City. The lodge at one time tendered him a reception. He died as the result of an accident while jumping from a moving train. His friends erected a miniature copy of Cleopatra's Needle over his grave.

Anonymous (1885). Henry H. Gorringe. *Trans. ASCE* 16: 875.
Anonymous (1890). Henry H. Gorringe. *Proc. ASCE* 16: 215-216.
Gorringe, H.H. (1885). *Egyptian obelisks*. Nimmo: London UK.
http://en.wikipedia.org/wiki/Henry_Honychurch_Gorringe *P*
http://www.pbase.com/jchiarella/gorringe_henry_honychurch&page=all *P*

GOTAAS

03.09. 1906 Melette SD/USA
24.08. 1977 Evanston IL/USA

Harold Benedict Gotaas received his BS degree in civil engineering in 1928 from the University of South Dakota, Vermillion SD, and his MS degree from Iowa State University in 1930. He returned then to his Alma Mater as assistant professor in sanitary engineering, continued from 1936 studies at Harvard Graduate School, receiving his MS title in 1937. He then served as assistant professor of public health, University of North Carolina, Chapel Hill NC, becoming there full professor in 1941, and receiving the DSc degree from Harvard University in 1942. In 1946 Gotaas joined the University of California, Berkeley CA, as professor of sanitary engineering, becoming in 1949 chairman of the Civil Engineering Division, and head of the Sanitary Research Laboratory. He finally served as Dean of the Northwestern University, Evanston IL.

Although continuously busy, Gotaas found time to write a book and papers covering the environment, water, waste treatment and control, engineering education, and economic development problems. Gotaas served on committees, and boards, notably membership of the Great Lakes Commission, Northeastern Illinois Planning Commission, the Inter-American Association of Sanitary Engineering, and the US Environmental Protection Agency. Awards and decorations came to him in many areas, including: US Legion of Merit; Order of Condor of the Andes, Bolivia; Order of Merit, Chile; Cross of Boyacá, Colombia; Harrison P. Eddy Medal, and Gordon Maskew Fair Medal, both of the Water Pollution Control Federation, and James R. Croes Medal and Rudolph Hering Medal, both of the American Society of Civil Engineers.

Anonymous (1957). Gotaas goes to Northwestern as dean. *Engineering News-Record* 158(Jan.3): 56. *P*
Anonymous (1958). Harold B. Gotaas. *Civil Engineering* 28(10): 777. *P*
Anonymous (1964). Gotaas, Harold B. *Who's who in engineering* 9: 689. Lewis: New York.
Anonymous (1977). Dr. Harold B. Gotaas dies. *Civil Engineering* 47(11): 87. *P*
Gotaas, H.B. (1956). *Composting*: Sanitary disposal and reclamation of organic wastes. Monograph Series 31: 1-205. World Health Organisation: Geneva, Switzerland.
Jennings, B.H., Berry, D.S. (1979). Harold B. Gotaas. *Memorial tributes* 1: 109-111. National Academy of Engineering: Washington DC. *P*
Wells, E.A., Gotaas, H.B. (1958). Design of Venturi flumes in circular conduits. *Trans. ASCE* 123: 749-771; 123: 774-775.

GOTTSCHALK

26.11. 1906 Lake Mills WI/USA
02.06. 1965 Falls Church VA/USA

Louis Christian Gottschalk was a 1931 graduate of the University of Wisconsin, Madison WI. After having been engaged with the Wisconsin Geological Survey and the West Survey, he joined the Soil Conservation Service SCS, headed until 1951 its Sedimentation Division, from when until retirement in 1964 he was SCS division engineer at Arlington VA. He was member of the American Geophysical Union AGU and the recipient of the superior service award by the Secretary of Agriculture in 1964.

Gottschalk became known through his contribution Reservoir sedimentation in the Handbook of applied hydrology, including 1. Introduction, 2. Problem, 3. Erosion, 4. Sediment from watersheds, 5. Sediment characteristics, 6. Trap efficiency of reservoirs, 7. Sediment distribution, 8. Sediment yield from watersheds, 9. Rates of reservoir sedimentation, and 10. Control of reservoir sedimentation. Dam construction results in reservoir sedimentation. The global replacement cost of storage lost to the sediment accumulation is tremendous, so that several dams were recently removed to free the river from the dam. The latter also changes the hydraulic flow characteristics and the sediment transport capacity, so that several sediment bypass tunnels were erected, by which a portion of the sediment inflow is flushed into the tailwater. Each dam erection has therefore, besides many advantages, this and other serious disadvantages, which must be critically considered prior to dam erection in terms of environmental impact and overall economy. Gottschalk's chapter is one of the early accounts on this currently widely-accepted problem.

Anonymous (1953). L.C. Gottschalk. Proc. 5th *Hydraulics Conference* Iowa: Frontispiece. *P*
Anonymous (1955). Gottschalk, L.C. *American men of science* 9: 719. Science Press: Lancaster.
Gottschalk, L.C. (1942). Sedimentation investigation of Carnegie Lake, Princeton NJ.
Gottschalk, L.C. (1948). Analysis and use of reservoir sedimentation data. Proc. Conf. *Federal Inter-Agency Sedimentation* Denver: 131-138.
Gottschalk, L.C., Brune, G.M. (1950). Sediment design criteria for the Missouri Basin loess hills. *Report* SCS-TP-97. US Soil Conservation Service: Milwaukee WI.
Gottschalk, L.C. (1952). Measurement of sedimentation in reservoirs. *Trans. ASCE* 117: 59-71.
Gottschalk, L.C. (1962). Effects of watershed protection measures on reduction of erosion and sediment damages in USA. *IASH Publication* 59: 426-450.
Gottschalk, L.C. (1964). Sedimentation 1: Reservoir sedimentation. *Handbook of applied hydrology* 17: 1-34, V.T. Chow, ed. McGraw-Hill: New York.

GOULD E.S.

13.08. 1837 New York NY/USA
24.01. 1905 New York NY/USA

Edward Sherman Gould was educated at private schools in New York, graduating then from *Ecole des Mines*, St. Etienne, France. After return to the USA he entered in the civil engineering profession in 1865, first as an assistant engineer to the US Corps of Engineers until 1876, then as an assistant engineer to the Croton Aqueduct, New York. From 1879 to 1886 Gould was then involved in a number of sub-projects for the New Croton Aqueduct. Then, as a consultant, he designed and constructed the water works of Scranton PA, among other similar projects. From 1890 to 1894 he was involved in the Havana water works, Cuba. He was member of the American Society of Civil Engineers ASCE, and received the Venezuelan decoration *El Busto del Libertador*.

Gould published a number of books, among which may be mentioned his Practical hydraulic formulae, Elements of water supply engineering, and Arithmetic of the steam engine. Further he contributed to several volumes of the Van Nostrand Science Series. In the first book, examples of both single pipelines and pipe networks are considered using Darcy's flow formula. The grade line is determined thereby accounting also for local head losses due to hydraulic elements. Typical dams of the United States are also shortly presented. The second book is essentially an updated version of the first, with additions on water flow through masonry conduits, methods of tunnel construction, pumps, and arch dams.

Anonymous (1905). Gould, Edward Sherman. *Who's who in America*: 582. Marquis: Chicago.
Anonymous (1905). Edward Sherman Gould. *Engineering News* 53(4): 107; 53(5): 124. *P*
Gould, E.S. (1894). *Practical hydraulic formulae* for the distribution of water through long pipes. Engineering News Publishing: New York.
Gould, E.S. (1894). The Dunning's Dam, near Scranton. *Trans. ASCE* 32: 389-420.
Gould, E.S. (1897). *The arithmetic of the steam engine*. van Nostrand: New York.
Gould, E.S. (1899). *The elements of water supply engineering*. Engineering News Publishing: New York.
Gould, E.S. (1902). Discussion of Experiments at Detroit MN on the effect of curvature upon the flow of water in pipes, by G.S. Williams, C.W. Hubbell, G.H. Fenkell. *Trans. ASCE* 47: 1-196.
Jacob, A., Gould, E.S. (1888). *The designing and construction of storage reservoirs*. van Nostrand: New York.

GOULD J.H.

21.10. 1844 Seneca Falls NY/USA
28.12. 1896 Seneca Falls NY/USA

James Henry Gould was a notable manufacturer of pumps in New York State in the second half of the 19th century. Pump manufacturing at Seneca Falls began in 1839 when wooden pumps in a cultivator shop were assembled. In 1848, Seneca Falls was a tiny village in upstate New York. Seabury S. Gould, the father of James, purchased the interests of the pump factory and started manufacturing cast-iron pumps, thereby creating the first all-metal pump in the world. Father Gould believed in the possibilities of the iron pump. He keenly watched as the first pump casting emerged from its mold of sand; this pump was believed to overcome all the disadvantages of wood, it would be strong and efficient and provide fresh flowing water for the pioneers who were opening the West, the farms in the East, and plantations down South.

In 1864 the company was incorporated in the State of New York. In 1869, its name was changed to Goulds Manufacturing Company. Seabury, the founder, ran the company until after the Civil War. In 1879 the entire plant, except the foundry, was destroyed by fire. James H. Gould took over as Gould's president from 1872 to 1896, the period of America's great industrial boom. The nation was rapidly expanding westward. Gould assumed the burden of reconstructing the ruined factory destroyed in the fire of 1879. In 1885 Goulds introduced its first fire engine with its new deluge suction and lift pump. It was then a period of great activity in the pump business because agriculture, lumbering, mining, and manufacturing were growing markets for a wide range of pumps. It was stated in 1938 that 'Goulds has become the world's largest manufacturer because it has refused to grow old, because we have continually looked forward, confident that we possessed in men, in management, in materials, and in money, resources sufficient to meet the needs and demands of each successive generation'. It is currently stated that this is as true as it was in 1938. Note that Henry R. Worthington (1817-1880) invented in 1840 the first direct-acting steam pump, and that Sulzer Brothers Co., Winterthur CH, the Swiss pump furnishers, was founded in 1834. The pump industry is currently a booming business, given the varied applications in modern technology and widespread infrastructure.

Anonymous (1904). James H. Gould. *Grip's historical souvenir of Seneca Falls* 2: 67. *P*
http://locator.gouldspumps.com/download_files/history/goulds_history.stm.
www.gouldspumps.com/.../Goulds_Pumps_History_... *P*

GOULD R.H.

14.06. 1889 Newton Upper Falls MA/USA
21.10. 1977 North Hempstead NY/USA

Richard Hartshorn Gould graduated as a sanitary engineer from Massachusetts Institute of Technology MIT in 1911. He then joined the Massachusetts State Board of Health as an inspector, and moved as sanitary engineer to *Emschergenossenschaft*, Essen, Germany, where important advances were made in his field of research. Upon return to the USA, he joined a consulting engineering firm at Poughkeepsie NY, was then a draftsman with Hazen & Whipple, Engineers, New York NY, worked later for Wards Island Sewage Treatment Works, New York, until becoming an engineer for the Sewage Disposal Department of Sanitation, New York. There he was until retirement in 1954 the director of the Division of Sewage Disposal of the New York City Department of Public Works.

Gould was interested throughout his career in elements of sewage treatment, including final settling tanks, by which the biological sludge should be separated from the purified water. Originally these large basins were designed using the particle settling velocity, but it was soon realized that this principle may only be applied if the particle is sufficiently large and dense. For biological sludge, these conditions are not satisfied, and other processes must be applied. Gould has contributed to this important task in environmental engineering mainly by prototype observations and proposals on how the processes are improved. His proposals on the activated sludge process including step aeration and high rate activated sludge have meant substantial plant-cost savings. He has been watched and adopted by sanitary engineers throughout the world. He was named Engineer of the Year in 1954 by the Metropolitan Section of New York, with the citation 'For outstanding creative contributions to the engineering profession by his inventive developments in the art of sewage treatment'.

Anonymous (1931). Gould, Richard H. *Who's who in engineering* 3: 502. Lewis: New York.
Anonymous (1954). Richard H. Gould. *Civil Engineering* 24(6): 402. *P*
Anonymous (1969). R.H. Gould. *Journal Water Pollution Control Federation* 41(10): 1816. *P*
Gould, R.H. (1922). The area of water surface as a controlling factor in the condition of polluted harbour waters. *Trans. ASCE* 85: 699-737.
Gould, R.H. (1943). Final settling tanks of novel design. *Water Works and Sewerage* 90(4): 133-136. *P*
Gould, R.H. (1953). Sewage disposal problems in the world's largest city. *Sewage and Industrial Wastes* 25(2): 155-160.

GOWEN

14.02. 1851 Barnstable NH/USA
19.10. 1909 Ossining NY/USA

After engineering education at the Massachusetts Institute of Technology MIT, Cambridge MA, until 1869, Charles Sewall Gowen became until 1876 assistant engineer to James B. Francis (1815-1892) of the Locks and Canal Company, Lowell MA, a stay which was of great value to his later work. He was engaged by the Boston water works until 1880, and the Boston Improved Sewerage, having charge of the Sudbury Supply until 1883. He then accepted as assistant engineer in charge of hydrography with the Philadelphia Water Department to study new sources of water supply from Schuylkill River, and was appointed in 1884 assistant engineer of the Aqueduct Commission, New York.

Gowen was engaged there on surveys for reservoir sites and on the location of the New Aqueduct from the Croton Lake to Harlem River, and on the construction of this great Aqueduct, normally a rock tunnel from the Old Croton Dam to Manhattan, over a total of 45 km. The Aqueduct drops below the hydraulic gradient near the city continuing for more than 10 km at a maximum depth of 100 m below tide water. It includes also a short inverted siphon. The horseshoe cross-sectional aqueduct shape is 4 m high and 4 m wide, designed for 12.5 m³/s discharge capacity. The accurate measuring of the line was made by a personally developed method. As division engineer from 1885, his work included a complicated gate-house at Old Croton Dam, and deep shafts. Alphonse Fteley (1837-1903), chief engineer of the works, stated of problems encountered by Gowen during the tunnel lining, which were however resolved. Once the Aqueduct was completed in 1890, Gowen was placed in charge of Reservoir M on Titicus River, near Purdy Station NY. Titicus Dam was completed in 1895, so that he took charge of the construction of the New Croton Dam, one of the largest dams of the early 20th century. Its maximum height is 90 m, including the notable curved stepped spillway. Changes imposed for parts of the dams were not accepted by Gowen, as described in the 1906 paper. The entire design was completed in 1906, marking a special moment for him. He had resigned from the service in 1905, becoming engaged in consulting work but soon became ill, which led to his early death.

Anonymous (1923). Charles S. Gowen. *Trans. ASCE* 86: 1670-1674.
Gowen, C.S. (1890). Discussion of Tunnel surveying on Division No. 6. *Trans. ASCE* 23: 32-33.
Gowen, C.S. (1900). The foundations of the New Croton Dam. *Trans. ASCE* 43: 469-565. (*P*)
Gowen, C.S. (1906). The changes at the New Croton Dam. *Trans. ASCE* 56: 32-72.

GRAFF F.

27.08. 1774 Philadelphia PA/USA
13.04. 1847 Philadelphia PA/USA

Frederick Graff was in his early life a carpenter and he acquired skill as a draftsman. At age 20 he was engaged by Benjamin H. Latrobe (1764-1820) as his assistant in erecting the first steam-powered waterworks of the USA at Philadelphia PA. In 1805 Graff was elected superintendent and engineer of these works, but they were found inadequate after some years of trial. In 1811 Graff recommended Fairmount PA as the best place for the waterworks. At this time the pipes were made of wood, but Graff devised an iron-pipe system to be used instead. He brought the work to perfection, and patterns of his fire plugs and stopcocks were sent to England. In 1818 two miles of 20- and 22-inch iron pipes was laid, with such an improvement in performance that at least two more miles of it per year was installed through the next decade. In 1820, the Schuylkill River was dammed to increase water supply. The associated feud between the domineering Schuylkill Navigation Company and the city administration was a significant sign of the times. In 1822, when the basic system was complete, the city water committee sent him to a resolution of thanks, and he was presented with a silver vase. His experience and ability became acknowledged all through the country, so that he supplied detailed information to other corporations in the USA, including New York City and Boston.

Graff was engaged for 42 years in the service of Philadelphia City, so that a monument to his memory was erected on the grounds of the Fairmount Water Works. In 1828 he received another from the water committee 'as a testimonial of respect for his talents and zeal effectually displaying in overcoming unforeseen difficulties encountered in the construction of the northeast reservoir at Fairmount'. At his death, his son Frederic Graff, Jr., took his place, combined the district and city waterworks, then planned and directed the construction of three additional reservoirs and a dam, thereby modernizing the Philadelphia water system, counting around 1850 to one of the best worldwide.

Anonymous (1963). Graff, F. *Who was who in America* 1607-1896: 213. Marquis: Chicago.
Egan, J.J., Granville, B. (1848). *To the memory of Frederick Graff*. Granville: Philadelphia.
Gibson, J.M. (1998). *The Fairmount Waterworks* 1812-1911. Philadelphia Museum of Arts. P
Graff, F. (1798). *Fred Graff's book*. Philadelphia.
Rouse, H. (1976). Frederick Graff. *Hydraulics in the United States* 1776-1976: 27. Iowa
        Institute of Hydraulic Research: Iowa City.
http://en.wikipedia.org/wiki/Frederick_Graff *P*

GRAFF, Jr. F.

23.05. 1817 Philadelphia PA/USA
30.03. 1890 Philadelphia PA/USA

Frederic Graff, son of Frederick Graff (1774-1847), was educated at Philadelphia PA, becoming assistant engineer in 1842 in the Water Department. In 1863 he designed and built the pipe bridge across River Wissahickon, in which the water pipe was used as compressive member. From 1873 to 1877 he was engaged on the design of water works in connection with Henry R. Worthington (1817-1880). Graff then had a private consulting office, notably for the water supply of Washington DC. However, he stayed all through his life essentially in his home town, adding to his city both in engineering and in its society, so that he was and still is considered one of the important citizens of Philadelphia.

When Graff was from 1851 chief engineer of the Philadelphia Water Department, he suggested to establish a park upon Schuylkill River. This was then developed to park improvements. He added considerably to his city such as near his home, where he was active in promoting engineering works and advanced arts, charity and science. He was member of the Franklin Institute from 1839, director for six years and vice-president for three years. He also was one of the founders of the Zoological Society and Gardens of Philadelphia, and acted as president from 1882. He presided over the Engineer's Club of Philadelphia in 1880. He was further one of the founders of the Photographical Society and its president during years; he was personally an excellent amateur photographer. His character was remarkably pure and all his actions were based of the highest nature. His habits were simple, his charities large but little known. Knowing for years that he was liable to sudden death, he continued in the constant unselfish performance of the duties and works, exhibiting always geniality and cheerfulness, which made his society and companionship always a pleasure. He was member and president of the American Society of Civil Engineers ASCE in 1885. His elevation to that distinguished office was accepted by him as the highest honour to which a civil engineer could aspire.

Anonymous (1891). Graff, Frederic Jr. *Proc. ASCE* 17(8): 247-250.
Graff, F., Worthen, W.E., Smith, E.W. (1862). *Reports* on trials of duty and capacity of the pumping engines No. 2, at Ridgewood, and of No. 1, at Prospect Hill. Trow: New York.
Graff, F. (1886). *Notes upon the early history of the employment of water power for supplying the city of Philadelphia with water, and the building and re-building of the dam at Fairmount.* Engineers' Club: Philadelphia.
http://pabook.libraries.psu.edu/palitmap/FairmountWW.html *P*

GRAY

.05. 1842 Andover MA/USA
05.11. 1921 Providence RI/USA

Samuel Merrill Gray was a recognized authority in the field of water works, sewerage, and hydraulics. While the later years of his professional experience were devoted to consulting work, he had served for thirteen years as city engineer of Providence RI, and was for thirty years a member of the State Board of Health for Rhode Island. His works covered not only cities in the USA, but also in Canada and in Mexico, the most important including the design of the reconstructed water works at Baltimore MD, the remodelling of the Worcester MA sewage treatment plant, as also the design of the complete system of intercepting sewers for the District of Columbia.

After leaving school, Gray began his engineering career on railroad construction in Connecticut. He went to Providence RI in the early 1860s, and secured a position as assistant to the chief engineer for the novel city water supply. His work received favourable notice so that he was elected city engineer of Providence. His duties also included supervision of the departments of water works, sewerage, bridges, and highways. While in that position, he was sent to Europe to study there novel sewage disposal methods, thereby visiting England, France, Germany, Italy, and the Netherlands. During these years, he was one of those who designed the modern toilet, including the midden closet, the excrement pail, the dry ash closet, and the night soil van. Gray resigned as city engineer in 1890 and opened a consulting office, specializing in municipal and sanitary work, and continuing in the field almost to his death. During these years he was retained on water and sewage works for various cities. In 1913, he became engineer for the study of a new water supply for Providence, keeping this position until the general plans for the new work were developed. He was then retained as chief consultant. Gray served in 1892 as vice-president of the American Society of Civil Engineers ASCE.

Anonymous (1921). Samuel M. Gray. *Engineering News-Record* 87(21): 872. *P*
Gray, S.M. (1884). *Proposed plan for a sewerage system, and for the disposal of the sewerage of the city of Providence*. Providence Press Co.: Providence.
Hering, R.G., Gray, S.M. (1889). *Plan of the city of Toronto*: Proposed intercepting sewers and outfall. City Engineer's Office: Toronto.
Nader, J., Gray, S.M. (1885). *Sewerage of Madison*: Report of the city engineer, transmitting the report of Mr. Samuel M. Gray, consulting engineer. Common Council: Madison WI.

GREELEY

20.08. 1882 Chicago IL/USA
03.02. 1968 Phoenix AZ/USA

Samuel Arnold Greeley obtained the BS degree in 1906 from Massachusetts Institute of Technology MIT, Cambridge MA. He was from 1904 to 1909 assistant engineer with Hering & Fuller, New York NY, led by Rudolph Hering (1847-1923) and George W. Fuller (1868-1934). Until 1911, after a study tour to Europe, Greeley was research engineer in charge of the Milwaukee Refuse Disposal Plant, and of a report on the water supply and sewage treatment of Caracas VE. From 1912 to 1915, he was engineer with the Sanitary District of Chicago, and until 1918 then supervising engineer at Camp Custer MI. He was member and honorary member of the American Society of Civil Engineers ASCE, recipient of its 1931 Thomas F. Rowland Prize, and its 1932 Rudolph Hering Medal; the American Water Works Association AWWA; and the Illinois Society of Engineers.

Greeley founded in 1914 the firm Greeley & Hansen G&H, Consulting Engineers, Chicago IL. Its history is largely the history of Greeley, who guided the firm for more than 50 years. He believed the most important and valuable contribution a practicing engineer has to offer is sound judgment based on experience. One of his earliest jobs was the design of the sewage treatment works for the North Shore Sanitary District in 1913, working for it still in 1963. As for experience, G&H had an impressive set of numbers. The firm's nine partners have been at G&H a combined total of 300 years. The firm was in the mid-1960s essentially at the size of the 1920s to permit the partners to keep in close touch with every job. At least one partner was familiar with each job maintaining contact with the client. Greeley considered him fortunate to have come in early contact with Hering, then one of the best-known sanitary engineers, including active practice from 1890 to 1920. The firm's first big foreign job comprised the study of sewage treatment on the Pacific and Atlantic sides of Panama Canal, the next was the design of the water supply works and sewage works of Barranquilla CO. G&H is currently a leader in developing innovative engineering solutions for water problems.

Anonymous (1948). Greeley, Samuel A. *Who's who in engineering* 6: 771. Lewis: New York.
Anonymous (1963). Where seniority is a tradition. *Engineering News-Record* 171(Nov.21): 53-54. *P*
Hering, R., Greeley, S.A. (1921). *Collection and disposal of municipal refuse.* McGraw-Hill: NY.
Greeley, S.A., Marston, F.A., Phelps, E.B. (1939). *Tentative plan for sewage disposal for the City of New York and the specific projects.* Empire State Printing: New York.

GREENE D.M.

08.07. 1832 Brunswick NY/USA
09.11. 1905 Adams NY/USA

David Maxson Greene entered in 1850 Rensselaer Polytechnic School, Troy NY, passing examinations for civil engineering in 1851. He was then appointed professor of mechanics and physics there until 1852, when being engaged as assistant engineer on the enlargement of Erie Canal. He rose from chainman to leveller, being involved in 1854 as engineer on railroads in Ohio and Indiana State. In 1855 he was appointed at his Alma Mater professor of geodesy and topographical drawing. His success there was immediate and he only left in 1861 because he was appointed assistant engineer in the US Navy. In 1862 he was detached and ordered to the US Naval Academy as senior assistant in the Department of Natural and Experimental Philosophy and instructor of steam engineering. After three years he was ordered to duty as assistant to the chief, the Bureau of Steam Engineering at the Navy Department, Washington DC, where he remained for another 3 years as member of a commission appointed by the US Treasury Department.

Greene was subsequently ordered to the US steamer Narragansett as chief engineer in the West India squadron. Because yellow fever broke out on board, the vessel was ordered north and went out of commission. Greene was then chief engineer of the Port Admiral's vessel in New York, but resigned in 1869 from the navy after almost nine years of service. He settled at Troy NY and began practice as civil, mechanical and hydraulic engineer. He was appointed chief engineer of the proposed Walloomsac railroad in 1872, and served as consulting engineer of the Ottawa City Water Works in Canada. In 1873 he became chief engineer of the Dansville Water Works, Livingstone County NY and in 1874 city surveyor of West Troy NY. He also served as engineer to the State Commission to examine plans for introducing steam on the canals. He was from 1875 also division engineer on the New York State canals. He served from 1872 to 1885 as engineer the Troy City Water Board, becoming in 1878 further director of the Rensselaer Polytechnic Institute.

Anonymous (1906). Greene, D.M. *Trans. ASCE* 56: 466-468.

Greene, D.M. (1888). *Notes on heat, steam and the steam engine for the use of students*. Lisk: Troy NY.

Nason, H.B., ed. (1887). Greene, D.M. *Biographical record of the officers and graduates of the Rensselaer Polytechnic Institute* 1824-1886. Young: Troy NY.

http://www.lib.rpi.edu/Archives/history/academic_heads/greene,dm.html *P*

GREENE G.S.

06.05. 1801 Warwick RI/USA
28.01. 1899 New York NY/USA

George Sears Greene entered in 1819 the US Military Academy, West Point NY, from where he graduated in 1823, becoming there then assistant professor of mathematics and engineering until 1827. He served in his artillery regiment in Massachusetts Maine, and Rhode Island until 1829, but resigned in 1836 his position entering a civil engineering practice in North Carolina. He was retained from 1856 by the Croton Aqueduct Department of New York City for six years to design and construct works for the extension of the water supply, comprising the large distributing reservoir in Central Park more than 10 m deep and covering almost 100 acres. Further he was responsible for the construction of the 2.5 m diameter and 400 m long wrought-iron pipe across Harlem River on the High Bridge, and the laying of the 1.5 m diameter and 1.2 km long cast-iron pipe across Manhattan Valley. At that time nothing comparable existed in the USA.

Greene served during the American Civil War as colonel from 1862 and later was commissioned brigadier general serving in Northern Virginia until 1865. Despite being severely wounded he rejoined his brigade in North Carolina. He was mustered out in 1866, and returned to New York City, where further reservoirs for the Croton Aqueduct had to be designed. The reservoirs at Boyd's Corner were the first, including a dam 25 m high and 200 m long at the crest. There were no American precedents of so high dams; the only available information was the 1853 paper of the French engineer Sazilly for the Furens Dam near St. Etienne. Upon the retirement of Alfred W. Craven (1810-1879) in 1867, Greene took over as chief engineer the Croton Aqueduct until 1871, when the project was transferred to the Department of Public Works. He then was engaged as chief engineer to devise the sewerage system of the capital. From 1873 he was one of the engineers to examine the cost of a ship canal from Lake Champlain to the St. Lawrence River. From 1874 he was engaged with the sewerage system of New York City, including the main outfalls into Harlem River. During his entire career Greene was progressive, keeping abreast of all advances in his profession. He served as president of the American Society of Civil Engineers ASCE from 1876 to 1877 and was elected its Honorary Member in 1888. He was one of the founding members of this society.

Anonymous (1902). Greene, G.S. *Trans. ASCE* 49: 335-340.
http://de.wikipedia.org/wiki/George_S._Greene *P*
http://en.wikipedia.org/wiki/George_S._Greene *P*

GREENE, Jr. G.S.

26.11. 1837 Lexington KY/USA
23.12. 1922 New York NY/USA

George Sears Greene, Jr., son of George S. (1801-1899), obtained his engineering training from his father. His early engagements were on the Croton Aqueduct of New York City, and copper mines in the Lake Superior District. He was engaged from 1875 to 1897 with the creation of the water front of Manhattan Island, as chief engineer of the NYC Department of Docks. Due to the small range of tides of up to 1.5 m the pier system was retained, with piers projecting out into the river with slips between; he added a bulkhead wall, which caused problems due to poor foundation but were finally resolved by the support of experienced men as John Newton (1823-1895) or William E. Worthen (1819-1897). The successful work placed Greene in prominence, at the head of American engineers in charge of dock and pier construction. The combination of pre-cast concrete blocks with both vertical and brace piles was a great concept. In 1895 commissioners of the Department of Docks appointed a Board of Consulting Engineers to review the water front improvements. In 1897 a stone was placed between Charles and West Street to commemorate the work of dock improvement made by Greene.

In 1897 Greene opened his engineering practice in New York. He became member of the Advisory Board of Engineers for the construction of the Barge Canal. John A. Bensel (1863-1922), another board member, sought a constant corporation of all members, in which Greene's long experience was of unusual value. This was during a period when the Scotia Dam, the Hinckley Dam and other large projects on the Barge Canal were built. Greene was member of the American Society of Civil Engineers, served as director from 1882 to 1886, as vice-president in 1885, and as treasurer from 1888 to 1890. He possessed the personal qualities of a gentleman, both in professional work and in his daily life, exemplifying markedly the fine and courtly ways of the professional man who never forgot himself under any circumstances.

Anonymous (1925). Greene, G.S., Jr. *Trans. ASCE* 88: 1392-1395.
Casella, R.A. (1989). *A brief history of the early years of the Department of Docks* of the City of New York including the outstanding engineering achievements of George S. Greene, Jr., and the construction of Pier A at the Battery. New York City.
King, M. (1899). George S. Greene, Jr.: Ex-engineer-in-chief, Department of Docks. *Notable New Yorkers* of 1896-1899: A companion volume to King's Handbook of New York City: 584. Orr: New York. *P*

GREGORY J.H.

07.08. 1874 Cambridge MA/USA
18.01. 1937 Baltimore OH/USA

John Herbert Gregory obtained in 1895 the degree of civil engineering from the Massachusetts Institute of Technology MIT, Cambridge MA. He was then employed by the Metropolitan Sewage Commission and the Metropolitan Water Board relating to the Wachusett Dam and Aqueduct, under Frederic P. Stearns (1851-1919). He also was associated with the early water purification in the USA. While in the employ with Allen Hazen (1869-1930) he was active on the design and construction of slow sand filter plants so that only after five years after graduation Gregory held a position of high responsibility. Until 1902 he was then assistant engineer of designs on the improvement of the water supply of Philadelphia PA, including the first water filter plants. From 1902 he was division engineer in charge of the 35 km long conduit from Boonton NJ to Jersey City NJ. In 1903 he collaborated with Rudolph Hering (1847-1923) and George W. Fuller (1868-1934) on a rapid sand filter plant at Milford NJ, the second of its kind in the USA.

The most noteworthy work of Gregory was for the Water Department of Columbus OH, which he served for 33 years. From 1904 he was successively design engineer, principal assistant engineer, and engineer in charge. He first designed the first large sewage test station. The treatment station included the largest water-softening plant; the trickling filters were the first large designed in the country. Gregory was awarded the Thomas Fitch Rowland Prize by the American Society of Civil Engineers ASCE for his 1910 paper. From 1920 to 1925 he was in charge of the construction of the O'Shaughnessy Dam and reservoir on Scioto River; his paper was awarded the ASCE James Laurie Prize. From 1926 he served Columbus with the design of relief and separate sewers. He was awarded the Rudolph Hering Medal in 1935 for the related paper. Gregory served for more than 40 years to advance the practice of sanitary engineering.

Anonymous (1931). J.H. Gregory: James Laurie Prize. *Engineering News-Record* 106(4): 164. *P*
Anonymous (1937). Death of John H. Gregory. *Engineering News-Record* 118(Jan.28): 144. *P*
Anonymous (1937). John H. Gregory. *Trans. ASCE* 102: 1552-1557.
Gregory, J.H. (1910). Improved water and sewage works of Columbus. *Trans. ASCE* 67: 206-425.
Gregory, J.H., Hoover, C.B., Cornell, C.B. (1929). The O'Shaughnessy Dam and reservoir. *Trans. ASCE* 93: 1428-1492; 1498-1504.
Gregory, J.H., Simpson, R.H., Allton, R.A. (1934). Intercepting sewers and storm stand-by tanks at Columbus OH. *Trans. ASCE* 99: 1295-1339.

GREGORY W.B.

13.03. 1871 Penn Yan NY/USA
29.01. 1945 New Orleans LA/USA

William Benjamin Gregory graduated from the Penn Yan Academy in 1890, obtaining the ME degree from Cornell University, Ithaca NY, in 1894, and made post-graduate works there in 1907. He was an instructor from 1894 to 1897, assistant and associate professor of experimental engineering until 1905, from when he took over as professor and dean the Engineering Department until 1938, of the Tulane University, New Orleans LA. From 1902, he also served the US Department of Agriculture USDA as irrigation engineer. He further was consultant of the Mississippi River Commission from 1903 mainly testing hydraulic dredges. From 1928 to 1929 he was a consultant to the US Army Corps of Engineers, conducting hydraulic tests for the Bonnet Carré Spillway, Lower Mississippi Valley. Gregory was president of the New Orleans Academy of Sciences in 1914. He was a member of the American Society of Civil Engineers ASCE, Council Member from 1916 to 1919, and vice-president in the term 1920-21.

Gregory was general engineer who served as teacher and consultant to the community. His professional interests included sanitary engineering, pumps and their development for engineering practice, irrigation and the optimum effect in the southern United States, as well as hydraulics. He proposed the so-called Tulane Pitot, an instrument to measure velocity and discharge in pipes and channels, which was less sensitive than the common Pitot Tube, a standard hydraulic instrument. He also developed a rice irrigation system 130 km long with canals and laterals in 1934.

Anonymous (1921). Prof. Gregory. *Jambalaya yearbook* 25: 4. Tulane University: New Orleans. *P*
Anonymous (1930). William B. Gregory. *Who's who in America* 16: 962. Marquis: Chicago.
Anonymous (1945). William Benjamin Gregory: Laboratory named for him dedicated at
        Tulane. *Civil Engineering* 15(9): 442.
Gregory, W.B. (1901). Tests of centrifugal pumps. *Trans. ASME* 22: 262-292.
Gregory, W.B. (1907). *Mechanical tests of pumps and pumping plants* used for irrigation and
        drainage in Louisiana in 1905 and 1906. Government Printing Office: Washington DC.
Gregory, W.B. (1908). *Cost of pumping from wells for the irrigation of rice in Louisiana and
        Arkansas.* Government Printing Office: Washington DC.
Gregory, W.B.. (1916). *Evolution of low-lift pumping plants in the Gulf Coast Country.* ASME.
Gregory, W.B. (1927). Pumping clay slurry. *Mechanical Engineering* 49(6): 609-616.
http://www.lahistory.org/site24.php

GREVE

28.07. 1885 Brooklyn NY/USA
13.06. 1950 West Lafayette IN/USA

Frederick William Greve was educated at University of Wisconsin, Madison WI, from where he obtained the ME degree in 1909. He then joined a pump company as draftsman, was instructor at Oregon State College, Corvallis OR, from 1910 to 1911, and then professor of hydraulic engineering at Purdue University, West Lafayette IN. He was involved in the Engineering Experiment Station, presenting a number of its Bulletins. Greve was a member of the American Society of Civil Engineers ASCE, and of the American Geophysical Union AGU.

Greve was an expert in fluid discharge measurement contributing with various devices to this technology for both hydraulic laboratory and prototype studies. He presented an improved Pitot tube, and introduced the parabolic weir. The latter was demonstrated to follow a head-discharge equation containing exactly the exponent of two, so in between the rectangular and the triangular weirs. He thereby pointed to the importance of the exact weir crest geometry. Other studies related to various basic weir types, thereby excluding the rectangular notch geometry for which numerous studies were available. Today, these discharge measurement devices are no more used in hydraulic laboratories because of accurate Inductive Discharge Measurement methods, providing discharge simpler than using these above mentioned. Greve also worked in questions relating to heating, piping, and air conditioning. During the 1930s he served on various committees concerned with the extensive drainage basin reports of the US Planning Board.

Anonymous (1933). Greve, Frederick William. *American men of science* 5: 440.
Anonymous (1941). Greve, Frederick W. *Who's who in engineering* 5: 701. Lewis: New York.
Anonymous (1942). F.W. Greve. Proc. 2nd *Hydraulics Conference* Iowa: Frontispiece. *P*
Anonymous (1950). Frederick W. Greve. *Mechanical Engineering* 72(12): 1037-1038.
Greve, F.W. (1921). Parabolic weirs. *Trans. ASCE* 84: 486-515.
Greve, F.W. (1924). Semi-circular weirs calibrated at Purdue University. *Engineering News-Record* 93(5): 182-183; 93(20): 804.
Greve, F.W. (1932). Flow of water through circular, parabolic and triangular vertical notch-weirs. *Research Bulletin* 40. Engineering Experiment Station, Purdue University: Lafayette IN.
Greve, F.W. (1945). Flow of liquids through circular, parabolic, and triangular vertical notch-weirs. *Engineering Bulletin* 16(2). Experiment Station, Purdue University: Lafayette IN.
Greve, F.W. (1945). Flow of liquids through vertical circular orifices and triangular weirs. *Engineering Bulletin* 29(3). Experiment Station, Purdue University: Lafayette IN.

GRIMM

12.01. 1886 Clear Lake IA/USA
01.12. 1942 Portland OR/USA

Claude Irving Grimm received education at Iowa State College, Ames IA, obtaining the degrees of B.CE in 1908 and CE in 1916. After having mainly worked for bridge construction, he was from 1924 to 1929 the principal assistant to the US Division Engineer at Cincinnati OH, and from then the chief designer of locks and dams on both the Ohio and Mississippi Rivers, and consulting engineer on the Mississippi River flood control. From 1930 to 1933 Grimm was principal engineer of the US Engineering Department, Pacific Division, San Francisco CA. He was further consulting engineer of Pacific streams for the development of water resources. From then until his premature death Grimm was the head of engineering, North Pacific Division, Portland OR, supervising the Bonneville project and flood control dams in Washington and Oregon.

Bonneville Dam was one of the ten dams developed for flood control, hydropower, navigation, and irrigation on Columbia River. In 1929, the US Corps of Engineers USCE published the '308 Report' proposing this scheme, but no action was taken until President Roosevelt adopted his New Deal. In 1934 two large dam projects were started with the Grand Coulee and the Bonneville Dams. To create the latter, USCE designed a new lock and a powerhouse on the Oregon side, and a spillway on the Washington side. Coffer dams had been built to block half of the river to allow for dam construction. These works were completed in 1937. The original Bonneville navigation lock was opened in 1938, then the world's largest single-lift lock. The dam produced commercial electricity also from then. The spillway includes eighteen gates over 440 m length, maintaining the 77 km long Bonneville Reservoir 18 m above the downstream river elevation. Grimm was therefore significantly involved in this large hydropower scheme, which was designated National Historic Landmark in 1987.

Anonymous (1941). Grimm, Claude I. *Who's who in engineering* 5: 704. Lewis: New York.
Anonymous (1943). Claude Irving Grimm. *Trans. ASCE* 108: 1585-1587.
Anonymous (2010). Claude Irving 'Pete' Grimm. *Civil Engineering* 80(6): 42. *P*
Grimm, C.I. (1928). Backwater slopes above dams. *Engineering News-Record* 100(23): 902.
Grimm, C.I. (1935). Cofferdams in swift water for Bonneville Dam. *Engineering News-Record* 115(Sep.5): 315-318.
Grimm, C.I. (1936). The Bonneville Project progress: Construction methods at Bonneville. *Civil Engineering* 6(10): 671-674.

GROAT

18.10. 1867 Hannibal MO/USA
16.06. 1949 Hightstown NJ/USA

Benjamin Feland Groat obtained his BSc degree in 1901 and the LL.M. degree in 1910 from University of Minnesota, Minneapolis MN. After early works with railroads, he went from instructor to professor in charge of mechanics and mathematics at the School of Mines at his Alma Mater from 1898 to 1910. In parallel he invented a lead pipe coupling with adjusted expansion properties in 1894. Groat, from then a consulting engineer, inventor and patent attorney, became interested in 'chemi-hydrometry' for discharge measurement applied to turbine tests. His 1916 paper was awarded the ASCE's Norman Medal. Further works related to the efficiency of screw current meters, as opposed to the cup current meter, and to the Pitot tube for velocity measurement, for which he was awarded the 1914 Silver Medal from the Engineering Society of Western Pennsylvania.

Groat then proceeded with research in hydraulic modeling. In the early 1920s, this technique was not generally accepted as an alternative to prototype observations and to hydraulic computations. Scale effects were considered the main limitation, preventing to up-scale laboratory observations to prototype scale. Groat also worked in this field with a final paper in 1932. He developed precise turbine engine tests and methods of measuring water discharges chemically. He authored many papers on water power. He was further interested in non-Newtonian models but recommended that Newtonian models should be adopted in hydraulic engineering. Following the proposal of John R. Freeman (1855-1932), Groat supported his National Hydraulic Laboratory, including thereby the experiences collected in Western Europe.

Anonymous (1941). Groat, Benjamin F. *Who's who in engineering* 5: 706. Lewis: New York.
Anonymous (1949). Benjamin F. Groat. *Civil Engineering* 19(8): 580. *P*
Groat, B.F. (1913). Characteristics of cup and screw current meters: Performance of these meters in tail-races and large mountain stream, statistical synthesis of discharge curves. *Trans. ASCE* 76: 819-870.
Groat, B.F. (1914). Pitot tube formulas. Proc. *Engineers' Society Western Pennsylvania* 30: 324.
Groat, B.F. (1914). Water-discharge measurement with chemicals. *Engineering Record* 70(8): 208-209; 70(9): 246-247.
Groat, B.F. (1916). Chemi-hydrometry and its application to the precise testing of hydro-electric generators. *Trans. ASCE* 80: 951-1305.
Groat, B.F. (1932). Theory of similitude and models. *Trans. ASCE* 96: 273-386.

GROVER

31.01. 1868 Bethel ME/USA
29.11. 1957 Washington DC/USA

Nathan Clifford Grover graduated with the BCE degree in civil engineering from the Maine State College in 1890, obtained the BS degree from the Massachusetts Institute of Technology, Cambridge MA in 1896, the civil engineering degree in 1897 and was awarded the D.Eng. degree in 1930. He was from 1891 to 1904 from instructor to professor in civil engineering at University of Maine, Orono ME, when joining until 1907 the US Geological Survey USGS. Grover was until 1911 a hydraulic, construction and irrigation engineer in New York City, then for two years chief engineer of the Land Classification Board, and from 1913 to 1939 chief hydraulic engineer in charge of the USGS Water Resources Branch. He was a member of the American Society of Civil Engineers ASCE, the Washington Society of Engineers, and the Washington Academy of Sciences.

The purpose of the 1907 book River discharge by John C. Hoyt (1874-1946) and Grover was to include everything that must be known in order to measure, analyse, and record the water discharge of the nation's streams. This book received wide recognition due to its federal background, and went through various editions, therefore. It was stated to be the most important American reference on hydrometry, which was widely used by engineers for generations.

Anonymous (1963). Grover, Nathan C. *Who was who in America* 3: 351. Marquis: Chicago.
Grover, N.C., Hoyt, J.C. (1906). *Report of progress of stream measurements*. Government
	Printing Office: Washington DC.
Grover, N.C. (1932). Contributions to the hydrology of the United States. *Water Supply Paper*
	W638. US Geological Survey: Washington DC.
Grover, N.C. (1935). Report of Committee on flood protection data. *Civil Engineering* 5(3): 170.
Grover, N.C., Harrington, W.H. (1943). *Stream flow*: Measurements, records and their uses.
	Wiley: New York.
Hall, M.R., Grover, N.C., Horton, A.H. (1907). *Surface water supply of the Ohio and Lower
	eastern Mississippi river drainages*. US Geological Survey. Washington DC.
Hoyt, J.C., Grover, N.C. (1907). *River discharge* prepared for the use of engineers and students.
	Wiley: New York.
Rouse, H. (1976). Nathan C. Grover. *Hydraulics in the United States* 1776-1976: 80-81. Iowa
	Institute of Hydraulic Research: Iowa City. *P*

GRUNSKY

04.04. 1855 San Joaquin CA/USA
09.06. 1934 Oakland CA/USA

Carl Ewald Grunsky obtained his higher education at Stuttgart *Polytechnikum*, Germany, graduating as a civil engineer in 1877. After having returned to the USA he worked as topographer for river surveys in the State Engineering Department of California, becoming assistant and chief assistant engineer to the State Engineer. From 1887 to 1899 he was a private consultant dealing mainly with irrigation, sewage and drainage works around Sacramento and San Francisco. He also contributed works to river rectifications and drainage problems as Member of the Examining Commission, Rivers and Harbours, California. He was from 1892 in parallel also a member of the San Francisco Sewerage Commission, and from 1900 to 1904 city engineer of San Francisco. During the next year Grunsky was a member of the Isthmian Canal Commission involved in Panama Canal. Since then he was a consultant for the US Reclamation Service, Washington DC. He was a member ASCE and its president in 1924, of the Technical Society of the Pacific Coast, and of the California Academy of Sciences, presiding it from 1912.

Grunsky authored various technical papers, including a work in the Transactions ASCE for which he was awarded the 1910 Norman Medal. He was also interested in all questions relating to water resources, including hydrology and methods of discharge measurement in rivers, specific rivers such as the Colorado River, or water supply. He further devised formulae for estimating rainfall intensities, which were applied to the design of sewer systems.

Anonymous (1921). Grunsky, Carl Ewald. *Who's who in America* 11: 1173. Marquis: Chicago.
Anonymous (1924). C.E. Grunsky. *Engineering News-Record* 92(3): 128. *P*
Anonymous (1934). Carl Ewald Grunsky. *Civil Engineering* 4(7): 373. *P*
Anonymous (1935). Carl Ewald Grunsky. *Trans. ASCE* 100: 1591-1595.
Grunsky, C.E. (1896). Method of approximate gauging of rivers. *Engineering Record* 33(14): 239.
Grunsky, C.E. (1907). *The problem of the Lower Colorado River*. Private printing.
Grunsky, C.E. (1917). *Valuation of depreciation and the rate-base*. Wiley: New York.
Grunsky, C.E. (1931). Simplified formulas for rainfall intensity. *Monthly Weather Review* AMS 59(2): 83.
Mead, E., Smythe, W.E., Manson, M., Wilson, J.M., Marx, C.D., Soule, F., Grunsky, C.E., Boggs, E.M., Schuyler, J.D. (1901). Report of irrigation investigations in California. *Bulletin* 100. US Dept. Agriculture: Washington DC.

GUMENSKY

26.10. 1897/RU
21.05. 1956 Berkeley CA/USA

D(i)mitry Benjamin Gumensky graduated as a civil engineer from the University of California in 1925. He then specialized in structural and hydraulic designs. During construction of the Metropolitan Water District of Southern California, he supervised design of the aqueduct structures. Later he served as consultant the US Engineer Department, Sacramento CA; he built hydroelectric works in China for the Chinese government; and he spent several years in Jerusalem laying out an irrigation system for the Israeli government. He returned in the early 1950s to the USA engaged on consulting work at Berkeley.

Gumensky worked in the field of water resources development. He was particularly engaged for the Colorado River Aqueduct by which drinking water is supplied to the large cities of Southern California. His 1935 paper reports on these works and the hydraulic problems that were solved. Gumensky was particularly attached to the design of almost 150 siphons of the 400 km long aqueduct. His 1938 paper lists the governing parameters to be considered in aqueduct design. First the water has to be lifted from the river to the required elevation, and second a sufficiently large conduit has to be provided in which the water flows to its destination. Another paper relating to pipeline design was a study on the head losses in pipe bends. The 1949 paper points to the fact that high-speed chute flow requires higher side walls than designed with conventional hydraulic formulae because of the air entrained, resulting in an air-water mixture. This effect was investigated scientifically only from the early 1960s.

Anonymous (1956). Dimitry Benjamin Gumensky. *Civil Engineering* 26(6): 93.
Anonymous (1956). D.B. Gumensky. *Engineering News-Record* 156(Jun.14): 90.
Gumensky, D.B. (1933). Fulton dam tainter gate discharge. *Civil Engineering* 3(11): 627-628.
Gumensky, D.B. (1935). Flow of water around bends in pipes. *Trans. ASCE* 100: 1033-1036.
Gumensky, D.B. (1935). Principles of siphon design for Colorado River Aqueduct. *Engineering News-Record* 114(Jun.27): 899-903. P
Gumensky, D.B. (1938). Governing factors in aqueduct design. *Engineering News-Record* 121(Nov.24): 653-658.
Gumensky, D.B. (1949). Air entrained in fast flowing water affects design of training walls and stilling basins. *Civil Engineering* 19(12): 831-833; 19(12): 889.
Gumensky, D.B. (1957). Earthquake and earthquake-resistant design. *American civil engineering practice* 3(34): 1-34, R.W. Abbett, ed. Wiley: New York.

GUY

05.08. 1922 Brighton IA/USA
21.01. 2011 Boiling Springs PA/USA

Harold Paul Guy was an US Army veteran serving in World War II in the Manhattan Project at Oak Ridge TN as technician for separation of uranium isotopes. He obtained in 1947 a BSc degree in civil, and the MS degree in agricultural engineering in 1951 from Iowa State University, Iowa City IA. He continued working with the US Geological Survey USGS, in its Water Resources Division, for 29 years by evaluating the movement of sediments in rivers. He lived during his career at Norton KS, Lincoln NE, Fort Collins CO and Reston VA. He prepared technical reports but relatively few journal papers. He was further involved in the American Society for Testing and Materials ASTM, including several years during retirement, becoming chairman of the section for evaluating the amount and movement of sediment in streams. He was a member of the American Society of Civil Engineers ASCE and the Soil Conservation Society SCS of America. In retirement, he and his wife enjoyed travel to visit friends and most of the National Parks. These resulted in the book published in 2002 on the Guy family.

Guy was a notable sediment specialists who particularly was interested in the large rivers in the United States. Based on his work during WWII he proposed in 1963 the use of radionuclides to study the entrainment and transport of sediments. The 1969 report discusses the laws of large falling particles in quiescent water, representing a basic parameter in sedimentological investigations. The 1975 Report was devoted to the large 1973 flood on Mississippi River.

Anonymous (1953). H.P. Guy. Proc. 5th *Hydraulics Conference* Iowa: Frontispiece. *P*
Chin, E.H., Skelton, J., Guy, H.P. (1975). 1973 *Mississippi River Basin flood*: Compilation and analyses of meteorologic, streamflow, and sediment data. US Government Printing Office Washington DC.
Guy, H.P. (1970). *Sediment problems in urban areas*. US Geological Survey: Washington DC.
Guy, H.P., Edwards, T.K., Glysson, G.D. (1999). *Field methods for measurement of fluvial sediment*. US Geological Survey: Denver CO.
Sayre, W.W., Guy, H.P., Chamberlain, A.R. (1963). *Uptake and transport of radionuclides by stream sediments*. US Government Printing Office: Washington DC.
Stringham, G.E., Simons, D.B., Guy, H.P. (1969). *The behaviour of large particles falling in quiescent liquids*. US Government Printing Office: Washington DC.
http://obitsforlife.com/obituary/343702/Guy-Harold.php

GUYTON B.

19.08. 1910 Durant MS/USA
14.06. 1996 Decatur AL/USA

Benson Guyton graduated from Mississippi State College, Starkville MS, with a BS degree in civil engineering in 1933. He was then engaged for two years in soil erosion control at the Forest Service, and another 2 years in the Soil Conservation Service SCS. Until 1944 he was in charge of model and prototype investigations of hydraulic structures with the US Corps of Engineers, Vicksburg MS. He was there from 1952 chief of the Prototype Test Section.

The 1953 paper co-authored by Frank B. Campbell (1904-1992) deals with air vents installed at gated outlet works. Bottom outlets are an important safety element, by which reservoir sedimentation is partially controlled. Due to the high flow velocities and the high sediment concentration, these elements undergo extreme loads. The paper summarizes field observations on the air demand to counter cavitation damage. The theoretical approach is based on the laws of turbulent flow. It was found that the roof drag and that the velocity distribution of the air flow must be considered. The data are compared with laboratory tests of Anton A. Kalinske (1911-1985) and James M. Robertson (1916-2012), indicating that the field data have a substantially higher air-water discharge ratio for a given Froude number. A design criterion for air vents counts to the first approach in this even currently not fully understood process. The 1954 Report deals with a complete piezometer system and air-demand measuring facilities for the outlet works of Pine Flat Dam. Field data were collected of both the vertical and flexural vibrations of a sluice slide gate, including the pressures at the sluice intake, on the gate leaf, in the conduit and on the tetrahedral deflector, using accelerometers attached to the gate leaf, indicating no dangerous flow conditions. The flow velocities in the air vent were also measured.

Anonymous (1956). B. Guyton. Proc. 6th *Hydraulics Conference* Iowa: Frontispiece. *P*
Campbell, F.B., Guyton, B. (1953). Air demand in gated outlet works. Proc. 5th *IAHR Congress* Minneapolis: 529-533.
Guyton, B. (1954). Vibration, pressure and air-demand tests in flood-control sluice, Pine Flat Dam, Kings River CA. *Report*. US Army Waterways Experiment Station: Vicksburg.
Guyton, B. (1959). Field investigations of spillways and outlet works. *Trans. ASCE* 124: 491-506.
US Army Corps of Engineers (1953). Pressure and air-demand tests in flood-control conduit. *Miscellaneous Paper* 2-31. Waterways Experiment Station: Vicksburg MS.
http://library.msstate.edu/FindingAid/MSS.345.html

GUYTON W.F.

15.10. 1917 Oxford MS/USA
02.03. 2013 Austin TX/USA

William Franklin Guyton obtained the BS and the MS degrees in 1938 from University of Mississippi, Rolla MS, and the civil engineering degree there in 1945. He was engaged as hydraulic engineer from 1939 to 1950 by the Groundwater Branch of the US Geological Survey USGS, becoming then consulting groundwater hydrologist at Austin TX. He also was in 1945 a consultant on groundwater supplies for the US Army on the Pacific Islands. He settled in 1951 at Austin TX, establishing a consulting firm which specialized in groundwater hydrology and general issues with groundwater flow. Guyton was a member of the American Society of Civil Engineers ASCE, the American Geophysical Union AGU, and of the American Water Works Association AWWA.

Guyton was an expert in groundwater hydrology. While working in Mississippi State he produced a study on the groundwater supply at Camp Van Dorn MI. Next to geological aspects of the Mississippi River Plain, the study is concerned with the water supply of Wilkinson County. The 1946 paper deals with cold water that was added to the principal aquifer through several supply wells. The large depression cone resulting from prior pumping was filled by artificial and natural recharge. The 1948 paper reviews methods used for measuring water levels to interprete changes in levels, and for using these measurements in aquifer tests.

Anonymous (1994). Guyton, William F. *American men and women of science* 3: 477.
Anonymous (2013): William Guyton. *Engineering News-Record* 270(9): 26. P
Guyton, W.F., Brown, G.F. (1943). *Geology and groundwater supply at Camp Van Dorn*. State Geological Survey, University of Mississippi: Rolla.
Guyton, W.F. (1944). *Progress report* on the groundwater resources of Louisville area, Kentucky.
Guyton, W.F. (1946). Artificial recharge of glacial sand and gravel with filtered river water at Louisville KY. *Economical Geology* 41(6): 644-658.
Guyton, W.F. (1948). *Fluctuations of water levels and artesian pressures in wells in the United States*: Their measurement and interpretation. Geological Survey: Washington.
Klaer, F.H., Guyton, W.F., Todd, D.K. (1948). *A preliminary list of references pertaining to artificial recharge of groundwater in the United States*. Geological Survey: Washington.
Lloyd, J.B., ed. (1981). *Lives of Mississippi authors* 1817-1967. University of Mississippi.
White, W.N., Guyton, W.F. (1951). *Ground water in Mimbres Valley NM*, with special reference to the available water supply in the Miesse District, east of Florida Mountains. Austin.

HAAS

28.08. 1870 Stockton CA/USA
13.04. 1939 San Francisco CA/USA

Edward Francis Haas graduated from University of California, Berkeley CA, with the BSc degree in civil engineering in 1892. After having been there instructor, he attended until 1894 Columbia School of Mines, New York NY, receiving the CE degree. He returned then to Stockton, where he had until 1897 his private practice. He became associated with the San Joaquin and Sacramento Rivers, and was retained for the reclamation of swamp land in their delta regions. He built 30 km of levees, 5 km of main canal, and large pumping plants for drainage. From 1898, Haas took residence at San Francisco CA, where he was associated with Carl E. Grunsky (1855-1934) in the design and the construction of the sewage system for Belvedere CA. In 1900 he was engaged by the Board of Engineers of San Francisco to design a comprehensive and modern sewage system. Haas was responsible for its design, so that he became assistant engineer in the newly erected Board of Public Works, with Grunsky as city engineer.

Haas entered in 1901 a partnership, which was engaged in dredging and reclamation work. In 1901 a dredging contract was received to deepen Pearl Harbor, Honolulu HI. In 1906 Haas organized the Union Dredging Company, and in 1909 then the Caledonia Dredging Company, acting as president of both. The work included river rectifications and reclamation work in the Sacramento and San Joaquin Deltas. Large amounts of money were spent on these projects which substantially added fertile land to the State. These works were done with floating dredgers, which were improved by Haas. From 1919 Haas turned his attention to dry land excavation, thereby forming a new partnership which lasted until his death. It was involved in the construction of the main canals and appurtenances of the Minidoka and Boise projects for the US Bureau of Reclamation. The firm was also engaged in large levees of the Sacramento Flood Control Project. The success of the firm rested mainly on the input of Haas, who had exceptional talents in work organization, and economical aspects of each project. He was described as a man with sympathetic manners and generous to the appeal of others. He was ASCE member.

Anonymous (1940). Edward F. Haas. *Trans. ASCE* 105: 1851-1854.
Haas, E.F., Morris, F.L. (1914). *Union Island reclamation*. San Francisco CA.
Haas, E.F. (1921). Discussion of Larger ships, deeper harbours, and better dredges. *Trans. ASCE* 84: 182-182.
Lewis, O. (1961). Edward F. Haas. *A family of builders*. Grabhorn Press: San Francisco. *P*

HABERMAN

04.05. 1922 Wien/A
27.12. 1996 Rockville MD/USA

William Lawrence Haberman received in 1949 the BS degree from the Cooper Union University, New York, the MS degree from University of Maryland, College Park, in 1952, and the PhD degree in 1956. He was from 1949 to 1950 a physicist at the Bureau of Ships, US Department of the Navy, until 1957 at the David Taylor Model Basin, Bethesda MD, then department director of the Gas Dynamics Division for two years, and then chief of research branch and director of advanced planning. From 1963 to 1971, he served as NASA senior staff scientist, and from 1973 to 1978 was professor of thermodynamics at Montgomery College, Montgomery MD. He was member of the American Institute of Aeronautics and Astronautics AAIA, the American Physical Society APS, the American Association for the Advancement of Science, and the American Geophysical Union AGU.

Haberman had research interests in potential flows, cavitation, theoretical mechanics, thermal radiation, research administration, and in viscous and two-phase flows. In the latter field, he contributed the significant 1956 paper with a colleague. Experiments were conducted to study the motion of air bubbles, thereby determining the drag and the shape of single-rising bubbles in various liquids. This fundamental hydraulic process is governed by viscosity, fluid density and surface tension. Three types of bubble shapes were observed, namely spherical, ellipsoidal and spherical cap, as the bubble dimension increases. For tiny spherical bubbles the drag coefficient equals that of the corresponding rigid sphere. As bubble size increases, the drag decreases as compared with the former case. Ellipsoidal bubbles occur at different ranges of Reynolds numbers, depending on the type of liquid. The drag coefficient of spherical cap bubbles was found constant. He authored three books in fluid mechanics, engineering thermodynamics and heat transfer.

Anonymous (1985). Haberman, W.L. *Who's who in engineering* 6: 256. AAES: New York.
Anonymous (1994). Haberman, William L. *American men and women in science* 3: 486.
Grossman, B.M. (2013). William L. Haberman. Personal communication. *P*
Haberman, W.L., Morton, R.K. (1956). An experimental study of bubbles moving in liquids. *Trans. ASCE* 121: 227-252.
Haberman, W.L., Sayre, R.M. (1958). Motion of rigid and fluid spheres in stationary and moving liquids inside cylindrical tubes. Dept. of Navy, *Report* 1143. David Taylor Model Basin.
John, J.E., Haberman, W.L. (1988). *Introduction to fluid mechanics*, 3rd ed. Prentice Hall: Englewood Cliffs NJ.

HALL B.M.

31.01. 1853 Wynnsboro SC/USA
19.11. 1929 Atlanta GA/USA

Benjamin Mortimer Hall was graduated with the BE degree from the University of Georgia, Atlanta GA, in 1876. He was then appointed at the North Georgia Agricultural College, Dahlonega GA, professor of mathematics, where he became interested in mining and hydraulic problems. From 1880 to 1890 he served as mining engineer in Georgia State, where he established a reputation for professionalism and initiative. As consulting hydrographic engineer for the US Geological Survey USGS from 1896 to 1903, he directed the organisation and active field operations for stream flow, run-off, and other water-power data of the main watersheds of Georgia, Alabama, and Tennessee States, and the Carolinas, resulting in various reports, written also with his brother, attesting their sound professional judgment and clarity of presentation.

From 1904 to 1907 Hall was a supervising engineer for the US Reclamation Service, negotiating for example the terms of the Mexico-Rio Grande Treaty at El Paso TX. He also prepared the original plans for the Elephant Butte Dam and Rio Grande Project, supervising the settlement of all water-right disputes arising from this large storage and irrigation enterprise. The lake formed by the Elephant Butte Dam was named Lake Hall in his honour, and in 1930 was the largest artificial reservoir of the USA. He was called from 1908 to 1910 as chief engineer to study, supervise and construct the widespread irrigation development for the Porto Rican Irrigation Service, a work which received distinct professional approval for its comprehensive planning, durability and efficiency. Returning to Atlanta in 1911 he became active as a consultant with various water power development and flooding problems. He also served in the arbitration of controversies in engineering issues, notably at Raleigh and Durham NC, in New York NY, and at Washington DC. Hall was member of the American Society of Civil Engineers ASCE, showing unswerving loyalty in its work. He was a kind gentleman who attracted and enjoyed friends; he was sagacious, clear-minded, energetic, thorough, and fair.

Anonymous (1931). Benjamin M. Hall. *Trans. ASCE* 95: 1493-1495.

Hall, B.M. (1903). A preliminary report on a part of the water powers of Alabama. *Bulletin* 7: Geological Survey of Alabama. Brown: Montgomery AL.

Hall, B.M., Hall, M.R. (1907). Water resources of Georgia. *Water Supply and Irrigation Paper* 197. Department of the Interior, US Geological Survey: Washington DC.

http://www.geni.com/people/Benjamin-Mortimer-Hall/6000000010697375864 *P*

HALL J.W.

26.02. 1885 Wellesley MA/USA
01.11. 1934 Glen Ferris WV/USA

John Wendell Hall graduated in 1911 with the BSc degree from Harvard University, Cambridge MA. He then went to Idaho, where he was employed as chief of party on a survey of an irrigation project for the Downey Irrigation Company. He entered then the employ of the Mississippi River Power Company at Keokuk IA, under Hugh L. Cooper (1865-1937). Hall there was first junior engineer, and later assistant to the chief engineer, in charge of power houses, and turbine machinery. Following the completion of this Plant, he was employed on a similar hydro-electrical development at Rutland VT.

In 1914 Hall entered the employ of Hugh L. Cooper & Company, in charge of field studies for a hydro-electric development at Davis WV. Later he was engineer on the East River Tunnels in New York NY for the Public Service Commission. From 1917 to 1920 Hall represented Cooper & Company as resident engineer on the construction of the Zumbro Dam for Rochester MN, and then was employed until 1926 as resident engineer in all matters pertaining to the construction of the Wilson Dam, power station, and navigation locks, Muscle Shoals AL. From 1927 to 1929 he represented Cooper & Company at Kichkas, Ukraine USSR, on the construction of the Dnieper River hydro-electric development as supervising engineer. This was a position of great responsibility because of the difficult foundation conditions. He returned then to the USA because of illness, but in 1930 re-entered the Company on the Hawks Nest-Gauley Junction power development of the New-Kanawha Power Company WV, serving as resident engineer until his unfortunate death, as a result of the failure of a plate in a penstock under test. An exceptional engineering career thus found its end. Hall had built up an experience as a field engineer difficult to replace in the field of hydro-electricity. He had a deep sense of loyalty, coupled with a thorough understanding of the work assigned to him, resulting in results of highest quality. He was member of the American Society of Civil Engineers ASCE from 1925.

Anonymous (1935). John W. Hall. *Trans. ASCE* 100: 1663-1665.
Cooper, H.L. (1912). The water power development of the Mississippi River Power Company at Keokuk. *Journal of the Western Society of Engineers* 17(3): 213-226.
Clary, M. (1924). *The facts about Muscle Shoals*. Ocean Publishing Co.: New York. (*P*)
Hall, J.W. (1926). *The control of mixtures and testing of Wilson Dam concrete*. American Concrete Institute: Chicago IL.

HALL L.S.

01.02. 1892 Corning NY/USA
05.07. 1947 Oakland CA/USA

Leslie Standish Hall graduated from Massachusetts Institute of Technology as a civil engineer in 1914. He started his professional career in 1915 as a junior hydraulic engineer with the US Office of Public Roads, joining Miami Conservation District, Dayton OH, in 1916 becoming later assistant engineer until 1924 and from then chief hydraulic engineer for the East Bay Municipal Utilities District, Oakland CA. From 1936 he was hydraulic engineer and later principal hydraulic engineer until his untimely death.

During his stay at San Francisco CA from 1917 to 1924, Hall was concerned with irrigation projects, including stream flow and location of storage reservoir sites, the design of diversion dams, and the location and design of canals. From 1924 Hall then served under Arthur Powell Davis (1861-1933) for the *Mokelumne* River water supply project. In 1926, once advanced to chief hydrographer, he determined the effect of the district's diversion from this river on the water supply of the respective river basin. From 1936 Hall was involved in hydrographic surveys, hydrology, river regulation and flood control, and hydraulic research. Hall is known for his excellent work on spillways and the flow of water through pipes and orifices. His 1943 paper deals with self-aeration of high-speed flow in chutes causing water bulking because of air presence. A paper on the yearly runoff variation was awarded the 1921 ASCE Collingwood Prize for Juniors, whereas the second paper was awarded the 1943 ASCE Karl Hilgard Prize. Hall also presented during the Third Hydraulics Conference at the University of Iowa a paper on the effects of air entrainment of chute design.

Anonymous (1941). Hall, L. Standish. *Who's who in engineering* 5: 729. Lewis: New York.
Anonymous (1946). L. Standish Hall. *Civil Engineering* 16(1): 36. *P*
Anonymous (1948). Leslie Standish Hall. *Trans. ASCE* 113: 1468-1470.
Hall, L.S. (1921). Probable variation of yearly runoff as determined by California streams. *Trans. ASCE* 84: 191-257.
Hall, L.S. (1924). Discharge of irrigation siphons in California. *Engineering News-Record* 93(26): 1024-1025.
Hall, L.S. (1931). Improving the accuracy of instruments. *Civil Engineering* 1(12): 1098-1101.
Hall, L.S. (1940). Silting of reservoirs. *Journal American Water Works Association* 32(1): 25-42.
Hall, L.S. (1943). Open channel flow at high velocities. *Trans. ASCE* 108: 1394-1434; 1494-1513.
Hall, L.S. (1947). The influence of air entrainment on flow in steep chutes. Proc. 3rd *Hydraulics Conference*: 298-314, J.W. Howe, J.S. McNown, eds. State University of Iowa: Iowa.

HALL W.M.

01.03. 1860 Fayetteville TN/USA
25.11. 1951 Parkersburg WV/USA

William McLaurine Hall graduated in 1881 from the US Military Academy, West Point NY. For a period he was then engineer and chief engineer of the New Croton Aqueduct for the water supply of New York City, and later he did construction work for railroad companies in New York, Pennsylvania and Virginia States. In 1893, he joined the US Engineers remaining there in federal employment until his retirement in 1932. On the Rough River in Kentucky, he built what was believed to be the first monolithic concrete river lock in the world.

Among the notable innovations introduced by Hall during the construction of Lock and Dam No. 18 on the Ohio River were the use of a core drilling to locate the bed rock, the anchoring of concrete masonry to the bed rock, the design of a novel bear-trap weir for controlling the pool stage above the dam, and the effecting of a new type of lock power plant. These methods were a standard practice after World War II in the USA. Hall also had full field charge of building sixteen Ohio River locks and dams. By an act of Congress he was retained in service beyond retirement age and served as chairman of a commission whose report was enacted into a law under which various government appropriations were made for flood control. While he was a member of the Parkersburg Water Works Commission WV in 1907, the controversial experimental river bed sand infiltration system was built, on which he wrote the 1917 paper. For notable service in the engineering field, Hall received in 1947 the honorary degree of Doctor of Science from Marietta College, Marietta OH. He was member of the American Society of Civil Engineers ASCE from 1893, becoming Life member in 1928.

Anonymous (1952). William M. Hall. *Trans. ASCE* 117: 1301-1302.

Hall, W.M. (1902). Discussion of Improvement of rivers. *Trans. ASCE* 49: 313-314.

Hall, W.M. (1917). The water supply of Parkersburg WV. *Trans. ASCE* 81: 749-787.

Hall, W.M. (1918). Discussion of Detention reservoirs with spillway outlets as an agency in flood control. *Trans. ASCE* 82: 1501-1503.

Hall, W.M. (1923). The location and construction of locks and movable dams on the Ohio River, with particular reference to Ohio River Dam No. 18. *Trans. ASCE* 86: 92-131.

Johnson, L.R. (1974). William M. Hall. *The Falls City engineers*: A history of the Louisville District and canalization. USACE: Louisville. *P*

http://www.brennancallan.com/BandR/NolinRiverKY/index.html

http://en.wikipedia.org/wiki/File:Mississippi_River_Lock_and_Dam_No._19_rear_of_dam.jpg (*P*)

HALMOS

06.12. 1877 Kassa/H
01.02. 1959 Manhasset NY/USA

Eugene (Jenö) Erwin Halmos graduated in 1899 as a civil engineer from the Royal Joseph Technical University, Budapest H. He moved in 1910 to the USA, starting as an engineering draftsman in New York NY. He was from 1912 to 1947 design engineer and associate of the New York engineering firm Parsons, Klapp, Brinkerhoff and Douglas and successors, thereby in charge as chief engineer of the Parklap Construction Corp., New York NY, from 1920 to 1932. He further was engaged on designs of the Sherman Island, Spier Falls, Salmon River and Feeder Dam hydro-electric plants, all in New York State, and of the Sutherland Power and Irrigation Project in Nebraska. Further projects during the 1920s and 1930s included twin graving docks of the Norfolk Navy Yard, or Bayboro Harbour at St. Petersburg FL. He was also in charge of the Cape Cod Canal hydraulic design, the Bayonne Ship Terminal, harbour developments at Ciudad Bolivar and *Los Morcos* in Venezuela. During World War II he acted as chief engineer of the naval dry docks at Portsmouth RI, Boston MA, and Brooklyn NY.

After 1947 Halmos directed the preparation of a Rapid Transit Report for Sao Paulo BZ, he studied the transit situation of Zurich CH, he prepared a Port Program for Turkey, but was chiefly engaged on the design of the *Seyhan, Hirfanli, Kemer*, and *Demirkopru* hydro-electric plants in Turkey. He authored numerous papers and reports mainly in the Engineering News-Record, and the ASCE journals. He became widely known in the hydraulic community also by his translation of the book on water hammer by the Italian Lorenzo Allievi (1856-1940), who led the mathematical bases for the fundamental wave phenomena exerted in unsteady pipe flow. He thereby was a participant of the 1933 Symposium on Water Hammer, organized by the ASME Hydraulic Division, and the ASCE Power Division in New York.

Abbett, W.A., Halmos, E.E. (1957). Harbor engineering. *American civil engineering practice* 2(21): 1-175, R.W. Abbett, ed. Wiley: New York.
Anonymous (1959). E.E. Halmos, Sr., dies: River, harbour designer. *Engineering New-Record* 162(Feb.19): 194. *P*
Anonymous (1959). Eugene E. Halmos. *Civil Engineering* 29(3): 220. *P*
Anonymous (1959). Halmos, Eugene E. *Who's who in engineering* 8: 1004. Lewis: New York.
Halmos, E.E. (1925). *Theory of water-hammer*. Garroni: Roma.
Halmos, E.E. (1945). *Report* on Soil compaction by vibroflotation. New York.

HAMILTON

14.02. 1911 Palmyra NJ/USA
02.05. 2001 Rapid City SD/USA

Wallis Sylvester Hamilton earned his BS degree in 1935 and the MS degree in 1939 from Carnegie Institute of Technology, Pittsburgh PA, and his PhD degree in hydraulics and mechanics from University of Iowa, Iowa City, in 1943. He moved then as Lecturer to Northwestern University, Evanston IL, where he became in 1953 full professor, remaining there until his retirement in 1977. He was awarded the ASCE Hydraulic Structures Medal in 1990, of which he was a member from the 1950s.

Hamilton's research interests included resistance of barges, velocity patterns around ship models, spray characteristics and directional stability of boats. He also studied wave effects, damage from cavitation and the body forces and properties of the earth crust and mantle. In 1937 he joined a group of engineers who were using a scale model to develop the design flood control dams above Pittsburgh PA. This work set the direction of his future researches. His PhD thesis was concerned with the measurement of the velocity distribution around a model hull in flowing water and the determination therefrom of the surface drag for comparison with the measured total drag. The prototype channel used at the David Taylor (1864-1940) Model Basin TMB at Carderock MD had a test section 18 m long, 7 m wide and 3 m deep with a maximum flow velocity of 5 m/s. His colleagues there included Charles A. Lee (1915-1995), and James M. Robertson (1916-2012). The Wallis S. Hamilton Papers fill eleven boxes spanning the period of 1935 to 1984, comprising education files, teaching files, as well as correspondence, research and consulting files, and publications. These were donated to the University Archives, the Northwestern University.

Boyce, D. (2011). *Retrospective* on the Civil Engineering Faculty at Northwestern University 1957-1961_rev.pdf. *P*

Dundurs, J., Hamilton, W.S. (1954). *Liquid jet disintegration*. Northwestern University.

Hamilton, W.S. (1953). Water tunnel teaches Bernoulli principle. *Civil Engineering* 23(12): 834. *P*

Hamilton, W.S., Lidell, J.E. (1971). Fluid force analysis and accelerating sphere tests. *Journal of the Hydraulics Division* ASCE 97(HY6): 805-817.

Hamilton, W.S. (1983). Preventing cavitation damage to hydraulic structures. *Water Power and Dam Construction* 35(11): 40-43; 35(12): 48-53; 36(1): 42-45.

Rouse, H. (1976). Wallis S. Hamilton. *Hydraulics in the United States* 1776-1976: 145-146. Iowa Institute of Hydraulic Research: Iowa City.

http://findingaids.library.northwestern.edu/catalog/inu-ead-nua-archon-947

HAMMITT

25.09. 1923 Trenton NJ/USA
29.06. 1989 Neptune NJ/USA

Frederick Gnichtel Hammitt obtained in 1944 his BS degree in mechanical engineering from Princeton University, Princeton NJ, and the MS degree in 1949 from University of Pennsylvania, Philadelphia PA. In 1958 he made his PhD degree in nuclear engineering at University of Michigan, Ann Arbor MI. In parallel Hammitt was engaged from 1946 to 1955 in private firms dealing with reaction motors, then joined University of Michigan as research associate, was there associate professor from 1959 and professor of nuclear engineering from 1961. He was also professor in charge of cavitation and multiphase flow there from 1967, the year he was a visiting scholar to *Electricité de France* EDF, Paris, France. In 1971 he visited *Société Grenobloise d'Hydraulique* SOGREAH, Grenoble, France, and there was a Fulbright senior lecturer in 1974. Further visits were made to the Polish Academy of Sciences, Gdansk in 1976, to the People's Republic of China in 1982, and to Japan in 1986.

Hammitt was one of the outstanding experts in questions relating to cavitation and cavitation damage of the USA. He has written more than 400 papers in journals and presented at conferences, and he is particularly known for his 1970 book written in collaboration with Robert T. Knapp (1899-1957) and James W. Daily (1913-1991) on cavitation. Hammitt made significant engineering achievements using theoretical and experimental works. He contributed to the understanding of the bubble dynamics, the bubble-bubble, and the bubble-surface interactions, leading to improvements in the design mainly of hydraulic machinery. His knowledge in bubble mechanics extended to study of nucleation in boiling of liquid metals, thus contributing to the advances in nuclear power. He was awarded fellowship of the Society of Mechanical Engineers ASME in 1973.

Anonymous (1965). Frederick G. Hammitt. *Mechanical Engineering* 87(1): 104-105. *P*
Anonymous (1973). Frederick G. Hammitt. *Mechanical Engineering* 95(10): 110.
Anonymous (1980). Frederick G. Hammitt. *Mechanical Engineering* 102(2): 116. *P*
Anonymous (1985). Hammitt, F.G. *Who's who in Engineering* 6: 262. AAES: Washington DC.
Hammitt, F.G. (1963). Observations on cavitation damage. *J. Basic Engng.* 85: 347-359.
Hammitt, F.G. (1964). Cavitation damage and performance research facilities. ASME Symp.
    Cavitation Research Facilities and Techniques: 175-184.
Knapp, R.T., Daily, J.W., Hammitt, F.G. (1970). *Cavitation*. McGraw-Hill: New York.

HAMMOND H.P.

21.12. 1884 Asbury Park NJ/USA
21.10. 1953 Bellefonte PA/USA

Harry Parker Hammond obtained the BS degree in 1909 from University of Pennsylvania, Philadelphia PA, the CE title in 1915, and the hon. D.Eng. title from Case School of Applied Science, Cleveland OH, in 1931. He was from 1909 to 1911 an instructor in civil engineering at his Alma Mater, then from 1913 to 1918 an assistant professor of civil engineering at Brooklyn Polytechnic Institute, professor of sanitary and hydraulic engineering until 1927, professor of civil engineering until 1937, and from then to 1951 dean of the School of Engineering, the Pennsylvania State College. Hammond was a member of the America Society of Engineering Education ASEE, and its president in the term 1936. He also was a *Lamme* Medallist in 1945, and a recipient of the J.H. McGraw Award in 1950 for his outstanding educational qualities.

Hammond's career included both scientific and practical works. He was for instance associated with the American Bridge Co., the New York City Board of Water Supply, and the Miami Conservancy District. Dean Hammond was a Honorary Member of the American Society of Civil Engineers ASCE. His teaching career lasted for more than forty years mainly at Penn State. Since retirement he was engaged in special consulting work at Wright-Patterson Air Force Base in Ohio. He has written an educational text on hydraulics, and a number of papers on engineering education, thereby contributing to the wealth of students.

Anonymous (1931). Hammond, Harry P. *Who's who in engineering* 3: 544. Lewis: New York.
Anonymous (1953). Harry Parker Hammond. *Civil Engineering* 23(12): 868. *P*
Anonymous (1953). Dean H.P. Hammond dies. *Engineering News-Record* 151(Oct.29): 24.
Cairns, W.D. (1931). The fifteenth summer meeting of the Mathematical Association. *American Mathematical Monthly* 38(9): 487-494. *P*
Hammond, H.P. (1925). The study of engineering education. *EICJ* (3): 106-110.
Hammond, H.P. (1940). Aims and scope of engineering curricula. *Journal of Engineering Education* 30(7): 555-566.
Hammond, H.P. (1956). General education in engineering. *Journal of Engineering Education* 46(4): 619-750.
Hammond, H.P. (1962). *Hydraulics*. Intl. Correspondence Schools: Scranton PA.
http://209.85.129.132/search?q=cache:nAsQAHuzlV4J:www.mne.psu.edu/History/chapter3.html+%22H.P.+Hammond%22&cd=10&hl=de&ct=clnk&gl=ch

HAMMOND J.J.

11.11. 1887 Troy NY/USA
18.10. 1971 Denver CO/USA

John Jacob Hammond was educated at the Colorado State A&M College, from where he graduated as a civil engineer in 1916. He then joined until 1920 the US Bureau of Reclamation USBR, being engaged with the Huntley Irrigation Project in Montana. He was from then to 1923 assistant design engineer for the Shoshone Project in Wyoming, thereby becoming acquainted with large dam projects, joining then the USBR Office of Engineering, Denver CO, where he mainly designed earthfill and concrete dams for the West, including outlet works, spillways and energy dissipators. Hammond was promoted in 1931 to a senior engineer in charge of concrete dams, and later to engineer in charge of all dam designs, continuing as senior engineer until 1955. He was also Head of the Concrete Dam Design Section, USBR, from 1952, retiring in 1955. From then, Hammond was a consultant. He was a member of the American Society of Civil Engineers ASCE from 1929, becoming an ASCE Fellow in 1959.

Hammond was for almost forty years with the Bureau, thereby having supervised the design of nearly each major of its concrete dams. This included structures of the calibre of the Grand Coulee Dam in Washington State, inaugurated in 1941, then the largest dam worldwide; Hoover Dam between Nevada and Arizona, formerly Boulder Dam, inaugurated in 1935 and until 1946 with 221 m the highest dam worldwide; or Shasta Dam in California, inaugurated in 1945 and one of the remarkable arch dams 155 m high and 1050 m long. Hammond was also a consultant for the governments of Taiwan with the *Wu-Sheh* Dam, and India with the *Bahkra* Dam, where similar large structures were later erected. He also directed dam designs in Panama Canal Zone and Australia. As a member of the International Commission of Large Dams ICOLD, he attended its 1955 Congress in Paris as an US delegate. He was also a member of the International Commission of Irrigation and Drainage ICID, thus stating his interest in combined dam structures for irrigation, hydropower, flood retention and drinking water supply.

Anonymous (1955). Designer of dams retires. *Engineering News-Record* 155(Sep.1): 59-60. *P*
Anonymous (1959). Hammond, John J. *Who's who in engineering* 8: 1012. Lewis: New York.
Anonymous (1972). John Jacob Hammond. *Trans. ASCE* 137: 1061-1062.
Hammond, J.J., Noonan, N.G. (1948). *Inspection program* for Bureau of Reclamation structures and facilities. US Dept. of Interior, Bureau of Reclamation: Denver.
Hammond, J.J. (1954). Design criteria for dams. *J. American Concrete Institute* 25(8): 657-668.

HANNA

16.09. 1867 Geneseo IL/USA
26.01. 1944 Webster City IA/USA

Frank Willard Hanna received education at Highland Park College, Des Moines IA, graduating as a civil engineer in 1894. He was from 1895 to 1902 there Dean, Civil Engineering Department, hydrographer of the US Geological Survey from 1903 to 1905 at Chicago IL, engineer of the US Reclamation Service at Washington DC from 1906 to 1908, then project engineer at Boise ID until 1912, when moving as a supervising engineer to Phoenix AZ until 1915, and as a consulting engineer to Ankeny IA until 1917. Hanna rejoined until 1921 the Reclamation Service, when becoming general manager of the Canada Land & Irrigation Company, Medicine Hat AB, supervising there both the design and the construction of irrigation works. From 1924 to 1928 he was a hydraulic engineer of the East Bay Municipal District, Oakland CA, and from then its chief engineer.

Hanna's career was devoted to irrigation of the West by designing dams and canals. He was the inventor of the angle multi-sector irrigation water meter, and the automatic stop and relief valves. He authored the books Measurement of irrigation water, Agriculture by irrigation, The agricultural value of peat soils, and The design of dams, written in collaboration with a colleague. He also presented a paper on the design of Pardee Dam on *Mokelumne* River, California. Pardee Dam was then with 100 m one of the highest of its type in the world. The water is transported from Pardee Reservoir across the Central Valley via three large pipe aqueducts to the hills east of San Francisco Bay to supply drinking water to the East Bay region. Hanna was a member of the American Water Works Association AWWA, and the American Society of Civil Engineers since 1913.

Anonymous (1931). Hanna, Frank W. *Who's who in engineering* 3: 547. Lewis: New York.
Anonymous (1946). Frank Willard Hanna. *Trans. ASCE* 111: 1474-1475.
Hanna, F.W. (1905). River discharge, mean velocity. *Engineering News* 53(12): 301-302.
Hanna, F.W., Hoyt, J.C. (1906). Report on progress of stream measurements: Hudson Bay and Upper Eastern and Western Mississippi drainages. *Water Supply and Irrigation Paper* 169. US Geological Survey: Washington DC.
Hanna, F.W. (1913). *Measurement of irrigation water*. US Bureau of Reclamation: Denver CO.
Hanna, F.W. (1928). Designing the high storage dam for the Mokelumne Project. *Engineering News-Record* 100(11): 444-446.
Hanna, F.W., Kennedy, R.C. (1938). *The design of dams*. McGraw-Hill: New York.
Kolupaila, S. (1961). F.W. Hanna. *Journal of the Hydraulics Division* ASCE 87(HY3): 179. *P*

HANTUSH

12.12. 1921 Hit/IQ
14.01. 1984 Kuwait City/KU

Mahdi Salih Hantush, born in Iraq, received in 1942 the degree of civil engineer from the American University in Beirut, Lebanon. After having been superintendant with the Irrigation Department in Baghdad, he came to the USA to pursue graduate studies. He received the MS degree in irrigation engineering in 1947 from University of California, Berkeley, and the PhD degree from the University of Utah, Salt Lake City UT, in 1949. Hantush then joined the Faculty of Baghdad University, serving as professor of irrigation and hydraulic engineering, and then as Dean of the College of Engineering until 1958. Hantush later joined the New Mexico Institute of Mining and Technology NMIMT as senior hydrologist and professor of hydrology, thereby establishing one of the nation's first graduate degree programs in groundwater hydrology.

Hantush was a devoted hydrologist, scientist and great teacher, who had specialized in the application of mathematics to the solution of transients groundwater flow problems. His particular expertise of well-flow equations led to refer to him as the Master of radial flow. His numerous publications contributed to the current theories of flow in leaky aquifers, unconfined aquifers, and anisotropic aquifers. He devised methods for the analysis of pumping-test data to determine their hydraulic properties. Hantush's treatise Hydraulics of wells became a must reference in groundwater hydrology. He was presented the 1968 Meinzer Award from the Geological Society of America GSA, and was a Fellow of the American Society of Civil Engineers ASCE.

Anonymous (1984). Mahdi S. Hantush. *Eos* 65(21): 361. *P*

Hantush, M.S., Jacob, C.E. (1960). Flow to an eccentric well in a leaky circular aquifer. *Journal of Geophysical Research* 65(10): 3425-3431.

Hantush, M.S. (1960). Modification of the theory of leaky aquifers. *Journal of Geophysical Research* 65(11): 3713-3725.

Hantush, M.S. (1962). Drawdown around a partially penetrating well. *Trans. ASCE* 127(1): 268-283.

Hantush, M.S. (1964). Hydraulics of wells. *Advances in hydroscience* 1: 281-432, V.T. Chow, ed.

Hantush, M.S. (1976). Pompage d'essai dans un puits à proximité d'une rivière colmatée.
    Bulletin Bureau Recherches Géologiques et Minières Ser. 23(3-4): 139-149 [in French].
http://209.85.129.132/search?q=cache:KRXroXMfKkoJ:www.ees.nmt.edu/hantush/+%22Mahd
i+Salih+Hantush%22+1921-&cd=1&hl=de&ct=clnk&gl=ch *P*

HARDER

02.12. 1926 Fullerton CA/USA
31.12. 2006 Talequah OK/USA

James Albert Harder graduated in 1948 with the BS, in 1952 with the MS and 1957 with the PhD degree from University of California, Berkeley CA. After having been design engineer with Soil Conservation Service SCS from 1948 to 1950, he was resident engineer at Berkeley CA from 1952 to 1957, from when he took over as assistant, in 1962 as associate, and from 1970 to 1991 there as professor of civil engineering. Harder was a Fellow of the American Association for the Advancement of Science AAAS, and of American Society of Civil Engineers ASCE.

Harder is known for his great hobby, the Unidentified Flying Objects UFOs, although his distinguished career as a hydraulic engineer and teacher. The latter included elementary fluid mechanics, advanced hydraulics, design of hydraulic structures and systems, computer programing, and hydraulic laboratory practice. His research in fluid mechanics was broad-ranging and innovative. His publications, though not numerous, were stated to be of high quality. Topics included river bends, sustained swimming of dolphins, analogue simulators for flood control systems, the identification of time-dependent non-linear hydraulic systems, and automatic control of irrigation control gates. He developed already in the 1950s electric analogue models simulating flows within San Francisco Bay, rating curves of rivers, non-linear storage effects within river levees, and reservoirs in which the water surface area was non-linearly related with the depth and from which programmable flow releases could be made. This development was a major advance in these years. He devised a system used by the US Corps of Engineers to control flooding of the Kansas River, and he further devised a system for the California Department of Water Resources for the Delta to combat salinity intrusions. His 1957 work received the Best Paper Award from the American Water Works Association.

Banks, H.O., Richter, R.C., Harder, J.A. (1957) Sea water intrusion in California. *Journal of the American Water Works Association* 49(1): 71-88.
Harder, J.A. (1957). Ground water development: Basin recharge. *Trans. ASCE* 122: 486-488.
Harder, J.A., Mockros, L., Nishizaki, R. (1960). Flood control analogs. Hydraulic Laboratory, *Water Resources Center Contributions* 24. University of California: Berkeley.
Harder, J.A. (1963). Analog models for flood control systems. *Trans. ASCE* 128(1): 993-1004.
Harder, J.A., Nelson, J.O. (1966). Analog modeling the California Delta Tidal System. *Journal of the Hydraulics Division* ASCE 92(HY4): 1-10; 93(HY2): 77-78.
http://www.universityofcalifornia.edu/senate/inmemoriam/jamesharder.html *P*

HARDESTY

05.04. 1865 Lincoln MO/USA
12.08. 1944 Portland OR/USA

William Preston Hardesty received his education at the University of Missouri, Columbia MO, obtaining the civil engineering degree in 1886. Until 1888 he was engaged with the Interstate Consolidated Rapid Transit Railway Company, Kansas City MO, from when he joined the Denver Tramway Co. until 1890. He was in 1890 assistant engineer on construction of the Colorado Canal for the Denver and Rio Grande Railway Co., in 1891 assistant engineer for the T.C. Henry Canal Co., San Luis Valley CO, and then was employed by a Gold Company in Idaho. From 1893 to 1903 Hardesty had a private practice at Salt Lake City UT; he was in 1903 appointed assistant engineer in the US Reclamation Service.

During his stay at Portland OR, Hardesty was, among many other involvements, also in charge of the design and construction of a new water works station. The 1908 paper describes earlier works for this oldest city in the Pacific Northwest, which was settled around 1850. The city acquired in the 1880s, when the population amounted to some 25,000 inhabitants, the entire system of the Portland Water Co. During the next decades various divisions of the water supply scheme were constructed, based on a standard gravity supply. Next, a number of reservoirs was erected, to distribute the water and to improve the scheme hydraulically. James D. Schuyler (1848-1912) was consultant for these works. On Hardesty's grave plate it is stated: 'He devoted his life to his work, and to the out-of-doors and the friends he made there'. He was indeed a great lover of nature, as is documented by various activities in the mountains of California and Oregon States. The so-called *Mazama* Club organized also lectures on topics including botany, geology, ornithology, and local history. The attendance during these activities was at times larger than 100 persons, with Hardesty as a lecturer on various topics. He also organized climbs to the nearby mountains with explorations of attractive spots.

Anonymous (1916). William P. Hardesty. *Engineering News* 76(2): 55. *P*
Hardesty, W.P. (1903). Utilization of Utah Lake as reservoir. *Engineering News* 49(21): 442-445.
Hardesty, W.P. (1906). Truckee-Carson irrigation project. *Engineering News* 56(16): 391-401.
Hardesty, W.P. (1907). The water and electric power systems of the Portland Railway Light and Power Company, Portland OR. *Engineering News* 57(26): 699-705.
Hardesty, W.P. (1908). Water-works of Portland OR. *Engineering News* 59(6): 137-141.
Hardesty, W.P. (1919). Mazama activities for the past year. *Sierra Club Bulletin* 10: 231-232.

HARDIN

17.06. 1897 Baltimore MD/USA
06.01. 1993 Talbot MD/USA

John Ray Hardin graduated from the US Military Academy, West Point NY, in 1919, and then was commissioned by the US Army Corps of Engineers. He was from 1934 to 1938 in charge of the spillway construction at Fort Peck Dam. In 1939 he began a tour in the Office of the Chief Engineer OCE serving as chief of the Rivers and Harbours Section, becoming in 1943 Deputy Chief Engineer. After the War he was associated with the Mississippi Valley Division. He was awarded the 1954 Toulmin Award of the Society of American Military Engineers SAME, then as Brigadier General.

Hardin became president of the Mississippi River Commission in 1953. He immediately faced the Old River Problem, and decided to do things quickly, because 'To postpone the task is to risk a greatly increased cost and maybe an impossible construction task in the eventual closure'. A Report was prepared in which the following three items were highlighted: 1. The cost and the extent of flowage easements over the entire Red River backwater area, exclusive the Tensas-Cocodrie area, that might be subject to overflow; 2. Engineering feasibility and economic justification of partially protecting additional lands within the backwater area; and 3. Advisability of designating some or all of the backwater area as a National Reservation, such as a Wild Life Preserve, or a National Forest. This Report was Hardin's highest priority. He developed a schedule, stressing that the report should assume that the control structures would be operated in a manner 'which will preserve as nearly as possible the present natural distribution of flow in the affected area'.

Fine, L., Remington, J.A. (2003). John R. Hardin. *The Corps of Engineers*: Construction in the United States. US Army: Washington DC. *P*

Hardin, J.R. (1937). Fort Peck Dam spillway. *The Military Engineer* 29(1/2): 24-28.

Hardin, J.R., Booth, Jr., W.H. (1953). Lake Michigan erosion studies. *Trans. ASCE* 118: 39-60.

Hardin, J.R. (1954). Mississippi-Atchafalaya diversion problem. *The Military Engineer* 46(3/4): 86-92.

Hardin, J.R. (1958). The general problem, in Old river diversion control: A symposium. *Trans. ASCE* 123: 1131-1141.

Reuss, M. (1998). *Designing the bayous*: The control of water in the Atchafalaya Basin 1800-1995. US Army Corps of Engineers: Alexandria VA. *P*

http://www.mvd.usace.army.mil/About/MississippiRiverCommission(MRC)/PastMRCPresidents.aspx *P*

HARLAN

22.11. 1806 Long Island NY/USA
06.02. 1883 Wilmington DE/USA

In 1837, the cabinetmaker Samuel Harlan joined the firm Betts, Pusey & Harlan, manufacturing freight cars for 2 years. In 1841 Pusey was bought out, and in 1849 Betts withdrew from the firm, which then became Harlan & Hollingsworth. Their experience with railcars and other ironwork led to become early experimenters in iron shipbuilding. In 1843, under the encouragement of Harlan, the company started engaging in marine engine building and repair. Their first ship-related project was repairing the steamboat *Sun*. This small step was the beginning of what would become one of the first iron shipyards in the United States.

Harlan & Hollingsworth's expanded steadily into iron shipbuilding. Only nine ships were built between 1841 and 1851. In 1843 the company leased a launching berth on the banks of Christiana River. Their facilities were limited, so all the work forming iron plates, bars, and fasteners was done at their main shop at Wilmington DE. The launch slipway was 60 m long accommodating only vessels of up to 600 tons, but this was deemed adequate for the needs of the time. The first two hulls built by the company, the *Ashland* and *Ocean*, were two of the earliest iron steamboats constructed in the USA, being delivered in 1844. That same year the company built *Bangor*, which is credited with the first seagoing iron propeller steamship built in the USA. By the early 1850s the company began to rely less on wood ship building for its income. Machine shops, office buildings, wharves, carpenter sheds, boiler shops, blacksmith shops and cranes were added in the first five years of the decade. As the firm's reputation grew, more orders for steamboats came in from across the country. A New York shipping magnate purchased his first ship from Harlan in 1856; he eventually would become one of the largest customers for Harlan & Hollingsworth, ordering over 31 vessels by 1878. During the Civil War the company won contracts for the construction of three monitors for the government. Hollingsworth's death greatly affected Harlan, so that the partnership was thereafter dissolved and the enterprise incorporated as The Harlan & Hollingsworth Company. Bethlehem Steel acquired the firm in 1904.

Anonymous (1898). *The Harlan & Hollingsworth Co.*, Ship and Car Builders. Philadelphia.
Hoffecker, C.E. (1974). *Wilmington DE*: Portrait of an industrial city 1830-1910. University
Press: Virginia. *P*
http://en.wikipedia.org/wiki/Harlan_and_Hollingsworth

2128

HARLEMAN

05.12. 1922 Palmerton PA/USA
28.09. 2005 Nantucket MA/USA

Donald Robert Fergusson Harleman obtained his MS degree from MIT in 1947, and the degree DSc. there in 1950, when he was a research assistant at the MIT Hydrodynamics Laboratory. From 1950 to 1956 he was assistant professor, then until 1962 an associate professor, and from then until retirement in 1991 professor of civil engineering first, later Ford Professor of Engineering. In parallel Harleman was from 1972 to 1983 head of the Water Resources & Environmental Engineering Division. He was from 1968 to 1969 senior visitor at Cambridge UK, and in 1977 visiting scientist of the International Institute of Applied Systems Analysis, Vienna, Austria. He was a member of the Water Pollution Control Federation WPCF, and of the International Association of Hydraulic Research IAHR. He was named in 1979 an outstanding alumnus of the Engineering College, Pennsylvania State University. He was awarded the 1960 ASCE Research Prize, the 1971 ASCE Karl Hilgard Hydraulic Prize, the 1973 ASCE J.C. Stevens Award, and ASCE's 1983 W.W. Horner Award. Harleman was elected to ASCE Honorary Member in 1989. The D.R.F. Harleman Lecture was established in 2001.

Harleman was a recognized international engineer, scientist and educator, whose research and innovations were directed toward improving water quality and making wastewater treatment available and affordable to all. He brought his expertise and enthusiasm to the next generation through leadership at the Institute. His is further known by the book Fluid dynamics, co-authored by James W. Daily (1913-1991). It may be compared with works of the calibre of Boundary layer theory by Hermann Schlichting (1907-1982).

Anonymous (1966). D.R.F. Harleman. *Civil Engineering* 36(6): 52. *P*
Anonymous (1982). First Hunter Rouse Hydraulic Engineering Lecture. *Journal of the Hydraulics Division* ASCE 108(HY3): 301. *P*
Anonymous (1995). Harleman, Donald R.F. *Who's who in America* 49: 1571. Marquis: Chicago.
Daily, J.W., Harleman, D.R.F. (1966). *Fluid dynamics*. Addison-Wesley: Reading MA.
Harleman, D.R.F. (1964). The significance of longitudinal dispersion in the analysis of pollution in estuaries. Proc. 2nd Intl. Conf. *Water Pollution Research*, Tokyo: 279-306.
Ippen, A.T., Harleman, D.R.F. (1956). Verification of theory of oblique standing waves. *Trans. ASCE* 121: 678-694.
http://209.85.129.132/search?q=cache:pig8ZWtMeeUJ:web.mit.edu/newsoffice/2005/obit-harleman.html+%22Donald+Harleman%22&cd=1&hl=de&ct=clnk&gl=ch *P*

HARPER J.L.

21.09. 1873 Harpersfield NY/USA
28.11. 1924 Niagara Falls NY/USA

John Lyell Harper graduated in 1897 from Cornell University, Ithaca NY, as mechanical engineer. In 1898 he became operating and construction engineer of the Twin City Rapid Transit Company, where he learned experimentation with electric and hydraulic machinery. From 1899 he was in charge of a project on the hydro-electric power plant of the St. Croix Power Company, Appleriver WI. In 1902 he moved to Buffalo NY, accepting a position with the Niagara Falls Hydraulic Power & Manufacturing Company as assistant to chief engineer Wallace C. Johnson (1859-1906). It was there where his principal work started, and the story of his life was intimately associated with the history of development of water power until his death.

This Company started in 1881 with the construction of power house Station 1. In 1902, when Harper joined the Company, it had a capacity of 2,000 HP, and power house Station 2 had partly been constructed, with a capacity of 20,000 HP, to be completed in 1904, so that Station 1 was abandoned but the construction of Station 3 was initiated for a capacity of 130,000 HP. However, in 1909, the Governments of Canada and the USA set a limit to the water discharge to be diverted to hydro-electric power production. In World War 1, the Company and the Cliff Electrical Distribution Company were merged under the name The Niagara Falls Power Company, owning all developments on the American side and controlling the Canadian Niagara Power Company on the other side. The combined plants had a total capacity of 350,000 HP, which was a fitting tribute to Harper's vision and ability, who was appointed vice-president in 1919 in addition to chief engineer. Harper was thereby especially interested to preserve the scenic beauty of the Falls. He built a model of the Niagara Falls and Rapids on a scale large enough to determine the effect of diversion and published the results of these studies. He was a man of high ideals and integrity, a loyal friend and was devoted to his family.

Anonymous (1925). John L. Harper. *Cornell Alumni News* 27(14): 174. *P*
Anonymous (1926). John L. Harper. *Trans. ASCE* 89: 1610-1613.
Harper, J.L. (1916). *The suicide of the Horseshoe Fall*. Courter: Niagara Falls NY.
Harper, J.L. (1920). Niagara Falls 100,000 HP development. *Electrical World* 16(12): 561-564.
Harper, J.L., Johnson, J.A. (1921). Hydroelectric development at Niagara Falls. 37th *Annual Convention of the American Institute Electrical Engineers* Salt Lake City: 881-923.
Mead, D.W. (1915). *Water power engineering*. McGraw-Hill: New York.

HARPER S.O.

06.06. 1883 Pacific Grove CA/USA
25.05. 1966 Alameda County CA/USA

Sinclair Ollason Harper graduated from University of California, Berkeley CA, with the BS degree in civil engineering in 1907. He was the recipient of a honorary degree ScD in 1940 from the University of Colorado, Denver CO. He was from 1907 engineer in charge of the design of the sewerage system of Montrose CO, remaining from 1907 all through his career with the US Bureau of Reclamation USBR. He served from 1908 to 1917 as assistant engineer of the Grand Valley Project in western Colorado, in charge of topographic and location surveys, designs and construction of its important features of this irrigation system. He was then until 1925 project manager of the Project in charge of its construction and operation, becoming until 1940 general superintendent and assistant chief engineer at Denver CO, in general charge of construction work of the USBR in 15 western States, including Boulder Dam, todays Hoover Dam, the All-American Canal in California, and Grand Coulee Dam on Columbia River. He was appointed in 1940 chief engineer of the USBR succeeding Raymond W. Walter (1873-1940). From 1935 to 1944 Harper was in parallel representative of the President, and chairman of the Commission negotiating the Compact between Colorado, New Mexico and Texas States for the water diversion of Rio Grande.

Since 1945 Harper was consulting engineer in private practice at Oakland CA, reporting on large hydropower and irrigation projects in India and Afghanistan. He also served on boards for the US Army Corps of Engineers, and on the International Board for the Egyptian Government, reporting on the hydro-electric power development of Aswan Dam on the Nile River. He was further involved in dam projects at the *Damodar* Valley, Calcutta IN, and on developments at Baghdad IQ. Harper was member of the American Society of Civil Engineers ASCE, the Colorado Society of Engineers, and the Society of Engineers, San Francisco CA.

Anonymous (1940). Harper named chief reclamation engineer. *Engineering News-Record* 125(Jul.18): 76. *P*
Anonymous (1944). S.O. Harper retires. *Engineering News-Record* 133(Dec.14): 735. *P*
Anonymous (1959). Harper, Sinclair O. *Who's who in engineering* 8: 1033. Lewis: New York.
Harper, S.O. (1966). *Sinclair O. Harper papers*. USBR: Denver CO.
Hundley, Jr., N. (1975). *Water and the West*: The Colorado River Compact and the politics of water in the American West. University of California Press: Berkeley CA.

HARRIS C.W.

17.11. 1880 Boisfort WA/USA
23.03. 1973 Seattle WA/USA

Charles William Harris obtained the BS degree from University of Washington, Seattle WA, in 1903, and the CE degree from Cornell University, Ithaca NY, in 1905. He was from 1906 the acting head of the Department of Civil Engineers of his Alma Mater, from 1915 to 1924 associate professor of hydraulic engineering, taking over as professor until 1951. He also was a member of its Engineering Experiment Station, and associated with local consultants such as a lumber company, or a fisheries commission.

Harris was interested in the questions of technical hydraulics, practical hydrodynamics and engineering economics. He published mainly in the Bulletins of the Engineering Experiment Station of Washington University, with a particular emphasis on weir and orifice flows. He also considered the effect of pipe thickness on the intake losses, a topic that received increased interest as larger hydraulic schemes were erected. He further presented three books in hydraulics, namely the Fundamentals, and two editions of Hydraulics. The 1936 edition is described as: 'The object of this book is to present the various principles of hydraulics in the relationship most commonly encountered in engineering practice. Earnest effort is made not only to promote the wholesome respect for a scientific and historic background, but also to encourage ample consideration of the needs of modern industry'.

Anonymous (1933). Harris, C.W. *American men of science* 5: 475. Science Press: New York.
Anonymous (1948). Harris, Charles W. *Who's who in engineering* 6: 844. Lewis: New York.
Harris, C.W. (1923). An analysis of the weir coefficient for suppressed weirs. *Bulletin* 22. Engineering Experiment Station, University of Washington: Seattle.
Harris, C.W. (1926). Pressure reduction on the face of orifice plates and weirs. *Bulletin* 35. Engineering Experiment Station, University of Washington: Seattle.
Harris, C.W. (1928). The influence of pipe thickness on re-entrant intake losses. *Bulletin* 48. Engineering Experiment Station, University of Washington: Seattle.
Harris, C.W. (1931). Constant flow characteristic of the plane orifice in proximity to side walls. *Bulletin* 56. Engineering Experiment Station, University of Washington: Seattle.
Harris, C.W. (1933). *Fundamentals of quantitative hydraulics*. University of Washington: Seattle.
Harris, C.W. (1936). *Hydraulics*. Wiley: New York.
Kramer, A. (1998). 1898-1998: *One hundred years of excellence in education*. UW: Seattle. *P*
Poggendorff, J.C. (1936). Harris, Charles William. *Biographisch-Literarisches Handwörterbuch* 6: 1030; 7b: 1856; 8: 1467. Verlag Chemie: Leipzig, Berlin, with bibliography.

HARRIS E.G.

27.06. 1861 Spartanburg SC/USA
21.12. 1944 Rolla MO/USA

Elmo Golightly Harris graduated from University of Virginia, Charlottesville VA, in 1882 with the civil engineering degree. He was in charge of surveys of the Missouri Pacific Railroad Company then, and later was employed by the Cotton Belt Railroad Co., in charge of rebuilding a bridge over Arkansas River at Pine Bluff AR. New piers had to be built under the old bridge while traffic was maintained. These were placed by the pneumatic process; in studying the process of blowing out the sand with compressed air, Harris discovered the principles of the 'air-lift pump'. He thus became interested in compressed air flow and later published a textbook on Compressed air. He also wrote in 1895 an article on the Theory of the air-lift pump. A patent was granted shortly later. He designed then the first water works and sewerage system of Rolla MO, with the air-lift pump used on the deep-well water supply installation. He later developed a return air pump which was widely used until the development of the electrically-driven centrifugal pump.

Harris joined in 1891 the Faculty of Missouri School of Mines and Metallurgy, Rolla MO, as civil engineering professor. He specialized in mining, hydraulics, and water supply. He published a treatise on centrifugal pumps, and developed a chart for the Manning formula, thereby replacing many tables by a simple diagram. He further published in the Bulletins of his School. He also called attention to the possibility of a hydro-electric power plant on Ouachita River in Arkansas. He was made emeritus professor in 1931; a new building on the university campus housing the civil engineering Department was named Harris Hall in his honour. He stated in his memoirs: 'The period of rugged individuality, wholesome achievement, peace and security, in which I have lived is past. We are at a turning point in human affairs'. He was member of the American Society of Civil Engineers ASCE from 1901, and became life member in 1932.

Anonymous (1939). Elmo G. Harris. *Civil Engineering* 9(1): 67. *P*
Anonymous (1945). Elmo G. Harris. *Trans. ASCE* 110: 1717-1720.
Harris, E.G. (1895). Theory of the air-lift pump. *Journal of the Franklin Institute* 140(1): 32-52.
Harris, E.G. (1903). Theory of centrifugal pumps and fans: Analysis of their action, with suggestions for designers. *Trans. ASCE* 51: 166-252.
Harris, E.G. (1910). *Compressed air*: Theory and computations. McGraw-Hill: New York.
Harris, E.G. (1915). Orifice measurement of air in large quantities. *Bulletin* 2(2). School of Mines and Metallurgy. University of Missouri: Rolla MO.

HARRIS J.H.

04.01. 1838 Troy NY/USA
22.01. 1894 New York NY/USA

John Henry Harris was educated at Springfield MA with the intention of entering Yale College, but this was abandoned, so that he went at age 18 to sea, making voyages to Japan and China. On his return, he was shipwrecked on the Nova Scotia coast, where he almost drowned. He arrived in New York in 1861, but his patriotism prompted him to enlist as Navy volunteer. He was assigned to USS *Albatross*, and immediately entered upon active service in the Civil War, thereby participating in the battle of Port Hudson. He was discharged in 1866, and decided to study law at Worcester University, but engaged soon in the mercantile business.

The first connection of Harris with the pump business was given when he was in charge of the Blake Manufacturing Co., New York. He there had for many years charge of the New York Company, and also went abroad to introduce the Blake pumps in Europe. Harris later organized the Harris Steam Pumping Co. In 1882 he became connected with the firm of Henry R. Worthington (1817-1880), who had patented the Worthington pump in 1850. In 1884, Harris went to London UK, to establish the Worthington Co. with branches throughout England, the English Colonies, and the East. In 1885 an order was awarded to the Worthington Pumping Engine Co. for high pressure pumps to supply the British Army in Sudan with water. This caused the outcry which brought the matter to the notice of another British Pumping Company, who had secured sole rights for manufacture the Worthington pumps. Finally, however, this led to an association of the two companies, which in 1903 led to the merger Worthington & Simpson. It should be noted that Worthington invented in 1845 the world's first direct-acting steam pumping engine, subsequently used to power canal boats and vessels. Due to his failing health, Harris had to return to the USA, becoming there chairman of the Worthington Co. He was member of the Institutions of Mechanical Engineers UK, the Naval Architects, and the Society of Arts, London, as also of the American Society of Mechanical Engineers ASME. He was described as a person earnest of purpose, honest of heart, strong in his convictions, sincere in his friendships, and manly in all his relations.

Anonymous (1894). John H. Harris. *Trans. ASME* 15: 1188-1189.
Anonymous (1894). John H. Harris. *Engineering Record* 29: 134.
Cronise, E.S. (1894). John H. Harris. *Cassier's Magazine* 6(31):84-86. *P*
http://www.gracesguide.co.uk/Worthington_Pumping_Engine_Co

HARRIS W.A.

02.03. 1835 Woodstock CT/USA
29.10. 1896 Providence RI/USA

William Andrew Harris came with his parents to Providence RI while being a child, and in 1840 they moved to North Adams MA. At age 11 he returned to Providence, where he since resided. In 1855, he engaged in the employ of the Providence Forge and Nut Company, later the Providence Tool Company, as draughtsman. The following year he accepted a similar position with the Corliss Steam Engine Co., founded by George H. Corliss (1817-1888). Here he remained eight years. From 1864, he manufactured the Corliss engine on his own account, paying the inventor Corliss a stipulated royalty. He occupied at first an old building carrying the business there four years. In 1869 he exhibited one of his 'Corliss Engines' at the American Institute in New York City. The 'New York Tribune' gave it the name 'Harris-Corliss Engine'. Since 1870, the date when the patent on the Corliss engine expired, Harris manufactured it, with his own and other patented improvements, under the name originally given it by the 'Tribune'.

Harris started his later extensive works west of the Union Railroad station in 1868. The buildings consisted of a machine shop, blacksmith shop, iron foundry, brass foundry, pattern shop, and pattern storehouse. A large force of skilled workmen was employed in the establishment, the most amicable relations existing between the employer and the employees, 'strikes' being an unheard of thing. A large part of the machinery and tools were invented and made especially for these works, the product of which consisted of stationary engines varying from 20 to 2,000 HP. The establishment was capable of turning out half a million dollars' worth of merchandise annually, which was shipped in addition to the USA also to Cuba, Mexico, and Spain. Harris made his works an industrial training school of the best, covering a period of 3 years as an apprenticeship. During this time the young person was taught to execute the complex work, so that he became a master of a good trade. These workmen have thus been instructed under direct supervision of his superintendent and foremen, thereby securing skilled mechanics. Each young person thoroughly understood what was expected of him, and upon compliance therewith merits and receiving the approbation of the proprietor.

Anonymous (1896). William A. Harris. *Power* 16(9): 23.
Anonymous (1896). William A. Harris. *Engineering Record* 34: 399.
Harris, W.A. (1869). Harris, builder of the Corliss Steam Engine. Hammond&Angell: Providence.
http://www.newsm.org/steam-engines/william_a_harris.html *P*

HARROD

19.02. 1837 New Orleans LA/USA
07.09. 1912 New Orleans LA /USA

Benjamin Morgan Harrod graduated with the BA degree from Harvard University, Cambridge MA, in 1856. He then studied engineering and architecture at New Orleans LA, becoming draftsman in the US Engineering Office in charge of constructions of lighthouses on the Gulf Coast from Mississippi to the Rio Grande Rivers. He began in 1859 practice in New Orleans, but in 1861 was enlisted as private in the Confederate Army. After Civil War service, he resumed professional activities at New Orleans until 1877. He was then appointed chief State engineer of Louisiana, designing the system of levees to protect the alluvial regions of the State from overflow until 1880. He was from 1879 to 1904 engineering member of the Mississippi River Commission, thereby surveying the Mississippi River and its tributaries, and improving the main streams from the junction of the Ohio River. He also was from 1888 to 1892 chief city engineer of New Orleans, including advisory engineer for the drainage, sewerage, and water works systems of the city. From 1897 to 1902 he was in charge of the design and construction of the drainage system. Harrod was appointed as one of the first US delegates to the Permanent Intl. Association of Navigation Congresses PIANC in 1903, but he was unable to attend the congress. He also was appointed member of the Panama Canal Commission by President Roosevelt from 1904 to 1907 to determine the type of canal. He was from 1906 again consulting engineer at New Orleans until his death.

Harrod was a member of the American Society of Civil Engineers ASCE, director from 1892 to 1894, vice-president from 1895 to 1896, and ASCE President in 1897. He was awarded the LL.D. degree from Tulane University, New Orleans LA, acted as vice-president of the Association of Harvard Engineers, and was member of the Louisiana Engineering Society. He also supported the conservation of natural resources through his work in the Audubon Society, and maintained a notable collection of art.

Anonymous (1912). Benjamin Morgan Harrod. *Engineering News* 68(12): 554. *P*
FitzSimons, N., ed. (1992). Benjamin M. Harrod. *A biographical dictionary of American civil engineers* 2: 47. ASCE: New York.
Harrod, B.M. (1903). The levee theory on the Mississippi River. *Trans. ASCE* 51: 331-344.
Shaw, A., ed. (1905). Mr. Benjamin M. Harrod. *The American Monthly Review of Reviews* 31: 519. *P*
http://en.wikipedia.org/wiki/Benjamin_Morgan_Harrod

HARROLD

09.04. 1905 Kokomo IN/USA
02.04. 1983 Cleveland OH/USA

John Coate Harrold obtained the BS degree in civil engineering from Harvard University in 1927. He further made a graduate course in fluid mechanics at Colorado State University in 1940. He was from 1927 to 1929 an instructor at Harvard University, and then until 1960 from junior engineer to project engineer with the Army Corps of Engineers, in the offices at Baltimore MD, St. Louis MO, Nashville KY, Iowa City IA, and Washington DC. He was from 1951 to 1954 also on the Panel of Oceanography of the National Defence Research and Development Board. He became a private consultant from 1960, dealing with general civil structures at Royal Oak MI. Harrold was a member of the American Society of Civil Engineers ASCE, the International Association of Hydraulic Research IAHR, and the Permanent International Association of Navigation Congresses.

Harrold had a 31-years' experience in the design of navigation locks and dams, flood control dams and channels, hydro-electric power plants, sewers and drainage works. He was a true designer of hydraulic structures in the Midwest of the USA, with experience mainly in details of engineering construction and development. In the Office of the Chief of Engineers, he also had to supervise the designs of the civil work structures executed by the Corps, including the direction of its hydraulic research programs. The 1947 paper deals with cavitation damage, which was discovered in 1935 on the Madden Dam, Panama Canal. Since 1941 the Corps of Engineers was engaged in the research of cavitation damage, thereby mainly studying rapidly varying pressure in high-speed flow. A particular attention was paid to gate locks, baffle blocks, siphons, and intakes. The effects of cavitation damage in hydraulic engineering were studied mainly in the 1960s.

Anonymous (1939). First Iowa *Conference of Fluid Mechanics*: Frontispiece. *P*
Anonymous (1964). Harrold, John C. *Who's who in engineering* 9: 772. Lewis: New York.
Harrold, J.C. (1935). Outline of the general principles used in the design of dams on the Upper Mississippi River. *Engineering as applied to the canalization of a river*. St. Louis.
Harrold, J.C. (1941). Slope-area method of determining flood discharge not exact. *Civil Engineering* 11(9): 551.
Harrold, J.C. (1947). Cavitation in hydraulic structures: Experiences of the Corps of Engineers. *Trans. ASCE* 112: 16-42; 112: 116-119.
Harrold, J.C. (1954). Discussion of Equation of the free-falling nappe. *Journal of the Hydraulics Division* ASCE 80(604): 16-19.

HARZA L.F.

06.02. 1882 Brookings County SD/USA
22.11. 1953 Chicago IL/USA

Leroy Francis Harza obtained the BS degree from University of Wisconsin, Madison WI, in 1906, and the civil engineering degree in 1908. After having been an instructor in mathematics at his Alma Mater until 1904, he was a hydraulic engineer there. From then Harza was a consulting engineer dealing mainly with the design and construction of dams, hydraulic works, and hydropower projects. He started practice with Daniel W. Mead (1862-1948) in 1906, opening his first office in 1912 at Portland OR there being involved in pioneering work on Bonneville and Dalles Dams on Columbia River. From 1916 to 1920 Harza was engaged in the design of hydroelectric plants for the Great Lakes Power Co., Sault Sainte Marie ON. He was from 1920 president of the Harza Engineering Company, Chicago IL. Harza was a member of the American Societies of Civil Engineers ASCE, Mechanical Engineers ASME, and the Western Society of Engineers.

Harza was an internationally known consulting engineer on dam and hydroelectric power plant engineering. He remained active in his profession until almost his very last day, after a career spanning almost half a century. He was considered in consulting the dean of the American hydraulic engineers. Many of the world's outstanding dam designs were the products of his judgement and originality of thought. His 1942 project for the Barnhart Island Powerhouse on the St. Lawrence River Project includes 36 units of 55 MW each, which was based on the New York State Power Authority. Further designs of Harza include the Dix River Dam in Kentucky, the Loup River plants in Nebraska, the power installations at Fort Peck on Missouri River, or international activities such as *Maithon* Dam in *Damodar* Valley, India, and *Ambuklao* Dam in the Philippines, in the 1940s the highest rockfill dam of the world. Harza's firm, the later Harza Engineering Co., was then carried on by Calvin V. Davis (1897-1981) and his son Richard (1923-).

Anonymous (1951). Harza, Leroy Francis. *Who's who in America* 26: 1167. Marquis: Chicago.
Anonymous (1953). L.F. Harza dies. *Engineering News-Record* 151(Dec.3): 25. *P*
Harza, L.F. (1907). An investigation of the hydraulic ram. *Bulletin* 205. Engineering Series 4(3). University of Wisconsin: Madison WI.
Harza, L.F., Reineking, V.H. (1916). The Columbia River power project near the Dalles OR. *Bulletin* 3, Oregon State engineer. Duniway: Salem OR.
Harza, L.F., Floor, E. (1932). 4,500 HP hydro turbine gives 94% efficiency. *Power* 76: 188-190.
Harza, L.F. (1935). Uplift and seepage under dams on sand. *Trans. ASCE* 100: 1352-1406.

HASKELL

10.05. 1855 Holland NY/USA
28.01. 1933 Holland NY/USA

Eugene Elwin Haskell was educated at the Cornell University, Ithaca NY, from where he received the BCE and the civil engineering degrees in 1879 and 1890, respectively. He initiated his career in 1879 with the US Lake Survey, Detroit, was then an engineer for the Sioux City and St. Pauls Railroads, and from 1880 to 1885 for the Mississippi River Commission at St. Louis. Until 1893 Haskell joined the US Coast and Geodetic Survey, Washington DC, and until 1906 was a staff member of the US Lake Survey, Detroit. From then he was the Dean of the College of Engineering, Cornell University, and professor of experimental hydraulics until retirement in 1921. Haskell was a member of the American Section of the International Waterways Commission from 1906 to 1915, a Fellow of the American Association for the Advancement of Science AAAS, and a member of the American Society of Civil Engineers ASCE.

Current meters are hydrodynamic instruments with rotating vanes or buckets. The speed of their rotation is proportional to the flow velocity. The forerunners of current meters were the paddle wheels developed in the early 18th century. These were applied by Francesco Domenico Michelotti (1710-1777) in 1767, or by Pierre-Louis Du Buat (1734-1809) in 1786. There are two principal types of current meters, namely the screw and the cup types. The first was conceived by the famous British engineer Robert Hooke in 1783 to measure wind velocity with four vanes similar to a windmill. This type was developed by Reinhard Woltman (1757-1837) in 1790, André Baumgarten (1808-1859), Albert Ott (1847-1895), Alphonse Fteley (1837-1903) and Haskell, among many others. The other type meter has several cups on spokes rotating around an axis oriented transverse to the current. These anemometers were first applied around 1850 to measure wind velocities, and then were developed by Theodore Gunville Ellis (1829-1883), or William G. Price (1853-1928) to record flow velocity in rivers.

Anonymous (1921). Dean Haskell's retirement. *Cornell Civil Engineer* 29(6): 109. *P*
Anonymous (1924). Dean E.E. Haskell. *Engineering News-Record* 93(24): 847. *P*
Anonymous (1925). Haskell, Eugene Elwin. *Who's who in America* 13: 1488. Marquis: Chicago.
Haskell, E.E. (1902). Current meter and weir discharges. *Trans. ASCE* 47: 387-389.
Haskell, E.E. (1902). Flow of water in pipes. *Trans. ASCE* 47: 283-389.
Haskell, E.E. (1913). Characteristics of cup and screw current meters. *Trans. ASCE* 76: 844-846.
Haskell, E.E. (1926). Reminiscences. *Cornell Civil Engineer* 34(9): 234-235; 34(10): 272-273.

HASWELL

22.05. 1809 New York NY/USA
12.05. 1907 New York NY/USA

Charles Haynes Haswell entered the shops of a well known furnisher of steam-engine builder. In 1836, he was commissioned to introduce these engines to navy ships, and thus became the first engineer in the US Navy. The double-engines installed in 1837 included cast-iron cranks and shafts, driving a side-wheel of almost 7 m in diameter. The boilers were designed by Charles W. Copeland (1815-1895). From 1839 to 1842, the two designed and built then the machinery of the vessels *Mississippi*, *Missouri* and *Michigan*. However, Haswell left this position in 1843 and designed the machinery of four cutters, but returned as chief engineer of the Navy in 1844. In 1845 he defined the duties and responsibilities of the engineer afloat, which was for more than fifty years the basis for the Navy's 'steam instructions'. In 1852, as a result of heavy overwork, controversy and chronic dyspepsia, he returned to civil life.

For almost fifty years, Haswell was then a consulting engineer for the city of New York. He designed commercial vessels, foundations for high buildings, harbour cribs and fills, was surveyor of steamers and a trustee for the New York and Brooklyn bridges. He also was a member and president of the New York City Council. Distinguished as he was in the history of steam navy, Haswell was best known for his Mechanic's and engineer's pocket book, first issued in 1842. This work, in a way the engineer's bible, carried through the seventy-fourth edition until 1913, with a total sale of almost 150,000 copies.

Anonymous (1904). Charles Haynes Haswell. *Engineering News* 51(22): 509-511. *P*
Anonymous (1931). Haswell, Charles Haynes. *Dictionary of American biography* 6.1: 391-392. Scribner's: New York.
Haswell, C.H. (1856). *Mechanics' tables*, containing areas and circumferences of circles, and sides of equal squares, circumferences of angled hoops, cutting boiler plates, covering of solids, &c, and weights of various metals. Harper: New York.
Haswell, C.H. (1896). *Reminiscences of an octogenarian of the city of New York* (1816 to 1860). Harper: New York.
Haswell, C.H. (1898). Reminiscences of early marine steam engine construction and steam navigation in the USA from 1807 to 1850. *Trans. Inst. Naval Architects* 40: 104-113.
Haswell, C.H. (1899). Pile-driving formulas: Their construction and factors of safety. *Trans. ASCE* 42: 267-287.
Haswell, C.H. (1908). *Mechanic's and engineer's pocket book*. Harper: New York.

HATHAWAY

11.10. 1895 Menomonie WI/USA
01.10. 1979 Washington DC/USA

Gail Abner Hathaway received a BS degree in civil engineering from Oregon State College in 1922, and the honorary D.Eng. degree from Drexel Institute of Technology, Philadelphia PA, in 1951. He started his career as a construction engineer in 1919, joining in 1928 the US Corps of Engineers until retirement in 1957. He was engaged with field assignments until 1937, was then chief of reservoir regulation and the hydrology section at the Office Chief of Engineers, Washington DC, until 1945, from when he was the special assistant to the chief of engineers, Dept. of Army, on flood control, river and harbour project planning and construction programs. After retirement, he was an engineering consultant for the World Bank.

Hathaway was recognized internationally for his work in hydrological engineering, and problems relating to the planning, design and operation of flood control, navigation and multi-purpose projects. He was particularly concerned with hydraulic matters on an international scale. He was in 1946 the engineering consultant to President Truman's Cabinet Committee on Palestine during meetings in London, and in 1947 consultant to the Venezuelan government on flood control and water resources development. Later he was involved in large hydropower schemes on Indus River in Pakistan, and the High Aswan Dam in Egypt. He served in 1951 as ASCE president, and from 1952 to 1958 as president of the International Commission of Large Dams ICOLD.

Anonymous (1957). Gail A. Hathaway. *Power Engineering* 61(1): 4. *P*
Anonymous (1957). River control expert to the World Bank. *Engineering News-Record* 158(Mar.28): 81. *P*
Anonymous (1958). Dam experts coming from 40 nations. *Engineering News-Record* 161(Aug.14): 27. *P*
Anonymous (1963). Hathaway, Gail A. *Who's who in America* 32: 1347. Marquis: Chicago.
Hathaway, G.A. (1939). The importance of meteorological studies in the design of flood control structures. *Bulletin American Meteorological Society* 20(6): 248-253.
Hathaway, G.A. (1945). Design of drainage facilities. *Trans. ASCE* 110: 697-730.
Hathaway, G.A. (1958). Dams: Their effect on some ancient civilizations. *Civil Engineering* 28(1): 58-63.
Vogel, H.H. (1984). Gail A. Hathaway. *Memorial tributes* 2: 107-109. National Academy of Engineering: Washington DC. *P*

HATTON

11.08. 1860 Avondale PA/USA
11.11. 1933 Milwaukee WI/USA

Thomas Chalkley Hatton was educated in public schools. He began his career in 1878 with railway companies, and in 1882 was a hydrographer in the Department of Public Works, Canada. From 1883 he was employed as assistant city engineer, Wilmington DE, and from 1900 was engineer of the Street and Sewer Department, in charge of the extension of the sewerage system. He was also private consultant for similar works in the Southeast, engaged on the construction of filtration plants and distribution systems for Ashbury Park NJ, or Austin PA. In 1914 he went to Milwaukee WI, as chief engineer of the then recently created Sewerage Commission, serving until 1927. From then until his death he was engaged in private practice, and appointed in 1930 consulting engineer to the Milwaukee Sewerage Commission. His death was caused by a car accident.

Hutton supervised as chief engineer of the Sewerage Commission the development of the activated sludge system in the USA. This work consisted of the design, construction, and operation of the intercepting sewerage system and the sewage treatment plant for the Metropolitan District of the County of Milwaukee. He was initiative and had the energy to advance the activated sludge practice of sewage disposal, which from 1915 was developed from its early experimental stages into a practical and efficient process, currently used extensively all over the world. He was in addition connected with the Great Lakes Water Diversion, involving the Sanitary District of Chicago IL. He further was engaged on the design of an activated sludge sewage treatment plant for East Chicago IL. He was president of the American Society of Municipal Engineers, member of the Institute of Consulting Engineers, the Institution of Civil Engineers ICE, London UK, and of the American Society of Civil Engineers ASCE. The Pennsylvania Military College, Chester PA, conferred upon him the Master Degree in civil engineering in recognition of his outstanding work in connection with the activated sludge process.

Anonymous (1935). Thomas C. Hatton. *Trans. ASCE* 100: 1665-1667.
Hatton, T.C. (1916). Activated sludge experiments at Milwaukee. *Engineering News* 74(3): 134-137; 75(6): 262-263; 75(7): 306-308.
Hatton, T.C. (1922). Deposition of sludges resulting from sewage disposal plants. *Trans. ASCE* 85: 448-450. *P*
Hatton, T.C. (1923). Milwaukee activated-sludge plant to use vacuum filters. *Engineering News-Record* 90(5): 203-204.

HAWLEY

19.11. 1889 Portland OR/USA
17.03. 1942 Sacramento CA/USA

George William Hawley obtained the BSc degree in civil engineering in 1913 from Stanford University, Stanford CA, and there in 1916 the CE degree. He joined the Oregon Electric Company, and in 1916 became construction engineer with the South San Joaquin Irrigation District, in charge of construction of the Woodward Dam. In 1917 he was engaged by the East Bay Water Company as chief engineer; the San Pablo Dam, and the Upper San Leandro Dam, both large hydraulic-fill structures, were completed. He remained in this position until 1928, from when he was assistant director of operations. From 1925 to 1929 he also maintained a private practice as consulting engineer on hydraulic projects.

In 1929, following the enactment by the Legislature placing all dams in California under the jurisdiction of the state engineer, Hawley was appointed deputy state engineer in direct charge of construction, maintenance, and repair of more than 800 dams. The duties of the office called for travel to all parts of the state, and contact with hydraulic engineers. Under his direction the Dam Department became a highly-efficient organization that had the full confidence of the general public and the US Forest Service or the Federal Power Commission, with which he had frequent contact. Hawley was described as person with an outstanding character. His example of loyalty to his superiors, fidelity to his job, and keen-mindedness and alertness to protect the interests of the state were well emulated by all. Because of his wide technical interest he held membership in the American Concrete Institute, the American Geophysical Union AGU, the American Water Works Association AWWA, and the American Society of Civil Engineers ASCE, whose member he was since 1929. He served therein in its Special Committee on Irrigation Hydraulics, and in the Committee of Earth Dams and Embankments.

Anonymous (1942). George W. Hawley. *Trans. ASCE* 107: 1774-1776.
Hawley, G.W. (1922). The problem of future water supply for the Eastbay Communities.
    Bubbles East Bay Water Company 5(10): 50-55. *P*
Hawley, G.W. (1932). *Supervision of dams* 4: Hydraulics of dams, spillways and reservoirs.
    Water Resources Center: Sacramento CA.
Hawley, G.W. (1933). *State supervision of dams in California*. ASCE: New York.
Hawley, G.W. (1935). Accomplishments in the State supervision of dams. *Journal of the
    American Water Works Association* 27(11): 1492-1503.
Marliave, C., Hawley, G.W. (1926). *Upper San Leandro Dam*. East Water Bay Co.: Berkeley.

HAZEN A.

28.08. 1869 Hartford VT/USA
26.07. 1930 Miles City MT/USA

Allen Hazen graduated in 1885 with the BS degree from New Hampshire College of Agriculture. After further graduation from Dartmouth College, Hanover NM, he continued studies at Massachusetts Institute of Technology MIT in sanitary chemistry, joining then the Massachusetts Board of Health at its newly erected station at Lawrence MA. He was concerned there with scientific research on water purification and sewage treatment. Hazen was appointed the first director of the station in 1888. For the first time in the USA, engineers, biologists and chemists were brought together to develop sanitary engineering. The 1890 report on typhoid remains the classic research in this field outlining the fundamentals on the biological action of filters and the grading and selection of material for water and sewage filters.

Hazen spent the year 1894 in Europe, mainly staying at Dresden Polytechnic Institute, Germany. He wrote during this period his first book on filtration practice, including a review of European practice. On his return to the United States he entered a private consulting office at Boston, and moved it in 1896 to New York. He took there charge of the design and construction of the first continuously operating slow-sand water-filter of the United States. Thereafter he served as consultant throughout North America. In 1904 Hazen formed a new partnership. The early work in water and sewage technology expanded soon into fields like hydrology, dam design, and piping systems. Hazen also studied with Gardner Williams (1866-1931) the flow of water in pipes, developing the popular Williams-Hazen formula and compiled the Hydraulic tables in 1905. Shortly later, Hazen published his popular book Clean water and how to get it. His last book was devoted to the concept of flood flows. Hazen died suddenly during a trip in the West.

Anonymous (1911). Allen Hazen. *New England Water Works Association* 25(1): Frontispiece. *P*
Anonymous (1930). Allen Hazen dies while on vacation. *Engineering News-Record* 105(5): 189. *P*
Anonymous (1944). Hazen, A. *Dictionary of American biography* 21: 389. Scribner's: New York.
Hazen, A. (1895). *The filtration of public water supplies*. Wiley: New York.
Hazen, A. (1907). *Clean water and how to get it*. Wiley: New York.
Hazen, A. (1913). Venturi meter coefficients. *Engineering News* 70(5): 198-202; 70(14): 674.
Hazen, A. (1919). Hydraulic-fill dams. *Trans. ASCE* 83: 1713-1821.
Hazen, A. (1930). *Flood flows*. Wiley: New York.
Williams, G.S., Hazen, A. (1905). *Hydraulic tables*. Wiley: New York.

HAZEN R.

05.08. 1911 Dobbs Ferry NY/USA
12.02. 1990 Dobbs Ferry NY/USA

Richard Hazen obtained in 1932 the AB degree from Dartmouth College, Hanover NH, in 1934 the BS degree from the Columbia University, New York NY, and in 1937 the MS degree in civil engineering from Harvard University, Cambridge MA. He was then until 1942 assistant and associate engineer at a private engineering company, and from then until his retirement in 1981 a partner within the Hazen & Sawyer Co., New York. Hazen was a member of the National Academy of Engineers NAE from 1974, the American Institute of Consulting Engineers and its president in 1968, the American Society of Civil Engineers ASCE, being its director from 1966 to 1969, and being awarded Honorary Membership in 1983, the American Water Works Association AWWA, from 1951 to 1956 being director and later also Honorary Member, among many other distinctions.

Hazen was a well known consultant in planning and design of water supply and waste disposal of municipalities and the industry. He made important accomplishments in water resources development and utilization, and the preparation of master plans for regional water supply and sewerage systems. Examples of the works made by Hazen & Sawyer include the integrated master plan for Sao Paulo, Brazil, or water supply planning for New York, Baltimore, and Washington DC. He also was a member of the Research Council Committee reviewing the Corps of Engineers' Metropolitan Washington Area water supply study. He stressed the need for effective utilization and protection of the nation's water resources. He contributed papers to these topics in professional journals and engineering handbooks.

Anonymous (1952). Richard Hazen joins Water Works School Faculty. *Water Works Engineering* 105(8): 750. *P*
Anonymous (1983). Richard Hazen. *Civil Engineering* 53(10): 83-84. *P*
Anonymous (1987). Hazen, Richard. *Who's who in America* 44: 1231. Marquis: Chicago.
Anonymous (1990). Richard Hazen. *Journal American Water Works Association* 82(7): 107.
Hazen, R. (1951). Elements of filter design. *Journal American Water Works Association* 43(3): 208-218.
Hudson, H.E., Hazen, R. (1964). Droughts and low streamflow. *Handbook of applied hydrology* 18: 1-26, V.T. Chow, ed. McGraw-Hill: New York.
Okun, D.A. (1993). Richard Hazen. *Memorial tributes* 6: 57-60. National Academy of Engineering: Washington DC. *P*

HEDGER

17.12. 1898 Riverside CA/USA
29.12. 1991 Irvine CA/USA

Harold Everett Hedger graduated in 1924 from the University of California, Berkeley CA, with a BSc degree in civil engineering. He had already been working before for the Flood Control District, so he continued his career there full time. He played a significant role in the successful activities of the Los Angeles County Flood Control District, serving as chief engineer from 1938 to 1959. His nomination followed a period when the agency gained an adverse reputation because of wrong-doings by contractor representatives and politicians. Hedger's honesty, determination, and professional attitude steered the agency through a period of professional growth, and in 1959 at his retirement, to the reputation of being a first-class organization. During his tenure as chief engineer, the national flood control works program of the US Army Corps of Engineers was being developed. Hedger made sure that the Los Angeles County's needs were well accounted for. Over one quarter billion federal dollars was obtained for the construction of major flood control dams and channels.

After World War II, it became evident that local drainage facilities were needed to gather and deliver run-off into the newly constructed flood control channels. Under Hedger's leadership, a large storm drain bond issue was conceived and presented to the Los Angeles County voters, which was largely approved. This program was based on a close cooperative arrangement with each of the more than fifty cities then controlling land use planning within their own jurisdictions. The cities provided design services for the storm drains in many cases. Construction management for each drain was handled by the LACFCD. The program was so successful that its pattern was used for three additional storm drain bond issue proposals over the next 20 years, aggregating almost one billion of US dollars worth of design and construction activity. Hedger joined the American Society of Civil Engineers in 1927, served as vice-president of the Los Angeles Section in 1941, and as president in 1942. He was in 1972 designated the Honorary Membership. He was accorded colonel of the US Army Corps of Engineers.

Anonymous (1992). Harold E. Hedger. *Trans. ASCE* 157: 505-506.
Hedger, H.E. (1960). *Lower San Joaquin River flood control project*. Reclamation Board: CA.
Hedger, H.E. (1962). *Cyprus Water Development program*. Agency for Intl. Dev.: Washington.
Newton, C.T., Hedger, H.E. (1959). LA County flood control and water conservation. Los Angeles.
http://74.54.153.80/CivilEngineeringLandmarksForum/history/ *P*

HELLAND

17.03. 1886 Black Earth WI/USA
07.09. 1958 Portland OR/USA

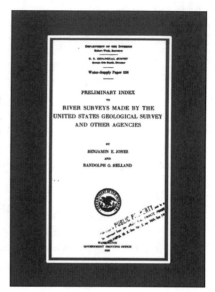

Randolph Olaf Helland graduated in 1913 from the University of Wisconsin, Madison WI, with the BA degree in science, and in 1916 with the BSc degree in agriculture. He also studied at George Washington University, Washington DC, to quantify later for an appointment with the US Geological Survey USGS. He rose subsequently through the grades of assistant classifier, assistant scientist, and further associate and hydraulic engineer, before retiring as regional hydraulic engineer with the Portland OR Office in 1955. He had entered USGS in 1917, and after a short absence while abroad, he rejoined it in 1923 in the Hydrographic Section of the Land Classification Branch, the later Water and Power Branch of the Conservation Division. He made numerous surveys of Western streams and prepared reports on the waterpower value of the adjoining lands. In 1943 he became regional hydraulic engineer for the Oregon-Idaho Region, including the State of Oregon and the Snake River Basin in Washington, Idaho, Nevada, Utah, and Wyoming.

Helland was an expert of water supply mainly in the West of his country. The 1926 paper written with a colleague deals with the determination and the extent of the natural resource water of the USA, which was previously collected by federal, state, and private agencies. These results were highly non-uniform from excellent surveys to difficulties in the assessment. The US Army Corps of Engineers have furnished surveys mainly on a large scale, with maps indicating the water depth, the location and character of any obstructions, and the general topography of banks. However, in many cases, the cross-sectional profiles of the water courses were not included, which are of prime interest for hydraulic purposes, but also for inland navigation. The 1926 compilation of data was mainly made for government bureaus to furnish improved information on the major US rivers. Helland was associate member of the American Society of Civil Engineers.

Anonymous (1960). Randolph O. Helland. *Trans. ASCE* 125: 1410-1411.

Helland, R.O. (1941). *Water utilization in the Nooksack River* WA. USGS: Washington DC.

Helland, R.O., Jones, B.E. (1948). *Index* to river surveys made by the USGS and other agencies, revised to 1947. USGS: Washington DC.

Helland, R.O. (1953). *Water power of the coast streams of Oregon*. US Geological Survey. US Department of the Interior: Washington DC.

Jones, B.E., Helland, R.O. (1926). *River surveys made by the US Geological Survey*. USGS: Washington DC. (*P*)

HEMPHILL

06.02. 1885 Abbeville SC/USA
19.06. 1930 Bexar TX/USA

Robert Grier Hemphill entered in 1903 the Clemson College, Clemson SC, yet failing health forced him to abandon the courses in civil engineering before the first year was completed. In 1906, after having regained health, he was engaged by Elwood Mead (1858-1936). It was at the base of the Rockies where the foundations between the plants and their water consumption was laid. If one cultural method seemed to be faulty, Hemphill tried another approach, always watching critically the effects on a plant. As years passed the scope of his work broadened. Clustering around the old Union Colony founded in 1870, amid cacti and jack rabbits were found to be the most productive lands of Colorado. The wealth derived from these soil products depended almost entirely on the mountain stream Cache la Poudre. Hemphill and his associates determined how the water supply was diverted or stored, distributed and used, in view of a further definition and protection of water rights to individuals and communities, to improve irrigation practice. In his Report Irrigation in Northern Colorado, he traces the water from the time it leaves its rocky canyon until it is transpired by the plant foliage. Carrying capacities, seepage losses, storage, exchange of water, distribution, and use, were all included in the account.

If Hemphill had nothing else done than to prepare this report, he would have received the plaudits of the rural population of the Rocky Mountain region, but the solution of still larger water problems was demanded. He thus considered the South Platte River: What is its return flow, and how can it be made to serve best the interests of farmers? Again, thousands of water measurements were made, resulting in a report written by his head Ralph L. Parshall (1881-1959). Based on these works, Hemphill was appointed in 1920 resident irrigation engineer for Western Texas. In these different climatic and plant conditions, he made similar studies on the water requirement of plants; he made there also a study on the river silt in Texas streams. Hemphill was described as a person with great perseverance and accuracy, combined with a cheerful disposition.

Anonymous (1931). Robert G. Hemphill. *Trans. ASCE* 95: 1639-1641.

Hemphill, R.G. (1922). *Irrigation in Northern Colorado.* US Dept. Agriculture: Washington DC.

Hemphill, R.G. (1930). Silting and life of southwestern reservoirs. *Proc. ASCE* 56(5): 967-979.

Marr, J.C., Hemphill, R.G. (1928). Irrigation of cotton. *Technical Bulletin* 72. US Department of Agriculture: Washington DC.

http://www.findagrave.com/cgi-bin/fg.cgi?page=pv&GRid=86446272&PIpi=89450882 (*P*)

HENNY

15.11. 1860 Arnhem/NL
14.07. 1935 Portland OR/USA

David Christiaan Henny was educated at the Delft Polytechnic School in the Netherlands, graduating as a civil engineer in 1881. In 1884 he emigrated to the United States and was there engaged until 1892 in general engineering work. He then joined the Excelsior Wooden Pipe Company as chief engineer and general manager, and from 1902 to 1905 was general manager of another wood company at San Francisco. From then he was a supervising engineer for the Pacific Division of the US Reclamation Service. His work included the extensive program of dam construction in the West.

In 1910 Henny began his long practice as a consultant and maintained an office at Portland OR for the remainder of his life. He was involved in a number of projects mainly in the United States in connection with land reclamation, power, and flood control. He first introduced the wooden stave pipe as a means of large pipeline for water transportation. He was an authority in dam design and contributed to the excellence of those built in the West. One of his important achievements was the invention of the Henny shear-joint, a special joint between the up- and downstream faces of concrete blocks used in the construction of massive concrete dams. This feature was successfully integrated in the Boulder Dam, then with 200 m the highest dam worldwide. It was later also incorporated in the Grand Coulee Dam on Columbia River, containing the largest amount of concrete then ever used for a dam. At the time of his death, Henny was a consultant to the Los Angeles County Flood Control District and also chairman of the Bonneville Dam Commission on the Lower Columbia River. He was a vice-president ASCE in 1932, and awarded the ASCE Norman Medal for his 1934 paper. In 1933 he was awarded Honorary Membership of the Royal Institute of Engineers, Holland.

Anonymous (1935). D.C. Henny dies in Portland OR. *Engineering News-Record* 115(3): 102. *P*
Anonymous (1936). David Christiaan Henny. *Trans. ASCE* 101: 1577-1580.
Anonymous (1944). Henny. *Dictionary of American biography* 21: 393. Scribner's: New York.
Griggs, F.E., ed. (1991). Henny, David Christiaan. *A biographical dictionary of American civil engineers* 2: 50. ASCE: New York.
Henny, D.C. (1898). Wooden-stave versus riveted pipe. *J. Assoc. Engineering Soc.* 21: 239-254.
Henny, D.C. (1927). New dam will double water supply of Portland OR. *Engineering News-Record* 98(21): 842-846.
Henny, D.C. (1934). Stability of straight concrete gravity dams. *Trans. ASCE* 99: 1041-1123.

HENRY D.F.

27.05. 1833 Detroit MI/USA
13.05. 1907 Detroit MI/USA

Daniel Farrand Henry graduated from the Sheffield Scientific School, Yale University, New Haven CT, in 1853. He then joined the US Lake Survey of the Northern Lakes continuing until 1871. From 1868 to 1871 he superintended the measurement of the outflow of the Lakes thereby accumulating valuable information. His observations of the overflow and on the sudden rise and fall of a reservoir were considered authoritatively. In connection with these observations he invented the telegraphic current meter used to measure velocity, and a flexible pipe inlet invented for the Detroit Water Works, for which he was awarded a medal at the 1876 Centennial Exhibition. From 1871 he was partner of a company at Chicago IL. He also served as chief engineer the Detroit Water Works from 1872 to 1878. Later he had a private practice as consulting engineer.

Counting the revolutions of paddle wheels over a certain time and then multiplying by the circumference was one method of measuring river velocity. Others included from measuring the rate at which rods weighted at one end moved, to determining how rapidly leaves were transported at the water surface from one end to the other. Around 1830 the modern current meter was introduced in America, based on long tests and perfections in Europe. This apparatus looks like a cross between a weather vane and an anemometer. Under the supervision of Andrew A. Humphreys (1810-1883), a civilian-military team surveying the Lower Mississippi developed in the 1850s a novel double-float system to measure river currents. It employed a weighted keg, its top and bottom removed, tied to a barrel floating at the water surface. Soon after the Civil War, Corps employees including Henry on the Great Lakes, and Theodore G. Ellis (1829-1883) on the Connecticut River considerably improved the original meters. Henry is credited having produced the first telegraphic current meter. An electrical contact occurs during each revolution of the meter cups. An operator either used headphones to count the clicks or employed an instrument that automatically totalled the electric impulses, so that the local water velocity could be determined.

Anonymous (1911). Henry, D.F. *Trans. ASCE* 71: 420-422.
Frazier, A.H. (1964). Daniel F. Henry's cup type 'telegraphic river current meter. *Technology and Culture* 5(4): 541-565. *P*
Henry, D.F. (1873). *Flow of water in rivers and canals*. Graham: Detroit.
http://www.usace.army.mil/About/History/HistoricalVignettes/CivilEngineering/013InCommon.aspx

HERBICH H.P.

15.01. 1896 Kalisz/PL
09.10. 1968 Sussex/UK

Henryk Pawl Herbich obtained his civil engineering diploma from the Technical University of Warsaw in 1923. He attended also courses at the University of Edinburg UK, receiving the DSc degree in 1943. He worked from 1924 for the Dept. of Public Works, and later for the Polish Dept. of Transportation as hydraulic engineer. He was from 1935 head of the Polish Water Power Bureau, and vice-president of the Inland Waterways Department until 1939. Then he directed the hydro-electric research division, the Polish Hydrological Institute, and he was part-time Lecturer at the Technical University of Warsaw.

Herbich escaped in 1939 with his family from Poland, settling first in France, where he was a consulting engineer to the French Department of Public Works. In 1940, after France was occupied by the Germans, he fled to the United Kingdom, becoming in 1941 a PhD student at Edinburgh University. From 1944 he was technical advisor of the Polish Exile Government, and professor and head of the Civil Engineering Department, Polish University College, University of London. He went in 1952 to the University of Minnesota, Minneapolis MN, as a professor of civil engineering until 1959. During this period he also was a consultant for Taconite Contracting Corporation, Cleveland OH. Then Herbich became professor as well as head of the Civil Engineering Department, University of Windsor, Windsor ON, until 1966. He then retired in England. Herbich had been responsible for the design and construction of twelve dams in Poland, France and the USA. He also was a permanent delegate of the Polish government to the World Power Conference and the International Committee of Large Dams. He was a Fellow of the Royal Meteorological Society RMS, London, the Institution of Civil Engineers ICE, and the International Association of Hydraulic Research IAHR.

Anonymous (1965). Henry P. Herbich. *Civil Engineering* 35(11): 14. *P*
Anonymous (1970). Henryk Pawel Herbich. *Trans. ASCE* 135: 1127-1128.
Anonymous (1970). Herbich, H.P. *Quarterly J. Royal Meteorological Society* 96(408): 156.
Herbich, H.P. (1933). Die Bestimmung der Einwirkung des Pflanzenwuchses auf den Wasserspiegelstand. 4. *Hydrologische Konferenz der Baltischen Staaten* Leningrad 103: 1-15.
Herbich, H.P. (1934). *Influence of vegetation on water stage in rivers*. Udsiał Centralnego Biura Hydrograficznego. Warszawa [in Polish].
Herbich, H.P., Brennan, L.M. (1967). *Prediction of scour at bridges*. Department of Civil Engineering, University of Windsor: Windsor Ont.

HERBICH J.B.

01.09. 1922 Warsaw/PL
19.06. 2008 College Station TX/USA

John Bronislaw Herbich, the son of Henry Pawel (1896-1968), came to the US in 1953. He graduated as a civil engineer from the University of Minnesota in 1957, and received the PhD degree in 1963 from Pennsylvania State University, University Park PA. He made post-graduate studies at the University of California, Berkeley, and at Utah State University, Logan UT. He was an assistant professor at Lehigh University from 1957 to 1960, then professor until 1967. Herbich was later professor of civil and ocean engineering, Texas A&M University, College Station TX until 1983, and there W.H. Bauer Professor of dredging engineering from 1989 to 1994. He was in 1972 on leave as project manager of the Central Water and Power Research Station, Government of India, Poona, and a Lecturer in Venezuela, India, China, among other countries. He further was Lecturer at US Waterways Experiment Station, Vicksburg MS. He was awarded the 1965 ASCE Hilgard Award, and ASCE's 1993 International Coastal Engineering Award. He was a member IAHR, the World Dredging Association, and the Marine Technology Society.

Throughout his distinguished career Herbich lectured and taught at many venues, in the USA and abroad. He served as consultant to numerous governments and international projects involving the engineering design of ports, harbours and coastlines. He authored more than 200 papers and several books, most notably editing a large Handbook, which is considered a standard in this field, as also the Handbook of dredging engineering. He was the recipient of prestigious awards.

Anonymous (1953). John B. Herbich. *Engineering Journal* 36(8): 1034. *P*
Anonymous (1960). John B. Herbich. *Civil Engineering* 30(8): 30. *P*
Anonymous (1966). John B. Herbich. *Civil Engineering* 36(10): 77. *P*
Anonymous (1995). Herbich, John B. *Who's who in America* 49: 1660. Marquis: Chicago.
Herbich, J.B., Shulits, S. (1964). Large-scale roughness. *Journal of the Hydraulics Division* ASCE 90(HY6): 203-230; 91(HY5): 242-262; 92(HY3): 75-79.
Herbich, J.B., ed. (1990). *Handbook of coastal and ocean engineering*. Gulf: Houston TX.
Herbich, J.B. (2000). *Handbook of dredging engineering*, 2nd ed. McGraw-Hill: New York.
Mariani, V.R., Herbich, J.B. (1966). Effect of viscosity of solid-liquid mixture on pump cavitation. *Modern Trends in Hydraulic Engineering Research*: 342-355. Golden Jubilee Symposia: Poona.
http://www.kbtx.com/obituaries/25335134.html

HERING

26.02. 1847 Philadelphia PA/USA
30.05. 1923 New York NY/USA

Rudolph Hering graduated in 1867 as a civil engineer from Dresden University, Germany. In 1907 he was awarded the title hon. D.Sc. from the University of Pennsylvania, Philadelphia PA. Upon return to the USA from Germany, Hering was assistant engineer at Philadelphia PA, and assistant city engineer there from 1872. After promotion to chief engineer, he dealt with both water supply and drainage works, and became a consulting engineer to the city of New York. He was concerned with both the design and construction of water supply and sewage works in 150 cities of the USA, Canada, Hawaii and Brazil. Hering was a member of the American Society of Civil Engineers ASCE, the American Society of Mechanical Engineers ASME, the Boston Society of Civil Engineers, the Western Society of Civil Engineers, and the Philadelphia Engineering Society. He was an Honorary Member of the American Water Works Association AWWA.

Hering was an outstanding figure in sanitary engineering in the United States. He translated with John C. Trautwine, Jr., (1850-1924) the work of Wilhelm Kutter (1818-1888) relating to uniform channel flow. He was then largely responsible for the adoption of plans for the construction of the Chicago drainage canal for removing the sewage of the city. He also was a member of the commission which in 1904 recommended that New York obtain its water from the Catskill Mountains, and was further responsible for the prominence of the Imhoff tank in sewage treatment proposed by Karl Imhoff (1876-1965). He was the acknowledged dean of sanitary engineering, maintaining a deep interest in all matters relating to sewerage, water quality and refuse disposal.

Anonymous (1918). Rudolph Hering. *Engineering News-Record* 81(6): 274-276. *P*
Anonymous (1922). Hering, Rudolph. *Who's who in engineering* 1: 589. Lewis: New York.
Anonymous (1923). Rudolph Hering, Dean of sanitary engineering, dies. *Mechanical Engineering* 45(7): 445.
Anonymous (1924). Rudolph Hering. *Trans. ASCE* 87: 1350-1354.
Hering, R. (1880). Report on the result of an examination made in 1880 of several sewerage works in Europe. *Bulletin* Suppl. 16. National Board of Health: Washington DC.
Hering, R., Trautwine, J.C., Jr. (1889). *A general formula for the flow of water.* Wiley: New York.
Hering, R. (1898). Dilution process of sewage disposal. *Engineering Magazine* 15(7): 575-583.
Hering, R., Greeley, S.A. (1921). *Collection and disposal of municipal refuse.* McGraw-Hill: New York.

HERMANY

09.10. 1830 Lehigh County PA/USA
18.01. 1908 Louisville KY/USA

Charles Hermany worked as young man on the family farm but showed an early aptitude for physics and mathematics. His aspirations turned into civil engineering. He was prepared for college at Minerva Seminary, Easton PA, but the financial difficulties made a formal education impossible. He returned to his family but continued studies in mathematics and engineering by his own. He went to Cleveland OH in 1853, securing a post in the city's engineering department, whose head recognized the unusual potential of the young assistant. When both went to Louisville KY in 1857 to design and supervise the construction of the city water works, the completed scheme was taken over by Hermany as chief engineer in 1861. Possessed of a keen mind and a fierce dedication to civil engineering, he had to solve the early problems of the water system. He was undoubtedly the one among the top officers who did most to shape the Louisville Water Company into a viable and efficient concern.

As early as in the 1870s Hermany was concerned with the problem of sediment in the water pumped directly from the river. He initiated a long series of experiments to solve the problem, a procedure that was adopted generally under these conditions. He designed Crescent Hill Reservoir, completed in 1879, building the company's second pumping station during the same year. He used his own labour force when private contractors turned down the job as too risky. He also had a hand in designing the huge steam pumping engine installed at the new station. An operating model of the Leavitt-Hermany engine, as introduced by Erasmus D. Leavitt (1836-1916) is on display at the Smithsonian Institute. Hermany was the founder, and in 1881 the first president, of the Louisville Engineers and Architects Club. The long experiments on filtering river water continued and were promising enough that work was begun in 1897 on the Crescent Hill Filtration Plant. In 1908 he was scheduled to attain the crowning recognition of his professional career, namely becoming president of the American Society of Civil Engineers ASCE, yet shortly before he contracted pneumonia and passed away.

Anonymous (1909). Charles Hermany. *Trans. ASCE* 65: 525-528.
Kleber, J.F., ed. (2001) Hermany, C. *Encyclopedia of Louisville*: 382. University Press: Kentucky.
Stoddard Johnson, J. (1896). *Memorial history of Louisville*. Am. Biographical Co.: Chicago.
Yater, G. (1996). *A history of the Louisville Water Company*. Water Company: Louisville.
http://www.google.ch/imgres?q=%22Charles+Hermany%22+1830- *P*

HERSCHEL

23.03. 1842 Boston MA/USA
01.03. 1930 Glen Ridge NJ/USA

Clemens Herschel graduated in 1860 from Lawrence Scientific School, Harvard MA. He studied further in France and Germany, graduating in 1863 as a civil engineering from Karlsruhe University. From 1879 to 1889 he was a hydraulic engineer with the Holyoke Water Power Co., thereby collaborating with James B. Francis (1815-1892), and consulting engineer of water power companies at Niagara Falls NY. He was until 1900 engineer and superintendant of the East Jersey Water Company. Later he was a consultant in New York. Herschel was a member of the Boston Society of Civil Engineers, the American Society of Civil Engineers ASCE, being its president in 1915 and elected to Honorary Membership in 1922, the American Water Works Association AWWA, as also of the Institution of Civil Engineers ICE, London UK.

During his stay at Holyoke, Herschel developed the Venturi meter, corresponding to a locally contracting pipe portion by which discharge can be measured using the pressure difference between the approach and the contracted pipe sections. He was awarded for this research ASCE's 1888 Rowland Prize and the Elliott Cresson Gold Medal by the Franklin Institute. During his stay in New Jersey Herschel provided a large additional water supply for Newark. These works included dams and reservoirs on the *Pequannock* River, riveted steel pipes and several Venturi meters. His many publications include a translation of the engineering classic text by the Roman hydraulic engineer Frontinus. Further notable work was devoted to an improved weir form for gaging discharge in channels. He was bestowed the honorary title Doctor of Engineering from Karlsruhe University in 1925, mainly under the influence then of Theodor Rehbock (1864-1950).

Anonymous (1922). Herschel, Clemens. *Who's who in engineering* 1: 591. Lewis: New York.
Anonymous (1930). Clemens Herschel. *Mechanical Engineering* 52(5): 571-572. *P*
Anonymous (1931). Clemens Herschel. *Trans. ASCE* 95: 1419-1423.
Herschel, C. (1885). The problem of the submerged weir. *Trans. ASCE* 14(5): 189-196.
Herschel, C. (1899). The two books on the *Water supply of the city of Rome* of Sextus Julius
        Frontinus, water commissioner A.D. 97. Dana Estes: Boston MA.
Herschel, C. (1899). The Venturi water meter. *Cassier's Magazine* 15(5): 411-421.
Herschel, C. (1921). The flow of liquids through short tubes. *Trans. ASCE* 84: 527-550.
Kent, W.G. (1912). An appreciation of two great workers in hydraulics: *Venturi and Herschel*.
        Blades: London. *P*

HERTZLER

06.09. 1906 Harrisburg PA/USA
21.05. 1990 Volusia FL/USA

Richard Adin Hertzler was educated at University of Michigan, Ann Arbor MI, from where he received the BS degree in civil engineering in 1934. He was then until 1939 a research engineer in hydrology for the Appalachian Forest Experiment Station, then to 1951 senior hydrologic engineer of the Flood Control Surveys FCS, and for the US Forest Production Laboratory, from when he served as a flood control survey officer at the US Department of Agriculture until 1953. Hertzler was from then until retirement Chief, the Office of Civil Functions, at the Office of the Secretary of the Army. He was a member of the American Geophysical Union AGU, and the American Association for the Advancement of Science AAAS.

Hertzler was both a notable hydraulic engineer and hydrologist. Early in his career he developed a formula for discharge measurement with a triangular weir, which is known for its accuracy mainly in hydraulic laboratories. Once working for the US Department of Agriculture he developed the Coweeta Experimental Forest in North Carolina as an hydrologic laboratory. This site was used from 1934 to advance watershed management and measurement of rainfall in forests. Observations indicated how natural and human disturbances to the watershed change the stream characteristics. Research at this site has provided important information on the effects of timber harvesting, road construction, and natural disturbance of watersheds. At the end of his career, Hertzler was member of the staff, Assistant Secretary of the Army for civil works and other civil functions. He has authored articles and pamphlets on the various aspects of hydrology and forest utilization, including flood control survey reports published as House Documents. He was from 1955 to 1957 a participant of the Presidential Advisory Commission on Water Resources Policy.

Anonymous (1964). Hertzler, Richard A. *Who's who in engineering* 9: 816. Lewis: New York.
Hertzler, R.A. (1936). *History of the Coweeta Experimental Forest*. USDA: Otto NC.
Hertzler, R.A. (1938). Determination of a formula for the 120° V-notch weir. *Civil Engineering* 8(11): 756-757.
Hertzler, R.A. (1939). Engineering aspects of the influence of forests on mountain streams. *Civil Engineering* 9(8): 487-489.
http://www.srs.fs.usda.gov/coweeta/about/history/
http://toto.lib.unca.edu/findingaids/photo/usfs/show.asp?PassVar=347683 *P*

HESSE

28.03. 1826 Treves/D
27.01. 1911 Oakland CA/USA

Frederick Godfrey Hesse was born in the Rhenish Prussia. After having received education at various German Polytechnics, he entered the Prussian Army and was engaged by the government with works in engineering. He left his native country during the 1848 wars moving to the USA. He there soon took up an engineering position, later receiving several engineering appointments under the US Government including a professorship of mathematics in the US Navy. He resided from 1864 at San Francisco and was there occupied in mechanical operations. He joined in 1875 the chair of mechanical engineering, the University of California, Berkeley CA. Its president noted that 'it is rare to find a man qualified to fill the duties of a chair of industrial mechanics, both by his scientific attainments and practical knowledge, yet Mr. Hesse is such a man'. He was described as a man beautiful and simple in his character, unsuspecting and credulous in his dealings with others, thoroughly scientific in the quality of his intellect, so that he was eminently suited for this position.

Hesse was actively involved in the development of the turbine proposed by Lester A. Pelton (1829-1909) by providing the original sketch of a split bucket. As stated by Rouse 'it is of note that whereas Pelton experimented first in this field of hydro-machinery, with buckets improvised from oyster cans, a model made from patterns supplied by Pelton had been tested in 1883 by Hesse in his hydraulic laboratory'. Pelton took out a second patent of his turbine in 1889 on small details, and this was followed by others on improvements of the bucket shape and in 1900 on the first needle nozzle. Hesse's laboratory appears to have been the first in the USA. Unfortunately, all that is known about it is found only in a printed report published in 1883, supplemented by a colleague of Hesse relating to the efficiency of a Pelton unit.

Anonymous (1911). Frederick G. Hesse. *The University of California Chronicle* 8: 117.
Greene, A.M. (1911). *Pumping machinery*. Wiley: New York.
Hesse, F.G. (1887). Hydraulic step. College of Engineering, *Bulletin* 2. University of California: Berkeley.
Hesse, F.G. (1911). *Hesse papers* 1861-1911. University of California: Berkeley.
Jones, W.C. (1895). Hesse, Frederick G. *Illustrated history of the University of California*: 113-114. Dukesmith: San Francisco. *P*
Rouse, H. (1976). Frederick G. Hesse. *Hydraulics in the United States* 1776-1976: 53-55. *P*

HEUER

02.03. 1843 St. Louis MO/USA
28.04. 1925 San Francisco CA/USA

William Henry Heuer was a graduate of the US Military Academy, West Point NY, commissioned first lieutenant of engineers in 1865. He was placed assistant engineer on the river and harbor work of the Pacific Coast until 1870, during which period he was in charge to remove Blossom Rock, a menace in San Francisco Harbour, to a depth of 7 m below low-water. From 1872 to 1873 he was placed in charge of the removal of Halletts Point in New York Harbour, known as the Hell Gate. In 1874 he joined a Commission of Military Engineers as junior officer to examine possible interoceanic canal routes at Darien, Nicaragua and Panama. From 1876 to 1880 he was on duty at Key West FL, as lighthouse engineer, and on the fortifications at the Dry Tortugas Islands. Until 1884 he was then in charge at New Orleans LA for river and harbor works, and instrumental for the jetties at Sabine Pass TX. He became major in 1884, stationed at Philadelphia PA, again for improvements of rivers and harbours. He also designed a lighthouse at Delaware Bay.

Heuer returned in 1887 to the Pacific Coast and remained in California until his death, except for a short stay on the Ohio River. He was made in 1900 lieutenant-colonel and continued river improvement works. During the Spanish-American War, he was in charge of the submarine defense of San Francisco Bay. After his retirement in 1907 he opened a consulting office at San Francisco CA, and was connected during the ten next years with important works along the Pacific Coast. In 1917 he was called to active duty taking charge of the office of division engineer for the districts San Francisco, Los Angeles, and Hawaii until 1919. Like many visitors of the Golden State, Heuer became infatuated with it, considering San Francisco his home. He became thoroughly identified with the people, and was widely and favorably known throughout the Pacific Coast.

Anonymous (1888). Report of Major W.H. Heuer, Corps of Engineers. *Annual Report* of the Chief of Engineers, US Army 4: 2690-2691. Government Printing Office: Washington.
Anonymous (1927). William H. Heuer. *Trans. ASCE* 90: 1168-1170.
Heuer, W.H. (1892). *Bar at entrance to harbor at Yaquina Bay, Oregon*. Committee on Rivers and Harbors: Washington DC.
Heuer, W.H., Handbury, T.H., Harts, W.W. (1905). *Sacramento Valley river improvement*: Government policy and works. Hall: San Francisco. (*P*)

HICKOX

01.03. 1903 Spokane WA/USA
11.12. 1986 Mount Vernon VA/USA

George Harold Hickox was educated at University of Iowa, from where he obtained in 1926 the MS degree in hydraulic engineering. He further received the PhD title from University of California in 1939. He was assistant hydro-electric engineer for the West Virginia Power and Transmission Co., Pittsburgh PA, from 1927 to 1928, hydraulic engineer at Memphis TE until 1930, junior hydraulic engineer for the US Corps of Engineers in charge of the Rock Island District until 1932, then instructor at the University of California until 1935, from when until 1947 he was in charge of the hydraulic laboratory, Tennessee Valley Authority TVA. Finally Hickox was director of engineering research at University of Tennessee, Knoxville TE, until his retirement.

Hickox has prepared a number of outstanding papers, among which is a work on duration curves, for which he was awarded ASCE's Collingwood Junior Prize in 1935, or the 1948 paper on friction coefficients in large tunnels, for which he was awarded ASCE's James Laurie Prize in 1950. During his stay with TVA, Hickox was in charge of all hydraulic model tests on TVA dams and hydraulic structures. He presented early papers on cavitation damage of spillways, and flow aeration due to high-velocity flow. He also has written in collaboration with Morrough P. O'Brien (1902-1988) Applied fluid mechanics, a basic text with a number of interesting additions relating to discharge measurement. He also contributed a chapter to the Handbook of applied hydraulics.

Anonymous (1935). G.H. Hickox. *Civil Engineering* 5(1): 43. *P*
Anonymous (1950). G.H. Hickox. *Civil Engineering* 20(1): 42-43. *P*
Anonymous (1954). Hickox George H. *Who's who in engineering* 7: 1098. Lewis: New York.
Anonymous (1987). George Harold Hickox. *Trans. ASCE* 152: 401-402.
Hickox, G.H. (1944). Aeration of spillways. *Trans. ASCE* 109: 537-556; 109: 565-566.
Hickox, G.H. (1946). Evaporation from a free water surface. *Trans. ASCE* 111: 1-66.
Hickox, G.H. (1947). Cavitation in hydraulic structures: Experiences of the Tennessee Valley Authority. *Trans. ASCE* 112: 59-67; 112: 123-124.
Hickox, G.H., Peterka, A.J., Elder, R.A. (1948). Friction coefficients in a large tunnel. *Trans. ASCE* 113: 1027-1046; 113: 1071-1076.
Hickox, G.H. (1952). Hydraulic models. *Handbook of applied hydraulics* 24: 1121-1158, C.V. Davis, ed. McGraw-Hill: New York.
O'Brien, M.P., Hickox, G.H. (1937). *Applied fluid mechanics*. McGraw-Hill: New York.

HIGGINS

19.11. 1926 Laconia NH/USA
22.06. 1975 Meredith NH/USA

George Richardson 'Dick' Higgins graduated from the University of New Hampshire, Durham NH, in 1948, with the BS degree in civil engineering. He was then employed by the Water Resources Branch of the US Geological Survey at Boston as junior hydraulic engineer. In 1949 he was hired as research assistant in the Hydrodynamics Laboratory of the Massachusetts Institute of Technology MIT, starting there his post-graduate study, obtaining in 1951 the SM degree in hydraulics and water power. He was engaged during the next two years as hydrographic engineer by the Safe Harbour Water Power Corp., Conestoga PA. Higgins launched his academic career as instructor of civil engineering in 1953 at Norwich University, Northfield VT. He was appointed assistant professor in civil engineering at MIT, Cambridge MA, in 1956. Although his teaching assignments were primarily in fluid mechanics, hydrology and hydraulic engineering, he was an enthusiastic teacher of a wide variety of courses. He was highly regarded by all, and his accomplishments were recognized by his promotion to associate professor in 1960, and full professor in 1973.

Higgins was a recognized authority of surface water hydrology in the North-east. The 1951 Report was concerned with an old hydraulic problem of mapping a water surface. Using gages had proven a lengthy process, so that the optical method was considered, once the adequate instrumentation became available. The accuracy of the latter method is certainly below that of traditional gaging, but a rough account of the often complex water surface is considered sufficient in many cases. From the 1960s, once Higgins was at MIT, he was concerned with hydrologic studies relating mainly to local questions of stream rehabilitation, and hydrologic factors determining watershed yield. He was a member of the American Society of Civil Engineers ASCE.

Anonymous (1948). George R. Higgins. *Granite yearbook*: 86. University of NH: Durham NH.
Anonymous (1976). George Richardson Higgins. *Trans. ASCE* 141: 543-544.
Higgins, G.R. (1951). *Stereoscopic measurement of water surfaces*. MIT: Cambridge MA.
Higgins, G.R. (1964). *Hydrology studies in Massachusetts*. Water Resources Commission: Boston.
Higgins, G.R. (1967). *Yield of streams in Massachusetts*: Interim report. Water Resources Research Center MIT: Cambridge MA.
Higgins, G.R., Colonell, J.M. (1971). *Hydrologic factors in the determination of watershed yields*. Water Resources Research Center MIT: Cambridge MA.

HILGARD J.E.

07.01. 1825 Zweibrücken/D
08.05. 1891 Washington DC/USA

Julius Erasmus Hilgard, the father of Karl E. (1858-1938) moved with his father in 1835 to the United States, studying civil engineering at Pennsylvania State University, Philadelphia PA. He joined in 1845 as assistant the Coastal Service. When coast-survey men were engaged in river and harbour works, he became principal executive engineer until 1881 when being appointed superintendent. He was in 1874 elected president of the American Association for the Advancement of Science. In his obituary it is stated that His character sometimes led him, if his sympathies were exited, into hasty and ill-advised participation in other people's controversies, and he made enemies, but his character was thoroughly kindly, charitable and self-sacrificing. He was an enthusiastic, indefatigable worker, quick to investigate and approve new methods and, from his linguistic proficiency and wide reading, thoroughly informed on engineering progress outside from the USA.

The work of Hilgard was directed to the improvement of methods in geodesy and in terrestrial physics. His publications on these topics are found in the Coast Survey Reports. His marked ability procured him immediate recognition so that he was already placed in 1846 at the head of a surveying party. In 1862, at the outbreak of the Civil War, he was placed in charge of the office and performed the onerous duties created by the demands of the Army and the Navy for assistance of the coast survey. He served the Union Army by the construction of maps and charts, and by tidal information. At the Centennial Exposition held at Philadelphia in 1876, he acted as a judge on scientific and mechanical apparatus. His work gained him favourable notice in Europe, and he was invited to the directorship of the International Bureau of Weights and Measure in Paris. Declining this honour, he was presented by the French President a beautiful Sèvres vase in recognition of his services.

Anonymous (1891). Prof. Julius E. Hilgard. *Engineering News* 25(May 16): 462. *P*
Anonymous (1892). J.E. Hilgard. *Trans. ASCE* 27: 189-190.
Anonymous (1895). J.E. Hilgard. *Biographical memoirs* 3: 327-338. National Academy of Sciences, Washington DC.
True, F.W. (1913). *A history on the first half-century of the National Academy of Sciences* 1863-1913. National Academy of Sciences: Washington DC.
http://en.wikipedia.org/wiki/Julius_Erasmus_Hilgard *P*

HILL J.W.

10.10. 1848 Covington KY/USA
22.06. 1930 Wyoming OH/USA

John Wilmuth Hill obtained his practical training in engineering in various machine shops from 1865 to 1874, during which period he was involved in a model shop and with water-wheels. He was selected to act as chairman of a commission to conduct tests on the Corliss engines for the Cincinnati Exposition, for which he was awarded a gold medal. In 1875 he was again selected to conduct further tests on other machinery. In 1880, then as consulting engineer, he patented a novel system of surface condensation for pumping engines, whereby the water pumped was used for condensing purposes. He also compiled a 200-page manual for steam engineers.

Hill was an ardent promoter of the purification of public water supplies; early in his career he advanced the theory that typhoid fever was essentially a water-borne disease, so that the typhoid rate of a city was a criterion of the purity of its drinking water. When the city of Cincinnati in 1896 planned its drinking water system, Hill in collaboration with two colleagues formed a board of consulting engineers, with him as chairman, to report on the best method of urban water supply. Many tests of water from various sources were made, resulting in the slow sand filtration as the optimum, which was then used in most cities. In 1897 Hill conducted a long series of tests on the bituminous coal for steam purposes at Cincinnati for the use of water works. In 1898 Hill compiled a more than 300 page volume on the Purification of public water supplies, which was considered an authoritative text in these days. In 1900 Hill became consulting engineer for the improvement and filtration of Philadelphia's water supply. He was also involved in similar projects in Washington DC, Knoxville KY, or Newport KY. On his return to Cincinnati in 1906, he was in charge to report on the cause and the prevention of floods of Miami River, which led to the first large flood prevention plan in the 1910s. Hill was an active promoter of the civil life, a member of the American Water Works Association AWWA and of the American Society of Civil Engineers ASCE. He also was member of the Franklin Institute; he presided over the Ohio State Board from 1915 to 1919.

Anonymous (1924). John W. Hill. *Engineering News-Record* 92(16): 647. *P*
Anonymous (1931). John W. Hill. *Trans. ASCE* 95: 1506-1509.
Hill, J.W. (1898). *Purification of public water supplies*. Van Nostrand: New York.
Hill, J.W. (1899). The accuracy and durability of water meters. *Trans. ASCE* 41: 326-409.
Hill, J.W. (1915). The Cincinnati water works. *Journal AWWA* 2(1): 42-60.

HILL L.C.

22.02. 1865 Ann Arbor MI/USA
05.11. 1938 Los Angeles CA/USA

Louis Clarence Hill graduated from the University of Michigan, Ann Arbor MI, with the BSc degree in civil engineering, receiving in 1890 there also the BSc degree in electrical engineering. In 1911 his Alma Mater conferred upon him a honorary degree of Master of Engineering. From 1890 to 1903, Hill was professor of hydraulics at the Colorado School of Mines, Golden CO. He then entered the US Reclamation Service as engineer, becoming later project engineer and supervising engineer, in charge of construction of the Salt River Project including Roosevelt Dam in Arizona, the Yuma Project, and projects in New Mexico and Texas; he was also in charge of the Colorado River Basin.

Hill entered in 1914 private practice, and was in parallel also retained by the US Bureau of Reclamation as consulting engineer. He was member of the engineering firm Quinton, Code & Hill-Leeds at Los Angeles CA. He was there connected with many engineering projects, among which were Gibraltar Dam on the Santa Ynez River in Santa Barbara County CA, the Pine Canyon Dam on the San Gabriel River, the Big Tujunga Dam on Big Tujunga River, or the Bouquet Dam, all of which are located in Los Angeles County CA. He had further been consulting engineer for the State of California on various other dams, and also was consulting engineer for the Yaqui River Project in Sonora, Mexico. In the mid-1930s he was consulting engineer to the US Engineers on various dams near Zanesville OH, on the Conchas Dam in New Mexico, the Tygart Dam in West Virginia, the Sardis Dam in Mississippi, and the Fort Peck on Missouri River, and Bonneville Dams on Columbia River. He also was then a member of the Board of Consulting Engineers on Boulder Dam, todays Hoover Dam on Colorado River, and on the All-American Canal Hill in south-eastern California. Hill was elected in 1937 president of the American Society of Civil Engineers.

Anonymous (1937). Louis C. Hill. *Engineering News-Record* 118(Jan.21): 108. *P*
Billington, D.P., Jackson, D.C., Melosi, M.V. (2005). *The history of large federal dams*: Planning, design, and construction. USBR: Denver CO.
Hill, L.C. (1937). The engineer, an employee-employer. *Trans. ASCE* 102: 1173-1178.
Maji, A.K., Lucero, J.L. (2001). Historic civil engineering landmarks in New Mexico. *Intl. Engineering History and Heritage*: 304-311, J.R. Rogers, ed.. ASCE: Reston VA.
Roosevelt, T., Newell, F.H., Pinchot, G., Kermit, E., Hill, L.C. (1928). *The Roosevelt Dam*. Roosevelt Memorial Association: Washington DC.

HILL R.A.

31.03. 1892 Golden CO/USA
06.04. 1973 Pasadena CA/USA

Raymond Alva Hill obtained in 1914 the BS degree in civil engineering from University of Michigan, Ann Arbor MI, and the civil engineering degree in 1922. He was until 1917 engineering assistant of the US Reclamation Service, joining after war service in 1919 as engineer the engineering firm Quinton, Code & Hill, Berkeley CA, directed then by John H. Quinton (1850-1939), William H. Code (1865-1951), and Louis C. Hill (1865-1938). He was member of this firm until 1940 when its name changed to Leeds, Hill, Barnard & Hill-Leeds, and from 1945 to Leeds, Hill & Jewett, Pasadena CA, now with Raymond A. Hill as partner. He was member and honorary member of the American Society of Civil Engineers ASCE serving as ASCE Director from 1936 to 1938, the American Water Works Association AWWA, and the American Geophysical Union AGU.

Hill was well-connected with the circles of power, and by the mid-1930s the youngest member of ASCE's Executive Board. His notoriety led to his appointment in 1938 to negotiate the Rio Grande Compact adjudicating the waters of the Rio Grande River between CO, NM, and TX. One of his firm's most unusual projects was the Camp San Luis Obispo Dam (renamed the Salinas Dam), which retains Santa Margarita Lake. It was the first arch dam to utilize thrust blocks on both abutments. Hill brought in Louis Hill's protégé Paul Baumann (1892-1982) as a consultant to oversee the dam design, because of its similarity to Gibraltar Dam. The project was completed in 1942. The firm was associated with ongoing improvements to the Ports of Los Angeles and Long Beach CA. Hill was an influential figure in water resources development in the 1950s and 1960s, advising California State as member of the Board of Engineering Consultants to the California Dept. of Water Resources during the planning, design, and construction of the largest non-federal public works project in American history, the California Water Project, between 1956-1971. After the death of Leeds in 1960, Hill moved the headquarters to San Francisco, which he believed to be a more lucrative marketplace for selling the firm's burgeoning expertise in water resources and dam engineering.

Anonymous (1959). Hill, Raymond A. *Who's who in engineering* 8: 1116. Lewis: New York.
Anonymous (1964). Conservative design entails innovation. *Engineering News-Record*
        172(Apr.30): 40-43. *P*
Hill, R.A. (1941). Salts in irrigation waters. *Proc. ASCE* 67(6): 975-990.
Rogers, J.D. (2013). http://web.mst.edu/~rogersda/Geotechnical-Practice/

HINDS

22.12. 1881 Warrenton AL/USA
15.07. 1977 Santa Barbara CA/USA

Julian Hinds obtained the civil engineering degree from University of Texas, Austin TX, in 1908. He was then there an instructor of civil engineering until 1911, when joining the design office of the US Bureau of Reclamation USBR, Denver CO, for the Elephant-Butte and the Yakima-Sunnyside projects. From 1926 to 1929 Hinds was a resident engineer for the Calles project, Aguascalientes MX, joining then the Department of Water & Power, Los Angeles CA as hydraulic design engineer. He was transferred to the Metropolitan Water District in 1931, becoming assistant chief engineer in 1933 and general manager and chief engineer in 1941 until retirement in 1951. From then to 1956 Hinds was chief engineer of the United Water Conservation District, then becoming a private consultant.

Hinds was the first to correctly formulate the equation of spatially-varied flow in his 1926 paper, for which he was awarded the ASCE Norman Medal. He was also a co-author of various important books in dam engineering, given his practical experiences in hydraulic structures. In parallel he also developed a number of hydraulic structures, including stilling basins, Venturi flumes or gate elements. Hinds was awarded Honorary Membership of ASCE in 1959 for his achievements in irrigation, reclamation and water supply, particularly the development of Southern California's Colorado River supply. The Julian Hinds Award, an annual ASCE distinction, is presented from 1974 for outstanding research in the field of water resources development.

Anonymous (1941). Julian Hinds. *Engineering News-Record* 146(Apr.3): 508. *P*
Anonymous (1954). James W. Rickey Medal: Julian Hinds. *Civil Engineering* 24(9): 610. *P*
Anonymous (1959). Hinds, Julian. *Who's who in engineering* 8: 1020. Lewis: New York.
Anonymous (1978). Julian Hinds. *Trans. ASCE* 143: 564-565.
Creager, W.P., Justin, J.D., Hinds, J. (1945). *Engineering of dams*. Wiley: New York.
Hinds, J. (1920). The hydraulic jump and the critical depth in the design of hydraulic structures. *Engineering News-Record* 85(22): 1034-1040; 86(3): 135; 86(4): 185-186; 86(6): 272.
Hinds, J. (1925). Automatic spillway gates of Black Canyon Dam. *Engineering News-Record* 94(26): 1046-1050.
Hinds, J. (1926). Side channel spillways: Hydraulic theory, economic factors, and experimental determination of losses. *Trans. ASCE* 89: 881-939.
Hinds, J. (1935). Design of hydraulic structures. *Civil Engineering* 5(2): 77-82.
Hinds, J. (1942). Canals. *Handbook of hydraulics*, C.V. Davis, ed. McGraw-Hill: New York.

HINTON

23.01. 1916 Lumberton MS/USA
07.02. 1985 Knoxville TN/USA

Julian Pitts Hinton was employed as rodman by the Tennessee Valley Authority TVA in 1934 following the construction of Norris Dam and Reservoir, the agency's first project. He graduated from University of Tennessee, Knoxville TN, in 1938 with the BSc degree in civil engineering, subsequently doing his graduate work there. He then went as full time TVA engineer, joining its Power Studies Branch, later a part of the Project Planning Branch in TVA's Water Control Planning Division. After war service during World War II, he returned to TVA's Project Planning Branch in 1946, and from 1954 until his retirement in 1971 served as head of the Power Studies and Hydrology Section. His duties there primarily involved leading a group of engineers engaged in preparing general plans of river development, and in making detailed power studies needed in planning for new and existing hydroelectric power projects.

TVA is a federally-owned corporation created by the Congress in 1933, providing for navigation, flood control, electricity generation, and the economic development of the Tennessee Valley, a region particularly affected by the Great Depression. TVA not only was envisioned as a provider, but also as a regional economic development agency that would use federal experts and electricity to rapidly modernize the region's economy and society. Its service area covers Tennessee State, portions of Alabama, Mississippi, and Kentucky, and small parts of Georgia, North Carolina, and Virginia. It was the first large regional planning agency of the federal government and remains the largest. TVA became a model for America's governmental efforts seeking to assist in modernizing agrarian societies in the developing world. TVA's power mix is 11 coal-powered plants, 29 hydroelectric dams, three nuclear power plants with six operating reactors, nine simple cycle natural gas combustion turbine plants and five combined cycle gas plants. TVA is the largest public power utility in the US, and one of the largest producers of electricity in the country. It acts as a regional grid reliability coordinator. TVA's Watts Bar reactor produces tritium as a by-product for the US National Nuclear Security Administration, which requires tritium for nuclear weapons. Hinton served TVA all through his professional career.

Anonymous (1985). Julian P. Hinton. *Trans. ASCE* 150: 362-363.
http://en.wikipedia.org/wiki/Tennessee_Valley_Authority
http://tngenweb.org/anderson/ (*P*)

HOGAN

12.06. 1881 Chicago IL/USA
09.06. 1961 Santa Barbara CA/USA

John Philip Hogan obtained in 1903 the AB degree from Harvard University, Cambridge MA, and the SB degree in 1904 from New York University, New York NY, from where he also was awarded in 1940 the honorary Dr. Engr. degree. He was until 1920 division engineer and acting deputy chief engineer of the Board of Water Supply, New York City, then until 1923 director of the New York Water Power Investigations, from when he was a member of the engineering firm Parsons, Brinkerhoff, Hogan & MacDonald, New York City. He was in addition employed by the US Government, municipalities, and the Army Corps of Engineers on projects in water supply, hydropower, navigation, and transmission lines in the USA, South America, and Europe. He also served as vice-president and chief engineer of the 1939 New York World's Fair. Hogan was chairman of the Construction League of the USA. He was member of the American Society of Civil Engineers ASCE, serving in 1940 as president; he was also vice-president of the Society of American Military Engineers SAME, and member of the American Society of Mechanical Engineers ASME. He further was decorated with various awards.

Hogan was an expert in water supply and hydropower engineering. As partner of the New York firm from 1926 to 1947, he served as consultant to many private utilities, to industry, to New York City, New York State, and the US Army. In the 1930s he was consultant to the US Army Corps of Engineers on the construction of the Bonneville Dam Project on Columbia River. His famous report on harnessing the potential of the Delaware River was also done for the Corps. When the United States built air bases in the Caribbean, Hogan's firm was selected to design and supervise their construction. He was in charge of this work as well as other US projects during World War II.

Anonymous (1935). Lieut. Col. John P. Hogan. *The Military Engineer* 27(7-8): 307. *P*
Anonymous (1948). Hogan, John P. *Who's who in engineering* 6: 926. Lewis: New York.
Anonymous (1961). John P. Hogan dies: Former ASCE President. *Engineering News-Record* 166(Jun.22): 56.
Anonymous (2011). John P. Hogan, chief engineer. *Newsletter* on the History and Heritage of American Civil Engineering 5(2): 6. *P*
Hogan, J.P. (1926). Proportioning of units in low-head plants. *Trans. ASCE* 89: 677-679.
Hogan, J.P. (1931). Record hydro-electric turbine installed. *Civil Engineering* 1(6): 519-523.
Hogan, J.P. (1940). Relation of the Society to national defense. *Trans. ASCE* 105: 1781-1784.

HOLLOWAY

18.01. 1825 Uniontown OH/USA
01.09. 1896 Cuyahoga Falls OH/USA

Joseph(us) Flavius Holloway attended for only few terms the settlement school, receiving elementary instruction from his father, a cabinet-maker. Joseph eventually became interested in mechanics through assisting a repairer of watches and clocks. Later, he served an apprenticeship with an engine builder, and at age twenty went to Cabotsville MA, where he worked for a year as a machinist. Returning to Ohio in 1847, he became associated with the *Cuyahoga* Steam Furnace Company, and designed within a year the machinery for the *Niagara* screw-propeller boat built at Cleveland OH for service on the Great Lakes. The design of this machinery, after receiving the approval of the dean of the country's mechanical engineers Horatio Allen (1802-1890), secured Holloway a position with a boat-building firm at Pittsburgh PA. He designed and constructed for this firm in 1850 the machinery for two boats, which he took down the Ohio and Mississippi Rivers and up the coast to New York.

At Wilmington DE, he next designed and built a side-wheel iron steamer for the Cuban service. The success of the steam equipment in these crafts made Holloway's name known among engine builders and created a demand for his services. He next went to Cumberland MD, as manager for the Cumberland Coal and Iron Company, and shortly after from there to Shawneetown IL, where he took a similar position with the iron and coal works. He returned in 1857 to Cleveland and became successively superintendent, manager and president of the *Cuyahoga* Steam Furnace Company. From 1887, when the company merged with the Cleveland Steamboat Company, to 1894, he was connected with Worthington, hydraulic engineers, New York, serving both as vice-president and adviser to the engineering branches of this business. Holloway was a member of the American Society of Mechanical Engineers ASME, being its president in 1884.

Anonymous (1897). Joseph Flavius Holloway. *Trans. ASME* 18: 612-643.
Anonymous (1932). Holloway, Joseph Flavius. *Dictionary of American biography* 9: 155.
    Scribner's: New York.
Anonymous (1963). Holloway, J.F. *Who was who in America* 1607-1896: 256. Marquis: Chicago.
Anonymous (1980). Holloway. *Mechanical engineers in America born prior* 1861: 175. ASME. *P*
Holloway, J.F. (1883). Cost of steam power. *Trans. ASCE* 12: 432-435.
Holloway, J.F. (1889). *Pumping machinery, ancient and modern*. Cornell University: Ithaca NY.
Holloway, J.F. (1891). Discussion of Proportional water meter. *Trans. ASCE* 24: 535-536.

HOLLY

08.08. 1822 Auburn NY/USA
27.04. 1894 Lockport NY/USA

Birdsill Holly learnt the trade of machinist as apprentice at Seneca Falls NY. He gradually rose to superintendent and proprietor of the Uniontown PA machine shop. He organized a firm at Seneca Falls for the manufacture of hydraulic machinery, thereby inventing the *Silsby* steam fire engine and its unorthodox rotary engine and pump. He founded in 1859 the Holly Manufacturing Co., Lockport NY, producing sewing machines, cistern pumps, and rotary pumps. His shop facilities doubled in size to build the machinery for the Lockport water-works, designed also by Holly to pump water directly into the city mains without a reservoir. This system was applied by him in over 2,000 cities of the United States and Canada.

By 1876, while business prospered, Holly's interests shifted to the problem of heating buildings by steam. An experimental steam heating system tested in his home convinced him and others of the viability of wide-scale central steam heating, resulting in the 1877 founding of the Holly Steam Combination Co., Lockport NY. Heating systems using wood-insulated pipes were installed in city businesses and eventually supplemented with a series of improvements covered by over 150 patents issued to Holly between 1876 and 1888. To avoid inefficiencies of heating buildings with individual small boilers, his system used a large central boiler furnishing a loop of supply and return mains. Each customer was charged for the steam consumed, determined by metering the water of condensation. Various steam regulating and measuring devices were developed to control and monitor the flow of steam in the system and to individual delivery points, all of which were designed and produced at his company. The central steam heating systems of Holly design spread to cities throughout the nation resulting in the 1880s a reorganisation of his firm into the American District Steam Co., within which Holly worked as principal and consulting engineer until his death. He was also associated with the invention of the fire hydrant.

Anonymous (1980). Holly, Birdsill. *Mechanical engineers in America born prior to* 1861: 175-176. ASME: New York. *P*

Hague, C.A. (1907). *Pumping engines for water works.* McGraw-Hill: New York. *P*

http://lockportcave.com/birdsill-holly/

http://www.firehydrant.org/pictures/holly.html

http://www.google.ch/imgres?q=%22Birdsill+Holly *P*

HOOPER E.G.

28.07. 1882 Lynn MA/USA
24.10. 1934 New York NY/USA

Elmer Guy Hooper graduated in 1907 with the BSc degree from the University of Maine, Orono ME, and there received the civil engineering degree in 1911. He became instructor in civil engineering at Washington University, St. Louis MO, in 1907, and there was also engaged in hydraulic research. From 1912 to 1919 he served as assistant engineer in the Department of Water Supply, Gas, and Electricity of New York NY, solving the problems of distribution and detection of waste, of which the solutions were published. In 1919 he was appointed to the Faculty of Civil Engineering, New York University, serving as assistant professor and associate professor of civil engineering until 1931, when becoming there professor of hydraulics, a position he held until his death. From 1921 to 1931 Hooper was also associated with a colleague, who formed a consulting office with projects in water supply, dam engineering, and highway construction.

Hooper conducted hydraulic research at New York University, with results published in the ASCE journals. He further edited the manuscripts of the monumental work Hydraulic Laboratory Practice issued by John R. Freeman (1855-1932). Hooper had a keen mind; his work was characterized by thoroughness of research coupled with high professional standards and personal ethics that won the admiration and appreciation of his associates. He was also of kindly nature, helpful to any who came to him for assistance. One of his associates summed up his qualifications as: 'It was Hooper's character that made significant his achievements as an engineer. The qualities that I came to recognize of him were his integrity and his considerations of others. He had worked through his personal problems to where he had effected a reconciliation with life. Whatever he did was expressive of the whole of him, clear and lucid'. He was member of the Society for the Promotion of Engineering Education, and the American Society of Civil Engineers.

Anonymous (1907). Elmer G. Hooper. *Prism yearbook*: 71. University of Maine: Orono. *P*
Anonymous (1935). Elmer G. Hooper. *Trans. ASCE* 100: 1673-1675.
Freeman, J.R. (1929). *Hydraulic laboratory practice*. ASME: New York.
Hooper, E.G. (1911). *Experiments with Pitot tubes*. University of Maine: Orono ME.
Hooper, E.G. (1915). The location of leaks in submarine pipe lines. *Journal of the New England Water Works Association* 29(4): 536-538.
Hooper, E.G. (1923). What is correct design for highway pavement? *Good Roads* 65(14): 105-106.
Hooper, E.G. (1929). Discussion of New theory for the centrifugal pump. *Trans. ASCE* 93: 44-46.

HOOPER L.J.

15.02. 1903 Essex MA/USA
09.04. 1977 Holden MA/USA

Leslie James Hooper was educated at the Worcester Polytechnic Institute, receiving the BS degree in 1924, and the MS degree in civil engineering in 1928. He was from 1924 to 1927 a hydraulic test engineer in Sao Paulo, Brazil, laboratory assistant at Alden Hydraulic Laboratory until 1932, then at his Alma Mater part-time teacher until 1938 and in parallel Freeman Scholar inspecting hydraulic labs in the USA and in Canada. He joined in 1938 the Alden Lab until retirement, taking over as director in 1952, and was further from assistant to professor of hydraulic engineering at Worcester Polytechnic. He was member of the American Societies of Civil and Mechanical Engineers ASCE and ASME, respectively, and a John R. Freeman Fellow from 1934 to 1936. He received the 1937 ASME Junior Award for his work on laboratory description.

Hooper collaborated in the 1930s with Charles M. Allen (1871-1950) on piezometers as used in hydraulic laboratories for pressure measurement. The 1940 paper deals with the salt-velocity method for determining the average velocity of water flow, later mainly used in small rivers. During World War II Hooper worked on secret projects for the Navy, developing the Navy's Underwater and Sound Laboratory at New London CT. In 1954 he made a tour through Europe visiting several hydraulic laboratories. He was awarded in 1964 an honorary doctorate in engineering from the Worcester Polytechnic Institute. He also served as director of the Boston Society of Civil Engineers, and was involved further in ASME, serving as secretary and then chairman of its Hydraulic Division. He was elected Fellow ASME in 1960, and received in 1976 the Goddard Award from his Alma Mater. He died while visiting friends at Millington MD.

Allen, C.M., Hooper, L.J. (1932). Piezometer investigation. *Trans. ASME* 54(Hyd 54-1): 1-16.
Allen, C.M., Hooper, L.J. (1935). Ten-foot weir and Venturi meter compared. *Civil Engineering* 5(4): 218-220.
Anonymous (1964). Hooper, Leslie James. *Who's who in engineering* 9: 860. Lewis: New York.
Hooper, L.J. (1940). Salt-velocity measurements at low velocities in pipes. *Trans. ASME* 62(11): 651-661.
Hooper, L.J. (1950). Calibrations of six Beth-flow meters. *Trans. ASME* 72(11): 1099-1110.
Hooper, L.J. (1960). Discharge measurements by the Allen salt-velocity method. Symposium *Flow measurement in closed conduits* Glasgow E(1): 1-17.
Rouse, H. (1976). Leslie Hooper. *Hydraulics in the United States* 1776-1976: 123. *P*

HORNE

12.10. 1873 Belmont MA/USA
23.05. 1928 Holden MA/USA

Harold Wellington Horne graduated from Harvard University, Cambridge MA, with the BA degree in 1894, and from Lawrence Scientific School, Harvard University, in 1896 with the civil engineering degree. He was first rodman with the Metropolitan Water Board of Massachusetts, advancing to engineer and being engaged on various structures for the water distribution to Boston MA. From 1906 to 1912 he served as assistant engineer the Board of Water Supply, New York City, first on the surveys for the Catskill Aqueduct, and then in the construction of a section including the Bull Hill Tunnel and Foundry Brook Pipe Siphon.

Horne became in 1912 associated with the Board of Water Commissioners of Hartford CT as division engineer, for the additional water supply. He was first engaged with topographical and subsoil surveys for reservoir and dam locations, then on constructing various structures including the *Nepaug* Reservoir and the large earth dam at Philips Brook, together with a 13 km long, 1.2 m diameter pipe. He also was involved in the construction of the dam at New Hartford CT until 1918. He was then assistant division engineer with the Miami Conservancy District of Ohio, in charge of the construction of the Englewood Dam for the flood control of the Miami River. Horne was for the next five years a private consultant, but he did not retire yet because he was convinced that his education and experience was still required. He became thus in 1926 assistant district engineer, the Metropolitan District Water Supply Commission of Massachusetts, and was promoted in 1927 to district engineer continuing in this position until his death. He conducted preliminary work for the aqueduct tunnel from Swift River to Wachusett Reservoir, and was subsequently in charge of the construction of a part of this tunnel. He lived to complete only the shafts. He left to his colleagues a memory of a kindly spirit, wide culture, and a soundly educated mind. They recalled the singular fidelity and painstaking devotion to whatever task was in his hand, and his true expression of his honourable nature. He was member of the American Society of Civil Engineers ASCE, and the New England Water Works Association.

Anonymous (1929). Harold W. Horne. *Trans. ASCE* 93: 1829-1830.
Horne, H.W. (1917). Engineering on the additional water supply for the city of Hartford. *Journal of the New England Water Works Association* 31(4): 588-594.
https://www.google.ch/search?q=%22Water+supply%22+Hartford+CT (*P*)

HORNER W.W.

22.09. 1883 Columbia MO/USA
23.09. 1958 St. Louis MO/USA

Wesley Winans Horner obtained in 1905 the BSE degree from Washington University, St. Louis MO. He was from then to 1919 assistant engineer, from when he was chief engineer for the city of St. Louis until 1932, becoming a consultant and senior partner of Horner & Shifrin, St. Louis MO, until retirement. He was in parallel from 1934 to 1942 lecturer and later professor of sanitary and hydraulic engineering at his Alma Mater, and water consultant and member of the national water commission within the National Resources Planning Board. Horner was a member of the American Geophysical Union, the American Public Works Association, presiding over it in 1923, the American Society of Civil Engineers ASCE, being its director in the term of 1933, and president in 1946. He was the recipient of the 1937 Rudolph Hering Medal for his 1936 ASCE paper, and the 1955 Gold Achievement Award Medal of the Engineers Club of St. Louis.

Horner was a notable civil engineer who devoted much activity to the city of St. Louis. It included the design and supervision of an extensive program of street widenings, traffic aids and sewer work. He played a major role in planning a comprehensive sewer system and in designing the East Side Levee and Sanitary District. His specialties included sanitary and municipal engineering, and water supply. Earlier in his career he developed the rational method for sewer design. He also worked in culvert hydraulics, an important hydraulic structure in flood defence that often fails under high discharges. In 1953, former president Hoover named him to a 26-man task force to study federal activities in the field of water resources and power development. He was awarded the honorary doctorate of engineering from his Alma Mater in 1952.

Anonymous (1946). Wesley W. Horner. *Engineering News-Record* 136(Jan.17): 85. *P*
Anonymous (1951). Horner, Wesley W. *Who's who in America* 26: 1294. Marquis: Chicago.
Anonymous (1958). W.W. Horner, former ASCE president, dies. *Civil Engineering* 28(11): 873. *P*
Anonymous (1959). Wesley Winans Horner. *Trans. ASCE* 124: 1049-1050.
Horner, W.W. (1913). A rational culvert formula. *Engineering News* 69(18): 912-913.
Horner, W.W., Flynt, F.L. (1936). Relation between rainfall and run-off from small urban areas. *Trans. ASCE* 101: 140-183.
Horner, W.W., Jens, S.W. (1942). Surface runoff determination from rainfall without using coefficients. *Trans. ASCE* 107: 1039-1117.
Horner, W.W. (1957). Hydrology. *American civil engng. practice* 3, R.W. Abbett, ed. Wiley: NY.

HORTON A.H.

24.07. 1875 Fairview PA/USA
04.03. 1945 Washington DC/USA

Albert Howard Horton obtained the civil engineering degree from Cornell University, Ithaca NY, in 1898. His first employment was with the US Board of Engineers on Deep Waterways, engaged on field surveys in New York State. From 1899 to 1903 he was with the US Lake Survey, conducting stream-flow measurements on the Niagara and St. Lawrence Rivers. He was then appointed assistant engineer in the US Geological Survey, engaged on surveys for the Shoshone and Utah Lake Projects. In 1905 he made discharge observations in the Central Atlantic States, and in 1906 he became district hydrographer in charge of the Upper Mississippi River District, with headquarters at Chicago IL. During this period he studied the errors resulting from the use of dams as gaging stations. In 1907 he was transferred to Washington DC and later to Newport KY, where he acted as district engineer in charge of surface water studies of the Ohio River.

In connection with the flood measurements of Ohio River, Horton developed techniques for velocity observations. He supervised the Ohio River work from Washington DC from 1913, thereby also making important contributions to the hydrologic literature through the analysis of fragmentary stage and discharge records of the Ohio River, which led to the compilation of long-time records of this important waterway. During World War I the Division of Power Resources was organized by the Geological Survey, with Horton as its chief from 1919. A collection of data pertaining to the installed capacities, and output of all electric power was started. These records were of inestimable value in connection with the prosecution of WWI. Horton continued this work until 1936, when these activities were transferred to the Federal Power Commission. Horton was given credit for the remarkable increase in the efficiency of power plants during these years. He had become district engineer in charge of surface water investigations in the Middle Atlantic District until 1940, with headquarters at Washington DC. Many perfections in the technique of stream gaging were brought about through his advice and suggestions. During his near 47 years in federal employment, he had trained many staff members so that he was highly respected among his organization.

Anonymous (1945). A.H. Horton. *Engineering News-Record* 134(Jan.04): 8; 134(Mar.8): 302. *P*
Anonymous (1946). Albert H. Horton. *Trans. ASCE* 111: 1589-1591.
Grover, N.C., Horton, A.H., Paulsen, C.G. (1922). Surface water supply of the United States 3: Ohio River Basin. *Water-Supply Paper* 473: USGS: Washington DC.

HORTON R.E.

18.05. 1875 Parma MI/USA
22.04. 1945 Voorheesville NY/USA

Robert Elmer Horton obtained the BSc degree from Albion College, Albion MI in 1897, and the ScD degree in 1932. He was until 1899 assistant of the US Deep Waterways Survey, then district engineer until 1906 with the US Geological Survey, engineer in charge of New York State Barge Canal until 1911 until becoming private consultant. He was hydraulic expert for the Department of Public Works in parallel until 1925, engineer in charge of Delaware River before the Supreme Court for New Jersey State, and consulting engineer for the Board of Water Supply, Albany NY. Other commitments included the Power Authority, State of New York, and chairman of the Board Consultants, Flood Control for the US Department of Agriculture from 1940. He was a member of the American Society of Civil Engineers ASCE, the American Water Works Association AWWA, and the American Geophysical Union AGU.

Horton authored some 150 papers and a number of books, including Turbine water wheel tests and power table in 1906, Weir experiments, and Water wheels in 1907, Determination of stream flow during the frozen season in 1907, Hydrography of New York State from 1900 to 1911, Hydrology of the Great Lakes in 1926, Rainfall, runoff and evaporation with Leroy K. Sherman (1869-1954) in 1933, and Surface runoff phenomena in 1935. The Robert E. Horton Medal, established in 1974, is annually presented for outstanding contributions to the geophysical aspects of hydrology.

Anonymous (1941). Horton, Robert E. *Who's who in engineering* 5: 856. Lewis: New York.

Anonymous (1945). R.E. Horton dies. *Engineering News-Record* 134(Apr.26): 597.

Anonymous (1984). The Robert E. Horton Medal. *Eos* 65(35): 520. *P*

Hall, F.R. (1987). Contributions of Robert E. Horton. *History of Geophysics* 3: 113-117.

Horton, R.E. (1907). Weir experiments, coefficients and formulas. *Water Supply and Irrigation Paper* 200. Department of the Interior, US Geological Survey. Government Printing Office: Washington DC.

Horton, R.E. (1918). Determining the regulating effect of a storage reservoir. *Engineering News-Record* 81(10): 455-458.

Horton, R.E. (1932). Drainage-basin characteristics. *Trans. AGU* 13: 350-361.

Horton, R.E. (1933). The role of infiltration in the hydrologic cycle. *Trans. AGU* 14: 446-460.

Horton, R.E. (1935). *Surface runoff phenomena*. Edward: Ann Arbor MI.

Horton, R.E. (1945). Erosional development of streams. *Bulletin* Geological Society 56: 275-370.

HOUK

18.02. 1888 Battle Creek IA/USA
10.08. 1972 Denver CO/USA

Ivan Edgar Houk obtained the BE degree in civil engineering from State University of Iowa in 1911. He was first a computer of an engineering office at Memphis TE, and then assistant engineer and flood forecaster of Miami Conservancy District, Dayton OH, until 1921. He further was in charge of stream measurements, rainfall investigations and open channel experimentation as city engineer of Dayton until 1923, when being appointed engineer of the US Bureau of Reclamation USBR, Denver CO, in charge of the Boulder Canyon Dam or the Grand Coulee Dam designs until 1946. After retirement he became a consultant at Denver CO. He was, among others, involved in the *Mascarehnas* Hydroelectric project near Rio de Janeiro, Brazil. He was a Fellow ASCE, and member of the American Meteorological Society AMS, the American Geophysical Union AGU, and the Colorado Historical Society, among other.

Houk became known in the USA after having published his 1918 report on open channel flow. This work includes the computation of backwater curves, then not yet standardized, and proceeding manually from one to the next cross-section. This report validated earlier approaches mainly developed in Europe, and successfully compared predictions with prototype observations. Another part of the Report relates to Rainfall and runoff in the Miami Valley. Once Houk had joined USBR he was involved in all aspects of dam design, including also hydraulic computations of spillways, intakes or conduits. He was known for the two book volumes Irrigation engineering, including (1) Agricultural and hydrological phases, and (2) Projects, conduits and structures.

Anonymous (1955). Ivan E. Houk. *The Engineers Bulletin* 39(2): 20. *P*
Anonymous (1959). Houk, Ivan E. *Who's who in engineering* 8: 1169. Lewis: New York.
Anonymous (1973). Ivan Edgar Houk. *Trans. ASCE* 138: 628-629.
Houk, I.E. (1918). Calculation of flow in open channels. *Technical Reports* 4. The Miami
    Conservancy District: Dayton OH.
Houk, I.E. (1922). Hydraulic design of bridge waterways. *Engineering News-Record* 88(26):
    1071-1075.
Houk, I.E. (1927). Evaporation on US Reclamation projects. *Trans. ASCE* 90: 266-378.
Houk, I.E. (1932). Model tests confirm design of Hoover Dam. *Engineering News-Record*
    108(Apr.7): 494-499.
Houk, I.E. (1951). *Irrigation engineering*. Wiley: New York.

HOWE

19.01. 1902 Omaha NE/USA
19.10. 1983 Iowa City IA/USA

Joseph Warner Howe obtained the BE degree in 1924, and the MSc degree in 1925 from Iowa State University. He joined the Mississippi River Power Co., Keokuk IA, until 1927, becoming instructor of theoretical and applied mechanics at the University of Illinois, Urbana IL, for the next year. From 1929, he was an assistant professor of mechanics and hydraulics at University of Iowa, then associate professor until 1942, from when he was head of department until retirement. After retirement Howe was a consulting engineer at Iowa City IA.

Howe was in charge of hydraulic tests on flow through rectangular notches for the US Department of Agriculture. He investigated also the precipitation of Iowa State. He co-authored the 1953 book with Hunter Rouse (1906-1996), and edited with John S. McNown (1916-1998) the third Hydraulics Conference at University of Iowa in 1947. He contributed to technical journals, including a review on the then current methods of open channel discharge measurement, or the flow features of weirs of circular sharp-crested plan arrangement. His work on fire monitors and nozzles as used for fire fighting was a significant contribution to this field, after the Pearl Harbour Disaster in 1941, which almost completely destroyed the Pacific Navy of the USA. Howe and his co-authors were able to demonstrate that a high turbulence level at the nozzle exit is the main cause for not having been able to extinguish the fire. Proposals were made on how these deficiencies could be improved, which were soon later adopted by fire brigades.

Anonymous (1942). Second Hydraulics Conference, Iowa State University: Frontispiece. *P*
Anonymous (1959). Howe, Joseph W. *Who's who in engineering* 8: 1176. Lewis: New York.
Anonymous (1983). Joseph W. Howe. *Eos* 64(46): 931.
Camp, C.S., Howe, J.W. (1939). Tests of circular weirs. *Civil Engineering* 9(4): 247-248.
Howe, J.W. (1950). Flow measurement. *Engineering hydraulics*: 177-229, H. Rouse, ed. Wiley: New York.
Howe, J.W., Shieh, G.C., Obadia, A.O. (1955). Aeration demand of a weir calculated. *Civil Engineering* 25(5): 289.
Rouse, H., Howe, J.W., Metzler, D.E. (1952). Experimental investigation of fire monitors and nozzles. *Trans. ASCE* 117: 1147-1175; 117: 1186-1188.
Rouse, H., Howe, J.W. (1953). *Basic mechanics of fluids*. Wiley: New York.
Yen, C.H., Howe, J.W. (1942). Effects of channel shape on losses in a canal bend. *Civil Engineering* 12(1): 28-29.

HOWELLS

26.03. 1859 Bowling Green OH/USA
22.04. 1927 San Francisco CA/USA

Julius Merriam Howells received his education at the Earlham College, Richmond IN. After work with various railway companies, he made from 1882 to 1884 studies of the tertiary history of Sierra Nevada at Quincy CA. Until 1888 he was then assistant and city engineer of Richmond IL, and practiced as a consultant at Chicago IL until 1892. Until 1894 he then was chief engineer of the Santa Fe NM Water Company, being involved in his first hydraulic work of hydraulic-fill dams, experimenting with hydraulic-fill processes. He shortly later built the first of these dams for the Water Company, Tyler TX. From 1895 to 1898 Howells was chief engineer of the San Diego Flume Co., building its La Mesa Dam. From 1899 to 1901 he was consulting engineer for the San Joaquin Electric Co., for which he erected the hydraulic-fill and rock-fill Crane Valley Dam in California.

Howells became associated in 1901 with James D. Schuyler (1848-1912). A number of dam projects resulted from this friendship, including the Lake Francis Dam in Eastern Texas for the Pacific Gas and Electric Company, or the Big Meadows and Feather River hydro-electric development in California. From 1905 to 1909 Howells was employed by the Kobe Syndicate to develop an Anglo-Japanese hydro-electric project, and then was for one year in Switzerland. He was from 1910 to 1912 chief engineer on an irrigation system near Barceloneta, Porto Rico, for which Schuyler was consulting engineer. Until 1927, Howells conducted a general private consulting office at San Francisco CA, with projects for the East Bay Water Company, the Western Canal Company, or the James Irvine Water Development. He designed and built also the Big Meadows Dam in 1913, and the Butte Valley Dam in 1923. Howells had a natural knowledge of proportion and logic, viewing all problems in their entirety without being confused by details. He had a rigid code of business ethics, and a fair but strict idea of justice. He was member ASCE.

Anonymous (1929). Julius M. Howells. *Trans. ASCE* 93: 1834-1836.

Howells, J.M. (1906). *Feather River power project*: Reports, photographs, maps. Stanford CA.

Jackson, D.C. (1984). *Great American bridges and dams*. Wiley: New York.

Schuyler, J.D. (1908). Reservoirs for irrigation, water power, and domestic water supply. Wiley: New York.

Schuyler, J.D., Howells, J.M. (1911). *Report* on a system of irrigation for the sugar cane lands, near Barceloneta, owned or leased by the Plazuela Sugar Company. Schuyler Papers.

http://www.wplives.com/frc/stairway_of_power.html (*P*)

HOWLAND

22.03. 1900 Athens NY/USA
15.10. 1980 Lafayette IN/USA

Warren Every Howland obtained the SB degree in 1922 from Massachusetts Institute of Technology, Cambridge MA, the MSc degree in 1929 from the Purdue University, West Lafayette IN, and the PhD degree in 1939 from Harvard University, Cambridge MA. He was from 1922 to 1923 junior assistant at Harvard University, until 1925 assistant engineer at the Sanitary District of Chicago IL, and from 1926 from instructor to professor of sanitary engineering at Purdue University. Howland was member of the American Water Works Association AWWA, and the American Society of Civil Engineers ASCE.

Howland was working in the 1920s to the 1940s mainly in pure hydraulics, and from then took interest in sanitary engineering. He also had a lifelong interest in questions of engineering education, and the history of technical sciences, mainly in hydraulics. The 1948 paper deals with orifice flow, which was usually solved by the use of the energy equation, or the Bernoulli equation given the potential flow. Few attempts were made to apply the momentum conservation principle because the pressure distribution is *a priori* unknown. This research attempts to discuss both methods, debating the pros and cons. The 1975 paper deals with the old problem of maximum discharge of a trapezoidal channel. Using the Manning formula for uniform flow, the result depends essentially on the hydraulic radius, and the bottom width of the section for a certain bottom slope and boundary roughness.

Anonymous (1935). Current research work at Purdue. *Engineering News-Record* 115(14): 456-458
Anonymous (1948). Howland, Warren E. *Who's who in engineering* 6: 961. Lewis: New York.
Howland, W.E., Richetta, J.D. (1937). Derivation of coefficients of orifices. *Journal of the Franklin Institute* 223(1): 83-94.
Howland, W.E. (1942). Interest in education. *School and Society* 56(1445): 173-178.
Howland, W.E. (1957). Flow over porous media as in a trickling filter. Proc. 12th Conference on *Industrial Waste* Lafayette: 435-465.
Howland, W.E. (1959). *The appeal of engineering teaching*. Purdue University: Lafayette.
Howland, W.E., Pohland, F.G., Bloodgood, D.E. (1960). *Kinetics in trickling filters*. Purdue University: Lafayette.
Howland, W.E., Howland, H.C. (1975). Economic shape of trapezoidal open channel sections with freeboard. *Journal of the Hydraulics Division* ASCE 101(HY5): 639-643.
Rouse, H. (1976). W.E. Howland. *Hydraulics in the USA* 1776-1976: Frontispiece. *P*

HOWSON

26.09. 1883 Sacramento CA/USA
07.08. 1952 Oakland CA/USA

George William Howson obtained in 1909 the BSc degree in civil engineering from the University of California, Berkeley CA. He was then civil engineer of the Sierra & San Francisco Power Co. As member of the Engineering Commission, he went to Greece to develop the municipal water supply of its capital Athens. On his return to the USA, he was appointed engineer of the Great Western Power Co., Oakland CA. In the 1920s Howson became research engineer of the Dix River Dam and Power Development in Kentucky. Later, Howson was travelling engineer for the Public Works Association PWA in Central California within the National Resources Planning Board, in charge of engineering in California, Idaho, Montana, and Oregon States. He was further design engineer on the Strawberry Dam in Utah after having joined the US Bureau of Reclamation USBR as engineer. He thereby was also coordinator on the 24 Central Valley Study Reports, composed of representatives from Federal, State, and local agencies, embracing many complicated problems such as allocation of costs, rates, and needed legislation.

The Strawberry Valley Project is centred around Spanish Fork UT; it provided the first large-scale trans-mountain diversion from Colorado River Basin to Bonneville Basin, and was an early USBR projects to develop hydro-electric energy. The original project included Strawberry Dam and Reservoir, Indian Creek Dike, Strawberry Tunnel, two diversion dams, three power plants, and a canal system. In 1974 Strawberry Reservoir was enlarged by the construction of the Soldier Creek Dam 10 km downstream of the former dam, to provide increased storage capacity. Water is now supplied through the Sixth Water Aqueduct and Syar Tunnel, which has replaced the old Strawberry Tunnel. The second large hydro-power project in which Howson was involved is Dix Dam. It was built to create a reservoir for operating a hydro-electric power plant on Kentucky River. Construction began in 1923, and the scheme was completed in 1927. It was then the largest rock-fill dam of the world of 87 m height and 330 m length.

Anonymous (1959). Howson, George W. *Who's who in engineering* 8: 962. Lewis: New York.
Anonymous (1952). George W. Howson. *Engineering News-Record* 149(Aug.21): 74.
Howson, G.W. (1916). Unique method used to build rock-fill dam. *Engng. News* 75(13): 604-606.
Howson, G.W. (1917). Discussion of Multiple-arch dams on Rush Creek, California. *Trans. ASCE* 81: 900-901. (*P*)
http://www.usbr.gov/projects/Project.jsp?proj_Name=Strawberry+Valley+Project

HOYT J.C.

10.06. 1874 Lafayette NY/USA
21.06. 1946 Paris VA/USA

John Clayton Hoyt graduated as a civil engineer from Cornell University, Ithaca NY in 1897, joining then the Cornell Hydraulic Laboratory Construction Company, and working for the US Deep Waterways Commission until 1899. From 1900 to 1902 he was engineer with the US Coast and Geodetic Survey, and to 1911 within the US Geological Survey, from when he was a hydraulic engineer in charge of the Surface Water Division. Later he directed the Union Trust Company, Washington DC. Hoyt was an US delegate to the International Navigation Congresses in London 1923, and Cairo 1926. He further was a delegate to the 1st World Power Congress, London, held in 1924. He was associated to ASCE, as a member, director from 1920 to 1922, and vice-president in 1927.

Hoyt did professional work for the geological survey in various States, including also Hawaii and Alaska. He has written a notable book with a colleague on River discharge, in which the various methods for river discharge measurement are described. This is a task still demanding a considerable effort, and is as relevant today as a century ago. Many of the riverine infrastructure must be designed based on the flood discharge, so that this is an essential basis for all works involving rivers. During his long service for purposes of water resources, Hoyt achieved a nation-wide reputation for his knowledge of hydrology and his improvements in equipment and methods for obtaining hydrologic data. He had a long-lasting influence on his colleagues particularly in the United States.

Anonymous (1939). Hoyt, John Clayton. *Who's who in America* 20: 1274. Marquis: Chicago.

Anonymous (1946). J.C. Hoyt, former ASCE vice-president, is dead. *Civil Engineering* 16: 366. P

Anonymous (1947). John Clayton Hoyt. *Trans. ASCE* 112: 1476-1478.

Hoyt, J.C. (1904). Methods of estimating stream flow. *Engineering News* 51(Jan.14): 47-48; 52(Aug.4): 104-105.

Hoyt, J.C., Grover, N.C. (1907). *River discharge*. Wiley: New York, 4th ed. in 1916.

Hoyt, J.C. (1910). The use and the care of the current meter. *Trans. ASCE* 66: 70-134.

Hoyt, J.C. (1914). Stream gaging stations as component parts of water-power and other hydraulic works. *Engineering News* 69(17): 834-835.

Hoyt, J.C. (1914). Artificial control sections for river measurement sections. *Cornell Civil Engineer* 23(5): 176-179.

Hoyt, J.C. (1939). Fifty years of water resources study. *Engineering News-Record* 122(Mar.16): 396-397.

HOYT W.G.

25.08. 1886 Lafayette NY/USA
25.08. 1971 Annandale VA/USA

William Glenn Hoyt, the brother of John Clayton (1874-1946), graduated as a civil engineer from Cornell University, Ithaca NY in 1909, joining then the US Geological Survey as a hydraulic engineer. He there was from 1921 to 1939 a district engineer of the Upper Mississippi River, from when he was principal hydraulic engineer. He was appointed in 1944 an executive officer of the Water Resources Commission, US Dept. of the Interior. Hoyt was a member of the American Society of Civil Engineers ASCE, and of the Washington Society of Engineers, serving as secretary from 1922 to 1928.

Hoyt was mainly interested in the run-off process related to hydrology. He was also working in aspects of power engineering, but again relating to the hydrologic design bases. The 1913 report on the effect of ice on stream flow presents a relationship between winter discharges of rivers and the meteorological factors precipitation and temperature. A method of stage correction is proposed. Sudden jumps due to ground ice are eliminated and it is maintained that actual measurements under the ice are the best guide for these corrections. Later in his career, Hoyt worked on floods, thereby collaborating with Walter B. Langbein (1907-1982). They for instance estimated that if floodplain development paralleled general development, the Ohio River floods would cause tremendous damages. Other works were directed to the relation between stream flow and forests, a topic that is still widely under research.

Anonymous (1939). Hoyt, William Glenn. *Who's who in engineering* 5: 868. Lewis: New York.
Anonymous (1971). Geophysicists: William G. Hoyt. *Trans. AGU* 53(9): 830.
Hoyt, W.G. (1912). Gaging Minnesota streams in winter. *Engineering News* 68(Sep.12): 499-502.
Hoyt, W.G. (1913). The effects of ice on stream flow. *Water Supply Paper* 337. Washington DC.
Hoyt, W.G. (1942). The runoff cycle. Physics of the Earth 9: *Hydrology*: 507-513, O.E.
        Meinzer, ed. McGraw-Hill: New York.
Hoyt, W.G., Langbein, W.B. (1955). *Floods*. Princeton University Press: Princeton.
Kolupaila, S. (1960). W.G. Hoyt: Early history of hydrometry in the United States. *Journal of the Hydraulics Division* ASCE 86(HY1): 39. *P*
Langbein, W.B., Hoyt, W.G. (1959). *Water facts for the nation's future*: Uses and benefits of hydrologic data programs. Ronald: New York.
Langbein, W.B., Hoyt, W.G. (1976). Hydrology and environmental aspects of Erie Canal (1817-1899). *Water Supply Paper* 2038. US Government Printing: Washington DC.

HUBBARD

04.03. 1921 Macon MO/USA
19.01. 2002 Iowa City IA/USA

Philip Gamaliel Hubbard obtained the MS and the PhD degrees from University of Iowa, Iowa City IA. He was research assistant in physics there from 1946 to 1949, then research associate at Iowa Institute of Hydraulic Research IIHR until 1954, from when he was assistant professor of mechanics and hydraulics there until 1959, finally taking over as full professor. In 1961 Hubbard was visiting professor at Colorado State University, Fort Collins CO. He was the first Afro-American professor at University of Iowa. He was appointed dean of academic affairs in 1966, a position he held until retirement in 1990. He then contributed to various academic and educational organisations.

Hubbard was known for excellent research in hydraulics and mechanics. He for instance co-edited the book Advanced mechanics of fluids with Hunter Rouse (1906-1996), then the head of IIHR and the tutor of Hubbard's PhD thesis. As an electrical engineer, he was well-qualified to submit a PhD thesis on the application of hot-wire anemometry in fluid flows, a method then not generally applied to problems in hydraulic engineering. Hubbard's research interests later included the study of the motion of matter in gas, liquid, plastic or plasma state. He pioneered turbulence measurements using again hot-film anemometry. He developed this type of instrumentation, corresponding to a device that has a length of very thin wire heated with an electrical current and placed in fluid flow. Measuring the rate at which the heat is dissipated from the wire leads to the measurement of fluid turbulence. The wire is usually only 50 μm in diameter. Such a thin wire allows for the device to respond quickly, so that fluid turbulence is assessed.

Anonymous (1964). Hubbard, Philip G. *Who's who in engineering* 9: 880. Lewis: New York.
Anonymous (1966). P.G. Hubbard. *Civil Engineering* 36(2): 10. *P*
Anonymous (1995). Hubbard. *Notable 20th century scientists* 2: 966-967. Gale: New York. *P*
Hubbard, P.G. (1949). Use of the electrical analogy in fluid mechanics research. *Review of Scientific Instruments* 20(11): 802-807.
Hubbard, P.G. (1954). Constant-temperature hot-wire anemometry with application to measurements in water. *PhD Thesis*. Iowa State University: Iowa IA.
Ling, S.C., Hubbard, P.G. (1956). The hot-film anemometer: A new device for fluid mechanics. *Journal of Applied Sciences* 23(9): 890-891.
Rouse, H., Hubbard, P.G., eds. (1959). *Advanced mechanics of fluids*. Wiley: New York.
http://www.lib.uiowa.edu/spec-coll/archives/guides/RG99.0248.htm *P*

HUBBELL

10.04. 1870 Cole County MO/USA
01.02. 1950 Detroit MI/USA

Clarence William Hubbell obtained the BS degree from the University of Michigan, Ann Arbor MI, in 1893, and the civil engineering degree in 1903. He was from 1893 to 1898 chief draughtsman of the Detroit Water Works Department, then until 1907 engineer in charge, from when he was until 1910 principal assistant engineer of water supply, and city engineer of Manila, the Philippine Islands. After having been chief engineer of public works there until 1914, he returned to Detroit MI, first as consulting engineer on Detroit pollution, then until 1922 as city engineer. He had founded in 1915 also his civil engineering company at Detroit, from where he retired in 1947. Hubbell was a member of the Board of consulting engineers, Detroit Sewage Collection and Treatment Project from 1925 to 1937. He also was a member of the American Society of Civil Engineers ASCE, and a recipient of its 1903 Norman Medal. He further was a member of the American Water Works Association AWWA, serving as president the North-east section.

Hubbell served as engineer on the design and the construction of large sewer systems of Detroit, involving an outlay of some 40 Million US$. He was also on the Board of the Sanitary District, Chicago IL, in the mid-1920s, and in the 1930s for Toronto, Ont. He had previously been involved in the 1902 paper dealing with the effect of curvature on pipe flow, one of the very early accounts on this topic, which was co-authored by Gardner S. Williams (1866-1931), and George H. Fenkell (1873-1949). The extremely long paper deals with the additional head loss in circular pipes curved by various relative radii and for a 90° flow deflection. The work includes a description of the Pitot tubes used, the experimental program, the distribution of velocities in curved pipes, and a discussion of the results. This paper attracted a large number of discussion, indicating the general interest in this topic. Hubbell was struck and killed by an automobile at age 79. The Hubbell Family Paper 1859-1983 are archived at the University of Michigan.

Anonymous (1924). Clarence W. Hubbell. *Engineering News-Record* 93(21): 846. *P*
Anonymous (1950). Clarence W. Hubbell. *Civil Engineering* 20(3): 226.
Anonymous (1960). Hubbell, Clarence W. *Who was who in America* 3: 424. Marquis: Chicago.
Hubbell, C.W. (1946). Sludge disposal practices at Detroit. *Sewage Work Journal* 18(3): 212-220
Williams, G.S., Hubbell, C.W., Fenkell, G.H. (1902). Experiments at Detroit MI on the effect
    of curvature upon the flow of water in pipes. *Trans. ASCE* 47: 1-369.

HUBER W.L.

04.01. 1883 San Francisco CA/USA
30.05. 1960 San Francisco CA/USA

Walter Leroy Huber obtained the BS degree in civil engineering from the College of Civil Engineering, University of California, Berkeley CA, in 1905. He was then assistant to John D. Galloway (1869-1943), from 1908 chief engineer and supervising architect at his Alma Mater, from 1910 then district engineer of the US Forest Service, when forming his private practice. Huber was member of the American Society of Civil Engineers, director from 1922 to 1924, vice-president from 1926 to 1927, and ASCE President in 1953. The Walter L. Huber Collection contains over 4,000 photographs of engineering and social interest. The Walter L. Huber Civil Engineering Research Prize recognizes members of the Society having demonstrated notable achievements in research related to civil engineering, with preference to members below 40 years of age.

During his years as a private consultant, Huber was engaged in hydro-electric, hydraulic and structural work, and valuation throughout the West. He was a consultant for dams, including Hogan, Big Tujunga, and Morris Dam in California; he also was consulting engineer for the California Debris Commission on arch dams; a member of the board of consulting engineers of the US War Department on flood control projects on the Los Angeles, the San Gabriel, and the Santa Ana Rivers, including the associated dams; member of the Governor's Advising Engineering Commission on the State Water Resources Development in 1931; special consultant of the War Department on water supplies; regional water consultant of the National Resources Planning Board from 1934 to 1943; and a special consultant on earthquake resistant design. He has written a large number of papers on these topics. He was named in 1954 by President Eisenhower adviser on the controversial study of the Arkansas-White-Red River Basins.

Anonymous (1948). Huber, Walter L. *Who's who in engineering* 6: 966. Lewis: New York.
Anonymous (1952). Huber nominated for highest ASCE post. *Engineering News-Record* 148(May8): 80. *P*
Huber, W.L. (1913). *Report* on the hydro-electric system of the Utica Gold Mining Company. Sierra and San Francisco Power Co.: San Francisco.
Huber, W.L. (1953). An engineering century in California. *Trans. ASCE* CT: 97-111.
Huber, W.L. (1960). *Scrapbooks* on hydroelectric and hydraulic projects, and water issues, 1913-1960. San Francisco.
http://library.ucr.edu/wrca/collections/huber/index.html *P*

HUGHES

23.10. 1871 Centralia PA/USA
01.03. 1930 Cambridge MA/USA

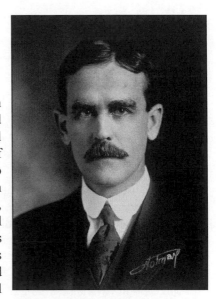

Hector James Hughes obtained the AB degree from Harvard University, Cambridge MA, in 1894, and the SB degree from its Lawrence Scientific School in 1899. He was from 1894 to 1898 in the office of the town engineer, Brookline MA, and from 1900 to 1902 resident engineer in charge of construction in Iowa. Further, from 1903 to 1913 he was associate, and from 1914 until his retirement professor of civil engineering at the Harvard University. He also was from 1914 to 1918 professor at the Massachusetts Institute of Technology MIT, Cambridge MA, and dean of the Harvard University Engineering School from 1920. Hughes was Fellow of the American Academy of Arts and Science AAAS, the American Society of Municipal Improvement ASMI, the American Society of Civil Engineers ASCE, the Boston Society of Civil Engineers, and the Society of Promotion of Engineering Education.

Around 1900, several textbooks were written in the United States, among which is also the 1911 book by Hughes and Arthur Truman Safford (1867-1951), consulting engineer at Boston MA, and a part-time lecturer at Harvard. While the book followed the pattern by then so well established, it also involves originality and perception, employing excellent flow diagrams, logarithmic plots, and a close tie between theory, including the Bernoulli equation, and engineering practice. Another, even better book was previously published by Georg Edmond Russell (1877-1953), also a Bostonian, who studied at MIT and in 1905 joined the MIT staff, and remained there all through his life. His 1909 text book on hydraulics went through many editions, and was praised for its simplicity in explanations rather than its originality.

Anonymous (1924). Hughes, Hector James. *Who's who in America* 13: 1663. Marquis: Chicago.
Anonymous (1943). Hughes, Hector J. *Who was who in America* 1: 603. Marquis: Chicago.
Hughes, H.J., Safford, A.T. (1911). *A treatise on hydraulics*. Macmillan: New York.
Hughes, H.J., Munro, W.B., Wilson, E.B. (1925). Minute on the life of Prof. George Chandler
    Whipple. *Science* 61(1572): 179-180.
Hughes, H.J.(1930). Engineering and other applied sciences in the Harvard Engineering School.
    The development of Harvard University 1869-1929, S.E. Morison, ed. Cambridge MA.
Rouse, H. (1976). Hector J. Hughes. *Hydraulics in the United States* 1776-1976: 83. Iowa
    Institute of Hydraulic Research: Iowa City.
http://ids.lib.harvard.edu/ids/view/686087?buttons=y *P*

HULBURD

02.06. 1879 Brasher Falls NY/USA
06.12. 1938 Poughkeepsie NY/USA

Lucius Sanford Hulburd graduated as civil engineer from Cornell University, Ithaca NY, in 1902. He was then engaged at York PA on the construction of a 15,000 HP hydro-electric development on River Chattahoochee near Atlanta GA. He returned to New York in 1904 as leveller in the Department of State Engineer and Surveyor, where his employment continued essentially until his death. The first four years were spent on the design of the improved State Canal System, mainly the slopes for the canalized sections of the system. In 1908 Hulburd was placed in charge of one of the first barge canal construction contracts. In 1909, he conducted surveys and made studies on the connection of the Canal with the Finger Lakes with the Improved Erie Canal. He was also assigned for the construction of the Cayuga and Seneca Canals, so that he was located for seven years at Seneca Falls NY, finally as resident engineer.

In 1916 he was transferred to Residency No. 8 on the Erie Canal, where he supervised 60 km of canal improvement through Wayne County NY. In 1918 he was commissioned captain of engineers at Camp A.A. Humphreys VA, from where he was discharged at the end of World War I. He returned to his work in the Engineering Department of the State, supervising canal improvement near Rochester NY, retracing surveys of old canal lands and investigating water supplies for State institutions. He was transferred in 1921 to the main office at Albany NY, to compile records for the canal improvement work. From 1923 he was also assigned for the construction of the vehicular bridge over the Hudson River at Poughkeepsie NY. Foundation problems were especially challenging, the main two piers being located in water 18 m deep with an additional 25 m of depth to the rock foundation. The bridge was taken into service in 1930. A heart affliction from which Hulburd never recovered developed during his training at Camp Humphreys, led finally to his death. His character and abilities were recognized by all who knew him. He possessed a fine moral fiber and so despised pretension as to appear at times modest. Hulburd was ASCE associate member from 1911.

Anonymous (1939). Lucius S. Hulburd. *Trans. ASCE* 104: 2040-2042.
Hulburd, L.S. (1903). *The development of water power in Fall Creek*. Cornell University: Ithaca.
Hulburd, L.S. (1930). *Lucius S. Hulburd papers* 1895-1930. Cornell University: Ithaca NY.
http://beta.worldcat.org/archivegrid/collection/data/64072997
http://en.wikipedia.org/wiki/File:Camillus-aqueduct1.JPG (*P*)

HUMPHREYS

02.11. 1810 Philadelphia PA/USA
27.12. 1883 Washington DC/USA

Andrew Atkinson Humphreys graduated from the US Military Academy in 1831, then serving in the academy and the 1835 Florida campaign. In 1836, he resigned and was employed as civil engineer by the US Government on lighthouses and breakwaters. He was reappointed two years later lieutenant in the Corps of Topographical Engineers. From 1848, then as captain, he was engaged in the hydrographical survey of the delta of Mississippi River, with a view on plans for securing it from inundations and for deepening its channel at the river mouth. He was however compelled by illness in 1851 and went to Europe, to examine the continent's river deltas, studying the means employed abroad for protection against inundation. Humphreys actively participated in the Civil War, finally as a major-general. In 1866 he was made brigadier-general and given command of the Corps of Engineers, the highest scientific appointment in the US Army.

On his return in 1854, Humphreys was given charge in connection with surveys for railways from Mississippi River to the Pacific. In 1857 he resumed his work on the Mississippi delta and in 1861 published with Henry Larcam Abbot (1831-1927) a notable work on the hydraulics of this large river. The report contains numerous velocity profiles measured with the double-float method, which was later not accepted as an accurate means, however. It also contains formulae for the velocity distribution and velocity under uniform flow conditions, yet these data have not been used after around 1900, mainly because of the immense problems associated with prototype observations.

Anonymous (1887). Humphreys, A.A. *Appletons' cyclopaedia of American biography* 3: 314. *P*
Humphreys, A.A. (1858). Method of obtaining the amount of water in rivers. 35th Congress, Senate, *Miscellaneous Documents* 49. Washington DC.
Humphreys, A.A., Abbot, H.L. (1861). *Report upon the physics and hydraulics of the Mississippi River*. Professional Papers, Corps of Topographical Engineers: Philadelphia.
Humphreys, A.A. (1874). *Regulations* respecting the navigation of channels at mouth of the Mississippi River. Office of the Chief of Engineers: Washington DC.
Humphreys, A.A., Abbot, H.L. (1878). Physics and hydraulics of the Mississippi. Reply to certain criticisms made by Dr. Hagen, Director of Public Works, Prussia. van Nostrand's *Eclectic Engineering Magazine* 18(109): 1-3.
Reuss, M. (1985). Humphreys and the development of hydraulic engineering: Politics and technology in the Army Corps of Engineers. *Technology and Culture* 26(1): 1-33.

HUNSAKER

26.08. 1886 Creston IA/USA
10.09. 1984 Boston MA/USA

Jerome Clarke Hunsaker graduated in 1908 from the US Naval Academy, receiving in 1912 the MS degree in naval architecture from Massachusetts Institute of Technology MIT, Cambridge MA. He was then an instructor at its Department of Naval Architecture, developing courses in aerodynamics, conducting also in Europe research in aeronautical engineering. He then built the first wind tunnel, earning for him in 1916 the first PhD degree in this research field at MIT. He headed the new Aircraft Division during World War I at Washington DC, and was in charge of building *Shenandoah*, the first rigid US airship to use helium.

Resigning from the Navy in 1926, Hunsaker joined the Bell Telephone Laboratories to develop improved communication for meteorological data transfer to in-flight aircraft. He returned to MIT in 1933 as head of the Department of Mechanical Engineering and in 1939 became head of the newly formed Department of Aeronautical Engineering. During World War II he was further engaged in wind tunnel experiments and naval officer training at MIT. From 1941 to 1945 he was coordinator and chairman of the Naval Research Development Board and chairman of the National Advisory Committee in Aeronautics NACA. After retirement from MIT in 1952, Hunsaker was the first president of the Institute of Aeronautical Services IAS. He had been an internationally known pioneer in aeronautical research and engineering during the first half of the 20th century. He developed flight systems, including the establishment of the scientific and mathematical basis for flight. Among the awards he received were the Navy Cross, the Presidential Medal for Merit, and the Gold Award of the Royal Aeronautical Society.

Anonymous (1921). Navy design competition for ship-plane. *Aviation* 10: 69-72. *P*
Anonymous (1984). Jerome C. Hunsaker. *Journal of the Society Naval Architects and Marine Engineers* 92: 387-388.
Hunsaker, J.C., Huff, T.H. (1916). Dynamical stability of aeroplanes. *Miscellaneous Collections* 62. Smithsonian Institution: Washington DC.
Hunsaker, J.C. (1935). Progress report on cavitation research at MIT. *Trans. ASME* 57: 211-216.
Hunsaker, J.C, Rightmire, B.G. (1947). *Engineering applications of fluid dynamics*. McGraw-Hill: New York.
Hunsaker, J.C. (1952). *Aeronautics at the mid-century*. Yale University Press: New Haven CT.
Launius, R.D. (1995). J.C. Hunsaker. *Notable* 20th *century scientists* 2: 980-981. Gale: New York.

HUNT

19.05. 1858 New York NY/USA
23.07. 1932 New York NY/USA

Charles Warren Hunt graduated with the BSc degree in civil engineering from New York University in 1876; he was conferred from his Alma Mater the honorary degree Doctor of Laws in 1907. After his graduation he entered into civil engineering practice including government service on river and harbour work, municipal employ in New York in water work engineering. He was thereby connected with offices of Charles B. Brush (1848-1897), and J. James Croes (1834-1906). The latter collaboration led to his 1892 appointment as assistant secretary of the American Society of Civil Engineers ASCE, taking over in 1895 as Secretary ASCE, thereby becoming ASCE executive head at age 37, a position he held until 1920. The membership increased from 1,900 to 9,400 in his 25 years term, during which he developed and sustained the ASCE practices, ideals and traditions.

It was not in the size of the Society only that Hunt's efforts were reflected, but also its resources and influence among engineers mounted. When he began his long work, the organization had headquarters in a small but ample building. Hunt advised the purchase of a larger building on West 57th Street, which was inaugurated in 1897. He further overlooked the development of the two Society's publications, namely the Proceedings and the Transactions ASCE, which both were edited by him, gaining an enviable place in engineering literature. He was also responsible for the Annual Conventions. He was further in charge of the Society's Library, which he improved at the Headquarters. His interest and pride in the Society was further attested by his historical writings, namely in 1897, and in 1902. Hunt was particularly fitted by heredity, education, experience, and training for his life work. He had a fine presence, a keen intellect, and a sterling character. He was made honorary member of the *Koninklijk Instituut van Ingenieurs* in Holland in 1922, and was ASCE member from 1884.

Anonymous (1932). Charles Warren Hunt dies. *Engineering News-Record* 109(4): 118. *P*
Anonymous (1933). Charles W. Hunt. *Trans. ASCE* 98: 1557-1562.
Hunt, C.W. (1897). *Historical sketch of the American Society of Civil Engineers*: 1852-1897. ASCE: New York. *P*
Hunt, C.W. (1902). The first fifty years of the American Society of Civil Engineers. *Trans. ASCE* 48: 220-226.
Hunt, C.W. (1918). The activities of the American Society of Civil Engineers during the past twenty-five years. *Trans. ASCE* 82: 1577-1652.

HUTTON F.R.

28.05. 1853 New York NY/USA
14.05. 1918 New York NY/USA

Frederick Remsen Hutton is not so much known for his scientific contributions to the profession, but more for his impact mainly into the American Society of Mechanical Engineers ASME. In 1873 and 1876, respectively, he obtained the AB and the EM degrees from the Columbia School of Mines, New York NY. His career was then always at his Alma Mater, namely instructor from 1877, 1881 PhD and adjunct professor, and from 1891 professor of mechanical engineering, acting from 1892 to his retirement in 1907 also as head of Department. He was from 1892 associate editor of the Engineering Magazine, from 1893 editor of Johnson's Universal Cyclopaedia; he published from 1897 the mechanic engineering of power plants, and from 1899 his book on Heat and heat engines.

Hutton may be considered a great developer of laboratories for mechanical engineering, similar to John R. Freeman (181855-1932) for hydraulic laboratories. He also had a significant influence on the educational advance in his field. He managed to edited various journals and writing books on gas engines, steam power plants and in particular authored the 1915 book on the history of ASME, including 20 chapters on aspects like the organisation, the principles, standing committees, ASME presidents including most portraits, the ASME Council, Honorary Members ASME, notable papers published, the ASME headquarters, trips and engineering congresses, ASME Library, and prizes and medals. He of course was a honorary member ASME, presiding the society in the term 1906-07. He also was member of the United Engineering Society. He was in addition a consulting engineer for the Department of Water, Gas and Electricity of New York City.

Anonymous (1920). Memorial to the late F.R. Hutton. *Mechanical Engineering* 42(1): 71. *P*

Anonymous (1980). Hutton, Frederick R. *Mechanical engineers in America born prior to* 1861: 188-189. ASME: New York. *P*

Hutton, F.R. (1898). *Heat and heat-engines*: A treatise on the internal combustion engine using gas, gasoline, kerosene, or other hydrocarbon as source of energy. Wiley: New York.

Hutton, F.R. (1903). *The gas-engine*. Wiley: New York.

Hutton, F.R. (1906). *The mechanical engineering of power plants*. Wiley: New York.

Hutton, F.R. (1908). *The mechanical engineering of steam power plants*. Wiley: New York.

Hutton, F.R. (1915). *A history of the American Society of Mechanical Engineers* from 1880 to 1915. ASME: New York. *P*

HUTTON W.R.

21.03. 1826 Washington DC/USA
11.12. 1901 Clopper MD/USA

William Rich Hutton received his first professional experience on survey works from the US Engineers. He acted as divisional engineer on the Washington Aqueduct from 1854 to 1861, was then promoted to chief engineer and in 1866 constructed the Annapolis waterworks. He completed a masonry dam across Potomac River as chief engineer of the Chesapeake and Ohio Canal in 1869, continuing as a consulting engineer until 1880. From 1876 to 1879 he was also in charge of the design and construction of locks and movable dams to improve *Kanawha* River.

In 1880 Hutton moved his offices to New York City, becoming consulting engineer for the New York Aqueduct, for the Colorado Midland Railroad, and for other important civil engineering projects. He was the chief engineer from 1886 to 1889 of the beautiful Harlem River Bridge, known as the Washington Bridge, and later he occupied a similar post on the Hudson River Tunnel serving on the commission for the improvement of the Columbia River. He was a member of the American Society of Civil Engineers ASCE, of which he was vice-president in the term 1896, of Société des Ingénieurs Civils de France, and of the Institution of Civil Engineers ICE, London UK. He was awarded the Diplôme d'Honneur at the 1878 Universal Exposition in Paris for his lock design on *Kanawha* River. Before having started his engineering career, Hutton left his home town for California, remaining there from 1847 to 1853. The book Glances at California chronicles his voyage via Panama and the life with the US Army occupation forces, his travels to Monterey, Santa Barbara and Los Angeles. In 1849 Hutton accompanied a surveying expedition to Los Angeles and later worked as a surveyor at San Luis Obispo. The book includes also notable paintings of the time of the Gold Rush in California.

Anonymous (1902). William Rich Hutton. *Minutes* Institution of Civil Engineers 150: 435-436.
Anonymous (1972). Hutton, William R. *A biographical dictionary of American civil engineers*: 64-65. ASCE: New York. *P*
Hutton, W.R. (1891). *The Washington Bridge over the Harlem River*: A description of its construction. von Rosenberg: New York.
Hutton, W.R. (1900). *Machinery to water supply of canals*. Intl. Navigation Congress, Paris.
Waters, W.O. (1942). *Glances at California* 1847-1853. Diaries and letters of William Rich Hutton, surveyor, with a brief memoir and notes. Huntington Library: San Marino CA.
Waters, W.O. (1942). Drawings by William Rich Hutton. Huntington Library: San Marino CA. *P*
http://en.wikipedia.org/wiki/William_Rich_Hutton *P*

INGERSOLL

08.06. 1920 Madison WI/USA
06.05. 1999 Crescent City CA/USA

Alfred Cajori Ingersoll received in 1942 the BS degree from University of Wisconsin, Madison WI, the MS degree in 1948, and the PhD title in 1950. He was associate professor of civil engineering until 1960 at California Institute of Technology, Berkeley CA, and guest professor of applied mechanics at the Bengal Engineering College, Calcutta, India, in 1954. From 1960 to 1970 he was dean of Caltech School of Engineering, and from then there associate dean of continuing education, the School of Engineering and Applied Sciences, until retirement in 1986. He was from 1982 a consultant for Bechtel Corp., San Francisco CA, and from 1987 president of Northop University, Los Angeles CA.

Engineering education was the main impact of Ingersoll. He was an educator who would drop all to help students, offer a compliment, and give a pat on the back of his students and colleagues. He had boundless energy and would rather run than walk or climb stairs than use an elevator. His enthusiasm affected all who were around him, and no task was too difficult to undertake in his professional and personal life. Ingersoll has added to the understanding of settling mechanisms in settling tanks used in wastewater treatment. He and his two co-authors were awarded the ASCE Rudolph Hering Medal in 1957. Ingersoll also collaborated with Robert L. Daugherty (1885-1978) on the fifth edition of the widely used text Fluid mechanics with engineering applications. His quality in writing technical work came first to attention in 1942, winning the Daniel W. Mead Prize for students with a paper on ethics. In 1969 Ingersoll was the winner of the ASCE Edmund Friedman Award, and in 1991 he was the recipient of the William H. Wisely American Civil Engineering Award for his services to ASCE.

Anonymous (1957). Alfred C. Ingersoll. *Civil Engineering* 27(10): 736. *P*
Anonymous (1969). Alfred C. Ingersoll. *Civil Engineering* 39(10): 88-89.
Anonymous (1991). Ingersoll, A.C. *Who's who in America* 46: 1613. Marquis: Chicago IL.
Anonymous (1991). Alfred C. Ingersoll. *Civil Engineering* 61(12): 90. *P*
Daugherty, R.L., Ingersoll, A.C. (1954). *Fluid mechanics with engineering applications*. McGraw-Hill: New York.
Ingersoll, A.C., McKee, J.E., Brooks, N.H. (1956). Fundamental concepts of rectangular settling tanks. *Trans. ASCE* 121: 1179-1204; 121: 1213-1218.
Ingersoll, A.C., Tanaka, H. (1958). Hydraulic model study. *Trans. ASCE* 123: 908-924.
http://www.universityofcalifornia.edu/senate/inmemoriam/AlfredC.Ingersoll.htm

INGRAM

16.06. 1908 Cleves OH/USA
10.10. 1991 Sonoma CA/USA

William Truitt Ingram obtained the BE degree in civil engineering from Stanford University in 1930 and the MPH degree in sanitary engineering from Johns Hopkins University, Baltimore MD, in 1942. He was until 1932 an office engineer of the Pacific Gas & Electro Company, San Francisco CA, then worked as surveyor until 1934 from when he was assistant company director for the California State Relief Administration. From 1935 to 1941 Ingram was sanitary engineer for the San Joaquin Local Health District. After war service he served as chief engineer the health division in Yugoslavia. From 1947 to 1949 he then was an engineer of the American Health Association when being appointed associate professor and from 1954 professor of public health engineering at the College of Engineering, New York University, until retirement. He in parallel was a private consultant. Ingram was a member of numerous professional societies and Fellow of the American Society of Civil Engineers ASCE, of the American Public Health Association APHA, and of the American Association for the Advancement of Science AAAS. He was the recipient of the 1959 Kenneth Allen Award for research in sewage treatment.

Ingram would have been a current environmental engineer, dealing with air pollution control, environmental engineering, septic tanks, sewage characteristics, sewage solids, sewage collection systems, sewage disposal, and sewage treatment in general. He has contributed to various sections of the 1960 Encyclopaedia and to the technical journals.

Anonymous (1964). Ingram, William T. *Who's who in engineering* 9: 906. Lewis: New York.
Anonymous (1971). Chairman William T. Ingram. *Civil Engineering* 41(4): 79. *P*
Diachishin, A.N., Hess, S.G., Ingram, W.T. (1954). Sewage disposal in tidal estuaries. *Trans. ASCE* 119: 434-452.
Ingram, W.T. (1952). Atmospheric pollution. *Science* 115(2995): 3.
Ingram, W.T., Dieringer, L.F. (1953). A critical examination of air sampling instrumentation methods. *American Industrial Hygienic Association* 14(2): 121-132.
Ingram, W.T. (1961). Trickling filter treatment of whey wastes. *Journal of the Water Pollution Control Federation* 33(8): 844-855.
Ingram, W.T. (1969). *Encyclopaedia of science and technology.* McGraw-Hill: New York.
Stone, R., Orford, H.E., Ingram, W.T. (1953). Land disposal of sewage and industrial wastes. *Sewage and Industrial Wastes* 25(4): 406-418.

ISAACSON

14.06. 1919 Brooklyn NY/USA
31.03. 2008 Queens NY/USA

Eugene (Gene) Isaacson was a graduate of the City
College, New York NY, as most Jewish students of
mathematics of his age were. He received the PhD
degree in 1949 under Kurt O. Friedrichs (1901-1982)
at New York University, where he spent his entire
professional career as professor of mathematical
sciences at the Courant Institute of Mathematical
Sciences, New York. Isaacson was the editor of
both the SIAM Journal of Numerical Analysis and
the journal Mathematical Tables and Aids to
computation, which later morphed into Mathematics
of Computation. Many of his fifteen PhD students
became prominent in research on numerical methods.

Isaacson's PhD thesis was on water waves on sloping beaches. He derived an explicit
formula for the case in which the angle of the slope is equal a rational multiple of pi.
Subsequently he was concerned with the numerical solution of differential equations.
He wrote in collaboration with a colleague his well-known book Analysis of numerical
methods. This was considered an excellent text with a distinctly modern flavour. While
most of the standard topics are covered, the emphasis is on proofs of convergence and
proper error estimates, rather than a full catalogue of procedures, as suggested in the
title. In the mid-1950s Isaacson worked on a problem related to the Grand Coulee Dam:
If the dam were breached by sabotage, would the resulting food wave on the Columbia
River inundate the Hanford reactor? In collaboration with James J. Stoker (1905-1992)
they did the mathematical formulation and the subsequent numerical solution of the
shallow-water equations. The answer was that the reactor was safe under the considered
flooding scenarios. The US Army Corps of Engineers USACE subsequently adopted the
numerical method developed by Isaacson.

Anonymous (1987). Isaacson and Saari step down as managing editors. *SIAM News* 20(1): 12. *P*
Isaacson, E. (1946). *Water particle paths near a vertical barrier*. Courant Institute: New York.
Isaacson, E., Stoker, J.J., Troesch, B.A. (1954). *Numerical solution of flood prediction and river
    regulation problems*. Institute of Mathematical Sciences, University: New York.
Isaacson, E., Stoker, J.J., Troesch, B.A. (1958). Numerical solution of flow problems in rivers.
    Journal of the Hydraulics Division ASCE 84(HY5, Paper 1810): 1-18.
Isaacson, E. (1959). Water waves over a sloping bottom. *Comm. Pure Applied Math*. 3(1): 11-31.
Isaacson, E., Keller, H.B. (1966). *Analysis of numerical methods*. Wiley: New York.
http://www.siam.org/news/news.php?id=1403 *P*

ISHERWOOD

06.10. 1822 New York NY/USA
19.06. 1915 New York NY/USA

Benjamin Franklin Isherwood was apprenticed from 1836 at the mechanical department of the Utica & Schenectady Railroad. He then worked for Croton Aqueduct and had an engineering job on the Erie Canal. At age 22, he was appointed first assistant engineer of the Navy. During the Mexican-American War, he served on a steam warship and a gunboat. From 1848 he was assigned to Washington Navy Yard, where he designed engines and conducted experiments with steam as the source of power for propelling ships. During the 1850s, he compiled then operational and performance data from steam engines to analyse the efficiency of engine types then in use. He thereby published more than 50 papers on steam engineering and vessel propulsion, including the 1859 paper on his thermodynamic tests. He thereby advanced to the moat prolific technical writer of the nation. This book includes a description to ascertain the distribution of energy and losses in engines and boilers.

Shortly after the outbreak of the Civil War Isherwood was appointed engineer-in-chief of the Navy, whereby he also founded the Bureau of Steam Engineering. When the war began, the Navy had 28 steam vessels, and this number increased to 600 during the war. Their design and construction was mainly accomplished by Isherwood, resulting in ships that were fast enough to pursue the blockade runners. The book Experimental researches in steam engineering was translated into six languages, becoming a standard engineering text. After the war, he was involved with the organisation of the US Naval Academy, Annapolis MD. By 1874 it served as the model for mechanical engineering education at most American universities. In 1870 he had conducted tests that resulted in a propeller used by the Navy for the next 27 years. Isherwood was a pioneer in the production of fast cruisers, and of the fastest vessel of the world. He became president of the Experimental Board under the Bureau of Steam Engineering until his retirement in 1884. He was appointed in 1915 with the rank of rear admiral.

Anonymous (1980). Isherwood. *Mechanical engineers in America born prior* 1861: 192. ASME. *P*
Bennett, F.M. (1896). *The steam Navy of the United States*: A history of the growth of the steam vessel of war in the US Navy, and of the Naval Engineer Corps. Nicholson: Pittsburgh.
Isherwood, B.F. (1859). *Engineering precedents for steam machinery*. Bailliere: New York.
Isherwood, B.F. (1863). *Experimental researches in steam engineering*. Hamilton: Philadelphia.
http://en.wikipedia.org/wiki/Benjamin_F._Isherwood *P*

IZZARD

26.11. 1904 San Mateo CA/USA
08.07. 1997 Loudoun VA/USA

Carl Frederick Izzard obtained the AB degree from Stanford University, Stanford CA, in 1930, and the MS degree in mechanics and hydraulics from Iowa State University, Iowa City IA, in 1940. He joined the Bureau of Public Roads in 1930, first for time and cost studies in highway construction, from 1933 to 1939 as federal aid inspector in Texas, and later until 1949 as drainage engineer. He was then until retirement chief of the Hydraulic Research Division, Bureau of Public Roads, Department of Commerce, Washington DC. From 1970 he was director, Office of development, Federal Highway Administration. He was a Fellow of the American Society of Civil Engineers ASCE, and a member of the American Geophysical Union AGU, and the Highway Research Board.

Izzard supervised hydraulic research of culverts, bridges and storm drains. He conducted research on peak discharge and frequency of runoff, developing methods of hydraulic designs for highway drainage structures. He was the recognized authority on the application of hydraulics on highway engineering, publishing reports and papers in professional journals. His 1954 paper is based on the use of stream flow records for determining the size of highway drainage structures. By developing regional flood curves, the peak discharge was estimated for a given frequency, whether or not stream gaging records are available for that particular site. His work also contributed to the improvement of hydraulic designs during construction of the interstate highway system, especially in reducing erosion and flood risks. Izzard realized that water was a concern for government agencies, such that he organized a Surface Drainage Committee for Highways under the aegis of the Transportation Research Board. Under his leadership, the group proved influential in collecting flood data to develop design procedures.

Anonymous (1964). Izzard, C.F. *Who's who in engineering* 9: 910. Lewis: New York.
Anonymous (1997). Carl F. Izzard. *Eos* 78(39): 419.
Anonymous (1998). Carl Izzard had major impact on hydraulics in highway design. *Civil Engineering* 68(1): 8. *P*
Izzard, C.F. (1946). Hydraulics of runoff from developed surfaces. Proc. *Highway Research Board* National Research Council, Washington DC 26: 129-150.
Izzard, C.F. (1953). Discussion of Importance of inlet design on culvert capacity. *Research Report* 15-B. Highway Research Board: Washington DC.
Izzard, C.F. (1954). Peak discharge for highway drainage design. *Trans. ASCE* 119: 1005-1015.

JABARA

27.06. 1910 Streator IL/USA
02.09. 1992 Denver CO/USA

Melvin (Mel) Andrew Jabara obtained the BS degree in civil engineering in 1931 from Oklahoma State University, Stillwater OK. He then made graduate work in hydraulics at the University of Colorado, Denver CO, in 1937, and further at the Graduate School of Business, Harvard University, Cambridge MA, in 1944. He became in the early 1930s a staff member of the US Bureau of Reclamation USBR, Denver CO, and there was from assistant to chief, Division of Design, head of the Earth Dam Section, and chief of the Hydraulic Structures Branch. He was thereby involved in almost 200 multi-purpose schemes, including the San Luis Pumped Storage Project as the USBR's largest earth dam. He was a member and later a Fellow of the American Society of Civil Engineers.

Jabara had experience in the layout, hydraulic and structural design of spillways, outlet works and other structures of earth dams. A number of factors influence the selection of spillway type, namely topography, geology, hydrology, and the purpose of the dam. The 1973 paper is based on USBR dam designs including both plunge pool stilling basins or flip bucket dissipators. Cavitation damage is reported and repairs and modification of spillway tunnels are described. Flow aeration was reported to be an effective means in the control of cavitation damage. The 1976 paper deals with the design and construction of Scoggins Dam in Oregon State. Aspects including climate, groundwater conditions, and marginal embankment material are considered. Pore-water pressure of almost the height of the completed embankment was observed, but no excessive pressures were recorded in the embankment during construction.

Jabara, M. (2014). Melvin Andrew Jabara. Personal communication. *P*

Jabara, M.A., Wagner, W.E. (1961). *Design and operating problems on the Glendo Dam high-head outlet works stilling basin*. US Bureau of Reclamation: Denver CO.

Jabara, M.A. (1969). *Tiber Dam auxiliary outlet works*. US Bureau of Reclamation: Denver.

Jabara, M.A., Legas, J. (1973). Selection of spillways, plunge pools and stilling basins for earth and concrete dams. 11th *ICOLD Congress* Madrid Q41(R17): 269-287.

Jabara, M.A., Harber, W.G. (1976). Problems associated with the construction of Scoggins Dam. 12th *ICOLD Congress* Mexico Q44(R28): 565-580.

Whinnerah, R.W., Jabara, M.A. (1961). *Whiskeytown Dam Spillway and outlet works*. US Bureau of Reclamation: Denver CO.

www.usbr.gov/history/.../HOFFMAN,CARLJ.pdf

JACKSON

18.01. 1874 Westmeath ON/CA
08.04. 1937 San Francisco CA/USA

Thomas Herbert Jackson was born in Canada. His family moved to the USA in 1880 and settled in Montague MI. He was appointed to the US Military Academy at West Point in 1895. After graduation in 1899, he was commissioned a second lieutenant of engineers. His first assignments were at Portland ME, and on the Philippine Islands. Upon returning to the USA in 1905, he served at Fort Leavenworth KA and then as captain at San Francisco CA. There he headed the planning effort for the innovative Sacramento River Flood Control Project. Following devastating floods in 1907, the Congress was asked for the authority to have the Corps of Engineers prepare a comprehensive plan for river rehabilitation. The Jackson Plan was submitted to the Congress in 1910, proposing the construction and enlargement of riverbank levees and bypasses to carry excess flood water. After acceptance at the national level, California added its endorsement and created the Reclamation Board to manage the involvement of the state in the joint federal-state effort. By utilizing this flood control system, the Jackson Plan served as a forerunner for the monumental plan to control the floods of Mississippi River.

From 1911 to 1917 Jackson was assigned district engineer, then he served as colonel in France. Upon return to his country, he took various positions in the Corps of Engineers. In 1928 he was appointed president of the Mississippi River Commission MRC in the rank of brigadier general. His tenure until 1932 marked a turning point in the Corps' involvement in the Lower Mississippi Valley. From its creation in 1879 the MRC had undertaken an improvement program in the alluvial valley, including both navigation and flood control. The large 1927 flood called for revision of the activities. The 1928 Flood Control Act consisted of improved levees to contain flood flows, floodways to bypass critical areas, bank and channel stabilization, and tributary basin improvements. Jackson was given the task of implementing this plan. The officer-engineer faced formidable challenges. The MRC headquarters were moved from St. Louis to Vicksburg MI, where he created a forerunner of the well-known Waterways Experiment Station.

Anonymous (1931). Jackson, Thomas H. *Who's who in engineering* 2: 667. Lewis: New York.
Jackson, T.H. (1935). *Bank protection on the Mississippi and Missouri Rivers*. US Army Corps of Engineers: Vicksburg.
Reuss, M. (2004). T.H. Jackson. *Designing the bayous*: 122. Texas A&M University Press. *P*
Robinson, M.C. (1983). Thomas H. Jackson. *APWA Reporter* (5): 6-7. *P*

JACOB C.E.

03.09. 1914 Mesa AZ/USA
30.01. 1970 Los Angeles CA/USA

Charles Edward Jacob received education in civil engineering at Utah University, Salt Lake City UT, obtaining the MS degree from Columbia University, New York, in 1936. Until 1947 he was associated with the US Geological Survey becoming chief of its Groundwater Hydraulics Section. He held various academic positions at University of Utah, Brigham Young University, and the New Mexico Institute of Mining and Technology, Socorro NM. Since 1947 he served as a groundwater consultant mainly in the USA, South America and the Middle East. His most notable student was Mahdi Hantush (1921-1984). Jacob was awarded the NSF Graduate Research Fellowship in 1952.

The untimely death of Jacob at age of 56 was a shock to his friends, colleagues and groundwater hydrologists. During the past three decades he had stimulated basic research. The pioneering efforts of Oscar E. Meinzer (1876-1948), Charles V. Theis (1900-1987), and Jacob formed the sound basis of the advances made in groundwater hydraulics. Jacob was one of the rare scientists who combined theoretical physics and mathematics with a real aptitude for practical engineering. He participated actively in the two main groundwater conferences, namely the Symposium on Transient Ground Water Hydraulics at Fort Collins CO in 1963, and the National Symposium on Ground Water Hydrology at San Francisco in 1967. He was an outstanding speaker, attracting attention and usually leaving the greatest impression with the participants. His analyses were outstanding on such controversial subjects as research versus engineering, pitfalls in routine investigations, or overlapping of disciplines at universities. Chapter 5 of the Engineering Hydraulics book was even in 1970 the best introduction for groundwater studies; it is clear and simple, yet rich and comprehensive.

Jacob, C.E. (1939). Fluctuations in artesian pressure by passing trains. *Trans. AGU* 20:666-674.
Jacob, C.E. (1940). On the flow of water in an elastic artesian aquifer. *Trans. AGU* 21: 574-586.
Jacob, C.E. (1945). *The water table in the western and central parts of Long Island NY.* US Dept. of the Interior: Washington DC.
Jacob, C.E. (1946). Slope-area measurements and the proper velocity-head coefficient. *Water Resources Bulletin* (Feb.10): 34-46; (May10): 124-125; (Aug.10): 172-181.
Jacob, C.E. (1947). Drawdown test to determine effective radius of artesian well. *Trans. ASCE* 112: 1047-1064; 112: 1068-1070.
Kashef, A.I. (1970). Charles E. Jacob. *J. American Water Resources Assoc.* 6(5): 841-843. *P*

JANSEN R.B.

14.12. 1922 Spokane WA/USA
14.12. 2011 Mead WA/USA

Robert Bruce Jansen obtained the BS degree from University of Denver in 1949, and the MS degree from University of Southern California, Los Angeles CA, in 1955. He was then a professional engineer in the West, from 1965 to 1968 chief of the California Division of Dam Safety, Sacramento CA, and until 1971 chief of operations, California Department of Water Resources, then its departmental director and its chief of design and construction until 1977. From then to 1980 Jansen was assistant commissioner of the US Bureau of Reclamation USBR, Denver CO, when becoming a consulting engineer at Bellingham WA. He was in charge of Tennessee Valley Authority TVA, Hydro-Quebec, Montreal, Canada, or the Alabama Power Co., Birmingham AL.

Jansen has authored and edited a number of important works. His 1951 paper deals with backwater curves in the pre-computer era. In contrast to other methods he made direct use of the governing differential equation in terms of free surface slope. He later edited the book Advanced dam engineering, a work summarizing recent techniques in dam engineering. The book includes 26 chapters, of which the main hydraulic interest is in earthfill dams, rockfill dams, gravity dams, arch dams, spillway design and construction, spillway performance and remedial measures, outlet design and construction, and also reservoirs. Other works are Dams and public safety, an important issue in hydro-electric schemes for the future, or Safety of existing dams, corresponding to a reassessment of the by then completed installations. Jansen was a member of the US Society on Dams, chairing it in the term 1979, and a member of the National Academy of Engineering.

Anonymous (1972). Robert B. Jansen. *Civil Engineering* 42(6): 84. *P*
Anonymous (1979). Robert B. Jansen. 13th *ICOLD Congress* Q49(GR 49): 611. *P*
Anonymous (2005). Jansen, Robert B. *Who's who in America* 59: 2292. Marquis: Chicago.
Jansen, R.B. (1951). Surface curves for steady non-uniform flow. *Proc. ASCE* 77(96): 1-7.
Jansen, R.B. (1972). Edmonston pumping plant: Nations mightiest. *Civil Engineering* 42(10): 67-71. *P*
Jansen, R.B. (1983). *Dams and public safety*. US Department of the Interior: Washington DC.
Jansen, R.B., ed. (1988). *Advanced dam engineering for design, construction, and rehabilitation*. van Nostrand Reinhold: New York.
Jansen, R.B. (1988). Dam safety engineering. *Development of dam engineering in the United States*: 1031-1050, E.B. Kollgaard, W.L. Chadwick, eds. Pergamon Press: New York.

JARVIS

17.02. 1880 St. George UT/USA
28.03. 1970 Watsonville CA/USA

Clarence Sylvester Jarvis obtained the BS degree in 1906, and in 1908 the civil engineering degree from the University of Missouri, Columbia MO. In 1925 he earned the MS degree from Michigan University, and the PhD degree in 1927 from the American University, Washington DC. He spent the year 1907 with the US Reclamation Service, and was then until 1917 irrigation engineer. From 1923 to 1929 he was a bridge engineer with the Bureau of Public Roads, and until 1934 principal hydraulic engineer at the US War Department. After a one-year stay with the US Geological Survey USGS, he became hydraulic engineer with the Soil Conservation Service SCS, at the US Department of Agriculture until 1942, from when he joined the Corps of Engineers, at the US War Department until retirement in 1946. He was a member of the American Society of Civil Engineers ASCE, awarded the 1926 Croes Medal. He also was a member of the Society of Military Engineers, and of the Washington Society of Engineers.

Jarvis was a known expert in flood control, the relation of rainfall to soils and run-off, the magnitude and frequency of floods in the USA, in forecasting run-off from index area data, and in hydrology in general. He was also active in questions of national river discharge, the design and construction of dams, and hydro-electric developments. He has written a number of papers, notably in the Transactions ASCE, including a paper on the hydrology of Nile River, and an inventory of unpublished hydrologic data. He also was the author of the floods chapter in the book edited by Oscar E. Meinzer (1876-1948). Another work relates to the Mississippi River.

Anonymous (1906). C.S. Jarvis. *Savitar yearbook*: 103. University of Missouri: Columbia. *P*
Anonymous (1944). Jarvis, C.S. *American men of science* 7: 902. Science Press: Lancaster.
Jarvis, C.S. (1926). Flood flow characteristics. *Trans. ASCE* 89: 985-1033; 89: 1091-1104.
Jarvis, C.S. (1927). Soils and erosional forms as affecting floods. *MS Thesis*. The American
         University: Washington DC.
Jarvis, C.S. (1936). Floods in the United States. *Water-Supply Paper* 771. USGS: Washington.
Jarvis, C.S. (1936). Flood-stage records on the river Nile. *Trans. ASCE* 101: 1012-1071.
Jarvis, C.S. (1942). Floods. *Hydrology*: 531-560, O.E. Meinzer, ed. McGraw-Hill: New York.
Jarvis, C.S. (1943). Gleanings from the field of hydrology. *Trans. ASCE* 108: 651-693.
Jarvis, C.S. (1943). *Supplementary gleanings from the field of hydrology*. Washington DC.
Jarvis, C.S. (1943). Early contributions to Mississippi River hydrology. *Trans. ASCE* 108: 605-655.

JENSEN

09.11. 1873 Summerville CA/USA
05.05. 1937 Fresno CA/USA

Christian Peter Jensen graduated from the Van der Naillen School of Engineering, Oakland CA, in 1895. He was then engaged in an engineering company at Fresno CA, forming a partnership which lasted until 1906. The engineering work attracting Jensen at that time was the construction of lumber flumes and irrigation systems. Lumbering in the Sierra Nevada Mountains was then a major industry, because the larger saw-mills found it economic to float their lumber to the valley by rapidly flowing water in V-shaped timber flumes. These clung to the sides of the canyons, crossing ravines and gullies on high trestles until they reached their destination in planing mills and lumber yards of the San Joaquin Valley. The location, design, and construction of these flumes called for engineering skill. Jensen had an important part in the engineering features of this early industrial and agricultural development of Fresno County, because he was also involved in irrigation engineering. In 1900 he was assistant engineer for an 80 km long lumber flume, whereas he was from 1901 to 1904 chief engineer for the location and construction of a logging railway. Until 1917 he was then consulting engineer for the Madera Canal and Irrigation Company, constructing canals and siphons.

From 1910 to 1919 Jensen was chief engineer of a Reclamation District on designs and the construction of levees on the San Joaquin River. From 1909 to 1913 he was city engineer at Fresno CA and shortly later consulting engineer at Bakersfield CA, and at Dinuba CA, on the design and construction of sewers and sewage disposal systems. He also served as engineer on the design and construction of water works for Strathmore CA. From 1917 to 1919 he was associated with his brother as engineering consultants. He then became County Surveyor of Fresno County. He developed the Jensenite paving process, which was widely used and eliminated the payment of royalties to private paving concerns. Jensen was active both in his civic and professional affairs. He was president of the Engineer's Club of Fresno; he was a member of the American Society of Civil Engineers ASCE since 1919. He may therefore be regarded an engineer who developed particularly the lumber transport by water, and who contributed to the wealth of Central California.

Anonymous (1937). Christian P. Jensen. *Civil Engineering* 7(7): 539.
Anonymous (1938). Christian P. Jensen. *Trans. ASCE* 103: 1817-1819.
https://www.google.ch/search?q=%22Christian+Peter+Jensen+%22+Fresno&source=lnms (*P*)

JERVIS

14.12. 1795 Huntington NY/USA
12.01. 1885 Rome NY/USA

John Bloomfield Jervis moved to Rome NY at age of three years, where he remained later all through his life. He assisted the construction of the Erie Canal, and conducted the survey and construction of the Delaware and Hudson Canals. He was then engaged as chief engineer of railroads, thereby inventing the locomotive truck, the principle of which was in use for a long time. In 1833, Jervis was appointed chief engineer of *Chenango* Canal, introducing then the method of providing artificial reservoirs for the water supply at its summit. In this connection Jervis did considerable original work to determine the percentage of total rainfall that could be used to replenish the water supply. His determination of some forty percents was higher than the constant used in Europe. Some of Jervis' constants for the computation of rainfall and runoff were cited in standard engineering handbooks as late as in 1900. In 1835 he was commissioned to make the surveys of the eastern section of the Erie Canal, in view of its enlargement.

In 1836 Jervis was the engineer in charge of the construction of the Croton Aqueduct, serving the water supply of New York City. He directed the completion of Croton Dam, the Ossining Bridge, the Harlem River Bridge as also the distributing reservoir. An extensive report on these activities was published in 1842. From 1846 to 1848 he was a consulting engineer for the waterworks of Boston MA. He was then engineer of the Chicago and Rock Island railroad, and was next engaged on the Pittsburgh, Fort Wayne and Chicago railroads until 1866. From 1868 until his death he was one of the trustees of Rome merchant-iron mill company. In 1878 Hamilton College conferred on him the honorary doctorate.

Anonymous (1885). The late John B. Jervis. *Railroad Gazette* 29(Jan.23): 49-50. *P*
Anonymous (1887). Jervis, John B. *Appletons' cyclopaedia of American biography* 3: 430. *P*
Cooper, L.G. (1987). *The Old Croton Aqueduct*. New York. *P*
Finch, J.K. (1931). John Bloomfield Jervis: Civil engineer. *Trans. Newcomen Society* 11(1): 109-120. *P*
Jervis, J.B. (1842). *Description of the Croton Aqueduct*. Slamm & Guion: New York.
Lankton, L.D. (1977). *The practicable engineer: John B. Jervis and the old Croton Aqueduct*. Public Works Historical Society: Chicago. *P*
http://209.85.129.132/search?q=cache:FAcyzC2LMMIJ:en.wikipedia.org/wiki/John_B._Jervis+Jervis+%22Croton+Aqueduct%22&cd=5&hl=de&ct=clnk&gl=ch *P*

JESSOP G.A.

04.05. 1883 York PA/USA
24.09. 1954 York PA/USA

George Augustus Jessop received the ME degree in 1909 from University of Michigan, Ann Arbor MI. He joined the S. Morgan Smith Company, York PA, a hydraulic laboratory established in 1921, becoming in 1923 chief engineer, a position he held until his retirement in 1949. He was killed in a car accident.

Jessop was a mechanical engineer within a notable furnisher and tester of turbines. He developed the Breakable link for turbine gates. His design was incorporated in the Bonneville Dam on Columbia River concerning further developments of Kaplan's Turbine as invented by Victor Kaplan (1876-1934) in Austria. In the 1925 New York Meeting of the ASCE Committee of Hydraulic Machinery, the first major American publication on low-head hydro-electric power plants was presented, based on the results of the 1924 London World Water Conference. The Meeting brought together the then leading American engineers on water turbines. The paper of Jessop describes tests on turbines made at the Hydraulic Laboratory, Holyoke MA, and in the specially-designed flume of the S. Morgan Smith Co., for whom he served then as hydraulic engineer. Although none of the tests was on Kaplan turbines, the Smith Flume provided a major new facility for testing the newer types of turbines for Bonneville. Its turbines were built in 1935 by the S. Morgan Co., which had already done this work for the Safe Harbour Project, with Jessop as its chief engineer, who was familiar with these designs. The turbines had more than 90% efficiency for a large range of power from 20,000 to 66,000 HP, resulting also from the 1931 and 1939 papers. The scale of the turbines demanded notable innovations in terms of discharge range and variation, and size, so that each detail had to be specially considered. In addition, concerns with cavitation damage had to be carefully tested due to the high turbine speeds.

Anonymous (1954). George A. Jessop. *Engineering News-Record* 153(Oct.14): 50.

Heslop, P.L., Jessop, G.A. (1931). Hydraulic and electrical possibilities of high speed low head developments. *Trans. AIEE* 50(4): 114-119.

Heslop, P.L., Jessop, G.A. (1939). Kaplan turbines at Bonneville. *Trans. ASME* 61(2): 97-108.

Jessop, G.A. (1925). High specific speed hydraulic turbines in their bearing on the proportioning of the number of units in low-head hydro-electric plants. *Trans. ASCE* 89: 659-665.

Liel, A.B., Billington, D.P. (2008). Engineering innovation at Bonneville Dam. *Technology and Culture* 49(3): 727-751.

Voaden, G.H. (1943). The Smith-Putnam wind turbine. *Turbine Topics* 1(3): 37-42. P

JEWELL

01.06. 1842 Wheaton IL/USA
19.05. 1931 Evanston IL/USA

Omar Hestrian Jewell, the first of the Jewell filter trio, was a master mechanic for grain elevators until he became interested in improving the quality of boiler-feed water taken from the notoriously foul Chicago River. He built his first filters in the 1870s, of which the first was located in an elevator of a company on the South Branch of Chicago River. Of the scores of filter patents taken out by the Jewell trio, the first ten were granted to Omar in 1888. It was significant for the time because it was intended to combine a filter and chlorine gas generator, the latter consisting of electrodes placed in the dome of the filter tank.

The Jewell Pure Water Co., Chicago IL, was formed by Omar and his son William in 1890. It was largely financed by well-known Chicago dealers in waterworks supplies. Son William was in Europe in 1887, erecting steel tank pressure filters shipped from Chicago, which served industrial plants. The Jewell Export Filter Co. built or equipped filters abroad. One for Warsaw, Poland, was said to be the largest mechanical filter plant in Europe. The first Jewell filters on a municipal water supply went into use at Rock Island IL in 1891; within five years Jewell filters were in use in twenty American cities. Impetus to the adoption of these filters was given by the Providence tests of 1893, in which the city was represented by Edmund B. Weston (1850-1916), and these of 1895 under the direction of George W. Fuller (1868-1934). An early demonstration in 1888 at Brockton MA would have resulted in an adoption, had it not been thwarted by the 'advice' of the Massachusetts State Board of Health, which had as strong aversion to the use of aluminium as a coagulant, as was previously also expressed in France. Omar and William entered the New York Continental Jewell Filtration Co. merger of 1900 under a 5-year contract. Then, William began his independent practice, centering in professional work. William's brother Ira joined the company in perfecting filters. Ira died in 1940, shortly before his brother William.

Anonymous (1913). *The New York Continental Jewell Filtration Company*. New York.
Baker, M.N. (1941). The three Jewells: Pioneers in mechanical filtration. *Engineering News-Record* 126(Jan.30): 179. *P*
Whipple, G.C. (1907). *The value of pure water*. Wiley: New York.
www.google.de/.../688312_STRAINER.pdf
http://www.google.es/patents/US377390

2206

JOHNSON C.F.

16.09. 1897 Mayfield KY/USA
05.05. 1988 Louisville KY/USA

Charles Franklin (Frank) Johnson was educated at University of Kentucky, Lexington KY, from where he received the BS degree in civil engineering in 1919, and the CE degree in 1938. He was until 1922 assistant engineer in charge of surveying and design at Union City TN, then until 1925 chief draftsman at Paducah KY, and at Grand Rapids MI, taking there over as project engineer the Office of Sanitary Engineering until 1930. Until his retirement Johnson was then from junior to senior project engineer at the Commissioners of Sewerage, Louisville KY. He was both member of the American Society of Civil Engineer ASCE, and the Kentucky Society of Professional Engineers.

Johnson worked in sewer hydraulics thereby contributing with his 1941 paper to the roughness estimation of free surface pipe flow. Historically, the so-called 'small formula' of Wilhelm R. Kutter (1818-1888), retaining the essence of the complete formula of Ganguillet and Kutter, had success in sewer hydraulics because of simplicity and the easily attributable Kutter roughness coefficient $1/n$, describing whether the boundary roughness was rather smooth or rough. It was however demonstrated toward the end of the 19th century that this approach was in disagreement with then updated experimental data. Yet, the concept of Kutter was retained mainly by the practitioners, because they 'knew' to apply the formula correctly for their purposes. Johnson in 1941 also stated that Kutter's formula is widely applied in the United States for sewers. The coefficient $1/n$ was observed to be determined only after actual discharge or velocity measurements. Johnson contributed therefore to this problem using clean water in circular pipes, but finally also stated that there were conceptual problems.

Anonymous (1941). C. Frank Johnson. *Who's who in engineering* 5: 917. Lewis: New York.
Johnson, C.F. (1941). Technique for a survey of old sewers. *Civil Engineering* 11(1): 27-30.
Johnson, C.F. (1941). Performance tests of large pumping plant. *Civil Engineering* 11(12): 703-706.
Johnson, C.F. (1943). Gaging and sampling Louisville's sewage. *Civil Engineering* 13(2): 97-100. *P*
Johnson, C.F. (1944). Determination of Kutter's *n* for sewers partly filled. *Trans. ASCE* 109: 223-247.
Johnson, C.F. (1958). Nation's capital enlarges its sewerage system. *Civil Engineering* 28(6): 428-431; 28(7): 502-505.

JOHNSON G.A.

26.05. 1874 Auburn ME/USA
31.03. 1934 New York NY/USA

George Arthur Johnson was from 1895 to 1898 member of the technical staff of Louisville KY and Cincinnati OH, where he conducted studies of water purification of the Ohio River. He was stationed in 1900 at St. Louis MO as associate director to study the effect of the Chicago Drainage Canal on the local water supply. From 1902 he was in charge of preliminary operations of the filter plant of the New Jersey Water Company, Little Falls NJ. In 1904 he studied the purification of sewage and the softening of water for Columbus OH. He made a trip around the world then to study the applicability of rapid filtration and coagulation in water supplies in Japan, China, India, and Egypt.

From 1906 to 1910 Johnson was principal assistant engineer of Hering & Fuller, New York NY, operating purification plants. His most valuable work was the application of lime hypochlorite for the sterilization of water supplies at Chicago IL. He then formed the partnership Johnson & Fuller, which continued work in water supply and sewage. After war service, Johnson returned to engineering practice in New York NY, and had his private practice from 1924. His work brought him engagements in nearly 100 cities, for which he served either as adviser, or as member of advisory boards. The most important included Cambridge MA, Trenton NJ, Jersey City NJ, or Columbus OH. He authored several technical papers and reports, including for instance Romance of water storage, The typhoid toll, Water filtration and sterilization, Municipal water works in the Far East, or Purification of public water supplies. Johnson was a man of marked and likeable personality, keen in his appreciation of human relations, and aggressive in advancing his views. He had remarkable ability, both in speaking and writing, for conciliating those holding opposite views, however. His achievements in advancing the cause of typhoid fever reduction through the elimination of bacteria by means of chlorination constitute one of the hot spots of water supply engineering during his era.

Anonymous (1935). George A. Johnson. *Trans. ASCE* 100: 1678-1681.
Johnson, G.A. (1907). Notes on oriental water-works. *Engineering News* 58(19): 481-484.
Johnson, G.A. (1913). The purification of public water supplies. *Water Supply Paper* 315. Government Printing Office: Washington DC.
Johnson, G.A. (1916). The typhoid toll. *Journal of the American Water Works Association* 3(2): 249-326; 3(3): 791-868.
nfocuslouisville.com *P*

JOHNSON H.

19.12. 1890 Willow Glen NY/USA
01.11. 1955 Watertown NY/USA

Hollister Johnson graduated in 1912 from Cornell University, Ithaca NY, with the civil engineering degree. After preliminary experience with the New York State Highway Commission, the New York State Conservation Commission, and the Board of Black River Regulating District, he joined the Water Resources Division, Surface Water Branch, Albany NY. He was from 1934 staff member of the US Geological Survey USGS, assigned from 1943 to the headquarters at Washington DC. Shortly before his retirement in 1952, Johnson was transferred to the Surface Water Branch, the Albany District.

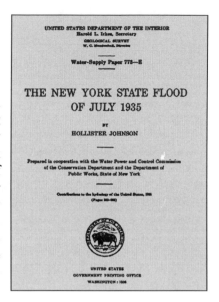

During his stay with the New York State Conservation Commission, Johnson presented reports on water power and drainage projects of the State. He was from 1912 to 1917 in charge of the construction and maintenance of the *Canaseraga* Creek improvements. After war service he returned to the USA in charge of surveys for the Intercoastal Canal in Louisiana until 1919. From 1921 to 1933 he was in charge again of New York State, making reports on the design and construction on operation studies of storage reservoirs. He further was assigned to prepare hydrologic investigations and studies. Once with the USGS, he was in charge of stream flow compilations and computations, as principal assistant to the District engineer at Albany NY, from where resulted a report on the main floods in the USA. The 1950 Report, compiled with two colleagues, deals with the relative accuracy and time saving in discharge computation with two methods. The study is based on 213 pairs of discharge measurements in rivers of different characters, from bridge, by wading, and from cableway or boat. He also supervised the preparation of two Special Reports on the water resources of Oklahoma State, and on general hydrologic studies and investigations. Johnson was member ASCE from 1917.

Anonymous (1957). Hollister Johnson. *Trans. ASCE* 122: 1244-1245.
Flynn, F.J., Johnson, H., McCall, J.E. (1950). *A comparative study of mean-section and mid-section methods for computation of discharge measurements.* Water Resources Division. USGS: Washington DC.
Harrington, A.W., Johnson, H. (1938). National aspects of flood control: The New York Floods of 1935 and 1936. *Trans. ASCE* 103: 613-621; 103: 714-719.
Johnson, H. (1938). The New York State flood of 1935. *Contributions to the hydrology of the United States* 1936, N.C. Grover, ed. USGS: Washington DC. (*P*)
Paulsen, C.G. (1933). *Floods of 1950 in SW Oregon.* US GPO: Washington DC.

JOHNSON J.W.

19.07. 1908 Pittsburgh KA/USA
11.04. 2002 Victoria BC/CA

Joe William Johnson was educated at the University of California, Berkeley CA, from where he obtained the BS degree in 1931, and the MS degree in civil engineering in 1934. He then conducted research on sediment transport of flowing water at Waterways Experiment Station, Vicksburg MI until 1935, was a hydraulic engineer at the Soil Conservation Service SCS until 1942, when taking over at his Alma Mater first as assistant, and later as full professor of hydraulic engineering until retirement in 1975. He was in parallel a consultant. He was a member of the National Academy of Science since 1976, the

American Society of Civil Engineers ASCE, of which he was from 1979 an Honorary Member, and of the Japan Society of Civil Engineers JSCE.

Johnson was a coastal engineering expert, with an early experimental work on impulse waves, in which the main wave characteristics were investigated as a function of the diameter and the weight of falling disks into a water tank. He was later also secretary of ASCE's Coastal Engineering Research Council, which organized notable international conferences. He was executive committee chairman of ASCE's Waterway and Harbour Division, and vice-president of the American Shore and Beach Preservation Association since 1974. Johnson also was the editor of its journal Shore and Beach. He was the recipient of the 1981 Harbor and Coastal Engineering Award 'for dedicated and untiring efforts in education, practical consulting activities around the world and for contribution to the advancement of coastal engineering through international conferences'.

Anonymous (1967). Joe W. Johnson. *Civil Engineering* 37(3): 87. *P*

Anonymous (1979). Joe W. Johnson. *Civil Engineering* 49(10): 89-90. *P*

Anonymous (1994). Johnson, J.W. *American men and women of science* 4: 115. Bowker: New Providence NJ.

Johnson, J.W., Bermel, K.J. (1949). Impulsive waves in shallow water as generated by falling weights. *Trans. AGU* 30(2): 223-230.

Johnson, J.W., Wiegel, R.L. (1959). *Investigation of current measurement in estuarine and coastal waters*. State Water Pollution Control Board: Sacramento CA.

Johnson, J.W., Eagleson, P.S. (1966). Sediment problems at coastal structures. *Estuary and coastline hydrodynamics*: 462-488, A.T. Ippen, ed. McGraw-Hill: New York.

O'Brien, M.P., Johnson, J.W. (1934). Velocity-head correction for hydraulic flow. *Engineering News-Record* 113(Aug.16): 214-216.

JOHNSON R.D.

09.01. 1874 Orchard Park NY/USA
28.06. 1949 Fort Lauderdale FL/USA

Raymond Deloraine Johnson obtained the ME degree from Cornell University in 1901. He was research engineer of Shawinigan Water & Power Company, Shawinigan Falls QC, until 1902, from where he moved as a hydraulic engineer to the Ontario Power Company, Niagara Falls. From 1908 Johnson was a hydraulic engineer of the New York State Water Power Commission, Albany NY, until becoming from 1914 to 1936 private consultant in New York City. He was a member of the American Society of Civil Engineers, and the recipient of the 1922 Elliot Cresson Gold Medal for meritorious inventions from the Franklin Institute for his pioneering works in hydraulic engineering. He was also awarded the 1947 Holley Medal from ASME.

Johnson's name is closely related with the surge tank, a term applied to a stand pipe or storage reservoir placed at the downstream end of a closed pipeline system to prevent undue pressure oscillations if discharge is suddenly reduced or increased. Surge tanks are a main element of all connections between reservoirs and turbines for hydro-electric power generation, because such an installation has to flexibly adapt to changes of users. Without a surge tank, such a scheme would either immediately fail, or the pipelines would be so massive that their cost would be highly excessive. Johnson proposed the differential surge tank as an improved version of the simple surge tank consisting only of the vertical stand pipe, as compared with a system of two or more tanks connected normally by overfall structures, to take up excessive discharge from the entire scheme, thus preventing the above mentioned pressure head oscillations. In the 1917 paper Johnson considered surges as appear on free surface channels, again due to a variation of typically the discharge at either channel end. These waves were originally investigated by Henry Bazin (1829-1917), providing information on the main wave characteristics.

Anonymous (1937). Johnson, Raymond D. *Who's who in engineering* 4: 715. Lewis: New York.
Anonymous (1948). Raymond D. Johnson: Holley Medal. *Mechanical Engineering* 70(1): 68. *P*
Anonymous (1949). Raymond D. Johnson. *Engineering News-Record* 143(Jul.14): 67.
Anonymous (1949). Raymond D. Johnson. *Mechanical Engineering* 71(9): 798.
Jaeger, C. (1977). *Fluid transients* in hydro-electric engineering practice. Blackie: Glasgow UK.
Johnson, R.D. (1908). The surge tank in water power plants. *Trans. ASME* 30: 443-474.
Johnson, R.D. (1915). The differential surge tank. *Trans. ASCE* 78: 760-805.
Johnson, R.D. (1917). Surges in an open canal. *Trans. ASCE* 81: 112-124.

JOHNSON W.C.

21.05. 1859 Granville MA/USA
15.12. 1906 Niagara Falls NY/USA

Wallace Clyde Johnson graduated in 1884 from the Worcester Polytechnic Institute, Worcester MA, and then accepted a position as assistant engineer with the Water Power Company, Holyoke MA. His energy, ability and engineering skill commanded soon attention. When Clemens Herschel (1842-1930) was requested to recommend an engineer to take charge of the development of the Niagara Falls Hydraulic Power and Manufacturing Company at Niagara Falls NY, Johnson was suggested in 1886. He remained chief engineer there until 1900, then becoming consulting engineer until his death.

The Niagara Falls Hydraulic Power and Manufacturing Company was during Johnson's stay there of outstanding importance for the entire country. He designed and executed the hydro-electric development of the Company on the banks of the Lower Niagara River, a pioneer accomplishment using turbines under a head of more than 60 m. This plant, as others which he designed, will stand as a monument to his memory. In 1900 Johnson also became chief engineer of the Shawinigan Water and Power Company, then the largest water-power development of Canada. The water wheels used there were the largest then ever built, and the transition line from Shawinigan Falls to Montreal the longest east of the Rocky Mountains. Johnson was appointed in 1906 member of the State Water Supply Commission. He was continually sought by other companies who were interested in his services. He made visits and reports on various proposals in the USA, Canada and Nicaragua. He was known for his genial disposition and cordial manner. Johnson was member of the American Society of Civil Engineers ASCE, the American Society of Mechanical Engineers ASME, and the Engineers' Society of Western New York, serving also as president. He died at early age, but his work bears testimony to his active life. He was constantly and effectively in the field of work, with his life devoted to his profession. His services were highly valued by many companies with which he was connected.

Anonymous (1907). Johnson, Wallace Clyde. *Trans. ASCE* 58: 538-539.
Anonymous (1907). Wallace Clyde Johnson. *Journal of the Association of Engineering Societies* AES 38(2): 100-101.
Johnson, W.C. (1894). A new development of power at Niagara. *Cassier's Mag.* 5(2): 326-330.
Johnson, W.C. (1899). Power development at Niagara Falls. *Journal AES* 23(2): 78-90.
Johnson, W.C. (1907). The Shawinigan Power Plant. *Cassier's Magazine* 31(5): 394-409. *P*

JOHNSON W.E.

23.09. 1910 Minneapolis MN/USA
26.02. 1982 McLean VA/USA

Wendell Eugene Johnson obtained his BS degree in 1931 from the University of Minnesota, Minneapolis MN. He joined the Minnesota Highway Department as a field engineer for two years, then was engineer until 1940 with the US Army Corps of Engineers for the Conchas Dam NM, and from 1946 to 1961 was chief division engineer at Omaha NE. From then until retirement in 1970, Johnson was chief of the engineering division for civil works, Office of Chief of Engineers, Washington DC. He continued as a private consultant on dams and water resources in McLean VA. He was Honorary Member of the American Society of Civil Engineers ASCE, the National Academy of Engineering NAE, and the Society of American Military Engineers SAME.

Johnson's career with the Corps of Engineers spanned 37 years and included work such as Conchas Dam, and the Missouri River Basin dams. He was there finally in charge of engineering for the Corps' water resources program. He had top-level responsibility for all engineering investigations and designs for major US water resources developments. He was as consultant active for large hydropower schemes both in the USA and abroad, including Canada, Greece, the Middle East, and South America. He was particularly a consultant of hydraulic engineering for *Tarbela* Dam, Pakistan, then one of the largest designs worldwide. Johnson had also been active in the affairs of the US Committee on Large Dams and its international organisation, the International Committee on Large Dams ICOLD. Being chairman of USCOLD in 1968 he had a major influence on dam safety, developing a model law for inspection of privately-owned dams in the USA.

Anonymous (1961). Wendell E. Johnson. *Engineering News-Record* 166(Mar.23): 155. *P*
Anonymous (1974). Wendell Johnson. *New Civil Engineer* 3(Oct.3): 7. *P*
Anonymous (1980). Wendell E. Johnson. *Civil Engineering* 50(11): 53. *P*
Anonymous (1981). Johnson, Wendell E. *Who's who in America* 41: 1730. Marquis: Chicago.
Anonymous (1982). Wendell E. Johnson, dam expert, dies. *Civil Engineering* 52(6): 88. *P*
Johnson, W.E., Breston, J.N. (1951). Directional permeability measurements on oil sandstones from various states. *Producers Monthly* 15(4): 10-19.
Johnson, W.E., Groot, C. (1963). Observations on the migration of young sockeye salmon through a large, complex lake system. *Journal of Fisheries Research Board* Canada 20: 919-938.
Johnson, W.E. (1977). Conservation tillage in western Canada. *Journal of Soil Water Conservation* 32(1): 61-65.

JOHNSTON C.T.

23.10. 1872 Near Denver CO/USA
15.01. 1970 Ann Arbor MI/USA

Clarence Thomas Johnston obtained the BS degree in 1895, and the civil engineering degree in 1899 from the University of Michigan, Ann Arbor MI. He was engaged as assistant engineer by Wyoming State from 1896 to 1898, then until 1903 was chief assistant engineer on irrigation investigations by the US Dept. of Agriculture, served as Wyoming State engineer until 1911, from when he took over until retirement in 1941 as professor of geology and surveys at the University of Michigan. He was a member of the American Society of Civil Engineers ASCE, and the Michigan Engineering Society.

Johnston had a research interest in all matters relating to irrigation techniques. He further was an expert in the irrigation and drainage laws, the flow of water in canals and ditches, the administration of water resources, the design of irrigation structures, the coefficient of friction and roughness coefficients. He also had a wide knowledge on the Egyptian irrigation. The 1901 paper deals with the discharge control in field conditions. The weir of Cesare Cippoletti (1843-1908) proposed for a linear increase of discharge with the head on the weir is presented. Further the effect of submergence on weir flow is discussed. Various types of discharge register apparatus are further described, among which that proposed by the English Gurley is particularly recommended. A standard register is also described. The 1903 paper describes discharge ratings made from a pontoon consisting of two boats towed across a pond at Cheyenne WY.

Anonymous (1941). Clarence T. Johnston. *Michiganensian yearbook*: 50. Ann Arbor. *P*
Anonymous (1955). Johnston, C.T. *American men of science* 9: 980. Science Press: Lancaster.
Johnston, C.T. (1899). Computation of discharge records and preparation of diagrams. *Bulletin* 86. Office of Experiment Station. US Department of Agriculture: Washington DC.
Johnston, C.T. (1901). Instructions for installing weirs, measuring flumes and water meters. *Engineering News* 46(9): 131-134.
Johnston, C.T. (1902). Report on US Irrigation Investigations for 1900. *Engineering News* 48(12): 208.
Johnston, C.T. (1903). Current meter ratings and observations by the US Irrigation Investigations Department. *Engineering News* 49(7): 158-160.
Johnston, C.T. (1914). Weir measurement of stream flow. *Trans. ASCE* 77: 1291-1294.
Johnston, C.T. (1915). Some principles relating to the administration of streams. *Trans. ASCE* 78: 630-648.

JOHNSTON T.T.

08.08. 1856 Piqua OH/USA
22.02. 1909 Evanston IL/USA

Thomas Taylor Johnston graduated in 1877 as civil engineer from Rensselear Polytechnic School, Troy NY. His career started with service on the Missouri and Mississippi Rivers within the US Army Corps of Engineers, where he made observations on the New Madrid Reach at Point Pleasant MO. Later he was concerned with gagings on the Mississippi River, including carefully collected data on the discharges and velocities, published in Reports of the Chief of Engineers. From 1886 he was involved in hydraulic studies of flow in regular channels in view of the application of the proposed Chicago Sanitary Canal. In the mid-1890s, Johnston took over as assistant chief and later as chief engineer of the Canal, preparing plans for structures, solving the hydraulic problems, and conducting the surveys necessary. He gave a special attention to the Desplaines River Diversion. He was in 1897 president of the Society of Western Engineers.

The 1896 paper deals with the classical weir arrangement composed of a rectangular weir crest, also referred to as the broad-crested weir. Due to the sharp front portion, the flow separates from the crest thereby generating a 'separation bubble' so that the critical flow depth does not appear. The paper also includes experimental data, but no attempt was made to propose a relationship for the discharge coefficient. The 1898 discussion relates to a paper written by James A. Seddon (1856-1921), who proposed headwater reservoirs to retain flood water and thus would allow for reduced dike heights along the river. Seddon was certainly an expert in this field, given his previous mathematical studies on the flood subsidence, so that he could determine the effect of the proposed reservoirs on the flooding scenarios.

Anonymous (1897). Johnston, Thos. T. *Journal of the Western Society of Engineers* 2(1): Frontispiece. *P*

Johnston, T.T., Cooley, E.L. (1896). New experimental data for flow over a broad crest dam. *Journal of the Western Society of Engineers* 1(1): 30-51.

Johnston, T.T. (1896). Data pertaining to rainfall and stream flow. *Journal of the Western Society of Engineers* 1(3): 297-404.

Johnston, T.T. (1897): The restoration of the water supply at Savannah, Georgia. *Journal of the Western Society of Engineers* 2(6): 711-723.

Johnston, T.T. (1900). Discussion of Control of Lower Mississippi, by James Seddon. *Journal of the Western Society of Engineers* 5(4): 310-315.

JONES B.E.

11.01. 1883 Conneaut OH/USA
03.07. 1961 Washington DC/USA

Benjamin Earl Jones was educated at the University
of Michigan, Ann Arbor MI, where he obtained the
BS degree in civil engineering in 1908. He then was
junior engineer of the US Office of Public Roads up
to 1910, and then from assistant engineer to chief,
the Water and Power Division of the US Geological
Survey USGS, Washington DC. He thereby directed
in the 1930s the Power Classification Division of
the USGS Conservation Branch. Jones was member
of the American Society of Civil Engineers ASCE,
of the Washington Society of Engineers, and of the
American Geophysical Union AGU.

Jones presented a number of reports and papers. His 1916 paper deals with the so-called
Jones-method of computing the change in slope through the change in stage of a flood
wave. The 1931 report deals with the Rogue River Basin in Oregon State, describing
power and reservoir sites, including maps and numerous potential dam sites. The 1953
bibliography co-authored by Jones describes water-supply papers of the USGS, dealing
with water power, floods, and droughts. It brings up the 1940 version of Jones on the
same subject. In total, hundreds of USGS related works are shortly reviewed. He had
been a nationally known consultant on water power and river measurements. From 1921
until his retirement in 1953, he made all revisions in water power estimates of the USA.
In 1932 he made the first geophysical investigations undertaken by the Survey for dam
site selection, supervising the surveys for Grand Coulee Dam and other large dams.
Jones was in 1945 president of the ASCE District of Columbia Section.

Anonymous (1930). Jones, Benjamin E. *Who's who in government* 1: 292. New York.
Anonymous (1945). B.E. Jones. *Engineering News-Record* 135(Aug.30): 267. *P*
Anonymous (1948). Jones, Benjamin E. *Who's who in engineering* 6: 1034. Lewis: New York.
Anonymous (1961). Benjamin E. Jones. *Engineering News-Record* 167(Jul.27): 53.
Jones, B.E. (1916). A method of correcting river discharge for changing stage. *Water Supply Paper* 375: 117-130.
Jones, B.E., Oakey, W., Stearns, H.T. (1932). Water-power resources of the Rogue River Drainage Basin, Oregon. *Water Supply Paper* 638-B. Washington DC.
Jones, B.E., Young, L.L. (1954). Developed and potential water power of the United States and other countries of the world. *Circular* 329. USGS: Washington DC.
Young, L.L., Jones, B.E. (1953). Annotated bibliography of US Geological Survey Reports on water-power resources, including floods and droughts. *Geological Survey Circular* 200.

JONES J.O.

10.09. 1884 Barrett KA/USA
02.02. 1982 Lawrence KS/USA

Jacob Oscar Jones was educated at the University of Kansas, Kansas KA, receiving the BA degree, and at Cornell University, Ithaca, obtaining from there the MS degree. He was from 1911 a hydrologist at the Idaho Irrigation Company, then in 1912 became instructor in civil engineering at his Alma Mater, was there from 1915 to 1922 associate and associate professor, moving then in this position to University of Minnesota, Minneapolis MN, becoming in 1928 professor of hydraulics there until his retirement in 1955. He also was the Acting Dean of its School of Engineering and Architecture from 1943 to 1947. Jones was a member of the American Society of Civil Engineers ASCE.

Jones has written several papers and a book in hydraulics. The 1917 paper deals with the effect of edge rounding on the outflow features from weirs and orifices. Currently, these edges are sharp-crested by standard, but at the time of Jones, this effect was not yet fully understood. Jones noted that the discharge coefficient increased by several per cents with the increase of edge rounding, yet he did not attempt to introduce a formula accounting for this effect, given the limited number of data collected. The described effect was found to be independent of the head on the orifice, but varies linearly with the ratio of the radius of rounding to the orifice diameter. The 1953 book is a basic account on hydraulics with usual chapters on pipe and open channel flows. In addition, the book includes chapters on centrifugal pumps, the impulse water wheel, and the reaction turbine.

Anonymous (1947). Dean Jones. *Jayhawker yearbook*: 21. University of Kansas: Lawrence. *P*
Anonymous (1964). Jones, Jacob O. *Who's who in engineering* 9: 950. Lewis: New York.
Anonymous (2001). Jacob O. Jones. *M.E. Vibrations* Alumni Newsletter, University of Kansas (9): 5-6.
Jones, J.O. (1917). The effect on orifice and weir flow of slight roundings of the upstream edge. *The Cornell Civil Engineer* 26(2): 108-117.
Jones, J.O. (1932). Inconsistent roughness coefficients. *Civil Engineering* 2(7): 454.
Jones, J.O. (1937). *Notes on the hydrology of Kansas*. Kansas State Printing: Topeka.
Jones, J.O. (1938). Discussion of Rainfall intensities and frequencies. *Trans. ASCE* 103: 372-379.
Jones, J.O. (1940). Discussion of Functional design of flood control reservoirs. *Trans. ASCE* 105: 1669.
Jones, J.O. (1953). *Introduction to hydraulics and fluid mechanics*. Harper: New York.

JORGENSEN

25.04. 1876 Saaborg/DK
08.05. 1938 Berkeley CA/USA

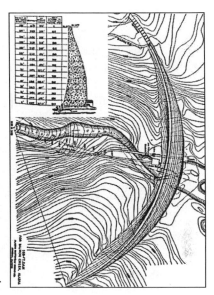

Lars Rasmus Jorgensen graduated as electrical and mechanical engineer from Charlottenburg University Berlin D, in 1900. He then served the Danish Army before coming to the USA in 1901. He was there draftsman and construction engineer with the General Electric Co., Schenectady NY, until 1903, the Los Angeles Edison Electric Company until 1905, from when he was until 1907 a hydro-electric engineer with the Pacific Gas and Electric Co., San Francisco CA. Until 1914 he was engaged in hydro-electricity at San Francisco, then founding his own firm there. He further was chief engineer and vice-president of Feather River Power Co., San Francisco.

Jorgensen has significantly contributed to the design and construction of arch dams. He invented the constant-angle arch dam, which was first built for a hydro-electric power plant on the Salmon Creek near Juneau AK. This dam in south-east Alaska is 50 m high and has a crest 200 m wide. It was completed in 1914. A curved arch dam withstands the water pressure better than a straight dam, because the pressures are better transferred to the valley sides. The ideal constant-angle arch dam in a V-shaped valley has a central angle of 133°. This design allows for a thin dam structure. Jorgensen evolved the first proposal, but the credit also went to his colleague Harry L. Wallenberg (1901-). Building this dam type allows for the economic use of construction material, which was first achieved by Jorgensen. He was a member of the American Society of Civil Engineers ASCE, and awarded its 1918 Norman Medal for the 1915 paper on the above topic. He also was member of the American Concrete Institute ACI, and the Engineers Club of San Francisco.

Anonymous (1929). Lars R. Jorgensen. *Who's who in California*: 43-44. San Francisco.
FitzSimons, N., ed. (1991). Lars R. Jorgensen. *A biographical dictionary of American civil engineers* 2: 59. ASCE: New York.
Jackson, D.C. (1995). *Building the ultimate dam*: Control of water in the West. Lawrence KS. (*P*)
Jorgensen, L.R. (1915). The constant angle arch dam. *Trans. ASCE* 78: 685-721.
Jorgensen, L.R. (1917). Multiple-arch dams on Rush Creek, Cal. *Trans. ASCE* 81: 850-906.
Jorgensen, L.R. (1933). Diablo constant angle arch dam. *Engineering* 136(Jul.07): 2-6.
Serafim, J.L., Clough, R.W., eds. (1990). L. Jorgensen. *Arch dams*: 533-535. Balkema: Rotterdam.
Wallenberg, H.L., Jorgensen, L.R. (1914). *Salmon Creek Dam and reservoir*. Alaska Gastineau Mining Company: Juneau AK.

JOSEPH

26.03. 1929 Chicago IL/USA
24.05. 2011 Minneapolis MN/USA

Daniel Donald Joseph earned the MA degree in sociology from the University of Chicago, Chicago IL in 1950. During the next years he worked as a semi-skilled machinist in various factories. He went back to school at the Illinois Institute of Technology IIT, earning his BS and MS degrees in mechanical engineering, and the PHD degree in 1963. He was since 1962 assistant professor at IIT, joining the faculty of University of Minnesota, Minneapolis in 1963 in the Aerospace Engineering and Mechanics Department, where he remained until retirement in 2009. He received awards including membership of the National Academy of Engineering NAE, the National Academy of Sciences NAS, and the American Academy of Arts and Sciences, among many others.

During his career, Joseph had ten patents, authored more than 400 journal articles and seven books, and was consultant for numerous companies mainly in the petroleum sector. He initiated his career with studies of fluid flows in geometries with permeable boundaries. He for instance proposed with a colleague a 'slip' boundary condition at the interface of a porous medium and a clear fluid, analogous to that in a rarefied gas flow, which is referred to as the Beavers-Joseph boundary condition. Around 1970 his work was more mathematically oriented, including research on fluid flow stability, and on the theory of bifurcation, two topics which were also dealt with in books and for which he remains known. In the late 1970s, Joseph developed interest in rheology, proposing an approach for analysing slow and slowly varying flows. He thereby developed a theory for the effect governing the rise of the free surface close to a rod rotating in a visco-elastic fluid. His research team then solved problems in which the governing equations involve a change of type from elliptic to hyperbolic, as in transonic flows. In the 1980s he worked on the water-lubricated transport of heavy viscous crude oil travelling within a sheath of water along a pipeline thereby reducing the power required for pumping.

Barenblatt, G., Iooss, G., Joseph, D.D., eds. (1983). *Nonlinear dynamics and turbulence.* Pitman: Boston MA.

Dafermos, C.M., Joseph, D.D., Leslie, F.M., Truesdell, C., Serrin, J., Ericksen, J.L. (1986). *The breadth and depth of continuum mechanics.* Springer: Berlin.

Joseph, D.D., Schaeffer, D.G. (1990). *Two-phase flows and waves.* Springer: New York.

www.aps.org/units/dfd/newsletters/.../fall11.pdf *P*

http://www.aem.umn.edu/people/faculty/bio/joseph.shtml *P*

JUDD

04.05. 1868 North Hatfield MA/USA
01.12. 1945 Columbus OH/USA

Horace Judd obtained the ME degree from the Ohio State University, Columbus OH, in 1897, and the MS degree from Cornell University, Ithaca NY, in 1899. After having been a graduate assistant at this university until 1899, he was appointed instructor in steam and machine design, Department of Science and Technology, Pratt Institute, New York NY, until 1902. Judd then joined as assistant professor of experimental engineering his Alma Mater until 1909, was associate professor there until 1920, and until retirement in 1939 hydraulic engineering professor.

Judd has written a number of papers relating mainly to fluid flow with a background in mechanical engineering. His first work relates to Pitot tubes as used then for point-wise velocity observations in either fluid or gas flows. Based on earlier tests made by John R. Freeman (1855-1932), the constants in the equation relating velocity to hydraulic head were verified. Pitot tubes are known for almost exact velocity determination provided the approach flow head is neither too small nor too large. A second research of Judd related to sharp-edged, thin-crested circular orifice flow; both the jet shape and the discharge coefficient were determined and compared to existing data. The distance from the orifice section to a point from where the jet thickness remains almost constant was also investigated. Still another research related again to orifices, yet with a cylindrical pipe addition producing low pressure and dependency on friction along the pipe. Diaphragms were also inserted in this pipe to detail their effect on the outflow features. Today, these classical studies have mainly historical value, but are no more relevant in hydraulic engineering practice. Judd also was interested in hydraulic machinery, namely centrifugal pumps for motor fire engines, and in the effect of turbulence on fluid flow.

Anonymous (1933). Judd, Horace. *American men of science* 5: 595. Science Press: New York.
Anonymous (1941). Judd, Horace. *Who's who in engineering* 5: 937. Lewis: New York.
Anonymous (1946). Judd, Horace. *Mechanical Engineering* 68(2): 188.
Anonymous (1946). Horace Judd. *Makio yearbook* 1946: 40. Ohio State University: Columbus. *P*
Boyd, J.E., Judd, H. (1904). Pitot tubes, with experimental determinations of the form and the velocity of jets. *Engineering News* 51(13): 318-320.
Judd, H., King, R.S. (1906). Some experiments on the frictionless orifice. *Engineering News* 56(13): 326-330; 60(2): 49-50.
Judd, H. (1916). Experiments on water flow through pipe orifices. *Trans. ASME* 38: 331-367.
Judd, H., Pheley, D.B. (1922). Effect of pulsations on flow of gases. *Trans. ASME* 44: 853-918.

JUDSON W.P.

20.05. 1849 Oswego NY/USA
12.02. 1925 Oswego NY/USA

William Pierson Judson was educated by his father and a captain of the US Army Corps of Engineers. From 1866 he was a civilian engineer within the Corps on fort, river and harbour surveys at Oswego Office. From 1874 he was appointed US engineer in charge of surveys, the design and construction of river and harbour works, and lighthouses on Lake Ontario and the St. Lawrence River, and surveying of projected works including 16 km of the Narrows of Lake Champlain. He reported on Niagara Ship Canal, advising on a barge canal from Oswego to the Hudson River instead of a ship canal in 1890. He made a report for the US Deep Waterway Commission proposing the canalization of these rivers in 1896. After works for the US Corps of Engineers at Buffalo NY, he entered a private practice, in which he was engaged on the construction of the Erie Canal at Buffalo in 1898. From 1899 to 1905 he was Deputy State Engineer of New York. He also collaborated with the US Geological Survey in the gauging of its streams. From 1905 to 1925 he was consulting engineer. He further served on the Commission on the Varick Water Power Canal at Oswego since 1876.

Judson's pamphlet on the best possible waterway from the West to the Sea was typical for his era. It is stated that the optimum is that by which the largest practicable lake steamers go nearest to the sea by deep-water navigation. Such a route could be imagined through the USA or Canada. The USA have constructed the Sault St. Mary Canal at the outlet of Lake Superior including large locks. A logical sequence would be a similar 6 m wide waterway from Lake Erie to Lake Ontario, including a ship canal around Niagara Falls. Judson proposed the best almost 30 km long route from Lockport NY to Olcott NY. The last idea of a Niagara Ship Canal was studied around 1910 when the US Army Corp of Engineers developed plans including a canal from Olcott Beach to the proposed Erie Barge Canal near Lockport. Despite of positive economic forecasts, the project was turned down, however.

Judson, W.P. (1888). *Maps of proposed Niagara Ship Canal*, with the connecting ways. Oswego.
Judson, W.P. (1890). *From the West and Northwest to the Sea, by way of the Niagara Ship Canal*. Oswego NY. (*P*)
Judson, W.P. (1901). *History of the various projects, reports, discussions and estimates for reaching the Great Lakes from tidewater*. Oswego NY.
Judson, W.P. (1902). *Lake Ontario harbors for canal commerce*. Oswego NY.

JUDSON W.V.

16.02. 1865 Indianapolis IN/USA
29.03. 1923 Winter Park FL/USA

William Voorhees Judson graduated in 1888 from the US Military Academy, serving then as officer in the US Corps of Engineers for the remainder of his life, finally from 1917 as colonel. From 1891 he was assistant engineer in charge of river and harbour improvements on Lake Erie, at Galveston TX and on the Upper Mississippi. From 1899 to 1900 he served as chief engineer and president of the Board of Public Works of Porto Rico under the military government, following the Spanish-American War. The excellence of his commitment was recognized in the Annual Report of the War Department. He then was until 1902 in charge of river and harbour improvements on the Gulf Coast, and then for three years at Washington DC as instructor at the Engineering School, acting as assistant to the chief of engineers.

Judson was sent in 1904 as military observer with the Russian Army to the Russian-Japanese War. He was captured in the battle of Mukden and sent back to the USA. He was placed in charge of the River and Harbour District, Milwaukee WI, where he had the opportunity to place concrete caissons which he had invented years before. This notable invention was a reason that he was awarded in 1911 the honorary MA degree from Harvard University. From 1909 he was for four years engineer commissioner of the District of Columbia, improving the lighting service, and the water distribution. He was then ordered by General George W. Goethals (1858-1928) to Panama as assistant division engineer of the Atlantic Division, which included the work at Gatun. After one year Judson was engaged with the River and Harbour District of Chicago IL, and in 1916 was ordered to Baltimore MD for a similar charge. After a stay in Russia during takeover of power by the Soviets, he returned after World War I to Chicago IL, working out a plan for the future harbour development which he called *Illiana* Harbor. By this time his health was ruined resulting in a final heart attack causing his death.

Anonymous (1925). William V. Judson. *Trans. ASCE* 88: 1405-1408.
Judson, W.V. (1909). Concrete-steel caissons: Their development and use for breakwaters, piers and revetments. *Journal of the Western Society of Engineers* 14(8): 533-608.
Judson, W.V. (1922). Illiana Harbour. *Journal of the Western Society of Engineers* 37: 201-212.
Salzman, N.V. (1998). *Russia in war and revolution*: General William Judson's accounts from Petrograd, 1917-1918. Kent State University Press: Kent OH. *P*
http://mms.newberry.org/html/Judson.html

JUSTIN J.B.

23.11. 1907 Buffalo NY/USA
10.08. 1995 Philadelphia PA/USA

Joel Bates Justin, son of Joel DeWitt (1881-1950), graduated from Cornell University, Ithaca NY, with a BS degree in civil engineering. He began his career with the construction of hydro-electric projects in New York State, Connecticut and Vermont. He joined in 1934 Tennessee Valley Authority TVA for construction, supervision and inspection of the *Chickamauga* Dam near Chattanooga TN and the Wheeler Dam on Tennessee River, in addition to general reservoir surveys. He was then from 1938 to 1948 manager of hydro-plants, the Appalachian Electric Power Company, Charleston WV. Justin joined his father as a partner in the consulting firm of Justin & Courtney, Philadelphia PA in 1948, assuming the role of president after incorporation of the firm in 1973. In his many years with this firm, Justin had supervised major water resources projects around the world. He retired as president from this firm in 1978, continuing working as an independent consultant for national and international projects.

Justin had also been active in a number of professional societies, including the American Society of Civil Engineers ASCE, the American Society of Mechanical Engineers, the American Consulting Engineers Council, the US Commission on Large Dams USCOLD, and the US Commission on Irrigation, Drainage and Flood Control USIDFC. He was chairman of the Water Resources Association of the Delaware River Basin in 1974. Justin had a fifty years experience in development, design, construction and management of both domestic and foreign water resources projects. These involved multi-purpose projects for water supply, irrigation, flood control, hydro-electric power generation and pumped-storage projects in the USA, Iran, India, Pakistan, Angola, Korea and the Dominican Republic.

Anonymous (1970). Joel B. Justin. 10th *ICOLD Congress* Montreal: 124. *P*
Anonymous (1978). Joel B. Justin. *Civil Engineering* 48(4): 100. *P*
Anonymous (1985). Justin, Joel B. *Who's who in engineering* 6: 334. American Association of Engineering Societies: Washington DC.
Anonymous (1996). Joel B. Justin. *Water International* 21(1): 61.
Justin, J.B. (1942). Discussion of Fort Peck slide. *Trans. ASCE* 107: 744-748.
Justin, J.B., Taleghani, K. (1955). Iran develops its rivers. *Civil Engineering* 25(3): 153-157.
Justin, J.B., Hough, T.C. (1985). Efficient small hydroelectric plant operation. Proc. *Waterpower* '85: 1979-1987, M.J. Roluti, ed. ASCE: New York.

JUSTIN J.D.

13.09. 1881 Syracuse NY/USA
21.02. 1950 Philadelphia PA/USA

Joel DeWitt Justin graduated in 1906 from Cornell University, Ithaca NY, as a civil engineer. He began his career with work for the *Cobbs* Hill Reservoir, Rochester NY, and was from 1907 to 1911 assistant engineer of the New York City Board of Water Supply. He was in charge of the Olive Bridge Dam until 1912, then resident engineer in New York in charge of the design and construction of dams until 1918. From 1919 to 1923 Justin was chief engineer, Department of Hydropower Plants at Winston-Salem NC, and consulting engineer for the New York State Board of Health, when joining until 1925 the Power Corporation of New York as hydraulic engineer. From then until 1928 Justin was a hydraulic engineer with a firm at Philadelphia PA, until finally joining a company at Watertown NY, and becoming a private consultant in 1932.

Justin was a hydraulic engineer who had specialized in dam engineering, hydropower and flood control. He was a consultant on the construction or rehabilitation of more than 120 dams, including Glenville and Nantahala Dams NC, *Arkabutla* Dam MI for the US Corps of Engineers USCE, Cooper River Dam NJ for the Camden Park Commission, *La Regadera* Dam, Bogota in Colombia, and many others mainly in the Southwest of the USA. He also was the chairman of consulting engineers that investigated slides at Fort Peck Dam. He was further involved in the Connecticut Flood Control Project with more than twenty dams, the Merrimack River Flood Control Project NH or the *Muskingum* Flood Control Project for USCE. Justin was co-author of the Hydroelectric handbook, and awarded the ASCE James R. Croes Medal in 1924, among others.

Anonymous (1941). Justin, Joel DeWitt. *Who's who in engineering* 5: 939. Lewis: New York.
Anonymous (1945). Joel D. Justin. *Engineering News-Record* 135(Aug.2): 139. *P*
Anonymous (1947). Joel D. Justin. *Civil Engineering* 17(12): 759. *P*
Anonymous (1950). Joel D. Justin dies. *Engineering News-Record* 144(Mar.2): 30.
Anonymous (1950). Director J.D. Justin, power expert, is dead. *Civil Engineering* 20(3): 199. *P*
Creager, W.P., Justin, J.D., Hinds, J. (1945). *Engineering for dams*. McGraw-Hill: New York.
Davis, C.V., Justin, J.D. (1942). *Handbook of applied hydraulics*. Wiley: New York.
Irwin, K.M., Justin, J.D. (1933). Economic balance between steam and hydro capacity. *Trans. ASME* 55(FSP-5): 63-71. *P*
Justin, J.D. (1914). Derivation of run-off from rainfall data. *Trans. ASCE* 77: 346-384.
Justin, J.D. (1924). The design of earth dams. *Trans. ASCE* 87: 1-141.

KALINSKE

02.09. 1911 Plymouth WI/USA
18.03. 1985 Walnut Creek CA/USA

Anton Adam Kalinske obtained in 1935 his MS degree in civil engineering from the University of Wisconsin, Madison WI. He was from 1936 to 1947 assistant, and associate professor of hydraulic and mechanical engineering at University of Iowa, Iowa City IA, and in parallel associate director of the Iowa Institute of Hydraulic Research from 1942 to 1946. He then joined until 1949 as chief hydraulic engineer the *Infilco* Inc., Tucson AZ, from when he there took over as research director. He became a private consultant in 1965 for an office at Salt Lake City UT. In 1969 he accepted the post of professor at the Utah Water Research Laboratory, Utah State University, Logan UT. He was later also associated with Camp, Dresser & McKee Inc., Boston MA.

Initiating research in turbulence theory, Kalinske gradually moved to the field of hydraulic and sanitary engineering. He was one of the main collaborators of the current Iowa Institute of Hydraulic Research IIHR founded and later directed by Hunter Rouse (1906-1996), yet left this Institute in 1947. During this period, he undertook noteworthy studies in collaboration with Edward R. van Driest (1913-2005) or James M. Robertson (1916-2012) on turbulence statistics and air-water flow in closed conduits. His research dealt with the application of fluid mechanics to problems of hydraulic engineering, sediment transport, and conduit flow. He was the recipient of the 1947 ASCE Karl Emil Hilgard Award for an excellent paper on energy conversion in expansions. Once he had left academia, he took interest in the movement of suspensions and their separation from water. He thereby invented five devices relative to water and sewage treatment equipment. He was a member ASCE, AWWA and AGU.

Anonymous (1948). A.A. Kalinske. *Civil Engineering* 18(1): 45. *P*
Anonymous (1959). Kalinske, A.A. *Who's who in engineering* 8: 1288. Lewis: New York.
Kalinske, A.A., van Driest, E.R. (1939). Application of statistical theory of turbulence to hydraulic problems. Proc. 5th *IUTAM Congress* Cambridge MA: 416-421.
Kalinske, A.A. (1941). Turbulence and energy dissipation. *Trans. ASME* 63(1): 41-48.
Kalinske, A.A., Robertson, J.M. (1943). Closed conduit flow. *Trans. ASCE* 108: 1435-1447.
Kalinske, A.A. (1953). Settling rate of suspensions in solids contact. *Proc. ASCE* 79(186): 1-8.
Kalinske, A.A. (1954). Flotation and sedimentation in treating wastes. *Water and Sewage Works* 101(3): 128-131. *P*
Kalinske, A.A. (1970). *Turbulence diffusivity in activated sludge aeration basins*. Pergamon: NY.

KAYS

22.01. 1881 Tonica IL/USA
13.03. 1946 Atlanta GA/USA

Marion Reed Kays started his studies at University of Arizona, Tucson AZ, but graduated in 1906 from University of Illinois, Urbana IL, as a civil engineer. He was then employed as assistant engineer by the US Reclamation Service on irrigation works on the North Platte Project in Wyoming and Nebraska, and later on the Idaho Irrigation Company, becoming in 1911 its vice-president and manager. During the next eight years he was responsible for all funds, water supply, management and operation, precipitation and stream flow measurements and records, incidental construction, and corporation with state authorities.

In 1920 Kays became associated with an engineering corporation of San Francisco CA, in charge of its Phoenix AZ office. He there acted as project engineer for the Paradise-Verde Irrigation District, and was in charge of the analyses of the design of reservoirs, dams, diversion works, and canals. In 1923 he became special field engineer for the Salt River Valley Water Users' Association, where he was engaged on the measurements on canal water loss, the quantity of water to be saved by lining the many canals, and the enlargement of the power plant at Roosevelt Dam. Kays was appointed manager of the Lake Worth Drainage District FL in 1924, which brought there a thorough, practical knowledge of the principles of the artificial control of water for agriculture. The relation between soil moisture and plant growth, the construction and operation of reclamation works, and the appreciation of public agencies engaged in this service were considered. The system of canals was incomplete, and the existing control works insufficient, and there were also no pumps when Kays started his work with the District. As manager, he studied these deficiencies and improved the conditions significantly. He resigned in 1938 the management by preparing a detailed report, making recommendations for the future. He was in 1939 appointed chief engineer of the West Palm Beach Housing Authority, and served from 1941 as resident manager the defence development during World War II. He was remembered for his enthusiastic handling of drainage problems in Florida, and for his tenacity in the accomplishment of long-time objectives. He was a member of the American Society of Civil Engineers ASCE from 1927.

Anonymous (1945). Marion R. Kays. *Engineering News-Record* 134(Feb.1): 152. *P*
Anonymous (1948). Marion R. Kays. *Trans. ASCE* 113: 1483-1487.
Harper, F., ed. (1913). Kays, Marion R. *Who's who on the Pacific Coast*: 312. Los Angeles.
http://www.google.ch/imgres?imgurl=http://o.mfcreative.com/f2/file08/objects/0/1/e/801e6c61-4483-40a2-9918 *P*

KEEFER

10.07. 1891 Baltimore MD/USA
02.09. 1974 Baltimore MD/USA

Clarence Edward Keefer was educated at the Johns Hopkins University, Baltimore MD, obtaining the BS degree in civil engineering. He was from 1919 to 1927 assistant civil engineer with the Bureau of Sewers, Baltimore MD, from when he was principal assistant engineer there of the Bureau of Sewers. From the 1940s Keefer acted as deputy sewerage engineer of this Bureau. He was also involved in lecturing sanitary engineering at Johns Hopkins. He was a member of the American Society of Civil Engineers ASCE, the British Institution of Sewage Purification, the American Society of Municipal Engineers, now the American Public Works Association APWA, and the American Public Health Association APHA.

Keefer was an authority in the sanitary engineering field. His main interest was the settling of particles in the separation process between sludge and pure water, as applied in settling tanks. He took also interest in the associated processes of sludge digestion. Authoring more than hundred articles dealing with sewerage practice or reporting the results of experimental and developmental work at the Baltimore sewage treatment plant, Keefer was all through his career closely related with his city of birth. He was the recipient of the 1947 Kenneth Allen Award, and the 1950 George B. Gascoigne Medal from the Federation of Sewage and Industrial Wastes Association. He also was awarded the 1952 Thomas Fitch Prize from ASCE for the paper dealing with the Back River Sewage Works. He was a member of the editorial board of Water and Sewage Works, and contributed equally to the Journal of Sewage Treatment Works. He also served the Maryland ASCE Section as secretary, treasurer and president.

Anonymous (1937). Keefer, Clarence E. *Who's who in engineering* 4: 736. Lewis: New York.
Anonymous (1952). C.E. Keefer, Thomas Fitch Rowland Prize. *Civil Engineering* 22(8): 587. *P*
Keefer, C.E., Kratz, H. (1929). Digesting sewage sludge at its optimum pH and temperature. *Engineering News-Record* 102(3): 103-105.
Keefer, C.E. (1934). The effect of sewage on cast-iron Venturi meter. *Engineering News-Record* 112(Jan.11): 46.
Keefer, C.E. (1940). *Sewage treatment works*. McGraw-Hill: New York.
Keefer, C.E. (1950). Settling tank design. *Water & Sewage Works* 97(10): 422-426.
Keefer, C.E. (1955). Venturi meter 42 years old still in good condition. *Journal of Water and Sewage Works* 102(4): 178-179.

KEENER

09.04. 1888 Moberly MO/USA
13.04. 1971 Denver CO/USA

Kenneth Bixby Keener graduated with a BS degree from Ohio Wesleyan University in 1910, and was in 1940 awarded the degree DSc from Colorado State University, Fort Collins CO, where he had formerly also extended studies to civil engineering in 1914. He was a staff member of the United States Bureau of Reclamation USBR all through his career, starting as chainman and rodman, then becoming a hydrographer on canal and drainage works, until finally chief design engineer when retiring in the 1950s. He was a member of the American Society of Civil Engineers ASCE, the National Geographical Society, and the American Concrete Institute.

Keener was from 1910 to 1915 involved in the Boise Irrigation Project, for which he made structural designs, water supply studies and field investigations. From then to 1926 he was active for the power plant construction and large concrete dams of this project. Once an associate engineer, he elaborated specifications and design work for major USBR dams at the Denver Office until 1930. Until 1936 he was a senior engineer, and principal assistant to the engineer in charge of dams, from when he was engineer in charge of the USBR Dam Division at Denver CO. Keener was strongly involved in the design and construction of Grand Coulee Dam on Columbia River, then one of the largest in the world, and Shasta Dam in California. Other dams where he was involved were the Anderson Ranch Dam on Boise River, and the Hungry Horse Dam in northwestern Montana. He was member, the US Committee of the International Commission on Large Dams ICOLD, attending various of its congresses, as also those of the World Power Conference. He had written a number of papers to professional journals mainly dealing with concrete dams.

Anonymous (1952). K.B. Keener. *Civil Engineering* 22(7): 22. *P*
Anonymous (1954). Keener, Kenneth B. *Who's who in engineering* 7: 1282. Lewis: New York.
Keener, K.B. (1935). Grand Coulee project and dam. *Engineering News-Record* 115(Aug.1): 141-143.
Keener, K.B. (1938). The Low Dam at Marshall Ford. *Engineering News-Record* 121(Dec.1): 697-699.
Keener, K.B. (1944). Spillway erosion at Grand Coulee Dam. *Engineering News-Record* 133(Jul.13): 95-101.
Keener, K.B. (1951). Uplift pressures in concrete dams. *Trans. ASCE* 116: 1218-1237; 1261-1264.

KEIM

22.04. 1902 Falls City NE/USA
03.10. 1990 Prescott AZ/USA

Paul Ferdinand Keim received in 1925 the BSc degree in civil engineering from the University of California, Berkeley CA, and the MSc degree in 1932 from the University of Nebraska, Lincoln NE. Keim started his professional career as consultant to the Platte Valley Public Power and Irrigation District NE. He moved back to California in 1936, working for the California Division of Highways. In 1937 he joined Los Angeles County Flood Control District, and from 1939 served as principal engineer the Federal Power Commission at Washington DC until WW2. After war service, he acted as consultant on water supply and transportation to the US Department of States Mission in Liberia. He was from 1948 associate of Tippetts, Albert, McCarthy, Stratton TAMS Engineers, New York City. Armed with a wealth of knowledge and practical experience, he was in 1952 appointed professor of civil engineering at the University of California, Berkeley CA. Upon leaving Berkeley in 1969, he served as chief staff engineer the Ralph M. Parsons Co., Pasadena CA, in Peru, Greece, Argentina, Mexico, Tunisia, and Morocco. In 1971 he acted as a consultant to the Organization of American States in Brazil.

Keim was an expert in water resources, hydrology, and engineering education. He has written on these topics, thereby participating in relevant aspects of engineering concern. Several deal also with irrigation, relating to the Platte-Missouri Rivers confluence, the irrigation efficiency and its management, or the irrigation demands in the Susquehanna Basin. Others deal with engineering education and the construction industry, changed conditions in construction contracts, quality control in construction, or construction personnel management. The 1965 discussion on groundwater development in Hawaii involves water resources and water supply aspects of Hawaii related to the groundwater characteristics.

Anonymous (1926). Paul F. Keim. *Blue and gold yearbook*: 87. Berkeley CA. *P*
Anonymous (1991). Paul F. Keim. *Trans. ASCE* 156: 503.
Keim, P.F. (1968). Discussion of Engineering geologic studies for sewer projects. *Journal of the Sanitary Engineering Division* ASCE 94(SA4): 760-761.
Keim, P.F. (1978). Discussion of Flood management for small urban streams. *Journal of the Water Resources Planning and Management Division* ASCE 104(WP1): 294.
Keim, P.F. (1966). Discussion of Development of ground water in Hawaii. *Journal of the Hydraulics Division* ASCE 91(HY2): 408-409.

KELLER H.B.

19.06. 1925 Mansfield OH/USA
26.01. 2008 Pasadena CA/USA

Herbert Bishop Keller was the brother of Joseph B. Keller (1923-). 'Herb' studied mathematics after war service at New York University, remaining there as faculty member, working there in numerical analysis and computing. He became the associate director of the Atomic Energy Commission's Computing and Applied Mathematics Center at the Courant Institute. In 1965, at the invitation of Gerald Whitham (1927–2014), he visited the Caltech Applied Mathematics Group, returning in 1966 to NYU, but accepting in 1967 an appointment at the Caltech as professor of applied mathematics, remaining there until his retirement.

It was during the Caltech years that Keller became a central figure in the numerical analysis and applied mathematics community. He was particularly known for numerical methods in boundary-value problems, and in numerical analysis of non-linear problems exhibiting folds and bifurcation phenomena. He also made extensive contributions to the scientific community as panel and committee member of the National Academy of Sciences, among others. He was in addition in the editorial boards of the SIAM Journal of Applied Mathematics and the SIAM Journal on Numerical Analysis. He further acted as editor of various book series. In addition to the classic textbook of Eugene Isaacson (1919-2008) and Keller, he authored books on two-point boundary-value problems, and on bifurcation and path-following problems. In 2000, after retirement from Caltech, Keller joined the Center for Computational Mathematics in the Math Department of UC San Diego as senior research scientist, splitting his time between his home at Pasadena and a condo at Leucadia CA.

Isaacson, E., Keller, H.B. (1966). *Analysis of numerical methods*. Wiley: New York.

Keller, H.B., Keller, J.B. (1951). *On systems of linear ordinary differential equations*. Research Report EM-33. New York University: New York.

Keller, H.B., Levine, D.A., Whitham, G.B. (1960). Motion of a bore on a sloping beach. *Journal of Fluid Mechanics* 7: 302-316.

Keller, H.B. (1976). *Numerical solution of two-point boundary value problems*. SIAM: New York.

Keller, H.B., Chorin, A.J., Dennis, S.C., Fornberg, B., Kreitzberg, C.W., eds. (1978). *Computational fluid dynamics*. American Mathematical Society: New York.

http://www.siam.org/news/news.php?id=1404

Keller_OHO.pdf *P*

http://oralhistories.library.caltech.edu/151/1/Keller_OHO.pdf *P*

KELLEY

17.04. 1839 Woodford/EI
24.01. 1914 Brooklyn NY/USA

John Carl Kelley was born in Irish Galway County. His parents emigrated to America soon after his birth, settling at Palmyra and Rochester NY. Fortified by a genuine Irish optimism and a rugged pertinacity, John came to New York City to make his fortune. He launched in 1870 the National Meter Company NMC, of which he was made president, remaining the active head until his death. Water meters were hardly used in the USA in 1870. In the early 1870s, Kelley was in addition to president also the chief engineer and the salesman. He had in this period no difficulty to sell meters to water works companies. The superintendent of the company of Concord NH was so delighted upon watching the movement of the recording dial, that he immediately ordered several meters without further test of their efficiency.

During the first ten years of its history, the NMC manufactured and sold on the Gem make, which was of velocity type, as invented and patented by Henry F. Read, Brooklyn NY, in 1869. The objections to early water meters were lack of accuracy, high cost, and their use for small pipes only. The first large-size Gem meters were two 4 inch meters ordered in the early 1870s by the Jersey City water works, and manufactured by Kelley. Although many patents were issued up to 1870, it was noted that The great need of water boards is a durable meter, registering with reasonable accuracy. The fortune of any inventor is secured who can introduce such a meter. The problems encountered with the rotary meters were to reduce the friction of the moving vanes, to direct the water upon the wheel that the latter moves with the least quantity of water, and to design the register that it would hardly retard the motion of the wheel. The NMC devoted the year 1878 to test various devices, resulting in its crowning effort with a meter named Crown Meter. This meter met with early and lasting success. In 1906 the NMC turned out its 500,000 meter, a number doubled by 1913. Kelley was a man of great character and of positive manner. He was intimately connected with the water works history and with the introduction of water meters.

Anonymous (1914). John C. Kelley and the American water meter industry. *Engineering News* 71(13): 674-675. *P*
Kelley, J.C. (1887). *Statistics, tables, and water rates of cities and towns*, together with facts about water meters. National Meter Company: New York.
Nash, L.H. (1879). Crown meter and its principles. *Engineering News* 6(Nov.29): 385.

KENNEDY J.F.

17.12. 1933 Farmington NM/USA
13.12. 1991 Iowa City IA/USA

John Fisher Kennedy obtained his BS degree in civil engineering from Notre Dam University, Notre Dame IN, in 1955. He then got the MS and PhD degrees from California Institute of Technology, in 1956 and 1960, respectively. He was from 1961 to 1964 assistant professor at Massachusetts Institute of Technology MIT, then there associate professor until 1966, from when he took over as professor of fluid mechanics at Iowa State University, Iowa City IA, and director of its Iowa Institute of Hydraulic Research IIHR until 1991. Kennedy was as Fulbright Scholar at Karlsruhe Technical University in 1972, as Erskine Fellow at the University of Canterbury, Christchurch NZ in 1976, and in 1985 as visiting professor at ETH Zurich. He was a Hunter Rouse Lecturer in 1981, a recipient of the 1959 J.C. Stevens Award, the 1964 W.L. Huber Prize, the 1974 Karl E. Hilgard Hydraulic Prize, all from ASCE.

Kennedy's first and primary love was sediment transport, to which he made significant contributions mainly relating to the understanding of alluvial processes. He in addition built and equipped IIHR with the first university ice laboratory. He also initiated research in cooling tower design and a new type of drop shaft for urban stormwater drainage. He was further involved in the design of vanes reducing erosion along alluvial river banks by reducing secondary currents. He was further an authority on management of waste heat from steam generation of electrical power, and on turbulent mixing of fluids. Kennedy was also involved in the International Association of Hydraulic Research, first as council member from 1971 to 1975, then vice-president until 1981, and president until 1985. He was awarded honorary IAHR membership in 1989.

Anonymous (1962). John F. Kennedy. *Civil Engineering* 32(10): 71. *P*
Anonymous (1985). Kennedy, John F. *Who's who in engineering* 6: 346. AAES: Washington DC.
Anonymous (1991). Kennedy, John F. *Who's who in America* 46: 346. AAES: Washington.
Anonymous (1992). John F. Kennedy. *IAHR Bulletin* 30(1): 9. *P*
Iwasa, Y., Kennedy, J.F. (1968). Free surface shear flow over a wavy bed. *Journal of the Hydraulics Division* ASCE 94(HY2): 431-454; 95(HY1): 524-530; 96(HY3): 837-841.
Kennedy, J.F., Froebel, R.A. (1965). Two-dimensional turbulent wakes in density-stratified liquids. ASME *Annual Meeting* New York: 1-8. ASME: New York.
Kennedy, J.F. (1987). Hydraulic trends towards the year 2000. *Hydraulics and hydraulic research*: 357-362, G. Garbrecht, ed. Balkema: Rotterdam.

KENNISON

06.05. 1886 Marysville NB/CA
30.04. 1977 Auburndale MA/USA

Karl Raymond Kennison obtained his BS degree from Massachusetts Institute of Technology MIT in 1908, and the honorary DSc degree from the Colby College, Waterville ME, in 1941. He was then from 1910 to 1915 the principal assistant of John R. Freeman (1855-1932) at Providence RI, dealing with water power and hydraulic structures projects. After some years as design engineer of dams in the North-East, Kennison founded in 1920 a consulting office at Boston MA. He was in addition chief engineer of the South Essex Sewerage District, deputy chief engineer of the Metropolitan District of the Water Supply Commission in Massachusetts, taking over as chief engineer in 1939. Kennison was in 1951 also a special lecturer at the Department of Civil Engineering, MIT, and from 1952 to 1956 chief engineer of the New York City Board of Water Supply. He was a Fellow of the American Society of Civil Engineers ASCE.

Kennison was involved both in hydraulic research and engineering practice. His 1916 paper on the direct hydraulic jump was the first dealing with high approach flow Froude numbers, attracting a large number of discussions, which indicated the interest in this topic. A simplified formula for the sequent depth ratio was derived in 1931. In the 1920s he was concerned with water power including knowledge on turbines. Toward the end of his career, questions of water supply were dealt with. He was considered an expert in this field stating that attention should be given to this problem because only the best sources should provide this resource. It was also his belief that the more you know the better paid you will be. He further stated that young engineers tend to need to much supervision instead of using their own resources to solve the problem at hand.

Anonymous (1952). Karl R. Kennison. *Engineering News-Record* 148(May 29): 68. *P*
Anonymous (1964). Kennison, Karl R. *Who's who in engineering* 9: 990. Lewis: New York.
Kennison, K.R. (1916). The hydraulic jump in open-channel flow at high velocity. *Trans. ASCE* 80: 338-420.
Kennison, K.R. (1920). Comprehensive plotting of water turbine characteristics. *Trans. ASCE* 83: 861-867.
Kennison, K.R. (1934). Ware River intake shaft and diversion works. *Civil Engineering* 4(8): 388-392.
Kennison, K.R. (1935). Quabbin Reservoir Work now in full swing. *Water Works Engineering* 88(18): 1062-1064. *P*

KERR

15.07. 1899 Philadelphia PA/USA
30.01. 1967 Flourtown PA/USA

Samuel Logan Kerr graduated from University of Pennsylvania, Philadelphia PA, as a civil engineer in 1924, after having obtained a professional certificate from *Ecole Speciale des Travaux Publics* ETP, Paris France, in 1919. He was from 1924 to 1927 assistant hydraulic engineer, then until 1929 assistant chief engineer, continuing as resident engineer until 1935. From 1937 to 1945 Kerr was a senior mechanical engineer of the US Engineering Office, Eastport ME, from when he joined large firms at Philadelphia and Washington DC. Finally, until retirement, he was a consulting hydraulic engineer. Kerr was a Fellow of the American Society of Mechanical Engineers ASME, a member of the American Society of Civil Engineers ASCE, and the American Water Works Association AWWA.

Kerr was widely known as excellent hydraulic engineer, particularly in the fields of water power, public water supply, pumping, water hammer, and surge control. During his early career he developed an automatic frequency and load control system for hydroelectric plants. Later he aided in the solution of water hammer problems in motor-driven water works pumping systems including the Croton Lake Pumping Plant for New York City, the city of Toledo OH, or the St. Louis County Water Company. In the mid-thirties he was in charge of the mechanical section of the Passamaquoddy Tidal Power Project in Maine. Later he also designed chemical plants, circulating water systems for steel mills and utility plants. Kerr was one of the organizers of the 1933 ASME Water Hammer Congress held at Chicago IL.

Anonymous (1938). S. Logan Kerr. *Mechanical Engineering* 60(12): 964. *P*
Anonymous (1953). S. Logan Kerr honoured by ASME. *Water Works Engineering* 106(3): 221. *P*
Anonymous (1954). Kerr, S. Logan. *Who's who in engineering* 7: 1302. Lewis: New York.
Anonymous (1968). S. Logan Kerr. *Mechanical Engineering* 90(4): 161. *P*
Kerr, S.L. (1929). New aspects of maximum pressure rise in closed conduits. *Trans. ASME* 51(3): 13-30.
Kerr, S.L. (1933). Committee report. *Symposium on water hammer*: 3-14. ASME: New York.
Kerr, S.L. (1933). Water hammer tests in Croton Lake Pumping Plant. *Symposium on water hammer*: 84-90. ASME: New York.
Kerr, S.L. (1935). Research investigation of current-meter behaviour in flowing water. *Trans. ASME* 57(HYD-9): 295-301.
Kerr, S.L. (1951). Water hammer control. *Journal AWWA* 43(12): 985-999.

KESSLER

12.03. 1900 South Haven MI/USA
19.03. 1974 Johnson KA/USA

Lewis Hanford Kessler was educated at University of Chicago, and University of Wisconsin, Madison WI, obtaining the degrees BS in 1922, and MS in 1928. He was instructor in hydraulic engineering at University of Wisconsin from 1922 to 1927, then from assistant to associate professor of hydraulics and sanitary engineering until 1941, and then until 1946 sanitary engineering professor at Northwestern University, Evanston IL. He was from 1925 to 1940 in parallel a consulting engineer, then chief of the water and sewage section, Office of the Chief of Engineers, War Dept., Washington DC. He was a member of the American Society of Civil Engineers, and the American Water Works Association. He received the Distinguished Service Citation of his Alma Mater in 1966.

Kessler is known for a number of papers in hydraulics. He was mainly interested in features of pipe flow and in hydraulic aspects of sanitary engineering. Early research included the hydraulics of air-lift pumps, in which the pumping action is generated by air flow injected in the riser pipe. This method is currently used mainly in the chemical and oil producing industries. A second notable work related to drop structures, which are often applied in hydraulic structures but may perform extremely poor because of air entrained by the water flow. These structures have received lots of attention in the 1960s and 1970s, based also on the early findings of Kessler. Still another research related to wrought-iron pipes and their particular friction behaviour in turbulent flow. He also invented a new type of mechanical-pneumatic water hammer arrestor.

Anonymous (1951). Kessler, Lewis H. *Who's who in America* 26: 1476. Marquis: Chicago IL.
Anonymous (1954). Kessler, Lewis H. *Who's who in engineering* 7: 970. Lewis: New York.
Anonymous (1966). Lewis H. Kessler. *Civil Engineering* 36(7): 13. *P*
Kerr, S.L., Kessler, L.H., Gamet, M.B. (1950). New method for bulk modulus determination. *Trans. ASME* 72(11): 1143-1154.
Kessler, L.H. (1934). Experimental investigation of the hydraulics of drop inlets and spillways for erosion control structures. Engineering Experiment Station, *Bulletin* 80. University of Wisconsin: Madison WI.
Kessler, L.H. (1935). Friction losses in wrought iron pipe when installed with couplings. Engineering Experiment Station, *Bulletin* 82. University of Wisconsin: Madison WI.
Ward, C.N., Kessler, L.H. (1924). Experimental study of the air-lift pumps and application of results to design. Engineering Series 9, *Bulletin* 1265. University of Wisconsin: Madison.

KILLEN

17.11. 1921 De Graff MN/USA
28.07. 2013 Olmito TX/USA

John Mark Killen received the BEE degree in 1944, and the MS degree in 1956 for a work entitled The measurement of sediment properties by scattering of ultrasonic radiation in water, from the University of Minnesota, Minneapolis MS. His PhD submitted in 1968 entitled The surface characteristics of self-aerated flow in steep channels, was initially tutored by Lorenz G. Straub (1901-1963); his advisor was Alvin G. Anderson (1911-1975), who continued the work of his predecessor. Killen remained during his professional career at St. Anthony Falls Hydraulic Laboratory, University of Minnesota, dealing with a variety of hydraulic questions, but with a particular interest into two-phase air-water high-speed flows.

The 1964 Report deals with a system for measuring the size and concentration of free air bubble nuclei in limited ranges of size and concentration in a water tunnel. This system was based on measuring the amplitude attenuation of an acoustic pulse of sound as it is propagated across the flow. An arbitrary method was proposed for converting the measured sound attenuation into size and concentration of air bubbles. It was this measuring system which led to the great advances in knowledge of air-water flows founded at the University of Minnesota.

Anonymous (1956). J.M. Killen. Proc. 6th *Hydraulics Conference* Iowa: Frontispiece. *P*

Killen, J.M., Ripken, J.F. (1964). A water tunnel air content meter. St. Anthony Falls Hydraulic Laboratory *Report* 70. University of Minnesota: Minneapolis.

Killen, J.M., Anderson, A.G. (1969). A study of the air-water interface in air-entrained flow in open channels. 13th *IAHR Congress* Kyoto 2(B36): 1-4.

Killen, J.M. (1974). *A buoyancy-propelled test body laboratory facility*. SAF Lab: Minneapolis.

Lamb, O.P., Killen, J.M. (1950). An electrical method for measuring air in flowing air-water mixture. *Technical Paper* 2, B. St. Anthony Falls Hydraulic Laboratory: Minneapolis.

Ripken, J.F., Killen, J.M. (1963). Gas bubbles: Their occurrence, measurement, and influence in cavitation testing. IAHR Symp. *Cavitation and Hydraulic Machinery* Sendai: 37-57.

Schiebe, F.R., Killen, J.M. (1968). New instrumentation for the investigation of transient cavitation in water tunnels. St. Anthony Falls Hydraulic Lab. *Memorandum* M-113. University of Minnesota: Minneapolis.

Stefan, H.G., Foufoula-Georgiou, E., Arndt, R.E.A. (2004). The St. Anthony Falls Laboratory: A rich history and a bright future. Proc. *Annual EWRI Meeting*: 185-197.

KINDSVATER

01.08. 1913 Hoisington KA/USA
15.11. 2002 Morrisville VT/USA

Carl Edward Kindsvater graduated as a civil engineer from State University Iowa, Iowa City IA, in 1937. He then joined the Tennessee Valley Authority TVA until 1941 as hydraulic engineer in its hydraulic laboratory, was in its Flood Control Section until 1943 when joining the US Army Corps of Engineers as hydraulic engineer for works in the District Little Rock AR. He was associate professor of hydraulics at the Georgia Institute of Technology from 1945 to 1948, then professor there until 1955 when becoming Regents Professor until retirement. Kindsvater was member of the American Society of Civil Engineers ASCE, the International Association of Hydraulic Research IAHR, and of the Georgia Engineering Society. He was awarded the 1945 ASCE Collingwood Prize, the 1954 ASCE Rickey Medal, the 1955 ASCE Norman Medal, and 1963 ASCE Honorary Membership.

Kindsvater was known for a number of outstanding researches in hydraulics and hydraulic engineering. His 1943 paper deals with hydraulic jumps on sloping channels in which the basic knowledge of classical hydraulic jumps is generalized by accounting for a slope parameter. In 1953 a report of peak discharges at river contractions was presented, followed in 1955 by a detailed analysis of the flow features at constrictions by bridges and abutments. Given the complex geometry of bridge elements and the approach flow conditions, a large number of parameters were considered whose effect was presented with a number of figures. The 1959 paper co-authored by Rolland Carter (1916-2011) adds to discharge measurement by thin-plate rectangular weirs, following the proposal of equivalent head and width by Theodor Rehbock (1864-1950).

Anonymous (1946). Carl E. Kindsvater. *Civil Engineering* 16(1): 36. *P*
Anonymous (1959). Kindsvater, Carl E. *Who's who in engineering* 8: 1344. Lewis: New York.
Anonymous (1960). Carl E. Kindsvater. *Civil Engineering* 30(10): 88. *P*
Harrison, E.S., Kindsvater, C.E. (1954). Dam modification checked by hydraulic models. *Trans. ASCE* 119: 73-92.
Kindsvater, C.E., Carter, R.W., Tracy, H.J. (1953). Computation of peak discharge at contractions. *Circular* 284. US Geological Survey: Washington DC.
Kindsvater, C.E., Carter, R.W. (1955). Tranquil flow through open-channel constrictions. *Trans. ASCE* 120: 955-980; 120: 991-992.
Kindsvater, C.E., Carter, R.W. (1959). Discharge characteristics of rectangular thin-plate weirs. *Trans. ASCE* 124: 772-801; 124: 818-822.

KING D.L.

13.08. 1938 Nampa ID/USA
10.11. 2013 Montrose CO/USA

Danny Lee King received the BS degree in 1959 in civil engineering from University of Idaho, Moscow ID, and the MS degree in hydraulic engineering in 1964 from the University of Colorado, Boulder, CO. He was from 1966 hydraulic engineer within the Hydraulics Branch, in 1971 head of the Applied Hydraulics Section, taking over in the mid-1970s as chief the Hydraulics Branch of the US Bureau of Reclamation USBR, Denver CO. In the 1970s he was a member of the ASCE Task Committee on Environmental Effects of Hydraulic Structures. He was involved in the organisation of the 1986 ASCE Water Forum. At USBR, King collaborated with Thomas J. Rhone (1921-1996), and Philip H. Burgi (1942-), among others.

The 1966 paper compares three transition designs in terms of flow appearance, head loss and velocity distribution in an open channel pump intake under subcritical approach flow. The symmetrical transition with the curved approach flow geometry was found to perform best. The final design included an angled transition for lower cost, however. The 1967 paper deals with the amplitude-frequency spectrum of pressure fluctuations caused by turbulence in stilling basins. Experiments with prototype structures indicated that fatigue failure was caused by vibration. The dynamic loads on the structures was determined. Results relative to design guidelines are discussed and possible sources of errors described. Other studies of the 1970s deal with the design of the approach flow channel to shaft spillways, environmental effects of hydraulic structures, cavitation control by aeration of high-velocity jets, or features in environmental engineering.

Anonymous (1965). Danny King. *Coloradan yearbook*: 225. University of Colorado: Boulder. *P*
King, D.L. (1966). Comparison of intake transition designs for a large pumping plant. ASCE
    Water Resources Engineering Conference Denver CO: 1-40.
King, D.L. (1967). Analysis of random pressure fluctuations in stilling basins. Proc. 12th *IAHR*
    Congress Fort Collins B(25): 210-217.
King, D.L., Johnson, P.L. (1971). Discussion of Mechanics of stratified flow through orifices.
    Journal of the Hydraulics Division ASCE 97(HY9): 1535-1541.
King, D.L. (1978). Environmental effects of hydraulic structures. *Journal of the Hydraulics
    Division* ASCE 104(HY2): 203-221.
King, D.L. (1979). *Bureau of Reclamation experience with destratification of reservoirs*.
    USBR: Denver CO.

KING H.W.

10.02. 1874 Big Rapids MI/USA
22.04. 1951 Pasadena CA/USA

Horace Williams King obtained his BS degree in civil engineering from University of Michigan, Ann Arbor MI in 1895. He was until 1898 city engineer of Big Rapids MI and then employed by the Isthmian Canal in charge of field works in Nicaragua until 1901. During 1902 King was assistant engineer for harbour works at Manila PH, and in 1903 resident engineer at Canton, China. Upon return to the USA he joined as engineer the US Reclamation Service in charge of works for the Umatilla Dam project in Oregon until 1909, from when he was a hydraulic engineer of a private consultant. King was appointed in 1912 professor of hydraulic engineering at University of Michigan, a position he held until his retirement in 1939. In parallel he also was a private consultant at Ann Arbor.

King is known for his Handbook of hydraulics, published in 1918 and revised in 1928. This was a standard reference of American hydraulic engineers for decades. The main book chapters include Hydrostatics, Orifices, Weirs, Pipe flow, and Open channel flow. He also was an associate editor of the American Civil Engineers' Handbook published in 1920 and revised in 1930, authoring chapters on Irrigation and drainage, and solution of the Manning formula for pipes. He advocated this formula as a substitute to that of Wilhelm Kutter (1818-1888) for both pipe and open channel flows. At this time, the formula of Robert Manning (1816-1897) was practically unknown in the USA but then almost universally adopted throughout the country. King was member of the American Society of Civil Engineers, elected to Honorary Member ASCE in 1945.

Anonymous (1941). King, Horace W. *Who's who in engineering* 5: 980. Lewis: New York.
Anonymous (1945). Horace Williams King. *Civil Engineering* 15(1): 46. *P*
Anonymous (1951). Horace King, Hon.M. ASCE, is dead. *Civil Engineering* 21(5): 293. *P*
King, H.W. (1918). *Handbook of hydraulics* for the solution of hydraulic problems. McGraw-Hill: New York, 7th ed. co-authored by E.F. Brater and E. Brater in 1976.
King, H.W., Wisler, C.O. (1922). *Hydraulics*. Wiley: New York.
King, H.W. (1937). *Manning formula tables* for the solution of pipe problems: Giving diameters in inches corresponding to different rates of loss of head. McGraw-Hill: New York.
King, H.W. (1944). Irrigation and drainage. *American civil engineers' handbook* 16: 1653-1714, T. Merriman, T.H. Wiggins, eds. Wiley: New York.
King, H.W., Wisler, C.O., Woodburn, J.G. (1948). *Hydraulics*. Wiley: New York.
King, H.W., Brater, E.F. (1959). *Handbook of hydraulics*, 4th ed. McGraw-Hill: New York.

KING J.S.

04.06. 1909 Monticello WI/USA
27.11. 1968 Warren IL/USA

James Sheldon King obtained in 1932 his BS degree in electrical engineering from University of Illinois, Champaign-Urbana IN. He pursued in 1940 studies in agriculture at Iowa State University, Iowa City IA. He also attended the Water Resources Seminar of Harvard University in 1957, receiving the degree of Master in Public Administration, Water Resources Planning. Following graduation, King worked until 1935 in various capacities when he commenced with the Federal Government. He was engaged until 1942 as engineer with the Soil Conservation Service SCS of the US Department of Agriculture, and then went to the US Army Corps of Engineers, the Chicago District. In 1952 he was transferred to the Great Lakes Division, and in 1963 selected to serve as chief of the River Basin Planning Branch. In 1966 King served as assistant chief of the Engineering Division, with responsibility in both civic and military works planning and design of the US Corps of Engineers, North Central Division, commencing then his tenure as chief of the newly formed Planning Division. From 1963 to 1965 he in addition lectured on water resources planning at the University of Illinois, Champaign-Urbana IL.

King was a notable hydraulic engineer in water resources planning and in agricultural engineering. He has served both as a designer and a teacher. He also aided on the ASCE Technical Committee of Navigation and Flood Control Facilities of the Waterways and Harbours Division, taking over as chairman from 1967 to 1968. His 1962 paper deals with a plan to develop the water resources of his District. He was instrumental with the 1967 Report on Flood proofing, by which flood damage to the structure and the contents of buildings in a flood-hazard area is prevented. This Report addresses public officials and building owners dealing with the principles of reducing flood damages. This issue is currently of concern, given the more intense floods, and the higher values of public infrastructure. King further was in his state registered engineer. He was member of the American Society of Civil Engineers ASCE, becoming ASCE Fellow in 1959.

Anonymous (1932). James S. King. *The Illio*: 558. University of Illinois: Urbana-Champaign. *P*
Anonymous (1969). James S. King. *Trans. ASCE* 134: 976-977.
King, J.S. (1962). *Comprehensive plan for the development of water and other resources*: Present and potential. US Army Engineer Division: Chicago IL.
Shaeffer, J.R., ed. (1967). *Introduction to flood proofing*: An outline of principles and methods. Center for Urban Studies, The University of Chicago: Chicago IL.

KING W.R.

29.04. 1888 Laramie WY/USA
26.07. 1934 Chattanooga TN/USA

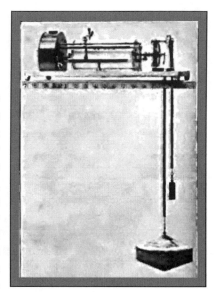

Warren Raymond King graduated from University of Utah, Salt Lake City UT, in 1911 with the degree of BSc in general engineering. He was appointed in 1912 junior engineer with the US Geological Survey USGS, Water Resources Branch, remaining there until 1916, with headquarters at Washington DC, Salt Lake City UT, and Denver CO. He then became assistant engineer with the Colorado Power Company until 1919, when returning to USGS, Denver CO. In 1920 he was appointed district engineer at Nashville TN, in charge of stream gauging investigations in the Southeastern States, and later that year moved to the District Office at Chattanooga TN, where he remained until his death.

King made an enviable record as engineer and as executive during his eighteen years of USGS. Based on convincing reasoning, and unfailing record of accomplishment, he built up from the small beginning one of the largest and most important districts in the USGS. His chief duty in Tennessee was the collection of the daily stream-flow records at various stations along the District rivers, and he was successful with an achievement of constant growth and improvement of stations. In 1920, when taking charge, there were 33 gauging stations, whereas in 1934 the District included 110 stations, of which 84 were equipped with the latest recording gauges, housed in a permanent structure. In addition to his contributions to the annual Water Supply Papers, he authored 2 Bulletins of Tennessee State and one Kentucky Bulletin. This information was basic for the water power development, flood control, water supply, and navigation, proving of immense value for the Tennessee River Valley. King was interested in addition in all things pertaining to this region, given that he was familiar with the various activities made by the Federal Government. He was an attractive personality, enjoying the society of others. He was the first president of the Tennessee Valley Section of the American Society of Civil Engineers ASCE, whose member he was since 1923.

Anonymous (1935). Warren R. King. *Trans. ASCE* 100: 1775-1776.
Grover, N.C., Hartwell, O.W., Stevens, G.C., Hall, W.E., King, W.R. (1924). Surface water supply of the United States 3: Ohio River Basin. *Water Supply Paper* 503. Washington DC. (*P*)
King, W.R. (1924). Record cloudburst flood in Carter County. *Monthly Weather Rec.* 52: 311-313.
King, W.R. (1925). *Water resources of Tennessee*: Compilation of existing data pertaining to the surface waters of Tennessee and their utilization. Division of Geology USGS: Nashville.
King, W.R. (1931). *Surface waters of Tennessee*. Division of Geology USGS: Nashville TN.

KINGMAN

06.03. 1852 Dover NH/USA
14.11. 1916 Atlantic City NJ/USA

Dan Christie Kingman graduated in 1875 from the US Military Academy, West Point NY, and was then assigned to duty with the engineer troops at Willets Point NY. He was commissioned first lieutenant in 1879 serving for three years as instructor in civil and military engineering at the Military Academy, and then was sent to Omaha NE, as engineer officer of the Platte Department, where he made explorations and surveys for military roads. In 1883 he was in charge of roads and bridges in the Yellowstone Park. In 1886 he was promoted to the rank of captain, and ordered to the Mississippi River Commission, New Orleans LA. He was given charge from the Vicksburg District to the head of passes at the river mouth. In 1890 he had charge of the protective works against the largest flood which had occurred up to that time, receiving thanks, by a joint resolution of the Senate and the House of Representatives, of Louisiana State.

Kingman was then ordered to Oswego NY, to take charge of harbour work on the south shore of Lake Ontario. He was placed in charge of improvements of Tennessee River in 1895. He thereby also supervised the construction of the great masonry lock at Colbert Shoals AL. In 1901 Major Kingman was transferred to Cleveland OH for improvement of harbour work on the south shore of Lake Erie. He was then commissioned lieutenant-colonel in 1905, and in 1906 ordered to Savannah GA, on duty of the Southeast Division, in charge of all coast work from Norfolk VA to the mouth of Suwanee River in Florida. He was promoted to colonel in 1908. In 1913 he succeeded William T. Rossell (1849-1919) as Chief of the Corps of Engineers, after 38 years of service in this organization. The Panama Canal was completed while he was Chief. Kingman retired in 1916, and was buried with high military honours at Arlington National Cemetery.

Anonymous (1913). Brigadier-General Dan C. Kingman, Chief of Engineers, USA. *Engineering News* 70(17): 834. *P*

Kingman, D.C. (1900). *Final Report* on Survey of Hiwassee River TN. US War Dept., US Army Corps of Engineers: Washington DC.

Kingman, D.C., ed. (1915). *Index* to the Reports of the Chief of Engineers, US Army, including the Reports of the Isthmian Canal Commissions. US Government Printing: Washington.

Whitford, N.E. (1921). *History of the Barge Canal of New York State*. Lyon: Albany NY.

http://www.cr.nps.gov/history/online_books/baldwin/chap6.htm *P*

http://www.arlingtoncemetery.net/dckingman.htm *P*

KIRKHAM

11.02. 1908 Provo UT/USA
07.03. 1998 Iowa City IA/USA

Don Kirkham made his studies at University of Utah and Columbia University, New York NY, graduating in 1934 as AM in physics and obtaining the PhD in 1938. He was then instructor and assistant professor there until 1940, civilian physicist at the Bureau of Ordnance, US Navy, until 1946, faculty member at Iowa State University, Iowa City IA until 1958, and until being retired in 1978 Curtiss Distinguished professor of agriculture. He was a visiting professor both at the Delft University, Delft NL, and at Ghent University, Belgium, from where he received in 1958 an honorary diploma. In 1958 also, Kirkham was a Lecturer at Vienna University, Austria; in 1974 at Göttingen University, Germany; in 1982 at Stuttgart University, Germany, or in 1985 in the People's Republic of China. He further was consultant for a large number of associations, including in 1959 the Turkish Government, the United Arab Republic UAR in the 1960s, or in Argentina in 1965. In 1994 then, Kirkham was awarded the honorary doctorate from Ohio State University, Columbus OH.

Kirkham was a well known scientist in soil physics, contributing thereby to knowledge in the field of environmental engineering. He authored numerous papers in technical journals of agricultural engineering and was an associate editor of the Water Resources Research from 1965 to 1971. He further was the Director of the Iowa Water Resources Research Council from 1964 to 1973. Kirkham was co-recipient of the International 1984 Wolf Prize in Agriculture, Israel, the recipient of the 1987 Merit Honor Award, University of Iowa, and a Fellow of the American Physical Society, among many other distinctions. He stated: My career has been based on these guidelines: 'Do don't stew, keep moving, don't take yourself too seriously'.

Anonymous (1995). Kirkham, Don. *Who's who in America* 49: 2023. Marquis: Chicago.
Anonymous (1998). Don Kirkham. *Eos* 79(40): 480.
Kirkham, D. (1950). Seepage into ditches in the case of a plane water table and an impervious substratum. *Trans. AGU* 31(3): 425-430.
Kirkham, D., Zeeuw de, J.W. (1952). Field measurements for tests of soil drainage theory. *Journal of the Soil Science Society* America 16(3): 286-293.
Kirkham, D., Powers, W.L. (1972). *Advanced soil physics*. Wiley-Interscience: New York.
http://www.lib.iastate.edu/arch/rgrp/9-9-60.html
http://ag.arizona.edu/kirkham/conference.htm *P*

KIRKPATRICK

25.02. 1915 Menlo IA/USA
27.05. 1995 Knoxville KY/USA

Kenneth William Kirkpatrick was educated at State University of Iowa, Iowa City IA, receiving the BS degree in civil engineering in 1937. Few information is available on his professional career, except that he moved to Tennessee Valley Authority TVA. He was there in the mid-1950s hydraulic engineer of its Hydraulic Laboratory, Norris TE, staying all through his career with TVA. He was junior member of the American Society of Civil Engineers ASCE from the late 1930s, and was finally a member.

Spillways are a basic hydraulic element by which an overflow is controlled. These structures form an essential part of each dam, so that their characteristics must be known for both so-called free, and submerged flow conditions. The weir shape is selected to provide a maximum discharge but that effects of cavitation damage, becoming significant under high overflow depths, are avoided. The 1957 paper of Kirkpatrick is a noteworthy addition to the engineering literature, dealing with the spillway discharge coefficients of TVA dams. Eleven dams were considered in the study including both free and submerged overflow conditions. Various overflow geometries were thereby also referred to. Discharge coefficients of Tainter gates as introduced by Jeremiah N. Tainter (1836-1920), but placed on curved spillway crests, are discussed. The data include further the effect of model scale on these coefficients, and the effect of closing adjacent spillway bays. These data were the basis of spillway design for decades in the USA. The model studies were made under the general direction of Albert S. Fry (1892-1974), then chief of the TVA Hydraulic Data Bank, under supervision of George H. Hickox (1903-1986), the former head of the TVA Hydraulic Laboratory, and Rex A. Elder (1917-), then head of the TVA Hydraulic Laboratory. Few additions have been made since for these hydraulic structures, mainly relating to the effects of vertical gates placed at the weir crest, and radial gates slightly up- and downstream of the weir crest. The number of parameters then becomes high, so that a general treatment of the hydraulic problem becomes more involved.

Anonymous (1956). K.W. Kirkpatrick. Proc. 6th *Hydraulics Conference* Iowa: Frontispiece. *P*
Bradley, J.N. (1952). *Discharge coefficients for irregular overfall spillways*. USBR: Denver.
Kirkpatrick, K.W. (1952). Graphical solution of hydraulic problems. *Proc. ASCE* 78(116).
Kirkpatrick, K.W. (1957). Discharge coefficients for spillways at TVA dams. *Trans. ASCE* 122: 190-210.
Rouse, H. (1976). K.W. Kirkpatrick. *Hydraulics in the USA* 1776-1976: Frontispiece. *P*

KIRPICH

01.11. 1914 Philadelphia PA/USA
27.08. 2008 Miami Beach FL/USA

Phillip Zalman Kirpich was educated at University of Pennsylvania, from where he received the BS degree in 1935. He then joined until 1942 two Engineering Departments of American universities, and then served in the US Army until 1946. From 1947 he was an associate with Tippetts-Abbett-McCarthy-Stratton TAMS, New York NY, at their Athens office until 1955, designing hydro, flood control and irrigation projects in Greece, Turkey and Iraq, from when he became chief engineer of the Cauca Valley project in Cali, Columbia, until 1961. From 1962 until retirement in 1990, Kirpich was a hydraulic and chief engineer within the Agriculture Division of the World Bank, concerned with irrigation and water resources projects. He was member of the American Society of Civil Engineers ASCE and its Fellow from 1959, the American Geophysical Union AGU, the International Association of Hydraulic Research IAHR, as well as the International Commission of Irrigation and Drainage ICID.

Kirpich was a hydraulic engineer mainly working for large irrigation and hydropower schemes, and an expert of hydrology, to which he contributed a popular note in 1940, relating to the determination of the time of concentration in the rainfall-runoff analysis. Once in Latin America, he studied the Cauca Valley scheme using integrated water resources development and administration. The flood control aspects were found to be greatly affected by power, irrigation and drainage. Kirpich also prepared the chapter Hydrology for the 1952 Handbook edited by Calvin V. Davis (1897-1981).

Anonymous (1964). Kirpich, Phillip Z. *Who's who in engineering* 9: 1012. Lewis: New York.
Anonymous (2009). Kirpich, Phillip Zalman. *Trans. ASCE* 174: 867-868.
Kirpich, P.Z. (1940). Time of concentration of small agricultural watersheds. *Civil Engineering* 10(6): 362; 10(8): 533.
Kirpich, P.Z. (1952). Hydrology. *Handbook of applied hydraulics*: 1159-1202. McGraw Hill.
Kirpich, P.Z., Ospina, C.S. (1959). Flood control aspects of Cauca Valley development. *Journal of the Hydraulics Division* ASCE 85(HY9): 1-34; 86(HY2): 123-124; 86(HY3): 63-65; 87(HY1): 131-132.
Kirpich, P.Z. (1999). *Water planning for food production in developing countries*. University Press of America: Lanham MD.
Kirpich, P.Z. (2000). Comments on Water crisis, current perceptions and future realities. *Water International* 25(3): 488-491.

KLEBANOFF

21.07. 1918 New York NY/USA
02.05. 1992 Iowa City IA/USA

Philip Samuel Klebanoff made post-graduate studies
at George Washington University, Washington DC,
from 1942 to 1945, and was then a staff member of
the National Bureau of Standards NBS, Gaithersburg
MD, all through his career until retirement in 1983.
He thereby was assistant chief of its Mechanics
Division from 1969 to 1978, and chief of the Fluid
Mechanics Section from 1975 to 1978. He also was
a member of the Boundary Layer Transition Study
Group, US Air Force, from 1970 to 1982. Klebanoff
was a member of the editorial board of the journal
Physics of Fluids from 1970 to 1973. He was the
recipient of the 1945 Naval Ordnance Award, and the 1975 Gold Medal from the US
Department of Commerce. He further was a Fellow of the American Association for the
Advancement of Science AAAS, the American Institute of Aeronautics and Astronautics
AIAA, and member of the Washington Academy of Sciences.

Klebanoff was a prominent researcher in turbulence. In his early career, he and his
colleagues used hot-wire anemometry to study both flows in the transitional and fully
turbulent regimes, resulting in the turbulence characteristics of a boundary layer. Later,
the development of waves in the laminar regime far beyond the linear range was
investigated, using the vibrating-ribbon technique. In the 1960s Klebanoff started a long
and fruitful collaboration with François Frenkiel (1910-1986), exploring the fundamental
statistical description of turbulence. He also contributed to studies of boundary layer
separation and the effect of roughness on the transition. He was in 1979 awarded the
Honorary Doctorate form Hokkaido University, Sapporo, Japan.

Anonymous (1981). Klebanoff. *Physics Today* 34(2): 89. *P*
Anonymous (1991). Klebanoff, Philip S. *Who's who in America* 46: 1807. Marquis: Chicago.
Anonymous (1993). Philip S. Klebanoff. *Physics Today* 46(3): 103-104.
Klebanoff, P.S. (1955). Characteristics of turbulence in a boundary layer with zero pressure
        gradient. *Report* 1247. NACA: Washington DC.
Klebanoff, P.S., Tidstrom, K.D., Sargent, L.H. (1962). The three-dimensional nature of
        boundary-layer instability. *Journal of Fluid Mechanics* 12: 1-34.
Klebanoff, P.S. (1966). The effect of a two-dimensional roughness element on boundary-layer
        transition. *Applied mechanics*: 803-805, H. Görtler, ed. Springer: Berlin.
Schubauer, G.B., Klebanoff, P.S. (1956). Contributions on the mechanics of boundary-layer
        transition. *Report* 1289. NACA: Washington DC.

KLEMIN

15.05. 1888 London/UK
13.03. 1950 Greenwich CT/USA

Alexander Klemin obtained his BS degree from the University of London UK, the MS degree in 1915 from Massachusetts Institute of Technology MIT, Cambridge MA, and the LLD degree in 1935 from Kenyon College, Gambier OH. He was Guggenheim research professor in aeronautical engineering at the Daniel Guggenheim School of Aeronautics, New York University, from 1925 to 1945, after having been officer in charge of the Research Department at MIT from 1917 to 1919. He further served as a consultant to the Navy Bureau of Aeronautics. He was a member of the Royal Aeronautical Society, and of the American Society of Mechanical Engineers ASME. He was awarded the honorary degree of doctor of engineering from New York University in 1950. The Dr. Alexander Klemin Award was established post-mortem.

Klemin was a notable aeronautical engineer who developed the helicopter and the gyroscope. He has written several textbooks mainly on aerodynamics. The 1918 book presents in Part 1 modern aeronautical laboratories, introduces the aerodynamical theory, compares various wing sections, and details the resistance features of airplanes. Part 2 relates then to the airplane design. Klemin also was concerned with the dynamic stability of the airplane. He served as the editor of the Aero Digest journal. He was president of the American Helicopter Society AHS when he died. He was also credited in 1921 for the design of the first amphibian landing gear in the USA, and was a winner of several Army and Navy airplane design competitions. He also contributed to various encyclopaedias and was an editor of the Scientific American.

Anonymous (1941). Alexander Klemin. *Mechanical Engineering* 63(3): 201. *P*
Anonymous (1944). Alexander Klemin. *American men of science* 7: 981. Science Press: Lancaster.
Anonymous (1950). Dr. Alexander Klemin. *IAS News* 9(5): 7. *P*
Klemin, A. (1918). *Aeronautical engineering and airplane design*. Gardner-Moffat: New York.
Klemin, A. (1924). Introduction to the helicopter. *Mechanical Engineering* 46(11a): 739-751.
Klemin, A., Huff, T.H., Dommett, W.E. (1925). *A text-book of aeronautical engineering*.
    Gardner-Moffat: New York.
Klemin, A. (1928). *Simplified aerodynamics*. Goodheart-Willcox: Chicago.
Klemin, A., Titterton, G.F. (1929). *Airplane stress analysis*. Ronald Press: New York.
Klemin, A. (1948). Sixteenth Annual Meeting of the Institute of the Aeronautical Sciences. IAS
    16th *Annual Meeting* 7: 22-33. *P*

KLINE

25.02. 1922 Los Angeles CA/USA
24.10. 1997 Stanford CA/USA

Stephen Jay Kline received in 1943 the BA degree
from Stanford University, Stanford CA, and the MS
and PhD titles in 1949 and 1952, respectively, from
the Massachusetts Institute of Technology MIT,
Cambridge MA. He was then a member of its faculty
until retirement in 1992, from 1961 as professor of
mechanical engineering. He was in parallel chairman
of the Thermodynamics Division and consultant to
General Electrics, General Motors, Brown Boveri,
or United Technology. Kline was the recipient of
the 1959 Melville Medal of the American Society of
Mechanical Engineers ASME, its Fluids Engineering
Award in 1975, and the 1977 George Stephenson Medal of the Institution of Mechanical
Engineers, London UK. He was a Fellow ASME, and member of the National Academy
of Engineering.

Kline was a pioneer in the field of fluid dynamics. In the 1940s and 1950s the nature of
turbulence was considered one of the most important unsolved problems in fluid
mechanics. Kline developed methods of scientific visualisation allowing him to take the
first detailed look at the layers of turbulent air flow that surround moving objects. This
turbulent layer is responsible for fifty percent drag force of an airplane, thus influencing
substantially the cost of flight. Later, Kline took interest in diffusor flow, a relevant
topic both in water and air flow and particularly associated to jet flow. He classified the
main flow types in expansions and proposed technical means to improve these elements.
In 1968, Kline organized a unique conference on turbulent boundary layer prediction at
Stanford University. According to ASME, this meeting may be considered 'a landmark
in the development of boundary layer technology'.

Anonymous (1960). S. Kline receives Melville Prize Medal. *Mechanical Engineering* 82(1): 91. *P*
Anonymous (1964). Kline, Stephen J. *Who's who in engineering* 9: 1018. Lewis: New York.
Anonymous (1965). S.J. Kline. *Civil Engineering* 35(10): 107. *P*
Anonymous (1995). Kline, Stephen Jay. *Who's who in America* 49: 2040. Marquis: Chicago.
Kline, S.J. (1965) *Similitude and approximation theory*. McGraw-Hill: New York, 2nd ed. 1986.
Kline, S.J. (1966). Some remarks on turbulent shear flows. *Proc. Institution of Mechanical
        Engineers* 180(3J): 222-244.
Kline, S.J., Sovran, G., Morkovin, M.V., Cockrell, D.J., eds. (1968). *Computation of turbulent
        boundary layers*. Mechanical Engineering Department. University: Stanford CA.
http://news.stanford.edu/pr/97/971028kline.html

KNAPP R.T.

05.01. 1899 Loveland CO/USA
07.11. 1957 Pasadena CA/USA

Robert Talbot Knapp was educated at Massachusetts Institute of Technology, Cambridge MA, from where he obtained in 1920 the degree SB in mechanical engineering. He moved in 1922 as instructor to the California Institute of Technology, Pasadena CA, from where he obtained the PhD degree in 1929. He was then for one year a Freeman Scholar, allowing him to travel to Europe for further education. Upon return to Pasadena, he was in charge of its hydraulic laboratory until 1942, directing it until 1951. Knapp there was until 1936 associate professor of hydraulic engineering, then associate professor until 1950, from when serving as professor until his premature death.

Knapp was in charge of special investigations of beach, harbour and channel problems for the Los Angeles Gas & Electric Corporation from 1932 to 1936, and a consulting engineer for the Water District of Southern California from 1933 to 1937. From 1935 to 1947 he acted as corporate agent for the Soil Conservation Service SCS within the US Department of Agriculture, and as consultant on pumping problems of Grand Coulee Dam. From 1942 to 1947 he was in charge of research in underwater projectiles for the US Navy, followed by harbour model studies until 1951. During these years he also took interest in the hydrodynamics of rotating channel flows, and in cavitation hydraulics. He authored research works in high-velocity open channel flows, collaborating with his PhD student Arthur T. Ippen (1907-1974), pipeline stream crossings, density currents, wave and surge studies, and particularly cavitation. He was the recipient of the 1956 Melville Prize Medal from the American Society of Mechanical Engineers ASME.

Anonymous (1954). Knapp, Robert T. *Who's who in engineering* 7: 1333. Lewis: New York.
Anonymous (1955). R.T. Knapp. *Mechanical Engineering* 77(12): 1131. *P*
Anonymous (1956). Robert T. Knapp. *Mechanical Engineering* 78(1): 116. *P*
Anonymous (1958). Le professeur Knapp. *La Houille Blanche* 13(2): 188. *P*
Knapp, R.T., Hollander, A. (1948). Laboratory investigations of the mechanism of cavitation. *Trans. ASME* 70(7): 419-435.
Knapp, R.T. (1952). Cavitation mechanics and its relation to the design of hydraulic equipment. *Proc. of the Institution of Mechanical Engineers* London UK 166: 150-163.
Knapp, R.T. (1955). Recent investigations of the mechanics of cavitation and cavitation damage. *Trans. ASME* 77(10): 1045-1054.
Knapp, R.T., Daily, J.W., Hammitt, F.G. (1970). *Cavitation*. McGraw-Hill: New York.

KNAPPEN

21.11. 1900 Minneapolis MN/USA
20.03. 1951 New York NY/USA

Theodore Temple Knappen graduated in 1920 from the US Military Academy, West Point NY, serving then as lieutenant in the US Corps of Engineers, and did graduate work at Rensselaer Polytechnic School, Troy NY, receiving the civil engineering degree in 1922. He moved to California as resident engineer then for the Nevada Irrigation District. After five years of valuable hydraulic design and construction experience, he rejoined the Corps as area engineer on the Mississippi River flood control, Memphis District, then as chief of the Engineering Division for the Muskingum Flood Control Project in Ohio, and from 1936 as chief of the Flood Control Section for the North Atlantic Division.

Knappen joined in 1937 a consulting engineering firm in New York NY, as engineer-manager of its Caracas, Venezuela, office, designing port, water supply, irrigation, and flood control projects. He became a partner in 1941, and then directed the design of US military bases in the Caribbean. He resigned in this position in 1942 to establish his own consulting office, which became later the firm Knappen, Tibbetts, Abbett & McCarthy Engineers, including Frederick H. Tibbetts (1882-1938), and Gerald T. McCarthy (1909-1990). This firms' numerous projects were mainly port improvements in Greece, Israel, and Sumatra, hydroelectric and irrigation projects in Turkey and Haiti, among many others, with Knappen as the guiding spirit. He was described as foresighted and brilliant. His writings on soil mechanics and flood control were widely read. He realized that American engineers should meet the challenge of engineering service abroad, so that he was called an engineering ambassador-at-large. His splendid personality and convincing expression of ideas, together with his business sense and tireless energy, were the secrets of his great professional success. He was member of the American Geophysical Union AGU, and the American Society of Civil Engineers ASCE from 1936.

Anonymous (1949). Theodore T. Knappen. *Engineering News-Record* 142(Apr.7): Frontispiece. *P*
Anonymous (1952). Theodore T. Knappen. *Trans. ASCE* 117: 1307-1308.
Knappen, T.T. (1931). New Madrid floodway levee. *Engineering News-Record* 106(16): 636-641.
Knappen, T.T., Philippe, R.R. (1936). Practical soil mechanics at Muskingum. *Engineering News-Record* 116(Mar.26): 453-455; 116(Apr.09): 532-535; 116(Apr.23): 595-598; 116(May07): 666-669; 116(May14): 711.
Knappen, T.T. (1942). Earth dams. *Handbook of applied hydraulics*: 261-288, C.V. Davis, ed. McGraw-Hill: New York.

KNEASS

05.11. 1806 Philadelphia PA/USA
15.02. 1858 Philadelphia PA/USA

Samuel Honeyman Kneass studied architecture and engineering in an office until 1824, designing then one of the triumphal arches at Philadelphia to honor General Lafayette. He was from 1825 an assistant engineer under William Strickland (1787-1854) for a survey of the Chesapeake and Delaware Canal. He also was commissioned by the Pennsylvania Society for the promotion of the internal improvements to report on public works in engineering. He was from 1825 to 1828 the principal assistant engineer in the Corps organized again by Strickland to construct the Susquehanna division of Pennsylvania State Canal. After having been active for several railroad companies, he was in charge of Delaware and Schuylkill Canal from 1832 to 1836. He visited Europe in 1840 thereby becoming familiarized with the technical improvements made there as compared with the United States. Upon return to his country he was in 1841 engaged on surveys and improvements for the southern district of Philadelphia.

Kneass had in the mid-1840s a contract to construct a canal at Cartagena, New Granada, currently in Colombia, up to Magdalena River. He took from 1846 to 1848 charge of the Wisconisco Canal in Pennsylvania and again collaborated for railroad projects in his State. From 1849 to 1853, he was Philadelphia City Surveyor thereby constructing the new bridge across Schuylkill River at Market Street carrying ordinary traffic and tracks of the Western Railroad; he was able to remove the old and to install the new bridge without interrupting the traffic. He was then from 1854 to 1858 chief engineer of the Northern Railroad of Pennsylvania. Although officially founded in 1852, the American Society of Civil Engineers ASCE was installed unofficially already in the 1840s, a time when Kneass was one of these who proposed this important professional association.

FitzSimons, N. (1972). Kneass. *Biographical dictionary of American civil engineers*: 75. ASCE.
Jackson, J. (1922). *William Strickland*: The first native American architect and engineer. Philadelphia. *P*
Kneass, S.H. (1844). *Report* of an Examination of the coal mines, lands, and estate of the Lykens Valley Coal Company, in Dauphin County PA. Philadelphia.
Kneass, S.H. (1853). *Report* on Drainage and sewerage made to the select and common councils of the City of Philadelphia. Crissy & Markley: Philadelphia PA.
Ports, M.A. (2001). Engineering intrigue at Barnum's Hotel. *International Engineering History and Heritage*: 94-99, J.R. Rogers, ed. ASCE: Reston VA.

KNOWLES

13.10. 1869 Lawrence MA/USA
08.11. 1932 Philadelphia PA/USA

Morris Knowles was apprenticed at the Essex Water Company, Lawrence MA. He was assistant engineer from 1893 for the East Jersey Water Co., and for the Massachusetts State Board of Health until 1897, from when he was water supply investigator for Boston MA, leading to the formation of Wachusett Reservoir and Aqueduct to supply the city with water. He also was member of the Water Board of Lawrence MA, and resident engineer of the Filtration Commission, Pittsburgh PA. From 1903 Knowles had his private consulting practice there until 1916, from when he was until his death president and chief engineer of Morris Knowles Inc., Pittsburgh and Canada. He was a member of leading engineering associations, including the American Society of Civil Engineers ASCE.

Knowles established at Pittsburgh notable laboratory and prototype filter experiments for the improvement of the city water supply. The art of filtration advanced from the first experiments conducted by Henry Darcy (1803-1858) to 1900, particularly in the methods of cleaning the slow sand filters. Knowles realized that these filters do not only remove gross particles but also organic content of the water, up to the bacterial action. He collaborated with John R. Freeman (1855-1932) presenting a report on the additional water supply for New York City. He advanced to the chief engineer of the Pittsburgh Bureau of Filtration in 1901, designing and supervising the construction of slow sand filters for the city's water supply on Allegheny River. From 1914, he in addition was a member of the Board of Advisory Engineers, the Miami Conservancy District in Ohio State. Knowles also served as member of the Engineering Board of Review, the Chicago Sanitary District in the 1920s. He was awarded the honorary engineering degree from the University of Pittsburgh in 1929.

Anonymous (1931). Morris Knowles, Director District 6. *Civil Engineering* 1(6): 559. *P*
FitzSimons, N., ed. (1972). Knowles, Morris. *Biographical dictionary of American civil engineers*: 61-62. ASCE: New York.
Knowles, M. (1899). *Report* of the Filtration Commission of the city of Pittsburgh PA. Filtration Commission: Pittsburgh.
Knowles, M. (1900). *Description of experimental filter plants at Pittsburgh, and results of experiments*. Pittsburgh.
Knowles, M. (1917). *Report* upon Metropolitan water and sewerage systems, to the Essex Border Utilities Commission, Ontario. Record Print: Windsor ON.

KOELZER

03.05. 1914 Seneca MO/USA
13.10. 1994 Fort Collins CO/USA

Victor Koelzer obtained the BS degree in 1937 from University of Kansas, Lawrence KS, and in 1939 the MS degree from Iowa State University, Iowa City IA. He was from 1938 to 1940 junior engineer of US Geological Survey USGS, junior assistant engineer of the Kentucky Corps of Engineers until 1942, and from 1946 to 1956 from senior engineer to head engineer, US Bureau of Reclamation USBR, Denver CO, and Washington DC. He then was until 1969 engaged from hydraulic engineer to vice-president at Harza Engineering Co., Chicago IL, from when he was chief engineer, National Water Commission, taking over as president the Engineering Farms Inc. until 1984. Concurrently he served both as professor at Colorado State University CSU, Fort Collins CO, and as consultant from 1972. Koelzer was member of the American Society of Civil Engineers ASCE, the American Geophysical Union AGU, and of the American Water Resources Association AWRA. He was the recipient of the 1975 Julian Hinds Award of ASCE.

Research interests of Koelzer included hydrology, sedimentation, hydropower, water resources, and the development and operation of irrigation land. He was involved as chief planner in the Guri Dam, Venezuela. This dam is currently one of the largest water power plants providing the lion's share of its country's electric power. A former CSU student stated that listening to Koelzer's experience on international projects stimulated for the same sort of activities, so that his international assignments had a profound effect on the student's life path. After his death he made a gift of 25,000 $ to the Kansas University Endowment Association, to be used in its Civil Engineering Department.

Anonymous (1937). Koelzer, Victor Alvin. *Jayhawker yearbook*: 219. University of Kansas: Lawrence KS. *P*
Anonymous (1965). Victor A. Koelzer. *Civil Engineering* 35(4): 50. *P*
Anonymous (1986). Koelzer, V. *American men and women of science* 16(4): 415. Bowker: NY.
Grigg, N.S. (2011). *Water finance*: Public responsibilities and private opportunities. Wiley: NY.
Koelzer, V.A. (1957). The use of statistics in reservoir operations. *Trans. ASCE* 122: 1187-1201.
Koelzer, V. (1972). Urban water management. *Journal AWWA* 64(9): 537-544.
Koelzer, V., ed. (1975). *Policies and organizations for urban water*. Water Resources Publications: Littleton CO.
Lane, E.W., Koelzer, V. (1943). Density of sediments deposited in reservoirs. *Report* 9, Hydraulic Laboratory. University of Iowa: Iowa City.

KOLIN

12.03. 1910 Odessa/RU
21.04. 1997 Los Angeles CA/USA

Alexander Kolin showed an early aptitude, doing at home experiments. Famine, revolution and difficult conditions in Russia induced the Kolin family to leave. They moved in 1922 to Berlin, Germany, where Kolin rapidly learned German, and received basic education. The study of physics became his passion, and he was inspired by the then famous physicists at Berlin, including Einstein, Planck or Hertz, who became his adviser for the PhD thesis while studying plasmas. In 1933, once this work was barely initiated, Kolin and his parents moved to Czechoslovakia, where Kolin continued studies at the German University of Prague. He then moved in 1935 to the USA to seek work. Through a mutual friend, he met there again Albert Einstein, who was able to find work for Kolin in a hospital at Chicago IL. He there worked during days, and did independent research at night.

During these years Kolin invented and developed the electromagnetic flow meter, initially applying it to the measurement of blood flow in animals. During the past decades, these discharge meters have been widely used in the industry and in medicine. Kolin held various research and teaching positions in Chicago and New York before becoming in 1946 assistant professor at the University of Chicago. There he wrote textbooks on physics, and invented isoelectric focusing, a laboratory technique widely used in biomedical research. He moved in 1956 to the University of California, Los Angeles UCLA, to do both teaching and research, retiring in 1977 as emeritus professor of biophysics. During this period he invented endless fluid belt electrophoresis, received the prestigious Albert F. Sperry Medal for flowmeter developments, and in 1977 received the Alexander von Humboldt Award from the Federal Republic of Germany. He was a strong candidate for the Noble Prize based on his inventions. The Alexander Kolin Papers, 1929-1990, include his correspondence, US patents and awards.

Kolin, A. (1953). An electro-magneto-kinetic phenomenon involving migration of neutral particles. *Science* 117(3032): 134-137.
Kolin, A., Archer, J.D., Ross, G. (1967). An electromagnetic catheter-flowmeter. *Circulation Research* 21(12): 889-899.
Kolin, A. (1968). Magnetic fields in biology. *Physics Today* 21(11): 39.
http://socialarchive.iath.virginia.edu/xtf/view?docId=kolin-alexander-1910-1997-cr.xml
http://ihm.nlm.nih.gov/luna/servlet/detail/NLMNLM~1~1~101420702~182030:-Alexander-Kolin- *P*

KOLOSEUS

24.09. 1919 New York NY/USA
16.05. 2004 Fort Collins CO/USA

Herman John (Ike) Koloseus graduated from the University of Iowa, Iowa City IA. He remained then there until 1962, teaching with Hunter Rouse (1906-1996) fluid mechanics for graduate students. He went in 1970 to the Colorado State University, Fort Collins CO, where he is remembered as a dedicated teacher. Further, it was stated that he always was a gentleman and was always considerate. He added to the Faculty as one who helped the department work in many ways, most of which were probably below the level of receiving accolades. He was considered as one who helped to give the Civil Engineering Faculty its stature. Koloseus was a member of the American Society of Civil Engineers ASCE, and the recipient of its J.C. Stevens Award in 1964.

Koloseus has had research interests mainly in open channel and pipe flow hydraulics. His PhD thesis was concerned with submerged weir flow, a topic much less studied than free surface flow, mainly because of the instability encountered. The submerged weir is however often a design concern, particularly under flood flow conditions. The 1984 paper deals with scour as occur below hydraulic structures. Riprap and filters were proposed to reduce scour. Attention was drawn to piezometric gradients originating from various water levels, and to scour which might result from these gradients. A flow scheme was considered in the paper demonstrating how these gradients are magnified in riprap vicinity, and the accompanying flow pattern that may also result in scour increase by dislodging riprap elements.

Anonymous (1959). H.J. Koloseus. Proc. 7th *Hydraulics Conference* Iowa: Frontispiece. *P*
Anonymous (2004). In memoriam: Herman J. Koloseus. *Civil Engineering News* 10(2): 5. *P*
Cermak, J.E., Koloseus, H.J. (1953). Lake Hefner model studies. *Final Report* 54JEC, Part 1. Department of Civil Engineering, Colorado A&M College: Fort Collins CO.
Koloseus, H.J. (1951). *Discharge characteristics of submerged spillways*. Colorado Agricultural and Mechanical College: Fort Collins CO.
Koloseus, H.J., Davidian, J. (1961). Flow in an artificially roughened channel. *Paper* 424-B. US Geological Survey: Washington DC.
Koloseus, H.J. (1971). Rigid boundary hydraulics for steady flow. *River mechanics* 1(3): 1-51, H.W. Shen, ed. Colorado State University: Fort Collins.
Koloseus, H.J. (1984). Scour due to riprap and improper filters. *Journal of Hydraulic Engineering* 110(10): 1315-1324.

KREUTZER

06.03. 1884 Douglas County CO/USA
23.11. 1929 New York NY/USA

George Charles Kreutzer was born on a farm some 15 km away from Sedalia CO. He graduated in 1908 with the BSc degree from the Colorado Agricultural College, Fort Collins CO, and with the MSc degree in 1922. He was employed from 1908 as a special field agent, the Bureau of Irrigation Investigations, US Department of Agriculture, and then was engaged until 1910 by the Colorado Experiment Station in studies of practical irrigation. Later, he entered the Bureau of Reclamation, as assistant superintendent of irrigation, on the Shoshone Project in Wyoming, supervising the operation of the Garland Division.

In 1910, Elwood Mead (1858-1936), then chairman of the State Rivers and Water Supply Commission of Victoria State, Australia, visited the USA. He asked the president of the Colorado Agricultural College to suggest a qualified person to serve as superintendent of an Australian project. Kreutzer was recommended, who accepted this offer and in 1911 was appointed superintendent of the *Cohuna* Irrigation District in Northern Victoria, marking the beginning of a long friendship among the two. Kreutzer's success in his mission was so marked that he was transferred to the Central Office at Melbourne Vic., serving as general economic adviser in the creation of new projects, and the reconstruction of existing works. He returned in 1916 to the USA, and there was appointed agricultural adviser of Kern County CA, in corporation with the US Dept. of Agriculture. In 1918 the State Land Settlement Act was accepted, to be under the control of Mead, who also had returned to the USA. Kreutzer was made superintendent of the Durham Settlement, the first to be established under the Act, remaining in this position until 1924. He was then appointed chief of the Division of Reclamation Economics, under Mead again, the Commissioner of Reclamation, remaining in this position until his death. He brought to all his work the influence of a remarkable personality, combined with warm human sympathy, a delightful sense of humour and of professionalism.

Anonymous (1931). George C. Kreutzer. *Trans. ASCE* 95: 1526-1528.
Kreutzer, G.C. (1928). *Organized rural communities*. US Congress, 70th Congress, First
    Session, HR 8221. House Committee on Irrigation and Reclamation. Washington DC.
Kreutzer, G.C. (1929). Soil survey, the foundation of successful reclamation development.
    Journal of the Soil Science Society of America B10: 1-8.
Packard, W.E. (1970). Kreutzer. *Land and power development in California*: 139. Berkeley. *P*
http://www.midvaleirrigation.net/History.aspx

KRUMBEIN

28.01. 1902 Beaver Falls PA/USA
18.08. 1979 Santa Monica CA/USA

William Christian Krumbein graduated with the Ph.B. degree in 1926, and the PhD degree in 1932 from University of Chicago. He was from 1932 to 1938 there an instructor in geology, then assistant professor until 1944, and associate professor until 1945 when being appointed professor of geology at Northwestern University, Evanston IL, until 1970. He was in parallel a research geologist for Gulf Research and a company at Pittsburgh PA until 1946, and a consultant of the Beach Erosion Board, US Corps of Engineers, from 1951. Krumbein was a John Simon Guggenheim Fellow from 1942, and a Fulbright Lecturer at Queen Mary University, London UK, from 1960. He was a Fellow of the American Academy of Arts and Sciences, a member of the Geological Society of America, of the American Geophysical Union, and of the International Association for Mathematical Geology IAMG, among other. He also was in 1950 the president of the Society of Economic Palaeontologists and Mineralogists.

Krumbein was the father of computer geology, pioneer in the application of quantitative methods, especially statistical techniques to sediments and sedimentary rocks. He explored sampling, textures, size distributions, transport, properties, and classification of sediments. He extensively applied statistical techniques to geological and hydraulic problems. With his analytical mind he could cut through all the extraneous material directly to the problem. In the mid-1950s, he transferred his statistical analyses to the computer. He was known as a masterful teacher using skill, patience, and humor to impart knowledge to his students. He was a notable geologist, after whom the Krumbein Medal of IAMG was established in 1976.

Anonymous (1975). Krumbein, William C. *Who's who in America* 38: 1767. Marquis: Chicago.
Anonymous (1979). William C. Krumbein. *Geotimes* 24(11): 28.
Croneis, C., Krumbein, W.C. (1936). *Down to earth*: An introduction to Geology. University of Chicago Press: Chicago. Excerpts also in *Groundwater* 24: 678-681.
Krumbein, W.C. (1932). The mechanical analysis of fine-grained sediments. *Journal of Sedimentology and Petrology* 2(3): 140-149.
Krumbein, W.C. (1941). Principles of sedimentation and the search for stratigraphic traps. *Economic Geology* 36(8): 786-810.
Merriam, D.F. (2004). Memorial to William Christian Krumbein. *Memorials* Geological Society of America 33(1): 27-30, with bibliography. *P*

KRUSKAL

28.09. 1925 New York NY/USA
26.12. 2006 Princeton NJ/USA

Martin David Kruskal obtained the MS degree from University of New York in 1948, and the PhD title in mathematics there in 1952. He was a research scientist in the Plasma Physics Laboratory, Princeton University, from 1951 to 1961, from when he took over there as professor of astrophysical sciences, and professor of mathematics in 1981 until retirement in 1989. He was then David Hilbert professor of mathematics at Rutgers University, New Brunswick NJ. He had been from 1985 to 1991 trustee of the Society of Industrial and Applied Mathematics SIAM, senior fellow of the Weizmann Institute of Sciences from 1973 to 1974, and 1979 Gibbs Lecturer of the American Mathematical Society AMS. Kruskal was the recipient of the 1983 Dannie Heineman Prize for mathematical physics, the 1986 Potts Gold Medal of the Franklin Institute, and the National Medal of Science of the National Science Foundation in 1993. He was member AMS, Fellow of the American Physical Society APS, and the National Academy of Sciences NAS.

Kruskal made fundamental contributions in mathematics and science, ranging from plasma physics to general relativity, and from non-linear analysis to asymptotic analysis. His single most celebrated contribution was the discovery and the theory of solitons. In the 1960s he discovered the integrability of certain non-linear partial differential equations involving functions of one spatial variable and time. These developments began with the pioneering computer simulations by Zabusky and Kruskal of the non-linear equation of Diederik Johannes Korteweg (1848-1941) and Gustav de Vries (1866-1934), the so-called KdV-equation. It is an asymptotic model for the propagation of non-linear dispersive waves. Kruskal and Zabusky made a startling discovery of a solitary wave solution of the KdV-equation propagating non-dispersively and even regains its shape after collision with other such waves. Because of the particle-like properties of such a wave, they named it soliton, a term that caught on almost immediately.

Anonymous (1987). Kruskal. *Physics Today* 40(4): 102. *P*
Anonymous (1993). Martin Kruskal receives National Medal of Science. *SIAM News* 26(7): 1. *P*
Anonymous (1995). Kruskal, Martin D. *Who's who in America* 49: 2100. Marquis: Chicago.
Miura, R.M., Gardner, C.S., Kruskal, M.D. (1968). KdV equation and generalisations: Existence of conservation laws and constants of motion. *J. Mathematical Physics* 9(8): 1204-1209.
Zabusky, N.J., Kruskal, M.D. (1965). Interaction of solitons in a collisionless plasma and the recurrence of initial states. *Physical Review Letters* 15(6): 240-243.
http://en.wikipedia.org/wiki/Martin_David_Kruskal *P*

KUETHE

24.08. 1905 Marshfield WI/USA
12.04. 2000 Ann Arbor MI/USA

Arnold Martin Kuethe graduated in 1926 from the
Ripon College, Ripon WI. He in 1933 obtained the
PhD degree from Caltech, under the guidance of
Theodor von Karman (1881-1963). Until 1938, he
was then a research engineer of the Guggenheim
Institute, Akron OH, then until 1940 lecturer at New
York University. During the next two years, Kuethe
then joined the National Advisory Committee of
Aeronautics NACA. From 1941, he was a faculty
member of the University of Michigan, Ann Arbor
MI, and there from 1943 professor of aeronautical
engineering. In parallel he was also a chief scientist
in the Office of Air Research, Wright Field OH, and a consultant in aerodynamics.
Kuethe was an Associate Editor of the Journal of Aeronautical Sciences, and a Fellow
of the Institute of Aeronautical Sciences.

Kuethe's PhD thesis deals with the mixing of two parallel turbulent jets. Whereas
Walter Tollmien (1900-1968) had considered two jets one of which is at rest, Kuethe
investigated their mixing if both have a finite speed. Tollmien's approach was thus
generalized using Prandtl's mixing concept. It was also observed, based both on a
theoretical approach and detailed laboratory observations, that the velocity profiles
beyond the potential jet core may be described with a unique function. Kuethe was thus
able to adequately define the jet characteristics for an important case in practice, namely
jet aviation. In the 1950s Kuethe considered the transition to turbulent flow in a tube,
thereby generalizing Poiseuille flow in the laminar flow regime.

Anonymous (1948). Arnold M. Kuethe. *Aeronautical Engineering Review* 7(2): 8. *P*
Anonymous (1959). Kuethe, Arnold Martin. *Who's who in America* 28: 1514-1515. Marquis:
        Chicago.
Dryden, H.L., Kuethe, A.M. (1930). The measurement of fluctuations of air speed by the hot-
        wire anemometer. NACA *Report* 320. Bureau of Standards: Washington DC.
Kuethe, A.M. (1935). Investigations of the turbulent mixing regions formed by jets. *Journal of
        Applied Mechanics* 2(A): 87-95.
Kuethe, A.M., Schetzer, J.D. (1950). *Foundations of aerodynamics*. Wiley: New York.
Kuethe, A.M. (1951). Some aspects of boundary-layer transition and flow separation on
        cylinders in yaw. Prof. 1st *Midwestern Conference on Fluid dynamics* 1: 44-55.
Kuethe, A.M., Chow, C.-Y. (1986). *Foundations of aerodynamic*s: Bases of aerodynamic
        design. Wiley: New York.

KURTZ

23.02. 1885 East Stroudsburg PA/USA
09.08. 1956 Newberry SC/USA

Ford Kurtz graduated in 1905 as a civil engineer from Cornell University, Ithaca NY. He was an aide of the US Coast and Geodetic Service from 1907 to 1909, assistant to the hydraulic engineer of J.G. White & Co., Inc., New York NY, until 1912, later there hydraulic design engineer, and successor of the firm from 1920 to 1922, chief hydraulic design engineer from 1923 to 1927, hydraulic engineer since 1928 and from 1940 to 1949 there engineering manager. From 1948 Kurtz was director of White & Co., and finally president. He was also in charge of a waterworks company in Cuba between 1916 and 1917, and then involved in Langley Field, Hampton VA. He was a member of the US Committee on Large Dams, was awarded the Fuertes Medal for the highest scholastic record in civil engineering by Cornell University, and was a member of the American Society of Civil Engineers ASCE, serving as secretary of its Power Division.

Kurtz devoted almost his entire career to an engineering company in New York, starting as hydraulic engineer and advancing to president. As an expert in hydraulic engineering, he was a consultant for the *Svirstroi* Dam in the Soviet Union, Fort Peck Dam in Montana, and other important projects in his country and abroad. He designed also various hydro-electric developments, including the Oak Grove plant in Oregon or the Saluda scheme in South Carolina. The scope of his endeavours further covered power plants, industrial set-ups, railroads and refineries. Kurtz introduced in 1925 the shaft spillway as an alternative to conventional chutes. His design includes a circular plan shape weir followed by a vertical shaft and an almost horizontal tunnel by which the discharge may be directed into the tailwater. The paper gives an almost final design with the head-discharge curve, and effects of valley topography. His paper received wide interest, and shaft spillways are currently a standard design.

Anonymous (1955). Kurtz, Ford. *Who's who in America* 28: 1518. Marquis: Chicago.
Anonymous (1956). Ford Kurtz. *Civil Engineering* 26(9): 636-637. *P*
Anonymous (1956). Ford Kurtz dies at 71, was head of J.G. White Corp. *Engineering News-Record* 157(Aug.23): 74.
Kurtz, F. (1920). Discussion to Pressures in penstocks caused by the gradual closing of turbine gates. *Trans. ASCE* 83: 741-775.
Kurtz, F. (1925). The hydraulic design of the shaft spillway for the Davis Bridge Dam, and hydraulic tests on working models. *Trans. ASCE* 88: 1-86.

LAITONE

06.09. 1915 San Francisco CA/USA
18.12. 2000 Berkeley CA/USA

Edmund (Ed) Victor Laitone obtained the BS degree from University of California, Berkeley CA in 1943, the MS degree in applied mathematics in 1944, and the PhD degree in 1962 from the University of California, Stanford CA. He had been from 1938 to 1945 aeronautical research engineer at Langley Field VA, and then for two years principal engineer at the Cornell Aeronautical Laboratory, Buffalo NY. From 1947 to 1954 Laitone was associate professor of mechanical engineering at Berkeley CA, and then continued as professor of aeronautical engineering there until retirement. He had in parallel a number of assignments in the industry relating to both aviation techniques and guided missiles research. He was a member of the American Physical Society APS, and the Institute of Aeronautical Sciences.

Laitone was a respected expert in aerodynamics with engagements in major engineering companies. He explored a large range of research topics, all within the realm of fluid mechanics and ranging from subsonic aerodynamic problems to papers on supersonic- and magneto-gasdynamics. He will be remembered particularly for his chapter Surface waves co-authored by John V. Wehausen (1913-2005), in which the mathematical theory of water waves is thoroughly discussed. Both two- and three-dimensional waves are considered, from the basic theory of sinusoidal waves to complicated waves in basins. A special chapter also includes steady two-dimensional waves known as shock waves, resulting from disturbances of a supercritical stream. This book is still of great value today because of its excellent mathematical approach. A Fellow of the American Institute for Aeronautics and Astronautics AIAA since 1988, he was also the US academic representative on the Flight Mechanics Panel of NATO's Advisory Group for Aerospace Research and Development AGARD from 1984 to 1988.

Anonymous (1959). Laitone, Edmund V. *Who's who in engineering* 8: 1413. Lewis: New York.
Anonymous (1962). Prof. Ed Laitone. *Mechanical Engineering* 84(8): 82. *P*
Laitone, E.V. (1947). Exact and approximate solutions of two-dimensional oblique shock flow. *Journal of the Aeronautical Sciences* 14(1): 25-41.
Laitone, E.V. (1952). A study of transonic gas dynamics by the hydraulic analogy. *Journal of the Aeronautical Sciences* 19(4): 265-272.
Wehausen, J.V., Laitone, E.V. (1960). Surface waves. *Encyclopaedia of physics* 9: 446-778, C.A. Truesdell, ed. Springer: Berlin.
http://www.me.berkeley.edu/announcements/laitone.html *P*

LAMB O.P.

11.12. 1922 Golden Valley MN/USA
08.08. 2008 Minneapolis MN/USA

Owen Peter Lamb made civil engineering studies at Minneapolis University, Minnesota MN, obtaining the BSc degree in 1943, and the MS degree in 1946, then joining in 1953 the Saint Anthony Falls SAF Hydraulic Laboratory as research fellow. He there remained all through his career having great success with researches in air-water flows. Besides, he was involved in studies with a sixty-inch water tunnel developed at SAF in the late 1940s, the set-up of an internationally known test channel allowing for the observation of air-water flows of bottom angles from 15° to 75° in 1952, velocity measurements of these flows shortly later, and two benchmark studies on the then final results of this test campaign. From 1956 Lamb took interest in aeronautical questions including for instance the lift effectiveness of spoilers.

The 1950 Report deals with an electrical method developed to measure concentrations of air of high-speed laboratory air-water flows. The method consists of the measurement of the difference between the conductivity of air-water flow, and water flow alone. A mechanical strut supporting a pair of electrical probes was combined with the electrical circuit so that air concentration measurements are recorded at a relatively small flow region. The instrument allowed for the detection of the air concentration both vertically and laterally of the flow field. It was found from further theoretical examinations that the recorded values were sufficiently accurate for engineering purposes. This instrument marked the initiation of detailed flow observations at the SAF Laboratory, culminating in the famous 1959 paper by Lorenz G. Straub (1901-1963) and Alvin G. Anderson (1911-1975).

Anonymous (2008). Owen Peter Lamb. Star Tribune: *Newspaper* Twin Cities (Aug. 11): 8. *P*
Lamb, O.P., Killen, J.M. (1950). An electrical method for measuring air in flowing air-water mixture. *Technical Paper* B2. St. Anthony Falls Hydraulic Laboratory: Minneapolis.
Stefan, H.G., Foufoula-Georgiu, E., Arndt, R.E.A. (2004). The St. Anthony Falls Laboratory: A rich history and a bright future. Proc. *EWRI Annual Meeting*: 185-197.
Straub, L.G., Killen, J.M., Lamb, O.P. (1954). Velocity measurements of air-water mixture. *Trans. ASCE* 119: 207-220.
Straub, L.G., Lamb, O.P. (1956). Experimental studies of air entrainment in open channel flow. *Trans. ASCE* 121: 30-44.
http://www.gps.caltech.edu/~mpl/owen_p.htm *P*

LANDWEBER

08.01. 1912 New York NY/USA
20.01. 1998 Iowa City IA/USA

Louis Landweber was educated at City College, New York NY, from where he obtained the BS degree in 1932, the MA degree from the George Washington University, Washington DC in 1935, and in 1951 the PhD title from University of Michigan, Ann Arbor MI. He was from 1932 to 1954 at David W. Taylor Naval Ship Research and Development Center, Bethesda MD, finally as head of the hydrodynamics division, then becoming principal investigator at the Office of Naval Research, Arlington VA, and from 1985 was a consultant of Mobil Oil Co. Landweber was awarded the Davidson Medal from the Society of Naval Architecture and Marine Engineers in 1978, he was the 1978 David Taylor Lecturer, and the 1981 Weinblum Memorial Lecturer. He was a member of the National Academy of Engineering from 1980 with the citation 'Research, design and educational contributions to modern naval architecture and marine engineering'. He was further a member of the Society of Naval Architects and Marine Engineers and a Fellow of the American Academy of Mechanics.

Landweber was a notable naval engineer. He conducted research about the flow of one or more bodies or ships moving through a fluid, their added masses and the forces and moments acting upon them. He wrote numerous papers in these fields and contributed also several books, among which his Investigation on components of ship resistance, published in 1974, may be mentioned. Another work entitled Irrotational flow about ship forms investigates the effect of ship shape on flow resistance. His prolific efforts and publications, paired by his humour and humanity, helped develop this research field into one of Iowa Institute of Hydraulic Research IIHR's major research activity.

Anonymous (1994). Landweber, Louis. *American men and women of science* 4: 715. Bowker: NJ.

Landweber, L. (1952). Der Reibungswiderstand der längsangeströmten ebenen Platte. *Jahrbuch* Schiffsbautechnische Gesellschaft 46: 137-150.

Landweber, L., Macagno, M. (1959). Added mass of a three-parameter family of two-dimensional forces oscillating in a free surface. *Journal of Ship Research* 2(1): 36-48.

Landweber, L., Macagno, M. (1969). Irrotational flow about ship forms. IIHR *Report* 123. Iowa Institute of Hydraulic Research: Iowa City.

Landweber, L. (1982). Irrotational flow within the boundary layer and wake. *Journal of Ship Research* 26(4): 219-228.

Rouse, H. (1976). Louis Landweber. *Hydraulics in the United States* 1776-1976: 189. Iowa. P
http://www.news-releases.uiowa.edu/1998/january/0121land.html

LANE E.W.

21.02. 1891 Lafayette IN/USA
10.08. 1963 Denver CO/USA

Emory Wilson Lane obtained his BS degree from Purdue University, West Lafayette IN in 1912, and his MS degree from Cornell University, Ithaca NY in 1914. He was a 'computer' and assistant engineer for the Miami Conservancy District from 1914 to 1919, manager of the China branch in Yangchow and advisor to Kiangsu Grand Canal Improvements until 1923, assistant engineer for flood control works at Pueblo CO until 1925, hydraulic engineer for the Mississippi River Spillway Board until 1928, then hydraulic engineer within the US Corps of Engineers at Baltimore MD, joining in 1929 the US Bureau of Reclamation USBR, Denver CO, thereby investigating the Boulder Canyon and the Grand Coulee Dams. In 1935 Lane was appointed professor of hydraulic engineering at the Iowa State University, Iowa City IA, where he was also in charge of its hydraulic laboratory. He returned in 1946 to USBR as hydraulic consultant of the chief engineer's staff. Lane was elected Honorary Member ASCE shortly before his death.

Lane was an internationally known hydraulic engineer. During his stay at USBR, he directed the pioneering model studies for two of the main dams in the West. He was further the Bureau representative in the program of correlating sedimentation work conducted by the agencies of the Department of the Interior and the Corps of Engineers in the Missouri River Basin. He was also for a number of years member of the Sediment Advisory Board of the Missouri River Division, Corps of Engineers. He completed in 1945 the final reports on the Imperial Dam and the All-American Canal Structures. Lane made major contributions to fluvial hydraulics, siltation, and sedimentation. His 1935 paper was a classic contribution for the design of hydraulic structures.

Anonymous (1946). Emory W. Lane. *Civil Engineering* 16(10): 474. *P*
Anonymous (1954). Lane, Emory W. *Who's who in engineering* 7: 1388. Lewis: New York.
Anonymous (1963). Emory W. Lane. *Civil Engineering* 33(10): 67-68. *P*
Lane, E.W. (1920). Experiments on the flow of water through contractions in an open channel. *Trans. ASCE* 83: 1149-1219.
Lane, E.W. (1935). Security from under-seepage masonry dams on earth foundations. *Trans. ASCE* 100: 1235-1351.
Lane, E.W. (1936). Flow conditions in steep chutes. *Engineering News-Record* 116(Jan.2): 5-7.
Lane, E.W., Lei, K.(1950). Stream flow variability. *Trans. ASCE* 115: 1084-1099; 115: 1130-1134.
Lane, E.W. (1955). Design of stable channels. *Trans. ASCE* 120: 1234-1260; 120: 1275-1279.

2264

LANE M.

16.11. 1823 Northfield VT/USA
25.01. 1882 Milwaukee WI/USA

Moses Lane graduated in 1845 from the University of Vermont UVM, Burlington VT, in engineering and began work as an engineer. In 1875 he received the PhD degree. He first helped to determine the routing of the Vermont Central Railroad, thereby being recognized as a good engineer. He worked on other railroad construction projects before migrating in 1857 to the State of New York, where he was in charge of the Brooklyn waterworks. He was there appointed to the principal assistant of James P. Kirkwood (1807-1877), and then succeeded him as chief engineer. In 1869 Lane became a partner of Ellis S. Chesbrough (1817-1886) at Chicago IL, coming thus into touch with sewerage works. His most important plans for sewers were the systems for Milwaukee WI and Buffalo NY, but he also furnished plans for a number of smaller schemes. When Lane died in 1882 he was serving as city engineer of Milwaukee, a place he had previously held from 1875 to 1878. While his prominence as a designer of water works overshadowed his sewerage engineering, he did excellent work of his time in the latter.

Lane served as chief engineer of the Brooklyn Water Works from 1862 to 1869. He later designed the Milwaukee Water Works and served there as city engineer. During his last years he was engaged in hydraulic and sanitary engineering throughout the country. He was for instance in charge of the new water supply system of New Orleans LA, or the sewerage system of Pittsfield MA. He was in addition a member of the commission appointed by the city of Memphis TN after the yellow fever scourge to perfect the drainage system. He therefore was one of the best known engineers of the country and was a recognized authority on waterworks and sewer construction.

Anonymous (1867). *The Brooklyn water works and sewers*: A descriptive memoir. Van Nostrand: New York.
Anonymous (1899). Lane, Moses. *The National cyclopaedia of American* 9: 34-35. White: New York.
Benzenberg, G.H. (1893). The sewerage system of Milwaukee and the Milwaukee River Flushing Works. *Trans. ASCE* 30: 368-385.
Floyd, B.L. (2004). *Images of America*: Toledo in the 19th century. Arcadia: Charlestown SC.
Lane, M. (1873). *Toledo water works*. Toledo OH.
Metcalf, L., Eddy, H.P. (1914). *Sewerage*. McGraw-Hill: New York.
http://vtbookofdays.com/months/november/november16.html *P*
http://www.ebooksread.com/authors-eng/th-ymbert/norwich-university-1819-1911-her-history-

LANGBEIN

17.10. 1907 Newark NJ/USA
10.12. 1982 Reston VA/USA

Walter Basil Langbein obtained the BS degree from Cooper Union School, New York NY, in 1931, and the ScD degree from Florida College in 1941. He was from 1935 hydrologist with the US Geological Survey USGS all through his career, finally at Washington DC. He was awarded, among other, membership of the National Academy of Sciences in 1970, and was the recipient of the 1969 William Bowie Medal from the American Geophysical Union AGU 'for outstanding contributions to fundamental geophysics and for unselfish corporation in research'. He also was awarded the 1976 Robert E. Horton Medal from the AGU 'for outstanding contributions to the geophysical aspects of hydrology'. He was further awarded the prestigious International Prize in Hydrology by the International Association of Hydrological Sciences in 1982. Langbein was member of AGU, ASCE, and the Geological Society of America GSA.

Langbein had distinguished himself in the field of geoscience. He learned others to understand the science of water, especially its role in water resources management and planning. Over his long professional career he became the hidden conscience and soul of the national hydrologic profession. He is particularly remembered for his studies on flood frequency, sediment transport, geochemistry, evaporation, reservoir storage, river meanders, and the hydraulic geometry of streams, but also for his work in floodplain regulation, water management, and hydrologic education, thereby establishing various national water policies and programs. He was the principal architect of AGU's journal Water Resources Research, serving also as one of its first two editors.

Anonymous (1969). William Bowie Medal to Walter B. Langbein. *Eos* 50(6): 446. *P*
Anonymous (1981). Langbein, Walter B. *Who's who in America* 41: 1942. Marquis: Chicago.
Anonymous (1996). Walter B. Langbein. *Water Resources Research* 32(10): 2969-2977. *P*
Langbein, W.B. (1938). Some channel-storage studies and their application to the determination of infiltration. *Trans. AGU* 19(1): 435-447.
Langbein, W.B., Leopold, L.B. (1964). Quasi-equilibrium states in channel morphology. *American Journal of Science* 262(6): 782-794.
Langbein, W.B. (1976). Hydrology and environmental aspects of the Erie Canal. US Geological Survey, *Water Supply Paper* 2038. US Government Printing: Washington DC.
Langbein, W.B. (1982). Dams, reservoirs and withdrawals for water supply: Historic trends. US Geological Survey, Open File *Report* 82-256. US Government Printing: Washington DC.

LANGHAAR

14.10. 1909 Bristol CT/USA
28.09. 1992 Corydon IN/USA

Henry Louis Langhaar was educated at the Lehigh University, Bethlehem PA, from where he obtained the MS degree in mechanical engineering in 1933, and the PhD degree in mathematics in 1940. He was until 1936, then test engineer at the Ingersoll-Rand Corporation, Phillipsburg NJ, further until 1937 in Tulsa OK assistant seismographer, from 1941 to 1947 structural engineer for an aircraft company at San Diego CA, from when being associate professor of theoretical and applied mechanics at University of Illinois, Urbana IL, taking there over in 1949 as full professor until his retirement.

Langhaar was internationally known for his works in solid mechanics and applied mathematics. During his career he was a Lecturer at various universities mainly in the Americas on the general theory of buckling, the theory of modulus, numerical methods for ordinary differential equations, energy principles and visco-elasticity. He is still remembered for the standard book Dimensional analysis and model theory, originally published in 1951. He was the recipient of the 1979 von Karman Medal just one year after retirement from University of Illinois. He won the medal for his 'Contributions to the mechanics of fluids and solids, and especially for his noteworthy works in dimensional analysis, energy principles and theories of shells'. He was a Fellow of the American Society of Mechanical Engineers ASME, a member of its Performance Test Code on model testing, and member of the American Association for the Advancement of Science AAAS.

Anonymous (1959). Langhaar, Henry L. *Who's who in engineering* 8: 1423. Lewis: New York.
Anonymous (1969). Prof. Henry L. Langhaar. *Engineering News-Record* 147(Nov.29): 57. *P*
Anonymous (1979). Henry L. Langhaar. *Civil Engineering* 49(10): 95. *P*
Langhaar, H.L. (1942). Steady flow in the transition length of a straight tube. *Journal of Applied Mechanics* 9(2): 55-58.
Langhaar, H.L. (1951). Wind tides in inland waters. Proc. 4th *Midwestern Conf. in Fluid Mechanics* Ann Arbor: 278-296.
Langhaar, H.L., Boresi, A.P. (1959). *Engineering mechanics*. McGraw-Hill: New York.
Langhaar, H.L. (1962). *Energy methods in applied mechanics*. Wiley: New York.
Langhaar, H.L. (1964). *Dimensional analysis and model theory*. Wiley: New York.
Langhaar, H.L. (1979). Geometry and stability theory. *Zeitschrift für Angewandte Mathematik und Physik* 29(4): 549-560.

LANGLEY

22.08. 1824 Roxbury MA/USA
22.02. 1906 Aiken SC/USA

Samuel Pierpont Langley completed in 1851 his formal education at Boston High School. He worked as a civil engineer until 1864 at Chicago, and St. Louis. After a tour in Europe he became assistant at the Harvard observatory. In 1866 he was appointed assistant professor of mathematics at the Naval Academy. From 1867 he was director of Allegheny Observatory and professor of physics at the current University of Pittsburgh PA. He was appointed in 1887 secretary of Smithsonian Institution, founding there the Astrophysical Observatory.

Langley began his research on aerodynamics at Pittsburgh in 1887, investigating the pressure on a plane surface inclined to its direction of travel. In 1891 he announced that 'mechanical flight is possible with engines we now possess', on the basis of his prediction that a twenty-pound, one HP steam engine could sustain the flight of a 200 pound airplane at a speed of 70 km/h. In 1896 Langley successfully flew models over the Potomac. Langley's experiments brought flight out of the stage of ridicule into a stage of science. The procedures of the Wright brothers were unlike those of Langley because they first mastered the art of flying using gliders before adding power. The main line of flight development followed their work and not that of Langley. It is difficult to agree on what constitutes successful flight of powered, man-carrying, heavier-than-air craft, and there appears to be a desire to improve Langley's popular image or to discount the contributions of the Wright brothers in favour of supposed precursors. In 1914, Glenn Curtiss, a patent opponent of the Wright brothers, barely flew Langley's airplane after significant but unreported modifications. The resulting controversy caused bitter feelings and did no service to Langley.

Anonymous (1906). Samuel P. Langley. *Illustrierte Aeronautische Mitteilungen* 10(5): 147. *P*
Anonymous (1906). The late Prof. S.P. Langley. *Aeronautical Journal* 10(2): 19-25.
Anonymous (1929). Samuel Pierpont Langley. *Mechanical Engineering* 51(2): 104. *P*
De Lalla, A. (1965). Samuel Pierpont Langley e i suoi esperimenti di volo meccanico. *Rivista Aeronautica* 41(2): 249-273. *P*
Langley, S.P., Manly, C. (1911). Langley memoir on flight. *Smithsonian Contributions to Knowledge* 27(1): 1-320.
Moyer, D.F. (1973). Langley, Samuel P. *Dictionary of scientific biography* 8: 19-21.
Soulé, H. (1967). NASA Langley Research Center: A great flight-research institution celebrates its fiftieth anniversary. *Aeronautics and Astronautics* 5(12): 50-53. *P*

LANSFORD

28.02. 1900 Shobonier IL/USA
01.01. 1996 Urbana IL/USA

Wallace Monroe Lansford received education from the University of Illinois, Urbana IL, from where he graduated with the BS degree in 1924, the MS degree in 1929, and the civil engineering degree in 1931. He had been from 1924 to 1927 assistant to the city engineer, Champaign IL, from 1928 to 1929 special research assistant at Engineering Experiment Station, the University of Illinois, from when he became staff member there. He was then appointed professor of theoretical and applied mechanics at its Talbot Laboratory. Lansford was a member of the American Society of Civil Engineers ASCE, serving in the 1950s within its Technical Division Executive Committees, and of the American Society of Engineering Education ASEE.

Lansford authored numerous papers in hydraulics and in engineering in general. He also published the notable sections Hydraulic machinery, and Diving in the 1944 edition of Encyclopaedia Britannica. Further, he presented chapters on Theoretical mechanics, Materials, or Mechanics of fluids in the 1947 edition of the Handbook of engineering. The 1936 paper deals with elbows as used to measure discharge in pipes, then a popular method due to low cost, small cost of upkeep and no major additional headloss in the piping system. Up to then, this 'discharge meter' had not received much attention, but it was later more widespread, although its accuracy has been considered average, as compared for instance with the weir discharge measurement. The measuring principle involves the pressure difference in front of and downstream the elbow, by which a unique relation for a certain element may be experimentally determined. The velocity coefficient was found to be constant for a certain design if the approach flow velocity was in excess of about 1 m/s, pointing at Reynolds effects. A second notable work of Lansford is concerned with backwater profiles of open channel flow, in which it was demonstrated that the theoretically predicted profiles agree well with laboratory data.

Anonymous (1953). W.M. Lansford. *Civil Engineering* 23(8): 560. *P*
Anonymous (1964). Lansford, W.M. *Who's who in engineering* 9: 1067. Lewis: New York.
Lansford, W.M. (1934). Discharge coefficients for pipe orifices. *Civil Engineering* 4(5): 245-247.
Lansford, W.M. (1936). The use of an elbow in a pipe line for determining the rate of flow in a pipe. *Bulletin* 289. Engineering Experiment Station EES, University of Illinois: Urbana.
Lansford, W.M., Mitchell, W.D. (1949). An investigation of the backwater profile for steady flow in prismatic channels. *Bulletin* 381. EES, University of Illinois: Urbana.

LARKIN

12.12. 1890 Louisburg WI/USA
17.10. 1968 Colusa CA/USA

Thomas Bernhard Larkin obtained the BA degree in 1910 from Gonzaga University, Spokane WA, and was awarded in 1936 the Dr.Sc. degree. He was from 1911 to 1915 at US Military Academy, West Point NY, and was sent to Mexico in 1916. After return to the USA in 1917, he graduated from the Engineer School at Washington Barracks, Washington DC, and then was sent to France where he was involved in the Second Battle of the Marne in 1918. Service with the US Army continued until the late 1920s. He was from 1929 to 1933 assistant district engineer at Vicksburg MA, and then until 1937 at the Fort Peck Project, when returning again to army service until after World War II. From 1939 to 1942 he was in charge of the construction of the Third Locks Project, Panama Canal. In 1943 he became commanding general for services of supply in the North African Theatre, and later commander for the communications zone in North Africa. In 1946 he became Quartermaster General of the US Army, serving until 1949; Larkin retired with the grade lieutenant general in 1952. He was awarded numerous medals and decorations among which the Distinguished Service Medal with two oak leaf clusters.

Larkin had a two-fold career, as member of the US Army, and also as successful civil engineer. In the latter field he was intimately connected with the Fort Peck Project. It was in the mid-1930s record-breaking in its proportions, namely a dam with 7×10^7 m^3 hydraulically placed earthfill, and the peculiar structure of the basic shale formation through which the diversion tunnels and the spillway were built, as described in the 1935 paper. The main purpose of this project was to improve the Missouri River for navigation from Sioux City IA to St. Louis MO, but it had and still has also functions in flood control, and was extensively used for irrigation purposes.

Anonymous (1935). Fort Peck dredges designed for high capacity on long pipe lines. *Engineering News-Record* 115(Dec.12): 810-813.
Anonymous (1940). Lieut. Col. T.B. Larkin. *Engineering News-Record* 125(Jul.11): 57. *P*
Anonymous (1948). Larkin, Thomas B. *Who's who in engineering* 6: 1155. Lewis: New York.
Larkin, T.B. (1933). Controlling floods along the Mississippi. *Civil Engineering* 3(10): 560-564.
Larkin, T.B. (1935). Fort Peck Project and Dam. *Engineering News-Record* 115(Aug.29): 279-282.
Larkin, T.B. (1937). Closing the Fort Peck Dam. *Civil Engineering* 7(9): 605-608.
http://en.wikipedia.org/wiki/Thomas_B._Larkin *P*

LARSON

21.09. 1909 Jamestown NY/USA
28.07. 1984 Crossville TN/USA

Floyd Clifford Larson graduated from the Rensselaer Polytechnic Institute, Troy NY, with the CE degree in 1934. He was then construction engineer first in New York, and from 1936 to 1939 employed at Bradford PA, joining then the Virginia Polytechnic Institute, Burruss Hall VA, as assistant professor. He served the US Navy during World War II, and in 1948 became instructor and assistant professor at the University of Oklahoma, Norman OK, until 1955, when appointed associate professor, and professor at the University of Tennessee, Knoxville TN. He was awarded in 1950 the MSc degree in sanitary engineering from VPI. After a short period as assistant director, he became director of the Water Resources Research Center WRRC, University of Tennessee, remaining in this position until his retirement in 1970.

Larson authored numerous papers in hydraulics and water resources. His particular research interest was water quality in rivers and reservoirs, and means of improvement. The 1973 paper presents a comprehensive surveillance of water quality conditions in the Forth Loudoun Reservoir on Tennessee River near Knoxville. From 1966 to 1973, the Knoxville Third Creek sewage treatment plant was upgraded from a primary plant to an activated sludge treatment plant. A comparison of the collected data was undertaken to elucidate the effect of these modifications on water quality conditions in the reservoir. A particular consideration was given to the improvements of water quality as related to the expenditure for modification of the treatment facilities. Comments are directed toward the public health significance of the water quality conditions. Larson was member of the American Academy of Environmental Engineers, a Fellow of the American Society of Civil Engineers ASCE, and member of the Water Pollution Control Federation WPCF.

Anonymous (1984). Floyd C. Larson. *Trans. ASCE* 149: 367-368.
Larson, F.C. (1968). *Water quality investigation* 1: Basic data, Fort Loudoun Reservoir. WRRC.
Larson, F.C. (1971). Changes in water quality parameters of reservoirs during regulated flow conditions. *Research Report* 23. WRRC: Knoxville TN.
Larson, F.C. (1978). *The impact of urban stormwater on the water quality standards of a regulated reservoir*. Water Resources Research Center: Knoxville TN.
Womack, J.D., Burdick, J.C., Larson, F.C. (1973). Impact of sewage treatment modifications on water quality of a reservoir. *Water Resources Bulletin* 9(1): 100-115.
http://www.google.ch/search?q=%22Floyd+C.+Larson%22+Knoxville *P*

LATROBE

01.05. 1764 Fulneck, Yorkshire/UK
03.09. 1820 New Orleans LA/USA

Benjamin Henry Latrobe was sent at age twelve to Saxony to complete education at Leipzig, Germany. In 1785 he entered the Prussian Army and was twice wounded severely. He returned in 1788 to England and there became an architect, was made in 1789 surveyor of the public offices and engineer of London. Influenced by his political ideas, he came to the United States in 1796. He was engineer of the James River and of Appomattox Canal first. In 1798 he moved to Philadelphia where he built the Bank of Pennsylvania, the Bank of the United States, and also the first water supply of the city. This was the first successful water supply in America, stipulated by the epidemic of yellow fever a few years before. The project advocated the raising of water from *Schuylkill* River, using pumps operated by steam-engines to an elevated reservoir on Center Square. This plan was so practical that it led to an immediate abandonment of the existing scheme. His 1799 book summarizes the concept adopted. President Thomas Jefferson appointed him surveyor of the public buildings in 1803, and Latrobe was then engaged with the architecture of the Capitol.

Among the engineers who were responsible for the very early water works of the young United States, Latrobe appears to have been the most skilled, as evidenced by his first water supply of Philadelphia. He was also involved in the construction of Chesapeake and Delaware Canals. In 1812 he became interested with Robert Fulton (1765-1815) in the introduction of steamboats on the western waters, and built the *Buffalo* at Pittsburgh, the forth steamer descending Ohio River. After the burning of the Capitol due to British war action, Latrobe was called to rebuild it. At his death he was engaged with works for the water supply of New Orleans LA.

Anonymous (1887). Latrobe, B.H. *Appletons' cyclopaedia of American biography* 3: 626-627.
Carter, E.C. (1976). *Latrobe and public works*. Public Works Historical Society: Washington DC.
Chase, E.S. (1967). American water supply engineers. *Water and Wastes Engineering* 4(4): 54-56. *P*
Donaldson, G.A. (1987). Bringing water to the crescent city: Benjamin Latrobe and the New Orleans waterworks system. *Louisiana History* 28(4): 381-396.
Hamlin, T. (1955). *Benjamin Henry Latrobe*. University Press: Oxford. *P*
Latrobe, B.H. (1799). *View of the practicability and means of supplying the city of Philadelphia with wholesome water*. Philadelphia.
http://en.wikipedia.org/wiki/Benjamin_Henry_Latrobe *P*

LAURSEN

24.01. 1919 Fairmount ND/USA
17.10. 2013 Tucson AZ/USA

Emmett Morton Laursen obtained the BCE degree from University of Minnesota, Minneapolis MN, in 1941, and the PhD degree from University of Iowa, Iowa City IA in 1958. He was from 1941 to 1942 an assistant at the Hydraulics Laboratory, University of Minnesota, junior engineer at a design company in Colorado until 1945, from when he joined the Iowa Institute of Hydraulic Research IIHR as assistant, research associate, and research engineer until 1958. He was then associate professor of civil engineering at Michigan State University, East Lansing MI until 1962, from when he was until retirement professor of civil engineering and head of department, University of Arizona, Tucson AZ. From 1985 Laursen was a private consultant. He was awarded the 1959 ASCE Hilgard Prize, and the 1961 ASCE Research Prize. Laursen also was Honorary Member ASCE, and a member of the International Association of Hydraulic Research IAHR, and of AGU.

Laursen was known for his research in sediment transport, fluid mechanics and its applications. He counted to the first Americans dealing with scour, an issue previously tackled mainly in Europe. Scour may be described as localized erosion mainly due to flow concentrations including jets and vortices. It may seriously damage structures erected in water, such as bridge piers and abutments. The 1962 paper attracted numerous discussions, stating a vital interest in this 'new topic' in the USA. During this period, mainly the maximum scour depth was considered, whereas the temporal scour hole advance was overlooked. The 1963 paper investigates scour in a contraction using mass and energy considerations coupled with a sediment transport index and a uniform flow equation. The result is a relation for scour depth as a function of channel geometry, discharge and sediment characteristics. The abutment as a special case was also studied.

Anonymous (1959). Emmett M. Laursen. *Civil Engineering* 29(10): 74-75. *P*
Anonymous (1962). Emmett M. Laursen. *Civil Engineering* 32(7): 65. *P*
Anonymous (1988). Laursen. *Civil Engineering* 58(12): 92. *P*
Anonymous (2000). Emmett M. Laursen. *ASCE News* (11): 2. *P*
Laursen, E.M. (1958). The total sediment load of streams. *Journal of the Hydraulics Division* ASCE 84(HY1): 1-36.
Laursen, E.M. (1962). Scour at bridge crossings. *Trans. ASCE* 127: 166-179; 127: 207-209.
Laursen, E.M. (1963). Analysis of relief bridge scour. *Journal of the Hydraulics Division* ASCE 89(HY3): 93-118; 90(HY1): 287; 90(HY4): 231.

LAUSHEY

13.05. 1917 Columbia PA/USA
02.06. 2001 Cincinnati OH/USA

Louis McNeal Laushey obtained the BS degree in
1942 from Pennsylvania State University, University
Park PA, and the MS and the PhD degrees from
Carnegie Institute of Technology, Pittsburgh PA, in
1947 and 1951, respectively. He was an assistant
and associate professor there from 1948 to 1954,
then professor and head of the Department of Civil
Engineering at Norwich University, Northfield VT,
until 1958, from when taking over as William Thoms
professor at the University of Cincinnati, Cincinnati
OH. In parallel Laushey was also private consultant.
He was a member of the American Society of Civil
Engineers ASCE, the International Association of Hydraulic Research IAHR, and the
American Geophysical Union AGU.

Laushey worked during his professional career on a wide range of hydraulic problems.
His PhD thesis was concerned with the measurement of turbulent fluctuations in flows
as the hydraulic jump, so that he may be considered a predecessor of the works made at
the Iowa Institute of Hydraulic Research then directed by Hunter Rouse (1906-1996).
The 1947 paper deals with similar aspects of tailwater characteristics beyond a hydraulic
jump to avoid erosion in the downstream river. The entrainment of sediment in the
fluvial environment, as first detailed by Albert Shields (1908-1974), was investigated in
the 1949 paper, co-authored by Frederick T. Mavis (1901-1983). The 1953 paper was
concerned with shaft hydraulics, a topic then almost unknown but important in many
application of the daily life.

Anonymous (1964). Laushey, Louis M. *Who's who in engineering* 9: 1078. Lewis: New York.
Laushey, L.M. (1947). Discussion of Tailwater erosion. *Trans. ASCE* 112: 1187-1189.
Laushey, L.M. (1951). Momentum and kinetic energy of turbulence, dispersion symmetry, and
        isotropy of the fluctuations. *PhD Thesis*. Carnegie Institute of Technology: Pittsburgh.
Laushey, L.M., Mavis, F.T. (1953). Air entrained by water flowing down vertical shafts. Proc.
        5th *IAHR Congress* Minneapolis: 483-487.
Mavis, F.T., Laushey, L.M. (1949). Formula for velocity at beginning of bed-load movement is
        reappraised. *Civil Engineering* 19(1): 26-27; 19(1): 60.
Mavis, F.T., Laushey, L.M. (1966). Discussion of Sediment transportation mechanics: Initiation
        of motion. *Journal of the Hydraulics Division* ASCE 92(HY5): 288-291.
Preul, H.C., Laushey, L.M. (1968). *Groundwater basin dynamics*. Water Resources Center,
        Ohio State University: Columbus OH.
http://libraries.uc.edu/libraries/arb/archives/collections/lmlaushey.html *P*

LAVERTY

09.07. 1901 Wellsville OH/USA
11.07. 1998 Pasadena CA/USA

Finley Burnap Laverty obtained in 1925 the BS degree in civil engineering from the Massachusetts Institute of Technology, Cambridge MA. He was then field engineer of the High Sierra power project construction in South California, becoming in 1928 engineer on sewers of the Flood Control District, Los Angeles CA. From 1935 he was chief hydraulic engineer of the Los Angeles Flood Control District, involved in the operation of 14 dams, and 8 km^2 water conservancy grounds. He also directed the District Hydraulic Division in accomplishment of its flood control and water conservancy research, including barrier to salt-water intrusion into groundwater basins through well recharge.

Laverty has written a number of papers and reports. The 1946 paper deals with water-spreading grounds developed in Los Angeles County, using the methods of ditch and furrow, basin, and regulation of flows in stream channels. This method is generally used in rough and sloping terrain, with canals and ditches laid out on the topographic contour with sufficient slope to prevent deposit of suspended material. A modification of this approach is to install broad, shallow main canals, from which smaller ditches at regular intervals spread the water over the area. The advantages of the ditch method are low cost of maintenance and operation, whereas the basin method involves the construction of dikes at regular intervals. Water is led to the upper basin by canal and spilled from basin to basin. From the last basin the waste water is returned to the flood channel. Laverty investigated both the advantages and disadvantages of these two approaches for the Los Angeles Coastal Plain. The 1952 paper is concerned with sea-water intrusion, controlled by injecting fresh water through a line of recharge wells. His conclusions include that the approach works if the injection rate is sufficient, if the height of the pressure mound is proportional to the input rate, and if the well spacing is 150 m.

Anonymous (1949). Laverty, chairman of Society's Committee. *Civil Engineering* 19(11): 846. *P*
Anonymous (1956). Finley B. Laverty, Director, District 11. *Civil Engineering* 26(10): 683. *P*
Anonymous (1959). Laverty, Finlay B. *Who's who in engineering* 8: 1440. Lewis: New York.
Anonymous (1998). Laverty, Finley B., Honorary Member ASCE. *Trans ASCE* 163: 630-631.
Laverty, F.B. (1946). Correlating flood control and water supply: Los Angeles Coastal Plain. *Trans. ASCE* 111: 1127-1144; 111: 1156-1159.
Laverty, F.B. (1952). Recharging wells expected to stem sea-water intrusion. *Civil Engineering* 22(5): 313-315.

LAWRENCE

09.01. 1903 Neodesha KS/USA
13.08. 1970 Kansas City MO/USA

Ray Ellsworth Lawrence graduated in 1925 with the BS degree in civil Engineering from University of Kansas, Lawrence KS. After a short stay with Black & Veatch, he was appointed engineer of the Kansas State Board of Health, and in parallel was member of the Civil Engineering Staff of his Alma Mater. He was from 1933 engineer examiner of the US Public Works Administration, and returned in 1937 as engineer to Black & Veatch, supervising various water supply, and sewerage projects in cities of the Midwest, including the new water supply system of Wichita KS. During his war service as civilian, he was attached to the US Army Corps of Engineers, responsible for the Water Supply and Sewerage Section, dealing with standards in this field and sanitary improvements.

After active war service in Europe, Lawrence returned in 1946 again to Black & Veatch at Kansas City MO, where he was responsible for the planning and prosecution of major engineering projects. He thereby contributed a number of technical papers and reports. Black & Veatch has designed many water works and sewerage projects for services in both World Wars. It was constituted only in 1956, but the partnership had then a 45 year old history. During these years of expansion, the firm had made a rigid rule of adhering to the letter of the code of professional practice, and regarded its professional integrity as its most valuable asset. Nathan T. Veatch (1886-1975), the sole owner of B&V, was then director. Lawrence was member of the American Society of Civil Engineers ASCE and member of its Sanitary Engineering Division, American Water Works Association AWWA, and of the Federation of Sewage and Industrial Wastes Association. He was elected in 1960 president of the Water Pollution Control Federation WPCF.

Anonymous (1959). Lawrence, Ray E. *Who's who in engineering* 8: 1443. Lewis: New York.
Anonymous (1960). WPCF moves consultant into its No. 1 job. *Engineering News-Record* 165(Sep.29): 52-53. *P*
Anonymous (1970). Raymond Lawrence dies. *Engineering News-Record* 185(Aug.20): 24.
Anonymous (1971). Lawrence, Ray E. *Trans. ASCE* 136: 1353.
Lawrence, R.E. (1940). A new water supply for Wichita. *Civil Engineering* 10(9): 555-558.
Lawrence, R.E. (1956). Planning sewerage services for suburban areas. *Proc. ASCE* 82(1): 1-15; 82(3): 25-26.
Lawrence, R.E., Hess, R.H. (1963). Wichita's past, present, and future water supply. *Journal of the American Water Works Association* 55(8): 1081-1092.

LAWSON

08.01. 1879 Washington DC/USA
21.12. 1963 El Paso TX/USA

Lawrence Milton Lawson studied engineering at the Leland Stanford University, Palo Alto CA. He was in 1901 assistant engineer of San Francisco water supply, then hydrologic aide of the US Geological Survey, from 1903 to 1904 in charge of topographic surveys of Colorado River, then from 1905 to 1911 assistant engineer of the Yuma Project, from 1912 to 1915 project engineer of the Rio Grande Project, El Paso TX, returning until 1917 as project engineer to the Yuma Project in charge of river control and canal construction, from when he was until 1926 in charge of the Rio Grande Federal Irrigation Project. He was appointed American Commissioner of the International Boundary Commission between Mexico and the USA in 1927, and there was an engineer of the US Department of State in negotiations and constructions of the Rio Grande International Rectification Project, the Nogales Flood Control Project, and the Lower Rio Grande Flood Control Project. He was instrumental in the US-Mexican Water Treaty providing for utilization of waters of boundary streams, construction of storage dams on Rio Grande, and flood control, channelization and drainage, which was ratified in 1944.

The 1934 paper of Lawson describes the flood control and river rectification operations along the boundary line that have been seriously menaced in the past by uncontrolled floodwater, then improved by the Intl. Boundary Commission. This Commission also formulated practical steps in constructive efforts between the two countries. The work was supported by Joseph B. Lippincott (1864-1942), among others. Lawson served with 27 years longest as US Section Commissioner; he found wide recognition as tamer of Rio Grande. In the early 1940s he provided a stable river channel, protecting residents from floods, thereby presiding over the construction of the American Dam at El Paso, diverting water from Rio Grande into the US canal. The 1944 Water Treaty certainly was his greatest accomplishment, leading to the completion of the Falcon Dam in 1953. Joseph F. Friedkin (1909-2008) stated: 'Lawson has left us with a great inspiration'.

Anonymous (1948). Lawson, Lawrence M. *Who's who in engineering* 6: 1167. Lewis: New York.
Lawson, L.M. (1929). Discussion of Silting of the lake at Austin TX. *Trans. ASCE* 93: 1690-1693.
Lawson, L.M. (1934). Mexico and United States join in border flood control. *Engineering News-Record* 113(Oct.04): 419-423.
Vigil, C. (2012). The canalization of the Rio Grande: A brief history. *New Mexico Journal of Science* 46(12): 249-259.
http://www.ibwc.state.gov/About_Us/Commish_History.html *P*

LEACH H.R.

14.07. 1891 Saginaw MI/USA
24.06. 1941 Fort Myer VA/USA

Harry Raymond Leach graduated as a civil engineer
from the University of Michigan, Ann Arbor MI, in
1916. He stayed there another year assisting Horace
Williams King (1874-1951) in preparing his notable
Handbook of hydraulics. He then was employed by
Robert Elmer Horton (1875-1945) at Albany NY. In
1918 Leach joined the US Army. He was trained in
the US Meteorological Section in the technique of
upper-air sounding in Missouri and North Carolina.
In 1919, Leach undertook a floods study of Saginaw
County and then also conducted a study of storage
possibilities in watersheds of Utah State. From 1920
to 1933, he collaborated again with Horton as principal assistant, conducting studies in
hydraulics and hydrology for hydroelectric and water supply works.

Leach was employed in 1934 by the State of New Hampshire, Concord NH, supervising
the establishment of recreational centers. In 1935 he became affiliated with the Soil
Conservation Service SCS of the US Department of Agriculture, devoting now his full
time to hydraulics and hydrology. He had hoped to bring into orderly array the
knowledge he had accumulated during the busy years of his professional practice,
contributing thereby a number of papers to the engineering community. A long illness
and his untimely death at age fifty prevented the attainment of this goal, however. The
'Leach method' for computing backwater curves was proposed in his 1919 paper. A
river reach is thereby divided into suitable portions, with its local conveyance consisting
of typical cross-sectional area times the roughness coefficient. Starting from a boundary
condition, and using flow diagrams, the water surface elevation of a neighbouring
section may be determined. Other works published around 1920 relate to devices of
irrigation engineering and the variation of the roughness coefficient in rivers.

Anonymous (1916). H.R. Leach. *Michiganenisan yearbook*: 165. Ann Arbour. *P*
Anonymous (1944). Harry Raymond Leach. *Trans. ASCE* 109: 1500-1502.
Jacob, C.C., Leach, H.R. (1931). Field experiments on a practical irrigation rating box.
        Engineering News-Record 88(13): 530-531.
Leach, H.R. (1919). New methods for the solution of backwater problems. *Engineering News-
        Record* 82(16): 768-770.
Leach, H.R. (1919). Variation of roughness coefficient in Manning and Kutter formulas.
        Engineering News-Record 82(11): 536-538.
Leach, H.R. (1931). Disc. on Effect of turbulence on current meters. *Trans. ASCE* 95: 816-826.

LEACH S.S.

27.04. 1851 New Carlisle IN/USA
16.10. 1909 Washington DC/USA

Smith Stallard Leach graduated in 1875 from the Military Academy, West Point NY. He was then stationed for 7 years at Willett's Point NY, when he was elected for duties at the Centennial Exposition at Philadelphia PA. In 1878 he was recorder of a Board of Engineers on the improvement of both the Mississippi and Missouri Rivers. He became in 1879 secretary of the Mississippi River Commission at St. Louis MO. He effected during the next 6 years a complete systematization of the survey work of the Commission. Until 1888 he was in charge of the improvement of the Second District from Cairo IL to the Passes. His reports indicated a thorough grasp of the entire problem and sound ideas as to its proper solution.

Leach was in 1888 transferred to the Memphis District. During the next two years he had an assignment to duty at Washington DC as assistant to the Engineer Commissioner. He was until 1892 military assistant in the Engineering District, Boston MA, remaining there until 1892, in charge of fortifications for the defence of Boston Harbour. He made a comprehensive study of the theory and practice of fortification, as developed in the USA and abroad, making himself a master in this field. From 1892 to 1896, Leach was in charge of the Burlington VT Engineering District, which comprised works of river and harbour improvement on Lake Champlain and St. Lawrence River. In 1896 Leach took charge of the Defences of the Long Island Sound, and of the river and harbour improvements in Connecticut State, stationed at New London CT. He proposed the idea there of protecting New York City by fortifying its entrance from the Sound by seacoast defences, and devised a type of battery suitable for use upon small islands at the western end of the Sound. After a short stay at Leavenworth KS from 1902 to 1904, he was attached to the General Staff at Washington DC, in charge of river and harbour works, and to finalize the Washington filtration work. From 1907 until his death, Colonel Leach was assistant to the chief of engineers, and was member of the River and Harbour Board. He passed away during an attack of uraemia, following poor health.

Anonymous (1909). Col. Smith S. Leach. *Engineering News* 62(18): 466. *P*
Anonymous (1911). Smith S. Leach. 42nd *Annual Reunion of the Association of Graduates*: 46-50. US Military Academy: West Point. *P*
Leach, S.S. (1890). *Mississippi River*: What it needs, and why it needs it. GPO: Washington DC.
Leach, S.S. (1912). *Engineer field manual* parts I – IV. Government Printing Office: Washington.

LEAVITT

27.10. 1836 Lowell MA/USA
11.03. 1916 Cambridge MA/USA

Erasmus Darwin Leavitt, Jr., conducted a three-years apprenticeship at Lowell Manufacturing Company. At age 20 he went to Boston MA where he designed the steam engine for the USS Hartford. He worked from 1859 to 1861 as chief draftsman for Thurston & Gardner Company, Providence RI, as a builder of steam engines. During the Civil War, he first served on various ships, ultimately becoming instructor in steam engineering at US Naval Academy, Annapolis MD. He was from 1867 consulting engineer.

Leavitt first achieved professional prominence in 1873 for his design of a novel pumping engine. From 1874 to 1880 he was a consultant for a mining company where he designed more than forty types of engines for a variety of uses. Each huge stationary steam engine was named, much like a steam locomotive or a ship, with names including Arcadia, Marquette or Superior. He also designed steam-powered water pumps for municipal water systems, including these of Louisville MA or Boston MA, and the power source for a large hydraulic forge at the Bethlehem Steel Company. He also designed and erected equipment for mines in Michigan for pumping, air compression, hoisting, stamping, and power. He was also a consultant for Henry R. Worthington (1817-1880). Leavitt was credited with doing more than other engineers to establish sound principles and propriety of design, and to appreciate the importance of weight in machinery. The Leavitt-Riedler Pumping Engine drew steam from a coal-fired boiler, and was installed in 1894 as one engine of the Chestnut Hill High Station, later the Boston Water Works. At normal speed of 50 revolutions per minute, it pumped almost 100 million litres of water within one day. He was awarded in 1884 the honorary doctorate from Stevens Institute of Technology, Hoboken NJ. He was a member of the American Society of Mechanical Engineers ASME, serving as president in 1893, and the Institution of Mechanical Engineers IME, London UK, among many others.

Anonymous (1924). Erasmus D. Leavitt. *Trans. ASCE* 87: 1376-1377.
Anonymous (1980). Leavitt, Erasmus D., Jr. *Mechanical engineers in America born prior to* 1861: 207-208. ASME: New York. *P*
Fitzgerald, D. (1895). *A short description of the Boston Water Works*. Rockwell: Boston MA.
Leavitt, E.D. (1917). *Erasmus D. Leavitt Papers*. Smithsonian Institution: Washington DC.
Peabody, C.H. (1909). *Thermodynamics of the steam engine*. Wiley: New York.
http://en.wikipedia.org/wiki/Erasmus_Darwin_Leavitt,_Jr.
http://en.wikipedia.org/wiki/Leavitt-Riedler_Pumping_Engine

LE CONTE

07.02. 1870 Oakland CA/USA
02.02. 1950 Carmel CA/USA

Joseph Nisbet Le Conte obtained his BS degree from the University of California, Berkeley CA, and the MME degree from Cornell University, Ithaca NY, in 1892. He was an assistant in mechanics at his Alma Mater until 1895, until 1903 instructor in mechanical engineering, assistant professor until 1912, when taking over as professor of engineering mechanics until 1930, and finally as professor of mechanical engineering until retirement in 1937. Le Conte was a member of the American Society of Mechanical Engineers ASME.

Le Conte was known for his technical publications and a number of books. These include the Elementary treatise with the main parts 1. Introductory, 2. Machinery of transmission, and 3. Mechanics of the steam-engine. His book Hydraulics published in 1926 represents an early American work in this field. He further made researches in pumping machinery with an industrial application at San Francisco, investigated resurge aspects of the water hammering, and contributed a short note to the main advances in hydraulics in the early 1930s. Le Conte was also a noted explorer of the Sierra Nevada. He went by 'Little Joe' among friends because of his short stature and as the son of his father, the famous geologist Joseph Le Conte. The son loved mountaineering and went all over the Sierra exploring and producing the first map of the central Sierra Nevada. He was an avid photographer such as from *Hetch Hetchy* Valley before it was flooded by the dam. The Le Conte Point there is named after him.

Anonymous (1937). Le Conte, Joseph N. *Who's who in engineering* 4: 810. Lewis: New York.
Anonymous (1950). Joseph Nisbet Le Conte. *Mechanical Engineering* 72(5): 450.
Anonymous (1951). Le Conte, Joseph N. *Who's who in America* 26: 1587. Marquis: Chicago.
Lage, A. (2000). The peaks and the professors. *Chronicle of University of California*: 91-98. P
Le Conte, J.N. (1902). *Elementary treatise on the mechanics of machinery* with a special reference to the mechanics of the steam-engine. MacMillan: London UK.
Le Conte, J.N., Tait, C.E. (1907). Mechanical tests of pumping plants in California. *Bulletin* 181. US Department of Agriculture: Washington DC.
Le Conte, J.N. (1926). *Hydraulics*. McGraw-Hill: New York.
Le Conte, J.N. (1930). Advance in the theory of hydraulics. *Mechanical Engineering* 52(4): 370-372. P
Le Conte, J.N. (1937). Experiments and calculations of resurge phase of water hammer. *Trans. ASME* 59(HYD-12): 691-694; 61(5): 440-445.

LEDOUX

28.08. 1860 St. Croix Falls WI/USA
07.11. 1932 Media PA/USA

John Walter Ledoux obtained the civil engineering degree from Lehigh University, Bethlehem PA, in 1887. From 1890 to 1920 he was a chief engineer for the American Pipe and Construction Company, thereby having designed and built more than 100 water plants in the USA and other countries, and having taken out more than 30 US patents. From 1920 he was vice-president of the Simplex Valve & Meter Company, Philadelphia PA, and in parallel was associated to an engineering consulting office, designing exceptional waterworks devices. He was a member of the American Society of Civil Engineers ASCE from 1895, the American Society of Mechanical Engineers ASME, the Franklin Institute, and the American Water Works Association AWWA. Ledoux was killed as a result of an automobile accident.

In the 1913 paper Ledoux proposes to use an immersed body shaped on a parabolic surface to transform stage into an exponential function of various powers. An universally applicable water meter is proposed in the 1924 paper within practical limits. A parabolic body transforms water height into discharge. The apparatus was equipped with an indicating, recording and totalizing device. The 1916 paper describes that velocity measurements may result in significant higher values if the flow is highly turbulent. He was awarded the 1919 *Longstreth* Medal of the Franklin Institute for water meter design.

Anonymous (1915). J.W. Ledoux: Committee of Engineers' Club of Philadelphia. *Engineering Record* 72(21): 646. *P*
Anonymous (1920). J.W. Ledoux. *Who's who in Philadelphia* 108. Stafford: Philadelphia. *P*
Anonymous (1929). Ledoux, John Walter. *Who's who in America* 15: 1273. Marquis: Chicago.
Anonymous (1939). John Walter Ledoux. *Trans. ASCE* 104: 1956-1960.
Kolupaila, S. (1960). John W. Ledoux. *Journal of the Hydraulics Division* ASCE 86(HY1: 42. *P*
Ledoux, J.W. (1913). A mechanism for metering and recording the flow of fluids through Venturi tubes, orifices, or conduits, by integrating velocity head. *Trans. ASCE* 76: 1148-1171.
Ledoux, J.W. (1914). The Pitot tube theory. *Journal AWWA* 1(3): 536-537.
Ledoux, J.W. (1916). Effect of stream turbulence on current meters. *Engineering News* 75(Jun.1): 1055.
Ledoux, J.W. (1924). Open-end flume water meter based on exponential equation. *Engineering News-Record* 93(13): 505-506.
Ledoux, J.W. (1927). Venturi tube characteristics. *Trans. ASCE* 91: 565-595.

LEE C.A.

16.11. 1915 Laramie WY/USA
20.06. 1995 Knoxville TN/USA

Charles Allen Lee obtained the BS degree in civil engineering from University of Wyoming, Laramie WY, and the MS degree from Lehigh University, Bethlehem PA, in 1940. He was then a hydraulic engineer at the David Taylor Model Basin, Navy Department, Washington DC, in the 1940s, later from 1948 to 1958 Chief of the Paper and Tissues Development Process, Neenah WI, and in 1959 he founded the Charles A. Lee Assocs. Inc., Knoxville TN, of which he was owner and president. Lee was member of the American Society of Civil Engineers, and winner of its 1950 Collingwood Prize. He was also recipient of the Award TAPPI, the leading association for the worldwide pulp, paper, packaging, and converting industries of paper.

The 1949 paper is concerned with model tests conducted at David Taylor Model Basin on the cross-sectional dimensions and the design of bends for Panama Canal. The study includes: (1) Effect of variable cross-sectional dimensions for both one-way and two-way traffic; (2) The comparative handling characteristics of various types of ships under specific conditions; (3) Effect of canal currents on the handling characteristics of ships; and (4) Comparison of bend types. The tests were conducted in a large laboratory facility with a scale factor of 1:50. The 1957 paper deals with the transition on a smooth flat plate in a zero pressure gradient using the then conventional surface tube, and hot-wire techniques. It was shown that the laminar oscillations occurring with the laminar layer stability play an important role in the transition phenomenon. It was demonstrated that the development of a turbulent wake from a vortex street governs the transition flow pattern. It was found that a higher turbulence level invokes no laminar oscillations. Lee also invented a dewatering system, separating solids from a slurry in 1980.

Anonymous (1950). C.A. Lee, Collingwood Prize winner. *Civil Engineering* 20(10): 675. *P*
Anonymous (1977). Lee, Charles A. *Who's who in engineering* 3: 314. Engineers Joint Council: New York.
Bennett, H.W., Lee, C.A. (1957). An experimental study of boundary-layer transition. *Trans. ASCE* 122: 307-329.
Lee, C.A. (1940). A study of the hydraulic characteristics of drop inlet spillways. *MS Thesis*. Lehigh University, Bethlehem PA.
Lee, C.A., Bowers, C.E. (1949). Panama Canal - The sea-level project: Ship performance in restricted channels. *Trans. ASCE* 114: 685-713; 114: 893-895.

LEE L.

02.01. 1887 Carbondale PA/USA
15.11. 1937 Columbus OH/USA

Lasley Lee obtained the BSc degree in 1910 from Massachusetts Institute of Technology, Cambridge MA. He was employed in 1911 as junior engineer in the Water Resources Branch, US Geological Survey USGS, San Francisco CA. He was there engaged in general stream-gaging work until 1914, from when he was connected until 1916 with studies of the Hetch Hetchy Valley, serving as additional water supply for San Francisco. In 1917 he moved as a surveyor to Tacoma WA, becoming in 1921 district engineer of the Ohio District. He there spent the rest of his life improving particularly the stream gauging stations and the quality of the stream-flow records.

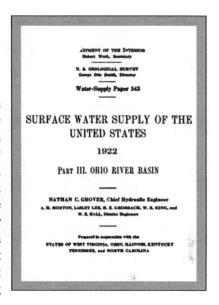

When Lee came to Ohio, only two gaging stations were maintained in corporation with USGS. This number was finally increased to over hundred, so that the problems relating to water flow were greatly improved until Lee's death. He was in addition interested in improving and developing the instrumentation and the methods in stream-gaging work, which were described in his Equipment for river measurements. In collaboration with colleagues, he developed and improved river-sounding reels, wire-weight gages, electric contact tape gages, streamline sounding weights, and other hydrographic instruments and appurtenances used in river engineering. These instruments were for a long time in general use all over the USA. Lee's most spectacular mechanical development were the power-driven rigs for obtaining discharge measurements of the Mississippi River at Vicksburg MS, allowing for the accurate determination of water discharge. He was member of the USGS Committee for Experimental Work at the National Hydraulic Lab. These projects included Studies of artificial controls, Investigations of current-meter performance in measurements of the velocity of water in shallow depths, or Calibration of a portable Parshall Flume. He was member, the US Geological Survey, the American Society of Civil Engineers ASCE, and president of the MIT Alumni Association.

Anonymous (1938). Lasley Lee. *Trans. ASCE* 103: 1830-1832.
Grover, N.C., Beckman, H.C., Burchard, E.D., Dirzulaitis, J.J., Grosbach, H.E., Harrington, A.W., Horton, A.H., Kessler, W., Lee, L., Morgan, J.H. (1939). Surface water supply of United States 3: Ohio River Basin. *Water Supply Paper* 823. USGS: Washington DC. (*P*)
Lee, L. (1927). *Equipment for gaging stations for measuring river discharge.* USGS: Washington.
Parker, G.L., Lee, L. (1923). Hydrometric data, 1878-1919. *Water Supply Paper* 492. USGS.
Soulé, S.B., Horton, A.H., Lee, L. (1925). St. Lawrence River. *Water Supply Paper* 544. USGS.

LEMEN

19.12. 1873 Collinsville IL/USA
07.04. 1935 Interlachen FL/USA

William Caswell Smith Lemen obtained in 1895 the BSc degree in civil engineering from the University of Illinois, Champaign IL. He was then recorder for the Mississippi River Commission, St. Louis MO. From 1899 he was topographer in the US Engineer Office, St. Paul MN, engaged in the survey of head-waters of the Mississippi River. In 1901 he moved to Paris IL as city engineer, but soon later returned to the Mississippi Basin as chief of reservoirs of the head-waters. In 1903 he became junior assistant engineer, the US Engineer Office, Savannah GA, having charge of surveys and construction work. From 1910, then as principal assistant engineer, he supervised all improvements of the District, including these conducted at Savannah Harbour and the Inside Waterway. This also included construction of bank-protections, stone jetties, wharves, and hydraulic dredges; these works were summarized in special reports.

In 1917, when the USA entered World War I, Lemen became captain of the US Army Corps of Engineers, and was promoted then to lieutenant colonel. He became the officer in charge of the Kearny US Engineer Depot, Newark NJ, and depot engineer of the Port of New York, of which deals his 1930 paper. From 1919 to 1920 he was commissioned, now as major in the regular army Corps of Engineers. From then to 1922 he acted as US district engineer at Jacksonville FL, for which position he fitted well from his earlier experience. He then attended the Engineer School at Fort Humphrey VA, serving until 1925 as engineer supplier at Fort Hayes OH. Until 1926 he was commanding engineer at Schofield Barracks HI. In 1928 he became at his old school assistant professor of Military Science and Tactics, University of Illinois, continuing until 1934, the year of his retirement due to illness. During his career Lemen was ever ready to lend his aid and to corporate to the actions in which he could be of assistance. He was described as most agreeable socially and a fine type of citizen, with his home always open to all friends. He was member of the American Association of Engineers, the American Society of Military Engineers, and from 1916 member of the American Society of Civil Engineers.

Anonymous (1936). William Caswell S. Lemen. *Trans. ASCE* 101: 1588-1590.
Crawford, W.G. (1997). Major William Lemen. The Atlantic Intercoastal Waterway from Jacksonville to Miami. *History of Florida's East Coast Canal*: 1-27. P
Lemen, W.C.S. (1930). Construction and reorganisation of the US Engineer Supply Depot, Kearny NJ. *The Military Engineer* 22(5/6): 262-265.

LENAU

14.11. 1936 Lubbock TX/USA
29.11. 2010 Columbia MO/USA

Charles Walter Lenau received both the BS and MS degrees from Texas Technological College, Lubbock TX, and the PhD degree from Stanford University, Stanford CA, in 1963. He then began teaching civil engineering with an emphasis in hydrology, at the University of Missouri, Columbia MO, and there retired in 2000 as emeritus professor. Lenau was a member of the American Society of Civil Engineers.

During his 40 years at 'Mizzou', Lenau advanced from assistant professor to full professor. He was voted in 1997 as Outstanding teacher in the College of Engineering. While teaching was his passion, he participated in various research projects including an automobile rollover project, building a hydraulic capsule pipeline prototype, including design of the jet pumps that make the system work, and designing computer-programming projects to assist students in their studies. The 1966 paper deals with the maximum amplitude of solitary waves, which was found as 83% of the still water depth, in excess of the accepted value of 78%. The 1969 paper deals with flow over spillway flip buckets as applied at large dams to dissipate excess energy. Both the free surface and the pressure head profiles along the bucket were determined by analytical and experimental methods. The 1972 paper deals with waste dispersion from a recharge well, located in a confined aquifer through which there is a natural uniform flow. Solving the convective-dispersion equation allows for the mathematical analysis of the problem. He also authored a computer-based refresher course for engineers who sought to renew their professional engineering licensure. He taught one summer in Brazil in 1972 and presented a research paper at a meeting in Australia in 1992.

Anonymous (1991). Dr. Charles Lenau. *Missouri Alumnus* (Spring): 24-25. P
Hjelmfelt, A.T., Lenau, C.W. (1970). Nonequilibrium transport of suspended sediment. *Journal of the Hydraulics Division* ASCE 96(HY7): 1567-1586; 97(HY5): 747-748.
Lenau, C.W. (1966). The solitary wave of maximum amplitude. *Journal of Fluid Mechanics* 26(2): 309-320.
Lenau, C.W., Cassidy, J.J. (1969). Flow through spillway flip bucket. *Journal of the Hydraulics Division* ASCE 95(HY2): 633-648.
Lenau, C.W. (1972). Dispersion from recharge well. *Journal of the Engineering Mechanics Division* ASCE 98(EM2): 331-344.
Lenau, D. (2014). Charles W. Lenau. Personal communication. P
http://www.columbiatribune.com/obituaries/charles-lenau/article_94ad2cc3-0885-5c27-9e64-4371212609a9.html

LENZ

22.09. 1906 Fond du Lac WI/USA
08.08. 1992 Madison WI/USA

Arno Thomas Lenz was educated at University of
Wisconsin, Madison WI, from where he graduated
with a BSc in 1928, the MSc in civil engineering in
1930, and the PhD in 1940. He there continued as
an instructor in hydraulics and sanitary engineering
until 1937, assistant professor in these fields there
until 1943, associate professor until 1948, from when
he was there a professor of civil engineering. He in
parallel directed from 1950 the hydraulic laboratory
of University of Wisconsin. He also was an official
TVA witness of the Wheeler Turbine Acceptance
Test in 1935, prepared the Wisconsin Water Plan
Report for the National Resources Commission in 1936, and was in charge of hydraulic
model tests for the Petenwell Dubay Dams, Wisconsin. Lenz was a member of the
American Society of Civil Engineers ASCE, the American Geophysical Union AGU,
the American Meteorological Society AMS, and the American Water Works Association.

Lenz was particularly known for his 1943 paper, presenting one of the most systematic
approaches to scale effects in hydraulic engineering. He considered the triangular notch
weir and determined its discharge coefficient for a large number of fluids, in which both
viscosity and surface tension were varied by the addition of chemicals, or by changing
temperatures of oils. The final proposal was also compared with the many existing data.
Limitations in terms of minimum Reynolds and Weber numbers were stated, above
which the conventional weir equation may be applied. The high alkalinity of rivers in
Wisconsin inspired the authors of the 1944 paper to establish a relationship between
alkalinity and river discharge and to use this for discharge determination by a natural
dilution method. The professional honors of Lenz include the Benjamin Smith Reynolds
Award for excellence in teaching, among other.

Anonymous (1954). Lenz, Arno Thomas. *Who's who in engineering* 7: 1433. Lewis: New York.
Anonymous (1956). Arno T. Lenz. *Civil Engineering* 26(12): 696. *P*
Anonymous (1963). Arno T. Lenz. *Civil Engineering* 33(6): 46. *P*
Anonymous (1966). Arno T. Lenz. *Civil Engineering* 36(12): 71. *P*
Anonymous (1992). Arno T. Lenz. *Eos* 73(49): 531.
Lenz, A.T. (1943). Viscosity and surface tension effects on V-notch weir coefficients. *Trans. ASCE* 108: 759-802.
Lenz, A.T. (1944). Estimation of stream-flow from alkalinity determinations. *Trans. AGU* 25(6): 1005-1011.

LEONARD

17.07. 1889 Lincoln NE/USA
23.06. 1969 Knoxville TN/USA

George Kinney Leonard received in 1912 the BSc degree in civil engineering from the University of Nebraska, Lincoln NE, earning there in 1939 the CE degree. He was employed after graduation by the State of Nebraska as resident engineer, working from 1916 to 1918 in its Public Works Department. From 1923 he was superintendent on the Starved Rock Lock and Dam on Illinois River, Ottawa IL, and other river control works including revetment work and channel improvements. During the years of the depression, he returned to the Nebraska State, serving until 1933 as engineer.

Leonard started in 1933 his work for the Tennessee Valley Authority TVA as assistant engineer on the Wheeler Dam in Alabama. When the works on the Guntersville Dam were initiated in 1936, he was transferred there as construction engineer. In 1939 he moved to Tennessee to start works of the Watts Bar Dam. In 1940 he was assigned to start the Cherokee Dam on Holston River, to meet the urgent need of additional power during war. In 1941, as TVA power needs increased further, the Appalachia, Chatuge, Ocoee No.3 and Nottely Dams on the Hiwassee, Ocoee and Nottely Rivers were started with Leonard as project manager, supervising all of them simultaneously. Within one year, storage water was released from Nottely Dam, and within two years, all four dams were contributing power to the war effort. From 1944 Leonard transferred to the TVA Water Control Planning Department, to assist in planning future projects. In 1946 he participated in the completion of the *Watauga* and South Holston Dams again as project manager. He became chief construction engineer in 1950 at the TVA headquarters, Knoxville TN. He was there associated with the expansion of the Tennessee River lock system, which was started at Wilson Dam with a new 35 m × 180 m lock having a single lift of 30 m. Leonard became in 1956 chief engineer of TVA, a position he held until his retirement in 1959. He was member of the US Committee on Large Dams, and the American Society of Civil Engineers, becoming ASCE Fellow in 1959.

Anonymous (1940). George K. Leonard. *Engineering News-Record* 125(Oct.17): 511. *P*
Anonymous (1971). George Kinney Leonard. *Trans. ASCE* 136: 1354-1356.
Leonard, G.K. (1959). *The economic justification for the navigation and flood control program of the Tennessee Valley Authority*. TVA: Knoxville.
Leonard, G.K., Raine, O.H. (1960). Rockfill dams: TVA central core dams. *Trans. ASCE* 125: 190-205.

LEOPOLD

08.10. 1915 Albuquerque NM/USA
23.02. 2006 Berkeley CA/USA

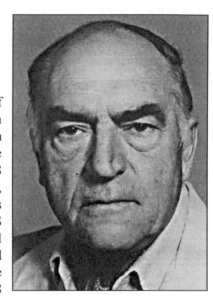

Luna Bergere Leopold graduated from University of California UCLA, Berkeley CA, with the MS in 1944, and received the PhD title in 1950 from Harvard University, Cambridge MA. The DSc. title was awarded in 1980. Leopold remained from 1938 to 1941 with the Soil Conservation Service SCS, joined the US Bureau of Reclamation in 1946, was from 1950 to 1971 hydraulic engineer with the US Geological Survey USGS, and finally became until retirement professor of geology at UCLA. Leopold was the recipient of the 1958 Distinguished Service Award from the Department of Interior, the 1968 *Cullum* Geographical Medal from the American Geographical Society, the 1983 Busk Medal of the Royal Geographical Society, the 1989 Linsley Award from the American Institute of Hydrology, or the 1991 National Medal of Science.

Leopold was the first to turn a scientific eye on rivers and streams drawing conclusions about their form and evolution. Long before fractals became an everyday word, he realized the similarity between large- and small-scale stream characteristics. In the 1950s he discovered with a colleague that river width, depth and velocity increase at a similar rate for nearly all rivers as the discharge increases downstream. This finding has been essential to theoretical and practical studies of rivers. He later travel led widely in Europe and Asia, writing on the nature of rivers, the problems of water and sediment supply, flood control, climate change, environmental planning, ecological restoration, scientific ethics, and the broader relationship between people and nature. This led to explanations for the natural form of rivers. He published some 200 papers and numerous books during a career spanning almost seventy years.

Anonymous (1972). Luna B. Leopold. *Eos* 53(1): 18. *P*
Anonymous (1975). Luna B. Leopold. *Water Resources Bulletin* 11(1): 200-201. *P*
Anonymous (1993). Luna B. Leopold. *Eos* 74(29): 323-324. *P*
Anonymous (2000). Leopold, Luna B. *Who's who in America* 54: 2880. Marquis: Chicago.
Leopold, L.B. (1953). Downstream change of velocity in rivers. *American Journal of Science* 251(8): 606-624.
Leopold, L.B., Maddock, Jr., T. (1954). *The flood control controversy*. Ronald: New York.
Leopold, L.B., Wolman, M.G., Miller, J.P. (1964). *Fluvial processes in geomorphology*. Dover: New York.
Leopold, L.B. (1994). *A view of the river*. Harvard University Press: Cambridge MA.

LEWIS

10.03. 1882 Ithaca NY/USA
12.07. 1948 Chevy Chase MD/USA

George William Lewis obtained the ME degree from Cornell University, Ithaca NY in 1908, and the ScD degree from Norwich University, Northfield VT. He was awarded in 1944 the degree D.Eng. from Illinois Institute of Technology, Chicago IL. He was from 1908 instructor in engineering at Cornell, from 1910 to 1917 then professor of engineering at Swarthmore College, Swarthmore PA, and from 1919 to 1924 an executive officer of National Advisory Committee for Aeronautics NACA, Washington DC, taking then over there as director of aeronautical research until 1947. He was member of the committee on power plants for air craft of NACA in 1918, chairman of the contest board of the National Aeronautical Association from 1925 to 1938, serving also for many other committees related to aeronautical research in the 1930s and 1940s. He was a Honorary Fellow of the Institute of Aeronautical Sciences IAS, which he presided in 1939. He also was a member of the American Philosophical Society, the National Academy of Sciences, and was awarded the Daniel Guggenheim Medal in 1936.

Lewis held the position of director of the Aeronautical Research from 1924 until his retirement in 1947. He was then stationed at Washington DC to handle NACA's political and bureaucratic issues. He in parallel oversaw research advancements and technical publications at the NACA's Langley Research Center, Hampton VA. Engineers at the Langley Facility were at these times responsible for numerous advancements in the understanding of aerodynamics, which earned Lewis international renown, and NACA expanded from a few employees to several thousand during his tenure.

Anonymous (1937). George W. Lewis. *Mechanical Engineering* 59(1): 34. *P*
Anonymous (1950). Lewis, George W. *Who was who in America* 2: 321. Marquis: Chicago.
Durand, W.F. (1949). George William Lewis. *Biographical memoirs* 25: 297-312. National
        Academy of Sciences. Washington DC. *P*
Hazen, H.C., Lewis, G.W. (1916). The measurement of viscosity and a new form of
        viscosimeter. *Trans. ASME* 38: 383-405.
Lewis. G.W. (1938). The contribution of the Wright Brothers to aeronautical science and
        engineering. *US Air Service* (5): 13-15.
Lewis, G.W. (1939). Wilbur Wright Memorial Lecture: Some modern methods of research in
        the problems of flight. *Journal of the Royal Aeronautical Society* 43(10): 771-800.
http://en.wikipedia.org/wiki/George_W._Lewis *P*

LEWY H.

20.10. 1904 Wroclaw/PL
23.08. 1988 Berkeley CA/USA

Hans Lewy was born in the former German city Breslau, today's Wroclaw in Poland. He completed his studies at Göttingen University, obtaining the PhD degree for a numerical boundary-value problem which was tutored by Richard Courant. He was in 1927 appointed *Privatdozent*, similar to a Reader, and in 1928 published with Courant and Kurt O. Friedrichs (1901-1982) the famous book on partial differential equations, including the fundamental CFL-condition as a numerical stability criterion for unsteady partial differential equations. Once the Nazis had taken power, Lewy was dismissed in 1933, so that he emigrated over France to the USA, where he was appointed professor of applied mathematics at the California Institute of Technology, Berkeley CA. Lewy was awarded the Leroy P. Steele Prize of the American Mathematical Society AMS in 1979, and the 1985 Wolf-Prize as a life-time achievement.

Lewy was concerned all through his career with the analysis of partial differential equations, partly with applications to hydrodynamic problems. Of relevance was his 1946 paper, in which the shallow-water equations are applied before computers were available. This hyperbolic type problem requires the application of the CFL-condition. Progressive waves were studied in two dimensions thereby neglecting compressibility and viscous effects. Other papers involve problems of water flow, mainly of unsteady type, so that the governing equations remain more or less identical. It was stated that 'Lewy's work is… extremely original and inventive: He created many ideas which have opened new fields and are still waiting to be developed or extended in the future'.

Courant, R., Friedrichs, K.O., Lewy, H. (1928). Über die partiellen Differentialgleichungen der mathematischen Physik. *Mathematische Annalen* 100(1): 32-74.

Friedrichs, K.O., Lewy, H. (1948). The dock problem. *Communications on Pure and Applied Mathematics* 1(2): 135-148.

Lewy, H. (1946). Water waves on sloping beaches. *Bulletin of the American Mathematical Society* 52(9): 737-832.

Lewy, H. (1952). A note on harmonic functions and a hydrodynamical application. *Proc. American Mathematical Society* 3(1): 111-113.

Lewy, H. (1952). On steady free surface flow in a gravity field. *Communications on Pure and Applied Mathematics* 5(4): 413-414.

http://de.wikipedia.org/wiki/Hans_Lewy *P*

LIEPMANN

03.07. 1914 Berlin/D
24.06. 2009 La Canada CA/USA

Following a PhD and a year as a research fellow at *Eidgenössische Technische Hochschule* ETH Zurich, Hans Wolfgang Liepmann joined in 1939 Caltech as a research fellow. He became in 1972 director of Caltech's Graduate Aeronautics Laboratory GALCIT thus following two leaders in aeronautics, Theodor von Karman (1881-1963), and Clark B. Millikan (1903-1966). Liepmann was awarded in 1969 the Ludwig-Prandtl-Ring, the primary decoration of the *Deutsche Gesellschaft für Luft- und Raumfahrt* DGLR. He was further awarded the National Medal of Science in 1986, and in 1993 the National Medal of Technology.

Liepmann was internationally known for his outstanding work in fluid mechanics and aerodynamics. He initially worked on laminar instability problems, the transition to turbulence and related topics. He then took interest in boundary layers along shock fronts and applied the results to high-speed aircraft design. Liepmann also investigated the direct measurement of local skin friction in both low- and high-speed flows. He did research in aerodynamic noise and turbulence, and in the dynamics of liquid helium. The latter work has furthered the understanding of fluid motion at low temperatures and thus opened possibilities for applications in energy storage and heat transfer. He also advanced a number of instruments to measure the characteristics of fluid flow.

Anonymous (1960). Chairman S.A. Schaaf with the speakers. *Aeronautics and Astronautics*: Frontispiece, N.J. Hoff, W.G. Vincenti, eds. Pergamon Press: Oxford. *P*

Anonymous (1972). Hans W. Liepmann. *Astronautics and Aeronautics* 10(6): 79. *P*

Benecke, T. (1969). Überreichung des Ludwig-Prandtl-Ringes an Prof. Dr.phil. Hans W. Liepmann. *DGLR-Mitteilungen* 2(2): 7-8. *P*

Coles, D., ed. (1988). *Perspectives in fluid mechanics*. Symposium held on the occasion of the 70th birthday of Hans Wolfgang Liepmann, Pasadena CA. Springer: Berlin.

Liepmann, H.W., Laufer, J. (1947). *Investigations of free turbulent mixing*. NACA TN 1257. Washington DC.

Liepmann, H.W., Dhawan, S. (1951). Direct measurements of local skin friction in low-speed and high-speed flow. Proc. 1st US Congress *Applied mechanics* Chicago: 869-874.

Liepmann, H.W. (1952). Aspects of the turbulence problem. *Zeitschrift für Angewandte Mathematik und Physik* 3(5): 321-342.

Liepmann, H.W., Roshko, A. (1957). *Elements of gas-dynamics*. Wiley: New York.

LINK J.W.

28.05. 1866 Richmond VA/USA
14.04. 1933 Evanston IL/USA

John William Link studied engineering at Virginia Mechanics' Institute Night School of Technology, Richmond VA. In 1884 he was employed as a clerk, and served from 1887 to 1890 as assistant engineer during the construction of the Richmond City Hall. From 1893 to 1896 he served as draftsman with the Missouri River Commission, St. Louis MO, and then was employed until 1900 as office assistant in the Metropolitan Water Board, Boston MA, during which time the Wachusett Aqueduct was constructed.

Link began his work on hydro-electric developments in 1901. He was until 1903 with the Niagara Falls Power Company. In 1904 he became principal assistant to William A. Brackenridge (1858-1929), engaged on water power studies including the Great Northern Development Company, Duluth MN, or the New York State Barge Canal. Until 1905 he was then at Columbus OH assistant engineer in charge of the design of water supply and sewerage disposal structures. He returned as assistant engineer to the field of hydro-electricity in 1908, collaborating with James W. Rickey (1871-1943), engaged on studies for the proposed water power development at Long Sault Rapids on St. Lawrence River. In 1910 Link became chief engineer with a firm in Chicago IL, in which company he remained until his death. The most notable of his achievements are the Rapidan, Coon Rapidan, and Chippewa Falls Development of the Northern States Power, or the Albany and Big Fork Power Developments and the Albany Filtration Plant of the Mountain States Power Company. Link was quiet and unassuming in manner, and somewhat reserved. In his uncompromising integrity and fine devotion to the interests of those he served, he was an outstanding example of what an engineer should be who desires to conform to the highest standards of his profession. He was a member of the Boston Society of Civil Engineers BSCE, the Western Society of Engineers, and since 1913 of the American Society of Civil Engineers ASCE.

Anonymous (1933). J.W. Link. *Engineering News-Record* 110(16): 514.

Anonymous (1933). John W. Link. *Trans. ASCE* 98: 1574-1575.

Link, J.W. (1914). Coon Rapids low-head hydro-electric development on the Mississippi River near Minneapolis. *Journal of the Western Society of Engineers* 19(10): 979-1015.

Martin, T.C., Coles, S.L., eds. (1919). *The story of electricity*. Marcy: New York.

Thayer, B.C., Allardice, E.R.B. (1915). The hydro-electric power plant at the Wachusett Dam, Clinton MA. *Journal Boston Society of Civil Engineers* 1(9): 481-548; 2(3): 109-119. (*P*)

publications.usace.army.mil/.../misc/.../c-11.pdf

LINSLEY

13.01. 1917 Hartford CT/USA
06.11. 1990 Santa Cruz CA/USA

Ray Keyes Linsley graduated in 1937 with the BS from the Worcester Polytechnic Institute, Worcester MA. He was engineer with the Tennessee Valley Authority TVA until 1940, chief hydrologist with the US Weather Bureau, Washington DC, until 1945, when being appointed in 1950 associate professor at Stanford University, Stanford CA, and professor and head of department from 1956 until his retirement in 1975. Linsley was a Fulbright professor at Imperial College, London UK, in 1958, vice-president of an engineering company from 1959 to 1967, president then of a hydraulic design company until 1978, until taking over as president Linsley-Kraeger Assoc. Ltd., Santa Cruz CA, from 1979 to 1985, and there continuing as chairman.

Linsley was an internationally well-known hydrologist and water resources expert. He advocated from the 1950s the need for water conservation, a time when the federal government was still constructing dam and irrigation projects. In the 1960s, he was a founder of the Engineering-Economic Planning program, which expanded the education of civil engineers beyond project design to include training in the social, political, and economic factors that affect their work. Linsley was also known for a number of books, including Applied hydrology, Elements of hydraulic engineering, becoming later Water resources engineering, written in collaboration with Joseph B. Franzini (1920-), or Hydrology for engineers. Linsley was awarded ASCE Honorary Membership in 1986, the 1943 Collingwood Prize, and the 1978 Julian Hinds Award. He was a member of the National Academy of Engineering, the American Meteorological Society, and Honorary Member of the Venezuelan Society of Hydraulic Engineers, among other.

Anonymous (1944). Ray K. Linsley. *Civil Engineering* 14(1): 35. *P*
Anonymous (1974). Ray K. Linsley. *Water Resources Bulletin* 10(1): 165-166. *P*
Anonymous (1977). Ray K. Linsley. *Civil Engineering* 37(7): 120. *P*
Anonymous (1991). Linsley, Ray Keyes. *Who's who in America* 46: 2007. Marquis: Chicago.
Anonymous (1991). Water expert Ray Linsley dies at 73. *Civil Engineering* 61(1): 77. *P*
Linsley, R.K., Kohler, M.A., Paulhus, J.L. (1949). *Applied hydrology*. McGraw-Hill: New York.
Linsley, R.K., Franzini, J.B. (1955). *Elements of hydraulic engineering*. McGraw-Hill: New York.
Linsley, R.K., Franzini, J.B. (1964). *Water resources engineering*. McGraw-Hill: New York.
Paulhus, J.L., Kohler, M.A., Linsley, R.K. (1981). *Hydrology for engineers*. McGraw-Hill: New York.

LIPPINCOTT

10.10. 1864 Scranton PA/USA
04.11. 1942 Los Angeles CA/USA

Joseph Barlow Lippincott graduated in 1887 from the University of Kansas, Lawrence KS, with the BSc degree in civil engineering, and was honored in 1914 there with the CE degree. From 1889 he was topographer for the US Geological Survey USGS in New Mexico and California. In 1893 he became an assistant engineer for the Bear Valley Irrigation Co., for the headwaters of Santa Ana River CA. In 1895 he became resident engineer for California State. He established with the USGS Hydrographic Branch gaging stations, collecting rainfall and stream-flow data. It was then when Lippincott decided to devote his career to water supply engineering. He prepared in 1899 an exhaustive report on the water supply, available reservoir sites, and irrigable areas on the Gila River AZ.

Lippincott became in 1902 hydraulic engineer of the newly created US Reclamation Service, taking over as supervising engineer all its activities in the Pacific Coast region from Klamath River OR to Colorado River in Arizona. In 1905 he was appointed with William Mulholland (1855-1935) to the board of engineers dealing with the water supply of Los Angeles CA. He left in 1906 the Service, however, becoming assistant chief engineer of the 400 km long Owens River Aqueduct, which should deliver Los Angeles from the Sierra Nevada with water. Once completed, this Aqueduct had a capacity of 11 m^3/s consisting of open-canal, covered conduit, 100 km of tunnels, five dams, storage reservoirs, and three hydro-electric power plants, so that this was a monumental work. Construction began in 1908, and the Aqueduct was completed in 1913 within the cost originally set. Lippincott was awarded the James R. Croes Medal from the American Society of Civil Engineers for a paper on this work. He entered in 1913 a private practice at Los Angeles, specializing in water supply, which he continued until his death. He was recognized as an expert on the design and construction of hydraulic fill dams. He was the first president of the ASCE Los Angeles Section, and greatly contributed to ASCE in various committees. He was ASCE member from 1899, and Honorary Member.

Anonymous (1943). Joseph B. Lippincott. *Trans. ASCE* 108: 1543-1550.
Hundley, Jr., N. (2001). *The great thirst*. University of California Press: Berkeley.
Lippincott, J.B. (1900). Storage of water on the Gila River. *Water Supply Paper* 33. USGS.
Lippincott, J.B. (1905). Water problems of Santa Barbara. *Water Supply Paper* 116. USGS.
http://waterandpower.org/museum/Early_DWP_Board_and_Management.html *P*
https://sunsite.berkeley.edu/WRCA/lipp.html *P*

LOCKETT

07.07. 1837 Mecklenburg County VA/USA
12.10. 1891 Bogota/CO

Samuel Henry Lockett graduated in 1859 from the US Military Academy, and was then commissioned a brevet second-lieutenant in the US Army Corps of Engineers. He was engaged in 1860 in engineering work in the 8th Lighthouse District, supervising the construction of a fort on the Florida Coast. During the Civil War, he served as captain and engineer, thereby designing and constructing the defenses of Vicksburg MS. He became in 1862 chief engineer of the Department of Mississippi and East Louisiana, supervising the fortification of Port Hudson LA. He was captured at Vicksburg in 1863, exchanged, and placed in charge of the defenses of Mobile AL, rising to the rank of captain.

After the War, he first was appointed professor of natural sciences at Judson Institute, Marion AL, and from 1868 to 1873 professor of engineering at the Louisiana State Seminary, Alexandria LA. During this period he wrote Louisiana as it is. During the next two years, Lockett in vain tried to open new schools, but he resigned due to economic reasons. In 1875 he moved to Egypt where he was commissioned a colonel in the Corps of Engineers within the Army of the Khedive. His construction of defenses during the Egyptian retreat from their near annihilation at the Plain of Dura was commendable. Following the collapse of the Egyptian economy, Lockett returned to the USA in 1877, serving then as professor of engineering and mathematics at the University of Tennessee, Knoxville TN. In 1880 he declined the presidency of the Louisiana State University LSU. From 1883, Lockett successfully engaged in consulting and published his Topographical Survey and Map of Louisiana. While returning from an engineering assignment in South America, he came down with fever and soon later died. The LSU Lockett Hall was named for him.

Anonymous (1891). Samuel H. Lockett. *Engineering Record* 24: 374.
Lockett, S.H. (1879). The valleys of the Nile and the Mississippi. *Trans. of the Wisconsin State Agricultural Society* 17(2): 380-394.
Post, L.C. (1964). *Samuel Henry Lockett*: A sketch of his life and work. Louisiana History: Baton Rouge LA.
Post, L.C., ed. (1969). *Louisiana as it is*: A geographical and topographical description of the State. Louisiana State University Press: Baton Rouge LA.
http://www.lahistory.org/site29.php
http://www.battleofchampionhill.org/lockett.htm *P*

LONG S.H.

30.12. 1784 Hopkinton NH/USA
04.09. 1864 Alton IL/USA

Stephen Harriman Long graduated from Dartmouth College, Hanover NH, in 1808. He was from 1814 a commissioned second lieutenant engineer of the US Army, became then for two years assistant professor of mathematics at the US Military Academy, served then as topographic engineer, and was sent by the War Department in 1817 to examine portages of the Fox and Wisconsin Rivers and to explore the Upper Mississippi River. He was from 1819 to 1823 an explorer of the Rocky Mountains, reaching Rockies in 1820. Longs Peak in northern Colorado, one of the tallest of the Rocky Mountains, was sighted in 1820 and named after Long. He later examined the sources of the St. Peter's River in Minnesota, and the northern boundary of the USA to the Great Lakes. Long became in 1823 a consulting engineer dealing mainly with railroad projects and was from 1837 to 1840 chief engineer of the Atlantic & Great Western RR, thereby mainly in charge of bridges. He was finally chief of the US Corps of Engineers and retired in 1863.

Along with his army duties, Long continued his consulting services until 1856, when he was put in charge of navigation improvements on the Mississippi River. He moved in 1858 to his home and headquarters to Alton IL. In 1861 he was promoted to colonel and called to Washington DC, to succeed as commander of the Topographical Engineers. Long there came into contact with the great works on Mississippi River relating to discharge measurement conducted by Andrew A. Humphries (1819-1883) and Henry L. Abbot (1831-1927). Long remained in his position until his retirement from the Army, three months after his Corps had been merged with the Corps of Engineers.

Anderson, H.A. (2002). Long, Stephen Harriman. *Handbook of Texas online*. Texas State Historical Association: Austin TX.

Anonymous (1963). Long, S.H. *Who was who in America* 1607-1896: 320. Marquis: Chicago.

James, E., Long, S.H. (1823). *Account of an expedition from Pittsburgh to the Rocky Mountains*. Carey: Philadelphia.

Kane, L.M., Holmquist, J.D., Gilman, C., eds. (1978). *The Northern Expeditions of Stephen H. Long*: The journals of 1817 and 1823 and related documents. Minnesota Historical Society Press: St. Paul.

Wood, R.G. (1966). *Stephen Harriman Long*: Army engineer, explorer, inventor. A.H. Clarke: Glendale CA.

http://en.wikipedia.org/wiki/Stephen_Harriman_Long *P*

LORING

26.12. 1828 Boston MA/USA
05.02. 1907 Hackettstown NJ/USA

Charles Harding Loring made an apprenticeship as machinist, becoming third assistant engineer of the US Navy in 1851. During the Civil War he was chief engineer and fleet engineer of the North Atlantic Station thereby supervising the construction of river, harbor and sea monitors at Cincinnati OH. Later he advanced to the general inspector of the iron-clad steamers west of Alleghenies. He was senior member of unused marine engines for studying compound marine engines, which recommended the adoption of the compound engine. He represented in 1874 the Navy in testing the relative economy of compound and simple engines. He was later fleet engineer of the Asiatic Station on the USS *Tennessee*, heading from 1886 the Steam Engineering Department, New York Navy Yard. He was from 1881 a member of the First Naval Advisory Board, participating in the decision to abandon wooden hulls. From 1884 he acted as engineer-in-chief of the Navy, resigning in 1887, but becoming senior member of the Experimental Board of Naval Officers. He thereby tested water-tube boilers, determined the economy of evaporation with various air pressures and rates of combustion. Loring retired in 1890 from the Navy, becoming a consultant for the US & Brazil Steamship Co. During the Spanish-American War he was inspector of engineering work in New York NY.

Loring conducted in 1874 a valuable series of experiments to ascertain the relative efficiency of given weights of steam when used under variable conditions in similar engines. The results constitute an important addition to the then available information on the essential economic conditions during the operation of steam engines. It was found that the simple engine, which cannot be considered as the best, may not only be economical, but so economical that the optimum performance of an excellent compound engine hardly beats it. Loring was a member of the American Society of Mechanical Engineers ASME, serving as president in 1891. He also was member of the Society of Naval Architects and Marine Engineers SNAME.

Anonymous (1907). Loring. *Journal of American Society for Naval Engineers* 19(1): 268-276.
Anonymous (1980). Loring. *Mechanical engineers in America born prior* 1861: 212. ASME: NY.
Clark, D.K. (1895). *The steam engine*: A treatise on steam engines and boilers. Blackie: London.
Hutton, F.R. (1915). Charles H. Loring. *History of ASME* 1880-1915: 108. ASME: New York. *P*
Loring, C.H., Emery, C.E. (1875). Report on the trials of the steam-machinery of the US
revenue steamers. *Journal of the Franklin Institute* 99(3): 197-205.

LOW

02.04. 1854 Pittsburgh PA/USA
25.05. 1927 Buffalo NY/USA

Emile Low was educated at the public schools of his native city. He became in 1871 assistant on the *Hiland* Avenue Reservoir of the Pittsburgh Water Works until 1875, from when he was city engineer. In 1882 he accompanied his father to Mexico as his assistant in railroad projects. From 1897 to 1903 Low was resident US assistant engineer in charge of the construction of the Buffalo Breakwater; the 1904 paper describes these works, and it was awarded the Norman Medal. From 1904 he was in the service of New York State as resident engineer in charge of its Barge Canal in Oneida County, thereby solving the difficult engineering problems. Out of this project resulted both the 1907 and the 1913 papers. In 1907, Low re-entered the Federal Services as assistant engineer at Buffalo NY, remaining until his retirement in 1924. He was during this period collaborating with a number of notable American engineers. Low was esteemed by his fellow engineers of Buffalo for his remarkable professional attainments. He was an indefatigable worker in engineering pursuits, and was an exemplary character.

The New York State Barge Canal is a successor to the Erie Canal within New York State. The current system is 850 km long and composed of the Erie, the Oswego, the Cayuga Canal, and the Champlain Canals. New York State authorized its construction in 1903. In 1905 the construction of the Barge Canal started, and was completed in 1918 with a cost of 100 million US$. Its new route followed rivers which were avoided by the original Erie Canal builders, thus bypassing cities as Syracuse or Rochester. Since the 1970s, the State has ceased modernizing the system due to the shift to truck transport. The Canal is preserved mainly for historical and recreational purposes, serving in addition for water supply and flood control. The system canals are 4 m deep, 36 m wide, with 57 electronically operated locks, allowing for up to 2,000 tons heavy vessels.

Anonymous (1916). Emile Low. *The Pittsburgh Press* (June 27): 9. *P*
Anonymous (1929). Low Emile. *Trans. ASCE* 93: 1874-1875.
Low, E. (1904). The breakwater at Buffalo NY. *Trans. ASCE* 52: 73-197.
Low, E. (1907). Wash-drill borings on the New York State Barge Canal. *Engineering News* 57(Jan.17): 54-57.
Low, E. (1913). Hydraulic dredging on New York State Barge Canal. *Engineering News* 69(Apr.10): 710-720.
http://en.wikipedia.org/wiki/New_York_State_Canal_System

LOWE

14.03. 1916 New York NY/USA
02.01. 2012 Seattle WA/USA

John Lowe III obtained the BS degree from the City College, New York NY, in 1936, and the MS degree from Massachusetts Institute of Technology MIT, Cambridge MA, in 1937. He was an instructor in civil engineering at University of Maryland, College Park MD, and at MIT until 1944, a physicist at David Taylor Model Basin, US Navy, in 1945, from when he joined Tippetts-Abbett-McCarthy-Stratton TAMS, New York NY, first as head of the Soil and Rock Engineering Department until 1956, then until 1962 as associate partner, and until 1983 as partner. From 1984 Lowe was an independent consultant on dam engineering in Yonkers NY. He was the eighth Terzaghi Lecturer of the Society of Soil Mechanics in 1971, the second USCOLD Lecturer of the US Committee on Large Dams in 1982, and keynote lecturer on roller compacted concrete in 1988. He was a Fellow of the American Society of Civil Engineers ASCE, and member of the National Academy of Engineering.

Lowe was a notable civil engineer with expert knowledge in soil mechanics and dam engineering. During his stay with TAMS he had been involved in major dam projects of the USA, Turkey, Taiwan, Morocco, Pakistan, Brazil, and Greece, such as *Tarbela* Dam counting among the largest worldwide, and being the recipient of the Order of Alouites from the King of Morocco for works there. He also contributed chapters to four books, including Dam engineering. He was further general reporter on Question 36 of the 10th ICOLD Congress held 1970 in Montreal.

Anonymous (1962). John Lowe III. *Civil Engineering* 32(9): 23. *P*
Anonymous (1970). John Lowe III. 10th *ICOLD Congress* Montreal 1: 126. *P*
Anonymous (1977). John Lowe III. *Water Power & Dam Construction* 29(8): 9. *P*
Anonymous (1994). John Lowe III. *American men and women in science* 4: 1079.
Anonymous (2005). Lowe, John III. *Who's who in America* 59: 2862.
Lowe, J. III, Knappen, T.T. (1969). Earth dams. *Handbook of applied hydraulics* 5: 149-206, C.V. Davis, ed. McGraw-Hill: New York.
Lowe, J. III (1978). Foundation design: Tarbela Dam. 4th Nabor Carillo Lecture, Mexican Society for Soil Mechanics. *Report* 39180-0631. USCE: Vicksburg MS.
Lowe, J. III, Zaccheo, P.F. (1991). Subsurface explorations and sampling. *Foundation engineering handbook* 1, H. Fang, ed. van Nostrand Reinhold: New York.
Lowe, J. (2008). *Filters and sinkholes*: Rapid drawdown stability. Homeland Security: New York.

LOWRY

26.01. 1900 Waxahachie TX/USA
30.05. 1959 Austin TX/USA

Robert Lee Lowry graduated in 1922 from Trinity University, San Antonio TX, with the BA degree, and in 1929 from the University of Texas, Austin TX, with the BSc degree in civil engineering. Early employment included experience with an engineering company, the Texas Board of Water Engineers, and the Texas Reclamation Department. In 1935 he was employed by the Brazos Conservancy District and the Texas Planning Board. He was in the mid-1930s a special consultant for the Colorado Conservation Board before joining the International Boundary and Water Commission in 1942. He also worked for the Mexican engineers on hydraulic studies of the Rio Grande and Pecos River Commission. Lowry opened in 1950 his consulting office. He was in 1951 appointed member of the international commission studying a water dispute between Afghanistan and Iran.

Lowry was a notable hydraulic engineer in water resources and water supply. He has written papers and reports mostly for regional problems, including works on the rainfall characteristics, both important for flood prevention and for agricultural purposes, and on the Brazos River Basin, with 2000 km one of the longest rivers in the USA from its headwaters in New Mexico to its mouth at the Gulf of Mexico. Lowry further made a study on the droughts in his state, which were particularly strong in 1950, resulting in a large water shortage mainly in 1951 and 1952, lasting until 1956, so that streams all over the State were mostly dried up. It only stopped once heavy spring rains started in 1956. Lowry described these hydrologic features and made proposals as to its reduction for all involved. The 1960 report deals with the reservoir evaporation in Texas, which is important in relation to floods and droughts. Lowry was ASCE member from 1940.

Anonymous (1959). Robert L. Lowry dies in Austin. *The Texas Engineer* 29(7): 8. *P*
Anonymous (1960). Robert L. Lowry. *Trans. ASCE* 125: 1416.
Lowry, R.L. (1934). *Excessive rainfall in Texas*. Texas Reclamation Board: Austin.
Lowry, R.L. (1956). Preliminary Plan, *Brazos River Basin*: Comprehensive development for water conservation. Texas Electric Service Company: Austin.
Lowry, R.L. (1958). *Surface water resources of Texas*. Technical Report. Texas Electric Service Company: Austin.
Lowry, R.L. (1959). *A study of droughts in Texas*. State Board of Water Engineers: Austin. (*P*)
Lowry, R.L. (1960). *Monthly reservoir evaporation rates for Texas*, 1940 through 1957.Texas Board of Water Engineers: Austin TX.

LUDLOW

27.11. 1843 Islip NY/USA
30.08. 1901 Convent Station NJ/USA

William Ludlow entered the US Military Academy in 1860, graduated in 1864 and was promoted to first lieutenant within the Corps of Engineers. After Civil War service, he was in 1867 engaged in river and harbor works at New York City and along the South Atlantic Coast until 1872, from when he took over as chief engineer in Dakota State until 1876. He also made surveys of the Yellowstone National Park and of the Black Hills County. From 1877 to 1882, he conducted general engineering services on the Delaware Bay and River, and river improvements. From 1882, he was major of the Corps, and secretary of the Lighthouse Board, Washington DC. From 1883 to 1886 he was chief engineer of the Philadelphia Water Department, reorganizing and improving the city's water system. He made from 1888 to 1893 river and lighthouse work on the Great Lakes, and then was military attaché of the US Embassy in London UK.

Ludlow inspected the canals of Suez, Corinth, Kiel and in the Netherlands, given that he chaired from 1895 the Nicaragua Canal Board. He was appointed lieutenant colonel of the Corps of Engineers in 1897, and was then again until 1898 in charge of harbor and river improvements in New York City, recommending that the East River channel be deepened. During the Spanish-American War he was transferred in 1898 to Santiago de Cuba, where he served as major general of the Volunteers. He also was president of the board to organize the Army Sea Transport Service during this year, becoming until 1900 Military Governor of Havana, Cuba. He inspected French and German military systems and methods of training then, and in 1901 had active duty in the Philippines but became then sick, which led to his premature death. Ludlow was a member of the American Society of Civil Engineers ASCE, and a Companion of the Military Order of the Loyal Legion of the US.

FitzSimons, N., ed. (1991). Ludlow, William. *A biographical dictionary of American civil engineers* 2: 65-66. ASCE: New York.
Ludlow, W. (1884). The present water supply: *Report* of chief engineer of the Philadelphia Water Department. Philadelphia.
Ludlow, W. (1899). *Report* of the US Nicaraguan Canal Commission. Washington DC.
McAndrews, E.V. (1973). William Ludlow: Engineer, governor, soldier. *PhD thesis*. Kansas State University: Manhattan KS.
http://en.wikipedia.org/wiki/William_Ludlow *P*

LYON

08.06. 1874 New York NY/USA
17.12. 1939 Parkersburg WV/USA

George John Lyon graduated in 1899 with the BSc degree from University of Nebraska, Lincoln NE, and received in 1904 the CE degree from School of Mines, Columbia University, New York NY. After work with railroad companies, he became in 1904 professor of civil engineering at Colorado College, Colorado Springs CO, installing there a hydraulic laboratory. He was appointed assistant engineer of the Water Resources Branch, US Geological Survey in 1910, studying the seepage and evaporation losses in the South Platte River, and reporting on irrigation projects in the San Juan River Valley CO. In 1912 he designed the Spiers Falls automatic gaging station on the Hudson River, and he later made hydrologic studies of the Hudson, Mohawk and Sacandaga Rivers in New York State, resulting in the 1913 Report. His book on this topic was published in 1915.

From 1916 Lyon was engaged by the Marseille Land and Water Power Company, Ottawa IL, on hydraulic studies of the Illinois River. In 1917 he was further consulted by the Imperial Russian Government relative to stream gagings in Turkestan, from which activities resulted the 1917 manual. After war service and other engineering contracts, he became in 1927 until his death hydraulic engineer of the National Board of Fire Underwriters, New York City. During this period he examined the water supplies of more than 100 cities with populations in excess of 50,000, to determine the availability of water for fire protection. Recommendations made in reports were then used for the determination of fire-insurance rates. Lyon was a member of the Colorado Polytechnic Society, the American Society for the Advancement of Science, and the Washington Society of Engineers. He also was member of the American Society of Civil Engineers ASCE since 1914.

Anonymous (1912). George John Lyon. *The Pike's Peak Nugget* 11: 24. Colorado Springs. *P*
Anonymous (1941). George J. Lyon. *Trans. ASCE* 106: 1609-1614.
Covert, C.C., Lyon, G.J., Pierce, C.H. (1913). *Plans and specifications for current-meter gaging stations*. USGS: Washington DC.
Lyon, G.J. (1915). Equipment for current-meter gaging stations. *Water Supply Paper* 371. USGS: Washington DC.
Lyon, G.J. (1916). Cable station equipment for stream gaging. *Stone & Webster Journal* 19(1): 40-44.
Lyon, G.J. (1917). *Manual of hydraulic engineering instruments*. Gurley: Troy NY. (*P*)

MADDOCK

06.04. 1907 Williams AZ/USA
07.03. 1991 Tucson AZ/USA

Thomas Maddock, Jr., joined the Soil Conservation Service in New Mexico before transferring to the US Bureau of Reclamation USBR as head of the Section of Sedimentation, Division of Hydrology. Later he became head of the Hydrology Section in the same Division before moving to Washington DC as assistant chief of the Division. He also served as chief of the Branch of Irrigation Operations, USBR. From 1957, he was with the US Geological Survey USGS, engaged in research as independent scientist in the fields of sediment transport, river mechanics, relations between soil and water flow and their uses, and other hydraulic problems. He also attempted to simplify problems associated with sediment transport and hydraulic roughness of alluvial channels. Maddock was presented with the Honorary Doctorate of Science by the University of Arizona in 1971, and was the recipient of the 1963 J.C. Stevens Award of ASCE. He was further awarded in 1975 with Luna B. Leopold (1915-1992) Honorary Membership of the American Water Resources Association AWRA.

Leopold and Maddock presented in 1953 a valuable study on the hydraulic geometry of stream channels. River width, average depth, and average velocity were investigated as functions of water discharge, and corresponding equations were derived. The change of this relationship along a river was also studied. The relationship to sediment load was further investigated. Similar results were presented as the regime equations by Gerald Lacey (1887-1979), and Thomas Blench (1905-1993). These equations indicate a trend of carefully selected data describing the main hydraulic features of a particular river.

Anonymous (1963). Thomas Maddock, Jr. *Civil Engineering* 33(10): 71. *P*
Anonymous (1975). Thomas Maddock, Jr. *Water Resources Bulletin* 11(1): 199-200. *P*
Leopold, L.B., Maddock, T., Jr. (1953). *The hydraulic geometry of stream channels and some physiographic implications*. US Government Printing Office: Washington DC.
Leopold, L.B., Maddock, T., Jr. (1955). The flood control controversy: Big dams, little dams, and land management. *Journal of Geology* 63(4): 395-396, also Ronald Press: New York.
Maddock, T., Jr. (1970). Indeterminate hydraulics of alluvial channels. *Journal of the Hydraulics Division* ASCE 96(HY11): 2309-2323; 97(HY5): 756-760; 97(HY7): 1149-1156; 97(HY9): 1533-1535; 98(HY3): 555-567.
Maddock, T., Jr. (1983). Discussion of Sediment transport and unit stream power function. *Journal of Hydraulic Engineering* 109(12): 1779-1781.

MAIN

16.02. 1856 Marblehead MA/USA
06.03. 1943 Winchester MA/USA

Charles Thomas Main graduated as a mechanical engineer in 1876 from the Massachusetts Institute of Technology MIT, Cambridge MA, and there was instructor until 1879. He was from 1881 to 1891 an engineer and superintendent with the Lower Pacific Mills, Lawrence MA, reorganizing its machinery and installing a new power plant, moving until 1893 to a similar plant in Providence RI. He formed then until 1907 a partnership at Boston MA, organizing the firm Charles T. Main, which was incorporated in 1926. The Charles T. Main Award was established in 1919, given annually to a student member of the American Society of Mechanical Engineers ASME, of which Main was member, served as president in 1918, and was recipient of its 1935 Gold Medal. He also was member of the American Society of Civil Engineers ASCE, the Boston Society of Civil Engineers and recipient of its 1913 Desmond Fitzgerald Medal, and was awarded the honorary degree from Northeastern University, Boston MA, in 1935, among many others.

Main designed and advised from 1900 on the building of steam and water power plants across the USA. He was involved in these projects particularly at the Conowingo Dam across Susquehanna River in Maryland, and the Keokuk Dam across the Mississippi River. His firm designed over eighty hydro-electric plants during the early 20th century. He was also one of the nine American engineers sent to France after World War I to consult the French authorities on reconstruction works in their devastated country. He was a delegate to the World Power Conference held in Tokyo, Japan, in 1929, during which international aspects of power engineering were particularly discussed. He also drafted the first code of ethics, as subsequently adopted by all American Engineering societies. Main may therefore be considered an outstanding expert in questions of power engineering, as also an individual who has significantly contributed to the engineering community by his lifelong service.

Anonymous (1916). Charles T. Main. *ASME yearbook*: Frontispiece. ASME: New York. *P*
Anonymous (1918). Charles T. Main. *Journal of the ASME* 40(1): Frontispiece. *P*
Main, C.T. (1886). *Relative cost of steam and water power*. MIT: Cambridge MA.
Main, C.T. (1907). Computation of the values of water power, and the damages caused by the diversion of water used for power. *J. New England Water Works Association* 21: 214-240.
Uhl, W.F. (1951). *Charles T. Main*: One of America's best. Newcomen Society: Boston.
http://en.wikipedia.org/wiki/Charles_T._Main *P*

MANSON

14.02. 1850 Leewood VA/USA
21.02. 1931 San Francisco CA/USA

Sweeping Back the Flood

Marsden Manson graduated in 1870 from Virginia Military Institute, Lexington VA, and obtained his first engineering experience on surveys along the Atlantic Ocean. He became in 1874 assistant in the US Army Corps of Engineers, on surveys for the extension of the James River and Kanawha Canal, and on making plans for the locks of this canal from Chesapeake Bay to the Ohio River. He moved to California in 1877, first as instructor at a School at San Mateo, and then with the newly erected State Engineer Department of California. He had to study the damage made to rivers and farm lands by debris from hydraulic mines, which were partly done at the University of California, from where he obtained the PhD degree in 1880. He was then chief engineer of the California State Board of Harbor Commissioners operating the harbor and water-front facilities of San Francisco CA. After eleven years in this position, he became member of the San Francisco Sewer Commission, and subsequently also had his private practice.

Manson was appointed in 1900 member of the San Francisco Board of Public Works. He took interest in the city's endeavor to secure the right to store water on Tuolumne River and its tributaries. He spent much time at Washington DC in the interest of his city, with the result that an adequate permit was granted in 1908, which finally led to the Hetch Hetchy Project. He was then city engineer serving in this capacity for four years, from when he had again his private practice until his death. He was called 'Dr.' Manson had a strong leaning to the study of scientific problems, and was particularly interested in meteorology, and in the earth's pre-glacial climate, which brought him in contact with scientists all over the world. He was described as a pleasant and charming personality. One of his friends has written: 'Maintaining his mental activity to the end, he loved to dwell upon the things that might be, and undoubtedly will be, done to make this Coast, especially California, with its marvellous scenic and climatic offerings the playground of the nation'.

Anonymous (1931). Marsden Manson. *Trans. ASCE* 95: 1554-1556.
Grunsky, C.E., Manson, M., Tilton, C.S. (1899). *System of sewerage for the city and county of San Francisco*. Britton & Rey: San Francisco.
Manson, M., Wilson, J.W. (1901). *Report of irrigation investigations in California*. Washington DC.
Manson, M. (1922). *The evolution of climates*. Lord Baltimore Press: Berkeley CA.
Unger, N.C. (2012). *Beyond nature's housekeepers*. Oxford University Press: New York. P

MANSUR

22.12. 1918 Kansas City MO/USA
29.12. 2010 Chesterfield MO/USA

Charles Isaiah Mansur was educated at Harvard University, Cambridge MA, from where he obtained the BS degree in 1939, and the MS degree in 1941. He was then until 1943 chief, Seepage Section, of the Waterways Experiment Station, Vicksburg MI, until 1954 chief of its Design and Analysis Section, assistant chief of its Embankment and Foundation Branch, when joining civil engineering contractors, becoming there vice-president and president until 1969, taking over as the general manager the newly established St. Louis office of McClelland Engineers Inc., a soils engineering firm. In 1980 he there took over as senior vice-president, and from 1984, Mansur was an independent consultant at St. Louis MO. He was recipient of the Thomas A. Middlebrooks Award, the Thomas Fitch Rowland Prize, the James R. Croes Medal, all sponsored by the American Society of Civil Engineers ASCE, of which he was a member.

Mansur made research in soil mechanics and foundation engineering, with a particular interest in dewatering construction sites. His early investigations dealt with particular site conditions at places where hydraulic structures were erected. Later, he considered with colleagues issues of foundation engineering and levee erection on the Mississippi River. He also added to the standardization of pile tests.

Anonymous (1959). Mansur, Charles I. *Who's who in engineering* 8: 1577. Lewis: New York.
Anonymous (1970). Charles Mansur. *Civil Engineering* 40(2): 10. *P*
Anonymous (1994). Mansur, Charles I. *American men and women in science* 5: 229.
Mansur, C.I., ed. (1943). *Waterworks engineering in disaster*. Office of Civilian Defense: Washington DC.
Mansur, C.I. (1947). *Malaria control of impounded water*. US Public Health Service and Tennessee Valley Authority, Health and Safety Department: Washington DC.
Mansur, C.I., Focht, J.A., Jr. (1956). Pile-loading tests, Morganza Floodway control structure. *Trans. ASCE* 121: 555-576; 121: 584-587.
Mansur, C.I., Kaufman, R.I. (1962). Dewatering. *Foundation engineering*: 241-350, G.A. Leonards, ed. McGraw-Hill: New York.
Mansur, C.I., Hunter, A.H. (1970). Pile tests: Arkansas River Project. *Journal of the Soil Mechanics and Foundation Division* ASCE 96(SM5): 1545-1582.
Turnbull, W.J., Mansur, C.I. (1961). Investigation of underseepage-Mississippi River Levees. *Trans. ASCE* 126: 1429-1539.

MARRIS

20.04. 1924 Lincoln/UK
07.09. 2005 Atlanta GA/USA

Andrew Wilfrid Marris obtained the BSc degree in physics from the University of London, London UK, and the PhD degree in mechanical engineering in 1952 from the University of New Zealand, Ardmore NZ. He moved in 1953 as assistant professor to the University of British Columbia, Vancouver BC, and was from 1960 to 1962 then associate professor of mechanical engineering at the University of Texas, Austin TX. He then moved to Georgia Institute of Technology, Atlanta GA, as professor of engineering mechanics. Marris was finally there until retirement in 1987 Regents' Professor of Engineering Science and Mechanics.

Marris worked in a wide range of engineering problems related to mechanical and hydraulic engineering, rheology, and the kinematics and dynamics of motion. His PhD thesis treated a problem of thermodynamics related to turbulent flow. After his move to the USA, he was initially interested in hydraulic engineering problems related to the design of hydropower installations. The 1965 Symposium on Rheology was one of the earlier meetings in this field. Questions including the viscoelastic behavior and lubricating properties of liquids, dragometers for elastic liquids, turbulent-flow rheometres, flow of solids-water mixtures in hydraulic dredging, among others. The 1967 paper deals with the classical bathtub vortex. The temporal growth of vorticity is analyzed. The results indicate both exponential growths of vorticity and the residual vorticity.

Marris, A.W. (1952). The heat transfer coefficient as a function of Reynolds and Prandtl moduli. *PhD Thesis*. Auckland University College, Auckland NZ.
Marris, A.W. (1959). Large water-level displacements in the simple surge tank. *Journal of Basic Engineering* 81(12): 446-454.
Marris, A.W. (1961). The phase-plane topology of the simple surge tank equation. *Journal of Basic Engineering* 83(12): 700-707.
Marris, A.W., Wang, J.T.-S., eds. (1965). *Symposium on rheology*. ASME Applied Mechanics and Engineering Conference. Catholic University of America: Washington DC.
Marris, A.W. (1967). Theory of the bathtub vortex. *Journal of Applied Mechanics* 34(1): 11-15.
Marris, A.W., Stoneking, C.E. (1967). *Advanced dynamics*. McGraw-Hill: New York.
Marris, A.W. (1974). *Kinematics in fluid motion*. School of Engineering Science and Mechanics, Georgia Institute of Technology: Atlanta.
http://www.legacy.com/obituaries/atlanta/obituary.aspx?n=andrew-marris&pid=15040257&fhid=5303 *P*

MARSHALL R.B.

22.01. 1869 Crows Landing CA/USA
21.06. 1949 San Francisco CA/USA

Robert Bradford Marshall went in 1891 as a chief topographer to California for the US Geological Survey. He there saw the paradox of millions of fertile acres lying parched in the summer, which might be saved by millions of acre feet of water flowing unused into the Pacific during winter and spring. He spent the next 30 years gathering the engineering data necessary to make real his plan of storing in dams the rain waters of the Sacramento, San Joaquin, and Santa Clara Valleys, and of South California.

In 1919 he turned his plan over to the State and resigned as chief hydrographer in the United States to devote his full time to fighting for his project. In face of scorn and ridicule, he succeeded in getting from the legislature a 200,000 US$ appropriation for the state engineer to make preliminary studies. During the next 12 years, more than one million $ was spent, investigating the feasibility of the project. Marshall proposed what became known as the Marshall Plan, one that called for a large dam and reservoir on the Sacramento River's northern reaches, along with two peripheral canals to help reclaim drier areas along both sides of the Central Valley. The plan also called for providing more water to growing cities as San Francisco, increasing flood control and navigability on the Sacramento River, and preventing salt intrusion into the Delta. The Marshall Plan also looked at diverting water to southern California from Kern River near Bakersfield via a tunnel under the Tehachapi Mountains. Revenue for this ambitious scheme would be produced through sales of water and electricity generated at state power plants. Finally in 1944, Col. Marshall saw water pouring through the giant Shasta Dam, which had been built basically as he had planned it. The Friant Dam also followed his original conception. In 1938 Marshall retired from the State Division of Highways for which he had worked for several years. Following his death, the California Assembly adopted a resolution acknowledging the state's debt to Col. Marshall 'for his vast contributions to the development of the wealth and heritage of this State'.

Anonymous (1949). Robert B. Marshall dies: Father of Central Valley. *Engineering News-Record* 143(Aug.4): 24.
Bailey, J. (2007). *Reclamation managing water in the West*: California's Central Valley Project, historic engineering features to 1956. USBR: Denver.
Marshall, R.B. (1921). *Irrigation of twelve million acres in the Valley of California*. Sacramento.
http://www.water.ca.gov/swp/historymarshall.cfm *P*

MARSHALL W.L.

11.06. 1846 Washington KY/USA
02.07. 1920 Washington DC/USA

William Louis Marshall graduated in 1868 from the US Military Academy, continuing there as assistant professor of natural and experimental philosophy until 1871. He was from 1872 to 1876 assistant in a party exploring the Rocky Mountains, discovering the Marshall Pass in 1873, and describing in 1875 the Marshall Basin of the San Miguel River CO. He was then until 1884 assistant engineer on projects in river improvements and levee construction in the States Alabama, Tennessee, Mississippi and Georgia. He was promoted to captain in 1882, and major in 1885. He worked then on governmental engineering projects in charge of the Chicago and Calumet Rivers, the construction of the Hennepin Canal connecting Illinois River at Lasalle IL with Mississippi River at Rock Island IL. He thereby patented improvements relating to a combined breakwater and beach, and an automated movable dam or sluiceway gate. He was from 1899 on the board to advise on the water supply of Washington DC, and on river and harbor improvement works for New York City. He then accomplished the Ambrose Channel there, and completed the extension of the coast defenses. Marshall was commissioned Chief of Engineers in 1908 with the rank of brigadier general, retiring from active service in 1910.

The Hennepin Canal is by now an abandoned waterway in northwest Illinois, listed on the National Register of Historic Places. It was opened in 1907 but soon abandoned because of railroad competition. The main canal is 121 km long with its feeder canal of 47 km length. The Canal was first conceived in 1834 but the pressure for transportation cheaper than the rail urged the Congress in 1871. Construction began in 1892 with the first boat traverse in 1907, reducing the distance from Chicago to Rock Island by 675 km. While the Canal was under construction, the Corps of Engineers widened its locks on both the Illinois and Mississippi Rivers. This Canal was the first in the USA built entirely of concrete. Although it enjoyed limited success as a waterway, engineering innovations employed in its construction were great, due to the additions of Marshall and his colleagues. From the 1930s the Canal was mainly used for recreational traffic.

FitzSimons, N., ed. (1991). Marshall, William L. *A biographical dictionary of American civil engineers* 2: 72-73. ASCE: New York.
Marshall, W.L. (1896). Marshall's bear-trap dams. *J. Assoc. Engng. Societies* 16(6): 218-226.
http://en.wikipedia.org/wiki/William_Louis_Marshall *P*
http://en.wikipedia.org/wiki/Hennepin_Canal_Parkway_State_Park

MARSLAND

05.10. 1829 Newcastle NY/USA
27.06. 1898 Osinning NY/USA

Edward Marsland, a distinguished master mechanic and steam engineer, was veteran of the Navy. His grandfather served as foreman Robert Fulton (1765-1815) while the famous and pioneer steamboat, the *Clermont*, was built. Marsland travelled in 1846 to New Orleans LA; returning to New York in 1847, he entered the iron works. In 1851 he then joined the Collins Line of ocean steamships, making his first sea voyage on the *Baltic*. During the next 6 years he rose from oiler to assistant engineer. In 1858, by the impact of the ship builder William H. Webb (1817-1899), who had just completed the construction of the steam corvette *Japanese* for the Russian Government, Marsland became second engineer to take the vessel to Nikolnefsk on Amur River in Siberia. Given his excellent services, he was commissioned chief engineer of the *Japanese*, and was in addition given charge in erecting complicated machinery at that port, so that Marsland was promoted to the rank of fleet engineer of the Russian Navy in Chinese waters.

In 1860, Marsland resigned his commission before returning to the USA, offering his services to the Navy Department in the event of the outbreak of hostilities. On his return to New York, he was appointed chief engineer of the Pacific Mail Steamship Company. He took the steamship *Constitution* to San Francisco, but on the outbreak of the Civil War in 1861, he was appointed first assistant engineer of the US Navy. In 1862 he was detached from the *Unadilla* and ordered to superintended duty on the steam sloop of war *Lackawanna* at the Brooklyn Navy Yard. As soon as she was ready for sea, he took her to Mobile AL, where he remained a year, then returning to Brooklyn. After a three years' service in the Navy he was obliged to resign on account of his failing health. He was then re-appointed chief engineer of the Pacific Mail Steamship Co. Shortly later, at the solicitation of Webb, he was given a leave of absence as first assistant engineer of the Italian Navy, to take a vessel to her destination in Italy. He entered In the early 1870s the service of the US Corps of Engineers, where he was in charge of dredging services on the Mississippi River. At the end of his career, he had been one of the last who was closely associated with the old steam navigation, leaving a lasting worth.

Anonymous (1888). The career of Edward Marsland, master mechanic and steam engineer. *The Newtown Register* (Feb.9): 8.

Anonymous (1892). Edward Marsland. *Engineer* 22: 99. *P*

http://en.wikipedia.org/wiki/User:Gatoclass/SB/Novelty_Iron_Works (*P*)

MARSTON

31.06. 1864 Seward IL/USA
21.10. 1949 Tama IA/USA

Anson Marston obtained the CE degree from Cornell University, Ithaca NY, in 1889, and the Dr. degree in 1925 from University of Nebraska, Lincoln NE. He was until 1892 resident engineer for the Pacific Railways, from when he was until 1920 professor and head of the Civil Engineering Department, Iowa State College, Ames IA. From 1904 to 1932 he there served as Dean of Engineering, and as Director of the Iowa Engineering Experiment Station, and was from then emeritus professor. Dean Marston served engineering on both statewide and nationwide basis. He published more than 200 papers and technical reports including Sewers and drains, and Engineering valuation, the latter with his successor as dean of engineering. He was killed in an automobile accident near Tama IA.

In parallel to his teaching activities, Marston was a private consultant. He was member of the Chicago Sanitary District from 1924 to 1929, served as consulting engineer at Miami FL from 1924 to 1927, and was member of the Florida Everglades Engineering Board of Review in 1927. From 1929 to 1932 he was member of the Interoceanic Canal Board, the Panama-Nicaragua Ship Canals. In 1933 he was member of the Mississippi River Engineering Board, contributing to flood control plans from Cairo IL to the Gulf of Mexico. Marston was closely associated with the American Society of Civil Engineers ASCE, serving as director from 1920 to 1923, vice-president until 1924, and ASCE President in 1929, becoming honorary member in 1939. He also presided over the Society for the Promotion of Engineering Education SPEE. He further was member and president of the Iowa Engineering Society. Marston was the recipient of the Fuertes Gold Medal from Cornell University in 1904, the 1904 Chanute Medal of the Western Society of Engineers, and the 1941 Lamme Medal from ASCE.

Anonymous (1948). Marston Anson. *Who's who in engineering* 6: 1280. Lewis: New York.
Anonymous (1949). Marston had notable career as a civil engineer. *Engineering News-Record* 143(Oct.27): 33; 143(Nov.3): 20.
Gilkey, H.J., Marston, A. (1916). *A preliminary investigation for an irrigation system for the Rogue Valley in Southwestern Oregon*. MIT: Cambridge MA.
Marston, A. (1935). *Water resources data for Iowa*. US Geological Survey: Iowa City IA.
Marston, A., Agg, T.R. (1936). *Engineering valuation*. McGraw-Hill: New York.
http://www.lib.iastate.edu/arch/rgrp/11-1-11.html
zihaoliu1990.files.wordpress.com/.../golden-shining... *P*

MARTIN H.M.

17.03. 1908 Argos IN/USA
03.03. 1997 Denver CO/USA

Harold Melville Martin received the BS degree in civil engineering from Purdue University in 1931, and the CE degree there in 1933. He then was engaged by the US Bureau of Reclamation USBR, Denver CO, as junior engineer until 1936, becoming associate engineer until 1941, when joining the US Army until 1945. He then returned to USBR as hydraulic research engineer conducting laboratory investigations on hydraulic structures and hydraulic machinery. From 1950 until retirement, Martin was chief of the USBR Hydraulic Laboratory, Denver CO. He was a Fellow of the American Society of Civil Engineers ASCE, thereby also served for its Hydraulics Division in the Executive Committee. He was a member of the International Association of Hydraulic Research.

Martin was a devoted experimenter contributing to a variety of aspects in hydraulic engineering. Both his 1939 and 1952 Reports were concerned with laboratory studies for dams then studied at the USBR Hydraulic Laboratory, Denver CO. The 1953 paper deals with the diversion of water from streams with alluvial beds into main irrigation canals to prevent sedimentation problems. It was proposed to apply both intermitting and continuous sluicing in combination with vortex tubes and short tunnels. It was also found that guide walls in front of the headworks and sluiceways may have a positive effect on the reduction of sediment deposition.

Anonymous (1967). Harold M. Martin. *Civil Engineering* 37(7): 66. *P*

Martin, H.M., Panuzio, F.L., Brewer, H.W. (1939). Hydraulic model studies for the design of the outlet works and spillway for the Caballo Dam. *Report* HYD 72. USBR: Denver.

Martin, H.M., Peterka, A.J. (1952). Hydraulic model tests on the moss-prevention devices for the Friant-Kern Canal. *Report* HYD 351. USBR: Denver.

Martin, H.M., Carlson, E.J. (1953). Model studies of sediment control structures on diversion dams. Proc. 5th *IAHR Congress* Minneapolis: 109-122.

Martin, H.M., Ball, J.W. (1955). Laboratory and prototype tests for the investigation and correction of excessive downpull forces of large cylinder gates under high heads. USBR: Denver.

Martin, H.M., Wagner, W.E. (1961). Experience in turbulence in hydraulic structures. Proc. 9th *IAHR Congress* Dubrovnik: 153-172.

Rouse, H. (1976). Harold Martin. *Hydraulics in the United States* 1776-1976: 185. The University of Iowa: Iowa City. *P*

MARX

10.10. 1857 Toledo OH/USA
31.12. 1939 Palo Alto CA/USA

Charles David Marx obtained his bachelor degree in civil engineering from Cornell University, Ithaca NY, in 1878, graduating as a civil engineer in 1881 from the Polytechnikum Karlsruhe, Germany. Upon return to the USA, he was an assistant engineer for river improvements in Montana State. He was from 1884 to 1890 assistant professor of civil engineering at Cornell University, then for one year in the same position at University of Wisconsin, and from 1891 professor at the Leland Stanford Junior University, today's Stanford University, Stanford CA, from where he retired in the mid 1920s. Marx was in 1925 awarded the honorary doctorate from Karlsruhe University.

Early in his career Marx made river and harbour works on the Missouri and later on the Mississippi River. During his stay at Cornell University he acted also as chief inspector of public improvements at Rochester NY, and was assistant engineer to Emil Kuichling (1848-1914). Marx was the first chairman of the California State Water Commission from 1911 once having moved to Stanford, framing to a large part the water laws of the state. He later served on many engineering boards and commissions, was consultant for the state on the Central Valley Project and on numerous dams, including San Gabriel No. 1, and under President Hoover was chairman of the Board of Advisory Engineers. The honorary degree of LL.D. was conferred on him by the University of California in 1918. He was an Honorary Member and past-president of the American Society of Civil Engineers ASCE. He was remembered affectionately for his kind and unfailing interest in all his associates, known also as 'Daddy Marx' to his students.

Anonymous (1915). Charles David Marx, president ASCE. *Engineering News* 73(8): 386-389. *P*
Anonymous (1929). Marx, Charles David. *Who's who in America* 15: 1389. Marquis: Chicago.
Anonymous (1940). Charles D. Marx dies. *Engineering News-Record* 124(Jan.4): 29. *P*
Anonymous (1940). Charles David Marx. *Civil Engineering* 10(2): 121. *P*
Anonymous (1940). Charles David Marx. *Trans. ASCE* 105: 1785-1789.
Marx, C.D., Wing, C.B., Hoskins, L.M. (1900). Experiments on the flow of water in the six-foot steel and wood pipe line. *Trans. ASCE* 44: 34-91.
Marx, C.D., Hyde, C.G., Grunsky, C.E. (1912). *San Francisco water supply*. Sacramento CA.
Marx, C.D. (1923). Flow in California streams. *Bulletin* 5, State of California: Sacramento CA.
Marx, C.D. (1929). *Causes of partial failure of the Lafayette Dam*. East Bay Municipal District.
http://histsoc.stanford.edu/pdfmem/MarxC.pdf

MASON M.A.

23.04. 1907 Washington DC/USA
12.01. 1982 Lorton VA/USA

Martin Alexander Mason obtained the BSE degree in geology from Washington University in 1931. He was awarded the sixth Freeman scholarship in 1937, thereby visiting the Grenoble Hydraulic Laboratory, then directed by Pierre Danel (1902-1966). He in 1938 obtained the PhD degree for having devised a method on measuring discharge using a salt tracer. He had previously served during six years on the staff of the National Hydraulics Laboratory NHL, the National Bureau of Standards NBS, Washington DC, joining in 1940 the Corps of Engineers Beach Erosion Board. In 1957, he left the Board to become dean of engineering at George Washington University, Washington DC.

During his stay at NHL Mason was concerned with current meters, a basic instrument in hydraulic research, with a paper published with Galen B. Schubauer (1904-1992), who later would greatly contribute to turbulence measurements. Mason then published two papers on the salt velocity method as proposed by Charles M. Allen (1871-1950), which was often applied in prototype experiments, to determine the discharge of turbines or that in mountainous rivers. Once with the Beach Erosion Board, Mason started with the publication of a series of papers, including for instance A study of progressive oscillatory waves in water, which was followed by A summary of the theory of oscillatory waves, co-authored by Morrough P. O'Brien (1902-1988).

Anonymous (1948). Mason, Martin A. *Who's who in engineering* 6: 1288. Lewis: New York.

Mason, M.A. (1938). Rapport sur la mesure des débits d'eau par la méthode Allen. *Rapport* 4. Commission des Mesures Hydrauliques. Grenoble.

Mason, M.A. (1940). Contribution to a study of the Allen salt velocity method of water measurement. *Journal of the Boston Society of Civil Engineers* 27(3): 207-241.

Mason, M.A. (1950). The transformation of waves in shallow water. Proc. 1st Conf. *Coastal Engineering* Long Beach CA: 22-32.

Mason, M.A. (1953). Surface water wave theories. *Trans. ASCE* 118: 546-574.

Mason, M.A., ed. (1957). Symposium *Saline water conversion*. National Research Council. National Academy of Sciences: Washington DC.

Rouse, H. (1976). Dr. M.A. Mason. *Hydraulics in the United States* 1776-1976: 182. *P*

Schubauer, G.B., Mason, M.A. (1937). Performance characteristics of a water current meter in water and in air. *Journal of Research* NBS 18(3): 351-360.

http://www.gogetpapers.com/Explore/Martin_Freeman_4_Lectures/

MASON W.P.

12.10. 1853 New York NY/USA
25.01. 1937 Little Boars Head NH/USA

William Pitt Mason graduated from the Rensselaer Polytechnic Institute RPI, Troy NY, in 1874 as civil engineer. He continued studies in Europe and in the chemical laboratory of Harvard University, becoming assistant in chemistry and natural sciences at RPI in 1875, from where he received the BS degree in 1877. He further received the MD degree from the Albany Medical College in 1881, and there became assistant professor of chemistry in 1882, and in 1885 professor of analytical chemistry. He studied the cholera epidemic at Messina, Sicily, and water supply systems of Europe in 1887. He also was in contact with the famous French Louis Pasteur, taking in addition courses at *Sorbonne*, *Ecole Centrale* and *Ecole Polytechnique* in Paris. In 1893 he studied the water supply systems of London, Glasgow, Paris, Vienna, Rome and the sewage farms of Paris, then leading institutions of the continent. Mason thus developed into the American expert of water chemistry and sanitary engineering, serving as consultant to cities and private institutions throughout his country. He was member of the US Assay Commission, and president of the Hygiene Division of the 8th Intl. Congress of Applied Chemistry held at Washington DC in 1912.

A pioneer in sanitary chemistry, Mason was an unusual combination of chemist, civil engineer and medical expert. He was teacher, scholar, and practising scientist at RPI almost through his entire life. Through his studies of water analysis and water supply, he became a major contributor to the knowledge and understanding of the need for pure municipal water supplies. His publications extended over a period of 40 years moving American cities toward pure water and better public health. He also offered at RPI opportunities for the training of chemists in water analysis and sanitation chemistry. He was member of the American Society of Civil Engineers ASCE, the American Chemical Society ACS, the American Water Works Society AWWS, serving as president in 1909, among many others. He was awarded the LL.D. degree from Lafayette College, Easton PA, and the ScD degree from the Union University, Jackson TE, in 1917.

Mason, W.P. (1890). *Examination of potable water*. Nims & Knight: Troy NY.
Mason, W.P. (1896). *Water supply* considered principally from a sanitary standpoint. Wiley: NY.
Mason, W.P. (1899). *Examination of water*: Chemical and bacteriological. Wiley: New York.
Mason, W.P. (1904). Water supply of Amsterdam, Holland. *Engineering News* 53(17): 437-438.
https://www.rpi.edu/about/alumni/inductees/mason.html *P*

MATTERN D.H.

19.07. 1905 Johnstown PA/USA
17.12. 1989 Knoxville TN/USA

Donald Heckman Mattern obtained the BS degree from Penn State University, University Park PA, in 1926, and the MS degree from Iowa State University, Iowa City IA, in 1928. He was until 1931 engineer at Jackson MI, until 1933 consulting engineer at Keyser WV, joining the Tennessee Valley Authority TVA, Knoxville TN, as design engineer, first as a project planning organiser, and from 1954 to 1970 as chief of the project planning branch, from when he was a water resources consultant. Mattern was elected in 1968 Honorary Member of the American Society of Civil Engineers ASCE and was recipient of its Professional Recognition Award in 1962. He was cited 'For his unselfish devotion to services in the interest of the engineering profession and active participation in all levels of Society affairs'. He served in 1949 as president of the Knoxville Branch, and in 1952 the Tennessee Valley Section. He further helped establish the District Council and served as managing editor of the Tennessee Valley Engineer. From 1954 to 1956 he chaired the 1956 National Convention Committee, and was member of the Committee on the Design of Hydraulic Structures, of ASCE's Hydraulics Division.

Mattern had spent the major part of his career with TVA. One of his first assignments was to investigate alternate sites for the Pickwick Landing Project. He moved through the ranks, gaining expertise, working on the design and construction of virtually all TVA's multiple-purpose hydro projects and steam generating plants. When he retired in 1970, he was head of TVA's project planning branch. His connection with ASCE activities began at Penn State and continued throughout his career. During his time on the Board of Direction, Mattern chaired national committees dealing with membership, employment conditions, and professional conduct.

Anonymous (1963). Donald H. Mattern. *Civil Engineering* 33(10): 72. *P*
Anonymous (1981). Mattern, Donald H. *Who's who in America* 41: 2180. Marquis: Chicago.
Anonymous (1990). D.H. Mattern, TVA engineer, dies. *Civil Engineering* 60(3): 89-90. *P*
Anonymous (1990). Donald H. Mattern. *Trans. ASCE* 155: 529.
Mattern, D.H. (1933). Discussion of Application of duration curves to hydro-electric studies. *Trans. ASCE* 98: 1291-1292.
Mattern, D.H. (1941). Discussion of Hydraulic turbine practice. *Trans. ASCE* 106: 369-371.
Mattern, D.H., Elliot, R.A. (1953). Record-size pump-turbine to be installed at TVA's Hiwassee project. *Civil Engineering* 23(3): 172-175.

MATZKE

28.08. 1910 New York City NY/USA
08.05. 1962 Huntington Station NY/USA

Few information is currently available on Arthur Edward Matzke. He made all studies at Columbia University, New York NY, obtaining the AB degree in 1928, the BS degree in 1929, and the CE degree in 1930. Its Civil Engineering Department then was directed by Boris A. Bakhmeteff (1880-1951), whom Matzke joined as research assistant in 1931. The two were recipients of the 1938 James Laurie Prize of the American Society of Civil Engineers ASCE. He was later chief engineer of the Aviation Products, Kenyon Instrument Co. Inc., Huntington Station NY.

Matzke was the last collaborator of Bakhmeteff. His first work relates to the classical hydraulic jump in rectangular channels. Whereas the so-called sequent depths received interest from the 19th century, this work particularly discusses its length characteristics, including both the lengths of the roller and the jump. It was further attempted to present a generalized free surface profile, based on the sequent depths and the roller length. The discussions significantly expanded the original scope of the paper, thereby including also studies made outside from the USA. The second study published in 1938 analyses gradually-varied open channel flow under adverse bottom slope, a topic considered hardly until then. The standard backwater equation for prismatic channel flow was integrated for hydraulic exponents ranging between 3 and 4, corresponding to cross-sectional channel shapes between rectangular and parabolic. The interest into this work was again reflected by a large number of discussions. As usual in this field until then, the approach was not verified with neither laboratory nor prototype experiments. The third paper was directed to the effect of the bottom slope in hydraulic jumps. The standard knowledge relating to both the sequent depth ratio, and the relative jump length, was expanded with a slope term, which was determined both analytically and experimentally. The approach was generalized only decades later, and still currently is under research.

Anonymous (1936). Arthur E. Matzke. *Civil Engineering* 6(2): 126. *P*
Anonymous (1938). Arthur E. Matzke. *Civil Engineering* 8(1): 51. *P*
Bakhmeteff, B.A., Matzke, A.E. (1936). The hydraulic jump in terms of hydraulic similarity. *Trans. ASCE* 101: 630-680.
Bakhmeteff, B.A., Matzke, A.E. (1938). The hydraulic jump in sloped channel. *Trans. ASME* 60(HYD-1): 111-118.
Matzke, A.E. (1937). Varied flow in open channels of adverse slope. *Trans. ASCE* 102: 651-677.

MAURY

14.01. 1806 Spotsylvania VA/USA
01.02. 1873 Lexington VA/USA

Matthew Fontaine Maury entered in 1822 Harpeth Academy, Franklin TN. In 1825 he was appointed midshipman in the US Navy, making his first cruise in the frigate Brandywine, on the coast of Europe, and in the Mediterranean. Upon return in 1826, he was transferred to the sloop-of-war *Vincennes* for a cruise around the world. He then passed the usual examinations, and in 1831 was appointed master of the sloop-of-war *Falmouth*, then fitting out for the Pacific. He was transferred to the schooner *Dolphin* then, serving as acting first lieutenant, until being again transferred to the frigate *Potomac*, in which he returned to the USA in 1834, and published his first work Maury's navigation, which was adopted as textbook in the Navy.

In 1837, after 13 years of service, Maury was promoted to lieutenant. He offered his knowledge in hydrography and astronomy to the exploring expedition to the South Seas, then preparing to sail, but he declined. In 1839 he met with a painful accident by which he was lamed for life. Being unable to perform the active duties of his profession for several years, Maury devoted the time to study the possible improvements of the Navy. His views were published mainly in the Southern Literary Messenger of Richmond VA, producing great reforms in the Navy, leading to the foundation of the Naval Academy. He further advocated the establishment of a navy-yard at Memphis TN, which was done under the Act of Congress. Under his direction, first series of observations were made on the flow of Mississippi River. Maury proposed a system of observations enabling to collect information, by telegraph, as to the state of the river, to the captains of steamers. He also advanced the enlargement of the Illinois and Michigan Canal, so that vessels of war might pass the Gulf and the Lakes. He further suggested to Congress plans for the disposition of the drowned lands along the Mississippi belonging to the US Government. In 1842 he was appointed superintendent of the later Hydrographical Office, Washington DC. In 1844 he proposed to respect the Gulf Stream, and ocean currents. His works on this topic culminated in the 1856 book. He is considered father of modern oceanography.

Anonymous (1888). Maury, Matthew F. *Appleton's cyclopaedia* 4: 264-266. New York. *P*
Maury, M.F. (1856). *The physical geography of the sea*. Harper: New York.
Maury, M.F. (1860). *Physical geography of the sea and its meteorology*. Harper: New York.
Williams, F.L. (1963). *Matthew Fontaine Maury*: Scientist of the sea. University Press: Rutgers.
https://en.wikipedia.org/wiki/Matthew_Fontaine_Maury *P*

MAVIS

07.02. 1901 Crocketts Bluff AR/USA
02.11. 1983 Macomb IL/USA

Frederic Theodore Mavis obtained the BS degree in
civil engineering from the University of Illinois in
1922, the MS degree in 1926, and the PhD degree in
1935. He was in 1928 a student at the *Technische
Hochschule* Karlsruhe, Germany. A junior engineer
at Chicago IL from 1922 to 1925, he was then a
research assistant at University of Illinois until 1926,
becoming then Freeman Scholar to Germany. Upon
return to the USA he became successively assistant
professor of mechanics and hydraulics, professor,
and head of department at State University of Iowa,
Iowa City IA, when joining the Pennsylvania State
College, University Park PA, in 1939, as professor and head of the Department of Civil
Engineering, from where he moved in 1944 in a similar position to Carnegie Institute of
Technology, Pittsburgh PA. From 1957 to 1967, Mavis was dean of the College of
Engineering and professor of civil engineering at University of Maryland, College Park
MD, retiring as a consultant. He was a member of the Division of Engineering and
Industrial Research NRC from 1955 to 1964, of ASCE and ASME, and of IAHR.

Mavis is known for a number of papers in professional journals, among which is an
account of the hydraulic laboratory at Iowa State University, already in the 1930s a
leading institution in hydraulic engineering in the USA, or research on the flow features
of submerged weirs, which are of relevance if the tailwater flow depth is relatively high.
His prime interest was weir flow for both free and submerged flow conditions. He also
authored and edited books in hydraulic engineering.

Anonymous (1944). Frederic T. Mavis. *Engineering News-Record* 133(Sep.21): 336. *P*
Anonymous (1957). Frederic T. Mavis. *Civil Engineering* 27(6): 29. *P*
Anonymous (1981). Mavis, Frederic T. *Who's who in America* 41: 2186. Marquis: Chicago.
Mavis, F.T. (1935). Hydraulic research at Iowa University. *Engineering News-Record*
    115(Sep.26): 433-437.
Mavis, F.T. (1942). The hydraulics of culverts. *Bulletin* 56. Engineering Experiment Station.
    Pennsylvania State College: University Park PA.
Mavis, F.T. (1949). Submerged thin-plate weirs. *Engineering News-Record* 143(Jul.7): 65-69.
Soucek, E., Howe, H.E., Mavis, F.T. (1936). Sutro weir investigations furnish discharge
    coefficients. *Engineering News-Record* 117(Nov.12): 679-680.
Stelson, T.E., Mavis, F.T. (1957). Virtual mass and acceleration in fluids. *Trans. ASCE* 122:
    518-525; 122: 529-530.

MAXIM

05.02. 1840 Sangerville ME/USA
24.11. 1916 Streatham/UK

Hiram Stevens Maxim studied whatever scientific books came in his way, with talent for drawing and painting and an uncanny facility in handling tools, quickly became an adept at several trades. First he went to Montreal and other cities in Canada to work as carriage painter, cabinet-maker and mechanic. After several years he returned home and secured employment in the engineering works of his uncle at Fitchburg MA. Maxim's genius for invention now came to fruition. During the next years he occupied himself with mainly machines to generate illuminating gas, but also invented a locomotive headlight. In 1878 he was appointed chief engineer of the US Electric Lighting Company, and turned his attention to the incandescent carbon lamp. In 1881 he went to the Paris Exposition to exhibit an electric pressure regulator. He then set up a laboratory close to London UK and remained there permanently. His Maxim Gun Co. eventually became Vickers Ltd., of which he was a director.

Maxim was the equal of any mechanician of his day. He could not use any machine or process without seeking to improve it. He took a total of 122 patents in the USA and 149 in the UK. His range of invention included an improved mouse-trap, automatic gas-generating plants, automatic sprinkling apparatus for extinguishing fires, automatic steam pumping-engines for supplying houses with water, feed-water heaters, steam and vacuum pumps, gas motors among many others. He also invented the automatic gun. Maxim is finally known in flying. His tests started in 1894 in Kent with a technically successful machine, since it lifted itself from the ground. However, he found no time to invent an internal combustion engine and therefore used steam. Despite the extremely light machinery, the weight of fuel and water made the machine impracticable.

Anonymous (1894). H.S. Maxim. *Industries and Iron* 16(Mar.23): 355-357; 20(Feb. 28): 172. *P*
Anonymous (1904). Hiram Maxim. *Illustrierte Aeronautische Mitteilungen* 8(8): 261-262. *P*
Anonymous (1933). Maxim, Hiram Stevens. *Dictionary of American biography* 12: 436-437.
        Scribner's: New York.
Maxim, H.S. (1908). *Artificial and natural flight*. Macmillan: New York.
Maxim, H.S. (1912). *A new system for preventing collisions at sea*. Cassell: London UK.
Maxim, H.S. (1915). *My life*. Methuen: London. *P*
Mottelay, P.F. (1920). *The life and work of Sir Hiram Maxim*. Lane: London. *P*
http://en.wikipedia.org/wiki/Hiram_Stevens_Maxim *P*

MAXWORTHY

21.05. 1933 London/UK
08.03. 2013 Los Angeles CA/USA

Tony Maxworthy obtained the BSc degree in 1954 from University of London, London UK, the MSE degree from Princeton University, Princeton NJ, in 1955, and the PhD degree in mechanical engineering from Harvard University, Cambridge MA, in 1960. He was then until 1967 a scientist and supervisor at the Jet Propulsion Labs, Pasadena CA, from when he was associate professor and professor of aerospace and mechanical engineering until retirement at the University of Southern California, Los Angeles CA. Maxworthy was a member of the NASA Committee of Fluid Mechanics in the 1960s, among many other commitments. He was on the editorial board of journals including Physics of Fluids, Geophysical Fluid Dynamics, and Dynamics of Oceans and Atmospheres. He was a Fellow of the American Physical Society APS and recipient of its 1990 Otto Laporte Award, and member of the American Geophysical Union AGU, and the American Meteorological Society AMS.

The research interests of Maxworthy included the application of basic principles of fluid mechanics to problems of technological and geophysical significance. Later he also worked on rotating stratified flows and unsteady aerodynamics. He was well known for the conception, design, and implementation of key laboratory experiments capturing the essence of important fluid flow phenomena. He was awarded the 2011 Fluid Dynamics Prize 'for outstanding and sustained contributions to fluid dynamics, elucidating stability of fluid interfaces, vortex dynamics, insect flight and, notable, to geophysical and environmental fluid dynamics, including stratified and rotating flows, gravity currents and convective processes'.

Anonymous (1972). Maxworthy, T. *American men and women of science* 12: 4162. Cattell: NY.
Anonymous (1995). Maxworthy, T. *Who's who in engineering* 9: 484. AAES: Washington DC.
Maxworthy, T. (1979). Experiments on the Weis-Fogh mechanism of lift generation by insects in hovering flight. *Journal of Fluid Mechanics* 93: 47-63.
Maxworthy, T. (1983). Experiments on solitary internal Kelvin waves. *JFM* 129: 365-383.
Monismith, S.G., Maxworthy, T. (1989). Selective withdrawal and spin-up of a rotating, stratified fluid. *Journal of Fluid Mechanics* 199: 377-401.
Spedding, G.R., Maxworthy, T. (1986). The generation of circulation and lift in a rigid-dimensional fling. *Journal of Fluid Mechanics* 165: 247-272.
http://news.usc.edu/#!/article/47837/in-memoriam-tony-maxworthy-79/ *P*

McAFEE

27.06. 1881 San Francisco CA/USA
01.01. 1942 San Francisco CA/USA

Lloyd Tevis McAfee began his civil engineering career in 1901, engaged successively from rodman to chief of party on surveys of Shasta County. In 1905 he became construction engineer for the Ocean Shore Railroad Co., Santa Cruz CA, and entered in 1909 the service of San Francisco CA, engaged on construction of cisterns after the 1906 earthquake, storm and sanitary sewers, the Fort Mason pumping station, and the high-pressure pipe system. In 1918 he was appointed engineer on O'Shaughnessy Dam for the Hetch Hetchy Water Supply, a project he was intimately identified for the rest of his career.

In 1913 the Congress had approved a grant of lands in the High Sierra to the city of San Francisco, to permit the development of an adequate water supply. The city constructed then a great system by which water is impounded and stored in reservoirs in the mountains and then conducted 200 km through tunnels and pipelines. Its principal elements include railways, a construction power plant, Lake Eleanor storage dam, Hetch Hetchy reservoir, aqueduct system, and Moccasin Creek power plant. McAfee as construction engineer until 1930 had direct charge of several project divisions. He was then promoted to chief assistant city engineer under Michael M. O'Shaughnessy (1864-1934), with the general supervision of all Hetch Hetchy work, from the mountains to the sea. In 1932 McAfee became chief engineer of the Hetch Hetchy Water Supply, serving also as chief of the Bureau of Engineering of the newly created Public Utilities Commission. McAfee was responsible after O'Shaughnessy's death for San Francisco Airport, and the Exposition building program on the Treasure Island. Until 1940, the work under his supervision excluded responsibility for operation of the electric power division of the Hetch Hetchy Program, but from then his authority was extended to its manager and chief engineer. He thus assumed both the construction and operation of this large water supply project.

Anonymous (1930). Lloyd T. McAfee. *The San Francisco Municipal Record* 4(4/5): 95-97. *P*
Anonymous (1944). Lloyd T. McAfee. *Trans. ASCE* 109: 1502-1504.
Anonymous (2005). *A history of the Municipal Water Department & Hetch Hetchy System*. San Francisco Public Utilities Commission: San Francisco.
McAfee, L.T. (1934). How the Hetch Hetchy Aqueduct was planned and built. *Engineering News-Record* 113(Aug.02): 134-141.
O'Shaughnessy, M.M. (1930). Hetch Hetchy Water Service Project. *The San Francisco Municipal Record* 4(4/5): 99-102. *P*

McALPINE W.H.

22.08. 1874 Lawrence MA/USA
01.11. 1956 Washington DC/USA

William Horatio McAlpine obtained the BS degree from Massachusetts Institute of Technology MIT, Cambridge MA, in 1896, and was then continuously employed in the US Engineering Department on the improvement of rivers and flood control structures. He was principal engineer in charge of construction of locks and dams in Ohio River from Louisville KY to its mouth from 1911 to 1929. Until 1933 he then was head engineer, the Division of Engineering Office, St. Louis MO, connected with the design and the construction of locks and dams along the Upper Mississippi River. Until 1940 he was in charge of the Engineering Section, the Chief of Engineers, Washington DC, from when he was until 1946 assistant to the chief of engineers, US Army Corps of Engineers. McAlpine was member of the American Society of Civil Engineers ASCE, the Society of American Military Engineers SAME, and the Engineers and Architects Club of Louisville KY.

McAlpine was in 1948 chief engineer at the Office of the Chief of Engineers. He was all through his professional career engaged in river regulation and control work for the US Army Corps of Engineers. In that period, he had served on many boards of consultants set up by the Corps, and in 1948 was member of the Board of Consultants for Panama Canal. Previously, he was engaged on the Ohio River, where the Corps of Engineers built major navigation improvements at the Falls in the 1920s. A lock, 180 m long by 33 m wide, was constructed along with a moveable wicket dam and a section of two bear-trap gates at the lower end. Between 1885 and 1929, the Corps built 51 locks and dams along Ohio River, each having a wicket dam. Also during the 1920s, a hydroelectric station was built. The locks and dams on Ohio River served ably during World War II, providing safe movement of crucial petroleum products and military equipment. In the 1960s the McAlpine Locks and Dam on Ohio River was named to honour his skills in river engineering. McAlpine was elected Honorary ASCE Member in 1947.

Anonymous (1948). William H. McAlpine. *Engineering News-Record* 140(Jan.22): 95. *P*
Anonymous (1948). McAlpine, W.H. *Who's who in engineering* 6: 1303. Lewis: New York.
Johnson, L.R., Parrish, C.E. (1999). *Kentucky River development*: The commonwealth's waterway. Louisville District Engineer, USACE, Louisville KY.
McAlpine, W.H. (1934). Roller gates in navigation dams. *The Military Engineer* 26: 419-423.
Parrish, C.E. (2010). *McAlpine locks and dam project history*. USACE: Louisville KY.
http://www.ket.org/cgi-local/fw_comment.exe/db/ket/dmps/Programs?do=topic&topicid=LOUL110025&id=LOUL

McALPINE W.J.

30.04. 1812 New York NY/USA
16.02. 1890 New Brighton RI/USA

William Jarvis McAlpine apprenticed to his father, a mechanical engineer. In 1836 he succeeded as a civil engineer and was involved in the eastern division of Erie Canal. He became chief engineer of the government dry dock at Brooklyn then, where the foundation had to be laid upon a deep layer of quicksand, 12 m below the tide level, posing great difficulty. McAlpine handled the construction and established himself as one of the leading engineers of his time. He served then as chief engineer for railroad projects and also prepared plans on water supply systems for Chicago, Brooklyn, Buffalo, Montreal, Philadelphia and San Francisco between 1850 and 1880. He was further active for a number of outstanding bridge projects, including the Eads Bridge over the Mississippi River at St. Louis or the Washington Bridge over the Harlem in New York.

McAlpine enjoyed wide professional recognition in England and Continental Europe, where he was consulted on many important projects, including the Manchester Ship Canal or the improvement of navigation on Danube River near the Iron Gate. He was the first American honoured by membership in the British Institution of Civil Engineers, was elected president of the American Society of Civil Engineers ASCE in 1870, and awarded honorary membership in 1889. McAlpine was a prolific writer on technical subjects. Contributing next to reports on the various projects, he was involved also in papers published in the Transactions ASCE.

Anonymous (1890). William J. McAlpine. *Minutes* Institution of Civil Engineers 100: 396-400.
Anonymous (1890). William Jarvis McAlpine. *Engineering News* 23(Mar.8): 223-225. *P*
Anonymous (1932). William Jarvis McAlpine. *Civil Engineering* 2(8): 522. *P*
Anonymous (1933). McAlpine, William Jarvis. *Dictionary of American biography* 11: 548-549. Scribner's: New York.
Anonymous (1936). William Jarvis McAlpine. *Civil Engineering* 6(6): 395-397. *P*
Anonymous (1963). McAlpine, W. *Who was who in America* 1607-1896: 342. Marquis: Chicago.
McAlpine, W.J. (1868). The supporting power of piles, and on the pneumatic process for sinking iron columns as practiced in America. *Minutes* Institution of Civil Engineers, London 27: 275-293.
McAlpine, W.J. (1872). Waves of translation in fresh water. *Trans. ASCE* 1: 333-343.
McAlpine, W.J. (1874). Foundation of the new capitol at Albany NY. *Trans ASCE* 2: 287-288.
McAlpine, W.J. (1874). *Modern engineering*. van Benthuysen: Albany NY.

McANEAR

08.09. 1929 Graford TX/USA
15.12. 1995 Vicksburg MS/USA

Clifford Leroy McAnear obtained the BSc degree in civil engineering in 1956 from the Texas Technical University, Lubbock TX. His education continued at the Mississippi State University, Starkville MS, from where he obtained the MS degree. He studied later soil mechanics and foundation engineering at the University of California, Berkeley CA. In 1956 he began his professional career with the US Army Corps of Engineers in the Fort Worth District, where he rose to the position of Chief of Soils, Design. He left in 1966 Fort Worth TX for a new assignment with the District of Jacksonville FL. As district soil engineer he distinguished himself in the study and reconstruction of the Panama Canal, known as Atlantic-Pacific Interoceanic Canal Studies.

McAnear became in 1969 chief of the Soil Mechanics Section, the US Army Engineer Waterways Experiment Station, Vicksburg MS. He was in 1972 promoted to chief of the Engineer Studies Branch, becoming chief of the Soil Mechanics Division in 1973. During his career he was primarily active in the design and construction of earth and rock-fill dams, deep excavated slopes, and foundations. He retired in 1988 from the Federal Government, and continued working as a project manager with a construction company at Vicksburg MS. McAnear was member of the American Society of Civil Engineers ASCE from 1956, and became ASCE Fellow in the 1990s. He served the Vicksburg Branch as president in 1991. He also was member of the US Committee on Large Dams USCOLD, the International Society of Soil Mechanics and Foundation Engineering ISSMFE, and the Society of American Military Engineers SAME. He was a notable civil engineer who greatly added to the knowledge in soil mechanics.

Anonymous (1996). Clifford L. McAnear. *Trans. ASCE* 161: 570.

Johnson, L.D., Anear, C.L. (1974). *Controlled field tests of expansive soils*. US Army Engineer Waterways Experiment Station: Vicksburg MS.

McAnear, C.L. (1969). *Route* 14 *excavation alternatives for Cerro Gordo* (Interoceanic Canal Studies). US Army Corps of Engineers. Jacksonville District: Jacksonville FL.

McAnear, C.L. (1972). *Three-dimensional seepage model study*, Oakley Dam, Sangamon River IL. Miscellaneous Papers. US Army Engineer Waterways Experiment Station: Vicksburg.

McAnear, C.L. (1975). *Evaluation of vertical sand drains* for St. Charles Parish Lakefront Levee, Lake Pontchartrain LA. US Army Waterways Experiment Station: Vicksburg.

http://gsl.erdc.usace.army.mil/gl-history/Chap10.htm *P*

McBIRNEY

14.03. 1884 Conrad IA/USA
14.11. 1975 Denver CO/USA

Harry Raymond McBirney studied civil engineering at the State University of Iowa, Iowa City IA, from 1904 to 1908. He was then draftsman and engineer of the US Reclamation Service at Boise ID until 1917, office engineer for the King Hill Project until 1920, from when he joined until his retirement the US Bureau of Reclamation USBR. He there was progressively in charge of the Sections of the Chief Engineers on standardisation of canal structures and equipment, the design of irrigation distribution and drainage systems including division dams and major bridges with the related field connection to surveys and construction, and specifications of materials. He was a member of the American Society of Civil Engineers ASCE.

McBirney retired in 1950 after nearly a quarter-century having been in charge of canal design for USBR. He was one of the engineering experts most responsible for the high degree of western water conservation. Walter E. Blomgren (1891-1974), then acting chief engineer of the Bureau, stated that 'McBirney, while a design engineer on the Boise Project, developed new and more economical types of canal structures and was largely responsible for the standardization of these designs for the entire Bureau by the chief engineer'. The 1931 Report deals with a then relatively new problem, namely the erosion of a concrete surface by water flow. Hoover Dam was then the highest in the world by which flow velocities increased to above 30 m/s, so that cavitation damage resulted. At that time, this work was premature, because the systematic investigation of these questions started only after World War II. McBirney's 1935 Memorandum contains hydraulic data pertaining to the design of Imperial Dam and Desilting Works of the All-American Canal.

Anonymous (1948). McBirney, Harry R. *Who's who in engineering* 6: 1304. Lewis: New York.
Anonymous (1950). Harry R. McBirney. *Engineering News-Record* 145(Oct.19): 63. *P*
McBirney, H.R., Crocker, E.R. (1931). *Erosion of concrete by clear water flowing at high velocities in open concrete channels and on concrete surface*: Hoover Dam Research, Boulder Canyon Project. USBR: Denver.
McBirney, H.R. (1935). *Memorandum* to chief designing engineer, H.R. McBirney. US Bureau of Reclamation: Denver CO.
McBirney, H.R. (1939). *Madera Canal*: Proposed construction to Fresno River, Friant Division. USBR: Denver CO.

McCARTHY

05.05. 1909 Dover NJ/USA
21.11. 1990 Summit NJ/USA

Gerald Timothy McCarthy graduated in 1930 from Pennsylvania State University, University Park PA, with a BS degree in civil engineering. After his stay with the US Army Corps of Engineers for 8 years, working on various flood control, navigation and power projects, he joined in 1938 an engineering company in New York until 1947, working in Latin America. He then joined the later Tippetts-Abbett-McCarthy-Stratton TAMS, where he was engaged in water resources projects in Burma, Greece, Morocco, the Philippines, and Turkey. He was responsible for design decisions on numerous company projects, was partner in charge of dams in South America, the Middle East, including the *Tarbela* Dam in Pakistan, the *Ziz* Dam in Morocco, or the *Demirkopru* Dam in Turkey. He also was associated with the *Shihmen* multipurpose dam and power programs in Taiwan. He retired from TAMS in 1974 after 24 years, being the acknowledged leader of the firm. He had the vision and imagination that brought the firm to be one of the foremost consultants in the world. He was quick to recognize and promoted talented people.

McCarthy was not only an outstanding civil engineer but also found time to provide leadership to the engineering profession. He presided over the International Commission of Large Dams ICOLD from 1964 to 1967, after having served as vice-president and chairman of the US Committee. He also presided over the American Institute of Consulting Engineers AICE in 1961. The American Society of Civil Engineers ASCE bestowed Honorary Membership on him in 1971. McCarthy was elected to the National Academy of Engineering NAE in 1973. His 1938 paper on the Muskingum Method is a notable contribution to the unit-hydrograph approach used in hydrology.

Anonymous (1960). McCarthy, Gerald T. *American men of science* 10: 2531. Cattell: Tempe AZ.
Anonymous (1961). Gerald T. McCarthy. *Engineering News-Record* 166(Mar.2): 45. *P*
Anonymous (1970). Gerald T. McCarthy. *Water Power* 22(5/6): 165. *P*
Anonymous (1991). Gerald T. McCarthy. *Civil Engineering* 61(2): 70.
Binger, W.V. (1993). Gerald T. McCarthy. *Memorial tributes* 6: 122-125. National Academy of Engineering: Washington DC. *P*
McCarthy, G.T. (1938). The unit hydrograph and flood routing. Conf. *North Atlantic Division* New London CT. US Army Corps of Engineers: Vicksburg.
McCarthy, G.T. (1940). *Engineering construction*: Flood control. The Engineer School: Fort Belvoir VA.

McCLURE

13.12. 1856 Perryville OH/USA
22.06. 1926 Berkeley CA/USA

Wilbur Fisk McClure made engineering studies, yet never obtained a degree. From 1879 to 1882 he was in charge of constructing railway lines in Missouri, Kansas and Arkansas. From 1883 to 1886 he was engaged in surveying oil claims for the Pacific Coast Oil Company, Los Angeles CA, acting additionally as engineer for the Mountain Water Company. He served then until 1893 as chief engineer of the Los Angeles Terminal Railway System, and went to Mexico to survey mining property, but there was contracted with malaria so that he returned to Los Angeles; the effects of this illness never left him. From 1900 to 1912 he was superintendent of the removal of rocks in San Francisco Bay, and city engineer, Berkeley CA. He also served as commissioner of public works from 1909.

McClure was appointed in 1912 to the State Engineering Office of California. From 1921 until his death he was in addition to state engineer also chief of the Division of Engineering and Irrigation, the State Department of Public Works, taking over in 1923 as director. Between 1911 and 1915 the California Irrigation District Act was perfected so that it became practical to proceed with a substantial irrigation development. Some 100 irrigation districts were organized involving some 120,000,000 US$ for construction works. This large scheme was passed under McClure, and it included 3,000,000 acres, so that it was larger than the 1926 Federal irrigation projects combined. The great increase in productivity resulting from successful irrigation made California to one of the top producers of agricultural goods. The 1923 Report on the water resources of the State contained the first complete inventory of Californian waters and future needs to accommodate the full development of its resources. Parallel to this immense program the reclamation of overflow lands in the Sacramento and San Joaquin Valleys progressed rapidly, so that all this work stands as a monument to McClure's administration.

Anonymous (1927). Wilbur F. McClure. *Trans. ASCE* 91: 1106-1109.
Anonymous (1927). Wilbur F. McClure. *Sierra Club Bulletin* 12: 429-436.
Adams, F. (1929). Irrigation Districts in California. *Bulletin* 21. Dept. Public Works: Sacramento.
McClure, W.F. (1918). *Report* on the utilization of Mojave River for irrigation in Victor Valley. California State Printing Office: Sacramento CA.
McClure, W.F. (1921). *Report* on Iron Canyon Project, California. Government Printing Office.
http://freepages.genealogy.rootsweb.ancestry.com/~npmelton/st52.jpg *P*.

McCONAUGHY

28.10. 1882 Atchinson KS/USA
21.10. 1957 Ohio WV/USA

David Charles McConaughy obtained the BS degree in civil engineering from the University of Kansas, Lawrence KS, in 1907. He was from 1908 to 1913 engineer of the Philippine Service, Manila PH, in charge of its water supply and that of Iloilo PH. On his return to the USA, he became design engineer of the US Reclamation Service at Phoenix AZ, and at Denver CO, in charge of various designs of irrigation systems. Until 1928, he then was engineer for the Edison Company in Southern California, in charge of designs of pipelines, tunnels, and other hydraulic structures. He was from 1929 section chief, and in the mid-1940s senior hydraulic engineer of US Bureau of Reclamation USBR, Denver CO, where he was in charge of spillway designs and other hydraulic works.

McConaughy was intimately related with the design of the spillways of Boulder Dam, todays Hoover Dam. The 1933 paper deals with its two side-channels, then a well-known hydraulic structure, but hardly used in so large dam structures. Their discharge capacity was 11,000 m³/s, 50% above the design discharge allowing for the handling of extreme floods. These channels have a particular design involving a diverging trapezoidal cross-section of relatively large bottom slope, which is abruptly broken shortly upstream the tunnel inlets 'to provide increased flow depth, reducing the disturbances caused by the cross flow'. This design basis is currently hardly applied, given the problems with the inlet, but McConaughy's design inspired future inlet works, and pointed at the problems related to high-speed air-water flows. The 1943 discussion deals with air entrainment in open channel chutes, causing flow bulking so that higher side walls are required. At the time, this topic was hardly explored and received fundamental treatment mainly by the research group of Lorenz G. Straub (1901-1963) at St. Anthony Falls SAF Laboratory.

Anonymous (1948). McConaughy, David C. *Who's who in engineering* 6: 1310. Lewis: NY.
McConaughy, D.C. (1933). Spillways in canyon walls to handle floodwaters. *Engineering News-Record* 111(Dec.21): 754-756; 113(Oct.25): 520-522; 114(Apr.4): 480-482.
McConaughy, D.C. (1943). Discussion of Open channel flow at high velocities. *Trans. ASCE* 108: 1484-1493.
McConaughy, D.C. (1944). Discussion Model-prototype conformity. *Trans. ASCE* 109: 148-150.
McConaughy, D.C. (1945). Discussion of Canal cross-sections. *Trans. ASCE* 110: 435-436.
McConaughy, D.C. (1945). Design of spillway chutes. *Civil Engineering* 15(11): 499-500.
http://www.usbr.gov/lc/hooverdam/History/essays/spillways.html (*P*)

McCORMICK

04.11. 1834 Tyrone PA/USA
21.08. 1924 West Mahoning PA/USA

John Buchanan McCormick was apprenticed by his uncle in an old-fashioned cabinet and chair shop. Later he helped with house painting and graining so that he made money which enabled him to develop his future position. In 1873 he went to Brookville PA and there was engaged by a turbine manufacturer. After the tests of the Hercules turbine at Holyoke by James B. Francis (1815-1892), McCormick and his colleague made an agreement with a company of Dayton OH, proving disastrous for them, however, so that McCormick re-joined in 1877 the Holyoke furnisher, remaining there for 10 years. The turbine perfection was a success, with turbine efficiencies above 80%, producing almost 20 different types in size, both right and left hand. Around 1890 McCormick designed and manufactured a turbine 25% stronger as to its diameter than the Hercules turbine, which was referred to as McCormick's Holyoke Turbine. It was also manufactured at York PA and at Dubuque IA. A famous turbine tester stated: 'Mr. McCormick as a designer and perfecter of hydraulic motors stands upon the top rung of the ladder, has stood there for 20 years, without a parallel, not in the United States alone, but upon this planet'.

The so-called American wheel was patented in 1858, and with the further blade-shape contributions of McCormick around 1870, it became the popular forerunner of the modern mixed-flow unit. Why the name of Francis continues to be associated with it presumably stemmed initially from the widespread attention attracted by his book, states Rouse, and then from the resulting adoption of this designation by the German and Swiss firms, which led in its scientific development later in the century.

Anonymous (1913). McCormick. *Indiana County PA*: Her people 2: 848-849. *P*
Francis, J.B. (1855). *Lowell hydraulic experiments*, being a selection from experiments on hydraulic motors on the flow of water over weirs, and in canals of uniform rectangular section and of short length made at Lowell. Little & Brown: New York.
Rouse, H. (1976). John B. McCormick. *Hydraulics in the United States* 1776-1976: 40-41. *P*
Safford, A.T., Hamilton, E.P. (1922). The American mixed-flow turbine and its setting. *Trans. ASCE* 85: 1237-1292.
Stewart, R.J. (1967). *John B. McCormick*: Pathfinder for a new age in water turbines. Indiana University of Pennsylvania: Philadelphia. *P*
Thurston, R.H. (1887). The systematic testing of turbine water-wheels in the United States. *Trans. ASME* 8: 359-420.

McELROY

04.10. 1825 Albany NY/USA
10.12. 1898 Brooklyn NY/USA

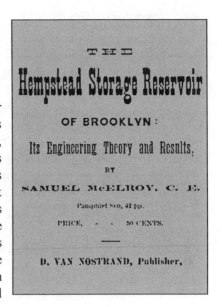

Samuel McElroy designed in 1850 the first water supply for the city of Brooklyn NY using reservoirs at Ridgewood, now at Queens NY. By the 1860s, thousands of houses were connected to sewers draining into the East River. One of his statements was that 'If New York is the active elephants trunk which ministers to a whole nation, these five towns of Kings County lay comparatively supinely on the Bay and Harbour in the form of a huge turtle'. In his later life McElroy served as chief engineer of the Bay Ridge Railways, and became known and an internationally accepted authority on the municipal water and sanitary engineering.

McElroy was a distinguished civil engineer, connected with the US Corps of Engineers. He was also the engineer of numerous water works, canals, harbour works and railways. He was a prominent expert in water power development, and was in his late years the oldest hydraulic engineer in practice in the USA. McElroy was also in charge of the Brooklyn Pumping Engine, based on the principles of the Cornish Engine. It was in 1860 the largest pumping engine of the world. The chief engineer of the water-works was James P. Kirkwood (1807-1877), whereas McElroy was his principal assistant engineer of the pumping machinery and the reservoir. Ridgewood Reservoir was built by Brooklyn which was rapidly outgrowing its local water supplies and a new reservoir at Mount Prospect. McElroy's original design called for three basins, but Kirkwood altered the designs and instead built a double basin, which was built on a hilltop near Ridgewood, Queens. Ground was broken for the reservoir in July 1856 and water was first raised on November 18, 1858. The reservoir was decommissioned in 1989.

McElroy, J.M. (1901). Samuel McElroy. *The Scotch-Irish McElroys in America*: 105-106. Fort Orange Press: Albany NY.

McElroy, S. (1866). Papers on hydraulic engineering. *Journal of the Franklin Institute* 81(3): 145-153.

McElroy, S. (1878). *The Hempstead Storage Reservoir of Brooklyn*: Its engineering theory and results. Van Nostrand: New York.

McElroy, S. (1885). Water-power at Niagara Falls. *Scientific American* Suppl. 20: 8217-8218.

http://www.nytimes.com/1860/02/01/news/scientific-notes-brooklyn-pumping-engine-built-messers-woodruff-beach-hartford.html?scp=32&sq=pumping&st=cse%3Fpagewanted%3D1&pagewanted=all

http://en.wikipedia.org/wiki/Ridgewood_Reservoir

McKEE

09.11. 1914 Pittsburgh PA/USA
22.10. 1979 Newport Beach CA/USA

Jack Edward McKee graduated with the CE degree in 1936 from the Carnegie Institute of Technology, Pittsburgh PA. He then joined the Tennessee Valley Authority TVA, where he became acquainted with Harold A. Thomas, Jr. (1913-2002), and his tutor Gordon M. Fair (1894-1970), whom McKee joined in 1937 at Harvard University, Cambridge MA. He obtained there the MSc degree in 1939, and the ScD degree in 1941, both in sanitary engineering. When McKee returned from war service, he joined the firm of Thomas R. Camp (1895-1971) at Boston MA, as associate, becoming a partner of Camp, Dresser & McKee in 1947.

McKee enjoyed practice and was very good at it. His 1946 study on the pollution control of the Merrimack River, prepared with Camp, was later published as Senate Document, becoming a guideline for the three decades program on this river. In 1949 he joined Cal Tech, Pasadena CA, creating new courses in sanitary engineering, with his research as an integral part of teaching, contributing not only to students' understanding of the environmental engineer's art, but also illuminating obscure areas. He explored the use of molecular filters for bacterial analysis, the fluoridation of drinking water, detection of trace contaminant in public water supplies, and the reclamation and re-use of waste waters as early as in the 1950s. Old-fashioned in some respects, his students knew him as strict disciplinarian, a stickler for detail, and a professor who insisted on the clear, correct use of language, oral and written. In 1970 he became senior vice-president of Camp, Dresser & McKee, and in 1977 was named honorary chairman. He had helped open the Pasadena office, participated in its activities, and continued to serve the firm as a consultant. He was member of the American Society of Civil Engineers ASCE, as Fellow from 1959. He was president of the Water Pollution Control Federation WPCF in 1962, and was in 1969 elected to the National Academy of Engineering NAE.

Anonymous (1962). We need researchers. *Engineering News-Record* 169(Oct.4): 101-102. *P*
Anonymous (1980). Jack E. McKee. *Trans. ASCE* 145: 751-753.
Cleary, E.J. (1984). Jack E. McKee. *Memorial tributes* 2: 205-208. NAE: Washington DC. *P*
McLaughlin, R.T., McKee, J.E. (1958). *On the mechanics of sedimentation*. Caltech: Pasadena.
McKee, J.E. (1961). 100 *problems in environmental health*: A collection of promising research problems. Jones: Washington DC.
McKee, J.E., Wolf, H.W. (1963). *Water quality criteria*. State Water Control Board: Sacramento.

McMATH

28.04. 1833 Varick NY/USA
31.05. 1918 Webster Groves MO/USA

Robert Emmet McMath graduated from Williams College, Williamstown MA, with the degree of AB at age twenty-four. He then went to St. Louis KA to survey the design and construction of improvements on Mississippi River. In 1862 he became assistant engineer of the US Coast Survey, making surveys in Nicaragua first in connection with the interoceanic canal. He then continued as an assistant engineer in the US Corps of Engineers from 1865, and was engaged during the next 20 years with navigation improvements of Mississippi, Arkansas and Illinois Rivers. In 1873 he was made principal civil assistant in charge of special physical investigations. From 1880 to 1883 he was employed by the Mississippi River Commission. From then he was appointed sewer commissioner of St. Louis, a position which he occupied for the next eight years. He was there from 1893 to 1901 president of the board of public improvements, and finally closed his official career as a consultant in Missouri.

McMath was in high esteem among engineers for his qualities. During his connection with the Mississippi River, he was looked upon as the best informed engineer on river hydraulics in the USA. He contributed to professional literature and was probably best known for a formula to determine the proper size of storm-water sewers. It relates the design discharge to rainfall intensity, catchment area, and its average slope. It was in use well until the 20th century, when formulae based on the rational approach were accepted. McMath's formula was also included in famous texts, including the American civil engineer's pocket book of Allen Hazen (1869-1930).

Anonymous (1901). Robert Emmet McMath. *Engineering News* 45(3): 44. *P*
Anonymous (1921). Robert Emmet McMath. *Trans. ASCE*: 83: 2268-2271.
Anonymous (1933). McMath, Robert Emmet. *Dictionary of American biography* 12: 142.
        Scribner's: New York.
McMath, R.E. (1874). *Discussion* of James B. Eads' project for construction of jetties at the
        mouth of the Mississippi River. Washington DC.
McMath, R.E. (1882). The mean velocity of streams flowing in natural channels. *Trans. ASCE*
        11: 186-218.
McMath, R.E. (1887). Determination of the size of sewers. *Trans. ASCE* 16: 179-190.
McMath, R.E. (1888). The waterway between Lake Michigan and the Mississippi River by way
        of the Illinois Canal. *Journal of the Association of Engineering Societies* 7(8): 313-329.

McPHERSON

11.10. 1919 Chicago IL/USA
20.08. 1981 Marblehead MA/USA

Murray Burns McPherson received education from Iowa State University, Iowa City IA, obtaining in 1947 the MS degree. He then continued until 1950 at Columbia University with graduate studies. In 1942 he was junior engineer for the US Engineering Department, Philadelphia PA, joined the US Corps of Engineers, obtained from University of Michigan, Ann Arbor MI a degree in public health. From 1947 to 1951 he was assistant and associate professor of civil engineering at Lehigh University, Bethlehem PA; from 1959 professor of hydraulic engineering at University of Illinois, Urbana IL, moving in 1966 as Director of the ASCE Urban Water Research Program to Marblehead MA. McPherson was in 1951 recipient of the Alfred Noble Robinson Faculty Award, and of the 1959 Distribution Division Award of the American Water Works Association AWWA. He was in the 1950s a member of the ASCE Hydraulics Division Task Force on Energy Dissipators for Spillways and Outlet Works, later taking over as chairman the ASCE Sanitary Engineering Division Task Committee on Water Hammer.

McPherson has written a number of papers in hydraulic structures and urban water resources. His first notable work dealt with the bucket-type energy dissipator, an alternative of the hydraulic jump stilling basin used to convert kinetic energy at the foot of a dam into heat. This type of structure is currently no more used because of cost and the complicated flow structure under variable tailwater conditions. He was interested also in the shaft spillway, which became an alternative spillway type in the 1950s.

Anonymous (1964). McPherson, M.B. *Who's who in engineering* 9: 1264. Lewis: New York.
Anonymous (1966). M.B. McPherson, Director for ASCE sewer research project. *Civil Engineering* 36(3): 90. *P*
Anonymous (1976). M.B. McPherson. *Eos* 57(11): 806. *P*
McPherson, M.B., Karr, M.H. (1957). A study of bucket-type energy dissipator characteristics. *Journal of the Hydraulics Division* 83(HY3, Paper 1266): 1-11; 83(HY4, Paper 1348): 57-64; 83(HY5, Paper 1417): 33-36; 84(HY1, Paper 1558): 43-45; 84(HY5); 41-48.
McPherson, M.B. (1970). *Prospects for metropolitan water management*. ASCE: New York.
McPherson, M.B. (1977). *Research on urban hydrology*. UNESCO: Paris.
McPherson, M.B. (1979). *Challenges in urban runoff control*. State University: Raleigh NC.
Wiseman, R.A., McPherson, M.B. (1965). *A study on the applicability of generalized distribution network head loss characteristics*. University of Illinois: Urbana IL.

MEAD D.W.

06.03. 1862 Fulton NY/USA
13.10. 1948 Madison WI/USA

Daniel Webster Mead received the BC degree in engineering from Cornell University, Ithaca NY, in 1884, and the honorary doctorate from University of Wisconsin, Madison WI, in 1931. He was until 1885 with the US Geological Survey, then until 1887 city engineer at Rockford IL, chief engineer and manager there of a construction company until 1896, from when he was a consultant for hydraulics works. Mead was appointed in 1904 professor of hydraulic and sanitary engineering, University of Wisconsin, retiring from there in 1932. He was a Fellow of the American Institute of Electrical Engineers AIEE, the American Public Health Association and a honorary member of the American Water Works Association AWWA, the American Society of Civil Engineers ASCE, taking over as president in 1936, among many other commitments.

Mead was known both as a practicing engineer and a scientist. He was involved in numerous designs of water works, such as in Rockford IL or Fort Worth TX, in hydro-electric power plants such as in Prairie du Sac WI. He also was a member of the Red Cross commission to China on the flood protection of *Huai* River in 1914, a consulting engineer of the Miami Conservancy District until 1920. He was appointed by President Coolidge to pass on the Boulder Canyon project in 1928. Mead was also a prolific writer of technical texts, including his Notes on hydrology, which was the first book in this field in the USA, or his Hydrology, first published in 1919 and revised in 1950. He also presented books on Water power engineering, and on Hydraulic machinery.

Anonymous (1914). Daniel W. Mead. *Engineering News* 73(Jun.4): 1273. *P*

Anonymous (1939). Mead, Daniel W. *Who's who in America* 20: 1719. Marquis: Chicago.

Anonymous (1948). Daniel W. Mead, educator and consultant, dies. *Engineering News-Record* 141(Nov.4): 12. *P*

Anonymous (1948). Dr. Daniel W. Mead. *The Engineering Journal* 31(12): 656-657. *P*

Mead, D.W. (1904). *Notes on hydrology* and the application of its laws to the problems of hydraulic engineering. Smith: Chicago.

Mead, D.W. (1913). The cause of floods and the factors that influence their intensity. *Journal of the Western Society of Engineers* 18(4): 239-289.

Mead, D.W. (1915). Four thousand years of practical engineering in China. *Wisconsin Engineer* 19(4): 153-181.

Mead, D.W. (1919). *Hydrology*: The fundamental basis of hydraulic engineering. McGraw-Hill.

MEAD E.

16.01. 1858 Patriot IN/USA
26.01. 1936 Washington DC/USA

Elwood Mead graduated from Purdue University, Lafayette IN, in 1882 and received from Iowa State College the degree of civil engineer in 1883. He was appointed instructor of mathematics at Colorado State University, Fort Collins CO, becoming there in 1885 professor of physics and engineering. When the college's Agricultural Experiment Station was established in 1887, Mead served as meteorologist and irrigation engineer. He went to Wyoming in 1888 as territorial engineer becoming widely known as authority on irrigation. He joined in 1899 the US Department of Agriculture as head of its Office of Irrigation Investigations. Mead organized the Department of Irrigation at University of California in 1901. In 1907, he was called to Australia as chairman of the State Rivers and Water Supply Commission of Victoria. Over the following six years he proposed a comprehensive plan of water conservation and reclamation. On his return to the USA in 1915 he rejoined the faculty of University of California as professor of rural institutions and in 1917 became in addition head of California's Land Settlement Board.

The irrigation development in the West raised legal problems, for which the old doctrine of riparian rights was inapplicable. Most western States had developed therefore the doctrine of prior appropriation, by which landowners could divert water for irrigation. But with increasing settlement a novel regulation system became necessary. Mead was instrumental in the State of Wyoming for the new water law, which became a model for the West. In 1924 he was appointed head of the US Bureau of Reclamation, where the design and construction of dams from then became a major activity. He there initiated the design of Boulder Dam on Colorado River. In his honour, the huge storage reservoir was named Lake Mead.

Anonymous (1936). Elwood Mead dies at 78. *Engineering News-Record* 116(Jan.30): 173. *P*
Anonymous (1937). Elwood Mead. *Trans. ASCE* 102: 1611-1618.
Anonymous (1958). Mead. *Dictionary of American biography* 22: 443-444. Scribner's: New York
Mead, E. (1900). The use of water in irrigation. *Bulletin* 86. US Dept. Agriculture: Washington.
Mead, E. (1900). Irrigation studies. *Trans. ASCE* 44: 149-180.
Mead, E. (1903). *Irrigation institutions*: A discussion of the economic and legal questions
          created by the growth of irrigated agriculture in the West. Macmillan: New York.
Mead, E. (1908). An Australian irrigation-ditch water meter. *Engineering News* 59(Mar.26): 346.
Mead, E. (1930). The Colorado River. *Engineering News-Record* 104(6): 240-246. *P*

MEINZER

28.11. 1876 Davis IL/USA
14.06. 1948 Washington DC/USA

Oscar Edward Meinzer received the BA degree from Beloit College in 1901. He became a teacher of physical science at Lenox College, Hopkinton IA, in 1903. There he began the study of geology at the University of Chicago. He joined the US Geological Survey in 1907 to investigate groundwater locating water resources in Utah, Arizona and New Mexico to irrigate previously arid valleys. He became chief of the Division of Ground Water in 1913, a position he held until retirement in 1946.

Meinzer transformed the previously neglected study of groundwater into a science. He realized that, as water became an important natural resource, it was necessary not only to discover underground reservoirs but also to measure their storage capacities and discharges, to arrive at a safe annual yield. He had an orderly mind and a capacity for hard work. He developed methods to estimate the groundwater basin yield, resulting in a standard reference published in 1923, for which he was awarded the PhD title in 1922. He also established a hydrologic laboratory to prove that as long as water flow through granular material is laminar, the velocity is proportional to the hydraulic gradient, conforming with the law of Henry Darcy (1803-1858). He developed geophysical methods for field investigations, including an automatic water-stage recorder on wells. He introduced the Ghijben-Herzberg formula to the USA to estimate the extent of saltwater encroachment in aquifers, in which fresh water is in dynamic equilibrium with seawater, as proposed by Willem Ghijben Badon (1845-1907). The 1930 droughts increased the demand for groundwater. Meinzer supervised geologists and engineers to develop sophisticated tools, methodology and techniques of modern groundwater hydrology, of which science he is the recognized father.

Anonymous (1974). Meinzer, Oscar Edward. *Dictionary of American biography* Suppl. 4: 567-568. Scribner's: New York.

Meinzer, O.E. (1923). The occurrence of groundwater in the United States. *Water Supply Paper* 489. US Geological Survey: Washington DC.

Meinzer, O.E. (1928). Compressibility and elasticity of artesian aquifers. *Journal of Economic Geology* 23(3): 263-291.

Meinzer, O.E., ed. (1942). *Hydrology*. McGraw-Hill: New York.

Meinzer, O.E. (1944). US ground water geologist warns against water diviners. *Water Works Engineering* 97(May31): 571. *P*

Sayre, N. (1948). Meinzer. *Proc. Geological Society of America* 3: 197-206. *P*

MELVILLE

10.01. 1841 New York NY/USA
17.03. 1912 Philadelphia PA/USA

After graduation from Brooklyn Polytechnic Institute George Wallace Melville entered the US Navy in 1861, becoming officer of the Engineering Corps. He finished the Civil War in Virginia, working with torpedo boats. After war he served as first assistant engineer on several ships. He undertook in 1879 an Arctic expedition, trying to find a quick way to the North Pole via Bering Strait, which failed, however. He was awarded the 1884 Congressional Gold Medal.

Melville was appointed by the US President in 1887 chief of the Bureau of Steam Engineering. In the rank of Commodore he was during 15 years responsible for the Navy's propulsion system during an era of force expansion, technological progress, and institutional change. He superintended the design of 120 ships of the 'New Navy'. Among the major technical innovations were the water-tube boiler, the triple screw propulsion system, vertical engines, floating repair ships, and a distilling ship. He was promoted in 1899 to rear admiral, and in 1900 engineer-in-chief. He reformed the service, putting Navy engineers on a professional footing. The Annapolis Laboratory was his brainchild, inaugurated in 1903 as Engineering Experiment Station EES. He thought that EES would increase the efficiency of the Navy. As other laboratories, he wanted to test improvement before installation in prototype ships; in addition the laboratory was a suitable installation for training of young marine engineers, so that it was located close to the Naval Academy, Annapolis MD. The US Navy has named two ships in honour of Melville, namely Destroyer Melville from 1915 to 1948, and in 1964 the oceanographic research ship Melville. The Navy's Melville Award recognizes outstanding engineering contributions in naval engineering, whereas the Melville Medal is awarded by the American Society of Mechanical Engineers ASME to honour the best paper from its Transactions.

Anonymous (1980). Melville, George W. *Mechanical engineers in America* born prior to 1861: 226-227. ASME: New York. *P*

Hirschfeld, F. (1980). George W. Melville. *Mechanical Engineering* 102(7): 24-27. *P*

Melville, G.W. (1884). *In the Lena Delta*: Narrative of the search for Lieut.-Commander DeLong and his companions, followed by proposed method of reaching the North Pole. Boston.

Melville, G.W. (1898). *Views of Commodore G.W. Melville*, chief engineer of the Navy, as to the strategic and commercial value of Nicaraguan Canal, the future control of Pacific Ocean, the strategic value of Hawaii, and its annexation to the United States. Washington DC.

http://en.wikipedia.org/wiki/George_W._Melville *P*

MENOCAL

01.09. 1836 Havana/CU
20.07. 1908 New York NY/USA

Aniceto Garcia Menocal was schooled at Havana, graduating in 1862. He was then assistant engineer for the *Vento* water works, supplying Havana with drinking water until 1865, and until 1869 chief engineer involved in the construction of similar schemes. He moved in 1870 to the USA where he was connected until 1872 with the Department of Public Works, New York NY. He was then until 1874 chief of a survey for the Nicaragua route of the proposed interoceanic canal. As chief engineer of this commission he proposed the construction of a lock canal in Nicaragua, based on various surveys, including the navigation of the Rio San Juan, a canal between Lakes Managua and Nicaragua, the so-called *Tipitapa* Route. He was in 1879 a delegate to the international conference under the auspices of the Geological Society of Paris for considering the proposals for a canal across the Isthmus, thereby made Chevalier of *Légion d'Honneur*, and becoming member of the Provisional Interoceanic Canal Society.

Menocal obtained from Nicaragua a concession for the construction of a canal in 1880. As chief engineer of the Maritime Canal Company of Nicaragua, plans allowed to lapse in 1884. He was charged by the Navy Department with surveying the Nicaragua route but in 1886 was engaged in the study of progress being made on Panama Canal. Despite these works were then under progress under French leadership, mainly by Ferdinand de Lesseps (1805-1894), Menocal obtained a new concession by Nicaragua State so that he began with the preliminary works in 1887. He proceeded with works for the Nicaraguan Canal until 1890, yet failed in efforts to induce the US Government to control assets, as also to obtain funds from Europe. He finally retired from the Navy but was ordered to Havana after the Spanish-American War to take over the Spanish naval properties.

FitzSimons, N. (1972). Menocal, Aniceto G. *Biographical dictionary of American civil engineers*: 90. ASCE: New York.
Menocal, A.G. (1879). *Inter-oceanic canal projects*. ASCE: New York.
Menocal, A.G. (1886). *Report of the US Nicaragua surveying party*. Government Printing: Washington DC.
Menocal, A.G. (1890). *Nicaragua Canal*: Its design, final location and works accomplished. Damrell & Upham: New York.
Menocal, A.G. (1906). The Panama Canal. *Trans. ASCE* 56: 197-218.
http://www.rpi.edu/magazine/fall2006/features/feature3-pg2.html *P*

MERMEL

12.09. 1907 Chicago IL/USA
07.10. 2001 Washington DC/USA

Thaddeus (Ted) Walter Mermel received in 1930 a degree in electrical engineering from the University of Illinois, Urbana IL. He was engaged in 1933 by the US Bureau of Reclamation at its engineering Center, Denver CO, where he participated as design engineer in the development of Hoover Dam first, then Grand Coulee Dam, Shasta Dam and Hungry Horse Dam in the 1940s. With large dams as his career focus, he also served as chairman of the Committee on the World Register of Dams for the International Commission of Large Dams ICOLD. Additional research interests included high-voltage underground cables, rights of way for utility services, and advanced tunnelling methods. In 1973, retiring as assistant to the commissioner for scientific affairs, he was awarded the Department of Interior's Gold Medal Award for distinguished service. Subsequently, the University of Wyoming's College of Engineering accepted his professional papers and records for their archive. In memory of his strong commitment to the engineering profession and his dedication to formal education, Mermel's family established the T.W. Mermel Engineering Scholarship Fund at the University of Wyoming.

Mermel played a major role in the Bureau's international activities in dam design and construction from the 1950s. He encouraged and sought recognition of individuals and groups making scientific and technological contributions to water resources development. During his distinguished career he collaborated with the main exponents of American dam engineers, including Wilbur A. Dexheimer (1901-1974), John C. Page (1887-1955) or Alfred (Al) R. Golzé (1905-1987). The interview with Mermel published in 2007 sheds an excellent light on USBR activities from the 1930s to the 1960s, including his portrait, the organisation of USBR, main dam projects from the perspective of one of its designers, his relations with ICOLD and other international organisations, trips to the Soviet Union and experiences with Russian dam designs, retirement, and memories on old engineering days.

Anonymous (2003). T.W. Mermel. 21st *ICOLD Congress* Montreal 4: 7. *P*
Babb, A.O., Mermel, T.W. (1968). *Catalog of dam disasters*. USBR: Washington DC.
Mermel, T.W. (1958). *Register of dams in the United States*. McGraw-Hill: New York.
Storey, B.A. (2007). *Oral history interviews*: Thaddeus W. Mermel: 1-331. USBR: Denver. *P*
http://www.uwyo.edu/ceas/scholarships/general.html *P*
http://www.uwyo.edu/ceas/scholarships/general.html

MERRICK

02.01. 1871 Wrightstown PA/USA
14.12. 1926 Ann Arbor MI/USA

Howard Benjamin Merrick in 1898 graduated from the College of Engineering, University of Michigan, Ann Arbor MI, as civil engineer. Until 1902 he was engaged by a railways company. He was called by his Alma Mater in 1903 to serve as instructor in surveying, advancing in 1906 to assistant professor and in 1918 to associate professor. From 1922 to his death, he was associate professor of geodesy and surveying.

In 1918 Merrick was among the three individuals selected from the University to go to China, to aid the Chinese Government in the improvement of the Grand Canal. Merrick was in charge of precise level work covering a distance of 500 km. He was transferred in 1919 to the Commission for the improvement of the river systems of *Chihli*, a large province of Northeast China. He there directed a force of some 400 men, among which were 90 engineers. He served there until 1922, yet the Chinese Government was anxious to retain his services. He was presented with a silver loving cup on his departure, engraved with the expression of 'Their high esteem and affection as their chief during three years'. He also was awarded the Fifth Class Chiao Hu Decoration, in recognition of his valuable services. Back to the University of Michigan, it was noted that Merrick was always in a close touch with the students and enjoyed their confidence. His equable temperament and his charming manners will be long remembered. A classmate reports: 'Howard was one of the most esteemed members of our class of '98 engineers'. A student noted: 'I gained Prof. Merrick's friendship while a student in his classes, and have valued it above most of my other friends'. Merrick was member of the Association of Chinese and American Engineers, the Michigan Engineering Society, and of the American Society of Civil Engineers ASCE. He became known as a painstaking and exact worker in his profession, a civil engineer of high standing; he was gentle in manner, but adamant in his advocacy of what he conceived to be right. He was described as one of the finest, best, and kindest ever known, or 'His consideration for others was outstanding'.

Anonymous (1928). Howard B. Merrick. *Trans. ASCE* 92: 1769-1772.
Davis, J.B., Merrick, H.B. (1910). *Direction of a line*. Wahr: Ann Arbor.
Merrick, H.B. (1906). *Plat of Judson's first addition to the city of Ann Arbor*. University of Michigan: Ann Arbor.
http://www.google.ch/imgres?q=%22Howard+B.+Merrick *P*

MERRILL

11.10. 1837 Fort Howard WI/USA
14.12. 1891 Shawneetown IL/USA

William Emery Merrill graduated in 1859 from the US Military Academy and was then assigned to the Corps of Engineers. He was involved in the Civil War, in which he was particularly charged with constructions and fortifications for the protection of railways supplying the army. His military services closed in 1870 as lieutenant colonel, and as chief engineer. As a military engineer he was excelled by none. He published in 1870 a book on iron truss bridges.

The second half of Merrill's career was devoted mainly to river and harbour improvement works carried out by the Corps of Engineers. He originated one of the greatest projects for the development of the American inland waterways, namely the canalization of the Ohio River from Pittsburgh to its mouth. In 1870 he was charged with these improvements, and in 1878, he was sent to Europe on his own request to study similar improvements of non-tidal rivers by means of locks and movable dams. On his return he advocated the method to improve the Ohio and in 1879 succeeded in securing from the Congress an appropriation for the Davis Island lock and dam below Pittsburgh PA. These were completed in 1885 and led to the approval of his project, for the entire river. He lived long enough to build only the first lock and dam. The entire project was not completed until 1929, when the President of the United States took part in the celebration. Merrill published in 1881 on the improvement works of rivers and inland navigation in the United States and in France. He was the US representative in the 1889 Paris meeting of the Permanent International Association of Navigation Congress PIANC.

Anonymous (1892). William Emery Merrill. *Proc. ASCE* 18: 90-93.
Anonymous (1933). Merrill, William Emery. *Dictionary of American biography* 12: 568-569. Scribner's: New York.
Anonymous (1963). Merrill, W.E. *Who was who in America* 1607-1896: 355. Marquis: Chicago.
Merrill, M.E. (1917). W.E. Merrill. *Professional memoirs* 9(48): 639-642. US Army Corps of Engineers: Washington DC. P
Merrill, S. (1917). *A Merrill memorial*: An account of the descendants of Nathaniel Merrill, an early settler of Newbury MA. Cambridge MA.
Merrill, W.E. (1870). *Iron truss bridges for railroads*. van Nostrand: New York.
Merrill, W.E. (1881). *Improvement of non-tidal rivers*. Gov. Printing Office: Washington DC.
USACE (1974). William E. Merrill. *Origins of Ohio River canalization*: 157. P

MERRIMAN M.

27.03. 1848 Southington CT/USA
07.06. 1925 New York NY/USA

Mansfield Merriman graduated as an engineer from Sheffield Scientific School, Yale CT, in 1871, there obtaining the PhD degree in 1876. He was from 1872 to 1873 assistant engineer with the US Corps of Engineers, a civil engineering instructor at his Alma Mata until 1878, and then assistant of the US Coast and Geodetic Survey. He had been appointed in parallel professor of civil engineering at Lehigh University, Bethlehem PA, in 1878, a post he held until his retirement in 1907, becoming from then a consulting engineer in New York City. Merriman was a member ASCE and member of the New York Academy of Sciences. He was awarded honorary PhD titles in 1906 from University of Pennsylvania, Philadelphia PA, and from Lehigh University, Bethlehem PA, in 1913.

Merriman was one of the prominent hydraulic engineers during his era in the United States, mainly known for his books. After his first books on the Method of least squares, and Mechanics of materials, he presented his Treatise on hydraulics, which appeared in a number of editions. The book chapters include: 1. Introduction, 2. Hydrostatics, 3. Theoretical hydraulics, 4. Flow through orifices, 5. Flow over weirs, 6. Flow through tubes, 7. Flow in pipes, 8. Flow in conduits and channels, 9. Flow in rivers, 10. Measurement of water power, 11. Dynamic pressure of flowing water, 12). Hydraulic motors, and 13. Naval hydromechanics. Further works relate to Roofs and bridges, to Higher mathematics, Strength of materials, Precise surveying and geodesy, and finally to Elements of sanitary engineering. Merriman served as the editor in chief of the American civil engineers' pocket book from 1911.

Anonymous (1909). Merriman, Mansfield. *Who's who in America* 6: 1288. Marquis: Chicago.
Anonymous (1915). Prof. Merriman elected president of the ASTM. *Engineering Record* 71(26): 818. *P*
Anonymous (1924). Mansfield Merriman. *Engineering News-Record* 92(16): 649-650. *P*
Anonymous (1926). Mansfield Merriman. *Trans. ASCE* 89: 1658-1662.
Merriman, M. (1884). *Method of least squares*. Wiley: New York.
Merriman, M. (1885). *Mechanics of materials*. Wiley: New York.
Merriman, M. (1889). *Treatise on hydraulics*. Wiley: New York.
Merriman, M. (1898). *Elements of sanitary engineering*. Wiley: New York.
Poggendorff, J.C. (1898). Merriman, Mansfield. *Biographisch-Literarisches Handwörterbuch* 3: 903-904; 4: 990-991; 5: 836; 6: 1707. Barth: Leipzig, with bibliography.

MERRIMAN T.

06.04. 1876 New Haven CT/USA
26.09. 1939 New York NY/USA

Thaddeus Merriman, son of Mansfield, graduated as a civil engineer from Lehigh University, Bethlehem PA, in 1897, there obtaining in 1930 the honorary PhD degree. He was from 1897 to 1899 draftsman on surveys in Nicaragua in charge of the US Canal Commission, then until 1900 assistant engineer of the US Isthmian Canal Commission, designer of the filtration plant for the Jersey Water Co. upon return until 1901, assistant engineer for the masonry dam of Boonton NJ until 1905, and from 1907 to 1918 assistant to the chief engineer for *Ashokan* Reservoir NY, continuing until retirement as deputy chief and chief engineer mainly for the Catskill Water System. He also was chairman of the Engineering Board of Review, Metropolitan Water District of South California for the 400 km long aqueduct from Colorado River to Los Angeles. He further was a Lecturer in hydraulics and water supply engineering at Lehigh University.

Merriman was known for his editorship of the American civil engineers' handbook, a work in collaboration with his father. He also published with his father the Treatise on hydraulics. From 1923 to 1927 he led negotiations with the States of Pennsylvania and New Jersey looking toward allocation of waters from Delaware River. In 1927 he prepared a plan for the development of 540 million gallons water daily for New York City from the *Rondout* and Delaware areas. He was from 1933 to 1936 a consultant engineer of the Board of Water Supply for New York City. He was director of the American Society of Civil Engineers ASCE, of the Franklin Institute, of the New England Water Works Association, and the American Water Works Association.

Anonymous (1922). Thaddeus Merriman. *Engineering News-Record* 88(21): 884. *P*
Anonymous (1929). Thaddeus Merriman. *Engineering News-Record* 103(25): 984. *P*
Anonymous (1932). Thaddeus Merriman. *Engineering News-Record* 108(24): 849. *P*
Anonymous (1939). Merriman, Thaddeus. *Who's who in America* 20: 1738. Marquis: Chicago.
Anonymous (1939). Thaddeus Merriman. *Civil Engineering* 9(11): 699. *P*
Anonymous (1942). Thaddeus Merriman. *Trans. ASCE* 107: 1799-1804.
Burr, W.H., Hering, R., Freeman, J.R. (1903). *Additional water supply for the city of New York*. New York.
Merriman, M., Merriman, T. (1916). *Treatise on hydraulics*, 10th ed. Wiley: New York.
Merriman, T. (1929). Dam construction and design. *Engineering News-Record* 102(6): 213-216.
Merriman, T. (1930). *The American civil engineers' handbook*, 5th ed. Wiley: New York.

METCALF

26.08. 1870 Galveston TX/USA
29.01. 1926 Concord MA/USA

Leonard Metcalf obtained his civil engineering degree from Massachusetts Institute of Technology MIT in 1892. He was first an engineer with a water works company until 1895, then became professor and meteorologist at Hatch Experimental Station, Massachusetts Agricultural College, Amherst MA, until 1897, from when he had a private consulting office at Boston MA, becoming in 1907 a partner of the firm Metcalf & Eddy, hydraulic and sanitary engineers.

The name Metcalf is known for both a successful engineering company within Metcalf & Eddy, as also for the books published. First, the American sewerage practice was issued as a three volumes work containing Design of sewers, Construction of sewers, and Disposal of sewage. This was followed by the American sewerage practice, abridged in 1923. He further published papers mainly in the New England Water Works Association, and in the Transactions of ASCE. Metcalf also was on a sub-committee on Emergency construction of buildings and engineering structures of the National Council for Defence, and advisor to the Construction Division of the USA. Metcalf's outstanding achievements were in the field of municipal water works. In the thirty years of his active career he designed and constructed a number of successful projects, including reservoirs, wells, pumping stations, filters and distribution systems. Metcalf was vice-president of the American Society of Civil Engineers ASCE, and president of the American Water Works Association AWWA, the New England Water Works Association NEWWA, and the Boston Society of Civil Engineers BSCE. He also was instrumental after World War I in the conception and organization of the Affiliated Engineering Societies of Boston.

Anonymous (1915). L. Metcalf. *J. New England Water Works Association* 29(1): Frontispiece. *P*
Anonymous (1923). Metcalf, Leonard. *Who's who in engineering* 1: 871. Lewis: New York.
Anonymous (1925). Metcalf, Leonard. *Who's who in America* 13: 2248. Marquis: Chicago.
Anonymous (1927). Leonard Metcalf. *Trans. ASCE* 91: 1110-1113.
Metcalf, L., Alvord, J.W. (1911). The going value of water-works. *Trans. ASCE* 73: 326-391.
Metcalf, L., Eddy, H.P. (1914). *American sewerage practice*. McGraw-Hill: New York.
Metcalf, L., Eddy, H.P. (1922). *Sewerage and sewage disposal*. McGraw-Hill: New York.
Metcalf, L. (1925). *Manual of American water-works practice*. Williams&Wilkins: Baltimore.
Metcalf, L., Eddy, H.P. (1928). *American sewerage practice*. McGraw-Hill: New York.
Metcalf, L., Eddy, H.P. (1972). *Wastewater engineering*. McGraw-Hill: New York.

MEYER A.F.

28.02. 1880 Cedarburg WI/USA
29.07. 1962 Minneapolis MN/USA

Adolph Frederick Meyer graduated from Wisconsin
University, Madison WI, in 1905 with the BS degree,
and the CE degree in 1909. He was from 1905 to
1912 an engineer of the US Army, then consulting
engineer until 1917, from when he was until 1919
associate professor of hydraulic engineering at the
University of Minnesota, Minneapolis MN, and then
from 1934 to 1935 professor of engineering at the
University of Iowa, Iowa City IA. However, he
remained a consulting hydraulic engineer all through
his career. Meyer was member of the Minneapolis
Engineers' Club.

Meyer was appointed consulting engineer in 1912 to the International Joint Commission
and at the same time opened a consulting engineering office at St. Paul MN, which he
maintained until his death. During his years at the University of Minnesota, he wrote a
textbook on hydrology. His numerous inventions include the Meyer Governor, a machine
controlling the speed of pulp girders. His clients included large industrial corporations,
agencies of federal, state, and local governments, as also individuals. He is remembered
as outstanding person who has particularly contributed to hydrology. His contribution to
the development of the State of Nebraska was acknowledged at a testimonial dinner in
1961. He also received in 1961 the Award of Merit for outstanding achievement and
lasting contributions to the science of water resources conservation from the American
Association for Conservation Information. He was member of the American Society of
Civil Engineers ASCE, of which he was a Fellow from 1959, the American Geophysical
Union AGU, and the American Society of Mechanical Engineers ASME.

Anonymous (1944). Meyer, A.F. *American men of science* 7: 1210. Science Press: Lancaster.
Anonymous (1964). Meyer, Adolph Frederick. *Trans. ASCE* 129: 945-946.
Barr, D.W. (1964). *Memoirs of Adolph F. Meyer*. Minneapolis. P
Meyer, A.F. (1913). Mississippi River high dam at St. Paul and Minneapolis. *Journal of the Engineering Societies* 50(5): 192-211.
Meyer, A.F. (1914). Power development at the high dam between Minneapolis and St. Paul. *Trans. ASME* 36: 255-281; 36: 305-315.
Meyer, A.F. (1915). Computing run-off from rainfall and other physical data. *Trans. ASCE* 79: 1056-1224.
Meyer, A.F. (1917). *The elements of hydrology*. Wiley: New York.
Meyer, A.F. (1940). Discussion of Hydrology of the Great Lakes. *Trans. ASCE* 105: 825-827.

MEYER R.E.

23.03. 1919 Berlin/D
06.01. 2008 Madison WI/USA

Richard Ernst Meyer received education from the Swiss Federal Institute of Technology, ETH Zurich, from where he obtained the mechanical engineering diploma in 1942, and the title Dr. sc. techn. in 1946. He had been from 1945 to 1946 junior scientific officer in mathematics at the British Ministry of Aircraft Production, then for one year lecturer at the University of Manchester UK, from 1947 to 1952 research fellow at Imperial Chemistry Industries, then for three years senior lecturer in aeronautics at the University of Sydney. Meyer was from 1957 to 1959 associate professor of applied mathematics at Brown University, Providence RI, and there professor until 1964. He then joined the Courant Institute of Mathematical Sciences, New York, from where he moved to the National Research Council until 1970. He was senior fellow of the Fluid Mechanics Research Institute, University of Essex UK, and in parallel professor of mathematics at University of Wisconsin, Madison WI, from 1964 until his retirement in 1994.

Meyer was known for his research in asymptotic analysis, partial differential equations, plasma physics, water waves, meteorology, and in gas dynamics. In the latter field he explored supercritical nozzle flows. In water waves theory, the fundamental hydraulics were studied in collaboration with Joseph B. Keller (1923-), including wave refraction and resonance, extending short wave asymptotics to obtain notable advances in the spatial theory of the classical water waves with applications to both coastal and shelf oceanography. He was a member of the Australian Academy of Sciences. Meyer was an individualist who marched to no one's drum but his own.

Anonymous (1994). Meyer, Richard E. *American men and women of science* 19(5): 437. Bowker: New Providence RI.

Meyer, R.E. (1956). Perturbations of supersonic nozzle flows. *Aeronautical Quarterly* 7: 71-84.

Meyer, R.E. (1962). A Liouville theorem in unsteady gas dynamics. *Applied mechanics*: 224-226, F. Rolla, W.T. Koiter, eds. Elsevier: Amsterdam.

Meyer, R.E. (1967). Note on the undular jump. *Journal of Fluid Mechanics* 28(2): 209-221.

Meyer, R.E. (1975). Theory of water-wave refraction. *Advances in Applied Mechanics* 19: 53-141.

Meyer, R.E., ed. (1981). *Transition and turbulence*. Academic Press: New York.

Meyer, R.E. (1982). *Introduction to mathematical fluid dynamics*. Dover: New York.

http://www.siam.org/news/news.php?id=1330

http://science.org.au/fellows/memoirs/meyer.html *P*

MIDDLEBROOKS

13.12. 1903 Yatesville GA/USA
03.02. 1954 Alexandria VA/USA

Thomas Alwyn Middlebrooks received the degree of BSc in 1928 in civil engineering from the Georgia Institute of Technology, Atlanta GA. He studied in 1930 under the direction of Karl Terzaghi (1883-1963) at Massachusetts Institute of Technology MIT, Cambridge MA, and thus was one of the pioneers in the application of modern soil mechanics in the US. He became then chief of the Soil Section, US Army Corps of Engineers, Vicksburg MS. In 1933 he was transferred to Fort Peck MT, responsible for the design and the construction of Fort Peck Dam, then by some standards the largest earth dam worldwide.

Among his many publications, the 1942 paper on the Fort Peck Slide was considered appropriate for being awarded by the James Laurie Medal of the American Society of Civil Engineers ASCE, whose associate member he was from 1935. During World War II, Middlebrooks served as a consultant on many military projects. He was an active member of technical societies, chairing the US delegation at the Second International Soil Mechanics Conference, held in 1948 at Rotterdam NL.

Middlebrooks' work included levee construction and surveys, soil studies relating to the design of levees, determination of the rate of erosion and silting in rivers; soil studies relative to the design of cutoffs, slides and seepage; control of hydraulic fill, rolled fills, and borrow pit selection; and the soil features of the third set of locks at Panama Canal. He received the Exceptionally Meritorious Award by the War Department in 1944. The ASCE Thomas A. Middlebrooks Award was established in 1955. It is made to authors of ASCE papers worthy of commendation for its merit as a contribution to geotechnical engineering. It consists of a certificate and cash prize. Awardees include Charles I. Mansur (1918-2010), John A. Focht, Jr. (1923-2010), or George F. Sowers (1921-1996).

Anonymous (1944). T.A. Middlebrooks, James Laurie Prize. *Civil Engineering* 14(1): 35. *P*
Anonymous (1955). Thomas A. Middlebrooks. *Trans. ASCE* 120: 1572-1573.
Billington, D.P., Jackson, D.C. (2006). *Big dams of the New Deal era*: A confluence of
    engineering and politics. University of Oklahoma Press: Norman OK.
Middlebrooks, T.A. (1942). Fort Peck Slide. *Trans. ASCE* 107: 723-764.
Middlebrooks, T.A. (1948). Seepage control for large earth dams. Proc. 3rd *ICOLD Congress*
    Stockholm Q10(R51): 1-16.
Middlebrooks, T.A. (1950). Earth dams. *Applied sedimentation*: 181-192. Wiley: New York.
http://gsl.erdc.usace.army.mil/gl-history/images/gl_img_8.jpg *P*

MILES J.B.

02.02. 1933 St. Louis MO/USA
19.01. 2012 Columbia MO/USA

John Bruce Miles obtained his BS and MS degrees in mechanical engineering from the University of Missouri-Rolla, and the PhD degree from University of Illinois, Urbana-Champaign. He was there an instructor of engineering mechanics from 1954 to 1958, then instructor of engineering mechanics at the Southern Illinois University, Carbondale IL until 1963, associate professor there until 1968, from when he took over as professor of engineering at the University of Missouri-Columbia, Columbia MO, until retirement in 1988. Miles was a member of the American Society of Mechanical Engineers ASME, the American Institute of Aeronautics and Astronautics AIAA, and the American Society for Engineering Education ASEE. He was recipient of the A.P. Green Award in 1955 from University of Missouri, or the 1985 SAE Teetor Award in aero-technology.

Miles performed the majority of his research in conjunction with NASA Ames space program at Moffett Field CA, continuing in 1970. Previously he had been involved in the design of steam turbines. The 1968 paper on the similarity parameter deals with two parallel co-flowing streams of unequal velocities, a problem of interest in separated flow phenomena. Assuming incompressible flow and equal density fluids with zero pressure gradient, the problem was first dealt with by Henry Görtler ((1909-1987) in 1942. He found the widely accepted error function velocity profile, which depends on a similarity parameter, and found to scatter significantly, so that detailed experiments were conducted, by which a better agreement resulted.

Anonymous (1955). Miles, John B. *Mechanical Engineering* 77(12): 1141. *P*
Anonymous (1985). Miles, John B. *Who's who in Engineering* 6: 452. AAES: Washington DC.
Anonymous (1992). Miles, John Bruce. *Men of achievement* 15: 534-535. IBC: Cambridge UK.
Miles, J.B., Shih, J.-S. (1968). Similarity parameter for two-stream turbulent jet-mixing region. *AIAA Journal* 6(7): 1429-1430.
Miles, J.B., Kim, J.H. (1968). Evaluation of Coles' turbulent compressible boundary-layer theory. *AIAA Journal* 6(6): 1187-1189.
Miles, J.B. (1968). Similarity parameter for two-stream turbulent jet-mixing region. *Journal of the American Institute Aeronautics and Asteroids* 6(7): 1429-1430.
Shi, Y., Isaac, K., Miles, J.B. (1997). Computational fluid dynamics simulation of turbulent waverider flow field with sideslip. *Journal of Spacecraft and Rockets* 34(1): 76-82.
http://www.columbiatribune.com/news/2012/jan/20/john-bruce-miles-1933-2012/ *P*

MILES J.W.

01.12. 1920 Cincinnati OH/USA
20.10. 2008 Santa Barbara CA/USA

John Wilder Miles obtained the BS, the MS, and the PhD degrees from California Institute of Technology, Pasadena CA, from 1942 to 1944. He joined then the Lockheed Aircraft Co., Burbank CA, until 1945, was assistant professor at UCLA until 1949, associate professor until 1955, and professor there until 1961. After having been professor of applied mathematics at Australian National University ANU, Canberra, he took over as professor of applied mechanics and geophysics at University of California, San Diego, in 1965. Miles was chairman of applied mechanics and engineering from 1968 to 1972, and vice-chancellor of academic affairs from 1980 to 1983. During his early career he was a consultant to Northrop Aviation Co., North American Aviation Co., the US Naval Ordnance, Douglas Aircraft Co., or the Aerospace Corp.

The real issue for the applied mathematician, as Miles considered to be himself instead of a fluid dynamicist, is the interaction between mathematical analysis and physical ideas. The primary job of the applied mathematician should be the generation of new physical ideas through mathematical investigations. Miles has indeed done this job excellently with dozens of papers published in top journals testifying his enormous output of physical ideas. He was an Associate Editor of the Journal of the Society of Industrial and Applied Mechanics SIAM, and from 1966 the Journal of Fluid Mechanics. He also served as co-editor of the Cambridge monographs on mechanics and applied mathematics. He was further involved as editor in Annual Review of Fluid Mechanics from 1967 to 1973, and Fluid mechanics – Soviet research. He was the recipient of the Fulbright Awards of the University of New Zealand in 1951, and of the University of Cambridge in 1969. Miles was awarded the 1982 Timoshenko Medal of ASME.

Anonymous (1983). John W. Miles. *Mechanical Engineering* 105(1): 65. *P*

Anonymous (1991). Miles, John W. *Who's who in America* 46: 2272. Marquis: Chicago.

Anonymous (2002). John W. Miles. *Eos* 83(31): 338. *P*

Miles, J.W. (1969). Wave and wave drag in stratified flows. *Applied mechanics*: 50-76, M. Hetényi, W.G. Vincenti, eds. Springer: Berlin.

Miles, J.W. (1974). Laplace's tidal equations revisited. Proc. 7th US Natl. Congress of *Applied mechanics*: 27-38. ASME: New York.

Miles, J.W. (1977). Obliquely interacting solitary waves. *Journal Fluid Mechanics* 79: 157-169.

http://scrippsnews.ucsd.edu/Releases/?releaseID=932 *P*

MILLER C.R.

29.12. 1920 Waltham MA/USA
02.11. 1964 Denver CO/USA

Carl Richard Miller originated from Massachusetts and was educated in New Hampshire. He took over the position of director at the USDA Sedimentation Laboratory, Oxford MS, in 1961, but was already replaced in 1963 obviously due to health problems, which also led to his untimely death at age 44.

Miller was a keen collaborator of Whitney Borland (1905-2001) when at the USBR. They for instance predicted the locations of sediment accumulations throughout the reservoir depth, as described in the 1960 paper. Therein the survey results from 30 US reservoirs were plotted as percentage depth versus percentage sediment deposited. The important factors affecting the deposition modes include the reservoir shape, sediment characteristics, the reservoir operation, the sediment-reservoir volume ratio, and inflow capacity relation. Another common project on the bank and levee stabilization of the Lower Colorado River on the border of Arizona and California dates from 1960. Dams on the Colorado had resulted in river problems caused by water regulation together with the changes in sediment loads. Along the 400 km from Davis Dam to the border with Mexico, the physical conditions of the river varied from a river channel to reservoir reaches and to a low-flow channel conveying essentially drainage water. After closure of Hoover Dam, sediment depositions or aggradations threatened the town of Needles and the Santa Fe Railroad. A Report on these problems was published in 1957, and an ASCE paper followed in 1960.

Anonymous (1960). Miller, Carl R. *American men of science* 10: 2780. Cattell: Tempe AZ.
Anonymous (1964). Carl R. Miller. *ASCE Panel*: Stable channel design. ASCE: New York. *P*
Borland, W.M., Miller, C.R. (1960). Distribution of sediment in large reservoirs. *Trans. ASCE* 125: 166-180.
Carlson, E.J., Miller, C.R. (1956). Research needs in sediment hydraulics. *Journal of the Hydraulics Division* ASCE 82(HY2): 1-33.
Miller, C.R. (1953). *Determination of the unit weight of sediment for use in sediment volume computations.* US Bureau of Reclamation: Denver.
Pemberton, E.L., Strand, R.I. (2005). Whitney M. Borland and the Bureau of Reclamation, 1930-1972. *Journal of Hydraulic Engineering* 131(5): 339-346.
Simons, D.B., Miller, C.R. (1965). *Sediment discharge in irrigation canals.* Fort Collins CO.
Woolhiser, D.A., Miller, C.R. (1963). *Case histories of gully control structures in southwestern Wisconsin.* US Dept. of Agriculture: Washington DC.

MILLIKAN

23.08. 1903 Chicago IL/USA
02.01. 1966 Pasadena CA/USA

Clark Blanchard Millikan graduated in 1924 from Yale College, earning in 1928 his PhD in physics at Caltech under Harry Bateman (1882-1946). He then joined the faculty, becoming in 1940 full professor, succeeding Theodor von Karman (1881-1963) in 1948 as director of the Guggenheim Aeronautical Laboratory at Caltech. Among the first Fellows of the Institute of Aeronautical Sciences IAS, he served as president in 1937, and delivered the third Wright Brothers Lecture at Columbia University, New York NY in 1939. He presented the 1957 Wilbur Wright Lecture at the Royal Aeronautical Society, London.

During his early career at CalTech, Pasadena CA, Millikan became one of the most outstanding experts in wind-tunnel experimentation. He directed design, construction, and operation of the CalTech Cooperative Wind Tunnel, supervising many scientific experiments that helped bring about greatly improved performance and efficiency of military and commercial aircraft. This activity earned him the Presidential Medal and the King's Medal of Great Britain. During Word War II and the following years, Millikan's interest turned to guided missiles and rockets. Both his theoretical and experimental work led to improvements in reaction propulsion and in the flight at supersonic speeds. Around 1950 he was chairman of the Guided Missiles Committee. He also served as chairman of the Space Systems Division Advisory Group of the Air Force Scientific Advisory Board, and was a member of the Army Ballistic Research Laboratory's Scientific Advisory Committee.

Anonymous (1949). Honorary American Fellow for 1948: Clark B. Millikan. *Aeronautical Engineering Review* 8(3): 22. *P*
Anonymous (1966). Clark B. Millikan. *Journal of Aeronautics and Astronautics* 4(1): 14-15. *P*
Karman von, T., Millikan, C.B. (1934). *On the theory of laminar boundary layers involving separation*. US Government Printing: Washington DC.
Millikan, C.B. (1931). *An extended theory of thin airfoils and its application to the biplane problem*. US Government Printing: Washington DC.
Millikan, C.B. (1941). *Aerodynamics of the airplane*. Wiley: New York.
Sechler, E.E. (1979). Clark Blanchard Millikan. *Memorial tributes* 1: 209-213. National Academy of Engineering: Washington DC. *P*
http://en.wikipedia.org/wiki/Clark_Blanchard_Millikan
http://calteches.library.caltech.edu/253/1/clark.pdf *P*

MILLS

01.11. 1836 Bangor ME/USA
04.10. 1921 Hingham MA/USA

Hiram Francis Mills graduated as a civil engineer from the Rensselaer Polytechnic Institute, Troy NY. During the next twelve years he was engaged with tunnel construction and dam erection on *Deerfield* River and *Penobscot* River. He conducted studies relating to water-power developments. In 1868 he opened his own office at Boston MA, soon attaining a high place in his profession. Although his principal work was done in Massachusetts, he was also a consultant in the USA and Mexico. He for instance was chief engineer of Essex Company, Lawrence, and of the Locks and Canals Company, Lowell MA.

At Lowell, Mills undertook a series of experiments on the water flow in natural and artificial channels, which led to the perfection of the piezometer and also advanced the knowledge on turbines. Due to lack of time, he had to abandon these activities, however, and his conclusions were only systematized posthumously in the 1923 book. As consultant to the Boston Metropolitan Water and Sewerage Board, he was responsible for the design of water supply, drainage, and sewer systems. When the Massachusetts State Board of Health was organized in 1886, he accepted the committee chair, a position which he held almost to his death. Much of the success of the Board was due to his efforts, because he not only standardized methods of sampling and analysis but also initiated tests on the purification of water and sewage. From these activities emerged the great Lawrence Experiment Station, which has long been recognized as the foremost in America. Mills also designed the slow-sand filter which marked the beginning of an new era in municipal engineering. He was a successful man of business, director and president, and accumulated a fortune which he used to found educational institutions.

Anonymous (1921). Hiram F. Mills dead. *Engineering News-Record* 87(15): 627-628. *P*
Anonymous (1924). Hiram F. Mills. *Memoirs* American Academy of Arts and Science 15: 1. *P*
Anonymous (1924). Hiram Francis Hills. *Trans. ASCE* 87: 1299-1302.
Anonymous (1934). Mills, H.F. *Dictionary of American biography* 13: 8. Scribner's: New York.
Mills, H.F. (1878). Experiments upon piezometers used in hydraulic investigations. *Proc. American Academy of Arts and Sciences* 14(1): 26-53.
Mills, H.F. (1893). Purification of sewage and of water by filtration. *Trans. ASCE* 30: 350-366.
Mills, H.F. (1923). Flow of water in pipes, with historical and personal note by John R. Freeman and introductionary outline by Karl R. Kennison. *Memoirs of the American Academy of Arts and Sciences* 15(2): 58-236. Lancaster. *P*

MILLSAPS

10.09. 1921 Birmingham AL/USA
19.12. 1989 Gainesville FL/USA

Knox Millsaps obtained in 1944 the PhD degree in mathematics and theoretical physics from California Institute of Technology, Berkeley CA. Previously he had earned a BS degree in English literature from the Auburn University, Auburn AL. He moved then through a number of appointments until arriving at the University of Florida UF, Gainesville FL, in the 1960s. He was in addition chief mathematician of Wright Air Development Center, and chief scientist of Air Force Missile Development Center, Holloman NM. He was further executive director of the Air Force Office of Scientific Research, Arlington VA, and chief scientist of its Office of Aerospace Research.

Millsaps was chief mathematician at the ARL Laboratory from 1952 to 1955. From 1956 to 1960 he held the post of chief scientist. During the term as executive director of the Air Force Office of Scientific Research from 1960 to 1963 he also served as chief scientist in the Office of Aerospace Research. His experiences in academe included professorships at MIT, Cambridge MA, at CSU, Fort Collins CO, and at Ohio State University, Columbus OH. At the time of death he was a professor at the Department of Aerospace Engineering, Mechanics, and Engineering Science at the Gainesville College of Engineering. His research interests included fluid mechanics and heat transfer, thereby also serving as chairman of the UF Engineering Sciences Department. He was in addition highly active in a number of professional organizations, including the American Institute of Aeronautics and Astronautics AIAA, the Society of Engineering Sciences, the American Mathematical Society AMS, and the American Physical Society APS.

Anonymous (1990). Millsaps, Knox. *SIAM News* 23(3): 2. *P*

Fearn, R.L., Shyy, W. (2003). *Aeronautical and aerospace engineering at the University of Florida*: 1-26. Gainesville. *P*

Millsaps, K., Pohlhausen, K. (1952). Heat transfer by laminar flow from a rotating plate. *Journal of Aeronautical Sciences* 19(2): 120-126.

Millsaps, K. (1963). A note on the statistical theory of turbulence. *Rider Anniversary Volume*: 149-157. Aeronautical Research Laboratories: Gainesville.

Millsaps, K., Soong, N.L. (1965). Thermal distribution in a round laminar jet. *Physics of Fluids* 8(1): 200-201.

http://archives.caltech.edu/search_catalog.cfm?results_file=Detail_View&recsPerPage=1&first RecToShow=17&search_field=Paul+B.+MacCready&entry_type=&photo_id=&cat_series= *P*

MITCHELL H.

16.09. 1830 Nantucket MA/USA
01.12. 1902 New York NY/USA

Henry Mitchell received his education in private schools; he entered in 1849 the US Coast Survey, where he was engaged by its Hydrographic Branch. In 1854, entrusted with carrying out a tidal survey of Nantucket and Vineyard Sounds, he devised a tide-gauge for use on open coast. During the next 36 years he was engaged in similar works and in studies in physical hydrography, the results of which were published in papers, mainly in the Annual Reports of the Coast Survey between 1854 and 1888. He also served as chief of the Department of Physical Hydrography.

In his harbour investigations Mitchell was concerned with tides and the current regime related to the hydrographic features. He elucidated successfully the complex forces at work in maintaining or changing the channels or shore lines of the harbours. His most important scientific works include Tides and tidal phenomena, Location of harbour lines, Physical survey of the Delaware River, or the Under-run of the Hudson. He became already in the 1860s a leading hydrographic engineer, collaborating with Julius E. Hilgard (1825-1891), then in charge of the Coast Survey Office. He also was member of the Advisory Council of the Massachusetts Board of Harbor Commissioners. In 1868 he was sent to Europe to study its hydrographic and engineering progress, and while there he made an inspection of the Suez Canal. In 1874 he was appointed member of the Commission on the Construction of the Oceanic Ship Canal proposing routes through Nicaragua and the Isthmus of Darien. He also was involved in the survey of the mouth of Mississippi River, and on the construction of the Mississippi Jetties. He received numerous awards from the scientific community, including his election to the National Academy of Sciences in 1875. He published shortly before his death a biographical sketch of Ferdinand de Lesseps (1805-1894), the builder of Suez Canal.

Anonymous (1931). Henry Mitchell. *Trans. ASCE* 95: 1560-1561.
Marmer, H.A. (1938). Henry Mitchell. *Biographical memoirs* 20: 141-150. National Academy
      of Sciences: Washington DC. *P*
Mitchell, H. (1867). *On recent soundings in the Gulf Stream*. National Academy of Sciences.
Mitchell, H. (1868). *Tides and tidal phenomena*: For use more particularly of US naval officers.
      US Navy Department, Bureau of Navigation: Washington DC.
Mitchell, H. (1869). *The coast of Egypt and the Suez Canal*. Fields, Osgood & Co.: Boston.
Mitchell, H., Mitchell, M. (1902). *Henry Mitchell Papers* 1864-1900. Nantucket MA.

MITCHELL W.S.

11.10. 1857 St. Louis MO/USA
27.01. 1932 St. Louis MO/USA

William Selby Mitchell studied engineering at the University of Charlottesville VA, but finally did not graduate. He entered in 1878 government service as rodman, and successively became instrumentman and recorder, and in 1881 assistant engineer of the US Engineer Office, St. Louis MO. He remained in this position for 23 years, when being appointed in 1904 principal assistant engineer of the St. Louis Office, being engaged there until 1917. During these years he had charge of all surveys and the collection of the physical and hydrographic data in St. Louis District. He was in charge of river improvements and the construction used for navigation, including dikes, dams, jetties and revetments. He further had the charge of the Mississippi River snag boats and snagging operations between the mouth of the Missouri River and New Orleans LA. He designed for the Mississippi River service a fleet of four standard all-steel towboats, and nineteen of the largest steel barges used then. He also served as engineer and member of the US Board on Experimental Tow Boats conducting hydrographic missions.

Mitchell authored several engineering reports to the War Department pertaining to the Mississippi River, including Measurement of stream flow by rod-floats and meters, or Design and construction of towboats and barges. In 1909 he prepared the Memorandum on a project for maintaining by hydraulic dredging alone a navigable channel 14 feet in depth by 500 feet in width in the Mississippi River between St. Louis MO, and the mouth of the Ohio River. It was stated that 'Mitchell had few equals and no superior in solving the intricate problems of regulation and dredging presented by the unruly middle Mississippi River. His greatest service to the government he served so long was in the design of hydraulic dredges and other floating plant, for which he had unusual talent'. Another of his head stated: 'His physical work on the Mississippi River was noteworthy and permanent. Still more noteworthy in my opinion was his influence upon his fellow-workers, particularly the officers of the Corps of Engineers, to whom he acted as a counsellor and a friend. So training the character of many men who have taken an influential part in the life of the United States, I believe that Mr. Mitchell's greatest work has been that of intellectual guidance and character training of the engineers'.

Anonymous (1933). William S. Mitchell. *Trans. ASCE* 98: 1591-1595.
Dobney, F.J. (1977). William S. Mitchell. *River engineers on the Middle Mississippi*: 82. P
www.dnr.mo.gov/shpo/nps-nr/85003102.pdf

MOCKMORE

07.11. 1891 Platte Center NE/USA
11.04. 1953 Corvallis OR/USA

Charles Arthur Mockmore was educated at State University of Iowa, Iowa City IA, from where he obtained the BE degree in 1920, in 1926 the CE degree and in 1935 the PhD degree. He was from 1921 to 1923 civil engineering instructor at Oregon State College, Corvallis OR, until 1934 associate professor, from when he continued as professor and head of the Department of Civil Engineering. He had been a research associate at University of Iowa from 1931 to 1932. He was in 1939 awarded the Croes Medal from the American Society of Civil Engineers ASCE, of which he was also a member.

Mockmore contributed a number of papers to journals, including his work on draft tubes, for which he was awarded in 1939. The 1943 paper was one of the first large researches on channel flow in bends. The test set-up consisted of two 180° bends in a rectangular channel in which the velocity distribution was measured using a Pitot tube. The results demonstrate the existence of secondary flow along with flow separation along the inner channel wall. This peculiar finding may be observed in river bends, in which there is a graduation of sediments along a curve, with large sediment sizes located at the outer curve side and smaller at the inner side. The helicoidal flow structure may be described with a surface current from the inner to the outer walls, and a bottom return flow along the slower paths due to roughness effects. The super-elevation of flow along the outer bend wall was also investigated. Another work of Mockmore analyzed the performance and efficiency of the Banki turbine, as proposed by Donat Banki (1859-1922) in 1918.

Anonymous (1924). Mockmore, Charles Arthur. *Who's who in America* 26: 1917. Marquis.
Anonymous (1940). Charles A. Mockmore. *Civil Engineering* 10(1): 55. P
Anonymous (1953). Charles A. Mockmore. *Engineering News-Record* 150(Apr.23): 74.
Anonymous (1954). Mockmore, C.A. *Who's who in engineering* 7: 1680. AAES: Washington DC.
Mockmore, C.A. (1935). *Hydraulic machinery*. Oregon State College: Corvallis OR.
Mockmore, C.A. (1944). Flow around bends in stable channels. *Trans. ASCE* 109: 593-628.
Mockmore, C.A. (1948). *Digest of Oregon land surveying laws*. Engineering Experiment Station, Oregon State College: Corvallis OR.
Mockmore, C.A., Merryfield, F. (1949). The Banki water turbine. *Bulletin* 25. Engineering Experiment Station, Oregon State College: Corvallis OR.

MOLITOR

16.08. 1866 Detroit MI/USA
08.09. 1939 Harlingen TX/USA

David Albert Molitor graduated as a civil engineer from Washington University, St. Louis MO, in 1887. He was then employed as an engineer on design and construction of strategic railways in Baden County, Germany, until 1890. Upon return to the USA he was an assistant engineer for the Mississippi Bridge, Memphis TN. He then entered the US Engineering Department serving in various capacities as design and superintending engineer from 1892 to 1898. These works were connected with the Sault Ste. Marie Falls Canal, and the canals through the Great Lakes. He conducted precise levelling operations for the US Board of Engineers on deep waterways, St. Lawrence River until 1899, when becoming a consultant until 1906. He was then a design engineer for the Panama Canal at Washington DC until 1908, thereby visiting the site on the Isthmus. From 1908 to 1911 Molitor was a professor of civil engineering at the Cornell University, Ithaca NY, from when he returned to his consulting office at Toronto ON. In parallel he was also involved in the design of Toronto Harbour. Molitor was a member of the American Society of Civil Engineers ASCE, and of the Washington Academy of Sciences.

Molitor is known for his 1908 book on weirs and sluices. This book is quite different from the usual textbooks then published in the USA, clearly reflecting his association with Germany. It is dedicated to Gustav von Wex (1811-1892), who had worked along similar lines as proposed by Molitor. The text is of old style using outdated equations but still attracted numerous engineers for which hydraulics was often not the main concern in their activities.

Anonymous (1921). Molitor, David Albert. *Who's who in America* 11: 2002. Marquis: Chicago.
Anonymous (1937). Molitor, David A. *Who's who in engineering* 4: 956. Lewis: New York.
Anonymous (1941). David Albert Molitor. *Trans. ASCE* 106: 1628-1632.
Molitor, D.A. (1908). *Hydraulics of rivers, weirs and sluices*: The derivation of new and more accurate formulae for discharge through rivers and canals obstructed by weirs, sluices, etc. according to the principles of Gustav Ritter von Wex. Wiley: New York.
Molitor, D.A. (1911). *Kinetic theory of engineering structures*. McGraw-Hill: New York.
Molitor, D.A. (1918). Discussion of Verification of the Bazin weir formula by hydro-chemical gaugings, by F.A. Nagler. *Trans. ASCE* 83: 164-168.
Molitor, D.A. (1935). Wave pressures on sea-walls and breakwaters. *Trans. ASCE* 100: 984-1017.
http://thm-a01.yimg.com/nimage/a6c0dcc6b8e4ca20 *P*

MONROE

13.05. 1889 Willow Ranch CA/USA
03.10. 1977 Knoxville TE/USA

Robert Ansley Monroe received education from the College of Civil Engineering, University of California San Francisco, with a BS degree in 1908. He was an engineer with the Pacific Gas & Electric Co. from 1912 to 1917, and there returned from 1919 to 1929 after war service, joining then until 1933 as engineer the Aluminium Co. of America, Pittsburgh PA. He was senior engineer until 1937 at the US Bureau of Reclamation USBR, Denver CO, and later assistant to the chief, Water Control Planning Engineers, of the Tennessee Valley Authority TVA, Knoxville TN, until his retirement in 1959. He was a member of the American Society of Civil Engineers ASCE, becoming ASCE Fellow in 1959. He also was president of the Knoxville Technical Society.

Monroe, finally chief design engineer of the TVA, had collaborated with this important authority for over twenty years. During this period he was involved in a large number of applied projects dealing with problems of hydraulic structures, dam engineering, water resources management and soil mechanics. He was particularly involved in the design of Norris Dam and Wheeler Dam. His 1929 paper co-authored by I. Cleveland Steele (1886-1973) may be considered one of the very first dealing with baffle piers as an appurtenance in stilling basins. In contrast to the Rehbock sill as proposed by Theodor Rehbock (1864-1950), Monroe's design relates to an element located at the basin center thereby actively reducing the tailwater elevation required and stabilizing the hydraulic jump under various flow conditions. As the approach flow velocity increases, cavitation damage has to be countered. Once having been appointed chief design engineer, he directed the engineering and architectural designs for all TVA construction projects during fourteen years. After retirement he served as consultant on a number of projects, notably the Wells Dam on the Columbia River in Washington State.

Anonymous (1941). Monroe, Robert A. *Who's who in engineering* 5: 1236. Lewis: New York.
Anonymous (1959). Robert A. Monroe. *Civil Engineering* 29(7): 23. *P*
Anonymous (1979). Robert Ansley Monroe. *Trans. ASCE* 144: 572-573.
Monroe, R.A., Templin, R.L. (1932). Vibration of overhead transmission lines. *Trans. AIEE* 51(12): 1050-1073.
Monroe, R.A. (1938). Disc. Economic diameter of steel penstocks. *Trans. ASCE* 103: 106-107.
Steele, I.C., Monroe, R.A. (1929). Baffle pier experiments on models of Pit River Dam. *Trans. ASCE* 93: 451-546.

MOODY

05.01. 1880 Philadelphia PA/USA
18.04. 1953 Plainfield NJ/USA

Lewis Ferry Moody graduated from the University of Pennsylvania, Philadelphia, in 1902 as mechanical engineer. He there was an instructor in mechanical engineering until 1904, joining the engineering staff of the hydraulic department of a firm in Philadelphia until 1908, when becoming assistant professor, later professor of mechanical engineering at Rensselaer Polytechnic Institute, Troy NY, until 1916. He was in parallel a consultant all through his career. He joined the faculty of Princeton University in 1930 as professor of hydraulic engineering. Moody was a member and Honorary Member ASME. He was awarded the Elliott Cresson Medal by the Franklin Institute in 1945.

Moody had been associated with various research projects in the field of hydraulic engineering. He conducted notable tests on Pitot tubes, current meters, cavitation and turbines, developing new types of turbines and pumps. He was granted over 90 patents for inventions in this field. He also served on the Advisory Committee of the National Hydraulic Laboratory for the US Bureau of Standards. Moody's name is related to a diagram showing the pipe friction coefficient versus the Reynolds number for a range of relative sand roughness heights. This relation was originally tested experimentally by Ludwig Prandtl (1875-1953) and Johann Nikuradse (1894-1979), and later analyzed by Cyril Frank Colebrook (1910-1997) and Cedric Masey White (1898-1993). Moody did not intend to offer a new relation, but his plot has become familiar among hydraulic engineers because of its simplicity in practical applications.

Anonymous (1918). L.F. Moody. *Journal ASME* 40(3): 253. *P*
Anonymous (1953). Lewis Ferry Moody. *La Houille Blanche* 8(8): 553-554. *P*
Anonymous (1954). Moody, Lewis F. *Who's who in engineering* 7: 1690. Lewis: New York.
Angus, R.W., Gibson, N.R., Kerr, S.L., Moody, L.F., Pirnie, M. (1937). Water hammer symposium. *Power* 81(14): 74-77. *P*
Moody, L.F. (1915). Wicket gates the logical development for hydraulic turbine regulation. *Engineering Record* 72(12): 358-360.
Moody, L.F. (1944). Friction factors for pipe flow. *Trans. ASME* 66(11): 671-684.
Moody, L.F. (1947). An approximate formula for pipe friction factors. *Mechanical Engineering* 69(12): 1005-1006.
Taylor, H.B., Moody, L.F. (1921). The hydraulic turbine in evolution. *Mechanical Engineering* 44(10): 633-640.

MOORE

12.03. 1916 Estrella CA/USA
24.03. 2009 Albuquerque NM/USA

Walter Leon Moore obtained the BS degree from California Institute of Technology in 1937, and the MS and PhD degrees from Iowa State University, Iowa City IA, in 1938 and 1951, respectively. He was from 1939 to 1940 a junior engineer with the Soil Conservation Service SCS, research analyst with the Lockheed Aviation Corp. until 1947, when joining University of Texas, Austin TX, becoming there professor of civil engineering in 1953 and department chairman from 1958 to 1965. He was from 1975 president of Moore & Sethness Inc. He was member of the American Society of Engineering Education ASEE, the American Geophysical Union AGU, the American Society of Civil Engineers ASCE, and the International Association of Hydraulic Research IAHR. Moore was the recipient of the 1945 ASCE Collingwood Prize for Juniors.

Moore presented papers which have become classic. The 1943 work relates to so-called free overfalls, resulting from an abruptly ending channel, with water flow discharging into the atmosphere. The original paper is one third of the entire publication including a large discussion section stating interest into this basic flow phenomenon but also its impact in advancing a certain question. The 1959 paper is another example in which the authors address a problem not having been advanced for years, receiving considerable interest. Hydraulic jumps may be fixed in their location by appurtenances, such as drops, sills, or blocks. Four types of jumps at abrupt drops are described, and the main features analyzed then based on simplifying assumptions relating to the pressure distribution at the jump vicinity. The results were validated using hydraulic modeling.

Anonymous (1945). Walter L. Moore. *Civil Engineering* 15(1): 44. *P*
Anonymous (1956). Walter L. Moore. *Civil Engineering* 26(12): 850. *P*
Anonymous (1959). Moore, Walter L. *Who's who in engineering* 8: 1749. Lewis: New York.
Anonymous (1981). Moore, Walter L. *Who's who in America* 41: 2361. Marquis: Chicago.
Masch, F.D., Moore, W.L. (1963). Drag forces in velocity gradient flow. *Trans. ASCE* 128(1): 48-58; 128(1): 63-64.
Moore, W.L. (1943). Energy loss at the base of a free overfall. *Trans. ASCE* 108: 1343-1392.
Moore, W.L., Morgan, C.W. (1959). Hydraulic jump at an abrupt drop. *Trans. ASCE* 124: 507-516; 124: 521-524.
Moore, W.L., Masch, F.D., Jr. (1962). Experiments on the scour resistance of cohesive sediments. *Journal of Geophysical Research* 67(4): 1437-1446.

MORGAN A.E.

20.06. 1878 Cincinnati OH/USA
16.11. 1975 Yellow Springs OH/USA

Arthur Ernest Morgan took up at St. Cloud MN civil engineering studies, qualifying to practice in 1902. His first work was as supervising engineer for the US Drainage Investigations in the south, designing reclamation projects. He formed his own engineering company in 1909, which reclaimed two million acres in several states. He went to Dayton OH to set up in 1913 the Miami River Conservancy District for flood prevention. In 1917 he founded there a school which joined Antioch College, Yellow Springs OH, of which he became president in 1920. He reorganized its educational structure known as the Antioch Plan, alternating study with work. When he left to head up the Tennessee Valley Authority TVA in 1933, the student body had increased to 560 with 150 cooperating employers.

Early in his career Morgan engineered the Miami Conservancy District, the first flood control program of its kind in the USA, and a cornerstone of hydrologic engineering. Five days after torrential rainfall which had devastated the valley, wrecking homes, and killing people, the waters reached its crest, rivers overtopped and Dayton being flooded by meters of water. The community moved quickly toward reconstruction and formed a flood prevention committee. Morgan was then only 35 but had developed remarkable genius for design of such plans. He approached the Miami Valley Project with the same thoroughness and foresight that had characterized all his personal and professional endeavors. He proposed a long-range, effective flood prevention program. To highlight the importance of community participation he generated the attitude of self-help. The entire project was funded from local sources without federal or state help. He pointed at the scarcity of adequate runoff and rainfall records. A detailed study indicated that flood control programs had to be conducted to reduce this natural hazard significantly.

Anonymous (1940). Arthur Morgan returns to Dayton-Morgan company. *Engineering News-Record* 124(Jun.6): 768. *P*

Anonymous (1953). Arthur E. Morgan. *Engineering News-Record* 151(Oct.22): 26. *P*

Anonymous (1969). Morgan, Arthur E. *Who's who in America* 35: 1555. Marquis: Chicago.

Anonymous (1976). Morgan: College innovator. *Civil Engineering* 46(1): 94.

Anonymous (1977). The genius of A.E. Morgan. *Civil Engineering* 47(10): 114-117. *P*

Morgan, A.E. (1922). Flood measures in the vicinity of Dayton, Ohio. *Journal of the Western Society of Engineers* 27(1): 1-12.

Morgan, A.E. (1929). Reservoirs for Mississippi flood control. *Trans. ASCE* 93: 737-754.

MORITZ

30.08. 1882 Sheboygan WI/USA
21.08. 1975 Boulder NV/USA

Ernest (Ernie) Anthony Moritz obtained the BS degree in 1904, and the CE degree in 1905 from University of Wisconsin, Madison WI. He was from 1907 to 1911 hydraulic engineer with the US Bureau of Reclamation USBR for the Yakima Project in Washington State; then he was transferred to Denver CO until 1919, from when he was a contracting engineer in Illinois State until 1934. He then joined the design of Parker Dam in California until 1938, continued in a similar position for the Marshall Ford Dam, Austin TX, and the Boulder Canyon Dam in Nevada, from when he acted as regional director of USBR. He retired in 1952, after having been 31 years in governmental service. He received the Department of Interior's highest honour, the Distinguished Service Award.

The 1915 book of Moritz deals with irrigation engineering in the US. Its seven chapters namely are: 1. Examination and reconnaissance, 2. Investigations and surveys, 3. Design of irrigation structures, 4. Hydraulic diagrams and tables, 5. Structural diagrams and tables, 6. Miscellaneous tables and data, and 7. Specifications. The book therefore mainly contains tables, which were popular among practising engineers in these days. Those books included often an accumulation of recipes of well-known and accepted engineers to be directly applied to engineering design. However, these books hardly contain any theoretical derivations, but rather examples of successfully made designs. It should be noted that most information of this and similar books originate from India, then a country well-developed in irrigation technologies. A similar book of Bernard A. Etcheverry (1881-1954) certainly more addressed basic hydraulic questions.

Anonymous (1954). Ernest A. Moritz. *Who's who in the West*: 483. Marquis: Chicago.
Anonymous (1974). Ernest A. Moritz celebrates 92nd birthday. *Henderson News* (9/10): 7. *P*
Anonymous (1981). Moritz, Ernest A. *Who was who in America* 7: 414. Marquis: Chicago.
Etcheverry, B.A. (1915). *Irrigation practice and engineering*. McGraw-Hill: New York.
Moritz, E.A. (1911). Experiments on the flow of water in wood stave pipe. *Trans. ASCE* 74: 411-482.
Moritz, E.A. (1914). Weir measurement of stream flow. *Trans. ASCE* 77: 1282-1285.
Moritz, E.A. (1915). *Working data for irrigation engineers*. Wiley: New York.
Moritz, E.A. (1918). Discussion of Wood stave pipe design. *Trans. ASCE* 82: 511-512.
Moritz, E.A. (1950). Power can be paying partner in Colorado River Development. *Civil Engineering* 20(5): 304-306.

MORRIS H.M.

06.10. 1918 Dallas TX/USA
25.02. 2006 Santee CA/USA

Henry Madison Morris graduated as a civil engineer from Rice University, Houston TX, in 1939. He then spent two decades as member of civil engineering faculties of four different universities, including the St. Anthony Falls SAF Hydraulic Laboratory of the University of Minnesota, Minneapolis MN, from where he earned the PhD degree in 1950. In 1951 he was appointed professor of civil engineering at the University of Louisiana, Lafayette LA, In 1959 he became head of department, Virginia Polytechnic Institute, Blacksburg VA, there teaching hydraulic engineering, including an accurate account on the history of hydraulics. After a dispute with the University administration in 1963, Morris resigned from this position in 1969. He then founded the Institute for Creation Research at Santee CA, which was taken over by his son when the father retired. Morris suffered a minor stroke, but shortly later died.

Morris is known for basic work in roughness of hydraulic engineering. Together with Lorenz G. Straub (1901-1963) he presented a report on the roughness effect of concrete and corrugated metal pipes as used in culverts. The 1955 paper is a noteworthy addition to the roughness problem in hydraulics, involving a new concept of rough boundaries, based on the effect of streamwise spacing of surface roughness elements. Three basic flow types were identified, namely isolated-roughness, wake-interference, and quasi-smooth flows. The friction factors were derived and the standard equations expanded by inclusion of the roughness height and its spacing. From 1970, Morris devoted his activities exclusively to the Creation Research Society, and he was considered later the father of modern creation science.

Morris, H.M. (1951). *The bible and modern science*. Moody Press: Chicago.
Morris, H.M. (1955). Flow in rough conduits. *Trans. ASCE* 120: 373-398; 120: 406-410.
Morris, H.M. (1961). Design methods for flow in rough conduits. *Trans.* ASCE 126: 454-490.
Morris, H.M. (1963). *Applied hydraulics in engineering*. Ronald Press: New York.
Morris, H.M. (1968). Hydraulics of energy dissipation in steep rough channels. Virginia Polytechnical *Bulletin* 19. Virginia Polytechnic Institute: Blacksburg.
Straub, L.G., Morris, H.M. (1951). Hydraulic data comparison of concrete and corrugated metal culvert pipes. *Technical Paper* 3, Series B. SAF Hydraulic Laboratory, University of Minnesota: Minneapolis.
http://en.wikipedia.org/wiki/Henry_M._Morris *P*

MORRIS S.B.

24.08. 1890 Los Angeles CA/USA
06.03. 1962 Los Angeles CA/USA

Samuel Brooks Morris obtained the AB degree in civil engineering from Stanford University, Stanford CA, in 1911. He joined the Water Department of the City of Pasadena CA until 1935, first as assistant engineer, then as chief engineer and from 1925 as general manager. He then was a professor of civil engineering at Stanford University until 1945, but returned in 1944 to the Department of Water and Power of the City of Los Angeles CA as general manager and chief engineer. He also was consultant of the Bonneville Power Administration from 1941 to 1944. He was awarded, among other, the John M. Diven Medal by the American Water Works Association AWWA in 1933, and its 1952 Fuller Award. He was elected in 1961 Honorary Member of the American Society of Civil Engineers ASCE.

During his tenure with the Department of Water and Power, a large number of reservoirs were constructed, dams rebuilt, and steam-generating plants undertaken. Outstanding projects include the Owens Gorge Power Project on the Upper Owens River in Eastern California. The Long Valley Dam at the head of the Owens River Gorge was completed in 1941. Morris was nationally known as consultant who had served to a number of government agencies. A member of the President's Water Resources Policy Commission in 1950 and 1951, he was then a member of the Colorado River Board of California. He had served further on the Committee on Geophysics and Geography of the Research and Development Board. He also had been AWWA president in 1943, and was ASCE vice-president in 1958.

Anonymous (1937). Samuel B. Morris. *Water Works Engineering* 90(May 26): 775. *P*
Anonymous (1942). Samuel B. Morris. *Engineering News-Record* 128(Mar.19): 447. *P*
Anonymous (1950). Truman appoints Water Resources Commission. *Civil Engineering* 20(2): 72.
Anonymous (1955). He did what he wanted to do, and he did it well. *Engineering News-Record* 155(Oct.13): 63-64. *P*
Anonymous (1959). Morris, Samuel B. *Who's who in America* 28: 1916. Marquis: Chicago.
Anonymous (1963). Samuel Brooks Morris. *Trans. ASCE* 128(5): 110-111.
Morris, S.B., Pearce, C.E. (1934). A concrete gravity dam for a faulted mountainous area. *Engineering News-Record* 113(Dec.27): 823-827.
Morris, S.B. (1955). Owens Gorge Project. *Journal Hydraulics Division* ASCE 81(738): 1-25.
Morris, S.B. (1963). Economic use of fresh water from the sea. *Trans. ASCE* 128(3): 130-141.

MORTIMER

27.02. 1911 Whitchurch/UK
11.05. 2010 Milwaukee WI/USA

Clifford Hiley Mortimer obtained the BS degree in 1932 and the DSc degree in 1946 from Manchester University, Manchester UK. He was an Alexander Humboldt Fellow at the University of Berlin from 1932 to 1935, thereby obtaining the PhD degree. From then to 1941 he served as scientific officer the Freshwater Biological Association in the UK, was then until 1946 at the Royal Naval Science Service, returned until 1956 to the former Association, from when he was until 1966 director of the Scottish Marine Biological Association, Oban UK. He was then appointed distinguished professor of zoology at the University of Wisconsin, Milwaukee WI, and in parallel was the director of its Center of Great Lakes Studies. Mortimer was a member of the American Society of Limnology and Oceanography ASLO, presiding it in 1970, of the Marine Biological Society MBS, and the International Association of Limnology, being the recipient of its 1965 Naumann Medal. He was a Fellow of the Royal Society since 1958.

The research interests of Mortimer included the physics, biology, and chemistry of lakes and oceans, in particular water motion in large basins and coastal marine waters, including internal waves. Almost all his research fell within the broad research field of limnology. In the 1930s his interests also included the control of chemical quantities within lakes, namely lake metabolism. After World War II he took interest in the physics of water movement in lakes and reservoirs. He observed strong wind-induced tilts of the deep temperature-density gradient followed by prolonged internal waves at the density interface. Internal *seiches* corresponding to temperature oscillations have implications on the water movement, which he analysed mathematically with Michael S. Longuet Higgins (1925-).

Anonymous (1972). Mortimer, C.H. *American men and women of science* 12: 4410. Cattell: NY.
Brooks, A.S., Lund, J.W.G., Talling, J.F. (2012). Clifford H. Mortimer. *Biographical Memoirs of the Fellows of the Royal Society* 57: 291-314. *P*
Mortimer, C.H., Longuet-Higgins, M.S. (1952). Water movements in lakes during summer stratification: Evidence from the distribution of temperature. *Phil. Trans.* 236: 355-398.
Mortimer, C.H. (1974). Lake hydrodynamics. *Mitteilung Intl. Verein Limnologie* 20: 124-197.
Mortimer, C.H. (1987). Fifty years of physical investigations and related limnological studies on Lake Erie. *Journal of Great Lakes Research* 13(4): 407-435.
Mortimer, C.H. (2004). *Lake Michigan in motion*. University of Wisconsin: Madison WI.

MORTON R.K.

03.12. 1925 Albemarle NC/USA
12.11. 1999 Washington DC/USA

Rose Katherine Morton graduated in 1945 from the University of North Carolina, Greensboro NC, as a mathematician. She then moved as mathematician to the David W. Taylor Model Basin, Navy Dept., Washington DC, where she was involved in model tests and computations of hydrodynamic problems. In a Report on the Hydrodynamics of slamming of ships, the author acknowledges her help in checking the manuscript and the computations. She then was in the research group of William L. Haberman (1922-1996), becoming known for her research on bubble flow in liquids. Morton married in 1953 and thus left the academic career in favour of her family.

The 1956 ASCE paper deals with a basic study of bubble motion. Experiments were conducted to determine the drag and shape of single air bubbles rising freely in various liquids. The results indicate that a complete description using dimensionless parameters containing viscous, surface tension, and density effects is impossible. In addition, three types of bubble shapes were observed, namely spherical, ellipsoidal, and spherical cap, in rising order of bubble diameters. For tiny spherical bubbles the drag coefficient is identical with that of the corresponding rigid sphere. As the bubble size increases, the drag reduction as compared with that of the rigid sphere occurs in some liquids, so that the drag curves of spherical bubbles fall between these of the rigid and the fluid spheres. The particular finding of this study is the so-called Morton number M by which the separate effects of fluid viscosity, surface tension, and density are included, forming a dimensionless number which has turned out important in relation with two-phase air-water flows. It defines the shape of bubbles or drops moving in a surrounding fluid. It may be expressed in terms of the Weber, the Froude, and the Reynolds numbers.

Anonymous (1945). Rose Morton. *Pine Needles*: 126. University of North Carolina: Greensboro NC. *P*

Haberman, W.L., Morton, R.K. (1953). An experimental investigation of the drag and shape of air bubbles rising freely in various liquids. *TMB Report* 802. Washington DC.

Haberman, W.L., Morton, R.K. (1956). An experimental study of bubbles moving in liquids. *Trans. ASCE* 121: 227-252; also *Proc. ASCE* 80(1, Separate 387): 1-25.

Pfister, M., Hager, W.H. (2014). The Morton number: History and significance in hydraulic engineering. *Journal of Hydraulic Engineering* 140(5): 02514001-1-6. *P*

Sayre, R.H. (2003). Rose Morton. *Sayre family*: Another 100 years. iUniverse: New York.

MULHOLLAND

11.09. 1855 Belfast/UK
22.07. 1935 Los Angeles CA/USA

William Mulholland was born in Northern Ireland. He spent from age 15 his life as a seaman primarily sailing Atlantic routes, arriving at Los Angeles in 1877. He there accepted a job digging a well by a hand-drill. After a brief stint in Arizona working on Colorado River he was engaged by Frederick Eaton (1855-1934) as deputy *zanjero* with the newly formed Los Angeles Water Company LAWC. During the Spanish and Mexican administrations, water was delivered to Los Angeles, then having a population of less than 10,000, in a large open ditch, with the man dealing with the ditch known as zanjero. In 1880 Mulholland oversaw the laying of the first iron water pipeline at Los Angeles. He was made in 1886 superintendent of LAWC, yet the city government decided in 1898 not to renew the contract with LAWC. In 1902, the Los Angeles Department of Water was established with Mulholland as its head.

Mulholland, a self-taught engineer, laid the foundations transforming Los Angeles into a modern metropolis. The city growth was limited because of the semi-arid climate, with its population reaching 50,000 in 1890, over 100,000 in 1900, and 300,000 in 1910. To provide water for this ever growing city, Mulholland oversaw the construction of the 375 km long Los Angeles Aqueduct, which opened in 1913, carrying water from Owens Valley in the Eastern Sierra to the San Fernando Valley. The project occupied more than 2,000 workers and involved 164 tunnels and 300 km of pipes and channels. This engineering work became possible by the rights obtained by the city to all surface flow water out of the city limits. Later in his career he provided technical assistance to the construction of the Panama Canal. His career ended in 1928 when the St. Francis Dam failed, whose construction he had supervised. 45 million m^3 of water discharged through the San Francisquito Canyon as a 40 m high surge at speed of 30 km/h. The final death toll was 600, until then the worst civil engineering disaster in the USA.

Davis, M.L. (1993). *Rivers in the desert*: William Mulholland and the inventing of Los Angeles. Harper Collins: New York. P
FitzSimons, N. (1972). Mulholland. *Biographical dictionary of American civil engineers*: 93-94. ASCE: New York.
Mulholland, C. (2000). *William Mulholland and the rise of Los Angeles*. UCal Press: Berkeley. P
Reisner, M. (1986). *Cadillac desert*: The American West and its disappearing waters. Viking: NY.
http://en.wikipedia.org/wiki/William_Mulholland P

MULLER

17.10. 1881 Stendal/D
04.06. 1944 New York NY/USA

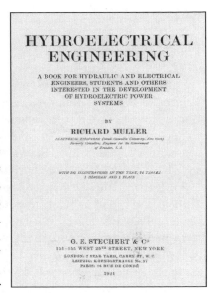

Few information is available on Richard O. Muller. He was a graduate of Hannover University, Germany, and associated with the steam turbine industry all through his life. He started at an early age with the British Thompson-Houston Co., and Westinghouse Co. He came to the USA and was employed as turbine pioneer at Brooklyn NY. He also worked as consultant at Washington DC. He further had been consulting engineer for the Government of Ecuador, and professor of electrical engineering at University of Quito. He joined the Terry Steam Turbine Co., Hartford CT, in 1915, and was made chief engineer in 1926. He had been in poor health and had retired from active work for some time, finally retiring from active work in 1943.

Muller was the true engineer who had proposed to design spillways based on the lower nappe profile of sharp-crested rectangular weirs. This idea is commonly attributed to William Pitcher Creager (1878-1953), yet Muller made this proposal as early as in 1908, when the first larger dams were erected. Creager, in turn, must be credited for the crest profile, which he proposed in 1917. Muller authored also a book on Hydro-electrical engineering, combing both the civil and electrical engineering aspects of these works. Its main chapters are Hydrology, Stream measurement, Canals, Pressure pipes, Dams, Turbines, Power house, Transmission lines, Water power projects, Hydroelectric plants, and Legislation. He had been an ASME member.

Anonymous (1911). Mr. Richard Muller. *Engineering Record* 63(21 Suppl.): 58.
Anonymous (1922). Muller. *Who's who in engineering* 1: 914. Leonard: New York
Anonymous (1944). Richard O. Muller. *Power Plant Engineering* 48(8): 132-134.
Muller, R. (1908). A formula for calculating flashboards for dams. *Engineering Record* 58(8): 208.
Muller, R. (1908). Development of a practical type of concrete spillway dam. *Engineering Record* 58(17): 461-462.
Muller, R. (1908). A diagram for calculating penstocks. *Engineering Record* 58(20): 549-550.
Muller, R. (1912). Sur les barrages-déversoirs: Détermination théorique d'un profil transversal minimum de barrage-déversoir. *La Technique Moderne* 5(4): 113-115; 5(5): 148-151.
Muller, R. (1914). Spacing of piers for reinforced-concrete pipes. *Engineering Record* 70(3): 71.
Muller, R. (1921). *Hydroelectrical engineering*: A book for hydraulic and electrical engineers, students and others interested in the development of hydroelectric power systems. Stechert: New York.

MUNSON

15.06. 1889 Columbus TX/USA
06.10. 1958 Freeport TX/USA

Thurmond Armour Munson received in 1910 the BSc degree from the Agricultural and Mechanical College, College Station TX, the civil engineering degree in 1924 from Iowa State College, Ames IA, and there the MSc degree in civil engineering in 1925. He was employed as assistant engineer from 1910 to 1911, and as sanitary engineer until 1912. He established a general consulting practice from 1913 to 1921 dealing with drainage projects, flood control, and surveys. Munson became in 1920 an associate professor and in 1926 then professor of hydraulic engineering at his Alma Mater, today's A&M College, accepting in 1946 the position of chief civil engineer for a chemical company at Freeport TX, continuing in this position until his retirement in 1955. During his professional career, he made surveys on public utilities locations. He was also known as hydraulic engineer, and an expert on land boundary suits. He further served during World War II.

Munson was rather a general civil engineer and educator than a hydraulic engineer advancing research. He was in the early 1920s a consulting engineer for waterworks and sewers, and in the mid-1920s a special engineer and advisor for high-pressure lines. In the late 1920s he made surveys and reports for the reclamation of state land in Texas from Brazos River floods. In the 1930s he made investigations, designs, and reports on rainfall, runoff, flood flows and flood frequencies, flood protection, and navigation improvements in the Missouri River Basin, with the US Engineers' Office, Kansas City MO. He was further involved in investigations and reports on the flood damage for two large railways companies from 1934 to 1940. Munson was member of the American Society of Civil Engineers from 1944, serving as chairman its Texas Section, and the Brazos County Branch from 1939 to 1940. He also was member of the Texas Water Conservation Association, the National Geographical Society, and the Brazos County Development Association.

Anonymous (1948). Munson, Thurmond A. *Who's who in engineering* 6: 1423-1424. Lewis: New York.
Anonymous (1959). Thurmond A. Munson. *Trans. ASCE* 124: 1072-1073.
Munson, T.A. (1924). *Investigation* of the effects of grade and types of road surface on the fuel consumption of motor vehicles. Iowa State College: Iowa City.
http://www.munsons-of-texas.net/pic-jwm-04.html *P*

MURPHY

17.06. 1859 Croydon ON/CA
18.09. 1934 Santa Monica CA/USA

Edward Charles Murphy obtained the BS degree in civil engineering in 1884, the MS degree in 1885, and the PhD degree in 1900. He was from 1887 to 1899 assistant professor of engineering at Kansas University, Lawrence KS, then instructor at Cornell University, Ithaca NY, and took over in 1903 until retirement in 1926 as engineer of the US Geological Survey. He was awarded the Fuertes Gold Medal.

Murphy was a successful hydrographer of the early 20th century, given that his experience was sought by individuals of the calibre of Robert E. Horton (1875-1945) or Grove K. Gilbert (1843-1918). The 1907 report summarizes the then available data on weir flow, including these of Henry Bazin (1829-1917), among many others, therefore providing a useful data-set to describe the main weir flow characteristics. The 1907 paper recommends the Stevens method for discharge extrapolation from available data of a certain river, which is discussed in most of the textbooks of the day. Murphy suggested that one should spend sufficient time to determine the accurate slope and cross-section. He was also interested in stream discharge measurement by the current meter, as evidenced by his discussion of a notable paper in the Trans. ASCE in 1910. Previously, he had presented in the 1902 report the then used methods of discharge measurement in the USA. The measurements with the meters proposed by William G. Price (1853-1928), Eugene E. Haskell (1855-1933), or even Alphonse Fteley (1837-1903) were compared with weir observations; the best results resulted from the Price meter.

Gilbert, G.K., Murphy, E.C. (1914). *The transportation of debris by running water*, based on experiments made with the assistance of Edward C. Murphy. Govt. Printing Office.

Horton, R.E., Murphy, E.C., Beardsley, R.C. (1907). On a suggested method for extrapolating values of stream discharge. *Engineering News* 58(8): 202-203.

Horton, R.E., Murphy, E.C. (1907). *Weir experiments, coefficients and formulas*. Washington.

Kolupaila, S. (1960). Edward C. Murphy: Early history of hydrometry in the United States. *Journal of the Hydraulics Division* ASCE 86(HY1): 34. *P*

Murphy, E.C. (1901). *The windmill*. Government Printing Office: Washington DC.

Murphy, E.C. (1902). *Accuracy of stream measurements*. Govt. Printing Office: Washington.

Murphy, E.C. (1902). Current meter and weir discharge comparisons. *Trans. ASCE* 47: 370.

Murphy, E.C. (1904). *Hydrographic manual of the United States Geological Survey*. Government Printing Office: Washington DC.

MYERS

09.02. 1903 Bethlehem PA/USA
31.01. 1994 Bethlehem PA/USA

Richmond Elmore Myers obtained the AB degree from Moravian College, Bethlehem PA, in 1925, the MA degree from the Penn State in 1929, as also the PhD degree in geology there in 1951. He served as department head at Muhlenberg College, Allentown PA, from 1938 to 1951, then continued as geologist at Pennsylvania State until 1953, from when he was director of research, the Department of Commerce, Commonwealth of Pennsylvania until 1954, then taking over as dean at Moravian College until 1959, and chairman of its Department of Geology until 1969, from when he was emeritus professor there. He was a member of the Geological Society of America GSA, and of the Mineralogical Society of America MSA.

The 1954 paper describes a river steamboat on Susquehanna River, which was identified as an important cog in the conservation scheme, reclaiming thousands of tons of coal from the river which otherwise would silt at the river bottom or wash on down into the Chesapeake, lost forever. The second 1954 paper describes the various types of bridges on Susquehanna River. It is stated that the ghosts of the old river bridges come to life again taking form over these abandoned skeletons of yesterday. The 1949 book contains 18 chapters, namely the geologist's river, Indian's river, trader's river, lumberman's river, ironmaster's river, anthracite's river, reel's river, politician's river, culture's river, New York's river and Maryland's river. The major emphasis is placed on the 16th and 17th centuries, yet with the story petering out in the 1800s, but the modern phase being hardly mentioned.

Anonymous (1956). R.E. Myers. Proc. 6th *Hydraulics Conference* Iowa: Frontispiece. *P*
Anonymous (1958). R.E. Myers. *Benigna yearbook*: 197. Moravian College: Bethlehem PA. *P*
Anonymous (1972). Myers, Richmond E. *American men and women of science* 12: 4483. Cattell: Tempe AZ.
Myers, R.E. (1949). *The long crooked river, the Susquehanna*. Christopher: Boston MA.
Myers, R.E. (1951). *The development of transportation in the Susquehanna Valley*: A geographical study. Ann Arbor.
Myers, R.E. (1954). Bottoms up on the Susquehanna. *Pennsylvania Angler* 23(1): 13-15.
Myers, E.E. (1954). Susquehanna bridges. *Pennsylvania Angler* 23(10): 2-5.
Myers, R.E. (1955). Middle-Atlantic geographic corridors. *Science* 80(5): 310-319.
Myers, R.E. (1963). *Floods at Des Moines IA*. US Geological Survey: Washington DC.

NAGHDI

29.03. 1924 Tehran/IR
09.07. 1994 Berkeley CA/USA

Paul Mansour Naghdi was naturalized in 1948. He made studies at University of Michigan, Ann Arbor MI, receiving the PhD degree in 1951. Until 1958, he was there assistant, associate and full professor. He moved to University of California, Berkeley CA as a professor of engineering science, chairing its Department of Applied Mechanics from 1964 to 1969. He was a member of the National Academy of Engineering NAE, the Society of Rheology, the American Society of Mechanical Engineers ASME, and ASME Fellow from 1969. He was the 1958 Guggenheim Fellow, recipient of the Timoshenko Medal in 1980, and awarded ASME Honorary Membership in 1983.

Naghdi's work on continuum mechanics extended over forty years encompassing most aspects of the mechanical behaviour of solids and fluids. He was strongly attracted by fundamental questions and always sought to treat these at the highest level of generality. He is known for his works in the areas of shell theory and plasticity, but also in visco-elasticity, fluid sheets and jets, continuum thermodynamics, and the mixture theory. His talents as a teacher were legendary. At Berkeley, he continually refined a magnificent series of courses on theoretical mechanics. These were well prepared, clearly delivered, original, and intellectually provocative. They also reflected his deep understanding of the history of mechanics and his encyclopaedic knowledge of literature.

Anonymous (1961). P.M. Naghdi. *Mechanical Engineering* 83(10): 124. *P*

Anonymous (1979). P.M. Naghdi. *Trans. ASME* 46(12): 734. *P*

Anonymous (1985). Naghdi Paul M. *Who's who in engineering* 6: 475. AAES: Washington DC.

Anonymous (1994). Paul M. Naghdi. *Journal of Applied Mechanics* 61(9): 509-510. *P*

Green, A.E., Laws, N., Naghdi, P.M. (1974). On the theory of water waves. *Proc. Royal Society* London A 338: 43-55.

Green, A.E., Naghdi, P.M. (1976). A derivation of equations for wave propagation in water of variable depth. *Journal of Fluid Mechanics* 78: 237-246.

Green, A.E., Naghdi, P.M. (1976). Directed fluid sheets. *Proc. Royal Society* London A 347: 447-473.

Green, A.E., Naghdi, P.M. (1986). A nonlinear theory of water waves for finite and infinite depths. *Philosophical Trans.* Royal Society London A 320: 37-70.

Green, A.E., Naghdi, P.M. (1995). A new thermoviscous theory for fluids. *Journal of Non-Newtonian Fluid Mechanics* 56(3): 289-306.

NAGLER F.A.

11.01. 1892 Howard City MI/USA
10.11. 1933 Iowa City IA/USA

Floyd August Nagler obtained the BS degree in 1914 in civil engineering from Michigan Agricultural College, East Lansing MI, the MS degree in 1915, and the PhD degree in 1917 from the University of Michigan, Ann Arbor MI. He was then an assistant engineer to Robert E. Horton (1875-1945) at Albany NY until 1920, when being appointed professor of hydraulic engineering at State University of Iowa, Iowa City IA. He founded its Hydraulic Laboratory. Nagler was consultant in hydraulic engineering too. He was a member of the American Water Works Association AWWA, and of the American Society of Civil Engineers ASCE, whose 1931 Normal Medal was awarded for the 1930 paper.

Nagler has authored a number of important papers, most of which were also awarded. Bridges are known to be a major cause for flooding during storms, such that the main effects of bridge elements protruding into the flow were investigated for a long time. Nagler's research particularly aimed to clarify the effect of obstruction by a variety of pier shapes and he proposed a formula allowing to account for these effects. The 1920 paper was a verification of the famous weir flow experiments conducted by Henry Bazin (1829-1917) using an alternative measurement approach. The 1930 paper was an early account on model observations in dam engineering, applied to Keokuk Dam, on the Upper Mississippi. This work includes crest velocity distributions, velocity contours in the crest section, and discharge coefficients by accounting for pier presence.

Anonymous (1931). Nagler Floyd A. *Who's who in engineering* 3: 944. Lewis: New York.
Anonymous (1932). Floyd A. Nagler. *Civil Engineering* 2(1): 57. *P*
Anonymous (1935). Floyd A. Nagler. *Trans. ASCE* 100: 1698-1700.
Mutel, C.F. (1999). A history of the Iowa Institute of Hydraulic Research published. *Journal of Hydraulic Engineering* 125(6): 558-561. *P*
Nagler, F.A. (1918). Obstruction of bridge piers to the flow of water. *Trans. ASCE* 82: 334-395.
Nagler, F.A. (1920). Verification of the Bazin weir formula. *Trans. ASCE* 83: 105-199.
Nagler, F.A. (1922). Utilization of surplus flood water to suppress backwater upon water power developments. *General Electric Review* 25(10): 598-602.
Nagler, F.A., Davis, A. (1930). Experiments on discharge over spillways and models, Keokuk Dam. *Trans. ASCE* 94: 777-844.
Yarnell, D.L., Nagler, F.A. (1931). Effect of turbulence on the registration of current meters. *Trans. ASCE* 95: 766-860.

NAGLER F.

21.04. 1885 Freeport MI/USA
01.04. 1952 Lake Forest IL/USA

Forrest Nagler received in 1905 his BS degree in mechanical engineering from University of Michigan Ann Arbor MI, specializing in hydraulic and marine engineering. In 1906, he joined the Allis-Chalmers Manufacturing Co., where he was employed in the foundry and machine shop. From then to 1930 he remained with that company, principally on work in hydraulic power stations. He then joined a company at Milwaukee WI in engineering and research fields. From 1933 to 1941 Nagler was chief engineer with the Canadian Allis-Chalmers, Ltd., Toronto CA. He returned to the USA in 1942 as chief mechanical engineer to this company at Milwaukee MI.

Nagler's principal assignments were the development of hydraulic machinery, and the production of heavy machinery. He was the inventor of the Nagler-type of axial-flow high-speed hydraulic turbine runner. The Nagler turbine is described in his 1923 paper. It essentially deals with a modification of a single wheel as compared to the then available Pelton, tangential, Girard, or Schwankrug wheels. Whereas the original wheels received the water on flat surfaces, the effect of reaction was improved by using curved wheel geometries. He also held a number of patents in this field. His other major work relating to turbines is described in the 1919 paper, for which he was awarded ASME life membership. Nagler was a member of the American Society of Mechanical Engineers ASME, there chaired the Executive Committee of its Hydraulic Division from 1939 to 1940 and was elected ASME Fellow in 1950. He also was a sponsor of the Committee on Hydraulic Prime Movers. In 1921 he was awarded the best paper presented before the Society, published in 1919.

Anonymous (1931). Nagler, Forrest. *Who's who in engineering* 3: 944. Lewis: New York.
Anonymous (1948). Forrest Nagler. *Mechanical Engineering* 70(8): 710-711. *P*
Anonymous (1952). Forrest Nagler. *Mechanical Engineering* 74(6): 538.
Brown, E.H., Nagler, F. (1914). Preliminary report of current meter investigations. *Proc. Engineers' Society of Western Pennsylvania* 30(5): 415-424.
Nagler, F. (1914). Plant tests of a low-head hydro-electric development. *Engineering Record* 72(25): 1193-1198.
Nagler, F. (1919). A new type of hydraulic-turbine runner. *Trans. ASME* 41: 829-853.
Nagler, F. (1923). The cross-flow impulse turbine. *Mechanical Engineering* 45(5): 275-281.
Nagler, F. (1929). Efficient hydro operation. *Electrical World* 94(18): 875-878.

NALDER

06.04. 1885 Waitsburg WA/USA
31.05. 1956 Denver CO/USA

William Henry Nalder obtained in 1909 the BSc degree in civil engineering from Washington State University, Pullman WA. He was then up until his retirement in 1952 always closely connected with the US Bureau of Reclamation USBR, serving as chief design engineer from 1924 to 1945, and as assistant chief engineer from 1946. It was stated from a colleague following retirement due to ill health: 'He is so much a part of this organization that it seems hard to imagine getting along without him. We all hope that his health will permit him to continue on a part-time basis as consultant'. Nalder was member ASCE, and of the Colorado Society of Engineers. He served in 1949 as staff adviser on irrigation to the United Nations for the Middle East.

Nalder was involved in the 1930s in the design of Hoover Dam as assistant to chief engineer Raymond F. Walter (1873-1940). His colleagues then were, among others, Sinclair O. Harper (1883-1966), John L. Savage (1879-1967), Erdman B. Debler (1885-1976), and Ivan E. Houk (1888-1972). The 1933 Report presents design fundamentals of this gigantic project and reviews the various procedures to be undertaken. The 1938 Report deals with the model studies relating to the penstocks and the outlet works, in which Nalder was involved as assistant chief design engineer. Except for few isolated works, this Report deals with the hydraulic losses in pipe junctions, and means to reduce these. So-called manifolds were also studied for the first time. A similar research was made in the early 1950s on the junction flow in rivers. Nalder was in the 1930s considered one of the nation's authorities on the design of water and power structures.

Anonymous (1948). Nalder, William H. *Who's who in engineering* 6: 1433. Lewis: New York.
Anonymous (1952). William H. Nalder. *Engineering News-Record* 149(Oct.30): 52. *P*
Nalder, W.H. (1952). Design of irrigation systems. *Trans. ASCE* 117: 230-241.
Nalder, W.H. (1953). Diversion structures and distribution systems for irrigation. *Trans. ASCE* CT: 437-450.
Page, J.C., Walter, R.F. (1938). *Boulder Canyon Project*, Final Reports 2: Model studies of penstocks and outlet works. USBR: Denver CO.
Wilbur, R.L., Mead, E. (1933). *The construction of Hoover Dam*: Preliminary investigations, design of dam, and progress of construction. Government Printing Office: Washington.
http://digitalcollections.uwyo.edu:8180/luna/servlet/detail/uwydbuwy~1~1~202973~156228:Group-photograph-of-the-Engineering *P*

NASON

31.12. 1815 Boston MA/USA
11.12. 1872 Montclair NJ/USA

Few is known on the young Joseph Nason. In 1841,
he founded with his brother-in-law James J. Walworth
(1807-1896) the Walworth, Nason & Co., Boston
MA, manufacturing in 1842 the first high-pressure
hot-water steam heating system in the USA. Nason
had previously spent time in England working on
this technique. However, its use appears to have been
limited until the 1880s, when suddenly becoming
popular. The radiator as known currently dates from
1863, when Nason and a colleague patented a design
featuring vertical wrought iron tubes screwed into a
cast iron base. Each tube was proportioned to have
exactly a surface of 1 ft^2 when screwed into the base, allowing standard radiator sizes to
be manufactured. The first popular cast-iron radiator was invented in 1874. Later, by the
1880s, cast-iron sectional radiators became popular. Competition between manufacturers
of boilers and radiators was intense. The late 19th century saw the rise of the Business
Trust, so that the heating industry was quick to use this business form to improve
competition. The most successful trust was the American Radiator Company, which
consolidated a number of the leading boiler and radiator manufacturers in 1891. Its
advertising boasted 'The largest makers of radiators in the world'.

The term 'radiator' is a misnomer since some 70% of the heat output is by convection.
The first patents appearing in the USA in 1841 referred to as heat distributors, a mix of
pipes and metal plates. Then came the vertical wrought-iron welded tubes fixed between
horizontal top and bottom headers, which were followed by the 'looped tube', an
inverted U fixed to a base plate, used for both steam and hot water. One of the early
radiators was developed in 1854. The Pin Radiator was developed in 1863 with the
attached pins increasing the heat transfer area. Nason patented with a collaborator in
1862 the steam radiator. Until 1890, Americans produced a variety of designs, many
highly ornamental, but the three principal manufacturers merged in 1892 to form the
American Radiator Co. It appears astonishing that the hydraulics of heating systems
were studied only from the 1950s, thereby using the manifold hydraulic system as the
fundamental notion.

http://www.hevac-heritage.org/pioneers/31-Mort-Nason.pdf *P*
http://www.achrnews.com/articles/87035
http://www.helm.org.uk/guidance-library/heating-ventilation/heatingventilation.pdf
http://www.heatinghelp.com/files/articles/1257/234.pdf

NEALE

23.09. 1918 Wayland MA/USA
24.10. 1984 St. Louis MO/USA

Lawrence Carlton Neale obtained in 1940 the BS degree from Worcester Polytechnic, Worcester MA, and graduated with an MS in 1957. He was from 1940 to 1950 instructor of mechanical engineering at his Alma Mater and there was associate professor of hydraulic engineering from 1958, taking over in 1965 as professor of mechanical engineering. He served in parallel as assistant director and later as director of Alden Hydraulic Research Laboratory, Worcester University, until 1975. He was member and Fellow, the American Society of Civil Engineers ASCE, and member of both the American Society of Mechanical Engineers ASME, and the International Association of Hydraulic Research IAHR. Neale passed away during a stay on St. Louis airport at age 66.

The research interests of Neale included hydraulic research in general, drag tests, smoke dispersion tests using model power plants, the missile development in water entry area, and salt velocity tests. He was attracted by the Alden Hydraulic Laboratory, given the practical background. Charles M. Allen (1871-1950), the founder of this institution, designed a so-called 'rotating boom' instead of a towing tank for calibrating velocity meters, given the reduced cost for construction, allowance for longer testing duration, and larger objects because boundary layer effects from the channel side walls were absent. The facility was used for current meter ratings, for studying aircraft propellers, ship logs, Pitot tubes, and Darrieus water turbines, as introduced by the Frenchman George Darrieus (1888-1979). Neale attempted to perfect the hydraulic techniques of the Alden Lab. He was awarded the 1954 Freeman Scholarship for an inspection tour of European hydraulic laboratories. He visited 64 sites in the UK, France, Switzerland, Italy, Austria, Germany, the Netherlands, Sweden, and Norway, reporting impressions in the 1957 paper.

Anonymous (1955). L.C. Neale. Proc. 6th *IAHR Congress* La Haye: Frontispiece. *P*
Anonymous (1972). Neale, L.C. *American men and women of science* 12: 3839. Cattell: Tempe.
Neale, L.C. (1957). Hydraulic laboratories in Europe. *Journal of the Boston Society of Civil Engineers* 44(10).
Neale, L.C. (1969). Alden Research Laboratories. *Worcester Polytechnic Journal* 73(1): 11-14. *P*
Neale, L.C. (1971). Chesapeake Bay Model study for Calvert Cliffs. *Journal of the Power Division* ASCE 97(PO4): 827-839.
files.asme.org/ASMEORG/.../History/.../5489.pd...

NELLES

15.04. 1856 Muscatine IA/USA
15.11. 1907 Cleveland OH/USA

George Thomas Nelles graduated from Rensselaer Polytechnic Institute, Troy NY, with the CE degree in 1877. After work with the US Corps of Engineers at Leavenworth KS, he became assistant engineer for improvements of Mississippi River, stationed at Atchison KS, St. Joseph KS, and Leavenworth KS until 1883, when there being elected city engineer. This city was then rapidly growing, so that much public work had to be done mainly in paving the streets and constructing sewers and bridges. He was then from 1889 engaged with general engineering projects, including the sewer design at Denver CO, and harbour improvements at St. Louis MO.

In 1895 Nelles returned to government service as assistant engineer on the Tennessee River improvements at Chattanooga TN. During the next six years he solved various difficult problems, including a study on the construction of locks and dams under fluctuating velocities which there prevail. He also studied the discharge of Tennessee River checking available formulae with his observations, and determining the Kutter roughness values. His solution of the effect of a dam on the submerged discharge was of particular note. He further investigated the flow conditions on the French Broad River, preparing plans for widening and deepening this river. A careful attention to details and the comprehensive consideration al all components are presented in his report. In 1901, he was transferred to Cleveland OH as US assistant engineer in charge of improvements of the harbours on Lake Erie at Cleveland. His health began to decline in 1903, so that he died four years later. He had been member and director of the Civil Engineers' Club of Cleveland, was from 1888 member of the American Society of Civil Engineers ASCE contributing discussions to various papers in his field, published in its Transactions.

Anonymous (1908). Nelles, George Thomas. *Trans. ASCE* 60: 586-587.

Fraser, C.B. (1986). *Nebraska City Bridge*, Fremont County IA. Historic American Engineering Record: Denver CO.

Horton, R.E. (1907). Weir experiments, coefficients, and formulas, including data collected by George T. Nelles. *Water Supply and Irrigation Paper* 200. USGS: Washington DC.

Nelles, G.T. (1900). Discussion of On the flow of water over dams. *Trans. ASCE* 44: 359-362.

Nelles, G.T. (1902). Discussion of Improvement of the Black Warrior, Warrior, and Tombigbee Rivers, in Alabama. *Trans. ASCE* 49: 284-295.

Ricketts, P.C. (1895). *History of the Rensselaer Polytechnic Institute*. Wiley: New York. *P*

NELSON M.L.

24.09. 1908 Cliff WA/USA
09.05. 1972 Boise ID/USA

Mark Lee Nelson received the BSc degree in civil engineering from Oregon State University, Corvallis OR, in 1930, later taking graduate work at University of Southern California, Los Angeles CA. His entire career was with the US Corps of Engineers, with earlier assignments in the field of water resources investigations. He performed and supervised many technical studies related to the planning and design of projects and river basin developments. He held responsible assignments in Corps offices, including at the Portland OR, Savannah GA, and Los Angeles Districts, covering a period of 42 years. At the time of his passing he was chief of the Water Control Branch, North Pacific Division, a position he held during 18 years. Under this assignment, he directed the operation of Corps and other reservoirs designated for operation by the Corps in the Columbia River System. He also developed and applied advanced system techniques in river regulation, whereby the operation was integrated for optimum functional utilization.

His assignment as chief of the Water Control Branch also involved direct supervision of hydraulic, hydrologic power, and water quality investigations required for water resource planning, design, and operation of projects in the Pacific Northwest and Alaska. Nelson's assignment involved dealings with engineers from other Federal or State agencies, as well as public and private utilities, to achieve his high management mission for the coordinated operation of the Columbia River System. Nelson was prominent as a water resource engineer, not only in the USA, but also internationally. He spent considerable time from 1962 to 1972 working with engineers in Southeast Asia and Brazil. His efforts led to applying advanced systems analysis techniques for studying the Mekong and the Upper Paraguay Rivers. He also was in charge of various studies relating to the planning of water resource developments for the Mekong River. Nelson also participated in local ASCE activities, particularly with the Oregon Section, and co-authored a paper on flood regulation. He was a member of the International Committee of Large Dams ICOLD.

Anonymous (1973). Mark L. Nelson. *Trans. ASCE* 138: 637-638.
Anonymous (1930). M.L. Nelson. *Beaver yearbook*: 243. Oregon State University: Corvallis. *P*
Lewis, D.J., Nelson, M.L. (1963). *Report* on computer application to system analysis: Lower Mekong River. US Army Corps of Engineers, North Pacific Division: Portland OR.
Nelson, M.L. (1971). *Report on Vientiane Laos flood control project*. US Army Engineer Division: North Pacific.

NELSON M.E.

20.01. 1898 Grantsburg WI/USA
15.07. 1987 Hennepin County MN/USA

Martin Emil Nelson obtained the BS degree in civil engineering in 1924 from College of Engineering, University of Minnesota, Minneapolis MN. He went as Fellow of the American-Scandinavian Foundation in 1927 to the Royal Institute of Technology KTH, Stockholm, Sweden, where the famous Wolmar K.A. Fellenius (1876-1957) was hydraulics professor. In 1928, upon return to the USA, he was in charge of the US Engineering Department, Hydraulic Lab of University of Iowa, Iowa City. From 1948 to 1962 Nelson was chief of the Hydraulics Lab Branch, Corps of Engineers, St. Paul's District, until 1962. He was a member of the American Society of Civil Engineers ASCE, the Permanent International Association of Navigation Congresses PIANC, and of the International Association of Hydraulic Research IAHR.

The work of Nelson involved the experimental analysis of hydraulic designs of flood control and navigation structures along the Mississippi River waterways. A number of designs was developed to improve the efficiency and safety of hydraulic structures along inland rivers. He has written technical papers and official reports on hydraulic research projects. The 1940 Report was sponsored by the Tennessee Valley Authority TVA, US Corps of Engineers, Department of Agricultural, Geological Survey, the US Bureau of Reclamation, and the Iowa Institute of Hydraulic Research, containing a description of 65 types of sediment samplers, their classification and evaluation. The bibliography includes 56 titles. This work was accomplished under the supervision of Emory W. Lane (1891-1963); regrettable this important work is without any authorship.

Anonymous (1964). Nelson, Martin E. *Who's who in engineering* 9: 1357. Lewis: New York.
Nelson, M.E. (1932). Laboratory tests on hydraulic models of the Hastings Dam. *Bulletin* 2. The University of Iowa: Iowa City.
Nelson, M.E., ed. (1940). Field practice and equipment used in sampling suspended sediment: A study of methods used in measurement and analysis of sediment loads in streams. *Report* 1. Iowa City.
Nelson, M.E., Hartigan, J.J. (1944). Conformity between model and prototype: Pickwick, Wheeler, and Guntersville Locks. *Trans. ASCE* 109: 35-58; 109: 177-185.
Nelson, M.E., Benedict, P.C. (1951). Measurement and analysis of suspended sediment loads in streams. *Trans. ASCE* 116: 891-918.
Rouse, H. (1976). Martin Nelson. *Hydraulic engineering in the United States* 1776-1976: 91. The University of Iowa: Iowa City. *P*

NELSON W.R.

21.11. 1897 Norwood CO/USA
21.04. 1998 Arlington VA/USA

Wesley Robert Nelson attended engineering schools at Colorado and Southern California, graduating in 1921 with the CE degree from the University of Southern California, Los Angeles CA. He served in the US Army during World War I, and then had an outstanding civil engineering career. He worked for the US Bureau of Reclamation USBR on projects including the construction of the Hoover Dam. In the 1930s and 1940s he worked as assistant engineer and associate engineer for USBR, becoming chief of the Engineering Division in the Office of the Commissioner of Reclamation, and assistant USBR commissioner. In the 1950s he developed the natural resources as an executive member of the Development Board in Iraq, as chief of the South Asia Division of the US International Corporation Administration, and as chief engineer and deputy general manager for a principal engineering firm advising the government of Burma. During the early 1960s, Nelson served then as director of foreign operations for an engineering consultant.

Nelson was transferred in 1952 from USBR to the Department of State, accepting an assignment with the Technical Corporation Administration. He had grown up 'on a ditch' in the irrigated area near Norwood CO. He entered USBR in 1922 as a surveyor and junior engineer on the Grand Valley Project at Grand Junction CO. He was in the 1930s concerned with the construction of the Hoover Dam, the great multipurpose scheme built then by USBR, becoming in 1936 chief of the USBR Engineering Division. He later organized and was director of the Amarillo TX Office of USBR. He was in 1952 recipient of the Distinguished Service Award. His main writings deal with the civil engineering projects relating to his era with USBR and beyond in Iraq. Nelson was a Fellow of the American Society of Civil Engineers ASCE.

Anonymous (1952). Wesley R. Nelson. *Engineering News-Record* 148(Feb.14): 163. *P*
Nelson, W.R. (1934). Construction of Boulder Dam: Government engineers and surveyors have made a notable record on perilous work. *Compressed Air Magazine* 39(11): 4585-4588.
Nelson, W.R. (1936). *Boulder Canyon Project*. US Government Printing Office: Washington.
Nelson, W.R. (1952). *Water and our future*. USBR: Washington DC.
Nelson, W.R. (1956). *Twin rivers, twin treasures*: A study of the control and use of the Tigris and Euphrates Rivers in Iraq. Development Board: Baghdad.
http://www.zoominfo.com/#!search/profile/person?personId=145253564&targetid=profile

NETTLETON

22.10. 1831 Medina OH/USA
22.04. 1901 Denver CO/USA

Edwin Shelton Nettleton moved to Pleasantville PA in 1865 serving there as county surveyor. He went West in 1870 joining the Union Colony on its way to Colorado, acting as colony engineer. He further built the Larimer and Weld Canal for a mortgage company, which was the largest irrigation system in Colorado at this time. He then proceeded to the High Line Canal, known as English Ditch, because it was under financial control of the English Co., thereby inventing a weir type. He was from 1883 to 1887 Colorado State Engineer, inaugurating projects of gauging streams and ditches. He also was engaged in the layout of various irrigation works in Wyoming and Idaho. From 1889 to 1893 he served as chief engineer in the diversion of Yaqui River in Mexico, and was appointed consulting engineer for the first investigation of irrigation by the US Government. He also was sent to study irrigation systems and methods for preventing forest destructions to Spain and Italy in 1892. Nettleton finally became the irrigation expert under Elwood Mead (1858-1936) of the US Department of Agriculture. He was in 1897 a founder and first trustee of the Colorado College, Colorado Springs CO, and established a weather bureau on Pike's Peak. He became wealthy in the real estate boom, but lost his money in the failure of a Denver bank.

Nettleton was among these who conducted significant hydrometric measurements in the 19th century. His observations of 1881 described in the 1885 Report involve the Colorado cup meter, which was attached to the end of a rod, allowing for velocity and discharge measurements in mountainous streams. The detailed methods and the equipment used were then described in the 1887 Report. His current meter consisted of substituting a vertical-axis rotor for the horizontal axis, screw type rotor. An original meter, along with similar designs of his era, is preserved at the Smithsonian Institution.

Frazier, A.H. (1974). Edwin S. Nettleton. *Water current meters* in the Smithsonian Collections of the National Museum of History and Technology: 75-77. Washington DC. *P*

Nettleton, E.S. (1885). *Report* of the State Engineer to the Government of Colorado for the years 1883 and 1884. Collier & Cleaveland: Denver.

Nettleton, E.S. (1887). *Third biennial report* of the State Engineer of the State of Colorado for the years 1885-1886. Collier & Cleaveland: Denver.

Nettleton, E.S. (1891). *Progress report on irrigation in the US*. Govt. Printing: Washington DC.

Nettleton, E.S. (1892). *Artesian and underflow investigation*. US Dept. Agriculture: Washington.

NEUMANN von

28.12. 1903 Budapest/H
08.02. 1957 Washington DC/USA

John (Janos) von Neumann was born in Hungary. After attending the University of Berlin from 1921 to 1923, he received the PhD title from ETH Zurich in 1926. von Neumann was a Rockefeller Fellow then at University of Göttingen, Germany, teaching mathematics at the University of Berlin, Germany. There his basic interest in quantum mechanics and mathematical logic was stimulated. In 1930, von Neumann spent one year as visiting professor of mathematical physics at Princeton University, and received in 1931 an appointment there. In 1933 he joined the staff of the Institute of Advanced Study, Princeton, of which he remained a member until his death.

Recognizing the inevitable armed conflict with German Nazi Regime, von Neumann contributed to the realization of this danger by the American colleagues and also to the defence of the USA. Beginning in 1940 he was a member of governmental agencies, including the technical advisory panel on atomic energy of the Department of Defence and its weapons evaluation group. He also served as chairman of the nuclear weapons committee of the US Air Force scientific advisory panel. His early interest resulted also in scientific understanding of the nature of shock waves. His realization of the practical importance of obtaining solutions to non-linear differential equations stimulated his interest in the currently accepted means of obtaining such solutions with computers. After the war von Neumann returned to the Institute of Advanced Study, where he was largely responsible for the development and construction of the mathematical analyzer, numerical integrator and the MANIAC computer completed in 1952. A colleague of his stated 'If he analyzed a problem, it was not necessary to discuss it further'.

Anonymous (1980). von Neumann, John. *Dictionary of American biography* Suppl. 6: 655-656. Scribner's: New York.

Bochner, S. (1958). John von Neumann. *Biographical memoirs* 32: 438-457. National Academy of Sciences: Washington DC. *P*

Neumann von, J., Richtmyer, R.D. (1950). A method for the numerical calculations of hydrodynamic shocks. *Journal of Applied Physics* 21(3): 232-237.

Neumann von, J. (1953). On ocean wave spectra. Beach Erosion Board, *Technical Memo* 43.

Neumann von, J. (1961). *Collected works*, A.H. Taub, ed. Pergamon Press: Oxford UK.

Poggendorff, J.C. (1953). von Neumann, Janos (John). *Biographisch-Literarisches Handwörterbuch* 6: 1845; 7b: 3592-3595, with bibliography.

NEWELL

05.03. 1862 Bradford PA/USA
05.07. 1932 Bradford PA/USA

Frederick Haynes Newell graduated in 1885 with the BA degree in mining engineering from the Massachusetts Institute of Technology MIT. He met in 1888 John W. Powell (1834-1902), then head of the US Geological Survey USGS at Boston MA, and Grove K. Gilbert (1843-1918). Powell organized the Irrigation Survey within USGS to map potential dams in the West, asking Newell to take charge of a group of young engineers to do the job.

Newell neither was a hydraulic engineer, nor ever designed any irrigation project, but knew lots about the nature of rivers. In addition, he had the right political and scientific connections. In 1890 he became member of a professional club discussing at Washington the critical natural resources issues facing then the nation. During the following years he became active in associations as the National Geographical Society NGS, lecturing before scientific and engineering public, more often on forestry than on hydrology. He became one of the close advisors of President Theodore Roosevelt on natural resources, and it was only natural that when the Reclamation Act passed Congress in 1902, the USGS would administer the new program with Newell as director. The start was excellent but there was an accumulation of projects resulting in 1915, when Newell left the Service, the mistakes became too obvious. Already by 1909 the Service had been bombarded with complaints so that Newell had become defensive and evasive. It was stated that his principal weakness was the inability to say 'No', and taking up too much work. Up to 1906 he had completed 28 government projects in Kansas, Nevada, Arizona. As head of the Reclamation Service he supervised the construction of 100 dams, 40 km of tunnels and 2000 km of canals and ditches, supplying water to 20,000 farmers. The Shoshone Dam in Wyoming completed in 1910 was with 100 m then the highest of the world. In 1915 Newell became head of the Civil Engineering Department, University of Illinois, Urbana IL. He was awarded in 1918 the NGS Cullum Gold Medal.

Anonymous (1921). Newell, F.H. *American men of science* 3: 504. Science Press: New York.
Newell, F.H. (1906). *Irrigation in the United States*. University of Wisconsin: Madison.
Newell, F.H. (1913). *Principles of irrigation engineering*. McGraw-Hill: New York.
Newell, F.H. (1920). *Water resources*: Present and future uses. Yale Univ. Press: New Haven.
Rouse, H. (1976). Frederick H. Newell. *Hydraulic engineering in the United States* 1776-1976: 65. The University of Iowa: Iowa City. *P*
http://www.waterhistory.org/histories/newell/. *P*

NEWTON I.

04.08. 1837 New York NY/USA
25.09. 1884 New York NY/USA

Isaac Newton graduated from New York University as an engineer in 1856. He was apprenticed by his father, who was related with steam navigation, in the machine shop of Cornelius H. Delamater (1821-1889), and on a North River steamboat. He was in 1858 engineer on steamers between New York and Liverpool, terminating in 1859 with a professional study tour through Europe.

Newton was commissioned during the Civil War as first assistant engineer by the US Navy, ordered on a steam frigate in the blockade of Charleston Harbor.

On the application of famous John Ericsson (1803-1889) he was assigned in 1861 to the construction of the then proposed ironclad *Monitor*, which was launched early in 1862. Newton was in charge of its engines during the historic combat following the rebel ram Merrimac, thereby marking a recognized era in the history of steam navies. In August 1862 he became superintendent of ironclad construction in New York for three years. He resigned from the Navy at the close of the war, taking then charge of a coal and iron company. In 1868 he was engaged in the reconstruction of Stevens' Steam Battery at Hoboken NJ. In the early 1870s Newton was involved in the design and construction of a 100 ton floating derrick. Until 1880 he was then engaged by the British Colony Prince Edward Island as a consulting engineer for railways projects, in locomotive engine construction, and in drainage works. He was in 1881 appointed chief engineer of the Department of Public Works, New York City, where he was involved in the water supply of the city. Within a short time he presented a project for this great engineering work, which was approved by the city council. His design was essentially the general outline of the new Croton Aqueduct which was then executed around 1900 by Edward Wegmann (1850-1935). His services to the city of his birth in the professional ability brought to this work, and the deep personal devotion given to it by him, remain a part of its history. Newton's attention to and thorough absorption in these studies undermined his health, however, thereby shorting his life considerably. He had been a member of the American Society of Civil Engineers ASCE from 1880. The Isaac Newton Papers are at the Mariners' Museum Library, Christopher Newport University, Newport News VA.

Anonymous (1885). Isaac Newton. *Proc. ASCE* 11: 128-129.
isaac_newton_papers.pdf
http://www.google.ch/imgres?q=%22Isaac+Newton%22+Monitor *P*

NEWTON J.

24.08. 1823 Norfolk VA/USA
01.05. 1895 New York NY/USA

John Newton graduated from US Military Academy in 1842 and there remained until 1846 as assistant professor of engineering. He was then until 1852 assistant engineer for construction of forts. In 1853 he was involved in the improvement of St. John's River in Florida, and on surveys for the breakwater at Owl's Head ME. From 1855 he was engaged as superintendent in the improvement of lighthouses on Savannah River, in the rank of captain. He was appointed in 1858 member of the Board to examine the floating dock at Washington Navy Yard, and until 1861 was superintending construction engineer of Fort Mifflin DE. He was in 1860 member of a special board on harbour defences in New York Harbour, and in 1861 chief engineer of the Department of Shenandoah, in the rank of brigadier general of Volunteers. In 1862 he served as engineer in defences of Washington DC. He left war service, once the Civil War had terminated, in 1865 in the rank of lieutenant colonel of engineers.

After the Civil War Newton was returned to fortification, river and harbour work, He was for example in charge for the removal of a large stone in the East River NY, which had caused many wrecks and previously had been unable to be removed to improve navigation. He was also in charge to blow up the Flood Rock or the Middle Reef at Hell Gate in 1885 on the same river. He had been promoted to colonel in 1879, becoming brigadier general and Chief of Engineers in 1884. After his successful works at Hell Gate he was retired at his own request in 1886. He then served as Commissioner on Public Works, New York City, until 1888, and declined a re-appointment. He was a Honorary Member of the American Society of Civil Engineers ASCE, of the National Academy of Sciences, and was awarded the Honorary Degree LL.D. of St. Francis Xavier College, New York NY, in 1886. He also acted as president of the Panama Railroad Company until his death.

Eads, J.B., Merrill, W.E., Newton, J., Mansfield, S.M. (1886). Discussion of The South Pass jetties: Ten years of practical teachings in river and harbour hydraulics, by E.L. Corthell. *Trans. ASCE* 15: 276-316.

FitzSimons, N., ed. (1991). John Newton. *A biographical dictionary of American civil engineers* 2: 81-82. ASCE: New York.

Newton, J. (1886). Improvement of East River. *Popular Science Monthly* 28(2): 433-448.

http://www.nndb.com/people/491/000103182/ P

NICHOLS

13.11. 1907 Cleveland OH/USA
21.02. 2000 Bethesda MD/USA

Major General Kenneth David Nichols was an US Army officer and an engineer. He worked both on the Manhattan Project during World War II, thereby responsible for both the uranium and plutonium productions. He became general manager of the Atomic Energy Commission in 1953, promoting the construction of nuclear power plants. In later life he became an engineering consultant. Nichols is here retained for his achievements before WWII.

Nichols graduated from the US Military Academy, West Point NY, in 1929, and then went to Nicaragua for conducting a survey of the Inter-Oceanic Nicaragua Canal. On return to the USA in 1931 he went to Cornell University, Ithaca NY, receiving the BS degree. In 1932 he was appointed assistant to the Director of the Waterways Experiment Station, Vicksburg MS. In 1934 he received a fellowship to study European hydraulic research methods at the *Technische Universität*, Berlin, Germany. The resulting 1935 paper won an award of the American Society of Civil Engineers ASCE. On returning to the USA he received another year posting at the Waterways Experiment Station. From 1936 to 1937 Nichols was a student officer at Fort Belvoir VA. He then became a student again, using his Berlin thesis as the basis for a PhD degree from the State University of Iowa. This study analyses 8 specific model studies for effects of geometric distortion. The analysis has indicated that a lesser degree of distortion may be required in movable bed models if light weight material is used to simulate the stream bed. This technique is currently used to avoid effects of sediment viscosity. Nichols made the following conclusions: 1. Geometrically distorted river models should be avoided, 2. Distorted models should be designed for specific purposes, 3. The effects of distortion should be kept in the eye in view of prototype up-scaling, and 4. particular details should not be studied using these models. This paper won the 1940 Collingwood Award. Nichols then became instructor at West Point, promoted to captain in 1939.

Nichols, K.D. (1935). Discussion of River laboratory hydraulics. *Trans. ASCE* 100: 159-163.
Nichols, K.D. (1937). Observed effects of geometric distortion in hydraulic models. *PhD Thesis*. The University of Iowa: Iowa City IA.
Nichols, K.D. (1938). Observed effects of geometric distortion in hydraulic models. *Proc. ASCE* 64(6): 1081-1102, also *Trans. ASCE* 104: 1488-1509.
http://en.wikipedia.org/wiki/Kenneth_Nichols *P*
http://chl.erdc.usace.army.mil/Media/8/5/5/Chap2.htm

NICOLLET

24.07. 1786 Cluses, Savoy/F
11.09. 1843 Washington DC/USA

Before having emigrated to the United States Joseph Nicolas Nicollet was a professor of mathematics at the Collège Louis-le-Grand, and astronomer at the Paris Observatory with Pierre-Simon Laplace (1749-1827). Political and academic changes in France led him to move in 1832 to the USA to do there work, strengthening his reputation among academics in Europe. He had the goal of using his expertise to map the Mississippi Valley. He travelled to New Orleans LA and there made travels to Baltimore, arriving at St. Louis MO in 1835.

Nicollet took from St. Louis a boat up the river to Fort Snelling MN, leading three expeditions to explore the Upper Mississippi mostly in the current States of Minnesota, North and South Dakota. The first expedition took place from 1836 to 1837, taking the group to the Mississippi source at Lake Itasca and the nearby Minnesota tributary, the St. Croix River. An error of an elder map was corrected which had placed the mouth of the Crow Wing River too far to the west, rendering all maps of this area inaccurate. Upon his return to Washington DC he was appointed head of the newly-formed Corps of Topographical Engineers and asked to lead his second expedition to map the area between the Mississippi and Missouri Rivers to correct the western maps. This tour started in 1838 at Traverse des Sioux MN, continued through Pipestone Quarry MN, from where the party proceeded along the Minnesota and Blue Earth Rivers toward Spirit Lake IA. During the third expedition in 1839, Nicollet made his own maps of the Missouri River basin, taking him and his party northwest from Iowa along the Missouri River toward Fort Pierre SD. His efforts were hampered by the sinking of his steamboat which was carrying supplies for the expedition. He then set out for Devil's Lake ND, from where he returned across the Coteau des Prairies to Fort Snelling. After return to Washington DC, his failing health led to his death, but he had been able to publish the results of his expeditions in his 1843 book. The maps of this book were highly accurate covering a region more than half of the size of Europe. His name was applied at places as the Nicollet Island, the Nicollet County, or Nicollet City, all in Minnesota.

Nicollet, J. (1843). *Joseph N. Nicollet on the plains and prairies*: The expeditions of 1838-39. Washington DC.
Nicollet, J., Fremont, J.C. (1843). *Map of the hydrographical basin of the Upper Mississippi*. Bureau of the Corps of Topographical Engineers: Washington DC.
http://en.wikipedia.org/wiki/Joseph_Nicollet *P*

NOBLE

07.08. 1844 Livonia MI/USA
19.04. 1914 New York NY/USA

Alfred Noble obtained in 1870 the civil engineering degree from the University of Michigan, Ann Arbor MI. He there was awarded the LLD degree in 1895, as also from the University of Wisconsin, Madison WI. He was from 1870 to 1872 engaged with surveys of harbours on Lake Michigan, and then with the improvements of St. Mary's River between Lakes Superior and Huron, and the enlargement of the St. Mary's Falls Canal at Sault St. Mary until 1882. He acted as assistant engineer under Godfrey Weitzel (1835-1884) in constructing the Weitzel Lock there. From 1886 he was resident engineer under William Hutton (1826-1901) for the Washington Bridge in New York City, and from then was involved in a number of prestigious bridge projects, among which were Cairo Bridge over Ohio River, or Memphis Bridge over the Mississippi River. He was from 1894 to 1897 consulting engineer at Chicago IL.

Noble was from 1895 member of the Nicaragua Canal Commission examining the lines of the Nicaragua and Panama Canals. He further was until 1900 member of the Deep Waterways Commission to study ship-canal routes from the Great Lakes to the sea. From 1899 to 1903 he was member of the Isthmian Canal Commission charged with determining the route of Panama Canal. He further served on the Board of Engineers to advise on the projected State Barge Canal in New York City. He was also associated with a bridge across Mississippi River at Thebes IL until 1905. As member of the Board of Consulting Engineers on Panama Canal he recommended a lock canal in 1905, which was finally executed. Other services from 1909 to 1914 included the Galveston Seawall design, the New Welland Canal, Catskill Aqueduct in New York State and the Pearl Harbour Dry Dock in Hawaii. He was awarded the John Fritz Medal of the American Society of Mechanical Engineers ASME in 1910, or the Elliott Cresson Medal from the Franklin Institute in 1912. He served as president of the Western Society of Engineers in 1897, and in 1903 of the American Society of Civil Engineers ASCE.

FitzSimons, N, ed. (1972) Noble, Alfred. *Biographical dictionary of American civil engineers*: 94-95. ASCE: New York. *P*

Noble, A. (1904). *The Isthmian Canal from an engineer's standpoint*. New York.

Noble, A. (1911). Dams. *American civil engineers' pocket book* 11: 1024-1126. Wiley: NY.

Riggs, H.E. (1948). Alfred Noble. *Michigan and the Cleveland Era*: 183-193, E.D. Babst, ed. University of Michigan Press: Ann Arbor. *P*

NORDIN

14.09. 1929 Albuquerque NM/USA
25.05. 1998 Fort Collins CO/USA

Carl Frederick Nordin, Jr., worked from 1954 to 1984 for the US Geological Survey USGS, including the direction of research at Colorado State University, Fort Collins CO. He in parallel was there a part-time professor. Since then, he was a consultant on sediment-related problems worldwide, including the Three Gorges Project on Yangtze River, China, or the *Xiaolangdi* Project on Yellow River, China. He served on committees of the American Geophysical Union AGU, and the National Research Council, and was member of the American Society of Civil Engineers ASCE, and the International Association of Hydraulic Research IAHR.

Nordin worked on research projects for the US Army Corps of Engineers, the Federal Highway Administration and was also project director for the Water Resources Division of USGS. His work took him around the world to China, Egypt, and the Amazon. The Nordin Collection of ColoState consists of photographs, reports, correspondence, and maps. His work on water in Colorado included the alluvial river flows of Colorado River. Further reports were made on Lake Nasser in Egypt, the Amazon River, and River Orinoco in Venezuela and Columbia. His papers include works in sediment hydraulics, of which the study on scour as related to bridges may be mentioned. In another work, he made a critique of the classical regime theory, in which relations among rivers parameters are given, mainly based on a semi-theoretical approach.

Anonymous (1972). Carl Nordin. *Eos* 53(12): 1152. *P*
Anonymous (1991). Carl F. Nordin. *Civil Engineering* 61(12): 82. *P*
Anonymous (1998). Carl F. Nordin. *Civil Engineering* 68(9): 88.
Lagasse, P.F., Nordin, C.F. (1991). Scour measuring and monitoring equipment for bridges. National Conf. *Hydraulic engineering*: 311-316, E.V. Richardson, ed. ASCE: New York.
Nordin, C.F., Algert, J.H. (1966). Spectral analysis of sand waves. *Journal of the Hydraulics Div.* ASCE 92(HY5): 95-114; 93(HY3): 228-229; 93(HY4): 310; 94(HY5): 1336-1338.
Nordin, C.F. (1971). Statistical properties of dune profiles. USGS *Professional Paper* 562-F.
Sabol, G.V., Nordin, C.F. (1978). Dispersion in rivers as related to storage zones. *Journal of the Hydraulics Division* ASCE 104(HY5): 695-708.
Stevens, M.A., Nordin, C.F. (1987). Critique of the regime theory for alluvial rivers. *Journal of Hydraulic Engineering* 113(11): 1359-1380.
http://comment.colostate.edu/index.asp?page=display_article&article_id=70995232

NORTH

16.07. 1835 Hartford CT/USA
20.07. 1911 New York NY/USA

Edward Payson North entered in 1854 the Union College but did not make a degree there. He joined in 1856 Ellis S. Chesbrough (1813-1886) as assistant engineer for Chicago's sewerage system. In 1864, after problems with his eyesight had been restored, he was appointed assistant engineer on a railways project. From 1873 to 1875 he served as principal assistant the US Corps of Engineers on improvement works of the Upper Mississippi between the Falls of St. Anthony and St. Cloud MN. His 1877 paper dates from these days. After works for New York City, he prepared from 1878 the contribution of the American Society of Civil Engineers ASCE to the *Exposition Universelle* in Paris. This exhibit consisted mainly of designs and photographs of the main American engineering works, notably dams, locks, bridges, hydraulic machinery, river and harbour improvements, and railroads; it was awarded at North's visit to Paris, during which he also studied the European techniques of road preparation. A paper published in 1878 on this topic was given the 1879 ASCE Norman Medal. He in addition wrote with Clemens Herschel (1842-1930) a notable book on the Science of road making.

North attended the 1892 Intl. Congress on Inland Navigation, held at Paris, presenting a work 'On the relations between railroads and waterways in the United States'. He also studied the canal systems in Holland, river control schemes in France, and railroads in Italy. From 1897 he was consulting engineer of the Department of Public Works, New York City. He also was consultant to the canal enlargement of New York State. He has written during these years papers 'characterized by earnestness, thorough study, and sincerity'. His technical education and the broad field of practical experiences led to outstanding papers in the then leading engineering journals, namely these of ASCE, but also in Engineering News, and in the Engineering Record. He contributed discussions to topics as varied as to the Inter-oceanic canal projects, Temperature of water at various depths in lakes and oceans, or Economic depth of canals. It was stated in his obituary 'That gentleness, which, when it mates with manhood, makes a man'.

Anonymous (1911). Edward P. North. *Engineering News* 66(4): 127. *P*
Anonymous (1912). North, E.P. *Trans. ASCE* 75: 1167-1176. *P*
Herschel, C., North, E.P. (1890). *The science of road making*. Engineering News Publ.: NY.
North, E.P. (1877). Wing dams in the Mississippi above the Falls of St. Anthony. *Trans. ASCE* 6: 268-276.
North, E.P. (1900). Canals from the Lakes to New York. *Proc. ASCE* 26: 1203-1223.

NOYES

10.07. 1850 South Boston MA/USA
12.10. 1896 Newton MA/USA

Albert Franklin Noyes was educated at Lawrence Scientific School, Harvard University, Cambridge MA. He was from 1871 employed in topographical work and landscape design. In 1874 he entered the office of the city engineer at Newton MA, and was in 1876 appointed there city engineer. He was one of the first to join the New England Waterworks Association, becoming member of its Executive Committee in 1887, and president in 1890. He also was until 1893 city engineer of Newton MA, being much involved with the extension and improvement of the water works. For the next two years he was assistant chief engineer of the Massachusetts State Board of Health, whereas he was during the time of his death consulting engineer, with a specialty in water works design.

Noyes presented various papers to the Journal of the New England Water Works, including Drive wells as a means of water supply, where he described an alternative approach to supply drinking water. Another paper deals with The Metropolitan water supply of Massachusetts, in which the outstanding role of water supply engineering in his State is highlighted, together with an outlook to future tasks to improve the quality of drinking water supply. Noyes was further a member of the Boston Society of Civil Engineers, for which he served also as vice-president and president, as of the American Public Health Association, and the American Society of Civil Engineers ASCE. He was described as a person with straightforward honesty and good will making him not only a favourite within professional associations, but also to all who knew him. While his advice on varied subjects was sought because of his wide experience, it was influenced by nothing but the best interests of those who sought his advice. He may be considered therefore as an early engineer both having developed water sciences, and his profession.

Anonymous (1897). Albert F. Noyes. *Journal of the New England Water Works Association* 11(3): 275-276. *P*
Anonymous (1897). Albert F. Noyes: A memoir. *Proc. of the Association of Engineering Societies* 18(1): 10-14. *P*
Noyes, A.F. (1886). Driven wells as a means of water supply. *Journal of the New England Water Works Association* 1(4): 19-27.
Noyes, A.F. (1891). Presidential annual address. *J. New England Water Works* 6(1): 5.
Noyes, A.F. (1895). The Metropolitan water supply of Massachusetts. *Journal of the New England Water Works* 10(2): 117-129.

OAKLEY

14.08. 1878 Roslyn NY/USA
06.11. 1936 Port Washington NY/USA

George Israel Oakley graduated with the BE degree in 1902 from Union College, Schenectady NY. He was then engaged mainly with railways projects. From 1905 to 1915 he was employed by the New York State Department on its Barge Canal. He was responsible for topographical surveys, rock borings, land and river drive-rod soundings through Mohawk River Valley. He also established measurements of stream flow data, and was in charge of pipe wells for studies on the groundwater movements affected by floods. He took an important part in the hydraulic and hydrological studies necessary for preparing the defence before the State Court of Claims.

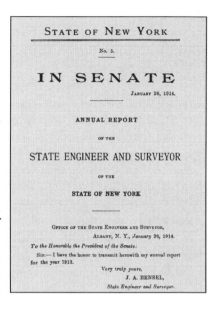

From 1915 Oakley was city engineer of Little Falls NY, executive officer of the Board of Public Works, and chief engineer of the Little Falls Water Works. Its plant consisted of four reservoirs, a slow sand filter, 30 km of transmission and distribution mains, a chlorine purification plant, and a metering system. Oakley was in charge of all design, construction and maintenance of bridges, sewers and the water works system. He recommended the Board of Public Works improvements including purchase of land for sanitary protection of the watershed; a general yearly plan for the reforestation of the cleared land on this watershed; rebuilding of a large sand filter; and a general plan for the conservation of the water supply. He previously collected data included the quantity of water from various springs and streams, and detailed rainfall records. From 1917 to 1919 he was responsible for the design and supervision of the reconstruction of a 0.5 m hydraulic dredge to be used for the Boston and New York Harbours. From 1922 to 1930 Oakley was employed by an engineering corporation as design engineer in its Hydraulic Department, and from then he served the Board of Water Supply, New York NY as an assistant engineer. He was associated with the American Society of Civil Engineers, as junior from 1903, associate member from 1907, and as member from 1922.

Anonymous (1937). George I. Oakley. *Trans. ASCE* 102: 1623-1626.

Bensel, J.A. (1914). *Annual Report* of the State Engineer and Surveyor of the State of New York. New York City. (*P*)

Oakley, G.I. (1902). The design of an interseptic sewer and sewage disposal plant for Schenectady. *Daily Times* (June 11): Troy NY.

Weidman, B.S., Martin, L.B. (1981). George I. Oakley. *Nassau County Long Island in early photographs* 1869-1940: 65. Dover: New York. *P*

OBER

20.05. 1871 Beverly MA/USA
30.08. 1931 Seattle WA/USA

Ralph Hadlock Ober received his education at the Massachusetts Institute of Technology, Cambridge MA. He was engaged in 1893 in surveying at his home town, but moved then to Pacific Northwest, serving from 1894 as instrumentman on land surveys in Washington State. In 1895 he was inspector of the construction of Cedar Lake storage dam. From 1897 he was employed as engineer with the US Army Corps of Engineers in Washington State, in charge of river surveys, of electric power houses at Flagler WA. From 1906 he served as consultant the Wenatchee High Line Canal Company in irrigation development, whereas in 1908 he was resident engineer in charge of construction of a bridge across Columbia River at Beverly WA. From 1908 to 1914 Ober was assistant engineer of Seattle WA, in charge of the study of the Cedar River Watershed for city water supply purposes, and for an additional hydro-electric power plant for the city. He was then engaged in private practice at Seattle, specializing in municipal, and river and harbour work.

Following the entry of the USA in World War I, Ober was commissioned a captain in the Engineers, stationed at Camp Humphreys VA, where he remained until 1920. He there served as instructor in civil engineering, receiving high praise from his colleagues and was awarded the Victory Medal. He then returned to the US Army Corps for three years with assignments on channel rectification, dredging, and construction of locks and dams along the 400 km stretch of the Ohio River, with headquarters at Louisville KY. He then returned to Seattle to resume his private practice. Among the projects received was an irrigation design, or the Skagit River power investigation for the city of Seattle. Ober was an able engineer and a gentleman of fine personality. His strict devotion to duty, and his capacity for thoroughness in all matters traced his character. He was president of the Engineers' Club of Seattle, and of the Pacific Northwest Society of Engineers, and from 1907 member of the American Society of Civil Engineers ASCE.

Anonymous (1931). Ralph H. Ober. *Engineering News-Record* 107(11): 430.
Anonymous (1940). Ralph H. Ober. *Trans. ASCE* 105: 1894-1896.
Ober, R.H., Johnson, H.C. (1913). *Report* on the reforestation of the lands in the Cedar River Watershed. Seattle.
http://www.flickr.com/photos/seattlemunicipalarchives/4907605377/ *P*
http://img9.fold3.com/img/thumbnail/150659369/278/0/0_0_284_421.jpg *P*

O'BRIEN

21.09. 1902 Hammond IN/USA
29.07. 1988 Cuernavaca/MX

Morrough Parker O'Brien obtained his engineering education at Massachusetts Institute of Technology MIT and from Purdue University, West Lafayette IN. He was an engineer of the Hudson River Regulating District from 1925 to 1927, then research assistant of the Engineering Experiment Station of Purdue University, when leaving for Europe as a Freeman Scholar. Upon return to the USA in 1928 he was appointed professor of engineering until retirement in 1959 at the University of California, Berkeley CA, and there acted also as director of research and engineering. He was in parallel from 1930 consultant to the Aircraft Gas Turbine Division, General Electric Co. O'Brien also was a civilian member of the US Beach Erosion Board from 1938, and a member of the US Army Scientific Advisory Panel from 1951. He was a member of the American Societies of Mechanical Engineers ASME, and Civil Engineers ASCE.

O'Brien's first assignment at University of California included the hydraulic laboratory and courses in hydraulics and hydraulic machinery. He had combined his qualities in teaching, research and professional engineering. In 1929 he initiated a research program on shoreline processes and coastal engineering. He later maintained an active interest in ocean waves and shoreline phenomena. His research in hydraulics led to practical applications by industry. Most of the jet pumps then sold in the USA followed his design. Application of the airfoil theory to the design of propeller pumps provided the basis for extensive production of low-head, high-capacity pumps for irrigation and drainage. He was elected Honorary Member ASCE in 1976.

Anonymous (1952). M.P. O'Brien. *Engineering News-Record* 149(Jul.17): 71. *P*
Anonymous (1961). M.P. O'Brien. *Mechanical Engineering* 83(3): 112-113.
Anonymous (1965). Morrough P. O'Brien. *Civil Engineering* 35(12): 73. *P*
Anonymous (1976). Morrough P. O'Brien. *Civil Engineering* 46(10): 95. *P*
Anonymous (1989). M.P. O'Brien dies at 85. *Civil Engineering* 59(2): 82.
O'Brien, M.P., Cherno, J. (1934). Model law for motion of salt water through fresh. *Trans. ASCE* 99: 576-609.
O'Brien, M.P., Johnson, J.W. (1934). Velocity-head correction for hydraulic flow. *Engineering News-Record* 113(7): 214-216.
O'Brien, M.P., Folsom, R.G. (1937). Modified ISA orifice with free discharge. *Trans. ASME* 59(RP-1): 61-64; 59(7): 756-757.

OCKERSON

04.03. 1848 Slattarod/S
22.03. 1924 St. Louis LA/USA

John Augustus Ockerson was born in the Province of *Skane*, South Sweden. At age twelve, his family emigrated to the United States. After participation in the Civil War, he graduated in 1873 as a civil engineer from the University of Illinois. He started as an assistant engineer in the Federal Great Lakes Survey and there was engaged as hydrographer. When in 1879 the Mississippi River Commission was established, he was appointed in charge of surveys from the river source to the Gulf. After a short pause, he rejoined his former commission from 1890 to 1898.

Ockerson's expert knowledge in all matters relating to river improvement made him a leading personality. One of his greatest individual achievements undertaken in 1910 was the construction of levees to control the flood waters of Colorado River. This was not only an engineering achievement, but since some constructions had to be done on Mexican territory, was also a task involving many delicate diplomatic problems. For the successful completion of this work, Ockerson received personal commendation from President Taft. He served in addition as US delegate for the Permanent International Association of Navigation Congress PIANC from 1900 to 1912. In 1912 he was elected president of the American Society of Civil Engineers ASCE. His numerous publications were mostly in the form of reports issued with his work on the Mississippi River Commission. Of particular relevance is his 1898 paper on dredges and dredging. Ockerson was an outstanding personality, possessing a powerful and commanding physique and charm of manner. He was a persuasive talker and a highly respected member in his profession. He died after an apoplectic stroke.

Anonymous (1934). Ockerson, John Augustus. *Dictionary of American biography* 13: 617-618.
        Scribner's: New York.
Griggs, F.E., Jr., ed. (1991). Ockerson, John A. *A biographical dictionary of American civil engineers* 2: 83-84. ASCE: New York. *P*
Ockerson, J.A. (1893). Erosion of river banks on the Mississippi and Missouri Rivers. *Trans. ASCE* 28: 396-424; 31: 1-28.
Ockerson, J.A. (1898). Dredges and dredging on the Mississippi and Missouri Rivers. *Trans. ASCE* 40: 215-354.
Ockerson, J.A. (1901). The Lower Mississippi River. *Proc. ASCE* 27: 139-162.
Ockerson, J.A. (1922). Flood control on the Mississippi River. *Trans ASCE* 85: 1461-1481.

ODONI

28.05. 1879 Luzern/CH
09.08. 1936 Los Angeles CA/USA

Vincent Phillip Odoni was born in Switzerland, and graduated as a civil engineer from the Swiss Federal Institute of Technology ETH, Zurich, in 1904. He spent his first years of practice at Lucerne, designing water supply systems, and hydro-electric plants, and was then in charge as construction engineer with the *Chemin de Fer Electric de la Gruyère*, Bulle FR, a railroad company.

Odoni came to the United States in 1907, finding employment at Los Angeles CA, in the design and construction of works for various water supply and irrigation projects. He moved in 1909 to Denver CO, where he was involved in the design of the Boulder and Shoshone Power Plants with the Central Colorado Power Co. From 1910 he was in charge of works for irrigation systems, dams, and hydro-electric power developments. He then moved to Tucson AZ as chief engineer of the Tucson Farms Co. from its project inception until its completion in 1917, including extensive studies of the movement and yield of groundwater supplies; the full development of this project saw more than eighty well-pumping plants and a successful irrigation enterprise in a semi-arid region. From 1918 Odoni was at Dayton OH in charge of the gigantic project of the Miami Conservancy District, from where he returned to Tucson as chief engineer and vice-president of the former Company, in charge of its irrigation system. He was called in 1923 as chief engineer to the Haitian-American Sugar Co, Port au Prince HT. This work included the drilling and the equipment of irrigation wells, the development of irrigation systems, the operation of pumps, the construction of pipelines, and the study of soils and drainage systems for the sugar plantations. He left the Sugar Company in 1931 because of poor health, and returned to California. He there studied the latest achievements of engineering science and enjoyed arts to ease his sufferings. He was an engineer of ability, with the capacity for original research and yet being able to adapt himself to circumstances, and master the many problems encountered as an engineer in remote places. He was there thrown on his own resources, responsible for the construction, maintenance and operation of the various plants and structures, called upon to meet emergencies with unskilled labour, requiring special ability to fill such a position. He was member of the American Society of Civil Engineers ASCE since 1916.

Anonymous (1937). Vincent P. Odoni. *Trans. ASCE* 102: 1626-1628.
http://search.ancestry.co.uk/cgi-bin/sse.dll?gl=43&rank=1&sbo=t&gsbco=Sweden&gsln=Odoni&gss=angs-d&uidh=000 *P*

OKEY

21.10. 1886 Corning IA/USA
04.11. 1971 Knoxville TN/USA

Charles Williams Okey graduated in 1909 from the Iowa State University, Iowa City IA, with the BSc degree in civil engineering. He was then employed by the US Department of Agriculture on studies of drainage and flood control projects in the Louisiana, Kansas, Oklahoma, and Florida States. His reports were employed for constructing drainage works. The 1915 Report deals with land drainage using pumps, written in collaboration with Sherman M. Woodward (1871-1953). Okey joined in 1916 an engineering company as superintendent of construction and as engineer of flood control. He was in charge as the

engineer on a pipeline construction at the Old Hickory Powder Plant, Nashville TN. He joined in 1919 another company at Memphis TN, as assistant chief engineer on reports on land drainage projects in the southeast. In 1926 he was employed at St. Louis MO.

Okey joined in 1935 the Tennessee Valley Authority TVA, Knoxville TN, remaining there for 21 years until his retirement. His duties included surveys and reports on flood control, and problems connected with the construction of TVA's reservoir system. One of the projects he was involved was the construction of Norris Dam on the Clinch River TN. Its construction in the late 1930s was the first major TVA project. The dam is a straight concrete gravity-type dam 570 m long and 81 m high. Norris Lake, the largest reservoir on a tributary of the Tennessee River, has 140 km^2 water surface and 1,300 km shoreline. Okey became in 1945 chief of the Flood Control Branch, responsible for the complicated studies of reservoir operations during floods, and for planning flood control projects throughout the Tennessee Valley. He was a Fellow ASCE from 1959.

Anonymous (1942). C.W. Okey. *Civil Engineering* 12(6): 340. *P*
Anonymous (1972). Charles W. Okey. *Trans. ASCE* 137: 1069-1070.
Crawford, C.W. (1970). *Oral history of the Tennessee Valley Authority*: Interview with Charles Okey. Memphis State University: Memphis TN.
Okey, C.W. (1914). The wet lands of southern Louisiana and their drainage. *Bulletin* 71. US Department of Agriculture: Washington DC.
Okey, C.W. (1918). The subsidence of muck and peat soils. *Trans. ASCE* 82: 396-432.
Okey, C.W. (1939). *Value of flood height reduction from Tennessee Valley Authority reservoirs to the alluvial valley of the Lower Mississippi River*. TVA: Knoxville TN.
Woodward, S.M., Okey, C.W. (1915). Land drainage by means of pumps. *Bulletin* 304. US Department of Agriculture: Washington DC.

OKUN

19.06. 1917 New York NY/USA
07.12. 2007 Chapel Hill NC/USA

Daniel Alexander Okun obtained the BS degree from the Cooper Union Institute, New York NY, in 1937, and the MS degree from the California Institute of Technology, Pasadena CA, in 1938. After war service he attended Harvard University, Cambridge MA, earning his PhD degree in sanitary engineering. His work led to the use of pure oxygen in wastewater treatment. Okun joined in 1952 University of North Carolina, Chapel Hill NC, where he remained until 1982, serving as the chairman of the Department of Environmental Sciences and Engineering from 1955 to 1973, and as Keenan Professor starting from 1973, during which period his Department grew from three to 25 faculty members. After his retirement, he remained actively involved in the profession through writing, lecturing and consulting. His post-retirement work included the campaign to build the Cane Creek Dam and Reservoir at Chapel Hill to ensure the most pristine water source possible for the city and the UNC campus. He had over 40 awards to his name, including one Fulbright, and was a member of the National Academy of Engineering NAE.

Okun was hailed worldwide for his groundbreaking discoveries and for the protection of the pristine water sources, water-resources management, water reclamation and reuse, watershed protection, and technologies and institutional solutions to water supply and wastewater management in developing countries. He worked in 89 countries as teacher and consultant. His work had influenced international policy making for organizations like the World Bank, the United Nations, or the World Health Organization WHO. He was named one of the 125 engineers who 'singularly and collectively helped shape this nation and the world'. He was awarded honorary membership of the American Water Works Association AWWA.

Anonymous (1960). Okun, Daniel A. *American men of science* 10: 3025. Cattell: Tempe AZ.
Anonymous (2008). Daniel A. Okun. *ASCE News* 33(2): 12.
Fair, G.M., Geyer, J.C., Okun, D.A. (1966). *Water and wastewater engineering*. Wiley: New York.
Okun, D.A. (1996). From cholera to cancer to cryptosporidiosis. *Journal of Environmental Engineering* 122(6): 453-458; 123(8): 819.
Singer, P.C. (2010). Daniel A. Okun. *Memorial tributes* 13: 178-182. National Academy of Engineering: Washington DC. *P*
http://www.sph.unc.edu/envr/daniel_a._okun_1917-2007_6498_1957.html *P*

OLBERG

19.08. 1875 St. Paul MN/USA
04.04. 1938 Washington DC/USA

Charles Real Olberg graduated in 1900 from today's George Washington University, Washington DC, with the BSC in civil engineering. He later took a post-graduate work in hydraulics at the University of California, Berkeley CA. From 1900, he was an assistant hydrographer for the US Geological Survey in connection with irrigation studies in Arizona. In 1902 he entered the US Reclamation Service, as an engineer in connection with the Roosevelt Project in Arizona. In 1907 he went to Los Angeles CA as a supervising engineer in charge of various irrigation works in Southern California. After war service, he was engaged as assistant engineer, US Indian Service, on the construction and design of irrigation structures, the most notable being the Coolidge Dam on Gila River, and the Roosevelt Dam on the Salt River, both in Arizona. The first dam was one of his great engineering works, creating a reservoir of one billion m^3; works started in 1924 and were finished in 1930. The dam height is 75 m, the crest width 170 m, with two Ogee spillways of 3,400 m^3/s design discharge. This structure is a reinforced concrete multiple dome and buttress dam, and represents a part of the San Carlos Irrigation Project.

Olberg held his position at Coolidge Dam from 1919 to 1929. He was then consultant for the Soviet Government on the design of dams and reclamation projects, including the *Mingichaur* Project, and various irrigation, hydro-electric, and drainage projects in Georgia, Armenia, and Azerbaijan. He also served during this period Peru, Argentina and Mexico as a consultant in irrigation work. He returned in 1933 to the USA, as engineer for the Public Works Administration. He had the rare combination of being both theoretical in investigation and design, yet extremely practical in construction and operation. He was known for his kindness to all who came into contact with him. He was in particular never too busy to assist those who had problems, whether they were personal or technical. He was member of the American Society of Civil Engineers.

Anonymous (1940). Charles R. Olberg. *Trans. ASCE* 105: 1896-1898.
Olberg, C.R. (1915). *Report* on Water rights for Gila River and feasibility of San Carlos Project.
US Indian Irrigation Service: Washington DC.
Olberg, C.R. (1915). *San Carlos Irrigation Project*. US Dept. of the Interior: Los Angeles. (*P*)
Olberg, C.R. (1932). Unique levelling method Irrigation Canal in America. *Civil Engng.* 2(2): 114.
Pfaff, C. (1996). *Historic American engineering record*: San Carlos Irrigation Project No. AZ-
50. Department of the Interior: San Francisco.

ORLOB

04.07. 1924 Seattle WA/USA
23.03. 2013 Poulsbo WA/USA

Gerald Thorwald Orlob obtained both his BS and MS degrees at the University of Washington, and the PhD degree from University of California, Berkeley CA, in 1959. He then worked on models describing the movement of salts and pollutants, and thermal pollution of lakes. He was associate professor of civil engineering at UC Berkeley from 1963, joining in 1968 UC Davis CA, chairing its Department of Civil and Environmental Engineering from 1986 to 1988. He retired in 1991. He was a member of the American Society of Civil Engineers ASCE and recipient of its 1963 Hilgard Prize and 1987 Julian Hinds Award. He was elected to the National Academy of Engineering in 1992.

The 1983 book includes the chapters: 1. Introduction, 2. Procedure for modelling, 3. General principles in deterministic water quality modelling, 4. Modeling the ecological processes, 5. Simulation of the thermal regime of rivers, 6. Stream quality modelling, 7. One-dimensional models for simulation of water quality in lakes and reservoirs, 8. Two- and three-dimensional mathematical models for lakes and reservoirs, 9. Ecological modelling of lakes, 10. Modeling the distribution and effect of toxic substances in rivers and lakes, 11. Sensitivity analysis, calibration, and validation, 12. Models for management applications, and 13. Future directions. This book therefore is an early attempt to highlight the basic principles in water quality modelling, and the description and modelling of water quality issues in rivers, lakes and reservoirs.

Krenkel, P.A., Orlob, G.T. (1963). Turbulent diffusion and the reaeration coefficient. *Trans. ASCE* 128(3): 293-334.

Orlob, G.T. (1959). Eddy diffusion in homogeneous turbulence. *Journal of the Hydraulics Division* ASCE 85(HY9): 75-101; 86(HY4): 95-109; 87(HY1): 137-143.

Orlob, G.T. (1961). Eddy diffusion in homogeneous turbulence. *Trans. ASCE* 126: 397-438.

Orlob, G.T., Selna, L. (1970). Temperature variations in deep reservoirs. *Journal of the Hydraulics Division* ASCE 96(HY2): 391-410; 96(HY10): 2154-2158.

Orlob, G.T. (1971). The urban water system: Technical aspects. *Treatise on urban water systems* I(D): 48-63, M.L. Albertson, ed. Colorado State University: Fort Collins.

Orlob, G.T. (1975). Present problems and future prospects of ecological modelling. *Ecological modelling in a resource management framework*: 283-312, C.S. Russell, ed. Baltimore.

Orlob, G.T., ed. (1983). *Mathematical modelling of water quality*. Wiley: New York.

http://www.kitsapsun.com/news/2013/apr/02/gerald-t-orlob-88/#axzz2StNwRCc5 *P*

ORROK

03.07. 1867 Dorchester MA/USA
06.04. 1944 Riverside CT/USA

George Alexander Orrok graduated in 1888 with the BS degree in mechanical engineering from the Massachusetts Institute of Technology MIT. After 2 years as teacher in public schools, he obtained in 1890 manual training at Ogden Military Academy, and was from 1892 to 1898 draftsman and engineer in New York City. He continued from then until 1916 as mechanical engineer with the Edison Co., New York, becoming there then consulting engineer. He was in parallel consulting professor at Brooklyn Polytechnic School from 1911 to 1914, Lecturer at Yale University, New Haven CT from 1920, and from 1927 Lecturer at the Engineering School of Harvard University, Cambridge MA. He was awarded in 1929 Honorary mechanical engineer at Stevens Institute of Technology, Hoboken NJ. He further was the recipient of the ASME Honorary Membership in 1936.

The American Society of Civil Engineers ASCE held in 1925 a Symposium on High specific speed hydraulic turbines in their bearing on the proportioning of the number of units in low-head hydro-electric plants. Contributors were, among others, George A. Orrok, Lewis F. Moody (1880-1953), Forrest Nagler (1885-1952), and George A. Jessop (1883-1954). Orrok reported the large developments and the variety of methods proposed since the War. He also reminded of the notable number of hydraulic laboratories all through the world, in which these processes were then studied. He particularly refers to the Chancy-Pougny Power-plant on the Rhone River at the Swiss-French border, whose modified Francis-type turbines were more than 5 m in diameter, under a head of 8 m, and with a characteristic speed of 120. He also stated the Lawaczeck wheels for the Lilla-Edet Power-plant in Sweden. He noted that Dieter Thoma (1881-1942), professor of hydraulic machinery at Munich Technical University, Germany, devoted a particular interest in the mathematical description of these turbines.

Anonymous (1924). George A. Orrok. *Power* 46(Aug.5): 201. *P*
Anonymous (1933). Orrok, George A. *American men of science* 5: 844. Science Press: New York.
Fernald, R.H., Orrok, G.A. (1927). *Engineering of power plants*. McGraw-Hill: New York.
Orrok, G.A. (1909). Small steam turbines. *Journal ASME* 31(5): 627-651.
Orrok, G.A. (1911). Experimental determination of the rate of heat transmission in surface
    condensers. *Engineering News* 65(6): 162-164.
Orrok, G.A. (1925). High specific speed turbines. *Proc. ASCE* 51(6): 1000-1008.
Orrok, G.A. (1937). *Engineering recollections* 1868-1898. Newcomen Society: Princeton NJ.

ORSZAG

27.02. 1943 New York NY/USA
01.05. 2011 New Haven CT/USA

Steven Alan Orszag received at age of 19 his BS degree in mathematics from Massachusetts Institute of Technology MIT, Cambridge MA, and his PhD degree from Princeton University. He then joined the MIT Faculty in 1967 as professor of applied mathematics, and from 1984 to 1998 the Princeton University as professor of engineering, from when he was on the faculty of the Yale University, New Haven CT. Orszag was awarded the 1986 Fluids and Plasma-dynamics Prize from the American Institute of Aeronautics and Astronautics AIAA, and was a Guggenheim Fellow from 1989, among others. He had been named a highly-cited author by the Web of Knowledge.

Orszag was a pioneer in applied and computational mathematics whose work had a deep influence in the field of fluid mechanics. He specialized in turbulence, computational physics, and mathematics. His work included the development of spectral methods, direct numerical solutions, renormalisation group methods for turbulence, and very-large-eddy-simulations. He achieved the first successful computer simulations of three-dimensional turbulent flows, and developed methods that provide a fundamental theory of turbulence. He also was interested in techniques for the simulation of electronic chip manufacturing processes. His accomplishments in the area of spectral methods include the introduction of fast surface harmonic transform methods for the global weather forecasting, and filtering techniques for shock wave problems. It was stated that 'the intrinsic difficulty of using these methods in nonlinear problems was known to fluid dynamicists and this was a major impediment to progress until Steve developed the transform methods still forming the core of many large-scale spectral computations. Understanding both the mathematics and the computational challenges of implementing them formed an enormous part of Steve's career and influence'. Orszag was the founder or chief scientific adviser to a number of companies, including Flow Research, Vector Technologies or Exa Corp. He was awarded six patents and wrote over 400 papers.

Bender, C.M., Orszag, S.A. (1978). *Advanced mathematical methods for scientists and engineers*: Asymptotic methods and perturbation theory. McGraw-Hill: New York.
Galperin, B., Orszag, S.A. (1994). *Large Eddy Simulation of complex engineering and geophysical flows*. Cambridge University Press: Cambridge UK.
www.aps.org/units/dfd/newsletters/.../fall11.pdf *P*
http://de.wikipedia.org/wiki/Steven_Orszag

OVERTON

04.06. 1938 Huntsville AL/USA
29.08. 2002 Louisville KY/USA

Donald Edward Overton was educated at University of Maryland, College Park MD, where he submitted his PhD thesis in 1972. He had previously been a hydraulic engineer of Soil and Water Conservation Research Division, and re-joined from 1974 as research hydraulic engineer again the Hydrograph Laboratory, US Department of Agriculture USDA, Beltsville MD, from where he moved to University of Tennessee, Knoxville TN, as professor of civil engineering. Overton was member of the American Society of Civil Engineers ASCE, as also of the International Association of Hydraulic Research.

Overton worked from the 1960s mainly in the numerical development of hydrologic, hydraulic, and environmental engineering. In his 1964 paper the runoff characteristics and the routing features are mathematically described. Three successive programs were used, namely the inflow hydrograph program, the flood routing program, and the summing and checking program, based on the storage equation for outflow to storage. The results from these computations were further developed for the Hurricane Creek Watershed in Arkansas. In 1966 the storage flood routing technique and its application to an experimental watershed in Wisconsin were highlighted. The Muskingum flood routing of upland streamflow was also considered in another 1966 paper. The results of flood routing trials in a small experimental water shed indicated that the routing coefficients varied for each storm. By approximating the observed inflow hydrographs by a triangle, a direct solution for these coefficients was possible. These are required to match the peak and time to peak of the outflow hydrograph. By examining various storms, the coefficients were related to the shape of the inflow hydrographs.

Anonymous (1962). Donald E. Overton. *Reveille yearbook*: 425. Univ. Maryland: College Park. *P*

Gburek, W.J., Overton, D.E. (1973). Subcritical kinematic flow in a stable stream. *Journal of the Hydraulics Division* ASCE 99(HY9): 1433-1447.

Overton, D.E. (1966). Muskingum flood routing of upland streamflow. *J. Hydrology* 4(3): 185-200.

Overton, D.E., Meadows, M.E. (1976). *Stormwater modelling*. Academic Press: New York.

Overton, D.E., Troxler, W.L., Bales, J.D. (1978). *A manual for simulating hydrographs in urban regions using kinematic wave theory and SCS curve numbers*. E.P.E. Inc.: Knoxville.

Overton, D.E. (1989). Comparison of Tenn-V and TR-55 on TVA watersheds. *Hydraulic Engineering* '89: 511-517, M.A. Ports, ed. ASCE: New York.

Overton, D.E. (1992). *Soil erosion and sediment yield*. Stormwater Publications: Knoxville.

OWEN W.M.

29.08. 1922 Cave City OH/USA
30.08. 1998 Decatur IL/USA

William Meredith Owen obtained the BS degree in 1943, and the MS degree in 1949 from University of Illinois, Urbana-Champaign IL. He started there in 1946 as a research assistant of theoretical and applied mechanics, and was there in 1948 assistant professor. In the 1950s he was development engineer, Airesearch Manufacturing Co., Los Angeles CA.

The 1950 discussion relates to a paper of Charles W. Harris (1880-1973) on the engineering concept for pipe flows. Based on the approaches proposed in the 19th and the early 20th century, it was realized that the friction factor varies essentially with the Reynolds number and the relative sand roughness height. The results of the paper do not really advance the topic, but the many discussions made allow for improved understanding. Halsey and Owen state the still currently most general Colebrook and White equation, and its particular forms for the turbulent rough and turbulent smooth regimes. The 1952 paper is concerned with the head loss and velocity distribution of annular pipes by centrally supporting pipe cores in outer pipes. The range of Reynolds numbers studied varied from 4,000 to 700,000. The results are compared with previous data, and conclusions relating to the flow conditions are drawn. The 1954 paper details the theoretical equations for laminar flow in long rectangular channels. Data taken at shallow depths on a 0.45 m wide and 6 m long channel show the relation between the friction factor and the Reynolds number. The point of departure of these data from the theoretical laminar flow line corresponds to the critical Reynolds number at which the flow changes from laminar to turbulent. A thread of die was injected into the flow to determine whether the flow was laminar, transitional or turbulent. The two methods resulted in similar findings.

Anonymous (1951). Rope lifts water: H.L. Langhaar, W.M. Owen. *Popular Science* 158(6): 77. *P*
Halsey, J.F., Owen, W.M. (1950). Discussion of Engineering concept of flow in pipes. *Trans. ASCE* 115: 938-941.
Owen, W.M. (1949). An experimental and analytical study of laminar flow and the critical velocity in rectangular open channels. *MS Thesis*. University of Illinois: Urbana.
Owen, W.M. (1952). Experimental study of water flow in annular pipes. *Trans. ASCE* 117: 485-496
Owen, W.M. (1953). Correlation between pipe flow and uniform flow in a triangular open channel. *Trans. AGU* 34(2): 213-219; 35(4): 659-660.
Owen, W.M. (1954). Laminar to turbulent flow in a wide open channel. *Trans. ASCE* 119: 1157-1164; 119: 1174-1175.

PADDOCK

15.04. 1884 Malone NY/USA
20.04. 1929 Great Falls MT/USA

Albert Edward Paddock started engineering studies at Syracuse University, Syracuse NY, but had to leave after two years because of his father's death. He was in 1906 accepted as foreman of construction of the water works at Dawson, Yukon Territory. In 1911 he joined the US Reclamation Service on the Tieton Project in Washington State, advancing soon to engineer. Next he was in charge of the Sunnyside Project there. When Yakima Project was completed in 1913, he was employed as superintendent of the large concrete diversion dam. He was assigned in 1915 to the Shoshone Project WY, engaged on the enlargement of the tunnel outlet and the installation of gates. In 1916 he accepted a position as superintendent of the Sweetwater Dam, San Diego CA.

Paddock's outstanding successes as superintendent of heavy construction brought him to the attention of the Utah Construction Company, so that he was transferred in 1920 to the Hetch Hetchy Dam. Until his premature death, he was in charge with the following assignments: Construction engineer on the Freeman Masonry Weir across Sacramento River above Woodland CA, superintendent of construction of double-tracking the main line of the Southern Pacific at Truckee NV, similar position for the construction of the American Falls Dam across Snake River ID. He also was assigned as superintendent of construction on the Gibson Masonry Dam, some 100 km west of Great Falls MT. The construction riggers of this dam fouled them against the high concrete hoist towers. Paddock, arriving on the scene and sensing the conditions and the consequent delay, jumped characteristically on top of the concrete bucket then at the base of the tower, signalling the engine-man to hoist him to where the chute was caught, some 30 m above the tower base. One of the riggers who had been fixing lines higher up in the tower also had noticed that the chute was caught, and without signalling for the bucket, and unknown to the hoisting engineer on the ground, came down the hoist cable. Thus he was inside the tower with his hand presumably on the hoist cable when the concrete bucket, with Paddock standing on top of it, was being hoisted. The rigger lost its hold on the hoist rope falling down inside the tower some 20 m so that Paddock was fatally injured and died. He was member of the American Society of Civil Engineers ASCE.

Anonymous (1921). Excavating the foundation for Hetch Hetchy Dam. *ENR* 87(6): 222-224. (*P*)
Anonymous (1930). Albert E. Paddock. *Trans. ASCE* 94: 1702-1704.
Paddock, A.E. (1923). O'Shaughnessy Dam completed and reservoir storage begun. *ENR* 90(17): 741-742.

PAGE

12.10. 1887 Syracuse NE/USA
23.03. 1955 Denver CO/USA

John Chatfield Page obtained the BS degree from University of Nebraska, Lincoln NE, and the MS degree in 1911 from Cornell University, Ithaca NY. After work as a topographer with the US Bureau of Reclamation USBR he became in 1909 city engineer of Grand Junction CO. From 1911 to 1925 he was junior engineer with USBR, Denver CO, then until 1930 superintendent of the Grand Valley Project in Colorado, involved in the design of Boulder Dam until 1935, from when he was acting commissioner until 1943, finally retiring as consulting engineer at Denver CO.

Page had been connected to irrigation work in Western US almost all through his professional career. In 1935 he was transferred to Washington DC to head up the USBR Engineering Division following Elwood Mead (1858-1936). Under Page's direction the Bureau carried forward the Colorado River Project as well as such large work as the Central Valley Project in California, and the Columbia Basin irrigation and power projects in Washington State. The main dams on which Page was involved were Hoover Dam, Grand Coulee Dam, and Marshall Ford Dam. Due to ill health, Page resigned in 1943 as commissioner, but continued as a consultant until 1947. He was elected in 1953 Honorary Member ASCE. A memorial to the late Page was unveiled at his namesake city Page AZ in 1964, located two miles from Glen Canyon Dam, consisting of a low concrete platform, a flagpole and a bronze mounted on the block of polished granite.

Anonymous (1937). John C. Page. *Civil Engineering* 7(3): 246. *P*

Anonymous (1947). John C. Page to retire from federal service. *Engineering News-Record* 139(Oct.16): 509. *P*

Anonymous (1951). Page, John C. *Who's who in America* 26: 2094-2095. Marquis: Chicago.

Anonymous (1953). John C. Page. *Engineering News-Record* 151(Oct.22): 26. *P*

Anonymous (1955). John C. Page. *Civil Engineering* 25(5): 303. *P*

Anonymous (1957). John Chatfield Page. *Trans. ASCE* 122: 1233.

Page, J.C. (1929). Producing Palisade peaches, Grand Valley Project, Colorado. *New Reclamation Era* 20(12): 188.

Page, J.C. (1935). Personnel building Boulder Dam. *Military Engineer* 27(7/8): 303-304.

Page, J.C. (1937). Water conservation and control. *Reclamation Era* 27(3): 46-50.

http://www.usbr.gov/history/CommissBios/page.html *P*

PALMER

20.01. 1878 Berkeley CA/USA
08.12. 1960 Los Angeles CA/USA

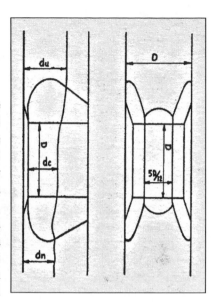

Harold King Palmer received his education from the University of California, Berkeley CA, with a BS degree in 1898, and the PhD degree in 1903. He was a fellow assistant of an observatory from 1898 to 1906, including two years in a branch observatory in Chile, and a computer of the San Joaquin Kings River Canal & Irrigation Co. from 1907 to 1908. From 1909 to 1925 he was a junior, assistant and engineer of the US Indian Irrigation Service, from when he was a designer and chief draftsman of its Office Engineer. Palmer was member of the American Society of Civil Engineers ASCE, and the National Geographical Society.

Palmer's name has been particularly connected to a design made to measure discharge in sewers of circular shape. Whereas the original design of Ralph Leroy Parshall (1881-1959) was applied in open channels, the Palmer-Bowlus design applies to closed-conduit conditions. Both are essentially based on the critical flow theorem, by which the transition from sub- to supercritical flow is forced by a channel contraction. This basic hydraulic principle allows for discharge determination with only one depth reading upstream of the contraction. The Palmer-Bowlus flume was perfected by Wells and Harold Benedict Gotaas (1906-1977), and an improved portable version was recently fabricated in Japan. The discharge coefficient varies only within a small range, and the modular limit undergoes an only small variation, rendering the device practical.

Anonymous (1941). Palmer Harold K. *Who's who in engineering* 5: 1343. Lewis: New York.
Anonymous (1961). Harold King Palmer. *Civil Engineering* 31(5): 100.
Palmer, H.K., Bowlus, F.D. (1936). Adaptation of Venturi flumes to flow measurements in conduits. *Trans. ASCE* 101: 1195-1239. (*P*)
Palmer, H.K., Bowlus, F.D. (1938). Discussion of Measurement of debris-laden stream flow with critical-depth flumes. *Trans. ASCE* 103: 1264-1266.
Palmer, H.K. (1939). Gaging stations for small streams. *Civil Engineering* 9(9): 497-498.
Perrine, C.D., Palmer, H.K., Moore, F.C. (1910). Determination of the solar parallax from photographs of Eros made with the Crossley reflector of the Lick Observatory, University of California. Carnegie Institution of Washington: Washington DC.
Rawn, A.M., Palmer, H.K. (1930). Predetermining the extent of a sewage field in sea water. *Trans. ASCE* 94: 1036-1081.
http://homepage.mac.com/jcrossley/wc/wc190/wc190_334.htm

PARKER E.S.

ca. 1828 Hasanoada (Indian Falls) NY/USA
31.08. 1895 Fairfield CT/USA

Ely Samuel Parker attended two academies before entering the civil engineering profession at age 21 as assistant engineer on the west end of the Erie Canal. In 1851 he was promoted to first assistant making improvements on the Canal until 1855. With this experience, Parker had received his first major appointment as chief engineer of Albemarle and Chesapeake Canal, a waterway commencing south of Norfolk VA, from where it went into *Currituck* Sound which, in turn, led to Albemarle Sound. This project was not far from the famous Dismal Swamp Canal, which was built 50 years earlier.

His next job took him back to the brisk climate of the Great Lakes, where he was the engineer of the Lighthouse District of Huron, Michigan and Superior. In 1857 he was in charge of the federal government at Galena IL. General U.S. Grant met Parker shortly later, and they became good friends. Grant was much interested in engineering and had been disappointed when he was rejected as country engineer at St. Louis in 1860. But upon the outbreak of the Civil War Grant rejoined the Army, whereas Parker as an Indian was rejected. Later he still joined the Army as brigadier general, remaining there with Grant, and after the War became Commissioner of Indian Affairs. During his term from 1869 to 1871, there was no significant hostile activity by the Indians. In 1871, Parker decided to return to the profession he had left. At age 43 only, he stated though 'Before the war I was in the successful practice of an honourable profession but this profession ran away from me; other and younger men had stepped in and filled my place, old men were discarded'. His later activities were not very successful and included no major projects. He made a small fortune and passed away at age 67. However, Do-Ne-Ho-Geh-Weh (Keeper of the western door) was interred in the land of the Senecas in the shadow of the monument of his idol, Red Jacket, at Buffalo NY.

Armstrong, W.H. (1990). *Warrior in two camps*: Ely S. Parker, Union General and Seneca Chief. Syracuse University Press: Syracuse NY.

Fitzsimons, N. (1973). Do-Ne-Ho-Geh-Weh: Seneca Sachem and civil engineer. *Civil Engineering* 43(6): 98. P

Parker, A.C. (1919). *The life of General Ely S. Parker*: Last grand sachem of the Iroquois and General Grant's military secretary. Historical Society: Buffalo.

http://en.wikipedia.org/wiki/Ely_S._Parker P

PARKER G.L.

23.12. 1884 Wickes MT/USA
12.02. 1946 Washington DC/USA

Glenn Lane Parker obtained the BS degree in civil engineering from Kansas University, Lawrence KS, in 1906. He was then instrumentman and assistant division engineer until 1907, and deck officer of the US Coast and Geological Survey in 1908. From 1909 he was from junior engineer to finally district engineer, the US Geological Survey USGS. He was member of the American Geophysical Union AGU, and the American Society of Civil Engineers ASCE.

Parker was hydraulic engineer within USGS almost all through his professional career. He had been district engineer at Tacoma WA from 1913 to 1939, and chief hydraulic engineer in charge of nation-wide water investigations until his death. The stream-gaging work was conducted by the USGS Water Resources Branch headed by Parker. During World War II he conducted more than 100 field offices, thereby contributing largely to solving the water problems relating to many phases of war activities. He became widely known through his works in water resources investigations, water power studies, and irrigation works. The 1943 Report, as an example, reports the data of measurements of stage and discharge made on streams, lakes, and reservoirs during a single year. This work was initiated in 1888 in connection with special studies relating to irrigation. In 1943 more than 5,000 gaging stations, including these of Hawaii, were maintained by USGS. The Parker Hall at the University of Kansas was dedicated in 1968 as an US Geological Survey facility. The USGS moved to other offices in 1989, and Parker now houses KGS offices including energy research and stratigraphic research. It is named for Glenn L. Parker, an alumnus and chief hydraulic engineer for the USGS from 1939 to 1946.

Anonymous (1933). Parker, Glenn L. *American men of science* 5: 856. Science Press: New York.
Anonymous (1939). G.L. Parker named chief hydraulic engineer. *Civil Engineering* 9(10): 633. *P*
Anonymous (1946). Glenn L. Parker dies, head hydraulic engineer. *Engineering News-Record* 136(Feb.21): 273.
Anonymous (1946). Death of Glenn L. Parker. *Civil Engineering* 16(3): 136. *P*
Parker, G.L., Storey, F.B. (1916). Water powers of the Cascade Range 3: Yakima River Basin. *Water-Supply Paper* 369. Government Printing Office: Washington DC.
Parker, G.L. (1941). *The Great Basin*. US Government Printing Office: Washington DC.
Parker, G.L. (1943). *Water resources in Alaska*. US Geological Survey: Washington DC.
Parker, G.L. (1945). Surface water supply of the United States 1943 7: Lower Mississippi River Basin. *Water-Supply Paper* 977: Washington DC.

PARKER P.M.

31.05. 1872 Greenwich/UK
04.08. 1920 Kuala Lumpur/MY

Philip à Morley Parker was educated at Cheltenham College, where his ability in mathematics enabled him to continue studies. Entering in 1889 the Central Technical College, the future Imperial College, London, he studied civil and mechanical engineering under William C. Unwin (1838-1933), but he left England before having graduated. He obtained the BCE degree from the University of Melbourne in 1894. He returned to Cambridge University taking the mathematical tripos in 1897. In 1898 he started work in London UK, constructing a reservoir for the city water supply. He entered in 1905 the Indian Irrigation Service rendering during the next five years valuable services as assistant engineer on the Upper Swat Canal, the Lower Jhelum Canal, and the Upper Chenab Canal, all located in Punjab, in today's Pakistan. He further served as officer in charge of hydraulic experiments on questions of weir and sluice discharges.

Parker started writing his book The control of water in 1910, and to complete this work including aspects of irrigation and general hydraulics, he visited the United States, Japan, Ceylon, Egypt and Italy. At the beginning of World War I he returned to London UK as a consulting engineer conducting irrigation projects in Jamaica. In 1915, before the Government system had been formed to deal with water questions, he prepared a water supply test facility including a 13 km long water main of 0.50 m diameter close to Dee River. He then left England for Sydney, and further went to the Federated Malay States, serving on irrigation projects, but he died in Kuala Lumpur from a heart failure. Parker's personality was described attractive and coupled with his great professional ability giving him a large circle of friends. He was well known by his 1913 book covering the then almost complete knowledge of practical hydraulics. Each subject is reviewed and examined in a masterly manner and with great completeness including references and cross-references, rendering the book to a monumental contribution of applied hydraulics. He had prepared other books at the time of his death. Parker was member of the Institution of Civil Engineers, London UK, and the American Society of Civil Engineers ASCE.

Anonymous (1920). Philip à Morley Parker. *Engineering News-Record* 85(19): 916.
Anonymous (1922). Parker, P.M. *Trans. ASCE* 85: 1706-1707.
Parker, P. à M. (1925). *The control of water* as applied to irrigation, power and town water supply purposes, ed. 2. Routledge: London. (*P*)

PARKER T.B.

20.08. 1889 Roxbury MA/USA
27.04. 1944 Wellesley MA/USA

Theodore Bissell Parker graduated in 1911 with the BSc degree from the Massachusetts Institute of Technology MIT, Cambridge MA. He then joined the staff of an engineering firm in New York NY, specializing in hydro-electric power plants. He was involved in the construction of new power plants on the Bear River in Idaho with the Utah Power and Light Company. After war service, he joined the Electric Bond and Share Co. as assistant hydraulic engineer, engaged in studies of the power sites on the Platte, Loup, and Niobrara Rivers in Nebraska. From 1922 he was a hydraulic engineer with an engineering corporation, active on hydro-electric developments of Bartlett's Ferry near Columbus GA, and the Rock Island Project on Columbia River in Washington State.

Parker joined in 1935 Tennessee Valley Authority TVA as chief construction engineer, responsible for the general supervision of lock, dam, powerhouse, and appurtenance work. At this time both the Norris and Wheeler Dam projects were under construction. In 1938 he was named chief engineer, in charge of all engineering and construction work of TVA, having general responsibility for the design and construction of many dams, among which was the Fontana Dam. Parker was devoted to the promise and completion of his work, and proud of the accomplishments of his men. In 1943 he left TVA, returning to MIT where he was appointed Head of the Department of Civil and Sanitary Engineering. Parker's career may be measured only partially by the notable engineering projects with which he was associated. The full measure of his success must include the encouragement he gave to younger members of the profession, the friendship he tendered his associates, and the high honour and esteem in which his colleagues held his memory. He was recipient of the Desmond Fitzgerald Medal for his 1926 paper from the Boston Society of Civil Engineers. He was a member of the American Society of Civil Engineers ASCE from 1931, and actively participated in the activities of its Tennessee Valley Section.

Anonymous (1938). Theodore B. Parker. *Engineering News-Record* 120(May19): 700. *P*
Anonymous (1945). Theodore B. Parker. *Trans. ASCE* 110: 1786-1790.
Hageman, H.A., Parker, T.B. (1926). The Bartlett's Ferry hydroelectric development. *Journal of the Boston Society of Civil Engineers* 13(3): 93-95.
Parker, T.B. (1943). Allocation of Tennessee Valley Authority projects. *Trans. ASCE* 108: 174-187
http://www.loc.gov/pictures/resource/hec.25510/ *P*

PARMAKIAN

09.09. 1908 Watertown MA/USA
15.08. 1994 Denver CO/USA

John Parmakian was educated at the Massachusetts Institute of Technology MIT, Cambridge MA, with a BS degree in civil engineering in 1930, and the MS degree in 1939 from the University of Colorado, Fort Collins CO. He was from 1930 to 1940 an engineer with the US Bureau of Reclamation USBR, and a field and test engineer of mechanical equipment there until 1946, from when he was a technical engineer of its Analytical Section. Parmakian was member of the American Societies of Mechanical Engineers ASME and Civil Engineers ASCE.

In his position as chief of the USBR Technical Engineering Analysis Branch Parmakian directed, coordinated and reviewed the activities of a group of highly-specialized engineers in the technical analysis of complex problems pertaining to the design and the construction of dams, hydro-electric plants and pumping schemes. He had published a number of papers in engineering and military journals in the fields of vibrations in hydraulic structures, hydraulic transients, and applied mathematics. He also authored the book Waterhammer analysis. After retirement from the USBR he was a consultant for a variety of international and domestic firms on aspects of hydropower plants, pumped storage, and general power plant engineering. In ASME he was elected to the Executive Committee of the Hydraulic Division in 1955, chairing it from 1958 to 1959. Later, he was chairman of the Waterhammer Sub-Committee and also a member of the Hydraulic Prime Movers and Cavitation Sub-Committees. Parmakian was Fellow of both ASME and ASCE. He was cited for superior accomplishment in solving complex stress and hydraulic design problems and for his work on the waterhammer problem in power penstocks.

Anonymous (1952). John Parmakian. *ASME News* 74(6): 530. *P*
Anonymous (1959). Parmakian, John. *Who's who in engineering* 8: 1894. Lewis: New York.
Anonymous (1961). John Parmakian. *Mechanical Engineering* 83(8): 116. *P*
Parmakian, J. (1950). Air-inlet valves for steel pipe lines. *Trans. ASCE* 115: 438-444.
Parmakian, J. (1953). Pressure surge control at Tracy pumping plant. *Proc. ASCE* 79(361): 1-29.
Parmakian, J. (1955). Pressure surges in pump installations. *Trans. ASCE* 120: 697-720.
Parmakian, J. (1955). *Waterhammer analysis*. Prentice-Hall: New York, 2nd ed. in 1963.
Parmakian, J. (1963). Hydraulic turbine operation difficulties. *Water Power* 15(10): 418-422.
Parmakian, J. (1980). The use of surge suppressors in pump discharge lines. *Water and Pollution Control* 118(8): 12. *P*

PARMLEY

08.12. 1862 Rock County WI/USA
19.02. 1934 Upper Montclair NJ/USA

Walter Camp Parmley was educated at University of Wisconsin, Madison WI, from where he obtained the MS degree in civil engineering in 1893. After graduation as a civil engineer in 1887, he went to California opening an office at San Bernardino, to practice in a partnership. The firm was immediately successful, with works including surveys and the construction of the *Jurupa* Canal and the *Vivienda* Pipeline, together with other irrigation systems. In 1889 Parmley became assistant engineer on the Bear River Canal Project in Utah, thereby being involved in the water works of Ogden UT. After graduation in 1893 he was assistant city engineer of Peoria IL until 1896, was a sanitation engineer for Cleveland OH until 1903, from when he had a private consulting office in New York City. He was a member of the American Society of Civil Engineers ASCE, and a president of the Cleveland Engineering Society. Parmley was remembered for his pleasing personality and for his dependability, both in business matter and socially. He was a man of fine character, unassuming, always cheerful, and always kind.

Parmley had specialized in large sewer designs during his stay in New York City as a consultant. He invented thereby reinforced concrete pipes and segmental construction and special arching reinforcing methods, particularly for tunnel construction. Parmley authored a number of articles in the then most important American civil engineering journals, including the ASCE journals or Engineering News-Record. These dealt with questions in sewerage engineering including the rainfall-runoff phenomena for sewer design purposes, and descriptions of particular wastewater treatment stations.

Anonymous (1931). Parmley, W.C. *Who's who in engineering* 3: 998. Lewis: New York.
Anonymous (1933). Parmley, Walter C. *Who's who in America* 17: 1789. Marquis: Chicago.
Anonymous (1934). Walter Camp Parmley. *Engineering News-Record* 112(Mar.1): 305.
Anonymous (1937). Walter Camp Parmley. *Trans. ASCE* 102: 1632-1635.
Parmley, W.C. (1898). Rainfall and run-off in relation to sewerage problems. *Journal of the Association of Engineering Societies* 20(3): 204-225.
Parmley, W.C. (1899). Discussion of Stream contamination and sewage purification. *Trans. ASCE* 42: 162-165.
Parmley, W.C. (1900). Discussion of On the flow of water over dams. *Trans. ASCE* 44: 346-359.
Parmley, W.C. (1905). The Walworth sewer, Cleveland Ohio. *Trans. ASCE* 55: 341-412.
http://search.ancestry.com/Walter+Camp+&gsfn_x=1&gsln=Parmley&gsln_x=1&msbdy=1862&msbdy_x=1&cpxt=0&Bucket *P*

PARR

20.03. 1880 Yonkers NY/USA
07.06. 1964 Yonkers NY/USA

Harry Lilienthal Parr obtained both the AB and ME degrees from Columbia University, New York NY, in 1902 and 1904, respectively. He then was there assistant of the Mechanical Engineering Department until 1906, joined until 1907 the Power Specialty Co., New York City, became until 1912 instructor of mechanical engineering at Columbia University, was there assistant professor until 1921, associate professor until 1928, from when he took over there until retirement in 1946 as professor of mechanical engineering. He was in parallel consultant in his professional field. He was a Fellow of the American Association for the Advancement of Science AAAS, and member of the American Society of Mechanical Engineers ASME, and the American Physical Society APS.

Parr was in charge of the mechanical laboratory of Columbus University. He spent the summer visiting engineering schools and industrial establishments in Europe, reporting that the Americans have learnt few until the start of World War I. However, the foreign methods were described not as useful as they might be by reason of limited American equipment compared to that abroad. Excellent European laboratories were described. His laboratory at Columbia was described especially poor in this regard, so that Parr intended to adopt German methods. The European instruction in engineering was considered particularly important with regard to mechanical engineering. In Germany again, he mentioned, they do not hesitate to state that mechanical engineering is the foundation of all their manufacturing success, a fact just as true in America, though then not really recognized and supported. Parr's ideas can be considered ahead these of John R. Freeman (1855-1932), who made visits to Europe in the 1920s, and consequently established the Freeman Scholarship allowing for gifted American students access to hydraulic studies at European universities.

Anonymous (1912). Prof. H.L. Parr. *Columbia University Quarterly* 15: 102-103.
Anonymous (1921). Parr, Harry L. *American men of science* 3: 527. Science Press: New York.
Anonymous (1959). Parr, Harry L. *Who's who in America* 28: 2070. Marquis: Chicago.
Anonymous (1964). Harry L. Parr. *Mechanical Engineering* 86(12): 106.
Parr, H.L., Thurston, Jr., E.D., Herrick, A.L. (1927). *Steam engines, boilers, pumps, turbines, gas engines*. Dept. Mechanical Engineering, Columbia University: New York.
Parr, H.L. (1937). Fluid flow analyzer. *Mechanical Engineering* 59(6): 437-439; 59(12): 960.
Parr, H.L. (1940). Smoke ribbons test streamline design. *Popular Science* 137(7): 99. *P*

PARSHALL

02.07. 1881 Golden CO/USA
29.12. 1959 Fort Collins CO/USA

Ralph Leroy Parshall obtained the BS degree from Colorado Agricultural College, Fort Collins CO, in 1904, specializing in civil and irrigation engineering. He was assistant at the Department of Physics there until 1906, assistant professor of civil engineering until 1913, assistant in irrigation investigations at the US Department of Agriculture until 1918, when taking over as laboratory head at the Experiment Station until his retirement in 1948.

Parshall is known for two devices used in irrigation engineering, namely the Parshall flume, and the Vortex sand trap. Following the common 'Venturi flume' introduced by Victor M. Cone (1883-1970) in 1916, Parshall presented an alterative structure for measurement of discharge under field conditions. His original device consisted of a short contraction from the approach to the tailwater canal width involving at the throat section a drop to avoid submergence effects. The device was laboratory-tested resulting in a discharge-head equation which points at certain scale effects. Irrigation engineers therefore had a simple and quite accurate means at hand to measure a basic quantity of fluid flow, rendering the device attractive despite certain disadvantages. Parshall flumes were developed from the 1960s and essentially may still be seen in the field. The Vortex Sand Trap was developed by Parshall in the early 1930s. A horizontal vortex is generated by setting a conduit diagonally across a flume, whose top is open. Due to turbulent water flow at the pipe crest, bedload sediment is entrained by the tube of which the sediment-water flow is thereby separated from the approach flow.

Anonymous (1922). Parshall, Ralph L. *Who's who in engineering* 1: 968. Lewis: New York.
Anonymous (1949). Council celebrates 30th anniversary. *Engineers' Bulletin* 33(10): 6. *P*
Anonymous (1960). Ralph Parshall. *Civil Engineering* 30(3): 230.
Parshall, R.L. (1925). The improved Venturi flume. *Trans. ASCE* 89: 841-880.
Parshall, R.L. (1929). Discussion of Experiments on models of Pit River Dams. *Trans. ASCE* 94: 507-511.
Parshall, R.L. (1931). Sand trap for canals works on vortex principle. *Civil Engineering* 1(5): 419-421.
Parshall, R.L. (1947). Riffle deflectors and vortex tubes remove suspended silt and sand. *Civil Engineering* 17(12): 740-741.
Parshall, R.L. (1948). The Parshall measuring flume. *Reclamation Era* 34(5): 97-98. *P*
Parshall, R.L. (1952). Model and prototype studies of sand traps. *Trans. ASCE* 117: 204-213.

PARSONS H.B.

06.01. 1862 New York NY/USA
26.01. 1935 New York NY/USA

Harry de Berkeley Parsons, the brother of William (1859-1932), graduated with the BSc degree from Columbia College, New York NY, in 1882. Further he graduated in 1884 as a mechanical engineer from the Stevens Institute of Technology, Hoboken NJ, which conferred on him in 1926 the honorary degree of Doctor of Engineering. Parsons established in 1885 an office in New York, which he maintained until his death. His earlier projects include railway works, and the water supply completion at Stevens Point WI. From 1892 to 1907, he was professor of steam engineering, Rensselaer Polytechnic Institute, Troy NY, from when he was emeritus professor of practical engineering. From 1901 to 1903 he designed the 500 m long and 50 m high Spiers Falls Dam on Hudson River. A similar structure, namely the Sherman Island Dam near Glens Falls NY, was designed from 1921 to 1923.

From 1908 to 1914 Parsons was a member of the Metropolitan Sewerage Commission, which prepared a monumental report on the Storm and sanitary drainage of New York City. His further activities included the design of water steam power plants, water supply and sewerage projects. He considered his most important works the dams stated above, as well as a study of tidal flows in New York Harbour for the Metropolitan Sewerage Commission. He was awarded by the American Society of Civil Engineers ASCE the Thomas Fitch Rowland Prize in 1925 for his paper on Sherman Island Dam, and in 1930 the J. James Croes Medal for the uplift paper. He further published three other noteworthy papers in the Transactions ASCE. He was member of the American Society of Mechanical Engineers ASME, the Society of Naval Architects and Marine Engineers SNAME, and ASCE member from 1897. Always courteous, his quiet refinement of manner, his sincerity of purpose, his modesty, and his grounding in the fundamentals of his profession impressed distinctly. He was indeed a fine person and a distinguished engineer, who has particularly contributed to his home city.

Anonymous (1926). H. de B. Parsons. *Engineering News-Record* 96(3): 132-133. *P*
Anonymous (1936). Harry de B. Parsons. *Trans. ASCE* 101: 1617-1620.
Parsons, H. de B. (1905). *Lectures on steam engines*. Worden: Troy NY.
Parsons, H. de B. (1907). *Steam boilers*: Their theory and design. Longmans & Green: New York.
Parsons, H. de B. (1925). Sherman Island Dam and power house. *Trans. ASCE* 88: 1257-1292.
Parsons, H. de B. (1929). Hydrostatic uplift in pervious soils. *Trans. ASCE* 93: 1317-1332.

PARSONS W.B.

15.04. 1859 New York NY/USA
09.05. 1932 New York NY/USA

William Barclay Parsons obtained the AB degree in 1879, and the CE degree in 1882 from the Columbia University, New York NY. He was awarded the LL.D degree from St. John's College, Annapolis MD in 1909, the ScD degree in 1920 from Princeton University, Princeton NJ, and the D.Eng. degree in 1921 from Stevens Institute of Technology, Hoboken NJ. Parsons was from 1885 a consulting engineer in New York City, was then from 1891 to 1904 deputy chief engineer for the Rapid Transit Commission, from when he was a member of the Isthmian Canal Commission, as also on the Board of Consulting Engineers for Panama Canal. Parsons was from 1905 to 1914 chief engineer of the Cape Code Canal, among other engineering projects. He was awarded the US Distinguished Service Medal, and the Order of the Crown from Belgium for distinguished war service. He was a Fellow of the American Academy of Arts and Sciences.

Parsons went to Panama as member of the Committee of Engineers reporting in favour of the sea-level canal, despite the final design was based upon a succession of locks to overcome the Isthmian height elevation. As chief engineer of Cape Cod Canal, he was instrumental for its opening in 1914, joining then the Massachusetts Bay with Buzzards Bay. It was demonstrated in the project, as before by the Frenchmen at Suez Canal in Egypt, that a canal without locks could be built between two large water bodies where considerable tidal differences exist. The 1920 book is a valuable and interesting record of the war activities in France and Belgium during World War I. The 1922 book on Robert Fulton (1765-1815) includes the chapters: 1. From art to engineering, 2. Early attempts at sub-surface navigation, 3. Fulton's first submarine, 4. Negotiations with France, 5. 'Drawings and descriptions', 6. The British contact, 7. Experience in England, 8. Negotiations with cabinet, 9. Further correspondence, 10. Failure of negotiations, 11. Return to America, and 12. Examination of Fulton's design.

Anonymous (1924). William B. Parsons. *Engineering News-Record* 92(16): 643-644. *P*
Anonymous (1932). Death of William Barclay Parsons. *Civil Engineering* 2(6): 401-402.
Anonymous (1943). Parsons, William B. *Who was who in America* 1: 940. Marquis: Chicago.
Parsons, W.B. (1900). *An American engineer in China*. McClure, Phillips & Co.: New York.
Parsons, W.B. (1920). *The American engineers in France*. Appleton: New York. *P*
Parsons, W.B. (1922). *Robert Fulton and the submarine*. Columbia University Press: New York.
http://en.wikipedia.org/wiki/William_Barclay_Parsons *P*

PARTRIDGE

07.04. 1891 San Francisco CA/USA
24.10. 1953 Knoxville TN/USA

John Frederick Partridge obtained in 1912 the AB degree in civil engineering from the Leland Stanford Junior University, Stanford CA. He then was until 1913 involved in the design and construction of the irrigation and drainage canals of the San Joaquin Irrigation District. Until 1917, he was chief engineer of the Pipe Department of a manufacturing company at San Francisco CA. After service in World War I, Partridge became research engineer on multiple arch dam construction at Sacramento CA, and then was until 1928 hydraulic engineer of the California Ore Power Co. From 1929 to 1931 Partridge was then assistant engineer of the American and Foreign Power Co in Brazil on investigations of hydro-electric power developments. After works for the Golden Gate Bridge, and the San Francisco-Oakland Bay Bridge until 1934, he was appointed application examiner of the Hiwassee Dam of the Tennessee Valley Authority TVA; he thereby also acted as assistant construction engineer of its Fort Loudoun Dam until 1943. After war service during World War II, he was a regional civil engineer in California. Finally, Partridge was the chief of TVA's Construction Branch at Knoxville TN.

Partridge had a notable engineering career both as irrigation engineer, and as designer of dams. His early years were spent in California and Oregon on irrigation and hydro-power construction, and he was also for several years in Brazil. He joined TVA in 1934, serving as construction engineer. Hiwassee Dam on Hiwassee River in North Carolina was built in the late 1930s to improve flood control and bring electricity to this region. With almost 100 m height, the dam is the third highest of the TVA system, behind the Fontana and Watauga Dams. Since 1947 Partridge directed construction plant planning for TVA's accelerated construction program. He was member of the American Society of Civil Engineers ASCE, and the Society of American Military Engineers SAME.

Anonymous (1948). Partridge, John F. *Who's who in engineering* 6: 1514. Lewis: New York.
Anonymous (1953). John F. Partridge. *Engineering News-Record* 151(Dec.24): 47.
Anonymous (1954). John F. Partridge. *Trans. ASCE* 119: 1356-1357.
Partridge, J.F., Bosworth, H.P. (1927). *Cost of Middle Fork-Red Blanket Diversion*. PacifiCorp Headquarters Archives: Jackson OR.
Partridge, J.F. (1940). Construction in rhyme. *Civil Engineering* 10(7): 467.
Partridge, J.F. (1948). Discussion of Tennessee Valley Authority. *Trans. ASCE* 113: 130-137.
http://www.pacificorp.com/content/dam/pacificorp/doc/Energy_Sources/Hydro/Hydro_Licensing/Rogue%20River/Prospect (*P*)

PATTERSON

23.12. 1914 Chicago IL/USA
20.04. 1982 Vicksburg MS/USA

Charles Bird Patterson received his education from University of Florida, Gainesville FL, obtaining in 1936 the BS degree in civil engineering. He was an assistant of water conservation to the National Resources Committee of the Drainage Basin Study first, joining then as hydraulic and civil engineer the Waterways Experiment Station WES of the Corps of Engineers, Vicksburg MI, all through his career. He was the chief of the Technical Service Division from 1951, retiring from WES in 1970. Patterson was member of the American Society of Civil Engineers ASCE.

Patterson devised the 'dynamic pressure meter' for measuring and recording rapid fluctuations of pressures within fluid bodies in 1941. He authored papers on the dry-dock suction chambers and was known for his chapter on applied sedimentation. Further works include his paper to the ASCE Symposium on the conformity between models and the prototype. In the early 1940s, hydraulic modeling was not generally accepted as an adequate alternative to prototype observations, because a scaling law could not always be identified, or scale effects did not allow for up-scaling results from standard laboratory observation, given their size was too small. The Symposium dealt with a number of typical problems of hydraulic engineering, and the results indicated that this approach could be indeed applied provided a number of basic limitations were respected.

Anonymous (1959). Patterson, C.B. *Who's who in engineering* 8: 1898. Lewis: New York.
Anonymous (1982). Charles B. Patterson. *Civil Engineering* 52(8): 78.
Brown, J. (2013). The Mississippi Basin in miniature. *Civil Engineering* 83(10): 42-45. (*P*)
Johnson, A.I., Patterson, C.B., eds. (1988). *Geotechnical applications of remote sensing and remote data transmission*. American Society for Testing and Materials: Philadelphia.
Patterson, C.B., Arnold, J.E. (1942). Model study of the pump suction chamber for dry dock No. 4, Puget Sound Navy Yard. *Technical Memorandum* 189-1. US Army Waterways Experiment Station: Vicksburg MI.
Patterson, C.B. (1944). Conformity between model and prototype: A dry-dock suction chamber. *Trans. ASCE* 109: 29-34.
Patterson, C.B., Brown, F.R. (1947). *Practical application of experimental hydraulics*. US Army Engineer Water Experiment Station: Vicksburg MI.
Patterson, C.B., Simmons, H.B. (1950). Contribution of model analysis to the solution of shoaling problems. *Applied Sedimentation* 66(8): 300-318. Wiley: New York.

PAUL

10.03. 1875 Rockport MA/USA
06.10. 1941 Dayton OH/USA

Charles Howard Paul started his studies in 1892 at Massachusetts Institute of Technology MIT, but his father's death stopped him from gaining a degree. He was employed by the Sewer Department, Boston MA. He was transferred to the Boston Metropolitan Water District in 1896, where he worked on designs of pipe lines, reservoirs, and pumping stations. From 1900 he was employed at the Bureau of Filtration, Philadelphia PA, assigned to the Belmont Plant, in charge of its sedimentation basin and sand filters. It was here where Paul made his first acquaintances with dam and reservoir designs and construction. In 1904 he became engaged by the newly organized US Bureau of Reclamation, to design irrigation works in the arid West. His first assignment was on the Lower Yellowstone project near Glendive MT. From 1910, he was placed in charge of the design and the construction of the Arrowrock Dam on Boise River ID, by then the highest dam of the world, with a height of 105 m, executed as a curved gravity type dam.

In 1915 Paul joined the Miami Conservancy District, a flood control organisation of Ohio State. The method of flood control involved a combination of retarding basins and channel improvements, thus different from all standard of that time. World War I delayed these works, so that it was decided to develop a construction organization by force account, with Paul as assistant chief engineer first, and from 1921 as chief engineer. He resigned in 1924, but was retained as a consultant until his death. Before he left the District, several floods occurred in the Miami Valley, demonstrating the efficiency of the works. Paul established then his consulting office at Dayton OH, where he remained until his death. In 1933 he was appointed member of the committee charged with the exploration of water problems of the Mississippi Valley. He was further appointed as a consulting engineer for the newly organized Tennessee Valley Authority TVA, and thus was connected with all dam projects. He was recipient of the Norman Medal for his 1922 paper published with the American Society of Civil Engineers, whose member he was.

Anonymous (1934). Charles H. Paul. *The Rotarian* 44(3): 48. *P*
Anonymous (1942). Charles H. Paul. *Trans. ASCE* 107: 1809-1813.
Morgan, A.E. (1951). *The Miami Conservancy District*. McGraw-Hill: New York.
Paul, C.H., Bennett, C.S. (1927). *Methods and plant for excavation and embankment*. McGraw-Hill: New York.
Paul, C.H. (1942). *The story of the Engineers' Club of Dayton*. Dayton OH.

PAULSEN

22.04. 1887 Chicago IL/USA
30.01. 1961 Caldwell ID/USA

Carl Gustav Paulsen obtained the BS degree in civil engineering from the University of Idaho, Moscow ID, in 1913. He was in 1914 employed by the US Geological Survey USGS within its Water Resources Branch. He was after war service from 1918 to 1930 USGS district engineer at Atlanta GA, taking over in 1931 as chief the Surface Water Division, first as assistant chief hydraulic engineer, and later as chief hydraulic engineer. Paulsen retired in 1957.

Paulsen has written numerous papers and reports, notably Water-Supply Papers, published by USGS. The 1938 paper describes USGS activities in measuring stream flows, involving also a discharge integrator for runoff computations. Paulsen was River Master on Delaware River since 1954, when the US Supreme Court established a new diversion formula for New York City. He also was chairman of the Arkansas River Compact Commission, and member of the International Joint Commission's engineering Boards for studying problems involving the Columbia and St. Johns rivers. He further was a member of the three-men United Nations Commission working on the development of Mekong River between Laos, Cambodia, and Thailand. The US Department of Interior awarded him in 1957 its highest honour, namely the Distinguished Service Medal, for his outstanding service during a long career. Paulsen was a member of the American Society of Civil Engineers ASCE, taking over as director from 1953 to 1955, and of the Washington Society of Engineers and past president of the Columbia District Section, among others. A notable hydraulic expert was lost with his death.

Anonymous (1946). Carl G. Paulsen. *Engineering News-Record* 136(May2): 144. *P*
Anonymous (1948). Paulsen, Carl G. *Who's who in engineering* 6: 1520. Lewis: New York.
Anonymous (1961). Carl G. Paulsen dies: Was hydraulics expert. *Engineering News-Record* 166(Feb.23): 61.
Anonymous (1961). Former ASCE Director Carl Paulsen dies. *Civil Engineering* 31(3): 88. *P*
Grover, N.C., Lamb, W.A., Parker, G.L., Paulsen, C.G. (1931). Surface water supply of the United States. *Water-Supply Paper* 672. Washington DC.
Paulsen, C.G. (1938). Geological Survey studies surface waters: Recent improvements in equipment aid in measuring stream flow. *Civil Engineering* 8(4): 247-250.
Paulsen, C.G. (1939). Measurement and computation of flood-discharge. *Trans. AGU* 20: 177-187.
Paulsen, C.G., Bigwood, B.L., Harrington, A.W., Hartwell, O.W., Kinnison, H.B. (1940). Hurricane floods of September 1938. *Water-Supply Paper* 867. USGS: Washington.

PAYNTER

11.08. 1923 Evanston IL/USA
14.06. 2002 Cambridge MA/USA

Henry Martyn Paynter received from Massachusetts Institute of Technology MIT, Cambridge MA, the SB, SM and PhD degrees, the latter in 1951. He was there assistant and professor of mechanics from 1954 to 1985, and head of the Systems Dynamics & Control Division from 1963 to 1966. In parallel, he was a consultant in general dynamics, gas dynamics from 1960 to 1966, propulsion from 1969 to 1972, and thermodynamics and fluid mechanics until 1974. A member of the American Societies of Mechanical Engineers ASME and Civil Engineers ASCE, he was the 1953 recipient of the Alfred Noble Prize for his paper on surges and water hammering.

Paynter was a leading expert in analogue computing. His early research led him to the formation of the Pi-Square Engineering Company, in which fast electronic computing was applied to industrial process control. Besides pioneering contributions to hybrid computation, he was known for his academic research culminating in the creation of Bond Graphing. In hydraulics he was known for a paper written in the early 1950s, in which unsteady flows are modelled with the Saint-Venant equations. These allow in the simplest case for one-dimensional modeling of the relevant features including flow depth and velocity, as functions of location and time. From the early 1960s Paynter considered systems engineering with an emphasis on system dynamics, modelling and simulation, machine computation and automatic control. He may be considered a hydraulic engineer who had left this field to improve the modern aspects of general engineering sciences. Paynter was a member of the National Academy of Engineering, Fellow of ASME, and senior member of the Institute of Electrical and Electronic Engineers IEEE.

Anonymous (1950). Henry M. Paynter, Jr. *Engineering News-Record* 144(Mar.9): 44. *P*
Anonymous (1953). Civil engineer wins Alfred Noble Prize. *Civil Engineering* 23(12): 849. *P*
Anonymous (1973). Prof. Henry M. Paynter. *Mechanical Engineering* 95(10): 103. *P*
Anonymous (1994). Paynter, H.M. *American men and women of science* 5: 1127. Bowker: New Providence NJ.
Hedrick, J.K., Paynter, H.M., eds. (1976). *Nonlinear system analysis and synthesis*. ASME: New York.
Paynter, H.M. (1953). Electrical analogies and electrical computers: Surge and water hammer problems. *Trans. ASCE* 118: 962-989; 118: 1004-1009.
Paynter, H.M. (1955). *A palimpsest of the electric analog art*. Philbrick: Boston.

PECK

23.06. 1912 Winnipeg MB/CA
18.03. 2008 Albuquerque NM/USA

Ralph Brazelton Peck obtained his civil engineering degree in 1937 from Rensselaer Polytechnic, Troy NY, from where he received the honorary degree of doctor in 1974; he was a post-graduate student at Harvard University in 1938, and assistant engineer for Chicago IL from 1939 to 1943, until 1948 then research assistant professor of soil mechanics at the University of Illinois, Urbana-Champaign IL, until 1957 research foundation engineering professor, and professor of foundation engineering until retirement in 1974. He was recipient of the 1974 National Medal of Science, the 1983 Golden Beaver Award, among others, Honorary Member ASCE, and Fellow of the Geological Society of America.

Peck was an eminent civil engineer who had specialized in soil mechanics. He had spent the majority of his professional career at the University of Illinois, initially in structures, but later focused on civil engineering under the influence of Karl Terzaghi (1883-1963). He was highly influential as a consulting engineer with more than 1,000 projects in foundations, ore storage facilities, tunnel projects, dams and dikes, including the dams in James Bay CA, the pending tower of Pisa, the Trans-Alaska Pipeline System, and the Dead Sea dikes. He also authored several books, including Soil mechanics in engineering. He was awarded the National Medal of Science by President Ford 'for his development of the science and art of subsurface engineering, combining the contributions of the sciences of geology and soil mechanics with the practical art of foundation design'.

Anonymous (1969). The Rankine Lecture 1969. *Geotechnique* 19(2): 169-170. *P*
Anonymous (1975). Peck receives Medal of Science. *Civil Engineering* 45(11): 93. *P*
Anonymous (1985). Ralph B. Peck. Proc. 11th Intl. Conf. *Soil Mechanics and Foundation Engineering* 5: xliii. *P*
Anonymous (1995). Ralph B. Peck. *Civil Engineering* 65(10): 96. *P*
Anonymous (2005). Peck, Ralph B. *Who's who in America* 59: 3609. Marquis: Chicago.
Peck, R.B. (1943). Earth-pressure measurements in open cuts, Chicago IL Subway. *Trans. ASCE* 108: 1008-1036; 108: 1097-1103.
Peck, R.B., Hanson, W.E., Thornburn, T.H. (1953). *Foundation engineering*. Wiley: New York.
Peck, R.B. (1969). Advantages and limitations of the observational method in applied soil mechanics. *Geotechnique* 19(2): 171-187.
Terzaghi, K., Peck, R.B., Mesri, G. (1948). *Soil mechanics in engineering practice*. Wiley: New York, 3rd ed. in 1996.

PELTON

05.09. 1829 Vermilion OH/USA
15.03. 1908 Oakland CA/USA

Lester Pelton immigrated in 1850 to Camptonville CA during the Gold Rush, where he made his living as carpenter and millwright. He realized that the then existing Hurdy-Gurdy wheels used for mining could be significantly improved. He introduced a novel turbine type based on impulse of water rather than reaction, revolutionizing turbines particularly for high hydraulic heads. Until then, turbines were reaction machines powered by hydraulic pressure. Pelton's invention was powered by the kinetic energy of a high-velocity jet. According to Durand's 1939 paper 'Pelton started from an accidental observation in the 1870s, watching a spinning water turbine when the key holding its wheel onto its shaft slipped, causing it to become misaligned. Instead of the jet hitting the cups in their center, the slippage made it impact near the edges. Rather than the water flow being stopped, it was now deflected into a half-circle, being reflected in the reversed direction. Surprisingly, the turbine now moved faster. This was Pelton's great discovery'.

Pelton produced from this observation his wheel. It improved upon earlier designs using two instead of only one cup. The nozzle of the old design sprayed water slightly off-center and at an angle into the cup to utilize the kinetic energy. However, using two cups and splitting the jet directly down the center, the Pelton wheel used more energy. In 1878, Pelton tested and manufactured the first Pelton Wheel at Miners Foundry, Nevada City CA. By 1879 he had tested a prototype Pelton wheel at the University of California. Pelton's design soon was accepted because of higher efficiency, as was proven in a head-to-head competition. In 1887, a miner attached Pelton's wheel to a dynamo producing the first hydro-electric power in the Sierra Nevada Mountains. A patent was granted in 1889 to Pelton, and he sold the rights to the Pelton Water Wheel Company of San Francisco. Pelton received the Elliott Cresson Medal from the Franklin Institute.

Anonymous (1892). La roué Pelton. *Le Génie Civil* 22(4): 49-52.
Anonymous (1895). The Pelton water wheel. *Journal of the Franklin Institute* 140(3): 160-197.
Anonymous (1959). Zum fünfzigjährigen Todestag von Lester Pelton. *Schweizerische Bauzeitung* 77(24): 384-385. P
Durand, W.F. (1939). The Pelton water wheel. *Mechanical Engineering* 61(6): 447-454; 61(7): 511-518; 61(10): 755-756.
Lescohier, R. (1992). *Lester Pelton and the Pelton water wheel*. Private printing: Grass Valley. P

PENFIELD

10.08. 1904 Battle Creek MI/USA
01.05. 1975 Santa Barbara CA/USA

Wallace Clay Penfield came with his family in 1907 to Southern California, and grew up at Pasadena CA. He received his BSc degree in civil engineering in 1926 from the California Institute of Technology, Pasadena CA. He moved in 1928 to Santa Barbara CA as the first county planning engineer and later director, a position he held until resigning, to found there in 1946 the firm Penfield & Smith Engineers. During his term as planning director, he prepared the first master shoreline plan. He represented the county on the State Park Committee, and was active in formulating and planning the *Cachuma* Water Project. He was interested in the harbour and beaches of his county, and represented it in the first beach erosion study made jointly by the government and the county from 1937 to 1946. Later, he prepared the first plan for the enlargement of Santa Barbara Harbor, a plan with dual provision for harbour facilities and prevention of beach erosion.

As president of Penfield & Smith, Penfield was responsible for small craft harbour plans, the Santa Barbara Marina, and the airport master plan of 1959. His firm was in charge for the plans of over 150 land development projects creating many homesites in Santa Barbara County, among others. Almost 200 km of city streets, sewer mains, and water supply mains were designed along with accompanying curbs, gutters, drains, and culverts. His publications include Oldest Beach Nourishment Project, relating to his activities with beach erosion, or Your planning commissioner, a long used handbook in Santa Barbara County. The 1939 paper describes the typical water problems of Santa Barbara County, including the demands of the increasing population, periodic damage due to floods, the loss of soil on agricultural land and the combined reservoir sedimentation, the forest destruction and the resulting accelerating runoff, and the confusion of effort toward the solution of these challenging problems. Penfield was Fellow of the American Society of Civil Engineers ASCE, director of the California Council of Civil Engineers and Land Surveyors, and president of the Tri-Counties Association of Civil Engineers.

Anonymous (1977). Wallace C. Penfield. *Trans. ASCE* 142: 565-566.
Fowler, M. (2013). Wallace C. Penfield. Personal communication. *P*
Penfield, W.C. (1939). Water problems of Santa Barbara County. *Water and Soil Conservation Conference* Los Angeles: 16-20.
Penfield, W.C. (1960). Oldest periodic beach nourishment project. *Shore and Beach* 28(1): 47-55.
Penfield, W.C. (1963). *Your planning commissioner*. Sacramento CA.

PENNIMAN

19.02. 1868 Winsor VT/USA
26.08. 1934 St. Louis MO/USA

William Merit Penniman graduated from Dartmouth College, Hanover NH, in 1893 with the BSc degree. He was then employed by the St. Louis District, US Engineer Department, on the improvement of the Mississippi River, remaining there in various grades until his death, finally as senior engineer. From 1907 to 1909 he was engaged as principal engineer on its survey between St. Louis and Cairo IL. His review on methods of the complete river regulation by bank protection, permeable dikes, and submerged dams, published as Appendix 5 to the 1911 Report, was so complete that it was in use for decades. Later, a formula for water flow in open channels was derived, but never published. It has been widely used in the District, where it was known as the Penniman Formula, dealing with the uniform flow in a natural river.

Because of his unusual knowledge and technical skill, he was employed during his later years on special investigations and reports. His services were utilized to aid the decisions reached by the Board of Engineers on problems beyond the bounds of St. Louis District. His training of younger engineers to produce accurate and reliable work proved to be of major value for the Engineering Department, since many of his former assistants held later responsible positions in the Districts. His service was characterized by his loyalty to the Department; his open-mindedness and ability to keep informed on the current developments in hydraulics and other engineering aspects made him a person whose counsel was frequently sought. As a tribute to his high standing, the District renamed a large steam towboat Penniman after his death. His long experience embraced a wide range of open-river improvements, including surveys, hydrography, bank protection, contraction work, the design and operation of hydraulic dredges, and other floating plant. He was trained as young engineer by William S. Mitchell (1857-1932). Penniman was fatally stricken two days before his death at age 66 and buried at St. Louis, on a hill overlooking the Mississippi River, which he loved so well. Penniman was member of the American Society of Civil Engineers ASCE since 1923.

Anonymous (1936). William M. Penniman. *Trans. ASCE* 101: 1621-1622.
Penniman, W.M. (1911). Review of methods of complete river regulations. House of
Representatives, Document No. 50. 61st Congress, 1st Session. Washington DC.
www.mvs.usace.army.mil/.../River_Engineers_on_the_Middle_Missis.... *P*
http://www.concordiasentinel.com/news.php?id=7328

PETERKA

14.01. 1911 Cleveland OH/USA
15.01. 1983 Denver CO/USA

Alvin Joseph Peterka graduated from Case Institute of Technology, Cleveland OH, in 1934, and added graduate work at the Universities of Tennessee and Colorado. From 1936 to 1946 he was in the staff of the Tennessee Valley Authority TVA, then joining the US Bureau of Reclamation USBR, Denver CO, eventually heading the Special Investigations Section of the USBR Hydraulic Laboratory. Peterka retired from this activity in 1968 to become a consultant.

Peterka reported in his 1946 on a tunnel spillway, which was later adopted also for the Glen Canyon Spillway, and on flip buckets. He developed into an expert on high-speed flows, co-authoring the 1948 paper on frictional effects of tunnel flows. This paper was awarded the 1949 ASCE James Laurie Prize. He also was the first to analyze the effect of air entrained on the reduction of cavitation damage in 1953. In 1954, now with USBR, he presented an important study on shaft spillways, an alternative to the standard spillway, thereby investigating both free and submerged flow conditions. At the vertical shaft base, the flow is horizontally deflected to the outlet structure. The so-called morning glory spillway is a standard today provided free surface flow is maintained from the inlet to the outlet. Peterka's name has become known for his 1958 publication on the design of stilling basins, a notable research campaign undertaken during the 1950s in collaboration with Joseph N. Bradley (1903-1993). Ten different basins were model-tested and designs were proposed that have remained a standard.

Anonymous (1950). A.J. Peterka. *Civil Engineering* 20(1): 42. *P*

Beichley, G.L., Peterka, A.J. (1961). Hydraulic design of hollow-jet valve stilling basins. *Journal of the Hydraulics Division* ASCE 87(HY5): 1-36.

Hager, W.H., Falvey, H.T. (2003). Alvin Peterka: Hydraulic engineer. *Journal of Hydraulic Engineering* 129(9): 657-659. *P*

Hickox, G.H., Peterka, A.J., Elder, R.A. (1948). Friction coefficients in a large tunnel. *Trans. ASCE* 113: 1027-1046; 113: 1071-1076.

Peterka, A.J. (1946). Model and prototype studies on unique spillway. *Civil Engineering* 16(6): 249-251; 16(8): 359; 16(10): 460-461.

Peterka, A.J. (1953). The effect of entrained air on cavitation pitting. 5th *IAHR Congress* Minneapolis: 507-518.

Peterka, A.J. (1958). Hydraulic design of stilling basins and energy dissipators. *Engineering Monograph* 25. US Dept. of the Interior: Washington DC.

PETERSEN M.S.

28.04. 1920 Rock Island IL/USA
19.01. 2013 Tucson AZ/USA

Margaret Sara Petersen obtained the BS degree in 1947, and the MS degree in 1953 from the University of Iowa, Iowa City IA. She began her employment with the US Corps of Engineers. After graduation she held various posts in research, the design of hydraulic structures, and water resources planning. She worked on some of the largest water projects, including the Mississippi River flood control and navigation effort, or the Missouri River multi-purpose storage reservoirs. Since 1981 she was faculty member of the College of Civil Engineering and Engineering Mechanics, University of Arizona, Tucson AZ. She developed new graduate-level courses in hydraulic engineering, and authored the widely-accepted 1984 book, because it draws attention to problems of lesser developed countries. In 1996 she completed a monograph on Inland navigation and canalization for the Chinese Intl. Research and Training Center of Erosion and Sedimentation. She lectured in South Africa, China and Morocco. Petersen was an active member of the American Society of Civil Engineers ASCE, elected in 1991 Honorary Member, and received the 2001 Hunter Rouse Hydraulic Engineering Award.

The 1994 paper is concerned with the stilling basin designs proposed by the US Bureau of Reclamation USBR in the 1950s. However, these design guidelines do not explicitly specify the required tailwater elevation for acceptable basin performance. Accordingly, USBR Basins III and IV were experimentally reconsidered in this regard by accounting for the resulting scour beyond the basins. It was found that the tailwater elevation is highly sensitive in this regard, particularly if it is below the standard elevation.

Hadjerioua, B., Laursen, E.M., Petersen, M.S. (1994). Behaviour of hydraulic jump basins. *Hydraulic Engineering* '94 Buffalo NY 1: 416-420. ASCE: New York.
Petersen, M.S. (1963). Hydraulic aspects of Arkansas River stabilization. *Journal of the Waterways and Harbors Division* ASCE 89(WW4): 29-65; 91(WW3): 139-140..
Petersen, M.S. (1984). *Water resource planning and development*. Prentice-Hall: Englewood Cl.
Petersen, M.S. (1986). *River engineering*. Prentice-Hall: Englewood Cliffs.
Petersen, M.S. (2003). How I became a hydraulic engineer. *Journal of Hydraulic Engineering* 129(5): 335-339.
Stephenson, D., Petersen, M.S. (1991). *Water resources development in developing countries*. Elsevier: Amsterdam.
http://www.engineering.uiowa.edu/honor-wall/alumni-academy/members/petersen.php *P*

PETERSON D.F.

03.06. 1913 Delta UT/USA
21.03. 1989 Logan UT/USA

Dean Freeman Peterson, Jr., was educated at State University of Utah, Salt Lake City UT, from where he obtained the BS degree in 1934, and the DSc hon. in 1978. The MS degree in civil engineering was obtained from the Rensselaer Polytechnic, Troy NY, and the Dr. title in 1939. Dean was a junior hydraulic engineer with the US Geological Survey from 1937 to 1939, project engineer for the Upper Potomac River until 1941, then progress engineer until being appointed in 1946 associate professor of civil engineering at Utah State University. He was there from 1957 until retirement in 1976 professor, dean of engineering, and vice-president of research.

Peterson was a teacher, researcher and consultant. He was for instance an US delegate for the Near East-South Asia Irrigation Seminar held in the 1960s, a consultant of water resources, Office of Science and Technology, presiding it in the mid-1960s, chief of the Agricultural Review Team for Afghanistan in 1967, or chairman of the Utah Advisory Council on Science and Technology from 1973. He was a Fellow of the American Geophysical Union AGU, member of the American Association for Advancement of Science AAAS, member of the American Society of Civil Engineers ASCE, acting from 1972 to 1974 as vice-president, and winner of the 1968 Royce J. Tipton Award. He also was a member of the American Academy of Engineering, and the American Academy of Arts and Sciences. His research interests were in irrigation and drainage mainly, and hydraulic engineering, with papers published in the ASCE and AGU journals.

Anonymous (1960). Peterson, Jr., D.F. *American men of science* 10: 3156. Cattell: Tempe AZ.
Anonymous (1968). D. Peterson, Jr., honoured for service to ASCE. *Civil Engineering* 38(3): 81. P
Peterson, Jr., D.F. (1955). Discussion of Effect of well screens on flow. *Trans. ASCE* 120: 597.
Peterson, Jr., D.F. (1957). Groundwater development: Hydraulics of wells. *Trans. ASCE* 122: 502-517.
Peterson, Jr., D.F., Mohanty, P.K. (1960). Flume studies on flow in steep, rough channels. *Proc. ASCE* 86(HY9): 55-75; 88(HY1): 83-87; 88(HY3): 199-202.
Peterson, Jr., D.F. (1962). Intercepting drainage wells in artesian aquifer. *Trans. ASCE* 127: 32-42.
Peterson, Jr., D.F. (1973). Some problems of on-farm water use in the Middle East and Asia. *Floods and Droughts*: 16-23, E.F. Schulz, ed. Water Resources Publ.: Fort Collins.
Zee, C.-H., Peterson, Jr., D.F., Bock, R.O. (1957). Flow into a well by electric and membrane analogy. *Trans. ASCE* 122: 1088-1105; 122: 1111-1112.

PHARR

08.05. 1875 La Grange AK/USA
03.11. 1947 Memphis TN/USA

Harry Nelson Pharr graduated from the University of Arkansas, Fayetteville AK, in 1893, with the BSc degree in civil engineering. After several years as assistant engineer of the St. Francis Levee District AK, he submitted a thesis to his Alma Mater, for which he was conferred the MSc degree. Pharr was made in 1897 chief engineer of the District, thereby following his father. In 1907 he became engaged in private practice, and served also as chief engineer of the Crawford County Levee District AK, designing and constructing levees along Arkansas River. He was then active in surveys on flood control for Obion River in Tennessee, Kansas River in Missouri, and Wabash River in Indiana. He also was consultant of flood control and drainage problems of various railroad companies.

In 1915, following the great floods of 1912 and 1913, when the existing levee system proved inadequate in protecting the St. Francis Basin, he was again appointed chief engineer of the District, planning and constructing extensive enlargements of the levee system. Pharr left this District in 1917 to engage again in private practice, but in 1918 returned to the District as chief engineer, remaining in that position until 1935. He planned and constructed in corporation with the federal government extensive levee enlargements and strengthenings, and was furthermore engaged in the relocation and rebuilding of parts of the entire levee system. He resigned in 1935 this position, accepting an appointment to become member of the Mississippi River Commission, on which he served until his death. This Commission had charge on the design and the construction of flood control projects, including levees, revetments, dikes, spillways, cutoffs, and outlets on the Mississippi River from Cairo IL to the Gulf of Mexico. Pharr was known as an engineer of highest integrity, soft spoken but firm in his convictions. He was member of the American Society of Civil Engineers ASCE from 1905, and was closely associated with the Engineers' Club of Memphis TN, where he lived from 1900.

Anonymous (1948). Harry N. Pharr. *Trans. ASCE* 113: 1529-1532.
Pharr, H.N. (1901). Brief history of levees in the Lower St. Francis Levee District. *Riparian lands of the Mississippi River*: Past, present, prospective, F.H. Tompkins, ed.: 314-317. *P*
Pharr, H.N. (1948). Discussion of Mississippi River Cutoffs. *Trans. ASCE* 113: 20-21.
Reuss, M. (2004). *Designing the Bayous*: The control of water in the Atchafalaya Basin 1800-1995. Texas A&M University Press: College Station TX.
http://www.mvd.usace.army.mil/About/MississippiRiverCommission(MRC)/PastMRCMembers.aspx *P*

PICKELS

05.09. 1883 Richmond KY/USA
02.12. 1944 Urbana IL/USA

George Wellington Pickels received in 1904 the BSc degree from University of Kentucky, Lexington KY, and in 1911 was then awarded the civil engineering degree by University of Illinois, Urbana IL. In 1907 he was there appointed civil engineering instructor, and there was teacher, author and investigator in the fields of flood control, hydrology, and land drainage during 37 years. His knowledge of the hydrological characteristics made him adept in selecting possible sites for reservoirs in various parts of the State. Crab Orchard Dam, built at Carbondale IL, then created the largest artificial water body in the State with an important economic potential. Pickels' report on this project formed the major part of a document issued by the University, the Illinois State Geological Survey, and the Illinois State Water Survey in 1935. A similar approach was later made for the Kaskaskia River Scheme.

Pickels published many other writings, including as co-author the 1929 Bulletin on Land drainage in Illinois. He also wrote two Bulletins on Run-off investigations, and on Floods on Illinois Streams. His most important contribution to the engineering literature was his 1941 book entitled Drainage and flood-control engineering. He was a member of the American Society of Civil Engineers ASCE, and past-president of the Illinois Society of Engineers. He also was member of the American Geophysical Union AGU, and the American Water Works Association AWWA. He was founder of the drainage conferences held at the University of Illinois from 1916 to 1921. He possessed an engaging personality, combining dignity with a ready wit and a warm disposition. He was keen, competent, and meticulous in his work, and tolerant and corporative with others. It was reported that it had been pleasant to work with him.

Anonymous (1945). George W. Pickels. *Trans. ASCE* 110: 1790-1792.
Pickels, G.W., Leonard, F.B. (1929). Engineering and legal aspects of land drainage in Illinois. *Bulletin* 42, ed. 2. Illinois State Geological Survey: Urbana IL.
Pickels, G.W. (1931). Run-off investigations in Central Illinois. *Bulletin* 232. Engineering Experiment Station, University of Illinois: Urbana IL.
Pickels, G.W. (1937). Magnitude and frequency of floods on Illinois streams. *Bulletin* 296. Engineering Experiment Station, University of Illinois: Urbana IL.
Pickels, G.W. (1941). *Drainage and flood-control engineering*. McGraw-Hill: New York.
www.isws.illinois.edu/pubdoc/cr/iswscr-162.pdf *P*

PIERSON

07.07. 1922 New York NY/USA
07.06. 2003 West Hempstead NY/USA

Willard James Pierson, Jr., started his professional career in meteorology at the University of Chicago, and made active service in China from 1943 to 1946. He then entered New York University NYU, from where he obtained the PhD degree in 1949. In spite of early works mainly in meteorology, he is remembered as an oceanographer, specifically for his wide-ranging studies of the surface waves from centimetres to kilometres, and his pioneering work on remote sensing of the ocean surface.

Pierson joined the NYU meteorology department in 1949 as instructor, achieving the rank of professor in 1961in the renamed Department of Meteorology and Oceanography, where he remained until it ceased to exist in 1973. In the first half of his professional career, he developed the spectral analysis of ocean waves, the systematic analysis of ship motion in a random sea, the generation and propagation of surface waves on the ocean, the methodology of numerical wave forecasting, and the remote sensing of ocean surface winds and sea state. His seminal 1952 research provides the underpinnings for research that continues up to now. Pierson was honoured on the 20th anniversary of this paper by the Society of Naval Architects and Marine Engineers SNAME for 'having changed our think about the ship, the sea, and the ship upon the sea'. Pierson knew already in the 1970s that the full realization of wave models would require satellite-borne remote sensing systems for providing marine winds to drive them, and wave information to validate and verify the models. He and a colleague conceived the idea of remote assessment of the winds in the marine boundary layer from the reflectivity to microwave radiation caused by the small waves that are in equilibrium with the wind in height and direction: the birth of scatterometry.

Anonymous (2003). Willard J. Pierson, Jr. *Eos* 84(42): 443-444. *P*

Neumann, G., Pierson, W.J., Jr. (1966). *Principles of physical oceanography*. Prentice-Hall: Englewood Cliffs NJ.

Pierson, W.J., Jr. (1949). The effects of eddy viscosity, Coriolis deflection and diurnal temperature fluctuations on the sea breeze as a function of time and height. *PhD Thesis*. University of New York.

Pierson, W.J., Jr., St. Denis, M. (1953). On the motion of ships in confused seas. *Trans. SNAME* 61(4): 280-357.

Pierson, W.J., Jr. (1972). The loss of two British trawlers: A study in wave refraction. *Journal of Navigation* 25(3): 291-304.

PILLSBURY

11.10. 1904 Hollywood CA/USA
12.04. 1991 Goleta CA/USA

Arthur Francis Pillsbury received the BA degree in 1928, and the civil engineering degree in 1930 from Stanford University, Stanford CA. He joined the staff of the University of California at Berkeley CA, in 1932, and moved to University of California, Los Angeles CA, UCLA, in 1939, serving as irrigation engineer until 1966, moving through all professorial ranks from 1940 until his retirement, retiring as full professor after 38 years on the UCLA Faculty, where he had headed the Irrigation and Soil Sciences Dept., the Engineering Systems Division, and the Water Resources Center. He was member of the American Society of Civil Engineers ASCE from 1930, receiving the 1977 Royce Tipton Award for outstanding teaching and research in hydrology, including soil science and drainage system engineering and environmental resource management. He also was member of the American Society of Agricultural Engineers ASAE, the American Geophysical Union AGU, and the US Committee of the International Commission of Irrigation and Drainage.

Pillsbury was a pioneer in the field of irrigation and water resources. He was among the first in California to warn of flood and erosion damage resulting from the loss of ground cover in the wake of disastrous brush fires. He served as consultant to a number of organizations in a variety of worldwide water rights, water quality, and delivery issues. These included the International Bank for Reconstruction and Development, and the Food and Agriculture Organization FAO, the United Nations. The 1968 Report presents pertinent information of engineering, agricultural and economic aspects of the then relevant sprinkler irrigation techniques. The selection, design and installation of the best irrigation system including soil, crop, climate, and water supply are detailed. He was a member of the US Panel, the International Boundary and Water Commission, addressing problems with the salinity of Colorado River.

Anonymous (1992). Arthur F. Pillsbury. *Trans. ASCE* 157: 509.
Pillsbury, A.F. (1933). *Erosion control in mountain meadows*. USDA: Pasadena CA.
Pillsbury, A.F., Degan, A. (1968). Sprinkler irrigation. *Agricultural Development Paper* 88. FAO: New York.
Pillsbury, A.F. (1973). Environmental quality in California. *Civil Engineering* 43(7): 50-52. *P*
Pillsbury, A.F. (1981). The salinity of rivers. *Scientific American* 245(1): 55-65.
http://texts.cdlib.org/view?docId=hb0h4n99rb&doc.view=frames&chunk.id=div00061&toc.depth=1&toc.id=
http://www.worthpoint.com/worthopedia/1920-photo-arthur-pillsbury-holding-149401532 *P*

PLATZMAN

19.04. 1920 Chicago IL/USA
02.08. 2008 Chicago IL/USA

George William Platzman received in 1940 the BS degree in mathematics and physics from University of Chicago, and the MS degree in 1941 from the University of Arizona, Tucson AZ. He then worked for the US Army Corps of Engineers at Portland OR, estimating the maximum possible precipitation in Willamette River Basin for spillway design. After WW II, he studied meteorology at his Alma Mater, joining in 1948 the Faculty of the University's Meteorology Department. In 1961 it formed the Department of Geographical Sciences, of which he served as chairman from 1971 to 1974. He retired as emeritus professor in 1990. Platzman received appointments as visiting scientist to the National Center for Atmospheric Research NCAR at Boulder CO from 1963 to 1983. As Guggenheim Fellow in 1967 at Imperial College, London UK, he conducted research on wind dynamics in the ocean and atmosphere. His honours included fellowships to the American Geophysical Union AGU, the American Meteorological Society AMS, and the American Association for the Advancement of Sciences.

Platzman pioneered the field of storm-surge forecasting. He had eventually specialized in dynamical meteorology and oceanography, including investigations of numerical weather prediction and storm surges. He became interested in these when a 2 m surge hit Chicago's Montrose Harbour in 1954, with fatal results, because high winds over large water bodies can cause these surges. 'George was one of the founders of modern meteorology who transformed weather forecasting from qualitative guesswork to quantitative science', said a colleague from the NCAR. He was also remembered as a man of gentle manners who maintained a sharp mind until his end. Platzman's seminal review of the Rossby waves, as introduced by Carl-Gustav Rossby (1898-1957), dealing with planetary-scale atmospheric oscillation that is critical to weather dynamics, as also his series of papers on ocean tides, was considered outstanding. Platzman also used computers to study the storm surge caused by Hurricane Carla, which stuck the Texas Coast in 1961. Later he dealt with natural oscillations of the world ocean and tides.

Platzman, G.W. (1968). The Rossby wave. *Quart. Journal Meteorological Society* 94: 225-248.
Platzman, G.W. (1970). Ocean tides and related waves. *Lectures in Applied Math.* 14: 239-291.
Platzman, G.W. (1972). Two-dimensional free oscillations in natural basins. *Journal of Physical Oceanography* 2(2): 117-138.
http://news.uchicago.edu/article/2008/08/18/george-w-platzman-meteorologist-1920-2008 *P*

PLESSET

07.02. 1907 Pittsburgh PA/USA
19.02. 1991 San Marino CA/USA

Milton Spinoza Plesset obtained the BS and MS degrees from the University of Pittsburgh, Pittsburgh PA, and the PhD degree from Yale University, New Haven CT, in 1932. He was an instructor in physics at University of Rochester, Rochester NY, from 1935 to 1940, until 1945 head of the analytical group, Douglas Aircraft Co., Santa Monica CA, taking then over as professor of engineering science at California Institute of Technology, Pasadena CA, till retirement in 1978. He was in parallel consultant to industrial firms, notably the Naval Ordnance Test Station, Pasadena CA, Electric Power Research, or General Motors. He was elected Fellow of the American Society of Mechanical Engineers ASME in 1967, and elected to the National Academy of Engineering in 1979, with the citation 'Pioneering in applied physics, including two-phase flows, and contributions to the understanding of the thermal hydraulics and safety of nuclear reactors'.

Plesset was a researcher and teacher concerned with various aspects of fluid dynamics, including cavitation mechanics and nuclear engineering. His investigations in cavitation dynamics are particularly important and led to the understanding of the most prominent physical features. The 1949 paper describes three regimes, namely the non-cavitating, incipient cavitating, and established cavitating flows. High-speed photographs illustrate the differences among these regimes. The cavitation number, previously introduced by Dieter Thoma (1881-1942), was used to describe the cavitation phenomena. Rayleigh's theory describing the movement of a cavitation bubble was successfully expanded, and validated with model observations. A review of the main aspects of cavitation in fluid dynamics was presented in 1977.

Anonymous (1972). M.S. Plesset. *Mechanical Engineering* 94(6): 90. *P*
Anonymous (1987). Plesset, Milton S. *Who's who in America* 43: 2227. Marquis: Chicago.
Plesset, M.S. (1949). The dynamics of cavitation bubbles. *Journal of Applied Mechanics* 16(9): 277-282; 17(3): 100-101.
Plesset, M.S., Mitchell, T.P. (1955). On the stability of the spherical shape of a vapour cavity in a liquid. *Quarterly Applied Mathematics* 13(4): 419-430.
Plesset, M.S., Chapman, R.B. (1971). Collapse of an initially spherical vapour cavity in the neighbourhood of a solid boundary. *Journal of Fluid Mechanics* 47: 283-290.
Plesset, M.S., Prosperetti, A. (1977). Bubble dynamics and cavitation. *Annual Review of Fluid Mechanics* 9: 145-185.

POE

07.03. 1832 Navarre OH/USA
02.10. 1895 Detroit MI/USA

Orlando Metcalfe Poe attended the US Military Academy, West Point, graduating in 1856, serving then until 1861 as assistant topographical engineer on the survey of the northern Great Lakes. During the Civil War he was involved in the defense of the capital, and then promoted to colonel, and to chief engineer in 1864. From 1865 he was engaged as the Lighthouse Board's chief engineer, promoted to the position of chief engineer of the Upper Great Lakes in 1870. He designed various 'Poe style lighthouses' which were unique in terms of location, construction, and cost. One was described as 'one of the greatest engineering features on the Great Lakes'. He solved the logistics problem of building a lighthouse on the remote Stannard Rock in Lake Superior with the proposal to use all the costly apparatus and machinery used to build the Spectacle Reef Light. Its exposed crib is rated in the top ten engineering feats of the United States. Many of these lights were of Italian architecture, a chief example being the Gross Point Light, Evanston IL.

Poe served from 1873 to 1883 as engineering aide-de-camp in the US Army. In 1883 he was made superintending engineer for the improvement of rivers and harbors on Lakes Superior and Huron, where he helped to develop the St. Marys Falls Canal. One of his crowning achievement was the design and implementation of the first Poe Lock in the American Soo Locks at Sault St. Marie, as it was instrumental in making possible the shipping industry, including steel craft freighters, in the Upper Great Lakes. His 240 m long and 30 m wide, then the largest of the world, creation was dismantled in the early 1960s with a larger and modern lock built on the same site. This new passage was renamed the Poe Lock and serves from then the largest of the Great Lakes freighters. He also was responsible for significant improvements made on Detroit River and the ship canals at Chicago, Duluth and Buffalo. Poe died of an infection following an on-duty accident at the Soo Locks, and was subsequently buried at the Arlington National Cemetery. Poe Reef and the Poe Reef Light in Lake Huron bear his name.

Pepper, T. (2011). *Seeing the light, Orlando Metcalfe Poe*: The great engineer of the western Great Lakes. GLLKA Mackinaw City. *P*
Poe, O.M. (1852). *Erie Canal Lock*. Military Academy: West Point.
Taylor, P. (2009). *Orlando M. Poe*: Civil War general and Great Lakes engineer. Kent University Press: Kent OH.
http://en.wikipedia.org/wiki/Orlando_Metcalfe_Poe *P*

POHLE

21.02. 1919 Buena Vista VA/USA
17.02. 1996 Nassau NY/USA

Frederick Valentine Pohle received education from Cooper Union Institute of Technology, New York NY, and from New York University, with the MS degree in 1943 and the PhD degree in mathematics in 1949. He was then from 1946 to 1948 research assistant at the Courant Institute, New York NY, and from 1950 to 1960 from assistant professor in mechanics to professor at the Polytechnic Institute of Brooklyn NY. He was in parallel in the 1950s researcher at the National Bureau of Standards and chairman of the Dept. of Mathematics at Adelphi University, New York, from 1962. He was member of the American Mathematical Society AMS, the Mathematical Association of America MAA, and the American Institute of Aeronautics and Astronautics AIAA.

Pohle had research interests in the theory of elasticity, in the vibration theory, and in non-linear mechanics. The 1952 paper describes the two-dimensional equations of the hydrodynamic motion, expressed in Lagrangian, as opposed to Cartesian coordinates. These were used to investigate the motion of an ideal fluid, a representation having far-reaching advantages in problems with time-dependant free boundaries, in which the independent space variables being the initial particle coordinates. The region occupied by the fluid is thus a fixed region independent of time. The front displacement and the pressure are expanded in powers of time. Equating to zero the coefficients of the powers relating to time leads to a systematic procedure for the determination of the successive terms, of which each is a solution of the Poisson equation. In all cases considered higher approximations require determination of a Green's function. This method was applied to the initial stages of a dam break. Pohle's approach was a singular work in the field of dambreak waves, but has certainly inspired the subsequent works of Robert F. Dressler (1920-1999) or Gerald B. Whitham (1927-) on the hydraulic resistance effect, or the wave front characteristics and the slope effect of dambreak waves.

Anonymous (1994). Pohle, Frederick V. *American men and women of science* 5: 1291. Bowker: New Providence NJ.
Dressler, R.F., Pohle, F.V. (1953). Resistance effects on hydraulic instability. *Communications in Pure and Applied Mathematics* 6(1): 93-96.
Pohle, F.V. (1952). Motion of water due to breaking of a dam and related problems. *Circular* 521: 47-53. US Bureau of Standards: Washington DC.
http://www.pohlecolloquium.org/ P

POMEROY

22.12. 1904 Burbank CA/USA
05.06. 1993 Twentynine Palms CA/USA

Richard Durant Pomeroy obtained the BS degree in 1926, and the PhD degree in 1931 from the Caltech, Pasadena CA. He was employed until 1940 by the Los Angeles City Sanitation District. Until 1973 he was founder and president of Pomeroy, Johnston & Bailey Inc., Pasadena CA. From 1978 he then joined another consulting firm at Pasadena CA. He was the recipient of the 1946 Harrison Prescott Eddy Award of the Water Pollution Control Federation WPCF, and the 1972 Rudolph Hering Award, the American Society of Civil Engineers ASCE. He was winner of the 1941 ASCE James Laurie Prize. Pomeroy was a member of the American Water Works Association AWWA.

Pomeroy was a notable American sanitary engineer who was both active in research and consulting. In the 1930s he developed the sludge digestion process. Later his research on hydrogen sulphide in sewers led him to introduce the method for sulphide testing, which subsequently became a standard in wastewater engineering. Once a consultant and president, the firm developed laboratory and design capabilities in both civil and chemical engineering. His research interests further included the design and operation of sludge digestion tanks, the design of ocean outfalls, the hydraulics of sewers, the corrosion of metals by water, and laboratory methods in environmental engineering in general. His 1983 paper deals with velocity formulae to be adopted in sewer systems.

Anonymous (1941). Richard Pomeroy. *Civil Engineering* 11(1): 54. *P*
Anonymous (1984). Richard D. Pomeroy. *Civil Engineering* 54(10): 80. *P*
Anonymous (1985). Pomeroy, R.D. *Who's who in engineering* 6: 524. AAES: Washington DC.
Anonymous (1994). Richard Durant Pomeroy. *Journal AWWA* 86(2): 141.
Parkhurst, J.D., Pomeroy, R.D. (1972). Oxygen absorption in streams. *Journal of the Sanitary Engineering Division* ASCE 98(SA1): 101-124.
Pomeroy, R.D. (1967). Flow velocities in small sewers. *Journal WPCF* 39(9): 1525-1548.
Pomeroy, Johnston & Bailey (1974). *Process design manual for sulphide control in sanitary sewage systems*. Pasadena CA.
Pomeroy, R.D. (1982). Biological treatment of odorous air. *Journal WPCF* 54(12): 1541-1545.
Pomeroy, R.D. (1983). Flow velocities in pipelines. *Journal of Hydraulic Engineering* 109(8): 1108-1117; 110(11): 1510-1514.
Rawn, A.M., Banta, A.P., Pomeroy, R. (1939). Multiple-stage sewage sludge digestion. *Trans. ASCE* 104: 93-132.

PORTER C.T.

18.01. 1826 Auburn NY/USA
28.08. 1910 Montclair NJ/USA

Charles Talbot Porter was an American mechanical engineer, manufacturer of machines, and the father of the high-speed steam machine. During the mid-1800s, steam power drove machinery in factories, drove the development of manufacturing systems, and revived dormant industries. Eventually, steam engines became powerful enough to supply power. Porter was a key inventor at the time, and the first to create high-speed steam engines. He graduated from Hamilton College, Clinton NY, and spent his early career as lawyer at Rochester NY, where he was introduced to inventing, so that he invested in what turned out to be a scam by a non-paying client. He purchased a patent for a steam-powered machine that cut, smoothed, and shaped stone blocks, and formed the Porter Stone Dressing Machine Company. His first prototype failed, but he built then machines that were simpler and contained fewer moving parts than the competition. It was stated in the 1855 *Scientific American* they cut 240 m of quarry stone/day, the work of 50 men.

Porter invented an isochronous centrifugal governor to control the temporal speed of an engine. According to his autobiography, the governor follows manufacturing techniques of machine tool producers, regulates machines more precisely. The "Porter Governor" won medals at the 1859 American Institute Fair. He became interested in steam power while locating people and potential machines that could benefit from his governor. He collaborated with steamboat engineer John Allen, who had patented ideas to overcome a defect in marine engines. The two thought the engines were suitable for high-speed operation, so that Porter visited engine rooms, designed and adapted overhead cranks, flywheels, and pulley borings, and eventually measured engine efficiency. Porter is best known as the first to realize high-rotative speeds in steam engines. Higher-speed pistons developed increased HP engines. He patented the high-speed engine, earning acclaim. He also modified and patented the Richards Steam Engine Indicator. It monitored the steam pressure inside cylinders and was a helpful tool for working with steam engines. Porter stated that this indicator made high-speed engineering possible. Although steamboats often used steam power in the 1850s, high-speed engines were not important until the 1880s when they directly drove electric generators. Porter installed the first electric-drive high-speed engine for Thomas Edison's laboratory at Menlo Park in 1880.

Porter, C.T. (1888). *Richards steam-engine indicator*. Van Nostrand: New York.
https://www.asme.org/engineering-topics/articles/energy/charles-talbot-porter *P*

PORTER D.

28.08. 1855 Hartford CT/USA
26.02. 1935 Malden MA/USA

Dwight Porter graduated from Sheffield Scientific School, Yale University, New Haven CT, obtaining in 1880 the PhB degree. After having served as instructor, he was appointed in 1883 professor of hydraulic engineering at Massachusetts Institute of Technology MIT, Cambridge MA. He was member of the American Society of Civil Engineers ASCE, the Boston Society of Civil Engineers, and the New England's Water Works Association.

Porter's compendium on stream gagings is almost unknown, despite it was a successful teaching basis. The contents includes Stream gaging, Discharge curves, Gage, Soundings, Water flow, Pulsations, Velocity distribution, Velocity measurement methods, Vertical velocity curves, Velocity at the surface, Mean velocity, Two-point or three-point methods, Rod or tube floats, Integration methods, Harlacher method, Computation of discharge, Discharge curve, Loops, Stout method, Floats, Current meters: Fteley, Price, Haskell, Discharge measurement by chemical means, Slope formulas, Accuracy of stream gagings, Gagings in ice-covered channels, Time required for gagings, Problems. It may be noted that all problems of theoretical and practical hydrometry are addressed in this book. In a later paper it was observed that the channel shape has a significant effect on accurate discharge measurement, and it was recommended that gagings should be made in artificial reaches of a river to eliminate abnormal discharge curves.

Anonymous (1899). Dwight Porter. *Technique yearbook*: 27. MIT: Cambridge MA. *P*
Anonymous (1905). Porter, Dwight. *Who's who in America* 3: 1181. Marquis: Chicago.
Porter, D. (1887). *Report on the water-power of the Ohio River basin and Ohio State canals*. Washington DC.
Porter, D. (1889). *Report* upon a sanitary inspection of certain tenement-house districts of Boston. Rockwell & Churchill: Boston.
Porter, D. (1899). Water-power streams of Maine. 19th *Annual Report* Part IV Hydrography: 34-111. US Geological Survey: Washington DC.
Porter, D. (1909). *Notes* on hydraulic measurements, prepared for the use of students in civil and sanitary engineering. Massachusetts Institute of Technology: Cambridge MA.
Porter, D. (1912). *Notes on stream gagings*, prepared for the use of students in civil and sanitary engineering at Massachusetts Institute of Technology. MIT: Cambridge MA.
Porter, D. (1917). Discussion of Effect of channel on stream flow, by N.C. Grover. *Journal of the Boston Society of Civil Engineers* 4(3): 123-131.

POSEY

12.06. 1906 Mankato MN/USA
30.08. 1991 Cedar Rapids IA/USA

Chesley Johnston Posey obtained the BS and MS degrees in civil engineering from Kansas University, and University of Illinois, Urbana IL, respectively. He was from 1929 to 1934 instructor at Iowa State University, Iowa City IA, spending thereby 1933 as assistant engineer at the Tennessee Valley Authority TVA. He was appointed assistant professor at State University of Iowa in 1934, promoted to associate professor in 1940, and was professor of hydraulics and structural engineering there, and Head of the Department of Civil Engineering from 1946 until 1962. He then moved to University of Connecticut, Storrs CT. Posey was in parallel consultant in civil engineering. He was member of the American Society of Civil Engineers ASCE, of the American Water Works Association AWWA, among others. He was awarded the 1958 ASCE James Laurie Prize.

Posey is known for his undergraduate textbook co-authored by Sherman M. Woodward (1871-1953). He also conducted research in collaboration with Ralph W. Powell (1889-1976) on large-scale open channel friction at the Rocky Mountain Hydraulic Laboratory, Fort Collins CO, which was founded by him in 1943, originally at Allenspark CO. The laboratory served as a summer residence for students and professionals interested in water experiments. The new name was adopted in 1991, and the Laboratory was taken over by the US Geological Survey to conduct a range of hydrologic and environmental science investigations. The 1974 paper investigates scour at bridge piers and means to protect these important infrastructural elements. The inverted filter as proposed by Karl Terzaghi (1883-1963) was recommended.

Anonymous (1958). Chesley J. Posey. *Civil Engineering* 28(10): 778. *P*
Anonymous (1964). Posey, C.J. *Who's who in engineering* 9: 1483. Lewis: New York.
Anonymous (1981). Chesley J. Posey. *Eos* 62(28): 577. *P*
Posey, C.J., Hsu, H.-c. (1950). How the vortex affects orifice discharge. *ENR* 144(Mar.9): 30.
Posey, C.J. (1957). Flood-erosion protection for highway fills. *Trans. ASCE* 122: 531-555.
Posey, C.J. (1974). Tests of scour protection for bridge piers. *Journal of the Hydraulics Division* ASCE 100(HY12): 1773-1783; 101(HY10): 1369-1371; 101(HY11): 1454-1455; 102(HY4): 531-532.
Powell, R.W., Posey, C.J. (1959). Resistance experiments in a triangular channel. *Journal of the Hydraulics Division* ASCE 85(HY5): 31-66.
Woodward, S.M., Posey, C.J. (1941). *Hydraulics of steady flow in open channels*. Wiley: NY.

POST J.C.

30.07. 1844 Newbury NY/USA
06.01. 1896 New York NY/USA

James Clarence Post entered in 1861 the US Military Academy at West Point NY, and on graduation four years later was appointed second lieutenant, and transferred to the Corps of Engineers, with which he was connected until his death. After his service as assistant engineer for improving government works in New England, he became assistant professor of mathematics at West Point, where he remained for four years. The next three years were spent with the engineer battalion at Willet's Point NY. He was promoted in 1871 to captain. From 1874 to 1882 he was assistant engineer, from when he was in charge of river improvements in several states. He was again promoted in 1886 to major.

Post was transferred in 1887 to Washington DC as assistant to the Chief of Engineers, and in 1889 became Military Attaché to the US Legation in London UK. In 1891 he was a delegate to the Geographical Congress at Bern, Switzerland, and in 1893 attended the Permanent International Maritime Congress in London. Shortly before he had married there an English lady, with whom he had a son. Major Post was from 1894 to 1895 in charge of important works in the Northwest. The large jetty at the mouth of Columbus River was completed under his supervision, but his work was in danger during a great flood in 1894, when the Cascades Canal had a serious injury. During this time he was also in charge of the construction and equipment of a boat railway from the foot of the Dallas Rapids to the head of Celilo Falls, and of many river and harbour works. He also served as board member on bridge construction and river and harbour improvements, and was in supervisory charge of the construction of bridges across various streams. In 1896 he was relieved from the duty at Portland OR to take over new duties at Detroit, when he died. He was a member of the American Society of Civil Engineers ASCE.

Anonymous (1896). Memoir of James C. Post. *Trans. ASCE* 36: 569.
Anonymous (1896). James Clarence Post. 27th Annual Reunion of the *Association of graduates*: 84-89. US Military Academy: West Point NY. *P*
Cunningham, J.H., Post, J.C. (1895). *The Upper Willamette*: Eugene to Portland. US Corps of Engineers: Washington DC.
Post, J.C. (1886). Discussion of South pass jetties. *Trans. ASCE* 15: 235-241.
Post, J.C. (1895). *Letter from the Secretary of War*, with reference to the continued improvement of the Columbia River between the mouth of the Willamette River and the city of Vancouver. Washington DC.

POST W.S.

21.08. 1871 Wien/A
24.01. 1945 Los Angeles CA/USA

William Schuyler Post was born in Austria, where his father served as US Consul General. The family returned in 1876 to the USA, settling at Galesburg IL. William graduated in 1891 from Johns Hopkins University, Baltimore MD, in electrical engineering. He then served as chief of party on topographic surveys for the US Geological Survey in Wyoming and South Dakota. He was topographer from 1898 on the exploration of southwestern Alaska. He then moved to Los Angeles CA, where he became his interest into hydraulics. He served as chief engineer for the *Guyamaca* Mutual Water Company on the reconstruction of flumes and siphons. He later was chief engineer of a water company at San Diego CA for the Lake Henshaw System in charge of hydraulic studies and water supply in Kings County CA, working thereby under Joseph B. Lippincott (1864-1942).

Post contributed much to the improvement of water affairs in California. From 1921 to 1923 he was one of the three principal assistants to the deputy state engineer on water resources investigation, in charge of reservoir studies and dam estimates. Under the same title, he was from 1927 to 1929 in charge of The Santa Ana River Investigation, authoring a Report on the flood control and water conservation of the Santa Ana River. Post was in 1930 resident engineer in charge of the construction of Hogan Dam on the *Calaveras* River for flood control purpose of Stockton CA. He was involved in the hydrographic surveys, and in association with Frederick H. Tibbetts (1882-1938), served as joint designer of this dam. In 1931 he was appointed director of irrigation, US Indian Service, in charge of 26 projects. In 1934 he entered the US Park Service, with headquarters at Death Valley CA. In 1935 he acted as resident engineer for the Hetch Hetchy Dam, the water supply for San Francisco CA. He was then also in charge of the water supply of Berkeley and Oakland CA. Finally, in 1940, he entered federal service as hydraulic engineer at Los Angeles CA, remaining at this position until his retirement. Post was described as a man with a wonderful memory, a great knowledge and deep understanding of general and current affairs. He was the courtesy of the old school, was kind, capable, and a loyal friend. He was a member of the American Society of Civil Engineers ASCE from 1901, and particularly of its Los Angeles Section.

Anonymous (1945). William S. Post. *Trans. ASCE* 110: 1899-1901.
Post, W.S. (1929). Santa Ana River Investigation. *Bulletin* 19. State of California: Sacramento.
http://farm7.staticflickr.com/6029/5940654637_3cba391155_z.jpg *P*

POTTER

24.01. 1864 Lisbon Falls ME/USA
06.08. 1928 St. Louis MO/USA

The career of General Charles Lewis Potter on the Mississippi River began in 1899 then as a young lieutenant, when he reported to Memphis from duty in the Philippines to take the charge of the Third Mississippi River Commission District. His work on river improvement and flood control from the mouth of the White River to Vicksburg MS added much to the District. From 1903 to 1910 he was on another work, returning to the river work then, taking charge of the St. Louis District. He served also as secretary of the Mississippi River Commission with charge of the dredging operations and engineering of the river upstream of Cairo. In 1912 he was transferred to St. Paul MN. In 1920 he returned to St Louis as president of the Commission, an office he held until 1928.

In 1922 and 1927 two great floods occurred on Mississippi River. The first came before the levees on the main river had been raised to grades established by the Commission, so that it was difficult to establish the levee quality adopted. As the 1927 flood reached heights indicating that the Commission's grades would have to be raised to record heights if the policy adopted was continued. Following this flood a revised plan for flood control based on the experience was considered. It abandoned the levee-only principle and adopted floodways and spillways as auxiliaries to levees. In 1928, Potter was retired and Thomas H. Jackson (1874-1937) took over his duties. The 1925 paper on the Mississippi River gives an impressive account on the works made from World War I, describing the methods adopted for the control of the then longest, fully-regulated river of the world. The differences between the control systems for the three distinct river sections are described. The paper is illustrated with excellent photographs showing wing dams, mattress laying and pile dikes, and typical levee sections.

Anonymous (1928). Retirement ends long service on the Mississippi River. *Engineering News-Record* 100(26): 1019-1021. *P*
Anonymous (1928). General Potter, Head of Mississippi work, dies. *ENR* 101(6): 223. *P*
Potter, C.L. (1925). How the Mississippi River is regulated. *Engineering News-Record* 94(13): 508-514; 94(14): 556-559.
Potter, C.L. (1927). Gen. Jadwin reports on flood protection system for Mississippi River. *Engineering News-Record* 99(24): 961-966.
Potter, C.L. (1928). Glossary of Mississippi River terms. *Congressional Digest* 7(2): 50-60.
hhttp://www.mvd.usace.army.mil/About/MississippiRiverCommission(MRC)/PastMRCPresidents.aspx *P*

POWELL J.W.

24.03. 1834 Mount Morris NY/USA
23.09. 1902 Haven ME/USA

John Wesley Powell studied at Illinois College, but took no degree. He joined the State Natural History Society in 1854 and was eventually elected secretary of its Illinois section. In parallel he made long boat trips on the Mississippi River and Ohio River. After service in the Civil War, during which he lost the right arm, Powell was then appointed professor of geology at Illinois Wesleyan College, Bloomington. Both in 1867 and 1868, he organized and conducted parties of students and naturalists across the plains to the Colorado mountains. He first saw the gorges of the Colorado River in 1868. In 1869 a party of eleven men and four boats crossed 1,500 km of the Grand Canyon within three months.

Powell continued his western explorations during the following years, from 1875 as the division director of the US Geological Survey. In 1880 he took over as director the Survey, and held the office with success until 1894. The results of the trips to the Colorado River were published in his 1875 report, and expanded twenty years later. In this report he made his bold appeal for immortality as a geologist by calling attention to the fact that these canyons were gorges of erosion and due to the action of rivers upon the rock which were undergoing gradual elevation. With this report, his geological work ceased. In 1883, Powell inaugurated the series of Bulletins which have continued until today. In 1890, he also started with the Monographs of the US Geological Survey, a way of publication later adopted by similar institutions. Powell was also interested in the native tribes with which he had come into contact during his voyages, and in 1877 inaugurated the publications series Contributions to North American ethnology.

Anonymous (1902). J. Wesley Powell. *The Engineering and Mining Journal* 74(13): 403-404. *P*

Anonymous (1935). Powell, John Wesley. *Dictionary of American biography* 15: 146-148. Scribner's: New York.

Newell, F.H. (1930). 40 years of research into water resources. *Engineering News-Record* 104(4): 132-136. *P*

Powell, J.W. (1875). *Explorations of the Colorado River of the west and its tributaries*. US Government Printing Office: Washington DC.

Powell, J.W. (1895). *Canyons of the Colorado*. Flood & Vincent: New York.

Stegner, W. (1954). *Beyond the hundredth meridian*: John Wesley Powell and the second opening of the west. Penguin: New York. *P*

http://en.wikipedia.org/wiki/John_Wesley_Powell *P*

POWELL R.W.

04.10. 1889 Ionia County MI/USA
30.01. 1976 Berkeley CA/USA

Ralph Waterbury Powell graduated from Michigan State College, East Lansing MI, with the BS degree in 1911, from Cornell University, Ithaca NY, with the CE degree in 1914, and the PhB degree from Yale University, New Haven NJ in 1916. He was an instructor in civil engineering from 1912 to 1914 at Cornell University, assistant in testing materials at Yale University until 1916, Head of the Department of Physics, College of Yale in China, Changsha until 1922, there appointed associate professor of applied mathematics until 1927, when taking over until 1934 as assistant professor of mechanics at the Ohio State University, Columbus OH, and as professor of mechanics until 1957. Powell was in 1955 visiting professor at State University of Iowa, Iowa City IA, and at University of Kansas, Lawrence KA, in 1957. He was a member of the American Society of Civil Engineers ASCE, and of the International Association of Hydraulic Research IAHR.

Powell was known for papers and books. His 1940 text deals with hydraulics in general for undergraduate students. The 1951 book is an update version, with the main chapters Hydrostatics, Fundamentals of fluid flow, Orifices, Tubes, Nozzles, Pipe flow, Flow in open channels, Effect of viscosity, Models, Properties of liquids, Dimensional analysis, and Rational basis of Nikuradse's formulas. He also co-authored the 1963 state-of-the-art paper. Powell collaborated for years with Chesley J. Posey (1906-1991) and was also involved as secretary of the Rocky Mountain Hydraulic Laboratory, Fort Collins CO.

Anonymous (1955). Ralph W. Powell. 6th *IAHR Congress* Den Haag: Frontispiece. *P*
Anonymous (1964). Powell, Ralph W. *Who's who in engineering* 9: 1487. Lewis: New York.
Anonymous (1976). Ralph W. Powell. *Eos* 57(10): 729.
Carter, R.W., Einstein, H.A., Hinds, J., Powell, R.W., Silberman, E. (1963). Friction factors in open channels. *Journal of the Hydraulics Division* ASCE 89(HY2): 97-143; 89(HY4): 283-293; 89(HY5): 169-170; 89(HY6): 265-276; 90(HY4): 223-227.
Powell, R.W. (1934). Dimensional analysis in model studies. *Civil Engineering* 4(11): 568-571.
Powell, R.W. (1940). *Mechanics of liquids*. MacMillan: New York.
Powell, R.W. (1946). Flow in a channel of definite roughness. *Trans. ASCE* 111: 531-566.
Powell, R.W. (1951). *Elementary text in hydraulics and fluid mechanics*. MacMillan: New York.
Powell, R.W., Posey, C.J. (1959). Resistance experiments in a triangular channel. *Journal of the Hydraulics Division* ASCE 85(HY5): 31-66.
Powell, R.W. (1968). Origin of Manning's formula. *J. Hydr. Div.* ASCE 94(HY4): 1179-1181.

PRATT

06.01. 1887 Flandreau SD/USA
12.07. 1961 New York NY/USA

Edmund Addison Pratt obtained in 1907 the BSc degree in civil engineering from Armour Institute of Technology, Chicago IL, and there received the CE degree in 1911. He was until 1910 assistant engineer of the Bureau of Public Works, joining then until 1913 as civil engineer an engineering company in Bombay IN, and was then assistant chief engineer of the Continental Jewell Filtration Co., New York NY, founded by Omar H. Jewell (1842-1931). From 1916 until 1942, Pratt served as design engineer an asphalt company at Barber NJ, taking then over as manager the Iroquois Works, Buffalo NY. From then until the early 1950s, he was engaged as project manager of the Pan American Airways for airport developments in Brazil, or as supervising engineer in New York for the preparation of technical reports on US Navy contracts. He also was involved in the exploration of asphalt deposits in Cuba in 1944, and from 1945 was the manager of the international relations of the American Standards Association ASA, New York NY.

Pratt had mainly a career as civil engineer, given his engagements in a variety of jobs, dealing with fluid flow, general engineering, water filtration, usage of asphalt in the building technology to finally management of a large association. He submitted in 1916 an appliance for purification of liquids. In 1914 he described the Sutro weir as proposed by Harry H. Sutro (1876-1913), corresponding to a thin-plate weir in a rectangular channel for which the approach flow depth is proportional to the discharge. This weir type was popular in the early 20th century particularly in relation to irrigation structures, to avoid both sedimentation and erosion by a proper selection of the flow velocity. The 1936 paper highlights these findings under the then available knowledge. Today, this technology is no more in use given alternative possibilities of flow control. Pratt was member of the American Society of Civil Engineers ASCE, the Newcomen Society, the Pan American Society, serving as its director from 1938 to 1943, and the Professional Engineering Club of New York.

Anonymous (1948). Pratt, Edmund A. *Who's who in engineering* 6: 1582. Lewis: New York.
Anonymous (1952). E.A. Pratt, consulting engineer. *Civil Engineering* 22(7): 502. *P*
Pratt, E.A. (1914). Another proportional-flow weir: Sutro weir. *Engineering News* 72(14): 462-463; 74(6): 277.
Pratt, E.A. (1936). Sutro weir formula. *Engineering News-Record* 117(26): 904.
Pratt, E.A. (1941). Automatically controlled mixing plant. *US Patent* 2,232,404.

PRESCOTT

03.10. 1755 Lancaster MA/USA
03.12. 1826 Waterford NY/USA

Few is known on the early life of Benjamin Prescott, which he had spent at Northampton MA. In 1793 he was one of the founders of an aqueduct company to bring a larger water supply to his city. Water was collected from local springs, carried through bored logs into a reservoir from where it was distributed with small log conduits to the individual residences. The lock and canal proprietors of the Connecticut River, among whom was Prescott, were authorized to construct bypass canals around the falls at South Hadley MA, and Montague MA. Initially the engineer Christopher Colles (1738-1816) was invited to do the surveys and the design, whereas Prescott was asked to do the engineering. Work commenced in 1793 with about 250 men excavating the canal and building a diversion dam. Locks were to accommodate boats, and the 330 m long and 3.5 m high dam raised the water level 1 m over some 16 km. Ague had been almost unknown before in this region but it became common, due to the swamps caused by the backwater of the dam.

The increasing expense of the construction caused Prescott to make a major deviation from the lock plan. He decided to install an inclined plane to overcome the 15 m lift. A carriage was built big enough to handle the canal boats. The idea was that a boat was floated over the submerged carriage in a lock chamber, the water was lowered until the boat settled on the carriage and the loaded carriage was then raised or lowered by means of chains by a pair of 5 m overshot waterwheels. In 1795 the 2 km long canal, the inclined plane and the dam were finished. For some ten years the drum hauled the boats but the objections of the anti-dam group grew so that in 1805, locks were substituted for the great Hampshire Machine. The canal was deepened, thus reducing the need for the dam, which was removed shortly later. Prescott then left Northampton accepting a position as superintendent of the famous US Armory at Springfield MA, where he remained in charge until 1815 when moving to Cohoes NY. Just before he died at age 72, Prescott was engaged in building a masonry dam across Hudson River at Waterford NY. His lasting work was the magnificent Hampshire Machine and the inclined plane.

Bacon, E.M. (1911). *The Connecticut River* and the Valley of the Connecticut: Three hundred and fifty miles from mountain to sea. Bangor ME.
Dwight, T. (1823). *Travels in New-England and New York*. Baynes & Son: London.
Fitzsimons, N. (1970). Benjamin Prescott and the Hampshire Machine. *Civil Engineering* 40(12): 68.
Love, N.D. (1903). *The navigation of the Connecticut River*. Hamilton: Worcester.
http://www.forgeofinnovation.org/Springfield_Armory_1794-1812/index.html *P*

PRESTON P.J.

23.12. 1871 Grinnell IA/USA
20.09. 1950 Denver CO/USA

Porter Johnstone Preston obtained the BS degree in civil engineering from Colorado State University, Fort Collins CO, in 1892. After having done survey work within Colorado State during the next two years, he became superintendent of the Fort Lyons Canal Co., Las Animas CO, where he stayed for five years. This was followed by a two-years stay as Colorado deputy state engineer. After four years of consulting work, he was retained by the US Bureau of Reclamation USBR, Denver CO. His work at the Bureau included preliminary investigation of the All-American Canal in California, the Uncompahgre Project in west Colorado, and the Yakima Irrigation Project in Washington State. From 1930 until his retirement in 1941, Preston was in charge of the Colorado River Project. He was further in charge of the USBR Colorado-Big Thompson Diversion Project as supervising engineer. He retired after 25 years of continuous service with the Bureau. He was member of the American Society of Civil Engineers ASCE.

Preston was an outstanding irrigation and hydraulic engineer whose work covered the western third of the USA. He was appointed in 1938 head of the Diversion Project by the Secretary of the Interior. It was through Preston's studies of irrigation possibilities in Colorado River Valley, made in 1935, that the project became reality. Preston was in collaboration with Raymond F. Walter (1873-1940), Sinclair O. Harper (1883-1966), or John L. Savage (1879-1967) also involved in the hydraulic experiments of the Boulder Canyon Dam. The Colorado-Big Thompson Project was a federal water diversion project collecting West Slope mountain water from the headwaters of the Colorado River and divert it to Colorado's Front Range and plains. Some 80% of the state's precipitation falls on the West Slope, while around 80% of the state's population lives along the East Slope. In search of a solution, farmers approached the Bureau. The water is diverted via a 21 km long tunnel under the Continental Divide and Rocky Mountain National Park.

Anonymous (1941). Preston retires from USBR. *Engineering News-Record* 125(Jan.9): 67. *P*
Hansen, C., Preston, P.J. (1936). *Water resources, Colorado's greatest wealth*: The Colorado-Big Thompson water tunnel plan would aid vast areas. Greeley: Denver CO.
Mead, E., Preston, P.J. (1920). *All-American Canal*. US Dept. of the Interior: Washington DC.
Preston, P.J., Engle, C.A. (1928). *Irrigation on Indian Reservations*. Dept. Interior: Washington.
Preston, P.J. (1938). Colorado-Big Thomson Project, Colorado. *Civil Engineering* 8(8): 517-519.
http://en.wikipedia.org/wiki/Colorado-Big_Thompson_Project

PRICE R.C.

17.05. 1911 Rio WI/USA
01.07. 1977 Bethesda MD/USA

Reginald Carrier Price obtained the BS degree in civil engineering from the University of Wisconsin, Madison WI, in 1935 and in 1943 the MA degree in economics from American University, Washington DC. He was from 1936 to 1939 district sanitary engineer of the Wisconsin State Health Board, then until 1941 assistant survey analyst, US Department of Agriculture, until 1943 then hydraulic engineer of the National Resources Planning Board, from when he was assistant professor of engineering until 1946 at the New York University, New York NY, continuing as engineering economist of US Bureau of Reclamation USBR. He also was secretary of the Commission on the Corporation with Technical Divisions of the American Society of Civil Engineers ASCE, whose member he was since the early 1940s.

Price was appointed in 1950 director of the US Department of the Interior's Division of Water and Power. He thus was in charge of project review and coordination for the then recently established Division, succeeding thereby William G. Hoyt (1886-1971). From 1954 to 1957 Price was an economic advisor for the United Nations' Bureau of Flood Control and Water Resources Development in Thailand. From 1961 to 1966 he was the director of California's Department of Water Resources, and from 1968 to 1972 took over as water resources advisor to the Mekong Project in Thailand sponsored by the United Nations. Price has published a number of reports on the economics of flood protection, on flood plain management, and on the beauties of waters in California State.

Anonymous (1948). Price, Reginald C. *Who's who in engineering* 6: 1588. Lewis: New York.
Anonymous (1950). R.C. Price will head Water and Power Division. *Engineering News-Record* 145(Oct.26): 31.
Anonymous (1961). Reginald C. Price. *Civil Engineering* 31(10): 36. *P*
Price, R.C. (1946). *Some problems in the economics of flood protection measures*. University of Wisconsin: Madison WI.
Price, R.C. (1964). *Statement on flood plain management*. California Dept. of Water Resources: Sacramento CA.
Price, R.C. (1965). *Economic analysis and water policy decisions*. California Dept. of Water Resources: Sacramento CA.
Price, R.C. (1966). *California beautiful*: Waters to enjoy. California Dept. Water Resources: Sacramento CA.

PRICE W.G.

06.07. 1853 Knoxville PA/USA
06.07. 1928 Detroit MI/USA

The aptitude for invention and engineering became evident quite early. After four years of instruction in both mathematics and engineering at Englewood NJ, William Gunn Price embarked on a brilliant career in these fields. Between 1879 and 1896 he was an assistant engineer within the Mississippi River Commission, measuring discharges of Mississippi, Ohio and Missouri Rivers. In 1882, after having installed a river gage on the Ohio River at Paducah KY, he conceived the design of the first current meter, whose circumstances are described in a letter written in 1927: In January 1882 I began measuring the discharge of the Ohio River. My equipment included a meter designed by Clemens Herschel (1842-1930). This meter was of propeller type, having a horizontal shaft, and there was no means for excluding water from the bearings. The meter of Theodore G. Ellis (1829-1883) had a coupled wheel revolving in a horizontal plane, but there was no provision for excluding water from its vertical-shaft bearings. I rated these meters in clear, still water and ratings indicated that they would give accurate measurements.

Price continues: 'I then began using them for river measurements. The water was very muddy, and boils in the swift current carried fine sands to the surface which I caught in my drinking cup. Neither would give discharge measurements which corresponded with the gradual gage increase. The idea then occurred to me that by using inverted cup bearings which would trap the air, I could exclude the water and grit. I made drawings for such a meter, and it was completed the next day. It was used for measuring the great flood of 1882. Among the many US patents that have been awarded to Price these relating to the meters range between 1885 and 1926'. Rouse states further that the Price water meter is hydrodynamically a monstrosity, for it would register even if simply moved up and down, but it was rugged, not easily fouled, and possessed bearings operating in air pockets and hence comparatively trouble-free in silty water. Price was the nation's foremost authority on current meters for a long period.

Frazier, A.H. (1967). *William Gunn Price and the Price current meters*. Smithsonian Press: Washington DC.
Price, W.G. (1885). *Current meter*. US Patent Office, Patent 325,011. Washington DC.
Rouse, H. (1976). Price current meter. *Hydraulics in the United States* 1776-1976: 66. Iowa Institute of Hydraulic Research: Iowa City. *P*
http://www.sil.si.edu/smithsoniancontributions/HistoryTechnology/text/SSHT-0028.txt *P*

PROCTOR

16.12. 1894 Grayslake IL/USA
12.10. 1962 Los Angeles CA/USA

Ralph Roscoe Proctor was educated at University of Southern California, Los Angeles CA, from 1914 to 1916. After war service in France, he returned to the employ of the Los Angeles Department of Water and Power. From 1933 until his retirement in 1959, he was in charge of the design, the construction, and maintenance of all dams in the Water System. He also served as consultant on dams and soil stability problems for the Metropolitan Water District of Southern California, the Los Angeles County Flood Control District, the Greybull Irrigation District, or the Tidewater Oil Company. He was a member of the American Water Works Association AWWA, and of the American Society of Civil Engineers ASCE, whose Fellow he was from 1959.

Proctor wrote a series of articles describing his theories on soil compaction control in 1933, to determine the water content-density relationship of soils. The principles set forth therein form the basis of the design and construction control methods used by the major agencies in dam building, airport runways, and highways. His principles address questions relating to the dam cross-section, its upstream slope, the selection of soils, and the rolling equipment. Proctor re-emphasized and expanded his principles in 1948. He evolved the Proctor Tests in the field of earth-fill dams, which are still widely used in soil engineering for compacting earth-fill materials in dam and highway construction.

Anonymous (1962). Foundation, dam expert, R.R. Proctor, dies. *Engineering News-Record* 169(Oct.25): 55.

Anonymous (1963). Ralph Roscoe Proctor, F. ASCE. *Trans. ASCE* 128(5): 139.

Proctor, R.R. (1933). Fundamental principles of soil compaction. *Engineering News-Record* 111(9): 245-248.

Proctor, R.R. (1933). Description of field and laboratory methods. *Engineering News-Record* 111(10): 286-289.

Proctor, R.R. (1933). Field and laboratory verification of soil suitability. *Engineering News-Record* 111(12): 348-351.

Proctor, R.R. (1933). New principles applied to actual dam-building. *Engineering News-Record* 111(13): 372-376.

Proctor, R.R. (1948). The relationship between the foot pounds per cubic foot of compactive effort and the shear strength of compacted soils. 2nd Intl. Conf. *Soil Mechanics and Foundation Engineering* Rotterdam 5: 219-223.

web.mst.edu/.../ge441/.../GE441-Lecture2-1.pdf *P*

PULS

30.01. 1897 Lowell MA/USA
11.05. 1992 Arvada CO/USA

Louis George Puls was a career engineer with the US Government, who worked on President Roosevelt's Tennessee Valley Authority TVA, becoming director of the US Bureau of Reclamation USBR of the Western USA. He was the successor of Kenneth B. Keener (1888-1971). Puls was involved in water projects and dams, not only in the West, but also in Honduras, Mexico, and the USSR. The Puls papers available at USBR Library contain personal files, an autobiography, water project papers, scrap-books, maps, photos, and engineering drawings.

Puls was the chief design engineer of Glen Canyon Dam. This project was authorized by Congress in 1956, and the first construction contract was awarded later that year. Work officially began when President Eisenhower pushed a button at Washington, and exploded by voice control a detonating cap, sending the first shower of rock into the canyon. Glen Canyon Dam is one of the engineering wonders and with 216 m the forth highest dam in the USA. Honoured upon its completion in 1964 as an outstanding civil engineering project, the thick-arch dam contains 1 million m³ of concrete. Lake Powell, the reservoir impounded by the dam, is the second largest in the country, storing enough water to meet the needs of 30 million families for a year. The power plant has eight generating units, with a total generating capacity of 1,320 MW. A enormous cavitation damage was experienced in 1983. After the flood event, the tunnels were inspected for constructional damages, confirming that large holes were excavated by the high-speed flow which almost were as deep as the tunnel diameter of 11 m, extending over a length of nearly 50 m. It was observed that these damages were avoided by an anti-cavitation system.

Anonymous (1958). Puls succeeds Keener as USBR's chief designer. *Engineering News-Record* 160(Jun.5): 78. *P*

Anonymous (1961). Puls leaves Bureau of Reclamation. *Engineering News-Record* 167(Oct.19): 78. *P*

Puls, L.G. (1928). Flood regulation of the Tennessee River. *House Document* 185, 70th Congress, 1st Session. Government Printing Office: Washington DC.

Puls, L.G. (1931). Spillway discharge capacity of Wilson Dam. *Trans. ASCE* 95: 316-333.

Puls, L.G. (1941). Mechanics of the hydraulic jump. *Technical Memorandum* 623. USBR: Denver.

Puls, L.G. (1950). Hungry Horse Dam, largest single contract of USBR now in force. *Civil Engineering* 20(12): 764-768.

PUTNAM

23.04. 1907 Sonoma CA/USA
23.10. 1966 Berkeley CA/USA

John Alpheus Putnam obtained from University of
California, Berkeley CA, the BS degree in 1930, the
MS degree in 1934, and the PhD degree in 1943, all
in mechanical engineering. He worked after his MS
degree as research engineer at Kettleman Hills CA
for the Standard Oil Company until 1939, applying
in the volumetric and phase behaviour of petroleum
hydrocarbons to reservoir analysis. This work on
the phase behaviour combined with the fundamentals
of flow in porous media was destined to become his
major professional interest.

Putnam's professional employment at the University of California was terminated by
the first of a long series of diseases and illness that would affect his entire professional
career. From 1944, he advanced through the academic ranks, becoming professor of
petroleum engineering in 1952, and professor emeritus on disability retirement in 1964.
He had served as assistant dean of the College of Engineering from 1954 to 1958, where
he was primarily concerned with the lower division program. His keen interest in the
fundamentals of two-phase fluid flow led to his appointment as director of the American
Petroleum Institute Research Project 47A for the period from 1947 to 1952. In parallel
he took interest in wave hydrodynamics including a paper with Joe W. Johnson (1908-
2002) on energy dissipation, or the 1949 paper on the loss of wave energy. In spite of
the many ailments he maintained his unfailing cheerful personality and uncomplaining
nature. He remarked at various occasions that much of his career was lived on borrowed
time. He was member of the American Institute of Mining Engineering AIME, and the
Society of Rheology.

Anonymous (1936). Putnam. *Blue and gold yearbook*: 173. Univ. California: Berkeley. *P*
Anonymous (1954). Putnam, John A. *Who's who in engineering* 7: 1942. Lewis: New York.
Laird, A.D.K., Putnam, J.A. (1951). Fluid saturation in porous media by X-ray technique.
    Journal of Petroleum Technology AIME 3(10): 275-284. (*P*)
Putnam, J.A., Johnson, J.W. (1949). The dissipation of wave energy by bottom friction. *Trans.
    AGU* 30(1): 67-74.
Putnam, J.A. (1949). Loss of wave energy due to percolation in a permeable sea bottom. *Trans.
    AGU* 30(3): 349-356.
Putnam, J.A., Munk, W.H., Traylor, M.A. (1949). The prediction of longshore currents. *Trans.
    AGU* 30(3): 337-345.
http://texts.cdlib.org/view?docId=hb238nb0d8&doc.view=frames&chunk.id=div00025&toc.depth=1&toc.id=.

QUINTON

19.10. 1850 Enniskillen/EI
10.05. 1939 Los Angeles CA/USA

John Henry Quinton was born in Ireland and educated at Queens University, Dublin EI. He moved in 1873 to the USA, one year after his graduation, as a civil engineer. He was then for a long time active in the water supply work on the Pacific Coast, having had an important part in the Owens River Aqueduct of Los Angeles CA, the San Gabriel Canyon power development, the Santa Ana River irrigation works, and many other reclamation projects.

During the early days of the Reclamation Service, its staff debated the need for uniformity in designs and specifications for constructions. When Frederick H. Newell (1862-1932) convened the first Conference of Engineers in 1903, his goal was to bring together his principal engineers to become better acquainted with each other, and to discuss the then newly organized Service. Before the meeting adjourned, Newell appointed a Committee on Standard Plans and Specifications to develop a proposal for the preparation of standard designs. This Committee consisted of three consulting engineers, namely Quinton, George M. Wisner (1870-1932), and Hiram N. Savage (1861-1934). In a letter to Quinton in 1904, Newell emphasized the immediate need for a general uniformity in all projects of the Reclamation Service. Newell expanded the Committee at the Second Conference of Engineers held at El Paso TX in 1904, consisting then of Quinton, Wisner, Arthur P. Davis (1861-1933), and another engineer, with Quinton as chairman. In 1909, when the preliminary edition of the Manual relating to the work of the US Reclamation Service was published, standardized plans were available for abutments of bridges, turnouts, drainage culverts, wooden drops, concrete flumes, cast-iron gates, retaining walls, concrete spillways, and buildings.

Anonymous (1939). Reclamation pioneer dies at eighty-nine: John H. Quinton. *Engineering News-Record* 122(May18): 689. *P*

Pfaff, C.E. (2007). *The Bureau of Reclamation's architectural legacy*: 1902 to 1955. USBR: Denver CO.

Quinton, J.H. (1905). *Experiments on steel-concrete pipes*. US Govt. Printing: Washington DC.

Rogers, J.D. (2013). *Threadlines* of geotechnical and engineering geology firms in the Greater Los Angeles Metro-Southern California Area. web.mst.edu/.../History-Geotech%20.doc

Wiltshire, R.L. (2004). Reclamation's 100 years of embankment dam design and construction. *Water resources and environmental history*:140-149, J.R. Rogers, G.O. Brown, J.D. Garbrecht, eds. ASCE: Reston VA.

RAFTER

09.12. 1851 Orleans NY/USA
29.12. 1907 Karlsbad/A

George Willson Rafter early lost his father. His mother sold the water-power which had been the source of income, but the purchasers had difficulties because of irregular water discharge. This may have influenced Rafter's later bent. He received education as civil engineer at Cornell University. In 1876, he became assistant engineer of Rochester waterworks, and then practiced as consulting engineer. Back in New York State in 1883, Rafter surveyed *Honeoye* Lake for use as a storage reservoir for water power of the Rochester mills. In 1890, while acting as water works chief engineer, he installed an original scheme for throttling down certain districts when a water famine threatened, distributing the insufficient supply to preserve public health and minimize discomfort.

Rafter also designed sewage disposal plants for various cities in New York State. He published his 1894 book on sewage, which became a standard text in this field. He was a sanitary expert to the Boston waterworks, thereby developing the Sedgwick-Rafter method for water analysis. Rafter was one of the first to use the microscope in water biology, and published a number of papers on this subject. In 1893 he was engaged by the State of New York to study river control. He devised a system of storage reservoirs to regulate river flow, and superintended a dam construction which formed Indian Lake in a tributary of the upper Hudson. Rafter considered his hydraulic investigation of the *Hemlock* Lake reservoir as most important work, because of the reforms it generated. His report on Hydrology is considered as authoritative on matters relating to stream flow. He authored a total of 170 books. Rafter died of pleurisy while on visit in Austria.

Anonymous (1935). Rafter, George W. *Dictionary of American biography* 15: 324-325.
        Scribner's: New York.
Griggs, F.E., ed. (1991). G.W. Rafter. *A biographical dictionary of American civil engineers*:
        92. ASCE: New York.
Rafter, G.W., Baker, M.N. (1894) *Sewage disposal in United States*. van Nostrand: New York.
Rafter, G.W. (1900). On the flow of water over dams. *Trans. ASCE* 44: 220-398.
Rafter, G.W. (1900). *Report on special water-supply investigation*. Washington DC.
Rafter, G.W. (1903). Discussion of Sewage purification. *Trans. ASCE* 51: 416-422.
Rafter, G.W. (1903). The relation of rainfall to run-off. *Water supply and irrigation papers* 80.
        US Geological Survey: Washington DC.
http://mcnygenealogy.com/bios/biographies030.htm *P*

RAMSER

01.11. 1885 Montezuma IA/USA
29.04. 1962 Washington DC/USA

Charles Ernest Ramser was educated at University of Illinois, Urbana IL, from where he obtained the BS degree in civil engineering. He was a hydraulic engineer at Polytechnic Institute, Brooklyn NY, in 1910, hydrographer at Alcoa TN until 1912, drainage engineer within the Bureau of Public Roads until 1917, when becoming senior drainage engineer at the Bureau of Agricultural Engineering until 1935. Ramser was until his retirement in 1947 senior soil conservationist and Head, the Division of Watershed Studies of the Soil Conservation Service SCS, Washington DC. He was finally consultant. Ramser was member of the American Society of Civil Engineers ASCE, the American Society of Agricultural Engineers ASAE, and the American Geophysical Union AGU.

Ramser was in charge of establishing and directing engineering experiments on soil erosion control on ten Soil Erosion Experiment Stations within the US Department of Agriculture. He devised and developed scientific methods of terracing farm lands to prevent soil erosion. His papers and reports include works on this topic as well as on drainage ditches, gullies, and how to control and reclaim them, flow of water in drainage channels, erosion and silting of dredged drainage ditches, among many others. His 1934 paper on dam outlets may be considered particularly important because the discharge characteristics of this intake were defined, based on the generalized Bernoulli equation. He was awarded for his outstanding work the John Deere Medal by ASAE.

Anonymous (1929). Charles E. Ramser. *Agricultural Engineering* 10(10): 335. P
Anonymous (1959). Ramser, Charles E. *Who's who in engineering* 8: 2003. Lewis: New York.
Anonymous (1962). Charles Ernest Ramser. *Trans. AGU* 43(3): 317.
Ramser, C.E. (1917). Studies of dredged drainage ditches before and after clearing. *Engineering News* 77(3): 104-105.
Ramser, C.E. (1919). Progressive erosion in a dredged drainage channel. *Engineering News-Record* 82(18): 876-877.
Ramser, C.E. (1923). Kutter's *n* for rough rock channel excavated by explosions. *Engineering News-Record* 91(23): 936.
Ramser, C.E. (1929). Terracing to combat soil erosion losses. *Engineering News-Record* 103(22): 848-851.
Ramser, C.E. (1934). Capacity of outlets for erosion-control dams. *Engineering News-Record* 112(May 10): 595-596.

RANDELL

08.05. 1890 Seattle WA/USA
ca. 1962 Washington DC/USA

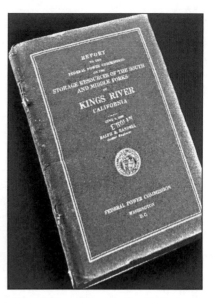

Ralph Reginald Randell received his education from University of Washington, Seattle WA, with the BS degree in civil engineering in 1911. He further was educated at Iowa State University in the 1930s. He was from 1911 to 1912 junior topographer of the Topographic Branch, US Geological Survey USGS, until 1914 junior engineer of its Water Resources Branch, joining then until 1920 as junior engineer the US Forest Service. From 1920 Randell acted as engineer of the Federal Power Commission, thereby representing the US Department of Agriculture on inter-departmental boards under the Federal Water Power Act. He was from 1933 to 1934 acting secretary of the Federal Power Commission. Randell was a member of the American Society of Civil Engineers ASCE.

Randell investigated around 1930 the storage resources of the South and Middle Forks of Kings River, California, to determine the value for power generation, irrigation needs, and other aspects. This study served as a guide in considering the proposal to transfer this region from a National Forest to Sequoia National Park. Randell invented an internal drainage system for dams. Embankment dams are currently the most common type with a vertical to horizontal side slope of one to two or flatter. Their capacity for water retention is due to low permeability of the entire dam mass or due to low permeability for zoned embankment dams. Both earthfill or rockfill dams exist, depending on whether they are made up of compacted earth or dumped pervious rock. Seepage under, through, or around a dam is a serious problem because of erosion of dam material. To reduce seepage internal impermeable barriers and internal filter systems are provided. It is here where Randell contributed to dam design, with the major advances in earthfill dams in the 1930s in the USA.

Anonymous (1937). Randell, Ralph R. *Who's who in engineering* 4: 1119. Lewis: New York.
Randell, R.R. (1931). *Report on the storage resources of the South and Middle Forks of Kings River, California*. Federal Power Commission: Washington DC. (*P*)
Randell, R.R. (1932). *Improvements* in the art of measuring fluid flow in open or non-pressure channels. US Court of Customs and Patent Appeals: Washington DC.
Randell, R.R. (1932). *Improvements in dams and other structures subject to fluid pressure*. US Court of Customs and Patent Appeals: Washington DC.
Randell, R.R. (1934). Discussion of Evaporation from reservoir surfaces, by R. Follansbee. *Trans. ASCE* 99: 719-722.

RANDOLPH E.S.

03.12. 1889 Bayou Goula LA/USA
16.10. 1968 East Baton Rouge LA/USA

Edward Sydney Randolph received education from Louisiana State University, Baton Rouge LA, with the BS degree in 1910. He was employed by the Isthmian Canal Commission until 1941, as draftsman first, then junior engineer, followed by design and office engineer, advancing to designer of Madden Dam in 1929, and its construction engineer until 1935, from when he was involved in the design of the Third Locks until 1941. From then Randolph was a private consultant, mainly for Panama Canal. He was a member of the American Society of Civil Engineers ASCE, the American Concrete Institute ACI, the Permanent International Association of Navigation Congresses PIANC, and of the International Association of Hydraulic Research IAHR.

Randolph was a leading engineer of Panama Canal from its opening in 1914 to the 1940s. He was involved in the design and construction of the ship piers, the dry docks, the coal and fueling plants, the marine shops, the bridges, the various reservoirs and dams, the hydro-electric power stations and locks, and the defence projects. He was in parallel also concerned with civil engineering works for the Republics of Panama, Colombia, and Nicaragua. He authored a number of reports and papers, including the Panama Canal lock gates. Various of his papers were published in the Military Engineer, a journal founded shortly after World War I.

Anonymous (1941). Randolph, Edward S. *Who's who in engineering* 5: 1440. Lewis: New York.
Anonymous (1962). Edward S. Randolph, top Canal engineer. *Panama Canal Review* 13(2): 12. *P*
Randolph, E.S. (1930). Overhaul of Panama Canal lock gates. *The Military Engineer* 22(5/6): 220-226.
Randolph, E.S. (1931). Enlargement of the Panama Canal facilities. *The Engineers Bulletin*, Colorado Society of Engineers 15(4): 9, 32.
Randolph, E.S. (1932). Model tests verify design of Madden Dam. *Engineering News-Record* 109(Jul.14): 42-44.
Randolph, E.S. (1936). Sealing reservoir leaks with clay grouting. *The Military Engineer* 28(5/6): 208-214.
Randolph, E.S. (1936). Parker dam operations center on diversion tunnel driving and plant. *Western Construction News* 11(2): 102-104.
Randolph, R.R. (1938). Hydraulic tests on the spillway of the Madden Dam. *Trans. ASCE* 103: 1080-1112.

RANDOLPH I.

25.03. 1848 New Market VA/USA
02.08. 1920 Chicago IL/USA

Isham Randolph's schooling was minimal because his family suffered greatly during the Civil War. In 1868, he decided to become civil engineer and took up railroad work, until becomirg resident engineer in 1873. After twenty years of railways engineering he was appointed chief engineer of the Sanitary District of Chicago in 1893, a position he held for fourteen years, during the entire construction period of Chicago Drainage Canal. This enterprise changing the direction of Chicago River to the Mississippi River instead of into Lake Michigan, was the largest artificial canal until completion of Panama Canal. Randolph was awarded for this achievement a gold medal by the 1900 Paris Exposition.

In 1905, Randolph was appointed by President Roosevelt a member of the board of consulting engineers to determine the type of Panama Canal. He was one of the few recommending the lock type which was adopted, as opposed to the sea-level canal, which was successfully implemented in the Suez Canal. Subsequent experience with the Culebra Cut proved this wisdom. In 1908, Randolph accompanied president-elect Taft to Panama to consider whether there was necessity for any fundamental change in the plans upon which the construction had begun. This board unanimously approved the lock canal across the Isthmus. Randolph then designed and constructed Obelisk Dam above the Horseshoe Falls, on Niagara River. Later he was involved in land reclamation work, serving as consultant for the Little River Drainage District of southeast Missouri. In 1913, the Franklin Institute of Philadelphia awarded Randolph the Elliott Cresson Medal in recognition of 'distinguished achievement in the field of civil engineering'.

Anonymous (1905). Isham Randolph. *Engineering News* 54(Sep.14): 263. *P*
Anonymous (1920). Isham Randolph dead. *Engineering News-Record* 85(6): 287. *P*
Anonymous (1921). I. Randolph. *Journal of the Western Society of Engineers* 26(1): 16-21. *P*
Anonymous (1935). Randolph, Isham. *Dictionary of American biography* 15: 359-360. Scribner's: New York.
Griggs, F.E., ed. (1991). Randolph. *A biographical dictionary of American civil engineers* 2: 93.
Randolph, I. (1895). Modern methods of canal excavation. *Engng. Record* 33(Oct.12): 348-360.
Randolph, I. (1895). The Chicago Sanitary District. *Journal of the Association of Engineering Societies* Boston 14(3), 1-9.
Randolph, I. (1909). The Sanitary District of Chicago, and the Chicago Drainage Canal: A review of 20 years of engineering work. *Engineering News* 62(4): 90-94.

RANDOLPH R.R.

11.01. 1900 Birmingham AL/USA
20.02. 1981 Birmingham AL/USA

Richard Rutherford Randolph, Jr, studied engineering at the University of Alabama, Birmingham AL. He joined the US Bureau of Reclamation, Fort Collins CO, in the late 1920s as hydraulic engineer in its Hydraulic Laboratories. He is seen on a photograph dated April 1931 among other colleagues of the early USBR Hydraulic Laboratory staff, including Emory W. Lane (1891-1963), Joseph N. Bradley (1903-1993), or Charles W. Thomas (1906-1979). The first project of Randolph dealt with Hoover Dam, then the largest dam in the world. In 1933 he moved to the Panama Canal Zone and there was involved in Madden Dam, the largest of the site. He re-joined the US Bureau of Reclamation USBR, Denver CO, from the late 1930s as hydraulic engineer and was from the mid-1950s hydraulic engineer and engineering manager of the Southern Services Inc., Birmingham AL. After retirement in 1965 he was there a private engineering consultant.

The 1938 paper deals with the spillways of Madden Dam, on the Panama Canal. This dam is 300 m long and 66 m high, built across Chagres River. Tests on a 1:72 model at the Hydraulic Laboratory of Denver were made in 1931. After dam construction, the prototype was checked by hydraulic measurements 800 m downstream of the dam. The discharge over the spillway obtained was considerably larger than deduced from the model tests, most probably due to scale effects. The 1954 paper is concerned with the Martin Dam power plant in Alabama. Following damages in the early 1950s, model studies were conducted at the hydraulic laboratory, the Georgia Institute of Technology, Atlanta GA. The results indicate that the hydraulic problems were solved.

Anonymous (1931). Model tests verify design of Madden Dam spillway. *ENR* 109(2): 42-44.

Anonymous (1981). Richard Rutherford Randolph, Jr. *Civil Engineering* 51(9): 62.

Kindsvater, C.E., Randolph, R.R. (1954). Hydraulic model studies of Martin Dam draft tubes. *Proc. ASCE* 80(382): 1-17; 81(748): 23-26.

Kindsvater, C.E., Randolph, R.R. (1955). Discharge characteristics of embankment-shaped weirs. *Water Supply Paper* 1617-A. US Geological Survey: Washington DC.

Randolph, R.R. (1931). *Report on Hoover Dam spillway model tests*, Fort Collins Laboratory. USBR: Denver CO.

Randolph, R.R. (1938). Hydraulic tests on the spillways of the Madden Dam. *Trans. ASCE* 103: 1080-1132.

http://www.usbr.gov/pmts/hydraulics_lab/history/75years2005.html *P*

RAWHOUSER

01.02. 1902 York PA/USA
31.03. 1991 Hawaii HI/USA

Clarence Rawhouser graduated in 1930 with a civil engineering degree from University of Cincinnati, Cincinnati OH, and the MSc degree in 1943 from University of Colorado, Boulder CO. He was from 1929 surveyor of the US Army Corps of Engineers, before becoming a dam engineer for the Design and Construction Branch, the US Bureau of Reclamation USBR, Denver CO, from 1930 to 1952. While with the Bureau, he was in the team for the planning and design of Hoover, Grand Coulee, and Hungry Horse dams, in the 1930s and 1940s the main structures built in dam engineering. From 1952 to 1955 he was a member of the Technical Corporation Mission to the Central Water and Power Commission of India, and then until 1959 served as dam design engineer and project engineer the International Corporation Administration, Washington DC. He also served as consulting dam design engineer Seattle WA from 1959 to 1960, and as hydro-electric engineer of the United Nations Technical Assistance Operation first in Colombo, Sri Lanka until 1962, and then in Taiwan.

Rawhouser was in 1948 the US delegate to the second International Conference on Soil Mechanics and Foundation Engineering, held at Rotterdam NL; the Third Congress on Large Dams ICOLD, held at Stockholm S; the International Commission on Irrigation and Drainage ICID; and the War Department Team for inspection of bombed hydro-electric structures. He was in 1963 and 1964 a special consultant to the Conservancy Bureau, Taiwan ROC. Rawhouser was a member and Fellow of the American Society of Civil Engineers ASCE, and the Colorado Society of Engineers. He was a registered engineers in Colorado. He contributed to the 1938 book Dams and control works, then a widely used book in dam engineering.

Anonymous (1948). C. Rawhouser. 3rd *IAHR Congress* Stockholm: Frontispiece. *P*
Anonymous (1992). Clarence Everett Rawhouser. *Trans. ASCE* 157: 509-510.
Rawhouser, C., Chapman, E.L., Spicer, C.B. (1936). *Charts and methods for making computations of cooling of concrete in Grand Coulee Dam*. USBR: Denver CO.
Rawhouser, C. (1938). Temperature control of mass concrete in large dams. *Dams and control works*: 246-250. USBR: Denver CO. (*P*)
Rawhouser, C. (1943). *Temperature control of mass concrete*. USBR: Denver.
Rawhouser, C. (1945). Cracking and temperature control of mass concrete. *Journal of the American Concrete Institute* 16(4): 305-346.

READ

02.07. 1759 Warren MA/USA
20.01. 1849 Belfast ME/USA

Nathan Read was the proper inventor of the high-pressure steam engine, twelve years before the steam engine was known to be used in the form of a high-pressure engine, leading to the great revolution in steam power to navigation and land-transport. His parents came from England. Read graduated in 1781 from Harvard University, Cambridge MA, in medicine, continuing there as tutor until 1787. He opened an apothecary store at Salem MA in 1788, but kept it only for one year.

From 1788 Read made a number of improvements of the steam engine. He first built the tubular boiler, a new kind of steam boiler. Then he made efforts to improve the function of steam cylinder, and placed it in a horizontal position. So the engine sustained much higher pressure than ever before, so that Read invented the high-pressure steam engine, a new kind steam engine which was different from James Watt's old engine. Read made the engine more convenient and portable, also much lighter and safer. The most important was that the new engine needed much less room and fuel than the old. He was succeeded in reconstructed steam engine, and modified the Watt engine to a high-pressure engine, widely used in new fields, including steamboat and locomotive. To prove this fact, Read manufactured several models of steamcar and steamboat in 1790. His experiments proved that these engines functioned well. He also devised as the true inventor of the chain-wheel to using paddle wheels for propelling the steamboat, and set up a shipbuilding factory with his friends in 1796. In agriculture, he invented the threshing machine, different forms of pumping engines, and a new kind of windmill. He came up with a plan for using the expansion and contraction of metals, multiplied by levers, widely used in winding up clocks and other purposes. He patented some of them, but mainly used them in the agricultural fields and never patented. Read from 1800 took also interest in politics, and was elected as Federalist to the Sixth Congress, and was re-elected to the Seventh Congress, serving from 1800 to 1803. In 1807 he moved with his family to Maine, acting there as judge. He was also instrumental in establishing the Belfast Academy, serving as its trustee for 40 years.

Read, D. (1870). *Nathan Read*: His invention of the multi-tubular boiler and portable high-pressure engine, and discovery of the true mode of applying steam-power to navigation and railways. Hurd & Houghton: New York. *P*
Read, N. (1801). *Nathan Read papers*: 1709-1914. Peabody Essex Museum: Salem MA.
http://en.wikipedia.org/wiki/Nathan_Read

REED

05.03. 1896 St. Paul IN/USA
29.10. 1953 Kingsport TN/USA

Oren Reed obtained in 1922 the BSc degree in civil engineering from Purdue University, Lafayette IN, and later also studied in Norway. He was in 1923 employed as assistant engineer on the hydro-electric power development for the San Joaquin Light and Power Corporation, Fresno CA; he was then sent to Germany as an inspector of steel pipes. Upon return he joined in 1934 the Tennessee Valley Authority TVA, Knoxville TN, where he was involved in its projects for the Pickwick Dam, the Hales Bar Dam, the Watts Bar Dam, Fontana Dam, and the Watauga Dam. He was in 1950 transferred as construction engineer to the Widows Creek Steam Plant, Stevenson AL. Reed was member of the American Society of Civil Engineers ASCE from 1936.

Reed made in 1946 several reports on the hydro-electric potential of Brazil, which was however only fully appreciated from the 1960s. He attended in 1951 the World Power Conference at Stockholm, Sweden, and then studied the hydro-electric developments in Europe. He was a frequent contributor to technical journals dealing with hydro-power engineering. The Watauga Dam, on which Reed had been associated, is a hydro-electric and flood control dam on the Watauga River TN, both owned and operated by TVA. The dam was built in the 1940s, its height is 100 m and thus the second highest of all TVA dams behind Fontana Dam. Watauga Dam impounds 26 km^2. It is controlled by a morning glory spillway of 2,000 m^3/s discharge capacity, one of the very first of these elegant hydraulic structures. Fontana Dam is on the Little Tennessee River, built in the early 1940s to accommodate the skyrocketing electricity demands in Tennessee Valley during World War II. With 150 m height, it is the tallest dam in the eastern USA, and it was the fourth tallest dam of the world at the time of construction.

Anonymous (1955). Oren Reed. *Trans. ASCE* 120: 1574-1575.

Jourdan, J.W., Reed, O. (1929). Hydraulic jump design for a 100 sec.-ft. conduit. *Engineering News-Record* 103(6): 224-225. *P*

Reed, O. (1927). Notes on European practice in penstock design. *Engineering News-Record* 98(18): 718-721; 98(19): 773-776. *P*

Reed, O. (1929). Maximum flood discharge in San Joaquin Valley, Calif. *Trans. ASCE* 93: 406-430; 93: 440-450.

Reed, O. (1951). *Watauga project construction*. Division of Construction, Upper Holston Project Branch. TVA: Johnson City TN.

REEVES

03.11. 1886 Ashton IA/USA
13.08. 1966 Denver CO/USA

Arthur Blaine Reeves obtained in 1910 the BS degree in civil engineering from Iowa State College, Iowa City IA, and the CE degree in 1920. He was from 1910 to 1926 engineer of the US Reclamation Service, in charge of the North Platte Project in the States of Wyoming and Nebraska. He was thereby concerned with its drainage and construction works until 1915, then until 1923 with canal location works and until 1926 then serving as assistant irrigation manager. From 1927 to 1930 Reeves was manager of the Goshen Irrigation District, Torrington WY, from when he was in the office of the chief engineer, US Bureau of Reclamation USBR, Denver CO. He there worked until 1934 on the design of canals and dams, was then until 1950 head of the Design Section including canals, tunnels, and canal structures, taking over until 1953 as chief of the USBR Canals Branch. From 1953 to 1956 Reeves was the chief of the USBR Irrigation Operations Division. He then joined as assistant construction engineer an engineering consultant at Denver CO. Reeves was member of the American Society of Civil Engineers ASCE, and of the Intl. Commission of Irrigation and Drainage ICID.

Reeves was a widely known authority in irrigation engineering. He ended in 1956 his 43-year career of government service. In his last position he supervised the technical activities of the Bureau's irrigation systems in 17 western states. He had earlier taken part in the design of Hoover Dam and supervised the design of some of the world's largest irrigation canals, including the Delta-Mendota Canal in Central California, the adjacent Friant-Kern Canal, and the Columbia Basin Canals in Oregon. He travelled abroad to advise on the Snowy Mountains Hydro-electric Project in Australia, and also advised the Intl. Bank for Reconstruction and Development on designs for an irrigation project in Peru. His papers deal with the design and construction of irrigation canals.

Anonymous (1956). Irrigation expert A.B. Reeves to retire from Reclamation. *Engineering News-Record* 157(Dec.6): 313. *P*
Anonymous (1959). Reeves, Arthur B. *Who's who in engineering* 8: 2022. Lewis: New York.
Reeves, A.B. (1920). *Location survey of the Fort Laramie Irrigation Canal*. Iowa City IA.
Reeves, A.B. (1955). Use and economy of concrete pipe in irrigation systems. *Proc. ASCE* 81(622): 1-10.
Rhoades, R., Reeves, A.B., Irwin, W.H. (1948). Discussion of Application of geology to tunneling problems. *Trans. ASCE* 113: 1345-1348.

RENWICK

30.05. 1792 Liverpool/UK
12.01. 1863 New York NY/USA

James Renwick was brought to New York City in 1794 by his father. A brilliant student, Renwick graduated first in his 1807 class at the Columbia University, where he spent most of his life teaching. His early engineering career included service with the US Topographical Engineers, where his duties included plans of all military positions to obtain intelligence of enemy movements, to keep a journal of the Army, noticing ground, buildings and culture, and to exhibit the positions of the contending armies. He was appointed professor of natural philosophy at Columbia University in 1820, a position he held until 1854. He wrote then widely read biographies on Robert Fulton, among others, and his 1830 Treatise on the steam engine. His textbooks Outlines of natural philosophy, and Outlines of geology were the first works of their kind published in the USA.

About this time he also made studies on a canal connecting the Delaware and Hudson Rivers. His suggestion to use inclined planes in lieu of locks was later adopted on the Morris Canal in New Jersey. Note that Robert Fulton (1765-1815) had suggested this system in England about 1794. Renwick, however, patented a stationary steam-engine-powered inclined plane in 1813. Years later, the first application of wire rope for cables on a similarly designed plane in Pennsylvania was proposed by John Roebling (1806-1869). Other engineering projects that claimed Renwick's attention were the Maine-Canada boundary survey, the Delaware and Hudson Railroad, the proposed Navy Yard at Bergen Point NJ, and the Hudson River Railroad. Early published works include his Outlines of natural philosophy. Renwick's sons all had distinguished careers in the fields of engineering and architecture. James Jr. was best known for his architectural works including St. Patrick's Cathedral in New York City, as also the original Smithsonian buildings at Washington DC.

Anonymous (1963). Renwick, J. *Who was who in America* 1607-1896: 438. Marquis: Chicago.
Fitzsimons, N. (1966). James Renwick. *Civil Engineering* 36(8): 78. *P*
Renwick, J. (1822). *Outlines of natural philosophy*, being the heads of a course of lectures, delivered in Columbia College, New York. Van Winkle: New York.
Renwick, J. (1825). *Lectures on geology*, being the outlines of the science, delivered at the New York Athenaeum. Bliss & White: New York.
Renwick, J. (1830). *Treatise on the steam engine.* Carvill: New York.
http://en.wikipedia.org/wiki/James_Renwick_(physicist)

RETTGER

06.04. 1871 Huntington IN/USA
09.10. 1938 Ithaca NY/USA

Ernest William Rettger graduated in 1893 with the AB degree from Indiana University, Bloomington IN, receiving the PhD degree from Clark University, Worcester MA, in 1898. He was assistant professor of applied mechanics at Cornell University, Ithaca NY, from 1908 to 1922, continuing there as professor until his retirement in 1936.

Rettger took interest in general mathematical and mechanical problems related to engineering. He for instance worked on Lie's theory of groups and applied it to the theory of least and of virtual work. He also determined the economic penstock diameter, then a problem of vital interest. He became particularly known for the so-called proportional weir, corresponding to a vertical plate-weir normally erected in a rectangular channel, by which the head on the weir is proportional to discharge. The weir shape may be determined by integration of the elementary weir equation, and corresponds essentially to an inverted twin parabola about the channel axis. These weirs have therefore a wide opening close to the channel bottom, reducing toward the channel top. A similar device proposed by Adolphe H. Sutro (1830-1898) had been taken into service in 1908. Whereas the latter has only one curved weir plate and the other being vertical, Sutro's design further limits the weir width close to the channel bottom. The Sutro and Rettger weirs were model-tested, resulting in a good data fit with the linear discharge-head curve. Both weirs are hardly employed currently, given that electronic devices are applied to control the flow depth in a channel.

Anonymous (1927). E.W. Rettger. *Engineering News-Record* 99(5): 195. *P*
Anonymous (1933). Rettger, E.W. *American men of science* 5: 924. Science Press: New York.
Anonymous (1938). E.W. Rettger. *Engineering News-Record* 121(Oct.13): 451.
Keshava Murthy, K., Rangaraj, C. (1998). Generalized design of single profile weirs. *Journal of Hydraulic Research* 36(4): 695-706.
Mavis, F.T. (1949). There's no mystery in weir-flow calculations. *ENR* 142(Jan.6): 76-78.
Rettger, E.W. (1914). A proportional-flow weir. *Engineering News* 71(26): 1409-1410; 72(3): 148-149; 72(9): 462-463; 74(22): 1018-1019.
Rettger, E.W. (1915). The inverted weir. *Engineering News* 73(2): 72-73; 74(6): 277; 74(22): 1018-1019.
Stout, O.v.P. (1914). The proportional-flow weir devised in 1896. *Engineering News* 72(3): 148-149.

REYNOLDS K.C.

28.05. 1897 Boston MA/USA
25.04. 1996 Orange CA/USA

Kenneth Cass Reynolds made his engineering studies at Massachusetts Institute of Technology, obtaining the MS degree in 1925. He was awarded the first Freeman Scholarship which led him to the hydraulic laboratories of the German universities of Danzig and Karlsruhe from 1927 to 1929. He then returned to MIT developing a small river hydraulics laboratory and obtained the PhD degree in 1937. He remained at MIT as professor and in the 1930s designed a tidal model of the Cape Cod Canal, featuring electric control and the continuous water level registration. He left MIT the for Cooper Union University, New York NY, in 1944, being replaced at MIT by Arthur T. Ippen (1907-1974).

Reynolds' travels as Freeman scholar are reported in his 1928 paper, giving an exciting view of a young American visiting Europe professionally. He first travelled to Berlin, taking lessons in German along with other Freeman scholars from the ASCE and the ASME. They were given a tour by John R. Freeman (1855-1932) himself to the Prussian Laboratory of Hydraulic Engineering and Shipbuilding (Preussische Versuchsanstalt für Wasserbau und Schiffbau), where important researches were conducted in the 1930s by later Freeman scholars. These then went to Danzig, attending lectures at its Technische Hochschule. In late 1927 Reynolds went with some of his colleagues to the Technische Hochschule Karlsruhe, Germany, whose director was Theodor Rehbock (1864-1950). They worked in his famous river hydraulics laboratory and attended lectures. Reynolds was in the 1930s instrumental for the development of hydraulic modelling standards.

Reynolds, K.C. (1925). A study of the value of the coefficient and exponents of a certain formula for the flow of water in asphalted cast-iron pipes. *MS Thesis*. MIT: Cambridge MA.
Reynolds, K.C. (1929). Notes on laws of hydraulic similitude as applied to experiments of models. *Hydraulic laboratory practice* App. 14: 759-773, J.R. Freeman, ed. ASME: New York.
Reynolds, K.C. (1934). Similitude in hydraulic models. *Engineering News-Record* 113(8): 238.
Reynolds, K.C. (1936). *Report* on model study of Cape Cod Canal and approaches. MIT: Cambridge MA.
Reynolds, K.C. (1946). *Report* on model study of spillway of Popolopen Water Supply Sam, West Point NY. US Army Corps of Engineers, New York District.
Reynolds, K.C. (1953). Wave-action on sea walls by the use of models. *Trans. AGU* 14: 512-519.
Rouse, H. (1976). Kenneth C. Reynolds. *Hydraulics in the United States* 1776-1976: 115. *P*
http://www.bsces.org/index.cfm/page/Chronology/pid/10713

REYNOLDS L.B.

29.12. 1884 Hillsdale MI/USA
26.09. 1955 Palo Alto CA/USA

Leon Benedict Reynolds obtained the AB degree in 1908 from Stanford University, Stanford CA, and the AB degree in civil engineering from the Cornell University, Ithaca NY. He was from 1909 engineer for a consultant at Kansas City MO, and from 1923 professor of hydraulics and sanitary engineering at Stanford University. Reynolds was member of the American Society of Civil Engineers ASCE, of the American Water Works Association AWWA, and the Californian Sewage Works Association. He also served from 1944 as executive head of the Stanford Civil Engineering Department, and was consulting engineer for San Francisco CA, and Santa Clara CA, on sewage disposal problems.

Reynolds' career was twofold: During the first part, he was a successful civil engineer and partner of a consultant. During his second part, he was civil engineering professor at a notable University. His research activities mainly concerned sewage treatment. The 1930 paper is concerned with the sanitary engineering aspects of Boulder Dam, then the largest remaining undeveloped asset of the US government within the public domain. With a length of 2,500 km and a huge drainage area, this river ranks as the third largest of the USA, so that the concerns of this large project were justified. In his 1933 paper, Reynolds deals with the so-called chemical precipitation, another word for sewage treatment. It consists in the addition of chemicals to the raw sewage, so by reaction of the two a flocculent is produced. Its purpose is coagulation as much as possible of the suspended, colloidal, and dissolved constituents, so that the sedimentation process is supported. The paper reports the various historical advances in this process, starting in England in the 18th century. The Leon B. Reynolds Memorial Scholarship at Stanford University was founded after Reynolds' death, and supports civil engineering students.

Anonymous (1948). Reynolds, Leon B. *Who's who in engineering* 6: 1636. Lewis: New York.
Anonymous (1955). Leon B. Reynolds. *Engineering News-Record* 155(Oct.13): 66. *P*
Reynolds, L.B. (1930). Boulder Dam Symposium: The sanitary engineering aspect. *American Journal of Public Health* 20(10): 1071-1079.
Reynolds, L.B. (1930). State sanitation in California. *Journal of the American Water Works Association* 22(4): 531-536.
Reynolds, L.B. (1933). History of chemical precipitation. *Sewage Works Journal* 5(4): 595-599.
Reynolds, L.B. (1951). *Project review of proposed sewage treatment plant to serve City of Sacramento and metropolitan area*. Sacramento CA.

REYNOLDS W.C.

16.03. 1933 Berkeley CA/USA
03.01. 2004 Los Altos CA/USA

William Craig Reynolds was educated at the Stanford University, Stanford CA, from where he obtained the BS degree in 1954, the MS degree in 1954, and the PhD degree in mechanical engineering in 1957. He joined until 1966 the Department of Mechanical Engineering of Stanford University as assistant and associate professor, from when he was professor of mechanical engineering there, from 1972 to 1982, and 1989 to 1993 department chair, and chairman of the Institute of Energy Studies from 1974 to 1982. He was the recipient of the 1972 G. Edwin Burks Award from the American Society of Engineering Education, and the 1989 Fluids Engineering Award from the American Society of Mechanical Engineers ASME.

Reynolds was an aeronautical research scientist, a nuclear engineer, and a consultant in fluid and applied mechanics. His research interests included blow-down phenomena in thermodynamics, non-isothermal heat transfer, zero-g fluid mechanics, the turbulence modeling, turbulent boundary layer flow structure, turbulence-wall interactions, stability of gas films, the stability of laminar and turbulent flows in general, boundary-layer calculation methods, surface-tension-driven flows, organized waves in turbulent shear flows, unsteady turbulent boundary layers, internal combustion engine cylinder flows, and unsteady jets and separating flows. He has written a large number of papers and books, including one in thermodynamics, and several chapters on the computation of turbulent flows. He was an outstanding teacher and excellent researcher who had a good relation within the faculty.

Anonymous (1994). Reynolds, W.C. *American men and women of science* 19(6): 175. Bowker: New Providence.
Anonymous (2000). Reynolds, W.C. *Who's who in America* 54: 4058. Marquis: Chicago.
Reynolds, W.C. (1965). *Engineering thermodynamics*, 2nd ed. McGraw-Hill: New York.
Reynolds, W.C. (1974). Recent advances in the computation of turbulent flows. *Advances in Chemical Engineering* 9: 193-246.
Reynolds, W.C., Cebeci, T. (1976). Calculation of turbulent flows. *Turbulence*: 193-229, P. Bradshaw, ed. Springer: Berlin.
Reynolds, W.C. (1977). Computation of turbulent flows. *Annual Review of Fluid Mechanics* 8: 183-208.
http://news.stanford.edu/news/2004/january21/reynoldsobit-121.html *P*

RHONE

02.04. 1921 Denver CO/USA
17.09. 1996 Denver CO/USA

Thomas (Tom) Joseph Rhone was educated as civil engineer. After graduation, he joined the US Bureau of Reclamation, Denver CO, where he stayed during his entire career as hydraulic research engineer in the Division of the Engineering Laboratory, dealing mainly with the design of dams in the West of the USA. He was winner of the 1985 ASCE Hydraulic Structures Award.

The Bureau of Reclamation was established in 1902. Since then USBR has designed and constructed more than 220 dams. Its main hydraulic structures are the spillway and the outlet works. Recently, many dams have been refurbished by adding stepped spillways, labyrinth spillways, traditional chutes and tunnel spillways, and energy dissipators, to improve safety aspects, cope with environmental issues or to obtain better efficiency.1938 was the birthdate of ASCE's Hydraulics Division, later publishing the Journal of the Hydraulics Division, followed from 1983 by the Journal of Hydraulic Engineering. Many hydraulic studies made at USBR were included in these journals, including the outstanding works on energy dissipators in the 1950s, the works on outlet structures conducted in the 1960s, or recommendations on cavitation damage in the 1970s. The 1988 paper summarizes the major USBR activities.

Anonymous (1975). Thomas J. Rhone. *Civil Engineering* 45(8): 69. *P*

Anonymous (1985). Thomas J. Rhone. *Civil Engineering* 55(10): 96. *P*

Pugh, C.A., Rhone, T.J. (1988). Cavitation in Bureau of Reclamation spillways. Intl. Symp. *Hydraulics for high dams* Beijing: 645-652.

Rhone, T.J., Peterka, A.J. (1959). Improved tunnel spillway flip buckets. *Journal of the Hydraulics Division* ASCE 85(HY12): 53-76.

Rhone, T.J. (1974). Hydraulic model studies for the penstock for the Grand Coulee third power plant. *Report* REC-ERC-74-14. US Bureau of Reclamation: Denver CO.

Rhone, T.J. (1977). Baffled apron as spillway energy dissipator. *Journal of the Hydraulics Division* ASCE 103(HY12): 1391-1401.

Rhone, T.J. (1988). Development of hydraulic structures. ASCE Conf. *Hydraulic Engineering* Colorado Springs: 132-147. ASCE: New York.

Rhone, T.J. (1990). 50th anniversary of the Hydraulics Division 1938-1988: 132-147, A.M. Alsaffar, ed. ASCE: New York.

Williamson, J.V., Rhone, T.J. (1973). Dividing flow in branches and wyes. *Journal of the Hydraulics Division* ASCE 99(HY5): 747-769.

RICE G.S.

28.02. 1849 Boston MA/USA
07.12. 1920 Montclair NJ/USA

George Staples Rice graduated in 1870 from Harvard University with the degree of SB. He started already in 1869 in the service of the Boston Water Works, assisting in the construction of the Chestnut Hill Reservoir. He became in 1870 assistant engineer of the water works, Lowell MA, and later an assistant division engineer of the Boston water works. From 1877 to 1880 he filled the position of both assistant engineer and principal assistant engineer in charge of the Boston Main Drainage Works, then one of the most important sanitary engineering projects ever undertaken in the USA. At this time James B. Francis (1815-1892), and Alphonse Fteley (1837-1903) were practising their profession at Boston. Daily contacts with these were an inspiration for young engineers as for Rice, and greatly influenced his entire career. In 1880 he went to Arizona and Colorado for seven years, where he was engaged in mining operations.

Rice became in 1887 deputy chief engineer of the Aqueduct Commission of New York City, and thus began his work for the city to which he was destined to give the greater part of his career. He returned in 1891 for nine years to Boston where he was in charge of the Boston Transit Commission, and with his consulting office. He received a call from his Alma Mater in developing its School of Engineering. From 1902 to 1910 he served Harvard University as instructor in sanitary engineering. He in parallel served New York City as deputy chief engineer of its Transit Commission taking over in 1905 as chief engineer the Commission. From 1910 he resigned all positions, engaging himself in private practice. He had in total served the cities of Boston and New York in a honourably and unselfish manner throughout thirty years of his professional career, devoting nearly two-thirds of his professional life to public service. His long and active life brought him many friends, to all of whom he was greatly endeared because of his lovable and unselfish character.

Anonymous (1904). Rice, George S. *Cyclopaedia of American biography* 12: 82. White: New York. *P*
Anonymous (1920). George S. Rice. *Engineering News-Record* 85(24): 1160; 85(25): 1207.
Anonymous (1922). George Staples Rice. *Trans. ASCE* 85: 1713-1715.
Rice, G.S. (1891). Discussion of Excessive rainfalls. *Trans. ASCE* 25: 111.
Rice, G.S. (1913). Suggested method of preventing rock slides. *Journal of the Western Society of Engineers* 18(7): 585-627.

RICE R.H.

09.01. 1863 Rockland ME/USA
10.02. 1922 Lynn MA/USA

Richard Henry Rice graduated in 1885 as mechanical engineer from the Stevens Institute of Technology, Hoboken NJ. In 1887 he accepted the position of designer and chief draftsman with a mining company at Cambridgeport MA, there gaining the reputation as an able machine designer. He was from 1891 to 1894 the general superintendent of a steam engine company at Providence RI, an office he resigned when organizing there the Rice & Sargent Engine Company, to manufacture steam engines. He held this office until it was merged with the Providence Engineering Company in 1899, and then continued there as treasurer. In 1903 he joined the General Electric Company GEC, Lynn MA, to direct work for fifteen years developing the steam turbine. In 1918 he was made general manager of the works, a position he held until his sudden death.

From the beginning of his career Rice demonstrated a market-inventive talent. During his life he received some fifty patents. He is remembered for his original design of the first turbo-blower for blast furnaces in America, though of equal value were the Rice-Sargent steam engines designed jointly with his partner. These were recognized as among the best slow and medium-speed engines in the country and were produced from 1894 to 1903. Earlier, Rice had designed the smaller ratings of Curtis turbines up to 5,000 HP. For the first two years of its existence Rice was president of the Associated Industries of Massachusetts. He found time to write a number of technical papers dealing with the steam turbine. Rice was therefore both a skilled engineer and an able executive and was widely recognized for fairness, justice and honesty of purpose, as well as a thorough understanding of technical problems. In 1924 GEC established and endowed in his memory the Richard H. Rice Scholarship at Stevens Institute.

Anonymous (1917). Rice, Richard Henry. *Who's who in America* 9: 2053. Marquis: Chicago.
Anonymous (1922). Richard Henry Rice. *Trans. ASME* 44: 1389.
Anonymous (1922). Death of Richard Rice, designer of steam engines and turbines. *Power* 55(Feb.21): 323. *P*
Anonymous (1935). Rice, Richard Henry. *Dictionary of American biography* 15: 544-545. Scribner's: New York.
Rice, R.H., Moss, S.A. (1917). Blast furnace and steel mill power plants. *Proc. Engineers' Society of Western Pennsylvania* 33(2): 81-130.
http://en.wikipedia.org/wiki/Richard_H._Rice

RICH

03.11. 1896 Worcester MA/USA
22.06. 1977 Boston MA/USA

George Rollo Rich obtained in 1919 the BS degree from Worcester Polytechnic Institute, Worcester MA and the honorary doctorate degree in 1948. He was a structural designer of firms in Boston and New York State until 1925, then hydraulic engineer at Boston MA until 1931, when joining as hydraulic engineer the US War Department until 1937, and becoming chief design engineer of Tennessee Valley Authority TVA until 1946. After further works as a consultant he became lecturer in hydraulic structures at the Graduate School, Columbia University, New York, becoming a visiting lecturer in civil engineering at Harvard University, Cambridge MA, from 1956. Rich was a Fellow of the American Society of Civil Engineers ASCE, and Honorary Member ASCE and ASME from 1974. He served the American Society of Mechanical Engineers ASME in works for the journal Applied Mechanics Review and a number of ASME committees. He received the ASCE Rickey Medal in 1968, among many other decorations.

Rich was considered an authority on the design of hydro-electric power plants, and overall hydraulic engineering. Over a career of almost sixty years he was in charge of design of major installations in the eastern USA. He was credited for designing such notable works as the *Conowingo* Hydropower Project, the Fort Peck Project, the Cape Cod Ship Canal, the *Marimbondo* Hydroelectric Project in Brazil, or the St. Lawrence Power Project. He was known for a number of textbooks and papers. These include Hydraulic transients, and a chapter in the Handbook of applied hydraulics, edited by Calvin Victor Davis (1877-1946). He was an expert of water hammer phenomena and related hydraulic transients.

Anonymous (1974). George R. Rich. *Mechanical Engineering* 96(6): 97. *P*
Anonymous (1974). George R. Rich. *Civil Engineering* 44(10): 93. *P*
Anonymous (1975). Rich, George R. *Who's who in America* 38: 2575. Marquis: Chicago.
Anonymous (1977). George R. Rich. *Civil Engineering* 47(10): 164. *P*
Anonymous (1979). George R. Rich. *Trans. ASCE* 144: 547-548.
Rich, G.R. (1945). Water-hammer analysis by the Laplace-Mellin transformation. *Trans. ASME* 67(7): 361-376.
Rich, G.R. (1951). *Hydraulic transients*. McGraw-Hill: New York.
Rich, G.R. (1952). Navigation locks. *Handbook of applied hydraulics* 16: 761-780, C.V. Davis, K.E. Sorensen, eds. McGraw-Hill: New York.

RICHARDSON E.V.

05.01. 1924 Scottsbluff NE/USA
06.08. 2013 Fort Collins CO/USA

Everett Vern Richardson received education as civil engineer from the Colorado State University, Fort Collins CO, with the BS degree in 1949, the MS degree in 1960, and the PhD in 1965. After having joined the US Geological Survey until 1953, he was with the Iowa Surface Water Brigade until 1956 and was at the Colorado State University from 1965 to 1968 associate professor first, and administrator of its Engineering Research Center until 1983, from when he was there professor of civil engineering. He was from 1988 consultant of the US Bureau of Public Roads. He was member and Fellow of the American Society of Civil Engineers ASCE, receiving its J.C. Stevens Award in 1961 for a discussion co-authored by Darryl B. Simons (1918-2005), and its 1996 Hans Einstein Award.

Richardson made significant contributions to the engineering profession in the areas of erosion control, sedimentation, and waterway development through teaching, research, design, and management. As project director for water use and irrigation projects in Egypt, he was responsible for improving its research and development capabilities in the Public Works Ministry. During his career, he developed a training manual and other publications for the Federal Highway Administration FHA. He chaired the ASCE Task Committee on Bridge Scour Evaluation, so that he received an excellence award from Colorado State's water resources alumni.

Anonymous (1961). Everett V. Richardson. *Civil Engineering* 31(10): 84. *P*
Anonymous (1987). Richardson, Everett V. *Who's who in America* 44: 2334. Marquis: Chicago.
Anonymous (1996). Everett V. Richardson. *ASCE News* 21(12): 5. *P*
Haushild, W.L., Simons, D.B., Richardson, E.V. (1961). The significance of the fall velocity and effective fall diameter of bed materials. USGS *Professional Paper* 424-d. US Dept. of the Interior: Washington DC.
Richardson, E.V., Harrison, L.J., Davis, S.R. (1991). Evaluating scour at bridges. *Publication* FHWA-IP-90-017. Hydraulic Engineering Circular 18. FHA: Washington DC.
Richardson, E.V. (1996). Historical development of bridge scour evaluations. Proc. *North American Water and Environment Congress* Anaheim CA: 3-27.
Richardson, E.V., Lagasse, P.F., eds. (1999). Stream stability and scour at highway bridges. Compendium of papers *Water Resources Engineering Conf.* 1991-1998. ASCE: Reston.
http://lib.colostate.edu/archives/findingaids/water/wevr.html

2478

RICKETTS

23.06. 1883 Baltimore MD/USA
31.10. 1944 Washington DC/USA

Allan Townshend Ricketts graduated in 1915 from the University of Michigan, Ann Arbor MI, with the BSc degree in civil engineering. His major subject was sanitary engineering. He was engaged until 1927 by an engineering company as assistant engineer, later as engineer on the design and the construction of water and sewer systems, including water treatment plants for both drinking water and sewage. He was appointed deputy chief engineer in 1927, the Public Works Engineering Corporation and its successor. These companies worked in engineering projects for the Federal Water Service Corporation. He thereby conducted investigations for new water supplies throughout the United States. He also supervised the design and construction of important municipal water supplies and distribution systems.

Ricketts entered in 1932 government service at Washington DC, remaining in federal employ until failing health compelled his retirement in 1943. From 1933 he was member of the Public Works Administration, serving successively as engineer examiner, assistant chief engineer, and chief engineer. After the Federal Works Agency was established, he became consultant in the office of the chief engineer. Millions of dollars of federally financed sanitary engineering construction felt Ricketts' sound and conservative effect when he was with the mentioned agencies. His broad engineering experience, ability, and knowledge were recognized of outstanding value. His booklet Diagram for the solution of problems by Williams and Hazen Formula for discharge of pipes provided a useful method for the solution of pipe flow problems in water works engineering. The equation proposed by Gardner S. Williams (1866-1931) and Allen Hazen (1869-1930) is currently no more in use, but was popular mainly in the USA in the early 20th century due to its simplicity. In addition to his membership within the American Society of Civil Engineers since 1919, he belonged to the American Society of Testing Materials, the American Public Health Association, and the Federal Sewage Research Association. He made and retained a host of friends who missed him after his death.

Anonymous (1915). Allan T. Ricketts, president Student Council. *Michiganensian yearbook*: 315. University of Michigan: Ann Arbor. *P.*
Anonymous (1945). Allan T. Ricketts. *Trans. ASCE* 110: 1810-1811.
Ricketts, A.T. (1934). *Diagram for the solution of problems by Williams and Hazen Formula for discharge of pipes*. Washington DC.

RICKEY

10.11. 1871 Dayton OH/USA
19.04. 1943 Washington DC/USA

James Walter Rickey graduated as a civil engineer from Rensselaer Polytechnic Institute, Troy NY, in 1894. He was from 1896 to 1897 principal assistant engineer for the Lake Superior Power Corporation, Sault Ste. Marie MI, joined the St. Anthony Falls Water Power Company, Minneapolis MN until 1907, becoming until retirement in 1938 chief hydraulic engineer at Pittsburgh PA. Four years after his death the American Society of Civil Engineers ASCE established the James W. Rickey Medal encouraging worthwhile studies in the field of hydro-electricity, to which Rickey had so much contributed. Papers in the Transactions ASCE originally, but now in the Journal of Hydraulic Engineering, are canvassed yearly for award winners. Both the value of technical issues, as also the presentation are considered. The design of the solid gold medal shows on the observe the portrait of Rickey, whereas an arch dam with an spillway operating is on the reverse. The statement on the medal reads 'To promote the art of hydroelectric engineering'.

Rickey had designed and constructed a number of dams in the States North Carolina, New York, and Tennessee. He was a member and director of the Engineering Society of Western Pennsylvania, member of the American Society of Civil Engineers ASCE, the Engineering Institution of Canada, and the International Commission of Large Dams ICOLD. His most dramatic manifestation occurred during the construction of the 60 m high *Chute à Caron* Dam on Saguenay River near Kenogami, 200 km north of Quebec City. It was necessary to place a 30 m high auxiliary dam across Saguenay River at a point where the turbulence and depth of flow made it almost impossible to accomplish the object by normal means. Rickey built a cofferdam standing on end at the river side and tipped it over with a charge of dynamite. These were indeed the days where the hydraulic engineer in charge of a structure was at place and thereby on full duty.

Anonymous (1938). James W. Rickey. *Engineering News-Record* 120(Jan.13): 87. *P*
Anonymous (1939). Rickey, James W. *Who's who in America* 20: 2098. Marquis: Chicago.
Anonymous (1943). James W. Rickey. *Engineering News-Record* 130(Apr.29): 644. *P*
Anonymous (1949). Rickey Gold Medal awaits its first award. *Civil Engineering* 19(9): 634. *P*
Rickey, J.W. (1899). Failure of dam at Minneapolis due to pervious weakening through ice pressure. *Engineering News* 41(May 11): 307-309.
Rickey, J.W. (1931). Chute-à-Charon hydro-electric development. *Journal of the Boston Engineering Society* 2(7): 9-34.

RIDGWAY

19.10. 1862 New York NY/USA
19.12. 1938 Fort Wayne IN/USA

Robert Ridgway never made studies beyond primary school. After work with several railroad companies, he was engaged in 1884 at Tarrytown NY on the building of the second aqueduct of the Croton water shed, meeting Alfred Craven (1846-1926) or Charles S. Gowen (1851-1909). The aqueduct alignment was fixed between Harlem River and Old Croton Dam, with Ridgway in charge of it. The aqueduct tunnel was driven from shafts 1.5 km apart; their sinking was on the division in which he was leveller. The New Croton Aqueduct was taken in service in 1890; Ridgway was thus sent to Purdy's Station NY, on the construction of Titicus Dam of the Croton System. Upon dam completion, he was transferred to New York City, to supervise the construction of the Jerome Park Reservoir.

In 1905 Ridgway was promoted to division engineer for the construction of the Catskill Aqueduct, under J. Waldo Smith (1861-1933). A large portion of this project was in a tunnel, following in the upper reach the hydraulic grade, while it was below it in the lower reach, notably below Hudson River. The conception and construction of this work for impounding Catskill and conveying the water to the city was monumental, of which Ridgway had a major part in it. From 1912 to 1918 he was engaged in New York's Rapid Transit Company as chief engineer. He stayed with the City until his retirement in 1932, becoming then consultant. In total, his journey with the city thus lasted 49 years. In 1929 he was sent to the World Engineering Congress in Tokyo, Japan, there representing the USA. He spent three months in the Orient, enjoying Japanese hospitality and the honour of the Emperor, who conferred on him the Order of the Rising Sun. Several American institutions recognized Ridgway's ability honouring him with awards. He was member of many societies, and Honorary Member of the American Society of Civil Engineers ASCE, serving as its director, vice-president, and president in 1925. The Robert Ridgway Award was established in 1965, which is presented annually to the top student ASCE Chapter of the USA.

Anonymous (1940). *Robert Ridgway*. Little & Ives: New York. *P*
Anonymous (1941). Robert Ridgway. *Trans. ASCE* 106: 1527-1538.
Diehl, G., Ridgway, R. (1938). *The consulting engineer*. American Inst. Consulting Engineers: NY.
Ridgway, R. (1938). My days of apprenticeship. *Civil Engineering* 8(9): 601-604. *P*
http://images.nycsubway.org/articles/engnewsdc0020.jpg. *P*
http://www.catskillarchive.com/rrextra/DNAQUE.Html

RIEGEL

31.10. 1881 Harrisburg PA/USA
20.07. 1966 Erie PA/USA

Ross Milton Riegel received education from Cornell University, Ithaca NY, from where he obtained the civil engineering degree in 1904. He was engaged then until 1912 with various engineering positions, joined as a water power engineer the Water Supply Commission of Pennsylvania WSCP until 1914, and then until 1920 the Miami Conservancy District as design engineer. He was hydraulic engineer with the West Pennsylvania Power Company, Pittsburgh PA, until 1926, then until 1933 was design engineer with the Pittsburgh Department of Public Works, joining then as head civil engineer the design department of the Tennessee Valley Authority TVA, Knoxville TN. Riegel was awarded the 1948 ASCE James Laurie Prize for the 1946 paper.

Riegel is known for his hydraulic works toward the Miami Conservancy District. He and his colleague John Cleaveland Beebe (1887-1954) were among the first to study hydraulic jumps from various aspects. They investigated the sequent depth ratio for relatively large approach flow Froude numbers, determining the most important length characteristics, the flow characteristics in expanding channels with a hydraulic jump, observing instability features, and proposing engineering means to improve these disadvantages. Later in his career, Riegel took interest in dam engineering describing hydraulic structures from the constructional point of view, providing recommendations on the optimum instrumentation for dam safety. He also discussed the then particular features of rock-fill dams, a technique that was mainly developed after World War II.

Anonymous (1937). Riegel, Ross M. *Who's who in engineering* 4: 1154. Lewis: New York.

Anonymous (1948). Ross M. Riegel. *Civil Engineering* 18(1): 45-47. *P*

Anonymous (1990). Miami, Ohio, Conservancy District. *SIA Newsletter* 19(1): 3.

Blee, C.E., Riegel, R.M. (1948). Methods and instruments for the measurement of performance of concrete dams of the TVA. 3rd *ICOLD Congress* Stockholm Q8(R45): 1-23.

Blee, C.E., Riegel, R.M. (1951). Rock fill dams. 4th *ICOLD Congress* New Delhi Q13(R22): 189-208.

Riegel, R.M. (1910). Paxton Creek flood controlling works, Harrisburg, Pennsylvania. *Engineering News* 63(7): 196-199.

Riegel, R.M., Beebe, J.C. (1917). Theory of the hydraulic jump as a means of dissipating energy. *Technical Report* 3. State of Ohio, The Miami Conservancy District: Dayton OH.

Riegel, R.M. (1946). Design developments: Structures of the Tennessee Valley Authority. *Trans. ASCE* 111: 1160-1174.

RIPKEN

26.02. 1914 Minneapolis MN/USA
30.11. 2004 Minneapolis MN/USA

John Frederick Ripken was educated at University of Minnesota, Minneapolis MN, obtaining the BS degree in civil engineering in 1934, and in 1941 the MS degree. He was there an instructor at the Dept. of Mathematics and Mechanics from 1937 to 1941, a research engineer at the Columbia University until 1945 within the Navy Underwater Sound Laboratory, and then until 1946 a hydraulic engineer of the Navy Department, at its David Taylor Model Basin, Bethesda MD. He then was from assistant professor to professor of hydrodynamics at his Alma Mater, thereby closely associated with its St. Anthony Falls SAF Hydraulic Laboratory. Ripken was a Fellow of the American Society of Civil Engineers ASCE, and a member of the American Society for Engineering Education ASEE.

Ripken was closely associated with the SAF Hydraulic Laboratory, founded by Lorenz Straub (1901-1963). Further collaborators of this facility were Fred W. Blaisdell (1911-1998), Edward Silberman (1914-2011), Alvin G. Anderson (1911-1975), and Ripken. The latter was particularly interested in cavitation as a physical process and means to reduce the cavitation damage in hydraulic applications. Cavitation damage had become a major issue in hydraulic engineering from World War II, once the hydraulic heads increased above, say 50 m, generating flow velocities on chutes or in tunnels of more than, say 20 to 30 m/s. Originally, it was thought to improve the boundary material of these hydraulic elements, but it soon became clear that the only technical means was flow aeration, by which these damages were relatively easily controlled. Ripken has added to these questions and therefore may be considered an eminent hydraulician.

Anonymous (1964). Ripken, John F. *Who's who in engineering* 9: 1555. Lewis: New York.

Ripken, J.F., Killen, J.M., Crist, S.D. (1965). A new facility for evaluation of materials subject to erosion and cavitation damage. SAF *Project Report* 77. University of Minnesota: Minneapolis.

Ripken, J.F. (1966). Reduction of cavitation damage by surface treatment. SAF *Project Report* 81. University of Minnesota: Minneapolis.

Ripken, J.F., Hayakawa, N. (1972). Cavitation in high-head conduit control dissipators. *Journal of the Hydraulics Division ASCE* 98(HY1): 239-256.

Rouse, H. (1976). John F. Ripken. *Hydraulics in the United States* 1776-1976: 164. The University of Iowa: Iowa City. *P*

ROACH

25.12. 1813 Mitchelstown/IE
10.01. 1887 New York NY/USA

John Roach was born in Cork County, Ireland. He came at age sixteen to the United States. He settled in Howell NJ and there learned the trade of an iron moulder. In 1840 he took a part of his savings and went to Illinois but had to give up because his plans failed. He therefore purchased in New York a small iron works. In 1856 he added land to this firm, but the shops were soon after wrecked by a strong boiler explosion. Roach now found himself penniless but his ability and integrity enabled him to borrow capital, and his business was resumed. In 1860 he obtained the contract for constructing an iron draw-bridge over Harlem River in New York City, and thereafter he prospered so that by the end of the Civil War his foundry and engine works was one of the best in the US.

Roach was one of the first to recognize the importance of the shift from wooden to iron vessels. He sent a representative to England to make a careful study of the methods of iron shipbuilding on Clyde River. In 1868, Roach purchased a number of smaller marine-engine plants in and near New York City, and three years later then transferred his headquarters to Chester PA. Among the iron vessels he built were the *City of Peking* and the *City of Tokio* in 1874, for the Pacific Mail Steamship Company. He was also authorized by the Navy Department to design compound engines in vessels. He in parallel built the dry-dock at Pensacola FL, and in 1876 received the contract for two monitors. In 1883 the construction of three cruisers was begun but this contract was later cancelled, so that Roach resigned all these activities by 1885. While not the first to build iron vessels in the US, Roach launched a total of 126 vessels between 1872 and 1886, and therefore deserves the title 'Father of the iron shipbuilding in America'.

Anonymous (1887). John Roach's life ended: Death of America's most noted shipbuilder. *New York Times* (Jan.11): 1-2.
Anonymous (1935). Roach, John. *Dictionary of American biography* 15: 639-640. Scribner's: New York.
Grose, H.B. (1895). *John Roach*. Atlantic: New York. *P*
Swann, L.A. (1980). *John Roach, maritime entrepreneur*. Arno Press: New York. *P*
Thayer, W.M. (1897). John Roach. *Men who win*: 164-182. Nelson & Sons: London. *P*
Williamson, L.M. (1895). *Prominent and progressive Pennsylvanians of the nineteenth century*. Record Publishing Company: Philadelphia. *P*
http://www.globalsecurity.org/military/facility/roach.htm *P*

ROBERTS N.S.

28.07. 1776 Piles Grove NJ/USA
24.11. 1852 Canastota NY/USA

Nathan Smith Roberts was appointed principal of
Whitesboro Academy in 1806. In 1816 he became
an assistant to Benjamin Wright (1770-1842), then
engineer in charge of the building of the middle Erie
Canal section. Until 1822 Roberts made surveys for
the Canal between Rome and Rochester NY with
plans for locks. He was until the completion of the
canal in 1825 in charge of the western section of the
canal, from Lockport to Buffalo NY. He designed
five pairs of locks at Lockport to overcome barriers
formed by a 20 m high rocky ridge, a more elaborate
lock scheme than had ever been built in America.

Upon the completion of Erie Canal, Roberts became a consultant for the Chesapeake &
Delaware Canal, and in 1826 for the New York State. He made a survey reporting on a
route for a ship canal around the Niagara Falls. He was then chief engineer of the
western end of the Pennsylvania State Canal, between Pittsburgh PA and Kiskimenetas
PA. He in parallel furnished a Report to the New York State Canal Board, on the
practicability of supplying the summit level of the projected Chenango Canal with water.
He reviewed with James Geddes (1763-1838) the estimates of the Chesapeake & Ohio
Canal, and examined the country between Johnstown PA and Franktown PA for a
possible route for either a railroad or a portage over the mountains to connect the eastern
and western sections of the Pennsylvania State Canal. Roberts was from 1828 member
of the board of engineers of the Chesapeake & Ohio Canal Company, engaged on its
location of extensions. As a notable canal engineer he was appointed by the federal
government to take charge of surveys for a ship canal around Muscle Shoals, Tennessee
River in Alabama. He made in 1835 with John B. Jervis (1795-1885) surveys for the
New York State Canal Board to enlarge the Erie Canal, beginning in 1839 as chief
engineer its enlargement between Rochester and Buffalo. Roberts retired in 1841 to
Madison County, where he spent the remaining years on his farm.

Levy, J. (2003). *The Erie Canal*: A primary source history of the canal that changed America.
    Rosen: New York.
Malone, D., ed. (1935). Roberts, Nathan S. *Dictionary of American biography* 16: 12-13.
    Scribner's: New York.
Roberts, N.S. (1830). *Chesapeake & Ohio Canal Company*. Georgetown.
http://lockportjournal.com/canaldiscovery/x212281772/ERIE-CANAL-DISCOVERY-Nathan-
Roberts-canal-engineer-Part-1 (*P*)

ROBERTS W.J.

13.05. 1860 Caroline Islands/KI
06.04. 1938 Tacoma WA/USA

William Jackson Roberts was born on the Caroline Islands, today a part of the Republic of Kiribati. His parents moved in 1861 to the USA; he graduated in 1891 with the BSc degree in civil engineering from Massachusetts Institute of Technology, Cambridge MA. He also received later the MA degree from the University of Oregon, Corvallis OR. He began his engineering work at Portland OR as topographer, becoming then engineer for the Hood River Irrigation Company. From 1895 he was associate professor of civil engineering, Washington State College, Pullman WA. He also served as consultant for cities in the Northwest in projects relating to water supply and irrigation. He was also consultant on the Washington State Board of Health.

In 1908, Roberts decided to devote his time exclusively to engineering consultancy. His first project was to build a modern water system for Medford OR. In 1913 the Kings and Pierce Counties in Washington State were confronted with a serious flood-control problem; they formed the Inter-County River Improvement, with the headquarters at Tacoma WA, of which Roberts became chief engineer. During the next nine years he straightened, widened and deepened the Puyallup River, a project of pioneer type, attracting the attention of engineers, because the flooding conditions were successfully removed. In 1917, when the USA entered World War I, Roberts again supervised the construction of the water and sewerage systems for the Army Cantonment at Camp Lewis WA, accommodating 55,000 troops. This facility consisted of 65 km water mains and 50 km of sewers. Later he built more than forty water and sewerage systems in Washington, Oregon and Idaho States. In 1936 he was bestowed the Honorary Life Membership of the Tacoma Engineers' Club, where he had been president and member, and whom he donated his engineering library. Roberts was not only engineer but also a gentleman, loved and esteemed by all who knew him. He also was an ASCE member.

Anonymous (1939). William J. Roberts. *Trans. ASCE* 104: 1995-1998.
Roberts, M. (1978). *Papers of William J. Roberts*. Washington State University: Pullman WA.
Roberts, W.J. (1916). Some of the problems in the flood control of the White-Stuck and
        Puyallup Rivers. *Proc. Pacific Northwest Society of Engineers* 15(3): 3-40.
Roberts, W.J. (1920). *Inter-County river improvement* on White-Stuck and Puyallup Rivers in
        King and Pierce Counties WA. Tacoma WA.
http://www.olsonengr.com/download/globios/robertswilliamjbio.pdf *P*

ROBERTS W.M.

12.02. 1810 Philadelphia PA/USA
14.07. 1881 Soledade/BR

William Milnor Roberts joined at age fifteen the engineering corps, to be engaged in the construction of the Union Canal in Pennsylvania. In 1826 he was a rodman for a road across Alleghany Mountains to connect the canals on either side of the hill. He then worked for the Lehigh Canal, helping to improve the inclined planes at *Mauch Chunk*. In 1831 he was appointed senior assistant engineer for a railroad. Upon the completion of this work in 1834, he served as general manager until he became chief engineer of the Lancaster & Harrisburg Railroad.

Roberts at only age of twenty-five had experience in canal and railroad construction. He possessed much mechanical ability and had developed a special aptitude for design. One of his greatest engineering feats at this time was the construction of a two-level lattice-truss bridge across *Susquehanna* River at Harrisburg in 1837, carrying a double-track railroad above and a double carriage-way below. He was given charge in 1838 of the extensions of the Pennsylvania State Canals. For the next eight years he was engaged in canal construction, enlarging the *Welland* Canal in Canada, directing the enlargements of the Erie Canal in Pennsylvania, and acting as chief engineer of the Sandy & Beaver Canal, Ohio. Later, Roberts was mainly engaged with railroad design and construction; in 1857 he went to Brazil, where he obtained the contract to build the *Dom Pedro Segunda* Railroad, a project involving considerable tunnelling and requiring eight years to be completed. He returned to the United States in 1866 proposing improvements of the Mississippi River at Keokuk IA. In 1869 he was appointed chief engineer of the Northern Pacific Railroad, a position he held for ten years, during which time he also served as a member of the Mississippi River Jetty Commission. In 1879 he accepted the appointment of chief engineer of all public works in Brazil, and for the remaining years of his life was occupied with the examination of rivers, harbours, and waterworks. Roberts was president of the American Society of Civil Engineers in 1878.

Anonymous (1935). Roberts, William Milnor. *Dictionary of American biography* 16: 18-19. Scribner's: New York.
Anonymous (1936). Early presidents of the Society: William Milnor Roberts. *Civil Engineering* 6(12): 833-834. *P*
Bogart, J. (1896). William Milnor Roberts. *Trans. ASCE* 36: 531-537.
Roberts, W.M. (1857). Practical views on the proposed improvement of the Ohio River. *Journal of the Franklin Institute* Ser.3 34(1): 2-82.

ROBERTSON

18.04. 1916 Champaign IL/USA
22.11. 2012 Golden IL/USA

James Mueller Robertson obtained the BS degree in 1938 in civil engineering from University of Illinois, Champaign IL, the MS degree from the University of Iowa, Iowa City IA in 1940, and the PhD degree there in 1941. He was an assistant physicist at the US Navy Department, David Taylor Model Basin, Bethesda MD until 1942, then associate professor until 1949, and professor of engineering until 1954, when joining as professor of theoretical and applied mechanics the University of Illinois until retirement. He was a member of the advisory commission of the Bureau of Ordnance, US Navy, consultant of the US Army Waterways Experiment Station, a visiting lecturer at Kansas State University, and at Colorado State University, Fort Collins CO. He was a Fellow of the American Society of Civil Engineers and its 1955 Hilgard Awardee. He served as chairman the Cavitation Commission of the American Society of Mechanical Engineers ASME, and the ASME Standards Committee on Measurement of Fluids in Closed Conduits.

Robertson was a specialist in fluid dynamics and numerical computations of unsteady, non-uniform viscous flow. He taught basic and advanced courses in fluid dynamics. He was known for a number of technical publications and particularly his excellent book Hydrodynamics in theory and application.

Anonymous (1966). Robertson, James M. *Mechanical Engineering* 88(7): 96. *P*
Anonymous (1979). James M. Robertson. *Mechanical Engineering* 101(5): 90.
Anonymous (2005). Robertson, James M. *Who's who in America* 59: 3927. Marquis: Chicago.
Pazwash, H., Robertson, J.M. (1975). Forces on bodies in Bingham fluids. *Journal of Hydraulic Research* 13(1): 35-55.
Robertson, J.M., Rouse, H. (1941). On the four regimes of open channel flow. *Civil Engineering* 11(3): 169-171; 11(5): 312; 11(7): 433.
Robertson, J.M., Clark, M.E. (1962). The drag of elongated bodies over a wide Reynolds number range. *Journal of Aerospace Science* 29(7): 842-847.
Robertson, J.M. (1965). *Hydrodynamics in theory and application*. Prentice-Hall: Englewood Cliffs NJ.
Robertson, J.M., Martin, J.D., Burkhart, T.H. (1968). Turbulent flow in rough pipes. *Industrial and Engineering Chemistry* Fundamentals 7(2): 253-265.
Robertson, J.M., Johnson, H.F. (1970). Turbulence structure in plane Couette flow. *Journal of the Engineering Mechanics Division* ASCE 96(EM6): 1171-1182.

ROBINSON A.R.

24.04. 1921 San Antonio TX/USA
19.09. 1990 Oxford MS/USA

August Robert Robinson began his career in federal service in 1947 with the US Geological Survey. He joined the US Department of Agriculture USDA in 1951. He held degrees in civil engineering from the University of Iowa, and Colorado State University. He also studied hydraulic engineering at University of Minnesota's St. Anthony Falls SAF Hydraulic Laboratory. Robinson was the director of the USDA Sedimentation Laboratory, Oxford MS, since 1969, from where he was transferred to the facilities at Beltsville MD, in 1974, where he was a specialist in erosion and sedimentation on the National Program Staff of Soil, Water and Air Service. While at Oxford, he was also a professor of civil engineering at the University of Mississippi. In 1979, Robinson became a consultant.

Robinson published a number of papers and reports, mainly in agricultural engineering, covering topics including seepage, irrigation, discharge measurement structures, sand traps, open channel flow, and sediment transport and yield. He also was interested in the advance of water in furrows and the corresponding roughness characteristics of these irrigation channels. He proposed a trapezoidal-shaped flume for discharge measurement, which is still popular in agricultural engineering. In contrast to other designs, his flume appears to be not really economic, yet it has a high modular limit. Robinson was the recipient of the 1968 USDA Superior Service Award, and the 1980 Hancor Soil and Water Conservation Engineering Award.

Anonymous (1974). Robinson, A.R. *Agricultural Engineering* 55(7): 30. *P*
Anonymous (1991). Robinson Jr., August Robert. *Trans. ASCE* 156: 505-506.
Kruse, E.G., Huntley, C.W., Robinson, A.R. (1965). Flow resistance in simulated irrigation borders and furrows. *Conservation Research Report* 3. USDA: Washington DC.
Robinson, A.R. (1951). Artificial roughness in open channels. *MS Thesis*. Colorado State University: Fort Collins.
Robinson, A.R. (1957). *Report on trapezoidal measuring flumes for determining discharges in steep ephemeral streams*. Colorado State University: Fort Collins CO.
Robinson, A.R., Rohwer, C. (1957). Measurement of canal seepage. *Trans. ASCE* 122: 347-373.
Robinson, A.R., Chamberlain, A.R. (1960). Trapezoidal flumes for open-channel flow measurement. *Trans. American Society of Agricultural Engineers* 3(2): 120-124.
Robinson, A.R. (1962). Vortex tube sand trap. *Trans. ASCE* 127(3): 391-433.
Robinson, A.R. (1971). Model study of scour from cantilevered outlets. *Trans. ASAE* 14: 571-581.

ROBINSON S.W.

06.03. 1838 South Reading VT/USA
31.10. 1910 Columbus OH/USA

Stillman Williams Robinson received apprenticeship in a machine shop until 1859, and obtained the civil engineering degree from the University of Michigan, Ann Arbor MI, in 1863. He worked as an instrument maker already during his education, inventing a machine for grading thermometers. He was until 1866 assistant engineer on US Lake surveys, was appointed instructor in civil engineering at his Alma Mater, continuing until 1870 as assistant professor of mining engineering and geodesy. Until 1878 he was professor of mechanical engineering and physics at the University of Illinois, Urbana IL, from when he continued in a similar position as professor at the Ohio State University, Columbus OH, until 1895, becoming professor emeritus in 1899. He was in the 1880s inspector of railways for Ohio State, and acted as consulting engineer for the Santa Fe Railway.

Robinson was a prolific inventor, having taken more than forty patents on a variety of subjects. He improved in 1892 the Pitot tube for measuring local gas and fluid flow velocities, an instrument that was employed in most hydraulic laboratories until late of the 20th century for the exact total head reading. He also improved the transmission dynameter in 1898, the Robinson-Delmers hypodermic syringe in 1898, the Robinson-Hitchcock automatic air brake mechanism in 1899, among many other. Robinson also was an active writer in scientific journals, and a book author. He contributed three volumes to the Van Nostrand's Science Series, and revised several of the then existing volumes. He was a member of the American Association for the Advancement of Science AAAS, acting as its vice-president. He was further member of the American Society of Civil Engineers ASCE, Mechanical Engineers ASME, and the Society of Naval Architects and Marine Engineers SNAME. He was recipient of the 1892 ASCE Thomas Fitch Rowland Prize.

Davis, J.B. (1912). *Stillman W. Robinson*: A memorial. Ohio State University: Columbus. *P*
FitzSimons, N., ed. (1991). Robinson, Stillman W. (1991). *A biographical dictionary of American civil engineers* 2: 98. ASCE: New York.
Robinson, S.W. (1876). Mathematical investigation of the use of floats in gaging rivers and streams. *Hydrographic Report*: 75-86. Water Commissioner: Detroit.
Robinson, S.W. (1881). The flow of gases through tubes. *Engineering Magazine* 24: 370-377.
Robinson, S.W. (1884). *Compound steam pumping engines*. Van Nostrand: NY.
http://umhistory.dc.umich.edu/history/Faculty_History/R/Robinson,_Stillman_Williams.html *P*

ROBY

09.01. 1883 Portland OR/USA
23.05. 1961 Alpena MI/USA

Harrison George Roby graduated from Dartmouth College, Hanover NH, with the BSc degree in 1904, and the CE degree in 1906. He had then for 22 years responsibility in many hydropower developments. In 1920 he worked as consultant on the improvement of navigation and power on the St. Lawrence River, and from 1922 to 1939 he was employed as assistant hydraulic engineer and chief hydraulic engineer by *Byllesby* Engineering and Management Company, Chicago IL, and its successors. He was manager of dam construction from 1939 to 1942 for the Great Lakes Dredge and Dock Company, Chicago IL, and represented in 1943 Harza Engineering Co., Chicago IL, at Fort Peck Dam. From 1944 to 1955 he headed the Hydro Power Branch, the Engineering Division, in the Office of the Chief of Engineers, Washington DC. Roby became consultant of hydro-electric power in 1957 with the Tudor Engineering Co., Oakland CA. He was a member of the Water Power Section of the National Electric Light Association, East Cleveland OH, and the American Society of Civil Engineers from 1924, becoming ASCE Fellow in 1959.

Roby was a successful hydraulic engineer dealing with river improvements, hydropower projects, and dam engineering. He was in charge of the St. Lawrence River in the early 1920s as principal assistant to the American Government Engineer. The total length of the river from Montreal to Lake Ontario is almost 300 km, over an elevation difference of 70 m. The work included removal of obstructions to ease river navigation, location of rock ledges within reasonable depths below the grades of the proposed canals for a large part of their total length, and to profit from the large amount of water power. Roby's work as consultant of the US Army Corps of Engineers included river developments of the Tennessee, the Red and the Missouri Rivers. He designed a 110,000 HP hydropower plant on the Ohio River at Louisville KY, as described in one of his papers.

Anonymous (1929). H.G. Roby. *Power* 69(9): 355. *P*
Anonymous (1948). Roby, Harrison G. *Who's who in engineering* 6: 1668. Lewis: New York.
Anonymous (1962). Harrison G. Roby. *Trans. ASCE* 127(5): 72.
Roby, H.G. (1927). Construction plan and plant at the falls of the Ohio. *ENR* 98(19): 762-769.
Roby, H.G. (1928). Ohio Falls hydro development at Louisville meets unusual high-water
        conditions. *Mechanical Engineering* 50(11): 828-832. (*P*)
Roby, H.G. (1957). Role of civil engineer in power. *J. Washington Acad. Sciences* 47: 238-240.
http://digital.library.louisville.edu/cdm/singleitem/collection/kyimages/id/6/rec/3 (*P*)

ROE

01.06. 1876 Afton MN/USA
27.11. 1962 Portland OR/USA

Harry Burgess Roe was educated at University of Minnesota, Minneapolis MN, from where he gained the BS degree in 1908 and the MS degree in 1934. He joined the Agricultural Engineering Division of the University of Minnesota as instructor until 1912, as assistant professor until 1918, until 1929 then as associate professor, from when he was professor of agriculture until his retirement. He was in parallel involved at the Agricultural Experiment Station, and at the College of Agriculture, in charge of all land reclamation activities. He further was acting chief agricultural engineer in 1938, and he organized the professional degree course in agricultural engineering at his Alma Mater. Following retirement from the University faculty he served consecutively as engineering specialist with the US Army Engineers, and the US Soil Conservation Service at Portland OR.

Roe lived a life for agricultural engineering, thereby contributing a particular attention to hydraulic problems. He authored a number of Bulletins of Minneapolis Experiment Station on topics such as Mathematics for agricultural engineers, Septic tanks for rural homes, Benefits of drainage, Causes and methods of control for soil erosion, or Farm drainage practice. He authored a remarkable paper on drop culverts much ahead of his time, in which anti-vortex devices, the head-loss characteristics, the tailwater effects, or discharge-head relations were investigated. These structures were developed mainly in the 1950s and 1960s, including aspects of air entrainment, scour in the tailwater zone and performance of the entire hydraulic structure. Roe was a member of the American Society of Agricultural Engineers ASAE, the Minneapolis Academy of Science, or of the American Society of Civil Engineers ASCE.

Anonymous (1954). Roe, Harry B. *Who's who in engineering* 7: 2033. Lewis: New York.
Anonymous (1963). Harry B. Roe. *The Senate* 1: 87-88. University of Minnesota: Minneapolis.
Anonymous (2009). Roe, Harry B. *The first century forward*: 14. Minneapolis. *P*
Roe, H.B. (1936). Experimental design of vertical drop culverts. *Agricultural Engineering* 17(10): 426-432; 17(11): 477-481.
Roe, H.B. (1936). A study of influence of depth of groundwater level on yields of crops on peat lands. *Bulletin* 330, Minnesota Agricultural Experiment Station: St. Paul MN.
Roe, H.B. (1950). *Moisture requirements in agriculture*: Farm irrigation. McGraw-Hill: New York.
Roe, H.B., Ayres, Q.C. (1954). *Engineering for agricultural drainage*. McGraw-Hill: New York.

ROGERS

07.08. 1890 Detroit MI/USA
05.06. 1957 Brooklyn NY/USA

Harry Stanley Rogers graduated from the University of Wyoming, Laramie WY, with the BSc degree in civil engineering in 1914, and he received in 1926 the degree of civil engineer. He ˙vas awarded in 1935 the DSc degree from Northeastern University, Boston MA, in 1942 the Doctor of Laws from the University of Wyoming, Laramie WY, and in 1950 the Honorary Doctor of Engineering degree from Rensselaer Polytechnic, Troy NY. Upon graduation, he served as instructor at the Universities of Iowa, Iowa City IA; Wyoming; Washington, Seattle WA; and at Lafayette College, Easton PA. He became in 1919 design engineer, but returned to education in 1920, as professor of hydraulics and irrigation engineering at Oregon State University, Corvallis OR, where he was in 1927 appointed Dean of Engineering, and in 1928 Director of the College's Engineering Experiment Station. He accepted in 1933 the appointment of president of the Polytechnic Institute of Brooklyn NY, which position he held until his death.

Rogers was involved in few hydraulic projects during his career, including a report on stream pollution control, and a report on the sanitary conditions of Willamette Valley. He further contributed a paper on the status of the Engineering School, which in the 1920s underwent notable changes from the old into the then modern style. Rogers was more the organizer of technical education, serving as dean and president of engineering colleges. He was member of the American Institute of Consulting Engineers AICE, the American Society of Mechanical Engineers ASME, and the American Society for Engineering Education ASEE, which awarded him the 1953 Lamme Award. He was a member of the American Society of Civil Engineers ASCE from 1927.

Anonymous (1957). Dr. Harry Stanley Rogers dies. *Electrical Engineering* 76(8):704. *P*
Anonymous (1958). Harry Stanley Rogers. *Trans. ASCE* 123: 1301-1302.
Langton, C.V., Rogers, H.S. (1929). *Preliminary report on the control of stream pollution in Oregon*. Oregon State Agricultural College: Corvallis.
Rogers, H.S. (1928). *The Engineering School*: Its opportunity and progress in state service. Oregon State College: Corvallis OR.
Rogers, H.S., Mockmore, C.A., Adams, C.D. (1930). *A sanitary survey of the Willamette Valley*. Engineering Experiment Station. Oregon State Agricultural College: Corvallis.
Rogers, H.S. (1933). Influences and trends in engineering education. *Engineering News-Record* 110(14): 438-439.

ROHWER

23.10. 1890 Fort Calhoun NE /USA
14.01. 1959 Fort Collins CO/USA

Carl Rohwer received in 1912 the BS degree in civil engineering from University of Nebraska, Lincoln NE, and the MS degree in 1913 from the Cornell University, Ithaca NY. He was laboratory assistant and assistant irrigation engineer at US Department of Agriculture USDA from 1914 to 1917, and after war service from 1920 to 1937 was there in various grades, from when he was an irrigation engineer, conducting research in irrigation techniques and in special studies dealing with the water supply for irrigation. He was also in charge of pump irrigation studies for USDA's Division of Irrigation. From 1937 Rohwer was further in charge of land and water use studies of Rio Grande in the San Luis Valley, Colorado. He was member of the American Society of Civil Engineers ASCE, the International Association of Hydraulic Research IAHR, and of the American Geophysical Union AGU.

Rohwer was finally senior irrigation engineer for USDA, Soil and Conservation Branch. His career spanned forty years, during which he had particularly contributed to the design and execution of numerous irrigation works. His research interests were aspects of evaporation particularly in the 1930s, later developing into groundwater flow features related to agricultural engineering. He authored two notable papers in well flow and on seepage hydraulics in the ASCE Transactions, thereby collaborating with Maurice L. Albertson (1918-2009), and August Robert Robinson (1926-1990). Further works relate to canal linings, irrigation wells and pumps, and a variety of flow measurement devices. He also conducted research on the end depth problem.

Anonymous (1954). Rohwer, Carl. *Who's who in engineering* 7: 2039. Lewis: New York.
Anonymous (1959). Carl Rohwer. *Civil Engineering* 29(3): 222-223.
Petersen, J.S., Rohwer, C., Albertson, M.L. (1955). Effect of well screens on flow into wells. *Trans. ASCE* 120: 563-585.
Robinson, A.R., Rohwer, C. (1957). Measurement of canal seepage. *Trans. ASCE* 122: 347-373.
Rohwer, C. (1931). Evaporation from free water surfaces. *Technical Bulletin* 271. US Department of Agriculture: Washington DC.
Rohwer, C. (1933). The rating and use of current meters. *Technical Bulletin* 3. Colorado Experiment Station: Fort Collins.
Rohwer, C. (1943). Discharge of pipes flowing partly full. *Civil Engineering* 13(10): 488-490.
Rohwer, C., Blaney, H.F. (1943). *Water supply of Mendota area, California*. Blaney Papers.

ROSS

07.05. 1865 Dunwhich ON/CA
22.06. 1935 Berkeley CA/USA

Douglas William Ross was born in Elgin County CA. He attended the University of Toronto, Toronto CA. He left Canada in 1886 working during 3 years for American railroad companies. In 1890 he became topographer and assistant to Arthur D. Foote (1849-1933) on surveys for the reclamation of desert lands in Idaho, under the US Geological Survey, during which Ross became interested in irrigation work. He was then in charge of the New York Canal, Boise Valley ID, 12 m wide and 160 km long, irrigating 800 km^2 land. In 1891 he became superintendent of the 90 km long Phyllis Canal of the Idaho Irrigation Co., gaining experience in the operation of irrigations works. During the following years, he was also in charge of irrigation works in Yakima County WA, the Riverside Canal in Idaho, and for the irrigation system in Payette Valley ID.

Ross was appointed state engineer of Idaho in 1899, a position held for four years. He drafted an Irrigation Code which was enacted into law by the Idaho State Legislature in 1903. He was then supervising engineer of the US Reclamation Service in charge of its work in Idaho, retaining this important executive position until 1908. It was during this period that the main features of the Great Minidoka and Boise Irrigation Projects were planned and constructed, including four dams, two power plants, and irrigation systems for some 1,200 km^2 of land. In 1908 he accepted a position as consulting engineer for the irrigation interests of the American Water Works and Guaranty Co., Pittsburgh PA, whose projects were located in Idaho and California. He was the manager and chief engineer of its land and irrigation project in the Sacramento Valley CA. From 1914 Ross was in charge of waste lands in the Gulf States, in charge of reclamation of lands, and from 1922 to 1926 was in charge of reporting on land colonization and irrigation projects in Mexico. Until 1932 he was chief engineer at Brownwood TX, in charge of the construction of Brownwood Dam, an earth-fill structure for flood control, irrigation, and municipal water supply. He was considered a pioneer in irrigation engineering.

Anonymous (1895). Douglas W. Ross. *The Irrigation Age* 8(4): 105. *P*
Anonymous (1937). Douglas W. Ross. *Trans. ASCE* 102: 1646-1651.
Ross, D.W., Noble, T.A., Whistler, J.T. (1906). Columbia River and Puget Sound drainages. *Report* 14 of Progress of stream measurements. US Government Printing: Washington.
Ross, D.W. (1914). *Failure of irrigation and land settlement policies of the Western States*: Address delivered before the 21st Meeting of the Intl. Irrigation Congress: Alberta CA.

ROSSELL

11.10. 1849 Mobile AL/USA
11.10. 1919 New Brighton NY/USA

William Trent Rossell was appointed a cadet at the US Military Academy, West Point NY, in 1869, and he graduated there in 1873, receiving his commission as second-lieutenant of the Corps of Engineers, in which he served all through his career. After serving at the Engineer School at Willets Point NY, he was in 1876 appointed assistant engineering professor at West Point. From 1880 to 1882 he was stationed at Portland ME, and then at Charleston SC, and Fort Monroe VA for river improvement work. In 1885, he was placed in charge on harbour improvement works at Key West FL, from where he went to the Mississippi River in charge of the Third District, for construction and repair of the levee system. In 1890 Captain Rossell was appointed assistant to the Engineer Commissioner of the District of Columbia, and in 1892 he was commissioned in this position by the US President. From 1893 to 1895 Rossell served with the engineer troops and as instructor at Willets Point, eventually becoming in command of that post.

From 1895 to 1901, Major Rossell was in charge of fortifications and harbour work in Alabama, Mississippi, and Louisiana. He was also a member of the Board of Promotion for the officers of the Corps. Until 1906 he then was engineer of the Third Lighthouse District, and also was in charge on the improvement of Connecticut River. During the next three years Rossell was in charge of Ohio River improvements, including design and construction of Lock and Dam No. 37. During this period he also was president of the Mississippi River Commission. He was appointed Engineer of the Third New York District in 1909, in charge of river and harbour improvements in New Jersey State. His appointment to Chief of Engineers followed in 1913, succeeding thereby William H. Bixby (1849-1928). Brig.-General Rossell retired in the same year, but was recalled to active service in 1917. He then led the Third New York and Puerto Rico Districts, and was Northeast Division Engineer, retiring again in 1918, and dying soon later. Dredging vessel William T. Rossell was built in 1925 and taken from service in 1957, following a collision on the Coos Bay bar with a freighter.

Anonymous (1913). William T. Rossell: New Chief of Engineers. *Engineering News* 70(7): 321.*P*
Rossell, W.T. (1889). *Improvement of Mississippi River*. Washington DC.
Rossell, W.T. (1896). *Preliminary examination* for canal to connect Black Warrior River and
        Five Mile Creek. Government Printing Office: Washington DC.
http://en.wikipedia.org/wiki/William_Trent_Rossell *P*

ROUSE

29.03. 1906 Toledo OH/USA
16.10. 1996 Sun City AZ/USA

Educated at Massachusetts Institute of Technology MIT, Hunter Rouse was a travelling hydraulics Fellow receiving the Dr.-Ing. title from Technische Universität Karlsruhe, Germany, in 1932. He was an instructor then at Columbia University, New York, until 1936, assistant professor of fluid mechanics at California Institute of Technology, Pasadena CA, until 1939, when taking over as professor of fluid mechanics at State University of Iowa, Iowa City IA, and as director of Iowa Institute of Hydraulic Research IIHR from 1944 until retirement in 1966. He was visiting professor at University of Grenoble, France, in the 1950s, among many other similar positions.

Rouse was a man whose name is synonymous with excellence in fluids engineering education, research, and application. His influence on fluids engineering was remarkable and it continues through his many milestones, still relevant publications, films, and the score of engineers who received advanced degrees under his supervision. He authored a number of successful books in hydraulic engineering and fluid mechanics, notably his Fluid mechanics in 1938, his Elementary mechanics of fluids in 1946, his Engineering hydraulics in 1950 as a summary of the 1949 Hydraulics Conference held at the State University of Iowa, History of hydraulics in collaboration with Simon Ince (1921-) as the far most cited of Rouse's books, Advanced mechanics of fluids in 1958 and his Hydraulics in the United States, 1976. Rouse was awarded IAHR honorary membership in 1985. The ASCE Hunter Rouse Annual Lecture was installed in 1979 as an award for distinguished hydraulic engineers.

Anonymous (1939). Hunter Rouse. *Civil Engineering* 9(1): 53. *P*
Anonymous (1964). Rouse, Hunter. *Who's who in engineering* 9: 1595. Lewis: New York.
Rouse, H. (1934). On the use of dimensionless numbers. *Civil Engineering* 4(11): 563-568.
Rouse, H. (1937). Modern conceptions of the mechanics of fluid turbulence. *Trans. ASCE* 102: 463-543.
Rouse, H. (1938). *Fluid mechanics for hydraulic engineers*. Dover: New York.
Rouse, H. (1946). *Elementary mechanics of fluids*. Wiley: New York.
Rouse, H., ed. (1950). *Engineering hydraulics*. Wiley: New York.
Rouse, H., Ince, S. (1957). *History of hydraulics*. Iowa Institute of Hydraulic Research: Iowa.
Rouse, H. (1959). *Advanced mechanics of fluids*. Wiley: New York.
Rouse, H. (1976). *Hydraulics in the United States* 1776-1976. University of Iowa: Iowa.

ROUSSEAU

19.04. 1870 Troy NY/USA
24.07. 1930 on Sea/USA

Harry Harwood Rousseau graduated in 1891 as civil engineer from the Rensselaer Polytechnic Institute, Troy NY. He served then as assistant engineer in New York. In 1898, after further studies, he was appointed lieutenant within the US Navy, remaining at the Bureau of Yards and Docks, stationed at Mare Island Navy Yard CA. He was appointed at age of only 36 chief of the Bureau, with the rank of rear-admiral. He subsequently became a Panama Canal Commissioner under George Goethals (1858-1928). Rousseau became head of the Departments Building Construction, Motive Power and Machinery, and of Municipal Engineering, with more than 10,000 employees under him. From 1908 he served as assistant to the chief until the Canal was completed in 1914. In addition to his other work, he was given charge in 1911 to design and build the canal terminals, with the dry docks, ship repair, piers, breakwaters, among others. Goethals noted: ' Rousseau was indispensable to the canal work', so that he remained there until these works were completed.

From 1915, in the rank of rear-admiral, he was appointed member of the Commission of Navy Yards and Naval Stations, including the three coasts of the USA. Particularly on the Pacific Coast, notably in San Francisco Bay, an exhaustive study was made for a main West Coast Base. During World War I, he became assistant general manager and head of the Shipyards Plant Division of the Emergency Fleet Corporation. This Division allocated and designed marine railways and dry docks to care for the ship-building program. For these services Rousseau was awarded the Navy Cross 'for exceptionally meritorious service in a duty of great responsibility'. In the 1920s, he was engaged with various government duties relating to naval engineering and management. Admiral Rousseau was ever ready to help; he carried the respect, the high regard of those who knew him in his work. With a keen sense of humour, he was both an enjoyer and a leader. He was member of the American Association of Port Authorities, the Permanent International Association of Navigation Congresses, and the American Society of Civil Engineers ASCE. He died on the sea on the way to the Canal Zone.

Anonymous (1931). Harry H. Rousseau. *Trans. ASCE* 95: 1597-1601.
Haskin, F.J. (1913). *The Panama Canal*. Doubleday: New York. *P*
Rousseau, H.H. (1909). *The Isthmian Canal*. US Government Printing Office: Washington DC.
http://www.google.ch/imgres?q=%22Harry+Harwood+Rousseau *P*

ROWLAND

15.03. 1831 New Haven CT/USA
13.12. 1907 New York NY/USA

Thomas Fitch Rowland entered at age of thirteen his father's grist mill as a miller's boy. Later Rowland was employed by New York & New Haven Railroad in their machine shop. In 1850 he accepted the post as assistant engineer of the steamboat *Connecticut*, but then was variously employed about New York in designing steamboat machinery. He established in 1859 at Greenpoint NY a business for iron works in association with a prominent builder of wooden vessels. The first contract was with Croton Aqueduct for a 400 m long wrought-iron water pipe of 2.2 m diameter. It was to be located on top of the high bridge over the Harlem River to carry water from the aqueduct to the new reservoir in Central Park. This work, requiring 450 tons of wrought-iron plates and some 400 tons of castings, was successfully carried out.

At the outbreak of the Civil War, Rowland began the manufacture of gun carriages and mortar beds for the Navy Department. He further fitted out most of the steamers purchased from the merchant service, which vessels took part in the capture of Port Royal and were known as the Porter Mortar Fleet. In 1861, Rowland contracted with John Ericsson (1803-1889) to construct an iron-clad floating battery, later known as the *Monitor*. He further designed a series of similar monitors and gunboats. In the 1870s he built a number of ferryboats for the Union Ferry Company, New York, and designed steamboats for Cuban waters. He then experimented in the art of iron and steel welding and also designed the process and apparatus used for his company in the manufacture of the Fox corrugated and Morison suspension furnaces then widely used. He incorporated later his business as Continental Iron Works, serving as president until his death. The Thomas Fitch Rowland Prize was instituted by the American Society of Civil Engineers ASCE in 1882. It is given to authors describing construction works with a valuable contribution to either construction management or construction engineering.

Anonymous (1908). Thomas Fitch Rowland. *Trans. ASME* 29: 1181-1183.
Anonymous (1935). Rowland, Thomas Fitch. *Dictionary of American biography* 16: 200-201.
    Scribner's: New York.
Rowland, T.F. (1888). Experiments on delivery of water from pipes. *Trans. ASCE* 19: 120-126.
http://209.85.129.132/search?q=cache:nTDRUIoc8cMJ:www.sonofthesouth.net/leefoundation/c
ivil-war/1862/iron-clad-ironsides.htm+%22Thomas+Fitch+Rowland%22+1831-
&cd=3&hl=de&ct=clnk&gl=ch *P*

RUBEY

19.12. 1898 Moberly MO/USA
12.04. 1974 Santa Monica CA/USA

William Walden Rubey obtained his AB degree from the University of Missouri, Columbia MO, and his D.Sc degree in 1953. He was an assistant geologist at US Geological Survey USGS until 1922, instructor in geology at Yale University until 1924, from when he rejoined USGS as associate geologist, geologist, and senior geologist until 1960. He was in 1954 guest geologist at the Institute of Geophysics, University of California, Los Angeles, and there professor of geology and geophysics from 1960. Rubey was a member, the Committee on Geophysics and Geology Research and Development Board from 1947 to 1950, and director and vice-president of the American Geological Institute AGI in the 1950s. He was recipient of the Award of Excellence from the US Department of Interior in 1943, member of the National Academy of Science, among many other distinctions.

Rubey was a notable and excellent geologist. He was interested in problems of earth science, including the origin and the evolution of mountain belts, the diversity of igneous, sedimentary and metamorphic rock, the growth of continents, the origin of ocean basins and of sea water, or the evolution of the terrestrial planets. He directed the studies by judicious questioning and with an open-minded, objective attack, paired with comprehensive appreciation of physics and chemistry coupled with broad background knowledge in geology. He is remembered for a formula to determine the sink velocity of sediments in still water thereby generalizing the approach of George G. Stokes (1819-1903) by accounting for the density difference between sediment and fluid, gravitational acceleration, fluid viscosity, and sediment diameter. His result may also be expressed as a relation between the particle Reynolds and the densimetric Froude number.

Anonymous (1969). Rubey, William W. *Who's who in America* 35: 1883. Marquis: Chicago.
Ernst, W.G. (1978). William W. Rubey. *Biographical memoirs* 49: 205-223. National Academy of Sciences: Washington DC. *P*
Rubey, W.W. (1933). Equilibrium conditions in debris-laden streams. *Trans. AGU* 14: 497-505.
Rubey, W.W. (1933). Settling velocity of gravel, sand and silt. *American Journal of Science* Ser. 5 25(148): 325-338.
Rubey, W.W. (1933). The size distribution of heavy minerals within a water-laid sandstone. *Journal of Sedimentary Petrology* 3(1): 3-29.
Rubey, W.W. (1938). The forces required to move particles on a stream bed. *Professional Paper* 189 E: 121-141. USGS: Washington DC.

RUDAVSKY

17.01. 1925 Jakubiw/PL
26.02. 2014 Santa Clara CA/USA

Alexander Bohdan Rudavsky was born in Poland. He immigrated to the United States in 1949, becoming US-citizen in 1954. In 1956, he gained the MSc degree in civil engineering from the University of Minnesota. Rudavsky received the Dr.-Ing. degree in 1966 from the Technical University, Hannover, Germany. During his early professional career, he was a research engineer at the Saint Anthony Falls SAF Hydraulic Laboratory and a design engineer at Philadelphia PA. For 25 years, he was a hydraulics professor at the San Jose State University, San Jose CA, from where he retired as emeritus in 1985. In parallel, he founded in 1963 Hydro Research Science HRS at Santa Clara CA, where he applied research in physical modeling and analysis of state-of-the-art techniques to hydraulic design. He was involved as HRS president in more than 200 hydraulic projects, including river and coastal engineering, sedimentation, dam failures, and sewage and water supply systems.

Rudavsky's PhD thesis was directed towards energy dissipators as used in hydraulic engineering. A number of energy dissipating structures was compared. In the same period, Rudavsky also investigated improvements for the *Escondido* Creek watershed, the energy dissipator of the Anderson Ranch development. Later, Rudavsky contributed to a number of dams in the USA, and to the rehabilitation of a large river in Iran. He also designed the spillways of two dams in Brazil, conducted the analysis of a disastrous flood in Spain, made extensive numerical and physical modeling of hawser forces on moored ships in Panama Canal, and major hydraulic projects in Greece, Thailand, and Venezuela.

Rudavsky, A.B. (1964). Escondido Creek watershed channel improvement – Hydraulic model investigations. Hydraulic Laboratory *Report* HY-0001-64. State College: San Jose CA.

Rudavsky, A.B. (1966). Energievernichter und Energieverzehrungsmethoden unter Überfallwehren. *Mitteilung* 26: 194-313. Franzius Institut, TU: Hannover.

Rudavsky, A.B. (1966). Albertson Ranch development, Potrero dam – Energy dissipator. Hydro Science Company *Report* HS-004-66. Santa Clara CA.

Rudavsky, A.B. (1976). Selection of spillways and energy dissipators in preliminary planning of dam developments. 12th *ICOLD Congress* Mexico Q46(R9): 143-180.

Rudavsky, J. (2003). Alexander B. Rudavsky. Personal communication. *P*

http://obits.dignitymemorial.com/dignity-memorial/obituary.aspx?n=Alexander-Rudavsky&lc=6922&pid=169965972&mid= *P*

RUMSEY

03. 1743 Bohemia Manor MD/USA
20.12. 1792 London/UK

James Rumsey was born in Cecil County, but little
is known until he was living at Bath VA in 1782. In
1784 when George Washington stayed there, Rumsey
showed him a model of a mechanical boat, which
had a bow-mounted paddle wheel that worked poles
to pull the boat upstream. Washington had plans to
make Potomac River navigable, so that a company
was formed. He recommended Rumsey in 1785 to
oversee the clearing of rocks at Harper's Ferry on
this river, while his brother-in-law built the pole-
boat at Shepherdstown WV, yet steam propulsion
was included.

The work on the hull of the boat had begun in 1785. Valve castings, cylinders and other
pieces were installed, so that the boat was tested on the river, yet the first trials were
unsatisfactory. The pole-boat mechanism caused the boat to yaw in the current, which
disabled the paddle-wheel and stopped the boat. Also the engine consumed too much
steam. Rumsey thus abandoned his work on the pole-boat mechanism, but tried a coil of
gorged iron pipe, which proved not only to be more efficient, but also smaller and
lighter. With a working steam engine, another problem was automatically resolved,
namely the single cylinder pump would draw liters of water from beneath the boat, send
it down a copper pipe to the stern. Because the water was drawn into the pump at the
same time as water was still flowing from it to the stern, the pump was working against
itself. This required replacing the copper pipe with a square wooden trunk with flatter
valves to allow water to enter from the river, so that the negative pressure at the pump
was relieved. The boat made its successful public demonstration on the Potomac in
1787. This was 20 years before Robert Fulton (1765-1815) constructed his *Clermont*,
but the idea of jet propulsion was also proposed by the Swiss Daniel Bernoulli. John
Fitch (1743-1798) demonstrated his steamboat at Philadelphia a year before. Although
he had no patents, he had exclusive rights to build steam boats, so that there started
disagreement with Rumsey. The latter moved in 1788 to England, taking four patents
before his death, of which the 1791 patent included the pumps, motors, and hydraulic
cylinders of fluid power engineering. One year later he had a true water turbine, some
40 years earlier that it would be invented in France. A monument was set up in 1906 in
a park of Shepherdstown WV, overlooking Potomac River.

Anonymous (1980). Rumsey. *Mechanical engineers in America*: 267. ASME: New York.
http://en.wikipedia.org/wiki/James_Rumsey *P*

RUSSELL G.E.

25.12. 1877 Boston MA/USA
04.12. 1953 Boston MA/USA

George Edmond Russell graduated in 1900 as civil engineer from Massachusetts Institute of Technology MIT, Cambridge. He served there as instructor until 1907, was then assistant and associate professor, and became professor of hydraulic engineering in 1921 until retirement in 1943. Russell was a Fellow of the American Society of Civil Engineers ASCE, Boston Society of Civil Engineers, the New England Water Works Association, and the American Academy of Arts and Sciences.

Russell was known for his textbook Hydraulics, first published in 1909. Its chapters are: 1. Introduction, 2. Hydrostatics, 3. Effects of translation and rotation upon bodies of water, 4. Fluid motion: Bernoulli's theorem, 5. Discharge from orifices, 6. Flow through mouthpieces, 7. Flow over weirs, 8. Flow through pipes, 9. Flow in open channels, 10. Dynamic action of jets and streams, and Appendix. The book is nothing extraordinary, but represents one of the early full texts in hydraulics of the United States. Outstanding contemporary texts including these of Alfred Flamant (1839-1915) or Philipp Forchheimer (1852-1933) have certainly had more impact, but Russell's book summarizes the then current knowledge well. He was the design engineer of the Charles River Basin Commission, State of Massachusetts, early in his career. Later he served in various municipalities of his State as advisor in water supply. He also was a member of the Advisory Committee of the US Coast Guard Academy, and he corporated with the US Navy Department in designing a new type of submarine. He will be remembered for his general hydraulic studies made in the early 20th century.

Anonymous (1933). Russell, George Edmond. *American men of science* 5: 962. Science Press: New York.
Anonymous (1944). Russell, G.E. *Water Works Engineering* 97(Dec.12): 1523. *P*
Anonymous (1951). Russell, George E. *Who's who in America* 26: 2380. Marquis: Chicago.
Anonymous (1965). George E. Russell. *Engineering News-Record* 152(Jan.7): 64.
Rouse, H. (1976). Russell, G.E. *Hydraulics in the United States*: 83. University of Iowa: Iowa. *P*
Russell, G.E. (1889). Thickness of water pipe. *Journal of the Association of Engineering Societies* 8(2): 100-113.
Russell, G.E. (1907). *Notes on hydraulics*, prepared for the use of students in the shorter courses of Massachusetts Institute of Technology. MIT: Boston.
Russell, G.E. (1909). *Textbook of hydraulics*. Holt: New York.
Russell, G.E. (1934). *Text-book on hydraulics*, 4th ed. Holt: New York.

SAFFMAN

19.03. 1931 Leeds/UK
17.08. 2008 Pasadena CA/USA

Philip Geoffrey Saffman graduated in mathematics from Cambridge University and there gained also his PhD title in 1956 to become there a Lecturer and from 1960 Reader. He joined until 1964 the Kings College, London University, moving then until 1969 to the California Institute of Technology, Pasadena CA, when taking over there as professor of applied mathematics. He was from 1970 to 1971 a visiting professor at Massachusetts Institute of Technology MIT and from 1995 Theodore von Karman professor of applied mathematics at Caltech. Saffman served as Associate Editor the Journal of Fluid Mechanics from 1958 to 1967, Physical Review Letters from 1979 to 1992, and Physics of Fluids from 1975 to 1977. He was a Fellow of the American Academy of Arts and Science, of the Royal Society, receiving the *Otto Laporte* Prize of the American Physical Society. He was awarded the 2000 *Ludwig-Prandtl-Gedächtnisvorlesung* by DGLR. Saffman was elected Fellow of the Royal Society in 1988.

Saffman had interests in turbulence, viscous flows, vortex motion and water waves. He made valuable theoretical contributions to different areas of low-Reynolds-number hydrodynamics. These included the lifting force on a sphere in a shear flow at small but finite Reynolds numbers, the Brownian motion in thin liquid films, and particle motion in rapidly rotating flows. Saffman's other contributions include dispersion in porous media, average velocity of sedimenting suspensions, and compressible low-Reynolds-number flows.

Pullin, D.I., Saffman, P.G. (1998). Vortex dynamics in turbulence. *Annual Review of Fluid Mechanics* 30: 31-51.

Saffman, P.G. (1962). The effect of wind shear on horizontal spread from an instantaneous ground source. *Quarterly Journal of the Royal Meteorological Society* 88(378): 382-393.

Saffman, P.G. (1967). The large-scale structure of homogeneous turbulence. *Journal of Fluid Mechanics* 27: 581-593.

Saffman, P.G. (1992). *Vortex dynamics*. University Press: Cambridge.

Saffman, P.G. (2003). Philip Geoffrey Saffman. Private communication. *P*

Taylor, G.I., Saffman, P.G. (1957). Effects of compressibility at low Reynolds number. *Journal Aeronautical Sciences* 24(8): 553-562.

Turner, J.S. (1997). G.I. Taylor in his later years. *Annual Review of Fluid Mechanics* 29: 1-25. *P*

http://www.physicstoday.org/obits/notice_310.shtml

SANGSTER

09.12. 1925 Austin MN/USA
22.08. 2000 Atlanta GA/USA

William McCoy Sangster was educated at the State University of Iowa, Iowa City IA, from where he obtained the BS degree in 1947, the MS degree in 1948, and the PhD in 1964. He was there assistant instructor in 1948, and from assistant professor to civil engineering professor until 1967 at University of Missouri, Columbia MO. From 1964 to 1967 he was in parallel associate dean of its College of Engineering, and associate director of its Engineering Experiment Station. From then he was professor of civil engineering at Georgia Institute of Technology, Atlanta GA. Later, he was there the director and Dean of the School of Engineering. He was closely related to the American Society of Civil Engineers ASCE, taking over as one of the youngest presidents in 1974. He also was a member of the American Society of Engineering Education. He was awarded a graduate fellowship to the State University of Iowa in 1954, and a National Science Foundation Science Faculty Fellowship for study there in 1962. He also served as chairman of the ASCE National Water Resources Meeting at Atlanta GA in 1972. He received the Linton E. Grinter Distinguished Service Award from the Accreditation Board for Engineering and Technology in 1991.

Sangster had research interests in the hydrodynamic stability of stratified flows, orbital mechanics, and general hydraulics. His PhD thesis was concerned with the head losses of junction manholes as used in sewer hydraulics. However, the manhole geometry considered did not account for the current standard, because no benches were included. He also provided the design basis for better storm drainage facilities under highways. During his presidency, he emphasized progress in ASCE's policing of the ethics of its members and its efforts to streamline administrative procedures.

Anonymous (1964). Sangster, W.M. *Who's who in engineering* 9: 1623. Lewis: New York.
Anonymous (1965). William M. Sangster. *Civil Engineering* 35(10): 89. *P*
Anonymous (1973). William M. Sangster. *Civil Engineering* 43(10): 88. *P*
Anonymous (1994). Sangster, W.M. *American men and women of science* 6: 525.
Sangster, W.M., Wood, H.W., Smerdon, E.T., Bossy, H.G. (1959). Pressure changes at open junctions in conduits. *Journal of the Hydraulics Division* ASCE 85(HY6): 13-42; 85(HY10): 157; 85(HY11): 153; 86(HY5): 117. Also available as *Engineering Series Bulletin* 6. Engineering Experiment Station, University of Missouri: Columbia.
http://www.engineering.uiowa.edu/alumni-friends/distinguished-engineering-alumni-academy-members/dr-william-m-sangster *P*

SAPH

26.10. 1871 San Jose CA/USA
13.02. 1920 Berkeley CA/USA

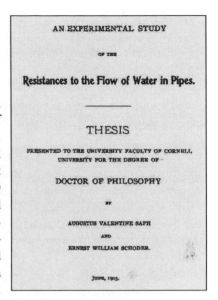

Augustus Valentine Saph graduated in 1894 from the College of Civil Engineering, the University of California, receiving there the MS degree in 1896, and then acted as an instructor there in drawing and mathematics until 1900. He continued his studies at Cornell University, Ithaca NY, receiving the PhD degree in 1902. From 1903 to 1905 he was engaged as a structural engineer at Reno NV for the US Reclamation Service, and then on the Truckee-Carson Project, designing the diversion dams and other structures. In 1906 he was employed by a structural engineer to design bridges and buildings. From 1908 to 1911 he acted as assistant engineer in the office of the City Engineer of San Francisco CA, designing the College Hill Viaduct, pumping stations and the Ashbury Heights Tank. From then to 1914, Saph was the assistant state engineer of California acting as chief engineer for the Board of State Harbor Commissioners at San Francisco CA, in charge of design and water front construction and maintenance. He became also engineer of the Spring Valley Water Company of San Francisco, in charge of the design of the water tower for the Calaveras Dam. He finally opened in 1914 his office at San Francisco, specializing in structural engineering. Saph suddenly passed away during an influenza epidemic.

Saph's name is known for a notable discussion submitted in collaboration with Ernest William Schoder (1879-1968) on the famous paper by Gardner Stewart Williams (1866-1931), C.W. Hubbell, and George Harrison Fenkell (1873-1949). Therein, the Pitot tube was used for discharge measurement in pipe flow, and its rating in a circular tank is discussed. The large number of discussers compared the current meter measurements with other recording methods. In 1903, the two discussers presented their own paper on pipe flow resistance based on a large experimental campaign. They first investigated the transition between laminar and turbulent flows, and then proposed an equation for the average velocity in terms of the hydraulic gradient and the hydraulic radius for a variety of pipe boundary material.

Anonymous (1921). Augustus V. Saph. *Trans. ASCE* 84: 905-906.
Saph, A.V., Schoder, E.W. (1902). Discussion of Experiments at Detroit MI, on the effect of curvature upon the flow of water in pipes. *Trans. ASCE* 47: 295-323.
Saph, A.V., Schoder, E.W. (1903). An experimental study of the resistances to the flow of water in pipes. *Trans. ASCE* 51: 253-330. (*P*)

SARGENT

29.09. 1872 Escanaba MI/USA
20.04. 1939 New York/USA

Joseph Andrews Sargent obtained the BSc degree in civil engineering in 1903 from the University of Nebraska, Lincoln NE, after having served during the Spanish-American War. He then joined the US Reclamation Service, where he was engaged on the design of the irrigation system of the Uncompahgre Project in Colorado. He returned in 1906 to Cuba as chief engineer, but came back to his country after the 1906 revolution. From 1909 to 1913 he had an engineering practice, dealing with irrigation projects in the West, followed by an engagement on the design of Medina Dam TX. In 1913 he was sent to Spain for a hydro-electric development near Barcelona, serving as chief engineer at *Seros* on Rio Segre and at *Tremp* on the Rio Noguera Pallaresa. The Seros development represented then the largest dam in Europe with a long, concrete-lined power canal.

After war service, Sargent was employed as consultant on hydraulic developments in Spain and Poland, returning to New York in 1920. His company obtained a contract with the Brazilian Government for the construction of five dams in the Ceara and Parahiba States, so that he was transferred in 1921, yet none of the dams was completed because of missing finances. In 1924 he made hydraulic studies for Poland, and in 1926 he had charge of irrigation exploration in Durango State, Mexico. In 1928 he joined a Brazilian hydro-power company, in charge of the power supply to *Curityba*. Although there were no flow records, no adequate maps, no roads in the heavily forested area, Sargent made his final selection at Rio Sao Joao, later known as *Chamine*. On completion of this work, he returned to New York in 1930. In 1936 he became supervising engineer of the Public Works Administration. Overall, Sargent's professional career was manifold. With a long and varied experience and a retentive memory, he was valuable in reporting on various phases of construction programs. He was described as being the product of an era having ended after World War I, when the use of the laboratory was replacing the judgment based on broad knowledge of engineering fundamentals. Sargent was member of the American Society of Civil Engineers ASCE since 1930.

Anonymous (1941). Joseph A. Sargent. *Trans. ASCE* 106: 1667-1871.
Sargent, J.A. (1910). Discussion of Water purification at Steelton PA. *Trans. ASCE* 66: 209-210.
Stevens, J.C. (1914). Seros Power Development, Lerida, Spain: Diversion dam and canals.
    Engineering News 72(Sep.03): 473-478; 72(Sep.10): 536-539.
http://www.rootsweb.ancestry.com/~nelancas/ww1/2.htm *P*

SAUNDERS

29.11. 1890 Washington DC/USA
11.11. 1961 Washington DC/USA

Harold Eugene Saunders obtained the BS degree in 1912 from the US Naval Academy, the MS degree in 1916 from Massachusetts Institute of Technology MIT, Cambridge MA, and the honorary ScD degree from Stevens Institute of Technology, Hoboken NJ. He was from 1912 commander ensign of the US Navy, advanced then through all grades to captain in 1939, and was from 1921 to 1924 in charge of submarine design at the Bureau of Construction and Repair, Washington DC, from where he proceeded to the US Navy Yard, Portsmouth NH. From 1929 to 1933 he was a model tester and engaged in full-scale ship trials at the Experimental Model Basin EMB, Washington DC, and was then in charge of the final designs at the Bureau of Construction and Repair until 1936. From 1937 to 1946 Saunders was technical director of the David Taylor Model Basin TMB, taking over as the director the following year. He was a member of the International Conference of Ship Tank Superintendents, and the American Towing Tank Conference. He was highly decorated with awards, including for instance the Legion of Merit, or the David W. Taylor Gold Medal.

In addition to the extensive ship model shops at EMB, its principal features were the three towing tanks, including one shallow of length, width and depth of 90×15×3 m, one of 300×15×7 m, and one for high speed of 340×6×3 m. The second and the third were so long that their tracks were but curved with the earth's surface. In 1940 the 12-in. and the 27-in. water tunnels for propeller testing, built at EMB respectively in 1930 and 1937, were moved to the new establishment TMB. Next, plans were initiated for a huge recirculating channel permitting models to be held stationary in water flow. A particular problem was the effect of free surface slope on the measurement of model drag, and possible means of minimizing this error. It was found that proper manipulation of flow depth and bed slope produce a level surface in the model vicinity.

Anonymous (1968). Saunders, Harold E. *Who was who in America* 4: 829. Marquis: Chicago.
Rouse, H. (1976). Harold E. Saunders. *Hydraulics in the United States* 1776-1976: 145-146. IIHR: Iowa. *P*
Saunders, H.E. (1933). *The prediction of speed and power of ships* by methods in use at the US Experimental Model Basin, Washington. Navy Bureau of Construction and Repair. Government Printing Office: Washington DC.
Saunders, H.E. (1957). *Hydrodynamics in ship design*. SNAME: New York.

SAVAGE H.N.

06.10. 1861 Lancaster NH/USA
24.06. 1934 San Diego CA/USA

Hiram Newton Savage graduated as a civil engineer from the New Hampshire College of Agriculture and Mechanic Arts, Durham NH, in 1887, and in 1891 from Thayer School of Civil Engineering, Dartmouth College, Hanover NH, from where he also obtained the DSc degree in 1913. From 1888 Savage was an engineer with the Nashville & Tellico Railways Co. From 1889 to 1890, he was a chief engineer of the Hydraulic Mining & Irrigation Co., New Mexico. Other engineering works were made for the White River Junction in Vermont, and for the sewerage system of West Randolph VT. From 1891 to 1903 Savage was a chief engineer for the Land & Town Company, San Diego CA, and in charge of maintenance and operation of the Sweetwater Dam, San Diego CA. In parallel he was a consulting engineer to the US Government for the jetty at the entrance to San Diego Bay. He was a member ASCE.

In 1903 Savage joined the US Reclamation Service, acting as supervising engineer its Northern Division. From 1904 to 1915 he was in charge of construction, operation, and maintenance of twelve reclamation projects. After retirement he was first consultant at Billings MO, and from 1916 at San Diego CA, thereby enlarging the Sweetwater Dam, and being the hydraulic city engineer from 1917. During this period falls also the design and construction of the Lower *Otay* Masonry Dam. Further works related to the Burrell Masonry Dam, and the Moring Dam. Savage made a world trip to inspect foreign dam designs in 1923. Two papers report on the works of the Sweetwater Dam in Southern California, a bold arch dam whose original construction started in 1886, with James D. Schuyler (1848-1912) then as designer. The 30 m high and 100 m long dam structure was made of uncoursed rubble masonry. It created a reservoir more than 6 km long. The third design included a siphon spillway, a side channel spillway allowing for a flood of 1200 m³/s to pass. In the tailwater, a stilling pool controls scour and energy dissipation.

Anonymous (1929). Savage, Hiram Newton. *Who's who in America* 15: 1835. Marquis: Chicago.
Anonymous (1938). Hiram Newton Savage. *Trans. ASCE* 103: 1873-1877.
Anonymous (1946). H.N. Savage, Reclamation Conference 1907. *Civil Engineering* 16(4): 188. *P*
Savage, H.N. (1895). Flood overflow at the Sweetwater Dam. *Engineering News* 34(Aug.15): 111-112.
Savage, H.N. (1919). Sweetwater Dam enlarged for the third time. *Engineering News-Record* 82(May15): 948-952.

SAVAGE J.L.

25.12. 1879 Cooksville WI/USA
28.12. 1967 Englewood CO/USA

John Lucian Savage obtained the BS degree in 1903 and the D.Sc degree in 1934 from the University of Wisconsin, Madison WI, and the D.Sc degree from University of Denver in 1946. He was until 1908 assistant engineer of the Idaho Division, Reclamation Service, predecessor organisation of the Bureau of Reclamation USBR, was then until 1916 associated with an irrigation consultant, and until 1924 design engineer for USBR, when taking over until 1945 as chief engineer in charge of all USBR designs, then becoming consultant for Tennessee Valley Authority TVA until retirement. Savage was awarded the 1937 Gold Medal from the Engineering Council, the 1945 John Fritz Medal, and the 1950 Gold Medal award from the Department of Interior, among many others. He was the American vice-president of the International Commission of Large Dams ICOLD, and a Honorary Member of the American Society of Civil Engineers ASCE.

Savage was an internationally known engineer and designer of more than forty major dams, including Hoover (Boulder) Dam on Colorado River, and Grand Coulee Dam on Columbia River. Known as 'engineer to the world' and the 'billion dollar engineer' during his twenty-one years of service as chief design engineer for USBR, he was described as the bureau's most distinguished alumnus upon retirement in 1945. He was also the designer of notable water resources projects such as California's Shasta, Parker and Imperial Dams and its All American Canal. Savage was further a private consultant to at least 20 foreign nations for dam projects such as the 280 m high *Grande Dixence* Dam in Switzerland, or dams in India, the Middle East, Mexico, Spain, and Australia.

Anonymous (1937). John L. Savage. *Civil Engineering* 7(2): 154. *P*
Anonymous (1937). John L. Savage. *Engineering News-Record* 118(Mar.04): 350. *P*
Anonymous (1959). Savage, John Lucian. *Who's who in America* 28: 2345. Marquis: Chicago.
Anonymous (1969). John Lucian Savage. *Trans. ASCE* 134: 968.
Munn, J., Savage, J.L. (1922). The flood of June 1921, in the Arkansas River, at Pueblo, Colorado. *Trans. ASCE* 85: 1-65.
Savage, J.L. (1928). Design of the Owyhee Irrigation Dam. *Engineering News-Record* 100(17): 663-667.
Savage, J.L., Houk, I.E. (1928). Tests on models at Boulder. *Proc. ASCE* 54(5): 215-218.
Savage, J.L. (1937). The Boulder Dam. *Journal of the Institution of Civil Engineers* 6: 161-180.
Wolman, A., Lyles, W.H. (1978). Savage. *Biographical memoirs* 49: 225-238. *P*

SAVILLE C.M.

27.05. 1865 Melrose MA/USA
14.02. 1960 Hartford CT/USA

Caleb Mills Saville, father of Thorndike (1892-1969), graduated in 1889 from the Harvard University, Cambridge MA, and continued during one year with post-graduate works at Lawrence Scientific School of Harvard University. He was Division engineer of the Metropolitan Water Board, Boston MA, from 1895 to 1905, engineer of the Isthmian Canal until 1912, conducting investigations on foundations of Gatun Dam, resulting in dam construction at the site proposed. He further investigated the hydrology and meteorology of Panama Canal. Upon returning to the USA he was manager and chief engineer of the Bureau of Water of the Metropolitan District, Hartford CT, installing and operating the new city water supply. Since 1948, Saville was a consultant at Hartford. He also was a member of the Commission on Regional Planning for Connecticut State.

Saville was a noted water supply authority of the United States, who had been active mainly in the Panama Canal Zone, and at Hartford CT, managing its Metropolitan District Water Commission during sixteen years. His 1935 paper describes the third enlargement of the water supply scheme, including a notable filtration plant. His 1937 paper compares the designs of earth dams erected in the 1910s and 1930s, with the differences mainly in design, construction methods, and cost. The conservative New England style includes earth embankments with concrete masonry core-walls built into base rock, a thick vertical section of impermeable material against the core-wall on the water side, a bearing layer of homogeneous material on the reverse side, banks of permeable material with flat slopes for weight and protection to the cutoff walls on both sides, and well-placed rip-rap on the water face. Saville was awarded the 1914 Norman Medal from the American Society of Civil Engineers ASCE for a paper on Panama Canal, the 1931 President's Premium from the Institution of Water Engineers, London UK, and in 1945 honorary membership from New England's Water Works Association.

Anonymous (1951). Saville, Caleb Mills. *Who's who in America* 26: 2411. Marquis: Chicago.
Anonymous (1960). Caleb M. Saville. *Water Works Engineering* 113(3): 229.
Saville, C.M. (1935). Third enlargement in 83 years under way in Hartford. *Engineering News-Record* 115(Sep.12): 351-356.
Saville, C.M. (1936). The underground water-index. *Trans. AGU* 17(2): 382-386.
Saville, C.M. (1937). Earth dams: Yesterday and today. *Water Works Engineering* 90(Sep.15): 1301-1307. *P*

SAVILLE T.

03.10. 1892 Malden MA/USA
21.02. 1969 Gainesville FL/USA

Thorndike Saville, son of Caleb Mills (1865-1960), graduated from Harvard University, Cambridge MA, with the MS degree in 1917, receiving honorary DE degrees from Syracuse University in 1951, and from New York University in 1957. He was then associate professor from 1919 and later professor of hydraulic and sanitary engineering at University of New York until his retirement in 1957. Saville further acted as Dean, the College of Engineering of his University, and served at University of Florida, Gainesville FL, as director of science and engineering studies from 1958 to 1960. In parallel he was chief engineer, the New York City Department of Conservation and Development. He was in 1961 elected Honorary Member of the American Society of Civil Engineers ASCE.

Saville was an outstanding leader in the advancement of the engineering profession and of engineering education in general. His academic programs have raised the fields of sanitary science, meteorology, oceanography, and engineering physics to a major status. Both as an engineer and a citizen, he has devoted a lifetime to the conservation and judicious use of the natural resources, particularly its water reserves. His early work in the statistical analysis of river flow have proved valuable in flood control and river pollution projects. Later, he described the power situation in South America including future possibilities of hydroelectric developments. In 1936, he published an inventory of the US water resources. He further conducted studies in stream gaging, beach erosion, underground waters, rainfall, floods, evaporation, and siltation. He was Director ASCE from 1945 to 1948 and awarded the 75th Anniversary Medal of the American Society of Mechanical Engineers ASME in 1955.

Anonymous (1937). Thorndike Saville. *Water Works Engineering* 90(May 26): 776. *P*
Anonymous (1942). Thorndike Saville. *Engineering News-Record* 128(Mar.5): 371. *P*
Anonymous (1948). Saville family established new record in ASCE. *Water Works Engineering* 101(11): 1037. *P*
Anonymous (1956). Thorndike Saville. *Engineering News-Record* 156(Mar.22): 67-72. *P*
Anonymous (1963). Saville, Thorndike. *Who's who in America* 32: 2734. Marquis: Chicago.
Saville, T. (1916). Rainfall data interpreted by laws of probability. *Engineering News* 76(26): 1208-1211.
Saville, T., Watson, J.D. (1933). An investigation of the flow-duration characteristics of North Carolina streams. *Trans. AGU* 14(1): 406-425.

SAXTON

22.03. 1799 Huntington PA/USA
26.10. 1873 Washington DC/USA

Joseph Saxton was a famous American inventor. He was a watchmaker at Philadelphia PA from 1817 to 1828, made then until 1837 the clock for the Belfry of Independence Hall in England, having invented in the meantime a magneto-electric machine as also a fountain pen, and a locomotive differential pulley. From 1838 to 1843 he was constructor and curator of the standard weighing apparatus of the US Mint, Philadelphia PA. He designed the standard balance used in government assays and coining offices and also was until his death the superintendent of weights and measures for the US Coast Survey, Washington DC. He further had invented a fusible metal seal and an ever-sharp pencil. Saxton was a member of the National Academy of Sciences, and the American Philosophical Society.

Saxton was particularly known for the current meter built and rated in 1832 at Adelaide Gallery flume, London UK. These meters provide the local river velocity and finally the river discharge at a certain section. Crude current meters resembling paddle wheels were invented centuries ago, but the meters received new importance in the 19th century because of scientific interest in river flow, and the associated hydraulic knowledge. The current meter was first designed in the USA by Saxton, based on the proposals of the German Reinhard Woltman (1757-1837). His apparatus appears like a cross between a weather vane and an anemometer, but it was not well-suited for application in rivers. Under supervision of Andrew A. Humphreys (1810-1883), a new double-float system was developed to measure river currents, yet also this method was erroneous. In the 1870s Theodore G. Ellis (1829-1883), among others, considerably improved the early meters. Saxton also invented in 1854 a self-registering tide gauge that is believed to have been the first gauge to record an earthquake. His papers and notebooks contain detailed information on his experiments and also the social life in London.

Anonymous (1963). Joseph Saxton. *Who was who in America* 1607-1896: 466. Marquis: Chicago.
Henry, J., Saxton Pendleton, J. (1935). *Joseph Saxton* 1799-1873. Pengelly: Reading PA. *P*
Kolupaila, S. (1961). Saxton. *Journal of the Hydraulics Division* ASCE 87(HY3): 178. *P*
Saxton, J. (1873). *Joseph Saxton Papers*. Record Unit 7056. Smithsonian Institution Archives: Washington DC.
http://museum.nist.gov/object.asp?ObjID=29 *P*
http://www.usace.army.mil/About/History/HistoricalVignettes/CivilEngineering/013InCommon.aspx

SAYRE A.N.

28.01. 1901 Granville OH/USA
12.10. 1967 Zürich/CH

Albert Nelson Sayre graduated in 1921 from Doane Academy, Granville OH, obtaining there the BS degree from Denison University in 1923, the MS degree from Kansas University, Lawrence KA, and the PhD degree in 1928 from University of Chicago, Chicago IL. He was assistant instructor in geology at Kansas University until 1924, from 1926 to 1929 instructor in geology at University of Pennsylvania, Philadelphia PA, from 1946 to 1959 from assistant geologist to chief geologist of the US Geological Survey USGS, acting as chief of its Groundwater Branch. He was in parallel from 1944 to 1945 water supply consultant for the US Army. He was awarded the Medal of Freedom for work in Leyte, the Philippines, after World War II, and obtained the degree D.Sc. from Denison University in 1949. He was Fellow of the Geological Society of America, and of the American Geophysical Union AGU, whose general secretary he was from 1956.

Sayre contributed to the science of hydrogeology, of which the foundations were laid in the mids of the 20th century. In the absence of infiltration, water flow would be reduced to the precipitation, overland runoff, and evapotranspiration. Subsurface geology would then have a small effect on hydrology and water resources, and the science of hydrogeology would not have been founded. Whereas Oscar E. Meinzer (1876-1948) may be considered the father of hydrogeology, at least in the USA, Sayre identified the origins and rise of qualitative groundwater hydrology. He was in 1967 stricken in a heart attack at Zürich, and died after his extended trip through Europe, Asia Minor, and Africa.

Anonymous (1963). Sayre, A. Nelson. *Who's who in America* 32: 2737. Marquis: Chicago.
Anonymous (1968). Dr. Nelson Sayre. *IAHS Bulletin* 13(1): 112-113.
Bennett, R.R., Sayre, A.N. (1962). *Geology and groundwater resources of Kinney County TX.* US Geological Survey: Washington DC.
Sayre, A.N. (1937). Geology and groundwater resources of Duval County TX. *Water Supply Paper* 776. USGS: Washington DC.
Sayre, A.N. (1942). Relation of ground water and surface runoff to total precipitation in the East Nueces River basin. *Water Resources Bulletin* (May 10): 100-102.
Sayre, A.N. (1954). *Water levels and artesian pressures in observation wells in the United States.* US Government Printing Office: Washington DC.
Smith, W.E. (1968). A. Nelson Sayre. *Trans. AGU* 49(2): 445-447. *P*
Smith, W.O., Sayre, A.N. (1964). *Turbulence in groundwater flow.* GPO: Washington DC.

SAYRE W.W.

31.08. 1927 Monroe County NY/USA
06.10. 1981 Lakewood CO/USA

William Whitaker Sayre received in 1949 his BS degree from Princeton University, Princeton NJ, and completed his graduate studies, leading to the PhD degree at Colorado State University, Fort Collins CO, in 1957. He spent his early career as a civil engineer with the US Bureau of Reclamation USBR, joining the US Geological Survey USGS in 1962, staying there until being appointed in 1968 to professor at University of Iowa, Iowa City IA. He also served as director of its Iowa Institute of Hydraulic Research IIHR until returning in 1980 to the USGS. He was the recipient of the 1966 Walter L. Huber Research Award, the 1967 Lorenz G. Straub Award, and the 1970 J.C. Stevens Award from the American Society of Civil Engineers ASCE.

Sayre first joined as hydraulic engineer USGS studying mixing processes for dissolved and particulate matter in open channel flow. Once having joined IIHR he did notable research on mixing processes in meandering channels. Other research efforts included boundary roughness and its spacing effect on open channel flow, and radioactive tracer techniques to investigate sediment transport and dispersion in rivers. He is remembered for his warmth and humanity. He was active in organizations which promoted world peace through non-violent means. His interactions with others was characterized by kindness and consideration. He was a gentle and compassionate man whose belief in peace was so strong that all who had the good fortune to know him found their own lives to be tranquil and harmonious. He died of a heart attack at age of only 54.

Anonymous (1970). William W. Sayre. *Civil Engineering* 40(10): 80-81. *P*

Anonymous (1982). Prof. W.W. Sayre. *Journal of Hydraulic Research* 20(1): 85-87. *P*

Jobson, H.E., Sayre, W.W. (1970). Predicting concentration profiles in open channels. *Journal of the Hydraulics Division* ASCE 96(HY10): 1983-1996; 97(HY5): 754-756; 97(HY12): 2087-2088.

Sayre, W.W., Albertson, M.L. (1963). Roughness spacing in rigid open channels. *Trans. ASCE* 128(1): 343-372; 128(1): 411-427.

Sayre, W.W., Chang, F.M. (1968). A laboratory investigation of open-channel dispersion processes for dissolved, suspended and floating dispersants. *Professional Paper* 433. US Geological Survey: Washington DC.

Yotsukura, N., Sayre, W.W. (1976). Transverse mixing in natural channels. *Water Resources Research* 12(4): 695-704.

SCHIEBER

28.08. 1888 Bucyrus OH/USA
23.06. 1944 San Marino CA/USA

Oliver Jay Schieber graduated from the University of Wisconsin, Madison WI, in 1912 as civil engineer. In 1913 he was engaged at Boston MA, as chief of party on the construction of two high-head hydro-electric power plants in the Sierra Nevada CA. He was there involved in field engineering of power-house and penstock construction. From 1915 to 1917 he was engaged on the Salt River Project of the US Reclamation Service at Phoenix AZ, with charge of the location and construction of irrigation canals and structures, and with preliminary works of Roosevelt Dam. After war service, Schieber returned to the US Reclamation Service, in charge of the King Hill Project in Idaho, as assistant engineer on the design of the irrigation structures. He accepted a position on the construction of the Big Creek hydro-electric development in 1920, demonstrating again his willingness to accept responsibility, and his capacity for hard work. The major work there included driving 40 km of hard-rock tunnel, the building of two concrete dams, the construction of three hydroelectric power stations, and the installation of power-generating equipment.

Schieber was in 1925 appointed technical assistant at Los Angeles CA, engaged on the layouts for construction plants, and arranging for construction material of another dam design. In 1931 Los Angeles had joined with other communities to form the Metropolitan Water District for constructing an aqueduct to bring drinking water from the Colorado River to Southern California, with Frank E. Weymouth (1874-1941) as chief engineer, and Schieber as his senior engineer. The almost 400 km route of the aqueduct is in mountainous and desert country. Schieber designed the district camps, the water supply facilities, and then superintended their construction in the desert. He acted as resident engineer then in charge of the aqueduct, including a 14 km 4.5 m diameter hard-rock tunnel, concrete conduits, and a long siphon. By 1938, when the aqueduct was nearly completed, Schieber served as construction superintendent of the Federal Works Project Administration in Southern California. He had charge of 1,500 men, engaged on the construction of a 3.5 m storm drain tunnel through Los Angeles, which is an integral part of the Arroyo Seco Parkway, connecting LA with Pasadena CA. He was involved also in a study of the hydrology of the entire Big Creek hydroelectric system. Schieber was ASCE member from 1933.

Anonymous (1913). O.J. Schieber. *Badger yearbook*: 509. University of Wisconsin: Madison. *P*
Anonymous (1947). Oliver J. Schieber. *Trans. ASCE* 112: 1512-1517.

SCHLEY

23.02. 1880 Savannah GA/USA
29.03. 1965 Washington DC/USA

Julian Larcombe Schley graduated in 1903 from the US Military Academy, West Point NY, then being on duty with engineering troops on the Philippine Islands. He returned in 1904 to the USA, and from 1905 to 1909 stayed in Cuba. He was then until 1912 instructor at his Alma Mater, and executive officer at Washington DC until 1916. He was in charge of the US Engineering District at New Orleans LA, for the improvement of rivers and harbours, including the Mississippi River mouths. After his war service until 1921 he returned to the US Engineer District at Nashville TN, where he improved and maintained Cumberland River, and Tennessee River downstream of Wilson Dam. From 1924 to 1928 he was in charge of the US Engineer District at Galveston TX, with works similar as previously along the Texas Gulf coast. He also served as board member to study and recommend spillways below the Red River to relieve floods on the Mississippi River. Until 1932 Schley was engineer on the Panama Canal, and until 1936 then Governor of Panama Canal. Upon return to the USA he was until 1941 major general, Chief of US Army Engineers. He partly continued work after 1941 on duty with the Office of Inter-American Affairs.

In choosing Schley as Chief of Engineers, President Roosevelt selected an officer who was familiar with the flood control problem on the Mississippi River, and with similar regions in the USA. He was for instance continuously on the Mississippi during the disastrous 1927 flood, having had the opportunity to observe that catastrophe at first hand. The 1938 paper describes that the large floods on Mississippi River along the lower river reach were successfully controlled and have demonstrated the advantage of cutoffs. The execution of the Houston Ship Canal was also under the personal direction of Schley, and he had built the first jetties at Port Isabel TX. He was member of the American Society of Civil Engineers ASCE, among other societies.

Anonymous (1937). Schley is appointed Chief of Engineers. *Engineering News-Record* 119(Sep.23): 409. *P*
Anonymous (1941). Retiring Chief of Engineers. *Engineering News-Record* 127(Oct.2): 465. *P*
Anonymous (1948). Schley, Julian L. *Who's who in engineering* 6: 1744. Lewis: New York.
Schley, J.L. (1938). Mississippi control works tested by flood. *Engineering News-Record* 120(Apr.14): 533-537.
http://de.wikipedia.org/wiki/Julian_Larcombe_Schley *P*

SCHMIDT, Jr., L.A.

16.11. 1900 Wauwautosa WI/USA
17.07. 1978 Chattanooga TN/USA

Lewis Adelbert Schmidt, Jr., obtained in 1923 the BSc degree from University of Wisconsin, Madison WI, and the MS degree from the University of Texas, Austin TX, in 1933. Between 1923 and 1936 he had acquired a broad experience in engineering. In 1936 he came to the Tennessee Valley Authority TVA at Knoxville TN, to work on the design of construction plants for Chickamauga, Guntersville, and Pickwick Dams. He received the civil engineering degree from University of Wisconsin in 1939, and from 1941 he became acting project engineer of the Hales Bar Dam below Chattanooga TN. In 1944 he formed the Schmidt Engineering Company, doing a wide variety of engineering work, including dams, water supply and wastewater disposal or highway plans. In 1963 he joined forces with a colleague, who did similar engineering work, forming therefore Hensley-Schmidt Consultants, doing planning, the design, and the management of projects. In the late 1970s they had offices at Atlanta GA, Charlotte NC, Jackson MS, Montgomery AL, Orlando FL, and Knoxville TN, in addition to Chattanooga.

Schmidt joined in 1930 the American Society of Civil Engineers ASCE, becoming in 1942 member, and in 1959 ASCE Fellow. He was secretary-treasurer of its Tennessee Valley Section, and president of both its Chattanooga and Tennessee Valley Sections. He was active on ASCE committees. He was in addition a member of the Permanent International Association of Navigation Congresses PIANC, the Society of American Military Engineers SAME, the American Water Works Association AWWA, the Water Pollution Control Federation WPCF, and the American Public Works Association APWA. He was awarded the 1946 ASCE James Laurie Prize for his 1943 paper. He authored a number of technical papers focusing on reservoirs and dams. He also authored the autobiography published in 1977.

Anonymous (1947). L.A. Schmidt, Jr.: James Laurie Prize. *Civil Engineering* 17(1): 41. *P*
Anonymous (1980). Lewis A. Schmidt. *Trans. ASCE* 145: 760-761.
Schmidt, L.A. (1939). Norris Dam construction. *CE Thesis*. University of Wisconsin: Madison.
Schmidt, L.A. (1943). Flowing water in underground channels: Hales Bar Dam. *Proc. ASCE* 69: 1417-1446.
Schmidt, L.A. (1977). *The engine ear*: 50 years in engineering. University of Wisconsin: Madison.
http://digicoll.library.wisc.edu/cgi-bin/UW/UW-idx?type=div&did=UW.WIEv51no5.RZirbel&isize=M *P*

SCHODER

17.08. 1879 Dewey, Fidalgo Island WA/USA
16.05. 1968 Ithaca NY/USA

Ernest William Schoder obtained the BS degree in 1900 in mining engineering from the University of Washington, Seattle WA, and the PhD degree from Cornell University, Ithaca NY, in 1903. He was in charge of the Hydraulic Laboratory of its latter School of Engineering in 1904, was until 1919 there then professor of experimental hydraulics, when taking over until retirement in 1947 as professor of hydraulics. Schoder was a Fellow of the American Association of the Advancement of Science AAAS, and the American Society of Civil Engineers ASCE.

Schoder was a notable hydraulic engineer and educator, known particularly for an early paper on the resistance of bend flow, and for his book written in collaboration with Francis M. Dawson (1889-1963). Until around 1900 few research was available on bend flow, a hydraulic element which may generate a considerable hydraulic loss, depending on its geometry, the flow regime and the pipe roughness. It was already observed in the 19th century that bends generate a spiral flow pattern as a secondary motion, which is superimposed to the primary curved flow pattern. This particular flow feature therefore complicates experimentation. Schoder's observations mainly relate to the headloss in the fully turbulent regime, which is even currently not fully assessed, given the many parameters influencing such flow. The second important paper relates to precise weir discharge measurement, then an important issue of many hydraulic laboratories. The corresponding paper of Theodor Rehbock (1864-1950) was also published in 1929, and the latter even discussed the paper of the two Americans. The long discussion section demonstrates the interest in this research field, resulting in an equation that was used for decades. The 1927 book may still be considered an outstanding and one of the earliest American examples in experimental hydraulics.

Anonymous (1942). Schoder, Ernest W. Proc. 2nd *Hydraulics Conference* State University of Iowa: Frontispiece. Iowa City IA. *P*

Anonymous (1963). Schoder, Ernest W. *Who's who in America* 32: 2761. Marquis: Chicago.

Gregory, W.B., Schoder, E.W. (1908). Some Pitot tube studies. *Trans. ASME* 30: 351-372.

Schoder, E.W. (1909). Curve resistance in water pipes. *Trans. ASCE* 62: 67-112.

Schoder, E.W. (1912). A method of plotting river stage-discharge data. *Engineering Record* 66(5): 138.

Schoder, E.W., Dawson, F.M. (1927). *Hydraulics*. McGraw-Hill: New York.

Schoder, E.W., Turner, K.B. (1929). Precise weir measurements. *Trans. ASCE* 93: 999-1190.

SCHOENBERGER

17.01. 1875 Buras LA/USA
24.09. 1928 New Orleans LA/USA

George Christian Schoenberger was born in the Plaquemines Parish of Louisiana State. He graduated from Louisiana State University, Baton Rouge LA, in 1898 with the BSc degree in civil and mechanical engineering. He then joined the State Board of Engineers, Louisiana, whose chief engineer he was later. He began his professional career as rodman, and was promoted to levee inspector on the Red River. He then accepted a position with the Houston Irrigation Company as assistant engineer for South-western Louisiana, and also was engaged with the Mississippi Construction Company. In 1902 he was appointed junior engineer with the US Engineer Office, New Orleans LA, in charge of levee construction. From then his work was exclusively related to river control, because he had realized that this was to be his life work, and that a broad training embodying all its branches was highly desirable.

Schoenberger was engaged from 1908 by the Mississippi River Commission at New Orleans, where he stayed until 1917, when being appointed member of the Board of State Engineers. During the high-water periods of the following decade, much of the responsibility for keeping the levee lines intact in the Louisiana and *Tensas* Districts was largely due to his management in rescue work after the break at Fagette LA, so that no lives were lost. From 1917 he was a member of the State Board of Engineers, and from 1925 he was chief engineer. The 1927 record flood of Mississippi River brought forth his qualities, upon his shoulders fell a large share for the protection of the levees and the property of thousands in the Lower Mississippi Valley. He took a leading part in the cutting of the Caernarvon Levee below New Orleans, creating the artificial crevasse to relieve the danger of a break at the city above. Supervising the stability of the levees in the upper section occupied all his attention, and he fought to the last ditch for every foot of levee. Only after everything possible was completed several disastrous crevasses occurred, yet this disaster opened the eyes for an adequate flood control. It was too late however, because Schoenberger died from a heart attack during the 1928 flood.

Anonymous (1929). George C. Schoenberger. *Trans. ASCE* 93: 1955-1957.
Barry, J.M. (1997). George C. Schoenberger. *Rising tide*: The great Mississippi flood of 1927 and how it changed America. Touchstone: New York. *P*
Kelman, A. (2003). *A river and its cities*: The nature of landscape in New Orleans. Berkeley CA.
http://www.zimbio.com/The+Mississippi+River/articles/QTp6mXs11GI/1927+Mississippi+River+Flood (*P*)

SCHORER

24.06. 1892 Wangen a. Aare/CH
20.12. 1950 Valhalla NY/USA

Herman (Hermann) Schorer studied at the Swiss Federal Institute of Technology, ETH Zurich, where he received the civil engineering degree in 1914. He was then engaged in hydro-electric surveys, field work, the hydropower station Lonza, Gampel VS, and designs with Schweizerische Wasserwirtschaft in Berne. He moved in 1920 to the USA, holding various positions in New York and Rochester NY, becoming then hydraulic engineer at Fresno and San Francisco CA. From 1926 to 1932 he was engaged in dam designs which allowed a full development of his talents, serving first as a hydraulic engineer and then as chief engineer for consultants at San Francisco. He prepared reports and designs for various hydroelectric and water supply projects, including the *Glines* Canyon arch dam and appurtenant structures near Port Angeles WA, or the *Guadaloupe* River Power Project near Medellin in Colombia.

For a short period Schorer was engineer of the US Bureau of Reclamation, Denver CO, becoming in 1934 manager and president of a tank corporation in New York NY, a position he held until his premature death. He there directed the design and construction of numerous concrete tank installations. Later, he formed a separate company for the utilisation of his patents for pre-stressed reinforced concrete structures. He published a number of papers in the Transactions ASCE, which were considered authoritative, becoming standards for engineering practice. He also published outstanding papers on general principles and special features of pre-stresses concrete structures. In addition to complete and original mastery of the theory, he demonstrated conspicuous ability in directing the design and construction activities, and the business management of large engineering firms. He was remembered for his loyalty to friends and for fairness and consideration in all his dealings with those who worked with him and for him. He was member of the *Gesellschaft Ehemaliger Polytechniker* GEP of ETH Zurich, and of the American Society of Civil Engineers ASCE from 1942.

Anonymous (1952). Herman Schorer. *Trans. ASCE* 117: 1319-1320.
Anonymous (1987). Hermann Schorer. Wissenschaftshistorische Sammlung: ETH Zürich.
Schorer, H. (1924). Discussion of Improved type of multiple-arch dam. *Trans. ASCE* 87: 388.
Schorer, H. (1932). The buttressed dam of uniform strength. *Trans. ASCE* 96: 681-683.
Schorer, H. (1933). Design of large pipe lines. *Trans. ASCE* 98: 101-119; 98: 178-191. *P*
USBR (1949). Welded steel pipes. *Engineering monograph* 3. US GPO: Washington. (*P*)

SCHUBAUER

07.07. 1904 Mechanicsburg PA/USA
24.11. 1992 Bethesda MD/USA

Galen Brandt Schubauer received the AB degree in 1928 from Pennsylvania State University, University Park PA, the MS in 1930 from California Institute of Technology, Pasadena CA, and the PhD degree in 1934 from Johns Hopkins University, Baltimore MD. He became in 1946 chief, the Aerodynamics Section of the National Bureau of Standards NBS, having previously been in its Electrical Division from 1929 to 1930. He and Harold K. Skramstad (1908-2000) were the winners of the 1948 S.A. Reed Award 'for contributions to the understanding of the mechanism of laminar to turbulent flow transition'. Schubauer became Fellow of the Aeronautical Society in 1949, and recipient of the Gold Medal for aerodynamic research in 1956, as the chief of the NBS Fluid Mechanics Section, in recognition of 'outstanding contributions to basic aerodynamics over the past 20 years'.

Schubauer's work on turbulence and airflow, and the development of instruments for measuring these phenomena, was vital to the development of the modern high-speed aircraft. In the 1950s, he studied the accuracy of the hot-wire anemometer at speeds up to twice the speed of sound. This instrument was previously a basis in aerodynamic research at subsonic speeds, but it was not known whether it could also be used at supersonic speeds. Schubauer was elected to the National Academy of Engineering in 1980 for the 'discovery of self-exited oscillations in laminar boundary layers, giving a new direction for further inquiry into the origin of turbulent flows'. He furthermore was the recipient of the 1988 Fluid Dynamics Prize from the American Physical Society APS.

Anonymous (1938). Schubauer, G.B. Proc. 5th Intl. *IUTAM Congress* Cambridge MA. *P*
Anonymous (1948). Galen B. Schubauer. *Aeronautical Engineering Review* 7(2): 7. *P*
Anonymous (1956). Schubauer wins Gold Medal. *Aeronautical Engineering Review* 15(7): 22. *P*
Schubauer, G.B., Skramstad, H.K. (1947). Laminar boundary-layer oscillations and stability of laminar flow. *Journal of the Aeronautical Sciences* 14(2): 69-78.
Schubauer, G.B. (1948). Aerodynamic characteristics of damping screens. Proc. 7th Intl. *IUTAM Congress* London 2(1): 620-622.
Schubauer, G.B. (1954). Turbulent processes as observed in boundary layer and pipe. *Journal of Applied Physics* 25(2): 188-196.
Schubauer, G.B., Tchen, C.M. (1959). *Turbulent flow*. Princeton University Press: Princeton.
http://nvl.nist.gov/pub/nistpubs/sp958-lide/html/046-048.html

SCHULEEN E.P.

25.10. 1897 Sioux City IA/USA
15.03. 1975 Allegheny PA/USA

Emil Philip Schuleen obtained the BS degree from State University of Iowa, Iowa City IA in 1926, and there the MS degree in engineering in 1927. He was then a hydrological assistant involved in the hydropower plant operation and maintenance at Keokuk IA for the Mississippi River Power Company until 1929, becoming until 1937 an associate engineer for flood control investigations and model tests. Until 1942 he was a senior engineer for investigations on navigation, flood control, hydro-electric power, and pollution abatement for the Upper Ohio River Basin. He was later concerned with the maintenance and operation of eight flood control and multi-purpose reservoirs until 1945, from when he prepared hydrologic studies, reservoir operation reports, hydraulic research and design for a variety of schemes. He was in charge of all engineering investigations, planning and design for navigation, flood control and multi-purpose projects of the Pittsburgh District of the US Corps of Engineers since 1951.

Schuleen was a typical hydraulic engineer during the Golden Years of hydropower development. He published few papers but submitted discussions to research journals of which he was interested, including on questions in hydrology such as the concept of the maximum probable flood, or model tests of hydraulic structures, or even the rainfall characteristics in the South of the USA. He published with a colleague a notable and one of the very first papers on cavitation damage in hydraulic engineering during the Cavitation Symposium in the early 1940s, for which he was awarded the 1943 Karl Emil Hilgard Prize from the American Society of Civil Engineers ASCE.

Anonymous (1926). Schuleen, E.P. *Hawkeye yearbook*: 467. University of Iowa: Iowa. *P*
Anonymous (1942). Schuleen, Emil P. Proc. 2nd *Hydraulics Conference* State University of Iowa: Frontispiece. Iowa City IA. *P*
Anonymous (1954). Schuleen, Emil P. *Who's who in engineering* 7: 2135. Lewis: New York.
Schuleen, E.P. (1941). Discussion of Maximum probable floods of Pennsylvania streams. *Trans. ASCE* 106: 1506-1508.
Schuleen, E.P. (1943). Discussion of Model tests on structures for hydro-electric developments. *Trans. ASCE* 108: 824-828.
Schuleen, E.P. (1956). Flood plain zoning as supplement to flood control. *Proc. ASCE* 82(954): 1-6.
Thomas, H.A., Schuleen, E.P. (1942). Discussion of Cavitation in outlet conduits of high dams. *Trans. ASCE* 107: 421-422.

SCHULEEN E.T.

29.10. 1902 Sioux City IA/USA
04.10. 1993 Lancaster PA/USA

Ernest Theodore Schuleen, the brother of Emil P. (1897-1975) obtained the BS degree from Iowa State University, Iowa City IA in 1926, and the MS degree in 1927. He was then until 1929 assistant hydraulic engineer of the West Pennsylvania Power Co., then another two years for the Aluminium Company of America, until joining the Pennsylvania Water & Power Co., Holtwood PA, as test engineer from 1931 to 1946, from when he was a hydrographic engineer until 1955. From then to retirement Schuleen served as assistant to the manager, the Safe Harbour Power Corp., Conestoga PA. He was a member of the American Society of Civil Engineers ASCE, the American Geophysical Union AGU, and the American Society of Professional Engineers ASPE.

Schuleen had research interests mainly in hydraulic turbines and hydrology, in a certain way two quite opposite fields of hydraulics, but in his case of almost equal relevance, given that turbines depend significantly on the load characteristics, the discharge and on the available head. Their effect under water flow depends solely from the product of velocity times discharge, indicating equal effects for low-head installations under large discharge but small velocity as high-head schemes with the opposite characteristics. Whereas his earlier works mainly dealt with the hydrologic descriptions of floods using various methods, he later advanced hydraulic engineering of his country by setting up a number of hydropower plants. He thereby was also concerned with aspects of reservoir sedimentation, as for Lake Clarke in Florida State.

Anonymous (1926). Schuleen, E.T. *Hawkeye yearbook*: 467. University of Iowa: Iowa. *P*
Anonymous (1942). Schuleen, Ernest T. Proc. 2nd *Hydraulics Conference* State University of Iowa: Frontispiece. Iowa City IA. *P*
Anonymous (1960). Schuleen, E.T. *American men of science* 10: 3609. Cattell: Tempe AZ.
Schuleen, E.T., Stewart, J.E. (1929). Flood predictions from storm paths, pre-flood river stages, precipitation data and peak river stages. *Monthly Weather Review* 57(5): 186-192.
Schuleen, E.T. (1940). Discussion of Flash-board pins. *Trans. ASCE* 105: 1447-1451.
Schuleen, E.T. (1949). Discussion of Conowingo hydroelectric plant. *Trans. ASCE* 114: 99-102.
Schuleen, E.T., Higgins, C.R. (1953). *Analysis of suspended sediment measurements for Lake Clarke, inflow and outflow*, 1948-53. Pennsylvania Power & Light Co.: Holtwood PA.
Schroyer, G.B., Schuleen, E.T. (1950). Discussion of Stream flow variability. *Trans. ASCE* 115: 1126-1130.

SCHULTZ

23.09. 1907 Brooklyn NY/USA
09.02. 1998 Denver CO/USA

Ernest Richard Schultz graduated from the Rutgers University, New Ark NJ, obtaining in 1930 the BSc degree in civil engineering. He then joined the US Bureau of Reclamation, Denver CO, staying there until his retirement in 1964, finally as Head of the Concrete Dams Section, the Division of Design. He worked on all phases of design and construction of various larger concrete dams both in the USA and abroad. Early work included the Hoover and the Grand Coulee Dams, which were the first large structures worldwide. When retiring, Schultz was awarded the Department of the Interior's highest honour, the Distinguished Service Award for 'outstanding contributions to engineering and the prestige of the Nation, the Department of the Interior, and the US Bureau of Reclamation'. He was ASCE Fellow and Life Member, and member of ICOLD.

The masterpiece of Schultz was the design of Glen Canyon Dam at Lake Powell on Colorado River, which received the 1963 Outstanding Civil Engineering Achievement Award from the American Society of Civil Engineers ASCE; this was the third time the Bureau had received this honour. Schultz visited in 1962 some of the Soviet Union's largest dams and hydro-electric installations as part of the 29th Executive Committee Meeting of the International Commission of Large Dams ICOLD. In 1963, he attended the Symposium on Concrete Dam Models, the National Civil Engineering Laboratories LNEC, Lisbon P. While there he was asked to inspect the *Vaiont* Dam Disaster in Italy, where a landslide had caused a flood killing some 2,000 people. From then, Schultz was used to rush off to survey disasters in dam engineering. He was asked in 1964 to survey three dams having overtopped during a flood near Glacier National Park MT. He also attended in 1963 the International Symposium on the Theory of Arch Dams in London UK, and the 8th ICOLD Congress in Scotland.

Schultz, E.R. (1947). *Preliminary flood routing studies for the Yangtze Gorge Dam*. MS Thesis. University of Colorado: Denver CO.
Schultz, E.R. (1951). *Final design report*: Angostura Dam. University of Colorado: Denver CO.
Schultz, E.R. (1961). *Study of a curved dam in a wide valley*. USBR: Denver CO.
Schultz, E.R. (1962). *Report* on participation in 29th executive committee meeting, ICOLD, Moscow. USBR: Denver CO.
Schultz, E.R. (1963). Design features of Glen Canyon Dam. *Trans. ASCE* 128(4): 236-265.
http://www.colorado.edu/engineering/deaa/cgi-bin/display.pl?id=170 *P*

SCHUMM

22.02. 1927 Kearney NJ/USA
10.04. 2011 Fort Collins CO/USA

Following service in the Navy during WWII Stanley Alfred Schumm received his BA degree in geology from the Upsala College, East Orange NJ, and later his PhD in geomorphology at Columbia University, New York. He moved then to the US Geological Survey USGS, until being appointed professor of geology at Colorado State University, Fort Collins CO. Schumm was an internationally acclaimed geomorphologist whose seminal research into earth surface processes was recognized worldwide. He received awards from the US National Academy of Sciences, the American Geophysical Union AGU, the Geological Society of America GSA, and of both the British and the Japanese Geomorphological Societies. He was well rounded and interested in all things, often surprising all around him with his wide knowledge.

The 1984 book on Incised channels deals with the geomorphic characteristics of three channelized streams in Mississippi, which was prepared for the Soil Conservation Service SCS. It applies a geomorphic approach to study these channels and determines if their future behaviour may be predicted. In spite of the extensive literature on the topic, no meaningful geomorphic synthesis on the problems had been completed until then, which may be considered a major reason for the lack of success of many channel control efforts. The book includes three main parts: 1. Literature survey to develop the necessary background, 2. Historical information on subject, and 3. Field investigations. The final chapter was an attempt to summarize the then available knowledge to the understanding and the control of incised channels.

Schumm, S.A. (1956). Evolution of drainage systems and slopes in badlands at Perth Amboy NJ. Geological Society of America *Bulletin* 67: 597-646.
Schumm, S.A. (1961). The effect of sediment characteristics on erosion and deposition in ephemeral stream channels. US Geological Survey *Professional Paper* 352C: 31-70.
Schumm, S.A., Khan, H.R. (1972). Experimental study of channel patterns. *Geological Society of America Bulletin* 83(6): 1755-1770.
Schumm, S.A. (1977). *The fluvial system*. Wiley: New York.
Schumm, S.A. (1980). Geomorphic thresholds: The concept and its applications. Institution of British Geographers *Trans*. 4(4): 485-515.
Schumm, S.A., Harvey, M.D., Watson, C.C. (1984). *Incised channels*. Water Res. Publ.: Littleton
http://www.tributes.com/show/Stanley-A.-Schumm-91254580 *P*

SCHUYLER

11.05. 1848 Ithaca NY/USA
13.09. 1912 Ocean Park CA/USA

James Dix Schuyler received most of his education by own study. He began in 1869 his professional career as assistant of the Kansas Pacific Railway in Western Kansas and Colorado. He continued these works for six years, serving the Denver & Rio Grande Railroad, the North Pacific Coast Railroad in California, and the Stockton & Ione Railroad, of which he was chief engineer.

Schuyler became chief assistant state engineer of California in 1877 and placed in charge of irrigation studies in Great Central Valley. From 1882, he was chief engineer as also the general superintendent of the Sinaloa & Durango Railroad in Mexico. Returning to California, he built a section of the sea wall on the waterfront of San Francisco, raised the Seawater Dam from 1887 to 1888, and constructed Hemet Dam, at the time the highest masonry dam in Western USA. He showed much skill and ability and therefore was prominent in his profession and enjoyed a wide and successful business as consultant, including also countries such as British Columbia, Japan, Brazil, and Hawaii. He was involved in water supply projects for Denver, Ogden UT, and Los Angeles. As one of the three consulting engineers for the Los Angeles Aqueduct, fed by Owens Rivers some 400 km distant, he was able to shorten the route by 10% resulting in considerable savings. He was also one of the commissioners for Panama Canal to judge the feasibility of *Gatun* Dam. Schuyler was awarded the Thomas Fitch Rowland Prize for the 1907 ASCE paper. He contributed extensively to technical journals but is known for his 1901 book on reservoirs. He was a prominent member of the American Society of Civil Engineers ASCE and known for his works in hydraulic fill dams.

Anonymous (1913). James Dix Schuyler. *Trans. ASCE* 76: 2243-2245.
Anonymous (1913). James Dix Schuyler. *Engineering News* 69(16): 785-788. *P*
Anonymous (1935). Schuyler, James Dix. *Dictionary of American biography* 16: 473-474. Scribner's: New York.
Schuyler, J.D. (1898). The failure of the Lynx Creek masonry dam, near Prescott AZ. *Engineering News* 39(Jun9): 362-363.
Schuyler, J.D. (1901). *Reservoirs for irrigation, water-supply, and domestic water supply*, 2nd ed. in 1908. Wiley: New York.
Schuyler, J.D. (1907). Practice in hydraulic-fill dam construction. *Trans. ASCE* 57: 196-277.
Schuyler, J.D. (1911). The extension of the Sweetwater Dam. *Engineering News* 65(Mar.30): 369-372.

SCOBEY

20.01. 1880 Greensburg IN/USA
19.07. 1962 Berkeley CA/USA

Fred Charles Scobey received in 1902 the BA degree in civil engineering from Stanford University, Stanford CA. After practical engineering work, he in 1909 became irrigation engineer of the Imperial Valley, but from 1911 spent all his career with the US Department of Agriculture USDA. He served as secretary and chairman of the Irrigation Division of the American Society of Civil Engineers ASCE. He also organized its Hydraulics Division in 1938 and from then was its chairman. He became Honorary Member ASCE in his retirement year 1960.

Scobey was an important irrigation engineer adding valuable knowledge to hydraulic engineering. During the first twenty years with USDA he was senior irrigation engineer specializing in the hydraulics of canals, pipes, and hydraulic structures. A main result were Bulletins dealing with the flow of water in irrigation and similar canals, in wood stave pipes, in concrete pipes, in steel and analogous pipes, and in flumes. This series of mainly experimental work detailed the roughness behaviour of prototype conveyance structures, with a particular focus on agricultural aspects. A similar work published with Samuel Fortier (1855-1933) in 1926 related to the 'permissible velocities' in loose channel hydraulics, for which Albert Shields (1908-1974) formulated a generalized concept a decade later. In the 1930s, Scobey planned and conducted field work in irrigation techniques, expanding interest also to fluvial hydraulics. He was further a consultant, either directly or through consulting engineers, to the War Department, to the Bureau of Reclamation, and to the cities of Tulsa OK or Denver CO in the matter of water supply and the conveyance of discharge increase from source to city.

Anonymous (1942). Fred C. Scobey. *Civil Engineering* 12(12): 696. *P*
Anonymous (1948). Fred C. Scobey. *Engineering News-Record* 141(Aug.26): 13. *P*
Anonymous (1954). Scobey, Fred C. *Who's who in engineering* 7: 2143. Lewis: New York.
Anonymous (1963). Fred C. Scobey. *Trans. ASCE* 128(5): 111-112.
Fortier, S., Scobey, F.C. (1926). Permissible canal velocities. *Trans. ASCE* 89: 940-984.
Scobey, F.C. (1914). Behaviour of cup current meters under conditions not covered by standard ratings. *Journal of Agricultural Research* 2(2): 77-83.
Scobey, F.C. (1916). Some better Kutter's formula coefficients. *Engineering News* 75(8): 373.
Scobey, F.C. (1920). The flow of water in concrete pipe. *Bulletin* 852. US Department of Agriculture: Washington DC.
Scobey, F.C. (1935). Water-conduit construction in the West. *Civil Engineering* 5(9): 569-571.

SCOTT

20.02. 1843 Dryden NY/USA
27.04. 1927 Charleston WV/USA

Addison Moffat Scott attended the Ithaca Academy, Ithaca NY, specializing in mathematics and survey. In 1866 he joined a party to the Northwest organized by Gouverneur K. Warren (1830-1882) for a survey of the Upper Mississippi. In 1867, Warren offered him a position as assistant engineer on improvement of Mississippi River. Scott remained there mainly on the construction of a large railroad bridge between Rock Island IL and Davenport IA. In 1873 the USA improved the Great Kanawha River, then in charge of William E. Merrill (1837-1891). Scott was placed at Charleston WV as chief assistant. During his long service there he devoted himself with zeal and energy to the duties of his professional position, mastering thoroughly the theory and the practical details of the movable dams as introduced by the Frenchman Chanoine, then the first dams of this type in the USA.

In 1901 Scott resigned from his position after 34 years as civil engineer, as member of the US Corps of Engineers. His last head noted in an article: 'I desire to call attention in the most emphatic way… to Mr. A.M. Scott, who was the principal engineer on this work when I took charge of it for the USA in 1874; he has remained on it up to the present time when it was completed. The conduct of the work has been most economical, wise and excellent in every way. Mr. Scott has exhibited an unusual degree of skill as a designing and constructing engineer in the management of the improvement and in dealing with the many perplexing problems which have presented themselves for solution in its progress. To him is due, more than to any other person, the success of this work. I say what I know, and it gives sincere pleasure thus to bear record to the merit of a most faithful and deserving man'. Scott retired to Charleston WV, serving there as a vice-president of the Charleston Chamber of Commerce. He was described as courteous and kind, while he possessed the unbounded confidence of his superiors. In his profession he achieved an enviable reputation for capacity, skill, and executive ability. He was a modest, unassuming gentleman thoroughly reliable in all engagements.

Anonymous (1929). Addison M. Scott. *Trans. ASCE* 93: 1897-1898.

Chanoine, J. (1839). Sur le barrage d'Epineau. *Annales des Ponts et Chaussées* 16: 238-280.

Chanoine, J., Lagrené, H.M. (1862). *Les barrages à hausses mobiles*. Dunod: Paris.

Kemp, E.L. (2000). Addison M. Scott. *The Great Kanawha navigation*: 43. University of Pittsburgh Press: Pittsburgh PA. *P*

Scott, A.M. (1894). Improvement of the Great Kanawha River. *Trans. ASCE* 31: 539-551.

SEARS F.W.

01.10. 1898 Plymouth MA/USA
12.11. 1975 Norwich VT/USA

Francis Weston Sears was a member of the physics staff at the Massachusetts Institute of Technology MIT, Cambridge MA, from 1925 to 1955, and then joined as professor of physics Dartmouth College, Hanover NH, until retirement in 1964. He influenced countless students in their preparation for careers in physics, chemistry, as also engineering. His lectures were models of clarity, and the demonstrations were meticulously prepared and performed. He was one of the few teachers who not only spoke slowly and loudly enough to be heard in the back of the lecture room but also wrote on the blackboard large enough symbols and figures. One of his students put it: 'He was a wonderful explainer'.

Sears' life was full of satisfaction and success. Stimulated by the physicist Debye on a visit to the MIT, Sears collaborated in a series of experiments showing that the density variations in a sinusoidal ultrasonic wave in a liquid serve to diffract light in a manner similar to that of a plane transmission grating. Their work was published in 1932, and the phenomenon is known as the Debye-Sears effect. In the mid 1940s Sears conceived the idea of a textbook covering the entire field of college physics, suitable for a two-years course in which elementary calculus would be used sparingly after the forth week but then more and more. This set of texts was called Principles of physics. Due to the success of these books, the printer Addison-Wesley of Cambridge MA has become an important publishing house. Other books that he published were University physics, or Thermodynamics. Sears was active on the American Association of Physics Teachers, serving as treasurer from 1950 to 1958, and as the president in 1959. He was awarded in 1962 its highest honour, the Oersted Medal. He is remembered by the large number of people whom he inspired and encouraged.

Anonymous (1959). Francis W. Sears. *Physics Today* 12(3): 60. *P*
Anonymous (1961). Francis W. Sears. *Physics Today* 14(4): 68. *P*
Anonymous (1976). Francis W. Sears. *Physics Today* 29(2): 65.
Anonymous (1976). Francis W. Sears. *American Journal of Physics* 44(1): 3. *P*
Sears, F.W. (1944). *Mechanics, heat and sound*. Addison-Wesley: Cambridge MA.
Sears, F.W. (1944). *Principles of physics*. Addison-Wesley: Cambridge MA.
Sears, F.W. (1950). *Thermodynamics, the kinetic theory of gases and statistical mechanics*. Addison-Wesley: Cambridge MA.
Sears, F.W., Zemansky, M.W. (1955). *University physics*. Addison-Wesley: Cambridge MA.

SEARS W.H.

08.12. 1847 Plymouth MA/USA
07.10. 1911 Plymouth MA/USA

Walter Herbert Sears graduated from Massachusetts Institute of Technology MIT, Cambridge MA, in 1868. He was then engaged for the construction of water works at Boston MA, and as chief engineer at Winchester MA until 1875, from when he held a similar position at Pawtucket RI. He then moved in 1880 to Stillwater MN where he was also in charge of the city water works. After return to the East he continued with comparable projects for the East Jersey Water Company, Paterson NJ. He was from 1892 to 1893 chief assistant engineer on additional water supply for Rochester NY, from when he was engaged until 1903 in general engineering practice.

Sears was appointed in 1904 division engineer of the Croton River Diversion of the Aqueduct Commission, New York City, with works near Katonah NY. He was until 1910 chief engineer in charge of the extensions of the Croton Water Supply project. The Cross River Reservoir was completed, and construction of the Croton Falls Reservoir started and carried out nearly to completion. However, Sears turned ill and was unable to return to active work. His professional work was marked with great thoroughness in which he studied each problem. He had the ability of foreseeing difficulties, and when they came up, he was ready with plans to surmount them. This preparedness resulted from constant observation of the nature, whereby he became familiar with the causes, effects, and processes which normally remain not noted by others. He thus had a keen appreciation of the beauties of natural sceneries, and when a Park Commission was created at Plymouth in 1895, he became member. The creation of an attractive park system in his native city was a source of absorbing interest to him, and the beauties are largely due to his artistic plans which he made, as commissioner, without any financial compensation. His unselfish civic interest was also displayed by presenting the city with an elaborate report on the improvement of its public water works, and these plans were duly realized. He was as a person modest, gentle, and lovable in a high degree so that he won the respect and sincere affection of many friends and associates.

Anonymous (1905). *New York City and vicinity*: Engng. works. McGraw-Hill: New York. (*P*)
Anonymous (1912). Sears, W.H. *Trans. ASCE* 75: 1180-1181.
Cowan, J.F., Walker, H.W., Sears, W.H. (1907). *Report to the aqueduct commissioners*. Mitchell & Sons: New York.
Sears, W.H. (1907). *Report of the chief engineer on the Jerome Park Reservoir*. New York.

SEARS W.R.

01.03. 1913 Minneapolis MN/USA
12.10. 2002 Tucson AZ/USA

William Rees Sears was educated at the University of Minnesota, Minneapolis MN, obtaining the degree BAeroE in 1934, and at the California Institute of Technology, Pasadena CA, receiving the PhD degree in 1938. He was there an instructor until 1941, chief aerodynamicist of Northrop Aircraft Inc. until 1946, taking over until 1964 as professor of aeronautical engineering, then as Director of the Graduate School of Aeronautical Engineering, at Cornell University, Ithaca NY. From 1974 to 1985 he was professor of aerospace and mechanical engineering at University of Arizona, Tucson AZ. He was awarded the 1992 Fluid Dynamics Prize for his 'outstanding contributions to both steady and unsteady aerodynamics, aero-acoustics, magneto-hydrodynamics and wind tunnel design; for his leadership in and devotion to aeronautical engineering education'.

As chief of Aerodynamics and Flight Testing of Northrop Sears headed the team that designed the first Flying Wing aircraft and the P-61, the so-called Black Widow. He later played a key role in the development of the XB-35 and YB-49 airplanes, which later provided the basic aerodynamic design for the Stealth bomber. Once at Cornell, he took interest in the physical and mathematical foundations of the airfoil theory, and introduced what is known as the Sears function, which gives the forces and moments on a wing due to a gust. Later he developed techniques for wind tunnel experimentation, especially of high-lift and high-drag configurations. Besides being a Honorary Fellow of the American Institute of Aeronautics and Astronautics AIAA, he was a member of the National Academy of Sciences NAS, the National Academy of Engineering NAE, and the American Academy of Arts and Sciences AAAS.

Anonymous (1955). William R. Sears. *Aeronautical Engineering Review* 14(12): 25. *P*
Anonymous (1974). William R. Sears. *Aeronautics and Astronautics* 12(5): 79. *P*
Anonymous (1992). William R. Sears. *Physics Today* 45(12): 105-106. *P*
Anonymous (1994). Sears, W.R. *American men and women in science* 6: 729. Bowker: NJ.
Rott, N. (2005). William Rees Sears. *Biographical memoirs* 86: 1-14. National Academy Press: Washington DC. *P*
Sears, W.R. (1940). Operational methods in the theory of airfoils in non-uniform motion. *Journal of Franklin Institute* 230: 95-111.
Sears, W.R. (1959). Magneto-hydrodynamic effects in aerodynamic flows. *ARS Journal* 29(6): 397-406. American Rocket Society: New York NY.

SEASTONE

18.04. 1872 New Boston IL/USA
26.09. 1940 Madison WI/USA

Charles Victor Seastone obtained the BSc degree in civil engineering in 1895 from University of Illinois, Urbana IL. He was already there interested in water flow, as is shown by his thesis. He was employed then by the Mississippi River Commission in charge of instrument work as assistant hydrographer. He was asked by Arthur N. Talbot (1857-1942) to return to his Alma Mater as his personal assistant in charge of laboratory work. Seastone went in 1900 to Purdue University, West Lafayette IN, as associate professor of sanitary engineering, but gave up his career as a teacher in 1907, despite the excellent records.

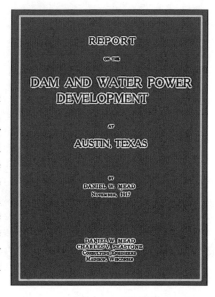

From 1898, Seastone served as assistant engineer Daniel W. Mead (1862-1948) on the reconstruction of the Rock Island Reservoir IL. This contact led later to an association. From 1907 Seastone was again engaged by Mead in his consulting office at Madison WI, entering a partnership in 1913, which lasted until 1933. Seastone was first in charge of the plans of the Kilbourn hydro-electric plant, on the Wisconsin River. In 1908, the Madison office under Seastone began plans for a hydro-electric plant at High Falls on the Peshtigo River WI, which was highly different from the previous installation. In 1911 work on the hydro-electric plant on the Wisconsin River upstream of Prairie du Sac started. The river bed there is sandy, 30 m deep, so that both the power house and the 300 m long dam were built on pile foundation. The next important plants included these at Twin Falls on the Menominee River near Iron Mountain MI, or on the Brule River near Florence MI. Many reports were furnished on the water power and water supply developments all over the USA, of which one of the largest possible was the hydro-electric development on the Missouri River for South Dakota at five successive locations within the State. One of the most interesting design was that on the Elwha River on the Olympic Peninsula WA. Seastone was credited for a successful completion of projects he was involved; he studied all of its aspects with unbiased judgment. He was described as kind person, utilizing with the fullest possible credit all sound suggestions that came to him. He was member of the American Society of Civil Engineers ASCE from 1916.

Anonymous (1942). Charles V. Seastone. *Trans. ASCE* 107: 1830-1835.
Ketchum, M.S., Seastone, C.V. (1895). *Flow of water through siphons*. University of Illinois.
Mead, D.W., Seastone, C.V. (1917). *Report* on the dam and water power development near
    Austin TX. Mead & Seastone, Consulting Engineers: Madison WI. (*P*)
Seastone, C.V. (1899). Measuring tank for a jet meter. *The Technograph* 14(2): 98-104.

SEDDON

1856 Richmond VA/USA
01.10. 1921 Chicago IL/USA

James Alexander Seddon was educated as a civil engineer. He soon joined as member the Engineers' Club of Saint Louis and from 1895 was US assistant engineer. He was stationed at Kansas City MO and there was in charge of works on Missouri River. Few details on his life are known, despite his name has been preserved until today, relating to unsteady flow in rivers, and the propagation of flood waves.

Seddon's 1900 paper was quoted by many texts because it describes the Seddon Law, apparently formulated independently, though first published by Charles Kleitz (1808-1886) in 1877. A basic application of the continuity equation, the law states that a moderate increase in discharge imposed on an approximately steady flow produces an increase in stage which travels downstream as a monoclinal wave at a celerity that is equal to the slope of the rating curve relating discharge to stage, divided by the surface width. This relation was checked experimentally during Seddon's study of flood movements on Mississippi and Missouri Rivers, and it marked the beginning of American analysis of unsteady flow. Monoclinal waves are an abstraction of real river flows, because their propagation depends exclusively on the local flow depth. More generally dynamical waves therefore are considered in current applications. The 1898 paper deals with the effect of storage on the hydrograph. The storage equation was integrated for a one cycle inflow function and for an outflow rating curve, which was assumed linear with respect to the reservoir level. Then, the reservoir level hydrograph was determined in which the reservoir area was assumed either constant or changing with the reservoir level. The storage effect on the amplitude of the crest flow and on the phase shifting was derived analytically for one or more reservoirs.

Anonymous (1922). James Alexander Seddon. *Proc. ASCE* 48(1): 2.

Rouse, H. (1976). James A. Seddon. *Hydraulics in the United States* 1776-1976: 78. IIHR: Iowa.

Seddon, J.A. (1886). Some considerations of the relation of bed to the variables in river hydraulics. *Journal of the Association of Engineering Societies* 5(4): 127-134.

Seddon, J.A. (1889). Clearing water by sedimentation: Observations and theory. *Journal of the Association of Engineering Societies* 8(10): 477-492.

Seddon, J.A. (1898). A mathematical analysis of the influence of reservoirs upon stream-flow. *Trans. ASCE* 40: 401-427.

Seddon, J.A. (1900). River hydraulics. *Trans. ASCE* 43: 179-243.

Seddon, J.A. (1902). Discussion of Flow of water in pipes. *Trans. ASCE* 47: 238-242.

A Deep Waterway from the Great Lakes to the Gulf of Mexico

PAPERS BEFORE
THE WESTERN SOCIETY OF ENGINEERS

By

JAMES A. SEDDON—Lower Mississippi River from the Gulf of Mexico to Cairo
LYMAN E. COOLEY—Cairo to the Sanitary and Ship Canal at Lockport
ISHAM RANDOLPH—The Sanitary and Ship Canal of Chicago

Discussions by

LYMAN E. COOLEY   ISHAM RANDOLPH
ROBERT E. McMATH   THOMAS T. JOHNSTON
C. H. TUTTON   JAMES A. SEDDON

Introduction by
The Legislative Committee

STANFORD LIBRARY

The Illinois River Valley Association

Republished from
Journal of Western Society of Engineers
August, 1909

SEELY

29.04. 1884 Chester NY/USA
15.09. 1968 San Mateo CA/USA

Fred (Frederick) B. Seely, Jr., received his education from the Worcester Polytechnic Institute, Worcester MA, from where he obtained the BS degree in 1907, and from University of Illinois, Urbana IL, receiving the MS degree in mechanical engineering in 1915. After service in 1908 as an instructor at Villanova College, King City Ont., Canada, he was from 1909 to retirement from associate professor to professor of theoretical and applied mechanics at University of Illinois, serving there also as Head of Department. Seely was a member of the American Society of Mechanical Engineers ASME, the Chicago Engineers Club, and the Champaign-Urbana Kiwanis.

Seely was known for several books and papers published in the then journals of the day. His Analytical mechanics for engineers was among the most read by students, given its basic but clear approach. The basic principles of mechanics are outlined, and were built as much as possible from common experience. A physical rather than a mathematical interpretation of these principles was attempted. Additional books included works in mechanics and material sciences, which were addressed to undergraduate and graduate students. The 1917 report deals with the outflow through submerged short pipes under various nozzle angles and lengths. The coefficient of discharge was found to vary with essentially the pipe velocity, becoming almost a constant for velocities in excess of 0.5 m/s, thereby pointing at viscous effects for small velocities. In addition, the angle of the mouthpiece was found to have a significant effect on the outflow features. The results were compared with classical data on this basic hydraulic problem. The 1918 report is a more general account on the topic of general pressurized pipe flow reflecting the experimental approaches made in the early 20th century.

Anonymous (1948). Seely, Fred B. *Who's who in engineering* 6: 1771. Lewis: New York.
Seely, F.B. (1915). *Flow of water through submerged orifices and short tubes with mouthpieces*. University of Illinois: Urbana-Champaign.
Seely, F.B. (1917). The effect of mouthpieces on the flow of water through a submerged short pipe. *Bulletin* 96. Engineering Experiment Station, University of Illinois: Urbana.
Seely, F.B., Ensign, N.E. (1921). *Analytical mechanics for engineers*, 2nd ed. in 1933. Wiley: NY.
Talbot, A.N., Seely, F.B., Fleming, V.R., Enger, M.L. (1918). *Hydraulic experiments with valves, orifices, hose, nozzles, and orifice buckets*. University of Illinois: Urbana.
http://archives.library.illinois.edu/archon/index.php?p=digitallibrary/digitalcontent&id=5995 *P*

SEERY

24.05. 1874 Waterbury CT/USA
27.07. 1947 Dover DE/USA

Francis Joseph Seery served from 1893 to 1900 his home city as engineering assistant on construction of the additional water supply. He was transitman in 1900 for the Isthmian Canal Commission, and then decided to continue his education. He received in 1905 the BSc degree in civil engineering from Tufts College, Medford MA, and then served as assistant engineer on the construction of the Barge Canal, New York State. He was in 1905 appointed instructor of hydraulic engineering, College of Civil Engineering, Cornell University, Ithaca NY, a teaching career that lasted until 1942. Seery was promoted to assistant professor in 1907, and full professor of hydraulic engineering in 1918. He taught courses in water supply, hydraulic construction, water power and pumping plants, hydraulic engineering, and reclamation problems.

Seery prepared numerous reports, including these on the hydroelectric development on *Geneganslet* Creek, and *Oswegatchie* River in New York State in 1909, the design of the Potters Falls Dam on the Six Mile Creek in Ithaca, or valuation appraisals for water plants. He also served as hydraulic engineer the US Geological Survey in 1918. After World War I, he made a design for the water supply of Moravia NY, a review of the *Jadwin* Plan for flood protection of the Mississippi River in 1925, an expertise on the hydraulics for the State of New York on various cases in 1936, or an expertise on the hydraulic meters for the New York State attorney general's department on water supply of Rochester NY in 1937. At Cornell University, Seery was convinced of the value of hydraulic experimentation. The educational worth of the most relevant problems found feedback by the students. This feature of Seery's teaching methods became incorporated in other fields in the curriculum of the School of Civil Engineering. In his work as teacher, Seery stressed conception, design, construction, and maintenance of hydraulic engineering structures. He was an expert in water works valuation and hydrology, and served as chairman of the Board of Public Works of Ithaca NY. During the first World War, he was superintendent of construction at Camp Dix NJ, in charge of water and sewage. Seery was a member of the American Society of Civil Engineers ASCE from 1921.

Anonymous (1947). Prof. Seery dies. *Cornell Alumni News* 50(2): 47. *P*
Anonymous (1948). Francis J. Seery. *Trans. ASCE* 113: 1541-1543.
Seery, F.J. (1906). Some features of Isthmian Canal Projects. *Cornell Civil Engineer* 14(Jan. 19).

SERRIN

01.11. 1926 Chicago IL/USA
23.08. 2012 Minneapolis MN/USA

James (Jim) Burton Serrin received the PhD degree in 1951 from Indiana University, Bloomington IN, under the supervision of David Gilbarg (1918-2001). He there met Clifford Truesdell (1919-2000), from whom he became his lifelong interest in thermodynamics. Serrin was appointed in 1952 mathematics instructor at the Princeton University, Princeton NJ, from where he moved to the Massachusetts Institute of Technology MIT, Cambridge MA. He joined in 1954 Minnesota University, Minneapolis MN, as assistant professor of mathematics, was promoted to associate professor in 1956, and to full professor in 1959 there. He served his university as head of the School of Mathematics in the 1960s, and was named Regents' Professor of Mathematics. In the early 1970s he served on the Council of the American Mathematical Society AMS. Serrin retired in 1995. He was elected a member of the National Academy of Sciences NAS in 1980.

Serrin is known for his contributions to continuum mechanics, nonlinear analysis, and partial differential equations. Before submitting his PhD thesis he already published a paper on Free boundaries and jets in the theory of cavitation, co-authored by Gilbarg. They discuss the theory of a finite cavity containing a backward jet behind a body in two-dimensional irrotational and incompressible flow. Conformal mapping is used to solve this problem; however, emphasis is placed on a generalization of the Schwarz transformation, in which branch points are arbitrarily placed occupying a Riemann surface. A set of conditions for the solution involving curved boundaries is given. The 1959 chapter in the *Handbuch der Physik*, possibly the last of these encyclopaedic series containing more than 250 volumes in all branches of the then known physics, has had a lasting impact in hydrodynamics..

Pucci, P. (1998). An appreciation of James Serrin. *Nonlinear analysis and continuum mechanics*: Papers for the 65th birthday of James Serrin, G. Buttazzo, ed. Springer: Berlin.

Serrin, J.B. (1952). Uniqueness theorems for two free boundary problems. *American Journal of Mathematics* 74(2): 492-506.

Serrin, J.B. (1959). Mathematical principles of classical fluid dynamics. *Handbuch der Physik* 8(1): 125-263, S. Flügge, ed. Springer: Berlin.

Serrin, J.B. (1962). On the interior regularity of weak solutions of the Navier-Stokes equations. *Archive for Rational Mechanics and Analysis* 9(1): 187-195.

http://www-history.mcs.st-andrews.ac.uk/Biographies/Serrin.html *P* http://www.webcitation.org/6575qI0Ae *P*

SHAPIRO

20.05. 1916 Brooklyn NY/USA
26.11. 2004 Boston MA/USA

Ascher Herman Shapiro received the BSc degree in 1938 from the Massachusetts Institute of Technology MIT, Cambridge MA, and the PhD degree there in 1946. He was from 1943 to 1947 assistant professor and until 1952 associate professor when taking over as professor of mechanical engineering at MIT until retirement. He was in 1955 a visiting professor at Cambridge University, England, and consultant for a number of industrial firms. He was the recipient of the 1960 Richards Memorial ASME Prize for his 'outstanding achievements in engineering'.

Shapiro was a prolific author of texts, including his two-volume treatise Dynamics, which is considered a classic. His 1961 book Shape and flow: The fluid dynamics of drag explains boundary layer phenomena and drag in simple, non-mathematical terms. In 1961 Shapiro founded the National Council for fluid mechanics films in corporation with the Educational Development Center. A series of 39 films was released which have been widely used in the teaching of fluid dynamics. He was elected to the American Academy of Arts and Sciences in 1952, the National Academy of Science in 1967, and the National Academy of Engineering in 1974. He also was the recipient of the 1977 Fluids Engineering Award and the 1999 Drucker Medal by the American Society of Mechanical Engineering ASME. This award was established in 1997 and is conferred in recognition to the field of applied mechanics and mechanical engineering through research, teaching and service to the community over a substantial period of time. The Ascher H. Shapiro Lecture in Fluid Mechanics was established in 1994.

Anonymous (1961). Ascher H. Shapiro. *Mechanical Engineering* 83(1): 112. *P*
Anonymous (1964). Shapiro, A.H. *Who's who in engineering* 9: 1684-1685. Lewis: New York.
Anonymous (1966). Ascher H. Shapiro. *Mechanical Engineering* 88(1): 101. *P*
Rivas, M.A., Shapiro, A.H. (1956). On the theory of discharge coefficients for rounded-entrance flowmeters and Venturis. *Trans. ASME* 78(4): 489-497.
Shapiro, A.H. (1953). *The dynamics and thermodynamics of compressible fluid flow*. Wiley: NY.
Shapiro, A.H., Siegel, R., Kline, S.J. (1954). Friction factor in the laminar entry region of a smooth tube. Proc. 2nd US Congress *Applied Mechanics* Ann Arbor, 733-741.
Shapiro, A.H. (1964). Educational films in fluid mechanics. *Proc. Institution of Mechanical Engineers* 178: 1187-1204.
Shapiro, A.H. (1970). *Shape and flow*: The fluid dynamics of drag. Heinemann: London UK.
http://en.wikipedia.org/wiki/Ascher_H._Shapiro

SHERARD

20.09. 1925 San Francisco CA/USA
31.07. 1987 La Jolla CA/USA

James Lewis Sherard graduated in 1946 with the BSc degree from the University of California, Berkeley CA, and from Harvard University, Cambridge MA, in 1948 with the MSc degree in soil and foundation engineering. His doctoral thesis was completed in 1952 under Karl Terzaghi (1883-1963), involving a data analysis on dam failures with the US Bureau of Reclamation; it became a reference on dam failures due to cracking, and from then led to an improved performance. Over the next 35 years, Sherard was resident engineer in Brazil, partner and principal with Woodward-Clyde-Sherard & Associates, and the balance of his career in his own consulting practice Sherard Overseas Consultants, Los Angeles. Until his health failed, he worked on 220 dam and reservoir projects all through the world, as consultant to private firms and government agencies. His work led him to countries as Venezuela, Brazil, Colombia, Ecuador, Peru, Chile, the Philippines, Greece, Spain, Israel or Iran. He also involved himself with innovative research with the Soil Conservation Service SCS on dam failures by piping, and research on filter criteria.

Sherard published over 40 journal papers of which more than half in ASCE journals. He also co-authored the 725-page technical book Earth and earth-rock dams. The 1992 book is dedicated to the memory of Sherard, who made numerous contributions to the embankment dam engineering. Twenty-one of his papers were selected for the memorial volume, of which a large portion was considered seminal and of outstanding value to the profession. These are preceded by an introduction prepared by either a co-author or someone closely associated with the work. Papers of lasting value as references on subjects such as seismic aspects, dispersive clays, hydraulic fracturing and filters are included. His last two papers on the concrete face rockfill dam are also in this volume.

Anonymous (1961). James L. Sherard. *Civil Engineering* 31(7): 68. *P*
Anonymous (1987). James L. Sherard. *Trans. ASCE* 152: 403-404.
Sherard, J.L., Woodward, R.J., Gizienski, S.F. (1963). *Earth and earth rock dams*: Engineering problems of design and construction. Wiley: New York.
Sherard, J.L., Dunnigan, L.P., Talbot, J.R. (1984). Filters for silts and clays. *Journal of Geotechnical Engineering* 110(6): 701-718.
Sherard, J.L. (1986). Hydraulic fracturing in embankment dams. *Journal of Geotechnical Engineering* 112(10): 905-927.
Singh, S., ed. (1992). *Embankment dams*: James L. Sherard contributions. ASCE: New York.

SHERMAN C.W.

26.09. 1870 Kingston MA/USA
17.01. 1958 Cambridge MA/USA

Charles Winslow Sherman graduated in 1890 from the Massachusetts Institute of Technology MIT, Cambridge MA, with the BS degree and later from Cornell University, Ithaca NY, with an MCE. From 1895 to 1904 he was assistant engineer and division engineer with the Boston Water Works. He became then associated with Leonard Metcalf (1870-1926) and continued as principal engineer with Metcalf & Eddy until 1913, when becoming a partner until his retirement in 1938.

Sherman's professional activities were particularly in the field of waterworks, including hydrology, design, and maintenance. He authored technical papers dealing with the rainfall intensity of New England, or partially-filled pipe flow as used in sewer hydraulics, for which he pointed at deficiencies for designing these conduits because an apparent variation of the roughness coefficient. He served as editor of the Journal of the New England Water Works Association during the first decade of the 20th century, and was the recipient of its Dexter Brackett Medal in 1922, a year after having served as its president. He was elected honorary member of the Association in 1931. Sherman further was a water commissioner at Belmont MA for 24 years, at the end as chairman of the committee. He had been associated with a number of professional organisations, including the American Society of Civil Engineers ASCE, the Boston Society of Civil Engineers BSCE, or the American Water Works Association AWWA. He was remembered for his genial and friendly presence, and the ease and fluency with which he joined in the discussion of papers.

Anonymous (1921). Charles W. Sherman. *New England Water Works Association* 35(1): Frontispiece. *P*
Anonymous (1938). Charles W. Sherman. *Water Works Engineering* 91(8): 1274. *P*
Chase, E.S. (1958). Charles W. Sherman. *Journal of the New England Water Works Association* 72(1): 74-75.
Sherman, C.W. (1905). Maximum rates of rainfall at Boston. *Trans. ASCE* 54: 173-212.
Sherman, C.W. (1928). Twenty years as a Water Commissioner in a New England town. *Engineering News-Record* 100(22): 848-849. *P*
Sherman, C.W. (1929). Hydraulic formulas incorrect for partly filled pipes. *Engineering News-Record* 103(7): 253-254.
Sherman, C.W. (1931). Frequency and intensity of excessive rainfalls at Boston, Massachusetts. *Trans. ASCE* 95: 951-968.

SHERMAN L.K.

20.07. 1869 Eastham MA/USA
04.01. 1954 San Diego CA/USA

LeRoy Kempton Sherman received the BS degree in 1892 from Massachusetts Institute of Technology MIT, Cambridge MA. He was from 1895 to 1900 involved in the construction of the Chicago Drainage Canal and its regulating works, then until 1905 in railroad works around Chicago IL, when joining the Sanitary District of Chicago in charge of construction the Lockport water power installation, and of the North Shore Channel. He became in 1912 contractor of the Illinois Rivers and Lakes Committee, and was from 1918 to 1921 assistant chief engineer of the US Housing Corporation, when taking over as the president an engineering company at Chicago IL. From 1933 Sherman was finally a consultant and also served the National Resources Committee.

Sherman was an expert in hydraulics and hydrology who significantly contributed to the unit-hydrograph approach. A hydrograph in general is a graph showing changes of discharge typically of river flow over a certain period of time. A unit hydrograph in particular represents the effect of rainfall in a particular basin. It is a hypothetical unit response of the watershed to a unit input of rainfall, allowing for an easy computation of the response to an arbitrary rainfall by simply performing a convolution between the rain input and the unit hydrograph output. Sherman in addition considered hydraulics of sewers. He was a Honorary Member of the American Society of Civil Engineers ASCE.

Anonymous (1941). Sherman, L.K. *Who's who in engineering* 5: 1611. Lewis: New York.
Anonymous (1947). L.K. Sherman. *Engineering News-Record* 138(Jan.16): 81. *P*
Anonymous (1954). L.K. Sherman, ASCE Honorary Member, is dead. *Civil Engineering* 24(3): 179. *P*
Sherman, L.K. (1912). Run-off from sewered areas. *Journal of the Western Society of Engineers* 17(4): 361-378.
Sherman, L.K. (1932). Streamflow from rainfall by unit-hydrograph method. *Engineering News-Record* 108(14): 501-505.
Sherman, L.K., Horton, R.E. (1933). Rainfall, runoff and evaporation. Association Internationale d'Hydrologie Scientifique, *Bulletin* 20: 17-24.
Sherman, L.K. (1936). The Horton method for determination of infiltration-rates. *Trans. AGU* 17(2): 312-314.
Sherman, L.K., Mayer, L.C. (1941). Application of the infiltration theory to engineering practice. *Trans. AGU* 22(3): 666-677.

SHULITS

07.12. 1902 Lawrence MA/USA
27.02. 1973 Ithan PA/USA

Samuel (Sam) Shulits graduated in 1924 with the BS degree from Massachusetts Institute of Technology MIT, Cambridge MA, receiving additional education from the *Technische Hochschule* Berlin, Germany in 1928, and from the Michigan College of Mines & Technology MCMT, Houghton MI. From 1933 he served first as an engineer with the US Waterways Experiment Station, Vicksburg MS, and with the US Bureau of Reclamation USBR, when joining Colorado School of Mines, Golden CO as assistant professor in charge of fluid mechanics from 1936 to 1939. Until 1940 he then contributed at the US Engineering Office, Louisville KY, to flood control investigations and in parallel was also involved in the activities of the Soil Conservation Service SCS. After war service Shulits was in 1947 appointed adjunct professor of fluid mechanics at New York City University, and later was associate professor of civil engineering at MCMT until 1953, when becoming professor of civil engineering at Penn State University, University Park PA, until his retirement.

Shulits was one of the first Freeman Scholars travelling from the USA to Germany. He then translated the two volumes Hydraulic structure of Armin Schoklitsch (1888-1969) from German into English. After work mainly in river engineering, he initiated a large research campaign on flow over large scale roughness elements in the 1960s. A paper written jointly with John B. Herbich (1922-2008) was awarded the Karl Emil Hilgard Hydraulic Prize of the American Society of Civil Engineers ASCE. He was a member of ASCE, the International Association of Hydraulic Research IAHR, and the Permanent International Association of Navigation Congresses PIANC.

Anonymous (1959). Shulits, Samuel. *Who's who in engineering* 8: 2242. Lewis: New York.
Anonymous (1963). Samuel Shulits. *Civil Engineering* 33(10): 73. *P*
Anonymous (1966). Sam Shulits. *Civil Engineering* 36(10): 77. *P*
Herbich, J.B., Shulits, S. (1964). Large-scale roughness in open-channel flow. *Journal of the Hydraulics Division* ASCE 90(HY6): 203-230; 91(HY5): 242-262; 92(HY3): 75-79.
Shulits, S. (1937). *Hydraulic structures*. ASME: New York.
Shulits, S. (1941). Rational equation of river-bed profile. *Trans. AGU* 22(3): 622-630.
Shulits, S. (1955). Graphical analysis of trend profile of a shortened section of river. *Trans. AGU* 36(4): 649-654.

SIBERT

12.10. 1860 Gadsden AL/USA
16.10. 1935 Bowling Green KY/USA

William Luther Sibert attended until 1880 University of Alabama, but graduated only in 1884 from the US Military Academy, and from the Engineering School of Application in 1887. He was for a year in charge of the Rivers and Harbours Districts of Louisville KY, and until 1892 of local improvement works on the Green and Barren Rivers, Bowling Green KY. He was engaged in construction works of the Sault Sainte Marie Canal, connecting the Great Lakes until 1894, taking over until 1896 command of the River and Harbour District at Little Rock AL. During the Spanish-American War he commanded the engineering troops in Manila as chief engineer, returning to his country in 1900. He was then district engineer at Louisville KY, and Pittsburgh PA until 1907.

Sibert was member of the Isthmian Canal Commission from 1907 to 1914, from 1909 in the rank of lieutenant colonel. He was responsible for the design and construction of the Gatun Locks and Dam, a hydraulic key element of Panama Canal, thereby collaborating with John F. Stevens (1853-1943). The 1915 book describes the main stages of this outstanding waterway from the Atlantic to the Pacific Oceans, including the Culebra Cut, the Gatun Dam with the spillway, and the devastating landslides. Sibert also was in charge of the West Breakwater, Colon Harbour, and of the channel excavations from Gatun to the Atlantic Ocean. He served in 1914 as chair of the Board of Engineers on flood prevention under the American Red Cross, and the Chinese Government in the *Huai* River Valley, China. In 1915 he was a board member on the Ohio River flood control. After war service in the Pacific Army, he retired as major general. He was appointed in 1928 chairman of the Boulder Dam Commission, of which one of the most notable dams erected in the 20th century resulted. He retired from engineer to his farm. Sibert was commander of the French *Legion d'Honneur*, he was awarded the LL.D. and the D.Engr. degrees from the University of Nebraska, Lincoln NE, in 1919. He was member of the American Association of Port Authorities, acting as its president from 1929 to 1930. He also was member of the American Society of Civil Engineers ASCE.

Clark, E.B. (1930). *William L. Sibert*, the army engineer. Dorrance: Philadelphia. *P*
Sibert, W.L. (1909). The improvement of the Ohio River. *Trans. ASCE* 63: 388-428.
Sibert, W.L. (1912). The Gatun Dam and Locks. *Scientific American* 107(19): 386-387.
Sibert, W.L., Stevens, J.F. (1915). *The construction of the Panama Canal*. Appleton: New York.
http://en.wikipedia.org/wiki/William_L._Sibert *P*

SICKELS

20.09. 1819 Camden NJ/USA
08.03. 1895 Kansas City MO/USA

Frederick Ellsworth Sickels was rodman first on the Harlem Railroad, and then apprenticed in New York, where he developed a new type of steam cut-off valve for steam engines, which was perfected until 1841, and patented in 1842. The novel valve was described as: 'It was a form which adapted it to use with the beam-engine used on the Eastern waters of the USA, and was modified to stationary engines by a ship builder in Providence RI, who made use of it before any other form of "drop cut-off" came into general use. Sickels' cut-off consisted of a set of steam-valves, usually independent of the exhaust-valves, and each raised by a catch, which could be thrown out at the proper moment by a wedge with which it came in contact as it rose with the opening valve. This wedge, or other equivalent device, was so adjusted that the valve should be detached and fall to its seat when the piston reached that point in its movement, after taking steam, at which expansion was to commence. From this point, no steam entering the cylinder, the piston was impelled by the expanding vapor. The valve was usually the double-poppet'.

In 1990 it was described as 'a quick-closing valve gear using poppet valves, and a trip gear to control the cut-off with gravity assisted closure. Sickels used a water-filled dashpot to decelerate the valve smoothly as it approached the end of its travel'. In 1843 and 1845 Sickels improved his design. His new type of valve made high-pressure steam engines possible, and it was copied extensively and appeared soon also in the Corliss engine. Sickels sued Corliss for patent infringement, but although he won these court cases, his modest fortune was ultimately consumed, because the patents had expired. He completed in 1854 the construction of a full-size steam steering unit, installing in 1858 the equipment on the steamer *Augusta*. He found no purchaser for the invention, and also no buyer during his trip to Europe. He was from 1890 a consulting engineer for the National Water Works Company of New York City, and from 1891 chief engineer of operations at Kansas City MO. Sickels was member of the American Society of Civil Engineers ASCE.

Anonymous (1853). *Frederick E. Sickels and Truman Cook against John F. Rodman*. New York.
Anonymous (1859). *Frederick E. Sickels*. Washington DC.
Haswell, C.H. (1844). *Engineers' and mechanics' pocket-book*. Harper: New York.
http://en.wikipedia.org/wiki/Frederick_Ellsworth_Sickels
http://www.findagrave.com/cgi-bin/fg.cgi?page=pv&GRid=39684214&PIpi=19726192 *P*

2544

SILBERMAN

08.02. 1914 Minneapolis MN/USA
05.07. 2011 Minneapolis MN/USA

Edward Silberman graduated with the BCE degree from University of Minnesota, Minneapolis MN, in 1935, and the MS degree in 1936. He was in 1937 a junior engineer on flood control at Tennessee Valley Authority TVA, an engineer constructor of the US Civil Aeronautical Administration until 1941, when joining the University of Minnesota from research associate to professor of civil engineering until his retirement. He also served as a commissioner of the Bassett Creek Water Management Commission in the 1970s, and chaired the Water Resources Planning and Management Division, ASCE in 1981. He was elected Honorary Member of the American Society of Civil Engineers ASCE, Fellow of the American Water Resources Association AWRA, which he presided in 1969, and member of the International Association of Hydraulic Research IAHR.

Silberman attained eminence through research in hydraulics, naval hydromechanics, and water resources management. His citation for the ASCE honorary membership states 'Unceasing dedication to engineering education'. He was largely involved in the Saint Anthony Falls SAF Hydraulic Laboratory of the University of Minnesota, serving as director from 1963 to 1974. He also produced the film Fluid mechanics – The boundary layer in 1961. His research interests included water resources management, model studies of hydraulic and fluid flow phenomena, super-cavitating flows, boundary layers, turbulence, air-water mixtures, flow losses in closed and open conduits, and underwater acoustics. His notable work on friction factors may be considered classic.

Anonymous (1991). Silberman. *ASCE News* (12): 75-76. *P*
Anonymous (1994). Silberman, E. *American men and women of science* 6: 913. Bowker: NJ.
Carter, R.W., Einstein, H.A., Hinds, J., Powell, R.W., Silberman, E. (1963). Friction factors in open channels. *Journal of the Hydraulics Division* ASCE 89(HY2): 97-143; 89(HY4): 283-293; 89(HY5): 169-170; 89(HY6): 265-276; 90(HY4): 223-227.
Silberman, E., Ripken, J.F. (1959). The St. Anthony Falls Hydraulic Laboratory gravity-flow free-jet water tunnel. *Technical Paper* 24B. SAF: Minneapolis.
Silberman, E. (1961). Instability of ventilated cavities. *Journal of Ship Research* 5(1): 13-33.
Silberman, E. (1980). Boundary layers in developing open channel flow. *Journal of the Hydraulics Division* ASCE 106(HY7): 1237-1241; 107(HY4): 527-528; 107(HY10): 1275.
Silberman, E. (1983). Discussion of The effect of drag-reducing additives on fluid flows and their industrial applications. *Journal of Hydraulic Research* 21(1): 72-73.

SIMONS

12.02. 1918 Payson UT/USA
03.03. 2005 Fort Collins CO/USA

Daryl Baldwin Simons received his BS degree from Utah State University, Logan UT, only in 1947 after war service. After having made also the MS, he joined the University of Wyoming, Laramie WY, as professor of civil engineering. In 1957 he earned the PhD degree at Colorado A&M College, Fort Collins, for a work on stable channels in alluvial material, guided by Maurice L. Albertson (1918-2009). The resulting ASCE papers became classic texts in the field of sediment transport, the 1964 paper being awarded the Croes Medal by the American Society of Civil Engineers ASCE.

From 1957 to 1963 Simons and Everett V. Richardson (1924-2012) conducted probably the most comprehensive laboratory and field experiments on sediment transport within the US Geological Survey. The objective of their research was to unravel the many complexities of surface water flow in alluvial streams. One of the results was the streambed classification for sand bed streams into a lower and an upper flow regime. In 1964 Simons joined Colorado State University, Fort Collins CO, as professor of civil engineering, teaching erosion, scour and sediment transport, river mechanics, hydraulic structures, soil mechanics, and fluid mechanics. During the 1970s he analyzed riprap to protect stream beds and banks from erosion. He was awarded the Hunter Rouse Lecture Award in 1991, and was named an outstanding professional engineer for Colorado.

Albertson, M.L., Barton, J.R., Simons, D.B. (1960). *Fluid mechanics for engineers*. Prentice-Hall: Englewood Cliffs NJ.
Anonymous (1960). Daryl B. Simons. *Civil Engineering* 30(12): 78. *P*
Anonymous (1979). Daryl B. Simons. *Civil Engineering* 49(10): 95. *P*
Anonymous (1992). Daryl B. Simons. *Journal of Hydraulic Engineering* 118(12): 1607. *P*
Bhowmik, N., Richardson, E.V., Julien, P.Y. (2008). Daryl B. Simons: Hydraulic engineer, researcher, and educator. *Journal of Hydraulic Engineering* 134(3): 287-294. *P*
Simons, D.B., Richardson, E.V. (1962). Resistance to flow in alluvial channels. *Trans. ASCE* 127: 927-954; 127: 954-1006.
Simons, D.B., Albertson, M.L. (1963). Uniform water conveyance channels in alluvial materials. *Trans. ASCE* 128(1): 65-107; 128(1): 153-167.
Simons, D.B., Richardson, E.V. (1963). Forms of bed roughness in alluvial channels. *Trans. ASCE* 128: 284-323.
Simons, D.B., Sentürk, F. (1992). *Sediment transport technology*: Water and sediment dynamics. Water Resources Publications: Fort Collins CO.

SKOGERBOE

01.04. 1935 Cresco IA/USA
20.04. 2005 Ogden UT/USA

Gaylord Vincent Skogerboe received the BS and the MS degrees in the early 1960s from University of Utah, Logan UT. He was then a resident engineer on the construction of Porcupine Dam in northern Utah until 1963, when joining his Alma Mater as a hydraulic engineer until 1968, then becoming until 1977 staff member of the Colorado State University, Fort Collins CO, during which period he developed best management practices for salinity control in the Grand Valley, western Colorado. He was in parallel concerned with the campus leadership for water management research projects both in Pakistan and Vietnam from 1974 to 1980, from when he presided over a small consulting firm at Fort Collins. Later, Skogerboe became professor in the Biological and Irrigation Department of the Utah State University, Logan UT, for a long term assignment as director of the International Irrigation Management Institute in Pakistan.

Skogerboe authored numerous papers, and had conducted laboratory and field research particularly on the discharge measurement, and hydraulics in general. He had trained hundreds in domestic and overseas courses, and served as advisor of dozens of graduate students. He is remembered for his Cut-throat Flume, corresponding to an open channel discharge measurement structure of limited length. Whereas the standard structures have a length of some ten approach flow widths, his structure is much shorter involving a polygonial plan shape. For free flow conditions, the measurement of the approach flow depth alone allows for discharge prediction, whereas the tailwater depths needs to be known in addition for submerged flow conditions.

Anonymous (1985). Skogerboe, G.V. *Who's who in engineering* 6: 611. American Association of Engineering Societies: Washington DC.

Skogerboe, G.V., Hyatt, M.L. (1967a). Analysis of submergence in flow measuring flumes. *Journal of the Hydraulics Division* ASCE 93(HY4): 183-200; 94(HY3): 774-794; 94(HY6): 1530-1531.

Skogerboe, G.V., Hyatt, M.L. (1967b). Rectangular cut-throat flow measuring flumes. *Journal of the Irrigation and Drainage Division* ASCE 93(IR4): 1-13; 94(IR3): 357-362; 94(IR4): 527-530; 95(IR3): 433-439.

Skogerboe, G.V., Merkley, G.P. (1996). *Irrigation maintenance and operations learning process.* Water Resources Publications: Littleton CO.

http://www.wrpllc.com/authors/imolauthors.html *P*

SKRAMSTAD

26.07. 1908 Tacoma WA/USA
17.10. 2000 Melbourne Beach FL/USA

Harold Kenneth Skramstad graduated from Puget
Sound College, Tacoma WA, with the BS degree in
1930, and the PhD degree in 1935 from University
of Washington, Seattle WA. He then joined the staff
of the Aerodynamics Section, National Bureau of
Standards NBS, engaged in wind tunnel turbulence.
During the war he worked on guided missiles and
played a key role in developing the Bat, the first
fully-automatic guided missile. For this work, he
received the Navy Bureau of Ordnance Exceptional
Service Award in 1945, and in 1991 was elected to
the Missile Technology Historical Association Hall
of Fame. He became in 1946 chief of the NBS Guided Missiles Section. Later, he
pioneered in digital computation and the development of computers. In 1967 he retired
from government service joining the University of Miami, Coral Gables FL, as scientific
director of the computing center and professor of industrial engineering. He then taught
until 1984 courses at Florida Institute of Technology, Melbourne FL.

Skramstad contributed to the stability problem of laminar boundary layer flow. The
original equations were published by William M. Orr (1866-1934) in 1906, who found
an exact solution for plane Couette flow, but was unable to find any unstable regions for
this particular case. In 1908, Arnold J. Sommerfeld (1868-1951) independently derived
these equations, yet without improving Orr's results. In 1924, Werner K. Heisenberg
(1901-1976) attempted a complicated asymptotic analysis which led to an erroneous
result. Walter G. Tollmien (1900-1968) was more successful in 1929 and developed a
stability curve. In 1944, Chia C. Lin (1916-2013) extended Tollmien's results clarifying
the behaviour of the solution. The 1947 paper of Galen B. Schubauer (1904-1992) and
Skramstad completed the analysis using experiments on the effects of disturbances of a
boundary layer, thereby confirming Lin's and Tollmien's analyses.

Anonymous (1946). Skramstad named chief of guided missiles section. *Aeronautical
    Engineering Review* 5(12): 8. *P*
Eckert, M. (2008). Turbulenz: Ein problemhistorischer Abriss. *Zeitschrift für Geschichte der
    Wissenschaften, Technik und Medizin* 16(1): 39-71.
Schubauer, G.B., Skramstad, H.K. (1947). Laminar boundary-layer oscillations and stability of
    laminar flow. *Journal of the Aeronautical Sciences* 14(2): 69-78.
Skramstad, H.K. (1962). Combined analog-digital techniques in simulation. *Advances in
    Computers* 3: 275-298.

SKRINDE R.A.

11.03. 1919 Stanwood WA/USA
07.12. 2009 Stanwood WA/USA

Raymond Arthur Skrinde, the brother of Rolf T. Skrinde (1928-), was educated at Washington State University, Pullman WA, graduating with the SB degree in civil engineering in 1945. He received the MS degree from University of Iowa, Iowa City IA, in 1947. After war service in the Pacific, he retired from the US Army as lieutenant colonel. After more than 30 years with the Seattle District of the US Army Corps of Engineers, he retired to his beloved McKee's Beach home, while wintering in Arizona. He was a strong supporter of his local chapter of the Sons of Norway. He was a member of the American Society of Civil Engineers, who awarded him in 1951 its Collingwood Prize.

Skrinde has written only one paper, published in 1950. This work made him famous in hydraulic structures, because it deals with a basic hydraulic problem. The hydraulic jump is a phenomenon dissipating a considerable amount of hydraulic energy. The application of this principle is used for the design of stilling basins, which are normally located downstream of a hydraulic structure, of which the flow has too much energy to be released into the tailwater, so that a stilling basin is often added to dissipate the excess hydraulic energy mainly into heat. In contrast to a hydraulic jump, whose stability depends significantly on the tailwater level, a stilling basin is designed so that a range of tailwater conditions is controlled. To achieve this goal, a stilling basin has well-designed appurtenances, including for instance sills, as were investigated in the 1950 paper. The approach of John W. Forster (1927-2007) and Skrinde is one of the very early accounts of this method by using the momentum equation. Depending on the sill height, its position from the toe of the hydraulic jump and the so-called approach flow Froude number, the sill may result in poor flow conditions, so that its proper design is essential. This paper has led many later studies on the performance of appurtenances in hydraulic jump stilling basins, and therefore is a worthwhile and relevant addition to hydraulic engineering.

Anonymous (1951). Raymond A. Skrinde. *Civil Engineering* 21(10): 605-607. *P*
Foster, J.W., Skrinde, R.A. (1950). Control of the hydraulic jump by sills. *Trans. ASCE* 115: 973-987; 115: 1007-1022.
Skrinde, R.A. (1985). *History of Agency and members*. ASCE Seattle Section: Seattle.
http://www.legacy.com/obituaries/seattletimes/obituary.aspx?n=raymond-a-skrinde&pid= 137223829 *P*

SLEIGHT

30.06. 1889 Laingsburg MI/USA
14.11. 1927 Montpellier VT/USA

Reuben Benjamin Sleight obtained in 1911 the civil engineering degree from the University of Michigan, Ann Arbor MI. He moved in 1912 to Denver CO in charge of irrigation works and was employed as construction foreman on irrigation structures near Tucson AZ, but shortly later was transferred to a duty-of-water study in the Mesilla Valley NM. He there had to design pumping plants, assisting also in tests on submerged orifices, which resulted in the 1925 Report, in collaboration with Fred L. Bixby (1880-1955). He was appointed assistant irrigation engineer of the US Department of Agriculture until 1917, assigned in 1914 to the Headquarters Office, Washington DC, as assistant to Fred C. Scobey (1880-1962) on hydraulic engineering research. He was in parallel in 1914 on leave of absence from his government position at the University of Michigan, from where he gained the BS degree in marine engineering in 1915. He then moved to Denver CO to study the evaporation from free water surfaces, soils, and river-beds, the results of which were published in the 1917 paper.

After war service, Sleight returned to Ann Arbor as appraisal engineer. He was involved in projects of the Kentucky and West Virginia Power Companies, and the Columbus Power Company. From 1922 he became engineer with the Minnesota Tax Commission supervising the valuation of public utilities; he placed the Department on an efficient and effective basis. Because of his ambition, he was recognized as an outstanding leader in engineering and scientific circles. He was member of the Minneapolis Engineer's Club, authored a number of technical papers, including a notable paper on engineering economics, so that he became aide to Secretary Herbert Hoover, Federal Department of Commerce. There he was in charge to detail work on special waterways problems. On the day of his death, he had left Washington DC in an airplane to gather advance information in the areas then flooded. The injuries from which he died were suffered during the landing at Montpellier VT. A much too young life had come to an end.

Anonymous (1912). Reuben Sleight. *Michiganensian yearbook*: 115. *P*
Anonymous (1928). Reuben B. Sleight. *Trans. ASCE* 92: 1775-1777.
Bixby, F.L. (1925). Tests on deep-well turbine pumps. *J. Agricultural Research* 31(8): 227-246.
Gregg, E.S., Cricher, A.L., Sleight, R.B., Titus, N.F. (1927). *Great Lakes-to-Ocean waterways*: Some economic aspects of the Great Lakes-St. Lawrence waterway project. Washington.
Sleight, R.B. (1917). Evaporation from the surfaces of water. *J. Agricultural Res.* 10(8): 209-262.

SLICHTER C.S.

16.04. 1864 St. Paul MN/USA
04.10. 1946 Madison WI/USA

Charles Sumner Slichter was the first American, who significantly contributed to groundwater flow. Graduating from Northwestern University, Evanston IL, he received the BS degree in 1885, and the MS degree in 1887. From 1889 as assistant professor of mathematics, he became in 1892 then professor of applied mathematics as also Dean of the Graduate School, University of Wisconsin, Madison WI, until his retirement in 1934. He was in parallel consulting engineer both for the US Geological Survey USGS, and the US Reclamation Service. He was a member of the Wisconsin Academy of Sciences, Arts and Letters presiding it from 1900 to 1903, the Mathematical Association of America MAA, the American Geophysical Union AGU, and USGS.

Slichter worked in engineering and applied mathematics, enjoying the challenge of applying abstract formulas to the solution of practical difficulties. As a member of USGS he began to specialize in problems of groundwater flow, developing an extensive consulting career dealing also with irrigation and urban water supply. He developed his 'electrical' method to determine the velocity of groundwater flows. Slichter focused also on enlarging the University's support for scholarly research of all types, and was a major figure in establishing the Wisconsin Alumni Research Fund.

Anonymous (1927). Charles S. Slichter. *Engineering News-Record* 99(5): 194. *P*
Anonymous (1939). Slichter, Charles S. *Who's who in America* 20: 2291. Marquis: Chicago.
Chamberlin, T.C., Moulton, F.R., Slichter, C.S., MacMillan, W.D., Lunn, A.C., Stieglitz, J. (1909). *The tidal and other problems*. Carnegie Institution: Washington DC.
Ingraham, M.H. (1972). *Charles Sumner Slichter*: The golden vector. University of Wisconsin Press: Madison. *P*
Slichter, C.S. (1898). Note on the pressure within the earth. *Journal of Geology* 6(1): 65-78.
Slichter, C.S. (1899). Theoretical investigation of the motion of ground waters. *Annual Report* 19(2): 195-284. US Geological Survey: Washington DC.
Slichter, C.S. (1902). Motions of underground waters. *Water Supply Paper* 67. US Geological Survey: Washington DC.
Slichter, C.S. (1904). Measurements of underflow streams in Southern California. *Journal of the Western Society of Engineers* 9(6): 632-653.
Slichter, C.S. (1905). Field measurements of the rate of movement of underground waters. *Water Supply and Irrigation Paper* 140. USGS: Washington DC.

SLICHTER F.B.

23.12. 1904 Kansas City MO/USA
12.01. 1992 Salt Lake City UT/USA

Francis Benjamin Slichter obtained in 1926 the BS degree in civil engineering from the University of Kansas, Lawrence KS. He was from 1926 junior test engineer at the Kansas City Power & Light Co., from 1928 junior assistant engineer in supervising drafting and preliminary designs of US Engineering Office, Kansas City MO, and there civil engineer from 1934 in charge of earth dam and foundation designs until 1941. He was involved in Fort Peck Dam in Montana, on the *Kanopolis* and Marshall Creek Dams in Kansas, the Colburn and Unionville Dams in Missouri, as also levees of the Kansas City Flood Protection Scheme. After war service until 1945, Slichter became assistant chief of the Engineering Division, involved in many flood control and navigation projects, including the Garrison Dam in North Dakota, the Fort Randall Dam in South Dakota, and the Cherry Creek Dam in Colorado. From 1947 he was secretary of the Missouri Basin Inter-Agency Commission, taking then over as chief engineer the Division of Civil Works, the Office of the Chief Engineers, supervising the design and construction of improvements for flood control, rivers and harbours, and hydropower.

Slichter, as chief of the Engineering Division, Civil Works, in the Office of the Chief of Army Engineers, retired in 1961 after 32 years of service. He then established a practice as consulting engineer at Burke VA. As chief of the engineering division at the Omaha headquarters of the Missouri River Division in the 1940s, he developed the Pick Sloan Plan. A series of droughts during the 1930s and severe flooding in 1943 along Missouri River prompted water planning for the Missouri River Basin. It was felt that increased management of the river and its tributaries would not only reduce flooding, but also benefit the region's economy through increased irrigation and hydropower production, enhanced navigation, and job opportunities for veterans returning from World War II. Following Congressional approval of that plan, Slichter was responsible for getting designs and construction, with the Garrison, the Oahe, and Fort Randall Dams as its key reservoir projects. Slichter was member of the American Society of Civil Engineers.

Anonymous (1959). Slichter, Francis B. *Who's who in engineering* 8: 2270. Lewis: New York.
Anonymous (1961). Slichter retires from the Corps. *Engineering News-Record* 166(Feb.9): 51. *P*
Bennett, P.T., Slichter, F.B. (1940). *Electrical conduction models for the solution of problems involving the flow of water through soils*. US Engineer Office: Kansas City MO.
Peck, R.B., Slichter, F.B. (1984). *Geological surprises at Itezhitezhi Dam*. St. Louis MO.

SLOCUM H.

23.10. 1887 National City CA/USA
10.11. 1961 Naya Nangal/IN

Harvey Slocum was a self-made construction man in the absolute sense of the term. He quit school after the eighth grade to become successively an apprentice and journeyman carpenter, an ironworker, squad boss on concrete work, and foreman by the time he was 25 years old. Five years later he was construction superintendant on Lake Hodges Dam, a multiple arch built for the city of San Diego CA. He never left any doubt on his jobs that he was the boss, which earned him the affection of his workmen, the fear of his foremen, and the deep respect of his employers.

Slocum completed in the early 1950s the Bull Shoals Dam in Arkansas, a Corps of Engineers flood control facility on White River. Before he was in charge of construction of the Bureau of Reclamation's Friant Dam in California, and on the Army's Madden Dam in Panama. Slocum's biggest job was that of general superintendant for Grand Coulee Dam on Columbia River. In 1952 he was engaged by the Government of Punjab State, India to construct *Bhakra* Dam, often referred to also as India's Hoover Dam, 210 m high on the Sutlej River in northwest India. Sutlej River has its source in the lake of demons on the western highlands of Tibet. During the 1950s, Bhakra Dam was the central unit in India's Bhakra-Nangal power and irrigation scheme. It was the second highest concrete gravity-type dam in the world, after the *Grand-Dixence* Dam in Switzerland. Its crest length is 500 m, it includes a central overflow-type spillway equipped with 12 m high gates. There are 16 irrigation and reservoir control outlets. The dam impounds a 170 km^2 reservoir, 100 km long. The scheme provides irrigation water for 10 million acres of land, of which two-thirds were never irrigated before. In 1963 India's president Jawaharlal Nehru inaugurated Bhakra Dam, which he described as 'something which shakes you up….the new temple of resurgent India'.

Anonymous (1952). Harvey Slocum to direct India dam building. *Engineering News-Record* 148(Apr.17): 45-46. *P*
Anonymous (1958). Harvey Slocum, chief of construction. *Engineering News-Record* 161(Dec.11): 42. *P*
Anonymous (1961). Dambuilder Harvey Slocum dies. *Engineering News-Record* 167(Nov.16): 28. *P*
http://www.life.com/image/50348437 *P*
http://www.life.com/image/50870298 *P*

SLOCUM S.E.

05.06. 1875 Glenville NY/USA
22.09. 1960 Ardmore PA/USA

Stephen Elmer Slocum graduated in 1897 from the Union College, Schenectady NY, continued studies at Clark University, Worcester MA, obtaining there in 1900 the PhD degree in mathematics and physics. He was until 1901 instructor of civil engineering, and of applied mathematics until 1904 at University of Cincinnati, Cincinnati OH, becoming until 1906 assistant professor of mathematics at University of Illinois, Urbana IL, professor of applied mathematics at University of Cincinnati until 1920, from when he was a consultant at Philadelphia PA, and Ardmore PA. He was a member of the American Society of Civil Engineers ASCE, the American Society of Naval Engineers, which awarded him the Gold Medal in 1927 for original research in modern hydromechanics.

Slocum was an expert in marine propulsion and in noise and vibration engineering. He has written the textbook Hydraulics, of which the main chapters are: 1. Hydrostatics, 2. Hydrokinetics, 3. Hydrodynamics, and 4. Hydraulic data and tables. In the Preface it is stated that hydraulics has had an remarkable impetus in the USA so that the new aspects of this technical branch have to be added to the literature, both for educational purposes and for research. The extent and the cheapness of hydropower energy was stated to greatly improve modern the society in both private and commercial domains. The text presented in the book should include a basic treatment of the principles of hydraulics with adequate recent important works made in the country, including the Catskill Aqueduct in New York State, or the power plants at Niagara Falls. The book includes also information on turbines, and typical examples are added at the end of each section.

Anonymous (1937). Slocum, Stephen E. *Who's who in engineering* 4: 1272. Lewis: New York.
Anonymous (1959). Slocum, S.E. *Who's who in America* 28: 2466. Marquis: Chicago.
Anonymous (1971). Stephen E. Slocum. *Aeronautics and Astronautics* 10(9): 71. *P*
Slocum, S.E. (1913). *Theory and practice of mechanics*. Holt: New York.
Slocum, S.E. (1915). *Elements of hydraulics*. McGraw-Hill: New York.
Slocum, S.E. (1927). Practical application of modern hydrodynamics to marine propulsion. *Journal of the American Society of Naval Engineers* 39(1): 1-38.
Slocum, S.E. (1931). Experimental research on vibration dampers and insulators. *Trans. ASCE* 95: 85-128.
Slocum, S.E. (1954). The specific speed method of propulsive analysis. *Journal of the American Society of Naval Engineers* 66(3): 727-736.

SMAGORINSKY

24.01. 1924 New York NY/USA
21.09. 2005 Hillsborough NJ/USA

The parents of Joseph Smagorinsky originated from Belarus, fleeing from there during the early 20th century's pogroms, settling finally in the US. The son earned his BS, MS and PhD degrees from New the York University respectively in 1947, 1948, and 1953. He then was educated dynamical meteorology at the Massachusetts Institute of Technology MIT, Cambridge MA, and was later invited to Princeton University NJ, to study at the Institute for Advanced Study the prediction of large-scale motions in the middle troposphere using the electronic computer designed by John von Neumann (1903-1957). Before 1950 weather forecasting was crude. George Platzman (1920-2009) felt that 'academic meteorology is still suffering from the trade-school blues'. The leaders of the American Meteorological Society AMS aspired to turn meteorology into a professional discipline given the same respect accorded to engineering or physics.

During his PhD Smagorinsky had developed a new theory for heat flow created by the thermal contrast between land and oceans disturbed the path of the jet stream. In 1953 he accepted a position at the US Weather Bureau becoming soon one of the pioneers of the Joint Numerical Weather Prediction Unit. In 1955, a General Circulation Research Section GCRS was there created under Smagorinsky's direction. He and his group attempted a three-dimensional, global, primitive-equation general circulation model of the atmosphere. GCRS was originally located at Suitland MD, but moved then renamed in 1959 as General Circulation Research Laboratory to Washington DC. From 1963 it was today's Geophysical Fluid Dynamics Laboratory GFDL, from 1968 at Princeton University, with Smagorinsky as director until his retirement in 1983. He was among the earliest researchers who sought to exploit novel methods of numerical weather prediction extending forecasting past one or two days. His seminal 1963 paper involves the primitive equations of atmospheric dynamics to simulate its circulation. He thereby developed with his colleagues one of the first successful approaches to large-eddy-simulation, providing a solution that is currently used in many fields of fluid mechanics.

Smagorinsky, J. (1963). General circulation experiments with the primitive equations 1: The basic experiment. *Monthly Weather Review* 91(3): 99-164.
Smagorinsky, J. (1972). The general circulation of the atmosphere. *Meteorological challenges*: A history: 3-42. Information Canada.
http://en.wikipedia.org/wiki/Joseph_Smagorinsky *P*

SMITH C.A.

01.10. 1846 St. Louis MO/USA
02.02. 1884 Newburyport MA/USA

Charles Augustus Smith inherited from his parents the instincts of a sailor, because the blood of several generations of his old family was of ship-masters. Though he never became a sailor, he always showed a sailor's fondness for 'fixing things, for having his hands, and for actual construction'. In 1864 he was involved in works for a railroad at Boston MA, and in 1865 became chief assistant of the city engineer's office, Springfield MA. He graduated in the pioneer engineering 1868 class from Massachusetts Institute of Technology MIT, Cambridge MA, conducting additional works for the famous James B. Francis

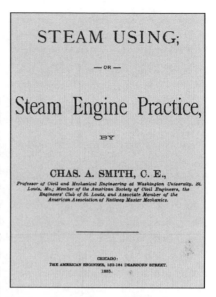

(1815-1892) at Lowell MA. After graduation he went into partnership with his former civil engineering professor, during which period he was again involved in railways projects.

Smith was also an instructor at MIT in civil and mechanical engineering, a position he held until 1883 when compelled by his last illness to resign. He was in parallel engaged as a consulting engineer for the construction of water works in Massachusetts and in Missouri. His last professional duties were in connection with these works. The pumping works of Richmond VA were designed by him, his plans being entered in competition and receiving the first prize. In 1879 he spent the summer vacations as resident engineer of the Baltimore Bridge Company, building piers in Mississippi River just below Minneapolis. The fatal malady in the shape of a cancerous tumor brought his life to an untimely end, at only 38 years of age. He had started from 1883 to work on his two books, lying on the bed, or reclining in an easy chair. Smith was considered a good example of a poor boy who made his own way, who fought his own battles, and who earned and honoured each position he took. As an engineer, he was trustworthy and bold. His professional confidence was based upon sound theory and careful practice. These accomplishments added greatly to his value as an instructor of young engineers.

Anonymous (1885). Charles A. Smith. *Proc. ASCE* 11(11): 122-123.
Smith, C.A. (1881). Discussion of Wind pressures upon bridges. *Trans. ASCE* 10: 164-165.
Smith, C.A. (1884). *Steam using or steam engine practice*. The American Engineer: Chicago. (*P*)
Smith, C.A. (1885). *Steam-boilers*. The American Engineer: Chicago.
Smith, C.A. (1886). Permanent transmitting dynamometer. *Trans. ASCE* 15: 357-358.
Woodward, C.M. (1884). *Preface* to Steam using or steam engine practice. The American Engineer: Chicago.

SMITH E.S.

28.03. 1897 Angola IN/USA
31.12. 1960 Baltimore MD/USA

Edward Sinclair Smith graduated from University of California, Berkeley CA, with the BS degree in 1919, and the ME degree in 1932. He was until 1923 a staff member of the Standard Oil Company, the Pacific Gas and Electric Company, and the Shell Petroleum Company, all in California. He moved to the East Coast as assistant chief engineer of Builders Iron Foundry, Providence RI, working as hydraulic engineer, and then as patent engineer. He spent a number of years with a manufacturer at Brooklyn NY, in charge of company patents, and as consultant on flow problems. He was employed from 1940 to 1948 as a research engineer of automatic control in Teterboro NJ, where his work included the development of stabilizing suction relief valves for fighters and super-chargers for bombers. In the 1950s he was concerned with the design of porosity meters for parachute fabric, scoops for bomber carburettors and for naval vessels, and essential hydraulic components for the Pearl Harbour dry dock. Finally, he was a scientific and technical adviser, Weapons Systems Laboratory, US Army Ordnance, Aberdeen MD.

Smith was a pioneer in fluid metering. One of his achievements was authorship of the book Automatic control engineering, the first of its kind in English. Much of his business activity was as liaison between the engineering and patent fields. He acquired experience with several companies until 1923, then used his knowledge in problems of fluid metering until the War started, when serving for war support. More than a dozen different devices were patented. He was elected Fellow of the American Society of Mechanical Engineers in 1946. He was one of the founders of the ASME Committee on Instruments and Regulators. Earlier, he was the winner of the 1931 ASME Junior Award. He was also a member of the Instrument Society of America.

Anonymous (1938). Ed S. Smith. *Mechanical Engineering* 60(12): 964. *P*
Anonymous (1946). Ed S. Smith. *Mechanical Engineering* 68(11): 1013. *P*
Anonymous (1961). Ed Sinclair Smith. *Mechanical Engineering* 83(4): 132.
Smith, E.S. (1923). The oil Venturi meter. *Trans. ASME* 45: 67-75.
Smith, E.S. (1930). Quantity-rate fluid meters. *Trans. ASME* 52(HYD 7b): 89-109.
Smith, E.S. (1930). Fluid metering. *Mechanical Engineering* 52(4): 372-374; 52(11): 968-970. *P*
Smith, E.S. (1934). The V-notch for hot water. *Trans. ASME* 56(9): 787-789; 57(3): 249-250.
Smith, E.S. (1939). Relations involved in metering rate-of-flow. *Instruments* 12(4): 115-126.
Smith, E.S. (1944). *Automatic control engineering*. McGraw-Hill: New York.

SMITH H.

05.07. 1840 Louisville KY/USA
04.07. 1900 Durham NH/USA

Hamilton Smith started his professional career with his father at age fourteen, who owned then a cotton factory and coal mines. Industrious, competent and with unusual aptitude for mathematics, he soon demonstrated his ability, and was recognized as chief engineer of the mines. During the 1860s, Smith developed other activities in Kentucky and Indiana, but was attracted to the Pacific Coast by its apparent great opportunities. He was engineer and manager first of a mine in Lower California, but his most notable were in Nevada County, which were worked by hydraulic methods, for which he became an authority. Attracting the favourable attention of Baron Rothschild, who made a visit to the properties, Smith became consulting mining engineer for the Rothschild interests.

In 1885 Smith opened a consulting office in London, and there in 1886 published his notable book Hydraulics. This classic text in hydraulics includes the following chapters: 1. Properties of water, 2. Theory of hydraulics, 3. Flow through orifices, 4. Velocity of approach, 5. and 6. Flow over weirs, 7. Flow through open conduits, 8. Flow through pipes, 9. to 11. Experiments. Besides this treatise, Smith also presented three additional papers published in the Transactions ASCE, dealing with pipe flow, water power, and temperature in lakes and oceans. His 1886 paper on the mining costs attracted wide attention, and his Exploration Company soon became an important factor in the development of mines throughout the world, notably in South Africa, where gold had been discovered in 1885. In 1895 Smith moved the headquarters of his office to New York. His death occurred through accidental drowning.

Anonymous (1900). Hamilton Smith. *Engineering News* 44(18): 300. *P*

Anonymous (1900). Hamilton Smith. *The Engineering and Mining Journal* 70(Jul.14): 34. *P*

Anonymous (1935). Smith, Hamilton. *Dictionary of American biography* 17: 273-274.
        Scribner's: New York.

Anonymous (1991). Smith, Hamilton. *A biographical dictionary of American civil engineers* 2:
        109-110. ASCE: New York.

Smith, H. (1883). The flow of water through pipes. *Trans. ASCE* 12(4): 119-125.

Smith, H. (1884). Temperature of water at various depths in lakes and oceans. *Trans. ASCE* 13:
        73-84.

Smith, H. (1886). *Hydraulics*: The flow of water through orifices, over weirs, and through open
        conduits and pipes. Wiley: New York.

SMITH J.W.

09.03. 1861 Lincoln MA/USA
14.10. 1933 New York NY/USA

Jonas Waldo Smith graduated in 1881 from Phillips Academy, Andover MA. He was then an assistant of the Essex Company at Lawrence MA, proprietors of the water power of *Merrimac* River. In 1887 he graduated as civil engineer from the Massachusetts Institute of Technology. He had then also spent two summers with the Holyoke Water Power Company, and served there after graduation until 1890. He became assistant engineer for the East Jersey Water Company and was engaged during the next twelve years in the construction and maintenance of water supply systems in northern New Jersey. He was appointed principal assistant engineer in 1891 and was ultimately chief engineer. He directed in 1901 the design and construction of the Little Falls mechanical filtration plant, a pioneer at this time, and the largest of its kind. He then supervised the water supply system for New Jersey City, including *Boonton* Dam, and a concrete aqueduct.

Smith accepted in 1903 the position of chief engineer of the Aqueduct Commission of New York City, taking charge of the construction of the New Croton Dam then under way, at the time the largest masonry dam in the world. He also made surveys for the Cross River and Croton Falls reservoirs, which were subsequently constructed. When the Board of Water Supply of the City of New York was created in 1905 to provide additional drinking water supply, Smith became its chief engineer, and began the most important work of his life as director of the *Catskill* Water Supply System. The initial design for this project included the *Ashokan* Reservoir, about 150 km from New York City, controlled by a masonry dam 75 m high and 300 m long, and *Catskill* Aqueduct, with a design capacity of some 20 m³/s. The Aqueduct, which is large enough for a railroad train to pass, terminated at *Kensico* Reservoir 50 km from the city, where another masonry dam was completed in 1918. Smith resigned his activities with the Board but continued as a consultant until his sudden death. He was awarded the John Fritz Medal in 1918, then the highest honour in the engineering profession.

Anonymous (1917). Catskill Aqueduct Builders. *Engineering News-Record* 79(15): 676-680. *P*
Anonymous (1925). Smith, Jonas Waldo. *Who's who in America* 13: 2962. Marquis: Chicago.
Anonymous (1931). J. Waldo Smith. *Engineering News-Record* 106(Mar.19): 498. *P*
Anonymous (1933). J. Waldo Smith. *Civil Engineering* 3(11): 642. *P*
Eliot, S.A. (1909). Smith, Jonas Waldo. *Biographical history of Massachusetts* 7: 387-391.
        Massachusetts Biographical Society: Boston. *P*

SMITH R.L.

31.10. 1923 Schaller IA/USA
09.12. 1995 Kansas City KS/USA

Robert Lee Smith graduated from the University of Iowa, Iowa City IA, and there received in 1948 the MSc degree. He became then assistant professor of applied mechanics at University of Kansas, Lawrence KS, being awarded from 1962 to 1966 Glen Parker Professor of water resources, and director of the Water Resources Institute; professor and chairman of the Department of Civil Engineering until 1972, being named in 1970 Deane Ackers professor until his retirement in 1989. He was member and Fellow of the American Society of Civil Engineers ASCE.

Smith had a lifelong career both as engineer and public servant. He helped shape the water policy for both Kansas and Iowa. He was special assistant on water resources in the Lyndon B. Johnson administration. His professional activity was in 42 states, the District of Columbia, Puerto Rico, and nine foreign countries. He received the 1993 distinguished engineering award from the School of Engineering: 'Smith's career has been a unique combination of public service, private sector practice, university teaching, and research on policy issues that involve water'. In 1975 he was the first member of Kansas University named to the National Academy of Engineering NAE. He received the 1988 Julian Hinds Award from ASCE, and was a recipient of the US Geological Survey Centennial Plaque. In 1990 he received the distinguished alumni achievement award from the University of Kansas, and the Ray K. Linsley Award from the American Institute of Hydrology. He further was a member of many honorary and professional organizations. 'One of the problem with water management', he said, 'is that it's both a chemical solvent and a social solvent. Some people irrigate, some generate power, some float barges, some even transport wastes. And some people use water just as a host to solitude'. He was posthumously awarded the Distinguished Engineering Alumni Award of his Alma Mater.

Anonymous (1996). Robert L. Smith. *Trans. ASCE* 161: 571-572.
McKinney, R.E. (1996). Robert L. Smith. *Memorial tributes* 8: 236-241. NAE: Washington. *P*
Smith, R.L. (1978). *Rural and urban stormwater runoff.* Water Quality Management: Topeka.
Smith, R.L., Carswell, Jr., W.J. (1984). Average annual fulfilment of instream uses. *Journal of Water Resources Planning and Management* 110(4): 497-510; 113(3): 446-448.
Smith, R.L. (1985). *Federal policies in water resources planning.* ASCE: New York.
http://www.engineering.uiowa.edu/alumni-friends/honor-wall/distinguished-engineering-alumni-academy-members/dr-robert-l-smith *P*

SMITH W.E.

20.08. 1900 New Hampton IA/USA
12.08. 1994 Washington DC/USA

Waldo Edward Smith graduated from the University of Iowa, Iowa City IA, with a MS degree in 1924. He then worked on hydraulic projects and was also involved in teaching activities at various universities. In 1936 he joined the American Geophysical Union AGU, taking an active engagement in its Hydrology Section. He moved in 1940 to Washington DC.

Up to the early 1940s, AGU with then some 2,200 members was mainly an association of volunteers in its administration. In 1944, with the title of Executive Secretary, Smith became AGU's first full-time staff member, and subsequently as Executive Director, he has presided over a period of 26 years during which AGU grew significantly. Smith presented a sound financial status, and managed AGU's books and journals. For many years he edited the Transactions AGU himself while operating the business affairs. Reviewing the accomplishments of Smith is to review the history of AGU during his term. In 1944 the Union published bi-months its Transactions. By 1960 the membership had grown to 6,100 and the Union published a series of books and the Journal of Geophysical Research in addition to the Transactions. In 1970, AGU counted 10,000 members and it was involved in three book series, a large translation program, three primary journals and EOS, the newsletter. With a firm but fatherly approach Smith directed a staff of then 40 people. On the international scene, with his hat on as Secretary of the US National Committee for the International Union of Geodesy and Geophysics IUGG, Smith provided a stimulus at home and a continuing contact for those abroad rendering the US National Committee particularly active. His position was though not an easy one, because he was pushed from all sides, but the record speaks for itself. His personality was a significant force within AGU. It was this personality that has brought matters back to an even keel, injected common sense and infused members with the sense of pride in the services and accomplishments of AGU. He was particularly known for his talent in organisation, his sharpness and his perseverance. In addition he had the capacity to recognize by name almost all of the AGU members, which by today has increased to 60,000 members. The Waldo E. Smith Medal created in 1982 recognizes individuals who have played a unique leadership.

Anonymous (1971). Waldo E. Smith. *Civil Engineering* 41(3): 6. *P*
Newell, H.E. (1970). Waldo E. Smith: A quarter century in service. *IAHS Bulletin* 15(4): 128. *P*
Smith, W.E. (1931). Byzantine aqueduct still in use. *Civil Engineering* 1(14): 1249-1254.
http://de.wikipedia.org/wiki/Waldo_E._Smith

SMITH W.M.

26.10. 1867 Newberry SC/USA
12.03. 1953 Glenn Springs SC/USA

Walter Mickle Smith received education from the Citadel of Charleston, Charleston SC, with the BSc degree in 1889, and the civil engineering degree in 1912. The title Doctor of Science was conferred on him there in 1933. He was in 1890 and 1891 engaged in railroad construction at Chattanooga TN. During the next ten years he was designer and consultant for the US Army Corps of Engineers, active on jetty constructions and coast fortifications at Charleston SC, and on fortification works at Portland ME. He became US assistant engineer again at Charleston in 1903. From 1905 to 1907 he was a consultant on the preliminaries to construction of the Panama Canal as division engineer. With designs so successful as the great Kensico Dam, and the Bronx earth dam for the New York Board of Water Supply, he earned the prestige which placed him among the great US engineers.

In 1914, Smith was called to Dayton OH, to design the flood control structures for the Miami Conservancy District, where his experience and application of engineering principles gained him admiration and cooperation. He was a strong advocate of the theory of 'least work', by use of which he designed large diameter conduits under heavy transport loads, and a canal lock wall containing a large diameter conduit. He resigned from the Conservancy District in 1919, initiating the plans for the Illinois Waterway, which was completed in 1933. He served during this period as chief designing engineer, assistant chief engineer, chief engineer, and consultant. In addition to this project, the plans of the State Department of Public Works included flood control and drainage works for many of the waterways of the state. The plans for the Illinois Waterway called for canalization of the Des Plaines and Illinois Rivers from Lockport IL, to the head of navigation at Starved Rock near Ottawa IL. Smith continued his position with the state agency until his retirement in 1947, acting in 1944 also as consultant for the Greater Chicago Lake Water Company and the Metropolitan Company, Chicago IL. He was remembered for his ability to sell the idea of an engineering project under trying and difficult circumstances. He was member of the American Society of Civil Engineers ASCE from 1906, becoming in 1936 Life Member.

Anonymous (1954). Walter M. Smith. *Trans. ASCE* 119: 1362-1363.
Smith, W.M. (1910). *Design of masonry dams*. Society of Civil Engineers: Albany NY.
Smith, W.M. (1931). Engineering features of Illinois Waterway. *Proc. ASCE* 57(8): 1189-1218.
http://en.wikipedia.org/wiki/Walter_Mickle_Smith
http://www.dnr.illinois.gov/WaterResources/Pages/Director.aspx *P*

SNOW

02.11. 1865 Providence RI/USA
29.10. 1942 Harrisburg PA/USA

Frank Herbert Snow started his career as surveyor at Brockton MA, becoming in 1890 city engineer. He designed a sewer system and a treatment plant. From 1896 to 1905 he was engaged in a private practice at Boston MA, and Columbus OH, conducting general engineering projects, including the sewerage system for the Metropolitan Water Board of Clinton MA, and water works at Peabody or Mansfield MA. He then moved to Pennsylvania where he lived until his death. He was appointed in 1905 chief engineer of the Pennsylvania Department of Health. First he was in charge to compile data of all public water works and sewerage systems of the State, realizing that the conditions were poor, and that a vast water supply network had to be designed to reduce waterborne disease. At this time, mainly filtration and chlorination were used.

In 1914 then, Snow was appointed chief engineer of the Public Service Commission of Pennsylvania, an office he held for 18 years. He supervised all engineering work; during these years the death rate from typhoid fever in Pennsylvania dropped drastically. After retirement in 1932, Snow organized an engineering firm, specializing in rehabilitation of industrial concerns. He was also elected secretary of the Pennsylvania Water Works Association, rendering a great service to this organization. 'In recognition of pioneer work and outstanding accomplishments in matters relating to public water supply', he was elected honorary member in 1940. He was proud as civil engineer, and had the burning passion to bring organization among engineers, for which he spent freely time and money. He was president of the Engineers Society of Pennsylvania, and member of the American Water Works Association AWWA, and of the American Society of Civil Engineers ASCE. The many reports that he wrote evidence his keen analytical mind, and the broad grasp of affairs. He had a wide acquaintanceship and numbered among his friends men prominent in the field of engineering, business, and municipal life. Snow carried on much pioneering in engineering work throughout his career.

Anonymous (1942). The late F. Herbert Snow. *Water Works Engineering* 95(25): 1501. *P*
Anonymous (1943). Frank H. Snow. *Trans. ASCE* 108: 1634-1638.
Snow, F.H. (1901). Discussion of The antecedents of the septic tank. *Trans. ASCE* 46: 472-476.
Snow, F.H. (1907). Discussion of Water supplies. *Proc. ASCE* 33(6): 688-696.
Venable, W.M. (1908). *Methods and devices for bacterial treatment of sewage*. Wiley: New York.

SONDEREGGER

14.04. 1875 Heiden AR/CH
03.09. 1964 Pasadena CA/USA

Arthur Ludwig Sonderegger graduated in 1900 as a civil engineer from the Swiss Federal Institute of Technology, ETH Zurich, Zürich, Switzerland. He was since 1904 consulting engineer at Los Angeles CA. During his career he was connected with some of the outstanding water and irrigation projects in the Southwest, namely the Imperial Valley, the Coachella Valley, the Pine Canyon Colorado River, or the Metropolitan Aqueduct for South California. He was member of the State Consulting Board for Southern California, the Water and Power Resources Committee, Los Angeles CA, the Metropolitan Water District, and of the Pasadena, San Marino and other water districts. He was member of the American Society of Civil Engineers ASCE, and in the 1930s president of its Los Angeles Section.

Sonderegger was a notable hydraulic engineer with interests in irrigation, water supply, and hydrologic engineering. The 1930 paper is concerned with the analysis of rainfall penetration on valley floors in semi-arid areas, including a discussion of methods to quantatively determine the resulting water supply, with a special reference to the conditions met in Southern California. Soil moisture tests allow for the determination of the rainfall penetration; a comparison with the mountain runoff gives the penetration on the valley floors, and the study of rainfall and irrigation allows for the characterization of penetration in addition. The 1933 paper deals with precipitation, erosion, sediment transport and deposition by stream flow, producing a so-called physiographical balance, which is undergoing in nature gradual changes. This work accounts for changes of the watershed, the regulation within a stream system, and the effect of debris barriers on the stream stability.

Anonymous (1964). *Los Angeles Section*: 50th anniversary 1914-1964. ASCE: New York.

Lätt, A., ed. (1931). A.L. Sonderegger. *Schweizer im Ausland*: 229. Sadag: Genf. *P*

Sonderegger, A.L. (1930). Water supply from rainfall on valley floors. *Trans. ASCE* 94: 1242-1311.

Sonderegger, A.L. (1933). Modifying the physiographical balance by conservation measures. *Proc. ASCE* 59(10): 1543-1563.

Sonderegger, A.L. (1941). Discussion of Transient flood peaks. *Trans. ASCE* 106: 234-239.

Sonderegger, A.L. (1946). Discussion of Correlating flood control and water supply, Los Angeles Coastal Plain. *Trans. ASCE* 111: 1145-1148.

SORENSEN

16.01. 1919 Minneapolis MN/USA
11.06. 1990 Oak Park IL/USA

Kenneth Edward Sorensen was, starting in 1946, for nearly forty years associated with Harza Engineering Company, Chicago IL. His career was charted in a number of increasingly important projects, including chief planning engineer, chief technical adviser, vice-president and, from 1983 to 1985, chairman of the board. He earned a BS degree in civil engineering from the University of Minnesota, Minneapolis MN in 1939, returning there to garner an MS degree. He was in World War II a lieutenant with the SeeBees in the Pacific.

Sorensen was a world-renowned water resources engineer. He was widely recognized as one of the world's outstanding planners of river development projects. One of his gifts, an approach that combined art, science and engineering, was his ability to conceive extensive river-basin projects, find ways to improve them, then conceive others of their worth. His successor at Harza said: 'Tens of millions of people throughout the world have a better life because of Sorensen's vision and ability'. His deep water resources experience included projects in seventeen river basins including more than twenty-five countries. Among them were several major hydroelectric projects, including the 2,800 MW *Inga* Project in Zaire or the 10,000 MW *Guri* Project in Venezuela. Sorensen also contributed to a master plan for irrigating the Jordan River Valley, as well as the Indus River Basin Project in Pakistan, where a 15-year plan was elaborated for developing agriculture, power and flood control in a 14 million hectare canal area. Sorensen was in addition co-editor of the Handbook on applied hydraulics, jointly with Calvin V. Davis (1897-1981). This successful work was based on two editions of 1942 and 1952 edited by Davis alone, and was in the 1970s one of the most cited text in hydraulic engineering.

Anonymous (1962). Kenneth E. Sorensen. *Engineering News-Record* 168(May10): 67. *P*
Anonymous (1985). Sorensen, K.E. *Who's who in engineering* 6: 623. AAES: Washington DC.
Anonymous (1990). Sorensen, water engineer, dies at 71. *ASCE News* (9): 3.
Davis, C.V., Sorensen, K.E., eds. (1969). *Handbook of applied hydraulics*, 3rd ed. McGraw-Hill: New York.
Sorensen, K.E. (1949). Curves solve reservoir flood-routing equations. *Civil Engineering* 19(11): 778-779.
Sorensen, K.E. (1953). Graphical solution of hydraulic problems. *Trans. ASCE* 118: 61-77.
Sorensen, K.E. (1971). *A program for preserving the quality of Lake Minnetonka*. Limnological Research Center, University of Minnesota: Minneapolis.

SORZANO

25.10. 1852 Santiago de Cuba/CU
25.06. 1923 Brooklyn NY/USA

Julio Federico Sorzano was educated in France and Belgium, graduating in 1872 as a civil engineer from *Institut de l'Etat*, Gembloux B. In 1874 he became assistant engineer in the Machinery and Industrial Department of the International Exhibition, Brussels. Until 1877, he then was engaged in the erection of machinery in a factory of Dorado, Puerto Rico. His greater part of work was done from then in Latin America, although he was established in New York NY. In the 1880s he was engaged with problems in the sugar producing industry. He was a pioneer and acknowledged expert in matters pertaining to both the cane and beet-root sugar industries, having added improvements in the agricultural and manufacturing phases. He thereby also did various engineering additions pertaining to foundations, ventilation, and drying work.

Sorzano took an active part in the Panama Canal controversy concerning the relative merits of the lock canal as compared with the sea level canal, and the water supply available for the former type. He was a member of the Committee appointed by the Chamber of Commerce of New York State to study this matter, writing various articles on the subject, presenting also the 1910 paper in the Transactions ASCE. This paper was considered interesting in the Discussion, yet the conclusions reached differ widely with these of others. Sorzano was besides a close student of Latin-American affairs and an ardent supporter of friendly and trade relations between these countries and the USA. He organized and was president of the Pan-American Chamber of Commerce in 1911, fostering relations between the two parties, and continued as president until his death. He was member of the American Society of Civil Engineers ASCE since 1884, of the American Society of Mechanical Engineers ASME, the Institution of Civil Engineers ICE, Great Britain, and the Société des Ingénieurs Civils de France, Paris.

Anonymous (1924). Julio F. Sorzano. *Trans. ASCE* 87: 1425.

Bishop, J.B. (1919). *A chronicle of one hundred and fifty years*: The Chamber of Commerce of the State of New York 1768-1918. New York.

Brodhead, M.J. (2012). *The Panama Canal*: Writings of the US Army Corps of Engineers officers who conceived and built it. US Army Corps of Engineers, Office of History: Alexandria VA.

Sorzano, J.F. (1910). Water supply for the lock canal at Panama. *Trans. ASCE* 67: 61-205.

http://numismatics.org/search/results?q=department_facet:LatinAmericanANDartist_facet:JulioF.Sorzano (*P*)

SOUCEK

15.08. 1909 Iowa City IA/USA
24.04. 1990 Iowa City IA/USA

Edward Soucek graduated from the State University
of Iowa, Iowa City IA, with the BS degree in 1932,
and the MS degree in 1934. He was engineer of the
Iowa State Planning Board and Iowa Conservation
Commission until 1935, research engineer at Iowa
Institute of Hydraulic Research until 1938, assistant
professor of civil engineering, University of Toledo,
Toledo OH, until 1939. He was assistant chief and
chief of hydraulic design for Third's Lock project of
Panama Canal until 1942, then until 1944 chief of
the civil engineering section, later assistant to chief
of the US Army Corps of Engineers, Wilmington
NC, from when he was hydraulic engineer at Omaha NE until 1949, and then there chief
and deputy chief of the Engineering Division, US Engineering Office, until retirement
in 1970. Soucek was a member of the American Society of Civil Engineers ASCE.

Soucek had a two-fold career, first in hydraulic research, later as a hydraulic engineer
dealing with large civil engineering projects. The 1936 paper relates to so-called Sutro-
weirs, as proposed by Harry H. Sutro (1876-1913), corresponding to plate weirs for
which the approach flow velocity remains constant. In other words, the discharge across
the weir is linearly related to the approach flow head. The first 1945 paper is an early
contribution to manifold flow, corresponding to elements connected with bottom-filling
locks. The performance of these elements was analyzed using model observations which
were then up-scaled to design the new Panama Canal locks. The other 1945 paper was
also directed to Panama Canal, but relating to surges generated by lock operations.

Anonymous (1944). Edward Soucek. *Engineering News-Record* 133(Aug.31): 257. *P*
Anonymous (1964). Soucek, Edward. *Who's who in engineering* 9: 1759. Lewis: New York.
Edwards, F.W., Soucek, E. (1945). Surges in Panama Canal reproduced in model. *Trans. ASCE*
        110: 345-355; 110: 359-362.
Mavis, F.T., Soucek, E. (1935). A summary of hydrologic data, Ralston Creek Watershed.
        Studies in Engineering, *Bulletin* 9. University of Iowa: Iowa City.
Soucek, E., Howe, H.E., Mavis, F.T. (1936). Sutro weir investigations furnish discharge
        coefficients. *Engineering News-Record* 117(Nov.12): 679-680; 117(Dec.24): 904.
Soucek, E. (1944). Meter measurements of discharge. *Trans. ASCE* 109: 86-99; 109: 185-191.
Soucek, E., Zelnick, E.W. (1945). Lock manifold experiments. *Trans. ASCE* 110: 1357-1400.
Soucek, E., Gau, J.N. (1967). Spillways and closures for the large earth dams on the Missouri
        River. 9th *ICOLD Congress* Istanbul Q33(R3): 29-55.

SOWERS

23.09. 1921 Cleveland OH/USA
23.10. 1996 Kennesaw GA/USA

George Frederick Sowers received education from Case Western Reserve University, Cleveland OH, graduating with a BS degree in 1942. He obtained the MS degree in 1947 from Harvard University, Cambridge MA. He joined from associate professor to professor of civil engineering Georgia Institute of Technology, Kennesaw GA until 1965, becoming then regents professor until 1984. He was in parallel consulting engineer. He was member of the National Academy of Engineering, the Geological Society of America GSA, and the International Society of Soil Mechanics and Foundation Engineers.

Sowers was an expert in soil mechanics, engineering geology and in rock mechanics. Although structures were built on the ground since the origins of mankind, it was only in 1935 when the theory of soil and rock mechanics was founded mainly by Karl Terzaghi (1883-1963). Sowers developed his ideas in the mechanics of soil, which was considered a dirty element, because he was fascinated to study the earth's strength and reactivity. He understood the critical interdependence between geology and geotechnical engineering, and solved problems that required both disciplines and hydraulics in addition. He had a rare combination of personal expertise, along with experience to apply his knowledge. Through his 50 years of teaching and consulting, Sowers became prominent in his field. He was remembered for wit, wisdom, and power of personality.

Anonymous (1946). George F. Sowers. *Engineering News-Record* 137(Aug.15): 212. *P*
Anonymous (1973). G.F. Sowers. *Civil Engineering* 43(2): 39. *P*
Anonymous (1994). Sowers, George F. *Men and women of science* 6: 1123. Bowker: NJ.
Fogle, G.H. (1999). Memorial to G.F. Sowers. *Journal Geological Society of America* 30: 37-41.
Sowers, G.B., Sowers, G.F. (1961). *Introductory soil mechanics and foundations*. MacMillan: New York.
Sowers, G.F. (1968). Foundation problems in sanitary landfills. *Journal of the Sanitary Engineering Division* ASCE 94(SA1): 103-116.
Sowers, G.F. (1973). Settlement of waste disposal fills. Proc. 8th Intl. Conf. *Soil Mechanics and Foundation Engineering* Moscow 4: 207-210.
Sowers, G.F. (1993). Human factors in civil and geotechnical engineering failures. *Journal of Geotechnical Engineering* 119(2): 238-256.
Sowers, G.F. (1996). *Buildings on sinkholes*: Design and construction of foundations in karst terrain. ASCE: New York.

SPANGLER H.W.

18.01. 1858 Carlisle PA/USA
17.03. 1912 Philadelphia PA/USA

Henry Wilson Spangler graduated in 1878 from the US Naval Academy, Annapolis MD. He was then until 1889 assistant engineer in the US Navy, and in parallel detached from 1881 as assistant professor of mechanical engineering to Pennsylvania University, Philadelphia PA, taking over in 1889 the Whitney Professorship in dynamical engineering. He served from 1887 until his death further as the head of the Mechanical and Electrical Department. The high standard of excellence achieved during this period was considered to be largely due to his remarkable talents as a teacher, and his pronounced ability as executive officer.

When Spangler took over at the University of Pennsylvania, there was only a small laboratory including just a 5 HP boiler, and a small engine connected by belt to two dynamos. In 1906, given the strong increase of students, the Engineering Building was erected, including an mechanical engineering laboratory which then was considered an adequate addition to the University. Spangler was the author of various textbooks, with topics ranging from Valve gears, Notes on thermodynamics, to Elements of steam engineering. His effort was to present the subject matters in the simplest and clearest manner, consisting with the intended scope of treatment, and to keep in view of practice. As teacher, Spangler was described as being lucid, stimulating, progressive, and always intensely practical. On no point he more insisted than that of individual responsibility, which his students were required to assume in each branch of their work. The University conferred in 1896 on him the honorary degree of Master of Science, and in 1906 the honorary PhD degree. He was member of the American Society of Mechanical Engineers ASME, the American Society of Naval Architects and Marine Engineers SNAME, and the Engineers' Club of Philadelphia, of which he was president in 1890 and 1908. He passed away from heart disease, after several months of illness. Spangler possessed to a remarkable degree the faculty of perceiving clearly the essential elements in his job.

Anonymous (1912). Henry Wilson Spangler. *Engineering News* 67(13): 616-617. *P*
Spangler, H.W. (1901). *Notes on thermodynamics*. Wiley: New York.
Spangler, H.W., Greene, A.M., Marshall, S.M. (1903). *Elements of steam engineering*. Wiley: New York.
Spangler, H.W. (1910). *Applied thermodynamics*. McVey: Philadelphia.
Spangler, H.W. (1910). *Valve gears*: Analysis of the Zeuner diagram. Wiley: New York.

SPEIR

13.07. 1892 Brooklyn NY/USA
06.07. 1964 Concord CA/USA

Oswald Speir graduated in 1915 from the University of California, Berkeley CA, with the BSc degree in civil engineering. He then worked as surveyor and junior engineer in Utah and California. From 1918 to 1919 he served in the US Army, and then was employed on the *Kerckhoff* Power Project at Fresno CA. He was from 1920 to 1923 resident engineer for the Lindsay Strathmore Irrigation District, serving as valuation engineer during the next year. From 1924 to 1934 he was engaged as design, hydraulic and civil engineer for private, district and municipal enterprises, including the Bucks Creek hydroelectric project, the Lindsay Strathmore Irrigation District, San Francisco's Hetch Hetchy Project, and the Trojan Engineering Corporation.

Speir joined in 1934 the US Army Corps of Engineers, Sacramento District. He was in 1940 transferred to the Pacific Division, San Francisco CA, and later to the South Pacific Division as staff engineer, where he supervised the reports on various flood control and multi-purpose water use projects for submission to the Chief of Engineers, US Army. From 1956 to 1962 he was supervising hydraulic engineer of the Department of Water Resources, Sacramento CA, serving in the development of the California Water Plan CWP. Speir retired in 1963 serving then as consultant the Corps, San Francisco, on their survey of San Francisco Bay and tributaries. He was member of the American Society of Military Engineers, the American Water Works Association AWWA, the American Geophysical Union AGU, and the American Society of Civil Engineers ASCE from 1922, becoming Fellow in 1959. He served as principal assistant, and as supervising hydraulic engineer in the 1957 CWP Report, directed by Arthur D. Edmonston (1886-1957). Speir was also involved in the 1963 Report of the Advisory Committee.

Anonymous (1915). Oswald Speir. *Blue and gold yearbook*: 337. Univ. California: Berkeley. *P*
Anonymous (1965). Oswald Speir. *Trans. ASCE* 130: 838.
Brown, E.G., Banks, H.O. (1959). *Report* on Observation and processing of basic water resource data in California. Dept. Water Resources, Division of Resources Planning: Sacramento.
Knight, G.J., Banks, H.O. (1957). The California Water Plan. *Bulletin* 3. Department of Water Resources, State of California: Sacramento.
McKee, J.E., Wolf, H.W., eds. (1963). *Water quality criteria*. California State Water Resources Control Board: Sacramento.
Speir, O. (1928). Building a penstock of 2,561 ft. head. *Engineering News-Record* 100(5): 191-195.

SPEZIALE

16.06. 1948 Newark NJ/USA
07.05. 1999 Newton MA/USA

Charles Gregory Speziale attended Newark College of Engineering, where he earned the BS degree in civil engineering and applied mathematics in 1970, and the MS degree in mechanics in 1972. He spent the next year at Rutgers University, New Brunswick NJ, studying mechanical and aerospace engineering. Continuing his education at Princeton University, Princeton NJ, he earned the MA and PhD degrees in aerospace and mechanical sciences in 1975 and in 1978, respectively. After graduation Speziale joined until 1985 the Engineering Faculty, Stevens Institute of Technology, Hoboken NJ, and then was until 1987 a member of the Engineering Faculty, the Georgia Institute of Technology, Atlanta GA. During these years he focused on non-Newtonian fluids and the kinetic theory of gases.

In 1987, Speziale moved to the Institute for Computer Applications in Science and Engineering, of NASA's Langley Research Center, Hampton VA. He began to devote his attention to fluid turbulence. He laid down important mathematical and physical milestones that continue to guide turbulence modellers. His paper on the extended Galilean invariance of the Reynolds stress, the Galilean invariance of the sub-grid stresses, and on the limiting behaviour of Reynolds stresses in flows subject to strong rotation or magnetic field have become classics in turbulence modeling. He developed also a nonlinear eddy viscosity model that is capable of predicting secondary internal flows. Speziale and his colleagues further advanced turbulence modeling by proposing a pressure-strain model and developing an algebraic model for Reynolds stress. He made contributions to the study of the final decay of turbulence and introduced nonlinear fixed-point analysis. In 1997, Speziale suffered a serious injury due to an accidental fall exacerbating some pre-existing health problems, which led ultimately to his death.

Speziale, C.G. (1983). Closure models for rotating two-dimensional turbulence. *Geophysical and Astrophysical Fluid Dynamics* 23(1): 69-84.

Speziale, C.G. (1985). Galilean invariance of subgrid scale models in the large-eddy simulation of turbulence. *Journal of Fluid Mechanics* 156: 55-62.

Speziale, C.G. (1987). On nonlinear k-l and k-ε models of turbulence. *Journal of Fluid Mechanics* 178: 459-475.

Speziale, C.G. (1991). Analytical models for the development of Reynolds-stress closures in turbulence. *Annual Review of Fluid Mechanics* 23: 107-157.

Zhou, Y., Girimaji, S. (1999). Charles G. Speziale. *Physics Today* 52(12): 77. P

SPIELMANN

20.09. 1847 Hoboken NJ/USA
29.11. 1883 New York NY/USA

SECTION OF TUNNELS UNDER THE HUDSON RIVER
WHEN COMPLETED

Arthur Spielmann graduated in 1867 from New York University with a BS degree in civil engineering, to be immediately engaged in this profession, forming a partnership with Charles B. Brush (1848-1897) in 1869. Both Spielmann and Brush were from 1874 until their deaths adjunct engineering professors at the College of Engineering, New York University. Spielmann obtained in 1878 the MS degree from his Alma Mater. He was thereby prominently identified with public works in Hudson County NJ in parallel, including street improvements and the sewerage of Hoboken NJ. The firm Spielmann & Brush was also connected with the Hudson River Tunnel, and with soundings for the new tunnel at Communipaw NJ. The water-works of the Hackensack Water Company completed in 1880, supplying Hoboken and adjacent cities, were also built by his firm. Spielmann devoted considerable time and money to the solution of proper drainage of the low lands of Hoboken.

In 1882, Spielmann went to Europe to study drainage engineering, visiting Holland for that purpose, and consulting distinguished engineers in England, France, and Germany. He was frequently called upon as an expert by public corporations and private parties, especially on matters relating to foundations. He also complied official maps of New York City which have been of invaluable aid in legal matters. His topographical and sanitary map of Hudson County for the National Board of Health is a monument to the ability, skill and knowledge of the subject which has claimed so much of his time. He died of typhoid pneumonia at age of only 37.

Anonymous (1884). Spielmann, Arthur. *Proc. ASCE* 10: 115-117.

Anonymous (1885). *Celebration* of the Centennial Anniversary of the University of the State of New York and the twenty-second University convocation held on July 8-10, 1884: 289. Weed, Parsons & Co.: Albany NY. (*P*)

Anonymous (1894). Arthur Spielmann. *Biographical catalogue* of the chancellors, professors and graduates of the Department of Arts and Science of the University of the City of New York: 120. Alumni Association: New York.

Jones, T.F., ed. (1933). Arthur Spielmann. *New York University* 1832-1932: 318. New York University Press: New York City.

Spielmann, A. (1880). Disc. of Construction and maintenance of roads. *Trans. ASCE* 8: 333.

Spielmann, A., Brush, C.B. (1880). The Hudson River Tunnel. *Trans. ASCE* 9: 259-272. (*P*)

SPRENKLE

21.01. 1895 Waynesboro PA/USA
07.12. 1986 East Cleveland OH/USA

Raymond (Ray) Eyler Sprenkle had earned by 1917 the BS degree in mechanical engineering from the Bucknell University, Lewisburg PA. He obtained the MS degree in mechanical engineering twenty years later. He joined the Bailey Meter Company, Cleveland OH in 1919, and there started selling and servicing its products. In 1921 he became head of its Meter Engineering Department, assuming all research on the flow primary elements within the company, as well as similar work conducted at Ohio State University. Sprenkle became in 1944 director of education of the company, thereby recruiting and training the hired engineers, and being responsible for the education of customers on instruments and automatic control. He also served as consultant on flow problems and flow primary element testing.

Sprenkle served the American Society of Mechanical Engineers ASME as member, vice-chairman of the Fluid Meters Committee, as also on the preparation of the ASME Report Fluid meters. He was ASME representative on the Cleveland Technical Societies Council, and an ASME member of the American delegation to Germany and France for the plenary meetings of the International Organisation for Standardization ISO of flow metering. His research was entirely devoted to discharge measurement using nozzles, and by developing the then current procedures. Bailey at the time was main furnisher of these instruments, of which Sprenkle was the main technical advisor. Currently, the discharge measurement is usually conducted using Inductive Discharge Measurement, at least for conduit flow, with an accuracy of 1% of the nominal discharge.

Anonymous (1917). Sprenkle. *L Agenda yearbook*: 95. Bucknell University: Lewisburg PA. *P*
Anonymous (1961). R.E. Sprenkle. *Mechanical Engineering* 83(3): 114.
Bean, H.S., Beitler, S.R., Sprenkle, R.E. (1941). Discharge coefficients of long-radius flow
    nozzles when used with pipe-wall pressure taps. *Trans. ASME* 63(7): 439-445.
Smith, R.B., Sprenkle, R.E., Pigott, R.J.S., Cooper, W.S. (1935). Discussion of Fluid-meter
    nozzles. *Trans. ASME* 57(5): 251-255.
Sprenkle, R.E. (1936). *Measurement of water flow to hydraulic turbines*. Bucknell University.
Sprenkle, R.E., Courtright, N.S. (1958). Straightening vanes for flow measurement. *Mechanical
    Engineering* 80(2): 71-73; 80(8): 92-95.
Tuve, G.L., Sprenkle, R.E. (1933). Orifice discharge coefficients for viscous liquids. *Instruments*
    6(11): 201-206; 8(8): 202-205; 8(8): 232-234.

STABLER

03.02. 1879 Sandy Spring MD/USA
24.11. 1942 Washington DC/USA

Herman Stabler graduated from Earlham College, Richmond IN, in 1899 with the BSc degree. He was then instructor in mathematics and engineering at the National Correspondence Institute, Washington DC. He was appointed in 1903 hydrographic aid of the US Geological Survey USGS to make studies on the economic aspects of water pollution at Schuylkill River PA. He was assigned in 1905 to investigate stream pollution by sewage. He investigated in 1906 for the State Board of Health, Rhode Island, various sources of pollution of rivers, reporting later 'the continued unsanitary condition of these streams offers olfactory evidence that my work there had little practical effect'. From 1907, then at Washington DC, he assisted in the nationwide study of the water quality, and devised a scheme for water analyses expressed in ionic form. In this period he also authored a number of USGS Water Supply Papers relating to stream pollution, and stream quality.

From 1908 Stabler was an irrigation engineer with the US Reclamation Service, but in 1910 returned to USGS engaged to study ground-water resources in southern California. From 1912 to 1922 he was involved in land classification studies, requiring for the determination of developed and undeveloped water power resources, land classification as to irrigable or non-irrigable, the feasibility of irrigation projects, and water availability. He was appointed chief of the USGS Land Classification Branch in 1922, making the first complete topographic surveys through Glen Canyon and the Grand Canyon of the Colorado River. From 1925 until his death he was chief of the USGS Conservation Branch. He was well known as hydraulic engineer, an expert on water-quality problems, and a land-use expert. He was member of many organisations, including the American Geographical Society, and the American Society of Civil Engineers ASCE, serving from 1935 to 1937 as director, and was also active in many ASCE Committees.

Anonymous (1936). Herman Stabler, Director District 5. *Civil Engineering* 6(3): 207. *P*
Anonymous (1943). Herman Stabler. *Trans. ASCE* 108: 1641-1645.
Mendenhall, W.C., Stabler, H., Dole, R.B. (1916). Ground water of San Joaquin Valley. *Water-Supply Paper* 398. USGS: Washington DC.
Stabler, H., Pratt, G.H. (1909). *Purification of some textile and other factory wastes*. USGS.
Stabler, H. (1911). Some stream waters of the Western United States. *Water-Supply Paper* 274. USGS: Washington DC.
http://archive.library.nau.edu/cdm/singleitem/collection/cpa/id/15286/rec/9 *P*

STANLEY C.M.

16.06. 1904 Corning IA/USA
20.09. 1984 Muscatine IA/USA

Claude Maxwell Stanley, Jr., obtained the BS degree from the State University of Iowa, Iowa City IA, in 1926, and the MS degree in hydraulic engineering there in 1930. He was from 1928 to 1932 hydraulic engineer at Dubuque IA, and Chicago IL, on hydro-electric and water works projects, until 1939 then consulting engineer for power plants, water works, electric systems and sewer projects, from when he was a partner in the Stanley Engineering Company, Iowa City IA, dealing with general engineering. He was a member of the American Society of Civil Engineers ASCE and was the recipient of its 1933 Alfred Noble Prize, and its 1935 Collingwood Prize.

The University of Iowa's UI hydraulic laboratory, a fixture on the Iowa River since 1932, was renamed C. Maxwell Stanley Hydraulics Laboratory in 2003. This renaming was a tribute to Stanley's wide-ranging contributions to UI during his lifetime and to the structure, a building he drafted in the late 1920s, following his graduation in 1926. The renaming also recognizes nearly 1 million US$ in commitments from Stanley's family to the Laboratory Renovation, which was completed in 2002. In the 1930s Stanley became co-owner of a two-man consulting firm in Muscatine IA, where he directed the firm into Stanley Consultants Inc., a nationally ranked multidisciplinary company. He later founded HON Industries, which he led to become a top manufacturer and marketer of office furniture and hearth products. In the 1950s he and his wife created the Stanley Foundation, a private foundation that promotes a secure peace with freedom and justice, through dialogue and education. He was inducted into the college's Distinguished Engineering Alumni Academy as a charter member, and extensively supported the university's educational mission through substantial contributions, service on volunteer boards, and the posthumous donation of a major African art collection.

Anonymous (1934). C. Maxwell Stanley. *Civil Engineering* 4(1): 51. *P*
Anonymous (1941). Stanley, C.M. *Who's who in engineering* 5: 1689. Lewis: New York.
Hudson, D. (2008). *The biographical dictionary of Iowa*: 485-486. University Press: Iowa.
Stanley, C.M. (1933). Discussion of Application of duration curves to hydro-electric studies. *Trans. ASCE* 98: 1300-1303.
Stanley, C.M. (1934). Study of stilling-basin design. *Trans. ASCE* 99: 490-523.
Stanley, C.M. (1934). Discussion of Improved type of flow meter for hydraulic turbines. *Trans. ASCE* 99: 867-872.

STANLEY O.G.

21.09. 1882 Humeston IA/USA
12.02. 1966 San Francisco CA/USA

Owen Garrett Stanley graduated in 1906 with a BS degree in civil engineering from Stanford University, Stanford CA. Soon later, he became associated with the US Army Corps of Engineers, an association that he maintained for 64 years. His work with the Corps encompassed engineering and administrative activities in both rivers and harbours, flood control, power, irrigation, water conservation, and military construction. While at the Sacramento CA District, he was involved in the design, the construction, and the operation of these important projects as debris storage dams on the Yuba and American Rivers, and the Stockton Deep Water Channel, the Sacramento River Flood Control Project, Folsom Dam on the American River, and the Pine Flat, Isabella, and Big Dry Creek Dams in the San Joaquin Valley.

In 1939, Stanley was transferred to the South Pacific Division at San Francisco CA, where he had to supervise all flood control work in California, Utah, and Nevada. In 1947 he became the senior civilian engineer in charge of the Engineering Division, responsible for all engineering activities in both military and civil work construction, receiving the Commendation for Meritorious Civilian Service from the Corps' chief of engineers. He was a special advisor to the South Pacific Engineer at the time of his retirement from the Corps in 1952, and thereafter was engaged in private practice as a consulting engineer. Stanley was considered a most valuable public servant and was recognized as such by the Assembly of the California Legislature in 1966. He was a member of the American Society of Civil Engineers ASCE, the Society of American Military Engineers SAME, the National Reclamation Association, and was a registered engineer in California State.

Anonymous (1952). Owen G. Stanley. *Engineering News-Record* 149(Oct.23): 74. *P*
Anonymous (1967). Owen Garrett Stanley. *Trans. ASCE* 132: 698-699.
Stanley, O.G., Lane, E.W. (1949). The importance of sediment control in the conservation and utilisation of water resources. UN Scientific Conference *Conservation and Utilisation of Resources* Lake Success NY. United Nations: New York.
Stanley, O.G. (1955). Discussion of Waterway economy. *Trans. ASCE* 120: 1532-1535.
Stanley, O.G. (1965). *Brief history* of hydraulic mining, gold dredging, creation of the California Debris Commission, and birth of the Sacramento District of the Corps of Engineers. Water Resources Center Archives, University of California: Berkeley CA.

STARLING

25.01. 1839 Columbus OH/USA
11.12. 1900 Greenville MS/USA

William Starling attended the University of New York, from where he graduated in 1856. He moved in 1859 to Frankfurt KY, serving during the Civil War the US Forces as first-lieutenant of a Kentucky Cavalry. He did the first engineering work during his military service, showing great aptitude so that he was soon made chief engineer. After the War he took interest in a cotton plantation in Arkansas, and sold it only in 1882. He moved to Greenville MS resuming his work as engineer. He was engaged in levee works and US Government surveys, thereby devoting time to study levees and the associated river hydraulics, a field in which he became an national and international authority. He was appointed chief engineer of the Mississippi Levee District, a position he held for ten years.

Starling was then appointed a member of the Southwest Pass Commission created to formulate a project for securing a permanent deep canal through Southwest Pass. This was his last, but his most important professional position, because he died of heart failure at Greenville MS shortly after the Board had completed its work. Starling wrote a number of technical papers, including on dikes in Holland, for which he was awarded the ASCE Norman Medal. In addition there are papers on dikes on the Mississippi River between Helena and Vicksburg, the Discharge of Mississippi River, or Improvements made on Mississippi River. Several of the papers were later issued in the pamphlet The floods of the Mississippi. Accordingly, Starling was a man related to one of the great rivers of the country, and he added with his services to the wealth of the nation, and to the population along the river. He was described as a man of great force of character. He was daring in his undertakings, and manly and frank in his intercourse with men. His culture and breeding impressed all who came into contact with him. He was cheerful and ever ready with anecdotes.

Anonymous (1901). William Starling. *Trans. ASCE* 46: 566-567.
Starling, W. (1889). The improvement of the Mississippi River. *Trans. ASCE* 20: 85-108.
Starling, W. (1889). On flood heights in the Mississippi River, with especial reference to the reach between Helena and Vicksburg. *Trans. ASCE* 20: 195-228.
Starling, W. (1892). Some notes on the Holland dykes. *Trans. ASCE* 26: 559-700.
Starling, W. (1901). River basins and the levees which protect them. *Riparian lands of the Mississippi River*: 75-80, F.H. Tompkins, ed. New Orleans. *P*

STEARNS

11.11. 1851 Calais ME/USA
01.12. 1919 Boston MA/USA

Frederic Pike Stearns went at the age of eighteen to Boston and there found a job with its engineering department. Here he came under the influence of James B. Francis (1815-1892) and Hiram F. Mills (1836-1921). In 1872 he was engaged on Sudbury River water supply of Boston, and became division engineer of the sewage tunnel under Dorchester Bay in 1880. In 1886 Stearns was called by the State Board of Health to become its chief engineer. This Board was in charge of the state's island waters, and empowered to advice the numerous municipalities with regard to their sanitation and water supply. The work required great sense in making decisions, and the influence the Board acquired was largely due to the excellence of its chief engineer. Stearns' exhaustive studies of water supplies and the means of controlling and improving them have become the basis for practice in many other states. He planned the improvement of the Charles River Basin, later carried out with his advice of consultant, by which the foul tidal estuary of the Charles was converted into a beautiful fresh-water basin.

Stearns' most notable work for the State Board of Health was the design of the Boston metropolitan water supply using *Nashua* River. When this plan was adopted in 1895, he became chief engineer of the new Board and completed the project in 1907. The water-works were recognized as examples. They included a provision by which the water fall into the aqueduct was utilized to develop water power. Stearns then became consultant, with projects for the water supply of New York, Baltimore, Los Angeles, Rochester or Winnipeg. He also did important work for dam design. In 1905 he was appointed by President Roosevelt Board Member of consulting engineers for Panama Canal, thereby advocating for the lock-type canal. Stearns published a number of important papers in hydraulics, including the 1883 paper with Alphonse Fteley (1837-1903).

Anonymous (1905). Frederic P. Stearns. *Engineering News* 54(11): 263-264. *P*
Anonymous (1920). Frederic Pike Stearns. *Trans. ASCE* 83: 2132-2138.
Anonymous (1935). Stearns. *Dictionary of American biography* 17: 542. Scribner's: New York.
Fteley, A., Stearns, F.P. (1883). Description of some experiments on the flow of water made during the construction of works for conveying the water of Sudbury River to Boston. *Trans. ASCE* 12: 1-118.
Stearns, F.P. (1885). Experiments on the flow of water in a 48-in pipe. *Trans. ASCE* 14: 1-18.
Stearns, F.P. (1900). The Wachusett Dam. *Engineering Record* 41: 50-51; 42: 218.

STEELE

07.11. 1886 Compton CA/USA
14.10. 1973 Oakland CA/USA

Isaac Cleveland Steele graduated from University of California, Berkeley CA, with a BS degree in civil engineering in 1909. His professional career was then essentially with the Pacific Gas & Electric Company, where he was employed from 1909 until his retirement in 1951. First as a labourer, he was rapidly promoted to foreman on projects of mountain construction and in 1911 was made a designer and in 1917 placed in charge of the Civil Engineering Group. In 1944 he became chief of the Division of Civil Engineering. He was appointed vice-president in 1947. Steele was a member of the US Committee on Large Dams, and the California Division of Water Resources Board.

Among the projects under the direction of Steele were the development of the Pit and Feather Rivers, and the Salt Springs Dam on *Makelumna* River in Central California. He made notable laboratory experiments on baffle piers as used in energy dissipators below dams to increase the efficiency of a hydraulic jump stilling basin, and to reduce its length. These appurtenances may significantly increase the dissipation effect, provided the approach flow velocities are not too large such that cavitation damage would result. After retirement, Steele was a consultant specializing in dam design throughout the world. These included the Upper Volta River project in Ghana, and the *Derbendi-Khan* Dam in Iraq. Also he was involved from 1952 to 1959 in the International Board of Consultants for the High Aswan Dam, Egypt. As a recognized authority on rock-filled dams, he wrote a section on that topic in the Handbook of applied hydraulics.

Anonymous (1947). I.C. Steele. *Engineering News-Record* 139(Aug.28): 297. *P*
Anonymous (1961). I.C. Steele. *Civil Engineering* 31(10): 83. *P*
Anonymous (1966). I.C. Steele. *Civil Engineering* 36(10): 74. *P*
Anonymous (1974). Isaac Cleveland Steele. *Trans. ASCE* 139: 559.
Steele, I.C. (1920). Making the final 15-ft. raise of the Spaulding Dam. *Engineering News-Record* 85(22): 1020-1024.
Steele, I.C. (1926). Baffle piers at toe of dam dispel energy of flow. *Engineering News-Record* 96(22): 886-889; 97(20): 800-802.
Steele, I.C., Monroe, R.A. (1929). Baffle-pier experiments on models of Pit River Dam. *Trans. ASCE* 94: 451-546.
Steele, I.C. (1952). Rock-fill dams. *Handbook of applied hydraulics* 6: 207-252, C.V. Davis, ed. McGraw-Hill: New York.

STEPANOFF

23.02. 1895/RU
01.08. 1973 Phillipsburg NJ/USA

Alexey Joakim Stepanoff obtained the BS degree in 1926 in mechanical engineering from University of California, Berkeley CA. He was design engineer in the 1930s, Byron Jackson Co., Huntington Park CA, who in 1901 developed the first deep-well vertical turbine pump. In the 1940s as development engineer for the Ingersoll-Rand Co., Phillipsburg NJ, he was closely associated with Richard G. Folsom (1907-1996). From the 1950s until retirement, Stepanoff had his consulting office in which he assessed large national and international pumping schemes, and wrote several of his books. He was a member of the American Society of Mechanical Engineers ASME. The Alexey J. Stepanoff Memorial Hydraulic Laboratory of the Ingersoll-Dresser Pump Co. is located at Phillipsburg NJ.

Stepanoff was and still is known for his outstanding books in the fields of pumps and blowers. In a review of the 1965 book, it is stated that during the past 17 years from the original publication of this book, no publication had appeared to rival its authority and comprehensiveness despite the uneven treatment of many aspects of pump design, so that this book was widely quoted. Aspects of cavitation in relation to the number and shape of the impeller blades and the degree of finish, the principle of least resistance, and the inlet guide vanes as a means to modify the pump performance are discussed. Mixed-flow blowers, then a comparatively recent introduction to engineering practice, are then dealt with. A construction is presented withstanding the bending stresses imposed on the blades by the high rotational speed enabling otherwise desirable pump characteristics. The second part of the book assembles and comments the results of others for transporting paper-pulp and solids in suspension. It is finally stated that the book is in the usual style expected including numerous graphs and diagrams.

Anonymous (1939). A.J. Stepanoff. *Mechanical Engineering* 61(4): 303. *P*
Stepanoff, A.J. (1929). Thermodynamic theory of the air-lift pump. *Trans. ASME* 51(5): 49-55.
Stepanoff, A.J. (1955). *Turboblowers*. Wiley: New York.
Stepanoff, A.J. (1957). *Centrifugal and axial flow pumps*: Theory, design, and application. Wiley: New York.
Stepanoff, A.J. (1964). Pumping solid-liquid mixture. *Mechanical Engineering* 86(9): 29-35.
Stepanoff, A.J. (1965). *Pumps and blowers*: Two-phase flow, ed. 2. Wiley: New York.
Stepanoff, A.J. (1970). *Gravity flow of bulk solids and transportation of solids in suspension*. Wiley: New York.

STEVENS J.

ca. 1749 New York NY/USA
06.03. 1838 Hoboken NJ/USA

John Stevens was the son of the ship owner John. Young John graduated in 1768 from King's College (now Columbia University), New York NY. With the outbreak of the Revolutionary War, Stevens offered his services to General Washington commissioning him as captain. He was appointed treasurer of New Jersey advancing thereby finally to colonel. In 1788 he drew attention to the work of John Fitch to develop steamboats and from then until his death, Stevens made a large fortune with the advancement of mechanical transport both on water and on land. He concentrated on the steam and boiler design and received for his vertical steam boiler and an improved Savery-type steam engine the patent in 1791. The experimental boat *Polacca* was tried on *Passaic* River, but proved unsuccessful. In 1800 Stevens became consulting engineer for the water supply of New York City. He convinced the directors that steam pumping should be used, but Boulton & Watt engines were adopted.

In 1803 Stevens had advanced sufficiently with his tests to secure a patent for a multi-tubular boiler. The following year, his small steamboat *Little Juliana* operated with a double screw was tested on Hudson River. His goal was to inaugurate a steam ferry across the Hudson between Hoboken and New York. In 1806 he began with the design of the 30 m long *Phoenix*. Before it was completed, however, Robert Fulton (1765-1815) made a successful voyage with his boat *Clermont* in 1807. The *Phoenix* was completed in 1808, made a sea trip and established itself the record of the first sea-going steamboat of the world. In 1810, Stevens began giving close attention to the adaption of the steam engine to the motive power for railways, leaving steam navigation in the hands of his sons. He argued the adoption of this transportation system rather than canals, and published a corresponding note in 1812. In his later years, he devoted his time to study.

Anonymous (1935). Stevens, John. *Dictionary of American biography* 17: 614-616. Scribner's: New York.
Stevens, J. (1812). *Documents* tending to prove the superior advantages of rail-ways and stream-carriages over canal navigation. New York.
Turnbull, A.D. (1928). *John Stevens, an American record*. The Century: New York. *P*
Turnbull, A.D. (1928). John Stevens and his sons. *Mechanical Engineering* 50(5): 353-357. *P*
http://en.wikipedia.org/wiki/John_Stevens_(inventor) *P*

STEVENS J.C.

09.01. 1876 Moline KS/USA
29.03. 1970 Portland OR/USA

John Cyprian Stevens graduated with a BS degree in 1905 from University of Nebraska, obtained the CE degree in 1928 and the Dr.Eng. degree in 1947. He was engineer of the US Geological Survey USGS until 1912 in charge of water supply investigations, and of the Columbia River Basin. He spent the next two years as an engineer with the Ebro Irrigation & Power Company, Barcelona Spain, and upon return to the USA became engineer of companies in New York and Washington State. He then had a private consulting office at Portland OR until 1920, from when being partner of Stevens & Koon, Consulting Engineers there until retirement in 1954. He was member of the American Society of Civil Engineers ASCE, and its president in 1945. He was elected ASCE Fellow in 1959. The J.C. Stevens Award is annually presented to the author of the best discussion of a paper published in the ASCE Journal of Hydraulic Engineering.

Stevens was a distinguished hydraulic engineer and inventor of water-metering devices. He had patented in 1911 the first continuous water stage recorder and made a contract to manufacture it. As consultant, he was involved in hydrographic studies in the Pacific Northwest. Later he contributed to Bonneville Navigation and Power Development, Mud Mountain flood control dam, Willamette Valley project for flood control, the Columbia River dams including McNary, and The Dalles Dams. He authored papers in technical journals, and the chapter Hydroelectric plants in the Handbook of applied hydraulics. He was awarded the 1938 ASCE Norman Medal for a paper on the silt problem.

Anonymous (1945). John Cyprian Stevens. *Civil Engineering* 15(2): 97-98. *P*
Anonymous (1959). Stevens, J.C. *Who's who in engineering* 8: 2359. Lewis: New York.
Anonymous (1970). John Cyprian Stevens. *Trans. ASCE* 135: 1113-1114.
Stevens, J.C. (1910). Experiments on small weirs and measuring modules. *Engineering News* 64(7): 171-177.
Stevens, J.C. (1922). Winter overflow from ice gorging on shallow streams. *Trans. ASCE* 85: 677-698.
Stevens, J.C. (1934). On the behaviour of siphons. *Trans. ASCE* 99: 986-1011.
Stevens, J.C. (1937). Scour prevention below Bonneville Dam. *Engineering News-Record* 118(Jan.14): 61-65.
Stevens, J.C. (1957). Flow through circular weirs. *Journal of the Hydraulics Division* ASCE 83(HY6, Paper 1455): 1-17.

2582

STEVENS J.F.

25.04. 1853 West Gardiner ME/USA
02.06. 1943 Southern Pines NC/USA

John Frank Stevens started his engineering career on field crews, surveying mills and industrial canals. In 1873 he went to Minneapolis MN and advanced in 1874 to assistant city engineer. After years as a railroad engineer he was appointed chief engineer of the Isthmian Canal Commission in 1904, giving him control over both construction and engineering phases of the Panama Canal. When Stevens arrived at the Canal Zone in 1906, he found poor conditions. Equipment, largely inherited from the French, was antiquated, housing and food were inadequate, and the 17,000 labourers were demoralized by frequent outbreaks of yellow fever and malaria. Stevens immediately reorganized the work force and the engineering staff. Recognizing that work progress on the canal depended upon efficient transportation, he gave most time to organizing and building an extensive system of railroads to carry out the soil and rock from the Culebra Cut at the inter-oceanic divide. Although he had at first favoured a sea-level canal as previously made for Suez Canal, he was soon convinced that only a locked canal was feasible. In his 1906 Report, he therefore concurred with the minority opinion in opposing a sea-level canal. To facilitate construction, he successfully reorganized the Canal Commission, giving the chief engineer complete control over the Canal Zone.

By the end of 1906 construction was under way but Stevens, frustrated by political manoeuvring at Washington, and eager to return to a less strenuous position, resigned a few months later. His successor, George Washington Goethals (1858-1928), later said of Stevens 'The Canal is his monument'. Upon return to the USA he was again in railroads but from 1911 was a private consultant. He served in 1927 as president of the American Society of Civil Engineers ASCE. In 1925 a statue was erected on Marias Pass where he had earlier provided the key passage across the Continental Divide.

Anonymous (1908). John F. Stevens. *Journal of the Franklin Institute* 165(1): 36. *P*
Anonymous (1925). Stevens receives J. Fritz Medal. *Mechanical Engineering* 47(5): 383-384. *P*
Anonymous (1939). J.F. Stevens to be given Hoover Medal. *Mechanical Engineering* 61(1): 97.
Anonymous (1973). Stevens, John Frank. *Dictionary of American biography* Suppl. 3: 735-737. Scribner's: New York.
Anonymous (2000). The engineering genius history forgot. *Civil Engineering* 70(3): 15. *P*
Stevens, J.F. (1927). The Panama Canal. *Trans. ASCE* 92: 946-967.
Stevens, J.F. (1936). *An engineer's recollections*. McGraw-Hill: New York. *P*

STEWART C.B.

08.03. 1868 Fairbury IL/USA
02.09. 1951 Anchor IL/USA

Clinton Brown Stewart obtained the CE degree from Cornell University, Ithaca NY, in 1890. He was then from 1893 to 1898 professor of civil engineering at the Colorado State School of Mines, Golden CO, from when he was US assistant engineer until 1900. From 1903 to 1908 he was researcher in hydraulics at the University of Wisconsin, Madison WI, from when he continued as a private consultant until his retirement at Madison WI. Stewart was recipient of the Octave Chanute Medal of the Western Society of Engineers, and of the Fuertes Medal of Cornell University. He published numerous books, reports and papers. He was member of the Western Society of Engineers, whose headquarters were in Chicago IL, and of the American Society of Civil Engineers ASCE.

During his stay at the University of Wisconsin, Stewart became known as excellent experimenter in hydraulics. His first work referred to centrifugal pumps, one of the pump types then under technical development. The second work describes submerged orifices and tubes, a classical topic in hydraulics, which has been investigated mainly in France during the 19th century. His fresh and accurate description of the flow processes added significantly to the engineering knowledge. It should here be stressed that the submergence effect of various hydraulic structures, including gates or weirs, is even currently not fully understood, mainly because of flow separation and the corresponding effects of turbulence. The 1916 paper deals with aspects of flood flows in rivers. This study includes the description of the failure of the Merrill Dam some 4 km upstream of Merrill City, which was to some 90% then submerged, but of which the discharge coefficient amounted to still almost 90% of the free overfall value. The increase of flood discharge due to the dam failure was also studied.

Anonymous (1948). Clinton B. Stewart. *Who's who in engineering* 6: 1906. Lewis: New York.
Anonymous (1951). Clinton B. Stewart. *Cornell Alumni News* 54(5):148.
Kolupaila, S. (1960). C.B. Stewart. *Journal of the Hydraulics Division* ASCE 86(HY1): 28. *P*
Stewart, C.B. (1907). Investigation of centrifugal pumps. *Bulletin* 173. University of Wisconsin: Madison WI.
Stewart, C.B. (1908). Investigation of flow through large submerged orifices and tubes. *Bulletin* 216. University of Wisconsin: Madison WI.
Stewart, C.B. (1916). Investigation of flood flow on the Wisconsin River at Merrill WI, July 23-24, 1912. *Journal of the Western Society of Engineers* 21(9): 717-745.

STEWART J.T.

13.01. 1868 Loda IL/USA
09.06. 1928 St. Paul MN/USA

John Truesdale Stewart obtained the BS degree from University of Illinois, Urbana IL, in 1893, and the civil engineering degree in 1909. He was from 1893 to 1897 engaged in general engineering practice in his State, and then was occupied with both drainage and railroad work. In 1899 he was appointed field assistant of the US Geological Survey USGS, and from 1904 he became drainage engineer within the US Department of Agriculture USDA, on drainage and irrigation work in South Dakota, North Dakota, Minnesota and Florida. Stewart was professor and chief of the Division of Agricultural Engineering at the University of Minnesota, Minneapolis MN, from 1908. During this era he conducted research on the drainage and development of peat lands, the durability of drain tiles, and the efficiency and maintenance of open ditches. Following his strong sense of loyalty and patriotism, he was active in the US Army, serving as lieutenant-colonel of Engineers in 1918. In 1922 he resumed his consulting practice as recognized authority on drainage and wetland development. He was member of the American Society of Civil Engineers ASCE, the Society of American Military Engineers, the Western Society of Engineers, and the American Society of Agricultural Engineers ASAE.

Stewart prepared a number of technical papers and reports, several of which were published by the USDA. He was an authority in all matters relating to land drainage, the drainage of farms, including drainage legislation. He also developed an experimental drainage system at the University of Minnesota. He was in addition the senior author of the book Engineering on the farm, whose main chapters are 1. Introductory, 2. Materials of construction, 3. Building equipment, and 4. Mechanical equipment. He was described as an always courteous, friendly and interested person, who had many friends in and out of the engineering fraternity. He was fond of outdoor life, enjoying hunting and fishing.

Anonymous (1929). John T. Stewart. *Trans. ASCE* 93: 1902-1903.
Robertson, W., Stewart, J.T. (1908). Installation of an experimental drainage system, Minnesota Experiment Station. *Bulletin* 110. USDA: Washington DC.
Stewart, J.T. (1907). Report on the drainage of the eastern parts of Cass, Traill, Grand Forks, Walsh, and Pembina Counties ND. Government Printing Office: Washington DC.
Stewart, J.T., Davenport, E. (1923). *Engineering on the farm*: A treatise on the application of engineering principles to agriculture. Rand, McNally: Chicago.
http://www.bookerworm.com/resources/authors/2427731-john-truesdale-stewart.html *P*

STICKNEY

24.01. 1869 St. Paul MN/USA
27.01. 1929 Albany NY/USA

George Fetter Stickney graduated in 1891 with the Ph.B. degree from the Sheffield Scientific School of Yale University, New Haven CT. He was an US assistant engineer on improvements of the rivers in Kentucky and of harbours in Northern New Jersey from 1893. He designed in 1901 a movable dam in Saint Mary's River, Ontario, and in 1903 piers for a bridge at Selma AL, and the East Viaduct approach to the Mississippi Bridge, Thebes IL. Later, from 1905 to 1914, employed by the Department of the State Engineer and Surveyor of New York, he then designed and supervised the construction of a large part of the New York Barge Canal, and from 1915 to 1917 he made preliminary investigations for the proposed Lake Erie & Ohio River Canal. Stickney also served as captain during the Spanish-American War. He was a member of the American Society of Civil Engineers ASCE since 1906.

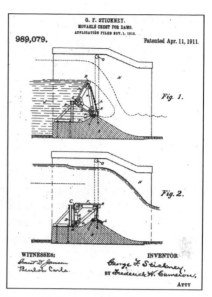

As a consulting engineer, Stickney maintained an office at Albany NY from 1917 to 1928, thereby being in charge of hydraulic investigations for waterpower developments and river improvements. He was the inventor of the siphon spillway for discharging water through dams, and of an automatic movable crest for dams. These elements are described in the 1922 paper. The various types of siphons are detailed including their principles of operation. The major siphon spillways of the USA are described and their use is illustrated. Other articles deal with projects in which he was active during this period. Stickney was described as a person of keen intellect, who had the faculty of quickly sifting out the essential elements of a problem, and the ability to work out in detail an accurate solution. He was methodical and kept his records in precise and complete form. He was self-reliant, and on occasion was tenacious in holding to his views. Always he was a thorough gentleman, with an ever-present quiet dignity. He passed away due to stomach cancer.

Anonymous (1922). Stickney, George F. *Who's who in engineering* 1: 1209. Lewis: New York.
Anonymous (1930). George Fetter Stickney. *Trans. ASCE* 94: 1634-1635.
Stickney, G.F. (1905). The compensating works of the Lake Superior Power Company. *Trans. ASCE* 54: 346-370.
Stickney, G.F. (1914). *The Stickney siphon spillway and the Stickney automatic crest for dams.* Hydraulic Specialty Comp.: Albany NY. (*P*)
Stickney, G.F. (1922). Siphon spillways. *Trans. ASCE* 85: 1098-1151.

STOKER

02.03. 1905 Pittsburgh PA/USA
19.10. 1992 New York NY/USA

James Johnston Stoker obtained the BS degree from Carnegie Institute of Technology, Pittsburgh PA, in 1927 and became instructor in technical mechanics. He obtained in 1931 there also the MS degree. In 1936 he moved to the Swiss Federal Institute of Technology, ETH Zurich, obtaining in 1936 the PhD degree. Stoker along with Kurt O. Friedrichs (1901-1982) joined the Department of Mathematics at the New York University, where the two developed the famous Courant Institute of Applied Mathematics and Physics. With Stoker's engineering background, and Friedrichs' mastery in mathematics, both greatly collaborated on many problems of applied mathematics and mechanics. On Courant's retirement in 1958, Stoker succeeded him as director, serving until 1966. During this period, the Institute acquired greater autonomy within the University, becoming the Courant Institute of Mathematical Sciences.

Stoker originally was interested in the elasticity theory, developing essentially into an expert of water waves. His 1947 paper deals with the effect of depth on the propagation of water waves. In 1957 Stoker published his book Water waves, one of the main works in this topic until the book chapter on Surface waves, by John V. Wehausen (1913-2005) and Edmund V. Laitone (1915-2000) published in 1960. Stoker's significant work summarizes the then current state of knowledge with a focus on the linear wave theory. In 1970 Stoker was awarded the Timoshenko Medal from the American Society of Mechanical Engineers ASME in recognition of distinguished contributions to the field of applied mechanics.

Anonymous (1971). Prof. James J. Stoker. *Mechanical Engineering* 93(1): 73. *P*
Anonymous (1985). J.J. Stoker. *Communications on Pure and Applied Mathematics* 38(1): 1. *P*
Anonymous (1992). J.J. Stoker. *Notices of the American Mathematical Society* 39(10): 1291.
Stoker, J.J. (1936). Über die Gestalt der positive gekrümmten offenen Flächen im drei-dimensionalen Raum. *Promotionsarbeit* 826. ETH: Zürich.
Stoker, J.J. (1947). Surface waves in water of variable depth. *Quarterly of Applied Mathematics* 5(1): 1-54.
Stoker, J.J. (1953). Numerical solution of flood prediction and river regulation problems 1: Derivation of basic theory and formulation of numerical method of attack. *Report* IMM-200. Institute of Mathematical Science. New York University: New York.
Stoker, J.J. (1957). *Water waves*: The mathematical theory with applications. Wiley: New York.

STORROW

25.03. 1809 Montreal/CA
30.04. 1904 Boston MA/USA

Charles Storer Storrow was born in Canada, but his family moved to Boston MA. Graduating in 1829 from the Harvard University, Cambridge MA, he moved to Paris, taking further courses at *Ecole des Ponts et Chaussees* studying then engineering works in France and Great Britain. Upon his return to Boston in 1832, he became a railroad engineer and directed the running of the first train from Boston to Lowell. He studied also the water quantity utilized by the Lowell mills, and published his pioneering 1835 treatise on water works, the first in the English language, making available to the English-speaking engineers the works of the French scientists Gaspard Riche de Prony (1755-1839), Jean-Baptiste Bélanger (1790-1874), or Henri Navier (1785-1836). Ten years later Storrow, became engineer for the Essex Company, Lawrence MA. He laid out the city, designing canals, designating mill sites, building mills, and designed a large masonry dam across *Merrimac* River, which still exists. From 1853, he was city mayor.

Storrow's work at Lawrence brought him into close contact with the president of the Essex Company. The latter tried to persuade Storrow to assume charge as professor of engineering at the newly erected Lawrence Scientific School at Harvard, yet Storrow declined this offer. In 1860 he established his home at Boston, and from 1861 served as engineer member of the state commission on the drainage of the Sudbury and Concord meadows. As a member of the *Hoosac* Tunnel Commission, he went in 1862 for a study tour to Europe and upon return advised the commission on plans and methods for tunnel construction. At age of eighty, Storrow resigned his position with the Essex Company. His eminent services to engineering were recognized by his professional brethren in the election to honorary membership of the American Society of Civil Engineers ASCE in 1893. He was also a Fellow of the American Academy of Arts and Sciences.

Anonymous (1904). Charles Storer Storrow. *Engineering News* 51(18): 422-423. *P*
Anonymous (1931). Charles Storer Storrow. *Engineering News-Record* 107(13): 476. *P*
Anonymous (1936). Storrow, Charles Storer. *Dictionary of American biography* 18: 98-99. Scribner's: New York.
Fitzsimons, N. (1968). Charles S. Storrow and the transition in American hydraulics. *Civil Engineering* 38(12): 81-82.
Storrow, C.S. (1835). *A treatise on water-works* for conveying and distributing supplies of water, with tables and examples. Hilliard, Gray and Co.: Boston.

STOUT

14.11. 1865 Jerseyville IL/USA
04.08. 1935 Denver CO/USA

Oscar van Pelt Stout obtained both his BCE and CE degrees in 1888 and 1897, respectively, from the University of Nebraska, Lincoln NE. He then joined various engineering departments until 1890, when becoming city engineer of Beatrice NE. From 1891 he taught at various grades civil engineering, and was from 1912 to 1920 the Dean of the College of Engineering, University of Nebraska. In parallel he was from 1894 to 1903 resident hydrographer of Nebraska and the adjoining States, and from then to 1913 irrigation expert and engineer, US Department of Agriculture. Later he joined as irrigation engineer the Nebraska State Board of Agriculture at the Agricultural Experiment Station, Lincoln NE. From 1922 Stout was employed on corporative irrigation investigations by the US Department of Agriculture, and the State of California.

Stout during his time as Dean at the University of Nebraska was mainly concerned with reclamation, drainage, and hydraulic engineering. He excelled in the combination of theoretical knowledge and simple and practical ways in doing. He devised for example an effective method to determine the daily discharges from a limited number of flow measurements on streams, referred to as 'Stout's method for shifting stream beds'. He further proposed the proportional weir, a plate weir of which the opening allowed for a constant approach flow velocity, as early as in 1896. He had a most subtle way of pointing out the human elements in engineering lessons not found in books, but of which he was a master. One of his epigrams has become famous, namely 'the engineer is a resourceful man'. Those who were graduated under him carried with them the conviction that engineering was an office of trust and a profession of honour, indeed a conviction so firmly instilled that it could never be forgotten in their profession.

Anonymous (1933). Stout, Oscar van Pelt. *Who's who in America* 17: 2208. Marquis: Chicago.
Anonymous (1935). Oscar van Pelt Stout. *Agricultural Engineering* 16(9): 373. *P*
Anonymous (1936). Oscar van Pelt Stour. *Trans. ASCE* 101: 1654-1657.
Kolupaila, S. (1960). Oscar P. Stout. *Journal of the Hydraulics Division* ASCE 86(HY1): 37. *P*
Stout, O. van P. (1897). A new form of weir notch. *Trans. Nebraska Engng. Society* 1(1): 13-16.
Stout, O. van P. (1904). Notes on the computation of stream gaugings. *Engineering News* 52(23): 521-522.
Stout, O. van P. (1914). The proportional-flow weir devised in 1896. *Engineering News* 72(3): 148-149.

STRAHLER

20.02. 1918 Kolhapur/IN
06.12. 2002 Santa Barbara CA/USA

Arthur Newell Strahler graduated in 1938 with an AB degree from the College of Wooster, Wooster OH, in 1940 with the AM degree from Columbia University, New York NY, and there with the PhD degree in 1944. He joined its Faculty in 1941, from 1958 to 1971 as a professor of geomorphology. He left University then to become a full-time academic author. He was a Fellow of the Geological Society of America, the American Geographical Society, the Association of American Geographers AAG, and the American Association for Advancement of Science.

In the mid-1940s Strahler's research consisted of descriptive studies linking landform development with structural and tectonic processes. He recognized the significance of the hydrologist Robert Horton (1875-1945). His 1945 paper was viewed 'a remarkable interdisciplinary information transfer from hydrology, a geophysical area of knowledge, to a geomorphology rooted in geological concepts'. Thus the transition from a qualitative geologist to a quantitative morphologist began. The importance of hydrology, hydraulics as also of soil mechanics became obvious, leading to erosion-control problems, channel stabilization and river engineering. In the early 1950s he used statistics to relate channel gradient to hillslope morphology. His simple modification of Horton's stream ordering system led to the application of this procedure to anatomical morphometrics. Between 1971 and 1998, Strahler produced 17 books dealing with topics ranging from basic physical geography and physical geology to the evolution-creation controversy. He had the ability to assimilate vast amounts of diverse material and to present it to audiences ranging from high school students to professional colleagues.

Anonymous (2000). Strahler, Arthur N. *Who's who in America* 54: 4767. Marquis: Chicago.
Schumm, S.A. (2004). Memorial to Arthur Newell Strahler. Geological Society of America *Memorials* 33(4): 53-55, with selected bibliography. *P*
Strahler, A.N. (1952). Dynamic basis of geomorphology. Geological Society of America *Bulletin* 63: 923-938.
Strahler, A.N. (1957). Quantitative analysis of watershed geomorphology. *Trans. AGU* 38(6): 913-920.
Strahler, A.N. (1964). Quantitative geomorphology of drainage basins and channel networks. *Handbook of applied hydrology* 4: 39-76, V.T. Chow, ed. McGraw-Hill: New York.
Strahler, A.N. (1987). *Science and earth history*: The evolution-creation controversy. Wiley: New York.

STRATTON

07.06. 1898 Stonington CT/USA
16.03. 1984 New York NY/USA

James Hobson Stratton obtained the BS degree from the US Military Academy in 1920, and in 1922 the civil engineering degree from Rensselaer Polytechnic Institute, Troy NY. He was from 1924 to 1927 at the Panama Canal Zone as supply officer, becoming from 1927 to 1931 then engineering instructor at the New Jersey National Guard, Englewood NJ, from when he went through several offices as district engineer at St. Paul MN until 1935. From then to 1937 he was chief of the Engineering District, the Tucumcari Engineering District, and of the Conchas Engineering District, Albuquerque NM. Until 1939 he stayed in a similar position at Boston MA, returning then to the Conchas District, and moving as district engineer to the Caddoa Engineer District in Colorado, where he was in charge of the Caddoa Reservoir Project. After war service in services of supply at the European Theatre of Operations, he returned to the USA in 1945 as assistant chief of the Engineers, Office of Engineers, Washington DC. In 1946 he was appointed director of Civil Works at this Office, in charge of all river, harbour and flood control works. In parallel he also was supervising engineer of the Panama Canal, in charge of Isthmian Canal Studies. He retired as brigadier general in 1949 after 30 years service, joining the engineering firm Tibbetts-Abbett-McCarthy-Stratton TAMS as partner until 1967.

Stratton was best known for his distinguished service in the war when he served as chief of the engineering branch of the Construction Division, US Army Corps of Engineers. Prior to the war he had many assignments on river and harbour work. He served as district engineer on the construction of the Conchas and Joe Martins Dams in New Mexico and Colorado for flood control, power generation, and irrigation. For a number of years he was also active in the canalization work on the Mississippi River. General Stratton was awarded in 1944 the Legion of Merit for his services with the Chief of Engineers. He was an expert in soil mechanics, construction of earth dams, and military airfields. He was awarded in 1981 membership of the National Academy of Engineering NAE. Stratton was member of the American Society of Civil Engineers ASCE.

Anonymous (1948). Stratton, James H. *Who's who in engineering* 6: 1924. Lewis: New York.
Anonymous (1949). Gen. J.H. Stratton. *Engineering News-Record* 143(Aug.11): 22. *P*
Anonymous (1984). Gen. J.H. Stratton dies. *Civil Engineering* 54(5): 68. *P*
Stratton, J.H. (1949). Panama Canal: The sea-level project. *Trans. ASCE* 114: 608-632.
Stratton, J.H. (1952). Uplift in masonry dams. *Trans. ASCE* 117: 1218-1252.

STRAUB

07.06. 1901 Kansas City MO/USA
27.10. 1963 Minneapolis MN/USA

Lorenz George Straub graduated from University of Illinois, Urbana-Champaign IL, with BCE. and MS degrees, and the PhD degree from three German Universities. After return from Germany, where he was one of the first Freeman Scholars, he joined the US Engineering Department of Kansas City District as head of special studies until 1930. He was then professor of hydraulics at University of Minnesota, Minneapolis MN, and director of its St. Anthony Falls Hydraulic Laboratory from 1938. In 1950 he was awarded the Order of Palms, France, and the 1958 ASCE Research Award. In parallel, Straub was involved in numerous projects as a consultant, and he was active in national and international associations, including president of the International Association of Hydraulic Research IAHR from 1948 to 1955, and honorary membership in 1959. He also received the 1961 Norman Medal for his 1960 paper. The Lorenz G. Straub Award is given to the most meritorious thesis in hydraulic engineering or ecohydraulics.

While still in Germany, Straub began the translation of *Verkehrswasserbau* authored by Otto Franzius (1877-1936), and later was the co-translator of *Wasserbau* authored by Armin Schoklitsch (1888-1969). He from the mid 1930s wrote frequently on sediment transport and related aspects of river engineering, and from the 1950s on the two-phase air-water flow in spillways, with the classic 1960 paper co-authored by Alvin G. Anderson (1911-1975). This work includes a set of concentration data used until now, to determine the main flow characteristics both of the uniform, and non-uniform reaches of chute flows in smooth, rectangular and straight chutes. He also made a noteworthy research with Edward Silberman (1914-2011) on small Reynolds number flows.

Anonymous (1949). Lorenz G. Straub. *Civil Engineering* 19(5): 328-329. *P*
Straub, L.G. (1934). Regelungsarbeiten am Missouri. *Zeitschrift Verein Deutscher Ingenieure* 78(39): 1144-1145.
Straub, L.G. (1936). Transportation of sediment in suspension. *Civil Engineering* 6(5): 321-323.
Straub, L.G. (1938). New hydraulic laboratory at St. Anthony Falls. *Engineering News-Record* 121(Dec.8): 725-726.
Straub, L.G., Silberman, E., Nelson, H.C. (1958). Open-channel flow at small Reynolds numbers. *Trans. ASCE* 123: 685-706; 123: 713-714.
Straub, L.G., Anderson, A.G. (1960). Self-aerated flow in open channels. *Trans. ASCE* 125: 456-481; 125: 485-486.

STREETER

21.11. 1909 Marcellus MI/USA
Ann Arbor MI/USA

Victor Lyle Streeter graduated from University of Michigan, Ann Arbor MI, with the ScD degree in 1934. He then moved to Germany to complete his education. Upon return to the USA in 1937 he joined the US Bureau of Reclamation USBR as associate engineer until 1939, becoming until 1941 associate hydraulic engineer within the Department of State, of the US-Mexican International Boundary Commission, from when he was associate hydraulics professor, Illinois Institute of Technology, Chicago IL until 1945, and later professor of fluid mechanics. He re-joined as hydraulics professor the University of Michigan in 1954, retiring in 1976. In parallel he was a consultant in fluid mechanics for the Armour Research Foundation. He also was a Fulbright visiting professor to two universities of New Zealand in the 1950s. Streeter was elected Fellow of the American Society of Mechanical Engineers ASME in 1979, and in 1982 was the third Hunter Rouse Hydraulic Engineering Lecturer, the American Society of Civil Engineers ASCE.

Streeter made a lasting impact on the field of computational fluid transients. With Evan B. Wylie (1931-) he wrote three books on this topic, namely Hydraulic transients in 1967, Fluid transients in 1982, and Fluid transients in systems in 1993. A paper dealing with computer analysis on water hammer was published as early as in 1963, thereby popularizing the computer-based method of characteristics combined with specific time intervals. At that time the predicted variables did not accurately attenuate because of numerical interpolations. In the mid-1960s, the effect of turbo-machinery on water hammer analysis was accounted for, given its relevance for the industry.

Anonymous (1938). Victor L. Streeter. *Civil Engineering* 8(1): 51. *P*
Anonymous (1954). Streeter, Victor L. *Who's who in engineering* 7: 2348. Lewis: New York.
Anonymous (1983). V.L. Streeter. *Journal of Hydraulic Engineering* ASCE 109(11): 1407. *P*
Streeter, V.L. (1936). Frictional resistance in artificially roughened pipes. *Trans. ASCE* 101: 681-713.
Streeter, V.L., ed. (1961). *Handbook of fluid dynamics*. McGraw-Hill: New York.
Streeter, V.L. (1964). Waterhammer analysis of pipelines. *Journal of the Hydraulics Division* ASCE 90(HY4): 151-172.
Wiggert, D.C., Wylie, E.B. (2003). A tribute to Victor L. Streeter. *Henry P.G. Darcy and other pioneers in hydraulics*: 160-173, G.O. Brown, ed. ASCE: Reston VA.

STREIFF

21.08. 1880 Java/ID
04.01. 1960 Jackson MI/USA

Abraham Streiff was born in the former Dutch Indies, todays Indonesia. He graduated in 1901 from the Technological Institute, Delft NL, taking then post-graduate studies at the Swiss Federal Institute, ETH Zurich, serving as hydraulic engineering assistant. He then worked as hydraulic engineer during eight years, joining in the 1910s Fargo Engineering Co., Jackson MI, and there was engaged in the design of major hydroelectric plants for 30 years. He also was associated with the Consumers Power Company, the Commonwealth and Southern Corporation, and the Ambursen Engineering Company as vice-president, then headed by Nils F. Ambursen (1876-1953). He conducted his own consulting office, and for the last eight years of his life served as hydraulic consultant to the American Smelting and Refining Company, Tucson AZ.

Streiff was a hydraulic engineer covering a wide field of tasks. He was connected with projects as chief engineer and consultant, including power plants in Texas, the *Rasgo* Project in Brazil, and the Narrows Dam and power project in Arkansas. He was a pioneer in the successful design of large Tainter Gates, as introduced by Jeremiah N. Tainter (1836-1920). He was in the 1920s particularly engaged with the relation between rainfall and the hydraulic design of flood-control structures. He also related sunspots to rainfall, and in 1951 therefore predicted: 'You won't need that raincoat much next year, but keep it handy for 1953'. His advice was based on a 30-years study of weather cycles. He further stated: 'In 1953 you'll see those big floods, but do not look for them before that'. His predictions were based on the recurrence of hydrological processes, which are governed by a systematic sequence that allows for a reasonably accurate forecasting. He was a member of the American Society of Civil Engineers ASCE from 1959.

Anonymous (1935). Brazos River act proponents. *The Meridian Tribune* 41(42): 1. *P*
Anonymous (1960). Abraham Streiff. *Trans. ASCE* 125: 1435.
Streiff, A. (1914). Testing low-head turbines by chemical method. *Engineering Record* 70(10): 276.
Streiff, A. (1927). Sunspots and rainfall. *Monthly Weather Review* 55(2): 69-71.
Streiff, A. (1928). Notes on estimating run-off. *Monthly Weather Review* 56(3): 98-99.
Streiff, A. (1935). *Control of flood waters of the Brazos River*. US Congress: Washington DC.
Streiff, A. (1946). Discussion of Design developments: Structures of the Tennessee Valley
    Authority. *Trans. ASCE* 111: 1216-1219.
Streiff, A. (1950). Discussion of Diffusion of submerged jets. *Trans. ASCE* 115: 684-687.

STRICKLAND

ca. 1787 Navesink NJ/USA
07.04. 1854 Nashville TN/USA

William Strickland was probably born around 1787 on a farm near Navesink NJ, from where his family moved to Philadelphia PA. William was apprenticed to Benjamin Latrobe (1764-1820). Strickland's first work was routine drawings for the US Capitol and surveying at Newcastle DE. From 1805 he devoted his time to artistic pursuits but in 1814 he then was involved in the defence of his town against the British. In 1818 Strickland bested his master Latrobe for the commission of the Second Bank of the USA. His engineering career began then only in 1821.

Strickland was appointed to find the best route for a canal between the Chesapeake and Delaware Bays. His next venture into engineering occurred in 1825 when he moved to Europe to observe the latest advances in transportation, construction as also industrial processes. His sponsor was the Pennsylvania Society for the Promotion of Internal Improvement, a group who clearly foresaw the commercial threat of New York City as completion of the Erie Canal neared. Although originally scheduled to tour the British Isles, France, Holland and Germany, Strickland apparently secured enough information in Britain alone. Upon completion of his famous report, his reputation as an engineer was greatly enhanced and he became active in canal and, later, railroad projects for Pennsylvania State. He also engineered in 1827 the Brandywine Shoal Lighthouse in New Jersey for the Federal Government, an industrial canal, and he was consulted on the Fairmont Dam of the Philadelphia water works. His best known engineering work was however the huge Delaware Breakwater whose construction took a decade until 1840 to extend the protection over 1 km out from Cape Henlopen DE. In the mid-1830s he discussed with colleagues the founding of an Institution of American Civil Engineers which occurred then in 1838. On his second journey to Europe, the British friends proposed him for membership in their Institution of Civil Engineers, which was founded in 1818, and Strickland was most probably the first American elected. His legacy were the men he trained to carry on the profession he fostered among which were John C. Trautwine (1810-1883) and a number of notable architects.

Anonymous (1972). Strickland, William. *A biographical dictionary of American civil engineers*: 113-114. ASCE: New York. *P*
Fitzsimons, N. (1969). William Strickland, architect-engineer. *Civil Engineering* 39(12): 84-85.
Gilchrest, A.A. (1950). *William Strickland*: Architect and engineer. University of Pennsylvania Press: Philadelphia.

STROWGER

09.10. 1895 Rochester NY/USA
06.11. 1992 Harper Woods MI/USA

Earl Byron Strowger was educated at University of Rochester, Rochester NY, from where he obtained the BS degree in mechanical engineering in 1918. He was engineer first with the US Shipping Board, from 1919 to 1924 engineer of the Niagara Falls Power Company, until 1929 then hydraulic engineer, joining until 1950 the Western Division of Niagara Hudson System, becoming chief engineer of the Niagara Mohawk Power Corp. in 1951, and finally serving there as systems project engineer until being retired. He was in parallel consultant for the Gibson Method in testing hydro-electric power plants.

Strowger was all through his career involved in testing hydropower installations in terms of discharge characteristics and relating to waterhammer effects. He also designed a large number of surge tanks, by which unsteady effects due to discharge variations or starting and shut-off processes in pipelines of hydraulic schemes may be significantly suppressed. He was also involved in the entire mechanical equipment of large schemes, including the Niagara Falls Power Company. He further popularized the Gibson Method for discharge measurement, as proposed by Norman R. Gibson (1880-1967). Most of his publications appeared in journals of the American Society of Mechanical Engineers ASME. Other works such as chapters on Waterhammer, or Surge tank computations are included in the Hydroelectric handbook of Creager and Justin. Strowger was an ASME member chairing the Hydro-Power Commission from 1937 to 1942. He was member of the American Society of Civil Engineers ASCE chairing its Hydraulic Division. He also was a member of the Advisory Committee on Steam and Hydraulic Turbines.

Anonymous (1954). E.B. Strowger, Niagara Mohawk Power Co. *Civil Engineering* 24(6): 301. *P*
Anonymous (1959). Strowger, Earl B. *Who's who in engineering* 8: 2390. Lewis: New York.
Creager, W.P., Strowger, E.B. (1957). Hydroelectric power. *American civil engineering practice* 2(16): 1-59, R.W. Abbett, ed. Wiley: New York.
Gibson, N.R., Strowger, E.B. (1935). Experimental and practical experience with the Gibson method of water measurement. *Trans. ASME* 57(HYD-5): 213-226.
Strowger, E.B. (1927). Niagara Falls Power Co. improves its Niagara plant by redesigning parts of the old waterwheels and re-building the generators. *Power* 66(25): 964-969.
Strowger, E.B. (1928). A water-level gage of the long-distance recording type. *Mechanical Engineering* 50(5): 365-367; 50(8): 614-617.
Strowger, E.B. (1934). How we raise hydro efficiencies. *Electrical World* 103(16): 535-538.

SUDLER

14.02. 1870 Anne Arundel County MD/USA
21.04. 1938 Washington DC/USA

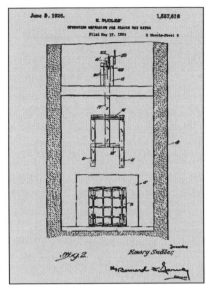

Emory Sudler attended the Baltimore City College, graduating in 1888. He was then first draftsman for Buena Vista VA. He returned in 1891 to Baltimore MD, remaining there until 1914. He served there as assistant engineer on river and harbour surveys to improve Baltimore's water shipping, and then was employed by the Baltimore City Water Department. Due to the growth of the city, many improvements were necessary, so that Sudler was engaged in these developments. The most outstanding related to the Roland-Park stand-pipe, along with an impounding reservoir on Gunpowder River, and the design and construction of Lake Ashburton. In 1910, Sudler was appointed division engineer in charge of these improvements, which were completed until 1914. In addition to the impounding dam, a gate house and a connecting pipeline system were installed. He then resigned from the Water Department.

From 1916 to 1918 Sudler was employed by the US Government Company as engineer on the design and construction of wharves, docks, bulkheads, and dredging operations at the Army Supply Base, Norfolk VA. He was from 1919 employed as chief engineer by the Ambursen Construction Company founded by Nils F. Ambursen (1876-1953), and there played an important part in the design and construction of dams and power houses. From 1930 until his death, Sudler prepared the data to be used in the legal case between New Jersey and New York States concerning the diversion of water from Delaware River. His outstanding work during his engineering career remained the design of dams and other hydraulic structures, however. During the ten years with Ambursen Company, he designed and supervised the installation of automatic crest gates for the *Patillas* Dam, Puerto Rico; he was involved in the reconstruction of the water-works reservoir of Oklahoma City OK, and made a preliminary design of the 75 m high Rodriguez Dam in Mexico, then the highest Ambursen dam, which was completed in 1938. Sudler was highly educated by years of study and reading, extending far beyond the engineering field. He was member of the American Society of Civil Engineers ASCE since 1925.

Anonymous (1939). Emory Sudler. *Trans. ASCE* 104: 2009-2011.
Beatty, P.A. (1913). Construction of a high-service reservoir at Baltimore MD. *Trans. ASCE* 76: 92-120.   http://www.google.com/patents/US1587616 (*P*)
Stokes, W.R., Hachtel, F.W. (1912). Some results of the treatment of the Baltimore drinking-water by calcium hypochlorite. *Trans. Preventive Medicine and Public Health*: 236-252.

SULLIVAN

23.05. 1876 Roxana MI/USA
12.07. 1938 El Paso TX/USA

Vernon Lyle Sullivan studied from 1898 to 1902 engineering under a private tutorage. He accepted in 1903 the position of chief engineer for the Pecos Irrigation Company NM, then the largest operated in the southwest. During 1905 and 1906 he designed and constructed the Public Utilities Hydro-Electric Power Plant at Carlsbad NM. He was appointed in 1907 territorial engineer of New Mexico, having charge of all water appropriations in the state, and thus working also for the Elephant Butte Project, the later Rio Grande Federal Irrigation Project. He loyally protected the government's water filings for the project, which was one of the great benefit to El Paso TX, and to the Rio Grande Valley. In 1910 he resigned as territorial engineer for New Mexico, accepting the position of engineer-manager of the irrigation projects of the Orient Railroad Company TX. From 1911 to 1913 he had charge of the construction, operation, and maintenance of all its irrigation projects.

Sullivan designed and constructed during the next four years irrigation projects in southwestern Texas, including these at Fort Stockton, Balmorhea, and Leon Springs. The first of these was by then probably one of the best smaller projects having concrete-lined canals provided with water tight turnout gates, and measuring devices. In 1918 he opened his private practice at El Paso TX, planning the Big Wichita Irrigation and City Water Supply Project near Wichita Falls TX. Between 1920 and 1933 he served as consulting engineer for the Pecos Valley Water Users Association in Texas, the State of Texas in connection with the New Mexico-Texas Pecos River Water Compact, and the *Basaseachic* Hydro-Electric Power Project in western Chihuahua MX. He purchased in 1927 the El Paso Testing Laboratories, and from that year until his death devoted much time and energies to their active management and operation. During the last years, he devoted time to the Red Bluff Project on the Pecos River, designing the Red Bluff Dam and the hydro-electric plant. Sullivan was a man of many interests an abilities, indeed an experimenter and inventor. He always had a smile on his face; he was ASCE member from 1913, and member of the American Association of Engineers.

Anonymous (1941). Vernon L. Sullivan. *Trans. ASCE* 106: 1680-1682.
Sullivan, V.L. (1909). *Irrigation in New Mexico.* US Office of Experiment Stations: Washington.
Sullivan, V.L. (1924). *Reclamation of the Middle Rio Grande Valley NM.* Albuquerque NM.
http://www.ose.state.nm.us/state_engineer_past_state_engineers.html *P*

SUTRO A.H.J.

29.04. 1830 Aachen/D
08.08. 1898 San Francisco CA/USA

Adolph Heinrich Joseph Sutro was born in Prussia. After the father had passed away in 1847, his mother emigrated in 1850 with the 11 children to the USA. Fired by Californian gold discoveries, Sutro arrived in 1851 at San Francisco CA, to be engaged for nine years in mercantile pursuits. In 1860 he established a quartz-reducing mill at East Dayton NV, working with a new process of amalgamation und thus laid the basis of his later fortune. Impressed by the old-fashioned mining methods, he drove a tunnel 3 m high and wide, and 5 km long into Mount Davidson from Carson River to Comstock Lode, providing the ventilation, drainage and ease of transportation. He formed the Sutro Tunnel Company then, obtained a charter in 1865, and persuaded mine owners to sign contracts to pay the company two dollars per ton for all ore mined after the opening of the tunnel for their use.

However, Sutro's California supporters turned against him, to get control of the tunnel and thereby reap the immense profits which were anticipated. Sutro in 1867 went to Europe to visit mines, studying their tunnels, consulting engineers, and obtaining the indorsement for his own plans. He published in 1868 his pamphlet on the Sutro Tunnel. After having secured money from Nevada and Europe, works began in 1869, the tunnel was completed in 1878, and a new era in western mining started. The project proved immediately and immensely profitable, so that Sutro sold his interests already in 1879, returning to San Francisco, where he invested into real estate. He at one time owned almost 10% of San Francisco city and county. In 1892 he began construction of the enormous Sutro salt-water baths for almost 1 million of US dollars, forming the finest bathing pavilion then in existence. In 1894 he was elected for two years mayor of San Francisco. He then collected rare books in technology and science, his library including finally more than 200,000 volumes, half of which were destroyed in the fire of 1906.

Anonymous (1936). Sutro, Adolph Heinrich Joseph. *Dictionary of American biography* 18: 223-224. Scribner's: New York.

Capen, C.H. (1943). History of Sutro weirs. *Engineering News-Record* 131(17): 585.

Sutro, A. (1868). *The mineral resources of the United States* and the importance and necessity of inaugurating a rational system of mining with special reference to the Comstock Lode and the Sutro Tunnel in Nevada. Murphy: Baltimore.

http://www.sfmuseum.org/sutro/bio.html *P*

SUTRO H.H.

02.01. 1877 New York NY/USA
21.01. 1913 New York NY/USA

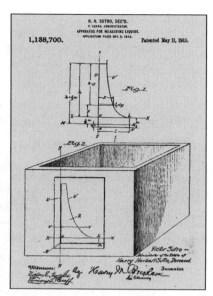

Harry Herbert Sutro graduated with a BS degree in chemistry from Columbia University, New York NY, in 1898. Few is known on his professional career, and whether he was associated to the family of A.H. Sutro (1830-1898). The Sutro weir was in any case described first in the 1912 paper, based on a patent submitted in 1907, and issued in 1911, referring to a drainage system. Sutro committed suicide jumping from the seventh floor of his apartment.

The Sutro weir is a liquid measuring notch shaped so that the quantity of water discharged always is in direct proportion to the approach flow head, rather than to an exponent differing from 1. The notch shape is parabolic, tending to infinite width at the base, which was modified however for practical purposes. The original Sutro weir had one vertical side and one curved side to approximately allow for the linear head-discharge relation. Ernest W. Rettger (1871-1938) was the first to devise a symmetrical proportional weir shape, as described in his 1914 paper. Edmund A. Pratt (1887-1961) developed the theory of the Sutro weir in 1914, which was in 1915 also experimentally investigated by a thesis. Clemens Herschel described the weir in articles published around 1920, as stated by Charles H. Capen (1895-1987) in his 1943 account. He further stated that 'there is little available at the moment to indicate just when someone conceived the idea of converting the Sutro weir into a symmetrical form similar to the Rettger type. It may be that the advantages of both types were recognized and combined'. These weirs were particularly used to control the approach flow to grit chambers.

Capen, C.H. (1943). History of Sutro weirs. *Engineering News-Record* 131(17): 585.

Dodge, M.G., ed. (1902). Harry Herbert Sutro. *Delta Upsilon Fraternity catalogue*: 827.

Pratt, E.A. (1914). Another proportional-flow weir: Sutro weir. *Engineering News* 72(14): 462-463.

Pratt, E.A. (1936). Sutro weir formula. *Engineering News-Record* 117(26): 904.

Rettger, E.W. (1914). A proportional-flow weir. *Engineering News* 71(26): 1409-1410; 72(9): 462-463.

Rettger, E.W. (1915). The inverted weir. *Engineering News* 73(2): 72-73; 74(6): 277; 74(22): 1018-1019.

http://www.google.de/patents?id=r6dCAAAAEBAJ&printsec=abstract&zoom=4&hl=de#v
http://patentimages.storage.googleapis.com/pages/US1138700-0.png (*P*)

SUTTON

26.01. 1877 Smyrna DE/USA
19.05. 1949 Lima/PE

Charles Wood Sutton graduated as a civil engineer in 1898 from University of Washington, Seattle WA. He was engaged by the US Geological Survey USGS working from 1902 as an assistant topographer. He went to Peru in 1904, in response to application for civil engineering from the Government of Peru to the US Government. There he was in charge of irrigation works, and general survey and boundary demarcation projects in *Callao*. Sutton returned to the USA in 1906, serving as assistant engineer the US Reclamation Service for the Yakima Project, and was in charge of mapping works for the Yakima River Valley in Washington State. In 1907 he was engaged by the US Navy Department to prepare a base map for development projects of the *Olongapo* Naval Reservation, the Philippine Islands, including design and construction of various hydraulic works.

Sutton returned in 1908 to Peru as chief of commission to investigate irrigation problems in the Central Coast Region, and from 1911 joined as chief engineer the Peruvian Irrigation Service. He was delegate of Peru to the 19th International Drainage Congress held at Chicago, and of the 1st International Drainage Congress, held at New Orleans. From 1914 to 1919 he was a consulting engineer both in New York, and of the Board of Water Supply at Lima, Peru. From then he was a consulting and construction engineer to the Department of Irrigation Works, Government of Peru. He was named Father of Peruvian irrigation for his outstanding services to this country. Sutton was a Member of the American Society of Civil Engineers ASCE, of the Peruvian Society of Engineers, and the Peruvian Geographical Society.

Anonymous (1917). Sutton, Charles W. *Who's who in America* 9: 2396. Marquis: Chicago.
Anonymous (1929). Sutton, Charles Wood. *Who's who in America* 15: 2016. Marquis: Chicago.
Mazuré, J.H. (2011). *Monumentos públicos en espacios urbanos de Lima*. Barcelona. P
Santisteban, V.P. (1980). *Sutton y la irrigación de Olmos*. Santisteban: Lima.
Sutton, C.W., Stiles, A.I. (1906). *Informes sobre aguas del departamento de Piura*. Prince: Lima.
Sutton, C.W. (1917). The relation of government to property and Enterprise in the Americas.
    Proc. New York Academy of Political Science 7(2): 502-513.
Sutton, C.W. (1921). *Irrigation and public policy in Peru*. Torres Aguirre: Lima.
Sutton, C.W. (1927). *Las obras de irrigación en el Departamento de Lambayeque*. Lima.
Sutton, C.W. (1927). Agriculture and irrigation in Peru. *Pan-America Union Bulletin* 61: 642-648.
www.imefen.uni.edu.pe/.../Lambayeque_1925_... P

SWAIN

02.03. 1857 San Francisco CA/USA
01.07. 1931 Holderness NH/USA

George Fillmore Swain graduated in 1877 from the Massachusetts Institute of Technology as a civil and topographical engineer. He went for three years to Europe, studying thereby at the Royal Polytechnic of Berlin. On his return to the USA he served for the Tenth Census to report on water power, with the Report published in 1881. He was then appointed an instructor of civil engineering at MIT, advancing to professor of engineering in 1887. During these years he spent one summer with the Office of Locks and Gates, Lowell MA, receiving further education by James Bicheno Francis (1815-1892). In 1909 he was appointed Gordon McKay professor of civil engineering Harvard University, Cambridge MA, a position he held until his retirement in 1929.

Swain was in parallel also a successful consultant on engineering projects. In this work he covered a wide field from transportation and structural engineering to hydraulics. He authored three volumes on Strength of materials, and a book Hydraulics. His greatest work was as a teacher: His influence on the teaching methods was profound. Familiarity with Continental methods of approach to problems of applied science enabled him to properly evaluate them and forcefully to discourage the emphasis on methodology that they frequently presented. He emphasized training in interpreting the data of science as distinguished from drill in the technique of assembling it. His books How to study, and The young man and civil engineering, summarize his particular approach. He was the first recipient of the *Lamme* Gold Medal of the Society for the Promotion of Engineering Education in 1928 'for accomplishment in technical teaching'. He served as president of the American Society of Civil Engineers in 1913.

Anonymous (1913). George F. Swain. *Engineering News* 69(5): 196-198. *P*
Anonymous (1924). George F. Swain. *Engineering News-Record* 92(16): 645-646. *P*
Anonymous (1933). George F. Swain. *Trans. ASCE* 98: 1476-1484.
Anonymous (1944). Swain, George Fillmore. *Dictionary of American biography* 21: 680. Scribner's: New York.
Swain, G.F. (1881). *Water-power of the Southern Atlantic water-shed*. US Census Report, Washington DC.
Swain, G.F. (1915). *Conservation of water by storage*. Yale University Press: New Haven.
Swain, G.F. (1917). *How to study?* McGraw-Hill: New York.
Swain, G.F. (1922). *The young man and civil engineering*. Macmillan: New York.

SWAN

17.08. 1842 Boston MA/USA
17.04. 1899 Boston MA/USA

Charles Herbert Swan graduated from the Lawrence Scientific School, Lawrence MC, in 1863 with the BSc degree. After service in the Civil War, he joined an engineering company dealing mainly with water works. He was also assigned to Charlestown City MA, in charge of laying the street mains. Later he was connected with the construction of the water works of Salem MA, and from there went in 1869 to Providence RI working on water and sewerage works. In 1881 he went to Boston, visiting Europe in 1884, to study the sewage systems of various cities. In 1886, Swan joined the engineering company of Rudolph Hering (1847-1923), then chief engineer of the Chicago Water Supply and Drainage Commission, to contribute to sewage disposal by land filtration. One year later, he studied the disposal scheme of the North Metropolitan Sewerage District by chemical precipitation for Massachusetts State Board of Health. He soon later was there appointed chief engineer and remained at Boston until his death.

While in Providence RI, Swan came into contact with the works of Wilhelm Rudolph Kutter (1818-1888). He applied his uniform flow formula originally developed for open channels to sewers, constructing a valuable set of sewer diagrams. He was in charge of the numerous investigations to develop water works and sewer systems, and his services were valuable and highly appreciated. In collaboration with a colleague the Kutter formula was further improved for direct engineering application, and published first mainly in tabular form in addition to diagrams in 1899. This work saw a number of re-editions and was in use well up to the 1950s. During his Boston years he was also engaged in siphon hydraulics as applied to the sewage treatment stations. He further conducted studies on pipe flow both for sewage and raw water.

Anonymous (1899). Charles Herbert Swan. *Proc. of the Association of Engineering Societies* 23(3): 6-8. *P*

Horton, T. (1901). Flow in the sewers of the North Metropolitan Sewerage System of Massachusetts. *Trans. ASCE* 46: 78-92.

Swan, C.H. (1880). Note on Kutter's diagram. *Trans. ASCE* 9: 326-328.

Swan, C.H. (1888). Notes on European practice in sewage disposal. *Journal of the Association of Engineering Societies* 7(7): 248-257.

Swan, C.H., Horton, T. (1899). *Hydraulic diagrams for the discharge of conduits and canals*, based upon the formula of Ganguillet and Kutter. Engng. News Publishing: New York.

SWEET

20.11. 1837 Cheshire MA/USA
26.01. 1903 Albany NY/USA

Elnathan Sweet graduated in 1859 from the Union College, Schenectady NY, as a civil engineer. His first work in the field was as deputy surveyor, but he soon returned to New York NY, married, and was employed as assistant engineer on railways projects. In 1864 he went to Franklin PA, being engaged as engineer on oil wells and coal mines, from where he moved in 1868 to Chicago IL where he was again in charge of railway construction. In 1872 he formed a partnership at Chicago. He was appointed in 1875 by the New York State Governor as expert engineer on the State canals. In 1876 he was promoted to a division engineer of the canals but resigned in 1880, returning to railways projects. In 1883 Sweet was elected state engineer; during his administration he made experiments on a large scale to define the laws governing the resistance of vessels propelled in narrow waterways, as affecting their best design, thereby accounting for optimum canal capacity and economy. The results of this study, including various made while having been division engineer, were published in the 1880 paper. As State Engineer, Sweet further made strenuous efforts to restore to the engineering department the control of all engineering matters in which the state was interested.

Upon retiring from Office in 1887, Sweet resumed the practice of his profession. He thereby introduced notable improvements in the design of movable bridges, and bridges of long span. His paper on The radical enlargement of the Erie Canal, published in 1884 was the first formal project for the construction of a deep waterway from the Lakes to Huron River. He urged that thorough surveys be made for the canal. He later took strong interest in the development of this canal, thereby favouring the project for a 6.3 m wide canal in preference to the only 4 m width. He stated that 'No waterway built to connect the Lakes and the Ocean can hope to secure great commercial importance unless it can accommodate the largest vessels navigating the Great Lakes'. Sweet also served as member of the New York Water Storage Commission; its purpose was to conserve the water supply for the public benefit. He was ASCE member since 1878.

Anonymous (1903). Elnathan Sweet. *Engineering News* 49(6): 124-125. *P*
Sweet, E. (1880). Proposed improvement of the Erie Canal. *Trans. ASCE* 9: 99-110.
Sweet, E. (1901). Discussion of Economic dimensions for a waterway from the Great Lakes to the Atlantic. *Trans. ASCE* 45: 282-284.
Sweet, E. (1902). Some important phases of canal navigation. *Trans. ASCE* 47: 435-444.

SWIFT

01.01. 1871 Cornwall CT/USA
28.02. 1953 Harwinton CT/USA

William Everett Swift was graduated in 1895 from the Massachusetts Institute of Technology MIT, Cambridge MA, with a degree in civil engineering. He was first responsible for planning of parts of the New York subway and then was in charge of special studies for construction of the Panama Canal. In the 1910s he was in charge of the location and then the construction of the Truckee-Carson Irrigation Project in Nevada, the first project of the US Reclamation Service. Other work in the western US included the design and construction of the Swift Dam, Valier MT, at the time the highest rock-fill dam worldwide. Swift was also involved in the location, design and construction of the Huntley Irrigation system near Billings MT, and he also located the Hudson River Division of the Catskill Viaduct for the New York Board of Water Supply. His work took him further to Alaska, Greece, The Dominican Republic, and to Costa Rica. He was member of the American Society of Civil Engineers ASCE from 1910, becoming Life Member in 1938.

The initial irrigation structures of the Truckee-Carson Project designed by reclamation engineers included the earthen Derby Diversion Dam on Truckee River, the 50 km long Truckee Canal, with a discharge capacity of 40 m^3/s, the Carson River Diversion Dam, and a distribution system of canals and laterals. The original plan for a storage dam on Lake Tahoe was halted by the property owners along the lake fearing the water levels would fluctuate, so that the engineers constructed the Lahontan Dam and Reservoir in 1911. A legal settlement was finally reached with the Truckee River Electric Company concerning storage of water at Lake Tahoe so that the engineers proceeded with dam construction. To insure the productivity of lands, the Reclamation Service constructed a network of drainage ditches to prevent alkali damage to the land and crops from the rising groundwater level. The 50 m high Swift Dam was originally constructed in 1910, but it collapsed in 1964 after heavy rains, resulting in at least 28 fatalities. The current concrete-arch dam was completed in 1967, with 60 m of height and 170 m crest length.

Anonymous (1955). William E. Swift. *Trans. ASCE* 120: 1577.
Swift, W.E., Williams, C.G. (1895). *Experiments on the loss of head of water flowing through diaphragms in pipes*. Department of Civil Engineering, BS Thesis. MIT: Cambridge.
Swift, W.E. (1910). Core drilling under Hudson River for Catskill Aqueduct. *Eng. News* 63(14):414 -415.
http://en.wikipedia.org/wiki/Swift_Dam_(Montana)
http://search.ancestry.com/William+Everett+&gsfn_x=1&gsln=Swift&gsln_x=1&msddy=1953 *P*

SWITZER

18.05. 1891 New York NY/USA
15.12. 1967 Honolulu HA/USA

Frederick George Switzer graduated in 1913 and 1914, respectively, with the ME and MME degrees from Cornell University, Ithaca NY. He then was a design engineer until 1916, assistant and professor of hydraulic engineering at Cornell University until 1940, from when he joined as division engineer the New York City Board of Water Supply until 1952. Until retirement he was a professor of mechanical engineering at Syracuse University, Syracuse NY. In parallel he was from 1916 a consulting engineer on hydro-electricity, centrifugal pumps, and general hydraulic problems. He was member and Fellow of the American Society of Civil Engineers ASCE.

Switzer was a leader in the field of pump design. He was known for various books, notably Hydraulics, Advanced mechanics, and General engineering. His 1915 paper is one of the few in which the temperature effect on the discharge coefficient on weir flow was investigated. For water flow, this effect is small, whereas it may become large when dealing with more viscous fluids, such as oils. He was a leader in the fields of pump design and applied his knowledge to both educational and technical purposes, while staying at Cornell University, and serving for New York City. He joined the American Society of Mechanical Engineers ASME in 1927. His many society activities included organizing the Ithaca Section in 1936, and serving as Secretary and member of the Executive Committee on the Hydraulics Division from 1935 to 1945, from when he was chairman of the PTC Committee No. 8 on centrifugal pumps until 1962.

Anonymous (1940). Frederick G. Switzer. *Engineering News-Record* 125(Dec.12): 792. *P*
Anonymous (1941). Switzer, F.G. *Who's who in engineering* 5: 1745. Lewis: New York.
Anonymous (1959). Switzer, Frederick G. *Who's who in America* 28: 2620. Marquis: Chicago.
Switzer, F.G. (1915). Tests on the effect of temperature on weir coefficients. *Engineering News* 73(13): 636-637.
Switzer, F.G., Miller, H.G. (1929). Floods. *Bulletin* 13. Engineering Experiment Station: Ithaca.
Switzer, F.G. (1930). *The centrifugal pump*: Theory, characteristics, operation and installation. Gould Pumps, Inc.: Seneca Falls NY.
Switzer, F.G. (1931). Discussion of Construction methods at Coolidge Dam. *Trans. ASCE* 95: 169.
Switzer, F.G. (1932). *Hydraulics*. Cornell University: Ithaca NY.
Switzer, F.G. (1937). *Pump fax bulletin* 400: A handbook of data and information on the characteristics, installation and operation of pumps. Gould Pumps Inc.: Seneca Falls NY.

TAINTER

06.01. 1836 Prairie du Chien WI/USA
05.02. 1920 Dunnville WI/USA

Jeremiah Burnham Tainter was the inventor of the Tainter Gate used in hydraulic engineering. He went to Menomonie WI to work in a lumber mill where his brother was manager, inventing there the gate. He began his work in hydrology also in 1862, with the modification of pre-existing mill pond dams at Menomonie. Tainter had talent for designing water control devices to be used on the Red Cedar River and tributaries within the territory of the company. In 1886 the company needed a water control device that would almost instantly release enough water from the mill pond to allow the ponderous 'Red River strings' of lumber to float down to Dunnville and the larger Chippewa River. Tainter redesigned a basic but clumsy water control gate first developed in the East. Installed at Menomonie Dam, it proved to be an efficient device relatively easy to manipulate when opened. A bank of six gates provided an almost instant rush of water sufficient to send the long river strings on their way to market.

Tainter took in 1885 patent no. 344, 878 for a sluiceway gate, as was announced in the Official Gazette in July 1886, the date of patent issue. Nothing further is known on Tainter according to Rouse except for the fact that he definitely did not spell his name with the 'o' sometimes found in handbooks. Rouse further states that the sector type of gate has been sketched by Leonardo da Vinci already in the 16th century. Tainter gates are hydraulic elements to control the flow of water at dams and locks. There are more than 300 of these gates on the Mississippi River Basin between Minneapolis and St. Louis alone. These standard hydraulic gates operate by using water power to support opening and closing manoeuvres. When a Tainter gate is closed water bears on the convex upstream side, but when the gate is rotated, the water flow below the gate helps to manoeuvre the gate. A critical factor in its design is the stress transferred from the skinplate to the radial arms, to the trunnion and the resulting friction.

Lynch, L., Russell, J. (1996). *Where the wild rice grows*: A sesquicentennial portrait of Menomonie 1846-1996. Dunn Co. Historical Society: Menomonie WI. *P*
Rouse, H. (1976). Jeremiah B. Tainter. *Hydraulics in the United States* 1776-1976: 55. Iowa Institute of Hydraulic Research: Iowa City.
http://en.wikipedia.org/wiki/Jeremiah_Burnham_Tainter
http://en.wikipedia.org/wiki/Tainter_gate
en.academic.ru.jpg *P*

TAIT C.E.

28.01. 1875 Vevay IN/USA
05.04. 1923 Los Angeles CA/USA

Clarence Everett Tait graduated in 1899 from Purdue University, West Lafayette IN, with the BS degree in mechanical engineering. He entered the service of the US Department of Agriculture at Cheyenne WY, in its Irrigation Investigations Division, serving as assistant engineer. He there was concerned with the design of current-meter rating stations, water-stage registers, measuring weirs, and rating flumes, of which position resulted a Report on water storage in Northern Colorado. In 1903 Tait was placed in charge of the work of the Central Division of the US Irrigation Investigations USII. He cooperated with the state engineers of the Rocky Mountain States, supervised the field work and designed and constructed works for the first rice irrigation experiment in Arkansas.

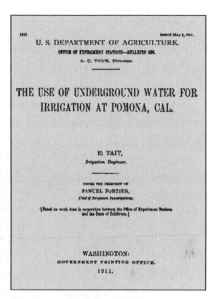

In account of failing health, Tait was transferred in 1905 to Southern California, and there was associated with Prof. Joseph N. Le Conte (1870-1950), in mechanical tests on pumping plants, the result being published in the 1907 Report. In 1906 and 1907, Tait studied the flow conditions of the Lower Colorado River, and wrote a Report on the Irrigation in Imperial Valley. In 1908 and 1909 he was in charge of the USII work in Arizona and Imperial Valley, studying silt problems and canal cleansing in the Southwest and on the Rio Grande River, with headquarters at Los Angeles CA. In 1912, in corporation with the State Conservation Commission, he supervised the field work in Southern California on the State irrigation resources. Another report followed in 1917 on the Conservation and control of flood water in Coachella Valley, and still another Report on the Utilization of Mojave River for irrigation in Victor Valley was published in 1918. At his stage, he was promoted to senior irrigation engineer. He was considered an authority in his profession.

Anonymous (1924). Clarence E. Tait. *Trans. ASCE* 87: 1433-1434.
Le Conte, J.N., Tait, C.E. (1907). Mechanical tests of pumping plants in California. US Office of Experiment Stations *Bulletin* 181. Dept. of Agriculture: Washington DC.
Tait, C.E. (1903). Storage of water on Cache La Poudre and Big Thompson Rivers. US Office of Experiment Stations *Bulletin* 134. Dept. of Agriculture: Washington DC.
Tait, C.E. (1908). *Irrigation in Imperial Valley, California*: Its problems and possibilities. US Government Printing Office: Washington DC.
Tait, C.E. (1911). The use of underground water for irrigation at Pomona CA. US Office of Experiment Stations *Bulletin* 236. Dept. of Agriculture: Washington DC. (*P*)

TALBOT

21.10. 1857 Cortland IL/USA
03.04. 1942 Chicago IL/USA

Arthur Newell Talbot graduated as a civil engineer in 1881 from Illinois Industrial University, the later University of Illinois. After four years of work for railroads, he returned to his Alma Mater as assistant professor of engineering and mathematics, and was promoted in 1890 to professor of municipal and sanitary engineering, a position he held until his retirement in 1926. Two of his earliest works were formulas describing the rates of maximum rainfall, and the size of waterways for bridge and culvert design. His interest in practical municipal problems such as septic tank design, led to the establishment of a research laboratory within the Engineering Experiment Station of the University of Illinois in 1903.

Talbot was one of the most distinguished members of the Joint Committee on Concrete of the American Society of Civil Engineers. A believer in professional associations as a 'powerful engine in technical affairs', he served in administrative capacities in many engineering societies. He was president of the Illinois Society of Engineers in 1890, of the Society for Promotion of Engineering Education in 1910, and of ASCE in 1918. His co-workers in research found him precise and argumentative, his professional colleagues enjoyed his wry sense of humor. Talbot was awarded the George Henderson Medal of the Franklin Institute in 1924, or the 1937 John Fritz Medal of the United Engineering Societies, the highest annual award to an engineer. In 1938, the University of Illinois named a laboratory for him, the first time a living individual had been so honoured. Talbot suffered a heart attack while attending a congress and died in a Chicago hospital.

Anonymous (1938). A tribute to Arthur Newell Talbot. *Bulletin* 62: 1-64. University of Illinois: Urbana IL.

Anonymous (1942). A.N. Talbot, research leader, dies. *Engineering News-Record* 128(Apr.9): 528. *P*

Anonymous (1973). Talbot, Arthur Newell. *Dictionary of American biography* Suppl. 3: 759-760. Scribner's: New York.

Talbot, A.N., Grover, A. (1893). Discussion of Flood waves in sewers. *Trans. ASCE* 28: 199-204.

Talbot, A.N., Seely, F.B., Fleming, V.R., Enger, M.L. (1918). Hydraulic experiments with valves, orifices, hose, nozzles and orifice buckets. *Bulletin* 105. Engineering Experiment Station: Urbana IL.

TALCOTT

07.04. 1809 Hebron CT/USA
08.12. 1868 Jersey City NJ/USA

William Hubbard Talcott was self-taught, mastering higher mathematics while tending a grist mill, and from 1830 to 1837 studied engineering with John B. Jervis (1795-1885) at Albany NY. He was engaged on surveys for railroads. In 1837, he entered upon the canal engineering, in which he was destined to become famous. He was a construction engineer of the Genesee Valley Canal for four years, and after its completion for another four years a resident engineer on the Erie Canal enlargement at Fort Plain NY. In 1845 he was called superintendent and engineer of the western division of Morris Canal in New Jersey, taking over the entire canal one year later as superintendent and chief engineer, staying in this position until his death. He was further made president of this company in 1864. He was in 1858 also called upon by the government of Nova Scotia to report on the practicability and probable cost of a ship canal between St. Peter's Bay and Bras d'Or Lake, on Cape Breton Island. During the Civil War, at the time of the threatened difficulties with Great Britain, he was appointed by the Governor of New York one of the engineers to devise means of defending the harbour of New York City against an attack by the British fleet.

Under Talcott's management of Morris Canal, the system of inclined planes in place of locks was installed and celebrated. The canal traffic increased from 58,000 to 825,000 tons per year within 22 years, in spite of a competing railroad on each side of the canal. Talcott's ability led to his wide reputation as an authority on matters both of hydraulic engineering and finances. He was called in 1857 to report on Shubenacadie Canal in Nova Scotia, and in 1858 on the practicability of a ship canal on Cape Breton Island. In 1860 he was offered the position of chief engineer of the James River and Kanawha Canal in Virginia, which he decline, however. During the Civil War he was one of a board of engineers appointed by the New York governor to devise means of protecting New York Harbour from attack by foreign fleets. Talcott counted among the founders of the American Society of Civil Engineers ASCE, staying on its Board until his death.

Anonymous (1893). William H. Talcott. *Proc. ASCE* 19: 97-99.
Anonymous (1899). Talcott, William H. *The National cyclopaedia of American* 9: 43. White: New York.
http://ronaldtalcott.com/Talcott%20Family/b2386.htm
wh_talcott.jpg *P*

TAYLOR D.W.

04.03. 1864 Louisa County VA/USA
28.07. 1940 Washington DC/USA

David Watson Taylor graduated in 1885 from the Naval Academy, Annapolis MD, and was sent to Greenwich UK, where he received highest honours of the Royal College. He was in 1907 the recipient of the honorary D.Eng. degree, Stevens University, Hoboken NJ, and in 1915 the DSc degree from the George Washington University, Washington DC. He was from 1904 captain of the US Navy, promoted to the rank of rear admiral in 1917. He retired in 1923. He was awarded from the French government the order *Légion d'Honneur*. He was vice-chairman of the National Advisory Committee of Aeronautics NACA, member of the Society of Naval Architects and Marine Engineers SNAME, presiding it from 1925 to 1927, and of the British Institution of Naval Architects INA, becoming in 1931 its honorary vice-president. He was recipient of the John Fritz Medal in 1931, the gold medallist of the British North-East Coast Institution of Engineers and Shipbuilders in 1931, and he was awarded the first David Watson Taylor Gold Medal established in his honour by the SNAME in 1936. The new US David Watson Taylor Model Basin was named in 1937 in his honour too.

Taylor designed and constructed in 1898 the first experimental tank for models of war vessels in the USA. The probably greatest achievement of his career was the Taylor Standard Series including 80 models by systematically varying vessel proportions and prismatic coefficients. The results of this series are still currently used for a preliminary determination of ship resistance for twin screw, moderate- to high-speed naval boats. The book was revised in 1933 with the addition of data on 40 new models. The series data were reanalysed using more recent methods in 1954. Both books Speed and power, and the mentioned Reanalysis were published in 1998 by SNAME, the centennial of the Experimental Model Basin.

Gertler, M. (1954). A reanalysis of the original test data for the Taylor Standard Series. *Report* 806. David Taylor Model Basin: Washington DC.
Hovgaard, W. (1941). David Watson Taylor. *Biographical memoir* 22: 134-153. *P*
Taylor, D.W. (1893). *Resistance of ships and screw propulsion*. Macmillan: New York.
Taylor, D.W. (1900). The US Experimental Model Basin. *Trans. SNAME* 8(1): 37-46.
Taylor, D.W. (1906). Model basin gleanings. *Trans. SNAME* 14(1): 65-79.
Taylor, D.W. (1910). *Speed and power of ships*. Wiley: New York.
http://en.wikipedia.org/wiki/David_W._Taylor *P*

TAYLOR E.H.

11.03. 1912 San Rafael CA/USA
11.03. 1988 Los Angeles CA/USA

Edward Holbrook Taylor attended the University of California, Berkeley CA, from where he received the BS degree in 1934 in civil engineering, and the MS degree in mechanical engineering in 1937. He there served as instructor of mechanical engineering then for five years. During World war II he was an associate engineer at Naval Ordnance Laboratory, Washington DC, and then a research engineer for the Ryan Aeronautical Company, San Diego CA. He was appointed in 1945 assistant professor of engineering at UCLA College of Engineering, there remaining until his retirement in 1979.

Taylor was a popular teacher, who received several teaching awards during the more than 20 years he spent in the classroom and in the laboratory. He taught and made research in fluid mechanics, irrigation engineering, and water distribution systems. His 1944 paper on subcritical open channel junction flow was the first in this field including a large experimental program; the data presented are still currently used for validation of more involved theoretical approaches than that of Taylor. He was known on the campus for his exceptional work in the guidance of foreign students, and he also was the faculty adviser who guided the Engineering Honorary Society. In 1960, he was appointed assistant dean for the undergraduate studies, holding this position until 1969, when the College of Engineering was reorganized as the School of Engineering and Applied Sciences. After his retirement he taught until his health began to fail. Dean Taylor will be remembered for his compassion and good humour by many students who sought his advice and guidance, and by all with whom he collaborated.

Anonymous (1951). Edward H. Taylor. *Southern Campus yearbook*: 61. UCLA: Los Angeles. *P*
Folsom, R.G., Taylor, E.H. (1939). *Laboratory manual*. Dept. Mechanical Engineering.
    University of California: Berkeley CA.
Taylor, E.A. (1938). Analysis of the positive surge in a rectangular open channel. *Civil*
    Engineering 8(10): 685-686.
Taylor, E.H. (1944). Flow characteristics at rectangular open channel junctions. *Trans. ASCE*
    109: 893-912.
Taylor, E.H. (1945). Discussion of Coefficients for velocity distribution in open-channel flow.
    Trans. ASCE 110: 646-648.
http://texts.cdlib.org/view?docId=hb4t1nb2bd&doc.view=frames&chunk.id=div00070&toc.dep
    th=1&toc.id=.

TAYLOR H.

26.06. 1862 Tilton NH/USA
27.01. 1930 Washington DC/USA

Harry Taylor was graduated from the US Military Academy in 1884, and appointed second lieutenant in the US Army Corps of Engineers. From 1887 he served as assistant to a brigadier in river and harbour improvements in the Carolinas. He was transferred in 1891 to the Pacific Coast in charge of construction of the lock at The Cascades OR, on Columbia River. He was promoted in 1896 to captain, and was in charge of the construction of fortifications on Puget Sound WA, remaining there until 1900, from when he was assigned to duty at Boston MA, in charge of river and harbour improvements in Boston District. From 1903 to 1905 he served on the Philippine Islands, where he was promoted to major. Back in the USA he was assigned to duty at New London CT, again in charge of river improvements, remaining there until 1911, finally as lieutenant-colonel. He was transferred to the Office of the Chief of Engineers, Washington DC, in charge of the River and Harbor Section, remaining until 1916, when being transferred to New York City in charge of river improvements. In 1917 he was appointed chief engineer of the American Expeditionary Forces, in the rank of brigadier-general, remaining in France until 1918, and there initiating large engineering works for the support of the American Army. On return to the USA he was appointed in 1920 Assistant Chief of Engineers, taking over in 1924 as Chief of Engineers with the rank of major general, serving until 1926. He was awarded the Distinguished Service Medal for his services in Europe. He also was made a Commander of *Légion d'Honneur* by the French Government.

During his 42 years of service within the Corps of Engineers, Taylor served on many official boards in connection with the defence of the coasts, and in the improvement and utilization of the American rivers and harbours. He also was member of the Advisory Board of Corps Officers to assist the Secretary of the Interior in granting the city and county of San Francisco to use the Hetch Hetchy Valley for a new water supply system. He had an intimate knowledge of fortification and river engineering, combined with a sound judgment, so that he was of great assistance to the government. He also was involved in the design of the Wilson Dam on Tennessee River at Muscle Shoals AL.

Anonymous (1931). Harry Taylor. *Trans. ASCE* 95: 1609-1612.
Taylor, H. (1906). *Norwalk Harbor* CT. Government Printing Office: Washington DC.
http://en.wikipedia.org/wiki/Harry_Taylor_(engineer) *P*
http://www.arlingtoncemetery.net/htaylor.htm *P*

TCHIKOFF

30.10. 1883 Vladikavkat/RU
07.10. 1945 Greenbelt MD/USA

Valentin(e) Vasilievich Tchikoff graduated in 1909 from the Institute of Ways and Communications, St. Petersburg, as civil engineer. He was then employed as assistant and later chief engineer in the Russian Ministry of Agriculture at Tashkent. He was sent by the government to Germany first, and then went to India, visiting the Punjab and Calcutta regions to study water power and reclamation projects. Upon return to his country, he became chief engineer at Kherson, and there published a Feasibility Report of irrigating a large area in the Dnieper River Valley. This publication includes also a complete description of the economic, hydrogeological, and agricultural factors of semi-arid regions.

Because of his interest in furthering his career, Tchikoff moved to the USA in 1916, as member of the Russian Military Purchasing Committee, being assigned chief inspector at Bridgeport CT. He then took a postgraduate at the University of California, Berkeley CA, studying water power, reclamation, and history of education. Following the 1917 revolution in Russia, he remained in the United States, entering a private practice for several years. He was also associated with an engineering firm, for which he interpreted plans and other data prepared in Russia for Hugh L. Cooper (1865-1937), who was then employed by the government of the Union of Soviet Socialist Republics for the design of the Dnieper River power development. In 1929 Tchikoff returned to his country for two years as first assistant engineer to Arthur P. Davis (1861-1933). This assignment consisted of a review of all important reclamation projects in the USSR, Turkestan, and Transcaucasia. In conjunction with this project Tchikoff wrote his 1932 book. In 1933 he went to France for a year, there studying water power projects, but then returned to the USA, writing on various engineering subjects. Following failing health, he moved to Greenbelt MD to work there as a consultant. A tireless and energetic worker, with a constant consideration for his fellow man, finally there passed away. He was member of the American Society of Civil Engineers since 1934.

Anonymous (1946). Valentin V. Tchikoff. *Trans. ASCE* 111: 1552-1555.
Tchikoff, V.V. (1916). *Irrigation of Taurida Region, South Russia*. Russian Department of Agriculture, Bureau of Soil Conservation: Petrograd.
Tchikoff, V.V. (1919). Irrigation opportunities in Russia are great. *Engineering News-Record* 82(4): 190-192.
Tchikoff, V.V. (1932). *Irrigation in agriculture*. Donskoy Institute: Novocherkassk (Russian). (*P*)

THAYER D.P.

13.12. 1902 Portland OR/USA
06.06. 1969 Sacramento CA/USA

Donald Packard Thayer studied from 1922 to 1925 at the Oregon State College, Corvallis OR, and then held various positions in the Pacific Northwest until 1930. He was from 1931 to 1933 engaged as assistant engineer on engineering work in Oregon, becoming from 1934 to 1936 assistant engineer on structural designs of Bonneville Dam, a major dam erected by the US Corps of Engineers, Portland OR. From 1939 to 1946 Thayer was in charge of the design of four major dams, miscellaneous hydraulic structures of the Corps program at Portland OR. Until 1949 he was then chief of the Design Branch, supervising military constructions in Hawaii, Guam, the Philippines for the Corps, and stationed at Sausalito CA. Until 1952 he then was assistant to the chief of the Military Branch, in planning and executing the military construction program in New England, at Boston MA. Until 1956 he was senior engineer of the Corps' Dam Design Branch, occupied with preliminary investigations and the design of the Feather River Project in California, including additional units of the California Water Plan, at Sacramento CA. From 1957 he was in charge of the design of all phases of water development projects as assistant Division engineer, including dams, power generation, aqueducts, and pumping plants of the Department of Water Resources, Sacramento CA. At the time of his death he was deputy division engineer of the California Dept. of Water Resources.

Thayer was credited as the one man with prime engineering responsibility for the design of Oroville Dam, thereby solving problems for an earth-fill dam far larger than built previously anywhere in the world. He received in 1962 the Governor's Award for superior accomplishment and a year later was given a special State citation for excellence as chief of the dams and plan design branch. He was member of the US Committee on Large Dams, and known for his work in hydraulics, geological exploration, and contract negotiation. He also had served as part-time instructor in engineering courses for the US Department of Education and the Oregon System of Higher Education from 1943 to 1946. Thayer was member of the American Society of Civil Engineers ASCE.

Anonymous (1959). Thayer, Donald P. *Who's who in engineering* 8: 2440. Lewis: New York.
Shultz, W.G., Thayer, D.P., Doody, J.J. (1961). Oroville Dam and appurtenant features. *Journal of the Power Division* ASCE 87(PO2): 29-40.
Thayer, D.P., Stroppini, E.W. (1965). *Hydraulic design for Oroville Dam spillway*. ASCE: NY.
http://en.wikipedia.org/wiki/File:OrovilleDam.jpg (*P*)

THAYER S.

09.06. 1785 Braintree MA/USA
07.09. 1872 South Braintree MA/USA

Sylvanus Thayer graduated in 1807 from Dartmouth College, Hanover NH. He graduated only one year later from the Military Academy, West Point NY. During the 1812 War he directed the fortifications and the defence of Norfolk VA, and was promoted to major. In 1815, he travelled to Europe studying at Ecole Polytechnique, Paris, for two years mainly mathematics. Upon return to the USA he became superintendent of the Military Academy, becoming successively America's first college of engineering. He ended this appointment in 1833 and was elected associate Fellow of the American Academy of Arts and Sciences in 1834. He returned to duty with the Army Corps of Engineers, spending the next 33 years as chief engineer for the Boston area. He oversaw the construction of both Fort Warren and Fort Independence to defend Boston Harbour. He retired from the Army in 1863 with the rank of colonel in the US Corps of Engineers, but in 1864 was awarded the rank of honorary brevet brigadier general. In 1867 Thayer donated 30,000 US$ to the trustees of Dartmouth College to create the Thayer School of Engineering. To honour his achievements, the Sylvanus Thayer Award was created in 1958 by the US Military Academy.

The Military Academy was founded in 1802 mainly under the strong sponsorship of President Jefferson. For the preparation of military engineers, the Academy eventually came to train as many man for civilian life as for the army. Thayer was sent to Europe by President Madison to study the theory and practice of fortification design. Thayer had in Paris the opportunity of observing instruction, where the mathematical training was and still is paramount, and his observations there were to play a considerable role in his development of West Point NY. During his stay as superintendent, both science and engineering were there greatly strengthened. Much of the instruction was patterned after the French system, in particular the strong emphasis on mathematics. Though there was periodic criticism of teaching civil as much as military engineering, and producing more engineers than the Army could absorb, it was decided that this policy was salutary rather than misguided.

Adams, C., ed. (1965). *The West Point Thayer papers* 1808-1872. West Point NY.
Kershner, J.W. (1982). *Sylvanus Thayer*: A biography. Arno Press: New York. *P*
Rouse, H. (1976). Thayer. *Hydraulics in the United States* 1776-1976: 23. IIHR: Iowa City IA. *P*
http://en.wikipedia.org/wiki/Sylvanus_Thayer *P*

THEIS

27.03. 1900 Newport KY/USA
31.07. 1987 Albuquerque NM/USA

Charles Vernon Theis received his BS degree in civil engineering in 1922, and in 1929 then the PhD degree in geology from University of Cincinnati, Cincinnati OH. In 1930 he joined the staff of the US Geological Survey USGS as assistant geologist in the Water Resources Division, where he became associated with Oscar E. Meinzer (1876-1948). He was assigned to investigate the Portales Valley NM groundwater resources. From 1934 he became head of the Public Works Administration to study the Southern High Plains, from Kansas south. In 1935 he published on the relation between the piezometric head and the discharge, which led the foundation upon which most of groundwater hydrology since has been built, including the Theis Equation. This is the mathematical solution allowing for time-dependent changes in water levels as a result of pumping.

Theis' contributions to groundwater hydrology shaped and reshaped the science as it is currently known. He has had a great impact in this field from the 1930s to the 1980s. There was a long debate between the pioneering work of Henry Darcy (1803-1858) in 1856 and the contributions of Adolf Thiem (1836-1908) and Charles S. Slichter (1864-1946) around 1900 relating to steady-state equilibrium flow. In the 1920s Meinzer and others became aware that steady-state was not adequate. However, it was Theis in 1935 by introducing the concept of analytical models to study hydrogeology, from when this field began to mature as a science. Later contributions in well hydraulics mainly by Charles E. Jacob (1914-1970), or Madhi S. Hantush (1921-1984), were based on the 1935 paper. The period from 1935 to 1962 was the golden era for transient groundwater hydrology, of which Theis' original work again was highly relevant.

Anonymous (1971). Charles V. Theis. *Groundwater* 9(1): 57. *P*
Anonymous (1971). C.V. Theis. *Journal of the American Water Works Association* 63(4): 32.
Anonymous (1984). 1984 Robert E. Horton Medal to Charles V. Theis. *Eos* 65(28): 436-437. *P*
Back, W., Herman, J.S. (1997). American hydrogeology at the millennium: An annotated chronology of 100 most influential papers. *Hydrogeology Journal* 5(4): 37-50. *P*
Theis, C.V. (1935). The relation between the lowering of the piezometric surface and the rate and duration of discharge of a well using groundwater storage. *Trans. AGU* 16: 519-524.
Theis, C.V. (1938). The significance and nature of the cone of depression in ground water bodies. *Economic Geology* 33(8): 889-902.
http://www.waterencyclopedia.com/St-Ts/Theis-Charles-Vernon.html

THOMAS B.F.

23.05. 1853 Ironton OH/USA
14.04. 1923 Catlettsburg KY/USA

Benjamin Franklin Thomas graduated in 1871 as a civil engineer from the National Normal University, Lebanon OH. He commenced his professional career as assistant to the county engineer, Lawrence County OH. From 1874 to 1877, he prepared the National Centennial Exposition held in 1876 at Philadelphia PA. From 1879 to 1883 he served on the construction of the Chattaroi Railway in Kentucky State. In 1883 he was appointed US assistant engineer in charge of the improvement of the Big Sandy River, where he stayed for the next following decades. One of the main work was the river control with three locks and movable dams. The needle-dam was the type selected, then the first built in the USA. His study relating also to existing European dams resulted in the 1898 paper, which was awarded the ASCE Norman Medal.

Thomas then worked on his book Improvement of rivers, published in conjunction with a colleague, which underwent several revisions and was later published in two volumes. It widely served as reference text for hydraulic engineering. Thomas also contributed various Discussions to journal articles on river improvements, including for instance artificial waterways, levees on the Mississippi River, or locks and movable dams on the Ohio River. Thomas was described as a thorough person, devoting himself to the close examination of similar works than he had conducted. He had a great love for the design, the planning, the supervision of large engineering projects. He was held in high regard and esteem by all who knew him for the unswerving integrity and as a conscientious worker with an unusual degree of energy. Thomas was a member of the American Society of Civil Engineers ASCE since 1887. He bequeathed his personal library to the US Engineer Office, Cincinnati OH.

Anonymous (1923). Benjamin F. Thomas. *Engineering News-Record* 90(17): 767.
Anonymous (1924). Benjamin F. Thomas. *Trans. ASCE* 87: 1435-1436.
Kemp, E.L. (2000). *The great Kanawha navigation*. University of Pittsburgh Press: Pittsburgh.
Thomas, B.F. (1898). Movable dams. *Trans. ASCE* 39: 431-615.
Thomas, B.F., Watt, D.A. (1905). *The improvement of rivers*: A treatise on the methods employed for improving streams for open navigation, and for navigation by means of locks and dams. Wiley: New York.
Thomas, B.F. (1923). Discussion of Locks and dams on Ohio River. *Trans. ASCE* 86: 151-154.
http://www.google.com/patents/US382395 (*P*)

THOMAS C.W.

11.03. 1906 Nepesta CO/USA
27.02. 1979 Denver CO/USA

Charles Walter Thomas obtained the BS degree in 1931 from Colorado State University, Fort Collins CO, and the MS degree in 1947 from the Colorado University, Boulder CO. He received in 1952 the PhD degree from Université de Grenoble, France. He was from 1931 to 1935 assistant in the Hydraulic Laboratory of the US Bureau of Reclamation USBR, heading then until 1936 the related lab in Montrose CO, from when he headed the USBR Hydraulics Field Investigation Group until 1940. After return from Grenoble he then headed the USBR Hydraulic Investigation Group at Denver until 1961. He was until 1962 hydraulic and sediment engineer in the Republic of China for the United Nations Special Fund. He was assigned in 1954 to the government of the Philippines as special consultant on hydraulics, taking over a similar position in 1957 for the government of Japan. Thomas was a Fellow of the American Society of Civil Engineers ASCE, and a member of the International Association of Hydraulic Research IAHR.

Thomas was interested all through his professional career in discharge measurement relating to prototype structures. His 1955 paper deals with the salt velocity method in which refinements in the test equipment and test techniques employed by the USBR are described. These include a brine injection device using compressed air, quick acting pop valves, two electrodes located in the conduit, and electric recorders. The 1959 paper on measurement errors discussed these due to constructional errors, transverse crest slope, incorrect head reading, incorrect zero setting, improper gage location, neglecting the approach flow velocity, effects of turbulence and surges, roughness effects, rounding of sharp crest, effect of submergence, nappe aeration, among other factors, including thus the significant sources of errors in free surface discharge measurement.

Anonymous (1957). Charles W. Thomas. 7th *IAHR Congress* Lisbon: Frontispiece. *P*
Anonymous (1964). Thomas, Charles W. *Who's who in engineering* 9: 1860. Lewis: New York.
Thomas, C.W., Dexter, R.B. (1955). Modern equipment for applications of salt velocity method
    of discharge measurement for performance tests. Proc. 6th *IAHR Congress* 2(B2): 1-10.
Thomas, C.W. (1959). Errors in measurement of irrigation water. *Trans. ASCE* 124: 319-332.
Thomas, C.W. (1959). Hydraulic performance of 96-inch regulating gates in closed conduits. 8th
    IAHR Congress Montreal A(8): 1-27.
Thomas, C.W. (1960). World practices in water measurement at turnouts. *Journal of the
    Irrigation and Drainage Division* ASCE 86(IR3): 29-52.

THOMAS F.

19.05. 1885 Red Oak IA/USA
27.08. 1952 Pasadena CA/USA

Franklin Thomas obtained the BE degree in 1908, and the CE degree in 1913 from the University of Iowa, Iowa City IA. He also studied from 1908 to 1909 at McGill University, Montreal CA, and then was employed until 1910 by the Mines Power Co., Cobalt Ont. Until 1912 he was then instructor at the Department of Engineering, University of Michigan, Ann Arbor MI, and until 1913 was in charge of the Alabama Power Co. for its developments at Lock 18, on Coosa River and Muscle Shoals, Tennessee River. Thomas was appointed in 1913 professor of civil engineering, California Institute of Technology, Pasadena CA, a position he held until his death. He thereby served as faculty chairman, and was chair of the Division of Engineering from 1926 to 1944, from when he took over as Dean. He was in parallel assistant engineer for the US Bureau of Reclamation USBR in 1919, and served as consultant on municipal projects. He was further member and vice-chairman of the Board of Directors, the Metropolitan Water District of Southern California from 1928. Thomas was member of the American Society of Civil Engineers ASCE, serving as president of the Los Angeles Section in 1924, and as member of the Committee on Irrigation Hydraulics from 1922 to 1930, taking over as chairman from 1930 to 1933, and as vice-president from 1944 to 1945. He was ASCE President in 1949. Thomas was awarded the Arthur Noble Gold medal by the City of Pasadena for notable service in promoting the city's welfare.

Thomas had a professional career both as a civil engineer in dam engineering, and as academic person mainly at Caltech. Next to his research, he served in many faculty administrative duties. In addition, water supply for California was his specialty. It was stated 'He was one of the planners and organizers of the Metropolitan Water District, who conceived the brilliant engineering project of bringing Colorado River water to the populous centers of the Southland by aqueduct'.

Anonymous (1948). Thomas, Franklin. *Who's who in engineering* 6: 1977. Lewis: New York.
Anonymous (1952). Franklin Thomas. *Engineering News-Record* 149(Sep.11): 54. *P*
Thomas, F. (1933). Value of water in Southern California. *Civil Engineering* 3(10): 555-559.
Thomas, F. (1941). Discussion of Transient flood peaks. *Trans. ASCE* 106: 262-263.
Thomas, F. (1950). Sanitary engineers face problem incident to rapid expansion in field of
        sanitation. *Civil Engineering* 20(2): 102.
Calteches.library.caltech.edu/318/01/tribute.pdf

THOMAS H.A.

24.07. 1885 Ann Arbor MI/USA
03.07. 1973 Murrysville PA/USA

Harold Allen Thomas was educated at the Columbia University, New York NY, receiving in 1906 the AB degree. He was instructor at the University of Washington, Seattle WA, from 1908 to 1910, until 1922 civil engineering professor at Rose Polytechnic Institute, Terre Haute IN, from when he joined in the same position Carnegie Institute of Technology, Pittsburgh PA, until 1950. He was in parallel active for the Miami Conservancy District, Dayton OH, in 1918, later was a hydraulic engineer for Pittsburgh, or worked for the US Navy at David Taylor Model Basin, Bethesda MD. The Harold A. Thomas Award consists of an honorarium and a certificate, intended to recognize graduating seniors who have provided exemplary services to the Department, the College, the University of the civil and environmental engineering profession at Columbia University.

Thomas' 1934 work deals with uniformly-progressive open channel waves, the dynamic propagation of stable wave forms, and the general laws of unsteady flow. The use of hydraulic models for investigating these flows was proposed, and approximate methods for flood routing investigated. Flood routing in natural rivers was also considered, at a time when no computers were available. The 1942 paper deals with outlets of high-speed flow. Erosion problems were then only dealt with by model observations, as were similar flows on ski jumps, the structure currently adopted for these flow conditions. The results were applied to the outlet structure of Tygart Dam in West Virginia. Cavitation damage was first experienced in the early 1940s, as occurred on Madden Dam, Panama Canal Zone. As in the 1939 paper, these problems develop under high-speed flow and may be removed by air supply to the approach flow. A cavitation-test facility was erected to study cavitation with model observations, so that Thomas may be considered one of the fathers in this research field.

Anonymous (1954). Thomas, Harold A. *Who's who in engineering* 7: 2414. Lewis: New York.
Rouse, H. (1976). Harold A. Thomas. *Hydraulics in the United States* 1776-1976: 140. *P*
Thomas, H.A. (1934). The hydraulics of flood movements in rivers. *Engineering Bulletin*.
    Carnegie Institute of Technology: Pittsburgh PA.
Thomas, H.A., Hamilton, W.S. (1939). Jet deflectors for high-dam outlet conduits. *Civil Engineering* 9(5): 297-300.
Thomas, H.A., Schuleen, E.P. (1942). Cavitation in outlet conduits of high dams. *Trans. ASCE* 107: 421-493.

THOMAS, Jr. H.A.

14.08. 1913 Terre Haute IN/USA
26.03. 2002 Cotuit MA/USA

Harold Allen Thomas Jr., son of Harold Allen (1885-1973), received education from Carnegie Institute of Technology, Pittsburgh PA, graduating in 1935 with a BS degree in civil engineering, and with an SD degree in 1938 from Harvard University, Cambridge MA. He was there a faculty member from instructor to associate professor in 1956 when being appointed until 1984 Gordon McKay professor of civil and sanitary engineering. He was in parallel consultant for the NRC of the National Academy of Sciences, for the Department of the Interior, or the scientific board for Middle East and Southeast Asia. Thomas was member of the American Geophysical Union AGU and recipient of its 1978 Robert E. Horton Medal, of the National Academy of Engineering NAE, and Fellow of the American Academy of Arts and Sciences.

Thomas followed his father in his professional career. He was one of the principal members of the Harvard Faculty who guided the Harvard Water Program from the late 1950s. For the first time, it established a working interdisciplinary approach to the development of water resources. A book summarizing the program was published in 1962. His most significant contribution to environmental management was the Animal Farm, in which he argued that environmental standards impose utility on human life, health, and well-being, and that we cannot set any quality criteria for control of the environment without making a value judgement that always takes the form of a cost-benefit ratio: 'To set a criterion is to impute a cost-benefit ratio', he wrote in 1964. For him hydrology was an integral part of the social fabric.

Anonymous (1954). Harold A. Thomas, Jr. *Civil Engineering* 24(9): 609-610. *P*
Anonymous (1978). Robert E. Horton Medal to Harold A. Thomas, Jr. *Eos* 59(8): 792-793. *P*
Anonymous (2000). Thomas, Jr., H.A. *Who's who in America* 54: 4892. Marquis: Chicago.
Anonymous (2003). Harold A. Thomas, Jr. *Eos* 84(13): 120. *P*
Maass, A., Hufschmidt, M.M., Dorfman, R., Thomas, Jr., H.A., Marglin, S.A. (1962). *Design of water-resource systems*: New techniques for relating economic objectives, engineering analysis, and governmental planning. University Press: Harvard.
Thomas, Jr., H.A., Archibald, R.S. (1952). Longitudinal mixing measured by radioactive tracers. *Trans. ASCE* 117: 839-856.
Thomas, Jr., H.A. (1964). The Animal farm: A mathematical model for the discussion of social standards for control of the environment. *Journal AWWA* 36(9): 1087-1091.

THOMPSON M.J.

28.07. 1904 Grand Rapids MI/USA
23.07. 1971 Austin TX/USA

Milton John Thompson received the BS degree in aeronautical engineering in 1925, and the MS degree in 1926 from the University of Michigan, Ann Arbor MI. He won in 1928 a Guggenheim Fellowship to study at the Polytechnical Institute, Warsaw, where he obtained the ScD degree in 1930, returning then to his Alma Mater as assistant professor and from 1937 as associate professor in the Department of Aeronautical Engineering. During these years he was active in wind tunnel projects and also became known as a consultant in aeronautical engineering.

He accepted in 1941 a professorship in aeronautics at the Department of Mechanical Engineering, University of Texas UT, Austin TX. Because of the importance of aviation in World War II, the UT Department of Aeronautical Engineering DAE was established in 1942, with Thompson as chairman. He toured often the USA, Europe and the Middle East as a Lecturer, predicting that 90% of ocean travel after WWII would be by air, and that jets would travel twice the speed of sound. For most of 1945 he was on leave from his university to serve as supervisor of aerodynamics at the Applied Physics Laboratory of Johns Hopkins University, Baltimore MD. After returning to University of Texas, he and a colleague established the Defence Research Laboratory, with Thompson serving as an associate director until shortly before his death. In 1958 the DAE became the Department of Aerospace Engineering, a change reflecting the emphasis on aerospace courses and the research in the travel outside the earth's atmosphere, with Thompson remaining as chairman. He co-authored several books and wrote many journal papers, particularly in the field of flight performance of aerospace vehicles. He received grants allowing for collaborations with the University of Göttingen, Germany, Imperial College, London UK, and Cambridge University, UK. He was member of the American Institute of Aeronautics and Astronautics AIAA, and the American Society of Engineering.

Anonymous (1925). Milton J. Thompson. *Michiganensian yearbook*: 122. Ann Arbor. *P*
Anonymous (1973). Milton J. Thompson. *Who was who in America* 5: 720. Marquis: Chicago.
Dodge, R.A., Thompson, M.J. (1937). *Fluid mechanics*. McGraw-Hill: New York.
Thompson, M.J., Wilson, R.E. (1951). Aerodynamic characteristics of nozzles and diffusors for supersonic wind tunnels. *Defence Research Laboratory* DRL-281. University of Texas.
Witoszynski, C., Thompson, M.J. (1935). The theory of single burbling. *Aerodynamic theory* 3: 1-33, W.F. Durand, ed. Springer: Berlin.
http://www.tshaonline.org/handbook/online/articles/fth23

THOMPSON P.W.

19.12. 1906 Alliance NE/USA
09.02. 1996 Daytona Beach FL/USA

Paul Williams Thompson graduated in 1929 from the US Military Academy, West Point NY, and then attended the State University of Iowa, Iowa City IA, obtaining the BS degree from University of Tulane, New Orleans LA. During the 1930s, after an early career in the US Army Corps of Engineers, he was awarded a Freeman Scholarship, to study waterways of European countries by attending graduate classes at the Berlin *Technische Hochschule*, Berlin D. His extensive engineering education helped prepare him for his assignment as director of the US Waterways Experiment Station at Vicksburg MS.

During his stay in Germany Thompson was detailed to various German army units and gained much valuable information and data 'upon which improvements in US military engineering practice were later based'. He eventually gained a knowledge of German forces which, together with his understanding of the capabilities of the Allied Forces to be assemble for D-Day, qualified him to command the European Theatre Assault Training Center in England. He there developed the amphibious assault techniques and tactics to be used at the Normandy attack, and trained the forces which made the initial attack. As D-Day approached, Thompson was appointed to command the 6th Engineer Special Brigade, which spearheaded the Normandy assault. For his leadership in the initial wave, he was awarded the Distinguished Service Cross. Later on D-Day he was seriously wounded and, after a lengthy recovery, was promoted to brigadier general and assigned Chief of Information and Education with responsibility for Stars and Stripes, the Army weekly magazine lank. Thompson retired from the Army in 1946. He was awarded the degree of Commander in the French Legion of Honor, the *Croix de Guerre* with Palms, and many other decorations. The Association of Graduates recognized in 1994 his enormous contribution to the nation by naming him a Distinguished Graduate.

Anonymous (1945). Paul W. Thompson. *Engineering News-Record* 135(Jul.12): 90. *P*
Thompson, P.W. (1935). Discussion of River laboratory hydraulics. *Trans. ASCE* 100: 151-155.
Thompson, P.W. (1938). The use and trustworthiness of small-scale hydraulic models. *Civil Engineering* 8(4): 255-257.
Thompson, P.W. (1938). Current waterway studies. *Engineering News-Record* 121(Dec.29): 822-824; 123(Aug.17): 223-224.
Vogel, H.D., Thompson, P.W. (1933). Flow in river bends. *Civil Engineering* 3(5): 266-268.
http://apps.westpointaog.org/Memorials/Article/8499/ *P*

THOMPSON W.L.

06.09. 1875 Chester MS/USA
23.02. 1932 Lanywa/MM

William Love Thompson graduated in 1897 with the BSc degree from Mississippi State College, Starkville MS. He was then maintenance inspector for the Mississippi Levee Board, Greenville MS, and there returned after participation in the Spanish-American War. He was engaged in the Culebra Division of Panama Canal in 1905, rose to junior engineer in 1907, and to senior engineer in 1911. He was in charge of hydraulic plant operation with the Pacific Division, and until 1913 of the lock construction at Pedro Miguel. When landslides at the Culebra Cut hindered work, he was placed in charge of design and operation of the Gold Hill sluicing project, under George W. Goethals (1858-1928).

In 1913 Thompson resigned his position, becoming chief engineer of the Mississippi Levee Board, Greenville MS. During his eight years in this office, he developed the District, lowering construction cost and raising the standards of safety. He was involved in the development of two tower excavators for levee construction, and was the author of the paper Backwater in the Yazoo Basin. He was often before the Rivers and Harbors Committee of Congress, attesting the need for flood-control appropriations. Thompson was engaged from 1922 to 1925 in consulting practice at Greenville, serving for various drainage projects, a flood control project, and general water-power investigations. He was then offered the position of superintending engineer for the Indo-Burma Petroleum Company at *Lanywa*, Upper Burma, where he was engaged for three years on its sand flat reclamation, which turned out to be the most interesting work of his career. He made in addition friends among the British engineers, but returned to the USA in 1928, taking a well-earned rest after the three years in the tropical East. In 1930 he entered the employ of a construction company at *Beauharnois* QC, in charge of dredging operations between Lakes St. Louis and St. Francis, but resigned to return to Burma for extending a guide-wall project, remaining there until his death, which was caused by a gastric haemorrhage. He was described as a cosmopolitan and broad-minded personality, but was uncompromising in his criticism. Thompson was elected ASCE member in 1911.

Anonymous (1933). William L. Thompson. *Trans. ASCE* 98: 1650-1652.
Thompson, W.L. (1918). Backwater in the Yazoo Basin. 67th Congress, 47th Session.
        Mississippi River Commission: Vicksburg MS.
Thompson, W.L. (1929). An estimate of tower machines for Mississippi levee construction.
        Engineering News-Record 102(17): 682-683.

THOMSON J.

25.10. 1853 Fochabers/UK
31.05. 1926 Brooklyn NY/USA

John Thomson was born in Scotland and brought to America when a child. He was educated in common schools in Wayne County NY, where his interest in mathematics was noted. He first was watchmaker, but later became engineer and inventor. For more than thirty-five years, he was engaged successfully in civil, mechanical and electrical engineering; also in the design and manufacture of water-meters, printing presses and electric furnaces. He formerly practised as solicitor before the US Patent Office, having been granted more than 200 patents in the USA and in Europe. He has also often been retained as expert in patent litigations before the Federal Courts and has made numerous studies with respect to the probable validity of patents and the merits of engineering and manufacturing enterprises. He was chief engineer of the Electrical Subway Commission, New York, 1886, which built the first underground conduit containing cabled telegraph and telephone wires. He was associated from the 1880s with the Colt's Patent Firearms Manufacturing Company, Hartford CT, in the design and manufacture of printing presses. He purchased their interests, so that the business was then conducted under the corporate title John Thomson Press Company, Long Island City NY.

Thomson was particularly known for the invention and manufacturing of water-meters. These devices are used to measure the water volume used by residential and commercial buildings supplied by a public water system. His first invention was the so-called Disc-Water-Meter made by the Thomson Meter Company. This method relies on the water to physically displace the moving measuring element in direct relation to the amount of water that passes through the meter. The disk moves a magnet that drives the register. Later, Thomson proceeded to the invention of the Trident and Crest water meters, which were manufactured by the Neptune Meter Company, Long Island City NY, by then the largest maker of these elements in the USA. At the end of his life Thomson was engaged in the design and manufacturing of the Zenith water meter, an extremely precise device also for small discharges. He served as president of the Engineer's Club, New York, and was a member of ASME and ASCE.

Anonymous (1927). John Thomson. *Trans. ASCE* 91: 1145.
Thomson, J. (1891). A memoir on water meters. *Trans. ASCE* 25: 41-69.
http://en.wikipedia.org/wiki/Water_meter
http://www.electricscotland.com/history/descendants/chap137.htm *P*

THOMSON R.H.

20.03. 1856 Hanover IN/USA
07.01. 1949 Seattle WA/USA

Reginald Heber Thomson graduated from Hanover College, Hanover ID, with the BA degree in 1877, and both the MA degree and honorary PhD degree in 1903. He made surveys in California, and taught at the Healdsburg Institute until 1877, then moved to the Northwest working as US mining engineer. He was assistant city engineer of Seattle WA from 1882 to 1884, city surveyor until 1886, and from 1892 to 1911 city engineer, from when he acted as First Engineer of the Port of Seattle. He conceived the layout of Strathcona Park on Vancouver Island in Canada in 1913, and in 1914 was consultant on the Lake Washington Pontoon Bridge, and the foundation of the Tacoma Narrows Bridge, Eugene OR. He further was engaged with the Rogue River Irrigation Canal, Prince Rupert BC, Canada. Thomson was member of the American Society of Civil Engineers ASCE, served as ASCE Director from 1917 to 1919, and elected Honorary Member in 1940. He also was member of the Canadian Society of Civil Engineers, a member of the Board of Managers, the University of Washington, Seattle WA, from 1905 to 1915, and in the Seattle City Council from 1916 to 1921. He was further member of the Pacific Northwest Society of Civil Engineers, taking over as president from 1902 to 1903.

Among the engineering achievements of Thomson were the railroad route through the Snoqualmie Pass, the Lake Washington Ship Canal, numerous bridges over rivers and valleys, and major improvements to Seattle's sewer system, as well as straightening and deepening the Duwamish River and developing the Cedar River watershed, now one of Seattle's major sources of drinking water. He was responsible for much of the regrading of Seattle, taking down hills and filling in the mudflats, and played a major role in the creation of Seattle City Light, the Port of Seattle, and the Hiram M. Chittenden Locks. Thompson was involved in the initial surveying and dredging of what would, years later, become the Montlake Cut of the Lake Washington Ship Canal. It was stated that 'Thomson probably did more than any other individual to change the face of Seattle'.

Anonymous (1948). R.H. Thomson, Honorary Member ASCE. *Civil Engineering* 18(9): 591. *P*
Thomson, R.H. (1950). *That man Thomson*. University of Washington Press: Seattle WA. *P*
Wilson, W.H. (2009). *Shaper of Seattle*: Reginald H. Thomson's Pacific Northwest. University
      Press: Washington State University, Seattle WA. *P*
http://en.wikipedia.org/wiki/Reginald_H._Thomson *P*

THORNDIKE

11.06. 1868 Beverly MA/USA
16.02. 1928 Weston MA/USA

Sturgis Hooper Thorndike received his BA degree in 1890 from Harvard College, Cambridge MA, and in 1895 the BS degree in civil engineering from the Massachusetts Institute of Technology, Cambridge MA. He was then city engineer of Boston MA with a notable activity in bridge design and in municipal engineering. In 1911, on the consolidation of the Engineering, Street, and Water Departments into the Department of Public Works, he became the design engineer of its Bridge and Ferry Division. He was connected with bridges over tide-water having an important span, including Cambridge Bridge across the Charles River between Boston and Cambridge, then the most notable bridge of Greater Boston. He also served from 1904 to 1906 as instructor of hydraulics, bridge design, and surveying at MIT. From 1911 to 1914 he had a private engineering practice at Boston MA.

In 1914 Thorndike joined a former colleague from the city of Boston as partner in an engineering office, a position he held until his death. During this period he designed and supervised the construction of the Boston Army Supply Base, one of the larger water-front terminals built during World War I for the US War Department, the Hampden County Memorial Bridge across the Connecticut River at Springfield MA, and the water supply and sewage disposal of Mariemont, in the suburbs of Cincinnati OH. In all these projects, both engineering and economic contributions were valued by reason of sound judgement and his analytical mind. He thereby advanced to an authority in engineering work, always interested in fair play and the advancement of ethical standards. He was an active member of the Committee of the Northeastern Section of the American Society of Civil Engineers ASCE, formulating the Code of Practice, adopted in 1927 by the Society. He also presided over this Section at the time of his death. He further was member of the Boston Society of Civil Engineers, serving as its Director from 1919 to 1920, member of the American Water Works Association AWWA, and the New England Water Works Association NEWWA.

Anonymous (1915). Sturgis H. Thorndike. 25th *anniversary report*: 250-252. Harvard
        University: Cambridge MA. *P*
Anonymous (1929). Sturgis H. Thorndike. *Trans. ASCE* 93: 1910-1912.
Fay, F., Spofford, C.M., Thorndike, S.H. (1919). *Report* upon the utilities at the Boston Army
        Supply Base made to the Construction Division of the US Army. Army Supply: Boston.

THURSTON R.H.

25.10. 1839 Providence RI/USA
25.10. 1903 Ithaca NY/USA

Robert Henry Thurston obtained the PhB degree from Brown University, Providence RI, in 1859 and a certificate as draftsman in civil engineering from his father's firm. During the Civil War he was third assistant engineer of the US Navy. After a visit to England, he became in 1865 assistant professor at the US Naval Academy. He then moved in 1871 as professor of mechanical engineering to the Stevens Institute of Technology, Hoboken NJ, leading to his creation of a new educational model and curricula for engineers. During this era engineering education was mainly a 'shop culture', where students with skill learned how to make mechanical machines. Thurston introduced novel methods for engineering education based on science and mathematics, which were widely published. He presented this novel curriculum to the US State Department, moving in 1873 to Europe visiting the then leading Berlin University and the Vienna Universal Exhibition, which led to the first mechanical laboratory in the USA at Brown University in 1875.

Thurston presented his educational model at the Philadelphia Centennial Exhibition in 1885, impressing the panel judge, who was the president of Cornell University, Ithaca NY. Thurston moved in 1885 to Ithaca, becoming first director of Cornell's Sibley College. He created a college of engineering, transforming the engineering study into one with a classroom culture and more laboratory tests and scientific principle-based classroom time. He introduced the idea that students should build upon work done in the classroom by testing theories in the laboratory, thereby establishing the first mechanical engineering laboratory, giving first-hand experience combined with applied engineering aspects. This idea was widely adopted at other universities after 1900. It was stated that 'Thurston was an intellectual and institutional visionary whose creative initiatives made him a strategic agent in the stabilization of engineering formation in schools'. Cornell University still has a collection of Thurston's models and instruments. He was the first president of the American Society of Mechanical Engineers from 1880 to 1882.

Anonymous (1929). A pioneer in engineering education. *Mechanical Engineering* 51(11): 805. *P*
Anonymous (1980). Thurston, Robert H. *Mechanical engineers in America born prior to* 1861: 297-298. ASME: New York. *P*
Thurston, R.H. (1878). *A history of the growth of the steam engine.* Appleton: New York.
Thurston, R.H. (1889). *A treatise on friction and lost work in machinery.* Wiley: New York.
http://www.asme.org/kb/news---articles/articles/energy/robert-henry-thurston *P*

THURSTON R.L.

13.12. 1800 Portsmouth RI/USA
13.01. 1874 Providence RI/USA

Robert Lawton Thurston, father of Robert H. (1839-1903), developed talent as a mechanic, so that he began to learn the trade of a machinist. He was apprenticed in the manufacture of an experimental steam engine which was placed in a small ferry-boat for use near Fall River MA. Its success led to the construction of engines for the *Rushlight* and the *Babcock*, which ran between Providence and New York. The engine was of 0.27 m cylinder diameter and of 1 m piston stroke. The boiler was a form of 'pipe-boiler'. The water was injected into the hot boiler as fast as required to furnish steam, no water being retained in the steam generator. The boat had 80 tons of weight, and steamed from Newport to Providence, a distance of 50 km, in only 3.5 hours, and to New York, a distance of 280 km, in 25 hours, using almost 2 cords of wood, corresponding to some 4 m^3. Thurston then entered the iron business at Fall River MA, but in 1830 returned to Providence, where he founded in 1834 the first steam-engine factory in New England, the Providence Steam-Engine Co. The patent for the drop cut-off for steam-engines was purchased from Frederick E. Sickles (1819-1895) to manufacture a standard form of expansion steam-engine. For years the company was engaged in litigation with George H. Corliss (1817-1888), against whom was brought suit for infringement of his patent.

Thurston introduced in 1863 the Greene Engine developed by David M. Greene (1832-1905), which was at the time considered to be one of the best steam engines. The unsettled condition of affairs resulting from the Civil War, with incidental lack of business, led to Thurston's withdrawal from his company. His decision was certainly influenced by his son, who did not want to pursue the affairs in the father's company, and made a different, yet highly successful career. The father Thurston was considered a practical, intelligent, far-seeing and enterprising machinist, whose energy and capacity developed successfully the long latent power of steam, and he applied it to navigation, railroad transportation, driving spindles, and to many other purposes which became familiar in the 19th century.

Miller, J. (1873). *Robert Lawton Thurston* 1800-1873. Providence RI.
Thurston, B. (1892). Robert L. Thurston. *Thurston genealogies* 1635-1892: 568-569. Thurston: Portland ME. *P*
Thurston, R.H. (1886). *A history of the growth of the steam-engine*. Appleton: New York.
https://familysearch.org/photos/images/2706519/robert-lawton-thurston *P*

TIBBETTS

28.04. 1882 Oshkosh WI/USA
02.08. 1938 Hollister CA/USA

Frederick Horace Tibbetts obtained the BSc degree in civil engineering from University of California, Berkeley CA, in 1904, and the MS degree in 1907. He was appointed instructor and associate professor in civil engineering there. He became partner with a colleague in Alameda County from 1909 to 1918. This firm designed and supervised the construction of projects in Central California, mainly relating to sewage disposal and a large reclamation project in Yolo Basin CA, where massive levees and drainage canals were erected. Tibbetts submitted in 1912 a report on the Knights Landing Ridge Cut, which had a major influence on the reclamation of the Upper Sacramento Valley. This Cut formed an artificial outlet for flood waters in Colusa Basin; it was completed right before the great 1915 flood and thus of immense benefit.

From 1918 Tibbetts had his own engineering practice at San Francisco. He continued as chief engineer the Colusa Basin Project, the Sacramento River West Side Levee District, and the Knights Landing Ridge Drainage District. These three Districts provided the complete flood protection for more than 400 km^2 of land which previously had been often flooded by both river overflow and foothill drainage. Tibbetts also became chief engineer of other reclamation districts, including the two largest in the Sacramento Valley, preparing reports and designs, but also actively supervising construction. These projects included large gravity intakes and pumping plants, screw pumps, irrigation and drainage canals with capacities of up to 45 m^3/s, and levees built by the world's largest clamshell dredges. In addition, a 50 m high rock-fill dam was erected including head gates, siphons, and flumes. Two other outstanding irrigation works were these of the Nevada Irrigation District in Nevada County CA, and of the Santa Clara Valley Water Conservation District, the latter for replenishment of the underground water supply. He was in parallel also active for civil duties, and chaired from 1925 to 1927 the Irrigation Section of the Commonwealth Club of California. He was ASCE member from 1917.

Anonymous (1929). Tibbetts, Frederick H. *Who's who in California*: 182-183. San Francisco. *P*
Anonymous (1940). Frederick H. Tibbetts. *Trans. ASCE* 105: 1924-1928.
Keyes, C.G. (2004). History of the Irrigation and Drainage Division to the Environmental and Water Resources Institute 1922-2004: 261-272. *Water Resources and Environmental History*: ASCE: Reston VA.
Tibbetts, F.H. (1929). Repair of breaks in outlet tunnel of rock-fill dam. *ENR* 102(23): 904-906.

TIDD

01.08. 1827 Woburn MA/USA
20.08. 1895 Woburn MA/USA

Marshall Martain Tidd lost the use of his right arm when he was a child. He began his engineering experience as assistant on the construction of a dam across the Merrimack River at Lawrence MA, under Charles S. Storrow (1809-1904). This was before photography was invented, so that Tidd had to make freehand sketches showing the work evolution. After dam completion he continued making plans of machines for more than two decades. In the 1860s he was connected with the construction of the dry docks at East Boston MA, and was the consulting engineer and designer of the dry dock at the Erie Basin, Brooklyn NY.

In 1872 Tidd was elected water commissioner of works for the water supply of various cities in Massachusetts, New Hampshire, and Maine. He was also employed to make improvements or additions to existing works in these cities. He designed the sewerage system at Marlboro MA. At the time of his death he was employed on the water works of Bath ME, the construction of the Woburn sewerage system, and as the consultant for various other engineering projects. As an engineer, Tidd had a remarkable mechanical ability, keen observation, fertility of resource, and entire honesty of purpose. He had in his private life a great interest in horticulture, and possessed a fine collection of shrubs and flowers; he was member of the Massachusetts Horticultural Society. Another of his hobbies was his workshop, which was a model in its equipment of tools and their neat arrangement. In his relations with the members of his profession, he was highly liberal and generous, ever ready to impart his great knowledge. He was a man of pronounced individuality, active and alert both in mind and body, an excellent companion indeed. He was a member of the American Society of Civil Engineers ASCE since 1878. His death was occasioned by heart failure, the result of a severe attack of a grippe two years earlier.

Anonymous (1897). Marshall M. Tidd. *Trans. ASCE* 37: 568-570.

Tidd, M.M. (1852). *View of Woburn*. Upham & Colburn: Woburn MA.

Tidd, M.M., Hobart, S.B., Simpson, J.E. (1860). *Simpson's patent drydock*, East Boston. Tidd's Lith: Boston MA.

Tidd, M.M. (1863). *View of the English steamer Caledonia wrecked on Cape Cod*. Tidd: Boston.

http://www.historicplaces.ca/en/rep-reg/place-lieu.aspx?id=5641

http://records.ancestry.com/Marshall_Martain_Tidd_records.ashx?pid=15995582 *P*

TIFFANY J.B.

12.04. 1909 Kansas City MI/USA
10.08. 1993 Vicksburg MS/USA

Joseph Ben Tiffany graduated as civil engineer from the University of Illinois, Urbana, in 1932, joining in 1933 the US Corps of Engineers at its Waterways Experiment Station WES, Vicksburg MS. Appointed executive assistant to the director WES in the early 1940s he was promoted to the rank of captain in the Corps of Engineers. In parallel he was in charge of the Station's Hydraulic Division. He was appointed in 1951 assistant director of WES. He remained all through his career with WES and organized in 1968 a workshop on the historical development of this important institution.

In his 1943 study, he reviewed the then current methods used in hydraulic laboratories. These included for instance investigations on wave forces by waves on breakwaters to develop vertical pressure curves for design purposes. Disks with sensitive pressure cells set flush to the face of the breakwater were proposed. The pressure of a wave striking against a pressure cell causes a change in the electrical output of a circuit, which was amplified and recorded, thus furnishing a continuous record of pressure. Another study reported in the 1943 paper relates to the power tunnel of Fort Peck Dam. It involved records of both surges and water hammer pressures from turbine and gate movements due to load changes. In still another study early observations of cavitation damage on baffle piers were conducted. Based on the pioneering work of Harold A. Thomas (1885-1973), a dynamic pressure meter was developed to measure the pressures accompanying cavitation. The first application concerned Bluestone Dam on the New River WV. The baffles were subjected with model velocities of up to 25 m/s. Based on observations, the original design was significantly improved to avoid cavitation damage while the energy dissipating action was preserved.

Anonymous (1944). J.B. Tiffany. *Engineering News-Record* 133(Aug.24): 207. *P*
Rouse, H. (1976). Tiffany. *Hydraulics in the United States*. The University of Iowa: Iowa City. *P*
Tiffany, J.B. (1935). Discussion of Sand movement in fluvial models. *Trans. ASCE* 100: 861-867.
Tiffany, J.B. (1943). Recent developments in hydraulic laboratory technique. Proc. 2nd
    Hydraulics Conf. Iowa. Studies in Engineering *Bulletin* 27: 31-50. University of Iowa.
Tiffany, J.B. (1945). Model study helps prevent Johnstown floods. *Civil Engineering* 15: 309-312.
Tiffany, J.B. (1963). Review of research on channel stabilization of the Mississippi River 1931-
    1962. *Technical Report* 2. US Army, Corps of Engineers: Vicksburg MS.
Tiffany, J.B., ed. (1968). *History of the Waterways Experiment Station*. WES: Vicksburg.

TIFFANY R.K.

11.06. 1879 Union IA/USA
01.06. 1939 Olympia WA/USA

Ross Kerr Tiffany graduated in 1900 from Cornell College, Mount Vernon IA, with the BSc degree in civil engineering. He was employed as engineer in 1901 by the Washington Irrigation Company on the Sunnyside Project, then the largest single irrigation project of Washington State, with headwater from Yakima River. He became chief engineer in 1906, and from 1910 supervised all land developments by his Company. In 1910 Tiffany was employed by the US Reclamation Service as superintendent of the Sunnyside unit, the Yakima Project. He was in 1914 advanced to its manager, in charge of operating and maintaining its irrigation units, and the extensive developments for storage purposes.

Tiffany left in 1920 federal service to engage in consulting engineering at Spokane WA. He there was project engineer and manager of the Spokane Valley Canal Company, supervising the Post Falls, Hayden Lake, and Dalton irrigation projects, reconstructing the Lewiston Orchards Canal System. An important engagement during this period was negotiation leading to a final settlement of questions between land owners and the Twin Falls North-Side Canal Company in Idaho. Tiffany was appointed in 1925 supervisor of hydraulics in the Washington State Department of Conservation and Development, thus representing the State on the Columbia River Allocation Board. He resigned in 1929, starting a private practice at Olympia WA. There he filled a wide range of engagements involving irrigation, drainage, diking, water supply, and water power. In 1934 he was appointed executive officer of the newly created Washington State Planning Council, an office he held until his death. He was qualified for this office because of his knowledge of the State and its varied and abundant natural resources. The traits of character were his geniality, his capacity to make and hold friends, his loyalty, his effectiveness, his liberal sense of humour, and his friendship. He was member of the American Society of Civil Engineers ASCE since 1920, and past-president of its Tacoma-Olympia Section.

Anonymous (1919). Ross K. Tiffany. *History of the Yakima Valley* 2: 729-730.
Anonymous (1938). Ross K. Tiffany. *Engineering News-Record* 120(Jan.27): 136. *P*
Anonymous (1940). Ross K. Tiffany. *Trans. ASCE* 105: 1928-1931.
Downs, L.V. (1993). *The mightiest of them all*: Grand Coulee Dam. ASCE: New York.
Ruettgers, A., Whitmore, A.A. (1930). Yakima Project. *New Reclamation Era* 21(10): 186-195.
Tiffany, R.K. (1925). Use of burned clay, concrete and wood pipe for irrigation. *ENR* 95(11): 419.
http://www.yakimamemory.org/cdm4/item_viewer.php?CISOROOT=%2 *P*

TINNEY

11.05. 1925 Man Ont./CA
10.12. 1974 Mexico City/MX

Edwy Roy Tinney made civil engineering studies at University of Washington, Pullman WA, receiving a MSc degree in 1950 and obtaining the PhD degree from St. Anthony Falls SAF Hydraulic Laboratory, University of Minnesota, Minneapolis MN in 1955. He then joined the Albrook Hydraulic Laboratory, Washington State University, serving as its first director. He worked tirelessly to build up the lab and the staff. He was responsible for planning the Central North American Water Project, a plan to irrigate the great plains from northern Canada to Mexico using the waters from the Arctic Ocean. He was in parallel professor there of civil engineering. In 1968 he moved as chief of the planning division of the Department of Energy, Mines and Resources to Canada.

Tinney has contributed to hydraulic engineering within a short period. Besides being laboratory director and responsible for its management, he published in various fields of hydraulics. His most notable research projects include fuse plugs corresponding to elements located on spillway crest that are washed out in excess of a certain discharge. Another topic of interest were advancing fluid flows in originally dry channels, as occur in irrigation technique. He also contributed to the hydraulics of penstock trifurcations as used in hydropower plants. He passed away at age 49 only from a heart failure following a tennis match in Mexico City, during which time he was adviser for the World Bank. He was a charter member of the American Water Resources Association AWRA.

Anonymous (1956). E. Roy Tinney. *Engineering News-Record* 157(Nov.22): 53-54. *P*
Anonymous (1968). E.R. Tinney. *Engineering Journal* 51(3): 54. *P*
Soliman, M.M., Tinney, E.R. (1968). Flow around 180° bends in open rectangular channels. *Journal of the Hydraulics Division* ASCE 94(HY4): 893-908; 95(HY2): 729-731; 95(HY3): 1064; 96(HY1): 257-258.
Tinney, E.R. (1957). Summary of the research program at R.L. Albrook Hydraulic Laboratory. *WSC Hydraulics Conference*, 99-110. State College of Washington: Pullman WA.
Tinney, E.R., Hsu, E.Y. (1961). Mechanics of washout of an erodible fuse plug. *Journal of the Hydraulics Division* ASCE 87(HY3): 1-29.
Tinney, E.R., Bassett, D.L. (1961). Terminal shape of a shallow liquid front. *Journal of the Hydraulics Division* ASCE 87(HY5): 117-133; 88(HY2): 183-186; 88(HY5): 277-280.
Tinney, E.R. (1962). The process of channel degradation. *Journal of Geophysical Research* 67(4): 1475-1480.

TIPTON

23.03. 1893 Litchfield IL/USA
23.12. 1967 Denver CO/USA

Royce Jay Tipton graduated from the Cripple Creek High School, Cripple Creek CO, in 1911. His career began in the San Luis Valley of Colorado where he worked as topographer and irrigation engineer before entering in 1915 University of Colorado, Fort Collins CO. After war service he entered private practice in irrigation engineering in the San Luis Valley. Except for brief assignments with an engineering company in Mexico, the US Bureau of Reclamation USBR in connection with Hoover Dam, and the State Engineer of Colorado in relation with interstate water matters, Tipton remained consultant until his death. He was a member of the American Society of Civil Engineers ASCE, serving from 1965 to 1967 as vice-president, and as member of executive committees of the Irrigation and Drainage Division, and the Power Division. He also was a member of the Colorado Society of Engineers, the Society of American Military Engineers SAME, the American Geophysical Union AGU, and the International Association of Hydraulic Research IAHR. He was president of the International Commission on Irrigation and Drainage ICID. He was recipient of the 1958 Norlin Medal, the University of Colorado.

Tipton's practice was exclusively devoted to water development projects, with emphasis on irrigation and reclamation, in the Western USA and overseas. Although the firms he headed were responsible for the design and the supervision of irrigation, flood control, and hydroelectric projects with construction cost in excess of one billion dollars, his most notable achievements were in his role as an individual consultant in connection with water resources investigations and interstate and international water problems. He played a prominent part in negotiation of many river compacts and treaties, including these involving Colorado River, Pecos River, Rio Grande, the Mexican Water Treaty, and the Indus Waters Treaty in the Indian Subcontinent in 1960.

Anonymous (1958). R.J. Tipton: Scholarship fund. *Engineering News-Record* 160(Jun.19): 177. *P*
Anonymous (1969). Royce J. Tipton. *Trans. ASCE* 134: 987-988.
Anonymous (1973). Tipton, Royce J. *Who was who in America* 5: 724. Marquis: Chicago.
Tipton, R.J. (1938). *Report on water supply of Colorado River and allied matters*. Denver.
Tipton, R.J. (1945). *The Six States Committee*: Arizona, Colorado, New Mexico, Texas, Utah
        and Wyoming. Denver.
Tipton, R.J. (1961). *A description of miscellaneous hydraulic structures*. Denver.
http://www.colorado.edu/engineering/deaa2/cgi-bin/display.pl?id=11 *P*

TOCH

23.01. 1922 Wien/A
19.07. 1962 Iowa City IA/USA

Arthur Toch graduated in 1948 from the University of Kansas, Lawrence KA, spent then two years with the US Bureau of Reclamation USBR, then moving to State University of Iowa, Iowa City IA, graduating there with the MS degree in hydraulic engineering in 1952. He became a research associate there on the staff of the Iowa Institute of Hydraulic Research IIHR, and in parallel was instructor at the Department of Mechanics and Hydraulics. Nothing is so far known on the further career of Toch.

During Toch's stay at University of Iowa, he dealt with a number of hydraulic problems. He collaborated with Emmett M. Laursen (1919-2013) as one of the first in the USA on bridge pier and abutment scour, a problem that had been investigated in Europe from the late 19th century but only in the 1950s was dealt with in America. A number of important bridges were damaged by scour, yet few means were then available to protect these important infrastructural elements. The complications of this two-phase flow problem involving both water flow and sediment, which is subjected by vortices mainly at the pier or abutment front, were investigated using simple models. Some preliminary results were established pointing at the large effect of the approach flow velocity, and proposals were made to reduce the scour action. The second basic study of Toch relates to so-called Tainter gates, a radial gate inserted in a rectangular channel. The discharge coefficient was determined in terms of relative gate radius, submerged outflow was considered, and limit conditions between free and submerged outflow were investigated.

Anonymous (1952). A. Toch. Proc. 5th *Hydraulics Conference* Iowa. Frontispiece *P*

Anonymous (1956). Arthur Toch. Proc. 6th *Hydraulics Conference* Iowa: Frontispiece. *P*

Laursen, E.M., Toch, A. (1953). A generalized model study of scour around bridge piers and abutments. 5th *IAHR Congress* Minneapolis: 123-131. *P*

Laursen, E.M., Toch, A. (1956). Scour around bridge piers and abutments. *Bulletin* 4. Iowa Road Research Board: Ames IA.

Toch, A. (1952). The effect of a lip angle upon flow under a Tainter gate. *MS Thesis*. State University of Iowa: Iowa City.

Toch, A., Moorman, R.W. (1953). Manifold efflux. Studies in Engineering *Bulletin* 35. State University of Iowa: Iowa City.

Toch, A. (1955). Discharge characteristics of Tainter gates. *Trans. ASCE* 120: 290-300.

Toch, A., Schneider, G.R., eds. (1958). Proc. 7th *Hydraulics Conference* Iowa City.

TODD

28.09. 1874 Rankin County MS/USA
11.09. 1929 Hot Springs AL/USA

Alexander Miller Todd obtained the BS degree in civil engineering in 1894 from Texas Agricultural and Mechanical College, todays Texas A&M. He then entered the service of the Mississippi River Commission, advancing rapidly from rodman to a superintendent of construction in 1900. His early survey work brought him into intimate field contact with Mississippi River, rendering him the practical background that proved invaluable later when he reached a position to study river control. He devoted his entire professional life to the Federal Service, rising steadily up to the position of senior engineer at the time of death. His studies of gauge relations were most exhaustive and used extensively to determine levee grade lines, and to predict flood peaks. He gave much time to the study of bank protection, so that he became an authority in levee building.

In 1913, when it became apparent that a much extended program for the Mississippi flood control would be necessary, Todd foresaw that the available supply of material for the older types of bank protection or revetments was insufficient to meet the demands. It was realized that the use of concrete offered a possible solution to the problem. In 1914 the Mississippi River Commission asked to increase the levee height and section so that an improved method of construction was necessary. Todd devoted much of his time to the study of more economical methods for placing the earthwork. A slack-line drag-bucket cableway was found successful, which had a capacity of nearly 4,000 m^3 earthwork per day. Todd was personally held in a high esteem; he was the quiet head in heated discussions bringing order and progress out of argument. He was appointed in 1925 president of the Mississippi River Commission. Death came to him suddenly, while on an inspection of the Carpenter Dam construction near Hot Springs AL, due to a stroke of apoplexy. He was a member of the American Society of Civil Engineers ASCE. He published a discussion of the famous paper by James A. Seddon (1856-1921) on the hydraulics of rivers, highlighting the relationship between the water stage, slope, and discharge for Mississippi River, and discussing experiences having gained during his long career and connection with this river. He was described as a man with sterling qualities, who was beloved by his fellow man.

Anonymous (1930). Alexander M. Todd. *Trans. ASCE* 94: 1640-1642.
Todd, A.M. (1900). Discussion of River hydraulics. *Trans. ASCE* 43: 230-233.
http://www.mvk.usace.army.mil/index.php?pID=cp *P*

TOFFALETI

25.10. 1906 Hillsborough FL/USA
05.04. 1982 Vicksburg MS/USA

Few is known on the education of Fred B. Toffaleti, He joined in any case the US Corps of Engineers of which he acted in 1960 as chief of the Hydraulic Model and Sedimentation Section, and in the late 1960s headed the US Army Engineering Division of the Lower Mississippi Valley. He was a member of the Mississippi River Commission, Vicksburg MS.

The 1963 paper reports a large number of velocity data collected at Atchafalaya River, Simmesport LA, and in the Mississippi River at Vicksburg MS, to define the vertical velocity distribution. These data indicate that the distributions of the two rivers are almost identical at the respective locations. However, a comparison with the standard formula derived from analytical sedimentation studies indicated a quite different result. Suspended sediment data were also collected indicating that the silt and clay portions were satisfactorily measured with a standard approach. The 1969 paper introduces the Toffaleti equation, corresponding to a procedure for the analytical determination of sand transport in rivers. His formulation was particularly adaptable to computational schemes. It was tested from small to large rivers, under the full range of flow conditions, and verified in the flume experiments conducted under widely-varied flow conditions whose beds also involved a large degree of variability. van Rijn in his famous 1984 paper series states that the Toffaleti approach is the best to deal with this problem. Nakato in 1990 found that the Toffaleti formula for sediment discharge computations performed well for all flow events investigated.

Rouse, H. (1976). F.B. Toffaleti. *Hydraulics in the United States* 1776-1976: 182. IIHR. *P*
Toffaleti, F.B. (1960). Sedimentation aspects in diversion at Old River. *Journal of the Hydraulics Division* ASCE 86(HY6): 37-46; 87(HY1): 259-261.
Toffaleti, F.B. (1963). Deep river velocity and sediment profiles and the suspended sand load. *Sedimentation Conference*: 207-228. Miscellaneous Publication 970. ARS: Washington.
Toffaleti, F.B. (1968). A procedure for computation of the total river sand discharge and detailed distribution, bed to surface. *Technical Report* 5. US Corps of Engineers.
Toffaleti, F.B. (1969). Definitive computation of sand discharge in rivers. *Journal of the Hydraulics Division* ASCE 95(HY1): 225-248; 95(HY6): 2185-2189; 96(HY6): 1352.
Vanoni, V.A. (1975). *Sedimentation engineering*. ASCE Manuals and reports on engineering practice 54. ASCE: New York.
van Rijn, L.C. (1984). Sediment transport 2: Suspended load transport. *Journal of Hydraulic Engineering* 110(11): 1613-1641.

TOLMAN C.F.

02.06. 1873 Chicago IL/USA
13.10. 1942 Spokane WA/USA

Cyrus Fisher Tolman graduated in 1896 with the BS degree from the University of Chicago, Chicago IL, continuing there with graduate studies in geology until 1899. He then was a consulting geologist and mining engineer until 1905, when being appointed professor of geology and mining until 1912 at the University of Arizona, Tucson AZ. He was further associate professor of economic geology at Leland Stanford University, Stanford CA, until 1919 when taking over there as professor until his retirement in 1938. He died suddenly during a professional trip. Tolman was a Fellow of the Geological Society of America GSA, as well as member of the American Institute of Mining Engineers, and the Seismological Society of America, serving there also as vice-president.

While staying in Arizona, Tolman took part in the active development of the southern copper district, learning mining geology by intimate field studies. For a decade he was a consulting geologist visiting mines in various parts of both the south-western USA and Mexico. At Stanford, Tolman recognized the importance of summer courses in the field. He was an effective teacher, clear and interesting in lectures and inspiring in field and laboratory research. He is remembered for the 1937 book on groundwater flow thereby emphasizing the need for geological investigations and engineering analysis to solve hydro-geological problems. Apart from the 1923 treatise of Oscar E. Meinzer (1876-1948), Tolman published the first text on groundwater in the USA. Tolman hoped that, besides students and engineers, his text would also assist in preparing attorneys for litigation involving subsurface water, and in the development of sound groundwater laws based on scientific principles.

Anonymous (1929). Tolman, Cyrus F., Jr. *Who's who in America* 15: 2072. Marquis: Chicago.
Meinzer, O.E. (1923). The occurrence of ground-water in the United Sates with a discussion of its principles. *Water Supply Paper* 489. US Geological Survey: Washington DC.
Remson, I. (2002). The history of hydrology at Stanford University. *Ground Water* 40(2): 205-206. *P*
Tolman, C.F. (1911). *Graphical solution of fault problems*. Mining and Scientific Press: San Francisco.
Tolman, C.F. (1937). *Ground water*. McGraw-Hill: New York.
Tolman, C.F., Poland, J.F. (1940). Ground water, salt-water infiltration, and ground-surface recession in Santa Clara Valley, Santa Clara County CA. *Trans. AGU* 21(1): 23-34.

TORPEN

27.08. 1886 Osseo WI/USA
29.01. 1970 Portland OR/USA

Bernhardt Enoch Torpen obtained the BSc degree from Washington State College, Pullman WA, and there obtained in 1924 a CE degree. From 1911 to 1913 he had been assistant engineer for the Calgary Power Company in Canada, and from 1914 locating engineer for the Cascade Irrigation District, Thorpe WA. From 1915 to 1917 he acted as field engineer for the municipal water supply, Aberdeen WA, and superintendent of construction of a copper mine in Peru. From 1921 to 1922 he was field engineer for the port of Aberdeen WA, and from 1923 hydraulic engineer of Cushman Power Project, Tacoma WA. From 1926 to 1929 he was then superintendent of design and construction of the Bull Run Dam, Portland OR, and then acted until 1930 as assistant hydraulic engineer on hydro properties at Guadalajara MX. In this period he was further engaged with layouts and estimates for the Ariel Hydro Project on Lewis River WA, the Z-Cannon Project at Clark's Fork on the Columbia River, and the Upper Salmon Falls on the Snake River.

Torpen was from 1930 to 1932 consulting engineer for the Soviet Government, Moscow, as member of the Board of Foreign Consulting Engineers. Upon completion of his work in Russia, he returned to the USA, visiting outstanding dams by then under construction or just having been completed. In 1933 he became engineer with the US Army Corps of Engineers, Portland OR, acting finally in 1939 as principal engineer. He was appointed in 1939 head engineer of the Corps at Portland OR, acting from 1946 to 1948 there as Assistant Chief of the North Pacific Division, and taking over in 1949 as its chief engineer. He was from 1951 to 1956 consulting engineer to this Division, and also a consultant to the governments of India and Pakistan on flood control of the Ganges and Brahmaputra rivers, and rivers in the Punjab. After retirement from the Corps he joined in 1956 a company at Portland for seven years, retiring from active service in 1969. He was both a member and Fellow of the American Society of Civil Engineers ASCE, and president of its Oregon Section, receiving its Engineer of the Year Award in 1969.

Anonymous (1956). B.E. Torpen. *Engineering News-Record* 156(May3): 79-80. *P*
Anonymous (1970). Bernhardt E. Torpen. *Trans. ASCE* 135: 1147-1149.
Torpen, B.E. (1924). *Cushman Power Project*. State College of Washington: Pullman.
Torpen, B.E. (1929). Bull Run Storage Dam for Portland OR. *ENR* 103(6): 204-208.
Torpen, B.E. (1951). *Power and Columbia River Storage*: Projects in region's potential. Seattle.
http://search.ancestry.co.uk/cgi-bin/sse.dll?db=mediaphotopublic&rank=1&sbo=t&gsbco=Sweden&gsln=Torpen *P*

TOTTEN

23.08. 1788 New Haven CT/USA
22.04. 1864 Washington DC/USA

Joseph Gilbert Totten graduated in 1805 from the US Military Academy and was commissioned a second lieutenant in the Corps of Engineers. He in 1806 resigned already, to assist in public land survey. He re-entered the Corps in 1808 and during the War of 1812 was chief engineer of the Niagara frontier. From 1825 to 1838 he oversaw the construction of Fort Adams, Newport RI, the second largest project of the Army in the 19th century exceeded only by Fort Monroe VA. Totten was appointed Army chief engineer in 1838, serving in this position until his death.

One of Totten's most significant achievements was the design and the construction of Minot's Ledge Light near Cohasset MA. Many ships had approached the Massachusetts coast but the location of Boston Harbour was misjudged. Cohasset Rock is a three km long geological formation similar to Eddystone Rocks off the Devon Coast in UK. The famous John Smeaton (1724-1792) built a notable lighthouse there which resisted all storms until today. A first lighthouse erected in 1849 failed in 1851 during a tremendous storm. In 1852 Totten designed an improved structure in collaboration with a colleague. The best site for the new lighthouse was determined based on a survey of the reef. It was decided that the location of the former lighthouse suited best for the purpose, but work could only proceed when wind and waves were low. The lighthouse rested on 7 interlocking foundation stones all below the water surface. By the fall 1856, 6 m tall permanent iron rods had been drilled , and wrought-iron frames connected the shafts. The iron framework would later reinforce the stone tower. Despite the rough working conditions no life was lost during construction. The work proceeded upward, layer by layer, until the stonework was finished in 1860. More than 1,000 granite blocks were required with a total weight of more than 2,000 tons. The lighthouse went in service late in that year. Today, it looks much like it did when erected, remaining an outstanding engineering feat. It was named a Heraldic Place in 1987.

Anonymous (1896). Joseph G. Totten. *Trans. ASCE* 36: 525-527.
Bernard, S., Totten, J.G., Humphreys, A.A. (1866). *Ohio and Mississippi Rivers*. Washington DC.
Brown, J.L. (2012). Braving the storm: Minot's Ledge Lighthouse. *Civil Engineering* 82(11): 46-49. *P*
Totten, J.G., Bache, A.D., Davis, C.H. (1862). *Surveys of Boston Harbour*. Farwell: Boston.
http://en.wikipedia.org/wiki/Joseph_Gilbert_Totten *P*

TOWER

12.01. 1819 Cohasset MA/USA
20.03. 1900 Cohasset MA/USA

Zealous Bates Tower was an American soldier and civil engineer who served as a general in the Union Army during the American Civil War. He was most noted for constructing the solid defenses of Federal-occupied Nashville TE, which proved to withstand repeated attacks by the Confederates. He graduated with first honors at West Point in 1841, serving then in the Mexican War. After the war, he served as an engineer, responsible for the initial construction of the Federal facilities on Alcatraz Island, and Fort Point, San Francisco CA. Following the acquisition of California by the USA as result of the Treaty of Guadalupe Hidalgo in 1848, ending the Mexican-American War, and the onset of the California Gold Rush the following year, the US Army began studying the suitability of Alcatraz Island for the positioning of coastal batteries to protect the approaches to the San Francisco Bay. In 1853, under the direction of Tower, the Army Corps of Engineers began fortifying the island until 1858, eventuating in Fortress Alcatraz. The island's first garrison at Camp Alcatraz, numbering 200 soldiers and 11 cannons, arrived at the end of that year. When the Civil War broke out in 1861 the island mounted 85 cannons in casemates around its perimeter, though the small size of the garrison meant that only a fraction of the guns could be used at one time.

Tower remained in the regular army after the Civil War. In 1865, he became lieutenant colonel in the US Army Corps of Engineers. He supervised the work of improving major harbors, both for commercial and military purposes. He was promoted to colonel in 1874, and retired from the service in 1883, returning to Cohasset MA, where he lived until his death. He was a notable military engineer and became known for river, harbour and fortification improvements. He received eight brevets for 'gallant and meritorious service'. He was one of the original members of the Aztec Club founded at Mexico City in 1847.

Anonymous ((1889). Tower, Zealous B. *Appleton's Cyclopaedia of American biography* 6: 146. Appleton: New York.

Anonymous (1900). Zealous B. Tower. *Engineering Record* 41: 281.

Tower, Z.B. (1871). An analytical investigation of the possible velocity of the Ice-Boat. *Van Nostrand's Engineering Magazine* 4: 55.

http://www.triposo.com/loc/San_Francisco/district_alcatraz_island

http://en.wikipedia.org/wiki/Zealous_Bates_Tower *P*

TRACY E.H.

31.03. 1817 Whitesboro NY/USA
28.08. 1875 Carmel NY/USA

Edward Huntington Tracy began his engineering career under John B. Jervis (1795-1885) as rodman on Chenango Canal NY. His work was thoroughly and satisfactorily accomplished so that he was in 1838 after the completion of the Canal engaged as assistant engineer on Croton Aqueduct, for which Jervis then served as chief engineer, and completed the work in 1842. Tracy constructed the portion of the aqueduct extending from Fordham Church to Manhattan, thereby superintending the entire works of the High Bridge over Harlem River. When the aqueduct was nearly completed and Jervis had stepped down from his position, Tracy took over in view of his capabilities, holding the position for a decade. In 1852, after having been actively connected with the Aqueduct for 14 years, he also resigned to become partner of iron works, but two years later was employed by the Canadian Government to make plans and surveys for a ship canal to connect the waters of the St. Lawrence River with Lake Champlain in the north of New York and Vermont States.

On accomplishment of this commission Tracy was appointed by the State of Iowa to superintend the design the slack-water navigation on the Des Moines River. He further made surveys on Mississippi River at Le Claire for the Rock Island, Mississippi and Missouri Railroad Company. Later he was appointed by New York State to superintend the removal of the Harlem River obstruction at McComb's Dam, and to erect Central Bridge on the site of the old dam. In 1870 Tracy was appointed chief engineer of Croton Aqueduct, which position his familiarity with the requirements of the aqueduct enabled him to fill excellently the office. His superior qualifications and extensive engineering operations won for him great confidence in his skill and capabilities, and conferred upon him a high and enduring reputation in his profession. He died of a heart disease, so that John C. Campbell (1817-1890), whose career had been similar to that of Tracy's, took over as chief engineer of the Croton Aqueduct.

Anonymous (1875). Edward H. Tracy. *The New York Times* (Aug. 29).
Anonymous (1878). Tracy, Edward H. *Proc. ASCE* 5: 337-338.
Burrows, E.G., Wallace, M. (1999). *Gotham*: A history of New York City to 1898. Oxford University Press: Oxford.
Wegmann, E. (1896). *Water supply of the city of New York* 1658-1895. Wiley: New York. (*P*)
New York NY Daily Graphic 1875 Aug-Jan 1876 Grayscale – 0180.pdf *P*

TRAUTWINE J.C.

30.03. 1810 Philadelphia PA/USA
14.09. 1883 Philadelphia PA/USA

John Cresson Trautwine entered at age of eighteen the office of William Strickland (1787-1854), the most prominent civil engineer then of Philadelphia. While receiving his technical training, Trautwine was engaged with the Delaware Breakwater and the erection of public buildings. In 1831, he secured a position with the Columbia Railroad, was appointed assistant engineer of the Philadelphia, Wilmington & Baltimore Railroad in 1835. In 1844 he sailed to New Granada in today's Columbia, being engaged for five years in the construction of *Canal del Dique* connecting the Magdalena River with the harbour of Cartagena on the Caribbean. Upon return to Philadelphia, he prepared for a stay in the Isthmus, where he made surveys for the Panama Railroad. A copy of his Isthmus map was published in 1871 in the Journal of the Franklin Institute. From 1851 he was asked to seek an inter-oceanic canal route. Crossing the Continental Divide, he descended San Juan River to the Pacific. He finally decided upon a canal route from the *Atrato* River to *Cupica* Bay as 'the least inadvisable'. However, he also stated 'I cannot entertain the slightest hope that a ship-canal will ever be found practicable across any part of the Isthmus'. In these times, the causes of malaria and yellow fever were unknown, and his work was therefore one of the most difficult and dangerous undertakings upon which an engineer could venture.

In the years to follow, Trautwine was engaged again in railroad design in Pennsylvania, he surveyed a route for an inter-oceanic railway in Honduras, planned a system of docks for Montreal, and a harbour for Big Glace Bay, Nova Scotia. After 1864, he took his life less strenuously, his health having undoubtedly been affected by previous works. In 1872 he published the first edition of his famous Civil engineer's pocket-book, a work that was immediately received with great favour by the profession. This book was later re-edited by his son John C., Jr. (1850-1924), with a total of 21 editions until 1937.

Anonymous (1883). John C. Trautwine. *Journal of the Franklin Institute* 116(2): 390-396.
Anonymous (1883). John C. Trautwine, C.E. *Engineering News* 10(Sep.22): 450-453. *P*
Anonymous (1936). Trautwine, John Cresson. *Dictionary of American biography* 18: 628-629. Scribner's: New York.
Anonymous (1965). John C. Trautwine. *Civil Engineering* 35(11): 102. *P*
Trautwine, J.C. (1872). *Civil engineer's pocket book*. Claxton, Remsen & Haffelfinger: Philadelphia.

TRAUTWINE, Jr. J.C.

17.03. 1850 Philadelphia PA/USA
04.07. 1924 Philadelphia PA/USA

John Cresson Trautwine, Jr., was the son of John C.
Trautwine (1810-1883). He became acquainted with
civil engineering in his father's design office, and
eventually became a partner. In addition to head his
office, he was from 1895 to 1899 chief of the Bureau
of Waters, Philadelphia PA. He was a life-member
of the Franklin Institute, Philadelphia, member and
president of the Philadelphia Club of Engineers, and
member of the Institution of Civil Engineers ICE,
London UK, and of ASCE. He served as Secretary
the Association of Engineering Societies, and was
business manager of its monthly journal.

Trautwine was from his father's death in 1883 proprietor of the office. He was an
advocate for water restriction by use of water meters, and proposed water purification
using adequate filter technology. He further advised New York City on the utilization of
Croton Watershed. He contributed a number of outstanding papers to the then leading
engineering journals on the above two topics. He further edited some of his father's
books, including the popular Civil engineer's pocket-book. He became also particularly
known in collaboration with Rudolph Hering (1847-1923) for translations of classical
hydraulic works, including the uniform flow formula of the two Swiss Emil Ganguillet
(1818-1894) and Wilhelm Rudolph Kutter (1818-1888), and the hydraulic observations
of Henry Emile Bazin (1829-1917) on orifice and weir flows.

Anonymous (1895). John C. Trautwine, Jr. *Engineering Record* 31(May 25): 453.
Anonymous (1917). Trautwine, Jr., John C. *Who's who in America* 9: 2482. Marquis: New York.
Anonymous (1924). John C. Trautwine, Jr. *Trans. ASME* 46: 1327.
Anonymous (1924). John C. Trautwine, Jr. *Engineering News-Record* 92(16): 647. *P*
Anonymous (1924). John C. Trautwine dies. *Engineering News-Record* 93(2): 77. *P*
Anonymous (1925). John Cresson Trautwine, Jr. *Trans. ASCE* 88: 1446-1447.
Hering, R., Trautwine, Jr., J.C. (1889). *A general formula for the uniform flow of water in rivers
        and other channels*, by E. Ganguillet, W.R. Kutter. Wiley: New York.
Trautwine, Jr., J.C. (1896). *Experiments* upon the contraction of the liquid vein issuing from an
        orifice, and upon the distribution of the velocities within it. Wiley: New York.
Trautwine, Jr., J.C., Trautwine, J.C. III (1906). *Civil engineer's pocket book*. Wiley: New York.
Trautwine, Jr., J.C. (1908). The water supply of Philadelphia, with special reference to the
        filtration works now under construction. *Journal of the Franklin Institute* 166: 363-394.
Trautwine, Jr., J.C., Trautwine, J.C. III. (1909). *Concrete*. Wiley: New York.

TRAUTWINE III J.C.

25.02. 1878 Philadelphia PA/USA
28.03. 1949 Ithaca NY/USA

John Cresson Trautwine, III, was the son of John C. Trautwine, Jr. (1850-1924). His grandfather John C. Trautwine (1810-1883) was the originator of the successful Trautwine's civil engineer's pocket-book, which was one of the frequently used texts during the 19th and early 20th century. John III graduated in 1900 from Cornell University, Ithaca NY, as civil engineer, and became soon associated as one of the editors of the mentioned family pocket-book. He was made editorial director in 1909, and after his father's death in 1924 took over the publication as both owner and editor, continuing in this position until his death. The book chapters were continuously revised and he planned extensive revisions of the text prior to his death. The 1930 version was the 21st edition of this book.

THE

CIVIL ENGINEER'S POCKET-BOOK

BY
JOHN C. TRAUTWINE
CIVIL ENGINEER

REVISED BY
JOHN C. TRAUTWINE, Jr.
AND
JOHN C. TRAUTWINE, 3D.
CIVIL ENGINEERS

EIGHTEENTH EDITION, NINETIETH THOUSAND

NEW YORK
JOHN WILEY & SONS
LONDON: CHAPMAN & HALL, LIMITED
1907

Besides, Trautwine III was engaged over the years in a diversity of engineering works. He was an investigator of water power possibilities in New Jersey and Pennsylvania, he was engineer with the Union Railroad Company at Port Perry PA, he was in charge of the water supply investigations for Philadelphia, and he was mine surveyor for the Glen Iron Furnace Company. From 1912 to 1915 he was a consultant at Philadelphia PA to study building vibrations and foundations from an engineering perspective, and he was in charge of the water power tests at Manayunk PA. In this period experimenting and inventing consumed a considerable time: He developed a monopod for dispensing with instrument tripods in mine surveying; and he experimented with a sounding balloon to carry sighting light into inaccessible portions of mines. For years, his home was at Philadelphia, and he held the membership of the Engineers' Club of that city. He moved in 1935 to Ithaca NY, so that he was able to carry on his editorial work near Cornell University. He also held for almost two decades membership in the Franklin Institute, Philadelphia, serving on its Science and Arts Committee. Trautwine III was a member of the American Society of Civil Engineers ASCE from 1920.

Anonymous (1949). John Cresson Trautwine, 3rd. *Civil Engineering* 19(7): 519.
Anonymous (1952). John C. Trautwine, III. *Trans. ASCE* 117: 1324-1325.
Trautwine, Jr., J.C., Trautwine, J.C., III (1919). *The civil engineer's reference book*. Trautwine Company: Philadelphia. (*P*)
Trautwine, Jr., J.C., Trautwine, J.C., III. (1909). *Concrete*. Wiley: New York.
Trautwine III, J.C. (1949). *Trautwine family papers* 1834-1947. Cornell University: Ithaca NY.

TRAVAINI

27.11. 1903 San Francisco CA/USA
10.01. 1974 Phoenix AZ/USA

Dario Travaini received in 1924 the BS degree in engineering from University of California, Berkeley CA, making in 1925 an extra year of graduate work. He first worked in the geophysical exploration for the Shell Oil Co. in California. He was employed next as sanitary engineer for the Grand Canyon Park and served for the Sanitary District of Indianapolis IN, returning in 1930 to Arizona as town engineer of Miami AZ. He was employed to supervise the first activated sludge sewerage treatment plant in 1931 at Phoenix AZ, remaining in charge of the plant after its completion. He was placed in charge

of the city's water system in 1941, and then was made engineer in charge of water and sewers in 1950, later Director of Water and Sewers Department, retiring in 1971.

Travaini received the Kenneth Allen Award of the Water Pollution Control Federation WPCF in 1945, sponsored by Kenneth Allen (1857-1930), for his meritorious service to the association. In 1951 he was honoured with the George W. Fuller Award, sponsored by George W. Fuller (1868-1934), of the American Water Works Association AWWA, for distinguished service in the water supply field. Travaini was named in 1957 Arizona Engineer of the Year. The American Public Works Association APWA presented him in 1962 the Samuel A. Greeley Award for outstanding community service. Travaini was a national director of AWWA, national director of WPCF, and president of the Arizona Society of Professional Engineers. He also was member of the American Academy of Sanitary Engineers, and the Arizona Water and Pollution Control Association. He was member of the American Society of Civil Engineers ASCE, attaining Life membership in 1972. He had understood that the city's growth in size and population required a similar growth in infrastructure. He helped plan and build the first three of the city's water treatment plants, a second wastewater treatment plant, and thousands of km of water supply and sewer mains. He has described these projects in his papers.

Anonymous (1954). Dario Travaini. *Civil Engineering* 24(4): 257. *P*
Anonymous (1974). Dario Travaini. *Trans. ASCE* 139: 577.
Harding, S.T. (1967). *A life in western water development*. University of California: Berkeley.
Travaini, D. (1947). Various applications of the autoxidation process. *Sewage Works Journal* 19(3): 478-482.
Travaini, D. (1960). *City of Phoenix water works*. Water and Sewers Department: Phoenix AZ.
www.azsce.org/.../history-150thAnniversaryBoo

TRIPP

05.08. 1888 Chicago IL/USA
26.07. 1971 Greensboro NC/USA

James (Jim) Gregory Tripp received the BSc degree in 1910 from Massachusetts Institute of Technology MIT, Cambridge MA. He moved to the West and then worked in the construction supply business of Southern California, forming his own engineering design service. From 1917 to 1930 he devoted his time to heavy construction as a contractor, and as a general superintendent of dams built in this time, which included the Palmdale and Creek Dams in California, the Lake Pleasant Dam and the Coolidge Dam in Arizona, and a section of the River Des Peres Sewer at St. Louis MO. In 1931 he joined a firm as construction manager, then as vice-president, handling the construction of the Hoover Dam, and Panama Canal work, or operations including the Mississippi River Lock and Dam Projects No. 5 and 15.

In 1937 Tripp reopened his own consulting office for contractors developing bids for clients for numerous dam projects, including the *Lackawak* Dam and Board of Water Supply tunnels in New York, and the Shasta Dam in California. During World War II he was again on the field, which he liked best, as operating manager for the Maritime Commission for the building of concrete tankers, and in the construction of the US Naval Base at Trinidad. After the war he went again into business for himself as a consultant which continued until his retirement in 1970. His clients numbered among the great contractors of the world, for whom he bid for work or provided on-site services, including for Folsom Dam, Hungry Horse Dam, Mount Morris Dam, Helena Dam, Niagara River power complex, Flaming Gorge Dam, Glen Canyon Dam, *Yanhee* Dam in Thailand, *Rihand* Dam in India, the water supply for Karachi, Pakistan, or the *Jacuara* Dam in Brazil. He was member of the Engineers Club, New York, and of the American Society of Civil Engineers ASCE, becoming ASCE Fellow in 1959.

Anonymous (1928). Construction of multiple-arch dam in Arizona: Lake Pleasant Dam. *Engineering News-Record* 100(5): 180-183.
Anonymous (1928). Construction features: Coolidge multiple-dome dam. *Engineering News-Record* 101(12): 438-442.
Anonymous (1951). James G. Tripp, Construction Engng. Prize. *Civil Engineering* 21(10): 607. *P*
Anonymous (1973). James G. Tripp. *Trans. ASCE* 138: 645-646.
Tripp, J.G. (1940). Discussion of Core drills at Chickamauga Dam. *Trans. ASCE* 105: 865-867.
Tripp, J.G. (1955). Construction engineer talks about tools of trade. *Civil Engng.* 25(10): 624-630.

TROESCH

02.03. 1920 Bern/CH
30.10. 2001 Beverly Hills CA/USA

B(eat) Andreas Troesch obtained in 1947 the degree of mathematician, and the PhD degree in 1952 from Swiss Federal Institute of Technology, ETH Zurich, Zürich. He moved to the United States as research associate of applied mathematics, joining the Institute of Mathematical Sciences, the New York University, serving from 1956 to 1958 as head of the applied mathematics section, the Computing Center of the Ramo-Woolridge Corp., Torrance CA, continuing then until 1961 at Space Technology Laboratories, Redondo Beach CA, from when he was the manager of the Computing Sciences Department, Aerospace Corp., San Jose CA. He was from 1966 professor of aerospace engineering, University of Southern California, Los Angeles. He was member of the American Mathematical Society AMS, the Society of Industrial and Applied Mathematics SIAM, and of the Association for Computing Machinery ACM.

The research interests of Troesch included applied mathematics and numerical analysis in hydrodynamics and gas dynamics, and both elliptic and hyperbolic partial differential equations. The 1954 Report considers problems of flood prediction on the Ohio and Mississippi Rivers by means of the finite-difference method, following James J. Stoker (1905-1992). The description includes the mesh system used. The junction problem is particularly dealt with. A theoretical analysis of steady flow and steady progressive wave features is also given. The 1956 report is concerned with the prediction of the 1945 and 1948 floods on the Ohio River, among others. The factors affecting the solution accuracy are discussed, and suggestions are provided for improvement of the numerical approach. The results are also compared with model studies. The 1958 work may be considered a synthesis of the two preceding studies. Examples indicate the flexibility of this approach, which may be counted to the very first in hydraulic engineering.

Anonymous (1955). A. Troesch. Proc. 6th *IAHR Congress* La Haye: Frontispiece. *P*
Anonymous (1960). Troesch, B. Andreas. *American men of science* 10: 4138. Cattell: Tempe AZ.
Isaacson, E., Stoker, J.J., Troesch, B.A. (1954). Numerical solution of flood prediction and river regulation problems. *Report* IMM-NYU-205. New York University: New York.
Isaacson, E., Stoker, J.J., Troesch, B.A. (1956). Numerical solution of flood prediction and river regulation problems. *Report* 6 IMM-NYU-235. New York University: New York.
Isaacson, E., Stoker, J.J., Troesch, B.A. (1958). Numerical solution of flow problems in rivers. *Proc. ASCE* 84(HY5, Paper 1810): 1-16.

TROWBRIDGE

25.05. 1828 Troy MI/USA
12.08. 1892 New Haven CT/USA

William Pettit Trowbridge graduated in 1848 first in class from the US Military Academy, West Point NY. He joined in 1850 the Corps of Topographical Engineers, surveying the coast of Maine; he was promoted in 1854 to first lieutenant. He was in 1856 appointed professor of mathematics at the University of Michigan, Ann Arbor MI, becoming in 1857 also scientific secretary of the Coast Survey. He installed in 1860 self-registering instrument in the permanent magnetic observatory, Key West FL, and there was also in charge of Gulf Stream observations. From 1862 he was in charge of the Ney York City branch office, constructing the NYC defences. He was professor of dynamic engineering from 1870 to 1877 at the Yale College, from 1872 to 1876 adjutant general, and from 1873 to 1878 was commissioned to build the State Capitol, Albany NY. He was from 1877 until his death professor of engineering at the School of Mines, Columbia College, New York NY. He was awarded in 1877 honorary membership of Yale University, New Haven CT, in 1879 the honorary PhD from Princeton University, Princeton NJ, among many others. He was elected a member of the National Academy in 1872.

Trowbridge has written books and papers; important in the present connection are his works on thermodynamics and on turbines. One of his colleagues at Columbia College noted: 'He had that rare combination of personal qualities of character with intellectual capacity and technical acquirement which made him an almost ideal selection for the position he was called upon to fill. With ability to appear before a broader public with credit, and appreciating the advantages which follow to an institution from such wider recognition of its professors, yet withal so modest and conscientious as never to be tempted to neglect a new duty because inconspicuous, his loss is made the more grevious by the very rarity of the union of these qualities in one person'.

Anonymous (1908). W.P. Trowbridge. *Trowbridge genealogy*: 608-609. Compiler: New Haven *P*

Comstock, C.B. (1895). William P. Trowbridge. *Biographical memoirs* 3: 363-367. National Academy of Sciences: Washington DC.

Trowbridge, W.P. (1874). *Heat as a source of power*, with applications of general principles to the construction of steam generators. Wiley: New York.

Trowbridge, W.P. (1879). *Turbine wheels*: On the inapplicability of the theoretical investigations of the turbine wheel, as given by Rankine, Weisbach, Bresse and others, to the modern constructions introduced by Boyden and Francis. Van Nostrand: New York.

TRUESDELL

18.02. 1919 Los Angeles CA/USA
14.01. 2000 Baltimore MD/USA

Clifford Ambrose Truesdell III obtained both the
BS and MS degrees in 1941 and 1942, respectively,
from California Institute of Technology, Pasadena
CA. He further received in 1942 the MS degree in
mechanics from the Brown University, Providence
RI, and the PhD degree in 1942 from Princeton
University, Princeton NJ. He worked then until
1946 at the Radiation Laboratory, Massachusetts
Institute of Technology MIT, Cambridge MA, from
when he joined the staff of the Naval Research
Laboratory, Washington DC, directing the group of
Theoretical Mechanics. He was appointed in 1950
professor of mechanics at Indiana University, Bloomington IN, taking over in 1961 as
professor of rational mechanics at Johns Hopkins University, Baltimore MD, where he
remained until his retirement. Truesdell was awarded the 1978 Birkhoff Prize, and in
1996 the Theodore von Karman Award.

Truesdell was a true leader in mechanics in general, and in the history of mechanics in
particular. He also was editor of two journals, namely Archive for Rational Mechanics
and Analysis from 1957, and the Archive for History of Exact Sciences from 1960. He
served as co-editor of various volumes of the *Handbuch für Physik* from 1956 to 1974.
He published 26 monographs, almost 300 scientific papers as also numerous Essays and
book reviews. He had a broad mind and a wide cultural background so that he was a
master in cultural history from early times up to the 18th century. He also edited the four
volumes on Leonhard Euler.

Truesdell, C. (1952). The mechanical foundations of elasticity and fluid dynamics. *Journal of
        Rational Mechanics and Analysis* 1(2): 125-291; 2(5): 593-616.
Truesdell, C. (1956). Experience, theory and experiment. Proc. 6th *Hydraulics Conf.* Iowa: 3-18.
Truesdell, C. (1960). A program toward rediscovering the rational mechanics of the age of
        reason. *Archive for History of Exact Sciences* 1(1): 1-36.
Truesdell, C.A., Noll, W. (1965). The non-linear field theories of mechanics. *Handbuch der
        Physik* 3(3): 1-602, S. Flügge, ed. Springer: Berlin.
Truesdell, C.A. (1968). *Essays in the history of mechanics*. Springer: Berlin.
Truesdell, C.A. (1980). *The tragicomic history of thermodynamics* 1822-1854. Springer: Berlin.
Truesdell, C.A. (1984). *An idiot's fugitive essays on science*: Methods, criticism, training,
        circumstances. Springer: Berlin.
http://de.wikipedia.org/wiki/Clifford_Truesdell *P*

TULTS

24.03. 1907 Tartus/EE
31.12. 1969 Santa Clara CA/USA

Harold (Harald) Tults was born in Estonia but he graduated from the University of Brno CR in 1932. He obtained in 1949 the PhD degree from Karlsruhe University, and moved to the United States in 1950. He was in the early 1950s then assistant hydraulic engineer within an engineering company of Chicago IL. He there also chaired the Estonian Society, and was member of the Estonian Students Corporation. He finally lived in California. Tults became ASCE Member in 1959.

Tults studied diffusor, water hammer, and side weir problems. His 1956 work is concerned with pressure recovery in expansions, including the flow separation from the walls. The diffusor is an important element in hydraulics and aerodynamics, to which research was devoted to improve the flow conditions. Tults measured the velocity profiles and the pressure distributions in a uni-laterally expanding plane rectangular test channel. Two characteristic phases, 'instant stop' and 'separation start', were recorded. The work resulted in information on the mechanics of separating flow. A simple relationship between the angle of divergence and the expansion rate at which the maximum pressure recovery occurs was established, permitting to determine the optimum divergence versus the required rate of gradual expansion. The pressure recovery can be significantly increased if the separation endangered area is connected to a bypass removing the retarded flow portion and moves it back into the approach flow portion. The 1956 paper on the flood protection is concerned with side weirs to laterally discharge a portion of the approach flow, thereby improving the discharge scenario in the tailwater. This approach is currently often applied if there is enough space to bypass temporally water into large natural ponds, and returned to the river after flooding. The third 1956 paper considers the design of penstocks using the Manning formula for pipe velocity, and a total annual cost relationship. If frictional effects are neglected, a simple solution results, which is improved by relaxing the latter condition.

Tults, H. (1955). Discussion of Diversion flow through Buford Dam conduits. *Proc. ASCE* 81(709): 7-10.
Tults, H. (1956). Flow expansion and pressure recovery in fluids. *Trans. ASCE* 121: 65-84.
Tults, H. (1956). Flood protection of canals by lateral spillways. *Journal of the Hydraulics Division* ASCE 82(HY5): 1-17; 83(1230): 47-49; 83(1417): 3-5.
Tults, H. (1956). Quicker design of penstocks. *Water Power* 8(8): 303-305.
Tults, J. (2013). Harold Tults. Personal communication. *P*

TURNBULL W.J.

19.03. 1903 Burchard NE/USA
28.09. 1997 Warren MS/USA

Willard Jay Turnbull obtained the BS degree in 1925 in civil engineering from University of Nebraska, Lincoln NE. He was from 1927 to 1928 a project engineer of the Nebraska State Highway Dept., and then until 1931 in charge of operations on farm land in Pawnee County NE, from when he was until 1935 assistant state test engineer at his Alma Mater. He was from 1935 to 1941 soils engineer and chief of the Laboratory, Central Nebraska Public Power and Irrigation District, from when he took over as chief the Embankment, Foundation and Pavement Division of the US Waterways Experiment Station, Vicksburg MS. Turnbull was member of the American Societies of Civil and Mechanical Engineers ASCE and ASME, and the Highway Research Board.

Turnbull was awarded in 1965 one of the eight Distinguished Civilian Service Awards of the Department of Defence in recognition of major achievements, then chief of the Soils Engineering Division of the US Army Corps of Engineers. The citation reads 'His broad engineering knowledge and exceptional managerial ability advanced the basic knowledge of soil engineering and the capacity of the Department to successfully support military operations under a wide range of different and difficult conditions'. In his career he had been responsible for the design and the evaluation of military airfields, studies for foundations of dams and levees, and research in soil dynamics, stabilization, and other aspects of soil mechanics. He was in 1969 recipient of the ASCE Terzaghi Award, which is presented for outstanding contributions to knowledge in the field of soil mechanics, subsurface and earthwork engineering, and construction.

Anonymous (1948). Turnbull, Willard J. *Who's who in engineering* 6: 2025. Lewis: New York.
Anonymous (1965). Top Pentagon Award to Turnbull. *Engineering News-Record* 174(Jun.17): 213. *P*
Turnbull, W.J., Woodland, G.S. (1958). Foundation design. *Trans. ASCE* 123: 1160-1171.
Turnbull, W.J., Krinitzsky, E.L., Weaver, F.J. (1966). Bank erosion in soils of Lower Mississippi Valley. *Journal of Soil Mechanics and Foundation Division* ASCE 92(SM1): 121-136.
Turnbull, W.J., Hvorslev, M. (1967). Special problems in slope stability. *Journal of Soil Mechanics and Foundation Division* ASCE 93(SM4): 499-525.
Turnbull, W.J., Mansur, C.L. (1973). Compaction of hydraulically placed fills. *Journal of Soil Mechanics and Foundation Division* ASCE 99(11): 939-955.
http://de.wikipedia.org/wiki/Willard_J._Turnbull

TURNBULL W.

09.10. 1800 Philadelphia PA/USA
09.12. 1857 Wilmington NC/USA

William Turnbull graduated in 1819 from the US Military Academy, entering as second lieutenant the Corps of Artillery. He was engaged from 1819 to 1831 as topographical engineer, finally in the rank of captain. He was from 1832 to 1843 assigned to the construction of the canal aqueducts across Potomac River at Georgetown DC. This work, one of the earliest of the important undertakings of American engineers, gave him a high rank among professional associates. The piers of the aqueduct were founded by coffer dams on rock, covered by mud and were nearly 15 m below the water surface. He was then in 1838 promoted to major, having charge of the repairs of the Potomac Long Bridge, connecting Washington DC and Alexandria VA.

Turnbull was subsequently in charge of harbour improvements at the Great Lakes and Lake Champlain. From 1846 to 1848 he was chief topographical engineer taking part in all operations from the siege of Vera Cruz to the capture of Mexico City. These services gained for him the brevet of lieutenant colonel, and colonel. From 1850 to 1852 he acted as assistant in the Topographical Bureau at Washington DC, and he investigated until 1854 the practicability of bridging the Susquehanna River at Havre de Grace MD, and the expediency of an additional canal around the Falls of Ohio. He also was from 1853 to 1855 light-house engineer for Oswego Harbour, New York, and in charge of harbour improvements of Lake Champlain, Lake Ontario, and the eastern part of Lake Erie. He finally worked on the improvement of the Cape Fear River in North Carolina until 1857. Among his various published government reports is his 1838 study on the Potomac Aqueduct, including 21 plates.

FitzSimons, N. (1972). Turnbull, William. *Biographical dictionary of American civil engineers*: 119. ASCE: New York.
Turnbull, W. (1838). On the survey and construction of the Potomac Aqueduct. House of Representatives, Document 459, 25th Congress, 2nd Session. Washington DC.
Turnbull, W. (1847). *An essay on the air-pump and atmospheric railway* containing formulae and rules for calculating the various quantities contained in Mr. Stephenson's report on atmospheric propulsion. Williams: London.
Turnbull, W. (1873). *Reports* on the construction of the piers of the aqueduct of the Alexandria Canal across the Potomac River at Georgetown, 1835-1840. Washington DC. (*P*)
http://www.topogs.org/b_wturnbull.html

TURNEAURE

30.07. 1866 Freeport IL/USA
31.03. 1951 Madison WI/USA

Frederick Eugene Turneaure was educated at Cornell University, Ithaca NY, receiving the CE degree in 1889. He was then an instructor in civil engineering at Washington University, St. Louis MO, professor of bridge and hydraulic engineering at the University of Wisconsin, Madison WI, from 1892 to 1903, and from then there dean of the College of Engineering until retirement in 1937. He was a member of the American Society of Civil Engineers and elected to its honorary membership in 1933. He also was a member of the Western Society of Engineers, and of the American Concrete Institute ACI. He further received the Turner Medal of this Institute in 1930.

Turneaure has written a number of civil engineering texts, including the Public water supplies. This book reflects well the engineering literature of its time, with a standard knowledge addressing civil engineers in general, without great details on the particular features of a certain phenomenon. The book is well illustrated to attract readers who are not familiar with the elements of water supply installations. Turneaure also wrote books on the Principles of mechanics, again a basic text, or his Hydraulic engineering, of which the title summarizes almost completely the book contents. Further works also relate to bridge design and to reinforced concrete, or to highway design and to sanitary engineering, covering almost completely all civil engineering fields.

Anonymous (1933). Turneaure elected to honorary membership. *Civil Engineering* 3(12): 696. *P*
Anonymous (1937). F.E. Turneaure receives Lamme Medal. *American Machinist* 81: 636c. *P*
Anonymous (1941). Turneaure, Frederick E. *Who's who in engineering* 5: 1810. Lewis: NY.
Anonymous (1951). Dean F.E. Turneaure dies. *Civil Engineering* 21(5): 291. *P*
Turneaure, F.E. (1900). *The principles of mechanics*: An elementary exposition for students of
    physics. Macmillan: New York.
Turneaure, F.E., Russell, H.L. (1901). *Public water supplies*. Wiley: New York.
Turneaure, F.E., Black, A. (1909). *Hydraulic engineering*: A practical treatise on the principles
    of water pressure and flow and their application to the development of water power,
    including the calculation, design, and construction of water wheels, turbines, and other
    details of hydraulic power plants. American School of Correspondence: Chicago.
Turneaure, F.E., Russell, H.L. (1933). Methods for ground-water supplies. *Water and Water
    Engineering* 34(411): 603-606; 35(413): 25-28.
Turneaure, F.E. (1935). Current work at the University of Wisconsin. *ENR* 114(Mar.28): 455-457.

TURNER H.M.

06.08. 1885 Wareham MA/USA
06.05. 1975 Marblehead MA/USA

Howard Moore Turner obtained the AB degree from Harvard University, Cambridge MA, in 1906, and the SB degree in civil engineering there in 1907. He was until 1910 with a construction company, was until 1911 research engineer with a gas company at Easthampton MA, returning until 1917 as hydraulic engineer to the Turners Falls Power & Electric Co., Turners Falls MA, taking there over as assistant of its president at Boston MA until 1918. After service in World War I, he became until 1923 member of an engineering firm at Boston MA, from when he had his private practice at Boston MA, dealing with works in hydraulic engineering and water power projects. He served in parallel as a professor on the practice of engineering at Harvard University. Turner was member of the Boston Society of Civil Engineers, taking over in 1943 as president. He also was member of the American Society of Civil Engineers ASCE, and the New England Water Works Association NEWWA.

Turner was in the 1940s identified with important public utility hydro-electric works and developments and other engineering projects. He had been for 20 years intimately associated with Harvard University, where he lectured on water power engineering from the practical view. From 1950, he was involved in the improvement of the Cambridge water system. After years of study under his guidance as president of the Cambridge Water Board, the city approved his plan. This provided for a new pumping station, extensions for water treatment facilities at the filtration plant, work on the sedimentation basins, and a new general shop building. He also was in charge of water projects and other works in hydraulics.

Anonymous (1943). Boston civil engineers elect Turner president. *Engineering News-Record* 130(Apr.15): 529. *P*
Anonymous (1948). Turner, Howard M. *Who's who in engineering* 6: 2026. Lewis: New York.
Turner, H.M. (1930). Discussion of Stream-flow data: Its collection and use. *Journal of the Boston Society of Civil Engineers* 17(5): 243-245.
Turner, H.M. (1948). 100 years of hydraulics. *Journal of the Boston Society of Civil Engineers* 35(3): 343-354.
Turner, H.M. (1949). *Water power storage in Maine.* Harvard University: Cambridge MA.
Turner, H.M. (1949). *Water power in New England.* New England Council: Boston MA.
www.locategrave.org/l/.../Howard-M-Turner-AZ

TUTTON

10.05. 1851 Tunkhannock PA/USA
19.06. 1908 Buffalo NY/USA

Charles Harold Tutton graduated from Rensselaer Polytechnic Institute, Troy NY, in 1872 as a civil engineer. After works with two railways companies until 1874, he was engaged until 1882 in mining and general engineering at Wilkes Barre PA and at Buffalo NY, where he stayed until his death. From 1882 to 1888 he was engaged in the Lehigh Valley Railroad, in connection with canals, coal trestles, and the shore protection along Lake Erie. From 1893 he was connected with the Engineering Bureau, Buffalo NY, where he had charge of city dredging operations, river improvements, the design of the new city water works, and designing the sewer system.

Tutton was a successful engineer who contributed a number of articles to the technical press. At his time, the flow of water in pipes was investigated by both researchers and engineers, given the importance in practice. Despite the effects of roughness and of viscosity on pipe discharge were known, no general formula allowed yet to estimate the main flow parameters. A final answer came only in the 1930s based on the experiments of Ludwig Prandtl (1875-1953) and Johann Nikuradse (1894-1979) in Germany, and the two Englishmen Cyril Frank Colebrook (1910-1997) and Cedric Masey White (1898-1993), who successfully identified the relative sand roughness height and the Reynolds number as the governing pipe flow parameters. In his discussion on the 1902 paper by Gardner S. Williams (1866-1931) and George H. Fenkell (1873-1949), Tutton presented the then very recent theory of Joseph V. Boussinesq (1842-1929) on turbulent flow of water. He erroneously deduced that the velocity profile of water flow in pipes should be of elliptical shape, however. Further developments led him to the definition of a flow formula, of which particular cases included these of Manning or Dupuit.

Anonymous (1909). Charles H. Tutton. *Journal of the Western Society of Engineers* 14: 112-113.
Anonymous (1909). Charles H. Tutton. *Trans. ASCE* 62: 560-561.
Harrison, W.S. (1905). Charles H. Tutton. *Political blue book*: 44. Dau: New York. *P*
Tutton, C.H. (1899). The flow of water in pipes. *Journal of the Association of Engineering Societies* 23(4): 151-165.
Tutton, C.H. (1900). Hydraulic theories on the flow of water. *Engineering Record* 42(26): 627-628.
Tutton, C.H. (1902). Discussion of Flow of water in pipes. *Trans. ASCE* 47: 215-221.
Tutton, C.H. (1902). A proposed solution of some hydraulic problems. *Trans. ASCE* 47: 392-425.

UHL

14.10. 1880 Sebewaing MI/USA
23.12. 1963 Wellesley Hills MA/USA

William Frank Uhl received education of Michigan State College, with a BSc degree in both civil and mechanical engineering in 1902. After having been a designer of construction equipment for an iron company at Dayton OH, he joined from 1904 to 1909 as design engineer and manager the Allis-Chalmers Company at Milwaukee WI, from when he was occupied until retirement in 1962 within the hydropower industry at Boston MA, serving finally as president of this firm. He has written in 1951 a pamphlet on his former head and colleague Charles T Main (1856-1943). Uhl received the honorary DEng degree in 1949 from Tufts University, Medford MA.

Uhl was considered one of the foremost hydraulic engineers in the USA. He participated in the design of over 50 hydroelectric schemes, notably the St. Lawrence Seaway, or the Niagara Power Project. He was a board member of consultants of the Tennessee Valley Authority TVA, and served as adviser for the US Army Corps of Engineers, Vicksburg MI. He has written a number of papers on hydraulic problems. The 1912 paper deals with speed regulations of hydraulic machinery, whereas the 1918 paper considers the industrial power problems in general. Uhl was also involved in the Mongaup River scheme in New York State. Following floods in the Eastern USA, he presented in 1937 a paper on the flood conditions in New England. The two papers on 100 years of water power in the 1950s are noteworthy because the early American hydropower industry is well presented, including for instance the breast wheel at Lowell MA, or the tub wheel as a particular Francis type reaction turbine developed in France. The historical water power scheme of Holyoke MA is also highlighted.

Anonymous (1913). W. Uhl. *Michiganensian yearbook*: 137. University of Michigan. *P*
Anonymous (1964). William F. Uhl. *Civil Engineering* 34(3): 105.
Uhl, W.F. (1912). Speed regulation in hydro-electric plants. *Trans. ASME* 34: 379-432.
Uhl, W.F. (1918). Industrial power problems. *Trans. ASME* 40: 537-578.
Uhl, W.F. (1937). Flood conditions in New England. *Proc. ASCE* 63(3): 449-483.
Uhl, W.F. (1951). *Charles T. Main*, 1856-1943, one of America's best. Newcomen Society in North America: New York.
Uhl, W.F. (1952). One hundred years of water power. *Civil Engineering* 22(9): 731-735.
Uhl, W.F. (1953). Water power over a century. *Trans. ASCE* CT: 451-460.
Uhl, W.F. (1954). Power resources. *Industrial and Engineering Chemistry* 44(11): 2538-2541.

VAN CLEVE

21.10. 1868 South Amboy NJ/USA
31.03. 1934 Fort Myers FL/USA

Aaron Howell van Cleve graduated in 1890 from the Lehigh University, Bethlehem PA, with the degree of civil engineering. He entered in 1892 the employ of the Niagara Falls Power Company as draftsman. He was appointed there in 1908 consulting engineer, having previously had the charges to maintain the Niagara Development Company including sewerage and the drainage system, to complete a water works system for some 10,000 persons, the installation of a hydro-electric plant with 110,000 HP, including a 2 km long discharge tunnel and two wheel-pits 50 m deep and 150 m long. After having been engaged in private practice, he became resident engineer and manager on the construction of a 6,000 HP hydro-electric plant at Cobalt ON. From 1911 to 1916, he was assistant to Charles W. Hunt (1858-1932), then secretary of the American Society of Civil Engineers ASCE. He then took charge of the design and construction of a large plant for the Sizer Forge Company, Buffalo NY, where he was in charge of heavy marine forgings for the Allies. He resigned in 1920 due to poor health, moving to Lynn Haven FL.

In 1921 van Cleve was consulted on the 275,000 HP development in the lower gorge of the Niagara River. From 1924 to 1928 he was with the US Engineers on a survey of Tennessee River as hydraulic expert on a dam project. He was co-author with Victor Gelpke of the book entitled Turbines, and he also authored a paper on the hydroelectric development. He was characterized by a deep and sincere interest in personal welfare of his friends. Modest and self-effacing himself, he was loud in the praise of his friends, and his loyalty to them was without bounds. Shortly before his death, he stated: 'As a money maker I have not been much of a success, but I have been wonderfully fortunate in my friends; it is a joy to reflect on them'. He died of heart trouble. He was ASCE member from 1904.

Anonymous (1935). Aaron H. van Cleve. *Trans. ASCE* 100: 1742-1746.

Gelpke, V., van Cleve, A.H. (1911). *Hydraulic turbines*: Their design and installation. McGraw-Hill: New York. (*P*)

van Cleve, A.H. (1903). Utilization of water power at Niagara Falls. *Bulletin* Buffalo Society of Natural Science 8(1): 3-20.

van Cleve, A.H. (1909). The hydro-electric development and transmission lines of the Canadian Niagara Power Company. *Trans. ASCE* 62: 199-237.

http://records.ancestry.com/Aaron_Howell_Van_Cleve_records.ashx?pid=89447618 (*P*)

VAN DRIEST

16.09. 1913 Cleveland OH/USA
01.01. 2005 Placerville CA/USA

Edward Reginald van Driest received a BS degree from Case Institute of Technology, Cleveland OH, the MS degree from University of Iowa in 1938 and the PhD degree from the California Institute of Technology in 1940. He had there studied under Theodor von Karman (1881-1963) in the areas of aerodynamics and boundary layer transition. During World War II he taught at Cornell University, the University of Connecticut, and at the Massachusetts Institute of Technology MIT. In 1947 he joined the Swiss Federal Institute of Technology, ETH Zurich, studying there aerodynamics under Jakob Ackeret (1898-1981), receiving the title Sc.D. He joined the North American Aviation's NAA Missile Division, Downey CA, where he directed the Space Sciences Laboratories, designing large rockets and missiles. Subsequently, he was employed by the Rand Corporation, and was in parallel a professor at the University of Southern California, Los Angeles CA, and at California State University, Long Beach CA, until retirement.

van Driest came into contact with the European hydraulicians through his stay at ETH Zurich, and got acquainted with the boundary layer theory as formulated by Ludwig Prandtl (1875-1953). van Driest extended this approach to compressible fluids by investigating the apparent turbulent stress, proposing a generalized formula for the skin friction, including the heat transfer to a flat plate. These observations were applied to supersonic flight during his stay at NAA. The energy equation for laminar boundary layers of variable Prandtl number was extended to turbulent flows by account for the von Karman analogy between heat transfer and fluid friction. He was a Fellow of AAS, and from 1971 of the American Institute of Aeronautics and Astronautics AIAA.

Anonymous (1957). Edward van Driest. *Aeronautical Engineering Review* 16(2): 47. *P*
Anonymous (1968). van Driest, Edward R. *Who's who in America* 34: 1709. Marquis: Chicago.
van Driest, E.R. (1949). Die linearisierte Theorie der dreidimensionalen kompressiblen Unter-
    schallströmung und die experimentelle Untersuchung von Rotationskörpern in einem
    geschlossenen Windkanal. Institut für Aerodynamik *Mitteilung* 16. Leemann: Zürich.
van Driest, E.R. (1951). Turbulent boundary layer in compressible fluids. *Journal of the
    Aeronautical Sciences* 18(3): 145-160.
van Driest, E.R. (1954). Boundary layer with variable Prandtl number. *Jahrbuch* WGL: 66-75. *P*
van Driest, E.R. (1956). On turbulent flow near a wall. *J. Aeronautical Sciences* 23(11): 1007-1011.
http://209.85.129.132/search?q=cache:SiM_KRD39NwJ:cletrac.org/newbb/viewtopic.php%3Ff
    %3D33%26t%3D921+%22Edward+R.+van+Driest%22&cd=2&hl=de&ct=clnk&gl=ch

VAN DYKE

01.08. 1922 Chicago IL/USA
11.05. 2010 Stanford CA/USA

Milton Denman van Dyke obtained his BS degree from Harvard University in 1943, his MS degree from the California Institute of Technology in 1947, and his PhD degree there in 1949. He was an aeronautical engineer with the National Advisory Committee for Aeronautics NACA from 1943 to 1946, and from 1950 to 1958. van Dyke joined the Douglas Aircraft Corporation in 1948, and from 1949 to 1950 was a consulting engineer to Rand Co. In parallel he was from 1950 at Stanford University a Lecturer in aeronautical engineering, a professor there from 1959, and from 1975 to 1992 professor of applied mechanics. He was a Guggenheim Fellow in 1954, a visiting professor to University of Paris, France in 1958, and an exchange visitor of the National Academy of Sciences to the USSR in 1965. He was in 1976 elected to the National Academy of Engineering, was a member of the American Physical Society APS, and the American Academy of Arts and Science AAAS.

van Dyke's research interests included the compressible and the viscous flow theories. His book Perturbation methods has led numerous students to analyze the structure of fluid flow. Chapters are: 1. Nature of perturbation theory, 2. Some regular perturbation problems, 3. Techniques of perturbation theory, 4. Some singular perturbation problems in airfoil theory, 5. Method of matched asymptotic expansions, 6. Method of strained coordinates, 7. Viscous flow at high Reynolds number, 8. Viscous flow at low Reynolds number, 9. Inviscid singular perturbation problems, 10 Other aspects of perturbation theory. His often cited Album of fluid motion presents a collection of some 400 black-and-white photographs allowing to visualize flows observed in laboratory experiments. van Dyke founded with William Rees Sears (1913-1989) the Annual Review of Fluid Mechanics in 1969, of which he was editor until 2000.

Anonymous (1998). Milton D. van Dyke. *AIAA Bulletin* 36(2): B6. *P*

Schwartz, L.W. (2002). Milton van Dyke: The man and his work. *Annual Review of Fluid Mechanics* 34: 1-18. *P*

van Dyke, M.D. (1954). Applications of hypersonic thin body theory. *Journal of Aeronautical Sciences* 21(3): 179-186.

van Dyke, M. (1964). Higher approximations in boundary layer theory. *Journal of Fluid Mechanics* 14: 161-177; 14: 481-495; 19: 145-159.

van Dyke, M.D. (1964). *Perturbation methods in fluid mechanics*. Academic Press: New York.

VAN LEER

16.08. 1893 Magnum OK/USA
23.01. 1956 Atlanta GA/USA

Blake Ragsdale van Leer graduated as an electrical engineer from Purdue University, West Lafayette, in 1915, as a mechanical engineer from University of California, Berkeley CA in 1920, and made further studies at Universities of Caen, France, and Munich, Germany. He taught hydraulics at Berkeley CA from 1915 to 1928, was then hydraulic engineer with a pump company and in parallel Lecturer in hydraulics at George Washington University, Washington DC, until 1932, when becoming dean of engineering at North Carolina State University, Raleigh NC, from 1937. In 1944 he became president of the Georgia School of Technology, Atlanta GA. Under his leadership Georgia Tech expanded not only in activities and services, but also in size and prestige. In addition to engineering and research administrative duties, he always had found time to serve as a consultant on government, local, civic, and professional engineering projects.

van Leer invented the California pipe method of water measurement, corresponding to depth measurement at the end of a short pipe portion discharging freely into the air. If the approach flow is subcritical, then the so-called end depth depends only on the pipe diameter and the discharge. This method has a number of limitations but is still used today at locations where a standard discharge measurement is not easily accessible. van Leer was awarded a Freeman Travelling Scholarship leading him to Germany in 1928, where he improved his knowledge in hydraulics. His 1929 article reviews the then taken actions in the main European hydraulics laboratories. He was also one who demanded a National Hydraulic Laboratory, which was then erected in the 1930s.

Anonymous (1941). van Leer, Blake R. *Who's who in engineering* 5: 1829. Lewis: New York.
Anonymous (1951). Blake Ragsdale van Leer. *Mechanical Engineering* 73(5): 447.
Anonymous (1956). B.R. van Leer dies. *Engineering News-Record* 156(Feb.2): 24.
Anonymous (1956). Blake R. van Leer. *Power Engineering* 60(1): 108-110. *P*
van Leer, B.R. (1922). The California pipe method of water measurement. *Engineering News-Record* 89(5): 190-192; 93(8): 293.
van Leer, B.R. (1929). Need of a National Hydraulic Research Laboratory. *Engineering News-Record* 102(2): 68-71.
van Leer, B.R. (1929). European hydraulics. *Mechanical Engineering* 51(3): 197-201. *P*
Moody, L.F., van Leer, B.R. (1930). Fifty years' progress in hydraulics. *Mechanical Engineering* 52(4): 366-367. *P*

VAN NORMAN

05.10. 1878 Victoria TX/USA
16.01. 1954 Los Angeles CA/USA

Harvey Arthur van Norman moved with his family in 1883 to California, settling in 1887 at Los Angeles CA, where he stayed all through his life. He secured a technical education through home study. He was first steam engineer at a pumping station near Yuma AZ, and from 1899 to 1906 in this position for the Los Angeles Railroad Company, and superintendent of construction for the Los Angeles Gas and Electric Corporation. He entered the LA Water Department in 1907, placed in the charge of constructing hydro-electric stations at Owens Valley CA, whose power was used on the construction of the 400 km long Los Angeles Aqueduct project. His association with this project ultimately brought van Norman wide recognition of his ability as builder. In 1909 he was in charge of the Owens Valley and Lone Pine aqueduct sections, and in 1912 he was placed in charge of the Mojave Division, where he was responsible for the difficult task of building several siphons. The Jawbone Siphon was one of the most imposing elements of the Aqueduct, and was considered at completion the most outstanding pipe construction in the USA.

When the Aqueduct was in 1913 put into operation, van Norman supervised its operation and maintenance until 1923, when given a leave of absence to design and construct the Los Angeles outfall sewer. He was in 1925 recalled by the Board of Water and Power Commissioners as assistant chief engineer of the water system, and in 1929 appointed chief engineer and general manager, the Bureau of Water Works and Supply. He assisted in surveys and planning the route of the Colorado River Aqueduct, ultimately built by the Metropolitan Water District, Southern California. He accomplished the construction of Bouquet Reservoir in 1934, and the Mono Basin extension in 1941. When in 1943 the separate Water and Power Bureaus of Los Angeles merged, van Norman was appointed general manager and chief engineer of the entire Department. From 1944 he was until his death there advisory engineer. He was a member and president of the California Section of the American Water Works Association AWWA, serving as national director from 1939 to 1941, and ASCE member from 1922, becoming life member in 1949.

Anonymous (1954). Harvey A. van Norman. *Trans. ASCE* 119: 1367-1368; 120: xv.
Davis, M.L. (1993). *Rivers in the desert*. Harper Collins: New York.
van Norman, H.A. (1925). Open sea construction of a concrete pipe sewer outfall. *ENR* 95(8): 292.
http://eng.lacity.org/aboutus/city_engineers_hist/VanNorman.htm *P*
http://www.scvhistory.com/scvhistory/lw2055.htm *P*

VANONI

30.08. 1904 Somis CA/USA
27.12. 1999 Pasadena CA/USA

Vito August Vanoni obtained the BS degree from California Institute of Technology, Pasadena CA, in 1926, the MS degree there in 1932, and the PhD degree in 1940. He stayed from 1935 to 1947 at the research laboratories of the US Soil Conservation Service SCS, was assistant professor of hydraulics at his Alma Mater from 1942 to 1949, then associate professor there until 1955, when taking over until retirement in 1974 as professor. He was a member of the American Society of Civil Engineers ASCE, he was awarded honorary membership in 1980, and was a member of the International Association of Hydraulic Research IAHR. He was the first recipient of the Hans Albert Einstein Award in 1989, and was from 1977 a member of the National Academy of Engineering NAE.

Vanoni was considered an authority on the mechanics of sediment transport by streams and rivers. He performed research on the hydrodynamics of sediment transport, coastal engineering, and hydraulic structures. He has written seminal papers particularly on sediment transport, including the bedload equations, distribution of suspended sediment, stream degradation, bed forms, and total sediment discharge relations. His 1946 work deals with the sediment distribution demonstrating that the then assumed values result in different than the actual momentum transfer coefficients. He also was able to show that suspended sediment tends to reduce the turbulent momentum transfer and hence the flow resistance. Suspended sediment may also cause secondary flow in a river. He was awarded for this paper the 1950 Karl Emil Hilgard Prize by ASCE.

Anonymous (1950). Vito A. Vanoni. *Civil Engineering* 20(1): 43-44. *P*
Anonymous (1980). Vito A. Vanoni. *Civil Engineering* 50(10): 143.
Anonymous (2000). Vito A. Vanoni. *Civil Engineering* 70(3): 13. *P*
Vanoni, V.A. (1946). Transportation of suspended sediment by water. *Trans. ASCE* 111: 67-133.
Vanoni, V.A, Hsu, E.-Y. (1952). Turbulence and diffusion as factors in sediment transportation. *Centennial Convocation* Chicago, Paper 67: 1-31. ASCE: New York.
Vanoni, V.A., Nomicos, G.N. (1960). Resistance properties of sediment-laden streams. *Trans. ASCE* 125: 1140-1166; 125: 1172-1175.
Vanoni, V.A. (1975). River dynamics. *Advances in applied mechanics* 15: 2-89, C.-S. Yih, ed. Academic Press: New York.
Vanoni, V.A. (1984). Fifty years of sedimentation. *Journal of Hydraulic Engineering* 110(8): 1022-1057.

VAN ORNUM

14.05. 1864 Hartford VT/USA
06.11. 1943 Clayton MO/USA

John Lane van Ornum graduated in 1888 with the BSc degree in civil engineering from University of Wisconsin, Madison WI. He began his professional career with the Municipal Engineering Department of Milwaukee WI, serving as surveyor and inspector of the US Harbor Works on the coasts of Georgia and Florida. In 1891, the International Boundary Commission between the United States and Mexico was organized, for which van Ornum became chief topographer. He was appointed in 1894 instructor in the Civil Engineering Department of Washington University, St. Louis MO, serving until 1897, and

THE REGULATION

OF RIVERS

BY

J. L. VAN ORNUM, C.E., M. AM. SOC. C.E.
PROFESSOR OF CIVIL ENGINEERING, WASHINGTON UNIVERSITY,
AND CONSULTING CIVIL ENGINEER
FORMERLY UNITED STATES ASSISTANT ENGINEER

FIRST EDITION

McGRAW-HILL BOOK COMPANY, Inc.
239 WEST 39TH STREET, NEW YORK
6 BOUVERIE STREET, LONDON, E. C.
Ulr 1914

then made a trip to Europe for studying foreign engineering works. After participation in the Spanish-American War, he remained in Cuba, there preparing his 1899 paper. He was appointed professor of civil engineering at Washington University, starting a 34 years career. He was there first in charge of the design of the University sewerage system, from which period resulted his 1900 paper. His services were often sought in major engineering problems of St. Louis. He took in 1903 charge of a flow study of the Illinois River from Chicago to Grafton IL, and its effect on the Mississippi.

van Ornum was a frequent contributor to technical journals, demonstrating a wide range of interest in the fields of economic problems, technical developments, and engineering education. Examples include his notes on the 1906 San Francisco disaster, Elements of effective education, and various papers on the behaviour of concrete. His 1914 textbook The regulation of rivers, which was widely used for classwork, was another example. Its reviews were favourable, notably 'It is a pleasure to note the sound literary style', or 'a noteworthy addition to the American engineering literature'. van Ornum withdrew from active teaching in 1933, because of poor health; he was retired in 1934, living quiet his final years, and then succumbed due to a severe heart attack.

Anonymous (1943). John Lane van Ornum. *Engineering News-Record* 131(Nov.18): 733.
Anonymous (1946). John L. van Ornum. *Trans. ASCE* 111: 1559-1563.
van Ornum, J.L. (1898). Disc. Geology and its relation to topography. *Trans. ASCE* 39: 90-92.
van Ornum, J.L. (1899). Work of the Third Regiment, US Volunteer Engineers in Cuba. *Engineering News* 41(25): 400-401.
van Ornum, J.L. (1900). The proposed septic tank and sewage purification studies at Washington University. *Engineering News* 44(20): 329-331.
van Ornum, J.L. (1914). *The regulation of rivers*. McGraw-Hill: New York. (*P*)

VAZSONYI

04.11. 1916 Budapest/H
13.11. 2003 Santa Rosa CA/USA

Andrew Vazsonyi made his PhD degree in 1938 at the Technical University of Hungary, Budapest. He left then his country because of Jewish background, staying for two years in Paris, France, from where he continued to the USA, where he had short stays at Harvard University and Brown University until 1942. He returned as a teaching fellow to Harvard until 1945, heading from 1946 to 1948 the Servo-mechanism Section, North American Aviation NAA Inc., Downey CA. He was engaged from then until 1951 by the Guidance and Control Division, US Naval Ordnance, Washington DC, and passed the next 3 years with Hughes Aircraft in California. From 1954, Vazsonyi was the head of the Management Sciences Research, Ramo-Woolridge Corp., Los Angeles CA, and in parallel a Lecturer of industrial mathematics at the California Institute of Technology, Pasadena CA.

Vazsonyi was the founder of the Institute of Management Sciences, and a pioneer in the application of OR/MS in industry and business. Although always a mathematician, he started his industrial work as an engineer, and later as a manager for a small pump manufacturer. Once with NAA he was concerned with a serious design problem for the P-51 fighter aircraft. At Hughes he became interested in computers and novel concepts of their management. He particularly developed Operations Research OR. In the early 1960s he founded with two colleagues The Institute of Management Sciences TIMS. In his early years he contributed to problems of gas dynamics, supersonic aerodynamics, and numerical solution of differential equations, so that he may be considered an early individual having numerically solved the equations of fluid mechanics. His 2002 autobiography gives an account of both his life and the passionate love of mathematics.

Anonymous (1955). Vazsonyi, A. *American men of science* 9: 1999. Science Press: Lancaster.
Gass, S.I. (2004). Andrew (Andy) Vazsonyi: Operations research/management science pioneer, educator, researcher, illustrator and author helped shape profession. OR/MS 31: 18-19.
Gass, S.I., Assad, A.A. (2005). *An annotated timeline of operations research*. Kluwer: Boston P
Vazsonyi, A. (1944). Pressure loss in elbows and duct branches. *Trans. ASME* 66(4): 177-183.
Vazsonyi, A. (1945). On rotational gas flow. *Quarterly Applied Mathematics* 3(1): 29-37.
Vazsonyi, A. (2002). *Which door has the Cadillac*: Adventures of a real-life mathematician. Writers Club Press: New York. P
http://www.orms-today.org/orms-2-04/frmemoriam.html

VEATCH

25.08. 1886 Rushville IL/USA
08.10. 1975 Kansas City MO/USA

Nathan Thomas Veatch graduated with a BS degree from University of Kansas, Lawrence KA, in 1909. He there was instructor first and was assistant to the State Board of Health. After work for the American Water Works and Guarantee Co. from 1912 to 1913, he founded Black & Veatch consulting engineers, Kansas City MO, in 1914. Veatch was a recipient of the award for distinguished service in engineering from the University of Missouri, Rolla MO in 1966, and there was given in 1973 the honorary doctorate. He further was Diplomate of the American Academy of Environmental Engineers, Fellow of the American Society of Mechanical Engineers ASME, and a member of the American Public Works Association APWA, serving there as director.

During his first years after graduation, Veatch became associated as resident engineer; he was working on the construction of water plants in Kansas with Worley & Black, predecessors of Black & Veatch. From 1914 Veatch was a partner in Black & Veatch, a firm which was involved in more than 500 water works engagements, varying from small developments to the major projects in the Mid-West. In 1947, Veatch was elected president of the American Water Works Association AWWA. He had a background of more than 30 years of work in this field, and his tenure came at a time when nationwide water work construction and expansion entered in an era of unprecedented activity. He was the recipient of the 1963 Harry E. Jordan Award, being cited for 'his outstanding contributions of talent and time to the benefit of his country and his community in a multitude of public services that have added to the stature of the industry with which he is identified'.

Anonymous (1942). N.T. Veatch, Jr. *Water Works Engineering* 95(Jun.17): 730. *P*
Anonymous (1947). N.T. Veatch. *Engineering News-Record* 139(Jul.17): 89. *P*
Anonymous (1963). N.T. Veatch. *Engineering News-Record* 171(Oct.17): 20. *P*
Anonymous (1976). N.T. Veatch. *Mechanical Engineering* 98(1): 109.
Veatch, N.T. (1927). Stream pollution and its effects. *Journal of the American Water Works Association* 17(1): 58-63.
Veatch, N.T. (1948). Industrial uses of reclaimed sewage effluents. *Sewage Works Journal* 20(1): 3-11.
Veatch, N.T. (1952). The Kansas flood of 1951. *Journal of the American Water Works Association* 44(9): 765-774.

VELTROP

09.03. 1922 Bussum/NL
24.03. 2007 Evanston IL/USA

Jan Adrianus Veltrop was a leading authority on large dams, designing projects to harness the power of rivers all over the world. He was born in the Netherlands, receiving in 1949 the BS degree from the University of Delft, the MS degree in 1951 from the Rensselaer Polytechnic Institute, Troy NY, and the PhD degree from the Massachusetts Institute of Technology MIT, Cambridge MA. He worked from 1954 to 1994 at Harza Engineering, Chicago IL, by now MWH, serving as chief engineer, executive vice-president, member of the Board of Directors, and retired as senior vice-president. He took three years off from 1964 to 1967 to become dean of the engineering school at the University of Nigeria, Nsukka. His tenure ended there with his family being evacuated before the Biafra War started.

Veltrop worked on hydroelectric dam projects around the world, including Iran, India, Pakistan, Taiwan, China, Israel, and the USA. He was co-author of the 1997 ASCE book dealing with the retirement of dams, in which the data to be considered for this action are identified, and the available engineering, environmental, and economic methods for assessing, quantifying and implementing the retirement are described. The book covers data collections and their analyses, studies of retirement, and sediment management, thereby reviewing selected case studies. He was chairman of the US Committee of Large Dams, and president of the Intl. Commission on Large Dams ICOLD. Referring on the role he was taking on as president, he said: 'Traditional emphasis on purely technical matters is no longer adequate. We must address the concerns of society about dam building'. He was awarded the ASCE 1997 Rickey Medal for his contributions in hydroelectric engineering. He also served as commissioner of the World Commission of Dams in 1998. He was elected to the National Academy of Engineering NAE in 1998.

Anonymous (2007). Jan A. Veltrop. *Hydropower and Dams* 14(3): 153. *P*
ASCE (1997). *Guidelines for retirement of dams and hydroelectric facilities*. ASCE Task
Committee for Retirement of Dams and Hydroelectric Facilities: New York.
Veltrop, J.A., Wengler, R.P. (1964). Design of Karadj arch dam. *Journal of the Power Division*
ASCE 90(PO1): 1-32.
Veltrop, J.A. (1992). *The role of dams in the 21st century*. USCOLD: Denver CO.
Veltrop, J.A. (2002). Future of dams. *IEEE Power Engineering Review* 22(3): 12-15.
http://www.highbeam.com/doc/1G1-161068468.html

VENNARD

24.06. 1909 Portsmouth NH/USA
27.12. 1969 Stanford CA/USA

John King Vennard graduated from Massachusetts
Institute of Technology MIT, with a BS degree in
civil engineering in 1930, the SM degree in 1932,
and there continued until 1935 as research engineer.
He joined then the faculty of the Civil Engineering
Department of New York University, serving for 11
years as professor in charge of the fluid mechanics
program. He joined the School of Engineering of
the California Institute of Technology, Stanford CA,
in 1946, as associate professor of fluid mechanics
and there was promoted to professor in 1951.

Vennard made an international impact on his profession. He assisted for instance the
Philippine Military Academy in setting up the laboratory facilities for teaching fluid
mechanics. He also lectured for young engineers at the Ministry of Public Works in
Venezuela, receiving high honor from the Venezuelan Society of Hydraulic Engineers
for the excellence of the Spanish edition of his book, which had served to educate many
engineers in Latin America. His textbook Elementary fluid mechanics, first published in
1940, and used by over a quarter of a million students all over the world, became classic
in its field. Its sections are 1. Fundamentals, 2. Fluid statics, 3. Flow of ideal fluid, 4.
Flow of real fluid, 5. Similarity and dimensional analysis, 6. Fluid flow in pipes, 7. Fluid
flow in open channels, 8. Fluid measurement, and 9. Flow about immersed bodies.
Vennard also served as a consultant to industry and engineering firms. He was active on
many committees of the American Society of Civil Engineers ASCE, and the American
Society of Engineering Education ASEE. He was awarded its 1949 Collingwood Prize
for Juniors for his 1947 paper.

Anonymous (1949). John K. Vennard. *Civil Engineering* 19(1): 36. *P*
Anonymous (1955). John K. Vennard. *Civil Engineering* 25(6): 374. *P*
Anonymous (1970). Prof. John K. Vennard. *Civil Engineering* 40(5): 14. *P*
Anonymous (1970). John K. Vennard. *Trans. ASCE* 135: 1150-1151.
Vennard, J.K. (1940). *Elementary fluid mechanics*. Braunworth: Bridgeport CT.
Vennard, J.K., Weston, R.F. (1943). Submergence effect on sharp-crested weirs. *Engineering
        News-Record* 130(Jun.3): 814-816.
Vennard, J.K. (1947). Nature of cavitation. *Trans. ASCE* 112: 2-15; 112: 115-116.
Vennard, J.K. (1961). One-dimensional flow. *Handbook of fluid dynamics* 3: 1-30, V. Streeter,
        ed. McGraw-Hill: New York.
Vennard, J.K., Street, R.L. (1976). *Elementary fluid mechanics*, 5th ed. Wiley: New York.

VETTER

15.12. 1892 Copenhagen/DK
30.04. 1973 Carmel CA/USA

Carl Peter Pedersen Vetter was educated at Danish Royal College of Engineering, Copenhagen, from where he obtained an MS degree in civil engineering. He had then various engineering appointments both in Denmark and in China, and was from 1924 to 1925 consulting engineer in Hong Kong. He moved then to San Francisco, joining until 1930 the Great Western Project Co., becoming until 1932 assistant engineer of the Pacific Gas and Electric Co., San Francisco CA. He was appointed senior engineer of the US Bureau of Reclamation USBR in 1934, in 1942 moved as consulting engineer to the Petroleum Administration of War, and then was consulting engineer to the Cuban Government, La Habana CU. In 1944 Vetter was appointed chief, the Office of River Control, USBR.

Vetter has seen his profession with an international view, but was essentially always attached to hydraulic engineering. His return to the USBR in 1944 as chief engineer was well planned; he became in the 1950s interested in aspects of reservoir sedimentation, as is also described in the 1954 Report. Previously, while engaged on the All-American Canal in California, he was interested in so-called desilting works. It was questioned whether the Boulder Reservoir with its enormous silt storage capacity will solve forever the silt problem on the Lower Colorado? The 'turquoise waters" of the lake forming above Boulder Dam, and the 'crystal clear' stream emerging below it were the topic of wide publicity. Vetter in his 1937 paper answers the question from where originate the 70,000 tons per day by which the desilting works at Imperial Dam were designed to be removed. In the mid-1950s he was research engineer for TAMS Engineers at Baghdad IQ. Vetter was member of the American Society of Civil Engineers.

Anonymous (1948). Vetter, Carl P. *Who's who in engineering* 6: 2058. Lewis: New York.
Smith, W.O., Vetter, C.P., Cummings, G.B. (1954). *Leak Mead comprehensive survey*. US Department of the Interior: Washington DC.
Vetter, C.P. (1937). Why desilting works for the All-American Canal? *Engineering News-Record* 118(Mar.04): 321-326.
Vetter, C.P. (1940). Rubber waterstops for dams. *Engineering News-Record* 124(Feb.1): 159-161.
Vetter, C.P. (1952). 20 years of sediment work on Colorado River. 5th *Hydraulics Conf.*: 5-32.
Wiltshire, R.L., Gilbert, D.R., Rogers, J.R., eds. (2011). Hoover Dam: 75th *Anniversary History Symposium*. ASCE: Reston VA.
http://archive.library.nau.edu/cdm/singleitem/collection/cpa/id/3873/rec/19 *P*

VIESSMAN

09.11. 1930 Baltimore MD/USA
08.04. 2010 Gainesville FA/USA

Warren Viessman, Jr. graduated from Johns Hopkins University, Baltimore MD, with the BE degree in 1952, the MSE degree in 1958, and the DEng degree in 1961. He was to 1966 from assistant to associate professor at University of New Mexico, Albuquerque NM, then until 1968 professor of civil engineering and director of the water resources centre, University of Maine, Orono ME, continuing at these positions at University of Nebraska, Lincoln NE, until 1975, from when he was a senior specialist engineer until 1983. Finally, Viessman was chairman and professor of environmental engineering at the University of Florida, Gainesville FL. He was presented the 1983 Icko Iben Award of the American Water Resources Association, and the 1989 Julian Hinds Prize, the American Society of Civil Engineers ASCE. He was further cited in 2001 'for his national leadership in civil and environmental engineering education and for developing state and national water resource management legislation and policies that integrated engineering and social and political sciences'. He was elected to ASCE Honorary Membership in 2001.

While at University of Nebraska, one of Viessman's responsibilities was administering funds provided by the Water Resources Research Act of 1964. He collaborated closely with faculty, state and federal agency personnel to ensure that the Institute's research program targeted both state and national concerns. One of his research highlights was heading a project studying the development of the Elkhorn River Basin. The water uses considered included irrigation, recreation, water supply, and flood control. This project began in 1973 and finished in 1974. The Missouri River Basin Commission used results in assessing options for the Elkhorn River Basin. This pioneering work was one of the first to incorporate simulation and optimization models simultaneously.

Anonymous (1989). Warren Viessman, Jr. *Civil Engineering* 59(12): 82. *P*
Anonymous (1994). Warren Viessman, Jr. *American men and women of science* 7: 405.
Anonymous (2001). Warren Viessman, Jr. *ASCE News* (11): 9. *P*
Viessman, Jr., W., Lewis, G.L., Knapp, J.W. (1989). *Introduction to hydrology*. Harper & Row: New York.
Viessman, Jr., W. (1990). Water management issues for the nineties. *Journal of the American Water Resources Association* 26(6): 883-891.
Viessman, Jr., W., Hammer, M.J., Perez, E.M., Chadik, P.A. (2008). *Water supply and pollution control*. Prentice Hall: Upper Saddle River NJ.

VILLEMONTE

11.05. 1912 Fennimore WI/USA
16.08. 1996 Madison WI/USA

James Richard Villemonte received education from the University of Wisconsin, Madison WI, with the BS degree in 1935, the MS degree in 1941, and the PhD degree in 1949. He was a research assistant there from 1941 to 1947, in parallel instructor and assistant professor until 1947 at Pennsylvania State College, University Park PA, in charge of its hydraulic laboratory, from when he continued from associate professor to professor of civil engineering at the University of Wisconsin. In the 1950s, he was twice guest professor of applied mechanics at the Bengal Engineering College, Calcutta, India, coordinating later the collaboration in hydraulics between India and the USA. Further international activities also included assistance to Nigeria and Indonesia.

Villemonte authored one of the successful papers on the submergence effect of weirs. Weirs may be employed to measure discharge by only reading the approach flow head, provided free flow conditions occur. If the tailwater level exceeds a certain height, referred to as the modular limit, then the discharge depends on both the approach flow and the tailwater elevations. Villemonte, based on classical flow studies, related the discharge of these flows to the free-flow discharge and a submergence parameter, which was demonstrated to depend exclusively on the weir geometry. Given this simple result, his 1947 became a classic paper in hydraulic engineering. Villemonte further was active in the Hydraulics Division Executive Committee of the American Society of Civil Engineers ASCE, chairing it in the mid-1960s. He also served as chairman his Civil and Environmental Engineering Department from 1972 to 1976, after having received the 1969 Polygon Engineering Council Outstanding Instructor Award. He in addition was presented the 1978 distinguished service award from the ASCE Wisconsin Section.

Anonymous (1964). Villemonte, J.R. *Who's who in engineering* 9: 1932. Lewis: New York.
Anonymous (1966). Prof. James R. Villemonte. *Civil Engineering* 36(12): 71. *P*
Anonymous (1996). James R. Villemonte. *Trans. ASCE* 161: 572-573.
Villemonte, J.R. (1943). New type gaging station for small streams. *Engineering News-Record* 131(Nov.18): 748-750.
Villemonte, J.R. (1947). Submerged-weir discharge studies. *Engineering News-Record* 139(Dec.25): 866-869.
Villemonte, J.R., Gunaji, V.N. (1953). Equation for submerged sharp-crested weirs found applicable to 6-in Parshall flume. *Civil Engineering* 23(6): 406-407.

VOETSCH

08.08. 1877 Waldenbuch/D
07.02. 1935 Denver CO/USA

Charles (Karl) Voetsch was born in Wurttemberg County, Germany. He received in the 1890s his degree of mechanical engineer from the Technical University of Stuttgart, and later then the degree of electrical engineering from Technical University of Darmstadt. He moved in 1904 to the USA, there employed as designer of hydraulic turbines in a company at Philadelphia PA. The first turbine had 10,500 HP made for the Shawinigan Water Power Co., Quebec QC. His next work was with the Allis-Chalmers Manufacturing Co., Milwaukee WI, where he was employed until 1906. There he designed the 13,500 HP turbines for the Great Northern Power Co. During the next year, he was engaged on hydro-electric power developments with a consultant of Philadelphia. From 1911 he served the *Falkenau* Electric Construction Co., Chicago IL, on the design of the hydro-electric power plant for the Davis and Weber County Canal Co., Ogden UT.

From 1913 to 1922 Voetsch served as electrical, hydraulic, and mechanical engineer the Utica Gas and Electric Co., Utica NY. He was in charge of reconstruction of steam and hydro-electric plants, and the design of power developments. He also prepared designs for the Trenton Falls Plant operating at 80 m head. In 1927 he entered government service, and from 1928 was employed in the US Engineer Office, Chattanooga TN, as hydro-electric engineer on the power surveys of Tennessee River. This survey included a comprehensive study of the hydro-electric possibilities on the River Basin, involving some 150 storage projects. He developed methods to determine the combined power of this complicated system with the most economical use of water for maximum power generation. In 1929 Voetsch was transferred to the Madden Dam Project, the Panama Canal Zone, where he prepared designs for the power plant. In 1931 he was transferred to the US Bureau of Reclamation, Denver CO, to assist the final design for Madden Dam. He there was finally in charge on the hydraulic designs of the Boulder Canyon, and Grand Coulee Dams. Voetsch was an indefatigable worker, devoting much time to the reading of technical papers. He was ASCE member since 1917.

Anonymous (1936). Charles Voetsch. *Trans. ASCE* 101: 1660-1664.
Voetsch, C. (1924). Redevelopment: Spiers Falls hydro-electric project. *Power* 60(25): 968-972. *P*
Voetsch, C. (1932). *Investigation* of discharge capacity, pressure rise and speed regulation at Hoover Dam, Lower Arizona Conduit. Bureau of Reclamation: Denver CO. (*P*)
Voetsch, C., Fresen, M.H. (1938). Economic diameter of steel penstocks. *Trans. ASCE* 103: 89-132.

WADDELL

01.05. 1877 Hillsboro NC/USA
20.04. 1945 Biltmore NC/USA

Charles Edward Waddell received his education from the Bingham Military Academy, Asheville NC, graduating in 1894. He became in 1901 chief engineer at Asheville, where he was engaged on the design of the largest electrical heating plant by then yet attempted. He became consulting engineer in 1903, and soon was one of the leading engineers in the Southeast for hydropower and steam-generated plants. His list of constructed works includes the North Carolina Electric Power Company's system with the Weaver and Marshall hydroelectric plants, the Sunburst Arch Dam, and a large filter plant for a fiber company at Canton NC, besides a number of wooden, earthen, gravity concrete, arch, and multiple arch dams.

Waddell served in 1913 a company at Boston MA, and he also made waterways studies for the Southern Railway Company at Washington DC. In 1917, he was involved in a similar study between Cincinnati OH and Harriman TN. In recognition of his work in developing hydroelectric power in North Carolina, the degree of Doctor of Science was conferred upon him in 1925 by the North Carolina State College of Agriculture and Engineering, Raleigh NC. He was a member and chairman of the NC State Board of Engineering Examiners from 1921 to 1926, of the NC Ship and Water Transportation Commission, consulting engineer with the Tennessee Valley Authority TVA from 1936 to 1938, among other engagements. He served as consulting engineer the Department of Antioquia in Columbia on the design of hydroelectric plants. He was identified as a great leader and teacher; to young engineers he was never too busy to impart to them the benefit of his vast store of knowledge and experience. He was from 1919 member of the American Society of Civil Engineers ASCE, organizing the North Carolina Section in 1923 at Durham NC, serving as its president until 1925.

Alexander, B. (2008). C.E. Waddell. *Around Biltmore Village*: 66. Arcadia: San Francisco. *P*
Anonymous (1947). Charles E. Waddell. *Trans. ASCE* 112: 1543-1545.
Waddell, C.E. (1905). The preservation of the Southern Appalachian streams: A forest problem. *Trans. AIEE* 24(8): 839-842.
Waddell, C.E. (1907). Southern Appalachian streams. *Journal of the Franklin Institute* 164(3): 161-175.
Waddell, C.E. (1930). Hydraulic-fill core control. *Engineering News-Record* 105(25): 958-961.
Waddell, C.E. (1936). *Charles E. Waddell Papers* 1914-1934. University of North Carolina.

WADSWORTH

13.01. 1865 New Haven CT/USA
07.07. 1923 San Francisco CA/USA

Henry Hayes Wadsworth graduated in 1886 as civil engineer from the Sheffield Scientific School, Yale University, New Haven CT. He was then engaged until 1890 in Minnesota and South Dakota on the Great Northern Railways. He further moved to Lake Superior until 1895 as principal assistant engineer of Superior WI in charge of many sewer and harbour works. From 1896 he served the US Engineer Office at Duluth MN, supervising and improving Duluth and other harbours of Lake Superior, initiating his 23 years' service for the Federal Government. As US assistant engineer he was in charge of dredging operations and the construction of breakwaters and jetties, together with hydrographic surveys. For the dredging works both hydraulic and dipper dredges of large capacity were used. He also had charge of several additions to the training walls of the Duluth Ship Canal.

In 1905 Wadsworth accepted a transfer to San Francisco CA, as principal assistant engineer of the California Debris Commission, a Board of Army engineer officers who controlled the hydraulic mining operations of the State. Its work was described in the 1906 paper. During this era, the Yuba River Barrier was raised, a restraining weir forming one unit of the extensive project for river improvement, whose supervision formed a large part of Wadsworth's duties. This project was done in connection with the Barrier until its failure in 1907, as described in the 1911 paper. He was later also in charge of the control and improvement of the Sacramento and San Joaquin Rivers in Central California. Inspections led him to remote places giving him a comprehensive grasp of stream-flow conditions in the Sierra Nevada, so that he became an expert in flood control of these rivers. He made several studies on a comprehensive flood-control plan for the Sacramento Valley. His 1910 Report served as basis for the rapid and continuous development of the project. In 1920 he resigned from these duties, opening his consulting office at San Francisco, but died 3 years later after a short illness.

Anonymous (1925). Henry H. Wadsworth. *Trans. ASCE* 88: 1449-1452.

Wadsworth, H.H. (1906). Discussion of The control of hydraulic mining in California by the Federal Government. *Trans. ASCE* 57: 31-33.

Wadsworth, H.H. (1911). The failure of the Yuba River Debris Barrier and the efforts made for its maintenance. *Trans. ASCE* 71: 217-232.

Way, H.W.L. (1912). Henry H. Wadsworth. *Round the world for gold*: 326. Sampson: London. *P*

WAGNER A.J.

12.01. 1912 Hillsboro WI /USA
14.07. 1990 Knoxville TN/USA

Aubrey Joseph Wagner received the BS degree in civil engineering from the University of Wisconsin, Madison WI, in 1933. He began in 1934 his long and distinguished career with the Tennessee Valley Authority TVA, which had been founded one year before. He first worked in the navigation program, being involved in the planning and the construction of the Tennessee River navigation facilities. In 1948 he was appointed chief of the TVA Navigation and Transportation Branch, where he was responsible for both engineering and economic studies with an account on the commercial use of the then improved Tennessee waterway. He was in 1951 appointed TVA assistant general manager, and general manager in 1954, so that he was the agency's chief administrative officer.

In 1961, President Kennedy appointed Wagner to the TVA Board of Directors, and next year to its chairman, a position he held until 1978, longer than anyone else who had been chairman. During his tenure, he gained a degree of respect and personal loyalty among the TVA employees; he was called 'Mr. TVA' by many. He is remembered for his tireless energy, never letting up until he accomplished what he was trying to do. Although he could be stubborn when he was convinced he was right, he would listen to others and worked hard to gain broader perspectives. He also saw TVA's projects as tools to create good jobs and build a better quality of life for the Tennessee Valley. Today people who travel to the Valley can see the legacy he left behind. He was elected to the National Academy of Engineering in 1973, and was the recipient of many awards.

Anonymous (1968). TVA chairman A.J. Wagner. *Engineering News-Record* 180(May23): 50. *P*
Hargrove, E.C. (1994). *The leadership of the Tennessee Valley Authority* 1933-1990. Princeton University Press: Princeton.
Wagner, A.J. (1961). *TVA tributary area development program*. TVA: Knoxville TN.
Wagner, A.J. (1965). *Natural resources*: A challenge for planning. The Tennessee Planner (3): 71-72.
Wagner, A.J. (1968). TVA looks at three decades of collective bargaining. *Industrial and Labor Relations Review* 22(1): 20-30.
Wagner, A.J. (1985). *Speeches and remarks*. TVA: Knoxville.
Wheeler, W.B., McDonald, M.J. (1986). *Prisoners of myth*: TVA and the Tellico Dam 1936-1979. University of Tennessee Press: Knoxville.
Willis, W.F. (1993). Aubrey J. Wagner. *Memorial tributes* 12: 238-242. NAE: Washington. *P*

WAGNER W.E.

24.01. 1915 Mancos CO/USA
25.07. 2000 Morrison CO/USA

William Emory Wagner joined the US Bureau of Reclamation USBR in 1946, after having obtained degrees from Colorado State College and University of Colorado, Fort Collins CO. He spent his career with USBR until retirement in 1974, finally as chief of the Hydraulics Branch of the General Research Division. He received then the Meritorious Service Award from the US Department of Interior.

Wagner contributed a number of outstanding papers to the knowledge of spillway flow. His 1956 paper deals with the so-called morning glory spillway, corresponding to a circular-crested intake connected with a vertical shaft and a 90° bend at its base, by which flood floods are discharged to an outlet structure. These spillways may be operated both free and submerged, but it was already found out in the 1940s that only free flow is adequate because of hydraulic control, cavitation damage, and unstable flow features. Despite the morning glory spillway was introduced around 1900, and there were notable research works for instance by Ford Kurtz (1885-1956) in 1925 or by William J.E. Binnie (1867-1949) in the 1930s, Wagner considered the flow features of the vertical sharp-crested circular pipe to find the lower and upper jet trajectories under fully-aerated conditions. He also determined the discharge coefficient versus the relative head on the weir, the effect of vacuum below the intake, and the effect of the approach flow velocity, resulting in an almost complete study of these flows, provided there is absolutely no rotational approach flow component. He then proceeded to the definition of the crest geometry, in analogy to the standard-crested weir with a straight weir crest. This paper was awarded the 1957 J. James Croes Medal from ASCE.

Anonymous (1957). William E. Wagner. *Civil Engineering* 27(10): 735. *P*
Anonymous (1974). William E. Wagner. *Civil Engineering* 44(10): 116.
Harleman, D.R.F., Wagner, W.E., Barnes, S.M. (1963). Bibliography on the hydraulic design of spillways. *Journal of the Hydraulics Division* ASCE 89(HY4): 117-139.
Wagner, W.E. (1951). Hydraulic model studies of the outlet works, Medicine Creek Dam, Frenchman-Cambridge Diversion, Missouri River Basin Project. Hydraulic Laboratory *Report* HYD-273. USBR: Denver CO.
Wagner, W.E. (1956). Morning glory shaft spillways: Determination of pressure-controlled profiles. *Trans. ASCE* 121: 345-368; 121: 380-384.
Wagner, W.E. (1967). Glen Canyon Dam diversion tunnel outlets. *Journal of the Hydraulics Division* ASCE 93(HY6): 113-134.

WALKER A.W.

17.03. 1882 Marlboro MA/USA
15.06. 1943 Fairfield MT/USA

Albert Willard Walker graduated in 1905 with the BSc sanitary engineering degree from Massachusetts Institute of Technology MIT, Cambridge MA. He was then engaged by the US Bureau of Reclamation USBR as engineering aide, staying continuously at Great Falls MT. He was first in charge to the Belle Fourche Project in South Dakota, and there served as inspector on the construction of Belle Fourche Dam, an earth-fill structure. He was transferred in 1912 to the Milk River Project in Montana, serving as assistant engineer on the construction of the office building at Malta MT, and on the design of Dodson North Canal. In 1913 he was placed to the Huntley Project MT, as engineer in charge on the design and construction of an extensive irrigation system consisting of deep tile drains, thereby entering the field of groundwater flow and soil mechanics, the topics to which he contributed significantly in the future.

In 1916 Walker was transferred to the Grand Valley Project, Grand Junction CO, where he prepared plans for the drainage of the Grand Valley Drainage District. Later, he conducted similar studies and supervised the construction of the drainage system for the Grand Valley Project. From 1921 to 1925 he was employed on the Newlands Project in Nevada, as engineer in charge of all phases of a drainage program involving deep open drains. He also served as special assistant on soil and drainage matters for the Minidoka, Boise, Klamath, and the Newlands projects. From 1926 to 1929 he directed drainage work, and was engaged in the reconstruction of the Truckee Canal. From 1929 he was construction engineer on the Sun River Project, Montana, and in 1931 was transferred to the Greenfields Irrigation District, serving as district manager under an arrangement with USBR. During this period the project was doubled in its irrigation area, the main canals were enlarged and improved, and the necessary drainage work was performed. Walker's sound judgment and solution of complex problems were admired. He was a man of simple tastes, having extreme patience and a calm demeanor under complicated situations. He was member of the American Society of Civil Engineers ASCE from 1910.

Anonymous (1944). Albert W. Walker. *Trans. ASCE* 109: 1580-1582.
Walker, A.W. (1924). Drainage construction on Newlands Reclamation Project. *Engineering News-Record* 93(10): 382-386. *P*
Walker, A.W. (1924). Rapid low-cost excavation of drain ditches. *Engineering News-Record* 93(11): 426-428.

WALKER H.B.

13.04. 1884 Macomb IL/USA
27.07. 1957 Davis CA/USA

Harry Bruce Walker graduated with the BSc degree in civil engineering in 1910, and in 1920 with the civil engineering degree from Iowa State College, Ames IA, completing additional graduate work at Kansas State College of Agriculture, Manhattan KS, in 1925. After initial experience as topographer, he served as drainage engineer for Humboldt County IA from 1909 to 1910. He was then drainage and irrigation engineer at the Kansas State College of Agriculture, and from 1913 state irrigation engineer of Kansas. Walker was appointed in 1921 professor of agricultural engineering at Kansas State College, from when he was professor of agricultural engineering until his retirement in 1950 at the Experiment Station, University of California, Davis CA. He directed from 1927 to 1928 a survey for the US Department of Agriculture. The US President appointed him in 1929 a delegate to the World Engineering Conference in Tokyo JP.

Walker distinguished himself as engineer, research worker, teacher, and administrator. He gained international reputation for his concept of the value of engineering and science in advancing agriculture. He demonstrated himself as leader with vision and unusual energy. To the minds of his students and his staff, he imparted the need for rigorous intellectual honesty, a reverence for truth, an open-minded approach to new situations, and a balanced judgment. During the twenty-one years at Davis CA, as chairman of the Committee on the Relation of Electricity to Agriculture, he saw the proportion of electrified farms in California increase from 38 to 96%. Electricity, besides raising the standard of living in the farm home, supplied one half the total energy requirement on the farm. In 1939 he was awarded the John Deere Medal by the American Society of Agricultural Engineers for his contribution The application of science and art to the soil. He twice served as president of the American Society of Agricultural Engineers ASAE, which elected him to the honorary membership in 1957. Appointed by the American Society for Engineering Education in 1951, he served on a board to make a study of engineering education in Japan. He was a member of the American Society of Civil Engineers ASCE from 1921, becoming Life Member in 1949.

Anonymous (1959). Harry B. Walker. *Trans. ASCE* 124: 1080.
Walker, H.B., Bainer, R. (1949). *Survey and problems in agricultural engineering*. Berkeley CA.
http://texts.cdlib.org/view?docId=hb3r29n8f4&doc.view=frames&chunk.id=div00026&toc.depth=1&toc.id=
http://sunsite.berkeley.edu/uchistory/archives_exhibits/in_memoriam/catalog/walker_harry_bruce.html *P*

WALLACE

10.09. 1852 Fall River MA/USA
03.07. 1921 Washington DC/USA

John Findlay Wallace received in 1882 the degree of engineering from Wooster College, Wooster OH. He then entered the US Corps of Engineers focusing on river and harbour projects on Upper Mississippi River. Over the next 15 years, he was promoted to chief engineer and finally general manager of the Illinois Central Railroad. He was in 1904 appointed by President Theodore Roosevelt chief engineer of the Panama Canal Project. This job earned him the second highest salary of 25,000 US$, the highest of any government employees right after the president. Wallace realized immediately after arrival on the Isthmus that the harsh terrain and the climate conditions were a serious obstacle to his project. The task ahead of him seemed almost impossible, to dig a canal 80 km long and 10 m below sea level stretching from the Atlantic to the Pacific Ocean. He would have to cut through dense jungle, control parts of the Chagres River, and haul away sections of the Culebra Mountain. Wallace asked for additional time to survey the area, but Roosevelt's directive to 'Make the dirt fly' prevented any further delay.

Wallace attempted to excavate the spoil as quickly as possible to meet the government's demands, but flooding and landslides caused repeated setbacks. The delays damaged the morale among workers, who already suffered from terrible food and living conditions. Logistical problems added to the inefficiency. At project start, the workers only had at their disposal the antiquated machines left behind by the Frenchmen. Wallace ordered new equipment from the USA, but the giant steam shovels excavated more spoil than the existing train infrastructure could remove, forcing him to operate them at 25% of their peak efficiency only. He was also faced with bureaucratic challenges from the Isthmus Canal Commission, who had to approve each of his decisions made in the Canal Zone. Wallace resigned abruptly in June 1905. His successor was John F. Stevens (1853-1943), who started his tenure that Wallace had failed to do: stop digging.

Anonymous (1921). John F. Wallace. *Engineering News-Record* 87(2): 83-84. *P*

Davis, G.W., Stevens, J.F., Wallace, J.F. (1906). *Report* of the Board of consulting engineers for the Panama Canal. Government Printing Office: Washington DC.

Navarro, B. (2010). *The emergence of power*. Xlibris Corp.: on-line.

Parker, M. (2008). *Hell's Gorge*: The battle to build the Panama Canal. Arrow Books: London.

Wallace, J.F. (1906). *Investigation of Panama Canal matters*. The Committee: Washington DC.

http://www.pbs.org/wgbh/americanexperience/features/general-article/panama-engineers/ *P*

WALTER D.S.

18.08. 1904 Greeley CO/USA
07.12. 1971 San Diego CA/USA

Donald Scott Walter was educated at the University of Colorado, Boulder CO, graduating with the BS degree in civil engineering in 1926, and obtaining the CE degree in 1957. He entered the US Bureau of Reclamation USBR in 1938, working as junior engineer and assistant engineer at Ellensburg WA until 1930. He was then from 1930 assistant chief inspector at the Boulder Dam Project until 1934, and continued in the same position until 1939 at the Coulee Dam Project. He was field engineer for the Friant Dam until 1941, returning in this position to the Boulder Dam until 1943, from when he was until 1948 consulting engineer for the Anderson Ranch Dam, and finally became regional engineer of USBR, stationed at Boise ID until his retirement in 1959.

Walter was for nearly 40 years in the service of the USBR. Since 1949 he was regional engineer for the Northwest District, directing the Columbia Basin and Yakima Projects in Washington State. Previously he had been involved in the Hungry Horse Dam in Montana, where the first large morning glory spillway had been successfully installed. Instead of a standard frontal spillway, this spillway type is based on a vertical shaft whose base is connected with an almost horizontal tunnel, discharging the water into the tailwater. A radial unsubmerged intake corresponding to the morning glory flower is thereby positioned at the shaft top resulting in water flow along the shaft walls, and an annular air core at its center. Walter was further involved in the design and construction of the Deschutes and Crooked River Project in Oregon State, and both the Palisades and Minidoka schemes in Idaho. He finally was construction engineer on Anderson Ranch Dam, a multipurpose structure providing benefits of irrigation, power, and flood and silt control. The 135 m high dam on the South Fork of Boise River was the world's highest earth-fill dam at the time of completion in 1950. The power plant has with two units an installed capacity of 27,000 kW.

Anonymous (1959). Walter, Donald S. *Who's who in engineering* 8: 2576. Lewis: New York.
Anonymous (1959). Donald S. Walter. *Civil Engineering* 29(11): 22. *P*
Walter, D.S. (1941). Engineering innovations at Friant Dam. *Reclamation Era* 31(12): 309-311.
Walter, D.S. (1957). *Rehabilitation and modification of spillway and outlet tunnels and river channel improvements at Hoover Dam*. University of Colorado: Boulder CO.
Wiltshire, R.L., Gilbert, D.R., Rogers, J.R., eds. (2010). *Hoover Dam*: 75th anniversary history symposium. ASCE: Reston VA.

WALTER R.F.

31.10. 1873 Chicago IL/USA
30.06. 1940 Fresno CA/USA

Raymond Fowler Walter graduated from Colorado State College of Agriculture and Mechanical Arts, Fort Collins CO, in 1893, with the BSc degree in civil engineering. He was engaged to enlarge Terry Lake first, and then designed and constructed the Independence Canal for the Big Horn Irrigation Co. He became in 1895 junior engineer at Greeley CO, where he was involved in the design of Reservoir 6 for the North Poudre Irrigation Company. From 1899 to 1901 he was there city engineer, and then was in charge of the Seven Lakes Reservoir at Loveland CO, and the Fossil Reservoir at Fort Collins CO. He designed many irrigation works, including the Union Reservoir at Longmont CO, and the McClellan and Law Reservoir at Windsor CO.

Walter was appointed assistant engineer of the US Reclamation Service, assigned to the Belle Fourche Project, South Dakota, in 1903. He then progressed to the position of supervising engineer at Denver CO, in charge of all projects of the US Bureau of Reclamation USBR in Colorado, Wyoming, Nebraska, and Utah. In 1915, then senior engineer at El Paso TX, he supervised the Rio Grande Project, returning in 1916 to Denver, where he concentrated on administrative duties and engineering matters. Due to his careful scrutiny of all operations, Walter earned the title 'watchdog of the treasury'. In 1925 he was appointed chief engineer, with supervision of all construction activities of USBR. To increase work efficiency and promote excellent design and construction, he organized the Construction Organization, which was recognized for the outstanding ability and for creation of new methods on great irrigation projects. During his service, he was instrumental in building the highest dams in the world, including Boulder, Shasta, and Grand Coulee. It was stated that Walter collected and opened the bids for the construction contract for Hoover Dam in 1931. He was by nature friendly and courteous, and he shielded his feelings by a shell of gruffness. Death by a heart disease ended the career of an outstanding public servant. He was ASCE member from 1909.

Anonymous (1924). Added reclamation responsibility. *Engineering News-Record* 93(25): 1011. *P*
Anonymous (1947). Raymond F. Walter. *Trans. ASCE* 112: 1546-1548.
USBR (2008). *Reclamation*: The Bureau of Reclamation, Centennial Symposium. Denver CO.
Walter, R.F. (1904). South Dakota investigations. 1st Conf. *Engineers of Reclamation Service*: 211-213, F.H. Newell, ed. Government Printing Office: Washington DC.
http://www.hooverdamstory.com/walter.htm *P*

WALWORTH

18.11. 1807 Canaan NH/USA
28.04. 1896 Boston MA/USA

James Jones Walworth was a brother of the famous inventor Caleb Clark. Walworth founded in 1841 with his brother-in-law Joseph Nason the Walworth, Nason & Co. at Boston MA, which became in 1872 the Walworth Manufacturing Co. In 1844, when the two partners installed a system for heating the Old Astor House on the Broadway of Boston, plumbers and fitters were the mechanic who not only drew a bath but also heated it. When going through the illustrated catalogues of Walworth, Nason & Co. of equipment for plumbers and steam fitters in 1870, then it will become evident that its connection with heating and cooking is almost as important as its work was for the baths. The daily life of the 19th century was significantly changed by the introduction of these benefits, which depended mainly on the availability of water and steam.

Nason and also Walworth are known as the fathers of steam heating. Their equipment was in the 1840s a novelty. Nason was patent holder of the globe and angle valve, the steam trap, cast and taper joint pipe fittings, and the "radiator," which name he created. Both men concentrated on a proposed line of iron fittings, a difficult and expensive process. In 1864, Nason and Walworth took over the Totoket Company of Branford CT. This Company was originally incorporated in 1854 for manufacturing of hardware and other goods from wrought, cast, and malleable iron. The plant was ideally located between New York and Boston on a busy waterfront and next to a newly built railroad. Walworth became company president, while Nason returned to New York, to run the Joseph Nason & Co., although he continued to cooperate with Walworth. It produced iron castings shipped to the Walworth Co. for finishing. The Walworth Co. continued as a separate company until 1962.

Anonymous (1895). Walworth, James J. *Engineering Record* 33: 380.
Romaine, L.B. (1960). *A guide to American trade* 1744-1900. Bowker: New York.
Walworth Manufacturing Co. (1870). *Catalogue and price list* of wrought and cast iron pipe, steam and gas pipe fittings, brass and iron steam valves and cocks, supplies, and other articles incidental to steam and gas engineering. James J. Walworth & Co.: Boston MA.
Walworth Manufacturing Co. (1878). *Illustrated catalogue and price list*. Gunn, Bliss & Co.: Boston MA.
http://doddcenter.uconn.edu/asc/findaids/MIF/MSS19820004.html
http://www.findagrave.com/cgi-bin/fg.cgi?page=gr&GRid=42973861 *P*

WARD

28.12. 1888 Belleville KS/USA
27.04. 1959 Riverside CA/USA

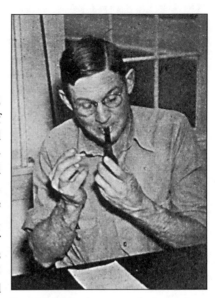

Richard Benskin Ward obtained in 1913 the BS degree in civil engineering from the University of Kansas, Lawrence KS. Until 1925 he was engaged in railroad projects in Alaska, construction of earth-fill dams, and irrigation schemes in Montana and Wyoming for the US Bureau of Reclamation USBR. Until 1931 he was then a design engineer with the County Sanitation District of Los Angeles County. He was in 1931 employed by the Metropolitan Water District of Southern California, preparing estimates for the Colorado River Aqueduct. He became in 1935 resident engineer on the construction of the dam and dike of the Cajalco Reservoir, todays Lake Mathews. In 1938 Ward accepted a position with the USBR at Denver CO. During the next eleven years he served as construction engineer on the Green Mountain Dam, Kremmling CO, location engineer on the San Diego Aqueduct, and construction engineer on four earth-fill dams for the Horsetooth Reservoir, Fort Collins CO. He was construction engineer from 1949 to 1954 on Falcon Dam and Power Plant on the Rio Grande River, Laredo TX, for the US International Boundary and Water Commission. In 1954 he returned to USBR as engineer advisor to the Snowy Mountains Hydro-Electric Authority in New South Wales, Australia. After return to the Metropolitan Water District in 1956, he was in charge of the construction work required to enlarge Lake Mathews.

Ward was all through his professional career intimately associated with dam design and construction, and thus may be considered among these engineers who have greatly added to the development of America's West in terms of water supply, irrigation, flood control, and energy supply. He was elected associate member of the American Society of Civil Engineers in 1933, and ASCE Member in 1938. He was presented a certificate of life membership at a meeting of the Los Angeles Section early in 1959. This was made by Julian Hinds (1881-1977), a friend and associate of long standing, who was assistant chief engineer of the Metropolitan Water District during Ward's assignment to the Cajalco Reservoir Project.

Anonymous (1941). R.B. Ward. *Engineering News-Record* 126(Jan.9): 65. *P*
Anonymous (1948). Ward, Richard B. *Who's who in engineering* 6: 2091. Lewis: New York.
Anonymous (1959). Richard B. Ward. *Civil Engineering* 29(7): 530.
Ward, R.B. (1921). *Clark Fork Secondary Project*, Montana-Wyoming. USBR: Denver.
http://www.ibwc.state.gov/Organization/Operations/Field_Offices/Falcon.html

WARING F.H.

05.10. 1890 Saratoga NY/USA
25.12. 1983 Columbus OH/USA

Frederick Holman Waring obtained the BSc degree in civil and sanitary engineering from the Worcester Polytechnic Institute, Worcester MA, in 1912. Until 1914 he joined the firm Metcalf & Eddy, Boston MA, as assistant engineer, from when he was until 1916 filtration chemist, the Municipal Engineering Department, the Panama Canal Zone. Upon return to the USA, Waring was then until 1926 assistant engineer, the Ohio Department of Health, Columbus OH, taking over until his retirement in the early 1960s as chief engineer. Waring was member of the American Society of Civil Engineers ASCE, and the American Water Works Association AWWA. He was Fellow of the American Public Health Association APHA.

The philosophy of Waring, and part of his legacy, was Reasonableness. This word was the cornerstone of his success in Ohio. The old-school sanitary engineer stepped into practical solutions to the problems offered during which environmental control has been greatly developed. His record of accomplishment, measurable in water plants and waste water treatment plants, and to a lesser degree in the installation of other public health projects, rated him to esteem of other state sanitary engineers. He was among the nation's pioneers in sanitary engineering. On Ohio's rivers, too many small cities still discharged in 1960 waste water to the next town downstream. Waring's philosophy of reasonableness was paying off, because in all Ohio there were in 1960 only 200,000 people not yet connected to a treatment plant. The Ohio River Valley Water Sanitation Commission was formed in 1948; it was and still is a promotional and coordinating agency, having legal authority. Waring described the regulatory process to prove the reasonableness of it. He stated how well the philosophy of pollution control was finally made understandable to the public. Waring seemed to know each sewage treatment plant of his state like the back off his hand. He kept his fingers on new processes, research, and development. Technical developments as the aerobic digestion plants gave him satisfaction. But his greatest satisfaction came from his engineering success.

Anonymous (1959). Waring, Frederick H. *Who's who in engineering* 8: 2585. Lewis: New York.
Anonymous (1960). How to stop water pollution. *Engineering News-Record* 164(Mar.3): 55-58. *P*
Waring, F.H. (1930). *Types of water supply and water treatment in Ohio cities*. Columbus OH.
Waring, F.H. (1955). *Public water supplies and control of water pollution in Ohio*. Ohio State University: Columbus OH.

WARING G.E.

04.07. 1833 Proud Ridge NY/USA
29.10. 1898 Havana/CU

George Edwin Waring was educated at College Hill, Poughkeepsie NY, and then studied agriculture. His first book The elements of agriculture was published in 1854. In 1857 he was appointed drainage engineer of Central Park, New York City. He served in the Civil War taking part in expeditions to Mississippi. Moving in 1867 to Newport RI he began in farm management but was drawn by sanitary engineering. His 1868 book Earth closets describes a system of handling human waste conveniently. He applied his ideas first at Ogdensburg NY. In 1879, as a member of the National Board of Health, he investigated the yellow fever epidemic of Memphis TN, which had caused 5,000 deaths.

Waring then built the first major sewerage system designed to exclude all but sewage from the conduits. The Memphis System became known, but it was controversial. Great sanitary engineers including Rudolph Hering (1847-1923) thought it a useful option, but maintained that the English Separate System or the Combined System allowing for rainfall water to tie into a sewer had also their place. Waring was careful to follow up on the performance of sewer systems. However collecting sewage was one problem, the other was its treatment and disposal. Waring in 1885 wrote '… the waste organic matter of our domestic life and of the industries must be withheld from the waters into which our sewers discharge'. Using a variety of mechanical devices and chemicals he explored precipitation, decantation, distillation, and disinfection, but became convinced that the disposal by surface or sub-surface must be the goal. His experiments started in 1869, yet his idea did not meet with the success of the Memphis System. In 1896 he finally stated that the salvation from sewage disposal is not in chemical disinfection, which should be avoided as a clog in the wheel of nature's beneficent processes. Even today, Waring's dream of natural recycling wastes is still under research.

Anonymous (1976). George Waring: Giving sanitation status. *Civil Engineering* 46(7): 81-82.
Fitzsimons, N. (1971). Pollution fighter: George Waring. *Civil Engineering* 41(7): 76-77.
Waring, G.E. (1854). *The elements of agriculture*. Appleton: New York.
Waring, G.E. (1867). *Draining for profit, and draining for health*. Orange & Judd: New York.
Waring, G.E. (1868). *Earth closets*: How to make them and how to use them. New York.
Waring, G.E. (1875). Sanitary drainage. *North American Review* 137(320): 57-67.
Waring, G.E. (1893). Memphis System of sewerage. *American Public Health Ass*. 18: 153-168.
http://popartmachine.com/item/pop_art/LOC+1142398/%5BCOL.-GEORGE-E.-WARING *P*

WARNER

21.02. 1858 New York NY/USA
17.06. 1927 Burlingame CA/USA

Edwin Hall Warner attended the University of the City of New York, the later New York University, registered in its Polytechnic Engineering Department of the School of Applied Science. From 1884 to 1890 he was engaged by the Mexican Central, and the Union Pacific Railways, from when he had his private practice at Seattle WA, developing mining properties, and hydraulic power investigations. From 1899 to 1903 he was engaged in a private practice at Republic OR, as engineer for the Republic Mining Co., from when he served as the principal assistant engineer for the Columbia Improvement Company on the Puyallup power development at Electron WA, installing a 20,000 HP unit. In 1904 Warner moved to California as chief engineer at Venice CA, supervising the construction of canals and bridges. In this period he was associated with James D. Schuyler (1848-1912), becoming in 1905 chief engineer of the Tri-State Land Company, Scottsbluff NE, a project for which Schuyler was consulting engineer. Warner was in charge of the construction of 100 miles of canal for a discharge of 50 m^3/s. From 1906 he was as assistant chief engineer in charge of the *Necaxa* Dam in Mexico, for the Mexican Light and Power Company, where he was responsible for the dam, the tunnel, conduits, and machinery.

From 1907 to 1910 Warner was then engaged in private practice at Los Angeles CA, designing irrigation projects and ocean piers. Until 1916 he then was the constructing engineer for the Southern California Edison Company, Los Angeles, including the construction of the 65,000 HP generating station at Long Beach CA and a large power tunnel on Kern River. After private practice, Warner served as construction engineer from 1919 to 1921 for the Kerckhoff Dam on the San Joaquin River, and the Snow Mountain Dam on Eel River, both early important concrete dams.

Anonymous (1908). *Construction of the Santa Monica Pier*. Public Library: Santa Monica. (*P*)
Anonymous (1929). Edwin H. Warner. *Trans. ASCE* 93: 1913-1915.
Warner, E.H. (1905). The hydraulic plant of the Puget Sound Power Company. *Trans. ASCE* 55: 228-261.
Warner, E.H. (1907). Discussion of The Necaxa Plant of the Mexican Light and Power Company. *Proc. ASCE* 33(1): 89-91.
Warner, E.H. (1923). Discussion of Tentative plan for the construction of a 780-foot rock-fill dam on the Colorado River, at Lee Ferry AZ. *Trans. ASCE* 86: 219-221.

WARNOCK

23.04. 1903 Honey Creek IN/USA
26.12. 1949 Denver CO/USA

Jacob Eugene Warnock graduated in 1925 with a BS degree in civil engineering and received in 1934 the CE degree from Purdue University, West Lafayette IN. In 1939 he graduated with the MS degree from University of Colorado, Fort Collins CO. He joined in 1931 the Bureau of Reclamation USBR, Denver CO, as an associate engineer and soon later headed its hydraulic laboratory, a position which he held until death. He was a member of the International Association of Hydraulic Research IAHR, and an active member of the American Society of Civil Engineers ASCE. He was appointed shortly before his untimely death to its National Executive Committee of the Hydraulics Division.

Warnock as laboratory head was involved in numerous dam projects of the 1930s and 1940s, including the Hoover, Grand Coulee, Shasta or the Friant Dams, as also large irrigation projects in the West. He had attained a national and international reputation as an authority in hydraulic engineering. He contributed articles on hydraulic modeling to books, including Hydraulic models, co-authored by a number of American experts, or the chapter Hydraulic similitude in the 1950 book edited by Hunter Rouse (1906-1996). It became evident in the 1930s that hydraulic laboratory models were an important addition to the then existing approaches, namely prototype structures and hydraulic computations. A large number of specific problems were amenable only by laboratory test observations, provided the fundamental model rules were satisfied.

Anonymous (1946). Jacob E. Warnock. *The Reclamation Era* 32(7): 163. *P*
Anonymous (1950). Hydraulic engineer Jacob Warnock dies. *The Reclamation Era* 36(2): 43. *P*
Anonymous (1950). Jacob Eugene Warnock. *Civil Engineering* 20(2): 150.
Stevens, J.C., Bardsley, C.E., Lane, E.W., Straub, L.G., Wright, C.A., Tiffany, J.A., Warnock, J.E. (1942). Hydraulic models. *Manuals of engineering practice* 25. ASCE: New York.
Warnock, J.E. (1936). Models guide work on Western dams: Experiments aid design of Grand Coulee. *Civil Engineering* 6(11): 737-741.
Warnock, J.E. (1938). *Report* on inspection trip to correlate present hydraulic design practice and the operations of structures in the field. USBR: Denver CO.
Warnock, J.E. (1947). Cavitation in hydraulic structures, a symposium: Experiences of the Bureau of Reclamation. *Trans. ASCE* 112: 43-58; 112: 119-123.
Warnock, J.E. (1950). Hydraulic similitude. *Engineering hydraulics*: 136-176, H. Rouse, ed. Wiley: New York.

WARREN G.K.

08.06. 1830 Cold Springs NY/USA
08.08. 1882 Newport RI/USA

Gouverneur Kemble Warren graduated from the US Military Academy in 1850, being appointed brevet lieutenant in the Corps of Topographical Engineers. He was assistant engineer on the delta survey of Mississippi River until 1854, becoming then board member of the canal improvement around the falls of Ohio River. He also served as head of surveys for the improvement of the Rock Island and the Des Moines Rapids. He was in 1855 chief topographical engineer of the Sioux Expedition, promoted to first lieutenant in 1856. Warren created in 1857 the first comprehensive map of the USA west of Mississippi River, which required extensive explorations of the vast Nebraska Territory, including Nebraska, North and South Dakota, and parts of Montana and Wyoming. Minnesota River Valley was found to be much larger than anticipated from low-flow river data. Warren first explained the hydrology of this region in 1868. He was from 1859 to 1861 assistant professor of mathematics at the Military Academy, from when he was in war service until 1865, finally in the rank of major, the US Army Corps of Engineers.

Warren was then member of the Board of Engineers to examine a canal at Washington DC, as superintending engineer of surveys and improvements of the Upper Mississippi. He supervised in 1869 the construction of the Rock Island Bridge across Mississippi River. He also was in charge of river and harbor works for the Corps of Engineers in the Upper Mississippi Valley, along the Atlantic Coast, and around the Great Lakes. He was promoted to lieutenant colonel of the Corps in 1879. A statue of Warren was unveiled at Gettysburg in 1888. He was member of the American Society of Civil Engineers ASCE, the American Association for the Advancement of Science, and the National Academy of Sciences. Its G.K. Warren Prize is awarded each four years.

Rhea, G.C. (2001). *Happiness is not my companion*: The life of General G.K. Warren. Indiana University Press: Bloomington. *P*

Warren, G.K. (1858). *Explorations in Nebraska*: Preliminary report of lieutenant G.K. Warren, topographical engineers to captain A.A. Humphreys. Harris: Washington DC.

Warren, G.K. (1874). *An essay concerning important physical features exhibited in the Valley of the Minnesota River*. Government Printing Office: Washington DC.

http://en.wikipedia.org/wiki/Gouverneur_K._Warren *P*

http://www.bridgemanart.com/asset/269487/American-School-19th-century/General-G.K.-Warren-at-the-signal-station-on-Littl?search_context. *P*

WARREN J.G.

12.09. 1858 Buffalo NY/USA
02.11. 1937 Buffalo NY/USA

James Goold Warren graduated from the US Military Academy, West Point NY, in 1881, being promoted to second lieutenant, US Army Corps of Engineers. He served then at Willets Point NY, graduating in 1884 to first lieutenant, eventually becoming assistant to the officer at Chattanooga TN. After service as instructor in civil and military engineering, he was appointed until 1891 secretary of the academic staff of the Engineer School, Willets Point NY. Warren became district engineer until 1898 at Louisville KY, and from then at Milwaukee WI until 1905. In 1906 he was appointed member of the Mississippi River Commission serving until 1919. In 1912, then lieutenant-colonel, he was appointed district engineer at Buffalo NY, having charge of both the Buffalo and Oswego Engineer Districts. In the same year he was appointed to the rank of colonel division engineer of the Great Lakes Division, supervising all river and harbour activities of the Great Lakes and the St. Lawrence River until 1919.

In 1917 Colonel Warren assumed charge of the Investigation of diversion of water from the Great Lakes and Niagara River, covering the phases of water diversion, sanitary and power purposes, and the preservation of the scenic beauty of Niagara Falls. His report was considered as authoritative text on these subjects. His superior stated: 'The report is the only comprehensive and thorough investigation of all these subjects ever made and possesses great value, not only from the technical, but also from the very full historic presentation'. After retirement in 1922, Warren continued his engineering practice at his home until his death. During his 45 years of active service with the Corps, Warren had earned the reputation of outstanding member of its organization. During his long term as engineer of the Great Lakes Division, or of the Mississippi River Commission, he gained experience and became an authority on inland navigation and inland waterways of the USA. He was member of the American Society of Civil Engineers ASCE.

Anonymous (1938). James G. Warren. *Trans. ASCE* 103: 1906-1908.

Warren, J.G. (1900). *Examination and survey of Sturgeon Bay and Lake Michigan Ship Canal Wisconsin*. Washington DC.

Warren, J.G. (1911). *Niagara River*. Committee on Rivers and Harbors: Washington DC.

Warren, J.G. (1913). *Compiled reports of the Intl. Waterways Commission*. Parmelee: Ottawa.

Warren, J.G. (1921). *Diversion of water from the Great Lakes and Niagara River*. US Army Corps of Engineers: Washington DC. (*P*)

WASHINGTON

24.09. 1883 Travis County TX/USA
04.07. 1954 Brownsville TX/USA

Walter Owen Washington graduated in 1904 from the University of Texas, Austin TX, with the BSc degree in civil engineering. He was then engaged in engineering works in connection with irrigation, land development, and large flood control projects. He was from 1920 to 1939 engineer of Cameron County TX, supervising the construction of its flood control system, and the Lower Rio Grande Valley. Until 1940 he was construction engineer for a large irrigation program in Willacy County TX, supervising in 1941 the building of the war-time facilities. From 1943 to 1945, he was engaged in similar work at Corpus Christi TX, becoming then until 1950 senior partner of the firm Washington and Ruff.

The flood control of the Lower Rio Grande resulted in a series of treaties between the USA and Mexico. The 1906 Treaty provided for the construction of Elephant Butte Reservoir in New Mexico. The Rio Grande Rectification Project of 1933 helped relieve the flood dangers in El Paso-Juarez Valley. The 1936 Act of Congress authorized a canalization project between El Paso and Caballo NM, the American Dam and Canal, which were completed by 1947. In the Lower Rio Grande valley considerable damage occurred from periodic floods. In 1924 and 1925 bond issues were voted to build levees from Donna to Brownsville TX, but the 1932 flood demonstrated that levees built on the American river side only could not give sufficient protection. In 1932, the International Boundary Commission recommended the construction of floodways on each river side; both countries agreed. American construction including 500 km of floodway and river levees, improvements, and control works was begun in 1933 and completed in 1951. The 1944 Treaty provided for plans of flood-control works from Fort Quitman TX to the Gulf of Mexico. Under the provisions of this treaty, construction of Falcon Dam was accomplished in 1954. Washington was involved in these activities both as a designer and a builder of an important flood relief scheme. He was a member of the American Society of Civil Engineers ASCE from 1919, and was active in its affairs by organizing the Lower Rio Grande Valley Branch, which he presided in 1941.

Anonymous (1936). Flood-control work in the Ro Grande Delta. *Engineering News-Record* 116(Mar.19): 407-411.
Anonymous (1954). Walter O. Washington. *Civil Engineering* 24(10): 646. *P*
Anonymous (1957). Walter O. Washington. *Trans. ASCE* 122: 1258-1259.
http://www.tshaonline.org/handbook/online/articles/mgr04

WATSON

14.03. 1827 Dumfries/UK
31.01. 1880 San Francisco CA/USA

William Stuart Watson, born in Scotland, came at age 12 with his family to the USA, settling in Miami County OH. He was from age 16 rodman on the Ohio Canals. After a visit to Scotland, he was until 1850 engaged on the Genesee Valley Canal, when moving to Buffalo NY as assistant engineer on the Erie Canal enlargement. After work for railways companies, Watson served as mining engineer in California. Many of his works exhibited originality of design and boldness of execution.

Watson raised the 8 km long North Fork Hydraulic Co. works in 1857. Some 3 km involving wrought-iron pipe of diameter 0.55 m were carried through the mountain gap 300 m deep, thereby conveying some 700 l/s water. The Cascade Canal Company's works, another of his successful designs, was almost 20 km long including 10 km of flumes. The canal was carried through one of the most formidable canyons of the State, with portions of the aqueduct and flumes suspended along the rocky sides of the stream, 100 m above the river. In 1858 he was engaged by the California Fluming Company to empty the bed of Feather River along almost 2 km and 40 m wide by means of dams and flumes. The main dam erected was 100 m long at top constructed by timber filled with stones. A total of 600 men were occupied for this notable hydraulic project. Watson's latest engineering exploit was completed in 1870; he furnished water from the Feather River for mining operations at Cherokee CA. He thereby employed a piping system to deliver water across the mountains with pipes of considerable wall thickness to withstand the pressure forces. To provide the escape of air due to over-pressure, a stand-pipe was erected including various 'blow-off' pipes controlling the pressure transients of the system. Although no novel hydraulic principles were involved in this project, it appears that the magnitude of the projects, at least in the USA, was formidable in terms of industrial water supply.

Stuart, C.B. (1871). William Stuart Watson. *Lives and works of civil and military engineers of America*: 293-300. Van Nostrand: New York.

Watson, W.S. (1854). *Preliminary report* of the Buffalo, Warren & St. Louis RR made to E.W. Cook, Esq., Springville, Erie County. Reese: Buffalo NY.

Watson, W.S. (1872). *Narrow gauge railroad system a complete success*: Its adaptability to the business of the Pacific Coast. San Francisco.

http://www.niceartgallery.com/William-Stuart-Watson/Shipping-Off-The-Coastof-Scotland-oil-painting.html (*P*)

http://publishing.cdlib.org/ucpressebooks/view?docId=ft758007r3;chunk.id=0;doc.view=print (*P*)

WATTENDORF

23.05. 1906 Boston MA/USA
11.06. 1986 Washington DC/USA

Frank Leslie Wattendorf majored in mathematics at Harvard University. After having received his BS degree in 1926, he enrolled at Massachusetts Institute of Technology MIT in aeronautical engineering. He there met Theodor von Karman (1881-1963). von Karman wrote 'He was greatly interested in my approach to the subject of aerodynamics. He also said that in America at that time there was only a very limited opportunity to learn the basic theory, and he asked me to recommend a school abroad where he might complete his studies for a master's degree. I recommended Göttingen and Aachen, and added that we did not have an American, and would be glad to have one. Wattendorf came to Aachen in 1927, and since that time I have considered him as a member of my family'. In 1930, Wattendorf accompanied von Karman to the California Institute of Technology, Pasadena CA, where he obtained the PhD degree, staying there until 1936, lastly as chief research engineer.

In 1936 Wattendorf accepted a professorship in aeronautical engineering at Tsing Hua University, Beijing China. There he designed and built a 5 m wind tunnel, but left China by the end of 1937 because of the war with Japan. Upon return to the USA he designed with von Karman a 6 m 40,000 HP wind tunnel at Wright Field OH, taking later over until 1946 as the civilian director. Toward the end of World War II, Wattendorf became a member of a group of scientists and engineers who were to visit Europe and the Far East to advise the Army Air Force Corps on future action in engineering developments. From 1946 to 1952 Wattendorf served as director of AGARD.

Anonymous (1986). Frank L. Wattendorf. *AIAA Bulletin* 24(10): B8. *P*

van der Bliek, J., ed. (1999). Dr. Frank L. Wattendorf. *AGARD*: The history 1952-1997. Advisory Group for Aeronautical Research and Development: Paris. *P*

Wattendorf, F. (1933). A study of the effect of curvature in fully-developed turbulent flow. *Proc. Royal Society* A 148: 565-598.

Wattendorf, F.L. (1948). High-speed flow through cambered rotating grids. *Journal of the Aeronautical Sciences* 15(4): 243-247.

Wattendorf, F.L., Noyes, J., Ponomareff, A.I. (1953). High-altitude and speed propulsion wind tunnel. *Mechanical Engineering* 75(10): 789-793.

Wattendorf, F.L., Ohain, H. von, Lawson, M. (1957). Factors influencing operating limits of high flux axial compressors. 9th *IUTAM Congress* Brussels 2: 309-319.

WEBB

19.06. 1816 New York NY/USA
30.10. 1899 New York NY/USA

William Henry Webb was trained by his father at the shipyard of New York, opening in 1818 his own yard near Corlears Hook, which eventually became Webb & Allen. At age 20, William was awarded his first commercial subcontract for the New York - Liverpool packet ship *Oxford*. He decided in 1840 to further his education by travelling to Scotland to visit the famous shipyards of the Clyde River near Glasgow. During this trip, his father died suddenly so that he returned home assuming the management of the shipyard. He began then turning out larger and more ambitious vessels, including clipper ships and sailing packets, for which the shipyard soon became famous. By 1849 his shipyard was at the cutting edge of sailing ship design. The California Gold Rush was by then in full swing, bringing with it a strong demand for new ships to convey the prospectors and supplies to and from the goldfields. Several of his ships set sailing speed records.

Webb's clipper designs employed the most judicious use of timber of all the major shipbuilders. For the *Challenge*, he relied on the hull planking as an integral part of the ship's structural strength, increasing the width between frames at the bow and the stern, thus using four fewer frames for a 60 m hull. In 1853 Webb built the 2,000 ton clipper *Young America*, considered by many to be the most beautiful clipper ever built, the 'acme of perfection' in clipper design. From the 1850s, Webb also designed steamboats and steamships, including *California*, the first steamer to enter the Golden Gate. By 1855, once the Gold Rush was over, Webb looked for new markets, completing in 1859 the steam frigate General Admiral for Imperial Russia, the fastest steam frigate then afloat. The most impressive warship built by the Webb shipyard was the giant ironclad USS *Dunderberg*, which was however completed only shortly after the Civil War. Later the American shipbuilding industry experienced a prolonged slump and Webb, having already made a large fortune, decided to close his shipyard and turn his energies toward philanthropic goals. He became a founding member of the Society of Naval Architects and Marine Engineers SNAME, establishing the Webb Academy, todays Webb Institute.

Anonymous (1899). Webb, William H. *American Machinist* 22: 1074.
Clark, A.H. (1910). *The clipper ship era*: An epitome of famous American and British clipper ships, their owners, builders, commanders and crews 1843-1869. Putnam's Sons: NY.
Dunbaugh, E.L., Thomas, W.B. (1989). William H. Webb, shipbuilder. Webb Institute: NY.
http://en.wikipedia.org/wiki/William_H._Webb *P*

WEBSTER

19.10. 1855 Philadelphia PA/USA
23.01. 1931 Philadelphia PA/USA

George Smedley Webster obtained in 1875 the civil engineering degree from Univerisy of Pennsylvania, Philadelphia PA, in 1909 the BSc degree, and in 1910 the honorary DS degree. He was engaged as surveyor in Eastern Pennsylvania after graduation. In 1877 he was employed for 44 years by the city of Philadelphia. Again, he served first as surveyor but then conducted engineering works of the sewerage systems, bridges and the parks. He was promoted in 1893 to chief engineer, which position he held until 1916, from when he was until 1920 director of the Department of Wharves, Docks and Ferries. As chief engineer, he was in charge of engineering projects including the construction of water purification, sewerage, port development, and city planning. Any history of his accomplishments is a history of the development of his city during this period.

More than 1,000 km of sewers were constructed during Webster's term as chief engineer. He conducted research work on the relation between the storm-water runoff and rainfall in various parts of the city, to allow for a better design of drainage systems. After a visit to Europe in 1908 he established a station for testing methods for sewage treatment, and designed and built the first sewage treatment plant of Philadelphia. In 1914 he prepared with colleagues a report on the collection and treatment of sewage, which was adopted. He was then appointed a member of the State Sanitary Water Board, to administer the statutes relative to sewerage and its discharge into streams. He was also involved in the design of slow sand filter plants. He was besides also active in the advancement of his profession, notably within the Engineers Club of Philadelphia, serving as director and president, the American Public Health Association, and the American Society of Civil Engineers ASCE, serving as director from 1904 to 1906, as vice-president from 1918 to 1919, and as ASCE President in 1920.

Anonymous (1895). George S. Webster. *Twenty years after*. University Archives. University of
        Pennsylvania: Philadelphia. *P*
Anonymous (1921). George S. Webster. *Engineering News-Record* 86(3): 139. *P*
Anonymous (1931). George S. Webster. *Trans. ASCE* 95: 1423-1428.
Webster, G.S. (1899). *Official handbook*, Department of Public Works. Philadelphia PA.
Webster, G.S., Wagner, S.T. (1900). History of the Pennsylvania Avenue Subway, Philadelphia,
        and sewer construction connected therewith. *Trans. ASCE* 44: 1-33.
http://freepages.history.rootsweb.ancestry.com/~oldnewspapers/phil_inq_misc_10_19_18.htm *P*

WEGMANN

27.11. 1850 Rio de Janeiro/BR
03.01. 1935 Yonkers NY/USA

Edward Wegmann was the son of Swiss parents. His early education was obtained in the public schools of Brooklyn NY and Zurich. He received the degree of civil engineering from New York University in 1871. After some years as a general civil engineer, he in 1884 began a thirty years service with the water supply system of New York City, as assistant engineer with the New York Aqueduct Commission. He first studied the design of profiles for the Quaker Bridge Dam to be built across Croton River, of which the height was 90 m, which was more than 30 m higher than any masonry dam built until then. He developed a simple formula for the dam section, which was adopted for the later Croton Dam, and other high dams of the USA. In 1885 Wegmann was appointed engineer of the Manhattan Division of the New Croton Aqueduct. The work included a tunnel under Harlem River and an aqueduct tunnel under the city. From 1893 to 1904 he was engineer in charge of the Croton River Division, supervising the construction of *Muscoot* Dam. From then to 1910 he continued until completion works for the Aqueduct Commission.

Wegmann then started as a consulting engineer. He was engaged from 1918 to 1920 by the New York and New Jersey Harbour Development Commission to estimate the cost and the construction of subways, elevated roads, and freight terminals. In 1920, he re-entered the service of New York City as consulting engineer and retired five years later. Wegmann authored the notable book Design and construction of masonry dams, which was originally published in 1888 and went through eight editions, its title eventually being changed to Design and construction of dams. It was widely used as a textbook both in the USA and abroad. Other notable works include his Water works of the city of New York, and Conveyance and distribution of water for water supply.

Anonymous (1944). Wegmann, Edward. *Dictionary of American biography* 21: 699-700. Scribner's: New York.

Hager, W.H. (2009). Edward Wegmann, sein Leben und Werk. *Wasser, Energie, Luft* 100(3): 235-240. *P*

Wegmann, E. (1888). *The design and construction of masonry dams*, giving the method employed in determining the profile of the Quaker Bridge Dam. Wiley: New York.

Wegmann, E. (1896). *The water-supply of the city of New York*. Wiley: New York.

Wegmann, E. (1911). The design and construction of dams including masonry, earth, rock-fill, timber, and steel structures, also the principal types of movable dams. Wiley: New York.

WEHAUSEN

23.09. 1913 Duluth MN/USA
06.10. 2005 Oakland CA/USA

John Vrooman Wehausen graduated with the BS degree in 1934, the MS degree in 1935, and the PhD degree in 1938 from the University of Michigan, Ann Arbor MN as mathematician. He was instructor at Columbia University, New York NY until 1940, at University of Missouri, Kansas City MO, until 1944, mathematician at David Taylor Model Basin from 1946 to 1949, then until 1956 executive editor of the Mathematical Reviews, from when he joined until retirement the Institute of Engineering Research University of California, Berkeley CA as head of its mechanics research group and from 1959 to 1984 as professor of engineering sciences. He was in 1960 Fulbright Lecturer at University of Hamburg, Germany, and in 1967 a visiting professor at Flinders University, Australia, among others. Wehausen was a Fellow of the Society of Naval Architects and Marine Engineers, and its recipient of the 1984 Davidson Medal, the American Mathematical Society AMS, and since 1980 member of the National Academy of Engineering NAE.

Wehausen contributed original scientific research in the areas of ship waves, ship manoeuvrability, floating systems in waves, and ship-generated solitary waves. He was the advocate of systematic theoretical analysis based on rational mechanics principles. In 1960 he published the 350 pages article Surface waves with Edmund V. Laitone (1915-2000). This synthesis has had such a long-lasting impact that it was republished in 2002. Wehausen also contributed review articles on the wave resistance of ships, and on the motion of floating bodies.

Anonymous (1962). Prof. J. Wehausen. *Mechanical Engineering* 84(8): 82. *P*
Anonymous (1987). J.V. Wehausen. *I have a photographic memory*: 76, P.R. Harmos, ed. American Mathematical Society: Providence RI. *P*
Anonymous (1991). Wehausen, John V. *Who's who in America* 46: 3418. Marquis: Chicago.
Wehausen, J.V., Laitone, E.V. (1960). Surface waves. *Handbuch der Physik* 9: 446-778. Springer: Berlin
Wehausen, J. (1964). Effect of the initial acceleration upon the wave resistance of ship models. *Journal of Ship Research* 7(3): 38-50.
Wehausen, J. (1973). The wave resistance of ships. *Advances in applied mechanics* 13: 93-245, C.-S. Yih, ed. Academic Press: New York.
http://www.universityofcalifornia.edu/senate/inmemoriam/johnwehausen.htm *P*
http://en.wikipedia.org/wiki/John_V._Wehausen *P*

WEIDNER

21.03. 1883 Buffalo NY/USA
20.08. 1941 Independence KS/USA

Carl Robert Weidner graduated as a civil engineer from the Cornell University, Ithaca NY, in 1904. He then was rodman for the Erie Railroad Company, and from 1904 worked as draftsman at the Bureau of Filtration, Philadelphia PA. Until 1908, he was assistant engineer for the Pipe Line Department for a gas company at Independence KS, in charge of construction of pipelines, pumping stations, tanks, and reservoirs. In 1908 he travelled and studied in Europe, and in 1909 he was engaged as instructor in civil engineering at Pennsylvania State College, State College PA, and then moved to the University of Wisconsin, Madison WI, where he taught hydraulic engineering during the next seven years. He there also conducted research of which resulted several Bulletins.

In 1916 Weidner left the academic field, and was engaged to make a property inventory of a pipeline company, which he had formed in 1915, and whose chief engineer he was. He had the charge of surveys for, and the design of many miles of pipelines, amounting in 1913 to some 25,000 km. Weidner also assisted in developing a pipeline accounting procedure. He was identified with oil pipeline transportation, thereby fostering technical advances including Welding in pipeline construction, or Prevention of corrosion of buried pipelines. He was active member of the American Petroleum Institute API, authoring a paper on the latter topic in 1927, and in 1931 on the Relation of pipe line currents and soil resistivity to corrosion. He organized tests on pipe coatings between 1930 and 1940 under the direction of the Bureau of Standards. In 1939 he edited a report on the History, construction, and operation of oil pipelines, thereby contributing to the standardization of this technique. In addition to his technical ability, he had talents in music and sports. He was described a pleasant companion with a wide range of interests. He was associate member of the American Society of Civil Engineers ASCE from 1910.

Anonymous (1941). Carl Weidner passes away. *The Petroleum Engineer* 12(13): 182. *P*
Anonymous (1944). Carl R. Weidner. *Trans. ASCE* 109: 1582-1584.
Davis, Jr., G.R., Weidner, C.R. (1911). An investigation of the air lift pump. *Bulletin* 450. University of Wisconsin: Madison.
Weidner, C.R. (1913). Theory and test of overshot water wheel. *Bulletin* 529. University of Wisconsin: Madison. (*P*)
Weidner, C.R. (1914). The diaphragm method for the measurement of water in open channels of uniform cross-section. *Bulletin* 672. University of Wisconsin: Madison.

WEIR P.

16.03. 1906 New York NY/USA
01.08. 1973 Atlanta GA/USA

Paul Weir received the CE degree in 1928 from the Georgia University of Technology, Atlanta GA, and was a registered engineer in Georgia. He joined the Atlanta Water Department all through his career and there served until 1942 as superintendent of the water purification plant, then until 1947 as assistant general manager from when he took over as general manager until retirement. Weir was in addition an active member since 1925 of the American Water Works Association AWWA, serving as chairman, national director from 1943 to 1946, and president in 1956. He was awarded the 1940 Fuller Prize and the 1941 Goodell Prize. He also served on the Editorial Board of Water and Wastes Engineering for 10 years. He was further a member of the American Society of Civil Engineers ASCE, and of the Georgia Engineering Society. He was named one of the US Top Ten Public Works Men of the year.

Weir was referred to as the famous water works superintendent of Atlanta. He was stated to be a synonym with the Atlanta water system. He was in 1969 singled out by a local newspaper for his outstanding achievement as city department head. The editorial said, among other, that 'Atlanta has been fortunate in the high calibre of some of the people who have been attracted by the place and have spent a lot of their life in public service. We have some fine examples in the city hall department heads and none finer than Paul Weir of the water department'. The occasion of the commendation was the award to Weir of a diamond pin for 40 years of service. The Atlanta Journal wrote on the 40th anniversary of his stay at the water department: 'The Weir career is remarkable for its longevity but even more remarkable for its progress. Mr. Weir keeps an eye on our waters and all things connected with it'.

Anonymous (1947). Paul Weir. *Engineering News-Record* 139(Jul.10): 170. *P*
Anonymous (1955). Paul Weir. *Water Works Engineering* 108(2): 124. *P*
Anonymous (1972). Paul Weir retires in Atlanta. *Mueller Record* (12): 7. *P*
Weir, P. (1940). The effect of internal pipe lining on water quality. *Journal of the American Water Works Association* 32(9): 1547-1576.
Weir, P. (1942). Use of bleaching clays in water purification. *Trans. AIME* 3(1): 167-177.
Weir, P. (1956). Public water supply and the future. *Journal of the American Water Works Association* 48(7): 755-760.
Weir, P. (1961). *The story of Atlanta's water*. Water Works Department: Atlanta GA.

WEIR W.V.

18.07. 1902 Warren IN/USA
19.11. 1959 St. Louis MO/USA

William Victor Weir graduated as a civil engineer from Washington University, with the BS degree in 1923. He was junior engineer at West St. Louis MO until 1926, then assistant manager of the St. Louis County Water Company until 1933, chief engineer and superintendent until 1945, and president and general manager since 1946. In parallel he also served the Missouri Water Company from 1943 in similar positions. He further was a member of the American Water Works Association AWWA and recipient of its 1940 Diven Memorial Medal, its 1943 Fuller Memorial Award, and its 1950 Goodell Prize. He also was its president in 1950, and elected to honorary membership in 1956. He further was member of the American Society of Civil Engineers ASCE. He was shot and fatally wounded in his office at University City, a suburb of St. Louis, and died soon after. His assailant attempted to shoot himself but was reported to have inflicted only a superficial wound. His act of violence resulted from the water company's foreclosure on property purchased by the individual and on which he was two years in arrears in payment on the deed of trust note.

Weir was a nationally-known water works authority, and the president of the St. Louis County Water Company and five other water utilities in various parts of the country. He was an expert in matters relating to water works design and construction. He also was a member of the Editorial Board of the Journal of Water Works Engineering. The first W. Victor Weir Award of the Missouri Water and Sewerage Conference was presented in 1968.

Anonymous (1947). W.V. Weir. *Journal of the American Water Works Association* 39(8): 10. *P*
Anonymous (1949). W. Victor Weir, vice-president AWWA. *Mueller Record* 35(3): 4. *P*
Anonymous (1959). Weir, W. Victor. *Who's who in America* 28: 2815-2816. Marquis: Chicago.
Anonymous (1959). W. Victor Weir shot. *Water Works Engineering* 112(12): 1115. *P*
Weir, W.V. (1949). Water main extension policy. *Journal of the American Water Works Association* 41(8): 729-741.
Weir, W.V. (1953). Consideration in supplying fluoridated water to industry. *Journal of the American Water Works Association* 45(4): 369-375.
Weir, W.V. (1959). 79th Annual Conference Civil Auditorium, San Francisco. *Journal of the American Water Works Association* 51(12): 1566-1578.
http://eece.wustl.edu/aboutthedepartment/Pages/history-env-eng.aspx

WEISS

18.07. 1867 Wels/A
02.09. 1951 Mexico DF/MX

Andrew Weiss came from Austria to St. Louis MO, completing his studies then at Colorado School of Mines, Golden CO, receiving the degree of engineer in 1899. He remained there as assistant professor of mathematics until 1903, from when he joined the newly created Reclamation Service. He first worked on the Salt River Project in Arizona, and the North Platte Project in Nebraska and Wyoming, serving as a project manager from 1907 to 1923. He was then appointed assistant director of reclamation economics to investigate projects experiencing difficulties in operation.

Weiss began in 1926 his work on reclamation in Mexico which he continued for the rest of his life. During the first six years, he was resident engineer on the Don Martin Project in Coahuila and Nueva Leon States, with partial responsibility for the Conchos Project in the Chihuahua State. He was appointed in 1932 consulting engineer for the National Commission of Irrigation, covering the technical and economic planning of all projects. From 1938, he was chief of the Technical Consulting Department of the commission. His projects included in addition to reclamation and irrigation also the flood control and sanitary water supplies, as well as multipurpose projects involving hydro-electric power plants. He was in addition consultant on the North Platte project in the USA. Weiss was remembered for having found a way to pass floods over earth dams under construction without problems in their stability, which was previously felt to be impracticable. He was awarded on the 50th anniversary of his graduation the medal for his distinguished achievements by his Alma Mater. He was a true member of the American Society of Agricultural Engineers ASAE. He also was member of the American Society of Civil Engineers ASCE from 1917, and became Honorary Member in 1948. He was buried in Mexico City after eulogies by leaders of engineering and government. It was noted that he had been a consultant for a quarter of a century, a most effective ambassador of good will from the USA, and an outstanding leader in his profession in Mexico.

Aboites Aguilar, L. (2012). The transnational dimensions of Mexican irrigation 1900-1950. *Journal of Political Ecology* 19(1): 70-80.
Anonymous (1949). Andrew Weiss. *Civil Engineering* 19(1): 43-44. *P*
Anonymous (1952). Andrew Weiss. *Trans. ASCE* 117: 1290-1291.
USBR (2002). *Reclamation*: History essays from the centennial symposium 1. USBR: Denver.
Weiss, A. (1921). *North Platte Project*. US Bureau of Reclamation: Denver.

WEITZEL

01.11. 1835 Cincinnati OH/USA
19.03. 1884 Philadelphia PA/USA

Godfrey Weitzel entered the United States Military Academy in 1851, graduating in 1855, and then was commissioned second lieutenant of engineers. His first duty was from 1855 to 1859 on fortifications of New Orleans LA. He then was assistant professor of engineering at the Military Academy until 1861, and then assigned to the engineer company on duty at Washington DC, taking part in the expedition to Pensacola FL. In 1862 he was made chief engineer in the operation of New Orleans and after surrender served as assistant military commandant of the city. In 1863 he became captain in the regular engineer corps. In 1864 he supervised the construction of the defences of Bermuda Hundred VA. During the final operations against Richmond, his command occupied the line between the James and the Appomattox Rivers. For his services in this campaign, Weitzel was promoted to major-general in the regular army. He was mustered out in 1866 returning returned to the duty with the Corps of Engineers.

Weitzel was from then until his death occupied notably with improvements of rivers and harbours. The most important were the ship canals at the falls of the Ohio, and at Sault Sainte Marie MI, and the lighthouse at Stannard's Rock, Lake Superior. The first project was completed in 1873. At Sault Sainte Marie he supervised the building of what was with 150 m long and 25 m wide at the time the largest in the world. The lighthouse, with a tower rising 35 m above the water, involved the construction below water, 20 m in diameter, on top of a rock located 50 km from the shore. During this time, Weitzel published also on hydraulic engineering and canal design. He was made a lieutenant-colonel in 1882 and shortly later, because of failing health, transferred from the Great Lakes to less arduous duty at Philadelphia, where he died at age of only 49.

Anonymous (1889). Weitzel, Godfrey. *Appletons' cyclopaedia of American biography* 6: 423. *P*
Anonymous (1936). Weitzel, Godfrey. *Dictionary of American biography* 19: 616-617.
     Scribner's: New York.
Brennan, J.F. (1879). *A biographical cyclopaedia of portrait gallery* of distinguished men, with
     a historical sketch of the State of Ohio. Yorston: Cincinnati.
Moore, C. (1907). *The Sainte Mary's Falls Canal*. Semi-Centennial Commission: Detroit.
Weitzel, G. (1878). *Report* of Board of Engineers on New Orleans Harbour. New Orleans.
Weitzel, G. (1880). *A lecture on the improvement of the Danube at Vienna*. Washington DC.
http://en.wikipedia.org/wiki/Godfrey_Weitzel *P*

WELCH

04.12. 1809 Nelson NY/USA
25.09. 1882 Lambertville NJ/USA

Ashbel Welch graduated from Albany Academy in 1826, starting the next year his engineering career at Lehigh Canal. He was in charge of the works of the Delaware and Raritan Canals from 1830, moving in 1832 to his home at Lambertville NJ. Next to canal works he was around 1840 also consulting engineer, collaborating with John Ericsson (1803-1889), and Horatio Allen (1802-1890). In 1844 he visited Europe travelling through England, Scotland, Ireland, France and Belgium. He returned to the Delaware Canal to build a wooden lock at Bordentown NJ. This was bold because its foundation was on quicksand, but the works were successful. In 1852 it was decided that the Delaware and Raritan Canal should enlarge its capacity, another work undertaken by Welch accomplished in 1853 for less money than estimated. A new tide-lock with walls 70 m long required the use of heavy coffer-dams. Its completion within 50 work days evidenced Welch's engineering ability, but his health was undermined. In 1854 he made another visit to Europe. After return he supervised the completion of the Chesapeake and Delaware Canal, which was opened in 1855. He was then engaged with railways projects.

Welch was appointed in 1882 commissioner to determine the storage of the New Jersey State waters for water supply. In developing improvements for freight transportation in canals, he succeeded to add significantly to the carrying capacity by introducing steam power to operate the locks and haul the vessels. He was interested in questions relating to navigation, including the Interoceanic Canal to be built in Panama. His character was of elevated tone. Great purity and entire disinterestedness lay at its foundation. These facts, combined with broad and generous views, were among the causes of the great moral influence that he exercised. He had a warm, sympathetic nature, and where best known, he was best loved. He was member of the American Society of Civil Engineers ASCE, serving as vice-president in 1880, and president in 1882.

Anonymous (1883). Ashbel Welch. *Proc. ASCE* 9: 137-144.
Anonymous (1937). Ashbel Welch. *Civil Engineering* 7(3): 236-237. *P*
Cohen, E. (1999). *Lambertville's legacy*: The Coryells, Ashbel Welch, and Fred Lewis.
Howell, J.R. (1972). *Ashbel Welch*: civil engineer. Historical Society: Lambertville. *P*
Welch, A. (1866). *Report* of C. & A. Railroad and Transportation Company. Lambertville.
Welch, A. (1880). Ship canal locks for operation by steam. *Trans. ASCE* 9: 293-314.
http://www.lindajbarth.com/TheDelawareandRaritanCanal.html

WELKER

01.06. 1857 Toledo OH/USA
24.12. 1926 Washington DC/USA

Philip Albert Welker graduated as a civil engineer in 1878 from Cornell University, Ithaca NY. He was appointed in 1879 to the US Coast and Geodetic Survey, and was engaged on the transcontinental triangulation of Missouri. From 1880 to 1890 he continued these works carrying out precise levels in Arkansas, Illinois and Tennessee. He also conducted hydrographic surveys at the harbour of Baltimore MD, San Francisco Bay and various rivers in Florida and Louisiana. From 1891 to 1910 he was in charge of hydrographic surveys along the Atlantic Coast as far north as Maine. He was placed in command in 1898 of the US Coast and Geodetic Survey Steamer *Bache*, making surveys of the harbour of Portsmouth NH and its approaches in New Hampshire and Maine.

In 1911 captain Welker took his charge of the sub-office of the US Coast and Geodetic Survey at Manila, the Philippine Islands, where he worked until 1914 as director of the coast surveys. He there served also as member of the Harbour Lines Commission. He was then appointed assistant in charge of the office, the US Coast and Geodetic Survey, Washington DC, from 1917 as hydrographic and geodetic engineer, and personnel officer from 1920 until his retirement from active duty in 1921. His surveys of the coasts of Porto Rico and off the Isthmus of Panama formed the basis for the first complete navigating charts of these waters prepared by the Federal Government. Through his systematic operations, Welker played an important role in the security and the extension of the American commerce. His results had always in evidence a distinctive mark of scientific originality, thereby increasing the accuracy and efficiency of the methods of the US Coast and Survey, and its reputation. His kind disposition gained for him the regard of all his associates in the Survey, and all who dealt with him could rely on him and his innate sense of justice. He was member of the Washington Society of Engineers, and of the American Society of Civil Engineers ASCE. Mount Welker is a geographical feature located in the Prince of Wales-Hyder Census Area AK, named in 1923 for captain Welker.

Anonymous (1916). *Centennial celebration of the United States Coast and Geodetic Survey*. US Dept. of Commerce: Washington DC.
Anonymous (1928). Philip A. Welker. *Trans. ASCE* 92: 1752-1753.
Baird, W.R. (1914). Philip A. Welker. *Betas of achievement*: 341. Beta Publishing Company: New York. *P*

WELLS

27.04. 1858 North Adams MA/USA
04.08. 1940 North Adams MA/USA

Charles Edwin Wells graduated from the Worcester Polytechnic Institute, Worcester MA, with the BSc degree in 1880. He was then employed by railroad companies, accepting in 1895 the position of division engineer, the Metropolitan Water Board of Boston MA. He was transferred to Clinton MA for raising the Wachusett Reservoir to augment the water supply for Boston MA. In 1899 he was transferred to the Reservoir Department in charge of the removal of topsoil from the 18 km^2 site. All Department work was placed from 1903 under his direction, from then as department engineer.

In 1905 Wells entered the US Reclamation Service as supervising engineer, in charge of the construction of irrigation works in southern Wyoming, Nebraska, and South Dakota. His first work was on the Pathfinder Dam and Reservoir, then one of the largest irrigation projects extending over 150 km in length. This dam is a masonry structure, 65 m high, located near Casper WY. The Belle Fourche Project SD lays within its territory, as did also the Gunnison Tunnel in Colorado. Wells resigned from the Service in 1907, moving as division engineer to the Board of Water Supply, New York NY, on its new Catskill water supply, then in its earliest stages. He was assigned to the 7 km long Hill View Division, located at Yonkers NY, where he spent the next nine years. It included a 2 km long steel pipe siphon 4 m in diameter, a pressurized aqueduct, and the 1 km long, 500 m wide and 10 m deep Hill View Reservoir; its filling began in 1915. In 1917 Wells became resident engineer on the construction of an embarkation camp, including water supply and sanitation works. In 1918 he was appointed supervising plant engineer for the US Shipping Board, San Diego CA, but retired in 1919, opening a private practice, serving finally from 1924 as city engineer of North Adams MA. It was stated: 'Wells had unusual execution ability. In carrying out the extensive work under my direction, I always felt the greatest confidence, because of his ability to get the work done… without friction'. He was ASCE member from 1892.

Anonymous (1906). Belle Fourche Irrigation Project. *The Irrigation Age* 21(5): 138-139.
Anonymous (1941). Charles E. Wells. *Trans. ASCE* 106: 1692-1696.
Anonymous (1946). C.E. Wells, with Reclamation Group, 1907. *Civil Engineering* 16(4): 188. *P*
Flinn, A.D., Wells, C.E. (1912). Protection of steel pipes. *The Canadian Engineer* 22: 205-208.
Wells, C.E. (1906). Operations in Nebraska. 4th *Annual Report* of the Reclamation Service:
    231-257, F.H. Newell, ed. Government Printing Office: Washington DC.

WESTON E.B.

25.03. 1850 Duxbury MA/USA
09.12. 1916 Boston MA/USA

Edmund Brownell Weston received education from the Highland Military Academy, Worcester MA. He was then a student in the office of the chief engineer of the Providence Water Works from 1871 to 1874. Until 1877 he served as an assistant engineer, from when he was a design engineer in charge of the Providence Water Department. From then he was a consulting engineer at Providence RI. Weston was a member of ASCE, the Institution of Civil Engineers London UK, the Boston Society of Civil Engineers, and the New England Meteorological Society. He also corporated closely with various departments of Brown University, Providence RI. He was a Fellow of Imperial College, London UK.

Weston conducted filter experiments for his city in 1893, which demonstrated the high efficiency of mechanical filtration for purification of municipal water supplies. He then designed filtration systems of up to 350 liters per second in many cities of the USA, India, and Austria. He also constructed an experimental plant for the city of Alexandria, Egypt. He has written several books next to journal papers, including Weston's friction of water in pipes, or Tables for estimating the cost of cast-iron water pipe. The first booklet was published first in 1893, and saw various re-editions. He devoted time to original research in hydraulics and water purification, and his foreign travels and studies of water supply and sanitary engineering were extensive. He was all through his career particularly interested in matters relating to water supply engineering.

Anonymous (1905). Weston, Edmund B. *Who's who in America*: 1585. Marquis: Chicago.
Anonymous (1917). Edmund Brownell Weston. *Trans. ASCE* 81: 1789-1790.
Weston, E.B. (1885). Description of some experiments made on the Providence RI water works, to ascertain the force of water ram in pipes. *Trans. ASCE* 14: 238-246.
Weston, E.B. (1890). Formulas for the flow of water in pipes. *Trans. ASCE* 22: 1-90.
Weston, E.B. (1896). *Tables* for estimating the cost of laying cast-iron water pipe of the 'Providence pattern'. Engineering News Publishing: New York.
Weston, E.B. (1896). *Report* of the results obtained with experimental filters at the Pettaconset pumping station of the Providence water works. Freeman: Providence RI.
Weston, E.B. (1903). *Tables* showing loss of head due to friction of water in pipes. Van Nostrand: New York.
Weston, E.B. (1916). *In memoriam my father and my mother* Hon. Gershom Bradford Weston, Deborah Brownell Weston of Duxbury MA. Providence RI. *P*

WESTON R.S.

01.08. 1869 Concord NH/USA
29.07. 1943 Wakefield RI/USA

Robert Spurr Weston graduated with a BS degree from the Amherst College, Amherst MA, continuing studies at the Massachusetts Institute of Technology MIT, Cambridge MA, and at University of Berlin, Germany. He was then an assistant at Louisville Water Company, and studied and travelled widely from 1893 to 1899. Upon return he joined a private consulting office at Boston MA, where his services were widely sought. He was in addition from 1912 to 1916 assistant professor of public health at MIT. He entered the consulting engineering partnership Weston & Sampson then, serving various cities and corporations in connection with sanitary engineering.

Weston's professional career began as a chemist, a subject that he greatly enjoyed, and a field in which he won distinction. One of his earliest engagements was concerned with water filtration experiments at Louisville KY, where his colleagues were Joseph W. Ellms (1867-1950) and George W. Fuller (1868-1934). Once at Boston with Weston & Sampson, where he remained until death, he dealt with projects in water supply, water purification, sewage treatment and disposal, industrial waste treatment, and stream and harbour pollution. He was engaged as sanitary expert in important litigations, including these of the Chicago Drainage Canal, Delaware River, and Connecticut River. He was a co-author of the Handbook and wrote a number of papers on public health subjects. He was an early advocate of the water supply filtration to improve public health, and its chemical treatment to prevent corrosion of the network. He was awarded the Dexter Brackett Memorial Medal by the New England Water Works Association for his works, and received the Desmond Fitzgerald Medal from the Boston Society of Civil Engineers. Weston was kind, understanding, generous to the fault, and helpful to young engineers. He also was fair and maintained always high ethical standards within the profession.

Anonymous (1930). Robert Spurr Weston. *New England Water Works Association* 44(1): 1. *P*
Anonymous (1941). Weston, Robert S. *Who's who in engineering* 5: 1904. Lewis: New York.
Anonymous (1944). Robert Spurr Weston. *Trans. ASCE* 109: 1542-1544.
Flinn, A.D., Weston, R.S., Bogert, C.L. (1916). *Water works handbook*. McGraw-Hill: New York.
Weston, R.S. (1937). Treating highly colored water. *Water Works Engineering* 90(3): 388-391.
Weston, R.S. (1941). 11-year operating experiences with zeolite softening plant. *Water Works Engineering* 94(2): 179-180. *P*

WESTON W.

ca. 1752 Oxford/UK
29.08. 1833 London/UK

William Weston was probably a pupil of the pioneer English canal engineer James Brindley (1716-1772). In 1792 Weston contracted first with the Schuylkill & Susquehanna Navigation Company, Pennsylvania, to serve for five years as engineer of its canal, then extending from Philadelphia up Schuylkill Valley to Reading and to Susquehanna, known later as Union Canal. He arrived at Philadelphia in 1793, serving the company for two years until becoming insolvent. He in parallel surveyed three canal projects. In 1794 he started the design of Middlesex Canal connecting Charlestown MA with the Merrimack River. In 1795 he examined the locks under construction at the Great Falls of Potomac River. The next two years were spent as engineer for the Western Inland Lock Navigation Company in New York State. Once Weston had severed his long connection with the Schuylkill & Susquehanna Company, he devoted himself for the New York State enterprise.

In 1799 Weston examined for New York State the possible sources for future water supply. He recommended damming Bronx River and regulating its flow by raising the level of the Rye Ponds, now a part of the *Kensico* Reservoir. He also proposed a dual distribution system, to apply after the water was brought to the reservoir near the City Hall Park. Among Weston's last American activities were these in connection with a 'permanent bridge' crossing the Schuylkill at Market Street, Philadelphia. As designer of the pier foundations, one extending to 12 m below the water surface, he remained in communication with the construction company for more than two years after his return to Great Britain in 1800. Little is known on Weston's subsequent activities. He was offered the position of chief engineer for the Erie Canal Project, but he declined in favour of his age. Weston's standing as an engineer in the United States may be judged by the obvious respect paid to his professional opinions by leading American men.

Anonymous (1936). Weston, W. *Dictionary of American biography* 20: 21-22. Scribner's: NY.
Fitzsimons, N. (1967). Weston: The seven American years. *Civil Engineering* 37(10): 86-87.
Kirby, R.S. (1936). William Weston and his contribution to early American engineering. *Trans. Newcomen Society* 16(1): 111-127.
Shallat, T. (1994). *Structures in the stream*: Water, science and rise of the US Army. Austin. (*P*)
Weston, W. (1794). *Schuylkill and Susquehanna navigation*. Philadelphia.
Weston, W. (1795). *An historical account of the rise, progress and present state of the Canal navigation in Pennsylvania*. Poulson: Philadelphia.

WEYMOUTH

02.06. 1874 Medford ME/USA
22.07. 1941 Los Angeles CA/USA

Frank Elwin Weymouth graduated in 1896 from the University of Maine, Orono ME, with a BSc degree in civil engineering, and was honoured in 1934 with the degree Doctor of Engineering. From 1897, he was engaged by Malden MA, and then with the Metropolitan Water Board MA on the construction of water works. In 1899 he joined the engineering organization of the Isthmian Canal Commission and there became engaged in hydraulic studies. In 1902, when the US Reclamation Service was created, he served in its Bureau. First he was in charge of studies for irrigation projects in Montana and North Dakota, becoming in 1908 supervising engineer of the Idaho District, in charge of the storage dam on the Snake River, Jackson Lake WY, and the Arrowrock Dam on Boise River ID, then with 105 m the highest of the world.

Weymouth served from 1916 to 1920 as chief engineer of construction in charge of all work undertaken by the Service, and in 1920 was made chief engineer of the Bureau. The Reclamation Service was under his leadership, pioneering in the development of novel dam designs, culminating some years later in the 220 m high Boulder Canyon Dam; the Weymouth Report furnished the basis upon which the final decision rested. In 1924 he took over as president the firm Brock & Weymouth, Philadelphia PA. From 1926 he worked with the National Irrigation Commission of Mexico, which involved extensive studies of irrigation possibilities, with the notable Don Martin Dam at Nuevo Laredo. In 1929 he was retained by Los Angeles CA as chief engineer of water works, and placed in charge of the Colorado River Aqueduct. Both its design and construction involved significant problems, including the optimum route across the desert, as also both its geology and topography. Construction started in 1931, and completion in 1941, shortly before Weymouth's death, involving a 390 km long main and 240 km laterals. He was an outstanding engineer and manager; he was ASCE member from 1907.

Anonymous (1929). F.E. Weymouth. *Engineering News-Record* 103(25): 984. *P*
Anonymous (1942). Frank E. Weymouth. *Trans. ASCE* 107: 1712-1716.
Weymouth, F.E. (1932). A great job looms in the Colorado River Aqueduct. *Engineering News-Record* 108(Jun.16): 847-850. *P*
Weymouth, F.E. (1940). *Report* of the Metropolitan Water District. District of Southern California: Los Angeles.
http://waterandpower.org/museum/Colorado%20River%20Aqueduct.html *P*

WHEELER E.S.

27.08. 1839 Buckingham PA/USA
05.01. 1913 Detroit MI/USA

Ebenezer Smith Wheeler moved with his parents in 1844 to Indiana State. He graduated in 1867 from the University of Michigan, Ann Arbor MI, and there received in 1897 the honorary MS degree. He entered in 1882 the US Corps of Engineers, was stationed at Detroit, and then at Sault Ste. Marie MI, where he was involved in the design and construction of the Hay Lake Channel. From 1897 he supervised the preliminary survey for the Nicaragua Canal Project in Central America, preparing a report to the Canal Commission, stating that the Nicaragua Canal was a feasible engineering project, yet this route was later abandoned in favour of the Panama Canal. On his return from Central America, he was assigned to duty in the US Engineer Office, Detroit MI, as chief assistant engineer, where he continued until his death. During this period he was in charge of the new St. Clair Flats Canal, and the breakwater at Mackinac Island. His engineering experience was also found to be invaluable when acting with other important works of the District.

While stationed at Detroit MI, Wheeler found time to invent the Wheeler bathometer, which was placed on a number of freight vessels on the Great Lakes to give warning when in shoal water. The device was operated successfully for several seasons, and it was considered to be a means of saving life and property on the Great Lakes. He was well informed on many subjects, was a modest, unobtrusive, cultured gentleman, gifted with a keen sense of humour. He had the rare talent to solve the difficulties of others and to illustrate his own thoughts with a good story well told. He was a most delightful companion, whom it was a genuine privilege to have known. He was a member of the Michigan Engineering Society, the Detroit Engineering Society, and of the American Society of Civil Engineers ASCE.

Anonymous (1913). Memoir of Ebenezer S. Wheeler. *Trans. ASCE* 76: 2249-2250.
Anonymous (1913). Ebenezer S. Wheeler. *Detroit Free Press*, January 06, 1913. *P*
Anonymous (1943). Wheeler, Ebenezer S. *Who was who in America* 1: 1328. Marquis: Chicago.
Libby, H.E. (1931). A graphic pressure sounder. *The Military Engineer* 23(9/10): 456-457.
Wheeler, E.S. (1889). *The old lock on the Canadian side of Sault Ste. Marie recently rediscovered*. Sault Ste. Marie.
Wheeler, E.S. (1899). *Report of the Nicaragua Canal Commission*. Friedenwald: Baltimore.
Wheeler, E.S. (1908). *Letterbook* concerning administration of river and harbour projects in the Great Lakes. Sault Ste. Marie.

WHEELER L.L.

05.02. 1851 Jackson MI/USA
13.03. 1927 Sterling IL/USA

Levi Lockwood Wheeler received in 1874 the degree of civil engineering from University of Michigan, Ann Arbor MI. He entered the US civil service as assistant of the US Lake Survey at Detroit MI until 1881, making precise levels of the elevations of the Great Lakes, resulting in the 1882 Report. In 1882, he was appointed assistant engineer, the Mississippi River Commission, St. Louis MO, remaining there until 1888. He there made hydrographic surveys from Vicksburg to Natchez MS. In 1888 he was placed in charge of a government survey for a 4 m waterway from the Lake Michigan to Joliet IL, the Report of which was used by the Sanitary District of Chicago IL in the construction of the main canal and the Calumet-Sag Canal. In 1890 Wheeler was appointed US assistant engineer in charge of construction of the Illinois and Mississippi Canal, extending from Illinois River near Hennepin to the Mississippi River near Rock Island IL. The main line has a surface width of 24 m and a depth of 2 m over a length of 110 km. Wheeler was in charge of the 45 km long Western Section. The Hennepin Canal includes 34 locks, 9 aqueduct bridges, 60 culverts, and 3 dams. Construction started in 1892 and the canal was opened in 1907. Wheeler was subsequently in charge of the canal operation and its maintenance until his retirement in 1921.

In addition to government work, Wheeler was also engaged in private practice, mainly consisting of hydraulic works. He designed and supervised the construction of water power dams across the Rock River at Sterling IL, Dixon, Oregon and Sears IL. After retirement he returned home from a visit to his son on the Hawaiian Islands, suffering a cerebral haemorrhage, from which he never recovered. He was extremely thorough and accurate in all his work. In connection with Hennepin Canal he established a record for economy and efficiency which seldom has been equalled. He was gifted with a mind and physique which allowed him to keep in contact with all details of his work. He seemed to be equally at home in the field, or in the office.

Anonymous (1928). Levi L. Wheeler. *Trans. ASCE* 92: 1754-1756.
Wheeler, L.L. (1882). Report on primary triangulation of US Lake Superior. Corps of Engineers *Professional Paper* 24(12). Government Printing Office: Washington DC.
Wheeler, L.L. (1890). *Map and profile of the proposed routes for a waterway between Lake Michigan and Mississippi River*. US Army Corps of Engineers: Washington DC.
http://www.tampicohistoricalsociety.com/Hennepin_Canal_Photos.html (*P*)

WHIPPLE

16.09. 1804 Hardwick MA/USA
15.03. 1888 Albany NY/USA

Squire Whipple attended from 1822 until 1828 the academic schools of the District, and then entered the senior class of Union College, Schenectady NY, graduating in 1829. He then began civil engineering practice, first as a rodman, then as leveller of the Baltimore and Ohio Railroad, and after two years was engaged on the surveys and computations for the enlargement of Erie Canal. From 1837 to 1850 he conducted numerous surveys of railroad and canal projects. His special talents laid, however, in the design and analysis of mechanical contrivances. In 1840, he designed a scale for weighing canal boats, which was adopted on the Erie Canal. In 1847 he published a treatise on bridge building, in which for the first time the fundamental laws of framed structures were elucidated. In 1869 he extended this treatise, adding more than 100 pages, for which he set the type and made the woodcuts, printing the book on a hand press at his own house. In 1847 he also had published a small book entitled The way to happiness, in which he advocated abstention from animal food. He had at his home a cabinet of mathematical and physical instruments of his own construction. He was a clear thinker, a forcible writer and socially, while retiring in his disposition, was a good companion. He was made a Honorary Member of the American Society of Civil Engineers ASCE in 1868.

Whipple was a most humble man, small in stature, but big in mind. In his alumni record sheets, he wrote the following on his moral essay published in 1847: 'A first-rate book, but not appreciated by the present generation. The author, nevertheless, glories in having made this effort'. It appears that there is no full-length biography on Whipple, but the Union College published an interesting pamphlet on him in 1949.

Anonymous (1896). Squire Whipple. *Trans. ASCE* 36: 527-530.
Anonymous (1899). Whipple, Squire. *The National cyclopaedia of American* 9: 35. White: New York.
Fitzsimons, N. (1966). Squire Whipple. *Civil Engineering* 36(5): 48. *P*
Hill, H.W. (1908). *An historical review of waterways and canal construction in New York State.* Buffalo Historical Society: Buffalo NY.
Whipple, S. (1847). *A work on bridge building*, consisting of two essays, the one elementary and general, the other giving original plans, and practical details for iron and wooden bridges. Curtiss: Utican NY.
http://en.wikipedia.org/wiki/Squire_Whipple *P*

WHISTLER

19.05. 1800 Fort Wayne IN/USA
07.04. 1849 St. Petersburg/RU

George Washington Whistler graduated from the US Military Academy in 1819, joining then the US Army until 1821. After having served as assistant drawing teacher the next year, he surveyed the international boundary between Lake Superior and Lake of the Woods. From 1826 to 1828 he served in the cabinet of the US President as commissioner to make surveys and plans. After having served to various railroad companies he resigned from the US Army in 1833, becoming then consulting and chief engineer in Massachusetts State. He initiated works for Russia in 1842, supervising the construction of fortifications and docks at *Kronstadt*, Russia, and designed a bridge there over Neva River. He was decorated with the Order of St. Anne by the Russian Emperor in 1847, but passed away only two years later. Whistler is buried at Stonington CT; there is a monument to him in Greenwood Cemetery, Brooklyn NY.

Fort Wayne was still a fort when Whistler was born, and his father was its commanding officer. After having been at West Point NY, his first engineering assignment was in the mid-1820s the survey of the highly irregular boundary from Minnesota to Canada. Until resigning from the Army he was put on 'detached' service for several railroads. His first civilian job was with the Middlesex Canal Company in England, but a few years later he returned to his country as a consultant. A seven-year contract to build the first major Russian railroads from Moscow to St. Petersburg was started in 1842, yet this was his last. Whistler did obviously excellent work in Russia; his older son was a capable civil engineer, whose career was curtailed by ill-health. Whistler was contracted by cholera during his works two years before the railroad line was completely. He is credited with selecting the five-foot rail gauge currently still used in Russia.

Anonymous (1893). Whistler, G.W. *Trans. Newcomen Society* 7(2): 126.
Anonymous (1963). Whistler, George W. *Who was who in America* 1607-1896: 574. Marquis: Chicago.
Fitzsimons, N. (1966). George Washington Whistler. *Civil Engineering* 36(7): 74. *P*
Vose, G.L. (1887). *A sketch of the life and works of George W. Whistler, civil engineer.* Lee & Shepard: New York. *P*
http://en.wikipedia.org/wiki/George_Washington_Whistler
http://penelope.uchicago.edu/Thayer/E/Gazetteer/Places/America/United_States/Army/USMA/ Cullums_Register/214*.html

WHITE C.

08.09. 1790 Whitesboro NY/USA
18.12. 1834 St. Augustine FL/USA

Canvass White shipped as supercargo on a merchant vessel for Russia in 1811, and returned to the USA in 1814, after having served as enlisted man during the 1812 War. He made in 1817 an extended trip to Great Britain examining canal construction. After return to the USA he patented a waterproof cement in 1820. He had worked then until 1825 on the East section of the Erie Canal, from when he was chief engineer of the Union Canal, Pennsylvania. Later, White became consulting engineer of the Schuylkill Navigation Company, designing thereby locks of the Connecticut River at Windsor, and at Farmington Canal. Finally he was chief engineer of the Delaware and Raritan Canal in New Jersey, and also in charge of the Lehigh Canal in Pennsylvania.

From the completion of the Erie Canal until his death, White was constantly employed in different parts of Montgomery County in public works, among which were these at the Susquehanna and Schuylkill Canal, the improvements of the *Scimnyikili* Navigation Company, the New Haven and Farmington Canal, and the Delaware breakwater. White was induced to take a contract for time completion of the latter structure, but by the mismanagement of others was a loser to a large amount. In 1834 his failing health compelled him to leave the business, so that he went to Florida, hoping that the climate would have a favourable effect upon his disease, but this step had been taken too late, and within a month after landing, he died. His remains were interred at Princeton NJ, where his family resided at the time. A colleague of White stated correctly: 'Get Mr. White, no man more competent, no man more capable. And while your faith in his ability and fidelity increases, your friendship will grow into affection'. An American engineer stated: 'As a civil engineer he had no superior; his genius and ability were of surpassing magnitude'. White's gentle disposition, and the kindly charm of his manner had earned him to all whom he chanced to meet, and his early death was mourned by a large circle of friends.

Anonymous (1963). White, Canvass. *Who was who in America* 1607-1896: 575. Marquis: Chicago.
Clarke, T.C. (1896). Waterways from the ocean to the Lakes. *Scribner's Mag.* 19(1): 103-114.
White, W.P. (1909). Canvass White's services. *Buffalo Historical Society* 13: 352-366. *P*
http://en.wikipedia.org/wiki/Canvass_White *P*
http://www.todayinsci.com/W/White_Canvass/WhiteCanvas-Obituary.htm *P*

WHITE W.M.

20.11. 1871 Valley Head AL/USA
09.02. 1949 Coral Gables FL/USA

William Monroe White received education at Tulane University, New Orleans LA, graduating with an ME degree in 1899. He was engineer until 1902 with the New Orleans Drainage Board, until 1911 hydraulic and chief engineer of a ship and engine building company at Philadelphia PA, then was in charge of hydraulic works to this company with designs and the construction of large hydro-electric installations, including at Niagara Falls. From 1911 until retirement White was the chief engineer and manager of the hydraulic department, Allis-Chalmers Manufacturing Co., Milwaukee WI. He was member and since 1943 Fellow of the American Society of Mechanical Engineers ASME, and the American Institute of Electrical Engineers AIEE. He was presented in 1930 the honorary doctorate of science from Tulane University. In 1940 he received a special Modern Pioneer award made by the distinguished committee of scientists for his invention of the *Hydraucone* draft tube regainer, an improved concentric-type draft tube for improved flow in hydraulic turbine draft tubes.

White authored numerous papers and was an authority on hydraulic turbines. Around 1900 he was also interested in the Pitot tube for point-wise velocity measurement, as used in flow elements such as pumps or turbines. Given that these tubes are particularly sensitive to the approach flow direction, they are hardly used currently for such tests. In ASME he served on the Power Test Codes Committee for centrifugal and rotary pumps, and was also involved in the Committee on Hydraulic Prime Movers.

Anonymous (1938). White, William M. *Who's who in America* 17: 2437. Marquis: Chicago.
Anonymous (1949). Dr. William M. White. *Power Generation* 53(4): 96.
Anonymous (1949). William Monroe White. *ASME News* 71(5): 460.
White, W.M. (1900). Water measurements in connection with a test of a centrifugal pump at Jourdan Avenue drainage station, New Orleans LA. *Journal of the Association of Engineering Societies* 25(4): 161-172.
White, W.M. (1901). The Pitot tube, its formula. *Journal of the Association of Engineering Societies* 27(2): 35-79.
White, W.M. (1920). Design in the Allis-Chalmers Unit. *ENR* 85(14): 650-653.
White, W.M. (1929). Hydraulic sessions feature power development. *Power* 70(24): 924. *P*
White, W.M., Rheingans, W.J. (1935). Photoflow method of water measurement. *Trans. ASME* 57(6): 273-280; 58(3): 156-165.

WHITHAM

13.12. 1927 Halifax/UK
26.01. 2014 Altadena CA/USA

Gerald Beresford Whitham gained his BSc in 1948, his MSc in 1949, and his PhD degree in 1953 from Manchester University, Manchester UK, where he stayed as Lecturer until 1956. He was then associate professor at the New York University until 1959. He spent the two next years as professor of applied mathematics, Massachusetts Institute of Technology MIT, Cambridge MA, joining in 1961 the California Institute of Technology Caltech, Pasadena CA, first as professor of aeronautics and mathematics, from 1967 as professor of applied mathematics. Whitham was elected Fellow of the American Academy of Arts and Science in 1959, of the Royal Society, London UK in 1965, and was awarded the Wiener Prize in applied mathematics in 1980.

Whitham was a scientist contributing significantly to water waves. His first important work relates to an extension of the classic dam-break wave as proposed in 1890 by August Ritter (1826-1908), thereby including the effect of wall friction. In collaboration with Michael J. Lighthill (1924-1998), Whitham introduced kinematic waves as simplest wave type in hydraulics, transporting information only in the flow direction. These waves have had a significant impact on hydrologic processes from the 1960s. From the 1970s Whitham took interest in water waves as occur in science and technology. His 1974 book has become classic in the field. He was also interested all through his career in shock waves, sonic booms, and gas dynamics in general.

Anonymous (1980). Citation for Wiener Prize: Gerald B. Whitham. *Notices of the American Mathematical Society* 27(6): 529-530. P

Anonymous (1994). Whitham, Gerald Beresford. *International who's who*: 1656. Europe Publications Ltd.: London.

Whitham, G.B. (1955). The effects of hydraulic resistance in the dam-break problem. *Proc. Royal Society* A 227: 399-407.

Whitham, G.B., Lighthill, M.J. (1955). On kinematic waves. *Proc. Royal Society* A 229: 281-345.

Whitham, G.B., Keller, H.B., Levine, D.A. (1960). Motion of a bore over a sloping beach. *Journal Fluid Mechanics* 7(2): 302-316.

Whitham, G.B. (1974). *Linear and nonlinear waves*. Wiley: New York.

Whitham, G.B. (1979). *Lectures on wave propagation*. Springer: Berlin.

http://www.caltech.edu/content/gerald-b-whitham-0 P

WICKER

14.01. 1907 Philadelphia PA/USA
29.04. 1994 Philadelphia PA/USA

Clarence Felton Wicker received his education from Pennsylvania State University, University PA, with the BS degree in civil engineering in 1929. He was engaged then until 1931 by the Philadelphia District, Corps of Engineers, taking over as section head in estuary research until 1940, from when he acted as assistant chief, and finally chief of the Engineering Division, Philadelphia Army Engineering District. Wicker was a member of the American Society of Civil Engineers ASCE, member and later chairman of its Hydraulics Division Commission on the Tidal Hydraulics and Research, member of the Society of the American Military Engineers SAME, of the Permanent International Association of Navigation Congress PIANC, and the International Association of Hydraulic Research IAHR. He was decorated for Outstanding Performance of duties of the Philadelphia District, and by the Philippine Islands in recognition of his services in 1946.

Wicker was interested in matters relating to coastal and tidal hydraulics. He advanced questions relating to tides, tidal currents, salinity, silt movement, bottom roughness, tidal inlets, and beach erosion. His 1965 Report details over 250 pages numerous factors to be considered in solving problems of practice. Its chapters are 1. Improvement of tidal waterways, 2. Computation of tides and currents, 3. Sedimentation of tidal waterways, 4. Salinity intrusions in estuaries, 5. Effects of density differences on estuarine hydraulics, 6. Effects of littoral processes on tidewater navigation channels, 7. Dredging and disposal practices in estuaries, 8. Dispersion and flushing of pollutants, 9. Hydraulic model studies of tidal waterway problems, and 10. Design of channels for navigation.

Anonymous (1955). C.F. Wicker. 6th *IAHR Congress* La Haye, Frontispiece. *P*
Anonymous (1994). Wicker, Clarence F. *Who's who in engineering* 9: 2012. Lewis: New York.
Wicker, C.F., Rosenzweig, O. (1950). Theories on tidal hydraulics. *Report* 1: 101-125. ASCE
        Commission on Tidal Hydraulics: New York.
Wicker, C.F., ed. (1965). Evaluation of present state of knowledge of factors affecting tidal
        hydraulics and related phenomena. *Report* 3. Committee on Tidal Hydraulics. US Corps
        of Engineers: Vicksburg MS.
Wicker, C.F. (1969). New horizons in the field of tidal hydraulics. *Journal of the Hydraulics
        Division* ASCE 95(HY1): 147-160.
Wicker, C.F. (1971). Economic channels and manoeuvring areas for ships. *Journal of the
        Waterways, Harbors and Coastal Division* ASCE 97(WW3): 443-454.

WILCOX

12.02. 1830 Westerly RI/USA
27.11. 1893 Brooklyn NY/USA

Stephen Wilcox seems to have followed his natural aptitude for mechanics without serving any regular apprenticeship. One of his early inventions was a practical hot-air engine for operating fog signals. In 1856 he invented then a safety water-tube boiler with inclined tubes, the germ of the later Babcock & Wilcox boiler. Around 1866, Wilcox with his partner Babcock designed a steam generator based on the principle of the early boiler. In 1867 the firm Babcock & Wilcox B&W, Providence RI was formed to manufacture this boiler. The concern was then incorporated in 1881, with Wilcox as vice-president until his death. The firm is still active, dealing currently mainly with nuclear power generation. During World War II, over half of the US Navy was powered by B&W boilers. The company has its headquarters now at Charlotte NC.

The B&W boiler and stationary steam-engine were used in the first central stations or power plants of the United States with a considerable importance in the development of electric lighting. B&W products were used all over the world with company offices also in Cuba or Puerto Rico. Wilcox was primarily both the inventor and mechanic of the combination, while Babcock was the executive. Thomas Edison stated that B&W had made 'the best boiler God has permitted man yet to make'. Wilcox continued his experimentation with engines and boilers until the end of his life. Much of his work was carried out on his yacht, which may have been responsible for the perfection of the marine form of the B&W boiler. Wilcox secured, alone or with others, forty-seven patents in forty years. He was handsome and popular, simple and unaffected by his rise to affluence. He presented to Westerly, his birthplace, a public library building, which after his death was enlarged by his widow, who also carried out their joint plans for other gifts to the town, including a park and a high-school building.

Anonymous (1894). Stephen Wilcox. *Trans. ASME* 15: 1188.

Anonymous (1931). *Fifty years of steam*: A brief history of the Babcock & Wilcox Company. B&W: New York.

Anonymous (1936). Wilcox, Stephen. *Dictionary of American biography* 20: 204-205. Scribner's: New York.

Babcock, G.H., Wilcox, S. (1894). *Steam*: Its generation and use. B&W: New York.

http://de.wikipedia.org/wiki/Stephen_Wilcox *P*

http://en.wikipedia.org/wiki/Babcock_and_Wilcox

WILEY A.J.

15.07. 1862 New Castle County DE/USA
08.10. 1931 Monrovia CA/USA

Andrew Jackson Wiley graduated from Delaware College with the civil engineering degree in 1882. From 1883 to 1886 he was rodman on irrigation work with the Idaho Mining and Irrigation Company, Boise ID. From 1888 to 1892 he was chief assistant engineer with this company for an irrigation project in Southern Idaho, and then until 1898 chief engineer and manager of the Owyhee Land and Irrigation Co., constructing a large project at Grandview ID. After difficult years in terms of finances, the turn came in 1900; the population had by then greatly increased and plans for the irrigation development were made.

Wiley worked during the next 30 years on a continuous improvement of great irrigation and power projects in Idaho, Oregon and California. His obituary contains a long list with all the major projects stated, including repairs, design, or extension of Milner Dam ID, Shoshone Falls power plant ID, Sweetwater Dam CA, or Bull River Dam OR. He also investigated the reasons for the St. Francis Dam disaster at Los Angeles CA, and proposed a flood control project for Orange County CA, or the San Gabriel Dam CA.

Wiley was in addition a private consultant from 1902, mainly for the US Bureau of Reclamation USBR. He was consulted with the design and the construction of major government dams including the Belle Fourche, the Shoshone, the Roosevelt, the Gibson, the Pathfinder, and the Hoover Dams. He had a close eye on questions of foundation, on material, and on the hydraulic aspects of the designs. He was of course also consulted on irrigation and power projects, mainly on aspects relating to the hydraulics of the diversion works, tunnels, flumes, siphons and spillways. He was further involved in the design and the construction of the Coolidge Dam AZ or the Tabor and Pablo Dams MT. His death marked the end of an exceptional engineering career because he held the highest respect of all who knew him. He was suddenly stricken with the illness that caused his death while engaged in the Los Angeles County Flood Control District CA. He was an engineer of exceptional ability having contributed to the better of the West.

Anonymous (1929). Andrew J. Wiley. *Engineering News-Record* 103(25): 984. *P*
Anonymous (1932). Andrew J. Wiley. *Trans. ASCE* 94: 1577-1584.
Wiley, A.J. (1894). Irrigation work in Idaho: A type of the difficulties encountered in building canals and dams. *The Irrigation Age* 7(6): 256-258.
Wiley, A.J. (1928). *Safety of water supply dams*, the city of Los Angeles. Los Angeles CA.
Wiley, A.J. (1931). *Safety of Hoover Dam*. US Bureau of Reclamation: Denver.

WILEY R.B.

05.03. 1884 Detroit MI/USA
14.05. 1963 Lafayette IN/USA

Ralph Benjamin Wiley obtained in 1906 the BS degree in civil engineering from the University of Michigan, Ann Arbor MI. He joined in 1908 Purdue University, West Lafayette IN, as instructor first, and then went through all positions, becoming from 1919 to 1937 professor of sanitary engineering, and then until 1954 professor of civil engineering, Head of the School of Engineering, and director of the Materials Testing Laboratory, from when he was an emeritus. He also served as chairman of the Indiana Stream Pollution Control Board, and was a member of the Indiana Food Control and Water Resources Commission. From 1954 he also was a private consultant at West Lafayette. Wiley was member of the American Society of Engineering Education ASEE, the American Water Works Association AWWA, and of the American Society of Civil Engineers ASCE, whose director he was from 1941 to 1943, and vice-president from 1947 to 1948. He was awarded Honorary ASCE Membership in 1956.

Wiley started his professional career with the city of Columbus OH, as design engineer on water purification and sewage disposal plants. He shortly later joined Purdue's staff, advancing within a short period to professor of sanitary engineering. To maintain close contact with engineering practice, he spent most of his summers working as consulting engineer for Detroit MI on the development of its sanitary system. He also served as consultant on sanitary drainage and water supply problems for various municipalities. He received a distinguished service award from the Indiana Sewage Works Association in 1947, and was named Engineer of the Year by the Indiana Society of Professional Engineers in 1955. Shortly before his death, the honorary civil engineering fraternity nominated him to its distinguished member.

Anonymous (1954). Prof. R.B. Wiley to leave Purdue Engineering School. *Engineering News-Record* 152(Jun.10): 69-70. P
Anonymous (1959). Wiley, Ralph B. *Who's who in engineering* 8: 2668. Lewis: New York.
Anonymous (1963). R.B. Wiley, Honorary Member of ASCE, dies. *Civil Engineering* 33(7): 65.
Howland, W.E., Wiley, R.B. (1941). Backsight at a turning point. *Civil Engineering* 11(3): 199.
Wiley, R.B., Greve, F.W., Zucrow, M.J. (1925). Characteristics of sewage-sprinkler nozzles. *Bulletin* 20. Engineering Experiment Station: Lafayette IN.
Wiley, R.B., Howland, W.E. (1947). Stream pollution abatement standards require economic justification. *Civil Engineering* 17(9): 539-543.

WILLEY

18.09. 1911 Climax KS/USA
12.06. 1980 Eureka KS/USA

Charles Keith Willey obtained in 1935 the BS degree in civil engineering from the University of Kansas, Lawrence KS, and in 1937 from the State University of Iowa, Iowa City IA, the MS degree in hydraulic engineering. He was from 1935 to 1937 draftsman at Topeka KS, assistant engineer of the Tennessee Valley Authority TVA, Knoxville TN, from 1937 to 1942, and then until 1946 member of the US Coast Artillery. From 1946 to 1947 Willey was hydraulic engineer of an engineering company at Knoxville TN joining then the Harza Engineering Co., Chicago IL, in charge of planning in its Power Studies Section. From 1952 he was there research engineer on the Box Canyon Project at Pende Oreille WA, representing Harza in the Pacific Northwest. From 1955 he served as the western manager of Harza Co., Chicago IL. Willey was an ASCE member.

Harza Engineering steadily rised to its position in the 1960s as one of the top international consulting firms in river development, an enlargement initiated by the founder Leroy F. Harza (1882-1953). The company was then taken over by Richard D. Harza (1923-), who stated that 'Several of my father's philosophies still help shape company policy today and as such are a highly respected heritage'. Leroy Harza established and maintained high technical standards so that three-fourths of the company's technical personnel was in the 1960s graduate engineers. Another Harza tradition was that the top management of the firm should be its principal technical individuals, but neither business managers nor administrators. In the 1960s the leaders of Harza next to Richard D. Harza included Willey, Calvin V. Davis (1897-1981), Kenneth E. Sorensen (1919-1990), and E. Montford Fucik (1914-2010). Charles K. Willey Fund of the American-Scandinavian Foundation supports graduate Scandinavian students in the USA for up to one year.

Anonymous (1933). Charles K. Willey. *Jayhawker yearbook*: 263. Univ. Kansas: Lawrence. *P*
Anonymous (1959). Willey, Charles K. *Who's who in engineering* 8: 2674. Lewis: New York.
Anonymous (1962). 'Harza' means river development in 25 countries. *Engineering News-Record* 168(May 10): 64-67. *P*
Willey, C.K. (1945). The engineer in foreign service: France. *Civil Engineering* 15(3): 152.
Willey, C.K. (1948). *First creek flood control Knoxville* TN. University of Kansas: Lawrence.
Willey, C.K. (1960). Wanapum hydroelectric development. *Civil Engineering* 30(9): 65-69.
Willey, C.K. (1970). *Some aspects of the design of ice passage facilities for the Burfell hydroelectric project*. IAHR: Delft.

WILLIAMS G.S.

22.10. 1866 Saginaw MI/USA
12.12. 1931 Ann Arbor MI/USA

Gardner Stewart Williams graduated in 1889 from University of Michigan, Ann Arbor MI. He was then first assistant engineer for water works construction at Bismarck ND, resident engineer for water works construction at Greenville MI, civil engineer for the Board of Water Commissioners, Detroit MI, from 1893 to 1898, engineer in charge of the hydraulic laboratory of Cornell University, Ithaca NY, until 1904, then professor of civil and sanitary engineering at University of Michigan. There he was from 1911 consultant dealing with problems in hydraulics and water power. He was a member of the International Commission on Navigation of Great Lakes, of ASCE, ASME, the North-Eastern Water Works Association, and the Detroit Engineering Society.

Williams is known for his 1905 booklet written jointly with Allen Hazen (1869-1930). After pipe flow had been thoroughly studied over a long time, with notable contributions mainly of Henry Darcy (1803-1858) in 1856, and his former collaborator Henry Bazin (1829-1917) in 1897, Williams and Hazen added new data to investigate mainly the roughness effect. In addition to straight pipes, they also analyzed the effects of local elements such as pipe bends. These books were then popular among hydraulic engineers. Currently, computers replace the numerical tables but use the identical mathematical background. The formula of Williams and Hazen was popular during the entire 20th century mainly in the USA. It particularly applies to design piping systems including sprinkler, water supply or irrigation networks. The formula relates only to water flow.

Anonymous (1909). Williams, Gardner S. *Who's who in America*: 2069. Marquis: New York.
Anonymous (1931). Gardner S. Williams. *Engineering News-Record* 107(27): 1053. *P*
Fenkell, G.H., Leisen, T.A., Hoad, W.C., Williams, G.S. (1924). *Reports on additional water supply for Detroit and environs*. Board of Water Commission: Detroit.
Williams, G.S., Hubbell, C.W., Fenkell, G.H. (1902). Experiments at Detroit MI on the effect of curvature upon the flow of water in pipes. *Trans. ASCE* 47: 1-369.
Williams, G.S. (1904). Advances in the design of high masonry dams. *Engineering Record* 50: 469-471.
Williams, G.S., Hazen, A. (1905). *Hydraulic tables* showing the loss of head due to friction of water flowing in pipes, aqueducts, sewers and discharge over weirs. Wiley: New York.
Williams, G.S. (1911). Hydraulics, pumping, water power. *American Civil Engineers' Pocket Book* 9: 830-912, M. Merriman, ed. Wiley: New York.

WILLIAMS G.R.

26.03. 1906 Newton MA/USA
20.02. 1999 Weston MA/USA

Gordon Ryerson Williams obtained the SB degree in 1929 from Massachusetts Institute of Technology MIT, Boston MA. He was from 1930 to 1934 junior engineer on stream-flow investigations for the US Geological Survey USGS, until 1939 then assistant and associate engineer, supervising flood studies for USGS at Washington DC, then until 1942 associate engineer for hydraulic and hydrologic designs at the US Engineers Office, Baltimore MD. He was then until 1943 involved in flood protection projects on Susquehanna River, Syracuse NY, taking over until 1945 as senior hydraulic engineer at the Civil Works Division, Office of Chief Engineers, US Army, Washington DC. Until 1946 Williams was principal hydraulic engineer at Providence RI and Boston MA, from when he was a principal hydraulic engineer for Tippetts-Abbett-McCarthy-Stratton TAMS Engineers, New York NY. He was from 1953 to 1960 professor of hydraulic engineering at MIT. From 1960 he was on planning and design of hydroelectric, irrigation, water supply and flood control projects in the USA and abroad. Williams was awarded the 1944 James Laurie Prize by the American Society of Civil Engineers ASCE, whose member he was from the 1930s, in addition to the American Geophysical Union AGU, the Boston Society of Civil Engineers BSCE, and the US Committee of Large Dams USCOLD.

Williams had seen through his career almost all aspects of hydraulic engineering, from detailed projects to surveys of large schemes, and education of young professionals during his years at MIT. He had written several papers in the Transactions ASCE as also three USGS Water Supply Papers. He was the author of the chapter Hydrology in the 1950 book of Hunter Rouse (1906-1996).

Anonymous (1922). Williams, Gordon R. *Who's who in engineering* 1: 2035. Lewis: New York.
Anonymous (1929). Gordon R. Williams. *Technique yearbook*: 384. MIT: Cambridge. *P*
Anonymous (1999). In memoriam Gordon Ryerson Williams. *ASCE News* 24(6): 6.
Williams, G.R. (1939). *Bibliography on limnology in the USA*. AIHS: Washington DC.
Williams, G.R., Crawford, L.C. (1940). Maximum discharges at stream-measurement stations. *Water Supply Paper* 847. Washington DC.
Williams, G.R. (1950). Hydrology. *Hydraulic engineering*: 229-320, H. Rouse, ed. Wiley: New York.
Williams, G.R. (1961). Cyclical variations in world-wide hydrologic data. *Journal of the Hydraulics Division* ASCE 87(HY6): 71-88.

WILLIAMS J.L.

06.05. 1807 Westfield NC/USA
09.10. 1886 Fort Wayne IN/USA

Jesse Lynch Williams moved with his family in 1814 to Cincinnati OH, and in 1819 to the Wayne County IN. He was student at Lancasterian Seminary, Cincinnati, and in 1828 joined the first survey of the Miami & Erie Canal in Ohio, from Cincinnati to Maumee Bay. He decided on the final canal location from Locking Summit to Chillicothe, constructing one of the divisions, including a dam and aqueduct across Scioto River. He was a member of the Board of Engineers deciding to use reservoirs rather than feeders from district streams to supply the water to the canal summit level. Williams advanced to chief engineer of the Wabash & Erie Canal, taking charge of all canals of Indiana in 1835, and was appointed in 1836 chief engineer of all canal routes. He was engaged from 1842 to 1847 in mercantile operations at Fort Wayne, and was chief engineer of the Wabash & Erie Canal until 1876. He presented in 1870 a report on the construction and the equipment of a road through the Rocky Mountains.

Begun at Fort Wayne in 1832, the Wabash & Erie was originally intended to connect the waters of the Wabash River with Lake Erie at Toledo OH. It reached Lafayette IN in 1841, and was extended to Terre Haute in 1849. Construction was then continued along the line of the uncompleted Cross-Cut Canal to Worthington IN and the uncompleted Southern Division of the Central Canal following that route to Evansville IN on the Ohio River, completed in 1853. With its 700 km between Toledo and Evansville, it was the longest canal in America. The Southern Division was never a financial success and was closed in 1860 after only seven years of fitful operation. The northern sections continued in operation until neglect and decay won out in the mid-1870s. Although the Wabash & Erie Canal was a financial loss to the State of Indiana, its contribution to the economic growth of the northern third of the state cannot be underestimated. In Indiana, there were 73 locks, 19 aqueducts, 16 dams, 239 culverts, 178 road bridges, and 15 waste weirs. Lockage from Fort Wayne eastward to the Ohio State Line was 6 m, from Fort Wayne to Terre Haute 60 m, Terre Haute to the Eel River summit 24 m, and from Eel River summit in Clay County to Evansville, the lockage was 50 m.

FitzSimons, N., ed. (1991). Williams, Jesse L. *A biographical dictionary of American civil engineers* 2: 125-126. ASCE: New York.
www.americancanals.org/.../Wabash%20&%20E...
http://trees.ancestry.com/tree/1587756/person/-1915861446 *P*

WILLIAMS R.B.

14.12. 1888 Beattyville KY/USA
08.11. 1973 Oceanside CA/USA

Roy Bruce Williams obtained the BS degree in civil engineering from Montana State College, Bozeman MT, in 1911. He was since 1912 engineer with the US Reclamation Service on a large number of dams, including the Flathead Dam and the Sun River Dam in Montana State, Kittitas Dam in Washington State, and St. Mary Storage Dam in Alberta CA. Later, he was associated with hydraulic structures of calibre as Gila Dam in Arizona, Boulder Canyon Dam on the Colorado River, and the All-American Canal in California. At the end of the 1940s, Williams was the assistant district manager of the Columbia River Basin Project, and also responsible for the Coulee Dam Project on Columbia River.

Williams was named in 1937, then construction engineer of the All-American Canal, assistant commissioner of the US Bureau of Reclamation USBR. His assignments in the Bureau have led him successively from instrumentman and surveyor to the position of senior engineer, with a notable experience on the many dam designs and constructions. The 1933 book co-authored by Elwood Mead (1858-1936) gives a detailed overview on the enormous work for Hoover Dam, in which Williams was one important element of the entire army of workers. Previously, he had been construction engineer of the All-American Canal. This is a 130 km long aqueduct in south-eastern California, conveying water from Colorado River into the Imperial Valley. The Canal was authorized along with the Hoover Dam in 1928, and built in the 1930s . Its design was supervised by the Bureau's then chief engineer John L. Savage (1879-1967), and completed in 1942. The Canal has a drop of 53 m, a width of 46 m to 210 m, and a depth of 2 m to 15 m. The canal size reduces toward west because less water has to be conveyed.

Anonymous (1937). Williams is assistant head of Reclamation Bureau. *Engineering News-Record* 119(Aug.05): 211.
Anonymous (1937). New appointments in the Bureau of Reclamation. *Reclamation Era* 27(9): 200-201; 27(9): 227. *P*
Anonymous (1948). Williams, Roy B. *Who's who in engineering* 6: 2181. Lewis: New York.
Anonymous (1949). Roy B. Williams retires. *Civil Engineering* 19(12): 878. *P*
Stene, E.A. (2009). *All-American Canal*: Boulder Canyon Project. USBR: Denver CO.
Wilbur, R.L., Mead, E. (1933). *The construction of Hoover Dam*: Preliminary investigations, design of dam, and progress of construction. US Dept. of the Interior: Washington DC.
Williams, R.B. (1942). General features of Friant Dam. *Civil Engineering* 12(2): 81-83.

WILLIAMSON S.B.

15.04. 1865 Lexington VA/USA
12.01. 1939 Lexington VA/USA

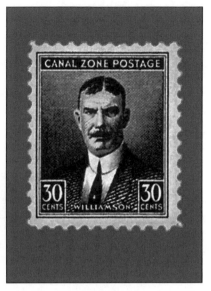

Sydney Bacon Williamson graduated in 1884 from the Virginia Military Institute, Lexington VA. He then accepted for two years an offer as instructor at King's Mountain Academy, York SC. From 1886 to 1887 he was staff member of the Engineering Depts. of the St. Paul and Duluth Railway, and then was in general engineering practice at Montgomery AL, in charge of sanitary engineering. From 1892 to 1898, Williamson became assistant engineer to George W. Goethals (1858-1928), in charge of improvements on the Tennessee River, including a canal lock at Riverton AL, and on Muscle Shoals Canal. After war service in Puerto Rico, he returned to Goethals in charge of fortifications at Newport RI, and from 1906 to 1914 was division engineer of the Panama Canal Pacific Division under Goethals, responsible for the excavation of two locks. Williamson also designed the Pacific terminal docks, the water supply, and sewers of Panama City.

Once Panama Canal was inaugurated Williamson became chief engineer of construction of an engineering company in London UK. He worked on a power station in Brazil, the water supply of Genoa, Italy, and served as chief of construction of the US Reclamation Service at Denver CO. From 1916 to 1924, he was consulting engineer for a company in New York NY, working on a copper mine in Chile. After war service during World War I, he returned to the company in New York. In 1924 he founded a civil engineering practice at Birmingham AL, and Charlottesville VA, and was consulting engineer with Goethals on the port of Palm Beach FL, and various municipal water supplies. In 1928, he was employed by the US District Engineer, Philadelphia PA, with projects on dikes and control of Delaware River. In 1929 Williamson became principal engineer of the US Engineering Department, occupied with the investigations on the feasibility of the Nicaragua Canal. In 1931 he was appointed by President Hoover as a member of the Interoceanic Canal Board, serving until 1935. In 1933 Williamson was engaged by the US Army Corps of Engineers to determine damages due to the large Mississippi floods.

Anonymous (1940). Sydney B. Williamson. *Trans. ASCE* 105: 1940-1954.
Sultan, D.I., Williamson, S.B. (1932). *Interoceanic Canal*. US Govt. Printing: Washington DC.
Williamson, S.B. (1948). *Sydney B. Williamson papers*. Archives of Virginia Military Institute: Lexington VA.
http://ead.lib.virginia.edu/vivaxtf/view?docId=vmi/vilxv00006.xml
http://picclick.com/Canal-Zone-113a-30-ct-380559305682.html *P*

WILSON E.K.

14.11. 1878 Southington CT/USA
31.03. 1943 Upper Montclair NJ/USA

Edgar Kennard Wilson received education from the Worcester Academy, Worcester MA, and from the University of Maine, Orono ME. From 1898 to 1905 he was engaged in surveying and construction, and advanced through the grades from chainman to chief of party, becoming in 1905 an assistant engineer at Philadelphia PA. He worked from 1906 to 1909 as transitman with the engineering force of the Panama Canal on the Culebra Division, returned to the USA, becoming assistant engineer of the Water District at Portland ME.

Wilson joined in 1912 the Pitometer Company, Chicago IL, founded by Edward S. Cole (1871-1950), continuing to serve this firm with skill and fidelity until his death. He there became eventually chief engineer and treasurer, and was actively engaged on hydraulic experiments to develop the Pitot tube, a standard device to measure velocity at a certain point of the flow field. He was also involved in the design of extensions for the water distribution systems of many American cities. In addition his work also included the flow analysis to improve the water supply mains in these cities, a service to which he contributed significantly in the conception of methods, and on the skill in organisation of these works. Wilson was widely known among water works men in the USA for his knowledge of water distribution problems. He authored a number of papers in this topic, which greatly added to his reputation. During his association with the Company he won the esteem of his associates for his sterling character and professional skill. He was later also vice-president and director of the Pitometer Log Corporation. He was member of the American Water Works Association AWWA, serving as chairman of the New York Section, of the New England Water Works Association, and of the American Society of Civil Engineers ASCE.

Anonymous (1942). E.K. Wilson, consulting engineer. *Journal AWWA* 95(12): 740. *P*
Anonymous (1944). Edgar K. Wilson. *Trans. ASCE* 109: 1551-1552.
Cole, J.A., Cole, E.S. (1909). *The portable test Pitometer*: Tables and directions. Chicago IL.
Wilson, E.K. (1930). Keeping water works maps and field records up to date. *Water Works Engineering* 30(16): 1143-1144; 30(17): 1177-1178.
Wilson, E.K. (1932). Trunk main surveys. *Journal AWWA* 24(5): 669-702.
Wilson, E.K. (1936). Conditions of mains in typical America cities. *Journal of the American Water Works Association* 28(9): 1304-1343.
Wilson, E.K. (1937). Finding lost water. *Engineering News-Record* 118(22): 819.

WILSON E.L.

05.09. 1931 Ferndale CA/USA
25.02. 2005 Haverhill MA/USA

Edward (Ed) Lawrence Wilson received education from University of California, Berkeley CA, from where he graduated with the BS, MS, and D.Eng. degrees in 1955, 1959 and 1963, respectively. He was research engineer with *Aerojet* General Corp., Sacramento CA, until 1965, returning then to his Alma Mater from assistant professor to professor of civil engineering until retirement in 1988. He was a member of the American Society of Civil Engineers ASCE and elected in 1985 to the National Academy of Engineering NAE. He also was the recipient of the 2003 John von Neumann Award with the full citation 'For his pioneering contributions to the Finite Element Method and the development of the SAP codes that were disseminated throughout the world, providing hundreds of institutions with their first finite element software'.

Wilson was a notable engineer developing the Finite Element Method for applications in hydraulic or structural engineering. In the early 1960s, computer programs were developed all over the Western World and Japan. The first widely accepted program was the Structural Analysis Package SAP of Wilson. Most programs were written in FORTRAN, the only suitable language at the time. Soon there was a veritable explosion in programs and currently there are sources of packages which are menu-driven and automated to the extent that, with minimal training, anybody can do a finite element analysis for better of for worse.

Anonymous (1994). Wilson, Edward L. *American men and women of science* 7: 810. Bowker: Providence NJ.

Anonymous (1996). Edward L. Wilson. *Who's who in science and engineering* 3:1087. Marquis.

Bathe, K.-J., Wilson, E.L. (1976). *Numerical methods in finite element analysis*. Prentice-Hall: Englewood Cliff NJ.

Ghaboussi, J., Wilson, E.L. (1972). Variational formulation of dynamics of fluid-saturated porous elastic solids. *J. Engineering Mechanics Division* ASCE 98(EM4): 947-963.

Sandhu, R.S., Wilson, E.L. (1969). Finite-element analysis of seepage in elastic media. *Journal of the Engineering Mechanics Division* ASCE 95(EM3): 641-652.

Wilson, E.L. (1963). Finite element analysis of two-dimensional structures. *PhD Thesis*. University of California: Berkeley CA.

http://en.wikipedia.org/wiki/Edward_L._Wilson

http://www.edwilson.org/Home%20Page/Professor%20Emeritus%20Edward%20L.htm *P*

WILSON H.M.

23.08. 1860 Glasgow/UK
25.11. 1920 Hartford CT/USA

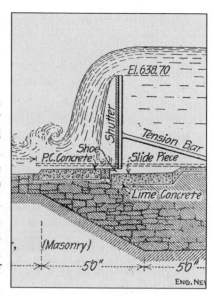

Herbert Michael Wilson obtained the degree in civil engineering in 1881 from Columbia University, New York NY. He then was levelman, transitman, and chief of party for railroad surveys in Mexico until 1882, taking over until 1888 as topographer of the US Geological Survey USGS, and then until 1890 being irrigation engineer, becoming geographer until 1906, from when he was chief engineer of structural and fuel testing until 1910. Until 1914 he was then engineer in charge of the US Bureau of Mines, and until 1918 served as director of the Department of Safety and Inspection of the Associated Companies, Hartford CT, from when he was its general manager. Wilson in parallel was Lecturer in irrigation at Columbia University from 1892 to 1893. He was member of the Washington Academy.

Wilson was a true expert in civil and mining engineering, specialising in geography, irrigation engineering, safety of mining, and of insurance compensation. He has written a number of papers and books, among which is the Manual of irrigation engineering. It contains the chapters 1. Introduction, 2. Precipitation, runoff, and stream flow, 3. Seepage, Evaporation and absorption, 4. Alkali, drainage, and sedimentation, 5. Quantity of water required, 6. Flow and measurement of water in open channels, 7. Subsurface water, 8. Classes of irrigation works, 9. Alignment, slope, and cross-section, 10. Headworks and diversion weirs, 11. Scouring sluices, regulators, and escapes, 12. Falls and drainage works, 13. Distributaries, 14. Application of water, and pipe irrigation, 15. Reservoirs, 16. Earth and loose rock dams, 17. Masonry dams, 18. Wasteways and outlet sluices, and 19. Pumping, tools and maintenance. The work therefore includes all important aspects of irrigation engineering; previous books of similar content include these of Patrick J. Flynn (1838-1893), and Robert B. Buckley (1847-1927). The 7th edition co-authored by Arthur P. Davis (1861-1933) includes more than 600 pages as compared to the original 500 pages book of the 1898 version.

Anonymous (1921). Wilson, H.M. *American men of science* 3: 749. Science Press: New York.
Davis, A.P., Wilson, H.M. (1919). *Irrigation engineering*, ed. 7. Wiley: New York.
Wilson, H.M. (1894). *Engineering results of irrigation survey*. Government Printing: Washington.
Wilson, H.M. (1898). *Manual of irrigation engineering*, ed. 2. Wiley: New York.
Wilson, H.M. (1903). *Irrigation in India*. Water Supply Paper 87. USGS: Washington DC.
https://archive.org/stream/irrigationengin01wilsgoog#page/n68/mode/2up (*P*)

WILSON P.S.

23.09. 1896 Montclair NJ/USA
13.07. 1992 Maplewood NJ/USA

Percy Suydam Wilson obtained the civil engineering degree in 1918, and the MS degree in 1920 from the Cornell University, Ithaca NY. He was employed then at a chemical engineering company at Buffalo NY, and was assistant engineer of the company of James H. Fuertes (1863-1932) on water supply and sewerage works notably for Harrisburg PA, and for Denver CO. In 1926 he stayed with an engineering company in New York NY, becoming in 1927 chief engineer, superintendent, and vice-president of the New Rochelle Water Supply Co., New Rochelle NY. Until 1932 he was in addition superintendent of 40 subsidiary operating water companies. Until 1936 Wilson was a private consultant of sanitary and hydraulic engineering, from when he became acting secretary and technical assistant of the American Water Works Association AWWA. Finally from 1940 he was vice-president of the Cathodic Protection Division of a company at Philadelphia PA, and thus had left the field of hydraulic engineering. He was a member of the New England Water Works Association, and of AWWA.

The 1920 paper deals with an experimental study of Venturi flumes as a discharge measurement structure. This flume was introduced in 1916 by Victor M. Cone (1883-1970), corresponding to a local flume contraction, by which the flow is forced by a suitable tailwater control from sub- to supercritical flow. The name is a misnomer, given that Gian B. Venturi (1746-1822) invented the Venturi pipe, involving a contraction. The pressure difference between the upstream reach and the contracted section may be used for discharge measurement, so that the basic principle differs from that of Venturi Flumes. The latter are still used mainly to determine discharge on wastewater stations, typically to ±5 to 10%. Wilson and his co-author observed that this Flume satisfies the requirements of practice, and that much less hydraulic losses are involved as compared to weirs. In addition, given the sediment-water flow with sewage, no serious depositions occur with these elements. The data also allowed for the definition of the head-discharge equation for the particular flume geometry studied. This study was supported by Ernest W. Schoder (1879-1968), then hydraulics professor at Cornell University.

Anonymous (1935). P.S. Wilson, Secretary AWWA. *Water Works Engineering* 88(23): 1318. *P*
Anonymous (1949). Wilson, Percy S. *Who's who in engineering* 6: 2193. Lewis: New York.
Wilson, P.S., Wright, C.A. (1920). A study of the Venturi Flume as a measuring device in open channels. *Engineering News-Record* 85(10): 452-457; 85(26): 1223-1224.

WILSON T.D.

11.05. 1840 Brooklyn NY/USA
29.06. 1896 Brooklyn NY/USA

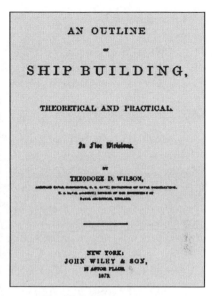

AN OUTLINE
-
SHIP BUILDING,

THEORETICAL AND PRACTICAL.

In Five Divisions.

BY

THEODORE D. WILSON,

NEW YORK:
JOHN WILEY & SON,
15 ASTOR PLACE.
1873.

Theodore Delavan Wilson was an American naval ship designer, and an instructor of naval architecture and shipbuilding. He was apprenticed at Brooklyn Navy Yard. At the outbreak of the American Civil War, he volunteered for the US Army service. Upon his return from the front in 1861, he was transferred to the US Navy and was appointed as a carpenter in the construction department. He served on the USS *Cambridge* until 1863, participating in the first day in the Battle of Hampton Roads against the ironclad CSS *Virginia*. In 1866, he was commissioned an assistant naval constructor, eventually serving at the naval facilities at Pensacola, Philadelphia, and Washington DC. Between 1869 and 1873 he taught naval architecture and shipbuilding at the US Naval Academy, Annapolis MD. In 1870, Wilson was sent on special duty to Great Britain and France, where he viewed vessels recently completed and under construction. In 1873, he was promoted to naval constructor.

Wilson served then at the Portsmouth Navy Yard, and in 1881 was appointed onto the first Naval Advisory Board for rehabilitating and modernizing the Navy. However, he, the two other naval constructors on the panel and chief engineer Benjamin Isherwood (1822-1915), dissented with the majority on recommendations for an improved warship construction. These would allow the Navy to take advantage of technological advances such as steel construction, revive the national economy, and develop a large domestic industrial base to support a large fleet. His dissent may have been due to the limitations of the domestic industry at the time, without taking into account that domestic industrial modernization had become a major goal of the Navy. After promotion to chief constructor of the Bureau of Construction and Repair in 1882, Wilson was placed in charge of naval design for all new warships. Among these were the pre-dreadnought battleship USS *Maine*, the protected cruisers USS *Boston*, USS *Chicago* and USS *San Francisco*, and the gunboats USS *Bennington*, USS *Concord*, and USS *Petrel*. Wilson's reputation in foreign naval circles was considerable, but he resigned from the Navy in 1893 due to failing health; he died of a heart stroke. Wilson was Honorary Member of the Royal Institution of Naval Architects in Great Britain, among others.

Wilson, T.D. (1872). *Lectures on the practice of US naval vessels*. Naval Academy: Annapolis.
Wilson, T.D. (1873). *An outline of ship building*: Theoretical and practical. Wiley: New York. (*P*)
http://en.wikipedia.org/wiki/Theodore_D._Wilson

WINSLOW

10.11. 1810 Bennington VT/USA
10.03. 1892 Poughkeepsie NY/USA

John Flack Winslow moved with his family at age five to Albany NY, where he was educated until age seventeen. He then entered a commercial house as a clerk. In 1832 he secured the management of the Boston agency of the New Jersey Iron Company. In 1833, he went into the iron industry on his own account and was engaged successfully in the pig iron production in Bergen and Sussex counties NJ. In 1837 then, Winslow joined an extensive hardware enterprise at Albany NY, forming the Corning & Winslow firm. It controlled both the Albany and the Rensselaer NY iron works, eventually becoming the largest producers of railroad and other iron in the United States.

Starting in 1861, Winslow and his business partner worked with John Ericsson (1803-1889) to build the USS *Monitor*. After the Naval Board had not approved Ericsson's proposal for the ironclad warship, Winslow and his partner met with President Lincoln personally to advocate for its construction. Lincoln then arranged a meeting with them in the office of the Secretary of the Navy. During the subsequent meetings, the Navy resisted the project but finally approved but without funding. Largely out of frustration, Winslow arranged to finance the project himself, at a cost of then 275,000 US$. Then construction began rapidly. Interim reimbursements were made by the Navy but by the time of the battle, the final payments had not yet been made. Therefore, the *Monitor* remained the property of Winslow. Some of the iron for the Monitor was produced at the Albany Iron Works at Troy NY. The boat was launched 101 days from the contract signing, in time to defend the Union blockade during the Battle of Hampton Roads. Along with Ericsson, Winslow and his partner received much praise for their efforts in producing the *Monitor*, and received additional contracts for ironclad warships. In 1865 Winslow was appointed president of the Rensselaer College, Troy NY. He also was the director of several banks and of a railroad.

Anonymous (1936). John Flack Winslow. *Dictionary of American biography* 1928-1936 20: 399-400. Scribner: New York.

Kuffner, W. (2012). John F. Winslow and the USS Monitor. *The Hudson River Valley Review* 28(2): 94-104. *P*

Weise, A.J. (1886). *The City of Troy and its vicinity*. Green: Troy.

http://www.rpi.edu/about/alumni/inductees/winslow.html *P*

http://www.johnflackwinslow.com/ *P*

WINSOR

16.11. 1870 Johnston RI/USA
30.01. 1939 West Newton MA/USA

Frank Edward Winsor graduated in 1891 with the BSc degree from Brown University, Providence RI, the CE degree in 1892, and he was awarded in 1929 the honorary degree of Doctor of Science. He began his engineering career with the Metropolitan Sewage Commission of Massachusetts as assistant engineer on the construction of the Shirley Gut Siphon and a pumping station. He then worked for the Wachusett Dam and Reservoir during two years, in charge of the Weston Aqueduct Department, Saxonville MA, including a masonry aqueduct and a 2 m steel pipe siphon across Happy Hollow. He was engaged from 1903 with the Commission of Additional Water Supply for New York City, in charge as principal assistant engineer of field investigations. His work covered more than 160 km of aqueduct line, but in the same year was called back to Massachusetts by the Charles River Basin Commission as division engineer in charge of the design of dam, locks, and flood-control structures. In 1906 he went back to the New York Board of Water Supply as senior division engineer for the construction of the 50 km long Catskill Aqueduct, and the Kensico and Hill View Reservoirs.

In 1915, the newly organized Water Supply Board of Providence RI accepted a report for a new water supply, and called Winsor for its chief engineer. The work included the construction of a storage reservoir and a 55 m high earth dam, 20 km of aqueduct and pipe lines, and a 5 km long tunnel. In 1926 he started work with the Metropolitan Water District involving the extension from Wachusett Reservoir. The new Quabbin Reservoir in the Swift River Valley had a capacity of 1.57 km^3 and is 50 m deep. The gravity connection to Wachusett Reservoir involves a 40 km long tunnel, completed in 1934. Two earth dams impound the waters of Quabbin Reservoir, which were completed by Winsor in 1937; the larger main dam is 780 m long, 90 m high from the rock, containing 3.6 million of m^3. It was nearly completed at Winsor' death, and was named Winsor Dam in his memory. He was ASCE member since 1905 and vice-president in 1930.

Anonymous (1940). Frank E. Winsor. *Trans. ASCE* 105: 1957-1962.
Updike, D.B. (1939). *A memorial to Frank Edward Winsor*, chief engineer. Boston.
Winsor, F.E. (1934). Boston's new Metropolitan Water Supply. *Civil Engineering* 4(6): 283-287.
Winsor, F.E. (1937). Quabbin Dams and Aqueduct. *Trans. ASCE* 102: 682-711.
http://en.wikipedia.org/wiki/Frank_E._Winsor
http://records.ancestry.com/Frank_Edward_Winsor_records.ashx?pid=144986632 *P*

WISLER

25.02. 1881 Nappanee IN/USA
24.12. 1961 Ann Arbor MI/USA

Chester Owen Wisler obtained his BCE in 1913 and the MSE in 1915 from the University of Michigan, Ann Arbor MI. He commenced an instructorship at his Alma Mater, later becoming a professor of civil engineering, which has been considered of notably high quality and had profoundly contributed to the Faculty. In the affairs of the University and of his College he has participated effectively through his membership on numerous committees, among which his service as a Board member of Directors of the Michigan Union, and as its financial secretary.

Wisler had been a loyal and valued faculty member. An excellent teacher, he maintained an active connection with the professional practice by serving as consultant on the water power and flood study problems to numerous organisations in the Michigan region. He was wider known in the US by several publications in the mentioned fields. The book Hydrology co-authored by Ernest Frederick Brater (1912-2003) includes chapters on runoff, floods, and streamflow records. Wisler also supported Horace Williams King (1874-1951) in writing the well-known book Hydraulics, one of the key publications in the early 20th century. The 1942 paper describes the successful use of a direct method in flood routing, which depends mainly on the availability of streamflow records during a flood at various points of the main stream. No cross-sections of stream channels or flow velocities are required, however, nor discharge records on all of the tributaries. An inflow hydrograph from the unmeasured area may directly be computed. This discharge and that at each of the upstream stations is then routed downstream. These discharges indicate the extent to which each of the upper tributaries contribute to the flood peak at each downstream point. A check on the accuracy of the results was provided by adding the routed discharges and comparing the resulting hydrograph with the actual records. Not only for his attainments as engineer and teacher of several generations of engineers, but also by its many excellent qualities of character and personality, Wisler merited the friendship and respect accorded him by his students and colleagues.

Anonymous (1951). Chester Owen Wisler. *Regents' Proceedings* (June 1): 1334. *P*
Jania, K. (2011). Chester Owen Wisler. Personal communication. *P*
King, H.W., Wisler, C.O. (1922). *Hydraulics*. Wiley: New York.
Wisler, C.O., Brater, E.F. (1942). A direct method of flood routing. *Trans. ASCE* 107: 1519-1562.
Wisler, C.O., Brater, E.F. (1949). *Hydrology*. Wiley: New York, 2nd ed. in 1959.
http://um2017.org/faculty-history/faculty/chester-owen-wisler/memoir *P*

WISNER G.M.

09.02. 1870 Detroit MI/USA
26.08. 1932 Chicago IL/USA

George Monroe Wisner graduated in 1892 with the BSc degree in civil engineering from University of Michigan, Ann Arbor MI. He joined in 1894 the Sanitary District of Chicago, having charge of the hydraulic and design work until 1897, dredging and docking of the Chicago River until 1901, directing then all the construction work including coffer dams and the water power development, becoming then in 1909 chief engineer. During his professional career, Wisner was connected throughout with the Sanitary District, from 1909 mainly with operation and work in maintenance.

When Chicago grew rapidly between 1900 and 1910, Wisner foresaw the necessity of sewage treatment to supplement the existing dilution system. He convinced the Board of Trustees to study a sanitary program, securing in 1909 a test station for domestic sewage. He also initiated the Sanitary Engineering Staff, equipped with a laboratory to handle the biological, bacterial, and chemical problems. He thereby developed a keen interest in sewage treatment, as was then practised in the USA and abroad. In 1910 he visited England, France and Germany to inspect their sewage works. In 1911 he reported to the Board of Trustees on sewage disposal works for the Sanitary District, so that the Des Plaines River Sewage Treatment Works and the Calumet Sewage Treatment Works were designed, the first units of the system. He also was one of the witnesses in the case of the USA versus the Sanitary District, and later in the hearings before the Rovers and Harbours Committee of Congress, becoming an expert in these matters. He resigned as chief engineer with the Sanitary District in 1920, entering then into private practice, but he was retained by the District as a consultant until his death. During this period he conducted harbour works in Labrador, dock work and shore protection in Florida, and drainage work in Cook County IL. His loyalty to his associates brought him enduring friendships. His ability to judge men quickly and strip matters presented for his decision to the essentials made him an effective executive. He possessed tact and an excellent business ability combined with engineering talent. He was member of the American Society of Mechanical Engineers ASME, and the Society of Civil Engineers ASCE.

Anonymous (1933). George M. Wisner. *Trans. ASCE* 98: 1662-1664.
Wisner, G.M. (1911). *Report on sewage disposal.* Sanitary District: Chicago IL.
Wisner, G.M. (1917). *Report relating to existing lake levels.* Sanitary District: Chicago IL.
http://www.historicbridges.org/bridges/browser/?bridgebrowser=illinois/evanstonrr/ *P*

WISNER G.Y.

11.07. 1841 West Dresden NY/USA
03.07. 1906 Detroit MI/USA

George Young Wisner graduated as civil engineer in 1865 from the University of Michigan, Ann Arbor MI. He then entered the employ of the government as US assistant engineer, being there employed for 27 years in government work and surveys of the Great Lakes, the Mississippi River, Des Plains and Illinois Rivers, and both the 10th and 11th Lighthouse Districts. In 1887 he resigned this employment to engage in private practice, from when he was in charge of a large number and important projects as a consulting engineer. He was chief engineer of the Brazos River Harbour improvement in Texas, then a consulting engineer to the Montreal, Ottawa & Georgian Bay Canal Co., and he had been during his last two years one of the consultants of the US Reclamation Service.

The most important work of Wisner was probably his service on the Board of Engineers on Deep Waterways between the Great Lakes and the Atlantic Seaboard. While the canal proposed by this Board was never executed, the investigations and the report made by the Board ranked as one of the most thorough and systematic pieces of engineering work undertaken in this field. Wisner was further involved in the design of the Pathfinder Dam on the North Platte River some 70 km southwest of Casper WY. It was originally constructed between 1905 and 1909 as cyclopean masonry dam, but has several times been modified since. The gravity dam is 65 m high, 132 m long, with a base width of 30 m and a crest width of 3.2 m. Wisner was at the time of his death also member of the International Commission on the Levels of the Great Lakes. He was a contributor to the professional literature, of which many papers were published in the *Engineering News* Journal. He was one of the best engineers in the Central West. He was member of the American Society of Mechanical Engineers ASME, and Civil Engineers ASCE.

Anonymous (1906). George Y. Wisner. *Engineering News* 56(2): 48.

Williams, B., Wisner, G.Y. (1902). *Report on the water supply of the city of St. Louis.* Commission of Hydraulic Engineers: St. Louis MO. (*P*)

Wisner, G.Y. (1901). The economic discussions for a waterway from the Great Lakes to the Atlantic. *Trans. ASCE* 45: 224-331.

Wisner, G.Y. (1902). *Montreal, Ottawa and Georgian Bay Canal.* Ottawa.

Wisner, G.Y. (1898). Discussion of Reservoir system of the Great Lakes of the St. Lawrence Basin: Its relation to the problem of improving the navigation. *Trans. ASCE* 40: 428-431.

WOLMAN A.

10.06. 1892 Baltimore MD/USA
22.02. 1989 Baltimore MD/USA

Abel Wolman received education at Johns Hopkins University, Baltimore MD, obtaining the AB, the BSE, and D.Eng. degrees in 1913, 1915 and 1937, respectively. He joined from 1914 to 1938 the US Public Health Service, conducting studies in stream pollution. He was involved already in 1914 in the construction of a sewage treatment station, promoted in 1915 to research engineer and in 1917 to division engineer, advancing in 1922 to chief engineer. He in parallel was from 1921 at his Alma Mater Lecturer in sanitary engineering, and from 1925 also at Harvard and Princeton Universities. From 1927 Wolman was a consultant at Baltimore MD, and the chairman of the Potomac River Flood Control Commission. From 1931 to 1941, he served as chairman the Maryland Water Resources Commission, acted as State Engineer and as director of Maryland Public Works Agency. He was from 1935 chairman of the National Water Resources Commission. He was in parallel professor of sanitary engineering and consulting engineer at Baltimore.

In the 1919 paper, Wolman and a colleague established standards for the application of chlorine to drinking water. The benefits of using hypochlorite salts to kill the bacteria in water were demonstrated as early as 1896 by George W. Fuller (1868-1934). However, because no method existed in determining the adsorption of chlorine into different kinds of water, it could not be applied to drinking water reliably. In this paper, a formula was devised for calculating the amount of chlorine based on particular water conditions and desired qualities. Their method soon gained universal acceptance. Wolman's expertise later guided the National Research Council in its efforts to improve sanitary engineering and environmental health issues.

Anonymous (1952). Wolman first to receive new AWWA award. *Municipal Utilities* 90(3): 18. *P*
Anonymous (1959). Wolman, Abel. *Who's who in engineering* 8: 2723. Lewis: New York.
Anonymous (1961). Abel Wolman Hon. M. ASCE. *Civil Engineering* 31(10): 80. *P*
Anonymous (1989). Abel Wolman. *Eos* 70(35): 801-806. *P*
Jacoby, H. (1957). Abel Wolman: Engineer extraordinary. *Engineering News-Record* 158(May9): 87-91. *P*
Wolman, A., Enslow, L.H. (1919). Chlorine absorption and the chlorination of water. *Journal of Industrial Engineering Chemistry* 11(3): 209-213.
Wolman, A., Gorman, A.E. (1931). *The significance of waterborne typhoid fever outbreaks* 1920-1930. Williams & Wilkins: Baltimore MD.

WOLMAN M.G.

16.08. 1924 Baltimore MD/USA
24.02. 2010 Baltimore MD/USA

Markley Gordon 'Reds' Wolman, the son of Abel (1892-1989), received the BA degree in geology in 1949 from the Johns Hopkins University, Baltimore MD, and the MS and PhD degrees from the Harvard University, Cambridge MA. He worked then for the US Geological Survey USGS from 1951 to 1958. Guided by his mentor Luna B. Leopold (1915-1992) he focused his research on the rigorous, quantitative description of rivers and their beds, and the relative effectiveness of geomorphic processes in shaping the landscape. His Brandywine Creek study became a classic example of a comprehensive geomorphic examination of a channel. The 1955 paper then generated the 'Wolman pebble count' technique in geomorphology for sampling how particles are distributed on river beds. His later research in collaboration with Leopold was mainly concerned with meandering.

In 1958, Reds was hired by the Johns Hopkins University as chairman of the Isaiah Bowman Department of Geography. A decade later, when Geography merged with the Sanitary and Water Resources Engineering Department to become the Department of Geography and Environmental Engineering, he chaired it too from 1970 to 1990. Like his father, Wolman had novel, innovative ideas about education and research: he was interdisciplinary long before it became fashionable. His research included the effects of land use and urbanization on the evolution of the landscape, along with issues including water quality, sustainability, and the relationship between public policy and science. Later, he served as director of the Center for Environmental Health Engineering, the Bloomberg School of Public Health, in addition to his faculty position. His list of honours, awards and other accomplishments is a small book in itself. Wolman was elected to the National Academy of Sciences INAS in 1988, and was recipient of the National Medal of Science.

Leopold, L.B., Wolman, M.G. (1957). River channel patterns: Braided, meandering and straight. *Professional Paper* 282-B. USGS: Washington DC.

Leopold, L.B., Wolman, M.G. (1957). River flood plains: Some observations on their formation. *Professional Paper* 282-C. USGS: Washington DC.

Leopold, L.B., Wolman, M.G., Miller, J.P. (1964). *Fluvial processes*. Freeman: San Francisco.

Wolman, M.G. (1955). The natural channel of Brandywine Creek PA. *Professional Paper* 271. USGS: Washington DC.

http://eng.jhu.edu/wse/Reds_Wolman/memory/pages/ *P*

http://webapps.jhu.edu/jhuniverse/featured/very_able_wolmans/ *P*

WOOD D.M.

22.04. 1884 St. Louis MO/USA
10.05. 1954 Wellesley MA/USA

Dana Melvin Wood obtained in 1906 the BSc degree in civil engineering from the Massachusetts Institute of Technology MIT, Cambridge MA. He began then his professional career as assistant engineer with service for the Water Resources Branch of the US Geological Survey USGS, joining in 1934 the Tennessee Valley Authority TVA, where he held the positions of senior and head hydraulic engineer, and later chief of the power studies branch in the Division of Water Control Planning, retiring in 1954. Wood was a member of the American Society of Civil Engineers ASCE from 1921, becoming Life Member in 1950. He was awarded the Desmond Fitzgerald Medal of the Boston Society of Civil Engineers in 1917.

The technical interests of Wood included rainfall and run-off as hydrological processes, power estimates for hydro-power facilities, and the design of multi-purpose reservoirs. The 1922 discussion of a paper of Carl E. Grunsky (1884-1934), studying the rainfall distribution of California, intends to clarify whether each watershed has its own specific characteristics, or if the rainfall pattern follows universal climatic relations. Wood recommends the use of flow-duration curves if the data are poor. It is further stated that working from the rainfall to the estimated runoff, the distribution follows in accordance with that in similar basins for which reliable data bases exist. The purpose of Wood's discussion offered additional suggestions proving helpful in difficult cases. The 1917 discussion on the effect of channel shape on stream flow results in the statement that the geometry, slope, and roughness of a channel have a significant effect on the discharge-head curve. Once with TVA, Wood developed plans for a unified development of the water resources of the Tennessee Valley, identifying sites which would satisfy the basic tenants of the TVA Act, namely the combined development for flood control, power and navigation.

Anonymous (1955). Dana M. Wood. *Trans. ASCE* 120: 1579-1580.
Wood, D.M. (1916). Power estimates from stream flow and rainfall data. *Journal of the Boston Society of Civil Engineers* 3(3): 77-120; 3(6): 324-328. (*P*)
Wood, D.M. (1917). Discussion of Effect of channel on stream flow. *Journal of the Boston Society of Civil Engineers* 4(3): 126-128.
Wood, D.M. (1922). Discussion of Rainfall and run-off studies. *Trans. ASCE* 85: 106-111.
Wood, D.M. (1950). Discussion of Multi-purpose reservoirs: Development of policy by the Bureau of Reclamation. *Trans. ASCE* 115: 905-906.

WOOD D.V.

01.06. 1832 Smyrna NY/USA
28.06. 1897 Hoboken NJ/USA

De Volson Wood graduated in 1857 from Rensselaer Polytechnic Institute as a civil engineer, and at once became assistant professor of civil engineering at the University of Michigan, Ann Arbor MI, promoted to professor two years later. In 1872 he was appointed professor of mathematics and mechanics at Stevens Institute of Technology, Hoboken NJ, and he was transferred to its chair of engineering in 1885. In 1881 he was elected president of the Board of Education at Boonton NJ. The honorary degree AM was conferred on him by Hamilton College, Clinton NY, and the MS degree by the University of Michigan, both in 1895. Wood was member of the American Society of Civil Engineers ASCE, and the American Association for the Advancement of Science.

Wood designed the ore-dock of Marquette MI in 1866 and invented a rock drill, a steam pump, and an air-compressor. He has published a number of papers and books, notably on the resistance of materials, on the theory of the construction of bridges and roofs, on analytical mechanics, on elementary mechanics, and on elements of coordinate geometry. Of particular relevance here are his books on thermodynamics, and on fluid mechanics. His influence as teacher was renowned. The American Mathematical Journal wrote the notable statement: 'The civil, mechanical and electrical engineers who formerly were Prof. Wood's students, and who are now scattered all over the world, would, if simultaneously rounded up, form the most intelligent army that ever moved on the face of this mundane sphere'.

Anonymous (1889). Wood, De Volson. *Appletons' cyclopaedia of American biography* 6: 591-592. D. Appleton & Co.: New York.
Anonymous (1897). De Volson Wood. *Trans. ASME* 18: 1106-1109.
Wood, V. de (1873). Backwater in streams as produced by dams. *Trans. ASCE* 2: 255-261.
Wood, V. de (1879). On the flow of water in rivers. *Trans. ASCE* 8: 173-178.
Wood, V. de (1884). *Mechanics of fluids*. Wiley: New York.
Wood, V. de (1888). *A treatise on civil engineering*. Wiley: New York.
Wood, V. de (1889). *Thermodynamics*: Heat motors and refrigeration machines. Wiley: New York.
Wood, V. de (1896). *Turbines, theoretical and practical*: With numerical examples and experimental results and many illustrations. Wiley: New York.
http://www.personal.psu.edu/gal4/AcademicLineage/AcademicLineage.html#anchorWood *P*

WOODBURN

30.11. 1894 Bloomington IN/USA
29.08. 1980 Madison WI/USA

James Gelston Woodburn obtained the BS degree in 1918 from Purdue University, West Lafayette IN, and the PhD degree in 1929 from the University of Michigan, Ann Arbor MI. He was from 1920 to 1923 associate engineer with the US Forest Products Laboratory, Madison WI, in 1924 a mathematics instructor at Indiana University, Bloomington IN, until 1927 an instructor in civil engineering at the State College of Washington, Pullman WA, until 1929 a research assistant at University of Michigan, and Freeman travelling scholar to Europe in 1930, from when he was associate professor of hydraulics at State College of Washington until 1937, taking then over until his retirement in 1965 as professor of hydraulic engineering at University of Michigan. He there served as chairman of the Department of Civil Engineering from 1949 to 1958. Woodburn was a member of the American Society of Civil Engineers ASCE, president of the Michigan Section in 1948, among many other commitments.

Woodburn is known for a classical paper on broad-crested weirs, a particular type of overflow structure used mainly for educational purposes. Whereas the sharp-crested weir serves for the precise discharge measurement, and the standard-shaped weir is employed in dam engineering, the broad-crested weir of standard shape has a horizontal crest of finite length, with vertical weir faces both up- and downstream of the crest. Woodburn observed the free surface profiles in the crest vicinity resulting in standing wave patterns if the relative weir length was large, and continuously decreasing flow depth as the relative weir length reduces. The hydraulic behaviour of broad-crested weirs with sloping aprons was furthermore analysed. The paper received a large number of discussions relating mainly to the effect of streamline curvature, the definition of critical depth, and the various flow types. Woodburn was also an author of the successful book Hydraulics by Horace W. King (1874-1951).

Anonymous (1944). James G. Woodburn. *Civil Engineering* 14(11): 489. *P*
Anonymous (1954). Woodburn, James G. *Who's who in engineering* 7: 2692. Lewis: New York.
Anonymous (1963). Woodburn, J.G. *Who's who in America* 32: 3447. Marquis: Chicago
King, H.W., Wisler, C.O., Woodburn, J.G. (1948). *Hydraulics*. Wiley: New York.
Woodburn, J.G. (1932). Tests of broad-crested weirs. *Trans. ASCE* 96: 387-453.
Woodburn, J.G. (1956). Hydraulics. *American civil engineering practice*. Wiley: New York.
http://www4.lib.purdue.edu/archon/?p=collections/findingaid&id=994&q=#boxfolder

WOODMAN

18.10. 1909 Orono ME/USA
14.03. 1999 Vicksburg MS/USA

Eugene Harvey Woodman graduated from Missouri School of Mines and Metallurgy, Rolla MO, with the BS degree in 1930. He joined in 1933 the US Waterways Experiment Station, Vicksburg, working on hydraulic models until 1934, was a designer of the Engineering Construction Division and a project engineer until 1939, when joining the US Army. In 1945 he returned to the Experiment Station, finally as senior engineer of the Instrument Branch, and as assistant chief of the Technical Sciences Division. All through his life Woodman was a devoted music lover, playing himself the piano in orchestras.

From the 1940s to the 1960s Woodman was responsible for the design and the operation of special control and measuring devices for hydraulic models and equipment in soil mechanics. He invented, among many other, the Woodman deflection gage for airport pavement tests. He also developed a wave height measuring apparatus for both model and field use, and an indicator to display the direction and velocity of currents in ship canals. He has written numerous papers both in technical journals and for conferences. A major involvement was the design of an automatic instrument for controlling the inflows and stages of the Mississippi Basin Model, then the world's largest hydraulic model covering an area of 3 km^2. This model, located at Clinton MS, was built between 1943 and 1966 simulating a large portion of the Mississippi watershed. The model was created during the paradigm shift in engineering, moving hydraulic research from on-site sampling and experience to a method of simulation and prototyping. The model was built with a 1:2000 horizontal scale, a 1:100 vertical scale and 1:200 time scale. Surface friction was adjusted using a variety of media, rocks adhered to the channel, wire mesh, lengths of metal posts, and metal grating. The size and scale of the model allowed for scientists and stakeholders to directly view the consequences of the various hydraulic control operations. The model was abandoned in the late 1980s because it was realized that it cannot adequately reproduce the major phenomena due to scale effects.

Anonymous (1964). Woodman, E.H. *Who's who in engineering* 9: 2071. Lewis: New York.
Rouse, H. (1976) Eugene Harvey Woodman. *Hydraulics in the United States* 1776-1976: 156-157. The University of Iowa: Iowa City. *P*
Woodman, E.H. (1952). *Instrumentation for dynamic measurements of vertical oscillations on hydraulic models of dam gates*. Waterways Experiment Station: Vicksburg MS.
Woodman, E.H. (1955). Pressure cells for field use. *Bulletin* 40. WES: Vicksburg MS.

WOODWARD

11.05. 1871 Richfield MN/USA
07.09. 1953 Knoxville TE/USA

Sherman Melville Woodward obtained in 1893 his MS degree from Washington University, Saint Louis MO, and the MA degree in 1896 from the Harvard University, Cambridge MA. He was a professor of mechanics and physics then until 1904 at University of Arizona, Tucson AZ, then irrigation and drainage engineer at the US Department of Agriculture until 1908, from when he was until 1934 professor of mechanics and hydraulics at the University of Iowa, Iowa City IA. He served in parallel as a consulting engineer the Miami Conservancy District, Dayton OH, from 1913 to 1920, and from 1925 to 1929 the Chicago Sanitary District. He also was a consultant planning engineer from 1933 for the Tennessee Valley Authority TVA. Woodward was a member of the American Society of Civil Engineers ASCE, and Mechanical Engineers ASME. He was elected Honorary Member ASCE in 1943.

Woodward was an expert in hydraulics mainly relating to flood flows. His 1941 book jointly written with Chesley J. Posey (1906-1991) includes the chapters: 1. Introduction, 2. Bernoulli's theorem, 3., 4. Steady hydraulic jumps, 5. Moving hydraulic jump, 6., 7. Backwater curves, 8., 9. Methods for backwater curves, 10. Bends, transitions and obstructions, 11. Slowly-varied flow. This work was mainly addressed to undergraduate students and professional engineers. During his time at Dayton OH, he took interest in hydraulic jumps and energy dissipators, extending the theory to expanding channels and applying as the first the phenomenon to large-scale tests. Later he was also involved in the designs of Norris and Wheeler Dams of TVA. His clear perception and integrity of purpose was invaluable in resolving conflicts. One of his Iowa students stated: 'He was one of the best, if not the best teacher I ever knew. He is a well-educated gentleman'.

Anonymous (1941). Woodward, S.M. *Who's who in engineering* 5: 1985. Lewis: New York.
Anonymous (1943). Sherman M. Woodward. *Engineering News-Record* 130(Jan.28): 108. *P*
Anonymous (1943). Sherman M. Woodward. *Civil Engineering* 13(1): 53-54. *P*
Anonymous (1951). Woodward, S.M. *Who's who in America* 26: 3019. Marquis: Chicago.
Anonymous (1953). ASCE Honorary Member Woodward dies. *Civil Engineering* 23(10): 709. *P*
Woodward, S.M., Nagler, F.A. (1929). The effect of agricultural drainage upon flood run-off. *Trans. ASCE* 93: 821-839.
Woodward, S.M., Posey, C.J. (1941). *Hydraulics of steady flow in open channels*. Wiley: NY.
Yarnell, D.L., Nagler, F.A., Woodward, S.M. (1926). Flow of water through culverts. Studies in Engineering *Bulletin* 1. University of Iowa: Iowa City IA.

WORTHEN

14.03. 1819 Amesbury MA/USA
02.04. 1897 Brooklyn NY/USA

William Ezra Worthen graduated in 1838 from the Harvard University, Cambridge MA. He began his professional career as an assistant for the water supply and hydraulic work at Lowell and Boston MA, then associated with James B. Francis (1815-1892). From 1840 to 1842 Worthen was engaged with a nearby railroad. Returning to Lowell MA, he designed and built with Francis a number of dams and mills. After a visit to Europe, Worthen settled in New York in 1849, engaging in building and mill construction. These included a dam across *Mohawk* River at Cohoes NY, and floating docks of Jersey City. Worthen was widely known as an expert in pumping machinery, so that he was called both upon design and to test such machinery in New York, Cincinnati, and St. Louis. He also selected pumping engines for Boston, and tested large pumping units at Brooklyn NY, or Philadelphia PA. He had profound practice in measuring the discharge of water in canals, reporting again for a number of cities.

From 1866 to 1869, Worthen was the sanitary engineer to the New York Metropolitan Board of Health. In Brooklyn NY he reported upon an extensive addition to the sewer system. With Francis and Theodore G. Ellis (1829-1883), he served upon a committee to report on the failure of the Mill River Dam at Williamsburg MA, which had occurred in 1874. Around 1890 Worthen was the chief engineer of the Chicago Main Drainage Canal. He was president of the American Society of Civil Engineers in 1887, and made honorary member ASCE in 1893.

Anonymous (1889). Worthen, William E. *Appletons' cyclopaedia of American biography* 6: 617.
Anonymous (1936). Worthen, William Ezra. *Dictionary of American biography* 20: 538-539. Scribner's: New York.
Francis, J.B., Ellis, T.G., Worthen, W.E. (1874). Failure of the dam on the Mill River. *Trans. ASCE* 3: 118-122.
Hunt, C.W. (1897) Worthen. *Historical sketch of ASCE*: 49-50 ASCE: New York. *P*
Worthen, W.E. (1848). *Report* upon the subject of introducing pure water into the city of Lowell. Norton: Lowell.
Worthen, W.E. (1862). *First lessons in mechanics*. Appleton: New York.
Worthen, W.E. (1869). *A practical treatise on architectural drawing and design*. Appleton: NY.
Worthen, W.E. (1896). *Report* on the future extension of water supply for the city of Brooklyn. Dept. City Works: Brooklyn.

WORTHINGTON

17.12. 1817 New York NY/USA
17.12. 1880 Brooklyn NY/USA

Henry Rossiter Worthington was already interested in mechanical problems as a boy. He concentrated his attention on problems of the city water supply once having left basic schooling, becoming familiar also with the steam engines and mechanical pumps. Canal navigation interested him too, and it was in this connection that he made his first invention. In 1840 he had an experimental steam canal-boat in operation which was fairly successful except that when the boat was stopped it became necessary to resort to a hand pump to keep the boiler supplied with water. Worthington invented an independent feeding pump to overcome this deficiency automatic in its action and controlled by the water level within the steam boiler.

After having pursued these tests, Worthington turned from 1845 his attention to pumping machinery, perfecting a series of inventions until 1855, which made him the first to propose and construct the direct steam-pump. In 1859, after having established a pump manufacturing plant in New York, he perfected his duplex steam feed pump, and in the following year built the first water-works engine of this kind. In the duplex system, one engine actuates the steam valves of the other, and a pause of the piston at the end of a stroke permits the water-valves to seat themselves quietly and preserve a uniform water pressure. A distinct improvement on the Cornish engines used then, Worthington's pump embodied most ingenious advances in engineering so that its principle was widely applied. Because of their reliability and low operating cost, these pumps were greatly used for waterworks and later for pumping oil through long pipelines in the oil fields. Worthington was founding member of the American Society of Mechanical Engineers.

Anonymous (1889). Worthington, Henry Rossiter. *Appletons' cyclopaedia of American biography* 6: 617. D. Appleton & Co.: New York.
Anonymous (1936). Worthington, Henry Rossiter. *Dictionary of American biography* 20: 539. Scribner's: New York.
Anonymous (1940). 100 *years of Worthington*. Worthington Pump and Machinery: New York.
Anonymous (1980). Worthington. *Mechanical engineers in America born prior to 1861*: 320-321. ASME: New York. *P*
Hirschfeld, F. (1980). Henry A. Worthington. *Mechanical Engineering* 102(3): 24-25. *P*
Worthington, H.R. (1876). *The Worthington steam pumping engine*. Worthington: New York.
http://en.wikipedia.org/wiki/Henry_Rossiter_Worthington *P*

WRIGHT

10.10. 1770 Wethersfield CT/USA
24.08. 1842 Brooklyn NY/USA

Benjamin Wright assisted the celebrated English engineer William Weston (1752-1833) in surveys of canals for what became later the Erie Canal. These works commenced in 1817, with Wright as engineer for the middle canal section first, and later as chief engineer, after Weston had declined to take charge of this project. The Erie Canal became the first school of civil engineering in the USA, 'professor Wright' being the best man available. Among his 'students' was for instance John B. Jervis (1795-1885). Once work on the Canal was well underway, Wright was prevailed upon to locate other canal routes. He was fifty-five years old when reaching the apex of acclamation of his career at the 'Great celebration of the opening of Erie Canal' on Nov. 7, 1825.

In 1827, Wright made his first studies on the use of steam locomotives. In conjunction with Jervis and his assistant Horatio Allen (1802-1890), the possibility of steam railways was investigated. He also was a consultant on the Blackstone Canal in Rhode Island and chief engineer of the Chesapeake and Ohio Canal, Maryland. In 1832 he was appointed chief engineer of the St. Lawrence Ship Canal, but later mainly worked on railways projects. Recognition of Wright is best described by citing three examples. In 1839 he prepared the constitution of a proposed society of civil engineers, which however came only into existence in 1852 as the American Society of Civil Engineers. It was only in 1968 that the professional legatees declared Wright as Father of the American Civil Engineering. On his bicentennial of birth in 1970, a bronze plaque was dedicated at his birthplace. His career included bridges, canals and railroads, with his assistants forming the nucleus of the American civil engineering profession during the early 19th century.

Anonymous (1970). Ceremony marks Connecticut birthplace of B. Wright. *Civil Engineering* 40(12): 65. *P*

Fitzsimons, N. (1970). Benjamin Wright: The Father of American Civil Engineering. *Civil Engineering* 40(9): 68. *P*

Langbein, W.B. (1975). Our Grand Erie Canal: A splendid project, a little short of madness. *Civil Engineering* Special Issue: 60-66.

Shank, W.H. (1982). *Towpaths to tugboats*. ASCE: New York. *P*

Stuart, C.B. (1871). *Lives and works of civil and military engineers of America*. Van Nostrand: New York.

Weingardt, R.G. (2005). *Engineering legends*: Great American civil engineers. ASCE: Reston.

YARNALL

28.06. 1878 Middletown PA/USA
11.09. 1967 Philadelphia PA/USA

David Robert Yarnall graduated from the University of Pennsylvania, Philadelphia PA, with a BS degree in 1901, and the MS degree in 1905 in mechanical engineering. He was recipient of the D.Eng. degrees from Lehigh University in 1942 and from Haverford College in 1947. He was design engineer from 1907 to 1912, vice-president of a valve company until 1918 at Philadelphia PA, director and president of Yarnall-Waring *Yarway* Company then until World War 2. He was member of the American Society of Mechanical Engineers ASME, recipient of its 1941 Hoover Medal Award, serving as its president in 1946 and awarded honorary membership in 1949. He also was a member of the Franklin Institute, and the Engineers Club of Philadelphia, presiding over it from 1926 to 1930.

Yarnall was a prominent industrialist, engineer, and civic leader. The Yarway company manufactured valves, steam traps, gauges and other engineering products for the power and processing industries. These devices were in use in steam-operated ships or in high-pressure steam-generating plants. He later served as director, president, and chairman of the board of another company, a manufacturer of scientific instruments. Besides, he authored papers including one which set forth his deep conviction on The engineer's responsibility in civic affairs. The technical papers considered essentially accuracy of the then current methods of discharge measurement, with a particular aspect to the V-notch weir. Before the inductive discharge measurement became common in hydraulic engineering, the standard V-notch was considered the most accurate to determine fluid discharges, typically to 1% of the test value.

Anonymous (1917). D. Robert Yarnall. *Journal ASCE* 39(10): 870. *P*
Anonymous (1939). D. Robert Yarnall. *Power* 83(4): 192. *P*
Anonymous (1945). D. Robert Yarnall. *Mechanical Engineering* 67(8): 551-552. *P*
Anonymous (1959). Yarnall, D.R. *Who's who in America* 28: 2942. Marquis: Chicago.
Anonymous (1967). D. Robert Yarnall, industrialist and past-president ASME. *Mechanical Engineering* 89(11): 121. *P*
Stuart, M.C., Yarnall, D.R. (1936). Fluid flow through two orifices in series. *Mechanical Engineering* 58(8): 479-484; 58(11): 744-746; 66(7): 387-397.
Yarnall, D.R. (1912). The V-notch weir method of measurement. *Trans. ASME* 34: 1055-1072.
Yarnall, D.R. (1926). Accuracy of the V-notch weir method of measurement. *Trans. ASME* 48: 939-964.

YARNELL

13.01. 1886 Storm Lake IA/USA
09.03. 1937 Iowa City IA/USA

David Leroy Yarnell graduated from the Iowa State College, Ames IA, with a BS degree in 1908, and the CE degree in 1916. He obtained the MSc degree from the State University of Iowa, Iowa City IA, in 1926. He was assistant drainage engineer with the US Department of Agriculture USDA from 1909 to 1911, then drainage engineer until 1916, taking over as senior drainage engineer until retirement. He was member of the American Society of Civil Engineers ASCE, and Agricultural Engineers ASAE. He was awarded the 1932 ASCE James R. Croes Medal for his 1931 paper.

Yarnell spent his early years in the field in connection with drainage engineering. From 1911 he was placed in charge of surveys and designs of drainage systems, technical investigations on runoff from drained land, and on groundwater of tiled land. He for instance prepared a plan for the Cypress Creek Drainage District in Arkansas. From 1916 he conducted systematic tests on the water flow in drain-tile and vitrified pipes, resulting in a widely known publication on open channel flow and its relation to surface roughness. Later he investigated in collaboration with Floyd A. Nagler and Sherman M. Woodward one of the very first hydraulic studies on culverts, an important element in flood control schemes. In 1929 he conducted hydraulic model tests to determine the benefits of proposed cut-offs of the Des Moines River at Ottumwa IA. This study was one of the first in the USA to predict the benefits to be derived from straightening rivers.

Anonymous (1933). David L. Yarnell. *Civil Engineering* 3(1): 47. *P*

Anonymous (1937). Yarnell, D.L. *Who's who in engineering* 4: 1553. Lewis: New York.

Anonymous (1938). David L. Yarnell. *Trans. ASCE* 103: 1922-1924.

Yarnell, D.L. (1920). The flow of water in drain tile. *Bulletin* 854. US Department of Agriculture: Washington DC.

Yarnell, D.L. (1924). *Comparison of discharge measurements by weir, Pitot tube, and current meter*. University of Iowa: Iowa City IA.

Yarnell, D.L., Nagler, F.A., Woodward, S.M. (1926). Flow of water through culverts. Studies in Engineering *Bulletin* 1. Iowa City IA.

Yarnell, D.L., Nagler, F.A. (1931). Effect of turbulence on the registration of current meters. *Trans. ASCE* 95: 766-860.

Yarnell, D.L. (1934). Bridge piers as channel obstructions. *Technical Bulletin* 442: 1-51. US Dept. of Agriculture: Washington DC.

YOUNG

07.05. 1885 Butler IN/USA
22.04. 1982 San Rafael CA/USA

Walker Rollo Young obtained in 1908 the BS degree in mining engineering from the University of Idaho, Moscow ID, and the honorary PhD degree there in 1935. He was surveyor from 1909 to 1911, assistant engineer until 1916 of the US Reclamation Service on Arrowrock Dam in Idaho, until 1920 engineer of the Chief Engineer's Office, Denver CO, and until 1924 engineer in charge of studies of Colorado River leading to the selection of the Boulder Dam site. He then was engineer in charge of investigations of the Salt Water Barrier in San Francisco Bay until 1926, advancing until 1935 to construction engineer of the Boulder Dam, taking over until 1938 as construction engineer the Central Valley Project in California. He there was assistant chief engineer, becoming in 1945 chief engineer.

Young worked for four years on the construction of the Arrowrock Dam at Boise ID. At that time, Arrowrock was the highest dam in the world. He then returned to Denver CO remaining at the USBR Office until selected to spearhead the study of a possible dam in Boulder Canyon on Colorado River. In 1921 he led a team of 58 men responsible for studying the feasibility of a dam site in either the Boulder or the Black Canyon. After completing the work in Boulder Canyon, the party moved downstream to Black Canyon in search of a more ideal dam site. By 1924 the project was completed and a detailed report submitted to the Secretary of the Interior. It was determined that Black Canyon was the better site. After demonstrating his capabilities in supervising other reclamation projects, he was assigned to be engineer in charge of the Boulder Canyon Project. This position included supervision of all construction, the contractors, and Boulder City. After completion of the dam, it was assumed that Young would head USBR. However, he did not get the nod from the government in 1936 when Elwood Mead (1858-1936) died, the position went to John C. Page (1887-1955) instead. Young moved on to the Central Valley Project. He rose to the rank of chief engineer before his retirement in 1948. He then became president of Thompson Pipe and Steel Company. He died in a California nursing home in 1982 at the ripe old age of 97.

Anonymous (1948). Young, Walker R. *Who's who in engineering* 6: 2246. Lewis: New York.
Young, W.R. (1935). Boulder Dam: Past construction and work yet to be done. *Engineering News-Record* 115(Dec.26): 878-883.
Young, W.R. (1939). Preserving the Central Valley. *Civil Engineering* 9(9): 543-546.
http://www.hooverdamstory.com/young.htm *P*

ZINGG

20.12. 1906 Winfield IA/USA
07.04. 1959 Fort Collins CO/USA

Austin Wesley Zingg received in 1930 a BS degree in agricultural engineering from Iowa State College, Iowa City IA. He joined in 1934 the US Department of Agriculture USDA, beginning his career in soil and water conservation researches at Bethany MO. For his later assignments, he lived subsequently at Columbia MO, Manhattan KS, Fort Collins CO and in the Washington DC area. He was the first USDA ARS supervisor of the Wind Erosion Research Unit, holding his post from 1947 to 1953. Previously he was with the Soil Conservation Service SCS, Bethany MO, finally heading the Watershed Technology Research Branch, Beltsville MD.

Zingg was an authority on the mechanics and control of both wind and water erosion. He devised in 1951 a method based on vacuum cleaner cloth infiltration to measure the total soil movement near a ground surface for use in wind tunnel studies of soil erosion by wind. This technique involving the simultaneous air intake at elevations of sampling was made equal to the air velocity at the sampling heights. The total movement resulted by integration of the product of concentration and velocity over the vertical distance sampled, as described in 1953. This method was unsuitable for field work, however, so that William S. Chepil (1904-1967) in 1957 proposed fine glass wool filters packed into small round tubes. Hoses connecting these to the suction units were lengthened so that the dust was collected up to 6 m. Air meters and barometers were added for accurate checks of the air volume intake. Again this method proved to be too complicated, so that the collector was modified, as proposed by Ralph A. Bagnold (1896-1990).

Anonymous (1959). Austin W. Zingg. *Agronomy Journal* 51(12): 767.

Smith, D.D., Whitt, D.M., Zingg, A.W., McCall, A.G., Bell, F.G. (1945). Investigations in erosion control and reclamation of eroded Shelby and related soils at the Conservation Experiment Station Bethany MO, 1930-42. *Technical Bulletin* 883. USDA: Washington.

Zingg, A.W. (1951). A portable wind tunnel and dust collector developed to evaluate the erodibility of field surfaces. *Agronomy Journal* 43(4): 189-191.

Zingg, A.W. (1953). Wind-tunnel studies of the movement of sedimentary material. Proc. 5th *Hydraulics Conference* Iowa: 111-135, J.S. McNown, M.C. Boyer, eds.

Zingg, A.W., Chepil, W.S., Woodruff, N.P. (1965). Sediment transportation mechanics: Wind erosion and transportation. *Journal of the Hydraulics Division* ASCE 91(HY2): 267-287.

http://trees.ancestry.com/tree/36177896/person/18895445739/photox/97ad2464-bbdd-4bc9-bcfd-69afaa69a4ad?src=search *P*

ZUCROW

15.12. 1899 Kiev/RU
05.06. 1975 Santa Barbara CA/USA

Maurice Joseph Zucrow was born in the Ukraine. He emigrated in 1900 with his parents to the United Kingdom, and to the USA in 1915. He began his engineering career in 1929, working mainly in the fields of gas turbines and jet propulsion. He was an early pioneer in one of the first rocket companies of the USA from 1942 to 1946. He held between 1946 and 1953 a joint appointment between aeronautical and mechanical engineering at Purdue University, West Lafayette IN. During this period Zucrow made tremendous progress in developing a top graduate and research program in rocket and jet propulsion. The first and second stages of the Rocket Engine Laboratory were completed at Purdue University under his direction. In 1954 he took all of the gas turbine and jet propulsion work to the new rocket Lab, so that the Purdue Aero School was left with no graduate program in this field.

Zucrow was an expert in fluid mechanics and gas dynamics. His 1929 paper deals with jets issued by submerged square-edges and chamfered short tubes. Using dimensional analysis and laboratory experiments, it is demonstrated that the coefficient of discharge for geometrically similar jets depends only from the discharge, the fluid viscosity, and the jet diameter. The effect of changing the chamfer angle was also studied, as well as the tube length-diameter ratio. His books Principles of gas dynamics, and Jet propulsion were the first texts in this field. The 1976 book includes the topics Compressible flows, Flow theory, Gas dynamics, Inviscid flow, Thermodynamics, Adiabatic conditions, Real Gases, Heat transfer, Ideal gas, Isentropic processes, Shock waves, and Small perturbation flow. It has become a classic text in this field of science. He was member of the International Academy of Astronautics since 1962. He also was a founder of the American Rocket Society, and recipient of many prestigious awards.

Anonymous (1975). Maurice J. Zucrow. *Bulletin* Chemical Propulsion Information 1(5): 6. *P*
Bonney, E.A., Zucrow, M.J. (1956). *Aerodynamics*: Propulsion. Van Nostrand: New York.
Dorman, B.L. (1976). Maurice Joseph Zucrow. *Acta Astronautica* 3(9-10): 20.
Grandt, Jr., A.F., Gustafson, W.A., Cargnino, L.T. (2010). *One small step*: The history of
        aerospace engineering at Purdue University, ed. 2. Purdue University: West Lafayette. *P*
Zucrow, M.J. (1929). Flow characteristics of submerged jets. *Trans. ASME* 51(APM-19): 213-218.
Zucrow, M.J. (1958). *Aircraft and missile propulsion*. Wiley: New York.
Zucrow, M.J., Hoffman, J.D. (1976). *Gas dynamics*. Wiley: New York.

Author index

Bold numbers refer to page of the individual mentioned, whereas other numbers refer to cites within the book.

A

Abbett, W.A. 2117
Abbot, C.G. 1760
Abbot, H.L. **1760**, 1809, 2187
Abernathy, F.H. 1886
Abert, S.T. **1761**
Aboites Aguilar, L. 2701
Ackerman, A.J. **1762**, 1829
Ackermann, W.C. **1763**
Adams, C. 2615
Adams, C.D. 2492
Adams, F. 1998, 2328
Adams, J.W. **1764**
Agg, T.R. 2311
Ahern, J. **1765**
Albertson, M.L. **1766**, 1824, 2493, 2514, 2545
Albright, J.J. **1767**
Alden, G.I. **1768**
Alexander, B. 2674
Alexander, L.J. **1769**
Alger, P.L. 1899
Algert, J.H. 2391
Allan, W. **1770**
Allardice, E.R.B. 2292
Allee, D.J. 1994
Allen, C.M. **1771**, 2170
Allen, H. **1772**
Allen, H.C. **1773**
Allen, J.R. **1774**
Allen, K. **1775**
Allen, R.S. 1775
Allen, Z. **1776**
Allison, J.C. **1777**
Allison, W.F. **1778**
Allton, R.A. 2100
Alvord, J.W. 1876, 2345
Ambrose, H.H. **1779**
Ambursen, N.F. **1780**
Amorocho, J. **1781**
Anderson, A.G. **1782**, 2235, 2591
Anderson, C.L. 1839

Anderson, G.G. **1783**
Anderson, H.A. 2296
Anderson, N.E. **1784**
Anderson, R.H. **1785**
Anderson, S. 1800
Anear, C.L. 2325
Angus, R.W. 2360
Archer, J.D. 2253
Archibald, R.S. 2621
Aref, H. **1786**
Arendt, J. 1795
Armstrong, E.L. **1787**
Armstrong, W.H. 2410
Arndt, R.E.A. 2235, 2261
Arnold, J.E. 2421
Arnoldt, G. **1788**
Arthur, H.G. **1789**
Ashley, C.M. 2079
Assad, A.A. 2666
Axtell, F.F. **1790**
Ayres, A.H. **1791**
Ayres, Q.C. 2491

B

Babb, A. 1781
Babb, A.O. 2340
Babb, C.C. **1792**
Babbitt, H.E. **1793**
Babcock, G.H. **1794**, 2718
Babcock, H.A. **1795**
Bache, A.D. 2641
Back, W. 1893, 2616
Bacon, E.M. 2450
Bailey, 2440
Bailey, E.G. **1796**
Bailey, G.I. **1797**
Bailey, J. 2308
Bainer, R. 2679
Baird, W.R. 2704
Baker, H.J.M. **1798**
Baker, M.N. 2205, 2458

Printed and bound by CPI Group (UK) Ltd, Croydon, CR0 4YY

24/10/2024

01778310-0001